ANNUAL REVIEW OF BIOCHEMISTRY

ANNUAL REVIEW OF BIOCHEMISTRY

VOLUME 65, 1996

CHARLES C. RICHARDSON, *Editor*
Harvard Medical School

JOHN N. ABELSON, *Associate Editor*
California Institute of Technology

CHRISTIAN R. H. RAETZ, *Associate Editor*
Duke University Medical Center

ANNUAL REVIEWS INC.   4139 EL CAMINO WAY   P.O. BOX 10139   PALO ALTO, CALIFORNIA   94303-0139

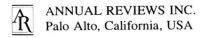

ANNUAL REVIEWS INC.
Palo Alto, California, USA

International Standard Serial Number: 0066–4154
International Standard Book Number: 0–8243–0865-4
Library of Congress Catalog Card Number: 32–25093

Annual Review and publication titles are registered trademarks of Annual Reviews Inc.

○ The paper used in this publication meets the minimum requirements of American National Standard for Information Sciences—Permanence of Paper for Printed Library Materials, ANSI Z39.48-1984.

Annual Reviews Inc. and the Editors of its publications assume no responsibility for the statements expressed by the contributors to this *Review*.

Typesetting by Kachina Typesetting Inc., Tempe, Arizona; John Olson, President; Jeannie Kaarle, Typesetting Coordinator; and by the Annual Reviews Inc. Editorial Staff

PRINTED AND BOUND IN THE UNITED STATES OF AMERICA

Dedicated in memory of Alton Meister

Associate Editor, 1965–1995

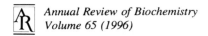

Annual Review of Biochemistry
Volume 65 (1996)

CONTENTS

How to Get Paid for Having Fun, *Daniel E. Koshland, Jr.*   1

Relationships Between DNA Repair and Transcription, *Errol C.
    Friedberg*   15

DNA Excision Repair, *Aziz Sancar*   43

Selenocysteine, *Thressa C. Stadtman*   83

Mismatch Repair in Replication, Fidelity, Genetic Recombination,
    and Cancer Biology, *Paul Modrich and Robert Lahue*   101

DNA Repair in Eukaryotes, *Richard D. Wood*   135

Mechanisms of Helicase-Catalyzed DNA Unwinding,
    Timothy M. Lohman and Keith P. Bjornson   169

Molecular Mechanisms of Drug Resistance in *Mycobacterium
    Tuberculosis*, John S. Blanchard   215

Protein Prenylation: Molecular Mechanisms and Functional
    Consequences, *Fang L. Zhang and Patrick J. Casey*   241

Protein Transport Across the Eukaryotic Endoplasmic Reticulum
    and Bacterial Inner Membranes, *Tom A. Rapoport, Berit
    Jungnickel, and Ulrike Kutay*   271

Molecular Biology of Mammalian Amino Acid Transporters,
    Marc S. Malandro and Michael S. Kilberg   305

Telomere Length Regulation, *Carol W. Greider*   337

The Structure and Function of Proteins Involved in Mammalian
    Pre-mRNA Splicing, *Angela Krämer*   367

Molecular Genetics of Signal Transduction in *Dictyostelium*,
    Carol A. Parent and Peter N. Devreotes   411

Structural Basis of Lectin-Carbohydrate Recognition, *William I.
    Weis and Kurt Drickamer*   441

Connexins, Connexons, and Intercellular Communication, *Daniel
    A. Goodenough, Jeffrey A. Goliger, and David L. Paul*   475

Rhizobium Lipo-Chitooligosaccharide Nodulation Factors:
    Signaling Molecules Mediating Recognition and
    Morphogenesis, *Jean Dénarié, Frédéric Debellé, and
    Jean-Claude Promé*   503

ELECTRON TRANSFER IN PROTEINS, *Harry B. Gray and Jay R. Winkler*   537

CROSSTALK BETWEEN NUCLEAR AND MITOCHONDRIAL GENOMES, *Robert O. Poyton and Joan E. McEwen*   563

HEMATOPOIETIC RECEPTOR COMPLEXES, *James A. Wells and Abraham M. de Vos*   609

DNA TOPOISOMERASES, *James C. Wang*   635

INTERRELATIONSHIPS OF THE PATHWAYS OF mRNA DECAY AND TRANSLATION IN EUKARYOTIC CELLS, *Allan Jacobson and Stuart W. Peltz*   693

RECODING: DYNAMIC REPROGRAMMING OF TRANSLATION, *Raymond F. Gesteland and John F. Atkins*   741

BIOCHEMISTRY AND STRUCTURAL BIOLOGY OF TRANSCRIPTION FACTOR IID (TFIID), *S. K. Burley and R. G. Roeder*   769

STRUCTURE AND FUNCTIONS OF THE 20S AND 26S PROTEASOMES, *Olivier Coux, Keiji Tanaka and Alfred L. Goldberg*   801

INDEXES

Author Index   849

Subject Index   901

Cumulative Index of Contributing Authors, Volumes 61–65   913

Cumulative Index of Chapter Titles, Volumes 61–65   915

SOME RELATED ARTICLES IN OTHER *ANNUAL REVIEWS*

From the *Annual Review of Biophysics and Biophysical Chemistry*, Volume 25 (1996)

Lipoxygenases: Structural Principles and Spectroscopy, B. Gaffney

Computational Studies of Protein Folding, R. A. Friesner and J. R. Gunn

Molding DNA in Aqueous Solutions: Theoretical and Computer Simulation Studies on the Ion Atmosphere of DNA, B. Jayaram and D. Beveridge

From the *Annual Review of Cell and Developmental Biology*, Volume 12 (1996)

Chromosome Condensation, D. Koshland

ATP Receptors, D. Julius

Cytokine Receptor Signal Transduction and the Control of Hematopoietic Cell Development, S. S. Watowich, H. Wu, M. Socolovsky, U. Klingmuller, S. N. Constantinescu, and H. Lodish

Structure-Function Analysis of the Motor Domain of Myosin, K. M. Ruppel and J. A. Spudich

From the *Annual Review of Genetics*, Volume 29 (1995)

Chromosome Partitioning in Bacteria, R. G. Wake and J. Errington

Molecular Genetic Aspects of Human Mitochondrial Disorders, N.-G. Larsson and D. A. Clayton

Yeast Transcriptional Regulatory Mechanisms, K. Struhl

Trinucleotide Repeat Expansion and Human Disease, C. T. Ashley Jr. and S. T. Warren

From the *Annual Review of Immunology*, Volume 14 (1996)

Altered Peptide Ligand-Induced Partial T Cell Activation: Molecular Mechanisms and Roles in T Cell Biology, J. Sloan-Lancaster and P. M. Allen

The Structure of the T Cell Antigen Receptor, G. A. Bentley and R. A. Mariuzza

Antigen Processing and Presentation by the Class-I Major Histocompatibility Complex, I. A. York, K. L. Rock

From the *Annual Review of Medicine*, Volume 47 (1996)

Gene Transfer into Hematopoietic Stem Cells: Implications for Gene Therapy of Human Disease, C. E. Dunbar

Trinucleotide Instability: A Repeating Theme in Human Inherited Disorders, J. F. Gusella, M. E. MacDonald

Nitric Oxide Synthase: Role as a Transmitter/Mediator in the Brain and Endocrine System, T. M. Dawson and V. L. Dawson

From the *Annual Review of Microbiology,* Volume 49 (1995)

   Mechanisms for the Prevention of Damage to DNA in Spores of Bacillus Species,
      P. Setlow
   The Structure and Replication of Kinetoplast DNA, T. A. Shapiro and P. T.
      Englund
   Viral Vectors in Gene Therapy, A. E. Smith

From the *Annual Review of Neuroscience,* Volume 19 (1996)

   RNA Editing, L. Simpson and R. B. Emeson
   Apolipoprotein E and Alzheimer's Disease, W. J. Strittmatter and A. D. Roses
   Trinucleotide Repeats in Neurogenetic Disorders, H. L. Paulson and K. H.
      Fischbeck
   Structure and Function of Cyclic Nucleotide-Gated Channels, W. N. Zagotta and
      S. A. Siegelbaum

From the *Annual Review of Nutrition,* Volume 16 (1996)

   *Structure, Function, and Regulation of the Mammalian Facilitative Glucose
      Transporter Gene Family,* A. L. Olson and J. E. Pessin

From the *Annual Review of Pharmacology and Toxicology,* Volume 36 (1996)

   Progress in Antisense Oligonucleotide Therapeutics, S. T. Crooke and C. F.
      Bennett
   Structural Mechanisms of HIV Drug Resistance, J. W. Erickson and S. K. Burt

From the *Annual Review of Physiology,* Volume 58 (1996)

   Molecular Mechanism of Growth Hormone Action, C. Carter-Su, J. Schwartz, and
      L. S. Smit
   Mutational Analysis of Motor Proteins, H. L. Sweeney and E. L. F. Holzbaur

From the *Annual Review of Plant Physiology and Plant Molecular Biology,* Volume
47 (1996)

   14-3-3 Proteins and Signal Transduction, R. J. Ferl
   Dioxygenases: Molecular Structure and Role in Plant Metabolism, A. G. Prescott
      and P. John
   The Functions and Regulations of Glutathione S-Transferases in Plants, K. A.
      Marrs
   *The Molecular-Genetics of Nitrogen Assimilation into Amino Acids in Higher
      Plants,* H.-M. Lam, K. T. Coschigano, I. C. Oliveira, R. Melo-Oliveira, G. M.
      Coruzzi
   DNA Damage and Repair in Plants, A. B. Britt

For the convenience of readers, a detachable order form/envelope is bound into the back of this volume

Dan Koshland

Annu. Rev. Biochem. 1996. 65:1–13

HOW TO GET PAID FOR HAVING FUN

Daniel E. Koshland, Jr.

University of California at Berkeley, Department of Molecular and Cell Biology, Berkeley, California 94720

CONTENTS

Introduction.   1
The Making of a Scientist.   1
Background.   2
The Manhattan Project.   3
Graduate School   5
Brookhaven National Laboratory.   7
Berkeley.   9
Science   10
Colleagues.   12
Summary   12

Introduction

When writing a prefatory chapter for Annual Reviews a scientist is confronted with the question of what in his or her life might be interesting to others. In my case I was appalled at the absence of material that generates good novels: no broken homes, no misunderstood childhood, no criminal youth gangs, no disastrous liaisons. A landscape of boredom from sea to shining sea. If there is one overlying theme it is that I got paid for doing what I enjoyed all my life. I wish I could say I had cleverly plotted to achieve this nirvana by a series of Machiavellian measures. The truth, however, is closer to the course of the Lord High Executioner in the Mikado: I was "...wafted by a favoring gale as one sometimes is in trances."

As I look back, each new chapter in my life seems to have been a mutation of Pasteur's phrase "chance to the prepared mind." Once I had decided to be a scientist, the events seemed to flow as if by accident. However, in retrospect I see that the experience of each phase of my life presaged the next "accidental happening." But I was surprised at the "random walk" nature of my life.

The Making of a Scientist

What started me on the course of being a scientist and why did I stay the course? I can retrospectively say I decided to become a scientist in the eighth

1

0066-4154/96/0701-0001$08.00

grade when in quick succession I read *Microbe Hunters* by Paul DeKruif and *Arrowsmith* by Sinclair Lewis. *Microbe Hunters* told the adventures of early microbiologists and their extraordinary success in curing diseases. *Arrowsmith* described what being a scientist really means and how a scientist has to make choices that differ from the practice of medicine. Being a scientist seemed highly desirable, so when I entered high school I enrolled in college preparatory courses and took all the math, physics, and chemistry I was allowed.

Today I am still a scientist. Am I typical or abnormal? In my job as a professor at the University of California, I make a point of asking students when they first decided to become scientists, and their stories are amazingly similar to mine. Though not all have read *Microbe Hunters* and *Arrowsmith*, some have. But additional patterns in their lives fit with my experience. All of these students got good grades in mathematics in elementary school and went on to do the same in high school. Not all were equally enamored of math by the time they were graduate students, but elementary school math was a common denominator of the elements of logic I think are essential for a scientist who does puzzle solving for its own sake.

This experience points out lessons that are sometimes forgotten in oratory regarding the supply of scientists. Almost all scientists decide on careers while very young, so if we are to have more scientists we had better entice them at the elementary and high school levels. Colleges are in the business of retaining scientists, but only rarely in the business of generating them. I have heard some speakers who wish to recruit minorities and women into science (a desirable goal in my opinion) who seem to have the idea that anybody who is interested in science is going to be good at it. In fairness to the recruits, recruiting should be targeted to those whose skills correlate with success in science. Exceptions will occur: Pasteur got a "C" in chemistry and Albert Einstein was renowned as a mediocre student, but most scientists like math and get good grades.

Background

From high school I went to the University of California. Tuition at that time was about $100 a semester for which one got free medical care and education. The school was a perfect opportunity for all, with no barriers of wealth or privilege; an egalitarian democracy. The college of chemistry, however, which I entered, was run by GN Lewis, who was an intellectual elitist and made no apologies for it. In freshman chemistry all students took an examination in the first week of classes. Although all students in chemistry had the same general lectures, assignments to lab sections were based on the rankings from the examination. At each hour there was a "number one" lab section composed of the top students, a number two section composed of the 30 next best students, a number three section composed of the next 30 and so on (chemistry was taught at 9:00, 10:00, 11:00, and 1:00 p.m.). The lab instructor for section one was a full professor of chemistry, something GN Lewis believed was important to keep his best students challenged and stimulated. The number one lab sections got no extra edge in course material, but a good deal of added stimulation. I also enjoyed the turbulent social and political life at Berkeley. Those early influences became important later.

Another event I thought minor at the time that turned out to be important was an encounter with Wendell Latimer, chairman of the department and one of the distinguished names at Berkeley. I took a summer course in which I was one of two students in Wendell Latimer's advanced inorganic chemistry. Both of us studied very diligently, and as I was walking out of a three-hour final exam, Wendell Latimer looked up at me, picked up my final exam, and asked "Would an 'A' be sufficient?" I laughingly said yes (I had gotten "A"s in three midterms), but then he started to tear up the exam implying that I had done so well in the course that he didn't need to read the final. With the typical impetuosity of youth I reacted angrily and said, "I spent three hours working on that exam; you owe it to me to correct it." In retrospect I wonder what was on my mind, but fortunately for me Wendell Latimer only laughed. He did correct the exam, and the incident passed without apparent further notice. When I recounted the incident to some classmates at the time they uniformly said that I had acted like an idiot. But shortly thereafter I received the James Monroe McDonnell Scholarship, which is given to the senior student deemed most likely to succeed, and I was later asked by Latimer to work with Glenn Seaborg on the Manhattan Project. I will of course never know whether my grade point average or some inner amusement of Latimer's at the antisocial behavior of a young student lead to those two appointments. But I concluded that being obnoxious was not all bad.

The Manhattan Project

When I graduated from Berkeley the war clouds were forming in Europe, and I tried to enlist in the Navy, but with an eyesight of 20/400 was told by the interviewing officer that, "As far as the Navy is concerned you are legally blind." As the war progressed Navy standards slackened, but at that time they wanted all recruits to be potential deck officers. I explained that my competence in mathematics and calculations would be useful below deck for this new discovery radar, even if I couldn't see the horizon, but I received no sympathy.

The rejection was fortuitous because shortly thereafter I received the call from Wendell Latimer saying that Glenn Seaborg was recruiting people for something called the Manhattan Project at the University of Chicago. By that time I had a job working on aviation fuels, and I told Latimer that my job was classified and getting approval for a transfer would be difficult. Latimer said, "Don't worry, this is the most important job in the world." Knowing Latimer's expertise and relationship to war efforts I accepted his word. A couple weeks later I was on my way to Chicago not knowing on what subject I was going to work, what I was going to be paid, or where I was to be located. I got one hint from Latimer who gave me Rassetti's book on nuclear physics and said to me slyly, "You might find it useful to read this on your way east." When I arrived in Chicago I was greeted by Glenn Seaborg who told me I would be working on an atomic bomb that could win the war, and he said the work was top secret and could not be mentioned to anyone. These were the unexpected events, far from Paul DeKruif and *Arrowsmith,* that started my research career.

Seaborg was not only a brilliant scientist but a superb organizer who could have been successful in any field (he became in fact a winner of the Nobel Prize, Chairman of the Atomic Energy Commission, Chancellor of the University of California, and scientific advisor to four presidents). As a bachelor student who had much to learn I could not have gone into a better training atmosphere. Seaborg's people all knew the job they had to do, and their assignments to projects made for interactions that maximized productivity. For example, we were using radioactive isotopes in an era in which simple Geiger counters were very untrustworthy and commercial instruments nonexistent. We sometimes went through an elaborate experiment to produce an isotope that had a half-life of hours only to have the Geiger counter break down, wasting the work of the experiment. Seaborg, whose primary goal was chemistry, hired instrument experts to develop his own set of reliable instruments that satisfied the needs of the chemists in his group.

Our major job was to purify plutonium from fission products and other impurities to obtain a product of unique purity that could be reduced to a metal and made part of a bomb. Since plutonium was a man-made element of unknown chemistry this charge was a tall order, and we scientists, under Seaborg's guidance, had to work out the chemistry of plutonium with amounts detectable only by their radioactivity. Moreover, this element whose chemistry was totally unknown had to be separated from many transition-state elements whose chemistry was largely ignored in classic textbooks and graduate research programs. Our success is an enormous feat in chemistry in which many young and totally untrained chemists like me were productive only because we were channeled into appropriate paths by an organizational genius.

When I was first hired it was explained to me that plutonium was extremely lethal, a conclusion based on calculations showing it to be much more dangerous than radium which had caused many deaths in the radium-dial industry. We were told to take precautions such as wearing gas masks and performing

elaborate pippetting procedures to protect ourselves from inhaling the plutonium. Almost unanimously, we young scientists discarded this advice because
we believed we were in a necessary war against an evil Hitler bent on global
domination. With our friends dying on battlefields, slowing research to be
extremely cautious about our own lives seemed inappropriate. We routinely
worked six and often seven days a week, spurred on by our reading of a
captured document in which German scientists speculated on the possibility
of a German bomb or nuclear power capability. Later we learned that the
German nuclear effort never really got going because leaders like Heisenberg
discounted these possibilities of nuclear potential. But fission had been discovered in Germany, and the secret weapon Hitler was bragging about could
have been the bomb as far as we knew.

Later, when I was at Berkeley in the 1960s, many discussions of "generation
gaps" filled the newspaper. Many were nonsense, but I did think there was
one real difference in generational attitudes. A generation that has gone through
a great "war for a noble cause" can never look on death in the same way as
one that has lived in an unthreatening peace. We had to realize that some things
are worth dying for, and pacifists in peacetime preaching that nothing is worse
than boys dying on the battlefield are not convincing.

Graduate School

When the war came to an end I decided to return to graduate school and was
tempted to stay in nuclear chemistry. Seaborg was a superb teacher, and I knew
I would benefit greatly by working with him. On the other hand my initial
attraction to biology was still very strong, and I decided that I should go back
to that subject with a chemical approach. Chemistry's application to biology
is so obvious today that one need not elaborate on it, but at that time it was a
frontier, so I chose to go to the University of Chicago. There, a young assistant
professor named Frank Westheimer was interested in applying chemistry to
biology. As a result of our mutual interest, I ended up in his laboratory, which
was enormously stimulating. A variety of graduate students were working on
classical problems of physical organic chemistry, whereas I was something of
the "odd man out" working on a chemical approach to biology.

My radioactive experience was valuable so I volunteered to apply carbon
14 (C^{14}) to the glycolytic pathway, a problem envisioned by Westheimer for
an incoming graduate student. The only available C^{14} at that time was supplied
by Oak Ridge National Laboratory in the form of barium carbonate. To obtain
useful C^{14} for my synthesis I needed to make hydrocyanic acid (HCN) which
I did by placing barium carbonate in a mixture of liquid ammonium and
metallic potassium. The result was a big explosion that resulted in a small
amount of potassium cyanide. That amount was sufficient to follow the clas-

sical Fisher -Kiliani synthesis of glucose 14 starting with arabinose and HCN. I used the glucose-1-C^{14} to trace metabolic pathways.

Getting a PhD was great fun for me, not only because of the excitement of Westheimer's laboratory, but also because the department of chemistry at the University of Chicago was a pioneering and intellectually stimulating group. Harold Urey was studying cosmology, Henry Taube was introducing a whole new approach to inorganic chemistry, and Bill Libby was starting his carbon-dating work.

I remember one Saturday afternoon when Frank Westheimer burst into the laboratory and said, "Come right away. We need you at a conference." I followed along dutifully to find Frank, Bill Libby, George Whelan, two other professors, and some assorted graduate students and postdocs assembled in a room. The problem put to us was that Libby wanted to know how to ash a penguin. Someone had told Libby that he should have a verified modern sample of carbon composition to compare with his ancient samples of carbon dating and that he should accumulate animals from the North Pole, South Pole, Equator, etc.

The Penguin had been flown from the Antarctic and we were charged with converting all the carbon in the flesh, beak, claws feathers, etc. to CO_2. The group started with obvious answers such as fuming sulfuric acid, aqua regia, fuming nitric acid, chromate solutions and so on. Each suggestion was discarded on the recommendation of someone whose experience showed it couldn't do the job. Finally, in frustration, the group dispersed for dinner. Several days later I happened to meet Libby, and I asked what had been decided. Libby said no chemical solution had been found, but he had mentioned the problem to his wife. She pointed out that all body materials were synthesized from a common source, and she therefore suggested that we cook the penguin and collect the grease, which of course could be easily oxidized to CO_2. We followed her advice and the problem was solved. Both this imaginative solution and the exchange of ideas among professors and students over the course of several hours are typical examples of what made the atmosphere at Chicago so exciting at that time. When I went to Harvard for my postdoctoral studies the competence level was the same but the mood was more sedate and regal.

Although my PhD program had only taken three years I was convinced I was a very old man because of the four-year war delay, so I wanted to skip a postdoc. But Frank Westheimer talked me into doing a postdoc and helped me get into Paul Bartlett's laboratory at Harvard. Bartlett's lab was at the forefront of organic chemistry mechanisms, and my years at Harvard were fascinating and challenging because I was focusing on enzymatic reactions, an area Bartlett's group respected but in which they had little interest.

Fritz Lipman, whom I met during my postdoctoral fellowship in Bartlett's laboratory, invited me to an important symposium because of my acetyl phosphate work and in spite of my youth and immaturity. He later offered to have me come to his laboratory for a few months to learn something about biology, and I suggested that I might like to crystallize an enzyme. This interest was a very logical step for me because crystallizing organic compounds was a routine matter; but it was a heroic project for enzymes at that time, and I did not realize the processes were so different. So word got around Massachusetts General Hospital that a young guy named Koshland was coming over to work with Lipman for a month and planned to crystallize an enzyme, a source of great humor to many. I did not succeed in crystallizing an enzyme in that short time, but I did learn an enormous amount of biology and fortunately was exposed to a lot of Lipman's philosophy, which was original and keenly perceptive.

Brookhaven National Laboratory

Up to this time I had been the typical happy-go-lucky graduate student and postdoc, but as happens to all such students it began to dawn on me that I would have to get a job. The job situation was very tight in the early 1950s. My varied background in chemistry and biology certainly didn't help, and I did not get an offer for an academic job. One interviewer from Columbia University looked at my record and said a man who had published little by the age of 31 would never amount to anything (all my war work was classified). So I went to Brookhaven National Laboratory somewhat reluctantly with the vague idea that I would stay a year or so and then go back to a university. Fourteen years later I was still at Brookhaven and very happy, and even then I probably would not have left if I had not had an attractive offer from the University of California at Berkeley.

Brookhaven National Laboratory turned out to be an excellent place for the beginning of my independent career. Because I was a chemist learning biology it was probably best that I didn't start teaching students immediately, and the position also gave me time to assimilate the knowledge in the new field. One lucky break during my years at Brookhaven was a rejection of one of my papers. The paper described a theory I had developed about the stereochemistry of enzymatic reactions which lumped group transfer reactions into two categories, single displacement reactions and double displacement reactions. Because the topic was chemistry I submitted the paper to the *Journal of American Chemical Society* using extensive references to the chemical literature for the stereochemistry of the substrates and products. I also chose this journal because they had recently invited theoretical articles. The editors promptly turned down the paper, saying they did not accept theories. When I pointed out that they had specifically asked for theory they told me that meant mathematical ideas. As a result of this rejection, I published in the review journal *Biological*

Reviews. Older biologists told me that lots of biologists read *Biological Reviews,* whereas most never looked at the *Journal of American Chemical Society,* and that I was extremely lucky to have picked that journal to bring chemistry to the biologists.

One of the great lessons of my life derived from the publication of the induced-fit theory. I was supposed to give a talk at a well-known symposium on my oxygen 18 (O^{18}) work, but decided to speak instead about new research on the specificity of proteins. As I was preparing the talk I was going through the classical explanation for the manner in which substrates are excluded from active sites and decided the "key-lock" or "template" theory of Emil-Fischer was too simple. The theory provided no explanation for the low enzymatic reactivity of water. When I reexamined the literature with this puzzling thought in mind, more and more anomalies arose. I postulated that the enzyme must be flexible and that a new structure of the protein would have to be induced by the binding substrate or activators. This deviation from the template hypothesis of Emil-Fischer met with no approval from distinguished journal editors but finally got published when DD Van Slyke offered to sponsor my paper for the *Proceedings of the National Academy of Sciences.*

Although we did many experiments that in my opinion could only be explained by the induced-fit theory, gaining acceptance for the theory was still an uphill fight. One referee wrote, "The Fischer Key-Lock theory has lasted 100 years and will not be overturned by speculation from an embryonic scientist." We did a lot of experiments, all of which supported the theory, and crystallographers saw small conformational changes that were enough in my opinion to validate the theory, but authors dismissed the changes as being too small. We did more experiments with protein reagents that supported the theory and later the crystallographers, Bill Lipscomb and Tom Steitz, found big conformational changes in proteins on binding ligands, which removed all doubt, and the theory became accepted. Textbooks now routinely include the induced-fit theory.

The episode made me sympathetic to novel theories and aware of the obstacles to their publication. New and tantalizing results require strong support to overturn well-established principles, but this experience also taught me that sometimes journals and literature can be too conservative. That attitude helped me later when I became editor of *Science.*

At Brookhaven we lived a fairly idyllic life. My wife Marian Koshland had her laboratory in the same institution as I did, simplifying our life. Our children grew up in a lovely small village with fine schools. They and we had great friends, and the laboratory had a stimulating intellectual atmosphere.

I got interested in muscular contraction and used lobster as a good source of muscle. We often ran what we called the "boiled control," and it was the culinary peak of my academic career. I had read that lobster muscle was essentially a pure ATPase, so I deduced that I could simply add an excess of H_2O^{18} to lobster muscle, let the H_2O^{18} exchange with the water in the muscle, and then let the reaction run in the intact lobster muscle. The experiment worked well and had the advantage that at the end I simply lifted out the intact muscle and had a pure water ATP solution without contaminants. I was then able to analyze the ADP and inorganic phosphate that was formed. I submitted the manuscript to the *Journal of Biological Chemistry* where it was accepted. Only later did Mildred Cohen tell me that the manuscript had come to Carl Cori who asked Mildred (an authority on O^{18}) what she thought of it, with the comment that "some nutty, young scientist at Brookhaven National Laboratory has added a whole lobster tail to a solution and thinks he is studying an enzymatic reaction." Mildred, with whom I later became good friends, told Carl Cori (with whom I also later became good friends) that in fact she could see nothing wrong with the experiment, and Cori accepted the manuscript.

Berkeley

While I was doing this work, I was offered and accepted a faculty position at Rockefeller University. Because moving five children into New York City did not seem practical I proposed that I come one day a week to teach a course and supervise a laboratory, and to my amazement this arrangement was accepted. I would have blithely continued for the rest of my life with that arrangement at Brookhaven and Rockefeller, but in 1964 I received a phone call from Horace Barker at the University of California asking if I would like to come to Berkeley. My wife, an Easterner, immediately pointed out how happy we were and how good our situation was, and I agreed intellectually that we could hardly better ourselves. One of my senior colleagues advised me, "If you are 95% happy at one institution never move to gain 5%, there are too many uncertainties."

Recalling all the arguments that were made at the time is difficult, but my great love for Berkeley and the campus in my early years cut through all the logic. At our family dinner table we had a brief discussion in which our five children and my wife voted "nay" on moving, and I quoted Lincoln to say the "ayes" have it. Actually, my wife made the decision saying, "Either we stay in the East and I spend the rest of my life making it up to you, or we move west and you spend the rest of your life making it up to me. We move." Berkeley lived up to its reputation of being turbulent as I moved in the 1960s when Berkeley was a center of uproar, but it also lived up to its reputation of being dynamic and exciting. In retrospect, I believe none of us regretted the move.

Being a professor at Berkeley added teaching to my job requirements, and I found it a great joy. Teaching is work, but very rewarding work. I enjoyed the large classes at Berkeley when I was an undergraduate, and since I am basically a ham I consider it a privilege to lecture to the 300 students who annually take our biochemistry major course. The students are very good and constantly keep a professor on his toes. The postdocs who made up my research laboratory at Brookhaven were augmented by graduate students and postdocs at Berkeley.

The pleasures of doing research and teaching made it difficult to pretend I was working instead of merely having a good time. Scientists are basically puzzle solvers, and they get hired to solve puzzles. So someone else is providing the capital for them to satisfy a lifelong desire. *The Organization Man,* a brilliant book by William Whyte explained that people in many professions love their work, but that scientists seem to be particularly lucky in that regard. Tenure can be given because those who love what they are doing work long hours even after such commitment is no longer necessary.

One Saturday morning I was working at the University when a student knocked on the door. When I answered the knock she said she had to ask about an anomaly. She asked, "You and Professor Snell, Professor Barker and Professor Hassid are all working on Saturday but you are members of the National Academy and have tenure so why are you working so hard?" I explained that we all enjoyed it. The incident made me remember a small conference during the hectic days of the atomic bomb research in which Enrico Fermi looked up from spirited argument and said humorously, "It's amazing how fascinating blowing up the world can be." Fermi was a kind and sensitive person and was well aware of the horror of a bomb and the importance of its achievement, but he was observing for a moment that developing a bomb entailed a massive puzzle of great difficulty in chemistry, physics, engineering, and mathematics.

Science

In 1984 I received a phone call from David Hamburg asking me to move to Washington to be full-time editor of *Science.* I answered without hesitation that I would not give up my laboratory but added laughingly, "If it was a part-time job I could be interested." I had just been chairman of the department, a time-consuming chore, and decided that one could do two things at once. Two months later I received a call from the same David Hamburg saying the American Association for the Advancement of Science (AAAS) (the owners of the journal *Science*) were willing to offer me the job part time. Then I got scared and asked myself if I could really do the job part time. I decided to visit the magazine and talk to the staff about it. After that initial appraisal I decided the arrangement was possible and said I would do the job with the condition that they would be honest with me if they thought I was doing a bad job and I would be honest with them if I thought I couldn't handle the burden.

But everything proceeded extremely well. Staff at the time were helpful and welcoming, and new staff, which I hired over the next couple of years, were equally enthusiastic and competent. The arrangement in which I spent one week a month at *Science* and the other three in Berkeley actually worked out quite well for both of us. My division of time between editing and research was close to 50-50, since I spent a great deal of *Science* journal time in Berkeley, not only phoning *Science* staff every day, but also soliciting articles and discussing policy from Berkeley.

When I was about to leave the editorship I recommended to the AAAS Board that they continue with the part-time editor for two reasons. First, I felt the editor of *Science* should be a continuing member of the scientific community. Not only because he or she would then speak with more credibility to the public, but also because he or she would be considered a colleague by fellow scientists and would understand policy issues such as funding that fellow scientists were experiencing. Second, this arrangement produced more independence and enjoyment for the staff. The editor was around enough so the staff could understand his wishes and get a feeling for his approach, which is important because the head of any large organization has to provide direction, enthusiasm, and motivation. But since I was not there most of the time, each editor, department, and staff member had an added degree of responsibility. Moreover, I deliberately selected people who could be more independent, and they in turn were the type of people who enjoyed the independence. When I started I was apprehensive that my part-time position would end badly and might cause great friction. When I left, the staff was in almost unanimous agreement that the part-time arrangement allowed the journal to select the editor-in-chief from a far more widespread pool and also to develop a more talented pool of staff members.

I had a wonderful time at *Science* and was tempted to go on forever. But as my ninth year was ending I noticed that I was beginning to lose some of my enthusiasm for new projects, and I was longing to get back to full-time research. So I notified the Board which then started a search for a successor, and I have returned to full-time research. The Board chose an excellent successor in Floyd Bloom. I miss the stimulation and excitement of the *Science* atmosphere and my many friends there, but I believe that any job in this world can only operate at full throttle. Once a manager starts thinking everything is going well, the staff is excellent, and there's nothing to fix up, he or she is on his way to disaster. The show business phrase "always leave them laughing" is a good one and is the indication of the time to move on.

Being able to do full-time research again provides compensation for the loss of the challenge of *Science*. When one is doing research half time, as I was for those 10 years, one can keep up if one stays in one's original field and knows the literature well, but starting an entirely new line of work is very difficult. Pursuing new areas of interest has been possible since I left *Science,* and I am extremely excited, feeling much as I did when I was a starting assistant professor and had to probe entirely new lines of research as part of my need to get a laboratory going.

Colleagues

One of the most easily forgotten ingredients of the life in science is its social aspect. The image of a reclusive curmudgeon solving esoteric puzzles in a badly heated attic is a movie director's idea of a scientist. The real life of a modern scientist involves interactions with many people: students, colleagues, editors, business staff, university administrators, visiting speakers, and so on. A very big social life is needed just to get the job done. That social life is very rewarding and in retrospect no minor part in the joy of science. There is a special camaraderie in science in which colleagues share the frustration of failure and the joy of success in the arduous battle to uncover nature's secrets.

That camaraderie is also part of editing where the challenges of producing a journal of highest quality require the great efforts and high abilities of many people. That enjoyment of a common endeavor is also shared with the business staff at universities, who rise to challenges such as getting the grant proposal in on time even though the principal investigator has left it until the last minute, and the university administrators who struggle to maintain the quality of the institution.

An inadequate step in this direction is to list the coauthors who contributed to the research in my laboratory (see Acknowledgments). Space does not allow me to add all those on the staffs of Brookhaven National Laboratory, Rockefeller University, the University of California, and *Science* magazine who were so helpful and fun to work with, but they are deeply and gratefully remembered for their effective and enjoyable contributions.

Summary

As I look back, I am still amazed that I was actually paid to do something I loved and others could describe as work. Yet my situation is no different from that of most scientists who find that they are asked to pursue their innate curiosity to solve puzzles, the solutions to which fortunately are of value to society. I enjoyed the beautiful logic of mathematics in elementary grades and was entranced by the exciting solution of puzzles described by DeKruif. So I drifted into the scientific profession without a clear idea of what to do or how to do it. Each experience prepared my mind and supplied the base for the next job, creating what was for me a smooth flow from scientist to professor to editor to scientist.

Fortunately for me and fellow scientists the problems of the world never disappear. "The one who rides the tiger can never get off" is an aphorism that expresses society's dependence on science. Automobiles improve transportation and create pollution, medical advantages prolong life and create overpopulation, pesticides bring cheaper food and create soil problems. Each advance brings on the need for more science to solve the new problems. Society, which likes to live well, is addicted to the products of science, and fortunately a peculiar set of humans are addicted to solving the problems. I am one of those typical addicts who finds the obstacle course fascinating and the endlessness of the quest utopia.

ACKNOWLEDGMENTS

The author owes too much to too many to acknowledge adequately the generators of his unbelievably enjoyable life. It all started with two exceptional parents who expected children to rise to challenges and provided the cushions of love and support that eliminated pain from failure. I had two outstanding scientific mentors in Glenn Seaborg and Frank Westheimer, and a stream of superb colleagues, graduate students, and postdoctoral fellows who not only provided scientific ideas and results but an atmosphere of enjoyment that made the trip as much fun as the destination. They were equaled by the staffs of Brookhaven, Rockefeller, Berkeley, and *Science* who were so good I would even inadvertently admit I liked administrators. Last but certainly the most important colleague and mentor is my wife, Marian Koshland, who has dispensed wisdom with amused compassion as a "shadow government" in all my endeavors.

Because space does not allow me to list all those collaborators who made research so much fun, I have settled for naming an illustrative sample chosen by chance to represent the totality whom I remember with affection. A Kowalsky, A Redfield, B Howlett, B Lynch, C Batt, C Blake, D Filmer, D Hoare, D Mulligan, D Phillips, E Herr Jr., E Kennedy, H Levy, H Weiner, J Spudich, J Thomas, M Erwin, M Kirtley, N Sharon, P Strange, R Macnab, R Weis, S Springhorn, S Strumeyer, W Ray Jr., W Springer, A Shiau, R Cook, R Zukim, R Yount, P McFadden, M Lee, T Ingolia, G Dafform, J Hurley, D Storm, W Stallcup, H Biemann, R Stroud, A Conway, C Jeffrey, K Carraway, R Cook, A Dean, A Cornish-Bowden, J Haber, D Saunders, C Long, M Shapiro, M Snyder, S Clarke, K Neet, R Bell, G Loudon, S Kim, M Brubaker, R Tjian, L Cheever, J Stock, S Mowbray, J Falke, K Walsh, A Goldbeter, G Bollag, A DeFranco, D LaPorte, R Dahlquist, G Moe, S Mockrin, T Terwilliger, B Stoddard, J Wang, D Mochley Rosen, M Smolarsky, S Parsons, A Russo, N Paoni, A Flint, D Aswad, P Thorsness, A Stock, D Clegg, M Fahnestock, A Newton, A Levitzki, and P Lovely.

Annu. Rev. Biochemistry 1996. 65:15–42

RELATIONSHIPS BETWEEN DNA REPAIR AND TRANSCRIPTION

Errol C. Friedberg

Laboratory of Molecular Pathology, Department of Pathology, The University of Texas Southwestern Medical Center, Dallas, Texas 75235

KEY WORDS: nucleotide excision repair, Cockayne syndrome, xeroderma pigmentosum, RNA polymerase II, repairosome

ABSTRACT

Multiple relationships have been noted between DNA repair and transcription in both prokaryotic and eukaryotic cells. First, in both prokaryotes and eukaryotes nucleotide excision repair of the template strand of transcriptionally active regions of the genome is faster than in the coding strand. In prokaryotes the biochemical basis for this kinetic difference appears to be related to the specific coupling of repair to arrested transcription by RNA polymerase. The biochemical basis for strand-specific repair in eukaryotes is unknown. Second, in eukaryotes some or all of the subunits of transcription factor IIH (TFIIH) are required for nucleotide excision repair. The biological significance of this dual function of TFIIH proteins is not obvious. Finally, there are indications that the genes *CSA* and *CSB,* which are implicated in the human hereditary disease Cockayne syndrome, may have a role in transcription.

CONTENTS

INTRODUCTION   16
GENERAL FEATURES OF BASE AND NUCLEOTIDE EXCISION REPAIR   17
   Base Excision Repair   17
   Nucleotide Excision Repair Pathways   17
GENERAL FEATURES OF RNA POLYMERASE II TRANSCRIPTION   19
   Initiation of Basal Transcription   19
   Promoter Clearance and Transcriptional Elongation   21
TRANSCRIPTIONAL ARREST IN TEMPLATES CONTAINING DAMAGED
   BASES   23
RELATIONSHIPS BETWEEN TRANSCRIPTION AND DNA REPAIR.   24
   Preferential Excision Repair of Transcriptionally Active Genes   24
   Strand-Specific Repair of Transcriptionally Active Genes   26
   Coupling of Transcription and Nucleotide Excision Repair in Escherichia coli   28
   Coupling of Transcription and Nucleotide Excision Repair in Eukaryotes   29

15

0066-4154/96/0701-0015$08.00

Some RNAP II Basal Transcription Proteins Are Required for Nucleotide Excision Repair in Eukaryotes   30

HOW DOES TFIIH OPERATE IN NUCLEOTIDE EXCISION REPAIR?   32
    TFIIH Is Required for Nucleotide Excision Repair in Transcriptionally Silent DNA .   32
    The Role of TFIIH in Nucleotide Excision Repair in Transcriptionally Active Genes   33
    Rad3 (XPD) and Ssl2 (XPB) Proteins Are DNA Helicases   34

THE TRANSCRIPTION HYPOTHESIS OF HUMAN HEREDITARY DISEASES ...   35

SUMMARY AND PERSPECTIVES   38

INTRODUCTION

Free-living forms have evolved manifold mechanisms for repairing or tolerating diverse types of DNA damage. Recent reviews of many of these repair and DNA damage–tolerance mechanisms provide a thorough discussion of DNA repair (1–17). Excision repair is one of the general strategies used by both prokaryotic and eukaryotic cells for repairing multiple types of base damage. During this process, chemically altered, mispaired, or inappropriate (such as uracil in DNA) bases are physically excised from the genome and replaced by bases with normal chemistry and sequence. Excision repair is classified into two mechanistically distinct modes called base excision repair (BER) and nucleotide excision repair (NER), on the basis of differences in the excised products. During BER, damaged bases are excised as free bases, whereas NER is characterized by the excision of offending bases as nucleotides, typically oligonucleotide fragments. NER operates most efficiently on chemically modified bases which promote helix distortion, such as cyclobutane pyrimidine dimers (CPD) and bases bearing adducts derived from interactions with chemicals. However, it is increasingly apparent that the excision of nucleotides occurs in the course of several enzymatically distinct pathways. For example, strand-directed long patch mismatch correction results in the excision of mispaired bases as nucleotides, yet the biochemical mechanism of this repair mode is substantially different from that of NER. Furthermore, the excision of CPD can also be effected by at least two other known pathways which differ from one another and from conventional NER. This diversity of mechanisms for the excision of photoproducts from DNA presumably reflects selection that operated during evolution as a defense against the effects of UV radiation from the sun, a prevalent natural genotoxic agent. New insights concerning the multiple mechanisms of excision repair call for a refinement of the classification of this repair mode to clarify the relationships between transcription and individual DNA repair pathways, the primary topic of this review. BER is always initiated by a DNA glycosylase and results in the excision of free bases. NER is always initiated by endonucleolytic incisions near sites of base damage. During general NER these incisions flank sites of base damage. Multiple specific forms of NER also exist. One such repair mode involves the action of a specific endonuclease which cleaves DNA exclusively 5' to certain photoproducts. A second specific mode of NER is initiated by a DNA glycosylase which only recognizes CPD, and so-called long patch mismatch repair represents a third form of specific NER.

GENERAL FEATURES OF BASE AND NUCLEOTIDE EXCISION REPAIR

Base Excision Repair

BER is initiated by a specific class of enzymes called DNA glycosylases. Each such enzyme recognizes a single or a relatively small number of damaged (e.g. following simple alkylation), inappropriate (e.g. uracil in DNA), or mispaired bases [reviewed in (17)]. DNA glycosylases catalyze the hydrolysis of the N-glycosyl bonds linking bases to the deoxyribose-phosphate backbone of DNA, thereby releasing free bases and generating sites of base loss (abasic sites). The hydrolysis of phosphodiester bonds 5' to abasic sites by apurinic/ apyrimidinic (AP) endonucleases creates suitable template-primers for DNA polymerases. Hence, concerted repair synthesis, nick translation, and excision reactions can result in repair patches that may be as small as a single nucleotide.

Nucleotide Excision Repair Pathways

GENERAL NUCLEOTIDE EXCISION REPAIR   NER repair pathways operate on many types of base damage caused by the interaction of DNA with physical and chemical agents. Such forms of base damage range from the simple covalent addition of methyl groups to complex chemical structures, such as covalently joined adjacent pyrimidines [CPD and so-called (6-4) photoproducts] and the covalent cross-linking of bases on opposite DNA strands through some sort of chemical bridge. Lesions that perturb the helical structure of the DNA duplex show indications of being processed more efficiently by NER than lesions that do not. So although nonhelix distortive lesions such as simple alkylated bases may be more efficiently repaired by the BER pathway, these lesions may on occasion be processed by NER.

General NER is characterized by two universal features. First, incision (nicking) of the damaged strand is effected by the action of two endonucleases, each of which operates exclusively on one side of the damaged base. Second, these incisions are located at rather precise distances from the damaged base. In *Escherichia coli* (*E. coli*) the 3' incision is located three or four nucleotides from the site of CPDs while the 5' incision is seven nucleotides removed. The distance relationship between the damaged base and the 3' nick is conserved in human (and presumably in other eukaryotic) cells. However, in such cells the 5' nick and the damaged base are separated by about 21 nucleotides (9). Similar results have been obtained with a reconstituted yeast NER system (17a).

These precise spatial relationships are consistent with the notion that during general NER a region of the DNA in the immediate vicinity of a damaged base undergoes specific conformational changes effected by interactions with the NER machinery. In the case of *E. coli* these interactions are understood in considerable detail (3, 18–20). A complex consisting of two molecules of UvrA protein and one molecule of UvrB protein ($UvrA_2B_1$ complex) binds to sites of base damage. The formation of the $UvrA_2B_1$-DNA complex is accompanied by conformational changes in the DNA and in UvrB protein. These changes are thought to promote the dissociation of UvrA protein and the tight binding of UvrB protein to DNA. The conformation of the $UvrB_1$-DNA complex facilitates the binding of UvrC protein, an event that results in further conformational changes and ultimately leads to incision of the DNA by UvrB protein 3' to the damaged base. This incision may further alter the conformation of the damaged DNA, resulting in a second nick catalyzed by UvrC protein 5' to the site of base damage (3, 18).

In eukaryotic cells the molecular events that are involved in the interaction of the NER machinery with damaged DNA and culminate in dual incisions flanking sites of base damage are less well understood. Nonetheless, recent studies in the yeast *Saccharomyces cerevisiae* and in human cells have led to the identification of candidate enzymes for such dual incisions and to the elucidation of their substrate specificities (21–23a).

SPECIFIC NUCLEOTIDE EXCISION REPAIR   Besides strand-directed long patch mismatch correction two other pathways have been identified for the excision of damaged bases as nucleotide structures. Both of these pathways are apparently specific for base damage produced by UV radiation. In *E. coli* infected with bacteriophage T4, and in the highly UV radiation-resistant organism *Micrococcus luteus,* excision repair of CPD can be initiated by hydrolysis of the 5' glycosyl bond in the dimerized pyrimidine pair by a pyrimidine dimer-specific DNA glycosylase [reviewed in (17)]. Since the 3' glycosyl bond remains intact, the dimerized pyrimidines are ultimately released as part of an oligonucleotide. Recent studies in the fission yeast *Schizosaccharomyces pombe* (24) and in the filamentous fungus *Neurospora crassa* (24a) have identified another enzyme that recognizes pyrimidine dimers. The substrate specificity of this endonuclease includes the other quantitatively major photoproduct in DNA, the (6-4) lesion (17). The endonuclease specifically catalyzes cleavage of a single phosphodiester bond 5' to these photoproducts in an ATP-independent fashion (24, 24a). How the damaged bases are ultimately excised is not clear.

GENERAL FEATURES OF RNA POLYMERASE II TRANSCRIPTION

Transcription by the eukaryotic RNA polymerase II has also been extensively reviewed in recent years (25–38). Hence, this discussion touches only on key features of this process to complete the stage for a more detailed analysis of the relationships between transcription and DNA repair.

Initiation of Basal Transcription

Purified RNA polymerase II (RNAP II) is inactive in initiating accurate transcription in vitro. The pioneering studies of Roeder and his colleagues [e.g. see (26)] demonstrated a requirement for multiple factors, designated basal or general transcription factors (TFs), for the initiation of transcription by this enzyme. Many of these transcription factors have been purified and characterized from a variety of lower and higher eukaryotic sources and include TFIIA, TFIIB, TFIID, TFIIE, TFIIF, TFIIH, TFIIJ (25–38), and more recently TFIIK (39–41). The prevailing view is that these transcription factors (most of which are comprised of more than one polypeptide in mammalian cells) assemble with RNAP II at promoter sites in an ordered, stepwise fashion. Support for this view comes from the demonstration that various transcription factors can be assembled onto promoter DNA in vitro. This ordered assembly is believed to be initiated by a complex of TFIIB and TATA binding protein (TBP) (a component of TFIID) to the promoter, followed by the binding of RNAP II in association with TFIIF. TFIIE and TFIIH (with bound TFIIK) are thought to be added to the initiation complex last. In vivo studies in yeast also indicate that a large RNA polymerase II holoenzyme comprising RNA polymerase II, TFIIF, TFIIB, TFIIH, and additional polypeptides involved in transcription initiation may preassemble in a DNA-independent fashion (42–45).

TRANSCRIPTION FACTOR IIH   The transcription factor TFIIH is particularly relevant to this review because it is required not only for RNAP II general transcription, but for NER as well (see section entitled "Some RNAP II Basal Transcription Proteins Are Required for Nucleotide Excision Repair in Eukaryotes"). In human cells TFIIH is thought to comprise nine subunits (46–51a; J-M Egly, personal communication). Five of these, with molecular masses of 89, 80, 62, 52, 44, and 34 kDa, are encoded by cloned genes designated *XPB* (*ERCC3*), *XPD* (*ERCC2*), *p62*, *p52*, *p44*, and *p34*, respectively (Table 1). Subunits of ~38 and ~37 kDa have also been identified (Table 1). Analysis of the ~38 kDa subunit showed that it is identical to a protein kinase called MO15 (49, 51a, 57a) and, more recently, Cdk7 (52). This kinase was originally identified as the catalytic subunit of cyclin-dependent kinase-activating kinase

Table 1 Composition of transcription factor IIH (TFIIH)[a]

Human		Yeast	
Subunit (kDa)	Gene	Subunit (kDa)	Gene
89	XPB	105	SSL2
80	XPD	85	RAD3
62	p62	73	TFB1
52	p52	55	TFB2
44	p44	50	SSL1
38	MO15 (CDK7)	33	KIN28
37	CYCH	47	CCL1
—	—	45	CCL1
—	MAT1	38	TFB3
34	p34	—	—

[a]Polypeptides encoded by homologous human and yeast genes are entered on the same line.

(CAK), an enzyme that phosphorylates the cyclin-dependent kinases $p34^{cdc2}$ and $p33^{cdk2}$ (53–57). Cyclin H, an ~37 kDa polypeptide that is associated with MO15 and is essential for kinase activity, has also been identified in TFIIH (49, 51a, 57a).

TFIIH from the yeast *S. cerevisiae* (designated as factor b in earlier literature) was originally purified using an in vitro transcription assay in which TFIID and TFIIH were inactivated by heat. Replenishment of inactivated TFIID with purified recombinant TATA-binding protein (TBP) facilitated the fractionation of TFIIH as a complex of three polypeptides of 85, 73, and 55 kDa (58). Cloning the *TFB1* gene encoding the 73 kDa subunit allowed the construction of yeast strains that encode histidine-tagged Tfb1 protein, and hence the use of high-resolution chromatography based on the affinity of histidine residues for a nickel-agarose matrix. These procedures revealed the presence of two other polypeptides of ~50 and ~38 kDa in purified TFIIH (59). This complex of five polypeptides is stable to extensive fractionation and has been designated as core TFIIH (41). When this form of TFIIH was tested in an in vitro transcription system reconstituted from purified transcription factors it was found to be inert. This observation led to the search for and discovery of a more complex form of TFIIH called holoTFIIH, which contains an additional four subunits of ~105, 47, 45, and 33 kDa (40). HoloTFIIH is the form of TFIIH that is believed to participate in RNAP II transcription initiation in yeast (41).

The genes encoding all of the nine subunits that have been identified in yeast holoTFIIH have been cloned and sequenced. The 105, 85, 73, and 50 kDa subunits are encoded by the *SSL2* (*RAD25*) (60, 61), *RAD3* (62), *TFB1* (63), and *SSL1* (64) genes, respectively, which are the yeast homologs of the human XPB, XPD, p62, and p44 subunits (Table 1). The 33 kDa subunit is encoded by the *KIN28* gene and is the yeast homolog of the MO15/Cdk7 kinase (39). In yeast, Kin28 protein interacts with a cyclin called Ccl1 (65). Kin28, together with the 47 and 45 kDa subunits of holoTFIIH, is designated TFIIK (39–41). Peptide and western analyses of the 45 and 47 kDa subunits of TFIIK indicate that they are both encoded by the *CCL1* gene (JQ Svejstrup, WJ Feaver & RD Kornberg, personal communication). The smaller subunit could be a degradation product of the larger polypeptide, or one of the proteins may be post-translationally modified. The genes that encode the remaining two subunits (55 and 38 kDa) of holoTFIIH are called *TFB2* and *TFB3* (Table 1) (WJ Feaver & RD Kornberg, personal communication). Recent studies suggest that the yeast *TFB2* and *TFB3* genes are represented in human cells as genes designated *p52* and *MAT1* respectively (Table 1) (J-M Egly, personal communication).

A recent study demonstrated that Kin28 protein is required for RNAP II transcription and that it phosphorylates the C-terminal domain (CTD) of RNAP II in vitro (65a). Consistent with these results, RNAP II transcription was dramatically reduced and phosphorylation of the largest subunit of RNAP II in a *kin28-ts* mutant at the restrictive temperature decreased (65b). CAK activity was not detected in the in vitro study just mentioned (65a). Similarly, CAK activity tested on appropriate yeast substrates has not been identified in purified preparations of holoTFIIH or TFIIK (WJ Feaver & RD Kornberg, personal communication). Hence, unlike its human counterpart, yeast TFIIH appears to be unendowed with CAK activity.

Promoter Clearance and Transcriptional Elongation

The biochemical events that ensue once a complete initiation complex is assembled at a eukaryotic promoter are less clear. In an attempt to determine the roles of various transcription factors in transcription initiation, promoter clearance, transcript elongation, and transcription termination in eukaryotic cells, Goodrich and Tjian (66) used an in vitro reconstituted human RNAP II transcription system. They defined transcription initiation as a state characterized by the assembly of the initiation complex at a promoter and the concomitant melting of the DNA surrounding the transcription start site. Promoter clearance was defined as a state in which the melted region (transcription bubble) begins to move away from the promoter and the first few ribonucleotides of the nascent mRNA are synthesized. Goodrich and Tjian (66) suggested that during this stage the composition of the initiation complex and its interaction with DNA are altered. Transcript elongation involves the passage of the transcription machinery along the template DNA strand and the progressive synthesis of a complete mRNA, and transcription termination involves the release of the completed transcript and RNAP II from the template. Using an abortive initiation assay that operationally defined the initiation stage, Goodrich & Tjian (66) observed a requirement for only TBP, TFIIB, TFIIF, and RNAP II. TFIIE and TFIIH were required in addition for promoter clearance, but not for transcription initiation or transcript elongation.

Similar results were obtained in an independent study using a defined reconstituted human transcription system and several experimental strategies to define the composition of the transcription complex at various stages of the transcription cycle (66a). This study showed that TFIID, TFIIB, TFIIF, TFIIE, and TFIIH are assembled in a promoter-bound initiation complex prior to the addition of nucleoside triphosphates. TFIID remained stably bound to the promoter, whereas TFIIB was released immediately upon addition of triphosphates. TFIIF was released from the initiation complex after the first 10 nucleotides were incorporated into the nascent RNA chain. TFIIF could reassociate with a stalled transcription elongation complex, but was released once the stalled complex re-entered productive elongation (66a). This finding is consistent with the results of other studies on transcription elongation (see below).

Both TFIIE and subsequently TFIIH were released from the initiation complex. TFIIE was released before the formation of the first 10 nucleotides in the RNA, whereas TFIIH was released after the first 30 nucleotides were synthesized but before the transcript was ~70 nucleotides in length (66a). Neither the p62, XPB, nor the XPD subunits of TFIIH were detected in isolated stalled elongation complexes (66a). As noted later in this review, the apparent absence of TFIIH from the transcription machinery during the elongation phase has important implications for understanding the apparent coupling of transcription by RNAP II to DNA repair.

Much remains to be learned about the composition and functional characteristics of the complex that promotes the elongation of a properly initiated transcript. Several proteins that interact directly with RNAP II and affect the rate of transcription have been identified from various eukaryotic sources. TFIIF has been reported to stimulate the rate of elongation of RNA chains and to promote read-through of RNAP II at pause sites (67). A 38 kDa polypeptide variously referred to as TFIIS, SII, RAP38, or p37 [quoted in (68)] and a factor called TFIIX (69, 70) have also been implicated in transcriptional elongation. Additionally, a factor called YES has been purified from *S. cerevisiae* and shown to stimulate the rate of RNA elongation by RNAP II on oligo (dC)-tailed DNA templates (71). More recently, an elongation factor called SIII has been isolated from rat liver extracts (72). SIII is comprised of three subunits of 110, 18, and 15 kDa (72) and is distinct from other known transcription factors.

Purified SIII stimulates the synthesis of accurately initiated RNAP II transcripts in vitro.

3' TRANSCRIPTION SHORTENING   A process that is intimately associated with transcript elongation by RNAP II is the shortening or retraction of the nascent transcript by hydrolysis of the 3' end (73). This hydrolytic activity is believed to be an intrinsic function of RNAP II (74), but is dependent on the presence of functional TFIIS (SII) (75). Shortening of the transcript by hydrolysis of its 3' end appears to be essential for TFIIS-mediated read-through past natural transcriptional pause sites. By direct analogy with the proof-reading function of the 3'->5' exonuclease function of many DNA polymerases, it has been suggested that hydrolysis of the 3' end of the emerging RNAP II transcript may promote the fidelity of transcription (75).

A final caveat of this brief overview of RNAP II transcription is the reminder that in living cells transcription operates on chromatin rather than on naked promoter DNA, and in addition to basal and transactivating factors, multiple factors that operate in relieving chromatin-mediated repression of transcription are important elements of transcriptional regulation. Experimental systems that apparently monitor both the antirepression of genes in the so-called inactive ground state and their concurrent activation by basal and transactivating factors are just beginning to emerge (29, 76).

TRANSCRIPTIONAL ARREST IN TEMPLATES CONTAINING DAMAGED BASES

The interpretation of early studies of base damage effects on in vitro transcription was complicated by the presence of multiple base adducts produced by some types of damage, particularly by their random distribution in both template and coding strands. The use of substrates containing chemically defined single base alterations in known locations led to the demonstration that various types of base damage, and even single strand breaks in the template strand (but not the coding strand) of DNA, can act as blocks to transcription catalyzed by various prokaryotic RNA polymerases in vitro (77–86). In all of these studies the efficiency of transcriptional arrest varied with the type of base damage, and transcriptional bypass, when observed, was frequently accompanied by misinsertion of bases, suggesting that bypass of unrepaired DNA damage may be a source of mutant proteins, at least in prokaryotic cells. An observation of particular interest with respect to T7 RNA polymerase is that single strand breaks or even large gaps with 3'OH and 5'P termini do not block transcription. However, breaks and gaps with 5'P and 3'P termini do block transcription, suggesting that charge repulsion between the two phosphoryl termini may perturb the catalytic function of the polymerase (87).

Transcriptional arrest of RNAP II on damaged templates has been less extensively studied. In recent experiments the effect of CPD on RNAP II-mediated transcription elongation was investigated using a reconstituted in vitro system (88). In these studies CPD were placed in known locations downstream of the adenovirus late major promoter. These lesions proved to be potent inhibitors of transcription when present on the template DNA strand. CPD at sites of arrested transcription were resistant to repair by DNA photolyase, suggesting that the arrested transcription complex was stably bound at sites of base damage and that it shielded the damage from this repair enzyme. TFIIS promoted transcript cleavage in this system. However, unlike the situation with natural pausing, this reaction was not accompanied by bypass of CPD sites.

RELATIONSHIPS BETWEEN TRANSCRIPTION AND DNA REPAIR

Preferential Excision Repair of Transcriptionally Active Genes

In 1985 Hanawalt and his colleagues (89) devised an elegant experimental technique in which they coupled the sensitivity of Southern hybridization for examining defined regions of the genome with the substrate specificity of certain DNA repair enzymes (e.g. the specificity of the pyrimidine dimer-DNA glycosylase for CPD in DNA), thereby devising a technical strategy for monitoring the kinetics of the loss of lesions such as CPD from defined genes. The essential elements of this technique are the following. The initial presence and persistence of lesions such as CPD in the DNA of a gene of interest results in sensitivity of the DNA to degradation by the damage-specific enzyme probe and hence a loss of the relative intensity of the hybridization signal during Southern analysis of that gene. In contrast, the progressive removal of CPD by NER as a function of the time of postirradiation incubation of cells protects the DNA against such degradation and restores the hybridization signal. This general technology has been refined to facilitate measurement of the repair of other types of base damage, including interstrand cross-links and base damage that yields alkali-labile sites that can be converted to strand breaks. Using this general technique Hanawalt and his colleagues observed that the rate of loss of CPD from the transcriptionally-active *DHFR* gene of Chinese hamster ovary (CHO) cells was about fivefold faster than in the genome overall. This phenomenon is referred to as the preferential repair of transcriptionally active genes, and it has been extensively documented in multiple genes for several types of base damage in both prokaryotic and eukaryotic cells (90–103).

A facile explanation for the preferential repair of transcriptionally active genes is that it reflects a more "open" configuration of the chromatin structure of such genes compared to genes that are transcriptionally silent, thereby providing preferential access of the DNA repair machinery to sites of base damage. In some experiments of this type this explanation may be correct. However, more refined analysis of the phenomenon using hybridization probes for each of the two DNA strands revealed that in many transcriptionally active eukaryotic genes the template strand is repaired more rapidly than the coding strand (103). Even this strand selectivity has been challenged with the caveat that arrest of the transcription elongation complex at lesions in the template strand might prolong the "open" conformation of that strand relative to the coding strand, thereby providing a kinetic advantage for its repair (104). But strand-specific repair has also been observed in the prokaryote *E. coli* (105) in which genomic organization is not believed to result in significant steric hindrance to the access of enzymes required for various metabolic transactions of DNA. Furthermore, the presence of stalled RNAP complexes at sites of base damage in the transcribed strand actually inhibited the repair of CPD by the *E. coli* UvrABC endonuclease in vitro (106). This result is consistent with the notion that an arrested transcription complex might actually protect lesions from repair (see "Coupling of Transcription and Nucleotide Excision Repair in *E. Coli*").

These observations, coupled with the observation that in mammalian cells both preferential repair and strand-specific repair of transcriptionally active genes are confined to genes which are transcribed by RNAP II, have led to the view that preferential repair and strand-specific repair are mechanistically related and reflect the operation of an NER mode that is somehow coupled to the process of transcription elongation by RNAP II at sites of base damage. Hence, the terms preferential repair, strand-specific repair, and transcription-ally-coupled repair are often used interchangeably in the literature.

The imprecision of this terminology is unfortunate because it has the potential to create considerable confusion. As already indicated, the preferential repair of transcriptionally active genes is not necessarily always predicated on a kinetic bias for the repair of the transcribed (template) strand relative to that of the nontranscribed (coding) strand. Furthermore, although the evidence that the *E. coli* NER machinery can be specifically coupled to arrested transcription in vitro is substantial (see "Coupling of Transcription and Nucleotide Excision Repair in *E. Coli*"), such coupling has not been definitively established as the biochemical correlate of all strand-specific repair in this prokaryote. In addition, the biochemical coupling of transcription and excision repair (of any type) has not been demonstrated in any eukaryotic system in vitro, and the molecular mechanism of strand-specific repair of transcriptionally active genes in higher organisms remains unknown.

Further confusion about the relationships between transcription and DNA repair stems from the evidence that many of the subunits of the RNAP II basal transcription factor TFIIH are also required for NER (see discussion below). The obligatory loading of TFIIH onto promoter sites during transcription initiation suggests an obvious mechanism for directly coupling RNAP II transcription to NER at sites of base damage in the template strand. However, as mentioned above, several studies (66, 66a) indicate that TFIIH is not associated with the transcription elongation complex when the nascent transcript becomes longer than about 30 nucleotides. Hence, the observation that some NER proteins are also essential for RNAP II transcription may have no direct bearing on the mechanism of strand-specific repair of transcriptionally active genes.

Strand-Specific Repair of Transcriptionally Active Genes

The strand-specific repair of transcriptionally active genes has been documented in many organisms, ranging from *E. coli* to mammalian cells (90–103). The phenomenon has also been observed with many but not all types of base damage. However, the extent of the kinetic differences observed between the repair of the template and coding strands varies considerably as a function of both the cell type and the type of base damage. For example, in CHO and human cells ~80% of the CPD were removed from the transcribed strand of the *DHFR* gene within 4 hr with little loss of these lesions from the nontranscribed strand of CHO cells during the same time period (107). This kinetic difference for the repair of CPD in the transcribed and nontranscribed strands is equally striking in the *lacZ* gene of *E. coli* (105). However, in human cells significant repair of CPD from the nontranscribed strand of the *DHFR* gene was observed in addition to the rapid repair of the transcribed strand (107).

With respect to the type of DNA damage, strand selectivity for the repair of CPD is more pronounced than for (6-4) photoproducts (97). The general observation is that lesions that are typically substrates for BER (such as base damage resulting from alkylation of DNA) are not repaired in a strand-specific manner (97). However, experimental evidence conflicts on this issue. No differences in the removal of alkylation damage (measured by the loss of alkali-labile sites in the gene of interest) were detected between the transcriptionally active *DHFR* and *HPRT* genes and the transcriptionally silent Duchenne muscular dystrophy gene of cultured T-lymphocytes following exposure to methylmethane sulfonate (108). In contrast, when cells were exposed to *N*-methyl-*N*-nitrosourea, the *DHFR* gene of CHO cells was cleared of ~60% of alkali-labile lesions during a 24-hr period, whereas the transcriptionally silent *c-fos* gene was not repaired at all (109).

In experiments in which the bias for repair of alkylation damage in the transcribed strand is statistically insignificant, such a bias is nonetheless highly reproducible (110, 111). This may reflect the strand-specific repair of some but not all lesions produced by alkylation treatment. More recent studies of the repair of both 3-methyladenine and 7-methylguanine from the amplified *DHFR* gene of CHO cells following exposure to several different alkylating agents showed no strand bias for the clearance of either lesion (112).

Strand-specific repair of transcribed genes is apparently confined to RNAP II. The repair of rDNA genes does not show a strand bias and is in fact reduced relative to genes transcribed by RNAP II (113–114). Strand-specific repair of genes transcribed by RNAP I has not been extensively investigated.

A strand bias for excision repair does in fact anticipate a bias for DNA damage–induced mutations. After cells were treated with UV radiation or the bulky chemical carcinogen benzo[a]pyrene, more mutations were observed in the slowly repaired nontranscribed strand of the human *HPRT* gene than in the more rapidly transcribed strand (115–118). Similar biases have been observed in the hamster *DHFR* gene (119).

Cells from individuals with Cockayne syndrome (CS) (discussed below) are defective in both the preferential repair of UV radiation damage and in the strand-specific repair of transcriptionally active genes (120, 121). Additionally, CS cells have been reported to be defective in strand-specific repair after exposure of cells to ionizing radiation (122). In contrast, cells from the human hereditary NER-defective disease xeroderma pigmentosum (XP) genetic complementation group C retain the capacity for the preferential repair of transcriptionally active genes [reviewed in (17)].

The evidence shows that strand-specific repair of transcriptionally active genes reflects a kinetic preference for the repair of the template compared to the coding strand of transcriptionally active genes. However, with the limited exceptions noted above, this kinetic preference is modest (two- to fivefold) and is most consistently associated with NER.

Unlike human CS cells, yeast mutants that are defective in genes called *RAD26* (the yeast homolog of the human Cockayne syndrome group B (*CSB*) gene) (123) and *RAD28* (the yeast homolog of the human Cockayne syndrome group A (*CSA*) gene) (P Bhatia & EC Friedberg, unpublished observations) are not abnormally sensitive to killing by UV radiation. These findings, and the general observation of the limited strand bias in human cells, have prompted questions as to what special advantages (if any) cells enjoy by repairing the template strand of transcriptionally active genes more rapidly than the coding strand. The most obvious advantage is the enhanced potential for remedying transcriptional arrest and hence completing the transcription of essential genes. This potential may be unlikely to provide a selective advantage on a population basis unless every cell in the population sustained "hits" in a particular gene. In any event the ability to bypass damage during semiconservative DNA synthesis may eventually yield lesion-free templates for transcription (100). However, if most cells in a population were at risk for hits in any essential gene at evolutionarily significant doses of DNA damage, those cells which could best cope with the potential for transcriptional failure attendant on such damage might indeed enjoy an immediate growth advantage.

Coupling of Transcription and Nucleotide Excision Repair in Escherichia coli

As early as 1987 Hanawalt (92) suggested that a plausible explanation for strand-specific repair is that "a repair complex is physically coupled to the transcription machinery." This notion was systematically explored in *E. coli*, an organism in which the biochemistry of NER is well characterized (see above). Using an in vitro system in which NER of the transcribed strand was inhibited relative to that of the coding strand, Selby & Sancar (124) screened extracts of *E. coli* for a fraction that could relieve such inhibition and also reverse the kinetic bias in favor of the transcribed strand. These studies led to the purification of a protein that, in the presence of UvrABC endonuclease, resulted in a fivefold faster rate of transcription of the template strand of a *tac* transcriptional unit carrying base damage produced by either UV radiation, cisplatin, or psoralen plus UV light (124). This protein is designated transcription repair coupling factor (TRCF). Subsequent studies showed that TRCF is encoded by a gene called *mfd* (for mutation frequency decline), a gene previously characterized by the isolation of a mutant allele (125).

Mfd (TRCF) protein is present in *E. coli* cells at a level of ~500 molecules/cell (126, 127). The predicted amino acid sequence of the cloned *mfd* gene (128) revealed the presence of consensus helicase motifs. However, the purified protein (~130 kDa) has no detectable DNA·DNA or DNA·RNA helicase activity in vitro and a weak DNA-independent ATPase activity (126, 127). This region of the polypeptide is required for binding of the TRCF protein to DNA, and ATP hydrolysis is required for the dissociation of DNA-TRCF complexes (129). The translated sequence of the cloned *mfd* gene also revealed a 140–amino acid domain near the N-terminus with significant homology to a region of *E. coli* UvrB protein. This region of the polypeptide has been implicated in the binding of Mfd protein to UvrA protein, since both Mfd and UvrB proteins have been shown to bind to UvrA in vitro. This implication has been directly confirmed experimentally (129).

Purified TRCF protein specifically interacts with *E. coli* RNAP that is stalled at a site of base damage, and it dissociates the ternary mRNA/DNA/RNAP complex. Amino acid residues 379–571 of TRCF are required for this binding (129). TRCF-mediated release of the RNAP and the truncated transcript also occurs at transcriptional blocks produced by sites of base loss and at sites where proteins are bound to DNA. However, the protein does not affect rho-dependent or rho-independent transcription termination (130).

Recent studies have provided insights into the possible mechanism by which Mfd (TRCF) protein stimulates the rate of repair of the transcribed strand. Purified UvrA protein binds to a Mfd affinity column (127). Additionally, when the UvrA$_2$UvrB$_1$ complex was applied to such a column only UvrA protein bound. These observations suggest that, after binding to stalled RNAP, Mfd protein releases the polymerase and the aborted transcript from the template DNA and recruits the UvrA$_2$UvrB$_1$ complex to sites of base damage by binding to the UvrA moiety of the complex. Mfd then dissociates this complex, allowing UvrB protein to bind tightly to the damaged DNA and to initiate strand-specific incision (127). Under conditions that permit transcription in vitro, stimulation of repair in the template strand affects only those lesions located 15 or more nucleotides downstream of the transcriptional start site, suggesting that TRCF only couples NER to transcription when the *E. coli* RNAP is in an elongation mode (127).

How these in vitro observations relate to strand-specific NER in vivo in *E. coli* remains to be established. Space constraints preclude a detailed discussion of mutation frequency decline, but the phenomenon is a decrease in UV radiation–induced mutation frequency observed under specific growth conditions that are believed to reflect enhanced NER of premutational damage of suppressor mutations in tRNA genes. However, the *mfd* mutant strain is about five times more mutable by UV radiation for both suppressor and nonsuppressor mutations.

Coupling of Transcription and Nucleotide Excision Repair in Eukaryotes

Two genetic complementation groups have been defined for CS: CS-A and CS-B (131). A human gene originally called *ERCC6* (132) was cloned by phenotypic correction of the UV radiation sensitivity of a rodent cell line from genetic complementation group 6 of a series of UV radiation–sensitive rodent lines. The *ERCC6* gene was shown to also correct the cellular phenotypes of CS-B cells (133). Additionally, mutations in the human *ERCC6* gene have been detected in CS-B individuals (133; D Mallery & AR Lehmann, unpublished data). Hence the *ERCC6* gene has been renamed *CSB* (134). Examination of the amino acid sequence of the translated *CSB* gene revealed some similarity to the *E. coli mfd* gene; in particular the presence of consensus ATPase/helicase motifs (128). This observation, coupled with the observation of defective strand-specific repair in CS cells, has led to the suggestion that CSB protein may function as a TRCF in human cells (11, 100, 103, 126, 127). However, no direct biochemical evidence supports this theory at present. Fibroblasts from the Li-Fraumeni syndrome that are homozygous for mutations in the *p53* gene have recently been reported to be defective in NER of bulk

DNA and in the coding strand of transcriptionally active genes, although they retain normal levels of repair of the transcribed strand of such genes (134a). This observation suggests that transcriptionally coupled NER is biochemically distinct from the mode that processes base damage in transcriptionally silent DNA.

In an attempt to demonstrate transcriptionally dependent NER in eukaryotic cells, transcription-competent extracts of human cells were incubated in the presence of randomly damaged plasmid DNA carrying the human CMV promoter. A small (~twofold) increase in repair synthesis was observed in a 650 bp fragment located immediately downstream of the promoter compared to the same fragment in a plasmid deleted for the promoter (G Dianov & EC Friedberg, unpublished observations). This increase was not observed in extracts of CS cells (G Dianov & EC Friedberg, unpublished observations). However, whether this observation specifically reflects repair of the template strand, or whether the increased repair is directly coupled to transcription, is not known.

Some RNAP II Basal Transcription Proteins Are Required For Nucleotide Excision Repair In Eukaryotes

In recent years studies in both human cells and *S. cerevisiae* have established another relationship between transcription and NER. Specifically, most if not all of the subunits of TFIIH are also indispensable for NER (27, 33, 51, 135–137). It is intuitively compelling to conceptually link this biochemical observation to the biological observation of strand-specific repair in transcriptionally active genes. However, as indicated above, there is no direct evidence that they should be linked, and what biological significance, if any, may be assigned to the fact that subunits of the transcription initiation factor TFIIH are also indispensable for NER is not clear.

The essential features of TFIIH were summarized above. The first documented indication that components of this multiprotein complex are required for NER came from the demonstration that the 89 kDa subunit of human TFIIH is the product of a previously cloned gene called *ERCC3* (46). Like the *ERCC6* gene mentioned above, *ERCC3* was isolated by functional complementation of the UV radiation sensitivity of a mutant rodent cell line known to be defective in NER, this time from genetic complementation group 3 (138). The *ERCC3* gene was shown to specifically and uniquely correct the NER-defective phenotype of cells from the XP-B genetic complementation group of the human hereditary disease xeroderma pigmentosum (XP) (139), a disease characterized by defective NER [reviewed in (17)]. The *ERCC3* gene is now called *XPB* (134). Direct evidence that XPB protein is required for NER came from the demonstration that microinjection of purified TFIIH into XP-B cells corrected defective NER in these cells in culture (140). Purified TFIIH also corrected defective NER in cell-free extracts of XP-B cells (140).

An independent study showed that the 85 kDa subunit of yeast TFIIH is the product of another NER gene called *RAD3* (59), the yeast homolog of the human *XPD* gene (141), which is also involved in XP [reviewed in (17)]. Direct evidence that Rad3 protein is required in NER came from the demonstration that purified yeast core TFIIH corrected defective NER in extracts of various *rad3* mutants (142). The additional demonstration that purified Rad3 protein alone did not correct defective NER in vitro indicates that this protein is stably bound in TFIIH (142). In an in vitro system for monitoring NER (143), the yeast Ssl2 (142) and Ssl1 and Tfb1 proteins (144) [all known subunits of TFIIH (see discussion above)], were shown to be necessary for NER. The Tfb2 and Tfb3 proteins of yeast core TFIIH have not been directly implicated in NER, but like the other four components of the core complex they are expected to be indispensable to this process. Related experiments in human cells have shown that the XPD (ERCC2), p44, and p34 protein subunits of TFIIH (see discussion above) are also required for NER (47, 48, 50).

Other RNAP II basal transcription factors show no indication of involvement in NER. However, to determine definitively whether these factors are involved in NER mutants each of the multiple genes that encode such proteins must be systematically evaluated for a NER-defective phenotype. TFIIE has been reported to negatively regulate the helicase activity of TFIIH in extracts of human cells through a direct interaction with the XPB (ERCC3) subunit (50). However, a conditional-lethal mutant of the large subunit of yeast TFIIE is not abnormally sensitive to UV radiation and is not defective in NER in vitro (Z Wang, M Sayre, X Wu & EC Friedberg, unpublished observations). This result is supported by studies showing that extracts of human cells depleted of TFIIE carry out NER at normal rates (145).

Microinjection of antibodies against the human MO15 kinase protein into normal human fibroblasts in culture caused a decrease in NER in these cells (49). However, this experiment does not provide direct evidence for a requirement of MO15 protein in NER because the antibody may have resulted in depletion of other proteins associated with MO15. Temperature-sensitive mutations in the homologous yeast *KIN28* gene yielded a very modest increase in UV sensitivity (Z Wang, G-J Valay, G Faye, X Wu & EC Friedberg, unpublished observations), which does not approach the sensitivity observed in strains carrying mutations in genes which encode proteins that are known to be required for NER. Furthermore, direct measurement of NER in vitro in several *kin28-ts* mutants at the restrictive temperature failed to reveal a significant defect (Z Wang, G-J Valay, G Faye, X Wu & EC Friedberg, unpublished observations). The *MO15* (*KIN28*) gene is unlikely to play an indispensable role in NER. However, the possibility that mutations in this and/or other components of the TFIIH complex that participate in transcription initiation may indirectly perturb the function of TFIIH in NER, thereby leading to modest NER-defective phenotypes, cannot be ruled out.

HOW DOES TFIIH OPERATE IN NUCLEOTIDE EXCISION REPAIR?

TFIIH Is Required for Nucleotide Excision Repair in Transcriptionally Silent DNA

The cell-free system used to characterize the NER function of TFIIH components of *S. cerevisiae* measures the repair of plasmid DNA containing defined types of DNA damage (143). This system operates in the apparent absence of active transcription as evidenced by the following criteria (Z Wang & EC Friedberg, unpublished observations). First, yeast extracts are extensively dialyzed and hence devoid of ribonucleoside triphosphates. Second, the plasmid DNA used as a substrate for repair does not contain yeast RNAP II promoters. Third, normal levels of NER were observed in the presence of RNAse A. Fourth, no incorporation of radiolabel into RNA was observed, and no transcripts were detected by gel electrophoresis and autoradiography when incubations were carried out in the presence of [^{32}P]UTP under conditions optimal for NER. Hence, this in vitro NER system is apparently not coupled to or dependent on RNAP II transcription, leading to the conclusion that TFIIH is required for NER in DNA that is transcriptionally silent. Whether an additional transcriptionally-dependent NER mode exists in yeast and higher eukaryotes remains to be determined.

Studies in yeast demonstrated a specific interaction between purified core TFIIH and in vitro–translated Rad2 and Rad4 proteins (147), both of which are indispensable for NER. Neither *RAD2* nor *RAD4* are essential yeast genes [(reviewed in (17)]. Hence, they are presumably not absolutely required for RNAP II transcription. Also, Rad2 protein has been shown to interact with both Tfb1 protein (the 73 kDa subunit of core TFIIH) and, through a different interacting domain, Ssl2 protein (the 105 kDa subunit of core TFIIH). Purified core TFIIH is typically depleted of Ssl2 protein. However, Ssl2 protein is required for NER in vitro (142) and in vivo (148) and is also essential for RNAP II transcription in a fully reconstituted system (40, 41). Furthermore, the human homolog of Ssl2 protein [XPB (ERCC3)] is an integral component of human TFIIH (46). Hence, Ssl2 protein likely dissociates from core TFIIH during purification, and in fact a form of core TFIIH that includes Ssl2 protein has been isolated (142).

Purified Rad1 and Rad10 proteins do not interact with purified core TFIIH in vitro. However, extensive purification of yeast extracts has yielded fractions that contain all the components of core TFIIH (including Ssl2 protein) as well as Rad1, Rad10, Rad2, Rad4, and Rad14 proteins (41). These observations suggest the presence of a preassembled repairosome complex in yeast comprising at least 11 polypeptides (41). Studies also suggest the existence of a repairosome complex that includes TFIIH in human cells (9, 49, 50, 149, 150), though it has not been established that such a complex is preassembled in the absence of active NER. Additionally, partially purified TFIIH from human cells has been shown to complement defective NER in extracts of XP-C cells (50). Specific interactions between human XPG protein (the human homolog of yeast Rad2 protein) and multiple components of TFIIH (N Iyer, MS Reagan, KJ Wu, B Canagarajah & EC Friedberg, unpublished observation), as well as between XPA protein (the human homolog of yeast Rad14 protein) and components of TFIIH (145; R Legerski, personal communication), have been demonstrated. Additionally, XPA protein has been shown to interact with ERCC1 protein (the human homolog of yeast Rad10) (150a).

The Role of TFIIH in Nucleotide Excision Repair in Transcriptionally Active Genes

The specific role(s) of the putative preassembled repairosome in NER in yeast is not known. No direct evidence shows that transcriptionally-independent NER detected in the cell-free system mentioned above is specifically mediated by this preassembled repairosome. An additional or alternative role for the repairosome is its participation in NER in the template strands of transcriptionally active units, a possibility equally lacking in experimental support, especially given the current absence of an in vitro system which specifically monitors such repair. Nonetheless, the possibility that a preassembled NER complex might be specifically recruited to sites of arrested transcription raises an interesting conundrum. If the dual function of TFIIH proteins in NER and transcription is indeed the biochemical correlate of strand-specific NER, one might anticipate that this pathway of excision repair exploits the obligatory loading of TFIIH onto the template strand during transcription initiation, i.e. TFIIH remains associated with RNAP II during transcriptional elongation. Such a scenario would place these NER proteins in the immediate proximity of sites of base damage when transcription was arrested, making them a potential nucleation site for the assembly of a complete functional NER complex.

This model would provide a good explanation for the more rapid repair of the template compared to the coding strand of transcriptionally active genes. However, as indicated above, several studies show that the components of TFIIH are not involved in transcript elongation (66, 66a). Hence, any model of transcriptionally coupled NER requires that TFIIH must somehow be re-recruited to the template strand following arrested transcription. The requirements for the resumption of transcription after stalling of RNAP II at sites of base damage may recapitulate some of the requirements for transcription initiation at natural promoters. Hence, TFIIH may have a special affinity for arrested RNAP II complexes, providing for the rapid delivery of NER proteins to such sites. TFIIH may be recruited to sites of arrested transcription as the core complex around which a repairosome is then assembled. However, the observation of a preassembled repairosome in extracts of yeast cells suggests the alternative possibility that following transcriptional arrest TFIIH returns to the template strand as part of a larger repairosome (135).

Several interesting ramifications of the latter model of strand-specific repair merit consideration. First, the model does not require (though it certainly does not exclude) additional components for the coupling of transcription to NER (as is the case in *E. coli.*), in which case human CSB (ERCC6) protein may not be the functional homolog of the *E. coli* Mfd protein. On the other hand, CSA protein has been shown to interact with the p44 subunit of TFIIH in vitro (150b), and CSB protein has been shown to interact with XPG protein in vitro (N Iyer, MS Reagan, KJ Wu, B Canagarajah & EC Friedberg, unpublished observations). Furthermore, CSA and CSB proteins have been shown to interact with each other both in vitro and in vivo (150b). These interactions have interesting potential implications for the regulation of transcription (see "The Transcription Hypothesis of Human Hereditary Diseases"). However, in the present context these interactions have equally interesting potential implications for the coupling of NER to sites of arrested transcription. Second, the observation that TFIIH is apparently shared by two different complexes, one of which (holoTFIIH) is dedicated to transcription initiation and the other (the repairosome) to NER, suggests a means for limiting transcription initiation in the presence of DNA damage (41, 135).

Rad3 (XPD) and Ssl2 (XPB) Proteins Are DNA Helicases

Of the six components of yeast core TFIIH, catalytic functions have been identified for Rad3 and Ssl2 proteins, both of which have been shown to be DNA-dependent ATPases with DNA helicase activity (151–153). The Rad3 (human XPD) helicase has a $5'->3$ directionality with respect to the strand on which it translocates (151, 152), whereas the Ssl2 (human XPB) helicase has the opposite directionality (153). Mutational inactivation of the Rad3 helicase activity is not lethal to yeast cells, but it is correlated with inactivation of NER. Hence, the helicase is presumably required for the repair function of this protein. Mutational inactivation of the helicase function of Ssl2 protein is lethal (154). This helicase is therefore almost certainly required for transcription, but may also participate in NER. Both helicases may participate in repair by facilitating unwinding of the duplex at sites of base damage. Ssl2, but not Rad3 protein, may play a similar role during transcription initiation and promoter clearance. The observation that TFIIE [which negatively regulates the helicase activity of TFIIH (50)] leaves the promoter prior to transcription elongation (66, 66a) is consistent with such a role.

The helicase function of Rad3 protein is inhibited by many types of base damage in the strand on which it translocates in vitro, but not by base damage on the opposite strand (155–157). This observation has led to the suggestion that Rad3 protein participates in the process of base damage recognition (155–157). This function may be restricted to the repair of the nontemplate strand of transcriptionally active genes and to transcriptionally silent DNA.

The biological implications of the dual function of the TFIIH proteins remain unclear. The most obvious explanation for this evolutionary development is that it directly couples NER to arrested transcription at sites of base damage, although this theory has not been verified experimentally. Given the uncertain biological significance of strand-specific NER in general, it is imperative to entertain the less interesting notion that eukaryotic cells simply utilize the same proteins in functionally distinct multi-protein complexes. Hence, these proteins may have evolved distinct functions that are appropriate to the particular complexes in which they are assembled.

THE TRANSCRIPTION HYPOTHESIS OF HUMAN HEREDITARY DISEASES

Regardless of whether the bifunctional role of proteins in RNAP II transcription and general NER has special utility for cells following exposure to DNA damage, this bifunctionality has led to interesting new insights about the possible pathogenesis of several hereditary human diseases in which NER genes are directly implicated. Noteworthy examples are the closely related syndromes referred to by the acronyms BIDS, IBIDS, and PIBIDS [158; reviewed in (17)]. BIDS is an acronym for a clinical state characterized by brittle hair, impaired intelligence, decreased fertility and short stature [158; reviewed in (17)]. IBIDS refers to the additional presence of ichthyosis (fish-like scaliness of the skin), PIBIDS the added feature of photosensitivity. In a clinical subtype of PIBIDS, often referred to as trichothiodystrophy (TTD), the features of brittle hair and photosensitivity are especially prominent.

Some (but not all) individuals with TTD are defective in NER, and this deficiency has been correlated at the molecular level with mutations in the NER/transcription genes *XPB* and *XPD* (159–160). Mutational inactivation of the NER function of XPD protein provides an explanation for the NER-defective phenotype. However, unlike patients with XP {the other known hereditary NER-defective disease [reviewed in (17)]}, TTD patients are not extremely prone to sunlight-induced skin cancer. Brittle hair, sterility, impaired intelligence, and growth retardation are difficult to rationalize on the basis of defective NER. This conundrum has led to the hypothesis that these and other clinical features of TTD (and perhaps of related syndromes) result from subtle defects in the transcription function of the XPB and XPD proteins and possibly other proteins endowed with bifunctional roles in RNAP II transcription and NER [161–163, reviewed in (163a)].

The *XPB* and *XPD* genes have also been implicated in a disease state in which the clinical features of XP are combined with those of Cockayne syndrome (CS). Like TTD, CS is characterized by diverse clinical features which are not obviously related to defective NER. Postnatal growth retardation leading to an appearance of so-called cachectic dwarfism is particularly prominent (164). The notion that defective transcriptional activity of the *XPB* and *XPD* genes may lead to the diverse clinical features of CS associated with XP and to features of TTD is attractive. However, the nature of such defects and their specific consequences for the biochemistry of transcription remain to be determined. Since the XPB and XPD proteins are clearly implicated in the initiation of RNAP II transcription, certain mutations in these proteins might alter the rate and/or extent of expression of a particular subset of genes that are critical for specific stages of development. Recent studies have shown that extracts of XP-B/CS cells are unable to support normal levels of RNAP II transcription from various promoters in vitro (G Dianov, N Iyer & EC Friedberg, unpublished observations).

In addition to the form of CS that is complicated by the clinical features of XP, certain individuals suffer from clinically "pure" CS [164; reviewed in (17)]. As indicated above, these individuals fall into two genetic complementation groups designated CS-A and CS-B. The *CSB* gene encodes a protein that resembles *E. coli* TRCF at the amino acid sequence level. CSB protein also bears a more pronounced resemblance to a family of proteins of which the yeast Snf2 protein and its human homolog Hsnf are notable examples (133). Yeast Snf2 is a subunit of the multi-protein SWI/SNF complex known to be involved in transcriptional activation (165–168). This scenario provides a possible explanation for how defective transcription could result in CS.

Consistent with this notion the *CSA* gene encodes a protein belonging to the WD-repeat family (150b), which comprises proteins required for diverse aspects of metabolism in eukaryotic cells, including transcription [reviewed in (170)]. The CSA and CSB proteins interact both in vitro and in the yeast two-hybrid system (150b). Additionally, CSA protein interacts with the p44 subunit of human TFIIH in vitro (150b) and CSB protein interacts with XPG protein (N Iyer, MS Reagan, KJ Wu, B Canagarajah & EC Friedberg, unpublished observations). As with extracts of XP-B/CS cells, those derived from

CS-A and CS-B cells are compromised with respect to RNAP II transcription in vitro (G Dianov, N Iyer & EC Friedberg, unpublished observations).

An observation that is more difficult to reconcile with the "transcription hypothesis" is that some patients with the combined clinical features of CS and XP belong to XP complementation group G. XPG protein encodes an endonuclease which is required for incision of DNA during general NER (22, 171). No experimental evidence supports a role for this protein in RNAP II transcription, however. Since yeast (and possibly human) TFIIH is common to both a repairosome complex and a complex required for RNAP II basal transcription (41), these complexes may exist in a regulated equilibrium, in which case mutations in any of the subunits of either complex might alter the function of the other. Thus, mutations in proteins that are functionally required for NER might result in secondary transcriptional defects and vice versa. By this argument one might anticipate that clinical features unrelated to defective NER might complicate most if not all cases of XP involving genes that encode components of the repairosome. Most cases of XP do in fact manifest with neurological complications that are not obviously related to defective NER [reviewed in (17)]. Further studies are required to establish whether this model accounts for the fact that mutations in the *XPG* gene can result in the clinical features of CS, whereas mutations in the *XPA, XPE,* and *XPF* genes result only (apparently) in neurological problems. Not all XP-D and XP-G patients exhibit the typical clinical features of CS. But examining such individuals, as well as patients from the XP-A, XP-E, and XP-F complementation groups, for subtle clinical features of CS that may have been ignored might be informative.

I have emphasized possible transcription defects as the underlying basis for the pathology of CS and TTD. However, CS cells are also abnormally sensitive to UV light and manifest a defect in strand-specific repair (11, 120, 121, 172, 173). Defective strand-specific repair has also been reported in CS cells exposed to ionizing radiation (122). Additionally, CS cells have been reported to be deficient in the repair of ribosomal RNA genes (174). These biochemical phenotypes are difficult to reconcile with the complex clinical features of CS, especially because the features of CS, unlike XP, are not linked to exposure to a known form of DNA damage. These features may result from defective strand-specific repair of spontaneous DNA damage such as that resulting from oxidative metabolism (11). The transcription and repair defects of CS cannot be simply reconciled at present. The CSA and CSB proteins, like some of the XP proteins, may have dual roles in transcription and DNA repair. Alternatively, transcriptional defects may result in secondary defects in repair when CS cells are exposed to agents such as UV radiation.

In addition to CS, other provocative relationships between transcription and repair are emerging in the literature. Recent studies (175, 176) have shown that p53 protein can bind to the XPB and XPD subunits of TFIIH. The protein has also been shown to bind to sites of insertion/deletion mismatches in DNA (177), and its binding to a p53 response element in supercoiled DNA is stimulated in the presence of short single-stranded regions (178). Cells from an individual with the cancer-prone hereditary Li-Fraumeni syndrome, carrying a heterozygous mutation in the *p53* gene, were found to have a significantly reduced rate of repair of the transcriptionally active *DHFR* gene relative to normal cells following exposure of the cells to UV radiation (175). NER is also reduced in vitro in p53 mutant cell lines (179). As indicated above, Li-Fraumeni cells carrying mutations in both copies of the *p53* gene are defective in global NER and in the repair of the coding strand of transcriptionally active genes, but they retain normal NER of the transcribed strand of such genes (134a).

SUMMARY AND PERSPECTIVES

Multiple distinct relationships are evident between excision repair and transcription by RNAP II. Regions of the genome in which the configuration of chromatin is in an "open" state because of active transcription are apparently more accessible to the repair machinery than regions that are in a more "closed" state. Specific mechanisms also appear to provide for more rapid repair of the template than does the coding strand of transcriptionally active genes. Further research is required to decipher the biochemical mechanism by which this strand-specific preference is effected. Arrested transcription at sites of base damage likely couples the repair machinery to the transcription elongation complex. Several mechanisms for such coupling have been postulated. A repair complex may be assembled in steps around a core unit that is integral to the elongation machinery. Alternatively, a preassembled repairosome may have special affinity for stalled RNAP II molecules, effectively recapitulating the process of transcription initiation. The coupling of a repair apparatus to sites of stalled transcription may alternatively or additionally involve one or more dedicated coupling factors for which there is good evidence in *E. coli.* The definition of a NER mode in eukaryotes that is specifically coupled to RNAP II transcription is an important biochemical challenge for the future. Finally, the discovery that components of TFIIH have an obligatory role in NER provides an important new dimension for investigating the molecular basis of hereditary diseases such as CS and TTD, for which the complex clinical phenotypes cannot be obviously reconciled with defective DNA repair. These diseases and possibly others require detailed exploration at the level of RNAP II transcription. The experimental validation of the transcription hypothesis may provide a new dimension for defining the molecular pathology of human hereditary diseases.

ACKNOWLEDGMENTS

I thank my laboratory colleagues as well as WJ Feaver for numerous helpful discussions and insightful review of the manuscript. Special thanks are due to RD Kornberg, WJ Feaver, and JQ Svejstrup for extensive experimental and intellectual collaboration and for permission to cite unpublished observations, and to Z Wang for his unique contributions to many of the studies reviewed here. I also wish to acknowledge the collaborative efforts of AE Tomkinson, RD Wood, CS West, G Faye, J-G Valay, M Sayre, and CC Harris. I apologize to those authors whose pertinent work may not have been cited due to space constraints. Studies from the author's laboratory were supported by research grants CA12428 and CA44247 from the United States Public Health Service.

Literature Cited

1. Sancar A, Sancar GB. 1988. *Annu. Rev. Biochem.* 57:29–67
2. Prakash S, Sung P, Prakash L. 1993. *Annu. Rev. Genet.* 27:33–70
3. Sancar A, Tang MS. 1993. *Photochem. Photobiol.* 57:905–21
4. Lindahl T. 1993. *Nature* 362:709–15
5. Barnes D, Lindahl T, Sedgwick B. 1993. *Curr. Opin. Cell Biol.* 5:424–33
6. Hoeijmakers JHJ. 1993. *Trends Genet.* 9:173–77
7. Hoeijmakers JHJ. 1993. *Trends Genet.* 9:211–17
8. Aboussekhra A, Wood RD. 1994. *Curr. Opin. Genet. Dev.* 4:212–20
9. Sancar A. 1994. *Science* 266:1954–56
10. Sweder K. 1994. *Curr. Genet.* 27:1–16
11. Hanawalt PC. 1994. *Science* 266:1957–58
12. Modrich P. 1994. *Science* 266:1959–60
13. Jeggo P, Carr AM, Lehmann AR. 1994. *Int. J. Radiat. Biol.* 66:573–77
14. Wallace SS. 1994. *Int. J. Radiat. Biol.* 66:579–89
15. Friedberg EC, Bardwell AJ, Bardwell L, Wang Z, Dianov G. 1994. *Mutat. Res.* 307:5–14
16. Friedberg EC. 1994. *BioEssays* 16:645–49
17. Friedberg EC, Walker GW, Siede W. 1995. *DNA Repair and Mutagenesis.* Washington, DC: Am. Soc. Microbiol. 648 pp.
17a. Guzder S, Habraken Y, Sung P, Prakash L, Prakash S. 1995. *J. Biol. Chem.* 270:12973–76
18. Lin JJ, Sancar A. 1992. *Mol. Microbiol.* 6:2219–24
19. Grossman L, Thiagalingam S. 1993. *J. Biol. Chem.* 268:16871–74
20. Van Houten B, Snowden A. 1993. *BioEssays* 15:51–59
21. Bardwell AJ, Bardwell L, Tomkinson AE, Friedberg EC. 1994. *Science* 265:2082–85
22. O'Donovan A, Davies AA, Moggs JG, West SC, Wood RD. 1994. *Nature* 371:432–35
23. Hoeijmakers JHJ, Bootsma D. 1994. *Nature* 371:654–55
23a. Davies AA, Friedberg EC, Tomkinson AE, Wood RD, West S. 1995. *J. Biol. Chem.* In press
24. Bowman KK, Sidik K, Smith CA, Taylor J-S, Doetsch PW Fryer GA. 1994. *Nucleic Acids Res.* 22:3026–32
24a. Yajima H, Takao M, Yasuhira S, Zhao JH, Ishii C, et al. 1995. *EMBO J.* 14:2393–99
25. Zawel L, Reinberg D. 1993. *Prog. Nucleic Acid Res. Mol. Biol.* 44:67–108
26. Conaway RC, Conaway JW. 1993. *Annu. Rev. Biochem.* 62:161–90
27. Drapkin R, Reinberg D. 1994. *Trends Biochem. Sci.* 19:504–8
28. Eick D, Wedel A, Heumann H. 1994. *Trends Genet.* 10:292–96

29. Paranjape SM, Kamakaka RT, Kadonaga J. 1994. *Annu. Rev. Biochem.* 63: 265–97
30. Dahmus ME. 1994. *Prog. Nucleic Acid Res. Mol. Biol.* 48:143–79
31. Buratowski S. 1994. *Cell* 77:1–3
32. Chalut C, Moncollin V, Egly J-M. 1994. *BioEssays* 16:651–55
33. Buratowski S. 1993. *Science* 260:37–38
34. Wolffe AP. 1994. *Cell* 77:13–16
35. Wolffe AP. 1994. *Trends Biochem. Sci.* 19:240–44
36. Tjian R, Maniatis T. 1994. *Cell* 77:5–8
37. Sheldon M, Reinberg D. 1995. *Curr. Biol.* 5:43–46
38. Zawel L, Reinberg D. 1995. *Annu. Rev. Biochem.* 64:533–61
39. Feaver WJ, Svejstrup JQ, Henry LH, Kornberg RD. 1994. *Cell* 79: 1103–9
40. Svejstrup JQ, Feaver WJ, Lapoint J, Kornberg RD. 1994. *J. Biol. Chem.* 269: 28044–48
41. Svejstrup JQ, Wang Z, Feaver WJ, Wu X, Bushnell DA, et al. 1995. *Cell* 80:21–28
42. Koleske AJ, Young RA. 1994. *Nature* 368:466–69
43. Carey M. 1994. *Nature* 368:402–3
44. Koleske AJ, Young RA. 1995. *Trends Biochem. Sci.* 20:113–16
45. Thompson CM, Young RA. 1995. *Proc. Natl. Acad. Sci. USA* 92:4587–90
46. Schaeffer L, Roy R, Humbert S, Moncollin V, Vermeulen W, et al. 1993. *Science* 260:58–63
47. Humbert S, van Vuuren H, Lutz Y, Hoeijmakers JHJ, Egly J-M, Moncollin V. 1994. *EMBO J.* 13:2393–98
48. Schaeffer L, Moncollin V, Roy R, Staub A, Mezzina M, et al. 1994. *EMBO J.* 13:2388–92
49. Roy R, Adamczewski JP, Seroz T, Vermeulen W, Tassan J-P, et al. 1994. *Cell* 79:1093–101
50. Drapkin R, Reardon JT, Ansari A, Huang JC, Zawel L, et al. 1994. *Nature* 368:769–72
51. Drapkin R, Sancar A, Reinberg D. 1994. *Cell* 77:9–12
51a. Shiekhattar R, Mermelstein F, Fisher RP, Drapkin R, Dynlacht B, et al. 1995. *Nature* 374:283–87
52. Morgan DO. 1995. *Nature* 374:131–34
53. Fesquest D, Labbé J-C, Derencourt J, Capony J-P, Galas S, et al. 1993. *EMBO J.* 12:3111–21
54. Poon RYC, Yamashita K, Adamczewski JP, Hunt T, Shuttleworth J. 1993. *EMBO J.* 12:3123–32
55. Fisher RP, Morgan DO. 1994. *Cell* 78: 713–24
56. Mäkelä TP, Tassan J-P, Nigg EA, Frutiger S, Hughes GJ, Weinberg RA. 1994. *Nature* 371:254–57
57. Tassan J-P, Schultz SJ, Bartek J, Nigg EA. 1994. *J. Cell Biol.* 127:467–78
57a. Serizawa H, Mäkelä TP, Conaway JC, Conaway RC, Weinberg RA, Young R. 1995. *Nature* 374:280–82
58. Feaver WJ, Gileadi O, Kornberg RD. 1991. *J. Biol. Chem.* 266:19000–5
59. Feaver WJ, Svejstrup JQ, Bardwell L, Bardwell AJ, Buratowski S, et al. 1993. *Cell* 75:1379–87
60. Gulyas KD, Donahue TF. 1992. *Cell* 69:1031–42
61. Park E, Guzder SN, Koken MHM, Jaspers-Dekker I, Weeda G, et al. 1992. *Proc. Natl. Acad. Sci. USA* 89: 11416–20
62. Naumovski L, Friedberg EC. 1983. *Proc. Natl. Acad. Sci. USA* 80:4818–21
63. Gileadi O, Feaver WJ, Kornberg RD. 1992. *Science* 257:1389–92
64. Yoon H, Miller SP, Pabich EK, Donahue TF. 1992. *Genes Dev.* 6:2463–77
65. Valay JG, Simon M, Faye G. 1993. *J. Mol. Biol.* 234:307–10
65a. Cismowski MJ, Laff GM, Solomon MJ, Reed SI. 1995. *Mol. Cell. Biol.* 15:2983–92
65b. Valay J-G, Simon M, Dubois M-F, Bensaude O, Facca C, Faye G. 1995. *J. Mol. Biol.* 249:535–44
66. Goodrich JA, Tjian R. 1994. *Cell* 77: 145–56
66a. Zawel L, Kumar KP, Reinberg D. 1995. *Genes Dev.* 9:1479–90
67. Flores O, Maldonado E, Reinberg D. 1989. *J. Biol. Chem.* 264:8913–21
68. Kerppola TK, Kane CM. 1991. *FASEB J.* 5:2833–42
69. Krauskopf A, Bengal E, Aloni Y. 1991. *Mol. Cell. Biol.* 11:3515–21
70. Bengal E, Flores O, Krauskopf A, Reinberg D, Aloni Y. 1991. *Mol. Cell. Biol.* 11:1195–206
71. Chafin DR, Claussen TJ, Price DH. 1991. *J. Biol. Chem.* 266:9256–62
72. Bradsher JN, Jackson KW, Conaway RC, Conaway JW. 1993. *J. Biol. Chem.* 268:25587–93
73. Kassavetis GA, Geiduschek EP. 1993. *Science* 259:944–45
74. Rudd MD, Izban MG, Luse DS. 1994. *Proc. Natl. Acad. Sci. USA* 91:8057–61
75. Reines D. 1992. *J. Biol. Chem.* 267: 3795–800
76. Lewin B. 1994. *Cell* 79:379–406
77. Shi Y-B, Gamper H, Hearst JE. 1988. *J. Biol. Chem.* 263:527–34
78. Selby CP, Sancar A. 1991. *Proc. Natl. Acad. Sci. USA* 88:11574–78
79. Nath ST, Romano LJ. 1991. *Carcinogenesis* 12:973–76

80. Sastry SS, Hearst JE. 1991. *J. Mol. Biol.* 221:1091–110
81. Sastry SS, Hearst JE. 1991. *J. Mol. Biol.* 221:1111–25
82. Chen Y-H, Bogenhagen DF. 1993. *J. Biol. Chem.* 268:5849–55
83. Zhou W, Doetsch PW. 1993. *Proc. Natl. Acad. Sci. USA* 90:6601–5
84. Masta A, Gray PJ, Phillips DR. 1994. *Nucleic Acids Res.* 22:3880–86
85. Sanchez G, Mamet-Bratley MD. 1994. *Environ. Mol. Mutagen.* 23:32–36
86. Choi D-J, Marino-Alessandri DJ, Geacintov NE, Scicchitano DA. 1994. *Biochemistry* 33:780–87
87. Zhou W, Doetsch PW. 1994. *Biochemistry* 33:14926–34
88. Donahue BA, Yin S, Taylor J-S, Reines D, Hanawalt PC. 1994. *Proc. Natl. Acad. Sci. USA* 91:8502–6
89. Bohr VA, Smith CA, Okumoto DS, Hanawalt PC. 1985. *Cell* 40:359–69
90. Hanawalt PC. 1986. In *Mechanisms of DNA Damage and Repair*, ed. L Grossman, AC Upton, pp. 489–98. New York: Plenum
91. Smith CA. 1987. *J. Cell Sci. Suppl.* 6:225–41
92. Hanawalt PC. 1987. *Environ. Health Perspect.* 76:9–14
93. Bohr VA, Phillips DH, Hanawalt PC. 1987. *Cancer Res.* 47:6426–36
94. Bohr VA. 1987. *Preferential DNA Repair in Active Genes.* Copenhagen: Laegeforeningens
95. Bohr VA, Wassermann K. 1988. *Trends Biochem. Sci.* 13:429–32
96. Bohr VA. 1988. *J. Cell Sci.* 90:175–78
97. Bohr VA. 1991. *Carcinogenesis* 12:1983–92
98. Terleth C, van de Putte P, Brouwer J. 1991. *Mutagenesis* 6:103–11
99. Bohr VA. 1992. See Ref. 180, pp. 217–27
100. Hanawalt PC. 1992. See Ref. 180, pp. 231–42
101. Brouwer J, Bang DD, Verhage R, van de Putte P. 1992. See Ref. 180, pp. 274–83
102. Downes CS, Ryan AJ, Johnson RT. 1993. *BioEssays* 15:209–16
103. Hanawalt PC, Mellon I. 1993. *Curr. Biol.* 3:67–69
104. Smerdon MJ, Thoma F. 1990. *Cell* 61:675–84
105. Mellon I, Hanawalt PC. 1989. *Nature* 342:95–98
106. Selby CP, Sancar A. 1990. *J. Biol. Chem.* 265:21330–36
107. Mellon I, Spivak G, Hanawalt PC. 1987. *Cell* 51:241–49
108. Bartlett JD, Scicchitano DA, Robison SH. 1991. *Mutat. Res.* 255:247–56
109. LeDoux SP, Thangada M, Bohr VA, Wilson GL. 1990. *Cancer Res.* 51:775–79
110. Scicchitano DA, Hanawalt PC. 1989. *Proc. Natl. Acad. Sci. USA* 86:3050–54
111. Scicchitano DA, Hanawalt PC. 1990. *Mutat. Res.* 236:31–37
112. Wang W, Sitaram A, Scicchitano DA. 1995. *Biochemistry* 34:1798–804
113. Vos J-M, Wauthier EL. 1991. *Mol. Cell. Biol.* 11:2245–52
114. Christians FC, Hanawalt PC. 1993. *Biochemistry* 32:10512–18
115. Vrieling H, van Rooijen ML, Groen NA, Zdzienicka MZ, Simons JWIM, et al. 1989. *Mol. Cell. Biol.* 9:1277–83
116. Chen R-H, Maher VM, McCormick JJ. 1990. *Proc. Natl. Acad. Sci. USA* 87:8680–84
117. McGregor WG, Chen R-H, Lukash L, Maher VM, McCormick JJ. 1991. *Mol. Cell. Biol.* 11:1927–34
118. Sage E, Drobetsky EA, Moustacchi E. 1993. *EMBO J.* 12:397–402
119. Carothers AM, Urlaub G, Steigerwalt RW, Chasin LA, Grunberger D. 1986. *Proc. Natl. Acad. Sci. USA* 83:6519–23
120. Hanawalt PC. 1991. *Mutat. Res.* 247:203–11
121. Venema J, Mullenders LHF, Natarajan AT, van Zeeland AA, Mayne LV. 1990. *Proc. Natl. Acad. Sci. USA* 87:4707–11
122. Leadon SA, Cooper PK. 1993. *Proc. Natl. Acad. Sci. USA* 90:10499–503
123. van Gool AJ, Verhage R, Swagemakers SMA, van de Putte P, Brouwer J, et al. 1994. *EMBO J.* 13:5361–69
124. Selby CP, Sancar A. 1991. *Proc. Natl. Acad. Sci. USA* 88:8232–36
125. Witkin EM. 1966. *Science* 152:1345–53
126. Selby CP, Sancar A. 1993. *J. Bacteriol.* 175:7509–14
127. Selby CP, Sancar A. 1994. *Microbiol. Rev.* 58:317–29
128. Selby CP, Sancar A. 1993. *Science* 260:53–58
129. Selby CP, Sancar A. 1995. *J. Biol. Chem.* 270:4882–89
130. Selby CP, Sancar A. 1995. *J. Biol. Chem.* 270:4890–95
131. Lehmann AR. 1982. *Mutat. Res.* 106:347–56
132. Troelstra C, Odijk H, de Wit J, Westerveld A, Thompson LH, et al. 1990. *Mol. Cell. Biol.* 10:5806–13
133. Troelstra C, van Gool A, de Wit J, Vermeulen W, Bootsma D, Hoeijmakers JHJ. 1992. *Cell* 71:939–53
134. Lehmann AR, Bootsma D, Clarkson SG, Cleaver JE, McAlpine PJ, et al. 1994. *Mutat. Res.* 315:41–42
134a. Ford JM, Hanawalt PC. 1995. *Proc. Natl. Acad. Sci. USA* 92:8876–80

135. Friedberg EC, Bardwell AJ, Bardwell L, Feaver WJ, Kornberg RD, et al. 1995. *Philos. Trans. R. Soc.* London Ser. B 347:63–68

136. Hanawalt PC, Donahue BA, Sweder K. 1994. *Curr. Biol.* 4:518–21

137. Cleaver JE. 1994. *Cell* 76:1–4

138. Weeda G, van Ham RCA, Masurel R, Westerveld A, Odijk H, et al. 1990. *Mol. Cell. Biol.* 10:2570–81

139. Weeda G, van Ham RCA, Vermeulen W, Bootsma D, van der Eb AJ, Hoeijmakers JHJ. 1990. *Cell* 62:777–91

140. van Vuuren AJ, Vermeulen W, Ma L, Weeda G, Appeldoorn E, et al. 1994. *EMBO J.* 13:1345–53

141. Weber CA, Salazar EP, Stewart SA, Thompson LH. 1990. *EMBO J.* 9:1437–47

142. Wang Z, Svejstrup JQ, Feaver WJ, Wu X, Kornberg RD, Friedberg EC. 1994. *Nature* 368:74–76

143. Wang Z, Wu X, Friedberg EC. 1993. *Proc. Natl. Acad. Sci. USA* 90:4907–11

144. Wang Z, Svejstrup JQ, Wu X, Feaver WJ, Kornberg RD, et al. 1995. *Mol. Cell. Biol.* 15:2288–93

145. Park C-H, Mu D, Reardon JT, Sancar A. 1995. *J. Biol. Chem.* 270:4896–902

146. Verhage R, Zeeman A-M, DeGroot N, Gleig F, Bang DD,et al. 1994. *Mol. Cell. Biol.* 14:6135–42

147. Bardwell AJ, Bardwell L, Iyer N, Svejstrup JQ, Feaver WJ, et al. 1994. *Mol. Cell. Biol.* 14:3569–76

148. Sweder KS, Hanawalt PC. 1994. *J. Biol. Chem.* 269:1852–57

149. Mu D, Park C-H, Matsunaga T, Hsu DS, Reardon JT, Sancar A. 1995. *J. Biol. Chem.* 270:2415–18

150. Aboussekhra A, Biggerstaff M, Shivji MKK, Vilpo JA, Moncollin V, et al. 1995. *Cell* 80:859–68

150a. Li L, Elledge SJ, Peterson CA, Bales ES, Legerski R. 1994. *Proc. Natl. Acad. Sci. USA* 91:5012–16

150b. Henning KA, Li L, Legerski R, Iyer N, McDaniel LD, et al. 1995. *Cell.* 82:555–64

151. Sung P, Prakash L, Matson SW, Prakash S. 1987. *Proc. Natl. Acad. Sci. USA* 84:8951–55

152. Harosh I, Naumovski L, Friedberg EC. 1989. *J. Biol. Chem.* 264:20532–39

153. Guzder SN, Sung P, Bailly V, Prakash L, Prakash S. 1994. *Nature* 369:578–81

154. Qui H, Park E, Prakash L, Prakash S. 1993. *Genes Dev.* 7:2161–71

155. Naegeli H, Bardwell L, Friedberg EC. 1992. *J. Biol. Chem.* 267:392–98

156. Naegeli H, Bardwell L, Friedberg EC. 1992. *J. Biol. Chem.* 267:7839–44

157. Naegeli H, Bardwell L, Friedberg EC. 1993. *Biochemistry* 32:613–21

158. Crovato F, Barrone C, Rebora A. 1983. *Br. J. Dermatol.* 108:247–53

159. Broughton BC, Steingrimsdotter H, Weber C, Lehmann AR. 1994. *Nat. Genet.* 7:189–94

160. Stefanini M, Vermeulen W, Weeda G, Giliani S, Nardo T, et al. 1993. *Am. J. Hum. Genet.* 53:817–21

161. Bootsma D, Hoeijmakers JHJ. 1993. *Nature* 363:114–15

162. Friedberg EC, Bardwell AJ, Bardwell L, Wang Z, Dianov G. 1994. *Mutat. Res.* 307:5–14

163. Vermeulen W, van Vuuren AJ, Chipoulet AJ, Schaeffer L, Appeldoorn E, et al. 1995. *Cold Spring Harbor Symp. Quant. Biol.* 59:317–29

163a. Cleaver JE, Hultner ML. 1995. *Am. J. Hum. Genet.* 56:1257–61

164. Nance MA, Berry SA. 1992. *Am. J. Med. Genet.* 42:68–84

165. Carlson M, Laurent BC. 1994. *Curr. Opin. Cell Biol.* 6:396–402

166. Peterson CL, Tamkun JW. 1995. *Trends Biochem. Sci.* 20:143–46

167. Winston F, Carlson M. 1992. *Trends Genet.* 8:387–91

168. Richard-Foy H. 1994. *Nature* 370:417–18

169. Deleted in proof

170. Neer EJ, Schmidt CJ, Nambudripad R, Smith TF. 1994. *Nature* 371:297–300

171. O'Donovan A, Scherly D, Clarkson SG, Wood RD. 1994. *J. Biol. Chem.* 269:15965–68

172. Mullenders LHF, Sakker RJ, van Hoffen A, Venema J, Natarajan AT, van Zeeland AA. 1992. See Ref. 180, pp. 247–54

173. van Hoffen A, Natarajan AT, Mayne LV, van Zeeland AA, Mullenders LHF, Venema J. 1993. *Nucleic Acids Res.* 21:5890–95

174. Christians FC, Hanawalt PC. 1994. *Mutat. Res.* 323:179–87

175. Wang XW, Yeh H, Schaeffer L, Roy R, Moncollin V, et al. 1995. *Nat. Genet.* 10:188–95

176. Wang XW, Forrester K, Yeh H, Feitelson MA, Gu J-R, Harris CC. 1994. *Proc. Natl. Acad. Sci. USA.* 91:2230–34

177. Lee S, Elenbaas B, Levine A, Griffith J. 1995. *Cell* 81:1013–20

178. Jayaraman L, Prives C. 1995. *Cell* 81:1021–29

179. Smith ML, Chen I-T, Zhan Q, O'Conner PM, Fornace AJ Jr. 1995. *Oncogene* 10:1053–59

180. Bohr VA, Wassermann K, Kraemer KH, eds. 1992. *DNA Repair Mechanisms.* Copenhagen: Munksgaard. 428 pp.

Annu. Rev. Biochem. 1996. 65:43–81
Copyright © 1996 by Annual Reviews Inc. All rights reserved

DNA EXCISION REPAIR

Aziz Sancar

Department of Biochemistry and Biophysics,University of North Carolina School of Medicine, Chapel Hill, North Carolina 27599

KEY WORDS: excision nuclease, molecular matchmakers, transcription-repair coupling, DNA damage and cancer, xeroderma pigmentosum, p53 and DNA repair

ABSTRACT

In nucleotide excision repair DNA damage is removed through incision of the damaged strand on both sides of the lesion, followed by repair synthesis, which fills the gap using the intact strand as a template, and finally ligation. In prokaryotes the damaged base is removed in a 12–13 nucleotide (nt)–long oligomer; in eukaryotes including humans the damage is excised in a 24–32 nt–long fragment. Excision in *Escherichia coli* is accomplished by three proteins designated UvrA, UvrB, and UvrC. In humans, by contrast, 16 polypeptides including seven xeroderma pigmentosum (XP) proteins, the trimeric replication protein A [RPA, human single-stranded DNA binding protein (HSSB)], and the multisubunit (7–10) general transcription factor TFIIH are required for the dual incisions. Transcribed strands are specifically targeted for excision repair by a transcription-repair coupling factor both in *E. coli* and in humans. In humans, excision repair is an important defense mechanism against the two major carcinogens, sunlight and cigarette smoke. Individuals defective in excision repair exhibit a high incidence of cancer while individuals with a defect in coupling transcription to repair suffer from neurological and skeletal abnormalities.

CONTENTS

INTRODUCTION 44
DNA REPAIR MECHANISMS 44
   Direct Repair 44
   Base Excision Repair 46
   Nucleotide Excision Repair 46
EXCISION REPAIR IN PROKARYOTES 47
   Damage Recognation by a Molecular Matchmaker 49
   Dual Incisions... 51
   Repair Synthesis 52
EXCISION REPAIR IN HUMANS 52
   Genetics of Human Excision Repair 53
   Structure and Function of Human Excinuclease.......................... 55
   Mechanism of Excision Repair... 61
TRANSCRIPTION-REPAIR COUPLING 67
   Transcription-Repair Coupling in Escherichia coli....................... 68
   Transcription-Repair Coupling in Humans 68

43

CELL CYCLE AND REPAIR... 69
   Excision Repair Potential as a Function of Cell Cycle 70
   Cyclin-Dependent Kinase-Activating Enzyme, Transcription Factor IIH, and
        Excision Repair 70
SOS RESPONSE IN *ESCHERICHIA COLI* AND IN MAN..................... 71
   SOS Regulation of Excision Repair in Escherichia coli 71
   Damage Response and Human Excision Repair............................. 72
SUMMARY AND CONCLUSIONS... 74
   Similarities Between Escherichia coli *and Human Excision Repair Systems*....... 74
   Differences in Excision Repair Between Escherichia coli *and Humans*........... 75

INTRODUCTION

A concept central to cancer biology is that mutations arising in oncogenes and tumor supressor genes as a result of replication errors or DNA damage lead to neoplastic transformation of cells (1). Base mismatches that result from replication errors or that occur during recombination are corrected by mismatch repair systems (2, 3). DNA lesions, which are noncoding or miscoding and include all types of base, deoxyribose, and phosphodiester bond modifications, are eliminated from the duplex by DNA damage–repair enzyme systems (3–5).

In general, mismatch and damage repair systems use the same overall strategy to maintain the integrity of genetic information: A mismatched or damaged nucleotide is removed and replaced by the correct and unmodified nucleotide using the intact strand of the duplex as a template. In addition to this basic mechanism, in direct damage repair the chemical bond(s) constituting the damage can be broken to restore the normal nucleotide. This review presents a brief survey of DNA damage repair mechanisms followed by a detailed analysis of nucleotide excision repair. The subject has been reviewed from different perspectives by several authors (5–11a).

DNA REPAIR MECHANISMS

There are three molecular mechanisms for repairing damaged DNA: direct repair, base excision repair, and nucleotide excision repair.

Direct Repair

In the direct repair mode, the abnormal chemical bonds between bases or between a nucleotide and an abnormal substituent are broken. The following are the currently known enzymes that catalyze direct repair.

1. DNA Photolyase (photoreactivating enzyme). This enzyme repairs cyclobutane pyrimidine dimers induced by ultraviolet light by splitting the cyclobutane ring using a light-initiated electron transfer reaction (12, 13). The crystal structure of the *E. coli* enzyme suggests that upon binding to DNA the enzyme flips the pyrimidine dimer out of the duplex into a hole that contains the catalytic flavin cofactor in the center of the protein (14). In

the absence of activating light, photolyase cannot catalyze the splitting reaction and remains stably bound to the damage. By an unknown mechanism this complex increases the rate of dimer removal by nucleotide excision repair both in *E. coli* (4) and in yeast (15, 16). Photolyase also binds to cisplatin-damaged DNA with relatively high affinity (16). However, the effects of these enzyme-substrate complexes on excision repair of cisplatin adducts in yeast and in *E. coli* are different. In yeast, photolyase inhibits excision repair and sensitizes cells to killing by cisplatin (16). In *E. coli*, photolyase stimulates excision repair and enhances resistance to killing by cisplatin (17). The distribution of photolyase in the biological world is erratic. Although photolyase is generally widespread in nature, many microorganisms, including *Bacillus subtilis* and placental mammals, lack photolyase. In contrast, photolyase is found in *E. coli* and other bacteria, and it is abundant in all tissues of some marsupial mammals such as *Monodelphus domesticus* (18–20). Why photolyase is present in internal animal organs has long been a mystery because the chances of dimer formation by exogenous ultraviolet radiation (UV) at these sites are essentially nil. The finding that *E. coli* photolyase stimulates excision of a nondimer lesion indicates that animal photolyases may play a similar accessory role in excision repair.

2. 6-4 Photoproduct Photolyase. This enzyme repairs the second major UV photoproduct by a light-initiated reaction (21). The enzyme has been found in Drosophila, silkworm, rattlesnake, and frog (21–23) but not in *E. coli*, yeast, or humans (21). During the formation of the 6-4 photoproduct, in addition to the C4-C6 bond formation between adjacent pyrmidines, the substitutent at C4 of one pyrimidine migrates to the C6 of the other pyrimidine. Interestingly, enzymatic photolysis reverses both reactions and hence restores the normal bases (21–23). The photochemistry by which this unusual reaction occurs remains to be elucidated.

3. Spore Photoproduct Lyase. UV irradiation of *B. subtilis* spores, which contain highly dehydrated DNA associated with small acid-soluble spore proteins, generates almost exclusively spore photoproduct (5-thyminyl-5,6-dihydrothymine) rather than cyclobutane dimers (24). This lesion is repaired by spore photoproduct lyase which breaks the C-C bond between the two thymines in a light-independent reaction. The enzyme is a 40 kDa protein with no apparent cofactor (25).

4. O^6-Methylguanine DNA Methyl Transferase. This enzyme is present in all species tested. It transfers the methyl group of O^6-methylguanine and (less efficiently) other alkyl groups at this position to a cysteine residue on the enzyme (26–28). The enzyme appears to play an important role in cellular defense against cancers induced by intrinsic and environmental alkylating agents (29, 30).

Base Excision Repair

In this mode of repair, usually, nonbulky DNA lesions such as uracil, thymine glycols and hydrates, N3-MeAde, and 8-oxo-guanine are removed from DNA in two steps. First, a DNA glycosylase releases the base by cleaving the glycosylic bond connecting the base to the deoxyribose. Second, the abasic sugar [apurinic/apyrimidinic (AP) site] is released by the combined actions of AP lyase and AP endonucleases (3, 5, 31, 32). In humans, there are uracil-, thymine glycol-, methylpurine-, and 8-oxoguanine DNA glycosylases with rather narrow substrate ranges and one major AP endonuclease (5, 31). Following removal of the AP sugar, the one-nucleotide gap is filled in to generate a 1–4-nt repair patch (3, 5, 31).

Nucleotide Excision Repair

The damaged base is removed by hydrolyzing phosphodiester bonds on both sides of the lesion. Two excision mechanisms could accomplish this removal. In removal by the endonuclease-exonuclease mechanism, an endonuclease makes an incision at a phosphodiester bond either 5' or 3' to the lesion, and then an exonuclease digests the damaged strand past the lesion. In removal by the excision nuclease (excinuclease) mechanism, an enzyme system incises phosphodiester bonds on either side of and at some distance from the lesion. The enzymes work in a concerted manner to excise the lesion in a fragment of relatively precise length.

REPAIR BY ENDONUCLEASE-EXONUCLEASE   Two such repair activities are known. In the first, 8-oxoguanine endonuclease (8-oxoG), which was partially purified from human cell free extracts, incises immediately 5' to 8-oxoG and then, presumably, the modified nucleotide is released by an exonuclease (33). The second involves *Schizosaccharomyces pombe* (*S. pombe*) DNA endonuclease (SPDE) or "UV-induced dimer endonculease," which was first detected in *S. pombe* (34, 35) and then in *Neurospora crassa* (36). The *S. pombe* enzyme is encoded by the *rad12* gene (35), and the partially purified protein can complement mutant cell-free extract in a repair synthesis assay. The *N. crassa* gene (*muts-18*) encoding the enzyme has been cloned and sequenced, and the en- zyme has been purified and characterized. The enzyme is a polypeptide of 74 kDa which cleaves the phosphodiester bond immediately 5' to both cyclobutane pyrimidine dimers and 6-4 photoproducts. Mutants lacking the enzyme are sensitive to UV but not to UV-mimetic chemicals, suggesting that the activity is specific to these two UV photoproducts (36).

REPAIR BY EXCISION NUCLEASE   Damage removal via concerted dual incisions on both sides of the lesion by an ATP-dependent enzyme system with an essentially infinite substrate range is the most universal form of nucleotide excision repair. This system has been found in all free-living species tested, from the smallest free-living life form *Mycoplasma genitalium* to humans (11). This enzyme system is called excision nuclease (excinuclease), a term that describes its mode of action and acknowledges its uniqueness to repair (37, 38). For historical as well as practical reasons, the nucleotide excision repair process initiated by excision nucleases is referred to as excision repair (31). Therefore, in this review the terms excision repair and nucleotide excision repair are used interchangeably.

Currently, two types of excision nucleases are known. The prokaryotic type removes damage by incising the 8th phosphodiester bond 5' and the 4th - 5th phosphodiester bond 3' to the lesion, and hence it excises lesions in 12–13 nt–long oligomers (37, 39). The eukaryotic type incises the 20-25th phosphodiester bond 5' and the 3rd - 8th phosphodiester bond 3' to the lesion and thus excises 24–32 nt–long oligomers (38, 40–45). Genetic and biochemical data show the prokaryotic pattern in *E. coli, B. subtilis, M. genitalium, Micrococcus luteus, Streptococcus pneumoniae,* and *Deinococcus radiodurans* [see (10)]. The eukaryotic pattern has been found in humans (38), *X. laevis* (43), *S. pombe* (11a), and *S. cerevisiae* (45). Thus, *S. pombe* has both the endonuclease-exonuclease and the excinuclease modes of excision repair, and mutants defective in general excision repair are still capable of removing both 6-4 and cyclobutane pyrimidine dimers by SPDE (45a,45b).

Following the dual incision by excision nuclease, a protein-free gap does not form. Rather, one or more of the repair proteins remain bound to DNA and are dissociated by replication proteins concomitant with repair synthesis to fill the gap. These processes are followed by ligation. Although the overall strategy of excision repair is quite similar in prokaryotes and eukaryotes, the prokaryotic excinuclease consists of 3 subunits whereas the human excinuclease results from the concerted action of 16 polypeptides. Furthermore, the subunits of prokaryotic excinuclease do not share significant homology with any of the eukaryotic excinuclease subunits. These two systems are discussed in more detail below.

EXCISION REPAIR IN PROKARYOTES

The *E. coli* excision nuclease has been extensively characterized (9, 10, 46). The activity results from the combined actions of three subunits, UvrA, UvrB, and UvrC, and the enzyme is referred to as (A)BC excinuclease. However, as discussed below, a multimeric complex containing all three subunits does not exist. Some of the properties of the three subunits are summarized in Table 1. The three formal steps of excision repair are damage recognition, dual incisions, and repair synthesis and ligation. A model for the entire excision repair reaction is shown in Figure 1.

Table 1 The six proteins required for excision repair in *E. coli*

Protein	Mr	Sequence motifs	Activity	Role in repair
a. Excision nuclease subunits				

I. UvrA	$(104)_2$	a) Walker ATPase (2)
		b) Zinc finger (2)
		c) Leucine zipper
		d) UvrA superfamily
			a) ATPase
			b) Damage-speci-
			   fic DNA binding
			c) UvrB binding
			d) TRCF binding
				a) Damage recog-
				   nition (Proximal)
				b) Molecular
				   matchmaker
				c) TRC

II. UvrB	78	a) Helicase motif
		b) Homology to
		   TRCF
			a) Latent ATPase
			b) Latent "helicase"
			c) Damage-specific
			   ssDNA binding
			d) Binds UvrA
			e) Binds UvrC
				a) Damage recogni-
				   tion (ultimate)
				b) Unwinding du-
				   plex
				c) 3′-incision

III. UvrC	69	a) Limited homology
		   to UvrB
		b) Limited (40 amino
		   acid) homology to
		   ERCC1
			a) Nonspecific DNA
			   binding
			b) UvrB binding
				a) Induces 3′-inci-
				   sion
				b) Makes 5′-inci-
				   sion

b. Repair synthesis				
IV. Helicase II (uvrD)	70	Helicase motif	True helicase	Releases UvrC and excised oligo
V. DNA Pol I	103		DNA synthesis	Repair synthesis Displaces UvrB
VI. Ligase	75		Ligase	Ligation

Figure 1 Model for excision repair in *E. coli*. Transcription-independent (left) and transcription-coupled (right) forms are shown: A, UvrA; B, UvrB; C, UvrC; RNAP, RNA polymerase; DNA Pol I, DNA polymerase I; helicase II, UvrD protein; TRCF, transcription-repair coupling factor. In the transcription-independent mode the A_2B_1 complex locates the lesion by tracking along DNA. Locating the lesion is a slow process and is the rate-limiting step of the overall reaction. Once A_2B_1 arrives at a lesion site the reaction proceeds as follows: (*a*) a transient A_2B_1-DNA complex is formed guided by UvrA in an ATP-independent reaction, (*b*) the DNA is kinked and unwound in an ATP-dependent reaction leading to the formation of intimate contacts between UvrB and the damaged strand, and (*c*) the molecular matchmaker (UvrA) dissociates, leaving a stable UvrB-DNA complex. In the transcription-coupled repair, RNA Pol is used as a surrogate damage recognition protein. Upon encountering a lesion RNA Pol stalls (Step 1) and makes a stable complex. Steps 2–5: The stalled complex is recognized by the TRCF, which releases RNA Pol and the truncated transcript while simultaneously recruiting the A_2B_1 repair proteins, helps UvrA load UvrB onto the lesion, and pulls UvrA off UvrB to accelerate the rate of formation of the preincision UvrB-DNA complex. These steps are highly concerted. The subsequent steps of excision repair are identical in the transcription-independent and transcription-coupled modes. Step 6: UvrC binds to the UvrB-DNA complex, and UvrB makes the 3′ incision. Step 7: The DNA straightens enabling UvrC to make the

Damage Recognition by a Molecular Matchmaker

Excision nucleases in general and (A)BC excinuclease in particular excise bulky adducts such as cisplatin-1,2-d(GpG) diadduct, psoralen-thymine monoadduct, and benzo[a]pyrene-guanine adduct. However, with varying efficiencies they also excise lesions with minor helical distortions, ranging from AP sites to O^6-MeGua (41, 47, 48). Clearly, complementary surfaces between the enzyme and substrate cannot be the basis for recognition of these lesions

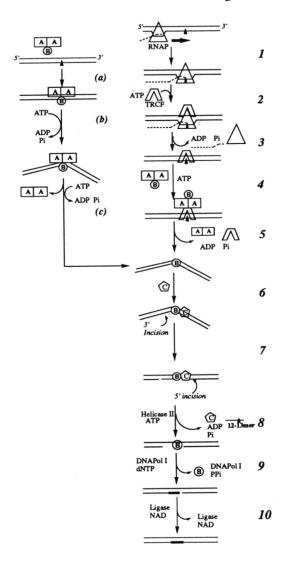

because the lesions have very few or no structural similarities. *E. coli* (A)BC excinuclease employs a molecular matchmaker, UvrA, to aid in recognition.

A molecular matchmaker (49, 50) is a protein that, in an ATP-dependent reaction, brings two compatible yet solitary macromolecules together, promotes their association, and then leaves the complex so it can engage in productive transactions. A molecular matchmaker must fulfill five criteria (50). First, in the absence of the matchmaker, the affinity of the matched protein for its binding site must be so low as to be physiologically insignificant. Second, the molecular matchmaker must promote stable complex formation between the matched components. Third, the matchmaker (or the matched protein) must be an ATPase, and ATP hydrolysis must be needed for association of the target molecules. Fourth, the matchmaker must make a complex with the matched components, causing a conformational change, but no covalent modification. Fifth, after stable complex formation the matchmaker must dissociate to allow the matched protein to carry out its effector function. UvrA meets all five criteria: It brings UvrB and damaged DNA together, promotes their association, and then leaves the complex in an ATP hydrolysis–dependent reaction.

Damage recognition proceeds as follows. UvrA dimerizes through a noncanonical leucine zipper (51) and makes an A_2B_1 complex with UvrB. This complex binds DNA nonspecifically with relatively high affinity ($K_{NS} \sim 10^{-7}$ M), and this binding activates the UvrB ATPase/helicase function (52) enabling the complex to probe the DNA for its propensity for local unwinding and bending. UvrA is a damage specific–DNA binding protein with specificity for damage in double-stranded DNA (53); UvrB is a damage specific–DNA binding protein with a specificity for single-stranded damaged DNA (54). The matchmaking step in which UvrA "loads" UvrB onto the damage exploits both the enhanced capacity of damaged regions to undergo deformation (bending and unwinding) and the intrinsic binding properties of UvrA and UvrB. Initial formation of the A_2B_1-DNA complex utilizes the damage-recognition specificity of UvrA. This process is followed by UvrB-dependent unwinding of 5 bp around the lesion (55, 56) and kinking of DNA by 130° into the major groove at phosphodiester bond 11 5′ to the lesion (57).

This unique conformation promotes extensive contacts between UvrB and the damaged strand. The initial contacts of UvrB with DNA are mainly ionic in nature and hence complex formation is sensitive to ionic stength. However, these interactions coupled to ATP hydrolysis by UvrB, which causes the local unwinding, expose the bases in the DNA and the hydrophobic core in UvrB, and they lead to a tight "hydrophobic bonding" between UvrB and the damaged strand, resulting in a salt-insensitive complex (49, 58). This model for damage recognition is in contradistinction to previous models (4, 9) which proposed that backbone deformity is the main determinant of recognition and that none of the subunits are in direct contact with the lesion. UvrB appears to have a hydrophobic binding pocket, and because of the lack of a requirement for specific H-bond donors and acceptors or for formation of ionic bonds of unique orientations, a vast number of chemical groups can be accomodated within this pocket (54).

Recent findings show that three repair enzymes with rather narrow substrate specificities, namely DNA photolyase (pyrimidine dimers), uracil glycosylase (uracil in DNA), and exonuclease III (AP site) "flip out" the lesion from the duplex into a "hole" within the enzyme to bring the active site cofactor or residues in close contacts with the target bonds (14, 59, 60). Thus, it is possible that UvrB employs a flip-out mechanism of substrate binding. Whether the excinuclease system flips out only the damaged nucleotides or the entire excised fragment remains to be seen.

To recapitulate, formation of the A_2B_1-DNA complex involves an ATP-independent step of "recognition" of any anomaly in DNA structure by UvrA, followed by "creation" of the ultimate recognition structure through UvrB-mediated helix unwinding, and consequent conformational change of UvrB and formation of intimate contacts between DNA and UvrB. Thus, the substrate structure cannot be considered independently of the binding reaction because the structure evolves in the process of recognition. In any event, formation of intimate contacts between UvrB and DNA weakens the contacts at the UvrA-UvrB interface, leading UvrA to dissociate and leaving behind a stable UvrB-DNA complex (49, 58, 61). Dissociation of UvrA is essential for binding of UvrC to the UvrB-DNA complex, a process that initiates the dual incisions (54, 62).

Dual Incisions

UvrB and UvrC carry out the excision reaction. The UvrB-DNA complex is recognized with high affinity and specificity by UvrC, and binding of UvrC to the complex leads to the dual incisions. Although contributions of amino acid residues from both subunits to both incision active sites cannot be eliminated, current evidence indicates that UvrB makes the 3' incision and UvrC makes the 5' incision (63). The two incisions are concerted but nonsynchronous. The 3' incision is made first, followed within a few seconds by the 5' incision (55). The 3' incision step requires ATP binding (but not hydrolysis) by UvrB. Because of its wide substrate range, a major challenge for (A)BC excinuclease is to discriminate between substrate and nonsubstrate DNA structures. The stepwise recognition and incision reactions help accomplish this goal. Thus, discrimination occurs at the following steps (46, 56): (a) binding of A_2B_1 to DNA, (b) dissociation of A_2 from A_2B_1-DNA complex, (c) dissociation of $UvrB_1$-DNA complex, (d) binding of UvrC to the $UvrB_1$-DNA complex, and (e) dissociation of the UvrB-3′-incised DNA complex. Such a stepwise mechanism amplifies the modest specificity of each step and safeguards against futile excision and resynthesis reactions on undamaged DNA.

Repair Synthesis

Following the dual incisions, UvrB, UvrC, and the "excised" oligomer remain in the postincision complex although the excised oligomer is no longer H-bonded (49). UvrC is not very stably bound in this complex; it dissociates slowly and this process is facilitated by helicase II (UvrD) which accomplishes its function by simply binding to the nicks, not by protein-protein interactions (64, 65). The remaining UvrB-gapped DNA complex is stable; however, following dissociation of UvrC the 3′-OH at the 5′ incision site becomes accessible to DNA polymerase I, which fills the gap and displaces UvrB. Under conditions approximating the physiological concentrations of DNA Pol I and ligase, virtually no nick translation occurs, and as a consequence more than 90% of the repair patches are 12–13 nt in length (66).

Since Pol I⁻ mutants are not as UV sensitive as Uvr⁻ mutants it has long been assumed that, in the absence of Pol I, either Pol II or Pol III can carry out repair synthesis, albeit less efficiently than Pol I (3). In a defined in vitro system all three polymerases are capable of repair synthesis (J Bouyer & A Sancar, submitted). However, Pol II and Pol III need accessory factors. In the absence of the polymerase β-clamp and the γ complex molecular matchmaker (68), Pol III is capable of limited repair synthesis, but Pol II is not, and the residual repair synthesis is inhibited by single-stranded DNA binding protein (SSB). In contrast, the β-clamp plus γ complex enables both polymerases to perform repair synthesis, and this repair synthesis is stimulated by SSB (J Bouyer & A Sancar, submitted). Thus, these polymerases require the same accessory proteins to fill a 12-nt gap as they do for semiconservative replication. Each cell has only about 20 Pol III holoenzyme molecules (68) and since, upon DNA damage, Pol II is induced to a level comparable to that of Pol I by the SOS response (69), Pol II likely does most of the repair synthesis in the absence of Pol I (J Bouyer & A Sancar, submitted).

EXCISION REPAIR IN HUMANS

Excision repair in humans is the prototype for excision repair in eukaryotes. The basic mechanism is similar to that of prokaryotes in that a multisubunit, ATP-dependent nuclease makes dual incisions on the damaged strand, excises an oligomer, and the resulting gap is filled and ligated. However, 13–16 polypeptides are needed to accomplish the task that is achieved by three polypeptides in E. coli. The two repair systems appear to represent convergent evolution. None of the excision repair proteins of humans shares significant sequence homolgy with *E. coli* excision repair proteins. In contrast, excision repair genes and proteins are conserved in eukaryotes ranging from *S. cerevisiae* to humans (3, 6, 70). As a consequence, the genetics and biochemistry of human excision repair is directly applicable to *S. cerevisiae* and other eukaryotes and vice versa. However, since the excision reaction entails coordinated action of 16 polypeptides and thus involves multiple protein-protein interactions, interspecies genetic or biochemical complementation is rare or absent.

Genetics of Human Excision Repair

Defective excision repair in humans is associated with three diseases: xeroderma pigmentosum, Cockayne's syndrome (CS), and trichothiodystrophy (TTD) (71).

XERODERMA PIGMENTOSUM   Xeroderma Pigmentosum (XP) is caused by an absence or greatly reduced level of excision repair (72). The disease is hereditary with autosomal recessive inheritance. The frequency of the disease is 10^{-6} in the United States and Europe and 10^{-5} in Japan. Symptoms fall into two groups: photodermatoses and neurological abnormalities. Photodermatoses include increased sensitivity to sunlight with manifestations ranging from erythema to xerosis and skin atrophy. Nearly 90% of these individuals develop basal cell and squamous cell carcinomas in their teens. Malignant melanomas also occur at high frequency. The overall rate of these three types of skin cancers is 2000-fold higher in XP individuals under the age of 20 than in the general population (71). Cancers of internal organs also occur at a 10–20-fold higher rate than in the general population. Most XP individuals suffer from neurological symptoms which include mental retardation, progressive ataxia, and deafness (71).

Somatic cell genetics revealed heterogeneity in XP individuals (73) and led to the identification of seven classic XP complementation groups named XP-A through XP-G (71). In addition, a group of individuals with near-normal UV resistance at the cellular level and normal levels of excision repair at the biochemical level exhibit the dermatological symptoms of XP, including skin cancer, but not the neurological symptoms (74, 75). These individuals are called XP variants (XP-V). The biochemical defect in XP-V is not known; however, XP-V cells have a reduced capacity to resume DNA replication after UV damage compared to normal cells. XP-V cells are said to be defective in postreplication repair (3). In eukaryotes, postreplication repair is an ill-defined phenomenon encompassing all molecular mechanisms enabling the cell to generate, through replication, two intact duplexes without actually removing the DNA damage (76–78). Clearly, the XP-V gene is not involved in excision repair.

XP complementation groups do not define all of the genes required for excision repair. Many repair defective–rodent cell lines have been isolated by screening mutagenized cell cultures for sensitivity to UV or chemotherapeutic agents such as mitomycin C (79, 80). These studies have resulted in the isolation of mutants falling into 11 complementation groups. Those in groups 2, 3, 4, 5, 6, and 8 are counterparts of human XP or CS mutants. The gene defined by complementation group 1 is required for excision repair but the genes in groups 6 to 11 are not. They either participate in transcription-repair coupling (6 and 8) or play accessory and as yet unknown roles in repair. Since all available data indicate a one-to-one correlation between the *S. cerevisiae* and human excision repair genes (6), yeast genetics has also aided in identifying human excision repair genes. The essential functions in excision repair of the transcription factor IIH (TFIIH) subunits p62 (hTFB1) and p44 (hSSL1) were revealed by the discovery that yeast *ssl1* (81) and *tfb1* (82) mutants are defective in excision repair. Human excision repair genes are called *XPA, XPB,* and so on or *ERCC1* (excision repair cross complementary group 1) and so on, depending on whether they were cloned by complementing human XP mutants or rodent UV-sensitive cell lines.

COCKAYNE'S SYNDROME   CS patients suffer from cachetic dwarfism, mental retardation, and progressive neurological symptoms caused by demyelination, and they are moderately sensitive to UV (83). Mutations in five genes cause CS. Two genes, *CSA* and *CSB,* are associated with "pure" CS. The corresponding rodent complementation groups are 8 and 6, respectively. Hence the genes are referred to by the names of *CSA(ERCC8)* and *CSB(ERCC6),* respectively (84, 85). In addition, some of the XPB, XPD, and XPG mutations give rise to XP/CS overlap syndrome (86). It has been suggested that CS is more of a transcription defect disease than a repair deficiency disease (87). CSA and CSB proteins are involved in coupling transcription to repair (88, 89), and XPB and XPD proteins are subunits of the transcription factor TFIIH (90–92); XPG is sometimes found to be associated with TFIIH (93).

TRICHOTHIODYSTROPHY   TTD individuals have sulfur-deficient brittle hair and suffer from dental caries, ichthyosis, skeletal abnormalities, and progressive mental retardation caused by demyelination. Mutations in three genes cause the disease: TTD-A, XPB, and XPD (94). Like CS, TTD caused by mutations in XPB and XPD exhibits the symptoms of both diseases. Reportedly, TTD is also mainly a transcription disease, because the repair defect in cell lines from all three complementation groups can be restored by microinjection of the transcription factor TFIIH, although none of the known TFIIH subunits is mutated in TTD-A (87).

Structure and Function of Human Excinuclease

The structure-function relationship of human excision nuclease is now understood in considerable detail. Sixteen polypeptides in six fractions are sufficient to reconstitute excision nuclease (93). The corresponding fractions of *S. cerevisiae* were also found to be necessary and sufficient for damage removal by dual incision in a defined system (45), further evidence of the striking similarities of the two systems. With the exception of XPA and XPG, all the reconstitution fractions contain 2–7 polypeptides in tight assemblies. Since the individual polypeptides are not present in free form in significant amounts in the cell, these six fractions may be justifiably considered the subunits of human excinuclease. The properties of each fraction are reviewed here to help explain the reaction mechanism of human excinuclease. Table 2 summarizes some of the properties of the six fractions. A more detailed account is given below.

XPA   The first of the six fractions is a zinc finger protein (95) with affinity for DNA and a marginally higher affinity for UV- or cisplatin-damaged DNA (96). In contrast, its yeast counterpart binds with high affinity and specificity to DNA containing 6-4 photoproducts (97). In fact, the real damage recognition entity of human excinuclease may be the XPA-RPA complex. XPA and RPA (HSSB) make a tight complex in vitro (98–100), and although each protein binds damaged DNA in isolation, when both are present increased amounts of both proteins are bound. However, whether enhanced binding results from complex formation is not clear. Even the increased affinity observed with the XPA-RPA complex is not sufficiently specific or avid enough to account for the ability of human excinuclease to locate rare lesions in the genome and excise them. Additional specificity may be conferred by interaction with other excision repair proteins. Indeed XPA binds to XPF-ERCC1 rather tightly (101, 102) and to TFIIH with modest affinity (103). However, standard gel retardation assays failed to reveal increased specificity upon association of these proteins with XPA (104).

RPA (HSSB)   The second fraction is a trimeric protein (p70, p34, p11) with high affinity for single-stranded DNA, and it performs an essential function in DNA replication analogous to that of *E. coli* single-stranded DNA binding protein (105–107). The large subunit binds DNA (108). The p34 subunit is involved in protein-protein interactions (109), undergoes phosphorylation/dephosphorylation reactions during the cell cycle (110, 111), and becomes hyperphosphorylated upon DNA damage by UV or ionizing radiation (112,

Table 2 The 25 polypeptides required for excision repair in humans

Fraction	Number	Proteins (yeast homolog)	Sequence motif	Activity	Role in repair
a. Excision nuclease subunits					
I. XPA	1	XPA/p31 (RAD14)	Zinc finger	DNA binding	Damage recognition
II. RPA	2	p70		DNA binding	Damage recognition
	3	p34			
	4	p11			
III. TFIIH	5	XPB/ERCC3/p89 (RAD25)	Helicase	a) DNA-dependent ATPase	a) Formation of preincision complex
	6	XPD/ERCC2/p80 (RAD3)	Helicase		
	7	p62 (TFB1)			
	8	p44 (SSL1)	Zinc finger	b) "Helicase"	b) Transcription-repair coupling
	9	Cdk7/p41 (KIN28)	S/T kinase		
	10	CycH/p38 (CCL1)	Cyclin	c) GTF	
	11	p34	Zinc finger	d) CAK	
IV. XPC	12	XPC/p125 (RAD4)		DNA binding	a) Stabilization of preincision complex
	13	HHR23B/p58 (RAD23)	Ubiquitin		b) Protection of preincision complex from degradation
V. XPF	14	XPF/ERCC4/p112 (RAD1)		Nuclease	5'-Incision
	15	ERCC1/p33 (RAD10)			
VI. XPG	16	XPG/ERCC5/p135 (RAD2)		Nuclease	3'-Incision
b. Repair synthesis and ligation					
I. RFC	17–21	$(p140)_1(p40)_4$	ATPase	ATPase	Molecular matchmaker
II. PCNA	22	$(p32)_3$			Polymerase clamp
III. RPA					
IV. Polε(δ)	23	p258	Polymerase	Replicase	Repair synthesis
	24	p55			
V. Ligase	25	p102			Ligation

113). Phosphorylation of HSSB occurs in two steps: first by cdk-cyclin A and then by Ku antigen–stimulated DNA-dependent protein kinase (114). Up to five serine and threonine residues become phosphorylated by the combined actions of these kinases. The role of the small (p11) subunit is unknown.

Depletion of cell-free extracts of RPA by antibodies or chromatography (115, 116) was found to severely inhibit repair synthesis *in vitro* and moderately reduce the damage-specific nicking activity. This finding led to the proposal that HSSB was essential for repair synthesis and in addition played a role in earlier steps of excision repair such as stabilizing the postincision complex (117). However, when the excision nuclease was reconstituted with purified proteins the incision step was found to be absolutely dependent on RPA (93). This conclusion was confirmed in the yeast reconstituted excision nuclease system (45). Hence, RPA is an essential subunit of human excinuclease. In addition to its interaction with XPA through both the p70 and p34 subunits (99, 100) it binds the XPG and XPF subunits of the excinuclease (98). Thus, the XPA-RPA complex with its multiple interactions with TFIIH, XPF-ERCC1, and XPG might constitute the nucleation component for the remaining subunits of the excinuclease.

TFIIH   The third fraction is one of the six general transcription factors (GTFs) (TFIID, TFIIA, TFIIB, TFIIF, TFIIE, and TFIIH) required for optimal transcription by RNA polymerase II (118–120). TFIIH has two enzymatic activities: helicase and CTD kinase, which phosphorylates the C-terminal domain (CTD) of the largest subunit of RNA Pol II. TFIIH is the last GTF to enter the initiation complex and is recruited to the complex by TFIIE. TFIIH is not required for transcription initiation but is required for promoter clearance which is the reaction encompassing the phosphorylation of CTD, the disruption of the initiation complex, and the synthesis of a transcript 30–50 nt in length (121, 122). After that reaction, RNA Pol II enters the elongation mode as TFIIH dissociates from the polymerase (122). The helicase activity of TFIIH is thought to be important for promoter clearance because in the absence of TFIIH aborted transcripts of less than 50 nt accumulate (121, 122a). Some genes such as the IgH gene can be transcribed without TFIIH. Furthermore, in in vitro systems with RNA Pol II lacking the CTD, a normal level of transcription occurs from several promoters. Even in TFIIH-dependent promoters this dependence can be abrogated by using a superhelical template (123). Nevertheless, TFIIH is an essential factor because mutations in its XPB (3, 6) and XPD (124, 125) homologs in yeast are lethal, indicating that TFIIH plays an essential role in transcription of genes important for normal cellular metabolism.

Depending on the purification scheme, TFIIH contains 5–10 subunits. Sequence analysis of its largest subunit (p89) revealed that it is identical to the

XPB/ERCC3 gene (91, 126). Concurrently, it was found that XP-B and XP-D human and rodent mutant cell-free extracts failed to complement in an excision assay (44), even though these cell lines complement upon cell fusion. The conclusion was that XPB and XPD make a tight complex and that the mutant subunits in these tight complexes exchange too inefficiently in vitro to complement each other. These observations, combined with the identification of the XPD homolog in yeast TFIIH (127), led to the eventual realization that TFIIH in its entirety is a repair factor (91, 92, 128). The highly purified TFIIH contains seven polypeptides (93, 129): XPB, XPD, p62, p44, p41 (cdk7), p38 (Cyclin H), and p34. Yeast genetics reveals that the first four are required for excision repair (119). Whether or not the other subunits are essential for excision repair is not known. However, the presence of the remaining three subunits in stoichiometric amounts in TFIIH does not interfere with excision (92, 129). Thus, the seven-subunit form of TFIIH is likely equally active in transcription and in repair.

However, the action mechanism of TFIIH in transcription differs in four important ways from that in repair. First, TFIIH is absolutely required for excising any type of lesion in any sequence context in both linear and supercoiled DNA. In contrast, TFIIH is not required for transcription from a subclass of RNA Pol II promoters or for transcription from any promoter in superhelical DNA. In this regard TFIIH is more of a repair factor than a transcription factor (93). Second, the XPA protein recruits TFIIH to the preexcision complex (103); in contrast TFIIE recruits TFIIH to the preinitiation complex (118, 120, 122a) and plays no role in general excision repair (103). Third, the ATPase ("helicase") activity of XPD is essential for excision repair (3, 6) but not for transcription. Finally, the CTD phosphorylating activity of TFIIH plays an important role in transcription initation but no protein is phosphorylated during excision repair (129). Anti-cyclin H antibodies which inhibit the cdk-activating kinase (CAK) activity of TFIIH (130–132) inhibit both transcription and excision repair. However, the inhibition of excision repair could simply result from steric hindrance by the antibody bound to a building block of TFIIH rather than from interference with the phosphorylating activity of CAK.

XPC-HHR23B   The fourth fraction contains two proteins; a 125 kDa and 58 kDa protein. The XPC gene, as defined by the XP-C complementation group, encodes a protein of 125 kDa with a modest degree of homology to the *S. cerevisiae* Rad4 protein (133, 134). Purification of XPC protein using an in vitro assay for repair yielded a fraction with stoichiometric amounts of two polypeptides. Sequence analysis revealed that p125 was the *XPC* gene product and p58 had a high degree of sequence homology to the *S. cerevisiae* RAD23 gene. The gene for the p58 was cloned. Humans have two RAD23 homologs called *HHR23A* and *B*. Of these homologs, only the protein encoded by

HHR23B is found in complex with XPC (13). There is no known human syndrome associated with mutations in HHR23A or B. In yeast, the *rad23Δ* mutant is not as UV sensitive as other yeast strains with mutations in the basal subunits of excinuclease (6). Neither p125 nor p58 (HHR23B) has any sequence signature revealing what function they may perform in excision repair. However, the p58 subunit has an interesting feature: the N-terminal 70 amino acids of both yeast RAD23 (135) and human HHR23A and B (134) show 25–31% sequence identity to ubiquitin and thus belong in the family of ubiquitin-fusion proteins. In some ubiquitin fusion proteins, the ubiquitin moiety is thought to function as a chaperone enabling proper folding and assembly in multiprotein complexes. In yeast, Rad23 has been found to help in stabilize the Rad14(XPA)-TFIIH complex (136).

The role of XPC protein in excision repair is rather interesting. The protein binds DNA with high affinity ($K_D \sim 10^{-9}$ M) and no specificity. XPC null mutants carry out normal strand-specific repair of transcribed genes but are defective in overall repair (137, 138). In contrast, yeast *rad4*(XPC) mutants are totally defective in excision, and *Rad7* and *Rad16* mutants behave like the human XPC mutant in that they repair the transcribed strand of a gene but not the lesions elsewhere (139). Whether these differences are real or apparent remains to be seen. However, in vitro experiments reveal that human excinuclease can be reconstituted in the absence of XPC and of RNA polymerase, and the partial nuclease reconstituted in this manner excises 27 nt–long fragments by incising at the appropriate 3′ and 5′ sites relative to the lesion (129). However, under these conditions both the excised oligomer and the damaged strand in the preincision complex were extensively degraded. XPC appears to bind to the damaged strand in the preincision complex and help target the nuclease subunits of the excision nuclease to the proper site while protecting the rest of the DNA in the preincision complex (which appears to be extensively single-stranded) from attacks by the two nuclease subunits, XPG and XPF-ERCC1. XPC, because of its high affinity to DNA, may help stabilize the preincision subassemblies on nucleosomal DNA and thus ensure proper assembly of the preincision complex and recruitment of the nuclease subunits. In transcribed DNA, an elongation complex stalled at a lesion apparently obviates the need for XPC.

XPG   The fifth fraction contains a solitary protein of 135 kDa (140–142). Its yeast homolog is RAD2 and both XPG and RAD2 show significant sequence homology to the human flap endonuclease (FEN1) which cleaves a DNA flap with a 5′-single stranded–end at the single-strand to double-strand DNA junction (143). XPG and Rad 2 proteins also have FEN activity (144) and thus XPG was predicted to be the 3′ nuclease of human excinuclease (143, 144). In fact XPG has three types of nuclease activities: (*a*) single-strand specific endonuclease (144, 145), (*b*) exonuclease activity of 5′ to 3′ directionality (145), and (*c*) FEN activity (144) which is stimulated by RPA (T Matsunaya & A Sancar, unpublished observation).

Direct evidence that XPG makes the 3′ incision comes from studies with XPG antibodies and from reconstitution experiments (146). Anti-XPG antibodies specifically changed the site and level of the 3′ incision in human cell-free extracts without affecting the 5′ incision strongly, suggesting that XPG makes the 3′ incision. Omission of the XPF-ERCC1 complex resulted in normal 3′ incision without any 5′ incision (129). Since XPG and XPF-ERCC1 are the only excinuclease subunits that have nuclease activity, these results established XPG as the 3′ nuclease (146). This fact may explain why XP-G mutants have some of the lowest residual repair activity of all of the XP mutants: XPB and XPD are essential genes and hence the mutants are always leaky; XPC protein is not required for gene-specific repair, and excision of nontranscribed sequences can occur without XPC when naked DNA is used as substrate. XPE protein is not required for excision; it may have a stimulatory effect (93). Mutants lacking XPF-ERCC1 do make the 3′ incision, which probably leads to some abnormal excision by a 3′ to 5′ exonuclease. As a consequence, XP-A and XP-G mutants have the lowest unscheduled DNA synthesis (UDS) of all the XP cell lines (71).

XPF-ERCC1   These two proteins (the sixth fraction) make a complex (44) of 1:1 stoichiometry (93, 147). The complex is a single strand–specific endonuclease which at the penultimate step of purification (or pure protein in the presence of RPA) also has junction endonucleolytic activity on a "bubble structure" on the strand which makes the transition from duplex to single-stranded DNA in the 5′ to 3′ direction (147). This activity is similar to that observed with the yeast Rad1-Rad10 complex which is the counterpart of XPF-ERCC1(148). These data, which are consistent with XPF-ERCC1 making the 5′ incision, were confirmed with antibody inhibition experiments. Anti-ERCC1 antibodies specifically inhibited the 5′ incision in a defined system giving rise to uncoupled 3′ incision (146). The same results were obtained by omission of XPF-ERCC1 in a reconstitution experiment (129). Moreover, these results showed that in the assembly of human excinuclease, XPF-ERCC1 is perhaps the last subunit to arrive and that in a normal excision reaction the 3′ incision may precede the 5′ incision even though the reaction is concerted.

Finally, XPF-ERCC1, like their yeast counterparts (149), are involved in recombinational repair as evidenced by the unusual sensitivity of these mutants to crosslinking agents such as mitomycin C (150).

XPE AND DDB   XP-E patients show mild symptoms of XP, and the XP-E cell lines are only moderately UV sensitive and have 50% of normal UDS activity

(71). XP-E cell-free extracts have reduced excision (44) and repair synthesis (134). Gel retardation assays using UV-irradiated DNA revealed that 2 out of 13 XP-E cell lines lacked a protein that specifically bound to damaged DNA. This activity was named damaged DNA binding protein (DDB) (151–153). The protein has been purified to homogeneity (154–156), and it is a heterodimer of $(p127)_1$ $(p48)_1$ composition (157). The genes for both subunits have been cloned and sequenced (158–160). The purified protein binds to 6-4 photoproducts with high affinity $(K_D \sim 10^{-10}$ M; 155) but not to pyrimidine cyclobutane dimers or to psoralen monoadducts (155). Microinjection of DDB into XPE-DDB⁻ cells has been reported to restore the UDS to normal level but has no effect on the UDS of XPE-DDB⁺ cells (161). However, DDB does not complement the excision activity of cell-free extracts from either DDB⁻ or DDB cells. In fact, DDB inhibits excision in vitro (162). Of the six fractions that are necessary and sufficient to reconstitute human excinuclease in vitro only RPA restores the excision activity of XP-E cell-free extract; yet no mutation was found in any of the three subunits of RPA in an XPE-DDB⁻ cell line (162). Clearly the relationship between the XP-E phenotype, DDB, and RPA remains to be elucidated.

Another class of proteins that binds to damaged DNA are high mobility group (HMG) domain proteins. Some members of this family bind to cisplatin-1,2-d(GpG) diadduct with high affinity (163–165). However this binding inhibits human excinuclease (42) and hence is not of direct relevance to excinuclease function. The binding may be relevant, however, to the tissue specifity of certain anticancer drugs.

In addition, human ribosomal protein S3 cleaves the phosphodiester bond of heavily UV-irradiated DNA at unknown lesions as well as at the intradimer phosphodiester bond. This activity, which has been referred to as AP endonuclease I or UV endonuclease III, is missing in some XP-D cell lines (166). The significance of these findings to XP pathogenesis is unknown (166).

Mechanism of Excision Repair

The three formal steps of excision repair are damage recognition, dual incisions, and repair synthesis and ligation. Figure 2 summarizes our current understanding of human excision repair which is based on (a) properties of individual components, (b) reactions with subsets of proteins, and (c) experiments with immobilized substrates. The three main steps are discussed below.

DAMAGE RECOGNITION   Although it has not been demonstrated experimentally, damage recognition is almost certainly the rate-limiting step of excision repair for two reasons. First, rare DNA lesions must be located among the 10^{10} bp present in the human genome. Second, lesions of infinite variety (Table 3) must be recognized by a small set of proteins (XPA, RPA, TFIIH) without recourse to the combinatorial recognition mechanism of transcription regulation which employs hundreds of proteins (transcription factors) in different combinations to activate specific genes or a small set of genes (167). Instead, damage recognition occurs in at least two stages: an ATP-independent step of low discrimination followed by an ATP-dependent step which leads to the

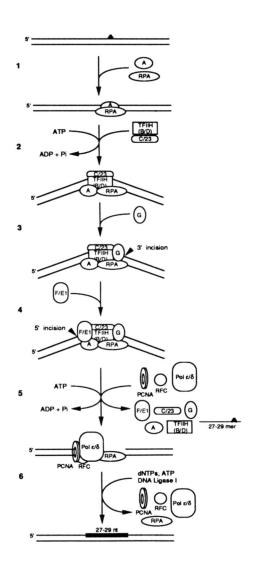

formation of a long-lived preincision complex. Although molecular match-making may be employed by human excinuclease to achieve high specificity, in contrast to *E. coli*, this possibility has not yet been experimentally demonstrated.

Binding of XPA-RPA   Both XPA (96) and RPA (168) have slightly higher affinity for damaged DNA than for undamaged DNA. The functional form of the recognition entity appears to be the XPA-RPA complex, because these two proteins associate tightly (98–100). Since RPA is quite abundant, in vivo all the XPA may be in XPA-RPA complex. This complex has higher affinity for damaged DNA than does either component alone (98–100). However, the binding data were qualitative and hence the level of improvement in damage-binding specificity by the complex compared to the individual components is not known at present. Theoretically, the occupancy of the target site in any DNA-protein interaction can be increased by two means. Either a protein of low abundance but high specificity finds its target because of its intrinsically high affinity for its site, or a protein of high abundance and low specificity occupies its target by the law of mass action while simultaneously occupying many nontarget sites as well. The second mechanism is utilized by RPA-XPA to find damage. RPA is one of the most abundant cellular proteins (106–108, 169) so it may occupy the damage sites and help recruit XPA to the lesion. In excinuclease reconstitution from purified components, 200–300 nM of RPA were needed for optimal activity as compared to picomolar amounts of XPF-ERCC1 (129, 170). In addition to XPA-RPA, XPC likely also contributes to damage recognition by stabilizing the XPA-RPA complex. This contribution may explain why XPC is not needed for transcription-coupled repair because in that case the stalled RNA polymerase is used as a surrogate damage-recognition protein.

◄ ───

Figure 2   Model for excision repair in humans. A-G, XPA through XPG (except XPE) proteins; C/23, XPC-HHR23B complex; F/E1, XPF-ERCC1 complex; Pol ε/δ, DNA pol ε or δ. Step 1: XPA-RPA recognizes damage in an ATP-independent reaction. Step 2: XPA-RPA recruits TFIIH to the lesion, and TFIIH unwinds DNA which leads to intimate DNA-protein contacts and makes the damaged strand accessible to XPC-HHR23B, which is recruited to the damage site through interactions with TFIIH and XPA. Step 3: XPG is recruited by RPA and TFIIH to the damage site and makes the 3' incision 3–5 nucleotides 3' to the lesion. Step 4: XPF-ERCC1 is recruited to the site by XPA, XPF, and TFIIH and makes the 5' incision 20–24 nucleotides 5' to the lesion. In vivo steps 3 and 4 are likely to be tightly coupled. Whether all proteins shown in steps 3 and 4 are simultaneously present in the incision and postincision complexes is not known. Step 5: The postincision complex is dissociated by the RFC molecular matchmaker, which loads the PCNA trimeric circle onto DNA and facilitates its (166a) association with Pol ε or δ, thus replacing excision proteins with repair synthesis proteins. Step 6: The gap is filled and the repair patch, corresponding in size to the excised fragment, is ligated. This is the transcription-independent mode of repair. In transcription-repair coupling, RNA Pol II, through its interaction with CSA/CSB proteins, is presumed to recruit XPA-RPA and TFIIH to the lesion that stalls transcription.

Table 3   Substrate spectrum of human excinuclease[a]

Bulky adducts
  Cholesterol
  Acetylaminofluorene
  Cisplatin-1,3-d(GpXpG)
  6 - 4 Photoproduct
  Cisplatin-1,2-d(GpG)
  Thymine dimer
  Texas Red
  Biotin
  Psoralen

Synthetic abasic analogs

Methylated nucleotides
  O^6-Methylguanine
  N^6-Methyladenine

Mismatches
  A : G
  G : G

[a] The lesions are listed in approximate order of catalytic efficiency. With the exception of N^6-MeAde all lesions are substrates for the *E. coli* excinuclease as well, although the order of preference is somewhat different.

Binding of TFIIH   This factor participates both in excision repair (92, 128) and in transcription initiation (118, 120). In transcription initiation, TFIIH is the last factor to enter the initiation complex and is recruited to the complex by TFIIE (121, 122, 122a). Binding of TFIIH sets in motion helix unwinding and CTD phosphorylation which result in promoter clearance and entry into elongation mode by RNA polymerase. In repair, the DNA-protein complex of (XPA-RPA)-DNA may be assumed to be the "closed form" preincision complex. The precise sequence of events leading to the formation of preincision "open complex" is not known at present. XPA is known to bind specifically to TFIIH, and therefore it must be involved in recruiting TFIIH to the damage site. Equally relevant is the fact that TFIIE, which specifically binds to TFIIH through XPB/ERCC3 (92, 122a), has no role in general (as opposed to transcription-coupled) excision repair. In *S. cerevisiae*, Rad23 stabilizes the TFIIH-Rad14(XPA) complex (97). Since XPC also associates with TFIIH (92) a reasonable assumption is that XPC-HHR23B binds to XPA-TFIIH in humans as well and aids in formation of a more stable preincision complex. However, this function is not essential for a functional pre-excision complex formation because excision can be achieved without XPC. In the absence of XPC, however, the reaction is not optimal and DNA is degraded extensively (129). XPG is also known to interact with RPA (98) and TFIIH specifically (93) and hence may play an important role in stabilizing the preincision complex.

No direct evidence is available on the nature of DNA conformational change caused by TFIIH in the preincision complex. However, based on its role in transcription (promoter clearance), and on the phenotype of yeast Rad3 (XPD) helicase active-site mutant (no repair; 6), TFIIH may reasonably be assumed to open the duplex at the lesion site, leading to evolution of new DNA-protein and protein-protein interactions that result in formation of a stable preincision complex. In fact, in reconstitution studies omission of XPC from the reaction makes the damaged strand uniquely susceptible to degradation by XPG in a TFIIH-dependent manner, suggesting that TFIIH creates a single-stranded region around the lesion which is attacked by this single-strand specific nuclease (129).

The composition of the preincision complex is not known at present, but XPA, RPA, TFIIH, and XPG are required for complex formation. The complex can form in the absence of XPC and XPF-ERCC1. However, since DNA degradation occurs in complexes without XPC, under physiological conditions XPC is likely also present. Analogous to the role of TFIIH in transcription— disengaging RNA Pol from other GTFs—TFIIH may, upon being recruited to the damage site, disengage some of the preincision (molecular matchmaker) proteins to prepare the preincision complex for the entry of the 5' nuclease.

DUAL INCISIONS/EXCISION   XPG makes the 3' incision and XPF-ERCC1 makes the 5' incision (143, 144, 146). The 3' incision is made first, followed within seconds by the 5' incision (129). A significant difference from the *E. coli* excinuclease is that the 3' incision is not triggered by the binding of the subunit that makes the 5' incision. In a reaction mixture lacking XPF-ERCC1, the 3' incision is made at almost normal levels by XPG (129). As a consequence both in cell-free extracts and in defined systems a significant amount of uncoupled 3' incision occurs even in the presence of XPF-ERCC1. In contrast to uncoupled 3' incision, uncoupled 5' incision is not observed in cell-free extracts or reconstituted systems, suggesting that XPG must be present in the preincision complex for XPF-ERCC1 to bind and make the 5' incision. However, with anti-XPG antibodies, which apparently do not interfere with assembly of XPG in the preincision complex but inhibit its nuclease activity, extensive uncoupled 5' incision was observed (146). This finding suggests that a conformational change caused by the 3' incision is not necessary to prime the DNA for the 5' incision by XPF-ERCC1.

The excised fragments range in size from 24 to 32 nt. The minimum size substrate for the excision reaction is about 100 bp (40). The incision sites are influenced by adduct type (41, 42, 93), and sequence context (40, 171) and reaction conditions. The 5' incision sites range from the 20 to 26 phosphodiester bond 5' (146) and from the 2nd to 10th phosphodiester bond 3' to the lesion (129, 146). However, in cell-free extracts the incisions occur over a more narrow range. Furthermore, both in cell-free extract and in defined systems the excised fragments are as a rule 24–32 nt with, in most cases, 27–29 mers being the dominant species (45, 93, 146). Thus, the nuclease appears to measure an "exact" distance. If the first 3' incision is made too far from the lesion, then the 5' incision is at the close end of the 5' incision range, which results in a narrow range of excised fragments.

The composition of the dual incision complex is not known. Naturally, the complex contains XPG and XPF-ERCC1. Considering that extensive degradation of DNA occurs in the absence of XPC, the complex most likely contains XPC as well. Whether XPA-RPA, which recruits XPG and XPF-ERCC1, and TFIIH, which forms the open complex, are present following the recruitment of the nuclease subunits is not clear. The composition of the postincision complex is currently not known. However, even though the excised oligomer is released (129), a single-stranded gap, free of DNA does not form. At least some of the excinuclease subunits remain bound in the postincision complex with the gap (93), and they must be displaced by the polymerase accessory factors RFC and PCNA before the polymerase can fill the gap.

REPAIR SYNTHESIS   Excision repair, in the strict sense, is the recognition and removal of damage from DNA. However, these two steps create a gap which must be filled to create a functional duplex. Of the five known human DNA polymerases (172–173) Pol δ and Pol ε carry out repair synthesis. The role of Pol ε in repair synthesis was discovered when a polymerase responsible for UV damage–induced DNA synthesis in permeabilized cells was purified (174) and later identified as Pol ε (175). A different set of in vivo experiments revealed that PCNA, which is the polymerase clamp of Pol δ and Pol ε, can be detected in association with chromatin of nonproliferating cells upon UV irradiation (176, 177), implicating the two PCNA-dependent polymerases, Pol δ and Pol ε, in repair synthesis. Indeed, depletion of cell-free extract of PCNA completely eliminates repair synthesis (116, 178), providing strong evidence that one or both of these polymerases carry out repair synthesis.

Studies using Pol δ antibodies indicate that nucleotide incorporation into UV-irradiated DNA in cell-free extract, which is commonly used as an assay for excision repair, prevents incorporation, and for this reason Pol δ was proposed as the repair polymerase (179). A different study, using gapped DNA as substrate and partially fractionated cell-free extracts, found that Pol ε was more efficient in generating ligatable products, suggesting that Pol ε was better suited to being a repair polymerase (180). As expected, in this latter system, gap filling was dependent on proliferating cell nuclear antigen

(PCNA) and was stimulated by the PCNA molecular matchmaker RFC (50) which loads the PCNA ring onto the primer-template. In fact a recent study with Pol δ and Pol ε yeast mutants showed that conditional mutations in either polymerase make cells UV sensitive (181). Thus the combination of available data is consistent with both polymerases being responsible for repair synthesis. Similarly, Drosophila (182) and yeast (183) conditional PCNA mutants are UV sensitive, providing in vivo support for PCNA in repair synthesis.

DNA Pol β, which is capable of filling in 25–30-nt gaps efficiently (172, 173), is the repair polymerase for certain types of base excision repair but apparently plays no role in nucleotide excision repair. In contrast, Pol δ and ε participate in filling gaps generated by either repair pathway (3, 31, 184). The inability of Pol β to carry out repair synthesis in nucleotide excision repair may stem from the fact that after excision of the oligomer the gap is occupied by RPA, which has a binding site of 30 nt (185), and other excision repair proteins. These proteins are apparently displaced by RFC-PCNA and, upon loading of PCNA at the 3'-OH of the gap, this primer terminus is no longer accessible to Pol β. Even though the main repair polymerases are quite different in prokaryotes (Pol I) and in eukaryotes (Pol δ and Pol ε), the E. coli replicase Pol III is also capable of repair synthesis, and when it carries out this function it is strongly dependent on the RFC-PCNA functional homologs δ and β proteins and on the RPA homolog SSB (J Bouyer & A Sancar, submitted).

The repair patch resulting from resynthesis was determined to be in the 30–50-nt range by a variety of in vivo methods (186–188). Considering the limitations of the in vivo methods, these studies gave remarkably accurate estimates. Determination of the repair patch size by the phosphorothioate method (66) revealed that within a resolution of 1–3 nt the 5' and 3' borders of the repair patch produced in cell-free extracts precisely matched the borders of the excision gap (38). Thus in humans, as in E. coli, the gap is filled without 5' enlargement or 3' nick translation.

The fact that the 3' end of the repair patch matches the 3' end of the gap means that DNA ligase is not a limiting factor in repair synthesis. Once the gap is filled, any of the four DNA ligases present in human cells (189–191) should be able to seal the patch.

TRANSCRIPTION-REPAIR COUPLING

Transcribed DNA is repaired faster than nontranscribed DNA both in humans (192) and in E. coli (193). Furthermore, the preferential repair is largely confined to the template strand (193, 194) and in humans only to genes transcribed by RNA Pol II (195). Several factors contribute to this phenomenon (195, 196): chromatin structure, topology of transcribed DNA, and the effects of lesions on the progression of RNA polymerase. The chromatin structure and topology of transcribed DNA undoubtedly play a role in repair (197); however, the major repair modulating factor appears to be the interaction of RNA polymerase with a lesion. Bulky lesions such as pyrimidine dimers in the template strand block transcription both in *E. coli* (198) and in humans (199); in contrast lesions in the coding strand have no effect on progression of the transcription complex. Such a stalled polymerase inhibits excision in a defined system of repair and transcription factors alone (198). This finding appears to be in conflict with the in vivo data indicating that stalled RNA polymerase enhances the repair rate of the lesion blocking transcription (198). A single protein in *E. coli* and a complex of at least two proteins in humans displace stalled RNA Pol and recruit the excision nuclease to the damage site and thus provide a solution for this apparent paradox. The mechanism of transcription-repair coupling (TRC) is well understood in *E. coli*; however, no in vitro system for coupling exists for humans and thus the mechanistic details remain unclear.

Transcription-Repair Coupling in Escherichia coli

In *E. coli*, enhanced repair of the transcribed strand is mediated by the transcription-repair coupling factor (TRCF) which is encoded by the *mfd* gene (200). The TRCF is a protein of 130 kDa with helicase motifs but no helicase activity (201). TRC in *E. coli* occurs as follows (Figure 1): TRCF specifically recognizes RNA pol stalled at a lesion, and then it dissociates the ternary complex, releasing RNA pol and the truncated transcript while simultaneously recruiting the A_2B_1 damage binding component of (A)BC excinuclease to the damage site by specifically binding to UvrA (201). The reaction is highly concerted such that the release of RNA polymerase and the delivery of A_2B_1 to the damage site occur simultaneously, and capturing an intermediate involving all these components has not been possible. The UvrA-binding domains of TRCF and UvrB partially overlap; hence after recruiting the A_2B_1 complex to the damage site TRCF helps UvrA dissociate from the A_2B_1-DNA complex, facilitating formation of the preincision B_1-DNA complex (202, Figure 1). The TRCF also dissociates RNA Pol stalled by nucleotide starvation or by a protein road block and hence it may play additional roles in transcription and repair (203). However, *mfd⁻* mutants are only moderately UV sensitive (204) and have about a threefold increase in spontaneous mutation rate (205, 206). Null mutants have normal growth properties, indicating that *mfd* is not an essential cellular gene under physiological conditions.

Transcription-Repair Coupling in Humans

In humans, individuals defective in TRC suffer from CS (88). The two genes necessary for TRC in humans, *CSA/ERCC8* and *CSB/ERCC6*, have been cloned (85, 207). The CS mutant cell lines are slow to resume RNA synthesis after DNA damage (208, 209) and show phenotypic properties similar to *E. coli mfd⁻* mutants (85, 196): They are moderately sensitive to UV and have an increased rate of UV-induced mutations, and the majority of these mutations are caused by lesions in the template strand, in contrast to normal cells in which most of the UV-induced mutations are caused by lesions in the coding strand (204, 210–212).

The mechanism of TRC in man is not known. The CSA protein is 46 kDa in size, and it has the WD motif (207) found in many proteins including those involved in skeletal assembly, membrane trafficking, and RNA metabolism. The motif may be used for protein-protein interactions (213). In contrast, the sequence of CSB/ERCC6 is rather revealing. CSB/ERCC6 is a protein of 160 kDa with "helicase motifs," and it almost certainly performs a function analogous to that of TRCF in *E. coli* (85, 201). Indeed, CSB/ERCC6 does bind to the proximal (XPA) and ultimate (TFIIH) damage recognition subunits of human excinuclease (119, 196) and to CSA (207). Based on the properties of CSA and CSB proteins and the known facts of transcription by RNA Pol II in humans (118, 120, 214), the following model has been proposed for transcription-repair coupling in humans (195, 196, 215). RNA Pol II stalls at a lesion, and the stalled complex is recognized by the CSA/CSB heterodimer which perhaps with the aid of TFIIS backs off RNA Pol without dissociating the ternary complex. CSA/CSB also recruit XPA and TFIIH to the lesion site and thus increase the rate of assembly of excinuclease. Following excision and repair synthesis, RNA Pol II resumes its transcription on the repaired template by elongating the truncated transcript.

The main difference between this model and the prokaryotic one is that in humans RNA Pol II is believed to back up rather than dissociate from the lesion site during repair. Currently, there is no experimental evidence to support this view. However, the argument has been made (195) that some human genes are so large that transcribing them without encountering a lesion is practically impossible (transcription may take up to 24 hr for the dystrophin gene). Had the transcripts been discarded, making full length proteins of such genes would never have been possible. However, this model is based on certain assumptions about the in vivo rates of transcription, damage formation, and repair, and some of these assumptions may not be entirely justified. Hence in humans, as in *E. coli*, the truncated transcript may be discarded during transcription-repair coupling.

CELL CYCLE AND REPAIR

Intuitively, one would predict that excision repair is tightly coupled to the cell cycle. Thus, it would appear that lesions present during a prolonged G2 phase have more time to be repaired. The lesions in G2 would also pose less of a threat to genomic integrity than lesions present during G1 or S phases, which would cause cellular death by blocking replication or by inducing mutation when bypassed by translesion synthesis. Studies on cell cycle and DNA repair have progressed along two lines of inquiry.

Excision Repair Potential as a Function of Cell Cycle

When the repair of pyrimidine dimer was measured in total genomic DNA in either CHO cells (216) or human fibroblasts (217) no significant differences were found except for an apparent decline during mitosis. These studies of low resolution were followed by high-resolution studies which analyzed gene-specific repair as a function of the cell cycle (218–221). Using flow cytometry to separate cells in various phases without using any synchronization procedure that might interfere with cell physiology, strand-specific repair of the actively transcribing CHO DHFR gene was found to be essentially constant during the entire cell cycle. This finding reveals that not only the overall repair but also the efficiency of TRC was constant during phases of the cell cycle (219). CHO cells were also obtained at high purity at various points of the cell cycle using a noninvasive synchronization method (treating cells with the plant amino acid mimosine) (220). Tests of these cells for gene-specific repair showed that the rate of excision of pyrimidine dimers was essentially constant throughout the cell cycle (221); however, DNA damage during S or G2 phases increased the length of G2 and hence allowed more time for repair, which explains the relatively high resistance of G2 cells to DNA damage compared to other phases. The consensus from these studies is that excision repair capacity of the cell does not change with cell cycle.

Cyclin-Dependent Kinase-Activating Enzyme, Transcription Factor IIH, and Excision Repair

Cyclin-dependent protein kinases such as cdk2 and cdc2 regulate the cell division cycle (222, 223). The activities of these kinases, in turn, are regulated by their associations with cyclins, which in general go through cyclic changes in concentration during the cell cycle, and by phosphorylation of serine or threonine residues of the kinases themselves. An activity that phosphorylates cyclin-complexed cdc2 and cdk is called cyclin-dependent kinase-activating enzyme (CAK). Recently CAK was purified from HeLa cells and found to be in two forms (222): a low molecular form containing cdk7 (41 kDa), cyclin H (37 kDa), and a third subunit (p36); and a large form of 300–400 kDa. Interestingly, the large form turned out to be TFIIH (130–132). Thus, two out of the seven subunits of TFIIH are cdk7 and cyclin H. The significance of this unexpected finding is unknown at present. However, Cdk7-Cyclin H needs to be phosphorylated by yet another kinase (CAKAK) to become CAK, and the levels of cyclin H and CAK remain constant during the cell cycle (223). Hence, whether or not CAK regulates excision repair or transcription-repair coupling, its potential effect is independent of the cell cycle. However, the full implications of CAK-TFIIH connection remain to be explored.

Another issue that remains to be explored is the effect of the replication complex on excision repair. Although the effect of a replication complex on a transcription complex moving in the same or in the opposite direction has been investigated (224), no direct evidence shows that a stalled replication complex affects excision repair. A stalled complex may hinder excision repair in a manner analogous to a stalled RNA Pol, or it may stimulate repair by increasing the local concentration of RPA, thereby facilitating the assembly of excinuclease. In vitro replication/repair experiments are needed to directly determine the effects of the replication complex on repair.

SOS RESPONSE IN *ESCHERICHIA COLI* AND IN MAN

The SOS response was originally defined in *E. coli* as a coordinated cellular response to DNA damage by UV and other agents that cause bulky lesions in DNA; a response that aids cell survival (225, 226). The response results from induction of about 30 cellular genes that have a common regulatory element called the SOS box with the consensus sequence of $CTG-N_{10}-CAG$ (225, 226). Attempts to find a similar response in human cells have revealed interesting damage response reactions which are often called SOS responses. The relation of these responses to excision repair will be discussed below after a brief review of the excision repair component of the SOS response in *E. coli*.

SOS Regulation of Excision Repair in Escherichia coli

The genes for rate-limiting subunits of (A)BC excinuclease, *UvrA* and *UvrB*, have SOS boxes (4) that are bound by the LexA repressor under physiological conditions. Upon DNA damage, the RecA protein binds to single-stranded DNA resulting from replication blocks and acts as a coprotease for autoproteolysis (and inactivation) of LexA. The levels of UvrA and UvrB increase, as does the cell's excision repair activity. In addition to *uvrA* and *uvrB,* other genes that play a role in excision repair, such as *uvrD* (helicase II) and *polB* (DNA Pol II), are induced by the SOS response and contribute to increased repair capacity. Upon completion of repair, the inducing signal disappears and the cell returns to preinduction conditions. A cardinal sign of the SOS response is increased repair capacity (4, 69), which led to Weigle's initial discovery of the response (226). Hence, increased repair capacity might be used as a reference point in describing a cellular response to damage as an SOS response.

Damage Response and Human Excision Repair

Ultraviolet and other DNA-damaging agents elicit a complex set of responses ranging from growth delay to apoptosis. Two response reactions are relatively well understood. One is mediated by the growth-stimulatory Ras signal trans-duction pathway, and the signal for this response is reactive oxygen species generated in the membrane (227). The other response reaction is mediated by the growth-inhibitory p53 pathway and the signal is DNA damage (228). Currently, no evidence shows convincingly that either of these response reactions has the *sine qua non* of SOS response (226): increased repair capacity. However, DNA damage does induce significant changes in cellular physiology, so these response reactions must be taken into account in any model of the role of excision repair in cellular survival of DNA damage.

EXCISION REPAIR IN INDUCED CELLS   Attempts have been made to detect Weigle Reactivation–like phenomena in human cells. In one such study UV irradiated–herpes virus had higher survival and mutation rates when plated on UV-irradiated cells than on untreated host (229). This finding was taken as an indication of increased repair capacity of UV-induced cells. However, repair was not measured directly in such cells. Considering the profound effect of UV on cellular physiology (230) alternative explanations are more likely.

p53 AND EXCISION REPAIR   *p53* is a tumor suppressor gene which plays an important role in the molecular pathogenesis of up to 50% of human cancers (231–233). The p53 protein is a transcriptional regulator and plays an important role in cell cycle regulation (234, 235). DNA damage by UV, ionizing radia-tion, and alkylating agents, which directly or indirectly cause single-strand breaks, results in increased levels of p53 by way of posttranslational modifi-cation and stabilization (236–238). p53 has several DNA binding properties that make it a candidate for a multifunctional DNA metabolism master regu-lator (239): (*a*) p53 binds to a specific sequence upstream of the target genes. (*b*) p53 binds to ends of DNA fragments and can promote strand exchange between single-stranded DNA and a homologous duplex. (*c*) p53 promotes the annealing of complementary single strands. (*d*) p53 binds single-stranded DNA 25–30 nt in length, and this binding causes a conformational change (allosteric regulation) that increases p53's affinity for its target sequences (240–242). (*e*) p53 binds to mismatches and bulges (243). In addition, p53 binds to certain proteins and protein assemblies: (*a*) p53 binds RPA and inhibits RPA-depend-ent replication initiation of SV40 in an in vitro system (244). (*b*) p53 binds XPB/ERCC3 (245). (*c*) p53 binds TFIIH and inhibits its helicase activity (246).

In light of these properties, suggestions have been made that, upon DNA damage, p53 induces several proteins that block the cell cycle at the G1/S boundary and perhaps increases the level of some excision repair proteins. It has been suggested that by binding to RPA, p53 may convert RPA itself from a "replication form" to a "repair form," and by binding to TFIIH, p53 may aid in transcription-repair coupling. Furthermore, the affinity of p53 for TFIIH may enable it to interact with TFIIH associated with RNA Pol II stalled at a lesion, become activated by phosphorylation by the CAK activity associated with TFIIH, and somehow couple transcription to repair (247).

This model predicts that p53-deficient cells would be more sensitive to UV and at least partially defective in excision repair and transcription-repair coupling. Indeed it has been reported that human p53 (–/–) cells have diminished capacity to excise pyrimidine dimers, suggesting that they have a direct role in excision repair (248). However, a comprehensive study which measured UV survival and pyrimidine-dimer and 6-4-photoproduct excision in p53(+/+), p53(+/–), and p53(–/–) mouse fibroblasts showed no difference between the three cell types (249). The nuclear accumulation of p53 in XP-A and CS-B cells that do not repair transcribed DNA, with lower doses of UV compared to normal cells (250, 251), may be a consequence of more replication gaps or stalled transcription bubbles in mutant cells. An attractive model is that binding of p53 to bulges and other lesions (243), or to the 27–29 mer excised by human excinuclease (241), activates the protein as a transcription factor by an allosteric mechanism, initiating the chain of events leading to G1/S arrest (241). Finally, even though p53 reportedly inhibits SV40 replication by binding to RPA (244) and inhibits the helicase activity of TFIIH (246), micromolar concentrations of p53 have no detectable effect on an in vitro excision repair system absolutely dependent on RPA and TFIIH (A Kazantsev & A Sancar, unpublished observations). Thus, no direct evidence shows that p53 has any direct effect on transcription-independent or -dependent repair. The reports regarding the p53 effect on excision repair activity through its regulatory function are discussed below.

p21, Gadd45, AND EXCISION REPAIR   These two genes are induced by p53 and contribute to G1/S arrest induced by p53 (234, 235, 252). The mechanism of cell cycle inhibition by p21 is well understood: It binds to Cdk/Cyclin complexes and inhibits kinase activity (234). p21 also binds PCNA with high affinity (253, 254). The PCNA-p21 complex reportedly inhibits replication but does not inhibit repair and thereby aids cell survival (255, 256). However, a comprehensive in vitro study with several different substrates did not confirm these preliminary results (257). The PCNA-p21 complex cannot participate either in replication or in repair, so increased p21 inhibits both cellular reactions and does not aid cell survival via differential effect on excision repair (257).

Similarly, it was reported that Gadd45 binds to PCNA and inhibits replication but stimulates excision repair either directly or in the form of a Gadd45-PCNA complex (252). However, this preliminary study has not been confirmed either. Up to micromolar concentrations of Gadd45 neither stimulated nor inhibited repair as measured by excision and repair synthesis assays (258). Thus, all available evidence is consistent with the following: (*a*) p53 does not directly participate in excision repair. (*b*) p53 does not induce the transcription of excision repair genes or modulate the activity of excision repair proteins by post-translational modification. (*c*) p53 does not upregulate proteins that stimulate excision repair. However, p53 undoubtedly plays a central role in cellular response to DNA damage, and signal transduction by p53 binding to nicks and abnormal DNA structures (243) or the excision product (241) remains an attractive possibility.

RPA (HSSB) PHOSPHORYLATION AND EXCISION REPAIR   RPA is phosphorylated as a result of DNA damage by UV and ionizing radiation (112, 113), and extracts from UV-irradiated cells that contain hyperphosphorylated HSSB are reportedly unable to sustain SV40 replication, leading to the suggestion that phosphorylation converts RPA from a replication form (RPA) to a repair form (RPA-P). However, subsequent studies revealed that inhibition of RPA phosphorylation did not affect replication (169, 259) and that unphosphorylated and hyperphosphorylated RPA were equally active in replication and in excision repair (170). Phosphorylation requires replication intermediates or single-stranded DNA to act as coactivators for DNA-PK, so phosphorylation of RPA during replication or repair may initiate a signaling pathway that prevents cell cycle progression while replication or repair intermediates exist (259). The significance of RPA phosphorylation in cellular physiology remains to be determined. However, RPA clearly plays no direct role in coordinating replication and repair.

SUMMARY AND CONCLUSIONS

Nucleotide excision repair is an important cellular defense mechanism in prokaryotes and eukaryotes. The basic mechanism is the removal of damage by dual incisions by an ATP-dependent multisubunit enyzme system called excision nuclease (excinuclease) followed by the filling and ligating of the single-stranded gap. Several similarities and differences between these two systems are enumerated below.

Similarities Between Escherichia coli and Human Excision Repair Systems

The excision repair systems of *E. coli* and humans share the following features. 1. They both have a wide (and essentially identical) substrate spectrum. 2. They are the sole repair mechanism for bulky adducts. 3. In general, they perform sequence-independent repair. 4. Damage recognition consists of an ATP-independent step followed by an ATP-dependent step. 5. DNA is unwound and kinked in the preincision complex. 6. The damage is removed by concerted but nonsynchronous dual incisions. 7. The 3' and 5' incisions are made by separate subunits. 8. The 3' incision precedes the 5' incision. 9. Following the dual incisions a subset of the subunits remains bound to DNA. 10. A helicase is required to dissociate the postincision complex. 11. The patch size equals the gap size. 12. Transcription is coupled to repair through a transcription-repair coupling factor.

Differences in Excision Repair Between Escherichia coli and Humans

The excision repair systems of *E. coli* and humans show the following differences. 1. Excision requires 3 polypeptides in *E. coli* and 16 polypeptides in humans. 2. The excinuclease subunits of the two systems show no sequence homology. 3. A replication protein, RPA(HSSB), is required for excision in humans but not in *E. coli*. 4. A transcription factor (TFIIH) is required for excision in humans but not in *E. coli*. 5. Although substrate spectra are similar the preferences are different, and sequence effect on excision affects the two systems differently. 6. Chromatin structure plays an important role in controlling excision in humans. 7. The nuclease subunits of *E. coli* do not show overt nuclease activity in isolation whereas the human nuclease subunits do. 8. In humans 3' incision can occur with a subassembly of the repair proteins; in *E. coli* all subunits are needed to elicit the nuclease activity. 9. *E. coli* excises the damage in 12–13 mers; humans excise 27–29 mers. The "excised" oligomer is released by the human but not by the *E. Coli* excinuclease. 10. Replication polymerases (Pol δ and Pol ε) are responsible for repair synthesis in humans; in *E. coli* the repair polymerase Pol I carries out repair synthesis. 11. *E. coli* excision repair proteins do not participate in recombination; human XPF-ERCC1 complex is involved in recombination. 12. *E. coli* excision nuclease is regulated by SOS response; human excinuclease is not.

The concepts and methodology of prokaryotic excision repair greatly aided in studies on eukaryotic repair. These studies have led to a detailed understanding of human excision repair. However, as indicated above, the two systems differ in significant ways. Furthermore, excision repair in humans transcends the realm of scientific curiosity: It is the most important defense mechanism against DNA damage caused by tobacco smoke, which accounts for more than 30% of cancer deaths worldwide (260). A concerted effort to improve understanding of human excision repair in relation to other cellular phenomena may lead to new ways of thinking about cancer prevention and treatment.

ACKNOWLEDGMENTS

I thank Drs. T Matsunaga and D Mu for providing the figures, Drs. JT Reardon and A Kazantsev for compiling the data in Table 3, and Drs. D Mu and GB Sancar for comments on the manuscript.

Literature Cited

1. Bishop JM. 1995. *Genes Dev.* 9:1309–15
2. Modrich P. 1994. *Science* 266:1959–60
3. Friedberg EC, Walker GC, Siede W. 1995. *DNA Repair and Mutagenesis.* Washington, DC: Am. Soc. Microbiol. 698 pp.
4. Sancar A, Sancar GB. 1988. *Annu. Rev. Biochem.* 57:29–67
5. Sancar A. 1995. *Annu. Rev. Genet.* 29: 69–105
6. Prakash S, Sung P, Prakash L. 1993. *Annu. Rev. Genet.* 27:33–70
7. Bootsma D, Hoeijmakers JH. 1994. *Mutat. Res.* 307:15–23
8. Tanaka K, Wood RD. 1994. *Trends Biochem. Sci.* 19:83–86
9. Grossman L, Thiagalingam S. 1993. *J. Biol. Chem.* 268:16871–74
10. Sancar A, Tang MS. 1993. *Photochem. Photobiol.* 57:905–21
11. Sancar A. 1994. *Science* 266:1954–56
11a. Sancar A. 1995. *J. Biol. Chem.* 270: 15915–18
12. Sancar GB. 1990. *Mutat. Res.* 236:137–60
13. Sancar A. 1994. *Biochemistry* 33:2–9
14. Park HW, Kim ST, Sancar A. 1995. *Science* 268:1866–72
15. Sancar GB, Smith FW. 1989. *Mol. Cell. Biol.* 9:4767–76
16. Fox ME, Feldman BJ, Chu G. 1994. *Mol. Cell. Biol.* 14:8071–77
17. Ozer Z, Reardon JT, Hsu DS, Malhotra K, Sancar A. 1995. *Biochemistry* 36. In press
18. Ley RD. 1993. *Proc. Natl. Acad. Sci. USA* 90:4337
19. Li YF, Kim ST, Sancar A. 1993. *Proc. Natl. Acad. Sci. USA* 90:4389–93
20. Kato T Jr, Todo T, Ayaki H, Ishizaki K, Morita T, et al. 1994. *Nucleic Acids Res.* 22:4119–24
21. Todo T, Takemori H, Ryo H, Ihara M, Matsunaga T, et al. 1993. *Nature* 361: 371–74
22. Kim ST, Malhotra K, Smith CA, Taylor JS, Sancar A. 1994. *J. Biol. Chem.* 269:8534–40
23. Kim ST, Malhotra K, Taylor JS, Sancar A. 1996. *Photochem. Photobiol.* In press
24. Setlow P. 1992. *J. Bacteriol.* 174:2737–41
25. Fajardo-Cavazos P, Salazar C, Nicholson WL. 1993. *J. Bacteriol.* 179:1735–44
26. Mitra S, Kaina B. 1993. *Prog. Nucleic Acid Res. Mol. Biol.* 44:109–42
27. Samson L. 1991. *Mol. Microbiol.* 6: 825–31
28. Lindahl T, Sedgwick B, Sekiguchi M, Nakabeppu Y. 1988. *Annu. Rev. Biochem.* 57:133–57
29. Dumenco LL, Allay E, Norton K, Gerson SL. 1993. *Science* 259:219–22
30. Nakatsuru Y, Matsukuma S, Nemoto N, Sugano H, Sekiguchi M, et al. 1993. *Proc. Natl. Acad. Sci. USA* 90:6468–72
31. Demple B, Harrison L. 1994. *Annu. Rev. Biochem.* 63:915–48
32. Dodson MC, Michaels ML, Lloyd RS. 1994. *J. Biol. Chem.* 269:32709–12
33. Bessho T, Tano K, Kasai H, Ohtsuka E, Nishimura S. 1993. *J. Biol. Chem.* 268:19416–21
34. Bowman KK, Sidik K, Smith CA, Taylor JS, Doetsch PW, Freyer GA. 1994. *Nucleic Acids Res.* 22:3026–32
35. Freyer G, Davey S, Ferrer JV, Martin AM, Beach D, Doetsch PW. 1995. *Mol. Cell. Biol.* 15:4572–77
36. Yajima H, Takao M, Yasuhira S, Zhao JH, Ishii C, et al. 1995. *EMBO J.* 14: 2393–99
37. Sancar A, Rupp WD. 1983. *Cell* 33: 249–60
38. Huang JC, Svoboda DL, Reardon JT, Sancar A. 1992. *Proc. Natl. Acad. Sci. USA* 89:3664–68
39. Yeung AT, Mattes WB, Oh EY, Grossman L. 1983. *Proc. Natl. Acad. Sci. USA* 80:6157–61

40. Huang JC, Sancar A. 1994. *J. Biol. Chem.* 269:19034–50
41. Huang JC, Hsu DS, Kazantsev A, Sancar A. 1994. *Proc. Natl. Acad. Sci. USA* 91:12213–17
42. Huang JC, Zamble DB, Reardon JT, Lippard SJ, Sancar A. 1994. *Proc. Natl. Acad. Sci. USA* 91:10394–98
43. Svoboda DL, Taylor JS, Hearst JE, Sancar A. 1993. *J. Biol. Chem.* 268:1431–36
44. Reardon JT, Thompson LH, Sancar A. 1993. *Cold Spring Harbor Symp. Quant. Biol.* 58:605–17
45. Guzder SN, Habraken Y, Sung P, Prakash L, Prakash S. 1995. *J. Biol. Chem.* 270:12973–76
45a. Birnboim HC, Nasim A. 1975. *Mol. Gen. Genet.* 136:1–8
45b. McCready S, Carr AM, Lehmann AR. 1993. *Mol. Microbiol.* 10:885–90
46. Lin JJ, Sancar A. 1992. *Mol. Microbiol.* 6:2219–24
47. Lin JJ, Sancar A. 1989. *Biochemistry* 28:7979–84
48. Snowden A, Kow YW, Van Houten B. 1990. *Biochemistry* 29:7251–59
49. Orren DK, Selby CP, Hearst JE, Sancar A. 1992. *J. Biol. Chem.* 267:780–88
50. Sancar A, Hearst JE. 1993. *Science* 259:1415–20
51. Sun Q. 1995. *Protein-protein Interactions Among the Subunits of E. coli: (A)BC Excinuclease.* MSc thesis. Univ. N. Carolina, Chapel Hill. 69 pp.
52. Oh EY, Grossman L. 1987. *Proc. Natl. Acad. Sci. USA* 84:3638–42
53. Seeberg E, Fuchs RPP. 1990. *Proc. Natl. Acad. Sci. USA* 87:191–94
54. Hsu DS, Kim ST, Sun Q, Sancar A. 1995. *J. Biol. Chem.* 270:8319–27
55. Lin JJ, Phillips AM, Hearst JE, Sancar A. 1992. *J. Biol. Chem.* 267:17693–700
56. Visse R, King A, Moolenaar GF, Goosen N, van de Putte P. 1994. *Biochemistry* 33:9881–88
57. Shi Q, Thresher R, Sancar A, Griffith J. 1992. *J. Mol. Biol.* 226:425–32
58. Orren DK, Sancar A. 1989. *Proc. Natl. Acad. Sci. USA* 86:5237–41
59. Savva R, McAuley-Hecht K, Brown T, Pearl L. 1995. *Nature* 373:487–93
60. Mol CD, Kuo CF, Thayer MM, Cunningham RP, Tainer JA. 1995. *Nature* 37:381–86
61. Visse R, de Ruijter M, Moolenaar GF, van de Putte P. 1992. *J. Biol. Chem.* 267:6736–42
62. Bertrand-Burggraf E, Selby CP, Hearst JE, Sancar A. 1991. *J. Mol. Biol.* 218:27–36
63. Lin JJ, Sancar A. 1990. *J. Biol. Chem.* 267:17688–92
64. Matson SW, Kaiser-Rogers KA. 1990. *Annu. Rev. Biochem.* 59:289–329
65. Lohman TM. 1992. *Mol. Microbiol.* 6:5–14
66. Sibghat-Ullah, Sancar A, Hearst JE. 1990. *Nucleic Acids Res.* 18:5051–53
67. Deleted in proof
68. Kelman Z, O'Donnell M. 1995. *Annu. Rev. Biochem.* 64:171–200
69. Bonner CA, Hays S, McEntee K, Goodman MF. 1990. *Proc. Natl. Acad. Sci. USA* 87:7663–67
70. van Duin M, de Wit J, Odijk H, Westerveld A, Yasui A, et al. 1986. *Cell* 44:913–23
71. Cleaver JE, Kraemer KH. 1989. In *The Metabolic Basis of Inherited Disease*, ed. CR Scriver, AL Beaudet, WS Sly, D Valle, 2:2949–71. New York: McGraw-Hill
72. Cleaver JE. 1968. *Nature* 218:652–56
73. De Weerd-Kastelein EA, Kleijzer W, Bootsma D. 1972. *Nature New Biol.* 238:80
74. Burk PG, Lutzner MA, Clarke DD, Robbins JH. 1971. *J. Lab. Clin. Med.* 77:759–65
75. Cleaver JE. 1971. *J. Invest. Dermatol.* 58:124
76. Kaufmann WK. 1989. *Carcinogenesis* 10:1–11
77. Cordeiro-Stone M, Boyer JC, Smith BA, Kaufmann WK. 1986. *Carcinogenesis* 7:1783–86
78. Naegeli H. 1994. *BioEssays* 16:557–64
79. Thompson LH, Busch DB, Brookman KW, Mooney CL, Glaser DA. 1981. *Proc. Natl. Acad. Sci. USA* 78:3734–37
80. Busch D, Greiner C, Rosenfeld KL, Ford R, de Wit J, et al. 1994. *Mutagenesis* 9:301–6
81. Yoon H, Miller SP, Pabich EK, Donahue TF. 1992. *Genes Dev.* 6:2463–77
82. Matsui P, DePaulo J, Buratowski S. 1995. *Nucleic Acids Res.* 23:767–72
83. Nance MA, Berry SA. 1992. *Am. J. Med. Genet.* 42:68–84
84. Shiomi T, Ito T, Yamaizumi M, Wakasugi M, Matsunaga T, et al. 1996. *Mutat. Res.* In press
85. Troelstra C, van Gool A, de Wit J, Vermeulen W, Bootsma D, Hoeijmakers JH. 1992. *Cell* 71:939–53
86. Vermeulen W, Jacken J, Jaspers NGJ, Bootsma D, Hoeijmakers JHJ. 1993. *Am. J. Hum. Genet.* 53:185–92
87. Vermeulen W, Van Vuuren AJ, Chipoulet M, Schaeffer L, Appeldoorn E, et al. 1994. *Cold Spring Harbor Symp. Quant. Biol.* 59:317–29
88. Venema J, Mullenders LHF, Natarajan AT, van Zeeland AA, Mayne LV. 1990. *Proc. Natl. Acad. Sci. USA* 87:4707–11

89. van Hoffen A, Natarajan AT, Mayne LV, van Zeeland AA, Mullenders LHF, Venema J. 1993. *Nucleic Acids Res.* 21:5890–95

90. Schaeffer L, Roy R, Humbert S, Moncollin V, Vermeulen W, et al. 1993. *Science* 260:58–63

91. Schaeffer L, Moncollin V, Roy R, Staub A, Mezzina M, et al. 1994. *EMBO J.* 13:2388–92

92. Drapkin R, Reardon JT, Ansari A, Huang JC, Zawel L, et al. 1994. *Nature* 368:769–72

93. Mu D, Park CH, Matsunaga T, Hsu DS, Reardon JT, Sancar A. 1995. *J. Biol. Chem.* 270:2415–18

94. Stefanini M, Vermeulen W, Weeda G, Giliani S, Nardo T, et al. 1993. *Am. J. Hum. Genet.* 53:817–21

95. Tanaka K, Miura N, Satokata I, Miyamoto I, Yoshida MC, et al. 1990. *Nature* 348:73–76

96. Jones CJ, Wood RD. 1993. *Biochemistry* 32:12096–104

97. Guzder SN, Sung P, Prakash L, Prakash S. 1993. *Proc. Natl. Acad. Sci. USA* 90:5433–37

98. He Z, Henricksen LA, Wold MS, Ingles CJ. 1995. *Nature* 374:566–68

99. Matsuda T, Saijo M, Kuraoka I, Kobayashi T, Nakatsu Y, et al. 1995. *J. Biol. Chem.* 270:4152–57

100. Li L, Lu X, Peterson CA, Legerski RJ. 1995. *Mol. Cell. Biol.* 15:5396–5402

101. Li L, Elledge SJ, Peterson CA, Bales ES, Legerski RJ. 1994. *Proc. Natl. Acad. Sci. USA* 91:5012–16

102. Park CH, Sancar A. 1994. *Proc. Natl. Acad. Sci. USA* 91:5017–21

103. Park CH, Mu D, Reardon JT, Sancar A. 1995. *J. Biol. Chem.* 270:4896–902

104. Park CH. 1995. *Protein-protein interactions in human nucleotide excision repair.* PhD thesis. Univ. N. Carolina, Chapel Hill. 116 pp.

105. Chalberg MD, Kelly TJ. 1989. *Annu. Rev. Biochem.* 58:671–717

106. Stillman B. 1989. *Annu. Rev. Cell. Biol.* 5:197–245

107. Hurwitz J, Dean FB, Kwong AD, Lee SH. 1990. *J. Biol. Chem.* 265:18043–46

108. Gomes XV, Wold MS. 1995. *J. Biol. Chem.* 270:4534–43

109. Lee SH, Kim DK. 1995. *J. Biol. Chem.* 270:12801–7

110. Din SU, Brill SJ, Fairman MP, Stillman B. 1990. *Genes Dev.* 4:968–77

111. Fotedar R, Roberts JM. 1992. *EMBO J.* 11:2177–87

112. Liu VF, Weaver DT. 1993. *Mol. Cell. Biol.* 13:7222–31

113. Carty MP, Zernik-Kobak M, McGrath S, Dixon K. 1994. *EMBO J.* 13:2114–23

114. Pan ZQ, Amin AA, Gibbs E, Niu H, Hurwitz J. 1994. *Proc. Natl. Acad. Sci. USA* 91:8343–47

115. Coverley D, Kenny MKR, Munn M, Rupp WD, Lane DP, Wood RD. 1991. *Nature* 349:538–41

116. Shivji MKK, Kenny MK, Wood RD. 1992. *Cell* 69:367–74

117. Aboussekhra A, Biggerstaff M, Shivji MKK, Vilpo JA, Moncollin V, et al. 1995. *Cell* 80:859–68

118. Conaway RC, Conaway JW. 1993. *Annu. Rev. Biochem.* 62:161–90

119. Drapkin R, Reinberg D. 1994. *Trends Biochem. Sci.* 19:504–8

120. Zawel L, Reinberg D. 1995. *Annu. Rev. Biochem.* 64:533–61

121. Goodrich JA, Tjian R. 1994. *Cell* 77:145–56

122. Zawel L, Kumar KP, Reinberg D. 1995. *Genes Dev.* 9:1479–90

122a. Maxon ME, Goodrich JA, Tjian R. 1994. *Genes Dev.* 8:515–24

123. Parvin JD, Sharp PA. 1993. *Cell* 73:533–40

124. Guzder SN, Sung P, Bailly V, Prakash L, Prakash S. 1994. *Nature* 369:578–81

125. Weber CA, Salazaar EP, Stewart SA, Thompson LH. 1990. *EMBO J.* 9:1437–47

126. Weeda G, van Ham RC, Vermeulen W, Bootsma D, van der Eb AJ, Hoeijmakers JHJ. 1990. *Cell* 62:777–91

127. Feaver WJ, Svejstrup JQ, Bardwell L, Bardwell AJ, Buratowski S, et al. 1993. *Cell* 75:1379–87

128. Wang Z, Svejstrup JQ, Feaver WJ, Wu X, Kornberg RD, Friedberg EC. 1994. *Nature* 368:74–76

129. Mu D, Hsu DS, Sancar A. 1995. *J. Biol. Chem.* 271. In press

130. Roy R, Adamczewski JP, Seroz T, Vermeulen W, Tassan JP, et al. 1994. *Cell* 79:1093–1101

131. Serizawa H, Makela TP, Conaway JW, Conaway RC, Weinberg RA, Young RA. 1995. *Nature* 374:280–82

132. Shiekhattar R, Mermelstein F, Fisher RP, Drapkin R, Dynlacht B, et al. 1995. *Nature* 374:283–87

133. Legerski R, Peterson C. 1992. *Nature* 359:70–73

134. Masutani C, Sugasawa K, Yangisawa J, Sonoyama T, Ui M, et al. 1994. *EMBO J.* 13:1831–43

135. Watkins JF, Sung P, Prakash L, Prakash S. 1993. *Mol. Cell. Biol.* 13:7757–65

136. Guzder SN, Bailly V, Sung P, Prakash L, Prakash S. 1995. *J. Biol. Chem.* 270:8385–88

137. Venema J, Van Hoffen A, Natarajan AT, van Zeeland AA, Mullenders LHF. 1990. *Nucleic Acids Res.* 18:443–48

138. van Hoffen A, Venema J, Meschini R, van Zeeland AA, Mullenders LHF. 1995. *EMBO J.* 14:360–67

139. Verhage R, Zeeman AM, deGroot N, Gleig F, Bang DD, et al. 1994. *Mol. Cell. Biol.* 14:6135–42

140. Scherly D, Nouspikel T, Corlet J, Ucla C, Bairoch A, Clarkson SG. 1993. *Nature* 363:182–85

141. MacInnes MA, Dickson JA, Hernandez RR, Learmonth D, Lin GY, et al. 1993. *Mol. Cell. Biol.* 13:6393–402

142. Shiomi T, Harada Y, Saito T, Shiomi N, Okuno Y, et al. 1994. *Mutat. Res.* 314:167–75

143. Harrington JJ, Lieber MR. 1994. *Genes Dev.* 8:1344–55

144. O'Donovan A, Davies AA, Moggs JG, West SC, Wood RD. 1994. *Nature* 371: 432–35

145. Habraken Y, Sung P, Prakash L, Prakash S. 1994. *J. Biol. Chem.* 269:31342–45

146. Matsunaga T, Mu D, Park CH, Reardon JT, Sancar A. 1995. *J. Biol. Chem.* 270:20862–69

147. Park CH, Bessho T, Matsunaga T, Sancar A. 1995. *J. Biol. Chem.* 270:22657–660

148. Bardwell AJ, Bardwell L, Tomkinson AE, Friedberg EC. 1994. *Science* 265: 2082–85

149. Shiestl RH, Prakash S. 1989. *Mol. Cell. Biol.* 8:3619–26

150. Hoy CA, Thompson LH, Mooney CL, Salazar EP. 1985. *Cancer Res.* 45:1737–43

151. Chu G, Chang E. 1988. *Science* 242: 564–67

152. Kataoka H, Fujiwara Y. 1991. *Biochem. Biophys. Res. Commun.* 175:1139–43

153. Keeney S, Wein H, Linn S. 1992. *Mutat. Res.* 273:49–56

154. Abramic M, Levine AS, Protic M. 1991. *J. Biol. Chem.* 266:22493–500

155. Reardon JT, Nichols AF, Keeney S, Smith CA, Taylor JS, et al. 1994. *J. Biol. Chem.* 268:27301–8

156. Hwang BJ, Chu G. 1993. *Biochemistry* 32:1657–66

157. Keeney S, Chang GJ, Linn S. 1993. *J. Biol. Chem.* 268:21293–300

158. Takao M, Abramic M, Moos M Jr, Otrin VR, Wooton JC. 1993. *Nucleic Acids Res.* 21:4111–18

159. Lee TH, Elledge SJ, Butel JS. 1995. *J. Virol.* 69:1107–14

160. Dualan R, Brody T, Keeney S, Nichols AF, Linn S. 1995. *Genomics.* In press

161. Keeney S, Eker APM, Brody T, Vermeulen W, Bootsma D, et al. 1994. *Proc. Natl. Acad. Sci. USA* 91:4053–56

162. Kazantsev A, Mu D, Zhao X, Nichols AF, Linn S, Sancar A. 1996. *Proc. Natl. Acad. Sci. USA* 93:In press

163. Bruhn SL, Pil PM, Essigmann JM, Housman DE, Lippard SJ. 1992. *Proc. Natl. Acad. Sci. USA* 89:2307–11

164. Pil PM, Lippard SJ. 1992. *Science* 256: 234–37

165. Hughes EN, Engelsberg BN, Billings PC. 1992. *J. Biol. Chem.* 267:13520–27

166. Kim J, Chubatsu LS, Admon A, Stahl J, Fellows R, Linn S. 1995. *J. Biol. Chem.* 270:13620–29

166a. O'Donnell M, Onrust R, Dean FB, Chen M, Hurwitz J. 1995. *Nucleic Acids Res.* 21:1–3

167. Johnson PF, McKnight SL. 1989. *Annu. Rev. Biochem.* 58:799–839

168. Clugston CK, McLaughlin K, Kenny MK, Brown R. 1992. *Cancer Res.* 52: 6375–79

169. Henricksen LA, Wold MS. 1994. *J. Biol. Chem.* 269:24203–8

170. Pan ZQ, Park CH, Amin AA, Hurwitz J, Sancar A. 1995. *Proc. Natl. Acad. Sci. USA* 92:4636–40

171. Mu D, Bertrand-Burggraf E, Huang JC, Fuchs RPP, Sancar A. 1994. *Nucleic Acids Res.* 22:4869–71

172. Bambara RA, Jessee CB. 1991. *Biochim. Biophys. Acta* 1088:11–24

173. Wang TSF. 1991. *Annu. Rev. Biochem.* 60:513–52

174. Nishida C, Reinhard P, Linn S. 1988. *J. Biol. Chem.* 263:501–10

175. Syvaoja J, Suomensaari S, Nishida C, Goldsmith JS, Chui GSJ, et al. 1990. *Proc. Natl. Acad. Sci. USA* 87:6664–68

176. Celis JE, Madsen P. 1986. *FEBS Lett.* 209:277–83

177. Toschi L, Bravo R. 1988. *J. Cell Biol.* 107:1623–28

178. Nichols AF, Sancar A. 1992. *Nucleic Acids Res.* 20:2441–46

179. Zeng XR, Jiang Y, Zhang SJ, Hao H, Lee MYWT. 1994. *J. Biol. Chem.* 269: 13748–51

180. Shivji MKK, Podust VN, Hubscher U, Wood RD. 1995. *Biochemistry* 34: 5011–17

181. Budd ME, Campbell JL. 1995. *Mol. Cell. Biol.* 15:2173–79

182. Henderson DS, Banga SS, Grigliatti TA, Boyd JB. 1994. *EMBO J.* 13:1450–59

183. Ayyagari R, Impellizzeri KJ, Yoder BL, Gary SL, Burgers P. 1995. *Mol. Cell. Biol.* 15:4420–29

184. Singhal RK, Prasad R, Wilson SH. 1995. *J. Biol. Chem.* 270:949–51

185. Kim C, Paulus BF, Wold MS. 1994. *Biochemistry* 33:14197–206

186. Edenberg H, Hanawalt PC. 1972. *Biochim. Biophys. Acta* 272:361–72

187. Regan JD, Setlow RB. 1974. *Cancer Res.* 34:3318–25
188. Cleaver JE, Jin J, Charles WC, Mitchell DL. 1991. *Photochem. Photobiol.* 54:393–402
189. Husain I, Tomkinson AE, Burkhart WA, Moyer MB, Ramos W, et al. 1995. *J. Biol. Chem.* 270:9683–90
190. Wei YF, Robins P, Carter K, Caldecott K, Pappin DJC, et al. 1995. *Mol. Cell. Biol.* 15:3206–16
191. Chen J, Tomkinson AE, Ramos W, Mackey ZB, Danehower S, et al. 1995. *Mol. Cell. Biol. 15*:5412–22
192. Bohr VA, Smith CA, Okumoto DS, Hanawalt PC. 1985. *Cell* 40:459–69
193. Mellon I, Hanawalt PC. 1989. *Nature* 342:95–98
194. Mellon I, Spivak G, Hanawalt PC. 1987. *Cell* 51:241–49
195. Hanawalt PC. 1994. *Science* 266:1457–58
196. Selby CP, Sancar A. 1994. *Microbiol. Rev.* 58:317–29
197. Smerdon MJ, Thoma F. 1990. *Cell* 61:675–84
198. Selby CP, Sancar A. 1990. *J. Biol. Chem.* 265:21330–36
199. Donahue BA, Yin S, Taylor JS, Reines D, Hanawalt PC. 1994. *Proc. Natl. Acad. Sci. USA* 91:8502–6
200. Selby CP, Witkin EM, Sancar A. 1991. *Proc. Natl. Acad. Sci. USA* 88:11574–78
201. Selby CP, Sancar A. 1993. *Science* 260:53–58
202. Selby CP, Sancar A. 1995. *J. Biol. Chem.* 270:4882–89
203. Selby CP, Sancar A. 1995. *J. Biol. Chem.* 270:4890–95
204. Oller AR, Fijalskowska IJ, Dunn RL, Schaaper RM. 1992. *Proc. Natl. Acad. Sci. USA* 89:11036–40
205. Witkin EM. 1966. *Science* 152:1345–53
206. Bockrath RC, Palmer JE. 1977. *Mol. Gen. Genet.* 156:133–40
207. Henning KA, Li L, Iyer N, McDaniel LD, Reagan MS, et al. 1995. *Cell* 82:555–64
208. Lehmann AR, Kirk-Bell S, Arlett CF, Patterson MC, Lohman PHM, et al. 1975. *Proc. Natl. Acad. Sci. USA* 72:219–23
209. Mayne LV, Lehmann AR. 1982. *Cancer Res.* 42:1473–78
210. Brash DE, Rudolph JA, Simon JA, Lin A, McKenna A, et al. 1994. *J. Biol. Chem.* 269:32672–77
211. Chen RH, Maher VM, McCormick JJ. 1990. *Proc. Natl. Acad. Sci. USA* 87:8680–84
212. Sato M, Nishigori C, Zghal M, Yagi T, Takebe H. 1993. *Cancer Res.* 53:2944–46
213. Doolittle RF. 1995. *Annu. Rev. Biochem.* 64:287–314
214. Kerppola TM, Kane CM. 1991. *FASEB J.* 5:2833–42
215. Drapkin R, Sancar A, Reinberg D. 1994. *Cell* 77:9–12
216. Collins ARS, Downes CS, Johnson RT. 1980. *J. Cell. Physiol.* 103:179–91
217. Kaufman WK, Wilson SJ. 1990. *Mutat. Res.* 236:107–17
218. Roussey G, Boulikas T. 1992. *Eur. J. Biochem.* 204:267–72
219. Lommel L, Carswell-Crumpton C, Hanawalt PC. 1995. *Mutat. Res.* 336:181–92
220. Orren DK, Petersen LN, Bohr VA. 1995. *Mol. Cell. Biol.* 15:3722–30
221. Petersen LN, Orren DK, Bohr VA. 1995. *Mol. Cell. Biol.* 15:3731–37
222. Fisher RP, Morgan DO. 1994. *Cell* 78:713–24
223. Solomon MJ. 1994. *Trends Biochem. Sci.* 19:496–500
224. Liu B, Alberts BM. 1995. *Science* 267:1131–37
225. Little JW, Mount DW. 1982. *Cell* 29:11–22
226. Walker GC. 1984. *Microbiol. Rev.* 48:60–93
227. Devary Y, Gottlieb RA, Smeal T, Karin M. 1992. *Cell* 71:1081–91
228. Kuerbitz SJ, Plunkett BS, Walsh WV, Kastan MB. 1992. *Proc. Natl. Acad. Sci. USA* 89:7491–95
229. Sarkar SN, Dasgupta UB, Summers WC. 1984. *Mol. Cell. Biol.* 4:2227–30
230. Fornace AJ Jr. 1992. *Annu. Rev. Genet.* 26:505–24
231. Levine AJ, Momand J, Finlay CA. 1991. *Nature* 351:453–56
232. Kastan MB, Onyekwere O, Sidransky D, Vogelstein B, Craig RW. 1991. *Cancer Res.* 51:6304–11
233. Harris CC. 1993. *Science* 262:1980–81
234. Harper JW, Adami GR, Wei N, Keyomarsi K, Elledge SJ. 1993. *Cell* 75:805–16
235. El-Deiry WS, Tokino T, Velculescu VE, Levy DB, Parsons R, et al. 1993. *Cell* 75:817–25
236. Maltzman W, Czyzyk L. 1984. *Mol. Cell. Biol.* 4:1689–94
237. Kastan MB, Zhan Q, El-Deiry WS, Carrier F, Jacks T, et al. 1992. *Cell* 71:587–97
238. Lu X, Lane DP. 1993. *Cell* 75:765–78
239. Hupp TR, Lane DP. 1995. *Curr. Biol.* 4:865–75
240. Bakalkin G, Selivanova G, Yakovleva T, Koseleva E, Kashuba E, et al. 1995. *Nucleic Acids Res.* 23:362–69
241. Jayaraman L, Prives C. 1995. *Cell* 81:1021–29

242. Bayle JH, Elenbaas B, Levine AJ. 1995. *Proc. Natl. Acad. Sci. USA* 92: 5729–33
243. Lee S, Elenbaas B, Levine A, Griffith J. 1995. *Cell* 81:1013–20
244. Dutta A, Ruppert JM, Aster JC, Winchester E. 1993. *Nature* 365:79–82
245. Wang YW, Forester K, Yeh H, Feitelson MA, Harris CC. 1994. *Proc. Natl. Acad. Sci. USA* 91:2230–34
246. Xiao H, Pearson A, Coulombe B, Truant R, Zhang S, et al. 1994. *Mol. Cell. Biol.* 14:7013–24
247. Jones CJ, Wynford-Thomas D. 1995. *Trends Genet.* 11:165–66
248. Wang XW, Yeh H, Schaeffer L, Roy R, Moncollin V, et al. 1995. *Nat Genet.* 10:188–95
249. Ishizaki K, Ejima Y, Matsunaga T, Hara R, Sakamoto A, et al. 1994. *Int. J. Cancer* 58:254–57
250. Jackson DA, Hassan AB, Errington RJ, Cook PR. 1994. *J. Cell. Sci.* 107:1753–60
251. Yamaizumi M, Sugano T. 1994. *Oncogene* 9:2775–84
252. Smith ML, Chen IT, Zhan Q, Bae I, Chen CY, et al. 1994. *Science* 266: 1376–79
253. Flores-Rozas H, Kelman Z, Dean FB, Pan Z, Harper JW, et al. 1994. *Proc. Natl. Acad. Sci. USA* 91:8655–59
254. Waga S, Hannon GJ, Beach D, Stillman B. 1994. *Nature* 369:574–78
255. Li R, Waga S, Hannon GJ, Beach D, Stillman B. 1994. *Nature* 371:534–37
256. Shivji MKK, Grey SJ, Strausfeld UP, Wood RD, Blow JJ. 1994. *Curr. Biol.* 4:1062–68
257. Pan ZQ, Reardon JT, Li L, Fores-Rozas H, Legerski R, et al. 1995. *J. Biol. Chem.* 270:22008–16
258. Kazantsev A, Sancar A. 1995. Science 270:1003–04
259. Brush GS, Anderson CW, Kelly TJ. 1994. *Proc. Natl. Acad. Sci. USA* 91: 12520–24
260. Ames BN, Gold LS, Willett WC. 1995. *Proc. Natl. Acad. Sci. USA* 92:5258–65

Annu. Rev. Biochemistry 1996. 65:83–100

SELENOCYSTEINE[1]

Thressa C. Stadtman

National Heart, Lung and Blood Institute, National Institutes of Health, Bethesda, Maryland 20892

KEY WORDS: selenophosphate, Se-dependent deiodinases, Se-Mo coordination, secys-thioredoxin reductase, selenocysteyl-tRNA^{Sec}, secys insertion elements, 21st amino acid

ABSTRACT

Selenocysteine is recognized as the 21st amino acid in ribosome-mediated protein synthesis and its specific incorporation is directed by the UGA codon. Unique tRNAs that have complementary UCA anticodons are aminoacylated with serine, the seryl-tRNA is converted to selenocysteyl-tRNA and the latter binds specifically to a special elongation factor and is delivered to the ribosome. Recognition elements within the mRNAs are essential for translation of UGA as selenocysteine. A reactive oxygen-labile compound, selenophosphate, is the selenium donor required for synthesis of selenocysteyl-tRNA. Selenophosphate synthetase, which forms selenophosphate from selenide and ATP, is found in various prokaryotes, eukaryotes, and archaebacteria. The distribution and properties of selenocysteine-containing enzymes and proteins that have been discovered to date are discussed. Artificial selenoenzymes such as selenosubtilisin have been produced by chemical modification. Genetic engineering techniques also have been used to replace cysteine residues in proteins with selenocysteine. The mechanistic roles of selenocysteine residues in the glutathione peroxidase family of enzymes, the 5′ deiodinases, formate dehydrogenases, glycine reductase, and a few hydrogenases are discussed. In some cases a marked decrease in catalytic activity of an enzyme is observed when a selenocysteine residue is replaced with cysteine. This substitution caused complete loss of glycine reductase selenoprotein A activity.

CONTENTS

INTRODUCTION 84
BIOSYNTHESIS AND SPECIFIC INSERTION OF SELENOCYSTEINE INTO
     PROTEINS 85
Selenocysteine Biosynthesis 85
A Special Translation Factor Is Required for Decoding UGA With
     Selenocysteyl-tRNA^{Sec} 87

[1]The US Government has the right to retain a nonexclusive royalty-free license in and to any copyright covering this paper.

Recognition Elements Within mRNAs Are Required for Decoding UGA as
        Selenocysteine Insertion Rather Than as Termination   87
The Biological Selenium Donor, Selenophosphate, Is Formed by the selD Gene
        Product, Selenophosphate Synthetase   89
Selenophosphate Synthetase Properties   90
Distribution of Selenophosphate Synthetase   91

ENZYMES AND PROTEINS THAT CONTAIN SPECIFIC SELENOCYSTEINE
        RESIDUES   91

SELENOCYSTEINE-CONTAINING PROTEINS PRODUCED BY CHEMICAL
        MODIFICATION OR GENETIC ENGINEERING TECHNIQUES   94

BIOCHEMICAL ROLES OF SELENOCYSTEINE RESIDUES IN ENZYMES   95

RELATIVE CATALYTIC ACTIVITIES OF CYSTEINE- VS
        SELENOCYSTEINE-CONTAINING ENZYMES   97

INTRODUCTION

The selenium biochemistry field has expanded rapidly since my previous reviews (1–3) and the 1991 reviews of August Böck and his associates (4, 5). During this period it has become well established in both eukaryotes and prokaryotes that selenocysteine is the 21st amino acid in ribosome-mediated protein synthesis. Our understanding of the numerous factors involved in the incorporation of this selenoamino acid in proteins in response to the UGA codon is based on the work of August Böck and his colleagues. Their analysis of a series of *Escherichia coli* mutants defective in their ability to synthesize selenium-dependent formate dehydrogenases revealed requirements for four new gene products that are directly involved in generation and insertion of selenocysteine into proteins. One of these products is a unique tRNA and two others are new enzymes that are essential for the formation of selenocysteyl-tRNA from seryl-tRNA. A fourth is a new elongation factor that specifically recognizes the selenocysteyl-tRNA. In eukaryotes an analogous specific tRNA recognizes the UGA codon (6), and selenocysteine is synthesized from serine esterified to this tRNA (7), but the lack of a corresponding series of mutants has prevented elucidation of the overall process.

New selenocysteine-containing enzymes and proteins are reported with greater frequency, and the specific roles of those with known catalytic functions illustrate the importance of the trace element selenium for mammalian and bacterial survival. For humans in particular, the key roles of the glutathione peroxidase family of enzymes in scavenging deleterious peroxides (8), and the synthesis and maintenance of optimum levels of essential thyroid hormones by the selenium-dependent deiodinase enzymes (9), serve as examples. The present review includes more recent information on the biosynthesis of seleno-cysteine, its presence in various newly discovered selenoproteins, and the types of catalytic roles of this amino acid in selenoenzymes. The biological importance of a key selenium donor compound, selenophosphate, and the enzyme, selenophosphate synthetase, responsible for its synthesis, is discussed.

BIOSYNTHESIS AND SPECIFIC INSERTION OF SELENOCYSTEINE INTO PROTEINS

Selenocysteine Biosynthesis

Important early findings in prokaryotes and eukaryotes established that seleno-cysteine incorporation into proteins is a cotranslational event directed by the UGA codon (10, 11). The impressive array of *E. coli* gene products required for selenocysteine incorporation are described in the review by Böck and associates in *Molecular Microbiology* (5). The genes are designated *sel*A, *sel*B, *sel*C, and *sel*D. Selenocysteine esterified to a unique tRNA^{Sec} [*sel*C gene product; anticodon UCA complimentary to UGA; (12)] is formed *in situ* from a serine residue initially ligated to this tRNA (Figure 1). A pyridoxal phosphate-dependent enzyme, selenocysteine synthase (*sel*A gene product), reacts with the seryl-tRNA^{Sec} forming aminoacrylyl-tRNA^{Sec} (13). Addition of a reactive selenium derivative across the double bond of the aminoacrylyl residue leads to the formation of selenocysteyl-tRNA^{Sec} (14). The reactive selenium donor compound, later identified as monoselenophosphate (15), is a product of selenophosphate synthetase, the *sel*D gene product (16–19). It is

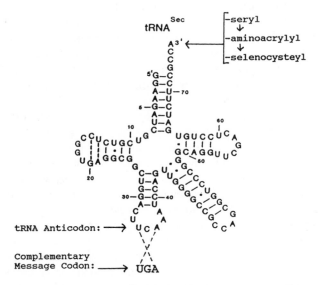

Figure 1   Nucleotide sequence of tRNA^{Sec} from a prokaryote. *E. coli* tRNA^{Sec} (4) is esterified with serine and converted to selenocysteyl-tRNA^{Sec} in a two-step process.

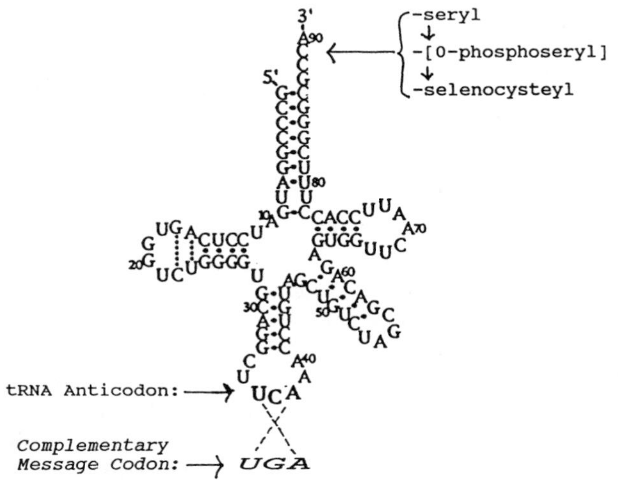

Figure 2 Nucleotide sequence of tRNA^{Sec} from a eukaryotic source. Bovine liver tRNA^{Sec} (6) containing an esterified serine is converted to selenocysteyl-tRNA^{Sec}.

important to emphasize that for specific insertion of selenocysteine into proteins the esterified form of this amino acid is derived from an L-serine residue previously ligated to tRNA^{Sec} by L-serine ligase, and the intermediate is also in ester linkage to the tRNA (Figure 1). Although free selenocysteine is synthesized by *E. coli* and *Salmonella typhimurium,* the free amino acid is not directly ligated to the *selC* tRNA. Instead, as observed earlier in *selD* mutants (20), this free selenocysteine can be esterified to tRNA^{Cys} (21) and then inserted randomly in many proteins in place of cysteine.

In eukaryotes the mechanism of synthesis of selenocysteyl-tRNA^{Sec} is not known in detail. Sunde's early discovery that the carbon skeleton of serine is a direct precursor of selenocysteine in glutathione peroxidase (22) provided an important clue concerning the mechanism of selenocysteine incorporation into mammalian proteins. The presence of a tRNA^{Sec} analogous to the *E. coli* *selC* gene product, the esterification of this tRNA with serine, and its conversion to selenocysteyl-tRNA^{Sec} are now well established [Figure 2; (7)]. However, except for the fact that 0-phosphoseryl-tRNA^{Sec}, along with seryl- and selenocysteyl-tRNA^{Sec}, was present in the aminoacyl-tRNA^{Sec} population isolated from a rat mammary tumor cell line (7), there is no direct evidence in support of an intermediate role for the phosphorylated compound. Suggestions that aminoacrylyl-tRNA^{Sec} is the intermediate in eukaryotes have not been substantiated. An enzyme analogous to the *selD* gene product has now been found in mammalian tissues (23), and genes exhibiting regions of homology with *E. coli selD* have been isolated from murine and human sources (24–27). Thus it appears that selenophosphate is the active selenium donor compound common to eukaryotes as well as prokaryotes. (See below for a discussion of selenophosphate and selenophosphate synthetase).

A Special Translation Factor Is Required for Decoding UGA with Selenocysteyl-tRNA*Sec*

Accumulation of selenocysteyl-tRNA[Sec] by the *E. coli selB* mutant led to the discovery of a new GTP-dependent translation factor that specifically interacts with the selenocysteine-charged tRNA (28, 29). This translation factor, the *selB* gene product, is alternate in its function to elongation factor EF-Tu, and it transports the selenocysteyl-tRNA[Sec] to the ribosome. The high affinity and specificity of binding of selenocysteyl-tRNA[Sec] to elongation factor SELB were demonstrated in RNase protection experiments and in direct binding assays. In contrast, little or no binding of seryl-tRNA[Sec] to SELB was detected. In the presence of SELB the rate of deacylation of selenocysteyl-tRNA[Sec] at alkaline pH decreased markedly, whereas deacylation of seryl-tRNA was unaffected by SELB. This difference suggests that binding of the ionized selenol group to SELB contributes to the stability of the ester bond. In similar experiments with EF-Tu, binding of tRNA[Sec] charged with either serine or selenocysteine was drastically reduced compared to the affinities of other charged aminoacyl tRNAs for EF-Tu. The extra base pair in the aminoacyl acceptor stem of tRNA[Sec] (vs the canonical seven base pair–tRNA stem) proved an important determinant in specificity of binding to elongation factors SELB and Tu (30). When the eight base pair–stem length of tRNA[Sec] was reduced to seven pairs, interaction with SELB was prevented, but binding to EF-Tu was now allowed. Similar effects were observed when the exceptionally long variable loop of tRNA[Sec] was replaced with the much shorter variable loop of tRNA[Ser]. The SELB protein, 68 kDa, although considerably larger than the 43 kDa EF-Tu, exhibits regions of sequence similarity in the N-terminal domain to regions of EF-Tu, including conserved GTP binding motifs.

Recognition Elements Within mRNAs Are Required for Decoding UGA as Selenocysteine Insertion Rather Than as Termination

The mechanism of distinguishing between selenocysteine insertion into a polypeptide and termination of polypeptide synthesis by release factors in response to the presence of a UGA codon in the open reading frame of an mRNA has been investigated in mutagenesis studies of prokaryotic and eukaryotic genes. Deletion of regions of the *E. coli* formate dehydrogenase gene (fdhF) from the

3' side of TGA 140 revealed that a 40-base sequence of the message immediately downstream from the UGA 140 must be present for efficient decoding with selenocysteine (31). There are a few bases within the loop of this putative stem-loop structure that are uniquely required (32). Positional effects of the putative stem-loop structure in this region in relation to the UGA codon strongly influenced efficiency of selenocysteine insertion. Thus increasing the distance between UGA and the unique secondary structure tended to decrease selenium-dependent translation, and nonspecific suppression of UGA was increasingly observed (33). The strong and overriding influence of the stem-loop structure is also illustrated by the reported ability of the two other termination codons, UAA and UAG, to direct selenocysteine insertion when they were substituted for UGA in the fdhF mRNA, provided the tRNA[Sec] anticodon was appropriately modified (32).

In eukaryotic selenoprotein messages there are now three examples of putative stem-loop sequences required for UGA decoding that are located in the 3' untranslated regions of the mRNAs, far removed from the UGA codon. These sequences, termed SECIS (selenocysteine insertion sequence) elements, were first detected in the 5' deiodinase (5' DI) and the glutathione peroxidase mRNAs (34) and were shown to be interchangeable. Later two such SECIS elements were identified in the 3' untranslated region of rat selenoprotein P mRNA, and these either together or singly could substitute for the rat 5' DI SECIS element (35). In fact the selenoprotein P mRNA element designated loop 1 was considerably more active than the original 5' DI SECIS element in allowing read-through of UGA. Although the glutathione peroxidase 3' untranslated region was shown to be sufficient to direct translation of a UGA codon inserted within the fused open reading frame of an unrelated nonselenoprotein as selenocysteine (36), it is clear that under normal conditions translation of authentic UGA termination codons must be very inefficient or entirely lacking. Investigation of predicted secondary structures within the coding regions of these three eukaryotic messages in the vicinity of the UGA codons revealed no stringent requirement for a particular context (35, 36). In this respect the requirements for selenocysteine insertion in eukaryotes differ from those operative in *E. coli* and closely related bacteria. In unrelated prokaryotes such as *Clostridium sticklandii* the nature of the mRNA element required for selenocysteine insertion is unknown. Although the glycine reductase selenoprotein A gene was expressed in *E. coli,* the UGA codon was translated primarily as cysteine, and only about 10% of the enzyme molecules were catalytically active owing to selenocysteine incorporation (37). The enzyme synthesized in media lacking selenium was catalytically inactive. The fact that enzyme of 10% normal activity also was produced in an *E. coli sel*D mutant in the presence of selenium was indicative of tRNA[Cys]-mediated nonspecific insertion of selenocysteine at UGA (37).

The Biological Selenium Donor, Selenophosphate, Is Formed by the selD Gene Product, Selenophosphate Synthetase

A mutation in *S. typhimurium* (38) and in *E. coli* (39) that prevented selenium incorporation in formate dehydrogenases and also in tRNAs was shown to be due to a defect in a single gene termed the *selD* gene (16, 20). Specifically these mutants were unable to (*a*) synthesize selenocysteyl-tRNA[Sec] from seryl-tRNA[Sec], and (*b*) convert a 2-thiouridine residue in the anticodons of certain tRNAs to 2-selenouridine (14, 17). The factor common to these two dissimilar processes proved to be a labile selenium compound formed from ATP and selenide. This labile selenium compound is required in an addition-type reaction for conversion of aminoacrylyl-tRNA[Sec], the intermediate formed from seryl-tRNA[Sec] by selenocysteine synthase, to selenocysteyl-tRNA[Sec] (Figure 1; 16). This compound is also required for a substitution-type reaction in which the sulfur of a 2-thiouridine residue in tRNAs is replaced with selenium (17). This labile, oxygen-sensitive compound was identified as a selenophosphate by ^{31}P NMR (17), and by comparison with the authentic synthetic compound the enzyme product was determined to be monoselenophosphate (15). The synthetic compound and the enzyme product were used interchangeably for the synthesis of a 2-selenouridine from a 2-thiouridine in tRNAs by a purified enzyme termed tRNA 2-selenouridine synthase (40). The only substrates required for the reaction catalyzed by the purified synthase are the thio-tRNA and selenophosphate. The mechanism of this substitution reaction, which is independent of ATP, appears to involve a direct attack on carbon-2 of the 2-thiouridine substrate by selenophosphate. Rearrangement of the intermediate followed by elimination of sulfur (as thiophosphate?) with retention of selenium could be envisioned as a possible mechanism.

The unique chemical properties of selenophosphate, particularly the relative weakness of the P-Se bond compared to a P-S or P-O bond (Figure 3) make it especially suited for its roles in the above addition and substitution reactions.

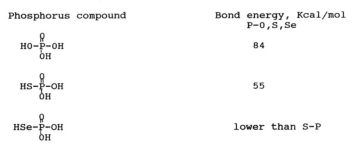

CHEMICAL BOND ENERGIES

Phosphorus compound	Bond energy, Kcal/mol P-O,S,Se
O‖ HO-P-OH OH	84
O‖ HS-P-OH OH	55
O‖ HSe-P-OH OH	lower than S-P

Figure 3   P-O, P-S, and P-Se bond energies. Data from (41).

The ability of selenophosphate to serve also as a phosphate donor producing phosphate esters of various alcohols (Glass, personal communication) suggests that this property may also be exploited in nature.

Selenophosphate Synthetase Properties

Selenophosphate synthetase, the *selD* gene product, is a 37 kDa protein (16) that is isolated from an overproducing *E. coli* strain (16, 18). In the reaction catalyzed by this enzyme, ATP and selenide are converted to equimolar amounts of selenophosphate, orthophosphate, and AMP (18). The γ-phosphoryl group of ATP is present in selenophosphate, and orthophosphate is derived from the β-phosphoryl group (18, 19). The fact that AMP is a competitive inhibitor of ATP in the catalytic reaction, whereas selenophosphate and orthophosphate inhibit significantly only at relatively high concentrations, indicates that the overall reaction is a multi-step process (18). Initial formation of an unstable enzyme-pyrophosphate intermediate with liberation of AMP could explain the observation that in the absence of selenide the enzyme can catalyze the complete conversion of ATP to AMP and orthophosphate (18). No pyrophosphate formation could be detected, and ADP is neither formed nor utilized by the highly purified enzyme. Attempts to detect the putative enzyme-PP by ^{31}P NMR using high enzyme levels have not been successful.

Analysis of mutant forms of selenophosphate synthetase produced by site-directed mutagenesis revealed that Cys-17 and Lys-20 located in a glycine-rich region near the amino-terminus are essential for catalytic activity, whereas substitution of Cys-19 with serine had no effect on enzyme activity (42). In experiments to define the ATP-binding site on the enzyme Mg-ATP, which is required for catalytic activity, was replaced with Mn-ATP. Replacement of either Cys-17 or Cys-19 with serine had little effect on ATP binding, but the double Cys-17,19 mutant and the Lys-20 mutants (K20R and K20Q) exhibited considerably lowered ATP-binding affinities (43). Labeling studies with [γ-^{32}P]8-azido-ATP, [^{32}P]ATP and [^{14}C]ATP provide further evidence in support of the importance of the two cysteine residues for ATP binding to the enzyme. A monovalent cation (K^+, NH_4^+, or Rb^+), which is essential for catalytic activity of selenophosphate synthetase (18) is also required for MnATP binding (44). Replacement of K^+ with Na^+ prevents MnATP binding, indicating a specific monovalent cation–induced conformational state of the enzyme. Low levels of zinc, which inhibit catalytic activity with MgATP, also inhibit MnATP binding to the enzyme. In view of the number of manipulatable variables affecting ATP binding and catalytic activities of selenophosphate synthetase, it seemed that conditions might be found that would favor stabilization of the putative enzyme-bound pyrophosphate intermediate. To date only very limited success has been realized, particularly in attempts to isolate derivatized peptides (SY Liu & TC Stadtman, unpublished observations).

Distribution of Selenophosphate Synthetase

Evidence of the widespread occurrence of selenophosphate synthetase, and by inference the general biological use of selenophosphate as selenium donor, comes from three types of current investigations. In our laboratory, using antibodies elicited to the overproduced *E. coli* enzyme in immunoblotting procedures, we detected the presence of selenophosphate synthetase in various mammalian tissues and in an archaean, *Methanococcus vannielii* (23; HL Kruschwitz & TC Stadtman, unpublished observations). Partially purified protein fractions from rat brain and from *M. vannielii* exhibited selenophosphate synthetase catalytic activity. A partial amino acid sequence of the *M. vannielii* enzyme (residues 1–25 from the amino terminus) did not include the catalytically important glycine-rich segment (HGAGCGCK) encompassing residues 13–20 of the *E. coli* enzyme, and the sequence otherwise differed significantly from the corresponding region of the *E. coli* enzyme. This finding is not surprising because in some respects Archaebacteria resemble eukaryotes more than prokaryotes.

In Marla Berry's laboratory (24, 25) mouse and human genes exhibiting regions of homology with the *E. coli* selenophosphate synthetase (*sel*D) gene were cloned and sequenced. The human gene sequence encoded a motif GTGCK (residues 28–32) resembling the *E. coli* 16–20 residue segment except that Cys-17 was replaced with threonine. Expression of the human gene was shown to regulate the synthesis of ^{75}Se-labeled 5′ deiodinase, providing evidence of the production of selenophosphate in the system. Especially interesting is the finding that an analogous gene detected in the mouse embryo at early stages of development contains an N-terminal sequence region corresponding to the *E. coli* residue 16–20 segment except that a selenocysteine residue is encoded in place of the essential Cys-17 residue (26, 27). This selenocysteine-containing gene product should be more active, and it will be interesting to compare the catalytic properties of the three purified enzymes when all are available. Both of the eukaryotic genes and the *E. coli sel*D gene contain a nucleotide sequence characteristic of a putative ATP binding domain, GXGXXG. This sequence, which corresponds to the amino acid sequence TGFGILGH located in the carboxy-terminal regions of the proteins, is completely conserved in the three genes.

ENZYMES AND PROTEINS THAT CONTAIN SPECIFIC SELENOCYSTEINE RESIDUES

For several years the only enzymes known to contain essential selenocysteine residues were some bacterial formate dehydrogenases, clostridial glycine reductase, a few hydrogenases of anaerobic bacteria, and mammalian glutathione peroxidases (1–5, 8). Recently the list of selenocysteine-containing enzymes and proteins has lengthened considerably. Among the new additions to the list is mammalian selenoprotein P, a glycoprotein of unknown function originally isolated from plasma (45, 46). The protein is expressed and secreted by many tissues in addition to liver and kidney (47). Selenoprotein P, a 42 kDa polypeptide, is unusual in that according to the gene sequence it should contain 10 selenocysteine residues although on average the isolated protein preparations contain about 7.5 selenocysteine residues per monomer (48). There is evidence that part of this discrepancy is the result of premature termination at one or more of the UGA codons, which results in a mixture of isoenzyme forms (47). In addition to selenocysteine the protein is rich in cysteine and histidine. An antioxidant role of the protein has been suggested but no regenerating system has been described. Although selenoprotein P turns over rapidly with an estimated half-life of 4 hr, this might be too slow to support its role as a single-turnover antioxidant.

Selenoprotein W, a low–molecular weight protein originally reported as a missing component in selenium-deficient animals suffering from white muscle disease, has recently been isolated from rat muscle (49). The partial amino acid sequence of the isolated protein (49), together with the deduced sequence determined from the cloned gene (50), shows that there is a single selenocysteine residue in the approximately 10 kDa protein. A TGA codon in the gene corresponds to the location of selenocysteine in the protein. In this case a second UGA codon in the message functions as a termination signal. As yet no function for selenoprotein W has been described. The protein is a component of muscle, including heart muscle, and it has been detected in many animal species (51).

The thyroid hormone deiodinases (9) consist of at least three types now designated Dl, D2, and D3. Mammalian Type I 5′ deiodinase (D1), an integral membrane protein of microsomes (52), was the first to be shown to be a selenoenzyme (53). The presence of selenocysteine in the enzyme is based on the deduced amino acid sequence and a selenium requirement for translation of the 5′ deiodinase message. In the reaction catalyzed by the D1 deiodinase the 5′ iodine is removed from the outer ring of 3,5,3′5′-tetraiodothyronine (thyroxine) thereby forming the active thyroid hormone 3,5,3′-triiodothyronine. The microsomal enzyme, a 55 kDa protein containing two 27 kDa identical subunits, occurs primarily in thyroid, liver, kidney, and pituitary. The gene encoding a Type III 5-deiodinase (D3) that removes the 5-iodine from the inner ring of 3,5,3′-triiodothyronine recently was isolated from *Xenopus laevis* (54). The product of this reaction, 3,3′-diiodothyronine, is an inactive metabolite. Sequence analysis of the cDNA revealed regions of homology to rat Type I 5′ deiodinase and a conserved in-frame TGA codon. Replacement of the UGA codon by site-directed mutagenesis with a UAA stop codon or with

UUA, a leucine codon, resulted in mutant enzymes that were inactive catalytically. Replacement with a UGU codon produced a cysteine mutant exhibiting reduced 5-deiodinase activity. These findings indicate that this deiodinase also is a selenocysteine-containing enzyme. In tadpoles this Type III 5-deiodinase is of critical importance in regulating the levels of the active hormone 3,5,3′-triiodothyronine, which is essential for metamorphosis yet highly toxic at elevated concentrations.

The sensitivity of the 5′ deiodinase (D1) to aurothioglucose in the nanomolar range and to propylthiouracil in the micromolar range was previously attributed to the presence of a reactive selenocysteine residue in the enzyme (53). However, the resistance of the Xenopus D3 enzyme to these inhibitors at similar concentrations indicates that the range of sensitivity to these compounds is not strictly diagnostic for the presence of selenocysteine (54). Sensitivity of the mammalian Type II 5′ deiodinase to propyl-thiouracil appears to depend on reaction conditions (55). Since none of the wild-type forms of the various deiodinases have been isolated in a highly purified state and characterized in detail, factors that determine sensitivity to inhibitors, precise catalytic activity, and other properties are difficult to establish.

A ^{75}Se-labeled 110 kDa flavoprotein isolated from cultured cells of a human lung adenocarcinoma cell line was shown by amino acid analysis to contain selenocysteine (85). The purified protein, a dimer of apparently identical 57 kDa subunits, contains FAD and exhibits NADPH-dependent thioredoxin reductase activity. The substrate specificity of this enzyme is comparable to that of mammalian liver thioredoxin reductase, but antibodies elicited to the liver enzyme do not crossreact with the selenoenzyme. The amino terminus of the isolated protein is blocked, and amino acid sequence data are not yet available. The distribution and possible unique biochemical roles of this new selenoenzyme are of special interest.

A gene detected at early stages of hematopoietic development in mouse embryos was cloned and sequenced (26, 27). Regions of sequence similarity with the *E. coli* selenophosphate synthetase (*sel*D) gene indicated this to be the mammalian equivalent of the bacterial gene (see above). Especially interesting was the presence of an in-frame TGA codon located in a motif corresponding to the active site of *E. coli* selenophosphate synthetase. The TGA codon was also present in a cloned human selenophosphate synthetase gene (26). The location of the putative encoded selenocysteine residue corresponds to the position of the essential Cys-17 residue in the *E. coli* enzyme, suggesting that this mammalian enzyme may be a more active catalyst.

According to the gene sequence, a cytochrome P-450 of human origin, designated hIIB3, should contain selenocysteine owing to the presence of an in-frame TGA codon (56). The corresponding mRNA was detected in normal human lung, as well as in a human lung adenocarcinoma cell line, suggesting that the cytochrome P-450 should be present (F Gonzalez, personal communication).

A 35–40 kDa selenocysteine-containing protein detected in extracts of ^{75}Se-labeled *M. vannielii* cells was purified and the amino acid sequence of the N-terminal residues 1–25 was determined (TC Stadtman, unpublished). A higher level of this protein was detected in extracts from a batch of cells cultured on formate with urate as the nitrogen source than in an extract from cells grown on formate plus ammonium chloride. The identity of this protein is unknown.

Numerous reports have shown that selenium is important for fertility of domestic animals, and the presence in sperm of a low–molecular weight selenoprotein has been suggested as the responsible entity (57, 58). This mitochondrial capsule protein contains high amounts of cysteine and proline. Structural analysis of a gene from mouse sperm that encoded a protein of similar composition showed the presence of three TGA codons in the region corresponding to residues 1–35 of the gene product (59). In subsequent studies, however, comparable genes from other animal species have been found to lack TGA codons (KC Kleene, personal communication). Thus the chemical identity of the selenium that has been detected in sperm mitochondrial capsule protein is unclear.

SELENOCYSTEINE-CONTAINING PROTEINS PRODUCED BY CHEMICAL MODIFICATION OR GENETIC ENGINEERING TECHNIQUES

There is one example of the direct synthesis of a selenocysteine-containing protein by the Merrifield method using an automated peptide synthesizer (60), but because of the lability of the selenocysteine derivatives during the ligation, deprotection, and purification steps this approach is seldom used. In view of these problems it is particularly impressive that the selenium analog of *Neurospora crassa* copper metallothionein could be produced because in this case direct insertion of seven selenocysteine residues in the 25–residue polypeptide was accomplished (60).

The selenium analog of subtilisin in which selenocysteine was substituted for the active-site serine residue was prepared from the native protease. Derivatization of the active-site serine with phenylmethylsulfonyl fluoride produced the sulfonyl ester of the serine hydroxyl group, and this, in the presence of selenide, was eliminated and replaced with selenium. The resulting seleno-subtilisin has been used as a glutathione peroxidase model in ^{77}Se NMR studies (61). The crystal structure of the selenoenzyme has been determined at 2.0Å-resolution (62). Selenosubtilisin, like glutathione peroxidase, reacts with a variety of hydroperoxides but, unlike glutathione peroxidase, tertiary-butyl peroxide is not a substrate of selenosubtilisin (63). Whereas glutathione and other alkyl thiols are relatively poor reducing agents for oxidized selenosubtilisin, the aromatic thiol 3-carboxy-4-nitrobenzenethiol allows efficient regeneration of the reduced enzyme (63). In a nonenzymic reaction, reduction of peroxides by 3-carboxy-4-nitrobenzenethiol involves free radicals, but the selenosubtilisin-mediated process was unaffected by the presence of a radical trap. A ping-pong type of reaction mechanism involving selenol (ESeH), selenenic acid (ESeOH), and selenenyl sulfide (ESeSR) has been implicated in the enzyme–catalyzed redox process analogous to that proposed for glutathione peroxidase.

Replacement of the two redox active cysteine residues in *E. coli* thioredoxin with selenocysteine was achieved by expression of the *E. coli* thioredoxin gene (trxA) in a cysteine auxotrophic strain of the organism in the presence of DL-selenocysteine (64). Because of the toxicity of the selenoamino acid at the concentration added (600 µM) the bacteria had to be cultured first in the presence of a limited amount of cysteine. After residual cysteine was washed out the thioredoxin gene was induced in cells resuspended in a medium containing selenocystine. The population of thioredoxin molecules in the isolated enzyme consisted of 75–80% containing two selenocysteine residues, 5–10% containing one, and 12–17% containing only cysteine. As would be expected, in vitro conversion of the diselenide form of thioredoxin to the diselenol form required a more powerful reducing agent than is needed to reduce the native disulfide form to the dithiol.

BIOCHEMICAL ROLES OF SELENOCYSTEINE RESIDUES IN ENZYMES

For a few selenocysteine-containing enzymes some information is available on the precise catalytic roles of the selenoamino acid residue in the catalytic processes. It is generally accepted that the mammalian glutathione peroxidases in the reduced form contain an ionized selenol that can react with an organic peroxide or H_2O_2 to form an enzyme-selenenic acid (RSeOH) product as shown in Figure 4. Regeneration of the reduced enzyme is a two-step process involving selenenylsulfide (Enz-SeS-G) as an intermediate with reduced glutathione (GSH) serving as reductant (65, 66). Reaction of the Enz-SeS-G with another equivalent of GSH regenerates Enz-SeH and forms GSSG. Support for this reaction scheme has been provided by ^{77}Se NMR studies on the model peroxidase, selenosubtilisin (61). From some of the kinetic studies there is suggestive evidence that reduction of the selenenylsulfide may be the slow step in the regeneration process.

In the reactions catalyzed by the selenium-dependent deiodinases, removal of iodine from a thyroid hormone substrate involves conversion of the ionized

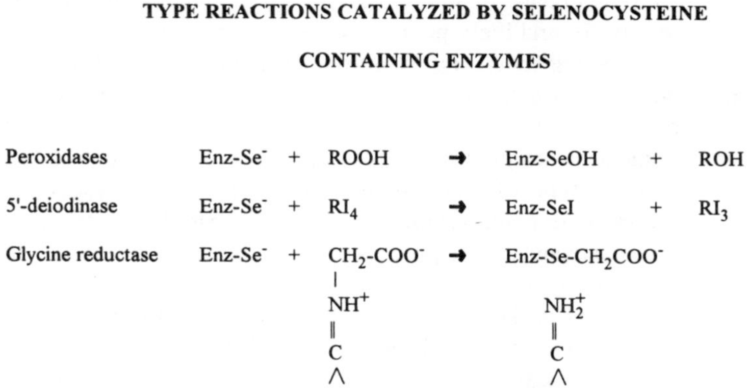

TYPE REACTIONS CATALYZED BY SELENOCYSTEINE

CONTAINING ENZYMES

Peroxidases	Enz-Se⁻	+	ROOH	→	Enz-SeOH	+ ROH
5'-deiodinase	Enz-Se⁻	+	RI₄	→	Enz-SeI	+ RI₃
Glycine reductase	Enz-Se⁻	+	CH₂-COO⁻	→	Enz-Se-CH₂COO⁻	

$$\text{Peroxidases} \quad \text{Enz-Se}^- + \text{ROOH} \rightarrow \text{Enz-SeOH} + \text{ROH}$$

$$\text{5'-deiodinase} \quad \text{Enz-Se}^- + \text{RI}_4 \rightarrow \text{Enz-SeI} + \text{RI}_3$$

$$\text{Glycine reductase} \quad \text{Enz-Se}^- + \underset{\underset{\text{C}}{\overset{\|}{\underset{\wedge}{\text{NH}^+}}}}{\text{CH}_2\text{-COO}^-} \rightarrow \text{Enz-Se-CH}_2\text{COO}^- \underset{\underset{\text{C}}{\overset{\|}{\text{NH}_2^+}}}{}$$

Figure 4   Some type reactions catalyzed by selenocysteine-containing enzymes.

selenol group of the enzyme to an RSeI derivative (Figure 4). Regeneration of the reduced enzyme by certain thiols, particularly DTT, is observed in vitro but the reductant that functions in vivo is uncertain. Sluggish reduction of Enz-SeI by GSH in vitro has led to the suggestion that additional factors are involved in the cell (67). With the potential availability of large quantities of an overexpressed Type I 5' deiodinase that contains cysteine in place of the selenocysteine residue, it may be possible to study in detail the requirements for the effective reduction of the RSI enzyme analogue.

A bacterial selenoenzyme, glycine reductase selenoprotein A, provides an example of a new type of catalytic role for a selenocysteine residue in an enzyme. This 18 kDa protein from *C. sticklandii* (37, 68–70) is part of an enzyme complex consisting of two other larger components, proteins B and C. In the overall glycine reductase reaction protein B forms a Schiff base with the substrate (71), and this intermediate is attacked by the ionized selenol of protein A, resulting in cleavage of the glycine carbon-nitrogen bond and transfer of the 2-carbon moiety to selenium (Figure 4; 72). The Se-carboxy-methyl derivative of selenoprotein A then is reductively cleaved and transferred to protein C, forming an acetylthiol ester derivative of protein C (73, 74). Transfer of the acetyl group to phosphate gives acetyl phosphate as the final product of the reaction or, in the presence of arsenate, free acetate is formed. The selenoether derivative of selenoprotein A, the Se-carboxymethy-lated selenocysteine residue, is not reduced by excess borohydride and is stable in the presence of DTT, but when incubated with protein C, DTT, and arsenate the carboxymethyl group is converted quantitatively to acetate (74). So far no analogous role of a selenocysteine residue in an enzyme has been found in eukaryotes, and in prokaryotes glycine reductase is still the only example.

Formate dehydrogenase H from *E. coli* contains a selenocysteine residue in the polypeptide (75), and the selenium atom of this amino acid is coordinated to molybdenum of the molybdopterin cofactor bound to the enzyme (76). In EPR studies the addition of formate to the native enzyme was shown to induce a signal typical of Mo(V) species (76). With enzyme containing ^{77}Se in place of normal abundance selenium, this signal was transformed, indicating direct coordination of selenium in the selenocysteine residue to the molybdenum atom. Mutant enzyme of much lower catalytic activity in which a cysteine residue replaces selenocysteine exhibited a different EPR signal. The appearance of a Mo(V) signal when the enzyme is reduced with formate under anaerobic conditions in the absence of an external electron acceptor may be explained as follows: The EPR-silent two-electron reduced Mo(IV) species initially formed by reaction with formate undergoes a one-electron oxidation by FeS centers in the enzyme giving rise to the EPR-detectible Mo(V) species which, under these conditions, is relatively stable (SV Kangulov, personal communication).

The large hydrogenase family of enzymes present in prokaryotes consists of those that contain iron sulfur centers and nickel and those that additionally contain selenium. One selenium-containing hydrogenase isolated from *M. vannielii* was shown by amino-acid analysis to contain selenocysteine (77) in addition to nickel, iron sulfur centers, and FAD (78). In some other hydrogenases a TGA codon identified in the gene indicates the presence of selenocysteine in the protein (79, 80). In one of these selenium-containing hydrogenases the nickel has been shown to be directly coordinated to selenium (81).

RELATIVE CATALYTIC ACTIVITIES OF CYSTEINE- VS SELENOCYSTEINE-CONTAINING ENZYMES

Except for detailed comparative studies (82) of highly purified wild-type and mutant forms of formate dehydrogenase H (hydrogenase linked) from *E. coli* little precise information is available on the effect of substitution of cysteine for selenocysteine on the catalytic activity of an enzyme. The relative activities of the selenium and sulfur containing formate dehydrogenases were compared using benzyl viologen as electron acceptor. The k_{cat}, sec^{-1} values determined were 2800 for the Se enzyme and 9 for the S enzyme, indicating that the selenocysteine-containing enzyme is 300 times more active catalytically. The Km value for formate exhibited by the selenocysteine enzyme was somewhat higher, 26 mM, than that for the cysteine enzyme, 9 mM. Inactivation of the two enzymes by alkylation was highly dependent on pH, reflecting the differences in pKa values of selenols and thiols.

In earlier studies on Type I 5' deiodinase (53) the selenocysteine-containing enzyme and the analogous mutant cysteine enzyme produced by replacement of the TGA codon with TGT were compared. Assays of enzyme activities in homogenized oocyte preparations showed that the cysteine-containing enzyme was only 10–20% as active as the native selenocysteine-containing enzyme. Replacement of TGA with TGT in the Xenopus Type III 5-deiodinase gene resulted in a 76% decrease in the activity of the gene product as measured in *X. laevis* oocytes (54). Detailed catalytic studies of these enzymes have not been carried out because of enzyme instability problems and lack of adequate amounts of purified enzymes.

Sulfur analogues of some of the glutathione peroxidases have been produced by various investigators, but the mutated genes have usually been expressed in systems such as reticulocyte lysates. Although much higher levels of the mutant gene products are obtained in these in vitro systems compared to the normal selenocysteine-containing enzyme products, the amounts are insufficient for isolation in active form and for detailed study. Recently in vitro synthesis of the cysteine analogue of rat phospholipid hydroperoxide glutathione peroxidase (83) and of the cysteine analogue of bovine cytosolic glutathione peroxidase (84) were reported. In instances where activities of the protein products have been estimated the cysteine-containing mutant enzymes apparently exhibit insignificant peroxidase activity (66).

Clostridial glycine reductase selenoprotein A is an example of a protein that has no detectable catalytic activity when the active site selenocysteine is replaced with cysteine (37). The gene encoding this protein contains the sequence TGC TTT GTC TGA, but when expressed in *E. coli* insertion of selenocysteine at UGA is nonspecific and occurs only to the extent of about 10%, resulting in a protein of 10% of normal activity. In the absence of selenium a protein product of comparable size and antigenic properties is synthesized in good yield but its activity as a glycine reductase component is undetectable.

From these examples the importance of an ionized selenol reaction center for the catalytic efficiency of an enzyme is clear. A selenol (pK_a = 5.2), in contrast to a normal thiol (pK_a = 8 or greater), is fully ionized in the normal physiological pH range and thus is fully active. To be of comparable reactivity a thiol must have an unusually low pK_a value owing to its location in a very specific environment. From recent studies it appears that more examples of effective selenium biological catalysts undoubtedly exist in nature and await discovery by persistent investigators.

Literature Cited

1. Stadtman TC. 1990. *Annu. Rev. Biochem.* 59:111–27
2. Stadtman TC. 1991. *J. Biol. Chem.* 266: 16257–60
3. Stadtman TC. 1994. In *Advances in Inorganic Biochemistry,* ed. GL Eichhorn, LG Marzilli, 10:157–75. Englewood Cliffs, NJ: Prentice-Hall
4. Böck A, Forchhammer K, Heider J, Baron C. 1991. *Trends Biochem. Sci.* 16:463–67
5. Böck A, Forchhammer K, Heider J, Leinfelder W, Sawers G, et al. 1991. *Mol. Microbiol.* 5:515–20
6. Hatfield D. 1985. *Trends Biochem. Sci.* 10:201–4
7. Lee BJ, Worland PJ, Davis JN, Stadtman TC, Hatfield DL. 1989. *J. Biol. Chem.* 264:9724–27
8. Schuckelt R, Brigelius-Flohé R, Maiorino M, Roveri A, Reumkens J, et al. 1991. *Free Radical Res. Commun.* 14: 343–61
9. Berry MJ, Larsen PR. 1995. *Annu. Rev. Nutr.* 15:323–52
10. Zinoni F, Birkmann A, Stadtman TC, Böck A. 1986. *Proc. Natl. Acad. Sci. USA* 83:4650–54
11. Chambers I, Frampton J, Goldfarb P, Affara N, McBain W, Harrison PP. 1986. *EMBO J.* 5:1221–27
12. Leinfelder W, Zehelin E, Mandrand-Berthelot M-A, Böck A. 1988. *Nature* 331:723–25
13. Forchhammer K, Leinfelder W, Boesmiller K, Veprek B, Böck A. 1991. *J. Biol. Chem.* 266:6318–23
14. Forchhammer K, Böck A. 1991. *J. Biol. Chem.* 266:6324–28
15. Glass RS, Singh WP, Jung W, Veres Z, Scholz TD, Stadtman TC. 1993. *Biochemistry* 32:12555–59
16. Leinfelder W, Forchhammer K, Veprek B, Zehelein E, Böck A. 1990. *Proc. Natl. Acad. Sci. USA* 87:543–47
17. Veres Z, Tsai L, Scholz TD, Politino M, Balaban RS, Stadtman TC. 1992. *Proc. Natl. Acad. Sci. USA* 89:2975–79
18. Veres Z, Kim IY, Scholz TD, Stadtman TC. 1994. *J. Biol. Chem.* 269:10597–603
19. Ehrenreich A, Forchhammer K, Tormay P, Veprek B, Böck A. 1992. *Eur. J. Biochem.* 206:767–73
20. Stadtman TC, Davis JN, Zehelein E, Böck A. 1989. *BioFactors* 2:35–44
21. Young PA, Kaiser II. 1975. *Arch. Biochem. Biophys.* 171:483–89
22. Sunde RA, Evenson JK. 1987. *J. Biol. Chem.* 262:933–37
23. Kim IY, Stadtman TC. 1995. *Proc. Natl. Acad. Sci. USA* 92:In press
24. Low SC, Harney JW, Bennett M, Reed R, Berry MJ. 1995. *BioFactors* 5:51
25. Low SC, Harney JW, Berry MJ. 1995. *J. Biol. Chem.* 270:21659–64
26. Zlotnik A, Bazan F, Cocks BG, McClanahan T, Wiles M, et al. 1995. *FASEB J.* 9:A833 (Abstr. 4832)
27. Guimaraes MJ, Lee F, Zlotnik A, Bazan F, McClanahan T. 1995. *FASEB J.* 9: A833 (Abstr. 4831)
28. Forchhammer K, Leinfelder W, Böck A. 1989. *Nature* 342:453–56
29. Forchhammer K, Rucknagel P, Böck A. 1990. *J. Biol. Chem.* 265:9346–50
30. Baron C, Böck A. 1991. *J. Biol. Chem.* 266:20375–79
31. Zinoni F, Heider J, Böck A. 1990. *Proc. Natl. Acad. Sci. USA* 87:4660–64
32. Heider J, Baron C, Böck A. 1992. *EMBO J.* 11:3759–66
33. Chen G-FT, Fang L, Inouye M. 1993. *J. Biol. Chem.* 268:23128–31
34. Berry MJ, Banu L, Chen Y, Mandel SJ, Kieffer JD, et al. 1991. *Nature* 353:273–76
35. Berry MJ, Banu L, Harney JW, Larsen PR. 1993. *EMBO J.* 12:3315–22
36. Shen Q, Chu F-F, Newburger PE. 1993. *J. Biol. Chem.* 268:11463–69
37. Garcia GE, Stadtman TC. 1992. *J. Bacteriol.* 174:7080–89
38. Kramer GF, Ames BN. 1988. *J. Bacteriol.* 170:736–43
39. Leinfelder W, Forchhammer K, Zinoni F, Sawyers G, Mandrand-Berthelot M-A, Böck A. 1988. *J. Bacteriol.* 170:540–46
40. Veres Z, Stadtman TC. 1994. *Proc. Natl. Acad. Sci. USA* 91:8092–96
41. Cotton FA, Wilkinson G. 1962. *Advances in Inorganic Chemistry.* New York: Interscience/Wiley. 88 pp.
42. Kim IY, Veres Z, Stadtman TC. 1992. *J. Biol. Chem.* 267:19650–54
43. Kim IY, Veres Z, Stadtman TC. 1993. *J. Biol. Chem.* 268:27020–25
44. Kim IY, Stadtman TC. 1994. *Proc. Natl. Acad. Sci. USA* 91:7326–29
45. Burk RF, Gregory PE. 1982. *Arch. Biochem. Biophys.* 213:73–80
46. Yang J-G, Morrison-Plummer J, Burk RF. 1987. *J. Biol. Chem.* 262:13372–75
47. Burk RF, Hill KE. 1994. *Proc. J. Nutr.* 124:1891–97
48. Hill KE, Lloyd RS, Yang J-G, Read R, Burk RF. 1991. *J. Biol. Chem.* 266: 10050–53
49. Vendeland SC, Beilstein MA, Chen CL,

Jensen ON, Barofsky E, Whanger PD. 1993. *J. Biol. Chem.* 268:17103–7

50. Vendeland SC, Beilstein MA, Yeh J-Y, Ream W, Whanger PD. 1995. *Proc. Natl. Acad. Sci. USA* 92:8749–53

51. Yeh J-Y, Beilstein MA, Andrews JS, Whanger PD. 1995. *FASEB J.* 9:392–96

52. Toyoda N, Berry MJ, Harney JW, Larsen PR. 1995. *J. Biol. Chem.* 270: 12310–18

53. Berry MJ, Banu L, Larsen PR. 1991. *Nature* 349:438–40

54. St. Germain DL, Schwartzman RA, Croteau W, Kanamori A, Wang Z, et al. 1994. *Proc. Natl. Acad. Sci. USA* 91: 7767–71

55. Safran M, Leonard JL. 1991. *J. Biol. Chem.* 266:3223–28

56. Yamano S, Nhamburo PT, Aoyama T, Meyer UA, Inaba T, et al. 1989. *Biochemistry* 28:7340–48

57. Calvin HI, Cooper GW, Wallace E. 1981. *Gamete Res.* 4:139–49

58. Wallace E, Calvin HI, Ploetz K, Cooper GW. 1984. In *Selenium in Biology and Medicine,* ed. GF Combs Jr, OA Levander, JA Oldfield, JE Spallholz, Part A, pp. 181–96. New York: AVI

59. Karimpour I, Cutler M, Shih D, Smith J, Kleene KC. 1992. *DNA Cell Biol.* 11:693–99

60. Oikawa T, Esaki N, Tanaka H, Soda K. 1991. *Proc. Natl. Acad. Sci. USA* 88: 3057–59

61. House KL, Dunlap RB, Odom JD, Wu Z-P, Hilvert D. 1992. *J. Am. Chem. Soc.* 114:8573–79

62. Syed R, Wu Z-P, Hogle JM, Hilvert D. 1993. *Biochemistry* 32:6157–64

63. Bell IM, Hilvert D. 1993. *Biochemistry* 32:13969–73

64. Muller S, Senn H, Gsell B, Vetter W, Baron C, Böck A. 1994. *Biochemistry* 33:3404–12

65. Flohé L. 1989. In *Glutathione:Chemical, Biochemical, and Medical Aspects,* ed. D Dolphin, R Poulson, O Avamovie, pp. 644–731. New York: Wiley

66. Ursini F, Maiorino M, Brigelius-Flohé R, Aumann KD, Roveri A, et al. 1995. *Methods Enzymol.* 252:38–53

67. Leonard JL, Visser TJ. 1986. In *Thyroid Hormone Metabolism,* ed. G Heinemann, pp. 189–229. New York: Dekker

68. Turner DC, Stadtman TC. 1973. *Arch. Biochem. Biophys.* 154:366–81

69. Cone JE, del Rio RM, Davis JN, Stadtman TC. 1976. *Proc. Natl. Acad. Sci. USA* 73:2659–63

70. Kimura Y, Stadtman TC. 1995. *Proc. Natl. Acad. Sci. USA* 92:2189–93

71. Tanaka H, Stadtman TC. 1979. *J. Biol. Chem.* 254:447–52

72. Arkowitz RA, Abeles RH. 1990. *J. Am. Chem. Soc.* 112:870–72

73. Arkowitz RA, Abeles RH. 1991. *Biochemistry* 30:4090–97

74. Stadtman TC, Davis JN. 1991. *J. Biol. Chem.* 266:22147–53

75. Stadtman TC, Davis JN, Ching W-M, Zinoni F, Böck A. 1991. *BioFactors* 3:21–27

76. Gladyshev VN, Khangulov SV, Axley MJ, Stadtman TC. 1994. *Biochemistry* 91:7708–11

77. Yamazaki S. 1982. *J. Biol. Chem.* 257: 7926–29

78. Yamazaki S. 1984. *Selenium in Biology and Medicine, 3rd Int. Symp., Beijing, China,* ed. GF Combs Jr, JE Spallholz, OA Levander, JE Oldfield, pp. 230–35. New York: AVI

79. Menon NK, Peck HD Jr, LeGall J, Pryzbyla AE. 1988. *J. Bacteriol.* 170: 4429

80. Sorgenfrei O, Linder D, Karas M, Klein A. 1993. *Eur. J. Biochem.* 213: 1355–58

81. He SH, Teixeira M, LeGall J, Patil DS, Moura I, et al. 1989. *J. Biol. Chem.* 264:2678–82

82. Axley MJ, Böck A, Stadtman TC. 1991. *Proc. Natl. Acad. Sci. USA* 88: 8450–54

83. Thimmalapura P-R, Burdsall A, Oleksa L, Chisolm G, Driscoll D. 1995. *BioFactors* 5:49–50

84. Jung J-E, Karoor V, Sandbaken MG, Lee BJ, Ohama T, et al. 1994. *J. Biol. Chem.* 269:29739–45

85. Tamura T, Stadtman TC. 1996. *Proc. Natl. Acad. Sci. USA*: In press

Annu. Rev. Biochem. 1996. 65:101–33

MISMATCH REPAIR IN REPLICATION FIDELITY, GENETIC RECOMBINATION, AND CANCER BIOLOGY

Paul Modrich

Howard Hughes Medical Institute and Department of Biochemistry, Duke
University Medical Center, Durham, North Carolina 27710

Robert Lahue

Department of Biochemistry and Molecular Biology, University of Massachusetts
Medical Center, Worcester, MA 01655

KEY WORDS: mutation, mutator, DNA repair, DNA replication, genetic recombination

ABSTRACT

Mismatch repair stabilizes the cellular genome by correcting DNA replication
errors and by blocking recombination events between divergent DNA sequences.
The reaction responsible for strand-specific correction of mispaired bases has
been highly conserved during evolution, and homologs of bacterial MutS and
MutL, which play key roles in mismatch recognition and initiation of repair,
have been identified in yeast and mammalian cells. Inactivation of genes encoding these activities results in a large increase in spontaneous mutability, and in
the case of mice and men, predisposition to tumor development.

CONTENTS

INTRODUCTION 102
OVERVIEW OF *ESCHERICHIA COLI* METHYL-DIRECTED MISMATCH REPAIR   102
LONG-PATCH MISMATCH REPAIR IN EUKARYOTIC CELLS 105
    Specificity, Mechanism, and Involvement of Mut Homologs 106
    Molecular Nature of MutS Homologs 110
    Molecular Nature of MutL Homologs 112
    The Question of Strand Signals 113
THE MITOTIC PHENOTYPE OF MISMATCH REPAIR MUTANTS. 114
    Nuclear and Mitochondrial Yeast Mutators 114
    Tolerance to DNA Methylation Damage 116
    Mismatch Repair Deficiency and Cancer Predisposition 118
MISMATCH REPAIR AND GENETIC RECOMBINATION 121
    Gene Conversion. 121
    Homeologous Recombination. 125

101

0066-4154/96/0701-0101$08.00

INTRODUCTION

Because mutations are rare events, the molecular origin of a mutation that arises spontaneously in a chromosome is typically impossible to specify. On the other hand, a growing wealth of information is available on mutation avoidance systems that ensure the integrity of a cell's genetic complement. Since defects in these pathways confer increases (in some cases very large) in spontaneous mutability, the substrates for these systems clearly represent immediate precursors to mutation. The importance of mismatch repair in stabilization of the bacterial genome was established more than 20 years ago with the demonstration that inactivation of the pathway increases the spontaneous mutation rate by several orders of magnitude (1, 2). We now know that eukaryotic cells also possess mismatch repair activities that are crucial in genetic stabilization. The importance of this function was dramatically illustrated by the recent demonstration that inactivation of the human pathway confers a strong predisposition to tumor development.

Bacteria and eukaryotic cells possess several distinct mismatch repair pathways, but we intend to focus here on the MutS- and MutL-dependent, so-called long-patch system. This pathway is characterized by broad mismatch specificity and is believed to be responsible for correcting DNA biosynthetic errors and processing recombination heteroduplexes that contain mismatched base pairs. Short-patch systems, which exhibit restricted mismatch specificity and whose primary function may be processing of particular chemically damaged base pairs (3–6), are discussed elsewhere (7). Since bacterial mismatch correction has been the subject of several reviews (8–10), our intent is to briefly discuss the prokaryotic system in order to emphasize results from higher cells. Recent alternate perspectives on the subject of this review are available (7, 11). The account in (7) is particularly comprehensive.

OVERVIEW OF *ESCHERICHIA COLI* METHYL-DIRECTED MISMATCH REPAIR

The prototypic long-patch mismatch correction system is the *E. coli* methyl-directed pathway, the inactivation of which results in a strong mutator phenotype (2, 12–14). Biological observations indicating action of the system on newly replicated DNA (15–20) confirmed the suspicion (21, 22) that this pathway stabilizes the bacterial genome by correcting DNA biosynthetic errors. Figure 1 illustrates the mechanism of the mismatch-provoked methyl-directed excision reaction.

The strand specificity necessary for repair of DNA biosynthetic errors is provided by patterns of adenine methylation in d(GATC) sequences (23). Since this is a postsynthetic modification, recently synthesized sequences exist in a

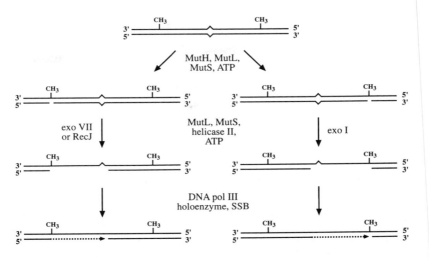

Figure 1   Methyl-directed mismatch repair in *E. coli*. Reproduced with permission from Grilley et al (45).

transiently unmodified state, and the absence of methylation on newly synthe-sized DNA targets correction to this strand (13, 14). The methyl-directed pathway has a broad mismatch specificity. Although the efficiency of repair of certain transversion mismatches can depend on sequence context, the only base-base mispair for which correction has not been reported is C-C [reviewed in (10)]. The system also repairs small insertion/deletion mismatches in which one strand contains one, two, or three extra nucleotides (24–26); heterodu-plexes with four extra nucleotides are weakly repaired, but larger heterologies do not appear to be recognized (26, 27). This mismatch specificity is consistent with the nature of mutations that accrue in *E. coli* strains deficient in mismatch repair (28, 29).

A single d(GATC) sequence is sufficient to direct mismatch repair (30–32). The distance separating the strand signal and the mismatch can be substantial: A d(GATC) site can direct correction of a mispair a kilobase (kb) away, but the strength of the strand signal is greatly reduced when separation distance exceeds two kb (30, 31). As shown in Figure 1, methyl-directed repair initiates via a mismatch-provoked incision of the unmethylated strand at a d(GATC) sequence in a reaction that requires the *mutS, mutL,* and *mutH* gene products and is dependent on ATP hydrolysis (33). MutS, which exists in solution as oligomers of a 95 kD polypeptide (34, 35), binds to the mismatch (26, 34, 36), and MutL, a homodimer of a 68 kD polypeptide, adds to this complex in a reaction that depends on ATP but not on ATP hydrolysis (37, 38). Interaction of MutS and MutL with the heteroduplex activates a latent endonuclease associated with the 25 kD MutH protein (39, 40), which cleaves the unmethylated strand at a d(GATC) site (33). The resulting strand break apparently serves as the primary signal that directs correction to the unmethylated strand (41, 42), because model heteroduplexes that contain a strand-specific incision, but lack a d(GATC) sequence, are subject to a nick-directed MutH-independent reaction that otherwise appears identical to methyl-directed repair (42). This finding suggests that strand discontinuities, other than those generated by d(GATC) cleavage, might serve to target mismatch repair to new DNA strands, and evidence for involvement of supplemental signals in mutation avoidance is available (32). The nature of the additional strand signal(s) has not been determined, but the 3'-terminus of the leading strand or discontinuities on the lagging strand might serve in this regard.

Excision, which is strictly exonucleolytic, initiates at the strand break and proceeds toward the mismatch to terminate at several discrete sites beyond the mispair (Figure 1). A surprising feature of methyl-directed repair is that the strand signal may reside on either side of the mismatch (31, 33, 43), reflecting a bidirectional capability of the system (44, 45). Excision initiating from either side of the mismatch requires MutS, MutL, and the *mutU/uvrD* gene product helicase II, but distinct exonucleases are required in the two cases (42, 44, 45). Excision from the 5' side of the mispair depends on RecJ exonuclease or exonuclease VII, both of which possess 5' → 3' hydrolytic activity (46, 47), whereas excision from the 3' side depends on exonuclease I, which hydrolyzes DNA with 3' → 5' directionality (48). Inasmuch as each of these exonucleases is single-strand specific, helicase II is thought to unwind the incised strand in order to render it sensitive to the appropriate exonuclease (45). Since helicase II is loaded into the heteroduplex at the site of the strand break in a reaction dependent on MutS and MutL (V Dao, M Yamaguchi & P Modrich, unpublished data), excision may involve concerted unwinding and hydrolysis.

Extracts of an *E. coli recJ xseA* double mutant (deficient in RecJ and exonuclease VII) do not support heteroduplex correction when the unmethylated d(GATC) sequence that directs the reaction resides 5' to the mismatch, but they do actively repair substrates in which the strand signal is located 3' to the mispair (44). A similar uncoupling of 3' and 5' excision modes has been observed in a reconstituted purified system upon omission of the appropriate exonuclease(s) (44, 45). Since the *recJ xseA* double mutant is not hypermutable (44), the independent 5' → 3' and 3' → 5' excision modes of the methyl-directed system may serve redundant functions in *E. coli* to ensure effective processing of mispaired nucleotides. In contrast to the *recJ xseA* extracts, in which an orientation-dependent repair defect was observed, fresh extracts derived from several exonuclease I-defective strains support repair of both heteroduplex orientations, although selective loss of the 3' → 5' repair mode has been observed after storage of exonuclease I-deficient extracts. The basis of this effect is not understood, but it may indicate the presence of a novel, labile activity that supports $3' \to 5'$ excision in lieu of exonuclease I (44).

The last step of the methyl-directed reaction is gap repair by DNA polymerase III holoenzyme and ligation of the repair product (42). The requirement for polymerase III holoenzyme in the purified system is quite specific because DNA polymerase I (42), T7 DNA polymerase, T4 DNA polymerase, and AMV reverse transcriptase will not substitute (J Smith & P Modrich, unpublished data). The molecular basis of this specificity for the replication polymerase is not understood but could be indicative of a special affiliation of the mismatch repair system with the replication apparatus.

LONG-PATCH MISMATCH REPAIR IN EUKARYOTIC CELLS

Homologs of MutS and MutL have been identified in yeast (49–55), mammals (56–70), *Xenopus laevis* (65), and *Drosophila melanogaster* (C Flores & W Engels, unpublished data, GenBank DMU17893). Yeast and mammalian MutS and MutL homologs for which biological functions have been identified are summarized in Table 1. In this section we consider the nature of mismatch repair in mitotic cells and the molecular properties of those Mut homologs that

Table 1 Eukaryotic MutS and MutL homologs[a,b]

S. cerevisiae		Mammalian	
MutS homolog	Function	MutS homolog	Function
MSH1	mitochondrial	—	
MSH2	mitotic & meiotic	MSH2[c]	mitotic & ?
MSH3	mitotic & ?	MRP1, REP-3	mitotic & ?
MSH4	meiotic	—	
MSH5	meiotic	—	
—		GTBP[c]	mitotic & ?
MutL homolog	Function	MutL homolog	Function
MLH1[d]	mitotic & meiotic	MLH1[d]	mitotic & ?
PMS1[d]	mitotic & meiotic	PMS2[d]	mitotic & meiotic[e]
—		PMS1	—

[a] References are provided in text.

[b] The indicated correspondence of yeast and mammalian activities is largely based on sequence homology, although evidence for functional homology is available in several cases.

[c] A heterodimer of hMSH2 and GTBP active in mismatch repair has been designated hMutSα.

[d] The yPMS1 and yMLH1 proteins interact, and the corresponding human proteins hPMS2 and hMLH1 form a heterodimer designated hMutLα that is active in mismatch repair.

[e] *PMS2*-deficient male mice are infertile and display abnormalities in chromosome synapsis during meiosis I.

have been implicated in the reaction. The mitotic and meiotic phenotypes of mutants deficient in these activities will be considered below.

Specificity, Mechanism, and Involvement of Mut Homologs

Direct evidence for mismatch correction in eukaryotes initially came from transfection of heteroduplex DNA into mammalian cells (71–74) and *S. cerevisiae* (75, 76). The most compelling of these experiments (73, 75, 76), which controlled for sequence context effects, demonstrated that individual mismatches are repaired with varying efficiency, indicating that mismatch recognition is involved in rectification. Repair in the yeast experiments was also dependent on a functional *PMS1* product (75, 76), implicating this MutL homolog (49) in mismatch correction (Table 1).

Because nearly all of the rectification events observed in the yeast transfection experiments of Kramer et al (76) were dependent on *PMS1* function, this study has provided the best assessment of the mismatch specificity of the yeast pathway. G-T, A-C, G-G, A-G, A-A, T-T, T-C, and a single nucleotide insertion/deletion mismatch were corrected with intermediate to high efficiency, whereas C-C and a 38-nucleotide insertion/deletion heteroduplex were subject to little if any repair, a specificity similar to that of the bacterial long-patch repair systems (10, 77). Some dependence on *PMS1* has also been observed for repair of heteroduplexes containing 8- or 12-nucleotide heterologies, indicating that insertion loops of this size are recognized to a significant degree by the yeast system (78).

Mammalian cell lines deficient in a MutS or MutL homolog have not been tested as recipients for heteroduplex transfection, thus complicating interpretation of mismatch specificity due to possible involvement of multiple repair systems. With this caveat in mind, application of the heteroduplex transfection assay to monkey kidney CV1 cells demonstrated intermediate to high efficiency rectification of each of the eight possible base-base mismatches (73). The most notable distinction between the CV1 results and those obtained in bacteria and yeast is that the mammalian cell line repaired C-C with good efficiency.

Strand-targeting signals have provided the conceptual basis for understanding the function of mismatch correction in bacterial replication fidelity (23). While the transfection studies summarized above established the mismatch specificity of eukaryotic repair, little strand bias was observed in these experiments, which did not address the question of strand signals (73, 75, 76). This issue has been explicitly examined by Hare & Taylor (71), who used transfection of CV1 cells to evaluate effects of DNA hemimethylation and single-strand breaks on repair of G-T and A-C mismatches in a CpG sequence environment. Although the data set was small and involvement of mismatch recognition was not tested, these experiments suggest that cytosine hemimethylation may direct correction so as to preserve CpG sequences in vertebrate cells and that single-strand breaks may serve as an even stronger determinant of strand repair bias.

The nature of eukaryotic mismatch correction has been clarified by cell-free systems that support the reaction. In vitro repair has been demonstrated in extracts prepared from *S. cerevisiae* mitotic cells (79), *Xenopus laevis* eggs (80), cultured *Drosophila melanogaster* (81), and human cell lines (81, 82). Correction efficiency in each of these systems depends on mismatch identity, implying a role for mispair recognition in provocation of repair. *Xenopus* extracts support repair of each of the base-base mispairs, albeit with differing efficiencies (80, 83). By contrast, all six transversion heteroduplexes have been found to be poor substrates for repair in yeast extracts, but G-T, A-C, and small insertion/deletion mispairs were efficiently corrected (79), a specificity somewhat different than that observed upon heteroduplex transfection (75, 76).

These yeast and *Xenopus* experiments employed covalently closed circular heteroduplexes, and in vitro repair displayed limited strand bias (79, 80, 83). Consequently, the role of DNA termini in directing correction in these systems is unclear. Repair in both in vitro systems is accompanied by DNA synthesis localized to the vicinity of the mispair (79, 83), but in the case of *Xenopus* extracts, the degree of repair synthesis did not always correlate with efficiency of correction, suggesting involvement of multiple repair systems and/or distinct modes of mismatch recognition (83). The dependence of yeast and *Xenopus* in vitro systems on MutS or MutL homologs has not been reported.

Use of circular heteroduplexes containing a site-specific strand-specific nick has demonstrated the striking effect of DNA termini on mismatch correction in extracts of *Drosophila* (81) and human cells (81, 82, 84, 85). Repair in extracts of both organisms is efficient and strongly biased to the incised strand of open circular substrates, but the two systems differ in their action on substrates lacking a strand break. Repair of covalently closed heteroduplexes has not been detected in *Drosophila* extracts (81). However, such molecules are repaired with reduced efficiency and without strand bias in human extracts (81, 82), findings similar to those obtained with *Xenopus* and yeast in vitro systems. The unbiased repair of closed circular heteroduplexes has been attributed to introduction of single-strand breaks by endonuclease activity present in human nuclear extracts (81), but other explanations have not been excluded.

While little is known about the reaction(s) responsible for unbiased correction of covalently continuous heteroduplexes, nick-directed mismatch repair in human extracts has been extensively characterized. This type of repair depends on MLH1 (67, 86), MSH2 (87), GTBP (68), and PMS2 (88), implicating these human MutS and MutL homologs in the reaction (Table 1). As discussed below, available evidence suggests that mismatch recognition in human cells is mediated by the MSH2•GTBP heterodimer (designated hMutSα), with the MLH1•PMS2 heterodimer (hMutLα) providing MutL function.

The mismatch specificity of nick-directed repair in extracts of human cells is similar to that of bacterial long-patch correction (10, 77) and *PMS1*-dependent repair in yeast (76). Human extracts support strand-specific correction of each of the eight base-base mismatches as well as insertion/deletion mispairs in which one strand of the helix contains one, two, three, or four extra nucleotides (81, 82, 84–86, 89). The most noteworthy specificity difference between the human and bacterial systems is the ability of the former to repair some C-C transversion mispairs (81, 85), a finding similar to that obtained with CV1 cells by transfection assay (73).

The nick-directed reaction in human extracts is dependent on the presence of ATP and the four dNTPs, and it is accompanied by repair DNA synthesis localized to the region spanning the mispair and the strand break that directs correction (81, 82). A single-strand break as much as a kilobase from the mismatch can direct repair, but the efficiency of correction has been shown to decline with increasing separation distance in the range of 100 to 1000 base pairs (84). This decrease is monotonic and independent of location of the nick 3' or 5' to the mismatch, suggesting that function of the system is independent of orientation of the two sites. A bidirectional excision capability was confirmed by visualization of mismatch-provoked excision tracts when repair DNA synthesis was blocked by aphidicolin or omission of exogenous dNTPs (84). With nicked circular heteroduplexes, mismatch-provoked excision removed that portion of the incised strand spanning the shorter path between the nick and the mispair, irrespective of the polarity of the DNA strand. The gaps so generated extended from the strand break to several discrete sites 90–170 nucleotides beyond the original location of the mismatch, similar to the gaps produced under similar conditions by the *E. coli* methyl-directed system (45). Since the eight base-base mispairs and a 2-nucleotide insertion/deletion mismatch provoke similar excision responses in human extracts (85, 86), their repair is now believed to be mediated by a single pathway.

As discussed below, several hypermutable human cell lines that contain mutations in genes encoding MutS and MutL homologs have been identified. Availability of these cell lines has provided additional support for the idea that repair of the eight base-base mismatches and of small insertion/deletion mispairs is largely mediated by a single system. Extracts of the hMLH1-deficient H6 tumor cell line (62) are defective in repair of the eight base-base mismatches and one-, two-, three-, or four-nucleotide insertion/deletion heterologies (86), but correction in all cases is restored upon addition of near homogeneous preparations of the hMLH1•hPMS2 hMutLα heterodimer (67).[1] Similar results implicating hMSH2 and hPMS2 in repair of base-base and insertion/deletion mismatches have been obtained with hMSH2-deficient LoVo cells (68, 87, 89) and hPMS2-deficient HEC-1A cells (88), but the set of heteroduplexes tested in these cases has not been as extensive. An apparent exception to this generalization has been observed with MT1 and HCT15 cell lines which are deficient in GTBP (68, 69, 90). Although extracts of these cell lines are defective in repair of the eight base mismatches and single nucleotide insertion/deletion mispairs, they are partially proficient in correction of two-, three-, and four-nucleotide insertion/deletion heterologies (68, 91). The selective nature of this repair defect has been attributed to a special role of the GTBP subunit of hMutSα in recognition of base-base and single-nucleotide insertion/deletion mismatches, and it has been suggested that mismatch recognition may be partially differentiated between the GTBP and hMSH2 subunits of the hMSH2•GTBP heterodimer (68).

Although the requirements for hMLH1, hPMS2, hMSH2, and GTBP in correction of base-base and 1-, 2-, 3-, or 4-nucleotide heterologies seem clear, their involvement in repair of larger insertion/deletion loops is not. Umar et al (89) have shown that 5-, 8-, and 16-nucleotide heterologies are subject to nick-directed repair in hMLH1-deficient extracts, implying that insertion/deletion loops of 5 nucleotides or larger are processed by a pathway distinct from that responsible for correction of base-base mispairs and smaller heterologies. However, repair of the 5-nucleotide loopout used in this study was dependent on hMSH2 function (89). This observation suggests that hMSH2 may function in the hMLH1-independent repair of larger loops, but the requirement for hMSH2 in 8- and 16-nucleotide loop repair was not assessed in these experiments. Assigning a special role to hMSH2 in the processing of such structures may therefore be premature.

Study of strand-specific mismatch repair in eukaryotic systems is at an early stage, but available information on the human pathway indicates several similarities between this type of repair and the bacterial methyl-directed reaction. The mismatch specificities of the two pathways are similar, a DNA terminus is sufficient to provide strand specificity in both cases, and both support bidirectional excision. The DNA polymerase responsible for repair synthesis in the human pathway has not been identified, but its sensitivity to aphidicolin suggests that it is one of the replication activities: α, ε, or δ (81, 82, 84). As discussed above, repair DNA synthesis in the methyl-directed system depends on DNA polymerase III holoenzyme, the replication enzyme in *E. coli*. These

---

[1]When not explicitly specified, the species of origin of MutS and MutL homologs is indicated using the single letter designation common in the literature; e.g. yMSH2 indicates yeast MSH2, hMLH1 human MLH1, etc.

similarities of specificity and mechanism, coupled with the homology of bacterial, yeast, and mammalian MutS and MutL activities implicated in the reaction, indicate that long-patch mismatch correction has been highly conserved during evolution.

Molecular Nature of MutS Homologs

Study of the mammalian *DHFR* gene led to the fortuitous identification of the first eukaryotic MutS homolog: transcripts encoding human MRP1 and mouse REP-3 are under the control of a divergent promoter that also regulates expression of dihydrofolate reductase (56, 57, 64, 92). Mammalian MRP1/REP-3 has not been studied biochemically, but its homology to bacterial MutS proteins has permitted design of PCR primers that have led to the identification of sequences encoding yeast MSH1, MSH2, and MSH3 (50, 52), and mammalian MSH2 (60, 65, 93). The degree of sequence similarity between individual members of the MutS family vary, but all members share a highly conserved carboxyl-terminal segment of about 100–150 amino acids (50, 52, 60, 93). This region contains a nucleotide binding consensus and a helix-turn-helix motif and, in the case of the bacterial protein, has been shown to include the active site of the MutS-associated ATPase (35).

Recombinant forms of yMSH1, yMSH2, and hMSH2 have been isolated and characterized. Yeast MSH1 is a 109 kD nuclear-encoded, mitochondrial protein that is involved in stabilization of the mitochondrial genome (51, 94). The protein binds to base-base and small insertion/deletion mismatches with high specificity and with a hierarchical preference for mispairs similar to that of bacterial MutS (94). Like bacterial MutS (33, 95), yMSH1 has a tightly associated, but nevertheless weak nucleoside triphosphatase activity that hydrolyzes ATP or dATP with a turnover number on the order of 1 min^{-1}. Chi and Kolodner (96) have demonstrated small but reproducible increases in basal ATP hydrolysis by yMSH1 in the presence of oligonucleotide homoduplex and heteroduplex substrates, with the degree of ATPase stimulation varying inversely with the intrinsic affinity of the protein for the DNA. Conversely, the presence of ATP resulted in a 60% increase in the ability of the protein to discriminate between G-T heteroduplex and G-C homoduplex DNAs, thus leading to the suggestion that the nucleotide acts as a positive effector of mismatch recognition (96). The effects of DNA on the basal rate of ATP hydrolysis by bacterial MutS have not been described, but the nucleotide has been found to decrease the protection afforded by MutS in footprinting assays (37) and to lead to mismatch-dependent formation of DNA loop structures with large heteroduplexes (97; M Grilley, P Modrich, & J Griffith, unpublished data). The relationship, if any, between these observations in the two systems is not clear.

Study of mismatch binding by band shift assay has revealed an activity in *S. cerevisiae* nuclear extracts that binds with high specificity to G-T, G-G,

A-C, T-C, and a single nucleotide insertion/deletion mismatch, but only weakly to C-C (98), a specificity similar to that of bacterial MutS and yMSH1. Since presence of this activity in extracts was dependent on a functional *MSH2* gene, binding was attributed to the yMSH2. However, the binding properties of purified, recombinant 109 kD yMSH2 differ markedly from those of bacterial MutS and yMSH1. In contrast to the bacterial protein, which does not appear to bind to insertion/deletion mispairs larger than 4 nucleotides (26), yMSH2 has been found to bind most avidly to heteroduplexes in which one strand contains a 12- or 14-nucleotide palindromic insert (99). Heteroduplexes with nonpalindromic inserts of 8–14 nucleotides are bound with intermediate affinity, and only weak binding with very modest specificity has been observed with a G-T mispair and insertion/deletion heteroduplexes containing 1- to 6-nucleotide heterologies. Formation of specific complexes between yMSH2 and a high affinity heteroduplex appears to be an extremely slow reaction, an effect postulated to reflect reiterative binding to and release from nonspecific sites until mismatch binding is achieved (99). Once formed, specific yMSH2 complexes have very long lifetimes.

In contrast to yMSH2, recombinant human MSH2 has been reported to bind with high specificity to a G-T heteroduplex (100). However, like the corresponding yeast homolog, hMSH2 binds with highest affinity to insertion/deletion oligonucleotide heteroduplexes containing 8- to 14-nucleotide heterologies (101). Electron microscopic visualization of hMSH2 complexes with 1 kb heteroduplexes localized the protein to the vicinity of the mismatch and has indicated that binding may involve multiple oligomeric states of the 105 kD protein. As judged by electron microscopy or band shift assay with oligonucleotide heteroduplexes, the presence of ATP appears to enhance affinity of hMSH2 for a mismatch (100, 101).

Although available information on yMSH2 and hMSH2 is largely based on work with the recombinant proteins, recent experiments indicate that human MSH2 functions as one subunit of a heterodimer, the other component of which is a 160 kD MutS homolog called GTBP (68, 69). GTBP (G-T binding protein) was initially isolated from HeLa cells by G-T mismatch binding assay, and purified preparations were found to contain a mixture of 160 kD and 100 kD polypeptides (58). The 100 kD protein was initially identified as a proteolytic product of the larger GTBP, but was subsequently shown to be hMSH2 (102). The formation of a functional complex between hMSH2 and GTBP has recently been demonstrated in two ways. Isolation of hMSH2 from HeLa nuclei by virtue of its ability to restore mismatch repair to extracts of an hMSH2-deficient tumor cell line yielded molar equivalents of the two proteins, which were shown to form a stable heterodimer designated hMutSα (68). Secondly, production of hMSH2 and GTBP by in vitro translation has demonstrated that both proteins are required for efficient binding to a G-T heteroduplex (69).

Monoclonal antibody assay has indicated that 80% of HeLa nuclear hMSH2 cofractionates with GTBP to homogeneity, so the hMutSα heterodimer appears to be the major mismatch recognition activity in HeLa cells (J Drummond & P Modrich, unpublished data).

The heteroduplex specificity of hMutSα has not been fully established, but the protein has been shown to bind with high specificity to G-T mispairs, as well as 1- and 3-nucleotide insertion/deletion mismatches, and to restore repair activity on various heteroduplexes to hMSH2 and GTBP-deficient cell extracts (68, 69). The effect of ATP on heteroduplex binding by hMutSα differs from that observed with recombinant hMSH2. In contrast to the increased binding affinity observed with the latter activity, the affinity of hMutSα for oligonucleotide heteroduplexes is greatly reduced in the presence of the nucleotide (58, 68), an effect similar to that observed with bacterial MutS (37). As noted above MutS and yMSH1 hydrolyze ATP with turnover numbers of about 1 min^{-1} (33, 95, 96). Preparations containing hMSH2 and GTBP have been reported to hydrolyze ATP and to possess DNA helicase activity (58), but given their extremely weak nature (< 1 ATP hydrolytic event and < 0.01 helicase event $mol^{-1}h^{-1}$), it is possible that these activities were due to contaminants in the preparation. Although ATP appears to have a large effect on mismatch affinity, its role in hMutSα function is therefore uncertain.

Molecular Nature of MutL Homologs

Isolation of *S. cerevisiae* mutants displaying elevated recombination between closely linked markers and increased incidence of postmeiotic segregation (103; and below) led to the initial identification of a eukaryotic MutL homolog, yPMS1 (49). Exploitation of the homology between yPMS1 and bacterial MutL proteins has led to identification of additional members of this family in both yeast and mammalian cells (53, 61–63, 66, 70). Table 1 summarizes those proteins for which functions have been identified.

The mitotic and meiotic phenotypes associated with inactivation of the yeast *MLH1* gene are similar to those of *pms1* mutations, and since the phenotypes of a *pms1 mlh1* double mutant are indistinguishable from either single mutant, the two genes have been assigned to a common epistasis group (53, 104). Additional support for the idea that yPMS1 and yMLH1 function in a common pathway has been provided by the demonstration that the two proteins interact physically in vivo and in vitro as judged by two-hybrid analysis and protein affinity chromatography (105). Neither protein was found to interact with yMSH2 by these criteria, but gel shift assays demonstrated that yPMS1 and yMLH1 bind to complexes between yMSH2 and homoduplex or heteroduplex DNA, with the highest yield of super-shifted complex observed with heteroduplex. Since both yPMS1 and yMLH1 were required for this effect, Prolla et al (105) postulated that the two proteins function as a heterodimer. In contrast to binding of the bacterial $(MutL)_2$ homodimer to the MutS•mismatch complex, which depends on the presence of ATP but not ATP hydrolysis (37), binding of the yPMS1•yMLH1 hetero-oligomer to the yMSH2•DNA complex does not require ATP (105).

Based on the degree of homology of their predicted products to yMLH1 and yPMS1, genes encoding mammalian MutL homologs have been designated *MLH1, PMS1,* and *PMS2* (61–63, 66, 70). As indicated in Table 1, human PMS2 is more homologous to yPMS1 than is human PMS1 (63). Furthermore, *hPMS2* appears to be only one member of a family of genes located on chromosome 7 (63, 66). Available cDNAs for the other members of this family (*hPMS3–hPMS8*) do not appear to be complete, but, surprisingly, all contain terminator codons at sites that, based on homology, would be internal to yPMS1, hPMS1, and hPMS2 polypeptides (66).

The complexity of the human MutL family suggests multiple functions, tissue-specific expression, or both. Unfortunately, biochemical information is available for only two of the homologs, and this information is limited. A HeLa cell activity that restores mismatch repair to extracts of hMLH1-deficient tumor cells has been isolated and shown to be a heterodimer of hMLH1 and hPMS2 (67). Although this heterodimer, designated hMutLα, restores repair of base-base and small insertion/deletion mispairs to hMLH1-deficient extracts, attempts to attribute simple biochemical activities to the complex were unsuccessful. This failure is not surprising, because negative results have also been obtained in attempts to assign simple activities to the bacterial $(MutL)_2$ homodimer (37), which is thought to function as an interface between MutS and other components of the repair system (10, 106). In view of the homology between yPMS1 and hPMS2, hMutLα may prove functionally equivalent to the yMLH1•yPMS1 complex described above; however, interaction of the human heterodimer with hMutSα•DNA complexes has not been evaluated.

The Question of Strand Signals

DNA termini can serve as strand signals for some *E. coli* mismatch repair events (32, 41, 42), but d(GATC) methylation also has a major role in directing correction in this organism (23). On the other hand, pre-existing DNA termini may represent the primary, if not the sole, basis for strand discrimination in some systems, as appears to be the case in *Streptococcus pneumoniae* (8). Whereas strand breaks suffice to direct mismatch repair in *Drosophila* and mammalian cells, the existence of alternate strand signals in eukaryotic systems has received little attention, but several possibilities are evident. Specific association of the mismatch repair system with the replication apparatus might serve to direct the correction of biosynthetic errors (21). However, since the biosynthetic activities involved in replication either act at pre-existing termini or generate new termini, association of repair components with biosynthetic activities is formally equivalent to correction directed by DNA ends. Some experiments (74) have suggested that DNA methylation might direct repair events in mammalian cells, but this possibility has not been pursued. Although methylation may provide the basis for strand discrimination in some eukaryotes, such signals are seemingly excluded in yeast and *Drosophila,* whose genomes appear to be largely devoid of this modification (107, 108). A third possibility, which has not been explored, envisions strand targeting based on noncovalent signals, for example, proteins that associate with individual strands of the helix and segregate with parental strands upon passage of a replication fork.

A related question concerns the nature of the undirected repair reaction that has been observed when covalently closed, circular heteroduplexes are transfected into yeast or mammalian cells, or introduced into extracts of yeast or human (but not *Drosophila*) cells. This effect can be explained in at least two ways. If such heteroduplexes are subject to adventitious introduction of strand signals in a manner that is random with respect to DNA strand, rectification by long-patch repair would yield the results observed. This possibility is analogous to the undirected correction that occurs when heteroduplexes lacking d(GATC) modification are processed by the bacterial methyl-directed pathway (23). A second possibility is that undirected repair occurs by a mechanism that is distinct from the long-patch reaction by virtue of its independence from pre-existing strand signals. Since strand-specific correction is the dominant mode of repair of incised heteroduplexes, and since undirected repair at the replication fork would seemingly lead to mutation fixation, undirected correction would presumably be restricted to regions of heteroduplex DNA devoid of functional signals for the long-patch system. Available evidence does not permit distinction between these possibilities.

THE MITOTIC PHENOTYPE OF MISMATCH REPAIR MUTANTS

Nuclear and Mitochondrial Yeast Mutators

Strand-specific mismatch repair has not been demonstrated in yeast, but inactivation of *S. cerevisiae PMS1, MLH1, MSH2,* or *MSH3* genes results in increased mitotic mutability (49, 51–53, 103, 104, 109) that has been attributed in several cases to failure to rectify replication errors (49, 104, 109). Deletion or insertional inactivation of *PMS1, MLH1,* or *MSH2* results in a 20- to 90-fold increase in the rate of mitotic mutation to canavanine resistance and a 1000-fold increase in the rate of reversion of the +1 *hom3-10* frameshift mutation (49,

51, 53). Mutations in *pms1*, *mlh1*, or *msh2* also result in a large destabilization of $(GT)_n$ repeat sequences, with observed increases in mutation rate of 85- to 800-fold for plasmid-borne repeats and 80-fold for a chromosomal repeat (104, 109). Analysis of $(GT)_n$ tract mutations that occurred in these experiments revealed that the majority were a result of the gain or loss of a single repeat element, with a modest bias toward deletions in an *msh2* background (109). Because recombination was excluded as a source of tract variation and simple repeat sequences are prone to replication errors due to strand misalignment (110, 111), $(GT)_n$ mutations were attributed to uncorrected biosynthetic errors in the repair-deficient background (104).

The mutator phenotype of *S. cerevisiae msh3* strains is distinct from that of *pms1*, *mlh1*, and *msh2* mutants. Disruption of *MSH3* has no effect on the rate of forward mutation to canavanine resistance and increases the rate of reversion of the +1 *hom3-10* frameshift mutation only tenfold (52). Deletion of *MSH3* also has a more modest effect on $(GT)_n$ mutability (30- to 60-fold increase) than does deletion of *MSH2*, and $(GT)_n$ tract mutations in an *msh3* background have been shown to be largely deletions of a single repeat element (109). These observations suggest that in contrast to PMS1, MLH1, and MSH2, the action of yeast MSH3 is largely restricted to processing of frameshift mismatches. However, the finding that the $(GT)_n$ mutation rates of *msh2 msh3* double mutants are similar to those of isogenic *msh2* single mutants rules out the possibility that the two MutS homologs function in distinct pathways that access a common substrate. To account for these observations, Strand et al (109) have suggested that MSH2 and MSH3 may cooperate in a single repair pathway to confer different repair specificities on the reaction, or alternatively may function in different pathways with mutually exclusive substrate access (e.g. on the leading and lagging strand).

Mutations in the *S. cerevisiae RTH1* gene, which encodes the homolog of a mammalian $5' \rightarrow 3'$ exonuclease that has been implicated in lagging strand DNA synthesis (112–115), also result in a large destabilization of $(GT)_n$ tracts (116). The mutation rate of plasmid-borne $(GT)_n$ tracts is elevated 100- to 300-fold, whereas that of a chromosomal tract is increased about 40-fold upon deletion of *RTH1*, values only slightly lower than those observed with *msh2*, *mlh1*, and *pms1* mutants (104, 109, 116). Furthermore, $(GT)_n$ mutability in double mutants between *rth1* and *msh2*, *mlh1*, or *pms1* is only threefold higher than that observed for single mutants. Based on these observations, Johnson et al (116) have suggested that RTH1 exonuclease functions in the MSH2, MLH1, PMS1-dependent mismatch repair system, and perhaps in a second minor repair pathway as well. As noted above, particular $5' \rightarrow 3'$ exonuclease activities have also been implicated in the bacterial reaction (44).

S. cerevisiae MSH1 is a nuclear gene, the product of which is targeted to the mitochondrion (50, 51, 94). The phenotype of *MSH1* disruptions is striking.

Reenan & Kolodner (51) have demonstrated a rapid and irreversible loss of mitochondrial function upon germination of *msh1* spores (derived from an *msh1/MSH1* diploid), and they have shown that this loss of function is accompanied by large changes in the intracellular distribution of mitochondrial DNA. Although *msh1* mutations have no effect on nuclear mutability, a sevenfold increase in the rate of mutation to mitochondrial erythromycin resistance has been demonstrated in an *msh1/MSH1* heterozygote (94). Coupled with the functional and morphological consequences of *msh1* mutations, this finding implies that MSH1 plays a major role in stabilization of the mitochondrial genome.

Tolerance to DNA Methylation Damage

Agents that methylate DNA by an $S_N 1$ mechanism, such as N-methyl-N'-nitro-N-nitrosoguanidine (MNNG) and N-methyl-N-nitrosourea (MNU), are mutagenic and cytotoxic. Both activities have been attributed to formation of O^6-methylguanine (O^6meG) because functional loss of the O^6-methylguanine-DNA methyltransferase, which repairs the O^6-modified base by abstraction of the methyl group, results in hypersensitivity to killing and mutagenesis by such agents [reviewed in (117, 118)]. Acquired resistance to the killing effects of $S_N 1$ methylators, which cannot be attributed to enhanced repair of the offending lesions, is referred to as methylation tolerance (119, 120).

Initial evidence implicating mismatch repair in the cytotoxic action of MNNG was provided by studies in bacteria (121, 122). *E. coli dam* mutants, which are deficient in the methylase that modifies the d(GATC) sequences responsible for strand directionality of bacterial repair, are hypersensitive to killing by MNNG. However, *dam mutL* and *dam mutS* double mutants are no more sensitive to the killing action of the alkylator than are *dam+* strains, implicating MutS and MutL in the enhanced killing observed with methylase-deficient cells. Despite their differential sensitivity to MNNG killing, *dam+*, *dam*, and *dam mutL* strains are subject to similar levels of mutagenesis by the alkylator (122), demonstrating that cytotoxic and mutagenic effects of MNNG can be uncoupled by mismatch repair mutations. These observations led to the suggestion that O^6meG lesions in the helix can provoke a mismatch repair response (122).

In vitro selection for MNNG or MNU resistance has led to identification of several methylation-tolerant mammalian cell lines (119, 123, 124).[2] Perhaps

---

[2]Since cell lines that express the O^6-methylguanine-DNA methyltransferase are resistant to relatively high doses of MNNG or MNU, mammalian alkylation-tolerant variants have been isolated from Mer⁻ (also called Mex⁻) cells that fail to produce this activity. The Mer⁻ (Mex⁻) phenotype, which is relatively common among cell lines derived from solid tumors (125), is apparently a consequence of transcriptional silencing (126).

the best characterized of these is the MT1 cell line, which was derived from human lymphoblastoid TK6 cells after single-step selection for high-level resistance to MNNG (119). Although MT1 cells are a hundred times more resistant than the parental line to the toxic effects of MNNG, they are nevertheless somewhat more sensitive to mutagenesis by the alkylator. Based on the latter finding and the demonstration of similar yields and kinetics of removal of various MNNG-induced lesions in the two cell lines, Goldmacher et al (119) concluded that MT1 cells tolerate adducts that are otherwise cytotoxic.

In addition to its methylation-tolerant phenotype, the MT1 cell line exhibits a cell cycle checkpoint defect after treatment with high doses of MNNG and is hypermutable in the absence of an alkylating agent. Spontaneous *HPRT* mutability is elevated about 60-fold in MT1 cells (119), and spontaneous mutations occurring in the line are largely transversions, A → G transitions, and single nucleotide frameshifts (90, 91). However, mutability of $(CA)_n$ dinucleotide repeat sequences appears to be enhanced only to a limited degree (87, 90). Karran and colleagues (123) have isolated alkylation-tolerant, hypermutable variants of CHO cells and the Burkitt's lymphoma RajiMex⁻ cell line whose mutation spectra appear to be distinct from that of MT1 cells. Although *APRT* and *HPRT* mutation rates are elevated only 2- and 4-fold in the CHO and RajiMex⁻ alkylation-tolerant variants, the mutability of $(CA)_n$ tracts is increased more than 25-fold in the CHO cell line (127).

Based on the phenotypic similarities of MT1 cells and bacterial mismatch repair mutants, Goldmacher et al (119) suggested that mismatch repair defects might also confer alkylation tolerance in mammalian cells. This possibility has been confirmed for all three of the cell lines described above. Alkylation-tolerant variants of CHO and RajiMex⁻ cells are deficient in a binding activity for G-T and small insertion/deletion heterologies (123, 127), whereas MT1 cells are defective in nick-directed repair of the eight base-base mismatches and single nucleotide frameshift heteroduplexes, but are proficient in repair of two-, three-, or four-nucleotide insertion/deletion mispairs (68, 91). MT1 cells are deficient in hMutSα activity (68) and harbor mutations in both alleles of the gene that encodes the GTBP subunit (90). Similar observations have been made with a mouse ES cell line containing *MSH2* knockout mutations: A homozygous defective is much more resistant to MNNG killing than heterozygotic or wild-type cells (128). Although all four of these alkylation-tolerant cell lines are defective in a mismatch recognition activity, loss of *hMLH1* function also results in alkylation tolerance. HCT116 tumor cells, which are defective in both alleles of *hMLH1* (62), are resistant to killing by MNNG (129), and like MT1 cells exhibit a G2 cell cycle checkpoint defect (130).

Mismatch repair mutations clearly confer methylation tolerance, but the mechanism responsible for this effect is still unclear. The favored explanation is similar to that originally invoked by Karran and Marinus (122), namely that O^6meG-T or O^6meG-C mispairs, which are produced upon encounter of template O^6meG by the biosynthetic apparatus, activate the mismatch repair system. Since action of this system is presumably restricted to the new strand, repair would be ineffectual and the methylated base would remain in the parental strand. Such abortive repair events are presumed to result in lethality, perhaps because of futile turnover of newly synthesized DNA (119).

An important corollary of the methylation tolerance studies is the demonstration that the long-patch mismatch repair system responds to genetic lesions other than mispaired Watson-Crick bases. Indeed, work in the bacterial system has shown that the methyl-directed pathway also responds to UV-induced lesions (131, 132). Such observations suggest that these systems may have a more general role as sensors of genetic damage (91).

Mismatch Repair Deficiency and Cancer Predisposition

Based on the thesis that cancer is a progressive genetic disease, Loeb (133) and Nowell (134) suggested 20 years ago that genetic destabilization would predispose a cell to malignant change. Progressive genetic change during the course of tumor development has subsequently been documented (135–138), and tumor-specific genetic instability has been observed in hereditary non-polyposis colorectal cancer (HNPCC) and certain sporadic cancers.[3] HNPCC (Lynch syndromes I and II) is a common cancer predisposition syndrome that is transmitted in an autosomal dominant fashion (139). Lynch I families are susceptible to early onset colorectal cancer, while Lynch II kindreds are also at risk for extracolonic epithelial tumors of the endometrium, ovary, stomach, small intestine, kidney, and ureter (140). In addition to the extracolonic cancers characteristic of Lynch II syndrome, Muir-Torre kindreds comprise a third, rare class of HNPCC patients that are also subject to sebaceous gland tumors (141, 142).

The initial clue linking increased mutability and cancer was provided by the finding that a subset of sporadic colon tumors (143–145) and the majority of tumors occurring in HNPCC patients (145) contain frequent mutations in the simple repeat sequences $(A)_n$, $(GGC)_n$, or $(CA)_n$. Mutations in these microsatellite repeat sequences are typically tumor-specific, indicative of a somatic origin, and their incidence is dramatic: Tumor cells with this characteristic contain thousands of microsatellite mutations (143, 145). Since the original description of the effect in HNPCC tumors and sporadic colon cancer, microsatellite instability has been described for a significant fraction of sporadic tumors (reviewed in 146), including colorectal (12–28%), endometrial (17–

---

[3]Sporadic refers to tumors in individuals who lack a family history of cancer.

23%), stomach (18–39%), ovarian (16%), cervical (15%), pancreatic (67%), esophogeal adenoma (22%), squamous cell skin (50%), and small-cell lung cancer (45%). Microsatellite instability has also been reported for prostatic cancers (20–38%) (147, 148), but reports of the phenotype in bladder (3–21%), breast (0–20%), testicular (0–18%), and nonsmall-cell lung cancer (2–34%) have differed (146).

The presence of numerous microsatellite mutations in tumors was postulated to reflect a breakdown in the fidelity of DNA replication or repair (143, 145), and the phenotype has been variously dubbed USM (ubiquitous somatic mutations), RER (replication errors), or MIN (microsatellite instability) (143–145). The MIN$^+$ and MIN$^-$ designation will be used here to refer to tumors that do, or do not, display microsatellite alterations, as these terms best describe the type of mutation scored by the microsatellite assay commonly used for tumor screening. Mutation rate analysis of established colon tumor cell lines has demonstrated that the MIN$^+$ phenotype is invariably accompanied by a large increase in the spontaneous mutability (86, 149–151), indicating that the frequent microsatellite mutations in MIN$^+$ tumors are almost certainly due to an increased mutation rate rather than selection. As anticipated by Loeb (133) and Nowell (134), establishment of the hypermutable phenotype is an early event in MIN$^+$ colorectal tumorigenesis, and genetic instability persists after transformation (149).

Hypermutable colon tumor cell lines fall into several distinct classes with respect to mutation spectrum. For example, $(CA)_n$ and *HPRT* mutation rates are both elevated several-100 fold in the MIN$^+$ HCT116 (also called H6) and RKO cell lines (86, 150, 151), whereas HCT-15 cells are characterized by a several–100 fold increase in *HPRT* mutability, but only a modest increase in dinucleotide repeat mutation rate (149, 150). This latter phenotype, which is reminiscent of that of MT1 alkylation-tolerant cells described above, will be referred to here as MIN$^\pm$. Eshleman et al (151) have identified a third class of colon tumor cell, typified by Vaco410 cell line, which although MIN$^-$, is nevertheless hypermutable at *HPRT*. Existence of the latter two classes suggests that screens for microsatellite mutations, while convenient, may underestimate the actual incidence of genetically unstable tumors.

The basis of the MIN$^+$ phenotype has been clarified by genetic analysis of HNPCC families and biochemical assay of established MIN$^+$ tumor cell lines. Four genes have been implicated to date in HNPCC: *MSH2* (59, 60, 142, 152–154), *MLH1* (61, 62, 155–157), *PMS1*, and *PMS2* (63). The majority of HNPCC kindreds examined to date harbor mutations in *MSH2* or *MLH1* (153, 158); *PMS1* and *PMS2* mutations appear to be responsible for only a small fraction of HNPCC cases (63). As expected from the dominant mode of HNPCC inheritance, normal cells of affected individuals are typically MIN$^-$ (145) and heterozygous for a germline defect in one of these four loci (60–63,

153, 157). By contrast, tumor cells have consistently been found to be defective in both alleles of the gene in question, with loss of the wild-type allele owing to somatic mutation (60, 63) or loss of heterozygosity (156, 159, 160). The simplest interpretation of these findings is that inactivation of the wild-type gene corresponds to the rate-limiting step for HNPCC tumor development.

Biochemical assay of extracts prepared from several tumor-derived MIN$^+$ cell lines has demonstrated deficiency of nick-directed mismatch repair (68, 86–88) and in some cases loss of mismatch binding activity (127). These cell lines define several in vitro complementation groups (68, 87). The most extensively characterized lines, H6 (also called HCT116), LoVo, and HEC-1-A are defective in both alleles of *MLH1* (62), *MSH2* (87, 160), or *PMS2* (88), respectively. Isolation of activities that restore repair to H6 and LoVo extracts have yielded hMLH1•hPMS2 (hMutLα) and hMSH2•GTBP (hMutSα) heterodimers (67, 68), implicating products of three of the HNPCC genes in mismatch repair. Biochemical involvement of the *PMS1* product in the reaction has not been assessed.

Mismatch binding assay and isolation of hMutSα (58, 68, 69) have implicated GTBP in mismatch repair, but mutations in the *GTBP* locus have not been identified in HNPCC kindreds (90). However, GTBP defects have been identified in several sporadic MIN$^\pm$ colorectal cancers (68, 69, 90). As noted above, *GTBP* mutants are selectively defective in repair of base-base and single nucleotide insertion/deletion mispairs, but are partially proficient in repair of two-, three-, and four-nucleotide heterologies (68), consistent with the MIN$^\pm$ phenotype of these tumors (90, 149, 150).

Mutations in *MSH2, GTBP, MLH1,* or *PMS2* result in mismatch repair deficiency and genetic destabilization, and *MSH2, GTBP, MLH1* defects confer alkylation tolerance. However, the relationship between these phenotypes and cancer predisposition remains to be established. The most obvious possibility is the one originally proposed by Nowell (134), namely that clonal expansion coupled with elevated mutability increases the probability of the multiple mutations necessary for tumor development. Parsons et al (161) have recently questioned the sufficiency of elevated mutation rate per se for tumor development, based on the identification of several HNPCC-affected patients with *hPMS2* and *hMLH1* mutations that act in a dominant fashion. Although heterozygous for the defect in question, normal lymphoblastoid and epithelial cells from these patients contain frequent microsatellite mutations and are defective in mismatch repair as judged by in vitro assay. Nevertheless, the age of onset and the incidence of colon tumors in these individuals were not unusual. Parsons et al (161) suggested that these observations may be indicative of an additional requirement for tissue damage, with the resulting mitogenic stimulus providing the proliferative capacity necessary to support the clonal expansion and selection required to produce a tumor cell (162). Given that production of replication errors is restricted to proliferating cells, a strong link between proliferative capacity and tumor development under conditions of mismatch repair deficiency would not be surprising. Another possibility is that the methylation-tolerant phenotype and the failure to respond to cell cycle checkpoints upon methylation damage may play a role in tumor development in repair-defective cells.

A related and particularly intriguing question concerns the tissue specificity of cancers that result from mismatch repair deficiency, an issue that has been rendered even more puzzling by the demonstration that knockout mice defective in both alleles of *PMS2* (70) or *MSH2* (128, 163) are viable and prone to lymphomas, but apparently not colon cancer. Such effects seemingly indicate distinct rate-limiting steps for tumor development in various tissues of a given species, or in similar tissues from different species. Although this paradox can be rationalized in several ways, one interesting possibility is suggested by the recent analysis of the type II TGF-β receptor gene in human colon tumors. Markowitz et al (164) have documented a strong correlation between defects in the type II TGF-β receptor gene and the MIN$^+$ phenotype in colorectal cancer cells. Though rare in MIN$^-$ tumor cell lines, defects in the type II TGF-β receptor gene are common in MIN$^+$ tumor cells, with the mutations responsible localized to an $(A)_{10}$ repeat in seven cases and a $(GT)_3$ repeat in one case (164). Since the failure to respond to TGF-β growth inhibition is a characteristic of certain epithelial tumors (165, 166), these findings directly link hypermutability and a mutational hotspot in a growth control locus.

In contrast to the human gene, the murine type II TGF-β receptor gene lacks the $(A)_{10}$ hotspot that is targeted for mutation in MIN$^+$ human cells (167), and the absence of this site provides a possible explanation for the failure of mouse *PMS2* or *MSH2* knockouts to develop colorectal tumors (S Markowitz, personal communication). A variation on this theme, which invokes mutational hotspots in growth regulatory genes whose function is restricted with respect to tissue, could in principle account for the tissue distribution of MIN$^+$ cancers.

MISMATCH REPAIR AND GENETIC RECOMBINATION

Gene Conversion

Several recent reviews and journal articles have examined genetic recombination in detail (168–173). This summary focuses on facets of recombination that affect the formation and processing of mispaired intermediates.

MISMATCH REPAIR AND GENE CONVERSION   Mismatch repair effects on genetic recombination have been inferred from certain inheritance patterns that deviate from the normal (4$^+$:4$^-$) Mendelian segregation in fungal systems. The consequence of one class of events, gene conversions, is an inheritance pattern of 6$^+$:2$^-$ or 6$^-$:2$^+$. For organisms like *S. cerevisiae* that yield four meiotic products, gene conversions are observed as three wild-type spores and one mutant spore or one wild-type spore and three mutant spores. Two mechanisms have been proposed for these events, which typically occur in 1–10% of unselected yeast tetrads. In one model (174–176), gene conversion results from correction of mispairs within the heteroduplex recombination intermediate. Mispairs are predicted to occur at positions of allelic difference when one or more strands are exchanged. Correction of the mispair in favor of the invading strand would result in gene conversion, whereas correction in favor of the recipient strand would restore the original markers and hence would not be observed in genetic experiments.

The second model (177) predicts that gene conversion is a consequence of processing double-strand breaks. In this model, repair of a gapped duplex occurs when broken strands invade the intact homolog and are used to prime DNA synthesis with the unbroken chromosome acting as template. In fact, numerous studies of double strand–break repair in *S. cerevisiae* indicate that breaks are processed by single-strand exonucleases to yield long single-stranded tails (178, 179). Strand invasion by the tails to yield mispaired intermediates (180–186) which then undergo repair is likely an important mechanism for gene conversion (summarized in 171). The other class of inheritance is postmeiotic segregation (PMS), which is manifested in *S. cerevisiae* as the presence of a sectored colony, reflecting the persistence of a mismatch resulting from an unrepaired recombination heteroduplex. PMS is relatively rare, occurring on average about tenfold less frequently than gene conversions in yeast [reviewed in (168)], implying that mismatch repair normally occurs with high efficiency.

EVIDENCE FOR MISMATCH CORRECTION OF RECOMBINATION INTERMEDIATES
Transformation experiments in *S. pneumoniae* [reviewed in (8)] and conjugational crosses in *E. coli* (187) provided some of the initial evidence for mismatch repair in recombination. More recent work has identified a requirement for mismatch correction in UV-induced recombination in *E. coli* (131, 132). These and other observations have lead to the proposal that the methyl-directed repair system processes UV-irradiated DNA to yield recombinagenic substrates, perhaps by generating single-strand gaps. Mismatch correction may thus function to recognize UV-damaged DNA as well as alkylation damage (described above) or alternative DNA structures (see description of *MSH4* and *MSH5* below).

The identification of yeast mutants that yield high levels of PMS with corresponding decreases in gene conversion (49, 51, 53, 103) was key evidence for correction of mispairs that arise from recombination. These mutations lead to PMS at many genetic loci, indicating their central role in correction. The first gene identified was *PMS1* (103). Disruption mutants of *PMS1* led to the loss of about 65% of gene conversion events, indicating that while *PMS1* is important for gene conversions, a significant fraction of such events can occur by a *PMS1*-independent mechanism (49). Subsequently, yeast cells with mutations in *MSH2* (51, 171) and *MLH1* (53) were shown to have phenotypes very similar to those in *PMS1*, consistent with the notion that the products of the three genes function together in mismatch correction. Of the other yeast MutS homologs, *msh3* mutants show a small increase in PMS (52). By contrast, *msh4* (54) and *msh5* (55) mutants reduce meiotic crossing over but are not mutators and do not alter the frequency of gene conversion. *MSH4* and *MSH5* products therefore play other roles during meiosis (see below). In *Drosophila*, *mei-9* mutants exhibit decreased gene conversion and increased PMS (180), suggesting that mismatch correction is also active during meiosis in this organism.

Another important proof of mismatch repair of recombination intermediates is the demonstration of allele specificity of conversion and PMS (reviewed in 168). Those alleles that yield well-repaired mismatches upon strand exchange, such as G-T or small (1–4 base) insertions, typically exhibit low PMS frequency. In contrast poorly repaired mismatches, such as C-C or insertions that can form stable hairpins, yield high PMS (51, 171, 181, 186, 188–192). Furthermore, poorly repaired mismatches will often undergo correction if placed near a well-repaired mismatch, an effect called co-correction (27, 193). This finding implies that poorly repaired mismatches simply escape correction rather than being intrinsically uncorrectable. Efficient processing of mispairs is also dependent on mismatch correction activities. In the case of *S. cerevisiae*, mutations in *PMS1*, *MSH2*, or *MLH1* increase the level of PMS at normally well-repaired mismatches to those of poorly repaired mismatches (49, 51, 53, 171, 186, 190), consistent with the notion that these gene products are directly involved in correction. In mouse cells, highly suggestive evidence for mispair formation and correction during recombination has also been obtained (194–196). With the isolation of mouse mismatch repair genes (65, 70) and cell lines harboring knockouts of these genes (70, 128, 163), direct tests of the role of mismatch repair in this process are now feasible.

The presence of mismatches in meiotic recombination intermediates has also been directly demonstrated by physical methods. Such experiments require that mismatch repair be obviated either through the use of mutants, the use of poorly repaired mismatches, or by testing for the presence of mismatches very shortly after their formation. Lichten et al (183) used markers that could generate either C-C or G-G mismatches. In wild-type cells the C-C mispair was observed in recombination intermediates, but both

C-C and G-G mispairs persisted in nearly equal levels in *pms1* mutants. Nag & Petes (185) observed persistent cruciform heteroduplexes arising from palindromic insertions during meiosis. In both cases the timing of mismatch formation was consistent with meiotic recombination. Haber et al (184) monitored formation and repair of a mitotic A-A mispair during mating type switching between wild-type and mutant *MAT* loci. Using PCR methodology, they demonstrated that the mismatch is formed as predicted, that correction of the mispair occurs rapidly (halflife of 6–10 minutes), and that repair is abolished in a *pms1* mutant. Furthermore, the invading strand, which contains a free DNA terminus eight nucleotides from the mismatch, is preferentially repaired.

INVOLVEMENT OF OTHER MUTS HOMOLOGS IN MEIOSIS   Genes encoding two novel MutS homologs have been identified in yeast. Mutations in *MSH4* (54), *MSH5*, or in both genes (55) reduce meiotic crossing over two- to threefold but do not affect gene conversion, PMS, or mitotic mutation rates. Thus in contrast to *MSH2* and *MSH3*, no role is apparent for these genes in strand-specific correction of mispairs; however, results to date do not rule out alternative functions involving mismatch recognition that may be essential for meiotic genetic stability. For example, one or both of these proteins might assist the search for homology in its final stages. This effect may not be apparent morphologically but might facilitate crossovers between homologous chromosomes and thus serve to prevent chromosome loss. This idea would account for the observation that *MSH4* (54) and *MSH5* (55) mutations result in decreased spore viability due to failure to segregate chromosomes during the first meiotic division, a phenotype frequently associated with low recombination rates. A mismatch recognition function for MSH4 and MSH5 is not mutually exclusive with existing hypotheses for the activity of these proteins (54, 55), such as recognition of alternative DNA structures. Biochemical assays for mismatch binding will help test these possibilities.

Transgenic mice harboring defects in *PMS2* (the homolog of yeast *PMS1*) exhibit, among other phenotypes, male infertility (70). Examination of meiotic chromosomes has yielded evidence that aberrant synapsis is responsible for the spermatogenesis defect in these animals (70), although inactivation of *mMSH2* does not result in male infertility (128). Based on these observations, Baker et al (70) and de Wind et al (128) have suggested that an as yet unidentified MutS homolog may function with PMS2 during meiotic recombination. Although the mouse homologs of yeast MSH4 and MSH5 are potential candidates, chromosome synapsis appears normal in yeast *msh4* mutants (54). Thus the identity of this putative MutS homolog remains to be established.

Homeologous Recombination

Homeologous recombination is defined as genetic exchanges between DNA sequences that are similar but not identical. The limits of homeology (in terms of percent sequence divergence) have not been defined precisely, in part because of the variable response to homeology observed in different experimental systems. For instance, conjugational crosses between *E. coli* and *Salmonella typhimurium* (whose sequences are approximately 16% divergent) are reduced five orders of magnitude relative to crosses within either species (197). Intrachromosomal gene conversion in mouse cells is also sensitive to homeology as 19% sequence divergence reduced recombination by at least 1000-fold (198) and as few as 2 mismatches in a 232-bp region was shown to result in a 20-fold reduction (199). Mitotic recombination in *S. cerevisiae* is less sensitive to homeology, as divergence in the range of 17–27% reduced recombination only 13–180-fold (171, 200–202). Mitotic gap repair in *Drosophila* was reduced about fourfold by the presence of 15 single heterologies over a 3455 span (203). Thus, variable extents of sequence divergence elicit attenuation of recombination. Exceptions to this idea have been reported, including a lack of detectable homeologous effects in *Drosophila* meiotic recombination (204).

Homeologous exchanges can occur because, in general, recombination strand transferases are able to proceed through regions of imperfect homology (205–207). Thus cells are believed to be capable of catalyzing genetic exchanges between imperfectly matched DNA partners. However, this capability presents a significant danger to genetic stability (9, 208). For example, inappropriate crossovers between widely dispersed homeologous sequences would lead to chromosomal rearrangements. Results from numerous studies support the idea that recombination intermediates containing multiple mismatches are targets for mismatch repair proteins, which act to prevent the completion of recombination events involving homeologous sequences (9, 128, 202, 208, 209) or to alter their outcome (171, 210–212). The distinction between these events and the gene conversion events described above is important. For gene conversion, typically only one or a few mispairs are present in the recombination heteroduplex, and their correction is an integral part of the conversion event. For homeologous recombination, a high density of mispairs is present and mismatch repair activities act to abort (or alter the outcome of) the process.

BACTERIAL HOMEOLOGOUS RECOMBINATION   Early work in *E. coli* proved that mismatch repair acts as a barrier to recombination between divergent sequences during conjugation (187, 209) or between phage and plasmids (213). Depending on the experiment, loss of mismatch repair function elevated homeologous recombination frequency up to 1000-fold when divergence was about 20% at the nucleotide level (209). Lower levels of divergence produced less dramatic effects (213). This work has been extended to include large chromosomal duplications (214). However, the rate of homeologous recombination was still several orders of magnitude below the rate of homologous recombination even in mismatch repair–deficient strains (209), raising questions concerning the basis of the residual impediment to homeologous exchange. This point has been clarified recently (197) by experiments proving that induction of the SOS system stimulates homeologous recombination, mainly through overproduction of RecA. The cumulative effect of SOS stimulation and loss of inhibition in mismatch repair mutants accelerated homeologous recombination to rates approaching those of homologous controls.

The availability of RecA-mediated strand exchange assays and purified MutS and MutL proteins allowed Worth et al (215) to conduct in vitro experiments that measured mismatch repair effects on homeologous strand transfer. Exchanges between sequences divergent by only 3% were dramatically inhibited by MutS, an effect that could be significantly enhanced by MutL. These proteins showed no significant inhibition of homologous control reactions. The major effect of MutS and MutL was to block branch migration during homeologous exchanges. Future experiments that include RuvA and RuvB proteins, which accelerate branch migration (206, 216), should further clarify the role of MutS and MutL in modulating branch migration in response to occurrence of mispairs within the heteroduplex.

YEAST HOMEOLOGOUS RECOMBINATION   Several studies have examined yeast homeologous mitotic recombination in transformed plasmids (171, 201), between transformed plasmids and chromosomes (217), or by intrachromosomal (202; E Selva & R Lahue, unpublished observations) or interchromosomal exchanges (200, 218–220). Estimates of the ratio of homologous to homeologous recombination range from 13- to 180-fold, depending on the experimental system and on the extent of sequence divergence. Thus yeast does not discriminate as much against homeologous recombination as other organisms.

The MutS homolog genes *MSH2* and *MSH3* play active roles in the control of homeologous exchanges (202). Since the *MSH2* and *MSH3* products appear to act independently, effects of these two proteins on homeologous exchanges may be a result of their involvement in distinct pathways. Although *PMS1* function affects the product distribution in homeologous recombination (217; E Selva and R Lahue, unpublished observations), specific enhancement of homeologous events in *pms1* mutants has not been reported (200, 202, 217, 219). Homeologous meiotic recombination events are unaffected in *MSH4* mutants (54).

Mismatch repair also exerts a regulatory influence over homeologous recombination in yeast meiosis. Borts and colleagues (N Hunter, SR Chambers, EJ Louis & RH Borts, unpublished observations) constructed a diploid hybrid between two species of *Saccharomyces* whose genomes are approximately 15% divergent. This diploid produced only 1% viable spores, and many of these spores exhibited aberrant chromosome segregation. These effects were substantially relieved in *pms1* mutants, a finding attributed to increased reciprocal exchanges between divergent chromosomes and to reduced nondisjunction. Thus, as in bacteria (197, 209), mismatch repair acts as a barrier to interspecies genetic exchange in yeast.

Mismatch repair may also play an alternate role in meiotic recombination when sequence divergence is limited [9 small heterologies in a 9-kb region (210, 211)]. Under these conditions, reciprocal exchanges are reduced by about one half with a concomitant increase in the number of exceptional recombination events. In *pms1* strains, the pattern of recombination was restored to nearly that seen without heterologies (211). This finding led to the hypothesis that mismatch repair of heteroduplexes containing multiple mispairs triggers a second recombination event, presumably because of double-strand breaks caused by overlapping excision tracts (210, 211).

HOMEOLOGOUS RECOMBINATION IN HIGHER EUKARYOTES   Mammalian cells also exhibit discrimination against homeologous recombination. Intrachromosomal gene conversions between sequences that have diverged by 19% are reduced at least 1000-fold relative to homologous controls (198). As few as 2 mispairs in a 232-bp stretch of homology have been found to reduce recombination by a factor of 20 (199). Targeted integration of plasmid DNA at the *Rb* locus in ES cells exhibits a 15–50-fold discrimination at the level of 0.6% divergence (128, 221), an effect that is relieved in *msh2* mutants (128). However, these effects appear to depend on the type of event being studied and the location (chromosomal vs plasmids) of the two DNA sequences. For example, when 19% diverged donor sequences were present on linear plasmids, the discrimination factor was reduced to only 3–15-fold (198). In some cases, the presence of homeology altered the product distribution rather than the frequency of recombination. For example, analysis of recombinational gap repair of LINE sequences revealed that half the events involved precise repair when sequence divergence was 5%, whereas no precise events were observed with 15% diverged sequences (212).

The role of homeology has not been resolved for recombination in *Drosophila*. Mitotic gap–repair experiments revealed sensitivity to homeology, with a reduction from 19% to 5% efficiency when 15 single base mispairs were present over a 3455-bp span (203); however in meiotic intragenic recombination similar amounts of homeology (11 single-nucleotide polymorphisms over 3.78 kb) did not detectably alter either the frequency or product distribution (204).

STRUCTURE OF HOMEOLOGOUS RECOMBINANTS   Analysis of genetic requirements and product structure of homeologous recombinants from *E. coli* and
yeast are consistent with the notion that homeologous recombination is mediated by the same (or similar) pathways as those responsible for homologous
events. Homeologous events in *E. coli* are dependent on *recA* and *recBC* (197,
222, 223), and homeologous gene conversions in yeast require *RAD52* (200).
Crossovers and gene conversions have been observed in yeast homeologous
experiments under circumstances in which both types of events could be
detected (220; E Selva & R Lahue, unpublished observations). The products
of homeologous and homologous recombination are therefore (to a first ap-
proximation) quite similar. Examination of recombination endpoints by DNA
sequencing indicated that exchanges were resolved within short stretches of
identity (201, 217, 220, 224; E Selva & R Lahue, unpublished observations).
Furthermore, information transfer was in continuous blocks (200, 201, 217,
220, 224, 225; E Selva & R Lahue, unpublished observations). Thus homeologous recombination resembles homologous recombination with mismatch repair superimposing an extra level of fidelity.

HOW DOES MISMATCH REPAIR INFLUENCE RECOMBINATION FIDELITY?   Although mismatch repair proteins clearly must access mispairs generated during
strand exchange and influence the subsequent course of the event, the underlying mechanisms of this action are not understood. One possibility is that
branch migration is halted or even reversed [heteroduplex rejection (9, 208)].
According to this hypothesis, mismatch repair proteins act to impede or even
halt the progress of strand exchange upon occurrence of mispairs within the
heteroduplex. Evidence to support this notion comes from biochemical (215)
and genetic experiments (171) indicating that MutS and MutL activities inhibit
branch migration when the heteroduplex contains mismatches. A second possibility is that mismatch recognition stimulates resolution of recombination
intermediates at or near the site of a mismatch and thus acts to preclude
exchange events containing additional downstream sequences (171). A third
possibility is that mismatch-stimulated excision occurs, leading to destruction
of the recombination event. This appears to occur in *E. coli* when the degree
of sequence divergence is small, as judged by dependence of the effect on
MutH and DNA helicase II, in addition to MutS and MutL (187). The requirement for MutH and helicase II, however, is less pronounced when divergence
is increased (209).

Although much evidence supports the idea that mismatch correction activities control the fidelity of genetic recombination, there clearly are exceptions,
and the mechanisms responsible for the different effects that have been observed are still obscure. Further clarification of the molecular mechanisms

responsible for genetic exchange and exploitation of the increasing availability of mismatch repair–deficient organisms should help resolve these issues.

ACKNOWLEDGMENTS

Work in our laboratories was supported by grants GM23719 and GM45190 (to P Modrich) and GM44824 (to R Lahue) from the National Insititute of General Medical Sciences. We thank Vickers Burdett, Jim Drummond, Matt Longley, Andy Pierce, and Erica Selva for comments on the manuscript.

Literature Cited

1. Tiraby J-G, Fox MS. 1973. *Proc. Natl. Acad. Sci. USA* 70:3541–45
2. Nevers P, Spatz H. 1975. *Mol. Gen. Genet.* 139:233–43
3. Lieb M, Allen E, Read D. 1986. *Genetics* 114:1041–60
4. Hennecke F, Kolmar H, Bründl K, Fritz H-J. 1991. *Nature* 353:776–78
5. Wiebauer K, Jiricny J. 1990. *Proc. Natl. Acad. Sci. USA* 87:5842–45
6. Michaels ML, Cruz C, Grollman AP, Miller JH. 1992. *Proc. Natl. Acad. Sci. USA* 89:7022–25
7. Friedberg EC, Walker GC, Siede W. 1995. *DNA Repair and Mutagenesis.* Washington, DC: ASM Press
8. Claverys J-P, Lacks SA. 1986. *Microbiol. Rev.* 50:133–65
9. Radman M. 1988. In *Genetic Recombination,* ed. R Kucherlapati, GR Smith, pp. 169–92. Washington, DC: Am. Soc. Microbiol.
10. Modrich P. 1991. *Annu. Rev. Genet.* 25:229–53
11. Kolodner RD, Alani E. 1994. *Curr. Opin. Biotechnol.* 5:585–94
12. Cox EC. 1973. *Genetics* 73:67–80 (Suppl.)
13. Pukkila PJ, Peterson J, Herman G, Modrich P, Meselson M. 1983. *Genetics* 104:571–82
14. Lu A-L, Clark S, Modrich P. 1983. *Proc. Natl. Acad. Sci. USA* 80:4639–43
15. Rydberg B. 1978. *Mutat. Res.* 52:11–24
16. Glickman B, van den Elsen P, Radman M. 1978. *Mol. Gen. Genet.* 163:307–12
17. Herman GE, Modrich P. 1981. *J. Bacteriol.* 145:644–46
18. Skopek TR, Hutchinson F. 1984. *Mol. Gen. Genet.* 195:418–23
19. Marinus MG, Poteete A, Arraj JA. 1984. *Gene* 28:123–25
20. Schaaper RM. 1989. *Genetics* 121:205–12
21. Wagner R, Meselson M. 1976. *Proc. Natl. Acad. Sci. USA* 73:4135–39
22. Radman M, Wagner RE, Glickman BW, Meselson M. 1980. In *Progress in Environmental Mutagenesis,* ed. M Alacevic, pp. 121–30. Amsterdam: Elsevier/North Holland Biomed.
23. Meselson M. 1988. In *Recombination of the Genetic Material,* ed. KB Low, pp. 91–113. San Diego: Academic
24. Dohet C, Wagner R, Radman M. 1986. *Proc. Natl. Acad. Sci. USA* 83:3395–97
25. Learn BA, Grafstrom RH. 1989. *J. Bacteriol.* 171:6473–81
26. Parker BO, Marinus MG. 1992. *Proc. Natl. Acad. Sci. USA* 89:1730–34
27. Carraway M, Marinus MG. 1993. *J. Bacteriol.* 175:3972–80
28. Schaaper RM, Dunn RL. 1987. *Proc. Natl. Acad. Sci. USA* 84:6220–24
29. Levinson G, Gutman GA. 1987. *Nucleic Acids Res.* 15:5323–38
30. Lahue RS, Su S-S, Modrich P. 1987. *Proc. Natl. Acad. Sci. USA* 84:1482–86
31. Bruni R, Martin D, Jiricny J. 1988. *Nucleic Acids Res.* 16:4875–90
32. Claverys JP, Mejean V. 1988. *Mol. Gen. Genet.* 214:574–78
33. Au KG, Welsh K, Modrich P. 1992. *J. Biol. Chem.* 267:12142–48
34. Su S-S, Modrich P. 1986. *Proc. Natl. Acad. Sci. USA* 83:5057–61
35. Haber LT, Pang PP, Sobell DI, Mankovich JA, Walker GC. 1988. *J. Bacteriol.* 170:197–202

130   MODRICH & LAHUE

36. Su S-S, Lahue RS, Au KG, Modrich P. 1988. *J. Biol. Chem.* 263:6829–35
37. Grilley M, Welsh KM, Su S-S, Modrich P. 1989. *J. Biol. Chem.* 264:1000–4
38. Mankovich JA, McIntyre CA, Walker GC. 1989. *J. Bacteriol.* 171:5325–31
39. Welsh KM, Lu A-L, Clark S, Modrich P. 1987. *J. Biol. Chem.* 262:15624–29
40. Grafstrom RH, Hoess RH. 1987. *Nucleic Acids Res.* 15:3073–84
41. Längle-Rouault F, Maenhaut MG, Radman M. 1987. *EMBO J.* 6:1121–27
42. Lahue RS, Au KG, Modrich P. 1989. *Science* 245:160–64
43. Lu A-L. 1987. *J. Bacteriol.* 169:1254–59
44. Cooper DL, Lahue RS, Modrich P. 1993. *J. Biol. Chem.* 268:11823–29
45. Grilley M, Griffith J, Modrich P. 1993. *J. Biol. Chem.* 268:11830–37
46. Lovett ST, Kolodner RD. 1989. *Proc. Natl. Acad. Sci. USA* 86:2627–31
47. Chase JW, Richardson CC. 1974. *J. Biol. Chem.* 249:4553–61
48. Lehman IR, Nussbaum AL. 1964. *J. Biol. Chem.* 239:2628–36
49. Kramer W, Kramer B, Williamson MS, Fogel S. 1989. *J. Bacteriol.* 171:5339–46
50. Reenan RA, Kolodner RD. 1992. *Genetics* 132:963–73
51. Reenan RA, Kolodner RD. 1992. *Genetics* 132:975–85
52. New L, Liu K, Crouse GF. 1993. *Mol. Gen. Genet.* 239:97–108
53. Prolla TA, Christie DM, Liskay RM. 1994. *Mol. Cell. Biol.* 14:407–15
54. Ross-Macdonald P, Roeder GS. 1994. *Cell* 79:1069–80
55. Hollingsworth NM, Ponte L, Halsey C. 1995. *Genes Dev.* 9:1728–39
56. Fujii H, Shimada T. 1989. *J. Biol. Chem.* 264:10057–64
57. Linton JP, Yen J-YJ, Selby E, Chen Z, Chinsky JM, et al. 1989. *Mol. Cell. Biol.* 9:3058–72
58. Hughes MJ, Jiricny J. 1992. *J. Biol. Chem.* 267:23876–82
59. Fishel R, Lescoe MK, Rao MR, Copeland NG, Jenkins NA, et al. 1993. *Cell* 75:1027–38
60. Leach FS, Nicolaides NC, Papadopoulos N, Liu B, Jen J, et al. 1993. *Cell* 75:1215–25
61. Bronner CE, Baker S, Morrison PT, Warren G, Smith LG, et al. 1994. *Nature* 368:258–61
62. Papadopoulos N, Nicolaides NC, Wei Y-F, Ruben SM, Carter KC, et al. 1994. *Science* 263:1625–29
63. Nicolaides NC, Papadopoulos N, Liu B, Wei Y-F, Carter KC, et al. 1994. *Nature* 371:75–80
64. Liu K, Niu LM, Linton JP, Crouse GF. 1994. *Gene* 147:169–77
65. Varlet I, Pallard C, Radman M, Moreau J, de Wind N. 1994. *Nucleic Acids Res.* 22:5723–28
66. Horii A, Han HJ, Sasaki S, Shimada M, Nakamura Y. 1994. *Biochem. Biophys. Res. Commun.* 204:1257–64
67. Li G-M, Modrich P. 1995. *Proc. Natl. Acad. Sci. USA* 92:1950–54
68. Drummond JT, Li G-M, Longley MJ, Modrich P. 1995. *Science* 268:1909–12
69. Palombo F, Gallinari P, Iaccarino I, Lettieri T, Hughes M, et al. 1995. *Science* 268:1912–14
70. Baker SM, Bronner CE, Zhang L, Plug A, Robatzek M, et al. 1995. *Cell* 82:309–19
71. Hare JT, Taylor JH. 1985. *Proc. Natl. Acad. Sci. USA* 82:7350–54
72. Folger KR, Thomas K, Capecchi MR. 1985. *Mol. Cell. Biol.* 5:70–74
73. Brown TC, Jiricny J. 1988. *Cell* 54:705–11
74. Hare JT, Taylor JH. 1988. *Gene* 74:159–61
75. Bishop DK, Andersen J, Kolodner RD. 1989. *Proc. Natl. Acad. Sci. USA* 86:3713–17
76. Kramer B, Kramer W, Williamson MS, Fogel S. 1989. *Mol. Cell. Biol.* 9:4432–40
77. Gasc AM, Sicard AM, Claverys JP. 1989. *Genetics* 121:29–36
78. Bishop DK, Williamson MS, Fogel S, Kolodner RD. 1987. *Nature* 328:362–64
79. Muster-Nassal C, Kolodner R. 1986. *Proc. Natl. Acad. Sci. USA* 83:7618–22
80. Brooks P, Dohet C, Almouzni G, Mechali M, Radman M. 1989. *Proc. Natl. Acad. Sci. USA* 86:4425–29
81. Holmes J, Clark S, Modrich P. 1990. *Proc. Natl. Acad. Sci. USA* 87:5837–41
82. Thomas DC, Roberts JD, Kunkel TA. 1991. *J. Biol. Chem.* 266:3744–51
83. Varlet I, Radman M, Brooks P. 1990. *Proc. Natl. Acad. Sci. USA* 87:7883–87
84. Fang W-H, Modrich P. 1993. *J. Biol. Chem.* 268:11838–44
85. Fang W-H, Li G-M, Longley M, Holmes J, Thilly W, Modrich P. 1993. *Cold Spring Harbor Symp. Quant. Biol.* 58:597–603
86. Parsons R, Li G-M, Longley MJ, Fang W-H, Papadopoulos N, et al. 1993. *Cell* 75:1227–36
87. Umar A, Boyer JC, Thomas DC, Nguyen DC, Risinger JI, et al. 1994. *J. Biol. Chem.* 269:14367–70
88. Risinger JI, Umar A, Barrett JC, Kunkel TA. 1995. *J. Biol. Chem.* 270:18183–86
89. Umar A, Boyer JC, Kunkel TA. 1994. *Science* 266:814–16

90. Papadopoulos N, Nicolaides NC, Liu B, Parsons R, Lengauer C, et al. 1995. *Science* 268:1915–17
91. Kat A, Thilly WG, Fang W-H, Longley MJ, Li G-M, Modrich P. 1993. *Proc. Natl. Acad. Sci. USA* 90:6424–28
92. Shinya E, Shimada T. 1994. *Nucleic Acids Res.* 22:2143–49
93. Fishel R, Lescoe MK, Rao MR, Copeland NG, Jenkins NA, et al. 1994. *Cell* 77:167
94. Chi NW, Kolodner RD. 1994. *J. Biol. Chem.* 269:29984–92
95. Haber LT, Walker GC. 1991. *EMBO J.* 10:2707–15
96. Chi NW, Kolodner RD. 1994. *J. Biol. Chem.* 269:29993–97
97. Grilley MM. 1992. *DNA and DNA-protein intermediates of methyl-directed mismatch correction in Escherichia coli.* PhD thesis. Duke Univ., Durham, NC. 159 pp.
98. Miret JJ, Milla MG, Lahue RS. 1993. *J. Biol. Chem.* 268:3507–13
99. Alani E, Chi NW, Kolodner R. 1995. *Genes Dev.* 9:234–47
100. Fishel R, Ewel A, Lescoe MK. 1994. *Cancer Res.* 54:5539–42
101. Fishel R, Ewel A, Lee S, Lescoe MK, Griffith J. 1994. *Science* 266:1403–5
102. Palombo F, Hughes M, Jiricny J, Truong O, Hsuan J. 1994. *Nature* 367:417
103. Williamson MS, Game JC, Fogel S. 1985. *Genetics* 110:609–46
104. Strand M, Prolla TA, Liskay RM, Petes TD. 1993. *Nature* 365:274–76
105. Prolla TA, Pang Q, Alani E, Kolodner RD, Liskay RM. 1994. *Science* 265:1091–93
106. Sancar A, Hearst JE. 1993. *Science* 259:1415–20
107. Proffitt JH, Davie JR, Swinton D, Hattman S. 1984. *Mol. Cell. Biol.* 4:985–88
108. Urieli-Shoval S, Gruenbaum Y, Sedat J, Razin A. 1982. *FEBS Lett.* 146:148–52
109. Strand M, Earley MC, Crouse GF, Petes TD. 1995. *Proc. Natl. Acad. Sci. USA* 92:10418–21
110. Kunkel TA. 1993. *Nature* 365:207–8
111. Ripley LS. 1990. *Annu. Rev. Genet.* 24:189–213
112. Ishimi Y, Claude A, Bullock P, Hurwitz J. 1988. *J. Biol. Chem.* 263:19723–33
113. Goulian M, Richards SH, Heard CJ, Bigsby BM. 1990. *J. Biol. Chem.* 265:18461–71
114. Turchi JJ, Bambara RA. 1993. *J. Biol. Chem.* 268:15136–41
115. Sommers CH, Miller EJ, Dujon B, Prakash S, Prakash L. 1995. *J. Biol. Chem.* 270:4193–96
116. Johnson RE, Kovvali GK, Prakash L, Prakash S. 1995. *Science* 269:238–40
117. Karran P, Bignami M. 1992. *Nucleic Acids Res.* 20:2933–40
118. Sibghat-Ullah S, Day RS. 1992. *Biochemistry* 31:7998–8008
119. Goldmacher VS, Cuzick RA, Thilly WG. 1986. *J. Biol. Chem.* 261:12462–71
120. Karran P, Bignami M. 1994. *BioEssays* 16:833–39
121. Jones M, Wagner R. 1981. *Mol. Gen. Genet.* 184:562–63
122. Karran P, Marinus M. 1982. *Nature* 296:868–69
123. Branch P, Aquilina G, Bignami M, Karran P. 1993. *Nature* 362:652–54
124. Aquilina G, Hess P, Fiumicino S, Ceccotti S, Bignami M. 1995. *Cancer Res.* 55:2569–75
125. Day RS, Ziolkowski CHJ, Scudiero DA, Meyer SA, Lubiniecki AS, et al. 1980. *Nature* 288:724–27
126. Pieper RO, Futscher BW, Dong Q, Ellis TM, Erickson L. 1990. *Cancer Commun.* 2:13–20
127. Aquilina G, Hess P, Branch P, MacGeoch C, Casciano I, et al. 1994. *Proc. Natl. Acad. Sci. USA* 91:8905–9
128. de Wind N, Dekker M, Berns A, Radman M, te Riele H. 1995. *Cell* 82:321–30
129. Koi M, Umar A, Chauhan DP, Cherian SP, Carethers JM, et al. 1994. *Cancer Res.* 54:4308–12
130. Hawn MT, Umar A, Carethers JM, Marra G, Kunkel TA, et al. 1995. *Cancer Res.* 55:3721–25
131. Feng W-Y, Lee EH, Hays JB. 1991. *Genetics* 129:1007–20
132. Feng W-Y, Hays JB. 1995. *Genetics* 140:1175–86
133. Loeb LA, Springate CF, Battula N. 1974. *Cancer Res.* 34:2311–21
134. Nowell PC. 1976. *Science* 194:23–28
135. Fearon ER, Vogelstein B. 1990. *Cell* 61:759–67
136. Stanbridge EJ. 1990. *Annu. Rev. Genet.* 24:615–57
137. Bishop JM. 1991. *Cell* 64:235–48
138. Vogelstein B, Kinzler KW. 1993. *Trends Genet.* 9:138–41
139. Lynch HT, Smyrk TC, Watson P, Lanspa SJ, Lynch JF, et al. 1993. *Gastroenterology* 104:1535–49
140. Watson P, Lynch HT. 1993. *Cancer* 71:677–85
141. Cohen PR, Kohn SR, Kurzrock R. 1991. *Am. J. Med.* 90:606–13
142. Kolodner RD, Hall NR, Lipford J, Kane MF, Rao MR, et al. 1994. *Genomics* 24:516–26
143. Ionov Y, Peinado MA, Malkhosyan S,

Shibata D, Perucho M. 1993. *Nature* 363:558–61

144. Thibodeau SN, Bren G, Schaid D. 1993. *Science* 260:816–19

145. Aaltonen LA, Peltomäki P, Leach FS, Sistonen P, Pylkkänen L, et al. 1993. *Science* 260:812–16

146. Eshleman JR, Markowitz SD. 1995. *Curr. Opin. Oncol.* 7:83–89

147. Uchida T, Wada C, Wang C, Ishida H, Egawa S, et al. 1995. *Oncogene* 10:1019–22

148. Egawa S, Uchida T, Suyama K, Wang C, Ohori M, et al. 1995. *Cancer Res.* 55:2418–21

149. Shibata D, Peinado MA, Ionov Y, Malkhosyan S, Perucho M. 1994. *Nat. Genet.* 6:273–81

150. Bhattacharyya NP, Skandalis A, Ganesh A, Groden J, Meuth M. 1994. *Proc. Natl. Acad. Sci. USA* 91:6319–23

151. Eshleman JR, Lang EZ, Bowerfind GK, Parsons R, Vogelstein B, et al. 1995. *Oncogene* 10:33–37

152. Hall NR, Taylor GR, Finan PJ, Kolodner RD, Bodmer WF, et al. 1994. *Eur. J. Cancer* 30A:1550–52

153. Liu B, Parsons RE, Hamilton SR, Peterson GM, Lynch HT, et al. 1994. *Cancer Res.* 54:4590–94

154. Mary JL, Bishop T, Kolodner R, Lipford JR, Kane M, et al. 1994. *Hum. Mol. Genet.* 3:2067–69

155. Lynch HT, Drouhard T, Lanspa S, Smyrk T, Lynch P, et al. 1994. *J. Natl. Cancer Inst.* 86:1417–19

156. Hemminki A, Peltomäki P, Mecklin J-P, Järvinen H, Salovaara R, et al. 1994. *Nat. Genet.* 8:405–10

157. Kolodner RD, Hall NR, Lipford J, Kane MF, Morrison PT, et al. 1995. *Cancer Res.* 55:242–48

158. Nystrom-Lahti M, Parsons R, Sistonen P, Pylkkanen L, Aaltonen LA, et al. 1994. *Am. J. Hum. Genet.* 55:659–65

159. Orth K, Hung J, Gazdar A, Bowcock A, Mathis JM, Sambrook J. 1994. *Proc. Natl. Acad. Sci. USA* 91:9495–99

160. Liu B, Nicolaides NC, Markowitz S, Willson JKV, Parsons RE, et al. 1995. *Nat. Genet.* 9:48–55

161. Parsons R, Li G-M, Longley M, Modrich P, Liu B, et al. 1995. *Science* 268:738–40

162. Ames BN, Gold LS. 1990. *Science* 249:970–71

163. Reitmair AH, Schmits R, Ewel A, Bapat B, Redston M, et al. 1995. *Nat. Genet.* 11:64–70

164. Markowitz S, Wang J, Myeroff L, Parsons R, Sun L, et al. 1995. *Science* 268:1336–38

165. Markowitz SD, Myeroff L, Cooper MJ,

Traicoff J, Kochera M, et al. 1994. *J. Clin. Invest.* 93:1005–13

166. Park K, Kim S-J, Bang Y-J, Park J-G, Kim NK, et al. 1994. *Proc. Natl. Acad. Sci. USA* 91:8772–76

167. Lawler S, Candia AF, Ebner R, Shum L, Lopez AR, et al. 1994. *Development* 120:165–75

168. Petes TD, Malone RE, Symington LS. 1991. In *The Molecular and Cellular Biology of the Yeast Saccharomyces*, ed. J Broach, E Jones, J Pringle, pp. 407–521. Cold Spring Harbor, NY: Cold Spring Harbor Lab.

169. Haber JE. 1992. *Curr. Opin. Cell Biol.* 4:401–12

170. Nicolas A, Petes TD. 1994. *Experimentia* 50:242–52

171. Alani E, Reenan RA, Kolodner RD. 1994. *Genetics* 137:19–39

172. West SC. 1994. *Cell* 76:9–15

173. Rattray AJ, Symington LS. 1995. *Genetics* 139:45–56

174. Holliday RA. 1964. *Genet. Res.* 5:282–304

175. Meselson MS, Radding CM. 1975. *Proc. Natl. Acad. Sci. USA* 72:358–61

176. Radding CM. 1982. *Annu. Rev. Genet.* 16:405–37

177. Szostak JW, Orr-Weaver TL, Rothstein RJ, Stahl FW. 1983. *Cell* 33:25–35

178. Cao L, Alani E, Kleckner N. 1990. *Cell* 61:1089–101

179. Sun H, Treco D, Szostak JW. 1991. *Cell* 64:1155–61

180. Carpenter ATC. 1982. *Proc. Natl. Acad. Sci. USA* 79:5961–65

181. White JH, Lusnak K, Fogel S. 1985. *Nature* 315:350–52

182. Ronne H, Rothstein R. 1988. *Proc. Natl. Acad. Sci. USA* 85:2696–700

183. Lichten M, Goyon C, Schultes NP, Treco D, Szostak JW, et al. 1990. *Proc. Natl. Acad. Sci. USA* 87:7653–57

184. Haber JE, Ray BL, Kolb JM, White CI. 1993. *Proc. Natl. Acad. Sci. USA* 90:3363–67

185. Nag DK, Petes TD. 1993. *Mol. Cell. Biol.* 13:2324–31

186. McDonald JP, Rothstein R. 1994. *Genetics* 137:393–405

187. Feinstein SI, Low KB. 1986. *Genetics* 113:13–33

188. Moore CW, Hampsey DM, Ernst JF, Sherman F. 1988. *Genetics* 119:21–34

189. Nag DK, White MA, Petes TD. 1989. *Nature* 340:318–20

190. Ray B, White CI, Haber JE. 1991. *Mol. Cell. Biol.* 11:5372–80

191. Detloff P, Sieber J, Petes TD. 1991. *Mol. Cell. Biol.* 11:737–45

192. Nag DK, Petes TD. 1991. *Genetics* 129:669–73

193. Detloff P, Petes TD. 1992. *Mol. Cell. Biol.* 12:1805–14
194. Steeg CM, Ellis J, Bernstein A. 1990. *Proc. Natl. Acad. Sci. USA* 87:4680–84
195. Bollag RJ, Elwood DR, Tobin ED, Godwin AR, Liskay RM. 1992. *Mol. Cell. Biol.* 12:1546–52
196. Deng WP, Nickoloff JA. 1994. *Mol. Cell. Biol.* 14:400–6
197. Matic I, Rayssiguier C, Radman M. 1995. *Cell* 80:507–15
198. Waldman AS, Liskay RM. 1987. *Proc. Natl. Acad. Sci.* 84:5340–44
199. Waldman AS, Liskay RM. 1988. *Mol. Cell. Biol.* 8:5350–57
200. Bailis AM, Rothstein R. 1990. *Genetics* 126:535–47
201. Mezard C, Pompon D, Nicolas A. 1992. *Cell* 70:659–70
202. Selva EM, New L, Crouse GF, Lahue RS. 1995. *Genetics* 139:1175–88
203. Nassif N, Engels W. 1993. *Proc. Natl. Acad. Sci. USA* 90:1262–66
204. Hilliker AJ, Clark SH, Chovnick A. 1991. *Genetics* 129:779–81
205. DasGupta C, Radding CM. 1982. *Proc. Natl. Acad. Sci. USA* 79:762–66
206. Muller B, Tsaneva IR, West SC. 1993. *J. Biol. Chem.* 268:17179–84
207. Iype LE, Wood EA, Inman RB, Cox MM. 1994. *J. Biol. Chem.* 269:24967–78
208. Radman M. 1989. *Genome* 31:68–73
209. Rayssiguier C, Thaler DS, Radman M. 1989. *Nature* 342:396–401
210. Borts RH, Haber JE. 1987. *Science* 237:1459–65
211. Borts RH, Leung WY, Kramer W, Kramer B, Williamson M, et al. 1990. *Genetics* 124:573–84
212. Belmaaza A, Milot E, Villemure J-F, Chartrand P. 1994. *EMBO J.* 13:5355–60
213. Shen P, Huang HV. 1989. *Mol. Gen. Genet.* 218:358–60
214. Petit M-A, Dimpfl J, Radman M, Echols H. 1991. *Genetics* 129:327–32
215. Worth L Jr, Clark S, Radman M, Modrich P. 1994. *Proc. Natl. Acad. Sci. USA* 91:3238–41
216. Tsaneva IR, Müller B, West SC. 1992. *Cell* 69:1171–80
217. Priebe SD, Westmoreland J, Nilsson-Tillgren T, Resnick MA. 1994. *Mol. Cell. Biol.* 14:4802–14
218. Resnick MA, Skaanild M, Nilsson-Tillgren T. 1989. *Proc. Natl. Acad. Sci. USA* 86:2276–80
219. Resnick MA, Zgaga Z, Hieter P, Westmoreland J, Fogel S, Nilsson-Tillgren T. 1992. *Mol. Gen. Genet.* 234:65–73
220. Harris S, Rudnicki KS, Haber JE. 1993. *Genetics* 135:5–16
221. te Riele H, Maandag ER, Berns A. 1992. *Proc. Natl. Acad. Sci. USA* 89:5128–32
222. Shen P, Huang HV. 1986. *Genetics* 112:441–57
223. Rayssiguier C, Dohet C, Radman M. 1991. *Biochemie* 73:371–74
224. Mezard C, Nicolas A. 1994. *Mol. Cell Biol.* 14:1278–92
225. Matic I, Radman M, Rayssiguier C. 1994. *Genetics* 136:17–26

Annu. Rev. Biochem. 1996. 65:135–67

DNA REPAIR IN EUKARYOTES

Richard D. Wood

Imperial Cancer Research Fund, Clare Hall Laboratories, South Mimms, Herts EN6 3LD, United Kingdom

KEY WORDS: base excision repair, nucleotide excision repair, UV photoproducts, DNA polymerases, strand break repair

ABSTRACT

Eukaryotic cells have multiple mechanisms for repairing damaged DNA. O^6-methylguanine-DNA methyltransferase directly reverses some simple alkylation adducts. However, most repair strategies excise lesions from DNA. Two major pathways are base excision repair (BER), which eliminates single damaged-base residues, and nucleotide excision repair (NER), which excises damage within oligomers that are 25–32 nucleotides long. The specialized DNA glycosylases and AP endonucleases of BER act on spontaneous and induced DNA alterations caused by hydrolysis, oxygen free radicals, and simple alkylating agents. NER utilizes many proteins (including the XP proteins in humans) to remove the major UV-induced photoproducts from DNA, as well as other types of modified nucleotides. Different DNA polymerases and ligases are used to complete the separate pathways. Some organisms have alternative schemes, which include the use of photolyases and a specific UV-endonuclease for repairing UV damage to DNA. Finally, double-strand breaks in DNA are repaired by mechanisms that involve recombination proteins and, in mammalian cells, a DNA protein kinase.

CONTENTS

INTRODUCTION 136
O^6-METHYLGUANINE-DNA METHYLTRANSFERASE 137
BASE EXCISION REPAIR.. 137
    DNA Glycosylases 138
    AP Endonucleases.. 142
    Pathways for Completion of Base Excision Repair......................... 142
    DNAse IV and Homologous Activities in Yeast............................. 143
    Poly(ADP-Ribose) Polymerase and Base Excision Repair.................... 144
NUCLEOTIDE EXCISION REPAIR 144
    Incision During Nucleotide Excision Repair............................... 146
    Proteins that Bind to DNA Lesions or Single-Stranded DNA 148
    Nucleotide Excision Repair Activities in TFIIH 150
    DNA Polymerases in Nucleotide Excision Repair........................... 151

0066-4154/96/0701-0135$08.00

    Purified Nucleotide Excision Repair Proteins and Interactions Between Them   152
    Regulation of Nucleotide Excision Repair.   153
    Other Pathways for Repair of UV Light-Induced Damage.   154
    DNA Photolyases   154
    (6-4) Photoproduct DNA Photolyase   155
    An Additional Excision Repair Pathway in Some Organisms   155
DNA LIGASES IN REPAIR   157
DOUBLE STRAND–BREAK REPAIR   158
CONCLUSION.   160

INTRODUCTION

This review covers the major mechanisms for removing damage from DNA in eukaryotes. Most of the relevant research has been done with mammalian cells and the yeast *Saccharomyces cerevisiae (S. cerevisiae)*, and the present coverage is biased towards the former. The past three years are emphasized because this period has seen enormous growth in the field. DNA mismatch repair is not discussed here, and in fact the main function of this process is not to repair damage but to correct errors of DNA replication and to act on heteroduplexes formed during recombination. Mismatch repair pathways in eukaryotes are reviewed in this volume by Modrich & Lahue and elsewhere (1). A valuable textbook on DNA repair was published in 1995 (2). It serves as the background to this summary, and is particularly informative on DNA repair in *Escherichia coli (E. coli)*, which has been so essential in setting paradigms in this field.

An outline of the pathway for direct reversal of DNA damage caused by simple alkylating agents is followed by a discussion of the two main mechanisms for excision of DNA damage. These mechanisms are base excision repair (BER) and nucleotide excision repair (NER). As an essential DNA maintenance pathway, BER acts continuously on both spontaneous and induced DNA damage caused by hydrolysis, oxygen free radicals, and simple alkylating agents. The most obvious function of NER in humans is to remove the major UV-induced photoproducts caused by sunlight from irradiated DNA. However, emerging evidence shows that NER in eukaryotes can excise nucleotides containing essentially any covalent adduct, although not all lesions are removed with equal efficiency. One aspect of NER not emphasized here is its connection with RNA pol II transcription and the fascinating implication for human disorders, including the so-called transcription syndromes. Many excellent summaries of these advances are available (e.g. 3, 4), and a relevant review by Friedberg appears in this volume. The remaining sections of this paper summarize the emerging variety of alternative mechanisms for repair of UV damage to DNA, the roles of different DNA polymerases and ligases, and some aspects of double strand–break repair.

O^6-METHYLGUANINE-DNA METHYLTRANSFERASE

The main function of O^6-methylguanine-DNA methyltransferase (MGMT) is to protect against the lethal and mutagenic effects of simple alkylating agents. Natural environmental sources of O^6-methylation in DNA may include atmospheric methyl halides and products of nitrite metabolism (5). MGMT enzymes exist in both yeast and mammalian cells. As in prokaryotes, the methyl group from O^6-methylguanine in DNA is transferred to a cysteine residue in the enzyme (2). Thus the adduct is directly reversed (rather than excised) in a reaction that inactivates the repair enzyme. cDNAs have been isolated that encode the *S. cerevisiae* enzyme MGT1 and several mammalian proteins [reviewed in (6)]; the human gene is near the end of chromosome 10q. Mammalian cell lines without MGMT activity (designated Mer⁻ or Mex⁻) are hypersensitive to methylating agents such as methylnitrosourea (7).

Variant Mex⁻ cells resistant to the toxic effects of simple N-nitrosating methylating agents can be readily selected in culture (7). One mechanism for methylation tolerance in such cells is the loss of mismatch repair capacity (8–10). This tolerance apparently occurs because O^6-methylguanine:C and O^6-methylguanine:T base pairs in DNA are recognized as mispairs by the mismatch correction system. The base opposite the methylated guanine is usually repaired, initiating a futile cycle that leads to toxicity in mismatch repair–proficient cells (7, 8). As predicted from studies of cell lines, double knockout mice defective in mismatch repair are anomalously resistant to the alkylating agent MNNG (11). The loss of mismatch repair that accompanies methylation tolerance leads in turn to a mutator phenotype. This sequence of events might contribute to oncogenesis, for instance in colon cancer, in that continual environmental exposure to alkylating agents might provide selective pressure for the loss of mismatch repair and the concomitant development of methylation tolerance (8, 12).

BASE EXCISION REPAIR

In general terms, BER removes DNA damage that can arise spontaneously in a cell from hydrolytic events such as deamination or base loss, oxygen free radical attack, or methylation of ring nitrogens by endogenous agents (13). Thus BER is a repair pathway essential for DNA maintenance. The key event of BER is the hydrolysis of the N-glycosyl bond linking a modified base to the deoxyribose-phosphate chain, which excises the base residue in the free form. The base release is catalyzed by the class of enzymes called DNA glycosylases, summarized in Table 1. Each DNA glycosylase is specific for a particular set of lesions, and each type of enzyme is discussed in turn. The resulting apurinic or apyrimidinic site is cleaved by an AP endonuclease

(Figure 1). Since complete loss of BER would be incompatible with life, a backup system or one of several alternate pathways can often repair a given lesion. Nevertheless, inherited syndromes may be found that have alterations in enzymes of the BER pathway.

DNA Glycosylases

URACIL-DNA GLYCOSYLASE   Uracil in DNA is formed continuously by the spontaneous or chemically induced deamination of cytosine (13), although it

Table 1   Eukaryotic DNA glycosylases and AP endonucleases for base excision repair

Enzyme	Mammalian gene	Human chromosome	Mammalian protein size[a]	S. cerevisiae gene	S. cerevisiae protein size	Substrates in DNA
Uracil-DNA glycosylase	UNG1	12	227 aa[b] 25.8 kDa[b]	UNG1	359 aa[c] 27–30 kDa[b]	uracil
3-Methyladenine-DNA glycosylase	AAG/MPG	16	294 aa 32 kDa	MAG	296 aa 34.3 kDa	3-methylpurines, 7-methylpurines, some ethylated bases, hypoxanthine, ethenoadenine
Pyrimidine hydrate-DNA glycosylase	—	—	~30 kDa	—	~40 kDa	pyrimidine residues altered by oxidative ring saturation, ring fragmentation or contraction
FaPy-DNA glycosylase	hMMH1	—		FPG1	~40 kDa	formamidopyrimidines, 7,8, dihydro 8-oxo-guanine
Thymine mismatch-DNA glycosylase	—	12q24	410 aa 46 kDa	—	—	thymine paired to G, uracil paired to G
AP endonuclease	HAP1, APE, (human) APEX (mouse) BAP1 (bovine)	14 q11.2-12	35.5 kDa	APN1	300 aa 40.5 kDa	apurinic and apyrimidinic sites; mammalian enzyme homologous to E. coli Xth, yeast enzyme homologous to E. coli Nfo

[a] Where the number of amino acids (aa) is shown, the size given in kDa is that predicted from the open reading frame of the cDNA sequence.
[b] Nuclear form.
[c] Full-length.

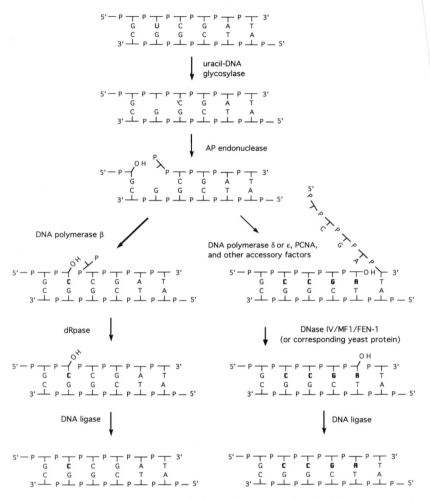

Figure 1 Model for base excision repair, similar to that suggested in (13a). The pathway may be envisioned as branched, with varying patch sizes resulting from distinct DNA polymerases.

can occasionally arise by the incorporation of dUMP from dUTP. Since deamination of cytosine in DNA generates a U•G mispair, excision of uracil by uracil-DNA glycosylase is needed to prevent GC → AT transition mutations during subsequent semiconservative DNA replication. The nuclear enzyme *UNG1* is the only identified physiologically relevant human uracil-DNA glycosylase (14) and it shares extensive amino acid–sequence identity with the

E. coli ung gene product. The biochemistry of uracil-DNA glycosylases has been reviewed (15).

Uracil-DNA glycosylases are small proteins (25–30 kDa) and the crystal structures for two of them have been determined. Binding sites for a uracil or a uracil analog were located for both the human (16) and Herpes simplex virus enzymes (17). The overall disposition of α-helices is similar in the two structures, but other features show some large differences. However, both enzymes have a binding pocket for the uracil substrate arranged at one end of a groove where the DNA backbone is expected to bind. The tight fit of uracil in the binding pocket has led to the proposition that the uracil base is flipped out from the double helix before hydrolysis of the glycosyl bond. Specific contacts in the binding site region explain why uracil in DNA is recognized as a substrate whereas uracil in RNA is not, and why the glycosylase removes uracil or 5-fluorouracil from DNA, but does not remove thymine or 5-bromouracil (16, 17).

A separate DNA glycosylase that eliminates formyluracil and hydroxy-methyluracil from DNA has been detected and partially purified from mammalian cells (18–20), but it has not been detected in yeast. The corresponding gene has not yet been found, but a mammalian cell line that lacks the activity exists (21). Formyluracil is generated by oxidation of the methyl group in thymine, whereas 5-hydroxymethyluracil might be formed in DNA after deamination of 5-hydroxymethylcytosine, an oxidation product of 5-methyl-cytosine. In *E. coli*, formyluracil and 5-hydroxymethyluracil are substrates for AlkA, a 3-methyladenine-DNA glycosylase, but the 3-methyladenine-DNA glycosylases from yeast and human cells do not show activity of this type (18).

3-METHYLADENINE-DNA GLYCOSYLASE   The most characterized substrates of 3-methyladenine-DNA glycosylases include major methylation adducts of purines at the N-3 and N-7 positions. These modifications can be continuously generated intracellularly from methyl donors such as S-adenosylmethionine (22), and studies in yeast show that such endogenous alkylation damage is a source of spontaneous mutation in eukaryotic cells (23). Some ethylation adducts are also removed (24). Other important substrates are cyclic adducts, such as $1,N^6$-ethenoadenine, which is formed by metabolites of vinyl chloride and related human carcinogens (25). However, a similar cyclic adduct of cytosine with p-benzoquinone appears to be repaired by a separate DNA glycosylase in mammalian cells (26). Several 3-methyladenine-DNA glycosy-lases and corresponding cDNAs have been isolated from mammalian cells (27–29) and *S. cerevisiae* (30, 31). The mammalian and yeast genes have little or no sequence homology. The yeast enzyme is transcriptionally induced by DNA-damaging agents (32).

Hypoxanthine can arise in DNA after deamination of adenine, and it is mutagenic. The eukaryotic 3-methyladenine-DNA glycosylases can remove hypoxanthine from DNA, and previously identified hypoxanthine-DNA glycosylase activities are likely to be attributable to 3-methyladenine-DNA glycosylase (33). As yet the rather broad substrate specificity of this enzyme has no clear structural explanation.

PYRIMIDINE HYDRATE-DNA GLYCOSYLASE   These enzymes act on oxidatively damaged pyrimidines with a saturated 5,6 double bond (such as thymine glycol and cytosine hydrate), as well as on ring-opened and degraded pyrimidine residues (34). The eukaryotic activity is a counterpart of the *E. coli* Nth and endonuclease VIII (35) proteins and is referred to variously as thymine glycol-DNA glycosylase, pyrimidine hydrate DNA glycosylase, urea-DNA glycosylase, and redoxyendonuclease (36–38). The activity has also been detected in mammalian mitochondria (39). A yeast enzyme of about 40 kDa with properties similar to Nth exists (40).

A physiological role for pyrimidine hydrate-DNA glycosylase in eukaryotes remains to be demonstrated and awaits isolation of a mutant defective in the activity. A proportion of pyrimidine oxidation products may be removed from nuclear DNA by NER.

FORMAMIDOPYRIMIDINE-DNA GLYCOSYLASE   Purines with oxidative damage are removed from DNA in *E. coli* by the Fpg (MutM) protein. Such damage includes formamidopyrimidine, other imidazole ring-opened residues, and 7,8-dihydro-8-oxoguanine (34). The latter common oxidation product of guanine is often referred to as 8-hydroxyguanine or 8-oxoguanine. 8-oxoguanine is mutagenic because it can pair with both A and C, and cells have several specialized ways to deal with this lesion. The mouse and human 3-methyladenine DNA glycosylases have been reported to act on 8-oxoguanine to some extent (41). An activity that more closely resembles *E. coli* Fpg is also present in human (42) and *S. cerevisiae* (43) cells, and characterization of the genes is underway.

Mammalian cells also have an activity, similar to *E. coli* MutT, that hydrolyzes the free nucleotide 8-oxo-dGTP to the monophosphate form, thus preventing its accidental and mutagenic incorporation during DNA synthesis. One human cDNA for 8-oxo-dGTPase activity maps to human chromosome 7p22 (44) and has been designated MTH1 (MutT homolog 1). This cDNA encodes a 156–amino acid protein of 17.9 kDa (45).

DNA GLYCOSYLASES THAT ACT ON MISMATCHES   A 46 kDa thymine mismatch-DNA glycosylase in human cells can catalyze the excision of T when paired with G (46; J Jiricny, personal communication). This mispair results from spontaneous deamination of 5-methylcytosine in DNA: It is a biological accident that the deaminated form of 5-methylcytosine is identical to thymine.

Hence, this form of BER represents one strategy for the repair of mismatched bases, although unlike general mismatch repair it shows no preference for a newly synthesized strand. The thymine mismatch-DNA glycosylase can also remove uracil when paired to G and may serve as a backup system for uracil-DNA glycosylase (47).

E. coli MutY is a DNA glycosylase that removes A when misincorporated opposite 8-oxoG, thus serving as another mechanism of defense against oxidative damage. A similar enzyme exists in human cells (47a) and the corresponding cDNA has been isolated (48).

AP Endonucleases

Apurinic or apyrimidinic (AP) sites are formed in DNA by the action of DNA glycosylases during the first step of BER (Figure 1) or by continuous slow spontaneous hydrolysis (13). The repair of these sites is initiated by AP endonucleases. *E. coli* has at least two such enzymes. However, studies of yeast and mammalian cells have thus far resulted in the purification and characterization of a single major enzyme in each organism that catalyzes the incision of phosphodiester linkages exclusively 5' to AP sites, leaving 5' deoxyribose-phosphate and 3' OH residues (Figure 1). In mammalian cells the gene that encodes this AP endonuclease is called *HAP1, BAP1, APE*, or *APEX* [reviewed in (49)]. The mammalian enzymes share significant sequence homology with the major AP endonuclease of *E. coli*, the Xth protein, for which the 3-D structure is known (50). HAP1 was also isolated as Ref-1, a protein that can modify the activity of Fos-Jun by reduction of a specific Cys residue in the transcription factor (51, 52). Thus, in addition to its role in BER, this protein might play a role in transducing cellular signals associated with oxidative stress.

In yeast, the major AP endonuclease activity is encoded by the *APN1* gene. This activity is unlike the major mammalian AP endonuclease in that it is homologous to the distinct *E. coli* AP endonuclease, Nfo (53). Yeast mutants lacking APN1 are sensitive to oxidative damage and alkylating agents (implying a severe defect in BER) and have an elevated rate of spontaneous mutation (54).

Pathways for Completion of Base Excision Repair

Completion of BER requires the removal of the 5' terminal deoxyribose-phosphate residue generated by the AP endonuclease, accompanied by repair DNA synthesis and ligation (Figure 1). An enzyme activity of ~50 kDa that can remove sugar-phosphate residues from duplex DNA (55) has been identified in human cells and is designated DNA deoxyribophosphodiesterase (dRPase).

This function may be used when a DNA polymerase generates a single-nucleotide repair patch.

There is good evidence that DNA polymerase β (pol β) is used in this very short–patch mode of repair synthesis during BER. This conclusion is based on studies of uracil repair in mammalian cell extracts in the presence of various DNA polymerase inhibitors (56), fractionation of *Xenopus* oocyte extracts to purify activities that repair natural AP sites (57), and the sensitivity of BER to anti-DNA pol β antibody (58). In a bovine testis nuclear extract, DNA pol β carried out repair synthesis during BER of a uracil residue (59). Pol β has an intrinsic dRp lyase activity in the small N-terminal domain of the enzyme, which might be employed for dRp removal as an alternative to the hydrolytic dRPase (60).

A double knockout mutation of pol β in mice causes embryonic lethality (61). This finding suggests either that the single-patch mode of BER is essential for maintaining normal viability (by eliminating products of hydrolytic and oxidative degradation) or that pol β has an additional essential role in mammalian cells, such as in chromosomal DNA replication. However, in *S. cerevisiae* the situation regarding DNA pol β is very different. A 67 kDa β-like DNA polymerase has been designated DNA polymerase IV (62, 63). Yeast strains harboring a disruption of *POL4* are viable and exhibit only a weak sensitivity to methylmethanesulfonate (MMS), a simple alkylating agent (63). Instead, either DNA pol δ (64) or pol ε (65) appears to participate in BER in *S. cerevisiae*. The function of yeast POL4 may be in a form of double strand–break repair (63).

Sometimes, BER is associated with the generation of longer repair patches, most likely generated by DNA pols δ or ε. Such repair shows a requirement for PCNA, in contrast to the short-patch pathway (57). These longer repair tracts may result from a nick translation reaction accompanied by strand displacement in the $5' \rightarrow 3'$ direction, thereby generating a flap type of structure (Figure 1). Removal of such an overhanging 5′ terminal single-stranded region of DNA can be effected by the 5′ single strand/duplex junction-specific enzyme originally designated DNase IV (66).

DNAse IV and Homologous Activities in Yeast

In an early survey of mammalian nucleases, DNAse IV was found as a $5' \rightarrow 3'$ exonuclease that released mononucleotides and short oligonucleotides from DNA (66). Given its currently known properties, DNAse IV might now be more properly termed a 5′ nuclease (67). This 43 kDa protein has since been rediscovered in several interesting contexts, notably as an exonuclease implicated in lagging strand DNA synthesis during semiconservative replication (68–70), as a "flap endonuclease" (71), and as the human homolog of *Schi-*

zosaccharomyces pombe rad2 (72). As a result this protein has acquired several additional names including factor pL, MF-1, and FEN-1. The predicted amino acid sequences of DNase IV and its yeast counterparts show regions of amino acid sequence similarity with the 5' → 3' exonuclease domains of prokaryotic DNA polymerases (73, 74).

In the fission yeast *S. pombe, rad2* mutants are hypersensitive to UV light and show a high rate of spontaneous chromosome loss, indicating a role in DNA repair, replication, and/or recombination (72). The corresponding gene for the nuclease in *S. cerevisiae* was first identified as an open reading frame with the temporary designation *YKL510* (75) and has since acquired the names *RAD27* (76), *RTH1* (77), and *ERC11* (78). Disruptions of the gene are temperature sensitive for growth (76, 77) and sensitive to MMS, and exhibit S-phase arrest or are inviable in combination with certain G1 cyclins (78). Certain DNA repeat structures show enhanced instability in an *rth1/rad27* mutant (79), confirming the DNA instability phenotype first found in *S. pombe*. All of these phenotypes suggest that the normal function of the enzyme is at the DNA replication fork, and perhaps additionally in mismatch repair and excision repair.

Poly(ADP-Ribose) Polymerase and Base Excision Repair

The enzyme poly (ADP-ribose) polymerase normally has a high affinity for strand breaks in DNA, but in the presence of NAD the DNA-bound enzyme undergoes extensive automodification which results in its dissociation. If the enzyme is not released from DNA, it actually inhibits BER (80, 81). The suggestion has been made that the binding of poly (ADP-ribose) polymerase to strand breaks introduced by AP endonucleases may signal a slowing or cessation of DNA replication while BER takes place. Alternatively, the binding of poly(ADP-ribose) polymerase to DNA may serve to reduce the potential for initiation of recombination at sites of strand breakage (82). Furthermore, ADP-ribosylation of histone and nonhistone chromosomal proteins changes chromatin structure, which may affect the rate or efficiency of BER (83). Nevertheless, poly(ADP-ribose) polymerase appears to have no essential role in BER, or even in essential transactions in chromatin. Mice with a homozygous knockout of the poly(ADP-ribose) polymerase gene develop normally and do not show unusual sensitivity to DNA-damaging agents (84). Instead, many develop epidermal hyperplasia, suggesting that the enzyme plays an auxiliary role in normal control of cell division. No poly(ADP-ribose) polymerase seems to exist in yeast.

NUCLEOTIDE EXCISION REPAIR

In addition to dipyrimidine photoproducts formed by UV light, NER can eliminate other lesions, including adducts of psoralen derivatives to pyrimidines (85, 86), various cisplatin-purine adducts (87, 88), and adducts formed by polycyclic carcinogens such as acetylaminofluorene (85, 89, 90). In vitro, NER can also act on less bulky alterations to some extent, including synthetic AP sites, 0^6-methylguanine, and mismatched base pairs (91). Even a cholesterol moiety can serve as a substrate when incorporated into the DNA backbone in place of a normal nucleotide (92). Which of these activities of NER are physiologically significant? Not all of the adducts are removed equally well. In general, the efficiency of repair may depend upon the degree of structural difference from a normal base pair (93). For example, the major lesions caused by 254-nm UV light, cyclobutane pyrimidine dimers, are removed from the bulk of the genome 5 to 10 times more slowly than the second most abundant lesions, (6-4) photoproducts (89, 94). The solution structure of a (6-4) photoproduct in DNA is significantly more distorted than that of a cyclobutane pyrimidine dimer (95). Nevertheless, the precise structural features that define the efficiency of recognition during NER are not known.

NER is not essential for viability, and some individuals with the inherited syndrome xeroderma pigmentosum (XP) have a total deficiency in the process (96). However, such patients usually have a roughly 1000-fold increased incidence of skin cancers (97). The increased incidence of internal cancers is not nearly as large, although not enough data exist to make a good estimate of the degree of risk (97). The hypersensitivity of XP patients to sunlight implies that in humans a major function of NER is to repair damage caused by UV radiation. However, some individuals with XP show progressive neurological degeneration over several decades, which may result from the slow accumulation of a class of oxygen free-radical–induced damage that is normally repaired by NER (98, 99).

The NER systems of the eukaryotes so far studied are very similar, from the yeasts *S. cerevisiae* and *S. pombe* through *Drosophila*, *Xenopus*, and mammalian cells. This high degree of conservation has facilitated rapid progress in the field. The following discussion considers biochemical features of the most studied systems, the budding yeast *S. cerevisiae* and mammals. Relevant proteins in mammalian cells and yeast are summarized in Table 2. In yeast, cells with defects in NER proteins were first isolated as UV-sensitive *rad* mutants. In mammalian cells, NER mutants include human cells from individuals with one of the seven repair-deficient complementation groups A through G of XP. Other important mutants are UV-sensitive rodent cell lines isolated in the laboratory, with defects in *ERCC* (excision repair cross-complementing) genes. Some of the *ERCC* genes are identical to *XP* genes, in which case the *XP* designation is preferred (100).

Hanawalt and coworkers discovered that UV-induced pyrimidine dimers are in general removed several-fold faster from transcribed genes than from nontranscribed genes and that most of this effect is due to preferential repair of the transcribed strand [reviewed in (101)]. This observation of transcription-coupled repair has been extended to many other DNA lesions and is discussed in other reviews in this volume.

Incision During Nucleotide Excision Repair

A key feature of NER is the introduction of two incisions into the damaged DNA strand, one on each side of a DNA lesion. The size of the repair patch formed during NER is about 25–30 nucleotides (nt) long, as shown by measurements in vivo (102, 103) and with in vitro systems (85, 104-106). This length is a consequence of the positions of the dual incisions. DNA substrates containing a single modification have been used to define the exact position of these incisions. For cyclobutane pyrimidine dimers and psoralen monoadducts, one incision is made 5 to 6 phosphodiester bonds 3′ to the lesion, and another incision 22 to 24 phosphodiester bonds away from the lesion on the 5′ side, giving rise to excision fragments 27–32 nt long (86). The exact positions of the principal incisions vary according to the adduct, but are 3 to 9 phosphodiester bonds from the lesion on the 3′ side and 16 to 25 phosphodiester bonds from the lesion on the 5′ side (106a).

Two different nucleases are used to create the dual incisions. In mammalian cells, these are the XPG protein (107), which makes the 3′ incision (108), and the ERCC1-containing complex, which makes the 5′ incision. In *S. cerevisiae* the corresponding nucleases are the RAD2 protein and the RAD1-RAD10 protein complex. RAD2 (109), XPG (110, 111), and RAD1-RAD10 (112, 113) were first thought to be single-stranded DNA endonucleases because they cleave M13 bacteriophage DNA. All of these proteins, however, have proven to be "structure-specific" DNA endonucleases that cut near the border of single-stranded and duplex DNA. Cleavage of M13 DNA may result from action at hairpin structures in the phage DNA. Members of the RAD2 protein family were found to cleave splayed-arm "Y" or flap structures at the junction between duplex DNA and single-stranded tails with 5′ ends (71). Independently, human XPG protein was shown to make endonucleolytic incisions in synthetic DNA substrates at the border between duplex and single-stranded regions when the single strand had a polarity 3′ to 5′ moving away from the duplex (108). These nucleases have two conserved domains of sequence similarity with DNAse IV and related enzymes (71, 72, 107).

In *S. cerevisiae*, Rad1 and Rad10 polypeptides associate tightly (112, 114, 115) to form a structure-specific endonuclease with a polarity opposite to that of XPG and Rad2. The Rad1-Rad10 complex cleaves model substrates near the border of duplex DNA and a single strand with a 3′ end (116). The human homolog of *S. cerevisiae* RAD10 is the 31-kDa ERCC1 protein (117), and the human homolog of Rad1 is the ERCC4 protein (L Thompson, personal communication). Together these human proteins are predicted to form a structure-specific nuclease. ERCC1 and ERCC4 form a tight protein complex 100–200 kDa in size that also includes correcting activity for XP-F cell extracts (118–120) and for the sole mutant in complementation group 11 of rodent cells (119). A polypeptide of about 110 kDa that binds tightly to ERCC1 (121, 122) is believed to be the XPF/ERCC4 protein, and *XPF* or *ERCC11* seem likely to be the same gene as *ERCC4*.

The exact mechanics of incision remain to be elucidated, but a partially unwound or open complex around a DNA lesion may be an intermediate in the NER process (Figure 2). Cleavage at the borders between duplex DNA and the opened region could then be performed by structure-specific nucleases of the appropriate polarity. This mechanism has been modelled with a synthetic DNA bubble structure with a 30-nucleotide unpaired region flanked by duplex regions of similar length. When the bubble substrate was incubated with XPG protein, cleavage on the 3′ side was observed (108), whereas cleavage on the 5′ side occurred upon incubation with Rad1-Rad10 protein complex, and both cleavage products were found when Rad1, Rad10, and XPG were incubated

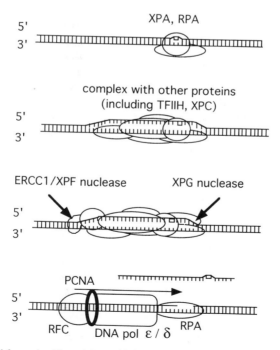

Figure 2   Model for nucleotide excision repair in mammalian cells showing (from top to bottom) recognition of damage, opening of a region around the lesion, dual incision, and repair synthesis.

together (123). A free single-stranded DNA end is not required for structure-specific cleavage by Rad1-Rad10 or XPG, suggesting that the duplex–single strand junction itself is recognized by the enzymes. Selective cleavage of the damaged DNA strand must result from the interaction of these endonucleases with other repair proteins.

Genetic evidence shows that in addition to a role in NER, Rad1 and Rad10 are involved in a pathway of mitotic recombination (124–131). The suggestion was recently made that Rad1 can act to resolve Holliday junctions (132). However, this result has not been confirmed with either Rad1 alone or with Rad10 in a complex (123). Instead, Rad1-Rad10 seems to play a role in recombination between direct repeats in a pathway mediated by homologous single-strand annealing (116, 131). ERCC1 and some ERCC4 mutants are much more sensitive to chemical DNA interstrand cross-linking agents than are mutants of other ERCC or XP groups (133). This may suggest an additional role of the complex in a form of DNA recombination, similar to the RAD1-RAD10 single-strand annealing pathway in yeast.

Mouse mutants with disruptions in *ERCC1* have been generated by gene targeting (134) (JHJ Hoeijmakers, personal communication). The ERCC1-deficient mice are small in size and die before weaning, primarily from liver failure.

Proteins That Bind to DNA Lesions or Single-Stranded DNA

XPA AND RAD14   The XPA protein and its *S. cerevisiae* homolog Rad14 are both absolutely essential for NER, as inactivating mutations completely eliminate the process in the respective organisms. Mammalian XPA protein shows a preference for binding damaged over nondamaged DNA, as demonstrated with UV-irradiated or cisplatin-treated DNA in filter-binding assays (135, 136) and by gel retardation assays (137). The relative preference for binding is not particularly large; on a per-nucleotide basis, XPA binds several hundred-fold more tightly to damaged than to nondamaged DNA (137). XPA also binds to single-stranded DNA fragments with an affinity similar to that found with damaged duplex DNA (137). Mobility shift experiments have likewise shown that *S. cerevisiae* RAD14 binds preferentially to UV-irradiated DNA (138). Both RAD14 (138) and XPA (136) contain a C_4-Zn finger motif and tightly bound zinc. A 122–amino acid (aa) core fragment of XPA containing the Zn-finger region retains the ability to bind damaged DNA preferentially (139), and a point mutation that disrupts the Zn-finger structure eliminates DNA binding (140).

Treatment of UV-irradiated probes with DNA photolyase to remove cyclobutane pyrimidine dimers leads to little or no reduction in binding of XPA (137) or RAD14 (138), indicating that these proteins bind significantly more tightly to other lesions, probably (6-4) photoproducts. Differential recognition by XPA/RAD14 may be an important factor in determining the faster repair of (6-4) photoproducts over cyclobutane pyrimidine dimers.

RPA AND ITS COMPLEX WITH XPA   The single-stranded binding protein RPA (also known in humans as HSSB) is required for NER in vitro in both mammalian and yeast systems. The heterotrimeric protein was first discovered as an essential factor for SV40 DNA replication in vitro [reviewed in (141)]. RPA was found to function in an early stage of repair (142, 143), and an assay for nicking of closed-circular UV-irradiated plasmid DNA showed that RPA was required for the introduction of NER incisions (104). RPA can also modulate the repair synthesis step (144). Some amino acid substitutions in yeast RPA lead to UV sensitivity, suggesting that RPA also functions in repair in vivo (145–147). In mammalian cells, NER is abolished by microinjection of neutralizing anti-RPA antibodies (W Vermeulen & JHJ Hoeijmakers, personal communication).

RPA binds single-stranded DNA tightly and duplex DNA only weakly, but some preferential binding to damaged DNA has been detected (148, 149). In human cells, the 34 kDa subunit of RPA interacts with XPA protein (149–151). This interaction seems to be functionally significant, since RPA and XPA together have a much higher affinity for damaged DNA than either protein alone (149).

UV-DDB AND XPE   Searches using crude extracts of primate cells for proteins that preferentially bind damaged DNA have regularly resulted in the discovery of an activity that binds damaged oligonucleotides in gel-retardation or filter-binding assays. The activity (DNA damage binding, DDB, or UV-DDB) contains a polypeptide of ~125 kDa that was purified originally in the mid-1970s (152). The protein has a particular affinity for (6-4) photoproducts in UV-irradiated DNA (153–155) but can also bind to other lesions with weaker affinity (156). The activity has been purified as a single subunit and as a complex between two proteins of 127 and 48 kDa (157). Damage-binding activity is missing in cells from two consanguineous XP-E patients, but present in other XP-E cell lines (158–161). The reduced DNA repair in the binding-defective XP-E cells can be corrected by microinjection of the two-protein complex (162). In a purified in vitro system, small amounts of UV-DDB preparations containing the p127 subunit (and undetectable p48) can stimulate repair synthesis in UV-irradiated DNA twofold (121), but larger amounts afford no further stimulation (A Aboussekhra & RD Wood, unpublished data) and can even inhibit incision. Reconstituted repair systems with purified proteins can carry out a core mammalian NER reaction in the absence of UV-DDB (92,

121), and a reconstituted system with *S. cerevisiae* proteins has no yeast UV-DDB equivalent (163).

The *p127* and *p48* genes have been cloned (164, 165). No mutations of these genes in XP-E patients have been reported, so designating one of the proteins as the XPE factor is not currently possible. None of the known *RAD* genes in yeast correspond to *p127* or *p48*, and so far DDB activity has been found only in primate cells.

XPC, RAD4, AND RAD23 HOMOLOGS   XP-C cells tested so far have mutations in the *XPC* gene, which shows some homology to *RAD4* of *S. cerevisiae* (166). XP-C cell extracts are corrected by a protein factor of about 160 kDa (167) that is a complex between two proteins, XPC and the 45 kDa HHR23B (168). The XPC complex binds firmly to single-stranded DNA cellulose (167, 168) and also has some affinity for damaged duplex over nondamaged duplex DNA (F Hanaoka, personal communication). In *S. cerevisiae*, Rad4 and Rad23 similarly form a complex (163). Yeasts with a disruption of *RAD23* are not as UV-sensitive as those with disrupted *RAD4*, suggesting that Rad23 is not absolutely required for NER in all situations (169). The N-termini of Rad23 and HHR23B have homology to ubiquitin (168). The function of the XPC/ HHR23B complex is unknown.

In human cells, repair of transcribed DNA strands can still take place in several genes that have been studied in at least some XP-C mutant cell lines (170, 171). In contrast, a *RAD4* deletion in *S. cerevisiae* eliminates all NER, whether transcription-coupled or not (172). Whether the transcription-coupled repair seen in the human system is actually XPC-independent or if all of the mutants studied so far are leaky is not clear. However, the complete deficiency in NER seen with XP-C cell extracts suggests that some alleles are not leaky (167, 173).

Nucleotide Excision Repair Activities in TFIIH

The multiprotein complex TFIIH participates in both basal transcription and in NER. TFIIH is normally found in the initiation complex at promoters transcribed by RNA polymerase II (174, 175). The finding that known NER proteins are components of TFIIH came as a surprise (176). Human TFIIH contains the XPB and XPD proteins (177, 178) and yeast TFIIH the corresponding Ssl2 and Rad3 (179). The core TFIIH subunits participate directly in NER as a unit and their role is discussed in detail in other articles in this volume. The 3' to 5' DNA helicase activity of XPB and Ssl2 and the 5' to 3' DNA helicase activity of XPD and Rad3 are essential for NER activity. The existence of multiple proteins with dual functions in both transcription and NER is leading to a deeper understanding of the multiple abnormalities in human syndromes associated with defects in nucleotide excision repair, including XP, Cockayne syndrome, and trichothiodystrophy. The profound implications for these human diseases and the concept of "transcription syndromes" are discussed in detail elsewhere (2–4).

TFIIH contains a closely associated kinase activity that can phosphorylate both the carboxy-terminal domain of RNA polymerase II and CDK2 kinases (174). A yeast TFIIH subassembly lacking the kinase activity can still function in NER (180).

DNA Polymerases in Nucleotide Excision Repair

NER in cells and in vitro is sensitive to aphidicolin (143), which inhibits DNA polymerases α, δ, and ϵ. Strong evidence from many sources shows that the replicative mammalian DNA pol α is not an NER polymerase, and many studies of cells and cell extracts have led to the conclusion that DNA polymerases δ or ϵ are responsible for NER synthesis (143, 181–184). A role for DNA pol ϵ or δ was strengthened by the finding that PCNA is required for NER by mammalian cell extracts (104, 185). PCNA forms part of a holoenzyme with DNA pols δ or ϵ, but not α (186, 187). A toroidal ring of three identical PCNA subunits encircles the DNA in both yeast and mammalian cells. The crystal structure of *S. cerevisiae* PCNA has been solved (188). PCNA functions in NER synthesis and in filling of short single-stranded gaps by assisting in the initiation of DNA synthesis (144, 189) and may serve as an anchoring clamp at the 3'-OH terminus of a nascent DNA strand to which the DNA polymerase can bind.

The evidence that PCNA also participates in mammalian NER in vivo is compelling. UV-irradiation of non–S phase human cells in culture rapidly increases the level of PCNA that is detectable in nuclei, by immunostaining after mild extraction and fixation, because PCNA is relocated in the nucleoplasm. This redistribution of PCNA probably reflects binding to sites of repair (190–194). Similarly, nonextractable PCNA arises in nonproliferating cells after irradiation of human skin (195). This association of PCNA with chromatin after irradiation is linked to NER, because rapid binding of PCNA is absent in UV-irradiated quiescent XP-A cells (196, 197). Association of PCNA with irradiated DNA is also absent in XP-G fibroblasts and partially defective in XP-B, C, D, and F (198).

The participation of PCNA in NER suggests that a further DNA polymerase accessory factor, RFC, is also involved. RFC is a multisubunit protein complex that loads PCNA onto DNA templates in an ATP-dependent manner, creating a sliding clamp for DNA polymerases δ and ϵ (186, 187, 198a). We have studied repair synthesis at gaps created in UV-irradiated DNA by dual incisions during NER (144). Either DNA polymerase δ or ϵ could carry out repair synthesis, and both polymerases were strictly dependent on both PCNA and RFC under the conditions used (circular DNA substrates, buffer containing ~70 mM KCl, and the presence of RPA in the reaction mixture). In the presence of DNA ligase I, polymerase δ gave rise to a low proportion of ligated products, because of limited strand displacement by pol δ. Ligation could be improved by including DNAse IV with pol δ but pol ε gave a higher proportion of ligated products.

Determining which polymerase normally participates in NER in cells is difficult, and perhaps either pol ε or pol δ can serve in this function. Nishida et al found that human DNA polymerase ε was needed for NER synthesis in permeabilized nuclei of UV-irradiated cells (182, 199). In contrast, antibodies against human pol δ were found to reduce synthesis mediated by nuclear extracts in UV-irradiated DNA (200). The damage-dependent DNA synthesis observed in that study may not have represented NER, and reinvestigating the activity of the antibody in an XP protein–dependent system would be useful. In yeast, studies of NER in DNA polymerase mutants led to the conclusion that either pol δ or pol ε can participate in repair of UV-damaged DNA when the other activity is absent (201).

Purified Nucleotide Excision Repair Proteins and Interactions Between Them

RECONSTITUTION OF NUCLEOTIDE EXCISION REPAIR WITH PURIFIED COMPO-NENTS   Progress in the isolation and purification of NER gene products (many in recombinant form) has allowed the core nucleotide excision process to be carried out with purified components. Aboussekhra et al (121) reconstituted the incision reaction with the purified mammalian proteins RPA, XPA, TFIIH (containing XPB and XPD), XPC, UV-DDB, XPG, partially purified ERCC1/XPF complex, and a factor designated IF7. UV-DDB (related to XPE protein) stimulated repair but was not essential. Complete repair synthesis was achieved by combining these factors with DNA polymerase ε, RFC, PCNA, and DNA ligase I. The reaction (involving both incision/excision and repair synthesis) involves about 30 polypeptides.

Mu et al used a combination of five fractions to reconstitute the dual incision reaction by assaying release of damage-containing oligonucleotides 24–30 nt in length (92). The components consisted of recombinant MBP-XPA fusion protein, purified RPA protein, a HeLa cell fraction containing both partially purified TFIIH and XPG, partially purified XP-C protein, and partially purified ERCC1/XPF complex. The oligomer was not displaced from DNA without further treatment such as heating at 95° C or incubation with SDS, suggesting that additional factors are needed for excision. However, the purified recom-binant yeast proteins used by Guzder et al (163) reconstituted the incision of

UV or AAAF-treated DNA as well as the release of 25–28 excision nucleotide fragments. The reaction required RAD14, RPA, RAD1, RAD10, RAD2, RAD4 (in a complex with RAD23), and TFIIH (with six identified subunits). No other factor was needed in yeast for this incision/excision reaction. This finding suggests that the IF7 factor identified in the reconstituted mammalian system may be specific to this system or may participate in a repair step other than dual incision.

INTERACTIONS BETWEEN NUCLEOTIDE EXCISION REPAIR PROTEINS Some tight complexes exist between NER proteins, for example within the ERCC1 complex, the XPC complex, and TFIIH. In addition, numerous weaker pair-wise interactions between various NER proteins have been identified using two-hybrid analysis in yeast, co-immunoprecipitation, or protein affinity schemes. For example, the mammalian XPA protein has been found to interact with ERCC1, RPA, and TFIIH (139, 149, 151, 202–204). In yeast, interactions of TFIIH have been found with RAD2, RAD4, and RAD23 (205, 206). Furthermore, in studies with yeast TFIIH containing His-tagged TFB1, some fractions eluted from nickel agarose were found to also contain RAD1, RAD10, RAD2, RAD4, and RAD14 (180). This finding suggested that a proportion of the NER proteins in yeast extracts are assembled into an active "repairosome" (180). Such an assembly would be reminiscent of the RNA polymerase II holoenzyme (207), a complex that contains a proportion of many general basic transcription factors and associated proteins and is active for transcription. It is not yet clear whether NER normally operates via the action of a repairosome or by the ordered assembly and action of DNA repair proteins on DNA.

Regulation of Nucleotide Excision Repair

In yeast, several NER genes including *RAD2* are subject to modest transcriptional induction after irradiation (2), and some evidence suggests that repair of pyrimidine dimers by NER has an inducible component (208). No transcriptional induction has been reported for the mammalian equivalents of these genes. Since NER is a complex process involving the coordinated action of many polypeptides, induction of any single component would only be effective if that component were rate-limiting for repair or if its expression in nondamaged cells were deleterious. However, in primate cells, UV-DDB binding activity is increased by UV irradiation, apparently by a post-transcriptional mechanism (209). Nevertheless, other proteins clearly must modulate the core NER proteins. Proteins must be in the proper phosphorylation state for optimal efficiency of NER, as shown by the reduction of the dual incision reaction after inhibition of a type II protein phosphatase in HeLa cell extracts (209a).

OTHER PATHWAYS FOR REPAIR OF UV LIGHT–INDUCED DAMAGE

DNA Photolyases

Enzymatic photoreactivation is a direct mechanism to repair UV-induced cyclobutane pyrimidine dimers (210, 211). Photolyases absorb blue or near-UV light via intrinsic chromophores. All characterized photolyases contain reduced FAD and a second chromophore that acts as an energy-harvesting antenna to transfer energy to the FAD cofactor. The excited FADH⁻ cofactor initiates splitting of the dimer by electron transfer, restoring the pyrimidines to the original monomeric form. Two types of second chromophore have been found: 5,10-methenyltetrahydrofolate (MTHF), which confers an absorption maximum of ~380 nm, and 8-hydroxy-5-deazariboflavin, which gives an absorption maximum of ~440 nm.

Photoreactivating activity has been demonstrated in many eukaryotes, including fungi, and most classes of vertebrates, including fish, reptiles, amphibians, and marsupials [reviewed in (2, 212)]. Genes for many of these DNA photolyases have been cloned. The photolyases from *S. cerevisiae* and the fungus *Neurospora crassa* contain MTHF cofactors (211, 213). The *S. cerevisiae* PHR1 gene encodes a 66.2 kDa protein of 565 amino acids that has extensive sequence similarity with the *E. coli* enzyme (211). The gene from *N. crassa* encodes a 615–amino acid protein that is similar to the *S. cerevisiae* enzyme (214). Transcription of the *S. cerevisiae* gene is inducible by UV light and other DNA-damaging agents (215).

The first example of a photolyase gene from a multicellular organism was from the goldfish *Carassius auratus* (216). This gene is induced by exposure of cells to visible light. The deduced amino sequence differed significantly from the bacterial and fungal photolyases. Four more metazoan photolyases have been cloned and with the goldfish gene they clearly form a new family of photolyases. These photolyases are from the killifish *Oryzias latipes* (217), the fruitfly *Drosophila melanogaster* (217, 218), and two marsupials, the opossum *Monodelphis domestica* (219) and the kangaroo rat *Potorous tridactylis*. The predicted molecular masses range from 55 to 74 kDa and all contain an FAD cofactor.

Photolyases are missing from many species in a manner that is so far unpredictable (210). For example, no photolyase has been found in *S. pombe*. Furthermore, no photolyase genes have been detected in placental mammals, including humans (219), and no photolyase activity has been detected in human cell extracts (212). In two plant species, *Arabidopsis thaliana* and the mustard *Sinapis alba*, genes were found with high sequence homology to microbial photolyases. However, both of these encode what are presumed to be blue-light photoreceptors and lack photolyase or DNA repair activity (220). The existence of a DNA photoreactivating enzyme in plants remains to be demonstrated.

(6-4) Photoproduct DNA Photolyase

Todo and coworkers isolated an enzyme from embryos of *Drosophila melanogaster* that caused a light-dependent removal of (6-4) photoproducts from DNA and restored biological activity to test DNA molecules (218). Further study confirmed this observation and strongly suggests that the products of the reaction are unmodified monomeric pyrimidines, arising through an oxetane intermediate (221). The efficiency of repair per incident photon is low in comparison with cyclobutane pyrimidine dimer photolyases (221). The biological role of this activity and its occurrence in other organisms require further investigation.

An Additional Excision Repair Pathway in Some Organisms

The fission yeast *S. pombe* has a system for NER that is analogous to the *S. cerevisiae* and human NER systems, and many of the genes have been identified (Table 2). However, *S. pombe* is not as UV-sensitive as *S. cerevisiae*, and none of the fission yeast NER mutants are as UV-sensitive as their budding yeast counterparts (73, 222). The existence of residual pyrimidine dimer removal in *S. pombe* NER mutants suggested that this organism might have several pathways to excise such damage (223). This idea was greatly solidified by McCready and coworkers who found that mutants with deletions in genes such as *rad13* (the *S. cerevisiae RAD2* equivalent) and *rad16* (the *RAD1* equivalent) still remove cyclobutane pyrimidine dimers and (6-4) photoproducts from their DNA, although at a reduced rate compared to wild-type cells (224). In marked contrast, removal of both photoproducts is completely abolished in *RAD2* or *RAD1* deletion mutants of *S. cerevisiae* (225). Freyer and colleagues found that cell-free extracts from *S. pombe* could nick UV-irradiated DNA in a Mg^{2+}-dependent and ATP-independent fashion, a process that apparently does not result from either BER or NER (226). Pursuing this observation led to the discovery of an ATP-independent enzymatic activity that introduces nicks 5' to both cyclobutane pyrimidine dimers and (6-4) photoproducts (227). The endonuclease is missing in an *S. pombe rad12* mutant (228), although *rad12* is not the structural gene.

In *N. crassa* more than 30 mutagen-sensitive mutants have been studied, but none so far corresponds to NER mutants of other organisms and *Neurospora* may conceivably be without an NER system (229). Only one known *Neurospora* mutant, *mus-18*, exhibits any defect in removal of pyrimidine dimers, and it is completely defective. *mus-18* mutant strains are UV-sensitive but are not cross-sensitive to other mutagens, including gamma rays, 4-nitro-

Table 2 Some eukaryotic nucleotide excision repair genes and proteins

Human gene	Rodent group	Human map position	S. cerevisiae homolog	S. pombe homolog	Molecular mass[a]	Comments
Dual incision						
XPA		9q34.1	RAD14		31 (40/42 on gels)	binds damaged DNA
XPC		3p25	RAD4		106 (125 on gels)	binds ss DNA
HHR23B		3p25	RAD23		43 (58 on gels)	associated with XPC
XPG (ERCC5)	5	13q33	RAD2	rad13+	133 (180–200 on gels)	DNA nuclease, for 3' side of lesion
ERCC4 (XPF?)	4	16p13.13	RAD1	rad16+	126 (S. cerevisiae)	components of DNA nuclease, for 5' side of lesion
ERCC1	1	19q13.2	RAD10	swi10+	31 (39 on gels)	
XPB (ERCC3)	3	2q21	SSL2 (RAD25)	rhp3+	89 (89 on gels)	3'–5' DNA helicase; in TFIIH
XPD (ERCC2)	2	19q13.2	RAD3	rad15+	87 (80 on gels)	5'–3' DNA helicase; in TFIIH
p44		5q13	SSL1		44 (50 in yeast)	TFIIH subunit
p62 (GTF2H1)		11p14.3–15.1	TFB1		62 (70–73 in yeast)	TFIIH subunit
p52			TFB2		41 (55 in yeast)	TFIIH subunit
p34			TFB4		34	"SSL1-like" TFIIH subunit
RPAp70		17p13	RFA1		68 (70 on gels)	binds to single-stranded DNA
RPAp32		1p35–36.1	RFA2		29 (34 on gels)	
RPAp14		7p21–22	RFA3		13.6	
DDB, UV-DDB (=XPE?)		11q12-13.2 / 11q12-p11.2			127 / 48	binds damaged DNA
Repair DNA synthesis						
PCNA		20p12	POL30	pcnt+	29 (36 on gels)	forms sliding clamp
hRFC140		4p13-p14	CDC44		128 (140 on gels)	RFC loads PCNA onto DNA in an ATP-dependent reaction
hRFC40		7q11.23	ScRFC4		39.1 (40 on gels)	
hRFC38		13q12.3-q13	ScRFC5		40.5 (38 on gels)	
hRFC37		3q27	ScRFC2		39.6 (37 on gels)	
hRFC36		12q24.2-q24.3	ScRFC3		38.5 (36 on gels)	
LIG1		19q13.2-3	CDC9	cdc17+	102 (125 on gels)	DNA ligase I
Transcription-coupled repair						
CSA	8	5	RAD28		44	WD-repeat protein
CSB (ERCC6)	6	10q11.2	RAD26		168	DNA helicase motifs

[a] kDa; human gene product unless indicated otherwise.

Table 3 Pathways for repair of cyclobutane pyrimidine dimers and (6 - 4) photoproducts in different eukaryotes

Organism	Nucleotide excision repair	Dimer/(6 - 4) endonuclease pathway	Pyrimidine dimer photolyase	(6 - 4) photoproduct photolyase
Saccharomyces cerevisiae	+	−	+	−
Schizosaccharomyces pombe	+	+	−	−
Neurospora crassa	−?	+	+	−
Drosophila melanogaster	+	?	+	+
Potorous tridactylis	+	−?	+	−?
Homo sapiens	+	−	−	−

quinoline-1-oxide, MNNG, mitomycin C, and MMS. Thus the normal function of Mus-18 appears to be exclusively in the repair of UV-induced damage (229). In a screen for *Neurospora* cDNAs that could complement the UV-sensitivity of an *E. coli* Δ*phr* Δ*uvrA*, Δ*recA* strain, Yasui and colleagues isolated a gene that encodes a 632–amino acid protein of 74.4 kDa. This cDNA proved to correspond to the *Mus-18* gene (230). Recombinant Mus-18 protein produced in *E. coli* was found to be a Mg^{2+}-dependent and ATP-independent endonuclease that cleaves UV-irradiated DNA 5′ to TC and TT sites of both cyclobutane pyrimidine dimers and (6-4) photoproducts (230). This enzyme seems likely to be the *Neurospora* equivalent of the *S. pombe* activity discussed above. Because this single polypeptide is a self-contained incision nuclease for pyrimidine dimers, heterologous expression of the *Mus-18* gene is able to partially alleviate the UV sensitivity of various cells, including *S. cerevisiae* *rad1* and *rad2* mutants and human XP-A cells (230).

Subsequent steps in this newly identified excision repair pathway in *Neurospora* and *S. pombe* remain to be characterized. Table 3 presents an overview of the two excision repair pathways and two types of photolyases known to remove cyclobutane pyrimidine dimers and (6-4) photoproducts in various eukaryotes. In general, organisms possess two or more mechanisms for dimer and (6-4) photoproduct removal. A notable exception appears to be placental mammals (including humans), as they can only eliminate such damage via NER.

DNA LIGASES IN REPAIR

Four distinct ATP-dependent DNA ligases have been identified in mammalian cells, but so far only one ligase has been found in each of the yeasts *S. cerevisiae* (Cdc9) and *S. pombe* (cdc17⁺). The single enzyme in yeasts seems to carry out all necessary replication and repair functions. In mammalian cells, there is strong evidence that DNA ligase I normally functions during semiconservative DNA replication [reviewed in (231)], and additional evidence implicates both DNA ligases I and III in DNA repair.

Point mutations in two conserved regions of the structural gene for DNA ligase I were located in human cells from an individual designated 46BR (232). This patient had retarded growth and severe immunodeficiencies. DNA ligase I activity was strongly reduced in 46BR cells, but not completely abolished, and this deficiency appears to be responsible for the clinical symptoms. 46BR cells are hypersensitive to MMS, and excessive DNA repair synthesis at repair patches, presumably the result of a malfunctioning DNA ligase I (233), has been observed.

Petrini et al (234) found that mouse embryonic stem cells with a homozygous knockout of the DNA ligase I gene were inviable. Rescue of viability was achieved by ectopic expression of intact DNA ligase I from a replicating plasmid. This result suggests that the enzyme is essential for growth, as might be expected for a core DNA synthesis enzyme. On the other hand, Melton and coworkers (D Bentley & D Melton, personal communication) found partial development of mouse embryos that have a homozygous C-terminal disruption of DNA ligase I, suggesting that other DNA ligases may sometimes compensate for a lack of DNA ligase I.

A clue to the function of DNA ligase III is its ability to form a tight complex with another nuclear protein, XRCC1 (235). XRCC1⁻ rodent cells have reduced single strand–break repair, a high level of sister chromatid exchange, lower levels of DNA ligase III activity, and are hypersensitive to simple alkylating agents (235, 236). The latter observation suggests that XRCC1 mutants may be defective in BER and implicates both XRCC1 and DNA ligase III in this process. Alternatively, DNA ligase III may be involved in recombination (237). cDNAs encoding DNA ligase III have been isolated (238, 239). An increased incidence of sister chromatid exchange is also a hallmark of the inherited human disorder Bloom's syndrome. This syndrome is caused by mutations in a gene on chromosome 15 that encodes a protein homologous to the *E. coli* RecQ DNA helicase (239a).

A cDNA encoding DNA ligase II has not been described, but the enzyme has been highly purified from bovine tissue, and some tryptic peptides have sequences identical to a region of DNA ligase III (240–242). The exact relationship of DNA ligase II to DNA ligase III remains to be determined. DNA ligase IV was discovered in a search for human DNA ligase cDNAs (238). The functions of DNA ligases II and IV are unknown.

DOUBLE STRAND–BREAK REPAIR

Ionizing radiation and some chemical agents cause single- and double-strand breaks. Failure to rejoin breaks properly can lead to loss of portions of chromosomes or to rearrangements. Few details of the biochemistry of radiation-induced double-strand break rejoining are truly understood in eukaryotes, but a good deal of genetic analysis in yeast continues to demonstrate which gene products are involved (2). Rejoining of double-strand breaks can occur in mammalian cell extracts, and some fractionation of such systems has been carried out (243, 244). End-joining can be nonhomologous, and study of the base changes occurring at the ends of breaks may give clues to the mechanism of rejoining (245, 246).

A notable development has been the realization that the DNA-dependent protein kinase (DNA-PK) is required for normal double-strand break rejoining activity in mammalian cells. DNA-PK consists of a huge 465 kDa catalytic subunit (DNA-PK$_{cs}$) and an associated DNA binding component, the Ku complex. Ku is composed of two subunits of ~70 and ~80 kDa, denoted Ku70 and Ku80. This DNA-PK holoenzyme is activated upon binding to DNA strand breaks and is able to phosphorylate many protein substrates in vitro, although its physiological targets are unclear (247, 247a). The involvement of DNA-PK in double strand–break repair was discovered by studying mutant mammalian cell lines with hypersensitivity to ionizing radiation. Such rodent mutants fall into at least nine complementation groups, some of which show defects in strand-break rejoining (248, 249). Cell lines from complementation group 5 are double-strand break–repair defective, defective in Ku–end binding activity, and are complemented by the Ku80 gene on human chromosome 2q33-35 (250, 251). Cell lines from complementation group 7, which include those from the scid (severe combined immunodeficiency) mouse, are also double-strand break–repair defective (252). Scid mutants and other cells from this group are defective in the 465 kDa DNA-PK$_{cs}$ and are complemented by the DNA-PKcs gene (XRCC7) on chromosome 8q11 (253–256).

In addition to having a defect in general double strand–break rejoining, mutants with defects in components of DNA-PK are also unable to perform correct recombinational V(D)J rejoining of maturing immunoglobulin genes, as assayed by transfection of test substrates. The immunodeficiency of the scid mouse directly arises from this rejoining defect. V(D)J recombination in developing lymphoid cells clearly uses part of the general cellular machinery for rejoining double-strand breaks (257). Many possible explanations have been put forward to account for the involvement of DNA-PK in strand-break rejoining and V(D)J recombination (249, 253). The end-binding activity of Ku may serve as an entry site for other proteins, and the phosphorylation activity of DNA-PK may activate or inhibit other repair factors.

In *S. cerevisiae,* double-strand break repair occurs through a pathway of homologous recombination that involves the products of the *RAD50–57* group of genes, reviewed extensively elsewhere (2, 258, 259). The *RAD50–57* genes have been cloned and sequenced, together with additional genes involved in the pathway, such as *MRE2, MRE11,* and *XRS2.* Some homologs in higher eukaryotes have been found (reviewed in 249). In yeast, the single-stranded binding protein RPA (encoded by the *RFA* genes) also participates in recombinational repair of double-strand breaks (146, 147) and activates strand-exchange reactions (260, 261). Several of the *RAD50–57* gene products show features of particular interest. The sequence of the *RAD54* gene shows that it belongs to the *SWI/SNF2* family of genes, which includes known transcriptional activators and helicases that can alter chromatin structure (249). Rad51 has significant homology with the *E. coli* RecA protein (reviewed in 2) and forms helical filaments on DNA similar to RecA (262), as does a homologous mammalian RAD51 protein (263). Furthermore, yeast Rad51 can perform a homologous pairing and strand-exchange reaction, aided by yeast RPA (261). Rad55 and Rad57 also have significant homology with RecA. Direct physical interactions have been detected between various proteins involved in double strand–break repair, including an association between Rad50 and Mre11 (264) and a complex that includes Rad51, Rad52, Rad55, Rad57, and RPA (265).

Individuals with the inherited syndrome ataxia telangiectasia (AT) display numerous clinical abnormalities including immunodeficiency, central nervous system problems, dilated blood vessels in some tissues, and a marked predisposition to cancer (266). Two features of AT have attracted the interest of DNA repair researchers: a marked radiosensitivity of AT cells and the phenomenon of "radioresistant DNA synthesis," whereby DNA synthesis in AT cells is not as suppressed by ionizing radiation as it is in normal cells. A DNA repair defect has often been postulated because of the perturbations of normal strand-break rejoining observed in AT cells (2, 266). The sequence of the ATM structural gene which is mutated in AT (267) strongly suggests that the protein is a kinase involved in a signaling pathway or in a checkpoint for cellular growth control. The carboxy-terminal portion of ATM is homologous to *S. pombe rad3*[+] and *S. cerevisiae MEC1* (*ESR1*), members of the PI-3 kinase family that are involved in cell-cycle checkpoint controls in the respective yeasts (268). The DNA-PK$_{cs}$ protein is also a member of this family (247a). Although *ATM* is not a DNA repair gene per se, mutations in the gene result in pleiotropic secondary effects, including alterations in NF-κB metabolism that seem to lead to radiation sensitivity (269) and possibly to alterations in repair. These effects will be better understood as the function of the gene product is studied further.

CONCLUSION

Recent years have been characterized by especially vigorous efforts to isolate eukaryotic DNA repair genes. At the beginning of 1990, no XP gene sequences had been reported. For *S. cerevisiae,* the gene identification stage will be finished rapidly, as the sequence of the yeast genome is completed. A catalog of human cDNAs will not be far behind. As these tasks are accomplished, the ingenuity of researchers will be challenged by attempts to discover the functions of genes involved in repair. More human syndromes and cancers will no doubt be found with mutations in DNA repair genes. One area of intense current activity is the construction of mice with knockouts or specific mutations in repair genes. Many such mice have been constructed, or are well under way, and much can be learned about physiological functions and any corresponding human syndromes (270–272). Careful interpretations are clearly required, because the genetic background of the animal can have profound influences. Another exploding area is the determination of the three-dimensional structures of repair enzymes and complexes with damaged DNA. As methods develop, the architecture of very large complexes such as the NER proteins will be revealed, as they work to repair DNA assembled into chromatin.

The field of DNA repair in eukaryotes now covers an enormous area and strong connections are being developed with many other areas including DNA replication, general transcription, cell-cycle control and checkpoints, and transduction of stress responses. This review, devoted largely to the DNA repair enzymes themselves, could not even touch on most of these subjects. The associations with each field will surely expand in the future.

ACKNOWLEDGMENTS

For useful comments on this review, I thank the members of my laboratory and Deborah Barnes, Stuart Clarkson, Ian Hickson, Alan Lehmann, and Tomas Lindahl. Steve Lacy and Herman Blount greatly assisted in the preparation of the manuscript.

Literature Cited

1. Karran P. 1996. *Semin. Cancer Biol.* 7:In press
2. Friedberg EC, Walker GC, Siede W. 1995. *DNA Repair and Mutagenesis.* Washington, DC: ASM Press
3. Bootsma D, Weeda G, Vermeulen W, van Vuuren H, Troelstra C, et al. 1995. *Philos. Trans. R. Soc. London Ser. B* 347:75–81
4. Lehmann AR. 1995. *Trends Biochem. Sci.* 20:402–05
5. Vaughan P, Lindahl T, Sedgwick B. 1993. *Mutat. Res.* 293:249–57
6. Mitra S, Kaina B. 1993. *Prog. Nucleic Acid Res. Mol. Biol.* 44:109–42
7. Karran P, Bignami M. 1992. *Nucleic Acids Res.* 20:2933–40
8. Karran P, Bignami M. 1994. *BioEssays* 16:833–39
9. Branch P, Aquilina G, Bignami M, Karran P. 1993. *Nature* 362:652–54
10. Kat A, Thilly WG, Fang WH, Longley

MJ, Li GM, Modrich P. 1993. *Proc. Natl. Acad. Sci. USA* 90:6424–28

11. de Wind N, Dekker M, Berns A, Radman M, te Riele H. 1995. *Cell* 82:321–30

12. Bodmer W, Bishop T, Karran P. 1994. *Nat. Genet.* 6:217–19

13. Lindahl T. 1993. *Nature* 362:709–15

13a. Lindahl T, Satoh MS, Dianov G. 1995. *Philos. Trans. R. Soc. London Ser. B* 247:57–62

14. Slupphaug G, Eftedal I, Kavli B, Bharati S, Helle NM, et al. 1995. *Biochemistry* 34:128–38

15. Mosbaugh DW, Bennett SE. 1994. *Prog. Nucleic Acid Res. Mol. Biol.* 48:315–70

16. Mol CD, Arvai AS, Slupphaug G, Kavli B, Alseth I, et al. 1995. *Cell* 80:869–78

17. Savva R, McAuleyhecht K, Brown T, Pearl L. 1995. *Nature* 373:487–93

18. Bjelland S, Birkeland N, Benneche T, Volden G, Seeberg E. 1994. *J. Biol. Chem.* 269:30489–95

19. Hollstein MC, Brooks P, Linn S, Ames BN. 1984. *Proc. Natl. Acad. Sci. USA* 81:4003–7

20. Cannon-Carlson S, Gokhale H, Teebor G. 1989. *J. Biol. Chem.* 264:13306–12

21. Boorstein RJ, Chiu LN, Teebor GW. 1992. *Mol. Cell. Biol.* 12:5536–40

22. Rydberg B, Lindahl T. 1982. *EMBO J.* 1:211–16

23. Xiao W, Samson L. 1993. *Proc. Natl. Acad. Sci. USA* 90:2117–21

24. O'Connor TR. 1993. *Nucleic Acids Res.* 21:5561–69

25. Dosanjh MK, Chenna A, Kim E, Fraenkel CH, Samson L, Singer B. 1994. *Proc. Natl. Acad. Sci. USA* 91:1024–28

26. Chenna A, Hang B, Rydberg B, Kim E, Pongracz K, et al. 1995. *Proc. Natl. Acad. Sci. USA* 92:5890–94

27. O'Connor TR, Laval F. 1990. *EMBO J.* 9:3337–42

28. Chakravarti D, Ibeanu GC, Tano K, Mitra S. 1991. *J. Biol. Chem.* 266:15710–15

29. Samson L, Derfler B, Boosalis M, Call K. 1991. *Proc. Natl. Acad. Sci. USA* 88:9127–31

30. Chen J, Derfler B, Samson L. 1990. *EMBO J.* 9:4569–75

31. Bjoras M, Klungland A, Johansen RF, Seeberg E. 1995. *Biochemistry* 34:4577–82

32. Chen J, Samson L. 1991. *Nucleic Acids Res.* 19:6427–32

33. Saparbaev M, Laval J. 1994. *Proc. Natl. Acad. Sci. USA* 91:5873–77

34. Demple B, Harrison L. 1994. *Annu. Rev. Biochem.* 63:915–48

35. Melamede RJ, Hatahet Z, Kow YW, Ide H, Wallace SS. 1994. *Biochemistry* 33:1255–64

36. Breimer LH. 1983. *Biochemistry* 22:4192–97

37. Doetsch PW, Henner WD, Cunningham RP, Toney JH, Helland DE. 1987. *Mol. Cell. Biol.* 7:26–32

38. Kim J, Linn S. 1989. *J. Biol. Chem.* 264:2739–45

39. Tomkinson AE, Bonk RT, Kim J, Bartfeld N, Linn S. 1990. *Nucleic Acids Res.* 18:929–35

40. Gossett J, Lee K, Cunningham RP, Doetsch PW. 1988. *Biochemistry* 27:2629–34

41. Bessho T, Roy R, Yamamoto K, Kasai H, Nishimura S, et al. 1993. *Proc. Natl. Acad. Sci. USA* 90:8901–4

42. Bessho T, Tano K, Kasai H, Ohtsuka E, Nishimura S. 1993. *J. Biol. Chem.* 268:19416–21

43. de Oliveira R, van der Kemp P, Thomas D, Geiger A, Nehls P, Boiteux S. 1994. *Nucleic Acids Res.* 22:3760–64

44. Furuichi M, Yoshida MC, Oda H, Tajiri T, Nakabeppu Y, et al. 1994. *Genomics* 24:485–90

45. Sakumi K, Furuichi M, Tsuzuki T, Kakuma T, Kawabata S, et al. 1993. *J. Biol. Chem.* 268:23524–30

46. Neddermann P, Jiricny J. 1993. *J. Biol. Chem.* 268:21218–24

47. Neddermann P, Jiricny J. 1994. *Proc. Natl. Acad. Sci. USA* 91:1642–46

47a. McGoldrick JP, Yeh Y-C, Solomon M, Essigmann JM, Lu A-L. 1995. *Mol. Cell Biol.* 15:989–96

48. Miller JH, Mroczkowska M, Scheimer C, Lloyd R. 1995. *J. Cell. Biochem.* 21A:268

49. Barzilay G, Hickson ID. 1995. *BioEssays* 17:713–19

50. Mol CD, Kuo CF, Thayer MM, Cunningham RP, Tainer JA. 1995. *Nature* 374:381–86

51. Walker LJ, Robson CN, Black E, Gillespie D, Hickson ID. 1993. *Mol. Cell. Biol.* 13:5370–76

52. Xanthoudakis S, Miao GG, Curran T. 1994. *Proc. Natl. Acad. Sci. USA* 91:23–27

53. Popoff SC, Spira AI, Johnson AW, Demple B. 1990. *Proc. Natl. Acad. Sci. USA* 87:4193–97

54. Ramotar D, Popoff SC, Gralla EB, Demple B. 1991. *Mol. Cell. Biol.* 11:4537–44

55. Price A, Lindahl T. 1991. *Biochemistry* 30:8631–37

56. Dianov G, Price A, Lindahl T. 1992. *Mol. Cell. Biol.* 12:1605–12

57. Matsumoto Y, Kim K, Bogenhagen DF. 1994. *Mol. Cell. Biol.* 14:6187–97

58. Wiebauer K, Jiricny J. 1990. *Proc. Natl. Acad. Sci. USA* 87:5842–45
59. Singhal RK, Prasad R, Wilson SH. 1995. *J. Biol. Chem.* 270:949–57
60. Matsumoto Y, Kim K. 1995. *Science* 269:699–702
61. Gu H, Marth JD, Orban PC, Mossmann H, Rajewsky K. 1994. *Science* 265:103–6
62. Prasad R, Widen SG, Singhal RK, Watkins J, Prakash L, Wilson SH. 1993. *Nucleic Acids Res.* 21:5301–7
63. Leem SH, Ropp PA, Sugino A. 1994. *Nucleic Acids Res.* 22:3011–17
64. Blank A, Kim B, Loeb LA. 1994. *Proc. Natl. Acad. Sci. USA* 91:9047–51
65. Wang ZG, Wu X, Friedberg EC. 1993. *Mol. Cell. Biol.* 13:1051–58
66. Lindahl T, Gally JA, Edelman GM. 1969. *Proc. Natl. Acad. Sci. USA* 62:597–603
67. Lyamichev V, Brow MAD, Dahlberg JE. 1993. *Science* 260:778–83
68. Ishimi Y, Claude A, Bullock P, Hurwitz J. 1988. *J. Biol. Chem.* 263:19723–33
69. Waga S, Bauer G, Stillman B. 1994. *J. Biol. Chem.* 269:10923–34
70. Turchi JJ, Huang L, Murante RS, Kim Y, Bambara RA. 1994. *Proc. Natl. Acad. Sci. USA* 91:9803–7
71. Harrington JJ, Lieber MR. 1994. *Genes Dev.* 8:1344–55
72. Murray JM, Tavassoli M, Al-Harithy R, Sheldrick KS, Lehmann AR, et al. 1994. *Mol. Cell Biol.* 14:4878–88
73. Carr AM, Sheldrick KS, Murray JM, Al-Harithy R, Watts FZ, Lehmann AR. 1993. *Nucleic Acids Res.* 21:1345–49
74. Robins P, Pappin DJC, Wood RD, Lindahl T. 1994. *J. Biol. Chem.* 269:28535–38
75. Jacquier A, Legrain P, Dujon B. 1992. *Yeast* 8:122–32
76. Reagan MS, Pittenger C, Siede W, Friedberg EC. 1995. *J. Bacteriol.* 177:4537–44
77. Sommers CH, Miller EJ, Dujon B, Prakash S, Prakash L. 1995. *J. Biol. Chem.* 270:4193–96
78. Vallen EA, Cross FR. 1995. *Mol. Cell. Biol.* 15:4291–302
79. Johnson RE, Kovvali GK, Prakash L, Prakash S. 1995. *Science* 269:238–40
80. Satoh MS, Poirier GG, Lindahl T. 1993. *J. Biol. Chem.* 268:5480–87
81. Molinete M, Vermeulen W, Burkle A, Menissier-De Murcia J, Kupper JH, et al. 1993. *EMBO J.* 12:2109–17
82. Lindahl T, Satoh MS, Poirier GG, Klungland A. 1995. *Trends Biochem. Sci.* 20:405–11
83. Ding R, Smulson M. 1994. *Cancer Res.* 54:4537–44
84. Wang Z-Q, Auer B, Stingl L, Berghammer H, Haidacher D, et al. 1995. *Genes Dev.* 9:509–20
85. Hansson J, Munn M, Rupp WD, Kahn R, Wood RD. 1989. *J. Biol. Chem.* 264:21788–92
86. Svoboda DL, Taylor JS, Hearst JE, Sancar A. 1993. *J. Biol. Chem.* 268:1931–36
87. Hansson J, Keyse SM, Lindahl T, Wood RD. 1991. *Cancer Res.* 51:3384–90
88. Huang JC, Zamble DB, Reardon JT, Lippard SJ, Sancar A. 1994. *Proc. Natl. Acad. Sci. USA* 91:10394–98
89. Szymkowski DE, Lawrence CW, Wood RD. 1993. *Proc. Natl. Acad. Sci. USA* 90:9823–27
90. Mu D, Bertrand-Burggraf E, Huang JC, Fuchs BPP, Sancar A. 1994. *Nucleic Acids Res.* 22:4869–71
91. Huang JC, Hsu DS, Kazantsev A, Sancar A. 1994. *Proc. Natl. Acad. Sci. USA* 91:12213–17
92. Mu D, Park CH, Matsunaga T, Hsu DS, Reardon JT, Sancar A. 1995. *J. Biol. Chem.* 270:2415–18
93. Hanawalt PC. 1994. *Environ. Mol. Mutagen.* 23:78–85
94. Mitchell DL, Nairn RS. 1989. *Photochem. Photobiol.* 49:805–19
95. Kim J-K, Choi B. 1995. *Eur. J. Biochem.* 228:849–54
96. Cleaver JE, Kraemer KH. 1989. In *The Metabolic Basis of Inherited Disease*, ed. CR Scriver, AL Beaudet, WS Sly, D Valle, 2:2949–71. New York: McGraw-Hill. 6th ed.
97. Kraemer KH, Lee MM, Andrews AD, Lambert WC. 1994. *Arch. Dermatol.* 130:1018–21
98. Robbins JH. 1988. *J. Am. Med. Assoc.* 260:384–88
99. Satoh MS, Jones CJ, Wood RD, Lindahl T. 1993. *Proc. Natl. Acad. Sci. USA* 90:6335–39
100. Lehmann AR, Bootsma D, Clarkson SG, Cleaver JE, McAlpine PJ, et al. 1994. *Mutat. Res.* 315:41–42
101. Hanawalt PC. 1994. *Science* 266:1957–58
102. Th'ng JPH, Walker IG. 1986. *Mutat. Res.* 165:139–50
103. Cleaver JE, Jen J, Charles WC, Mitchell DL. 1991. *Photochem. Photobiol.* 54:393–402
104. Shivji MKK, Kenny MK, Wood RD. 1992. *Cell* 69:367–74
105. Szymkowski DE, Hajibagheri MAN, Wood RD. 1993. *J. Mol. Biol.* 231:251–60
106. Huang JC, Svoboda DL, Reardon JT, Sancar A. 1992. *Proc. Natl. Acad. Sci. USA* 89:3664–68

106a. Moggs JG, Yarema KJ, Essigmann JM, Wood RD. 1996. *J. Biol. Chem.* 271: In press
107. Scherly D, Nouspikel T, Corlet J, Ucla C, Bairoch A, Clarkson SG. 1993. *Nature* 363:182–85
108. O'Donovan A, Davies AA, Moggs JG, West SC, Wood RD. 1994. *Nature* 371: 432–35
109. Habraken Y, Sung P, Prakash L, Prakash S. 1993. *Nature* 366:365–68
110. O'Donovan A, Scherly D, Clarkson SG, Wood RD. 1994. *J. Biol. Chem.* 269: 15965–68
111. Habraken Y, Sung P, Prakash L, Prakash S. 1994. *Nucleic Acids Res.* 22:3312–16
112. Bardwell AJ, Bardwell L, Johnson DK, Friedberg EC. 1993. *Mol. Microbiol.* 8:1177–88
113. Sung P, Reynolds P, Prakash L, Prakash S. 1993. *J. Biol. Chem.* 268:26391–99
114. Bardwell L, Cooper AJ, Friedberg EC. 1992. *Mol. Cell. Biol.* 12:3041–49
115. Bailly V, Sommers CH, Sung P, Prakash L, Prakash S. 1992. *Proc. Natl. Acad. Sci. USA* 89:8273–77
116. Bardwell AJ, Bardwell L, Tomkinson AE, Friedberg EC. 1994. *Science* 265: 2082–85
117. van Duin M, de Wit J, Odijk H, Westerveld A, Yasui A, et al. 1986. *Cell* 44:913–23
118. Biggerstaff M, Szymkowski DE, Wood RD. 1993. *EMBO J.* 12:3685–92
119. van Vuuren AJ, Appeldoorn E, Odijk H, Yasui A, Jaspers NGJ, et al. 1993. *EMBO J.* 12:3693–701
120. Thompson LH, Brookman KW, Weber CA, Salazar EP, Reardon JT, et al. 1994. *Proc. Natl. Acad. Sci. USA* 91:6855–59
121. Aboussekhra A, Biggerstaff M, Shivji MKK, Vilpo JA, Moncollin V, et al. 1995. *Cell* 80:859–68
122. van Vuuren AJ, Appeldoorn E, Odijk H, Humbert S,Moncollin V, et al. 1995. *Mutat. Res.* 337:25–39
123. Davies AA, Friedberg EC, Tomkinson AE, Wood RD, West SC. 1995. *J. Biol. Chem.* 270:24638–41
124. Walworth N, Davey S, Beach D. 1993. *Nature* 363:368–71
125. Craig RJ, Arraj JA, Marinus MG. 1984. *Mol. Gen. Genet.* 194:539–40
126. Gao Q, Williams LD, Egli M, Rabinovich D, Chen SL, et al. 1991. *Proc. Natl. Acad. Sci. USA* 88:2422–26
127. McBride TJ, Preston BD, Loeb LA. 1991. *Biochemistry* 30:207–13
128. Steck TR, Drlica K. 1984. *Cell* 36:1081–88
129. Schaaper RM. 1993. *J. Biol. Chem.* 268: 23762–65
130. Saffran WA, Greenberg RB, Thaler-scheer MS, Jones MM. 1994. *Nucleic Acids Res.* 22:2823–29
131. Ivanov EL, Haber JE. 1995. *Mol. Cell. Biol.* 15:2245–51
132. Habraken Y, Sung P, Prakash L, Prakash S. 1994. *Nature* 371:531–34
133. Hoy C, Thompson L, Mooney C, Salazar E. 1985. *Cancer Res.* 45:1737–43
134. McWhir J, Selfridge J, Harrison DJ, Squires S, Melton DW. 1993. *Nat. Genet.* 5:217–24
135. Robins P, Jones CJ, Biggerstaff M, Lindahl T, Wood RD. 1991. *EMBO J.* 10:3913–21
136. Asahina H, Kuraoka I, Shirakawa M, Morita EH, Miura N, et al. 1994. *Mutat. Res.* 315:229–37
137. Jones CJ, Wood RD. 1993. *Biochemistry* 32:12096–104
138. Guzder SN, Sung P, Prakash L, Prakash S. 1993. *Proc. Natl. Acad. Sci. USA* 90:5433–37
139. Nagai A, Saijo M, Kuraoka I, Matsuda T, Kodo N, et al. 1995. *Biochem. Biophys. Res. Commun.* 211:960–66
140. Miyamoto I, Miura N, Niwa H, Miyazaki J, Tanaka K. 1992. *J. Biol. Chem.* 267:12182–87
141. Hurwitz J, Dean FB, Kwong AD, Lee S-H. 1990. *J. Biol. Chem.* 265:18043–46
142. Coverley D, Kenny MK, Munn M, Rupp WD, Lane DP, Wood RD. 1991. *Nature* 349:538–41
143. Coverley D, Kenny MK, Lane DP, Wood RD. 1992. *Nucleic Acids Res.* 20:3873–80
144. Shivji MKK, Podust VN, Hübscher U, Wood RD. 1995. *Biochemistry* 34: 5011–17
145. Longhese MP, Plevani P, Lucchini G. 1994. *Mol. Cell. Biol.* 14:7884–90
146. Firmenich AA, Elias-Arnanz M, Berg P. 1995. *Mol. Cell. Biol.* 15:1620–31
147. Smith J, Rothstein R. 1995. *Mol. Cell. Biol.* 15:1632–41
148. Clugston CK, McLaughlin K, Kenny MK, Brown R. 1992. *Cancer Res.* 52: 6375–79
149. He Z, Henricksen LA, Wold MS, Ingles CJ. 1995. *Nature* 374:566–69
150. Li L, Lu XY, Peterson CA, Legerski RJ. 1995. *Mol. Cell Biol.* 15:5396–402
151. Matsuda T, Saijo M, Kuraoka I, Kobayashi T, Nakatsu Y, et al. 1995. *J. Biol. Chem.* 270:4152–57
152. Feldberg RS, Grossman L. 1976. *Biochemistry* 15:2402–8
153. Treiber DK, Chen ZH, Essigmann JM. 1992. *Nucleic Acids Res.* 20:5805–10
154. Reardon JT, Nichols AF, Keeney S, Smith CA, Taylor JS, et al. 1993. *J. Biol. Chem.* 268:21301–8

155. Hwang BJ, Chu G. 1993. *Biochemistry* 32:1657–66
156. Payne A, Chu G. 1994. *Mutat. Res.* 310:89–102
157. Keeney S, Chang GJ, Linn S. 1993. *J. Biol. Chem.* 268:21293–300
158. Chu G, Chang E. 1988. *Science* 242: 564–67
159. Hirschfeld S, Levine AS, Ozato K, Protic_ M. 1990. *Mol. Cell. Biol.* 10: 2041–48
160. Kataoka H, Fujiwara Y. 1991. *Biochem. Biophys. Res. Commun.* 175:1139–43
161. Keeney S, Wein H, Linn S. 1992. *Mutat. Res.* 273:49–56
162. Keeney S, Eker APM, Brody T, Vermeulen W, Bootsma D, et al. 1994. *Proc. Natl. Acad. Sci. USA* 91:4053–56
163. Guzder SN, Habraken Y, Sung P, Prakash L, Prakash S. 1995. *J. Biol. Chem.* 270:12973–76
164. Takao M, Abramic M, Moos M, Otrin VR, Wootton JC, et al. 1993. *Nucleic Acids Res.* 21:4111–18
165. Dualan R, Brody T, Keeney S, Nichols AF, Admon A, Linn S. 1995. *Genomics* 29:62–69
166. Li L, Bales ES, Peterson C, Legerski R. 1993. *Nat. Genet.* 5:413–17
167. Shivji MKK, Eker APM, Wood RD. 1994. *J. Biol. Chem.* 269:224729–57
168. Masutani C, Sugasawa K, Yanagisawa J, Sonoyama T, Ui M, et al. 1994. *EMBO J.* 13:1831–43
169. Watkins JF, Sung P, Prakash L, Prakash S. 1993. *Mol. Cell. Biol.* 13:7757–65
170. Venema J, van Hoffen A, Karcagi V, Natarajan AT, van Zeeland AA, Mullenders LHF. 1991. *Mol. Cell. Biol.* 11: 4128–34
171. Evans MK, Robbins JH, Ganges MB, Tarone RE, Nairn RS, Bohr VA. 1993. *J. Biol. Chem.* 268:4839–47
172. Verhage R, Zeeman A, Degroot N, Gleig F, Bang D, et al. 1994. *Mol. Cell. Biol.* 14:6135–42
173. Reardon JT, Thompson LH, Sancar A. 1993. *Cold Spring Harbor Symp. Quant. Biol.* 58:605–17
174. Conaway RC, Conaway JW. 1993. *Annu. Rev. Biochem.* 62:161–90
175. Chalut C, Moncollin V, Egly JM. 1994. *BioEssays* 16:651–55
176. Schaeffer L, Roy R, Humbert S, Moncollin V, Vermeulen W, et al. 1993. *Science* 260:58–63
177. van Vuuren AJ, Vermeulen W, Weeda G, Appeldoorn E, Jaspers NGJ, et al. 1994. *EMBO J.* 13:1645–53
178. Drapkin R, Reardon JT, Ansari A, Huang JC, Zawel L, et al. 1994. *Nature* 368:769–72
179. Feaver WJ, Svejstrup JQ, Bardwell L, Bardwell AJ, Buratowski S, et al. 1993. *Cell* 75:1379–87
180. Svejstrup JQ, Wang Z, Feaver WJ, Wu X, Bushnell DA, et al. 1995. *Cell* 80:21–28
181. Dresler SL, Frattini MK. 1986. *Nucleic Acids Res.* 14:7093–102
182. Nishida C, Reinhard P, Linn S. 1988. *J. Biol. Chem.* 263:501–10
183. Hunting DJ, Gowans BJ, Dresler SL. 1991. *Biochem. Cell Biol.* 69:303–8
184. Popanda O, Thielmann HW. 1992. *Biochim. Biophys. Acta* 1129:155–60
185. Nichols AF, Sancar A. 1992. *Nucleic Acids Res.* 20:3559–64
186. Kelman Z, O'Donnell M. 1994. *Curr. Opin. Genet. Dev.* 4:185–95
187. Hübscher U, Spadari S. 1994. *Physiol. Rev.* 74:259–304
188. Krishna TSR, Kong X-P, Gary S, Burgers PM, Kuriyan J. 1994. *Cell* 79: 1233–43
189. Podust LM, Podust VN, Floth C, Hübscher U. 1994. *Nucleic Acids Res.* 22: 2970–75
190. Celis JE, Madsen P. 1986. *FEBS Lett.* 209:277–83
191. Toschi L, Bravo R. 1988. *J. Cell Biol.* 107:1623–28
192. Prosperi E, Stivala LA, Sala E, Scovassi AI, Bianchi L. 1993. *Exp. Cell. Res.* 205:320–25
193. Stivala LA, Prosperi E, Rossi R, Bianchi L. 1993. *Carcinogenesis* 14:2569–73
194. Jackson DA, Hassan AB, Errington RJ, Cook PR. 1994. *J. Cell Sci.* 107:1753–60
195. Hall PA, McKee PH, Menage H, Dover R, Lane DP. 1993. *Oncogene* 8:203–7
196. Miura M, Domon M, Sasaki T, Takasaki Y. 1992. *J. Cell. Physiol.* 150:370–76
197. Miura M, Domon M, Sasaki T, Kondo S, Takasaki Y. 1992. *Exp. Cell. Res.* 201:541–44
198. Aboussekhra A, Wood RD. 1995. *Exp. Cell Res.* In press
198a. Cullmann G, Fien K, Kobayashi R, Stillman B. 1995. *Mol. Cell Biol.* 15:4661–71
199. Syväoja J, Suomensaari S, Nishida C, Goldsmith JS, Chui GSJ, et al. 1990. *Proc. Natl. Acad. Sci. USA* 87:6664–68
200. Zeng XR, Jiang YQ, Zhang SJ, Hao HL, Lee MYWT. 1994. *J. Biol. Chem.* 269:13748–51
201. Budd ME, Campbell JL. 1995. *Mol. Cell. Biol.* 15:2173–79
202. Lee SH, Kim DK, Drissi R. 1995. *J. Biol. Chem.* 270:21800–05
203. Li L, Elledge SJ, Peterson CA, Bales ES, Legerski RJ. 1994. *Proc. Natl. Acad. Sci. USA* 91:5012–16

204. Park CH, Mu D, Reardon JT, Sancar A. 1995. *J. Biol. Chem.* 270:4896–902
205. Bardwell AJ, Bardwell L, Iyer N, Svejstrup JQ, Feaver WJ, et al. 1994. *Mol. Cell. Biol.* 14:3569–76
206. Guzder SN, Bailly V, Sung P, Prakash L, Prakash S. 1995. *J. Biol. Chem.* 270:8385–88
207. Koleske AJ, Young RA. 1994. *Nature* 368:466–69
208. Waters R, Rong Z, Jones NJ. 1993. *Mol. Gen. Genet.* 239:28–32
209. Abramic M, Levine AS, Protic' M. 1991. *J. Biol. Chem.* 266:22493–500
209a. Ariza RR, Keyse SM, Moggs JG, Wood RD. 1996. *Nucleic Acids Res.* 24:In press
210. Sancar A. 1994. *Biochemistry* 33:2–9
211. Sancar GB. 1990. *Mutat. Res.* 236:147–60
212. Li YF, Kim ST, Sancar A. 1993. *Proc. Natl. Acad. Sci. USA* 90:4389–93
213. Eker AP, Yajima H, Yasui A. 1994. *Photochem. Photobiol.* 60:125–33
214. Yajima H, Inoue H, Oikawa A, Yasui A. 1991. *Nucleic Acids Res.* 19:5359–62
215. Sebastian J, Sancar GB. 1991. *Proc. Natl. Acad. Sci. USA* 88:11251–55
216. Yasuhira S, Yasui A. 1992. *J. Biol. Chem.* 267:25644–47
217. Yasui A, Eker APM, Yasuhira S, Yajima H, Kobayashi T, et al. 1994. *EMBO J.* 13:6143–51
218. Todo T, Takemori H, Ryo H, Ihara M, Matsunaga T, et al. 1993. *Nature* 361:371–74
219. Kato TJ, Todo T, Ayaki H, Ishizaki K, Morita T, et al. 1994. *Nucleic Acids Res.* 22:4119–24
220. Malhotra K, Kim ST, Batschauer A, Dawut L, Sancar A. 1995. *Biochemistry* 34:6892–99
221. Kim ST, Malhotra K, Smith CA, Taylor JS, Sancar A. 1994. *J. Biol. Chem.* 269:8535–40
222. Carr A, Schmidt H, Kirchhoff S, Muriel W, Sheldrick K, et al. 1994. *Mol. Cell. Biol.* 14:2029–40
223. Birnboim HC, Nasim A. 1975. *Mol. Gen. Genet.* 136:1–8
224. McCready SJ, Carr AM, Lehmann AR. 1993. *Mol. Microbiol.* 10:885–90
225. McCready S. 1994. *Mutat. Res.* 315:261–73
226. Sidik K, Lieberman HB, Freyer GA. 1992. *Proc. Natl. Acad. Sci. USA* 89:12112–16
227. Bowman KK, Sidik K, Smith CA, Taylor JS, Doetsch PW, Freyer GA. 1994. *Nucleic Acids Res.* 22:3026–32
228. Freyer G, Davey S, Ferrer JV, Martin AM, Beach D, Doetsch P. 1995. *Mol. Cell. Biol.* 15:4572–77
229. Ishii C, Nakamura K, Inoue H. 1991. *Mol. Gen. Genet.* 228:33–39
230. Yajima H, Takao M, Yasuhira S, Zhao JH, Ishii C, et al. 1995. *EMBO J.* 14:2393–99
231. Lindahl T, Barnes DE. 1992. *Annu. Rev. Biochem.* 61:251–81
232. Barnes DE, Tomkinson AE, Lehmann AR, Webster ADB, Lindahl T. 1992. *Cell* 69:495–503
233. Prigent C, Satoh MS, Daly G, Barnes DE, Lindahl T. 1994. *Mol. Cell. Biol.* 14:310–17
234. Petrini JHJ, Xiao Y, Weaver DT. 1995. *Mol. Cell. Biol.* 15:4303–8
235. Caldecott KW, McKeown CK, Tucker JD, Ljungquist S, Thompson LH. 1994. *Mol. Cell. Biol.* 14:68–76
236. Thompson LH, Brookman KW, Jones NJ, Allen SA, Carrano AV. 1990. *Mol. Cell. Biol.* 10:6160–71
237. Jessberger R, Podust V, Hübscher U, Berg P. 1993. *J. Biol. Chem.* 268:15070–79
238. Wei YF, Robins P, Carter K, Caldecott K, Pappin DJC, et al. 1995. *Mol. Cell. Biol.* 15:3206–16
239. Chen JW, Tomkinson AE, Ramos W, Mackey ZB, Danehower S, et al. 1995. *Mol. Cell. Biol.* 15:5412–22
239a. Ellis NA, Groden J, Ye T-Z, Straughen J, Lennon DJ, et al. 1995. *Cell.* 83:655–66
240. Roberts E, Nash RA, Robins P, Lindahl T. 1994. *J. Biol. Chem.* 269:3789–92
241. Wang YC, Burkhart WA, Mackey ZB, Moyer MB, Ramos W, et al. 1994. *J. Biol. Chem.* 269:31923–28
242. Husain I, Tomkinson AE, Burkhart WA, Moyer MB, Ramos W, et al. 1995. *J. Biol. Chem.* 270:9683–90
243. Fairman MP, Johnson AP, Thacker J. 1992. *Nucleic Acids Res.* 20:4145–52
244. Derbyshire MK, Epstein LH, Young CS, Munz PL, Fishel R. 1994. *Mol. Cell. Biol.* 14:156–69
245. Ganesh A, North P, Thacker J. 1993. *Mutat. Res.* 299:251–59
246. Nicolas AL, Young CS. 1994. *Mol. Cell. Biol.* 14:170–80
247. Anderson CW. 1994. *Semin. Cell Biol.* 5:427–36
247a. Hartley KO, Gell D, Smith GCM, Zhang H, Divecha N, et al. 1995. *Cell* 82:849–56
248. Jeggo PA, Tesmer J, Chen DJ. 1991. *Mutat. Res.* 254:3171–74
249. Troelstra C, Jaspers NG. 1994. *Curr. Biol.* 4:1149–51
250. Taccioli GE, Gottlieb TM, Blunt T, Priestley A, Demengeot J, et al. 1994. *Science* 265:1442–45
251. Smider V, Rathmell WK, Lieber MR, Chu G. 1994. *Science* 266:288–91

252. Biedermann KA, Sun JR, Giaccia AJ, Tosto LM, Brown JM. 1991. *Proc. Natl. Acad. Sci. USA* 88:1394–97

253. Blunt T, Finnie NJ, Taccioli GE, Smith GCM, Demengeot J, et al. 1995. *Cell* 80:813–23

254. Kirchgessner CU, Patil CK, Evans JW, Cuomo CA, Fried LM, et al. 1995. *Science* 267:1178–83

255. Lees-Miller SP, Godbout R, Chan DW, Weinfeld M, Day RSI, et al. 1995. *Science* 267:1183–85

256. Peterson SR, Kurimasa A, Oshimura M, Dynan WS, Bradbury EM, Chen DJ. 1995. *Proc. Natl. Acad. Sci. USA* 92:3171–74

257. Vangent D, McBlane J, Ramsden D, Sadofsky M, Hesse J, Gellert M. 1995. *Cell* 81:925–34

258. Game JC. 1993. *Semin. Cancer Biol.* 4:73–83

259. Heyer WD, Kolodner RD. 1993. *Prog. Nucleic Acid Res. Mol. Biol.* 46:221–71

260. Alani E, Thresher R, Griffith JD, Kolodner RD. 1992. *J. Mol. Biol.* 227:54–71

261. Sung P. 1994. *Science* 265:1241–43

262. Ogawa T, Yu X, Shinohara A, Egelman EH. 1993. *Science* 259:1896–99

263. Benson FE, Stasiak A, West SC. 1994. *EMBO J.* 13:5764–71

264. Johzuka K, Ogawa H. 1995. *Genetics* 139:1521–32

265. Hays SL, Firmenich AA, Berg P. 1995. *Proc. Natl. Acad. Sci. USA* 92:6925–29

266. Murnane JP, Kapp LN. 1993. *Semin. Cancer Biol.* 4:93–104

267. Savitsky K, Bar-Shira A, Gilad S, Rotman G, Ziv Y, et al. 1995. *Science* 268:1749–53

268. Lehmann AR, Carr AM. 1995. *Trends Genet.* 11:375–77

269. Jung M, Zhang Y, Lee S, Dritschilo A. 1995. *Science* 268:1619–21

270. Sands AT, Abuin A, Sanchez A, Conti CJ, Bradley A. 1995. *Nature* 377:162–65

271. Nakane H, Takeuchi S, Yuba S, Saijo M, Nakatsu Y, et al. 1995. *Nature* 377:165–68

272. de Vries A, van Oostrom CTM, Hofhuis FMA, Dortant PM, Berg RJW, et al. 1995. *Nature* 377:169–73

Annu. Rev. Biochem. 1996. 65:169–214
Copyright © 1996 by Annual Reviews Inc. All rights reserved

MECHANISMS OF HELICASE-CATALYZED DNA UNWINDING

Timothy M. Lohman and Keith P. Bjornson

Department of Biochemistry and Molecular Biophysics, Washington University School of Medicine, St. Louis, Missouri 63110

KEY WORDS: energy transduction, motor proteins, mechanism, kinetics, DNA-protein interactions, allosterism, replication

ABSTRACT

DNA helicases are essential motor proteins that function to unwind duplex DNA to yield the transient single-stranded DNA intermediates required for replication, recombination, and repair. These enzymes unwind duplex DNA and translocate along DNA in reactions that are coupled to the binding and hydrolysis of 5'-nucleoside triphosphates (NTP). Although these enzymes are essential for DNA metabolism, the molecular details of their mechanisms are only beginning to emerge. This review discusses mechanistic aspects of helicase-catalyzed DNA unwinding and translocation with a focus on energetic (thermodynamic), kinetic, and structural studies of the few DNA helicases for which such information is available. Recent studies of DNA and NTP binding and DNA unwinding by the *Escherichia coli* (E. coli) Rep helicase suggest that the Rep helicase dimer unwinds DNA by an active, rolling mechanism. In fact, DNA helicases appear to be generally oligomeric (usually dimers or hexamers), which provides the helicase with multiple DNA binding sites. The apparent mechanistic similarities and differences among these DNA helicases are discussed.

CONTENTS

PERSPECTIVES AND OVERVIEW 170
PROPERTIES OF DNA... 172
  Single-Stranded vs Duplex DNA 172
  Effects of Solution Conditions on the Energetics and Kinetics of Protein-DNA
    Interactions 173
STRUCTURAL FEATURES OF DNA HELICASES 174
  Oligomeric Nature of Helicases... 174
  Structural Features of Hexameric Helicases............................. 177
  Primary Structures 179

169

DNA BINDING   180
  Polarity of Binding to Single-Stranded DNA.   181
  Stoichiometries and Energetics of Single-Stranded and Double-Stranded DNA
          Binding   181
  Allosteric Effects of Nucleotides on DNA Binding to the Dimeric E. coli Rep
          Helicase   185
  Helicase Interactions at Single-Stranded/Double-Stranded DNA Junctions (Forks). .   186
NUCLEOTIDE BINDING AND NTP HYDROLYSIS   188
  Stoichiometry and Equilibrium Binding.   188
  Steady-State NTP Hydrolysis and the Effects of Protein Oligomerization.   190
  Mechanism of ATP Binding and Hydrolysis by E. coli Rep Monomer   191
  Mutations Within the Nucleotide Binding Site   192
PHENOMENOLOGICAL FEATURES OF DNA UNWINDING AND
          TRANSLOCATION.   192
  Initiation of DNA Unwinding in Vitro   192
  "Polarity" of Duplex DNA Unwinding and the Role of Flanking Single-Stranded
          DNA   193
  Translocation Along Single-Stranded DNA.   196
  Rates and Processivities of DNA Unwinding   198
  Thermodynamic Efficiency of DNA Unwinding.   200
MECHANISMS OF DNA UNWINDING AND TRANSLOCATION   201
  Active vs Passive Mechanisms of DNA Unwinding.   202
  E. coli Rep Helicase.   203
  E. coli UvrD (Helicase II)   206
  Hexameric DNA Helicases.   207
  E. coli Rho Protein.   208
SUMMARY   209

PERSPECTIVES AND OVERVIEW

Double helical DNA is the stable form of most DNA in vivo; however, the DNA duplex must be unwound transiently and the two complementary strands must be separated, at least partially, in order to form the single-stranded (ss) DNA intermediates required for DNA replication, repair, recombination, and DNA transfer during conjugation. In each of these processes, duplex DNA unwinding is catalyzed by enzymes known as DNA helicases, which function to destabilize the hydrogen bonds between the complementary base pairs (bp) in duplex DNA in reactions that are coupled to the binding and hydrolysis of nucleoside 5'-triphosphates (NTP). Intimately linked to the DNA unwinding reaction is the requirement for helicases to translocate along the DNA filament in order to unwind the DNA duplex processively at rates that can be as fast as 500–1000 bp s^{-1}. Since DNA helicases transduce the chemical free energy change associated with NTP hydrolysis into mechanical energy to unwind DNA and also translocate along DNA, they are members of the general class of "motor proteins" with which they have several similarities (1).

DNA helicases appear to be ubiquitous, having been identified in various prokaryotes and eukaryotes as well as in bacteriophages and viruses [for reviews see (2–6)]. Most organisms also encode multiple helicases; for example, E. coli encodes at least 12 different helicases (7) and S. cerevisiae encodes at least six (8). We estimate that DNA helicase activity has been demonstrated in vitro for more than 60 enzymes since their discovery in 1976 (9), although this number is increasing rapidly. In addition to functioning in DNA replication, recombination, repair, and conjugation, enzymes with DNA helicase activity are components of eukaryotic transcription complexes and are important in coupling transcription to DNA repair (10, 11). Several human diseases, including xeroderma pigmentosum and Cockayne's syndrome, involve defects in proteins involved in nucleotide excision repair that possess helicase activity [for reviews see (10–13)]. A number of enzymes with demonstrated RNA helicase activity have also been identified (14–17), one of which functions in eukaryotic translation initiation (18). Some helicases also unwind RNA•DNA duplexes (19–23). Many putative RNA helicases have also been identified (24), although most of these do not possess helicase activity in vitro.

Since DNA helicases are essential enzymes in all aspects of DNA metabolism a detailed understanding of the mechanism(s) by which helicases function at the molecular level is important. We have been selective in this review in order to focus on those helicases for which significant mechanistic information is available, although most studies of helicases are still at an early stage and complete mechanistic details of DNA unwinding and translocation are not yet available for any helicase. The thermodynamic and kinetic considerations of energy transduction and the coupling of NTP binding and hydrolysis to vectorial processes in general have been discussed previously (25–27), and these apply to helicases. To catalyze the unwinding of duplex DNA, a helicase must cycle, vectorially, through a series of energetic (conformational) states, driven by the binding and/or hydrolysis of NTP and subsequent release of products $(NDP + PO_4^=)$ (6, 25, 28). Therefore, understanding helicase-catalyzed DNA unwinding at the molecular level requires information on the coupling of NTP binding and hydrolysis to DNA unwinding as well as the identification of the intermediate helicase-DNA states that occur during unwinding. Such an understanding requires quantitative studies of the energetics (thermodynamics) and kinetics of helicase binding to DNA and nucleotide cofactors (NTP, NDP, P_i), as well as structural information. Most postulated mechanisms for helicase function require the helicase to possess multiple DNA binding sites, and this requirement appears to be have been satisfied by the fact that the functionally active forms of helicases are oligomeric (6, 29). Therefore, the energetics and kinetics of assembly of the active helicase also must be understood. Furthermore, all of these interactions (protein assembly, DNA binding, NTP binding, and hydrolysis) are coupled (i.e. one process influences the other), making an examination of their linkage essential. Transient, pre-steady-state kinetic methods are also required to measure the elementary kinetic steps of NTP hydrolysis and DNA unwinding, which will ultimately be needed for an understanding of the mechanism(s) of these molecular motors. These topics are discussed along with the current mechanistic models of how some helicases carry out this important process. Most previous reviews have considered mainly the biological functions, genetics, and biochemical characterizations of DNA helicases (2–5, 7, 30) and RNA helicases (24), although recent reviews have emphasized the mechanistic aspects of DNA helicases (6, 29).

PROPERTIES OF DNA

We briefly discuss some properties of DNA that are relevant to a mechanistic understanding of helicase function.

Single-Stranded vs Duplex DNA

Due to the chemical nature of its sugar-phosphate backbone, ss DNA has a chemical and structural polarity, designated as either 5'-to-3' or 3'-to-5' with respect to the orientation of the sugar. Duplex DNA, being composed of two complementary antiparallel single strands, does not possess such polarity. Therefore, any helicase property that is sensitive to DNA strand polarity must reflect an interaction of the helicase with ss DNA. Single-stranded DNA is also substantially more flexible than duplex DNA. The flexibility of any linear polymer can be described quantitatively by a statistical quantity referred to as its persistence length, P_∞ (31) (stiffer molecules have longer persistence lengths). The average value of P_∞ for B-form duplex DNA is ~150 base pairs (bp) at moderate salt concentrations ([NaCl] $\geq$ 10 mM or [MgCl$_2$] $\geq$ 0.5 mM), although P_∞ does increase at lower salt concentrations, reflecting its polyelectrolyte nature (31). Considerably less energy is required to bend ss nucleic acids; estimates of P_∞ for the ss homopolynucleotides poly(U) and poly(A) are ~10 and ~60 nucleotides, respectively (32).

Natural ss DNA (e.g. phage DNA from ϕX174, fd, or M13) contains significant, but generally unknown, amounts of duplex DNA owing to the occurrence of intramolecular base pairing. Furthermore, the fraction of nucleotides present as ss DNA is also dependent upon solution conditions (especially salt concentration, temperature, and pH). Thus, the results of experiments that examine the effects of ss DNA on helicase activities (e.g. NTPase) are less ambiguous if ss homopolynucleotides [e.g. poly(dT) or poly(dA)] are used since intramolecular base pairing does not occur under most conditions (at pH $\geq$7).

B-form duplex DNA is more stable, relative to ss DNA, by ~1.5–2 kcal/mol bp at 37°C (pH 7); however, this relative stability depends upon solution conditions, especially salt concentration and type (33, 34). This dependence on salt concentration is due to the fact that DNA is a highly negatively charged linear polyelectrolyte (polyanion); the B-form duplex has two phosphates every 3.4 Å along its contour length. Single-stranded DNA has one phosphate every 3.4–4.5 Å along its contour length, depending on pH (35, 36). As a result, cations such as K^+, Na^+, Mg^{2+}, and Ca^{2+} and polyamines bind strongly to both duplex and ss DNA, although they generally bind with higher affinity to duplex DNA owing to its higher linear charge density. In the presence of a monovalent salt only (e.g. NaCl or KCl), B-form duplex DNA can be viewed thermodynamically as having 0.88 monovalent cations (M^+) associated thermodynamically per phosphate, whereas only ~0.7 M^+ are thermodynamically associated per ss DNA phosphate. Thus, a net release of cations occurs upon melting duplex DNA (35, 37, 38). As a result, the relative duplex stability, as reflected by its melting temperature (T_m), increases with increasing salt (cation) concentration, with multivalent cations exhibiting a much larger effect than monovalent cations (35, 38). Salt concentration also influences the kinetics of denaturation and renaturation (39–41). The extent of cation association per DNA phosphate is also lower for linear oligonucleotides than for polynucleotides (42, 43). The electrostatic and thus cation binding properties of ss/ds DNA junctions are also different than for either ss or ds DNA alone (44), and this difference has the potential to influence binding of a protein to the junction.

Effects of Solution Conditions on the Energetics and Kinetics of Protein-DNA Interactions

Changes in solution conditions (salt concentration and type, pH, temperature, etc) can and generally will influence the energetics (stability) and kinetics, as well as the specificity, of protein-DNA complexes. Due to the polyelectrolyte nature of DNA, these properties are influenced most dramatically by salt concentration (34, 45–47) as a result of differential cation or anion binding to the complex vs the free protein and DNA. Since the DNA binding sites of most proteins are positively charged, protein binding to DNA generally results in partial neutralization of DNA phosphates with concomitant release of cations from the DNA, although differential ion binding to the protein can also occur (48). In fact, the increase in entropy accompanying cation release from the DNA provides a major favorable contribution to the stability of most protein-DNA complexes (34, 45, 46, 49). As a result of the release of cations from the DNA (and potentially ions from the protein), an increase in salt concentration will generally lower the observed association equilibrium constant (34, 45, 46, 49) and also influence the kinetic rate constants (45, 47) for protein-DNA complex formation. For example, the equilibrium association constant for E. coli SSB tetramer binding to ss nucleic acids decreases by a factor of ~100 upon raising the [NaCl] from 0.1 to 0.2 M (50). Thus, meaningful comparisons of helicase activities (DNA binding, unwinding rates,

NTPase rates, processivity, etc) must be made under identical solution conditions. These general effects of monovalent cations and Mg^{2+} need to be considered along with the role of Mg^{2+} as a cofactor in the hydrolysis of NTP by helicases.

STRUCTURAL FEATURES OF DNA HELICASES

Information about the self-assembly properties and functionally active forms of DNA helicases is not available for most DNA helicases, although investigations of these properties and the influence of DNA and nucleotides are essential for interpreting biochemical and functional studies of these enzymes. Although it is too early to form many general conclusions, the active forms of most helicases appear to be oligomeric (generally dimeric or hexameric) (6). Atomic resolution structural information is not yet available for any helicase; however, recent electron microscopic studies have provided interesting low resolution structural information for a few hexameric helicases (51–56).

Oligomeric Nature of Helicases

Most proposed mechanisms for helicase-catalyzed DNA unwinding, such as rolling or inchworm mechanisms, require the functional helicase to possess at least two DNA binding sites (6, 29). These two sites would accommodate intermediates that require simultaneous binding of the helicase to either two ss DNA regions or to both ss and duplex DNA at an unwinding junction (6, 28, 29, 57). In fact, evidence for ternary complexes of a helicase with both ss and ds DNA exists for both the *E. coli* Rep (28, 58–60) and DnaB helicases (61). Even translocation of a helicase along ss DNA by an inchworm or rolling mechanism (28, 62) requires intermediates in which two regions of ss DNA are bound simultaneously to the helicase. Although multiple DNA binding sites can potentially exist within a single polypeptide, all helicases for which the assembly state has been examined in detail form oligomeric structures (6, 29, 57), generally dimers or hexamers. The immediate consequence of such oligomeric structures is that they provide a simple mechanism for helicases to acquire multiple DNA binding sites.

Table 1 lists DNA helicases for which the assembly state has been characterized in some detail, as well as those for which oligomerization has been demonstrated, but the form that is active in DNA unwinding is not yet known. With the apparent exception of the RecBCD enzyme, these helicases form either dimers or hexamers. Of course, under some conditions, these helicases exist as a mixture of oligomeric states. Since the assembly states of the large majority of known helicases have not yet been characterized, it is too early to conclude whether there is a predominant oligomeric form of DNA helicases.

Table 1 Oligomeric nature of helicases

Helicase	Assembly state[a]	Unwinding polarity
E. coli DnaB	hexamer (65, 66)	5′ to 3′
SV40 large T antigen	hexamer (51)	3′ to 5′
T4 phage gene *41*	hexamer (requires GTP) (55)	5′ to 3′
T7 phage gene *4*	hexamer (56, 67)	5′ to 3′
E. coli RuvB	hexamer (54)	5′ to 3′
E. coli Rho	hexamer (53, 82)	5′ to 3′
E. coli Helicase III	oligomer (182)	5′ to 3′
E. coli TraI	oligomer (146, 183)	5′ to 3′
E. coli RecBCD	hetero-trimer (or hexamer) (177)	undefined[b]
E. coli RecB	oligomer (dimer-tetramer) (184)	3′ to 5′
E. coli Rep	dimer (DNA-induced) (28, 57, 59)	3′ to 5′
E. coli Helicase II (UvrD)	dimer (possibly larger oligomers) (69)	3′ to 5′ [c]
E. coli Rep/UvrD	hetero-dimer (68)	undetermined
E. coli UvrAB	A_2B hetero-trimer (185)	5′ to 3′
HeLa Helicase	dimer (186)	3′ to 5′
Herpes (HSV-1) Origin Binding Protein (UL-9)	dimer (187)	3′ to 5′
HSV-1 helicase/primase	hetero-dimer (UL5/52) (188)	5′ to 3′

[a] Reference numbers are given in parentheses following column entries.

[b] Rec BCD preferentially unwinds blunt-ended DNA (119).

[c] At high protein to DNA ratios, helicase II initiates unwinding from blunt ends and nicks (129–131).

For example, even though *E. coli* Helicase III, *E. coli* UvrD (Helicase II), *E. coli* RecBCD, F factor TraI (Helicase I), HeLa Helicase, HSV-1 Origin Binding Protein (UL9), and HSV-1 Helicase/primase are known to self-assemble, the active forms of these helicases have not been determined. Furthermore, DNA helicase activity may not be unique to one particular oligomeric form of a DNA helicase (e.g. dimeric and hexameric forms of the same helicase may both be active). Of particular interest in this regard is the fact that the SV40 T antigen also can function as an RNA helicase, although this activity is not stimulated by ATP or dATP, but by a different set of NTPs (GTP, UTP, or CTP) (63). Since NTPs other than ATP or dATP do not favor hexamer formation, the suggestion has been made that SV40 T antigen may not function as a hexamer when unwinding RNA (64).

There is strong evidence that the functionally active forms of a number of DNA helicases are oligomeric; *E. coli* DnaB (65, 66), *E. coli* RuvB (54), phage T7 gene 4 helicase/primase (56, 67), phage T4 gene 41 helicase (55), SV40 T antigen (51), and the *E. coli* Rho protein, an RNA and RNA/DNA helicase (62), all can assemble to form ring-like toroidal hexamers. The *E. coli* Rep protein is induced to dimerize upon binding ss or duplex DNA and a chemically cross-linked Rep dimer retains both ss DNA–dependent ATPase and DNA helicase activities (28, 57, 59, 60). Furthermore, since no evidence for a Rep oligomeric state larger than a dimer has been found, even under conditions (e.g. ATP-γ-S) that promote hexamer formation for other DNA helicases (I Wong & T Lohman, unpublished results), the Rep homo-dimer appears to be the active form of the helicase. In this regard, the ss DNA-stimulated ATPase activity of the Rep protein is also enhanced significantly (by a factor of 8–10) upon dimerization (68). The *E. coli* UvrD protein (Helicase II) also forms dimers and its ss DNA-stimulated ATPase activity is also enhanced upon dimerization (69). The *E. coli* Rep and UvrD (Helicase II) proteins, which share ~40% sequence similarity, can also form heterodimers in vitro (68). Although a Rep/UvrD heterodimer has no known function, the formation of heterodimers is interesting in light of the fact that a *rep/uvrD* double mutant is lethal (70), although neither Rep nor UvrD are essential in *E. coli*. Furthermore, whereas a complete deletion of the *uvrD* gene is not lethal in *E. coli*, overproduction of a UvrD mutant (UvrD-K35M) in a wild-type *rep* background is lethal (71), suggesting that UvrD and Rep can form hetero-oligomers in vivo and that a wild-type Rep/UvrD-K35M hetero-oligomer is inactive (68).

Oligomerization of several helicases is modulated by interactions with other ligands. The most dramatic case is the *E. coli* Rep helicase which exists as a stable monomer (M_r=76,400) up to concentrations of at least 12 μM in the absence of DNA (72) (I Wong & T Lohman, unpublished data), indicating a maximum dimerization constant of ~10^3 M^{-1}; however, binding of either ss or ds DNA induces Rep to dimerize (57), with dimerization constant L~2×10^8 M^{-1} (59). Thus the Rep dimerization constant increases by at least a factor of 10^5 upon binding DNA (59). The formation of stable hexamers of the phage T4 gene *41* protein (helicase/primase) is facilitated by the binding of GTP-γ-S as well as GTP or ATP (55, 73). Formation of hexamers of the SV40 T antigen is facilitated by Mg^{2+} and ATP (51, 74), whereas tetramers are formed in the absence of ATP (51). Formation of hexamers of the T7 gene *4* protein is also facilitated by Mg^{2+} and a nonhydrolyzable analogue of dTTP (56, 67). Dna B hexamer formation is stabilized by Mg^{2+} (66). These examples suggest that other helicases that appear to be monomeric should be examined to determine whether an oligomeric form is stabilized upon binding DNA, nucleotides, or divalent cations.

Although some researchers (75) have recently claimed to be able to differentiate between helicases that function as monomers as opposed to oligomers, the assembly states of the "monomeric" RNA helicases that are discussed (human RNA helicases p68 and RNA helicase A, vaccinia NPH-II) have not been characterized in sufficient detail to conclude that they function as monomers. In fact, the assembly states of these proteins when bound to nucleic acid have not been characterized.

Structural Features of Hexameric Helicases

Low resolution structural information has been obtained for some hexameric DNA helicases, SV40 large T antigen (51), *E. coli* RuvB protein (54), phage T4 gene 41 protein (55) and phage T7 gene 4 proteins (56), *E. coli* DnaB protein (66), and *E. coli* Rho protein, a hexameric RNA-RNA and RNA-DNA helicase (53). In their hexameric forms, these helicases form very similar toroidal or ring-like structures, with outer diameters ranging from 100 to 130 Å and 20–30 Å holes through the center of the hexamer. This strong similarity is somewhat surprising since the subunit molecular weights range from 37 kDa for RuvB to 92 kDa for SV40 T antigen. Furthermore, evidence shows that when T antigen (51) and RuvB double hexamers (54) are bound to duplex DNA, the DNA can pass through the hole in the hexamer. The T7 gene 4 hexamer can bind to ss DNA with the DNA passing through the center of the hexamer (56). In contrast, ss RNA can bind to Rho by wrapping around the hexamer (53, 62). Such ring-like structures have not been observed for the *E. coli* Rep or UvrD helicases using these same cryo-EM approaches (X Yu & E Egelman, personal communication), consistent with the conclusion that Rep helicase functions as a dimer rather than a hexamer (28, 57).

SV40 LARGE T ANTIGEN   Scanning transmission EM (51) and atomic force microscopy (52) have shown that in the presence of ATP, Mg^{2+}, and DNA containing the SV40 origin of replication, the SV40 large T antigen (monomer M_r= ~92,000) assembles into bi-lobed double hexamers, and these hexamers appear to assemble around the DNA (51, 52). In the absence of DNA, T antigen forms single hexamers that have low affinity for DNA and do not support DNA replication, although these hexamers still retain helicase activity in vitro (74). However, conditions that favor dissociation of hexamers to monomers (37° C, no ATP) can reactivate the T antigen so that it functions in DNA binding and replication (74). These studies suggest that the pathway for formation of the T antigen double hexamer at the replication origin occurs through assembly of monomers, rather than binding of preformed hexamers, consistent with the proposal that T antigen hexamers encircle the DNA at the origin (74). However, preformed T antigen hexamers retain helicase activity on partial duplex DNA substrates, hence helicase function may not require the hexamer to encircle the DNA (74).

E. coli RuvB PROTEIN   The RuvB protein (monomer M_r=37,177) functions in recombination to promote branch migration of Holliday junctions in a reaction that is coupled to ATP hydrolysis (76). RuvB, in the presence of RuvA protein, can also unwind short duplex oligodeoxynucleotides annealed to ss DNA, with apparent 5′-to-3′ polarity (77). EM studies indicate that RuvB, in reactions requiring ATP binding, can assemble onto covalently closed circular duplex DNA to form double hexameric rings possessing D_6 symmetry (54). Averaging of ~800 EM images suggests that these double hexamers are in the form of a ring with an outer diameter of ~120 Å and a 20–25 Å–diameter hole through the center of the ring (54). Binding of RuvB hexamers does not change the contour length of the DNA, suggesting that the DNA passes through the center of the double hexamers (54).

PHAGE T4 GENE 41 PROTEIN   Cryogenic EM studies of the phage T4 gene 41 helicase (monomer M_r= ~53,000), which is involved in phage replication, indicate that this helicase can also form a stable hexamer upon binding the nonhydrolyzable nucleotides, ATP-γ-S or GTP-γ-S (55). This conclusion is also supported by sedimentation velocity, protein crosslinking, and gel filtration studies. In the absence of nucleotides, the T4 gene 41 protein is in a monomer-dimer equilibrium, with a dimerization constant near 10^6 M^{-1}. However, these dimers assemble further to form hexamers upon binding ATP or GTP. The EM images indicate a toroidal structure for the T4 gene 41 hexamers with outer diameters of ~100 Å and a region of low density in its center, suggesting a hole (55). The hexamers dissociate to dimers upon hydrolysis of ATP or GTP, so further studies will be required to determine whether the hexamer is the functionally active form during DNA unwinding. Another possibility is that dimers and hexamers both possess helicase activity, but the hexamer is required for processive unwinding.

PHAGE T7 GENE 4 PROTEINS   Three-dimensional reconstruction of EM images of the phage T7 gene 4 proteins indicates that these proteins can form toroidal hexamers with a diameter of ~130 Å and an ~25–30 Å–diameter central hole (56). Both the short form of the protein (4B′, M_r=56,000), which possesses only helicase activity (78, 79), and the long form (4A′, M_r= 63,000), which possesses both helicase and primase activity (80), can form hexamers. The hexamer has apparent D_6 symmetry, and each subunit possesses two lobes, so the hexamer has the appearance of being two-tiered, with a small ring stacked upon a larger ring. The gross features of this structure are very similar to those of the RuvB hexamer, including the bilobal structure of the subunits (54), but they differ from the D_3 symmetry suggested for the Rho (81) and phage T4 gene 41 hexamers (55). The T7 gene 4 hexamers can assemble on covalently closed circular ss phage M13 DNA, with the ss DNA passing through the center of the hexamer; binding is polar, with the smaller ring oriented toward the 5′ end of the ss DNA (56).

E. coli Rho PROTEIN   The Rho protein (monomer M_r=46,000), a 5′-to-3′ RNA and RNA/DNA helicase (19), also forms hexameric structures with D_3 symmetry (82). Although hexamer formation does not require nucleotide binding, it is promoted by the binding of ATP and/or RNA (82). Low-resolution structural information has been obtained by cryo-EM (53) and X-ray and neutron scattering (81, 83), which complements the extensive biochemical and biophysical studies of this protein (84–86). These studies have been summarized and a mechanism for Rho translocation along ss RNA has been proposed (62). Equilibrium binding of oligo- and polynucleotides indicates that ss RNA wraps around the hexamer, with ~78±6 nucleotides required to occupy all six subunits (87). Image reconstruction analysis of cryo-EM pictures of the Rho hexamer indicates a ring-like structure with an outer diameter of 125 Å and a separation of 45 Å between the center of mass of each subunit (53). At this resolution, this structure is remarkably similar to those observed for the RuvB (54), T7 gene 4 (56), and T4 gene 41 (55) hexamers; the center of the Rho hexamer even has a similar region of low density. However, models based on X-ray and neutron-scattering studies of Rho hexamers bound to RNA (rC_{70}) suggest that the RNA is bound at the periphery of the hexamer (83).

E. coli DnaB PROTEIN   The *E. coli* DnaB helicase also forms hexamers as shown by protein crosslinking (65) and sedimentation studies (66). In the presence of 5 mM $MgCl_2$, hexamers are observed at protein concentrations as low as 100 nM hexamers (66). Formation of the DnaB hexamers does not appear to require nucleotide binding, in contrast to RuvB (54), T7 gene 4 (56), T4 gene 41 (55), and T antigen (51). Sedimentation studies indicate that hexamers dissociate to form trimers in the absence of Mg^{2+}; in fact, a net binding of 4 Mg^{2+} accompanies formation of the DnaB hexamer from two trimers (66). The sedimentation studies of the DnaB hexamer also suggest a ring-like or cyclic structure with each protomer contacting only its two nearest neighbors (66).

Primary Structures

The literature on the use of comparative studies of protein primary structures to identify and classify proteins as putative helicases is extensive (88, 89) [see (90) for a recent review]. Several different sequence patterns are conserved in some helicases, with each sequence pattern defining a family or superfamily. The two largest superfamilies, SF-1 (containing *E. coli* Rep, UvrD, RecB, RecD, TraI, HSV-1 UL5, etc) and SF-2 (containing *E. coli* RecQ, PriA, UvrB, yeast RAD3, eIF-4A, etc), are defined by seven conserved regions of sequence homology, whereas SF-3, which includes SV40 T antigen, is defined by only three conserved regions. A smaller family, F4, defined by five conserved regions, includes *E. coli* DnaB, T7 gene 4, and T4 gene 41, all of which are hexameric helicases that are associated with or (as in the case of T7 gene 4 proteins) possess primase activity. The *E. coli* Rho protein, a hexameric RNA and RNA/DNA helicase, falls within a separate family that includes proton-translocating ATPases (90); the RuvB protein is also in a separate superfamily (90).

The only regions of sequence similarity that are shared uniformly among all of the helicase families are the "A" and "B" motifs of the "Walker Box" that have been shown to be useful predictors of a nucleoside-5′-triphosphate binding site (91). Region I (GXGXGK[T/S]) represents the Walker "A-type" consensus sequence that forms the "P-loop" within the NTP binding site, and region II contains an aspartate (D) that interacts with NTP via Mg^{2+} (92). These two regions are necessary but not sufficient for a protein to have helicase activity. Region II is equivalent to the so-called DEAD-box or DExH-box that has been used as a predictor of putative RNA helicases (24); however, most RNA binding proteins containing a DEAD-box do not possess helicase activity in vitro; thus this region is useful only as a predictor of NTP binding.

Site-directed mutagenesis studies of several helicases indicate that the Walker A and B regions are both important for NTP binding (93–97). However, no functions have been identified with any of the remaining "conserved" regions, although site-directed mutagenesis studies have been done with eIF-4A (98), HSV-1 UL5 (99), and HSV-1 UL9 (100). Mutations within each of the six conserved sequences of HSV-1 UL5 and five of the six conserved sequences of HSV-1 UL9 result in loss of function in viral DNA replication, although helicase activity has not been examined. However, not all of these so-called helicase motifs are needed for helicase activity, because region VI can be deleted from the *E. coli* Rep protein while helicase activity in vitro is retained (101). In interpreting such results, one must also keep in mind that the assembly state of a helicase influences its activity and that an association with another protein may be required for helicase activity, as in the case of eIF-4A which requires eIF-4B (18).

None of the known hexameric helicases are grouped in the SF1 or SF2 superfamilies, so some of the "super families" may therefore reflect differences in the assembly state of the active forms of these helicases. In general, it is still too early to judge the utility of primary structure as a predictor of "putative" helicases, since far more putative helicases have been identified through these computer searches than have been characterized biochemically. Furthermore, since not all proteins that have been identified as putative helicases possess helicase activity in vitro (102), these approaches should be used with caution.

DNA BINDING

An understanding of helicase mechanisms requires studies of the interactions of helicase with its DNA substrate, specifically the functional energetics and kinetics of binding and the influence of nucleotide cofactors. The following questions are of central importance to developing this understanding. Does the helicase interact with both duplex and ss DNA or only ss DNA? Does the helicase interact with only one or both single strands at a ss/ds DNA junction? Does a helicase recognize a ss/ds DNA junction per se, and/or do separate sites allow simultaneous binding to ss and ds DNA? Of course, there will be a set of answers to these questions reflecting the transient formation of multiple intermediates during DNA unwinding and translocation. The relative energetics of these interactions and their allosteric responses to nucleotide cofactors must also be examined in order to determine their functional significance. Kinetic studies are also essential to elucidate the pathway and mechanism of translocation and DNA unwinding. In addressing such questions one must consider the oligomeric nature of the helicase and the fact that DNA and nucleotide binding will influence the energetics of protein assembly and thus the distribution of assembly states (103).

Polarity of Binding to Single-Stranded DNA

The ss DNA binding site of a helicase is expected to be polar and thus bind ss DNA with a unique orientation with respect to the polarity of the sugar-phosphate backbone. However, duplex DNA should be capable of binding in either orientation if binding is nonspecific. In fact, direct evidence shows polar binding of ss DNA to the subunits of the *E. coli* Rep dimer. Two types of $(dT)_{16}$ molecules containing the fluorescent base etheno-adenosine (dεA) on either the 5′ end or the 3′ end bind Rep with the same affinities but with very different extents of fluorescence enhancement, indicating that ss DNA binds with polarity (M Amaratunga & T Lohman, unpublished data). Similar studies have demonstrated polar binding of ss DNA to the phage T4 gene 32 protein (104). The affinity of Rep protein for ss oligodeoxynucleotides also shows little dependence on DNA base composition or sequence (59), suggesting that Rep recognizes primarily the sugar-phosphate backbone in ss DNA. The hexameric T7 gene 4 protein also binds with polarity to ss DNA while encircling the DNA (56) (see section on Structural Features of DNA Helicases). Although polar binding of ss DNA has only been demonstrated for these two helicases, this finding is likely to be true for all helicases. In fact, how helicases could display a "polarity" or "directionality" of DNA unwinding in vitro without this property is difficult to imagine (see section on DNA Binding).

Stoichiometries and Energetics of Single-Stranded and Double-Stranded DNA Binding

Qualitative studies of nonspecific binding to polymeric DNA have shown that helicases generally bind with higher affinity to ss DNA than to ds DNA (58,

105, 106) [for reviews of early studies see (2, 3, 30)]. Although quantitative estimates of equilibrium binding constants and the energetics of DNA binding are needed to interpret ATPase and DNA unwinding studies, these estimates are complicated by the fact that most helicases are oligomeric and thus possess multiple DNA binding sites. Furthermore, at the protein concentrations generally used in early experiments, these proteins exist as an equilibrium mixture of oligomeric forms, and DNA affinity is influenced by the assembly state of the protein. In addition, multiple helicases can bind to long DNA molecules. Therefore, we restrict our discussion to quantitative studies of DNA binding that have been performed under conditions for which the assembly state of the helicase is defined or in which the linkage of DNA binding to helicase assembly has been accounted for explicitly in the analysis.

E. coli Rep PROTEIN   One approach to resolve the energetics of DNA binding from protein oligomerization is to use oligodeoxynucleotides that are short enough that only one protein monomer or subunit can bind to each oligodeoxynucleotide (28, 59, 103). The use of such short DNA molecules is also needed to determine if cooperativity exists among the DNA binding sites of an oligomeric helicase as well as to resolve the energetic contributions resulting from the simultaneous binding of ss DNA and ds DNA to separate subunits of an oligomeric helicase. These latter effects will almost certainly contribute to the energetics of binding of helicases to a replication fork.

The Rep protein undergoes a DNA-induced dimerization, and the dimer appears to be the active form of the helicase (28, 57, 60). Using a modified nitrocellulose filter–binding method (107), Wong et al (59) and Wong & Lohman (28) examined the energetics of DNA binding and dimerization of the Rep protein using oligodeoxynucleotides that are short enough (≤ 16 nucleotides or bp) to insure that only one Rep monomer binds to each oligodeoxynucleotide. Figure 1 depicts the five DNA-ligation states that a Rep dimer can form in the presence of both ss and ds oligodeoxynucleotides; each subunit of a Rep dimer can bind either ss DNA (S) or ds DNA (D) to form Rep dimers that are either half saturated (P_2S and P_2D) or fully saturated (P_2S_2, P_2D_2, and P_2SD). Seven independent equilibrium constants are required to describe the equilibrium binding of Rep to ss DNA and ds DNA, and these were determined from three sets of equilibrium titrations: One set was performed with ss DNA alone, a second set was performed with ds DNA alone, and a third set of competition experiments was performed with both ss DNA and ds DNA (59, 103).

The equilibrium constant for Rep monomer binding to ss oligodeoxynucleotides, $K_{1S} = 4.5 \pm 1.0 \times 10^6$ M^{-1} (4° C, 6 mM NaCl, pH 7.5), shows relatively little dependence on base composition or sequence (59). The dependence of the binding constant on the length of a series of $(dT)_N$ molecules indicates that

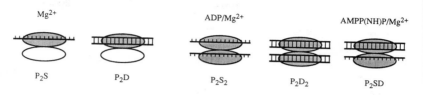

Figure 1   Representation of the five different DNA ligation states of the Rep dimer that can form in the presence of ss- and ds-oligodeoxynucleotides that are short enough to allow only one Rep monomer to bind per oligodeoxynucleotide. At equimolar concentrations of ss DNA, duplex DNA, and Rep monomers, the P_2S dimer is favored in the presence of Mg^{2+}, the P_2S_2 dimer is favored in the presence of ADP/Mg^{2+}, and the P_2SD dimer is favored in the presence of $AMPP(NH)P/Mg^{2+}$. Rep monomers can also bind ss DNA or duplex DNA to form the monomer species PS and PD, respectively. Modified from Wong & Lohman (28).

~13–14 nucleotides are needed for optimal binding (59), which is consistent with an occluded site size of 16 ± 2 nucleotides per monomer on ss DNA (57). Either ss (S) or ds (D) oligodeoxynucleotides can bind to Rep monomers (P) to form PS or PD, with DNA binding inducing Rep monomers (P) to dimerize, with a dimerization equilibrium constant ~1–2×10^8 M^{-1} (59). Except in the presence of ss DNA and a nonhydrolyzable ATP analogue (see below), the equilibrium constant for binding a 16 bp DNA hairpin to the Rep monomer and the second subunit of a Rep dimer is a factor of ~100 lower than for $(dT)_{16}$ (59). These results indicate that all five DNA ligation states of the Rep dimer shown in Figure 1 can form in solution, including the P_2SD complex in which ss and ds DNA bind simultaneously to each subunit of the Rep dimer (28, 59). An early study concluded that Rep could bind ss and ds polynucleotides simultaneously, but to separate sites (58). However, Rep monomers and individual dimer subunits bind ss and ds DNA competitively under all conditions examined (28, 59), suggesting that the same DNA binding site can be occupied by either ss or ds DNA or that binding of one DNA conformation precludes binding of the other.

Of particular interest is the observation that there is communication (cooperativity) between the two Rep dimer DNA binding sites. DNA binds with overall negative cooperativity to the Rep dimer since the affinity of DNA for the second subunit is dramatically lower (by $\geq \sim10^4$) than for the first subunit (59). However, the affinities for binding either ss DNA or ds DNA to the second subunit of a half-saturated Rep dimer (P_2S or P_2D) are also influenced by the conformation of DNA (ss vs ds) bound to the first subunit. A most interesting finding is that this cooperativity, and thus the relative stabilities of the P_2S_2 vs P_2SD complexes, is affected by nucleotides (ATP, ADP). Thus, nucleotide binding influences primarily the energetics of DNA binding to the second site of the Rep dimer (28). This latter effect appears to be important functionally for the mechanism of Rep-catalyzed DNA unwinding (28) (see section on Mechanisms of DNA Unwinding and Translocation).

PHAGE T7 GENE 4 HELICASE   Hingorani & Patel (108) have examined the equilibrium binding of the phage T7 gene 4A′ helicase/primase to mixed sequence ss oligodeoxynucleotides 10, 30, and 60 nucleotides long, and to a 20 bp hairpin duplex using nitrocellulose filter binding. Although relatively little binding is observed in the absence of nucleotides, binding is optimal in the presence of Mg^{2+} and thymidine 5′-(β, γ-methylenetriphosphate) (dTMP-PCP), a nonhydrolyzable dTTP analogue, which stabilizes the gene 4A′ hexamer. Therefore, DNA appears to bind primarily to the hexameric form of the enzyme, which is believed to be the oligomeric state of the functional helicase, although some evidence for binding to dodecamers was observed (108). High-affinity binding of one molecule of either ss $(dN)_{10}$ or ss $(dN)_{30}$ was observed, although a second molecule of DNA binds with ~50-fold lower affinity. A 60-nucleotide ss DNA can bind two hexamers, suggesting that each hexamer interacts with ~30 nucleotides of DNA, a finding consistent with nuclease protection studies of gene 4A′ protein bound to poly(dT) (108). This length of ss DNA does not seem long enough to wrap around the perimeter of the hexamer and interact with all six subunits simultaneously. Thus the hexamer may interact with ss DNA using only a subset of its subunits at any time. Protein cross-linking and EM studies also suggest that the ring-like T7 gene 4A′ hexamer binds long ss DNA by encircling the ss DNA, rather than by wrapping the ss DNA around the hexamer (56). Binding is considerably weaker to a 20 bp duplex hairpin, but simultaneous binding of both ss and ds DNA to hexamers was observed (108). The ability to bind both ss and ds DNA simultaneously may be important during one stage of DNA unwinding by these helicases as suggested for DNA unwinding by the *E. coli* Rep dimer (28, 59).

E. coli DnaB   Bujalowski & Jezewska (109) have examined the stoichiometry of DnaB hexamer binding to ss DNA based on equilibrium titrations by monitoring the fluorescence increase associated with DnaB binding to the fluorescent ss homopolydeoxynucleotide, poly(dεA). A stoichiometry of 20±3 nucleotides per DnaB hexamer (pH 8.1, 10° C, 1 mM AMPP(NH)P, 5 mM $MgCl2$, 50 mM NaCl, 10% glycerol) is found upon saturation of poly(dεA). Consistent with this finding, binding experiments with ss oligodeoxynucleotides that were 20 and 40 nucleotides long showed stoichiometries of one and two hexamers per oligodeoxynucleotide, respectively. Furthermore, photo-crosslinking experiments suggest that only one subunit of the hexamer cross-links strongly to $(dT)_{20}$ (109). Based on these results, Bujalowski & Jezewska (109) conclude that ss DNA does not wrap around the Dna B hexamer, but rather interacts with one or at most two subunits of the hexamer under these conditions. These conclusions are similar to those reached for the phage T7 gene 4A' hexamer (56, 108). However, this stoichiometry is considerably lower than the value of 78±6 nucleotides determined for ss RNA binding to the *E. coli* Rho hexamer (87), which is similar in size to the DnaB and T7 gene 4A' hexamers. Binding of the DnaB hexamer to long ss DNA shows only weak positive cooperativity indicating that hexamers do not form long clusters when bound to ss DNA (109).

E. coli Rho   The binding of a series of oligonucleotides to the Rho protein, a hexameric RNA and RNA/DNA helicase, has been investigated using electrophoretic band shift and ultrafiltration binding assays (110). Octanucleotides containing mixed sequences of cytidine and uridine were examined. Each Rho hexamer can bind six molecules of $(rC)_{10}$, with three high-affinity and three low-affinity binding sites differing in affinity by approximately tenfold. This is consistent with the conclusion based on neutron scattering (83), RNase A digestion studies (111), and fluorescence titrations (87) that long ss RNA can wrap around the hexamer, interact with all six subunits, and occlude ~78±6 nucleotides per hexamer. The affinity of a given oligonucleotide depends primarily on its cytosine content, but not the sequence. In contrast the base sequence of these same oligonucleotides has a substantial effect on the ATPase activity of Rho (112). Rho hexamers that bind to long ss RNA do show some ability to form small clusters of hexamers (111).

Allosteric Effects of Nucleotides on DNA Binding to the Dimeric E. coli Rep Helicase

Processive DNA unwinding by a helicase requires the binding of nucleoside-5'-triphosphate(s) (NTP), their subsequent hydrolysis, and release of products (NDP and inorganic phosphate, P_i). This cycle of binding, hydrolysis, and product release must drive the helicase through energetic states that effect DNA unwinding and helicase translocation (25). This concept is supported by observations of nucleotide- and DNA-dependent changes in the conformational states of DnaB (113), Rho (114), Helicase II, and Rep (101, 115, 116). However, which step(s) in the NTP cycle is coupled to DNA unwinding is not yet known; in principle, it could be any step.

As discussed below (see section on Nucleotide Binding and NTP Hydrolysis), helicases generally possess one potential ATP binding site per subunit within the oligomeric helicase. Early qualitative studies with Rep (58), DnaB (105), and Helicase III (106) indicate that the relative affinities of these helicases for ss and ds DNA are influenced by the type of nucleotide cofactor (ATP or ADP) that is bound to the helicase. DnaB can bind both ss and ds DNA, and its affinity for ss poly(dT) increases upon binding nonhydrolyzable

ATP analogues (61). The Rep protein binds to both ss and ds DNA (59, 106), and both the equilibrium affinity (28) and dissociation rate constant for ss DNA are affected differentially by ATP, ADP, and nonhydrolyzable ATP analogues (58). Thus ATP and its hydrolysis products are allosteric effectors of DNA binding in these and likely all helicases. In contrast, effects of nucleotide cofactors (ATP, ADP, AMPPCP) on the binding affinities of oligonucleotides for the Rho hexamer have not been observed (110).

E. coli Rep DIMER   Wong & Lohman (28) have examined the allosteric effects of nucleotide cofactors on the energetics of DNA binding to the *E. coli* Rep helicase. Equilibrium binding studies of short ss and ds oligodeoxynucleotides to Rep monomers and dimers were analyzed to determine all seven independent equilibrium constants under three sets of conditions (4° C, pH 7.5, 6 mM NaCl): (*a*) 5 mM Mg^{2+}, (*b*) 5 mM Mg^{2+} plus ADP (2 mM), and (*c*) 5 mM Mg^{2+} plus AMPP(NH)P (β–γ-imidoadenosine-5'-triphosphate), a nonhydrolyzable ATP analogue. Whereas these nucleotides show no major effect on Rep dimerization or ss or ds DNA binding to Rep monomers, AMPP(NH)P and ADP show a dramatic modulation of the affinity of ss DNA and ds DNA for the second subunit of the half-saturated Rep dimer species, P_2S and P_2D. The interesting result is that binding of both ADP and AMPP(NH)P influences the ability of a Rep dimer to form the different DNA ligation states depicted in Figure 1. Binding of ADP stabilizes the P_2S_2 state, in which ss DNA is bound to both Rep dimer subunits, whereas AMPP(NH)P stabilizes the P_2SD state in which ss and ds DNA are bound simultaneously, one to each subunit of a Rep dimer. In the presence of excess Mg^{2+}-AMPP(NH)P, the P_2SD state is favored over the P_2S_2 state by $\Delta G° = - 2.7$ kcal mol^{-1}, whereas in the presence of excess Mg^{2+}-ADP, the P_2SD state is disfavored over the P_2S_2 state by $\Delta G° = + 1.4$ kcal mol^{-1} (28). This allosteric effect of nucleotides strongly supports the proposal that a P_2SD complex, in which a Rep dimer binds simultaneously to duplex DNA and the 3'–ss DNA tail, is an important intermediate in the Rep-catalyzed DNA unwinding cycle and is stabilized by ATP binding (28). These results also suggest a cycling scheme that might be used by Rep dimers to unwind duplex DNA (6, 28, 29) (see section on Mechanisms of DNA Unwinding and Translocation).

Helicase Interactions at Single-Stranded/Double-Stranded DNA Junctions (Forks)

The regions of the DNA contacted by a helicase can, in principle, be detected by DNaseI and chemical protection (footprinting) methods, but unless there is high specificity for helicase binding to a DNA junction, differentiating productive from nonproductive complexes and thereby determining functionally important contacts is difficult. Furthermore, such studies are often not sensitive to the energetics of these interactions; hence energetically different complexes may appear indistinguishable using footprinting approaches.

SV40 LARGE T ANTIGEN   The binding of SV40 large T antigen with a synthetic DNA fork (twin-tailed) has been examined by two groups using DNA footprinting techniques; however, the conclusions from these studies are somewhat different. Sen Gupta & Borowiec (117) used P1 nuclease to probe ss DNA regions and DNase I to probe ds DNA regions, whereas Wessel et al (118) used P1 nuclease to probe ss DNA regions and exonuclease III to probe ds DNA regions. In both cases, significant protection of the ss/ds DNA junction was observed in the presence of T antigen and ATP. Sen Gupta & Borowiec (117) report that protection was asymmetric in that ~10 nucleotides of the 3'–ss DNA tail were protected, whereas the 5'–ss DNA tail was relatively unprotected (117). On the other hand, Wessel et al (118) observed protection of ~10 nucleotides on the 3' single strand and ~16 nucleotides on the 5' single strand, as well as ~24 bp of duplex DNA. Subtle changes in the pattern of P1 nuclease protection of the 3'–ss DNA region were observed as a function of nucleotide cofactor (117). Some protection from DNase I of ~4 bp at the junction was observed in the presence of T antigen and ATP or AMPP(NH)P, whereas this region was not protected in the presence of ADP (117). This apparent preference for SV40 T antigen binding to the duplex region in the presence of AMPP(NH)P is similar to the preference for simultaneous binding of ss and ds DNA by the *E. coli* Rep dimer to form a P$_2$SD complex in the presence of AMPP(NH)P (28). Base-specific modification of the ss DNA regions of the synthetic DNA forks showed only minor effects on T antigen binding, whereas ethylation of the phosphate backbone by treatment with ethylnitrosourea inhibits T antigen binding, suggesting recognition of the sugar-phosphate backbone (117). Again, this conclusion is similar to that reached for *E. coli* Rep helicase on the basis of equilibrium studies of Rep binding to ss oligodeoxynucleotides of varying base composition (59). Sen Gupta & Borowiec (117) conclude that T antigen binds the 3'-ss/ds DNA junction asymmetrically primarily by recognizing the sugar-phosphate backbone. They further conclude that T antigen does not encircle the DNA but rather binds to the outside of the DNA, and they suggest a "rolling" mechanism for the hexameric T antigen–catalyzed DNA unwinding (117). However, Wessel et al (118) conclude that duplex DNA threads through a T antigen double hexamer, with extrusion of the unwound single-stranded DNA loops.

E. coli RecBCD   The *E. coli* Rec BCD helicase initiates DNA unwinding at the ends of fully duplex DNA (119). The complex of RecBCD enzyme bound to duplex DNA ends in the absence of ATP has been examined (120) using DNase I cleavage of end-labeled DNA. In this complex, 16–17 nucleotides of the 3′ terminated strand and 20–21 nucleotides of the 5′ terminated strand are protected from DNase I cleavage. In addition, UV-treatment of the RecBCD-DNA complex resulted in crosslinking of the RecB subunit to the 3′ terminated strand and the RecC subunit to the 5′ terminated strand. These results suggest that the RecBCD enzyme interacts with both strands of the DNA duplex, at least to initiate DNA unwinding. Based on these studies, a DNA unwinding model has been proposed (120) that is a modified version of the active, rolling model proposed for the dimeric Rep helicase (28) (see section on Mechanisms of DNA Unwinding and Translocation).

NUCLEOTIDE BINDING AND NTP HYDROLYSIS

All DNA helicases appear to possess a consensus NTP binding site, as indicated by the presence of the conserved Walker A and B consensus motifs (motifs I and II) (90). Therefore, homo-oligomeric helicases possess at least one potential NTP binding site per subunit. This appears to be the case for the few helicases for which direct studies of nucleotide binding have been performed; however, nucleotides appear to bind nonequivalently to the oligomer, displaying negative cooperativity or possibly two classes of "high" and "low" affinity sites (1, 84, 121–123).

Stoichiometry and Equilibrium Binding

E. coli Rep   The active form of the Rep helicase is a homodimer which forms only upon binding DNA (28, 57, 59), and the steady-state ATPase activity of the dimer bound to ss DNA (P2S) is ~8–10-fold higher than the monomer bound to ss DNA (PS), which in turn is ~10^3-fold higher than Rep monomer in the absence of DNA (115). As a prelude to studying the Rep dimers, Moore & Lohman (115, 116) have investigated the equilibrium binding affinities and mechanism of binding of various nucleotides (ATP, ADP, ATP-γ-S, AMPP (NH)P, AMP) to the Rep monomer (in the absence of DNA). The Rep monomer has one site for nucleotide binding and ATP, ADP, AMPPNP and ATP-γ-S bind competitively (58, 115, 116). Therefore, dimer formation is not required for nucleotide binding. All nucleotides bind to the monomer by a two-step mechanism (115, 116), such that binding is followed by a protein conformational change. Independent evidence for nucleotide-induced changes in Rep protein conformation also comes from studies of the influence of nucleotides on the sensitivity and pattern of proteolysis of Rep protein (101). Rep monomer binds ATP tightly in the presence of Mg^{2+} ($K_{overall}$=1.3 × 10^8 M^{-1} at 4° C, pH 7.5, 6 mM NaCl, 10% (v/v) glycerol, 5 mM MgCl2); ATP can also bind in the absence of Mg^{2+}, although with ~10^3-fold lower affinity (~8 × 10^4 M^{-1}) (116). ADP binds competitively with ATP, but with significantly lower affinity

($K_{overall}=1.1 \times 10^6$ M^{-1} (5 mM Mg^{2+}). The affinity of Rep monomer for ATP-γ-S is $2 \pm 0.1 \times 10^7$ M^{-1}, ~5-fold lower than for ATP, whereas the overall binding constant for AMPPNP is only $1.4 \pm 0.1 \times 10^6$ M^{-1}. The nucleotide binding constants decrease with both increasing temperature and salt concentration (115, 116).

E. coli DnaB   Equilibrium binding of nucleotides to the DnaB protein indicates a stoichiometry of 1 nucleotide per DnaB monomer (105, 123, 124). The equilibrium binding of fluorescent nucleotide analogues to the hexameric DnaB protein has been examined (123) under conditions in which the DnaB is fully hexameric (66). Three high-affinity and three low-affinity nucleotide binding sites are observed per hexamer, which likely reflects a negative cooperativity for nucleotide binding. A statistical thermodynamic model was used to analyze the binding isotherms quantitatively in terms of an intrinsic equilibrium constant, K, for nucleotide binding per site on the hexamer and a cooperativity constant, σ ($\sigma<1$ indicates negative cooperativity), reflecting communication among the nucleotide binding sites. At 10° C (pH 8.1, 20 mM NaCl), $K=5.9\pm1 \times 10^5$ M^{-1}, and $\sigma=0.55\pm0.05$ for the binding of 2'(3')-O-(2, 4, 6-trinitrophenyl)adenosine 5'-triphosphate (TNP-ATP) (123). This negative cooperativity becomes more pronounced at higher temperatures, making it more difficult to bind nucleotides to the last three sites on the DnaB hexamer. The nucleotide binding sites of the DnaB hexamer have also been characterized using fluorescence approaches (125, 126).

PHAGE T7 GENE 4 PROTEINS   Binding of dTTP and thymidine 5'-(β, γ-methylenetriphosphate) (dTMP-PCP) to the phage T7 gene 4A' protein, the helicase-primase, has been examined using nitrocellulose filter binding (122). Three nucleotides bind per hexamer with high affinity (50 mM Tris-acetate, pH 7.5, 3 mM $Mg(CH_3CO_2)_2$, 50 mM $NaCH_3CO_2$, 10% glycerol, 23° C). The same stoichiometry was observed in the presence and absence of a 30 nucleotide-long ss DNA. This result is qualitatively consistent with the observation that the DnaB hexamer has three high-affinity sites and three lower-affinity sites for nucleotides (123). Any lower-affinity nucleotide binding sites on the T7 gene 4A' hexamer may not be detectable by a nitrocellulose filter–binding assay, and thus one cannot rule out the presence and possible importance of such lower-affinity sites.

E. coli Rho   The binding of nucleotides to the hexameric Rho helicase has been examined in the absence of its RNA cofactor (84, 121). Geiselmann & von Hippel (84) showed that the Rho hexamer binds a maximum of six nucleotides. However, the six sites are not equivalent: Three sites possess high affinity ($K= ~3 \times 10^6$ M^{-1}) and three sites possess 30-fold lower affinity

(0.1 M KCl, 10 mM MgCl$_2$, pH 7.8, unspecified temperature). The two classes of ATP binding sites may reflect the same negative cooperativity observed for nucleotide binding to the *E. coli* DnaB hexamer (123). Rho monomers and dimers can also bind ATP, the monomeric form having a relatively higher affinity for nucleotide (84); thus each Rho protomer contains an ATP binding site. The high-affinity site on a Rho hexamer binds ATP with ~100-fold higher affinity than either ADP or AMPPNP (121). These affinities are similar to the relative affinities measured for the Rep monomer (58, 115, 116). However, both inorganic phosphate and AMP bind more weakly to Rho [$K_{Pi} \approx 30$ M^{-1}; $K_{AMP} \ll 10^3$ M^{-1} (121)] than to Rep [$K_{Pi} \approx 10^3$ M^{-1}; $K_{AMP} \approx 10^4$ M^{-1} (116)]. Nucleotide binding to Rho was not detectable in the absence of Mg^{2+} with spin column methods (121). Similarly, nucleotide binding to Rep (58) and DnaB (105) was not detectable in the absence of Mg^{2+} with nitrocellulose filter binding. However, stopped-flow fluorescence studies have shown that nucleotides can bind to the Rep monomer in the absence of Mg^{2+}, although with significantly faster rate constants (115, 116). Therefore, spin-column and filter-binding techniques apparently do not detect binding owing to the higher dissociation rate constants of nucleotides in the absence of Mg^{2+}.

Steady-State NTP Hydrolysis and Effects of Protein Oligomerization

Most studies of NTP hydrolysis by helicases have been performed under steady-state conditions and many of the earlier results have been reviewed previously (2, 3). Although such steady-state experiments yield information on the order of binding and can provide constraints on the elementary rate constants, they provide little direct mechanistic information. Furthermore, most studies of NTP hydrolysis have been performed under conditions in which the assembly state(s) of the helicase is unknown or exists as a mixture of oligomeric states. Since the helicase assembly state influences the kinetics of ATP binding and hydrolysis it is essential to know the distribution of assembly states under the conditions of the experiment. For example, in the presence of excess ss oligodeoxynucleotide, (dT)$_8$, the steady-state ATP hydrolysis of Rep monomer (PS) increases ~8-fold (68) upon dimerization. Similarly, dimerization of *E. coli* Helicase II (UvrD) protein increases its steady-state DNA-stimulated ATPase by ~2.5–4-fold, depending on the DNA cofactor and solution conditions (68, 69). At 4° C (pH 7.5, 10% (v/v) glycerol, 6 mM NaCl, 5 mM MgCl$_2$), the free Rep monomer (P) has a $k_{cat}=0.002$ s^{-1}; the Rep monomer bound to (dT)$_{16}$ (PS) has $k_{cat}=2\pm0.5$ s^{-1}; and the Rep dimer with one subunit bound to (dT)$_{16}$ (P$_2$S) has $k_{cat}=18\pm2$ s^{-1} (1).

Mechanism of ATP Binding and Hydrolysis by E. coli Rep Monomer

Pre-steady-state kinetic studies are essential for a determination of the mechanism of NTP hydrolysis (127). However, studies of the detailed kinetic mechanism(s) of NTP binding and hydrolysis by DNA helicases are in their early stages. Extensive pre-steady-state kinetic studies of nucleotide binding and ATP hydrolysis have been performed with the *E. coli* Rep monomer. Although the Rep monomer is not the active form of the helicase, pre-steady-state studies of the monomer were performed as a necessary prelude to studies of the active Rep dimers, and they provide fundamental information about the basic steps in nucleotide binding to a helicase. Moore & Lohman (115, 116) used stopped-flow fluorescence to study the transient binding kinetics of nucleotides (ATP, ADP, AMPPNP, ATP-γ-S and others) to the *E. coli* Rep monomer (the stable form of the protein in the absence of DNA) as well as the single turnover kinetics of ATP hydrolysis by the Rep monomer. Binding of the fluorescent ATP analogue 2′(3′)-[*O*-(N-methylanthraniloyl]-ATP (mantATP) was monitored by the enhancement of its fluorescence upon Rep binding due to energy transfer from tryptophan (115); binding of the nonfluorescent parent nucleotides was studied by competition methods (116). A minimal mechanism for the DNA-independent binding and hydrolysis of ATP by the Rep monomer is shown in the Scheme below (4° C, pH 7.5, 10% (v/v) glycerol, 6 mM NaCl, 5 mM $MgCl_2$). Binding of ATP (T) to the Rep monomer (P) occurs by a two-step process (binding plus isomerization), followed by slow hydrolysis at 0.002 s^{-1} (k_{cat}), followed by a two-step mechanism for ADP (D) dissociation.

$$
\begin{array}{ccccccccccc}
 & \overset{1}{\underset{}{}} & & \overset{2}{\underset{}{}} & & \overset{3}{\underset{}{}} & & \overset{4}{\underset{}{}} & & \overset{5}{\underset{}{}} & \overset{6}{\underset{}{}} \\
 & 12\ \mu M^{-1} s^{-1} & & 15\ s^{-1} & & 0.002^{-1} & & >0.002\ s^{-1} & & 0.067\ s^{-1} & 720\ s^{-1} \\
P+T & \underset{1.5\ s^{-1}}{\overset{\rightarrow}{\leftarrow}} & PT & \underset{0.9\ s^{-1}}{\overset{\rightarrow}{\leftarrow}} & (PT)^* & \overset{\rightarrow}{} & (PD{\cdot}Pi)^* & \underset{}{\overset{\rightarrow}{\leftarrow}} & (PD)^* & \underset{8s^{-1}}{\overset{\rightarrow}{\leftarrow}} PD & \underset{6.7\ \mu M^{-1} s^{-1}}{\overset{\rightarrow}{\leftarrow}} P+D
\end{array}
$$

The only information missing from this scheme is the rate constant for phosphate release from the Rep-ADP-P_i complex. However, the absence of a burst of ADP formation in multiple turnover (steady-state) experiments indicates that P_i release occurs with a rate constant > 0.002 s^{-1}. Under these conditions, ATP binding to the Rep monomer is tighter than ADP, primarily as a result of differences in the dissociation rate constants. Similar rate and equilibrium constants are obtained with ATPγS, whereas AMPPNP binding is 15-fold slower and equilibrium binding is ~100-fold weaker (115, 116). Although the Rep monomer does bind mantATP in the absence of Mg^{2+} ($K_{overall}$ ~5 x 10^5 M^{-1}), all four rate constants (steps 1 and 2 in the above scheme) increase substantially. For example, the bimolecular rate constant (step 1) increases to 80 μM^{-1} s^{-1} (115). Two-step binding of nucleotides is an intrinsic property of nucleotide binding to the Rep monomer. Similar information on the elementary rate constants for nucleotide binding and NTP hydrolysis by the active dimeric form of the Rep helicase in its different ss and ds DNA ligation states will be needed in order to understand the coupling of NTP binding and hydrolysis to helicase translocation and DNA unwinding.

Mutations Within the Nucleotide Binding Site

An examination of the cooperativity among the multiple NTP sites within an oligomeric helicase is required to understand the coupling of NTP binding and hydrolysis to DNA unwinding. Several studies of helicases containing site-directed mutations within the nucleotide binding site have been performed (93, 94, 96, 97). These mutant helicases generally display dramatic decreases in the steady-state rate of NTP hydrolysis as well as in helicase activity. Mixing experiments with wild-type enzyme provide further evidence for the functional importance of the oligomeric nature of helicases because mixed oligomeric species possess decreased NTPase and helicase activity (97).

PHENOMENOLOGICAL FEATURES OF DNA UNWINDING AND TRANSLOCATION

Before considering molecular mechanisms of helicase-catalyzed DNA unwinding and helicase translocation, we first consider the macroscopic or phenomenological features of these processes. These features include the types of nucleic acid substrates required for initiation of unwinding in vitro, the "polarity" of the unwinding reaction, and the rates, processivities, and efficiencies of NTP hydrolysis during translocation and DNA unwinding. Such information may provide constraints on possible mechanisms; however, inferences about mechanisms from such information can be misleading.

Initiation of DNA Unwinding in Vitro

Most DNA helicases show definite preferences for unwinding particular types of DNA substrates in vitro. Some helicases, such as the SV40 large T antigen (128) can initiate DNA unwinding by binding to fully duplex DNA containing the SV40 origin of replication and "melting" the duplex to generate the single strands. Others, such as E. coli RecBCD (119), E. coli UvrD (Helicase II) (129–131), and E. coli RecQ (132), can initiate DNA unwinding at the ends of fully duplex DNA; in fact, initiation of DNA unwinding by RecBCD is inhibited if the DNA duplex has ss DNA flanking regions that are unequal in length (119). E. coli UvrD (Helicase II) can also initiate DNA unwinding at a nick (129, 130, 133), which is the biologically important site for initiation of unwinding in its roles in methyl-directed mismatch repair (134, 135) and excision repair (10). However, initiation of DNA unwinding in vitro by most DNA helicases requires or is strongly stimulated by a ss DNA covalently attached to the duplex DNA. These flanking ss DNA regions provide high-affinity sites for binding and possibly promote further assembly of the helicase to form an oligomeric initiation complex. Such helicases likely either initiate unwinding in vivo at preformed ss/ds DNA junctions or require accessory proteins to promote initiation of unwinding on blunt-ended, nicked, or fully duplex DNA.

"Polarity" of Duplex DNA Unwinding and the Role of Flanking Single-Stranded DNA

Most DNA helicases display a preference for unwinding duplex DNA containing a ss DNA region flanking the duplex. Some helicases require ss DNA attached to a 3' end of the duplex (a 3'-to-5' helicase), whereas others require ss DNA attached to a 5' end of the duplex (a 5'-to-3' helicase). Determination of this preference has generally been made using a partially duplex DNA substrate consisting of a linear ss DNA with complementary oligodeoxynucleotides annealed to each end as shown in Figure 2 (136–138). The "polarity" of unwinding is operationally defined by the backbone polarity of the flanking ss DNA that facilitates initiation of DNA unwinding in vitro. For example, if a helicase preferentially unwinds duplex A in Figure 2, it unwinds with 5'-to-3' polarity (a 5'-to-3' helicase), with respect to the polarity of the internal region of ss DNA to which it presumably binds. If duplex B is unwound preferentially, it is a 3'-to-5' helicase. Such studies indicate that the polarity of the ss DNA backbone flanking the duplex is an important determinant for initiation of unwinding, and the polarity can differ for different helicases. The polarity of unwinding is undefined for helicases that initiate unwinding only on blunt-ended duplex DNA or "forked" duplex DNA.

A helicase that displays a polarity in DNA unwinding is also often assumed to translocate uni-directionally along ss DNA with that same polarity. However, experimental tests of DNA unwinding polarity do not provide information about unidirectional translocation along ss DNA. In fact, polarity in a DNA unwinding reaction may not reflect uni-directional translocation of the helicase

Figure 2   Schematic of a DNA substrate used to investigate the macroscopic "polarity" of DNA unwinding by a helicase.

along ss DNA, but rather a requirement for a particular orientation of the flanking ss DNA with respect to the ss/ds DNA junction that enables the helicase to form a proper initiation complex. Convincing evidence that any helicase translocates uni-directionally along ss DNA is still lacking.

To initiate DNA unwinding in vitro, the *E. coli* DnaB protein (139), phage T4 gene 41 protein (140), and phage T7 gene 4 proteins (138), all of which are 5'-to-3' hexameric helicases, appear to require a forked DNA substrate, i.e. a duplex possessing noncomplementary ss DNA tails attached to both the 3' and 5' ends of each complementary DNA strand. This requirement may reflect a need for these helicases to interact with the ss DNA regions of both complementary strands during DNA unwinding, although the possibility that the 3'–ss DNA tail functions to facilitate binding, rather than being directly involved in unwinding, has not been tested. These three helicases all function in DNA replication and also associate with their respective primase. In the case of the T7 gene 4 protein, the long (63 kDa) form of the protein (4A) possesses both helicase and primase activities (141). It is interesting that DNA unwinding by the SV40 T antigen, a hexameric helicase with opposite polarity (3' to 5'), does not appear to require a 5'–ss DNA tail (142), although whether a 5'–ss DNA tail stimulates helicase activity is not known. Although these helicases have been denoted 5'-to-3' helicases, a polarity of unwinding cannot be determined unambiguously for such helicases that require both a 3'– and a 5'–ss DNA flanking region to initiate unwinding.

Efforts have been made to determine which DNA conformations (ss vs ds) and which strands are contacted by a helicase at an unwinding fork using synthetic DNA substrates to determine the influence of 3'– and 5'–ss DNA tail lengths on DNA unwinding. Although these methods do not address the questions directly, they can be used to determine what characteristics of the DNA substrate are important for formation of complexes that are productive in DNA unwinding. However, the results of such experiments can be ambiguous if DNA unwinding is measured at steady state ([DNA] >> [Helicase]). For example, an additional ss DNA tail could stimulate steady-state unwinding by providing a high-affinity binding site thus facilitating loading of the helicase onto the correct DNA strand, but not participate directly in the DNA unwinding reaction.

In principle, single turnover experiments ([Helicase] >> [DNA]) (60, 143) can be used to separate the DNA binding step from DNA unwinding and thus address these questions directly. Such experiments, in which rapid quench-flow and stopped-flow techniques were used, have provided information about the DNA substrate requirements for initiation of DNA unwinding by the *E. coli* Rep helicase (60, 143). In these experiments, the helicase is pre-incubated with the DNA substrate and unwinding is initiated by addition of ATP. The interpretation of such studies can be further simplified by including a protein "trap"

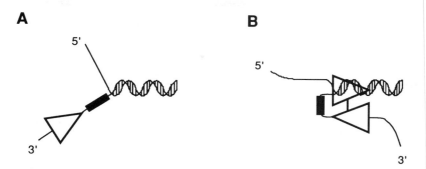

Figure 3 (*A*) Schematic representation of "blocked" DNA substrates used to test a "passive" mechanism for helicase-catalyzed DNA unwinding. A 3'-to-5' helicase (triangle) that unwinds by a "passive" mechanism cannot unwind the duplex owing to the presence of a "block" in the ss DNA flanking the duplex (rectangle) that can be either a segment of non-DNA or a segment of ss DNA with reversed (5' to 3') polarity. (*B*) In an active rolling mechanism for DNA unwinding, a dimeric Rep helicase (triangular protomers assumed to be related by C_2 symmetry) is postulated to form a P_2SD complex in which a Rep dimer is bound simultaneously to the 3'-ss DNA and the ds DNA (28) such that a segment of the 3'-ss DNA containing a "block" can be looped out and thus bypassed (reproduced from 60).

(e.g. excess ss DNA or heparin) along with the ATP, which serves to prevent re-initiation by the helicase (143). Such studies with Rep indicate that a 3'–ss DNA flanking region is required for unwinding, whereas a 5'–ss DNA flanking region is neither needed for nor stimulates unwinding of a 24–bp duplex (60). The same results are observed for UvrD-catalyzed DNA unwinding (J Ali & T Lohman, unpublished observations).

Experiments with the *E. coli* Rep helicase (a 3'-to-5' helicase) indicate that Rep does not unwind duplex DNA by translocating continuously along the 3' flanking–ss DNA until the duplex is encountered, but that a ss DNA region can facilitate unwinding even when removed some distance from the duplex DNA (60). To examine this possibility, a series of novel DNA substrates were synthesized that possessed a 3' flanking–ss DNA within which was embedded either a segment of ss DNA with reversed backbone polarity (e.g. 5' to 3') or a segment of non-DNA (poly-ethylene glycol) (60), as depicted in Figure 3A. Single turnover rapid quench-flow experiments show that these "reverse polarity" substrates are unwound by both the *E. coli* Rep helicase (60) and *E. coli* UvrD (Helicase II) (J Ali & T Lohman, unpublished observations) as long as the flanking ss DNA is covalently attached in the correct 3' orientation. Thus, both Rep (60) and Helicase II can unwind duplex DNA without translocating into the duplex, ruling out a "passive" mechanism (see section on Mechanisms of DNA Unwinding and Translocation). However, the results are consistent with an "active" rolling mechanism as proposed for the Rep helicase (6, 28, 29) in which an essential intermediate is a P_2SD complex in which one subunit of the Rep dimer binds to duplex DNA and the other subunit is bound to the 3'–ss DNA flanking region (see section on Mechanisms of DNA Unwinding and Translocation). Such a complex could circumvent the polyethylene glycol or the reverse polarity DNA by "looping out" these regions as depicted in Figure 3B.

Further support for this conclusion comes from the finding that the *E. coli* Rep helicase can unwind short regions of duplex DNA (18 bp) possessing only a 5'–ss DNA flanking region, which indicates that uni-directional translocation along ss DNA is not required for unwinding by this helicase (60). Therefore, the observation of polarity in a helicase-catalyzed DNA unwinding reaction does not indicate that the helicase translocates uni-directionally along ss DNA alone.

Translocation Along Single-Stranded DNA

The molecular details of translocation of a DNA helicase along ss DNA and its coupling to NTP binding and hydrolysis are not known for any helicase. Even the phenomenological aspects of this reaction have not been characterized for most helicases owing mainly to the absence of direct assays to monitor translocation. As discussed above, although helicase translocation along ss DNA has often been assumed to be uni-directional, this has not yet been demonstrated unambiguously for any helicase. However, this assumption has biased many postulated mechanisms of helicase-catalyzed DNA unwinding.

Lee & Marians (144) have characterized indirectly the translocation along ss DNA of the *E. coli* PriA protein, a 3'-to-5' helicase involved in primasome assembly. ATP-dependent assembly of PriA onto SSB protein-coated φX174 ss DNA occurs only at a specific primasome assembly site (PAS) (144). Therefore, the average net rate of translocation was determined by monitoring the time required for PriA to displace an oligodeoxynucleotide (24 bp) annealed at a known distance from the PAS initiation site. The net rate of PriA translocation along ss DNA increases hyperbolically with [ATP] reaching a plateau above 1 mM ATP at a net rate of ~90 nucleotides s^{-1} [30° C, pH 8.0, 5 mM $Mg(CH_3CO_2)_2$] (144). Furthermore, the amount of ATP required increases as the length of the duplex region to be unwound increases (144).

The effects of ss DNA length on the steady-state kinetics of NTP hydrolysis by a helicase have been examined in attempts to study translocation of DNA helicases (73, 138, 145–148). This approach is based on the idea that if helicase translocation is driven by NTP hydrolysis, then the steady-state kinetic parameters for NTP hydrolysis should be influenced by the length of the ss DNA effector if the ss DNA is shorter than the average distance over which the helicase translocates before dissociation. Young et al (147) show that, in principle, the quantitative dependencies on ss DNA length of V_{max} and K_{act} (the DNA concentration required to reach half-maximal velocity) can be sensitive to whether a helicase translocates uni-directionally or bi-directionally. However, such steady-state approaches are complicated by the linkage between DNA binding and helicase assembly, by preferential binding of helicases to the ends of the DNA, and by the influence of intramolecular base pairing within natural ss DNA.

This approach has been used (73) to show that the GTPase activity of the phage T4 gene 41 protein increases with the average length of ss DNA. In a slightly different experiment, stimulation of the TTPase activity of the T7 gene 4 proteins was shown to occur at a ~20-fold lower concentration when circular M13 ss DNA is used as the effector rather than linear M13 ss DNA (138). Similar results have been reported for *E. coli* Helicase II (145) and were interpreted as an indication that translocation along the ss DNA is coupled to NTP hydrolysis. The most dramatic difference in the ATPase activity of a helicase on linear vs circular ss DNA was observed with *E. coli* Helicase I (146). These results suggest that Helicase I translocates along ss DNA with high processivity, but dissociates slowly from the end of a linear ss DNA or reassociates slowly to ss DNA (146).

Young et al (148) examined the steady-state GTPase activity of T4 gene 41 protein as a function of the length of a series of ss oligo(dT) lattices and on polymeric poly(dT) bound with varying amounts of T4 gene 32 protein, a cooperative SSB protein. The latter approach takes advantage of the fact that at sub-saturating amounts of T4 gene 32 protein, with respect to poly(dT) lattice, cooperative clusters of T4 gene 32 protein will form an equilibrium distribution of ss DNA gaps that may mimic ss DNA molecules of differing lengths. The average length distribution of these ss DNA gaps can be calculated based on the known equilibrium binding properties of the T4 gene 32 protein for poly(dT) (149–152). These results demonstrate that the T4 gene 41 protein translocates along ss DNA; however, an unambiguous demonstration of uni-directional translocation is still lacking.

Several studies suggest that when a helicase is prevented from translocating along ss DNA, NTP hydrolysis is inhibited (153–155). Upon encountering a benzo[a]pyrene-DNA adduct, the phage T7 gene 4 helicase is apparently sequestered by the adduct, blocking further translocation and strongly inhibiting its dTTPase activity (153). The ATPase activity of the *S. cerevisiae* RAD3 helicase is also inhibited upon encountering UV damage (155). Similarly, the phage T4 Dda helicase is blocked and sequestered, with a concomitant reduction in ATPase activity, upon encountering a yeast GAL4 protein-DNA complex (154). These results have led to the suggestion that translocation, or at least transient dissociation of DNA from one site or subunit on the helicase, is linked to multiple NTP turnover. However, the molecular interpretation of these results is complicated by the fact that helicases possess multiple DNA and ATP binding sites.

Rates and Processivities of DNA Unwinding

The macroscopic rate and processivity are important characteristics of helicase-catalyzed DNA unwinding. Although these characteristics provide few mechanistic details, a processive helicase must clearly remain bound to the DNA during the time required for translocation. Processivity is a measure of the number of bp unwound before the helicase dissociates from the DNA. A DNA helicase involved in DNA replication is expected to have high processivity, whereas a helicase involved in repair of short patches of DNA may not require high processivity. However, as discussed above, the rates and processivities of helicase-catalyzed DNA unwinding measured in vitro are sensitive to solution conditions (salt type and concentration, pH, temperature, etc) and thus meaningful comparisons can only be made under identical solution conditions. These kinetic properties can also be influenced through interactions with accessory proteins.

Quantitative measurements of the macroscopic rates of DNA unwinding on long duplex DNA have been reported for only a few helicases and are dependent upon solution conditions, especially the salt concentration. Some DNA helicases can unwind duplex DNA at rates approaching 500–1000 bp s^{-1}, which are comparable to replication rates in prokaryotes (139, 156–158). The most detailed studies of DNA unwinding rates have been performed with the RecBCD helicase (156, 158). Under optimal salt concentrations (0.1 M NaCl), RecBCD can unwind duplex DNA at rates of 470 ± 80 bp s^{-1} at 25° C; this rate increases to ~900 bp s^{-1} at 37° C. These rates decrease both above and below 0.1 M NaCl (158). The rate of DNA unwinding by E. coli DnaB in vitro is ~30–40 bp s^{-1} (30° C, pH 7.6, 11 mM $Mg(CH_3CO_2)_2$) (139) and ~60 bp s^{-1} in the presence of DNA gyrase [37° C, pH 7.6, 30% glycerol, 6 mM $CaCl_2$, 10 mM $Mg(CH_3CO_2)_2$] (157); however, this increases substantially to ~700 bp s^{-1} in the presence of DNA polymerase III (30° C, pH 8, 10 mM $Mg(CH_3CO_2)_2$) (159). The rate of unwinding of short duplex DNA (18 or 24 bp) by E. coli Rep in vitro is 23 ± 3 bp s^{-1} (25° C) and 8 ± 2 bp s^{-1} (15° C) (pH 7.5, 6 mM NaCl, 1.7 mM $MgCl_2$, 10% glycerol, 1.5 mM ATP) (143).

The processivity of DNA unwinding is related to the probability that the helicase will translocate to and unwind the next bp (or n bps if the helicase unwinds multiple bp per catalytic event; i.e. with a "step size" of n bp) relative to the probability that the helicase will dissociate from the DNA. Processivity can be described quantitatively in either of two related ways. Processivity can be measured as the average number of bp, N, unwound per helicase binding event, or as the probability, P, that the helicase will unwind the next bp (or n bp) rather than dissociate from the DNA; P has limits of $0 \leq P \leq 1$ (160, 161). A distributive helicase ($P=0$ or $N=n$) unwinds only n bp per binding event, whereas a hypothetical helicase with $P=1$ would never dissociate and thus would be capable of unwinding an infinitely long duplex. For a helicase that unwinds with a step size of n bp, the processivity, P, can be written in terms of k_u, the macroscopic rate constant (bp s^{-1}) for unwinding n bp, and k_d (s^{-1}), the macroscopic rate constant for helicase dissociation from the DNA, and is related to N as in equation (1), such that $k_u/k_d =(N-n)$.

$$P = (N-n)/N = k_u/(nk_d + k_u).$$ 1.

If only 1 bp is unwound in each step (i.e. $n =1$ bp), equation (1) simplifies to equation (2):

$$P = k_u/(k_d + k_u) = (N-1)/N.$$ 2.

The only quantitative study of DNA unwinding processivity has been carried out for *E. coli* RecBCD (162). DNA unwinding was examined in the presence of excess *E. coli* SSB protein, by monitoring the quenching of the SSB tryptophan fluorescence upon its binding to the ss DNA formed during unwinding. The effects of temperature, NaCl, Mg^{2+}, Ca^{2+}, and ATP concentrations were examined. Although the quantitative effects of solution conditions on RecBCD processivity will likely differ from those found for other helicases, the qualitative effects on processivity are likely to be general. The RecBCD helicase has a highly processive DNA unwinding reaction. At 25° C (pH 7.5, 1 mM Mg^{2+}, 30 mM NaCl, 1 mM ATP), RecBCD unwinds an average of $N=30\pm3$ kilobase pairs (kbp) before dissociating, corresponding to $P= 0.99997$ (assuming an unwinding step size of $n=1$). Above 100 mM NaCl, processivity decreases with increasing [NaCl], such that $N=15$ kbp ($P=0.99993$) at 0.275 M NaCl (25° C). As the [Mg^{2+}] is raised from 1 to 10 mM, N also decreases from ~30 kbp to ~18 kbp (162). The decrease in processivity with increasing salt concentration results, at least in part, from an increase in the protein-DNA dissociation rate constant, k_d, since increased salt concentration increases k_d generally for protein–nucleic acid interactions (47). The processivity of RecBCD unwinding is also influenced by [ATP], with N decreasing hyperbolically for [ATP] below 0.2 mM; N is half maximum at 41 ± 9 μM ATP (162). Whether these effects reflect changes in the rate constants for DNA unwinding, helicase translocation, or dissociation from the DNA is not known.

The *E. coli* DnaB helicase (139, 157, 159) and Helicase I (*E. coli* F factor TraI) also unwind DNA with relatively high processivities, although these have not been examined quantitatively (146, 163, 164). *E. coli* Helicase II (165) and the phage T4 Dda helicase (164, 165) appear to unwind DNA with considerably lower processivities in vitro. The *E. coli* Rep helicase is an example of a helicase whose processivity seems to be influenced by interactions with other proteins. The Rep protein, in the presence of the phage ϕX174 gene A protein or the phage f1 gene II protein, can unwind duplex DNA as long as 7000 bp and thus must possess high processivity (136, 166). However, in the absence of either of these accessory proteins, the Rep's DNA unwinding processivity is significantly lower (167, 168). In fact, the ϕX174 gene A protein, which introduces a nick into the plus strand of the ϕX174 duplex RF DNA and remains covalently attached to the 5' end of the nick (166), appears to interact directly with the Rep protein (169).

Thermodynamic Efficiency of DNA Unwinding

Any biological process that is vectorial or directional in nature must be coupled to a net expenditure of energy. For helicase-catalyzed DNA unwinding, this coupling occurs in the form of NTP binding and hydrolysis; nonhydrolyzable NTP analogues (e.g. AMPPNP), although capable of binding to the helicase, do not support continuous DNA unwinding, but partial unwinding of only a few bp that is coupled only to NTP binding has not been ruled out. A fundamental characteristic of such processes is their thermodynamic efficiency, which for helicase-catalyzed DNA unwinding is defined as the net free energy change needed to unwind one bp, and it can be related to the average number of ATP molecules hydrolyzed per bp unwound. The general aspects of free energy transduction in vectorial biological processes have been discussed (26, 27, 170, 171) and Hill & Tsuchiya (25) provide a useful discussion of vectorial coupling of NTP hydrolysis to DNA unwinding and helicase translocation.

For helicase-catalyzed DNA unwinding, two cases need to be considered. The first is a processive helicase functioning in the absence of a helix-destabilizing protein (HDP). The second is a helicase that functions in the presence of a HDP (or excess helicase) such that the resulting single strands become bound with protein. In the absence of a HDP, the free energy change associated with ATP hydrolysis ($\Delta G_{ATP} < 0$) provides the sole thermodynamic force for unwinding and translocation. That is, after the helicase undergoes a full cycle of unwinding, the only net change in the energetics of the system is that r ATP molecules have been hydrolyzed and N bp have been separated. If ΔG_{bp} is the free energy change associated with melting one bp ($\Delta G_{bp} > 0$), then the thermodynamic constraint for the net unwinding of N bp is: $r(-\Delta G_{ATP}) > N\Delta G_{bp}$. Under conditions relevant to *E. coli* (pH 7, 3 mM Mg^{2+}, 0.2 M K^+, 8 mM ATP, 1 mM ADP, 8 mM P_i), $\Delta G_{ATP} \sim -13$ kcal mol^{-1} at 25° C (171, 172). Since the average ΔG_{bp} for B-form DNA is ~1–1.5 kcal mol^{-1} bp (33), then hydrolysis of one ATP could "fuel" the unwinding of up to ~9–12 bp if coupling is 100% efficient; this efficiency will be reduced if "slippage" occurs [i.e. ATP hydrolysis that is uncoupled to DNA unwinding (25, 27)]. However, in the presence of a HDP (or nonspecific binding of another helicase molecule), an additional favorable free energy change associated with the binding of x proteins to the ss DNA will be contributed, and the thermodynamic constraint becomes: $x(-\Delta G_{HDP}) + r(-\Delta G_{ATP}) > N(\Delta G_{bp})$, where ΔG_{HDP} is the free energy change per mole of HDP bound to the ss DNA. As emphasized above, these ΔG_i are dependent upon solution conditions and concentrations.

Only a few estimates of the efficiency of DNA unwinding have been reported. For the E. coli Rep helicase in complex with the phage ϕX174 gene A protein and in the presence of the E. coli single stranded DNA binding (SSB) protein, an HDP (173), the thermodynamic efficiency was estimated to be 1–2.6 ATP hydrolyzed per bp unwound (136, 174, 175). Similar estimates of 2–3 ATP bp^{-1} (176) and 1.7–2.3 ATP bp^{-1} (95) have been made for E. coli RecBCD, also in the presence of excess E. coli SSB protein. The efficiency of ATP hydrolysis has been estimated to be 1.3–1.5 ATP bp^{-1} for RecBC enzyme (without the RecD subunit) and 1.1–1.2 ATP bp^{-1} for RecBCD-K177Q, which contains a mutation in the putative ATP binding site of the RecD subunit (95). These values differ considerably from the maximum theoretical efficiencies of ~0.083–0.11 ATP per bp discussed above. Although a molecular interpretation is not possible at this time, these results suggest that substantial uncoupled ATP hydrolysis accompanies DNA unwinding by these enzymes under these conditions in vitro.

MECHANISMS OF DNA UNWINDING AND TRANSLOCATION

Although all DNA helicases do not appear to unwind DNA by precisely the same mechanism, certain features of DNA unwinding and translocation are likely to be common to all helicases. One major similarity is that the functionally active forms of the DNA helicases appear generally to be oligomeric, with each subunit possessing a potential DNA binding site. In fact, the proposal has been made that DNA helicases are likely to function as oligomeric enzymes (6, 29, 57) since most molecular mechanisms of DNA unwinding and translocation that have been proposed require the active helicase to possess at least two DNA binding sites (6, 29).

Only two molecular mechanisms for helicase-catalyzed duplex nucleic acid unwinding and/or protein translocation have been proposed at this time: one for E. coli Rep, a dimeric DNA helicase (28), and the other for E. coli Rho, a hexameric RNA/DNA helicase (62). Much of the mechanistic information on helicase-catalyzed DNA unwinding has been obtained from studies of the dimeric Rep helicase, and Wong & Lohman (28) have proposed a detailed mechanism for how this helicase may function. Although low-resolution structural information is available for some of the hexameric helicases, the kinetic and thermodynamic studies needed to derive mechanistic information is only beginning to emerge. However, at least some of these hexamers appear to be formed as trimers of dimers, thus the information obtained from studies of the functional Rep dimer may also facilitate mechanistic studies of the hexamers. The other helicase for which equilibrium and kinetic information has been used to draw mechanistic inferences is the *E. coli* Rho RNA helicase and Geiselmann et al (62) have proposed a mechanism for translocation of this hexameric helicase along ss RNA. However, since the function of Rho is very different from that of the hexameric DNA helicases involved in DNA metabolism, it is too early to judge the generality of the conclusions reached with Rho.

Active vs Passive Mechanisms of DNA Unwinding

In general, mechanisms of helicase-catalyzed DNA unwinding can be classified as either "active" or "passive." In an active mechanism, the helicase plays a direct role in destabilizing the duplex DNA. However, a helicase that functions by a passive mechanism would facilitate unwinding indirectly by binding to the ss DNA that becomes available through transient fraying of the duplex caused by thermal fluctuations at the ss /ds-DNA junction. In a passive mechanism, as depicted in Figure 4, ss DNA that is formed transiently at the ss/ds DNA junction is trapped by the translocating helicase; net unwinding would be coupled to translocation with a step size of one or at most a few bp. A passive mechanism would require the helicase to interact only with ss DNA

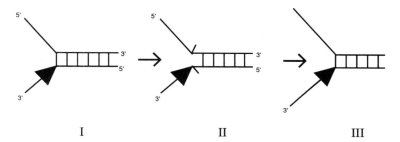

I   II   III

Figure 4 Passive mechanism for helicase-catalyzed DNA unwinding. In this mechanism the helicase, represented by a triangle, interacts only with ss DNA and translocates uni-directionally or with biased directionality in the 3′-to-5′ direction along ss DNA. The helicase does not interact with the duplex, but rather waits at the ss/ds-DNA junction until a bp becomes unpaired transiently because of thermal fluctuations and then translocates onto the newly formed ss DNA.

using a single site, but would require the helicase to translocate unidirectionally along ss DNA towards the duplex. However, a helicase with multiple DNA binding sites used for translocation along ss DNA could still unwind DNA using a passive mechanism. Direct evidence does not support a passive mechanism, although direct evidence does indicate that both the *E. coli* Rep (60) and the *E. coli* UvrD (Helicase II) helicases (J Ali & T Lohman, unpublished observations) do not unwind DNA by passive mechanisms (see below).

An active unwinding mechanism requires the functional helicase to possess at least two DNA binding sites, and in principle it can be of two types. In one type, the helicase would interact directly with the ds DNA at the junction (at least transiently) and actively destabilize some number of bp, presumably through conformational changes in the helicase that are triggered by NTP binding, hydrolysis, or product release. This type of active mechanism requires the helicase to bind to both ss and ds DNA and to both simultaneously during at least one intermediate stage. Sub-classes of this type of active mechanism are the "inch-worm" (25, 136) and "rolling" models (28) (see below). A second type of active mechanism is a "torsional" model in which the helicase does not interact with duplex DNA, but binds simultaneously to both of the single strands at the ss/ds-DNA junction and unwinds by distorting the adjacent duplex region through an NTP-induced conformational change (6).

An active, inch-worm mechanism was proposed on the basis of studies of *E. coli* Rep and Helicase III (136). This mechanism requires that the functional helicase possess two non-identical DNA binding sites that bind with polarity to ss DNA; the leading site (H) which will interact with the duplex to be unwound can bind both duplex and ss DNA, whereas the tail site (T) need bind only ss DNA. In the inch-worm model, the same site on the protein (H-site) always interacts with the duplex DNA during successive unwinding cycles. The inchworm model was proposed for Rep-catalyzed DNA unwinding (136) before it was recognized that Rep forms a homo-dimer upon binding DNA.

E. coli Rep Helicase

The best studied DNA helicase, with regard to its mechanism of DNA unwinding, is the homo-dimeric *E. coli* Rep helicase (6, 28, 29, 57, 59, 60, 115, 116, 143). Wong & Lohman (28) have proposed an "active, rolling" mechanism for DNA unwinding and translocation by the Rep dimer, shown schematically in Figure 5. This mechanism is based on the observations that DNA binding induces the Rep monomer to dimerize, that a chemically cross-linked dimer is functionally active (57), and that nucleotides influence allosterically the ss and ds DNA binding affinities and thus the five different DNA ligation states of the Rep dimer (28, 59). This mechanism requires at least two identical

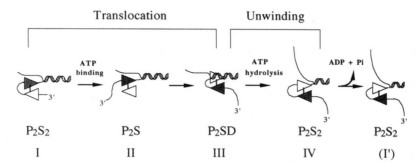

Figure 5 "Active, rolling" mechanism for dimeric Rep-catalyzed DNA unwinding. The dimeric Rep helicase is shown with triangular subunits assumed to be related by C_2 symmetry. The two Rep protomers are distinguished (open vs stippled) in order to indicate how the positioning of each protomer changes during the unwinding cycle. Modified from (28).

DNA binding sites, both of which must be able to bind ss DNA as well as ds DNA as is observed for Rep (28, 59), and could apply to any homo-oligomeric helicase with these properties. The model assumes that the Rep dimer possesses C_2 symmetry and incorporates the observation that each subunit of the Rep dimer binds ss DNA with defined polarity with respect to its sugar-phosphate backbone (M Amaratunga & T Lohman, unpublished observations). In the case of Rep, a non–base paired 5' single strand is neither necessary nor does it facilitate DNA unwinding in vitro (60); therefore only intermediates in which Rep is bound to the 3'-ss DNA are functionally relevant for DNA unwinding.

In an active, rolling mechanism, at least one subunit of the Rep dimer is always bound to the 3'-ss DNA at the fork, while the other subunit can be bound either to the same single strand or to the duplex region ahead of the fork. However, each subunit of the dimer alternates binding to ss DNA and ds DNA, as controlled allosterically by ATP and ADP binding. Thus, the active, rolling model differs from an inch-worm model, which also requires two DNA binding sites, but one binding site always binds ss DNA while the other always binds ds DNA (136). Figure 5 starts with intermediate I in which both subunits of the Rep dimer are bound to the 3' single strand in a P_2S_2 complex. On binding ATP, the affinity of one subunit of the dimer decreases for ss DNA, whereas its affinity increases for duplex DNA (28), resulting in formation of intermediate III (P_2SD) [by way of a P_2S complex (intermediate II)]. In the P_2SD complex, the Rep dimer is bound simultaneously to the duplex ahead of the fork as well as the 3'-ss DNA strand. In the next step (leading to intermediate IV), hydrolysis of ATP by the P_2SD complex induces protein conformational changes that destabilize multiple bp within the duplex DNA, which results in displacement of the 5' single strand while the Rep subunit remains bound to the 3' strand, yielding a P_2S_2 complex, after having unwound with a step size of $n \gg 1$. Subsequent release of ADP and inorganic phosphate completes the cycle by forming intermediate I', which is functionally equivalent to intermediate I because Rep is a homodimer.

In this model, translocation of the Rep dimer (steps I to III) is coupled to ATP binding and occurs by a rolling mechanism, whereas DNA unwinding (steps III to IV) is coupled to ATP hydrolysis. Furthermore, since the functional Rep dimer always remains bound to the 3'-ss DNA through at least one subunit, this mechanism provides a simple means for maintaining highly processive unwinding. The model also predicts that translocation and unwinding occur with a step size that is comparable to the site size of the protein (~14–16 nt) (57, 59) (rather than one nucleotide at a time) and consequently that multiple bp are unwound per catalytic event (i.e. an unwinding step size, $n > 1$). Although this prediction appears to be in conflict with estimates that 1–3 ATP molecules are hydrolyzed per bp unwound (136, 174, 175), it can be explained if coupling of ATP hydrolysis to DNA unwinding is less than 100% efficient. Such lower efficiency could result from a rolling mechanism because Rep dimers do hydrolyze ATP efficiently when bound as a P_2S complex (intermediate II); hence uncoupled ATP hydrolysis could occur at this stage of the cycle.

One important aspect of the rolling model is that Rep dimer translocation along ss DNA alone is predicted to occur without directional bias owing to the fact that the ss DNA regions on either side of a bound Rep dimer are equivalent. In a rolling model for a dimeric helicase, biased directional movement is predicted to occur only during duplex DNA unwinding owing to the presence of a duplex region on one side of the helicase which breaks the symmetry on either side of the helicase (6, 28, 29). One observation that argues strongly that Rep does not unwind DNA by simply binding to ss DNA and translocating unidirectionally in a 3'-to-5' direction is the fact that Rep can unwind an 18 bp duplex DNA possessing only a $5'-(dT)_{20}$ flanking region (60).

The kinetics of Rep-catalyzed DNA unwinding have been studied using both rapid quench-flow (60) and stopped-flow fluorescence (143) approaches. Oligodeoxynucleotide substrates were used that possessed duplex regions of either 18 or 24 bp along with 3'- and/or $5'-(dT)_{20}$ ss DNA tails. Under single turnover conditions ([Rep] $\gg$ [DNA]), the kinetics of unwinding are observed to be biphasic, the first phase representing rapid unwinding of pre-formed Rep-DNA complexes and the second reflecting slower unwinding that is limited by the rate of dimerization of the Rep protein on the DNA substrate (60, 143). However, for substrates possessing a 24 bp duplex, only those with a 3' flanking–$(dT)_{20}$ tail display the rapid phase of unwinding reflecting Rep's requirement for a 3'–ss DNA flanking region; a $5'-(dT)_{20}$ tail neither supports nor facilitates unwinding, even in the presence of a 3'-ss DNA tail (60). An apparent unwinding rate of 1.5 s^{-1} (23±3 bp s^{-1}) was measured for the unwinding of both an 18 bp and a 24 bp duplex possessing a 3'-(dT)$_{20}$ tail (25° C, 20 mM Tris, pH 7.5, 1.7 mM Mg^{2+}, 10% glycerol) (143). The rate of DNA unwinding at 25°C [~1.5 s^{-1} (143)] is comparable to the k_{cat} for ATP hydrolysis by P$_2$D but much slower than the ATPase rate for P$_2$S (190 s^{-1} per dimer). However, the apparent K_m for ATP during unwinding (30 µM at 25°C) is comparable to that observed with P$_2$S under the same conditions (~20 µM). This finding suggests that while binding (and possibly hydrolysis) of ATP in both subunits of a P$_2$SD complex (see Figures 2 and 5) is required for unwinding, the kinetics of unwinding may be limited by the rate of ATP hydrolysis by the Rep subunit bound to the duplex.

Rep's ability to unwind DNA by a passive mechanism has also been tested. In these experiments, a series of novel, nonnatural duplex DNA substrates (18 or 24 bp) possessing 3' flanking–ss DNA within which was embedded either a segment of ss DNA (four or five nucleotides) in which the backbone polarity was reversed or a non-DNA (poly-ethylene glycol) spacer as shown in Figure 4. Either of these segments should block unwinding by a 3'-to-5' helicase that functions by simply translocating unidirectionally along ss DNA in the 3'-to-5' direction and unwinding the duplex using a passive mechanism. The E. coli Rep helicase effectively unwinds these DNA substrates, ruling out a strictly passive mechanism of unwinding. However, the results are consistent with an active, rolling mechanism during which a P$_2$SD intermediate can form in such a way that Rep binds to ss and duplex DNA simultaneously, forming a looped region as in Figure 4B, which can bypass the "block" (60). These studies provide support for the active, rolling mechanism proposed for the dimeric Rep helicase (28).

In principle, any oligomeric helicase, displaying either 3'-to-5' or 5'-to-3' polarity, could operate by a rolling mechanism as long as the helicase possesses at least two DNA binding sites that can alternate binding of ss and ds DNA. A modification of the rolling model has also been proposed for the RecBCD helicase to account for the interaction of the enzyme with both DNA strands (120). That model assumes that the active form of RecBCD is a hexamer [(BCD)$_2$], an assumption for which there is some experimental support (177).

E. coli UvrD (Helicase II)

The oligomerization and DNA binding properties of E. coli UvrD protein (Helicase II) have not been examined in as much detail as they have for the Rep helicase. However, UvrD can form at least a dimer, and dimerization increases its specific steady-state ATPase activity (68, 69). Single turnover kinetics, similar to those performed with the Rep helicase (60, 143), have been used to examine the mechanism of UvrD-catalyzed unwinding of short duplex

DNA possessing 3'- and/or 5'-$(dT)_N$ tails (J Ali & T Lohman, unpublished experiments). Upon addition of ATP to pre-formed UvrD-DNA complexes, unwinding of a duplex DNA possessing a 3'–ss DNA tail occurs in a rapid phase reflecting unwinding by productive UvrD-DNA complexes. As with Rep helicase, a 5'-ss DNA tail is not sufficient for unwinding, and it does not enhance the rate of unwinding of duplex DNA that possesses a 3'–ss DNA tail. UvrD is also able to unwind "reverse polarity" DNA substrates similar to those used with Rep (60) (see Figure 4), indicating that UvrD does not unwind DNA by a strictly passive mechanism (J Ali & T Lohman, unpublished observations). These results support a mechanism in which the functional form of UvrD is an oligomer and a model in which unwinding occurs by an active, rolling mechanism similar to that proposed for the Rep dimer (28, 60).

Hexameric DNA Helicases

The gross structural similarities among the ring-like hexameric forms of SV40 T antigen, RuvB, T7 gene 4, T4 gene 41, DnaB, and Rho helicases, along with the evidence that DNA can bind through the center of some of these hexamers, has led to the suggestion that these hexamers may encircle the nucleic acid while functioning as helicases. Although such a binding mechanism would increase the processivity of the helicase by decreasing its rate of dissociation from the DNA, whether DNA is encircled by the hexamers as they are actively unwinding DNA is not known. Furthermore, no consensus has yet emerged on whether all of these hexamers encircle DNA or even on which form of DNA is encircled. For example, although the RuvB (54) and large T antigen (74) hexamers can encircle duplex DNA, the T7 gene 4A' hexamer appears able to encircle only ss DNA (56). In contrast, ss RNA appears to interact with all subunits and wrap around the perimeter of the Rho hexamer (62). An alternative possibility is that during DNA unwinding, the DNA is encircled by the helicase, while the unwound ss DNA (or RNA) also wraps around the hexamer. However, even if hexameric helicases do encircle the DNA during helicase activity, how the helicase unwinds DNA or translocates remains unexplained.

Due to the lack of quantitative information about stoichiometries, energetics, and kinetics of both DNA binding and DNA unwinding, detailed mechanisms have not been proposed for the hexameric DNA helicases. However, the following features may be common to such mechanisms. First, the hexameric "ring" helicases possess multiple nucleotide binding sites (84, 122, 123) and, in principle, multiple DNA binding sites. Second, there is evidence that DNA (ss and/or ds) can pass through the center of some of these hexamers (at least under some circumstances). Based on these features and the assumption that some of the mechanistic conclusions drawn from studies of the Rep dimer helicase can be applied to the hexamers, an active, rolling mechanism similar to that suggested for Rep (28) can be proposed. In a 3′-to-5′ hexameric helicase, one subunit (A) of the hexamer could interact with the 3′ ss DNA, while a second subunit (B) could bind simultaneously to the ds DNA to form the equivalent of the P_2SD complex as proposed for the Rep dimer (28). ATP hydrolysis (or ATP binding) could be coupled to unwinding of the duplex resulting in a P_2S_2 complex with the 3′–ss DNA tail bound to both subunits A and B. ADP release and subsequent ATP binding would then result in release of ss DNA from subunit A and concomitant binding of a third subunit (C) to the duplex DNA to reform a P_2SD complex. Therefore, each dimer within the hexamer functions as a helicase in much the same manner as proposed for the Rep dimer. The main difference is that for the Rep dimer, the same subunit that dissociates from ss DNA then binds to the duplex, whereas for a hexameric helicase, a third subunit (C) binds the duplex after subunit A dissociates from the ss DNA.

Another possibility is that one strand of the DNA is encircled by the hexamer, increasing the processivity of unwinding by inhibiting dissociation of the hexamer, but that the unwinding activity occurs through separate interactions with the complementary single strand and the duplex. Although speculative, this model incorporates most of the observations that have been made for hexameric DNA helicases as well as some features of the Rep dimer. Such a model may also explain why initiation of unwinding by many hexameric helicases, including DnaB (139), T7 gene 4 (138), and T4 gene 41 (140), appears to be facilitated by DNA possessing a forked DNA substrate, i.e. DNA possessing both a flanking 3′– and a 5′–ss DNA tail. On the other hand, the SV40 large T antigen appears not to require a 5′–ss DNA tail to unwind DNA efficiently (142). If any helicase uses such a mechanism or any other wherein it interacts with both single strands during unwinding, then the directionality of unwinding is ambiguous. Such a mechanism also precludes an unambiguous determination of whether the helicase functions on either the leading or lagging strand.

E. coli Rho Protein

Although *E. coli* Rho protein is not a DNA helicase, we discuss this protein because it unwinds duplex RNA and RNA-DNA hybrids (19), it appears to function as a hexamer, and its physical properties have been examined in detail, along with its interaction with ss RNA and nucleotides. Furthermore, von Hippel and colleagues (62) have proposed a detailed "rolling" mechanism to explain how a hexameric helicase might utilize ATP binding and hydrolysis to translocate with biased directionality along ss RNA in the 5′-to-3′ direction, although evidence for uni-directional translocation is lacking. However, since

Rho is only required to unwind short stretches of RNA/DNA duplexes, its helicase mechanism may differ from those of DNA helicases that function in replication with high processivity.

The Rho hexamer is arranged as a trimer of asymmetric dimers with D_3 symmetry, and ss RNA appears able to interact with all six subunits and thus "wrap" around the hexamer. Geiselmann et al (62) have proposed a model in which the functional unit of the hexamer is the "dimer" that undergoes conformational transitions driven by ATP binding and hydrolysis. In this model ATP hydrolysis drives release of the 5'-ss RNA from one dimer, followed by binding of the 3' ss RNA to the same dimer to yield 5'-to-3' translocation. This cycle is then repeated for the next "dimer". This type of rolling model is similar in some respects to the rolling mechanism of translocation along ss DNA proposed for the *E. coli* Rep dimer (28). However, the Rep dimer would translocate without directional bias along ss DNA (6, 29, 60). In principle, a helicase needs at least three nucleic acid binding sites (trimeric) in order to translocate with biased directionality by the mechanism proposed by Geiselmann et al (62), and this may provide one explanation for why some helicases are hexameric. In this model, RNA does not go through the center of the hexameric ring, although such a feature could easily be accommodated if the RNA binding sites were on the inside of the hexamer. Although Geiselmann et al (62) propose this model only to explain a biased (5' to 3') directional translocation of Rho along ss RNA, they argue that the helicase activity of Rho could result if Rho binds to a segment of ss RNA that becomes transiently exposed owing to thermal fluctuations in the duplex (a passive mechanism in our nomenclature).

SUMMARY

An understanding of how ATP binding and hydrolysis are coupled to DNA unwinding and helicase translocation at the molecular level requires quantitative studies of the energetics and kinetics of binding of DNA and nucleotides (and their hydrolysis) in addition to structural information. In particular, transient kinetic studies are needed to examine the intermediate helicase-DNA-nucleotide complexes that are functionally important for DNA unwinding. Such mechanistic studies of DNA helicases are still at an early stage, and current information is insufficient to define the detailed mechanism of DNA unwinding for any one DNA helicase. Based on current information it seems unlikely that all helicases function by the same mechanism. In fact, whether helicases involved in replication use different mechanisms than repair or recombinational helicases and what mechanistic aspects are shared between dimeric and hexameric helicases remains to be seen.

Even though detailed studies of several helicases will be needed to determine which mechanistic aspects are general, several features appear to be common to the few helicases that have been examined. First, they are all oligomeric, which results in the potential for multiple nucleotide and DNA binding sites on the functional helicase. Both nucleotide and DNA binding also display a functional negative cooperativity. Some helicases also bind duplex DNA as well as ss DNA suggesting that these may function by active mechanisms. Evidence is also sufficient to suggest that rolling mechanisms are likely to be involved in helicase translocation and DNA unwinding, although not all sub-units of the helicase need to contact the DNA simultaneously. Although the elementary steps of DNA unwinding are still to be mapped out, the most challenging problem remains: that of deciphering the energetic communications that are central to the energy transduction process. Answers to these questions are also being pursued with the classical motor proteins, such as myosin, kinesin (178–180), and dynein (181). The ability to begin to address these questions with DNA motor proteins using well defined and more easily manipulatable DNA substrates should greatly facilitate progress in this exciting area.

Acknowledgments

The preparation of this review was supported in part by grants from the American Cancer Society (NP-756) and the NIH (GM45948). We thank all those who shared manuscripts prior to publication and the past and present members of the laboratory for their many contributions and insights into our studies of DNA helicases. We particularly thank Janid Ali, Razmic Gregorian, John Hsieh, Keith Moore, and Isaac Wong for their comments on this review. We also thank Wlodek Bujalowski and Smita Patel for their comments on this review.

Literature Cited

1. Moore KJM, Lohman TM. 1995. *Biophys. J.* 68:S180–85
2. Matson SW, Kaiser-Rogers KA. 1990. *Annu. Rev. Biochem.* 59:289–329
3. Matson SW. 1991. *Prog. Nucleic Acids Res. Mol. Biol.* 40:289–326
4. Thommes P, Hubscher U. 1990. *FEBS Lett.* 268:325–28
5. Thommes P, Hubscher U. 1992. *Chromosoma* 101:467–73
6. Lohman TM. 1992. *Mol. Microbiol.* 6: 5–14
7. Matson SW, Bean DW, George JW. 1994. *BioEssays* 16:13–22
8. Li X, Yoder BL, Burgers PMJ. 1992. *Chromosoma* 102:S93–S99
9. Abdel-Monem M, Durwald H, Hoffmann-Berling H. 1976. *Eur. J. Biochem.* 65:441–49
10. Sancar A. 1994. *Science* 266:1954–56

11. Hanawalt PC. 1994. *Science* 266:1957–58
12. Friedberg EC. 1992. *Cell* 71:887–89
13. Tanaka K, Wood RD. 1994. *Trends Biochem. Sci.* 19:83–86
14. Flores-Rozas H, Hurwitz J. 1993. *J. Biol. Chem.* 268:21372–83
15. Lee C-G, Hurwitz J. 1993. *J. Biol. Chem.* 268:16822–30
16. Lee C-G, Hurwitz J. 1992. *J. Biol. Chem.* 267:4398–407
17. Hirling H, Scheffner M, Restle T, Stahl H. 1989. *Nature* 339:562–64
18. Ray BK, Lawson TG, Kramer JC, Cladaras MH, Grifo JA, et al. 1985. *J. Biol. Chem.* 260:7651–58
19. Brennan CA, Dombroski AJ, Platt T. 1987. *Cell* 48:945–52
20. Matson SW. 1989. *Proc. Natl. Acad. Sci. USA* 86:4430
21. Bailly V, Sung P, Prakash L, Prakash S. 1991. *Proc. Natl. Acad. Sci. USA* 88:9712–16
22. Naegeli H, Bardwell L, Harosh I, Friedberg EC. 1992. *J. Biol. Chem.* 267:7839–44
23. Scheffner M, Knippers R, Stahl H. 1991. *Eur. J. Biochem.* 195:49–54
24. Schmid SR, Linder P. 1992. *Mol. Microbiol.* 6:283–92
25. Hill TL, Tsuchiya T. 1981. *Proc. Natl. Acad. Sci. USA* 78:4796–800
26. Jencks WP. 1980. *Adv. Enzymol. Relat. Areas Mol. Biol.* 51:75–106
27. Jencks WP. 1989. *J. Biol. Chem.* 264:18855–58
28. Wong I, Lohman TM. 1992. *Science* 256:350–55
29. Lohman TM. 1993. *J. Biol. Chem.* 268:2269–72
30. Geider K, Hoffmann-Berling H. 1981. *Annu. Rev. Biochem.* 50:233–60
31. Hagerman PJ. 1988. *Annu. Rev. Biophys. Biophys. Chem.* 17:265–86
32. Bloomfield VA, Crothers D, Tinoco I Jr. 1974. *Physical Chemistry of Nucleic Acids.* New York: Harper & Row
33. Record MT Jr, Mazur SJ, Melancon P, Roe J-H, Shaner SL, Unger L. 1981. *Annu. Rev. Biochem.* 50:997–1024
34. Record MT Jr. Anderson CF, Lohman TM. 1978. *Q. Rev. Biophys.* 11:103–78
35. Record MT Jr. 1975. *Biopolymers* 14:2137–58
36. Record MT Jr, Woodbury CP, Lohman TM. 1976. *Biopolymers* 15:893–915
37. Manning GS. 1972. *Biopolymers* 11:937
38. Manning GS. 1972. *Biopolymers* 11:951
39. Record MT Jr. 1972. *Biopolymers* 11:1435–84
40. Wetmur JG, Davidson N. 1968. *J. Mol. Biol.* 31:349–70
41. Wetmur JG. 1991. *Crit. Rev. Biochem. Mol. Biol.* 26:227–59
42. Record MT Jr, Lohman TM. 1978. *Biopolymers* 17:159–66
43. Olmsted MC, Anderson CF, Record MT Jr. 1989. *Proc. Natl. Acad. Sci. USA* 86:7766–70
44. Blake RD, Delcourt SG. 1990. *Biopolymers* 29:393–405
45. Record MT Jr, Ha J-H, Fisher MA. 1991. *Methods Enzymol.* 208:291–343
46. Lohman TM, Mascotti DP. 1992. *Methods Enzymol.* 212:400–24
47. Lohman TM. 1986. *CRC Crit. Rev. Biochem.* 19:191–245
48. Overman LB, Lohman TM. 1994. *J. Mol. Biol.* 236:165–78
49. Record MT Jr. Lohman TM, de Haseth PL. 1976. *J. Mol. Biol.* 107:145–58
50. Overman LB, Bujalowski W, Lohman TM. 1988. *Biochemistry* 27:456–71
51. Mastrangelo IA, Hough PVC, Wall JS, Dodson M, Dean FB, Hurwitz J. 1989. *Nature* 338:658–62
52. Mastrangelo IA, Bezanilla M, Hansma PK, Hough PVC, Hansma HG. 1994. *Biophys. J.* 66:293–98
53. Gogol EP, Seifried SE, von Hippel PH. 1991. *J. Mol. Biol.* 221:1127–38
54. Stasiak A, Tsaneva IR, West SC, Benson CJB, Yu X, Egelman EH. 1994. *Proc. Natl. Acad. Sci. USA* 91:7618–22
55. Dong F, Gogol EP, von Hippel PH. 1995. *J. Biol. Chem.* 270:7462–73
56. Egelman EH, Yu X, Wild R, Hingorani MM, Patel SS. 1995. *Proc. Natl. Acad. Sci. USA* 92:3869–73
57. Chao K, Lohman TM. 1991. *J. Mol. Biol.* 221:1165–81
58. Arai N, Arai KI, Kornberg A. 1981. *J. Biol. Chem.* 256:5287–93
59. Wong I, Chao KL, Bujalowski W, Lohman TM. 1992. *J. Biol. Chem.* 267:7596–610
60. Amaratunga M, Lohman TM. 1993. *Biochemistry* 32:6815–20
61. Arai K, Kornberg A. 1981. *J. Biol. Chem.* 256:5253–59
62. Geiselmann J, Wang Y, Seifried SE, von Hippel PH. 1993. *Proc. Natl. Acad. Sci. USA* 90:7754–58
63. Scheffner M, Knippers R, Stahl H. 1989. *Cell* 57:955–63
64. SenGupta DJ, Blackwell LJ, Gillette T, Borowiec JA. 1992. *Chromosoma* 102:S46–S51
65. Reha-Krantz LJ, Hurwitz J. 1978. *J. Biol. Chem.* 253:4043–50
66. Bujalowski W, Klonowska MM, Jezewska MJ. 1994. *J. Biol. Chem.* 269:31350–58
67. Patel SS, Hingorani MM. 1993. *J. Biol. Chem.* 268:10668–75

212   LOHMAN & BJORNSON

68. Wong I, Amaratunga M, Lohman TM. 1993. *J. Biol. Chem.* 268:20386–91
69. Runyon GT, Wong I, Lohman TM. 1993. *Biochemistry* 32:602–12
70. Washburn BK, Kushner SR. 1991. *J. Bacteriol.* 173:2569–75
71. George JW, Brosh RM Jr, Matson SW. 1994. *J. Mol. Biol.* 235:424–35
72. Lohman TM, Chao K, Green JM, Sage S, Runyon G. 1989. *J. Biol. Chem.* 264:10139–47
73. Liu CC, Alberts B. 1981. *J. Biol. Chem.* 256:2813–20
74. Dean FB, Borowiec JA, Eki T, Hurwitz J. 1992. *J. Biol. Chem.* 267:14129–37
75. Gibson TJ, Thompson JD. 1994. *Nucleic Acids Res.* 22:2552–56
76. West SC. 1994. *Cell* 76:9–15
77. Tsaneva IR, Muller B, West SC. 1993. *Proc. Natl. Acad. Sci. USA* 90:1315–19
78. Bernstein JA, Richardson CC. 1988. *J. Biol. Chem.* 263:14891–99
79. Bernstein JA, Richardson CC. 1988. *Proc. Natl. Acad. Sci. USA* 85:396–400
80. Bernstein JA, Richardson CC. 1989. *J. Biol. Chem.* 264:13066–73
81. Geiselmann J, Seifried SE, Yager TD, Liang C, von Hippel PH. 1992. *Biochemistry* 31:121–32
82. Finger LR, Richardson JP. 1982. *J. Mol. Biol.* 156:203–19
83. Geiselmann J, Yager TD, Gill SC, Calmettes P, von Hippel PH. 1992. *Biochemistry* 31:111–21
84. Geiselmann J, von Hippel PH. 1992. *Protein Sci.* 1:850–60
85. Geiselmann J, Yager TD, von Hippel PH. 1992. *Protein Sci.* 1:861–73
86. Seifried SE, Bjornson KP, von Hippel PH. 1991. *J. Mol. Biol.* 221:1139–51
87. McSwiggen JA, Bear DG, von Hippel PH. 1988. *J. Mol. Biol.* 199:609–22
88. Hodgman TC. 1988. *Nature* 333:22–23
89. Lane D. 1988. *Nature* 334:478
90. Gorbalenya AE, Koonin EV. 1993. *Curr. Opin. Struct. Biol.* 3:419–29
91. Walker JE, Saraste M, Runswick MJ, Gay NJ. 1982. *EMBO J.* 1:945–51
92. Schulz GE. 1992. *Curr. Opin. Struct. Biol.* 2:61–67
93. Sung P, Higgins D, Prakash L, Prakash S. 1988. *EMBO J.* 7:3263–69
94. Korangy F, Julin DA. 1992. *J. Biol. Chem.* 267:1727–32
95. Korangy F, Julin DA. 1994. *Biochemistry* 33:9552–60
96. Notarnicola SM, Richardson CC. 1993. *J. Biol. Chem.* 268:27198–207
97. Patel SS, Hingorani MM, Ng WM. 1994. *Biochemistry* 33:7857–68
98. Pause A, Sonenberg N. 1992. *EMBO J.* 11:2643–54
99. Zhu L, Weller SK. 1992. *J. Virol.* 66:469–79
100. Martinez R, Shao L, Weller SK. 1992. *J. Virol.* 66:6735–46
101. Chao K, Lohman TM. 1990. *J. Biol. Chem.* 265:1067–76
102. Kim S-H, Smith J, Claude A, Lin R-J. 1992. *EMBO J.* 11:2319–26
103. Wong I, Lohman TM. 1996. *Methods Enzymol.* 259:95–127
104. Giedroc DP, Khan R, Barnhart K. 1991. *Biochemistry* 30:8230–42
105. Arai K, Kornberg A. 1981. *J. Biol. Chem.* 256:5260–66
106. Das RH, Yarranton GT, Gefter ML. 1980. *J. Biol. Chem.* 255:8069–73
107. Wong I, Lohman TM. 1993. *Proc. Natl. Acad. Sci. USA* 90:5428–32
108. Hingorani MM, Patel SS. 1993. *Biochemistry* 32:12478–87
109. Bujalowski W, Jezewska MJ. 1995. *Biochemistry* 34:8513–19
110. Wang Y, von Hippel PH. 1993. *J. Biol. Chem.* 268:13947–55
111. Bear DG, Hicks PS, Ecudero KW, Andrews CL, McSwiggen JA, von Hippel PH. 1988. *J. Mol. Biol.* 199:623–35
112. Wang Y, von Hippel PH. 1993. *J. Biol. Chem.* 268:13940–46
113. Nakayama N, Arai N, Kaziro Y, Arai K. 1984. *J. Biol. Chem.* 259:88–96
114. Bear DG, Andrews CL, Singer JD, Morgan WD, Grant RA, et al. 1985. *Proc. Natl. Acad. Sci. USA* 82:1911–15
115. Moore KJM, Lohman TM. 1994. *Biochemistry* 33:14550–64
116. Moore KJM, Lohman TM. 1994. *Biochemistry* 33:14565–78
117. Sen Gupta DJ, Borowiec JA. 1992. *Science* 256:1656–61
118. Wessel R, Schweizer J, Stahl H. 1992. *J. Virol.* 66:804–15
119. Taylor AF, Smith GR. 1985. *J. Mol. Biol.* 185:431–43
120. Ganesan S, Smith GR. 1993. *J. Mol. Biol.* 229:67–78
121. Stitt BL. 1988. *J. Biol. Chem.* 263:11130–37
122. Patel SS, Hingorani MM. 1995. *Biophys. J.* 68:S186–90
123. Bujalowski W, Klonowska MM. 1993. *Biochemistry* 32:5888–900
124. Biswas EE, Biswas SB, Bishop JE. 1986. *Biochemistry* 25:7368–74
125. Bujalowski W, Klonowska MM. 1994. *J. Biol. Chem.* 269:31359–71
126. Bujalowski W, Klonowska MM. 1994. *Biochemistry* 33:4682–94
127. Johnson KA. 1992. *Enzymes* 20:1–61
128. Wold MS, Li JJ, Kelly TJ. 1987. *Proc. Natl. Acad. Sci. USA* 84:3643–47
129. Runyon GT, Lohman TM. 1989. *J. Biol. Chem.* 264:17502–12

130. Runyon GT, Bear DG, Lohman TM. 1990. *Proc. Natl. Acad. Sci. USA* 87: 6383–87
131. Runyon GT, Lohman TM. 1993. *Biochemistry* 32:4128–38
132. Umezu K, Nakayama K, Nakayama H. 1990. *Proc. Natl. Acad. Sci. USA* 87: 5363–67
133. Washburn BK, Kushner SR. 1993. *J. Bacteriol.* 175:341–50
134. Modrich P. 1989. *J. Biol. Chem.* 264: 6597–600
135. Modrich P. 1994. *Science* 266:1959–60
136. Yarranton GT, Gefter ML. 1979. *Proc. Natl. Acad. Sci. USA* 76:1658–62
137. Venkatesan M, Silver LL, Nossal NG. 1982. *J. Biol. Chem.* 257:12426–34
138. Matson SW, Tabor S, Richardson CC. 1983. *J. Biol. Chem.* 258:14017–24
139. Lebowitz JH, McMacken RM. 1986. *J. Biol. Chem.* 261:4738–48
140. Richardson RW, Ellis RL, Nossal NG. 1990. *Molecular Mechanisms in DNA Replication and Recombination*, ed. C Richardson, IR Lehman, pp. 247–59. UCLA Symp. Mol. Cell. Biol. New York: Liss
141. Mendelman LV, Notarnicola SM, Richardson CC. 1993. *J. Biol. Chem.* 268: 27208–13
142. Wiekowski M, Schwarz MW, Stahl H. 1988. *J. Biol. Chem.* 263:436–42
143. Bjornson KP, Amaratunga M, Moore KJM, Lohman TM. 1994. *Biochemistry* 33:14306–16
144. Lee MS, Marians KJ. 1990. *J. Biol. Chem.* 265:17078–83
145. Matson SW, George JW. 1987. *J. Biol. Chem.* 262:2066–76
146. Lahue EE, Matson SW. 1988. *J. Biol. Chem.* 263:3208–15
147. Young MC, Kuhl SB, von Hippel PH. 1994. *J. Mol. Biol.* 235:1436–46
148. Young MC, Schultz DE, Ring D, von Hippel PH. 1994. *J. Mol. Biol.* 235: 1447–58
149. McGhee JD, von Hippel PH. 1974. *J. Mol. Biol.* 86:469–89
150. McGhee JD, von Hippel PH. 1976. *J. Mol. Biol.* 103:679
151. Lohman TM. 1983. *Biopolymers* 22: 1697–713
152. Kowalczykowski SC, Paul LS, Lonberg N, Newport JW, McSwiggen JA, von Hippel PH. 1986. *Biochemistry* 25: 1226–40
153. Brown WC, Romano LJ. 1989. *J. Biol. Chem.* 264:6748–54
154. Maine IP, Kodadek T. 1994. *Biochem. Biophys. Res. Commun.* 198: 1070–77
155. Naegeli H, Bardwell L, Friedberg EC. 1992. *J. Biol. Chem.* 267:392–98
156. Taylor A, Smith GR. 1980. *Cell* 22:447–57
157. Baker TA, Funnell BE, Kornberg A. 1987. *J. Biol. Chem.* 262:6877–85
158. Roman LJ, Kowalczykowski SC. 1989. *Biochemistry* 28:2863–73
159. Mok M, Marians KJ. 1987. *J. Biol. Chem.* 262:16644–54
160. McClure WR, Chow Y. 1980. *Methods Enzymol.* 64:277–97
161. Newport JW, Kowalczykowski SC, Lonberg N, Paul L, von Hippel PH. 1980. *Mechanistic Studies of DNA Replication and Genetic Recombination*, ed. B Alberts, pp. 485–505. New York: Academic
162. Roman LJ, Eggleston AK, Kowalczykowski SC. 1992. *J. Biol. Chem.* 267: 4207–14
163. Abdel-Monem M, Lauppe HF, Kartenbeck J, Durwald H, Hoffmann-Berling H. 1977. *J. Mol. Biol.* 110: 667–85
164. Kuhn B, Abdel-Monem M, Krell H, Hoffmann-Berling H. 1979. *J. Biol. Chem.* 254:11343–50
165. Abdel-Monem M, Durwald H, Hoffmann-Berling H. 1977. *Eur. J. Biochem.* 79:39–45
166. Eisenberg S, Griffith JD, Kornberg A. 1977. *Proc. Natl. Acad. Sci. USA* 74: 3198–202
167. Smith KR, Yancey JE, Matson SW. 1989. *J. Biol. Chem.* 264:6119–26
168. Yancey JE, Matson SW. 1991. *Nucleic Acids Res.* 19:3943–51
169. Sumida-Yasumoto C, Ikeda JE, Benz E, Marians KJ, Vicuna R, et al. 1978. *Cold Spring Harbor Symp. Quant. Biol.* 43: 311–29
170. Hill TL. 1977. *Free Energy Transduction in Biology.* New York: Academic
171. Simmons RM, Hill TL. 1976. *Nature* 263:615–18
172. Alberty RA. 1969. *J. Biol. Chem.* 244: 3290–302
173. Lohman TM, Ferrari ME. 1994. *Annu. Rev. Biochem.* 63:527–70
174. Kornberg A, Scott JF, Bertsch LL. 1978. *J. Biol. Chem.* 253:3298–304
175. Arai N, Kornberg A. 1981. *J. Biol. Chem.* 256:5294–98
176. Roman LJ, Kowalczykowski SC. 1989. *Biochemistry* 28:2873–81
177. Dykstra CC, Palas KM, Kushner SR. 1984. *Cold Spring Harbor Symp. Quant. Biol.* 49:463–67
178. Vale RD, Reese TS, Sheetz MP. 1985. *Cell* 42:39–50
179. Gilbert S, Webb MR, Brune M, Johnson KA. 1995. *Nature* 373:671–76
180. Hackney DD. 1994. *Proc. Natl. Acad. Sci. USA* 91:6865–69

181. Johnson KA. 1985. *Annu. Rev. Biophys. Biophys. Chem.* 14:161–88
182. Yarranton GT, Das RH, Gefter ML. 1979. *J. Biol. Chem.* 254:12002–6
183. Dash PK, Traxler BA, Panicker MM, Hackney DD, Minkley EG Jr. 1995. *Mol. Microbiol.* 6:1163–72
184. Boehmer PE, Emmerson PT. 1992. *J. Biol. Chem.* 267:4981–87
185. Oh EY, Grossman L. 1989. *J. Biol. Chem.* 264:1336–43
186. Seo Y-S, Lee S-H, Hurwitz J. 1991. *J. Biol. Chem.* 266:13161–70
187. Bruckner RC, Crute JJ, Dodson MS, Lehman IR. 1991. *J. Biol. Chem.* 266: 2669–74
188. Dodson MS, Lehman IR. 1991. *Proc. Natl. Acad. Sci. USA* 88:1105–9

Annu. Rev. Biochem. 1996. 65:215–39

MOLECULAR MECHANISMS OF DRUG RESISTANCE IN *MYCOBACTERIUM TUBERCULOSIS*

John S. Blanchard

Department of Biochemistry, Albert Einstein College of Medicine, Bronx, New York 10461

KEY WORDS: drug resistance, tuberculosis, antibacterial, drug action

ABSTRACT

In spite of forty years of effective chemotherapy for tuberculosis, the molecular mechanisms of antibacterial compounds in *Mycobacterium tuberculosis* have only recently been revealed. Broad spectrum antibacterials, including streptomycin, rifampicin, and fluoroquinolones have been demonstrated to act on the same targets in *M. tuberculosis* as they do in *E. coli*. Resistance to these agents results from single mutagenic events that lead to amino acid substitutions in their target proteins. The mechanisms of action of the unique antitubercular drugs, including isoniazid, ethambutol, and pyrazinamide have also recently been defined. Resistance to isoniazid can be caused either by mutations in the *katG*-encoded catalase-peroxidase, the enzyme responsible for drug activation, or by the molecular target, the *inhA*-encoded long chain enoyl-ACP reductase. Ethambutol appears to block specifically the biosynthesis of the arabinogalactan component of the mycobacterial cell envelope, and pyrazinamide has no known target. With the resurgence of tuberculosis and the appearance of strains which are multiply resistant to the above compounds, present tuberculosis chemotherapies are threatened. New approaches to the treatment of multi drug–resistant tuberculosis are needed.

CONTENTS

INTRODUCTION   216
RESISTANCE TO INHIBITORS OF PROTEIN SYNTHESIS   219
   Streptomycin and Related Inhibitors   219
RESISTANCE TO INHIBITORS OF NUCLEIC ACID SYNTHESIS.   221
   Rifampicin.   221
   Fluoroquinolones   223

215

RESISTANCE TO INHIBITORS OF CELL WALL SYNTHESIS 225
   Isoniazid and Ethionamide... 225
   Ethambutol 230
RESISTANCE TO OTHER ANTITUBERCULARS 232
   Pyrazinamide 232
ANTIBIOTIC RESISTANCE AND TUBERCULOSIS TREATMENT 233
   Future Studies and Solutions 235

INTRODUCTION

Tuberculosis is the world's leading cause of mortality owing to an infectious bacterial agent, *Mycobacterium tuberculosis*. Between 2 and 3 million people will die from the disease this year, some 8–10 million new cases will be reported, and estimates put the total number of infected individuals at 1700 million (1). Tuberculosis has an extraordinary impact on the economies of the developing world because the disease generally strikes individuals in their prime working years. Prior to World War II and the development of modern antibiotic therapy, the disease was fatal in 50–60% of all cases. However, the discovery of the antibacterial and antitubercular properties of streptomycin in 1944 (2), and both isoniazid and pyrazinamide in 1952 (3, 4), led to effective chemotherapies that decreased tuberculosis mortality rates both in the United States and worldwide. This trend continued from 1953, when national surveillance programs were instituted, to the mid-1980s, when the number of new reported cases began to increase in the United States. From 1985 to 1992, 52,000 excess cases of tuberculosis have been reported nationally. What is the cause of the resurgence of this ancient disease? Several explanations have been offered (1): The appearance of the AIDS virus in the early 1980s is certainly a contributing factor. The prevalence of tuberculosis in HIV-infected individuals is high, and the largest increases in tuberculosis nationally and internationally have occurred in urban areas that have a high percentage of HIV-infected individuals. Correspondingly, tuberculosis is one of the most common opportunistic infections in AIDS patients. Another reason for the increase is the deteriorating social structure within major national urban areas. Increases in the homeless population and declining health care structures and national surveillance have also contributed to the rise of tuberculosis. In many cases physician complacency in patient monitoring of tuberculosis chemotherapy can result in the appearance of drug-resistant tuberculosis.

Although the emergence of drug resistance is a serious health concern (5, 6), the appearance of multi drug–resistant strains of *M. tuberculosis* is particularly disturbing (1, 7), since few drugs are effective against tuberculosis. Among the broad spectrum antibacterials used against bacterial infection, only streptomycin and related protein-synthesis inhibitors, rifampicin, and fluoroquinolones are effective against *M. tuberculosis* in vitro and in vivo. Resistance to each of these agents has been reported, and drug resistance is encountered even in patients who have never been treated with the drug. Such primary resistance, as opposed to resistance acquired as a result of incomplete or ineffective drug therapy, is particularly disturbing, as it suggests that nosocomial (hospital-acquired) infection with MDR-TB may be increasingly common. Resistance to other atypical antibacterials, including ethambutol, pyrazinamide, and especially isoniazid, has also been reported. No doubt the extremely long period of tuberculosis chemotherapy is a major element in the initial acquisition of resistance to antitubercular drugs. The modern, standard "short-course" therapy (8, 9) for tuberculosis involves the treatment of patients with a four-drug combination of isoniazid, rifampicin, ethambutol, and pyrazinamide for two months, followed by treatment with a combination of isoniazid and rifampicin for an additional four months. This combination therapy must be strictly followed to prevent drug resistance and relapse, and direct observation of patient compliance is the most reliable way to ensure effective treatment and prevent the acquisition of resistance (10).

Tuberculosis chemotherapy is complicated by the slow-growing nature of the bacillus. *M. tuberculosis* has a doubling time of ~ 24 hr (11), compared to 2–3 hr for the fast-growing saprophytic mycobacteria, *Mycobacterium smegmatis*. This growing time increases the length of time required for chemotherapy and makes drug susceptibility determinations time-consuming (6–8 weeks). Tuberculosis chemotherapy is also complicated by the metabolic activity of the bacteria and its cellular localization. Mitchison (8) has suggested that pathogenic mycobacterial populations be divided into four components: actively metabolizing and rapidly growing, semidormant in an acidic intracellular environment, semidormant in a nonacidic intracellular environment, and dormant. The latter categories appear to be unique properties of mycobacterial infections, with the organisms able to remain quiescent for years, or decades. Effective tuberculosis chemotherapy must include early bactericidal action against rapidly growing organisms (to both reduce the time of infectiousness and shorten the treatment period) and subsequent sterilization of the semidormant and dormant populations of bacilli. Certain drugs are clearly more effective against some of these subpopulations. Isoniazid, rifampicin, streptomycin, and ethambutol all individually exhibit rapid bactericidal action against actively metabolizing organisms (12). Pyrazinamide appears to be most effective against semidormant bacilli in acidic intracellular environments, and rifampicin can be effective against semidormant bacilli in nonacidic environments. Whether any single drug is effective against dormant bacilli is not clear (for exception, see 13, 14), but combinations of the above-mentioned drugs appear to result in effective sterilization of even dormant populations when treatment is continued for the suggested six-month regimen.

An additional level of complexity in antimycobacterial drug treatment results from the extraordinary difference in susceptibility of various mycobacterial species to specific drugs. *M. tuberculosis* and *M. bovis* BCG (an attenuated strain used for vaccination against tuberculosis worldwide) are exquisitely susceptible to isoniazid, whereas *M. smegmatis* and the *Mycobacterium avium-intracellulare* complex (MAC) are 100–1000 times less susceptible to isoniazid. These problems are considered in detail in a recent review (12). Throughout the remainder of this article, the focus is on mechanisms of action and resistance to clinically useful antitubercular drugs.

Considering the enormous impact tuberculosis has had on humanity for the past five millenia and the forty years since chemotherapy was shown to be effective, the fact that so little was known about the mechanism of action of any antitubercular drug is surprising. The process of antibacterial drug discovery has been intimately linked to tuberculosis chemotherapy: Examples are the discovery of prontosil, and the subsequent identification of sulfanilamide antibiotics in 1936, and the discovery of streptomycin in 1944. In spite of this long association, only in the past several years have researchers begun to understand the molecular mechanisms of action and resistance of antitubercular compounds. Certainly some of the delay has been due to the highly pathogenic nature of the organism and the lack of basic biochemical knowledge of the mycobacteria. Mycobacteria are members of the order *Actinomycetales* which includes the genera *Rhodococcus, Corynebacteria,* and *Nocardia.* They are gram-positive eubacteria that are distinguished by the high G+C content of their DNA (60–70%) and unusual cell walls. In addition to a slightly atypical peptidoglycan (15, 16), these bacteria have a unique polysaccharide component, arabinogalactan (17). To this polysaccharide are covalently attached long, branched lipids, termed mycolic acids, which can be from 40 to 90 carbons long. The structure of the cell wall of the mycobacteria has been recently reviewed in this series (18), and this subject will be discussed below when cell wall inhibitors are presented.

Bacteria develop resistance to drugs via a limited number of mechanisms (19), some of which apply to mycobacterial drug resistance (20–23). Mutations in the enzymes that either activate antimycobacterial drugs or are themselves the target of drug action are most commonly observed in *M. tuberculosis.* Drug inactivation mechanisms that result in resistance have only been described for β-lactam antibiotics in fast-growing mycobacteria (24, 25), and this has been of limited clinical interest because β-lactams are not used in the treatment of tuberculosis. No drug efflux mechanisms have yet been described that can account for drug resistance in *M. tuberculosis,* although diffusion and transport into mycobacterial cells is an extremely important variable in drug activity. Lastly, episomal or transposon-mediated transfer of resistance genes into *M. tuberculosis* has never been demonstrated, although this is a common mechanism for the acquisition of drug resistance in other bacteria. Although both *in vivo–* and *in vitro*–developed resistance to antimycobacterial drugs has been documented since the introduction of these drugs, their targets and mechanisms of resistance in *M. tuberculosis* at the molecular level have all been described in the past several years (26).

RESISTANCE TO INHIBITORS OF PROTEIN SYNTHESIS

Streptomycin and Related Inhibitors

Streptomycin (Figure 1) was first shown to be an effective antitubercular drug in 1944 (2). The liquid Minimum Inhibitory Concentration (MIC) of streptomycin against *M. tuberculosis* has been reported to be 0.4–1.5 µg/ml (12), making it one of the most effective early antitubercular drugs. The antibacterial activities of streptomycin, and related aminoglycosides, are due to the inhibition of prokaryotic protein translation. Specifically, initiation of mRNA translation appears to be inhibited, although translational accuracy is also affected (19). A common mechanism of resistance to aminoglycoside antibiotics in other bacteria is drug inactivation via acetylation (19). However, this mechanism of resistance has not been reported in *Mycobacterium tuberculosis*. Instead two classes of mutations account for some 80% of the high-level streptomycin resistance in *M. tuberculosis* (27). The first consists of point mutations in the ribosomal S12 protein (28, 29), encoded by the *rpsL* gene, resulting in single–amino acid replacements (30–32). These mutants account for two thirds of the resistant mutations. Mapping of these mutations revealed that all mutations occurred in highly conserved regions of the gene encoding one of two critical lysine residues (K43 and K88). In all cases, either K88 was converted to an arginine residue or K43 was converted to either an arginine or threonine residue (31). Corresponding mutations result in streptomycin-resistant *E. coli* (33).

The second class of mutations, which account for the remaining third of the streptomycin-resistance conferring mutations, occur on the 16S rRNA, are encoded in the *rrs* locus, and are thought to interact with the ribosomal S12 protein. Mutations in *M. tuberculosis* have been mapped to two regions, the 530 loop and the 915 region. In the 530 loop, C>T transitions at positions 491, 512, and 516 are observed, as is an A>C transversion at position 513 (these correspond to positions 501, 522, 526, and 523 in *E. coli* 16S rRNA). Mutations at corresponding positions have been mapped in streptomycin-resistant *E. coli* mutants, including the invariant C513 residue. A single A>G mutant has been mapped to position 913 (30). In contrast to *E. coli*, little is known of the biochemical basis of the interaction between streptomycin and either the S12 protein or the 16S rRNA of *M. tuberculosis*. However, it seems likely that the mutations result in similar changes in the *M. tuberculosis* ribosome and that

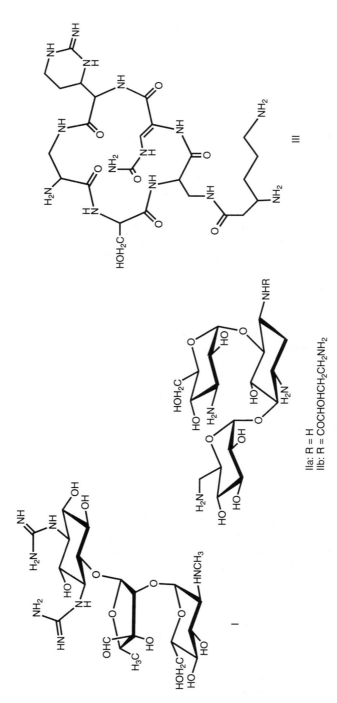

Figure 1 The structures of streptomycin (I), kanamycin (IIa) and amikacin (IIb), and capreomycin (III).

IIa: R = H
IIb: R = COCHOHCH₂CH₂NH₂

S12 mutants decrease the binding of the drug. Several recent reviews expand on this discussion (27, 33).

Resistance to streptomycin appears in 5.7% of all cases reported in the United States (5), making streptomycin resistance the third most commonly encountered drug resistance after isoniazid (9.1%) and pyrazinamide (5.8%). Cross-resistance in *M. tuberculosis* to either related aminoglycosides, kanamycin and amikacin, or cyclic peptide protein synthesis inhibitors, such as capreomycin, (Figure 1) has never been observed (34–38). This lack of cross-resistance suggests that these compounds may be substituted for streptomycin in those cases where resistance has been determined. Since all of these compounds must be administered by injection, whether streptomycin and related inhibitors of bacterial protein synthesis will continue to be considered as first-line antituberculars in short-course chemotherapies is not clear.

RESISTANCE TO INHIBITORS OF NUCLEIC ACID SYNTHESIS

Rifampicin

Rifampicin (Figure 2), a semi-synthetic derivative of the natural product rifamycin, obtained from culture filtrates of *Streptomyces mediterranei*, was introduced in 1972 as an antitubercular drug (39). Rifampicin is extremely effective against *M. tuberculosis*, (MIC-0.1-0.2 µg/ml,) and its rapid bactericidal activity (8, 12) helped to shorten the course of treatment against drug-susceptible infections. Currently rifampicin is a key component of all short-course multi drug–treatment regimens (8, 9) of tuberculosis and leprosy. For more than 20 years the target of rifampicin action in *M. tuberculosis* has been assumed to be the mycobacterial RNA polymerase (40, 41). In 1993 this target

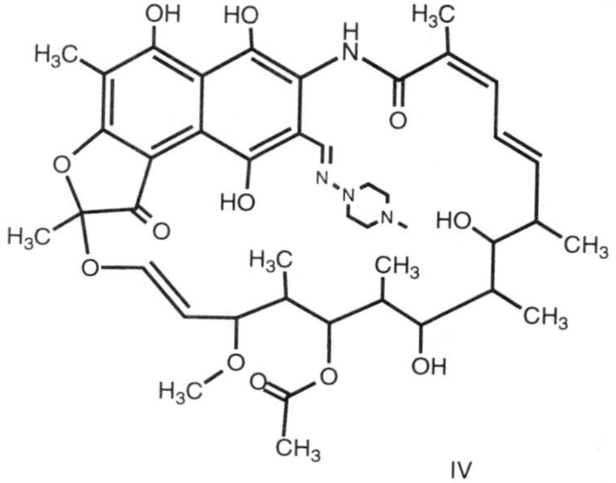

Figure 2   The structure of rifampicin (iv).

Figure 3 Mutations in the *rpoB* gene that confer rifampicin resistance and the corresponding amino acid substitutions in the β subunit of *M. tuberculosis* RNA polymerase. The circled substitutions are the most commonly observed mutations. Adapted from (47) and (48).

was confirmed and the mechanism of action described (42). These studies, using purified RNA polymerase from *M. smegmatis*, strain mc²155, demonstrated that rifampicin specifically inhibits the elongation of full-length transcripts, but has little or no effect on transcription initiation.

Resistance to rifampicin is increasing rapidly as a result of its widespread use. Rifampicin-resistant tuberculosis, often observed in conjunction with isoniazid resistance, leads to a longer treatment period and significantly poorer chemo-therapeutic outcomes. Resistance has been observed in 3.9% of all cases nationally, but in 9.0% of patients who have been previously treated for tuberculosis (5). Resistance is acquired at a rate of ~ 10^{-8} in cultures of *M. tuberculosis*, evidence of a single-step mutational event (41). The genes encoding the β subunit of the mycobacterial RNA polymerase, *rpoB*, have been sequenced from both *M. tuberculosis* (43, 44) and *M. leprae* (45), and the mutations that result in resistance have been identified (43, 45–47). These findings now allow for the rapid determination of the drug-susceptibility characterization of clinical isolates of *M. tuberculosis* using PCR-SSCP or automated DNA sequencing methods (46, 48).

The vast majority of rifampicin resistance–conferring mutations in the mycobacterial *rpoB*-encoded RNA polymerase are single nucleotide changes that result in single amino acid substitutions (93%; 45). The remaining mutations are insertions (3%) and deletions (4%), but all mutations map to the presumed rifampicin binding site (Figure 3) between amino acid positions 511 and 533 (*E. coli* numbering system). In two recent studies, the most commonly encountered amino acid substitutions occurred at His526 and Ser531 [~ 60% of all mutations; (47, 48)]. Additional mutations (43) have been mapped to positions 511, 512, 513, 516, and 533 (~ 20%) as seen in Figure 3. These mutations have been found in clinical isolates from geographically diverse regions and do not correlate with the level of resistance. Nor do they correlate with multidrug resistance phenotypes, suggesting the independent acquisition of rifampicin resistance unrelated to other drug resistance. A very small percentage of rifampicin-resistant clinical *M. tuberculosis* isolates do not map to the 511–533 region of RNA polymerase, but may be present in the carboxy terminal region of the protein.

The increasing occurrence of rifampicin-resistant *M. tuberculosis*, especially in individuals not previously treated for tuberculosis (5), is cause for significant concern. Few other antitubercular compounds are as rapidly effective as rifampicin. Newer synthetic derivatives (rifabutin and rifapentine MICs = 0.03–0.06 and 0.01–0.06) do not appear to be substantially more effective than rifampicin (49) and exhibit cross-resistance with rifampicin. Although the target and mechanisms of action and resistance of rifampicin in *M. tuberculosis* have been well documented in the past several years, more potent new analogs to replace rifampicin have not been developed.

Fluoroquinolones

The quinolone antibacterials, and the newer fluoroquinolones, are synthetic derivatives of nalidixic acid (Figure 4), a natural product whose antibacterial action was described more than 30 years ago. The antimycobacterial activity of fluoroquinolones was described in 1984 (50), and numerous reports of their efficacy have since appeared [reviewed in (51, 52)]. These compounds are bactericidal against *M. tuberculosis*, and newer fluoroquinolones (53–57), such as ciprofloxacin (Figure 4), have MICs less than 1 μg/ml (12). There does not appear to be any synergism between fluoroquinolones and other antitubercular drugs (58), and their activity is independent of resistance to other antitubercular drugs. Because of their relatively recent introduction, fluoroquinolone therapy for tuberculosis is predominantly used in patients who are infected with multi drug–resistant organisms. In spite of these precautions in the use of fluoroquinolones as antituberculars, fluoroquinolone resistance is emerging via primary resistance mechanisms as well as via nosocomial infection with fluoroquinolone-resistant organisms (59).

The target of fluoroquinolone action is the bacterial DNA gyrase, an ATP-dependent Type II DNA topoisomerase that catalyzes the negative supercoiling of DNA (60). The enzyme is a heterotetramer composed of two A and two B subunits (A_2B_2), encoded by the *gyrA* and *gyrB* genes, respectively. Fluoro-

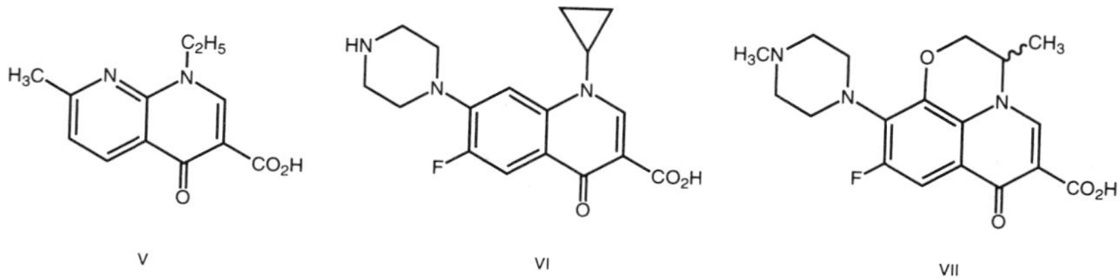

Figure 4   The structures of nalidixic acid (V), ciprofloxacin (VI) and ofloxacin (VII).

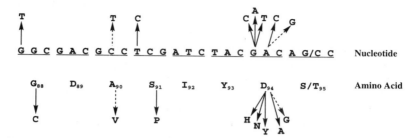

Figure 5   Mutations in the *gyrA* gene that confer fluoroquinolone resistance and the corresponding amino acid substitutions in the DNA gyrase. Solid lines represent mutations observed in *M. tuberculosis* and dotted lines represent mutations mapped in both *M. tuberculosis* and *M. smegmatis*. Adapted from (61) and (62).

quinolones bind to the gyrase, inhibiting supercoiling and subsequent processes dependent on DNA topology such as replication and transcription. The cloning and sequencing of the *gyrA* and *gyrB* genes from *M. tuberculosis* has allowed the quinolone binding site to be identified and the mutations that confer resistance to be mapped (61). As in other bacteria, these mutations cluster in a small region ~ 40 residues amino-terminal to the catalytic tyrosine (Y122 in *E. coli*) involved in DNA strand scission. In both *in vitro*–selected ciprofloxacin-resistant strains, and resistant clinical isolates, single–amino acid substitutions for residues 88–94 (equivalent to residues 81–87 in *E. coli*) were identified in these strains (Figure 5). These single–amino acid substitutions lead to a ~ 10-fold increase in the MIC for ciprofloxacin (104).

In a parallel study, mutations in the *gyrA* gene of *M. smegmatis* were obtained by selection for ofloxacin-resistant strains (62). Two amino acid substitutions were observed in these studies, A90V and D94G, which could account for low level resistance, comparable to the levels of resistance observed in *M. tuberculosis*. However, a second round of selection at higher ofloxacin levels yielded double mutants that displayed high-level resistance to fluoroquinolones. These double mutants contained substitutions at both positions 90 and 94, A90V/D94G, suggesting that the accumulation of specific mutations in the presumptive quinolone binding site had cumulative effects on the MIC for fluoroquinolones (62). These results suggest that fluoroquinolone susceptibility must be continuously monitored in the treated patient population to prevent low-level fluoroquinolone-resistant strains from acquiring additional mutations that will result in high-level resistance.

Other mechanisms of fluoroquinolone resistance in mycobacteria have been proposed in the literature to quantitatively account for MIC values including non-*gyrA* mutations (59). Additional mechanisms of resistance include changes in cell wall permeability or active quinolone efflux pumping. The detailed interactions between target and drug remain unknown; however, newer fluoroquinolone derivatives such as sparfloxacin (MIC = 0.2 µg/ml; 63) appear to be even more potent antimycobacterial compounds than ciprofloxacin and ofloxacin, promising better therapeutic results against multi drug–resistant tuberculosis.

RESISTANCE TO INHIBITORS OF CELL WALL SYNTHESIS

Isoniazid and Ethionamide

Isoniazid (INH, isonicotinic acid hydrozide) was first reported to be effective in the treatment of tuberculosis in 1952 (4, 64). Both *M. tuberculosis* and *M. bovis* BCG are susceptible to isoniazid in the range of 0.02–0.2 mg/ml (12, 65). Isoniazid is bactericidal and is both the oldest synthetic antitubercular and the most commonly prescribed drug for active infection and prophylaxis. Neither isoniazid nor ethionamide are broad spectrum antibacterials, and both are tolerated in a majority of patients even for prolonged treatment periods. Isoniazid (Figure 6) is rapidly transported into actively growing bacteria in a cyanide-inhibitable oxygen-dependent process (65). The discovery of the potent antitubercular activity of isoniazid was followed by a search for structural analogs, of which ethionamide (Figure 6) was found to be the most potent. Both drugs are presently assumed to act in similar ways, although there are some significant differences in their activation mechanisms (66). Although the subject of biochemical investigations for more than 40 years (65), the mechanism of action of and resistance to isoniazid has been clarified only in the past several years. The remainder of this section deals primarily with isoniazid, with occasional reference to ethionamide.

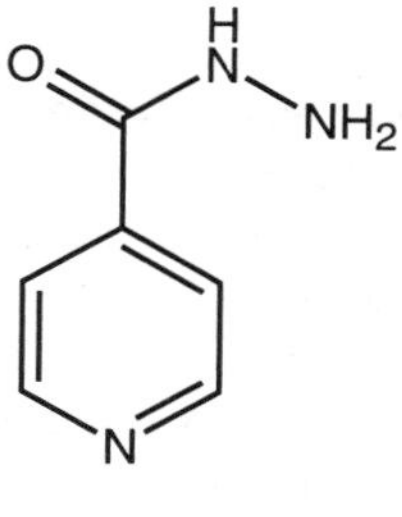

VIII 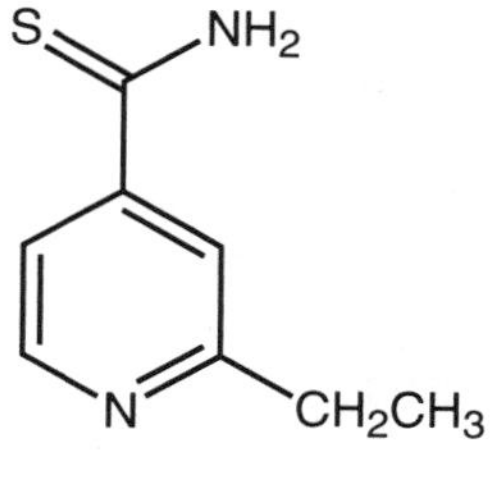   IX

Figure 6   Structures of isoniazid (VIII) and ethionamide (IX).

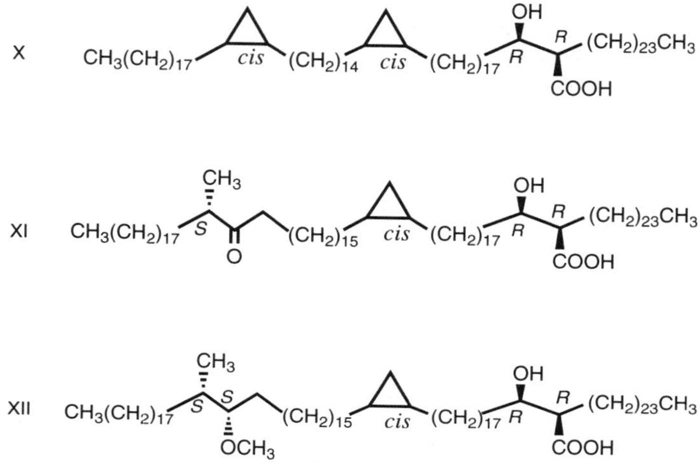

Figure 7 Structures of α-mycolates (X), ketomycolates (XI), and methoxymycolates (XII).

Some of the earliest studies of isoniazid-resistant clinical isolates of *M. tuberculosis* noted a correlation between resistance and attenuated catalase-peroxidase activity (68–70). A second effect relevant to the mechanism of action of isoniazid was the observation that acid-fastness, an important clinical diagnostic for *M. tuberculosis* infection, was rapidly lost after treatment with isoniazid (71). The acid-fast staining procedure takes advantage of the unique dye-binding properties of mycobacteria, which are related to the presence of covalently attached mycolic acids to the 5′-hydroxyl groups of the arabi-nogalactan polymer in the cell wall (72). Mycolic acids (Figure 7) are α-branched lipids, with a species-dependent saturated "short" arm of 20–24 carbon atoms and a "long" meromycolic acid arm of 50–60 carbon atoms, functionalized at regular intervals by cyclopropyl (α-mycolates), α-methyl ketone (ketomycolates), or α-methyl methylethers (methoxymycolates) groups. The presence of the mycolic acids in the outer envelope of the myco-bacterial cell wall provides an impressive permeability barrier to hydrophilic solutes. In 1970, the demonstration that isoniazid inhibited the synthesis of mycolic acids in *M. tuberculosis* (73, 74) led to the correlation of biosynthesis of mycolic acids and viability (75). A specific inhibitory effect of isoniazid was reported on the synthesis of saturated fatty acids greater than 26 carbons (76–79) and on the ability of a cell-free system to synthesize mycolic acids from radiolabelled precursors (80), implicating a target for the drug's action in the elongation of fatty acids.

Genetic and molecular biological approaches have been recently employed to identify the mechanism of activation of isoniazid, its enzymatic target, and mechanisms of resistance. As described above, the attenuation of catalase activity was correlated with resistance to isoniazid in *M. tuberculosis*. In 1992, the *katG* gene, encoding the mycobacterial catalase-peroxidase, was cloned from *M. tuberculosis* (81, 81a). Transformation of the wild-type gene into either *M. smegmatis* or *E. coli*, which are naturally less susceptible to isoniazid (82) or isoniazid-resistant strains of *M. tuberculosis*, sensitized these organisms to the drug (83). Additional evidence shows that ~ 50% of isoniazid-resistant clinical isolates of *M. tuberculosis* had deletions or missense mutations within the *katG* gene (84–86). These results suggested that isoniazid was a prodrug that requires the *katG* gene product for activation. Earlier studies had shown that peroxidases could react with isoniazid to generate a number of oxidized products similar to those observed *in vivo* (87), and the *M. tuberculosis* catalase-peroxidase has recently been shown to be capable of oxidizing isoniazid to an electrophilic species (88). The target for the activated form of isoniazid remained unknown.

Selection for single-step spontaneous mutants of *M. smegmatis* mc^{2}155 resistant to isoniazid yielded a strain, mc^{2}651, which was an order of magnitude less susceptible to isoniazid, but retained wild-type catalase-peroxidase activity. This strain was similarly resistant to ethionamide, but exhibited wild-type susceptibility to other antitubercular drugs. Subcloning and sequencing of a 3 kilobase DNA fragment, which could be transfected into wild-type strains to yield isoniazid- and ethionamide-resistant transfectants, revealed two open reading frames, one of which was identified as *inhA* (89). Comparison of wild-type and mutant genes revealed a single nucleotide difference, in which a T→G transversion resulted in the substitution of an alanine residue for serine 94. Using the cell-free mycolic acid synthesizing system, extracts of mc^{2}651 were shown to exhibit substantially reduced sensitivity to isoniazid, suggesting the *inhA*-encoded protein was involved in fatty acid elongation and mycolic acid biosynthesis (89).

Expression of the wild-type *M. tuberculosis* H37Rv *inhA*-encoded protein in *E. coli* allowed the facile purification of the protein. The purified, recombinant protein exhibited an appropriate molecular weight, and the amino terminal sequence exactly matched that predicted by the gene sequence. The protein bound NADH tightly (K_d= 2 μM) and stoichiometrically and was shown to catalyze the reduction of Δ^2-*trans*-enoyl thioesters of either CoA, or preferentially, acyl carrier protein (ACP). In a homologous series of enoyl-CoA substrates, the enzyme exhibited a marked preference for long chain (C_{16}–C_{20}) substrates (90). The S94A mutant was similarly expressed, purified, and characterized. The enzyme exhibited statistically indistinguishable maximum velocities and K_m values for fatty acyl substrates compared to the wild-type enzyme, but a 5–8 fold higher K_m value for NADH. Kinetic studies supported a random kinetic mechanism and a rate-limiting hydride transfer of the 4S hydrogen of NADH to the C3 position of the bound enoyl thioester.

Both the wild-type and S94A mutant of the *M. tuberculosis* H37Rv enoyl reductase–NADH complexes were crystallized, and their three-dimensional structures were determined to 2.2 and 2.7 Å resolution, respectively (91). The overall fold of the enzyme is reminiscent of a "Rossmann-fold" of alternating parallel β strands and α helices (92). The β sheet is composed of two half sheets: a very regular first half sheet composed of three β strands connected by three α helices and an unusual second half sheet composed of two elongated (β4 and β5) and two short β strands and two elongated α helices (α4 and α5). These elements of the second half sheet generated a large cavity over the β sheet fold which is lined with hydrophobic and aromatic side chains. NADH is bound at the carboxyl termini of the β strands, and the nicotinamide ring of NADH penetrates the hydrophobic cavity, which is the presumed fatty acid binding site. Serine 94 is positioned near the P_N atom of bound NADH, and the side chain hydroxyl is hydrogen bonded to a water molecule, which is in turn hydrogen bonded to one of the oxygen atoms of the P_N atom of NADH and the backbone carbonyl of Gly14. These interactions are disrupted in the S94A mutant and can account for the observed higher K_m value for NADH exhibited by the mutant enoyl reductase.

Together, the studies on the *M. tuberculosis* *katG*-encoded catalase-peroxidase and *inhA*-encoded enoyl reductase provide a revealing, but complicated, picture of the mechanism of action of (Figure 8) and mechanisms of resistance to isoniazid (65, 93, 94). Isoniazid is a prodrug that must be activated by reaction with the mycobacterial catalase-peroxidase. Mutations in the catalase-peroxidase which generate inactive enzyme will fail to activate the prodrug and lead to high-level isoniazid resistance. This mechanism is the one most commonly encountered for the acquisition of isoniazid resistance in clinical isolates of *M. tuberculosis* (95, 96). Isoniazid is also a potent inhibitor of the mycobacterial catalase-peroxidase (97), and a secondary effect of drug administration may be to enhance the susceptibility of bacteria to the toxic effects of reactive oxygen, especially H_2O_2 (98, 99) generated in the macrophage phagolysosomal compartments.

Although isoniazid clearly does not bind to the *inhA*-encoded enoyl reductase (90), the catalase-peroxidase–activated drug does (66), and its binding is correlated with inhibition of the reductase activity. Drug binding occurs predominantly, or exclusively, to the reductase-NADH binary complex, providing a reasonable explanation for the resistance to isoniazid exhibited by organisms expressing the S94A mutant enzyme. The intracellular levels of NADH in *M. tuberculosis* are low [estimated at < 10 μM; (67)], and thus the wild-type

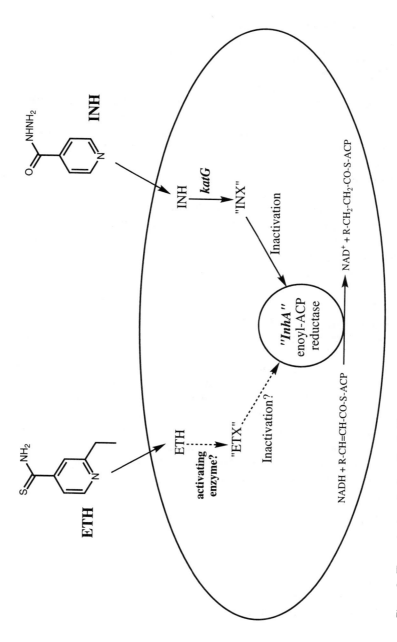

Figure 8   The mechanism of action of isoniazid.

enzyme is present as the drug inhibitable–binary complex. The mutant enzyme, because of its lower affinity for NADH, would be only partially saturated with NADH at these concentrations, and would not bind or be inhibited by the drug. Although the residue(s) of the enoyl reductase that interact, or react, with the activated form of isoniazid are not known with certainty, the protective effects of added substrates (66) suggest that residues in the active site are most likely modified by the drug.

The mechanism of action of ethionamide is almost certainly similar to that of isoniazid, including a requirement for prodrug activation. Whereas mutations in the enoyl reductase generate resistance to ethionamide, isoniazid-resistant strains containing mutations in the catalase-peroxidase gene remain susceptible to ethionamide. A second enzyme that activates ethionamide must be present in mycobacteria to convert this prodrug into a form capable of binding to, and presumably inhibiting, the *inhA*-encoded enoyl reductase, because ethionamide does not bind to the reductase (90).

Lastly, resistance due to the overexpression of the enoyl reductase has been demonstrated *in vitro* (89), and the majority of isoniazid-resistant clinical isolates of *M. tuberculosis* that map to the *inhA* locus have mutations in the promoter region of the gene (21). This finding suggests that this common bacterial-resistance mechanism is operative in the mycobacteria.

Ethambutol

Ethambutol is a specific bactericidal agent used in most modern combination antitubercular therapies. Its powerful antimycobacterial effects were first described in 1961 (100), and the *S,S* stereochemistry has been shown to be essential for activity. The compound's simple molecular structure (Figure 9) gives few clues to its biochemical target, although its specific inhibition of mycobacteria suggest that its target might be involved in the construction of the outer cell wall structures of mycobacteria. Support for this theory comes from the demonstration that coadministration of ethambutol and other drugs produces a synergistic effect, presumed to result from a decreased mycobacterial permeability barrier and resultant increased drug uptake (8, 101, 102). While neither the mechanism of action of (103, 104), nor resistance to (105),

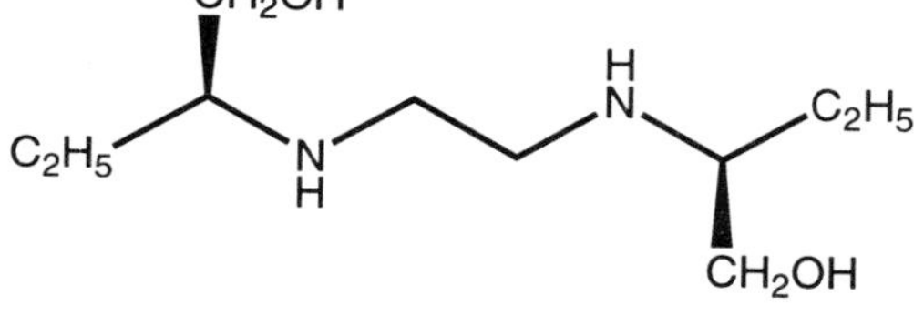

Figure 9   The structure of ethambutol (XIII).

ethambutol are understood in detail, the past five years have brought us closer to a biochemical description of both.

Early studies on the effect of ethambutol showed that transfer of mycolic acids into the cell wall was inhibited as early as fifteen minutes after drug administration (106). Even more rapid effects were subsequently demonstrated in the accumulation of trehalose mono- and dimycolates (107), suggesting that the biochemical target was indeed involved in the construction of the outer envelope of the mycobacterial cell wall (108). The structure and composition of the outer cell wall of *M. tuberculosis* is still under active investigation, although substantial progress has been made in the past decade [reviewed in (18)]. The evidence shows clearly that mycolic acids are specifically attached to the 5′-hydroxyl groups of the terminal D-arabinose residues of the arabinogalactan polymer of the cell wall. The effects of ethambutol could be due to the inhibition of the transfer of the active mycolyl groups onto the arabinogalactan or to steps involving arabinogalactan synthesis. The latter possibility was shown to be correct when the rapid inhibitory effect of ethambutol was observed on the synthesis of arabinogalactan in a drug-susceptible strain of *M. smegmatis*, which was not observed in an ethambutol-resistant strain (109). The inhibition of arabinogalactan biosynthesis by ethambutol could account for the accumulation of mycolic acids and their trehalose esters, as well as the effects on cell wall permeability.

More recent results have clarified and extended these studies. The identification of β-D-arabinofuranosyl-1-monophosphodecaprenol as the major intermediate in the biosynthesis of arabinogalactan, and the rapid accumulation (<2 min) of this intermediate following ethambutol administration, suggested that the target of ethambutol was either the arabinosyl transferase or some enzyme involved in the synthesis of the arabinose acceptor (104). The recent synthesis of the decaprenylphosphoarabinose intermediate, and development of an arabinosyltransferase assay, has allowed the specific inhibitory effect of ethambutol on the transfer reaction to be demonstrated (Besra & Brennan, personal communication). These studies should allow for the further biochemical description of the mechanism of action of ethambutol.

Single-step mutants of *M. tuberculosis* H_{37}Ra resistant to ethambutol have been described previously (108), although the genetic basis for the resistance has not been demonstrated. Similar strategies of mutant selection in *M. smegmatis* followed by cloning and sequencing have now revealed two open reaching frames in which single base changes are observed in the mutant genotypes (WR Jacobs Jr, personal communication). The level of resistance to ethambutol appears to correlate with the accumulation of mutations in these genes, as has been observed with fluoroquinolone resistance in mycobacteria (*vide supra*). These recent reports suggest that the mechanism of resistance to ethambutol in mycobacteria will soon be made clear with the aid of this powerful combination of genetic and biochemical methods.

RESISTANCE TO OTHER ANTITUBERCULARS

Pyrazinamide

The weak antitubercular activity of nicotinamide began a search for more powerful analogs and resulted in the discovery of pyrazinamide (Figure 10). Although shown to have substantial activity in 1952 (3), pyrazinamide was not used extensively in the treatment of tuberculosis until the mid-1980s. Pyrazinamide is now recommended in essentially every combination therapy for the treatment of the disease because of its strong synergistic and accelerating effect in combination with isoniazid and rifampicin (8, 12). The introduction of pyrazinamide combination chemotherapy allowed treatment regimens to be reduced from 9–12 months to 6 months. The MIC for pyrazinamide varies from 8 to 60 μg/ml depending on the assay method and media, and the drug is most active against cultures of *M. tuberculosis* at pH values below 6 (12, 110). Pyrazinamide does not appear to be bactericidal, even at concentrations significantly greater than its MIC, and its *in vitro* effect is presently termed "sterilizing" (111) to distinguish it from the effects of other drugs such as rifampicin. Much of the confusion about, and clinical utility of, pyrazinamide is due to its unique activity *in vitro* at low pH values, conditions which naturally inhibit the growth of mycobacteria (110). The drug appears to exhibit its *in vivo* sterilizing activity on the semidormant populations of *M. tuberculosis* in acidic intracellular compartments, such as the macrophage phagolyosomes. Although pyrazinamide is effective against *M. tuberculosis* infection, the compound is not effective in the treatment of other mycobacterial infections, in particular *M. bovis* and fast-growing mycobacteria.

Susceptibility of mycobacteria to pyrazinamide correlates with the presence of a specific amidase, which hydrolyzes both pyrazinamide and nicotinamide, its presumed physiological substrate (112–114). This finding suggests that pyrazinamide, like isoniazid, is a prodrug, transported or diffused as a neutral species through the mycobacterial cell wall and converted into pyrazinoic acid, the presumed active drug form. This proposal is supported by the finding that pyrazinoic acid is active *in vitro* against pyrazinamide- and nicotinamide-resistant strains of *M. tuberculosis*, as well as the naturally resistant *M. bovis*, which lacks pyrazinamidase (113). Pyrazinoic acid is ineffective in treating *M. tuberculosis*–infected mice, although whether the acid can be transported to the site of infection, or whether it is too rapidly cleared from the mouse to attain pharmacologically significant concentrations, is not clear. Pyrazinoic acid also appears to reduce the pH of the media of cultures of *M. tuberculosis*

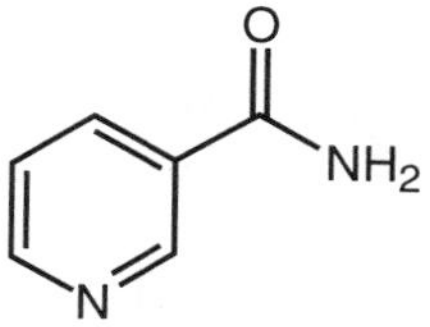

Figure 10   The structures of nicotinamide (XIV) and pyrazinamide (XV).

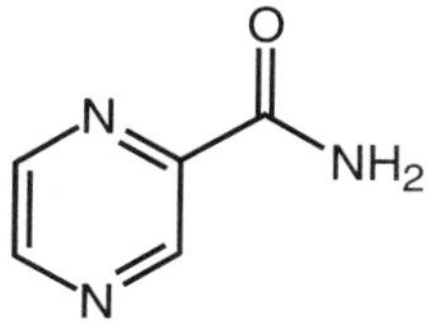

(115), but whether this is a primary or secondary effect is unclear. Very recently, a series of pyrazinoic acid esters have been shown to exhibit better *in vitro* activity against *M. tuberculosis* than pyrazinamide (116), supporting the view that the acid form is the active component.

Resistance to pyrazinamide has been observed *in vitro*, and pyrazinamide-resistant cultures are cross-resistant to nicotinamide (117). Resistance to pyrazinamide is the second–most commonly observed drug-resistant phenotype in clinical isolates, accounting for 5.8% of all cases and up to 17.6% of recurrent cases (5). These phenotypes are probably the result of mutations that reduce or abolish the activity of the amidase, although no detailed enzymological analysis to test this theory has been performed. The target of pyrazinamide, or pyrazinoic acid, is unknown, although the obvious structural similarity to nicotinamide suggests that the compound could interfere with pyridine nucleotide biosynthesis and turnover, and enzymes in these pathways would be logical candidates for targets. As for isoniazid, a combination of genetic and biochemical approaches will be helpful to elucidate the mechanisms of activation and action of this important antitubercular drug.

ANTIBIOTIC RESISTANCE AND TUBERCULOSIS TREATMENT

Tremendous concern has been voiced about the increasing incidence of antibiotic resistance in bacterial infections that appeared to be of little threat as recently as a decade ago. The emergence of single and especially multi drug–resistant strains of enteric bacteria and *Staphylococcus aureus* presents an enormous problem in confined populations, including hospitals, where nosocomial infection can have disastrous consequences. In addition to the problem associated with primary resistance as a result of noncompliance with chemotherapy, the transfer of genes encoding drug-inactivating enzymes between bacterial populations is a significant and demonstrated concern. An important example of the general problem is the recent appearance of vancomycin resistance in clinical isolates of methicillin-resistant enterococci (118).

The appearance of multi drug–resistant strains of *M. tuberculosis* shares many of the societal and clinical problems of general bacterial antibiotic resistance and generates some unique concerns. Drug-resistant *M. tuberculosis* strains are initially the result of noncompliance with chemotherapeutic regimens and selective genetic pressure. Because of the extremely infectious nature of the organism and its ability to be transferred via aerosols from infected to noninfected individuals, multi drug–resistant organisms are now being detected in previously untreated tuberculosis patients (5). Patients are often treated with ineffective drug combinations for long periods before susceptibility screens can be analyzed, further complicating treatment and jeopardizing therapeutic outcomes.

Because of the unique nature and antibacterial properties of antitubercular drugs, and the lack of evidence to date for episomal transfer of resistance-conferring genes from the general bacterial population to *M. tuberculosis*, multi drug–resistant tuberculosis remains a treatable infectious disease for the majority of cases. In large part this high level of treatability is due to the early recognition that combinations of at least three or four drugs were required for both effective sterilization and the prevention of the acquisition of resistance commonly observed in monotherapy. Unfortunately, the most commonly encountered resistance is against the two most effective antitubercular compounds, isoniazid and rifampicin. As discussed above, these are the two drugs which are most bactericidal against rapidly growing organisms, and rifampicin appears to be effective against semi-dormant organisms that reactivate and become metabolically active. Rifampicin has no effective homologs for which cross-resistance is not encountered, and ethionamide is only a modestly effective therapeutic substitute for isoniazid in cases in which isoniazid-resistance is due to mutations in the catalase-peroxidase gene.

Other broad spectrum antibacterials, which have significant activity against mycobacteria, can be substituted for isoniazid or rifampicin. The fluoroquinolones, including ciprofloxacin, ofloxacin, and the newer sparfloxacin, are extremely effective antituberculars. Although fluoroquinolone resistance has been reported in clinical isolates of *M. tuberculosis*, the compounds remain effective in the treatment of isoniazid- and rifampicin-resistant tuberculosis. The macrolide antibiotics such as erythromycin and clarithromycin exhibit antimycobacterial activity.The latter has significant activity against *M. avium* infections in HIV-infected individuals (119), and ethambutol has a pronounced synergistic effect on clarithromycin administration (102).

New drug discovery is hampered by the fact that although tuberculosis remains the world's leading cause of human mortality among infectious diseases, the vast majority of disease occurs in the undeveloped nations which have limited resources with which to address these health concerns. In contrast, newly diagnosed cases of tuberculosis in the developed nations number ~ 25,000, and chemotherapy is successful in 95% of compliant cases. By far the most effective way to ensure a decrease in primary resistance, acquired resistance, and relapse is to institute directly observed combination chemotherapy (10), even amongst individuals harboring multi drug–resistant strains. Drug-resistant phenotypes of *M. tuberculosis* appear to be uncorrelated with virulence (120). Thus the solution to the problem of drug-resistant tuberculosis is clear, but the disease remains both a societal and scientific problem.

Future Studies and Solutions

The challenge to the scientific and pharmacologic community is to eliminate the comparative lack of modern biochemical and genetic information about mycobacteria in general. *M. tuberculosis*, in particular, is not an organism whose large scale culture is achievable in any but a handful of laboratories. The recent development of strains of the nonpathogenic fast-growing *M. smegmatis* that can be genetically manipulated (121, 122) has allowed researchers to begin classical genetic studies in this organism. These approaches have clarified the mechanism of action of, and resistance to, both rifampicin (42) and isoniazid (89). The biochemical transformations occurring in mycobacteria during the acquisition of drug resistance are generally inferred, rather than demonstrated, and tremendous progress should be made in this area in the next decade.

The serendipitous discovery of isoniazid, ethionamide, and ethambutol has now provided clues into critical and unique biosynthetic pathways in mycobacteria. The chemical simplicity of these molecules, and a decade of synthetic endeavors after their initial discovery, suggests that more potent analogs, prepared by classic organic synthesis or novel combinatorial synthetic methods, will be hard to find. Given what is now known about the mechanism of action of isoniazid, the discovery of homologs that are both actively accumulated and oxidatively activated, and still inhibit long chain fatty acid elongation processes involved in outer envelope biosynthesis, seems unlikely. However, as the mechanism of action of these drugs is clarified, and their molecular targets identified, more rational mechanism-based and structure-based approaches to inhibitor design will be possible. The synthesis and clinical evaluation of any lead compound is a long and expensive process, and the worldwide distribution of tuberculosis in the developing nations may discourage such investments.

Alternative solutions that should be examined include the revaluation of existing antibacterials. The example of fluoroquinolone inhibitors of DNA

236 BLANCHARD

gyrase was discussed above. A second example could include the β-lactam inhibitors of peptidoglycan biosynthesis. Mycobacteria are naturally insensitive to β-lactams, because of their extremely hydrophobic cell wall (123) and the presence of both periplasmic penicillin–binding proteins (124) and an active β-lactamase (25, 123). However, the combined administration of β-lactams and β-lactamase inhibitors has recently been shown to be effective in inhibiting the growth of mycobacteria (125–127). Given their oral availability and favorable toxicology profile, the thousands of β-lactams that have been synthesized seem worthy of re-examination, in combination with inhibitors of the mycobacterial β-lactamases. The recently demonstrated bactericidal activity of nitroimidazoles such as metronidazole (13, 14) against dormant populations, which are poorly treated in present drug regimens, represents a potential new chemotherapeutic addition. Ultimately, the only solution to the problems with tuberculosis chemotherapy and the explosion of multi-drug resistance is patient surveillance and compliance. We have beaten the scourge of tuberculosis once before, but we must remain vigilant in this present day cat-and-mouse game.

ACKNOWLEDGMENTS

This work was supported by NIH grants GM-33449 and AI-33696. I would like to thank Drs. Lincoln Miller, Vern Schramm, and Thomas Shrader for helpful comments and suggestions on the manuscript, and Ms. Lisa Idi for wordprocessing assistance.

Literature Cited

1. Bloom BR, Murray CJL. 1992. *Science* 257:1055–64
2. Schatz A, Waksman SA. 1944. *Proc. Soc. Exp. Biol. Med.* 57:244–48
3. Kushner S, Dalalian H, Sanjuro JL, Bach FL, Safir SR, et al. 1952. *J. Am. Chem. Soc.* 74:3617–26
4. Middlebrook G. 1952. *Am. Rev. Tuberc.* 65:765–67
5. Bloch AB, Cauthen GM, Onorato IM, Dansbury KG, Kelly GD, et al. 1994. *J. Am. Med. Assoc.* 271:665–71
6. Frieden TR, Sterling T, Pablos-Mendez A, Kilburn JO, Cauthen GM, et al. 1993. *N. Engl. J. Med.* 328:521–26
7. Heym B, Honore N, Truffot-Pernot C, Banerjee A, Schurra C, et al. 1994. *Lancet* 344:293–98
8. Mitchison DA. 1985. *Tubercle* 66:219–26
9. Stratton MA, Reed MT. 1986. *Clin. Pharm.* 5:977–87
10. Weis SE, Slocum PC, Blais FX, King B, Nunn M, et al. 1994. *N. Engl. J. Med.* 330:1229–30
11. Hiriyanna KT, Ramakrishnan T. 1986. *Arch. Microbiol.* 144:105–9
12. Heifets LB. 1994. *Semin. Respir. Infect.* 9:84–103
13. Ashtekar DR, Costa-Perira R, Nagrajan K, Vishvanathan N, Bhatt AD, Rittel W. 1993. *Antimicrob. Agents Chemother.* 37:183–86
14. Wayne LG, Sramek HA. 1994. *Antimicrob. Agents Chemother.* 38:2054–58

15. Rastoggi N. 1991. *Res. Microbiol.* 142: 464–76

16. Rastoggi N, Barrow WW. 1994. *Res. Microbiol.* 145:243–52

17. Daffe M, McNeil M, Brennan PJ. 1993. *Carbohydr. Res.* 249:383–98

18. Brennan PJ, Nikaido H. 1995. *Annu. Rev. Biochem.* 64:29–63

19. Benveniste R, Davies J. 1973. *Annu. Rev. Biochem.* 42:471–506

20. Cole ST. 1994. *Immunobiology* 191: 584–85

21. Morris S, Bai GH, Suffys P, Portillo-Gomez L, Fairchok M, et al. 1995. *J. Infect. Dis.* 171:954–60

22. Rastoggi N, David HL. 1993. *Res. Microbiol.* 144:133–43

23. Suzuki AE, Inamine JM. 1994. *Res. Microbiol.* 145:210–13

24. Eum HM, Yapo A, Petit JF. 1978. *Eur. J. Biochem.* 86:97–103

25. Fattorini L, Orefici G, Jin SH, Scardaci G, Amicosante G, et al. 1992. *Antimicrob. Agents Chemother.* 36:1068–72

26. Zhang Y, Young D. 1994. *J. Antimicrob. Chemother.* 34:313–19

27. Honore N, Cole ST. 1994. *Antimicrob. Agents Chemother.* 38:238–42

28. Douglass J, Steyn LM. 1993. *J. Infect. Dis.* 167:1505–6

29. Yamada T, Nagata A, Ono Y, Suzuki Y, Yamanouchi T. 1985. *Antimicrob. Agents Chemother.* 27:921–24

30. Finken M, Kirschner P, Meier A, Wrede A, Bottger EC. 1993. *Mol. Microbiol.* 9:1239–46

31. Meier A, Kirschner P, Bange FC, Vogel U, Bottger EC. 1994. *Antimicrob. Agents Chemother.* 38:228–33

32. Nair J, Rouse DA, Bai G-H, Morris SL. 1993. *Mol. Microbiol.* 10:521–27

33. Bottger EC. 1994. *Trends Microbiol.* 2:416–21

34. Heifets LB, Lindholm-Levy PJ. 1989. *Antimicrob. Agents Chemother.* 33: 1298–301

35. Hoffner SE, Kallenius G. 1988. *Eur. J. Clin. Microbiol. Infect. Dis.* 7:188–90

36. McClatchy JK, Kanes W, Davidson PT, Moulding TS. 1977. *Tubercle* 58:29–34

37. Tsukamura M, Mizuno S. 1975. *J. Gen. Microbiol.* 88:269–74

38. Tsukamura M, Mizuno S. 1980. *Microbiol. Immunol.* 24:777–87

39. Woodley CL, Kilburn JO, David HL, Silcox VA. 1972. *Antimicrob. Agents Chemother.* 2:245–49

40. Siddiqi SH, Aziz A, Reggiardo Z, Middlebrook G. 1981. *J. Clin. Pathol.* 34: 927–29

41. Tsukamura M. 1972. *Tubercle* 53:111–17

42. Levin ME, Hatfull GF. 1993. *Mol. Microbiol.* 8:277–85

43. Donnabella MV, Martiniuk F, Kinney D, Bacerdo M, Bonk S, et al. 1994. *Am. J. Respir. Cell Mol. Biol.* 11:639–43

44. Miller LP, Crawford JT, Shinnick TM. 1994. *Antimicrob. Agents Chemother.* 38:805–11

45. Honore N, Cole ST. 1993. *Antimicrob. Agents Chemother.* 37:414–18

46. Telenti A, Imboden P, Marchesi F, Lowrie D, Cole S, et al. 1993. *Lancet* 341:647–50

47. Williams DL, Waguespack C, Eisenach K, Crawford JT, Portaels F, et al. 1994. *Antimicrob. Agents Chemother.* 38: 2380–86

48. Kapur V, Li L-L, Iordanescu S, Hamrick MR, Wanger A, et al. 1994. *J. Clin. Microbiol.* 32:1095–98

49. Heifets LB, Lindholm-Levy PJ, Flory MA. 1990. *Am. Rev. Respir. Dis.* 141: 626–30

50. Gay JD, DeYoung DR, Roberts GD. 1984. *Antimicrob. Agents Chemother.* 26:94–96

51. Leysen DC, Haemers A, Pattyn SR. 1989. *Antimicrob. Agents Chemother.* 33:1–5

52. Stratton C. 1992. *Clin. Ther.* 14:348–75

53. Klopman G, Wang S, Jacobs MR, Bajaksouzian S, Edmonds K, et al. 1993. *Antimicrob. Agents Chemother.* 37: 1799–806

54. Klopman G, Wang S, Jacobs MR, Ellner JJ. 1993. *Antimicrob. Agents Chemother.* 37:1807–15

55. Klopman G, Li JY, Wang S, Pearson AJ, Chang K, et al. 1994. *Antimicrob. Agents Chemother.* 38:1794–802

56. Piersimoni C, Morbiducci V, Bornigia S, DeSio G, Scalise G. 1992. *Am. Rev. Respir. Dis.* 146:1445–47

57. Truffot-Pernot C, Ji B, Grosset J. 1991. *Tubercle* 72:57–64

58. Marinis E, Legakis NJ. 1985. *J. Antimicrob. Chemother.* 16:527–30

59. Sullivan EA, Kreiswirth BN, Palumbo L, Kapur V, Musser JM, et al. 1995. *Lancet* 345:1148–50

60. Wang JC. 1991. *J. Biol. Chem.* 266: 6659–62

61. Takiff HE, Salazar L, Guerrero C, Philipp W, Huang WM, et al. 1994. *Antimicrob. Agents Chemother.* 38:773–80

62. Revel V, Cambau E, Jarlier V, Sougakoff W. 1994. *Antimicrob. Agents Chemother.* 38:1991–96

63. Lalande V, Truffot-Pernot C, Paccaly-Moulin A, Grosset J, Ji B. 1993. *Antimicrob. Agents Chemother.* 37:407–13

64. Bernstein J, Lott WA, Steinberg BA,

Yale HL. 1952. *Am. Rev. Tuberc.* 65: 357–64

65. Youatt J. 1969. *Am. Rev. Respir. Dis.* 99:729–49

66. Johnsson K, King DS, Schultz PG. 1995. *J. Am. Chem. Soc.* 117:5009–10

67. Gopinathan KP, Sirsi M, Ramakrishnan T. 1963. *Biochem. J.* 87:444–48

68. Middlebrook G. 1954. *Am. Rev. Tuberc.* 69:471–72

69. Middlebrook G, Cohn ML, Schaefer WB. 1954. *Am. Rev. Tuberc.* 70:852–72

70. Winder FG. 1960. *Am. Rev. Respir. Dis.* 81:68–78

71. Koch-Weser D, Ebert RH, Barclay WR, Lee VS. 1953. *J. Lab. Clin. Med.* 42: 828–29

72. McNeil M, Daffe M, Brennan PJ. 1991. *J. Biol. Chem.* 266:13217–23

73. Winder FG, Collins PB. 1970. *J. Gen. Microbiol.* 63:41–48

74. Winder FG, Collins PB, Rooney SA. 1970. *Biochem. J.* 117:P27

75. Takayama K, Wang L, David HL. 1972. *Antimicrob. Agents Chemother.* 2:29–35

76. Davidson LA, Takayama K. 1979. *Antimicrob. Agents Chemother.* 16: 104–5

77. Kikuchi S, Takeuchi T, Yasui M, Kasaka T, Kolaitukudy PE. 1989. *Agric. Biol. Chem.* 53:1689–98

78. Takayama K, Schnoes HK, Armstrong EL, Boyle RW. 1975. *J. Lipid Res.* 16:308–17

79. Wang L, Takayama K. 1972. *Antimicrob. Agents Chemother.* 2:438–41

80. Quemard A, Lacave C, Laneelle G. 1991. *Antimicrob. Agents Chemother.* 35:1035–39

81. Zhang Y, Heym B, Allen B, Young D, Cole S. 1992. *Nature* 358:591–93

81a. Heym B, Zhang Y, Poulet S, Young D, Cole ST. 1993. *J. Bacteriol.* 175:4255–59

82. Rosner JL. 1993. *Antimicrob. Agents Chemother.* 37:2251–53

83. Zhang Y, Garbe T, Young D. 1993. *Mol. Microbiol.* 8:521–24

84. Cockerill FR, Uhl JR, Temesgen Z, Zhang Y, Stockman L, et al. 1995. *J. Infect. Dis.* 171:240–45

85. Heym B, Cole ST. 1992. *Res. Microbiol.* 143:721–30

86. Heym B, Alzari PM, Honore N, Cole ST. 1995. *Mol. Microbiol.* 15:235–45

87. Shoeb HA, Bowman BU, Ottolenghi AC, Merola AJ. 1985. *Antimicrob. Agents Chemother.* 27:399–403

88. Johnsson K, Schultz PG. 1994. *J. Am. Chem. Soc.* 116:7425–26

89. Banerjee A, Dubnau E, Quemard A, Balasubramanian V, Um KS, et al. 1994. *Science* 263:227–30

90. Quemard A, Sacchettini JC, Dessen A, Jacobs WR Jr, Blanchard JS. 1995. *Biochemistry* 34:8235–41

91. Dessen A, Quemard A, Blanchard JS, Jacobs WR Jr, Sacchettini JC. 1995. *Science* 24:1638–41

92. Rossmann MG, Liljas A, Branden C-I, Banaszak LJ. 1975. *Enzymes* 11A:61–102

93. Rouse DA, Morris SL. 1995. *Infect. Immun.* 63:1427–33

94. Zhang Y, Young D. 1993. *Trends Microbiol.* 1:109–13

95. Altamirano M, Marostenmaki J, Wong A, FitzGerald M, Black WA, Smith JA. 1994. *J. Infect. Dis.* 169:1162–65

96. Stoeckle MY, Guan L, Riegler N, Weitzman I, Kreiswirth B. 1993. *J. Infect. Dis.* 168:1063–65

97. Marcinkeviciene J, Magliozzo RS, Blanchard JS. 1995. *J. Biol. Chem.* 270: 22290–95

98. Jackett PS, Aber VR, Lowrie DB. 1978. *J. Gen. Microbiol.* 104:37–45

99. Jackett PS, Aber VR, Mitchison DA, Lowrie DB. 1981. *Br. J. Exp. Pathol.* 62:34–40

100. Thomas JP, Baughn CO, Wilkinson RG, Shepherd RG. 1961. *Am. Rev. Respir. Dis.* 83:891–93

101. Rastoggi N, Goh KS, David HL. 1990. *Antimicrob. Agents Chemother.* 34: 2061–64

102. Rastoggi N, Goh KS, Labrousse V. 1992. *Antimicrob. Agents Chemother.* 36:2843–46

103. Silve G, Valero-Guillen P, Quemard A, DuPont M-A, Daffe M, Laneelle G. 1993. *Antimicrob. Agents Chemother.* 37:1536–38

104. Wolucka BA, McNeil MR, de Hoffmann E, Chojnacki T, Brennan PJ. 1994. *J. Biol. Chem.* 269:23228–335

105. Schroder KH, Hensel I. 1970. *Antibiot. Chemother.* 16:302–4

106. Takayama K, Armstrong EL, Kunugi KA, Kilburn JO. 1979. *Antimicrob. Agents Chemother.* 16:240–42

107. Kilburn JO, Takayama K. 1981. *Antimicrob. Agents Chemother.* 20:401–4

108. Sareen M, Khuller GK. 1990. *Antimicrob. Agents Chemother.* 34:1773–76

109. Takayama K, Kilburn JO. 1989. *Antimicrob. Agents Chemother.* 33:1493–99

110. Salfinger M, Heifets LB. 1988. *Antimicrob. Agents Chemother.* 32:1002–4

111. Heifets LB, Lindholm-Levy PJ. 1992. *Am. Rev. Respir. Dis.* 145:1223–25

112. Butler WR, Kilburn JO. 1983. *Antimicrob. Agents Chemother.* 24:600–01

113. Konno K, Feldmann FM, McDermott W. 1967. *Am. Rev. Respir. Dis.* 95:461–69

114. Machaness GB. 1956. *Am. Rev. Tuberc.* 74:718–28
115. Heifets LB, Flory MA, Lindholm-Levy PJ. 1989. *Antimicrob. Agents Chemother.* 33:1252–54
116. Yamamoto S, Toida I, Watanabe N, Ura T. 1995. *Antimicrob. Agents Chemother.* 39:2988–91
117. Havel A, Trnka L, Kuska J. 1964–1965. *Chemotherapy* 9:168–75
118. Walsh CT. 1993. *Science* 261:308–9
119. Ruf B, Schurmann D, Mauch H, Jautzke G, Fehrenbach FJ, et al. 1992. *Infection* 20:267–72
120. Ordway DJ, Sonnenberg MG, Donahue SA, Belisle JT, Orme IM. 1995. *Infect. Immun.* 63:741–43
121. Shinnick TM, King CH, Quinn FD. 1995. *Am. J. Med. Sci.* 309:92–98
122. Snapper SB, Melton RE, Mustafa S, Kieser T, Jacobs WR. 1990. *Mol. Microbiol.* 4:1911–19
123. Jarlier VL, Gutmann L, Nikaido H. 1991. *Antimicrob. Agents Chemother.* 35:1937–39
124. Basu J, Chattopadhyay R, Kundu M, Chakrabati P. 1992. *J. Bacteriol.* 174:4829–32
125. Prabhakaran K, Harris EB, Randhawa B, Hastings RC. 1992. *Microbios* 72:137–42
126. Prabhakaran K, Harris EB, Randhawa B, Adams LB, Williams DL, et al. 1993. *Microbios* 76:251–61
127. Zhang Y, Steingrube VA, Wallace RJ. 1992. *Am. Rev. Respir. Dis.* 145:657–60

Annu. Rev. Biochem. 1996. 65:241–69

PROTEIN PRENYLATION:
Molecular Mechanisms and Functional Consequences

Fang L. Zhang and Patrick J. Casey

Departments of Molecular Cancer Biology and Biochemistry, Duke University Medical Center, Durham, North Carolina 27710-3686

KEY WORDS: prenylation, isoprenylation, farnesylation, lipidation, Ras, G proteins

ABSTRACT

Prenylation is a class of lipid modification involving covalent addition of either farnesyl (15-carbon) or geranylgeranyl (20-carbon) isoprenoids to conserved cysteine residues at or near the C-terminus of proteins. Known prenylated proteins include fungal mating factors, nuclear lamins, Ras and Ras-related GTP-binding proteins (G proteins), the subunits of trimeric G proteins, protein kinases, and at least one viral protein. Prenylation promotes membrane interactions of most of these proteins, which is not surprising given the hydrophobicity of the lipids involved. In addition, however, prenylation appears to play a major role in several protein-protein interactions involving these species. The emphasis in this review is on the enzymology of prenyl protein processing and the functional significance of prenylation in cellular events. Several other recent reviews provide more detailed coverage of aspects of prenylation that receive limited attention here owing to length restrictions (1–4).

CONTENTS

INTRODUCTION AND HISTORY 242
ENZYMOLOGY OF PRENYL PROTEIN PROCESSING....................... 244
  General Features 244
  Prenylation of CaaX Proteins—Farnesyltransferase and Geranylgeranyltransferase
    Type I 245
  Prenylation of Rab Proteins—Geranylgeranyltransferase Type II............... 250
  Proteolytic Maturation of Prenylated C-Termini 252
  Methylation of C-Terminal Prenylcysteines................................ 253
FUNCTIONAL CONSEQUENCES OF PROTEIN PRENYLATION 255
  Membrane Targeting and Cellular Localization............................. 255
  Protein-Protein Interactions.. 258
  Information from Inhibitor Studies 259
  Metabolism of Prenylated Proteins 262
CONCLUSION.. 264

241

INTRODUCTION AND HISTORY

Isoprenoids constitute an array of compounds formed from isopentenyl pyrophosphate, which contains the basic five-carbon building block termed an isoprene unit. Early studies of isoprenoid metabolism focused on the major end-product of the pathway: cholesterol (5). The first evidence that isoprenoids can modify polypeptides came from studies in the late 1970s and early 1980s on the structures of certain fungal mating factors. These mating factors were found to consist of a short peptide terminating in a Cys residue to which a farnesyl group or, in one instance, an oxidized farnesyl group, was covalently linked (6, 7).

The discovery that mammalian proteins could be subject to prenylation took much longer, partly because precise structures of proteins cannot be determined as easily as that of peptides. The elucidation of the cholesterol biosynthesis pathway laid the groundwork for this discovery. The synthesis of mevalonate by the enzyme HMG CoA reductase is the committed step in cholesterol formation. Identification of compactin, a specific inhibitor of HMG CoA reductase, made possible studies designed to follow the metabolism of mevalonate in cells (8). In a series of studies in the early 1980s, a product of mevalonate metabolism other than cholesterol was found to be required for entry of cells into the S phase of the cell cycle (9, 10). The search for the required metabolite of mevalonate, which involved following the fate of exogenously added [³H]mevalonate to compactin-treated cells, revealed that metabolites of mevalonic acid are incorporated into proteins (11–13). Subsequently, one of the proteins was identified as the nuclear envelope–protein lamin B (14, 15), and the modifying species was shown to be a farnesyl group (16).

At about the same time as the lamin B discoveries, independent studies demonstrated that Ras proteins are farnesylated (17–19). Ras proteins are small GTP-binding proteins (G proteins) that play crucial roles in signaling pathways controlling cell growth and differentiation (20, 21). Yeast genetics provided important clues for the discovery that Ras is modified by an isoprenoid lipid. Genes required for the post-translational maturation of both Ras and the peptide-mating pheromone a-factor in S. cerevisiae were discovered, indicating that both polypeptides are processed via a common route (2, 22). The finding that, like the fungal mating factors mentioned above, a-factor contained a farnesylated cysteine at its carboxyl terminus (23) prompted investigators to examine whether Ras proteins were also subject to this modification. This search led to the discovery that Ras proteins are farnesylated and that the modification is required for oncogenic forms of Ras to transform cells (17–19). These studies dramatically stimulated interest in the field of protein prenylation (24).

The coding sequences of lamin B, Ras proteins, and a-factor indicated that the C-termini of all these proteins contain a Cys residue fourth from the end

(16, 22, 25), the so-called CaaX motif. The identification of the CaaX motif as that which directs prenylation united the field from lower eukaryotes to mammalian systems (1, 2). Searches of sequence databases revealed that a variety of proteins contain the CaaX motif (1, 26). Comparison of the CaaX motif with the C-termini of mature proteins also indicated that the processing of CaaX-containing proteins consists of at least three steps because, in addition to being prenylated, the mature proteins lacked the three C-terminal residues (i.e. the 'aaX') and contained a carboxymethyl group on the prenylcysteine (1) (see Figure 1).

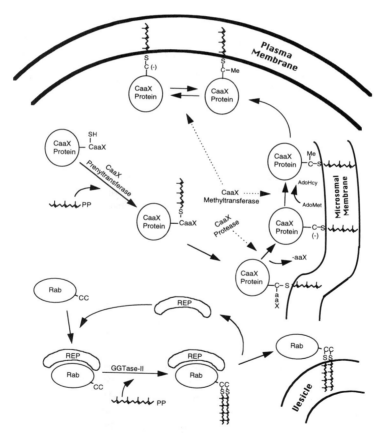

Figure 1 Overview of protein prenylation in cells. Modification of C-terminal Cys residues (designated "C") for both the CaaX motif and Rab proteins are depicted. In the case of the CaaX motif proteins, farnesylation is the modification shown. The minus sign (−) on particular forms designates those containing a free carboxyl group on the Cys residue and "C-Me" indicates that the carboxyl group has been methylated. See text for further details.

Two subsequent findings finished the framework for the protein prenylation field. The first was that, in addition to farnesyl, the 20-carbon geranylgeranyl isoprenoid can be attached to proteins. In fact, geranylgeranyl is the predominant isoprenoid found on cellular proteins (27, 28). The second finding was that prenylation was not confined to proteins containing the CaaX motif, but also occurred on members of the Rab family of Ras-related G proteins, which lack this motif (29, 30). Most of these proteins contain two Cys residues at, or very near, their C-termini, and they are processed in a fashion distinct from that of the CaaX-containing proteins (see Figure 1).

Tremendous progress has been made in this field in the past five years. Prenylation has been shown to exert profound effects on a host of processes involving signal transduction and intracellular trafficking pathways, and in many cases the specific prenylated proteins involved have been identified and the functional consequences of their processing at least in part defined. The enzymes responsible for isoprenoid addition to proteins have been identified and characterized at a molecular level both in mammalian systems and in lower eukaryotes. The requirement for prenylation in transformation by oncogenic Ras has made the enzyme responsible for this modification an important target for anti-tumor drug design. Additionally, a defect in one of the molecules required for Rab prenylation has been identified as responsible for a genetic form of retinal degeneration in humans. Clinical as well as biological implications of protein prenylation will clearly continue to drive the effort in this field.

ENZYMOLOGY OF PRENYL PROTEIN PROCESSING

General Features

Prenylated proteins can be grouped into two major classes: those containing the CaaX motif and the so-called CC- or CxC-containing proteins (3). The former class contains a diverse group of proteins (1, 26), whereas the latter is almost exclusively composed of members of the Rab family of small GTP-binding proteins that participate in intracellular membrane trafficking (31). Three known enzymes catalyze isoprenoid addition to proteins; these are termed protein farnesyltransferase (FTase), protein geranylgeranyltransferase type I (GGTase-I), and protein geranylgeranyltransferase type II (GGTase-II). FTase and GGTase-I are closely related and transfer a farnesyl group or a geranylgeranyl group from farnesyldiphosphate (FPP) and geranylgeranyl-diphosphate (GGPP), respectively, to the cysteine residue of CaaX-containing proteins (32–35). The carboxyl-terminal residue of the CaaX motif (i.e. the "X") in general determines which isoprenoid will be added to a protein. When "X" is serine, methionine, or glutamine, proteins are recognized by FTase, whereas a leucine at this position results in modification by GGTase-I (33, 34). Following prenylation, the aaX residues are cleaved by an endoprotease and the carboxyl group of the modified cysteine is methylated by a specific methyltransferase.

GGTase-II transfers geranylgeranyl groups from GGPP to both cysteine residues of CC- or CxC-containing proteins in a process mechanistically distinct from that of the CaaX proteins (36–38). Additionally, proteins containing the CxC motif are methylated at the C-terminal prenylcysteine, whereas CC-containing proteins are not (38–40).

Prenylation of CaaX Proteins—Farnesyltransferase and Geranylgeranyltransferase Type I

Mammalian protein farnesyltransferase (FTase) was first identified and purified to homogeneity from rat brain cytosol (32). FTase is a heterodimer consisting of 48 kDa ($\alpha_{F/GGI}$)– and 45 kDa (β_F)–subunit polypeptides (32); the nomenclature $\alpha_{F/GGI}$ is chosen because the α subunit is also a component of GGTase-I (see below). Substrates for FTase in mammalian cells include all known Ras proteins, nuclear lamins A and B, the γ subunit of the retinal trimeric G protein transducin, rhodopsin kinase, and a peroxisomal protein of unknown function termed PxF (1, 26, 42, 43). One important property of FTase is that it can recognize short peptides containing appropriate CaaX motifs as substrates (32, 44, 45). Specificity in recognition of Ca_1a_2X sequences by FTase indicates that the a_1 position has a relaxed amino acid specificity, whereas variability at a_2 and X are more restricted. Basic and aromatic side chains are tolerated at a_1 but not at a_2, whereas acidic residues are not well tolerated at either position (35, 46). Moreover, substitution at the a_2 position by an aromatic residue in the context of a tetrapeptide creates a molecule which is not a substrate for FTase but rather is a competitive inhibitor (44, 45). One such peptide, CVFM, has served as the basis for design of peptidomimetic inhibitors of FTase (see below).

cDNA clones encoding mammalian FTase have been isolated from rat, bovine, and human libraries (47–50). The mammalian $\alpha_{F/GGI}$ subunit has 377 amino acid residues and a calculated molecular weight is 44 kDa (48). A string of nine consecutive proline residues near the amino terminus is responsible for the apparent molecular weight of 48 kDa observed on SDS-PAGE (51). The mammalian β_F subunit consists of 437 residues with a calculated molecular weight of 48.6 kDa (47, 50). The $\alpha_{F/GGI}$ and β_F subunits of mammalian FTase show about 30% and 37% identity with the proteins encoded by the S. cerevisiae genes RAM2 and RAM1 (also known as DPR1), respectively (47–49). The two yeast genes were originally identified in a genetic screen of suppressors of RAS2[val19], a mutationally activated RAS allele (22, 52). RAM1 was also identified based on its involvement in a-factor processing and as a suppressor of G protein function (53, 54). Mutations in either *RAM1* or *RAM2* abolish FTase activity, and coexpression of the two genes in *E. coli* produces FTase activity that can farnesylate a-factor peptide and Ras protein substrates (55–57). Similarly, coexpression of mammalian FTase subunits in HEK 293 cells, *E. coli*, and Sf9 cells yields functional enzyme (47, 50, 58). The baculovirus-mediated Sf9 cell expression is particularly efficient (58) and is capable of producing the quantity of enzyme required for mechanistic and structural studies.

FTase is a zinc metalloenzyme. Prolonged dialysis against chelating reagents such as EDTA completely inactivates the enzyme. Neither Zn^{2+} nor Mg^{2+} alone restores the activity of metal-depleted FTase, but addition of both Zn^{2+} and Mg^{2+} fully restores activity (59, 60). The dependence on millimolar levels of Mg^{2+} for full activity indicates that it is probably not an integral component of FTase (59, 60). Measurement of the zinc content of recombinant FTase has confirmed that FTase contains one mole of zinc per mole of enzyme (58). The zinc is not required for FPP binding, but is required for protein substrate binding (60). Whether the zinc plays a structural role or whether it is directly involved in catalysis is not yet known. One possibility for a catalytic role of the zinc is that the metal could activate the sulfhydryl of the substrate protein cysteine residue and make it more nucleophilic. Evidence for such a "metalloactivation of Cys" mechanism has been found in a DNA repair enzyme termed Ada, which catalyzes a reaction chemically similar to that of FTase (61). Metal substitution studies combined with spectroscopic analysis could provide evidence for such a mechanism.

FTase can bind either substrate independently (62). Binding of peptide substrate has been closely examined by NMR using a transferred NOE approach, revealing that the CaaX sequence of a peptide substrate adopts a Type I β-turn conformation when bound to the enzyme (63). A similar study of binding of a peptidomimetic inhibitor of FTase termed L-739,787 revealed a slightly different conformation most closely approximating a Type III β-turn (64). Binding of FPP by FTase is of such high affinity that the complex can be isolated by gel filtration. No covalent adduct is involved, however, because FPP can be released intact by denaturing the enzyme (62). Although kinetic analysis indicates that FTase has distinct binding sites for its two substrates, direct evidence for the formation of a ternary complex has not been obtained.

The binding sites for both substrates seem to lie on the β subunit of the enzyme. Both protein and peptide substrates can be crosslinked to $β_F$ (62, 65), whereas divalent affinity-labeled short peptide substrates are crosslinked preferentially to both the α and β subunits upon photoactivation (65). Similarly, a photoactivated FPP analog is crosslinked to the $β_F$ subunit upon activation (50). Therefore, these results suggest that both substrates are bound to the β subunit of FTase, and the binding site for the peptide substrate may be near the interface of the two subunits. However, because in these experiments the affinity group was attached to the amino terminus of peptide substrate or the C-15 end of the FPP substrate (i.e. some distance from the site of chemistry), one should be cautious in interpreting the results.

Steady-state kinetic data have been interpreted to indicate that FTase proceeds through a random sequential mechanism, meaning that either substrate can bind to the free enzyme first, but both substrates must bind before either one can go on to form product (65a). This interpretation is also consistent with the substrate binding studies. However, isotope partitioning studies have indicated that the preferred catalytic pathway is through the FTase•FPP binary complex, whereas the pathway through FTase•RasCVLM binary complex is slower (66). Presteady-state kinetic analysis has further confirmed this ordered sequential mechanism. In these studies, FTase was shown to bind FPP in a two-step process to form FTase•FPP*, with the second step presumably involving a conformational change in the enzyme-substrate complex. FTase•FPP* then rapidly reacts with the peptide substrate to form product, and product release is the rate-limiting step (67). Therefore, the random-sequential model is only an approximate description of the true ordered-sequential mechanism. The precise chemical mechanism of the reaction is still undefined. A recent study using fluoro-substituted FPP analogs has been interpreted to indicate that the mechanism is electrophilic in nature (68), whereas analysis of the stereochemical course of the reaction using FPP with chiral deuterium-for-hydrogen substitutions at the C-1 carbon indicate that the reaction proceeds with inversion, suggesting a more nucleophilic mechanism (206). Additional studies will be required to firmly establish the chemical mechanism.

Molecular biological approaches are just beginning to be used to study the structure and function of FTase. Multiple alignment analysis of mammalian and yeast $\alpha_{F/GGI}$ subunits revealed that the sequences contain five tandem repeats of a specific pattern. The positions of four residues—glutamate, asparagine, tryptophan, and arginine—are highly conserved in the repeats (69). Mutation of mammalian $\alpha_{F/GGI}$ in four of these positions (i.e. R172E, N199D, N199K, N203H) impairs FTase activity; however, enzyme heterodimers still form from these mutants (50, 51). Therefore, these residues may be involved in a conformational requirement of FTase rather than subunit interactions. Deletion of 51 residues from the N-terminus of the rat $\alpha_{F/GGI}$ subunit does not affect enzyme activity, but deletion of 106 residues from the N-terminus or 5 residues from the C-terminus abolishes FTase activity (51). Mutation of Lys164 to Asn in $\alpha_{F/GGI}$ produced a polypeptide that still dimerized with the β subunit, and the resulting FTase retained its ability to bind substrates, but the mutant enzyme had no activity (51). These data are probably the best evidence to date that the α subunit has a direct role in the catalysis by FTase.

Point mutations in conserved residues of the human β_F subunit that correspond to known yeast mutants impair FTase activity, but still form dimers with the α subunit (50). The residues involved (D200, G249, G349) are highly conserved among all protein prenyltransferases.

Mammalian protein geranylgeranyltransferase Type I (GGTase-I) was also first identified as a cytosolic enzyme in bovine brain and subsequently purified to homogeneity from this tissue (35, 59, 70). Like FTase, it consists of two subunits, the 48 kDa α subunit shared with FTase ($\alpha_{F/GGI}$) and a 43 kDa polypeptide (β_{GGI}) (59). Known targets of GGTase-I include the γ subunit of brain heterotrimeric G proteins and Ras-related small GTP-binding proteins such as Rac, Rho, and Rap (33, 34). Substrate binding properties of GGTase-I are quite similar to FTase, in that short peptides containing the CaaX motif of its substrates (i.e. those containing Leu at the C-terminus) can be recognized by GGTase-I and the isoprenoid substrate GGPP binds quite tightly to the enzyme (33, 34, 70).

cDNA clones encoding the β subunit of mammalian GGTase-I have been isolated from rat and human libraries (71). The cDNAs encode polypeptides of 377 residues that are homologous to a yeast gene known as *CDC43/CAL1*. The *CAL1* gene was isolated based on a Ca^{2+}-dependent phenotype; it grows well in Ca^{2+}-rich media but not in Ca^{2+}-deficient media (72). *CDC43* was isolated based on a temperature-sensitive defect in cell polarity and also exhibits defects in localization of budding and secretion (73). Cells containing a temperature-sensitive cdc43 mutation exhibit greatly reduced GGTase-I activity and introduction of *CDC43* gene into the mutant restores enzyme activity (74).

It is now quite clear that the α subunits of GGTase-I and FTase are identical. Antibodies raised against the α subunit of FTase also recognize this α subunit of GGTase-I (75). Additionally, peptide sequences obtained from the α subunit of bovine GGTase-I are identical to regions of the α subunit of bovine FTase (71). Finally, coexpression of human FTase α with the β subunit of human GGTase or of *S. cerevisiae* RAM2 and CAL1/CDC43 generates GGTase-I activity (71, 76).

GGTase-I, like FTase, is a zinc metalloenzyme (59, 77). However, activity of metal-depleted GGTase-I can be restored with Zn^{2+} alone, unlike FTase which requires both Zn^{2+} and Mg^{2+} (78). As noted earlier, Mg^{2+} is absolutely required for FTase activity (59, 60, 78). The lack of a requirement for Mg^{2+} exhibited by GGTase-I is surprising (since other properties of FTase and GGTase-I are so similar) and highlights the importance of determining the precise role(s) for the metal ions in the function of both enzymes. The role of zinc is likely same in both FTase and GGTase-I. As with FTase, the zinc of GGTase-I is required for the binding of peptide substrate but not for isoprenoid binding (78, 79). The zinc of GGTase-I can be replaced by Cd^{2+} with retention of enzymatic activity (78, 79), but the Cd^{2+}-substituted enzyme does exhibit somewhat altered specificity for substrates (78). The ability to substitute Cd^{2+} for Zn^{2+} in this enzyme opens the door to the use of spectroscopic techniques (e.g. ^{113}Cd-NMR) to study the role of the metal in catalysis.

GGTase-I also binds GGPP and peptide substrates independently. The GGPP/GGTase-I complex can be isolated by gel filtration, and the isoprenoid can be released intact upon denaturation of the enzyme (70). A photoactivatable GGPP analog has been shown to crosslink to β_{GGI} (79, 80). Thus, as with FTase, it is reasonable to assume that both substrate binding sites are localized predominately in the β subunit of GGTase-I. Steady-state kinetic analysis of GGTase-I is consistent with a random-sequential mechanism (77) but, as with FTase, isoprenoid substrate binding first to the free enzyme seems the preferred kinetic pathway (79).

The β subunits of rat GGTase-I and FTase share about 30% identity. The central regions are very homologous, but the C- and N-termini are more divergent (71). The ram2 mutant of the $\alpha_{F/GGI}$ subunit in *S. cerevisiae* has totally lost its FTase activity whereas its GGTase-I is only reduced twofold, indicating that either the regions interacting with β_F and β_{GGI} are different or the ram2 mutation is tolerated in GGTase-I (35). In vitro mutagenesis has also identified a mutation in *RAM1* (S159N) based on suppression of the call phenotype. The mutant enzyme has increased activity to farnesylate a substrate for GGTase-I, but the ability to farnesylate its own substrate is reduced (81). Therefore, the S159N mutant may be in the peptide substrate binding site.

FTase and GGTase-I are generally quite selective for their substrates. However, cross-specificity has been observed (34, 35), and such ability to modify alternate substrates may be of biological significance. In *S. cerevisiae RAM1* null mutants are viable, although they exhibit growth defects (22). Overexpression of CDC43 (the β subunit of GGTase-I) can partially suppress the growth defect of *RAM1* null mutants, suggesting that GGTase-I can partially prenylate substrates of FTase (82). *CDC43* null mutants are not viable, but overexpression of two essential substrates of this enzyme, Rho1 and Cdc42, allows growth in a *RAM1*-dependent manner, suggesting that FTase can prenylate these substrates of GGTase-I when they are overproduced (82, 82a). A specific form of mammalian Ras—K-RasB—can serve as a substrate for both FTase and GGTase-I, although the K_M is much lower for the former enzyme (83). RhoB, a Ras-related G protein implicated in organization of the actin cytoskeleton, can be modified by either farnesyl or geranylgeranyl (84), and it has been shown that both isoprenoids can be transferred to the protein by GGTase-I (85).

In summary, FTase and GGTase-I have very similar properties. Both are zinc metalloenzymes that require this metal for binding of protein but not isoprenoid substrate. While either substrate can bind independently to these enzymes, isoprenoid binding first is the preferred kinetic pathway in catalysis. The major difference detected to date between these enzymes, apart from their ability to recognize distinct forms of protein and isoprenoid substrates, is that FTase requires Mg^{2+} for its activity, whereas GGTase-I does not. Remaining questions on these enzymes primarily pertain to the precise chemical mechanism of the transferase reaction and the roles of the metal ions in catalysis. Additionally, active site residues and binding sites for substrates have not been localized. Further structure-function studies and elucidation of structures of FTase and GGTase-I should provide insights into these important questions.

Prenylation of Rab Proteins—Geranylgeranyltransferase Type II

The discovery that Rab proteins, most of which do not contain a CaaX motif, were modified by the geranylgeranyl isoprenoid provided the first hint of the existence of a protein prenyltransferase distinct from FTase or GGTase-I (29, 30). The enzyme responsible, protein geranylgeranyltransferase type II (GGTase-II), was identified using Rab proteins as substrates (35, 37). Purification of GGTase-II from rat brain cytosol revealed that three polypeptides were required for activity—a 95 kDa protein originally termed component A but recently renamed Rab Escort Protein 1 (Rep1) and an $\alpha\beta$ heterodimer of 50 kDa (α_{GGII}) and 38 kDa (β_{GGII}) polypeptides (36, 86). cDNA clones of the α and β subunits have been isolated from rat cDNA libraries (87), and they encode products of 567 (α_{GGII}) and 331 (β_{GGII}) amino acids that each share about 30% identity with their counterparts in FTase and GGTase-I. The actual enzymatic function of GGTase-II is considered to reside in the $\alpha\beta$ dimer component, whereas the involvement of Rep1 in the reaction seems primarily limited to its binding to Rab protein substrates in such a way as to form a complex which can be acted upon by the enzymatic component (see below).

Rep1 binds both unprenylated and prenylated Rab proteins, leading to the suggestion that Rep1 plays an escort role in the GGTase-II reaction (88). In this model, Rep1 binds unprenylated Rab, presents it to the catalytic heterodimer of GGTase-II, and remains bound to the Rab after the geranylgeranyl transfer reaction. In the absence of detergent, the reaction terminates after all Rep1 is occupied by prenylated Rab. Certain detergents can dissociate this product from Rep1, allowing multiple cycles of catalysis. In cells, another Rab binding protein termed GDI may substitute for detergents, since GDI also binds prenylated Rab proteins (88, 89).

The finding that the Rab-Rep complex is the actual substrate for GGTase-II explains why GGTase-II, in contrast to FTase and GGTase-I, does not recognize short peptides containing the appropriate prenylation motif (e.g. CC and CxC) (30, 89a). Several studies have examined the requirement Rab proteins have for upstream sequences in order to serve as GGTase-II substrates. Examples include point mutations in the effector domain of Rab1B which abolish its prenylation (90) and analysis of Ras/Rab6 chimeras showing that the loop 3/β3 and hypervariable regions in the C-terminus of Rab6 are required for efficient processing of this protein by GGTase-II (91). The effects of these mutations on processing are likely due at least in part to disruption of Rab-Rep complex formation.

At least in the case of the CC- and CxC-containing proteins, GGTase-II apparently modifies both Cys residues with geranylgeranyl groups (38). The precise mechanism through which this occurs is not clear, but both residues are likely prenylated in a single cycle of the reaction, i.e. the mono-prenylated species does not dissociate and then re-bind for addition of the second isoprenoid (38). GGTase-II is generally assumed to prenylate all CC- or CxC-containing Rab proteins, as well as other Rab proteins with paired Cys residues near the C-terminus (e.g. the CCSN motif of Rab5), since only one enzyme has been identified so far. GGTase-II requires Mg^{2+} for activity (36), but whether this enzyme contains Zn^{2+} as do FTase and GGTase-I is not clear.

The elucidation of protein sequences from Rep1 and its subsequent cloning from a rat cDNA library revealed that the corresponding gene was the counterpart of a human gene identified by positional cloning as that responsible for choroideremia, a disease of retinal degeneration (88, 92). Extracts from lymphoblasts from patients with choroideremia (Rab3A was used as a substrate) were in fact found to be deficient in GGTase-II activity (92). Activity could be restored, however, by addition of purified Rep1 protein. Therefore, both the genetics and biochemistry indicate that a defect in Rep1 function is responsible for human choroideremia (92). However, these findings also indicated the existence of another form of Rep, because cells from the patients retained some functional GGTase-II activity and furthermore the effects of choroideremia are limited to the retina. These issues have been at least partially resolved with the discovery that the product of a gene designated "choroideremia-like," owing to its homology to the choroideremia gene, possessed Rep activity. Furthermore, this protein, now termed Rep2, has somewhat altered specificities in supporting prenylation of distinct Rab isotypes compared to Rep1 (93). In this regard, prenylation of a protein identified as Ram/Rab27 has been shown to be preferentially reduced in choroideremia, and Rep1 is much better at supporting prenylation of this protein than Rep2 (94). Furthermore, Ram/Rab27 is expressed at high levels in the retinal cell layers that degenerate earliest in the progression of this disease (94). Taken together, these data provide a potential molecular mechanism for how loss of Rep1 activity can selectively lead to retinal defects.

Yeast genes encoding the α_{GGII} and β_{GGII} polypeptides have been identified as *BET4* (formerly known as *MAD2*) and *BET2*, respectively (95, 96). *BET4* was originally isolated from budding yeast based on its sensitivity to the anti-microtubule drug benomyl (95), accounting for the original MAD nomenclature. However, the true *MAD* gene was later found to be the one next to *BET4* (97). *BET2* was initially identified as a gene required for vesicular transport from endoplasmic reticulum to Golgi apparatus in yeast (96). The products of the *BET4* and *BET2* genes form a complex, and mutants in either gene are deficient in geranylgeranylation of Ypt1 and Sec4, which are the yeast Rab proteins (95, 98). The yeast homolog of the mammalian Rep component has been identified as the product of the *MRS6/MSI4* gene (99, 100). *MRS6* was isolated as a high copy suppressor of the respiration-deficient phenotype of a mrs2 mutant, which is deficient in mitochondrial intron splicing (101). *MSI4* was isolated as a multicopy suppressor of ira1 (an inhibitory regulator of Ras) (99). The *MRS6* gene is essential for growth, and mutants in this gene are defective in both protein trafficking and in localization of Ypt1 (99, 101). Extracts from cells coexpressing *BET4* and *BET2* plus yeast Rep component can transfer a geranylgeranyl group to Ypt1 (102). Therefore, *S. cerevisiae BET4, BET2,* and *MSI4/MRS6* correspond to mammalian GGTase II α, β, and Rep, respectively.

In summary, GGTase-II has been characterized at a molecular level as a dimeric enzyme in both mammalian and yeast systems. Unlike the CaaX prenyltransferases, GGTase-II requires a third component, termed Rep, to function. Rep binds an unprenylated Rab protein and presents it to the catalytic dimer of GGTase-II. The enzymology of GGTase-II has not yet been thoroughly studied. Determining whether its catalytic mechanism is similar to that of the CaaX prenyltransferases and exactly how the di-geranylgeranylation of Rab proteins is accomplished is particularly important. Furthermore, the question of whether the enzyme exists solely for the modification of Rab proteins is still open, because evidence has been presented that casein kinase I in yeast may be modified by this enzyme (103). If so, this finding raises the question of whether this kinase also requires a Rep-type escort protein for its modification.

Proteolytic Maturation of Prenylated C-Termini

The cleavage of the three C-terminal amino acid residues of prenylated CaaX proteins was first observed in the maturation process of mammalian and yeast Ras proteins (52, 104). An enzymatic activity has been identified in canine pancreatic microsomal membranes using in vitro–expressed Ras as a substrate (105) and in bovine and rat microsomal membranes through use of short prenylated peptides as substrates (106–108). The endoprotease activity that is being studied by most groups has the following properties: (*a*) the activity is stereospecific, and D-amino acid substitutions in the prenylated peptide abolish substrate activity; (b) the activity recognizes both farnesyl- and geranylger-anyl-containing peptides; and (c) the activity is not inhibited by either nonfar-nesylated peptides or common protease inhibitors, including those for serine proteases, cysteine proteases, metalloproteases, and the aspartyl proteases (107–110). Prenylated tetrapeptides with acetylated N-termini are cleaved by the enzyme at the modified cysteine to generate a tripeptide product. Although the corresponding farnesylated tripeptides and dipeptides are also substrates of the enzyme, they exhibit V_{max} values only 10% of that of the corresponding tetrapeptide (106). All these properties are consistent with those expected for a protease specific for prenylated proteins and have led to a general consensus that this is the biological activity for processing prenylated proteins.

A prenylpeptide endoprotease activity has also been identified in the membrane fraction of S. cerevisiae (108, 111). Like the activity in mammalian cells, this enzyme cleaves farnesylated a-factor peptide at the prenyl cysteine to release a tripeptide, and unfarnesylated a-factor peptide is not a competitor. This activity is sensitive to sulfhydryl reagents and $ZnCl_2$, and these properties are also observed for the mammalian activity (108, 111). Given these similarities, this activity seems to be the yeast counterpart of the mammalian protease involved in processing prenylated proteins.

No molecular information is yet available on the prenyl protein C-terminal protease in any organism, which in large part explains the paucity of mechanistic information available on the process. Some quite potent inhibitors of the protease have been designed based on the structure of the prenylpeptide substrate (110), but to date have not been applied either in mechanistic studies or in biological systems. Identification of the gene(s) encoding this enzyme will probably be required before such crucial questions as whether there is but a single enzyme for all CaaX-type proteins [there is some evidence for substrate-selective isoforms (112)], and what the functional consequences of its action are, can be addressed. Identification of the yeast gene(s) involved will be particularly useful for such studies.

Methylation of C-Terminal Prenylcysteines

The final step in the C-terminal processing of CaaX-containing proteins is methylation of the carboxyl group of the prenylated cysteine residue that is exposed by the proteolytic step. Methylation of H-Ras was first observed by incorporation of radiolabel from [3H-Methyl] methionine into the protein, and studies indicated that the site of methylation was the C-terminal cysteine residue (25). Subsequent studies have established that essentially all CaaX-type prenyl proteins, and a subset of Rab proteins (see below), are subject to such C-terminal methylation (1, 4).

A methyltransferase activity capable of specifically modifying prenylcysteines has been identified in several mammalian tissues, including bovine retina, rat liver, rabbit brain, and human neutrophils (113–115). The activity is enriched in the endoplasmic reticulum fraction of cells, but significant activity is also found in plasma membranes (113, 114). The enzyme uses S-adenosyl-L-methionine (AdoMet) as a methyl donor, and the simple prenylated amino acids N-acetyl-S-farnesyl-L-cysteine (AFC) and N-acetyl-S-geranylgeranyl-L-cysteine (AGGC) can serve as substrates (114, 116, 117). The smallest recognition unit identified is S-farnesylthiopropionic acid (FTP) (118). The structure of the isoprenoid attached to the Cys residue is quite important for substrate recognition. Substrates containing either the farnesyl or geranylgeranyl are recognized essentially equally well, but replacement by the 10-carbon geranyl isoprenoid or removal of the double bonds from the prenyl group abolishes substrate activity (118). As with the prenyl protein endopeptidase, a single enzyme appears to be responsible for the methylation of both farnesylated and geranylgeranylated CaaX-type substrates (116, 117). Kinetic analysis has indicated that the methyltransferase proceeds with an ordered BiBi mechanism with AdoMet binding first and its homocysteine form (AdoHcy) departing last (119). Inhibitors of this enzyme have been designed based on the structure of the prenylcysteine substrate and characterized extensively both in vitro and in their effects in cells (120–123) (and see below).

Carboxyl methylation of prenylated proteins in cell membranes has been reported to be stimulated by GTP or GTPγS (124). Since these nucleotides affect the conformation of G proteins, many of which are prenylated proteins, these findings may reflect an increased accessibility of the methylation site upon the activation of these G proteins rather than any modulation of the enzyme's activity. An activity capable of demethylating AFC has also been observed in the membranes (125). This finding indicates that the methylation step may be reversible and may play a regulatory role in signal transduction. Such a process has been found in biological systems before. In bacterial chemotaxis, for example, reversible carboxyl methylation of proteins is an important component of the signaling pathway (126, 127).

For proteins processed by GGTase-II, those containing the CxC motif—but not those with the CC motif—are subject to methylation (38, 40, 128). Replacement of the CC motif on a Rab protein with CxC produces one that is also methylated, indicating that the ability of the proteins to be methylated is dependent only on the carboxyl-terminal motif (40). The activity responsible for this class of methylation has been identified in bovine brain membranes (40, 129), and it appears to be a distinct enzyme from that which methylates CaaX-type proteins (129).

The gene encoding the methyltransferase for CaaX proteins in *S. cerevisiae* has been identified as *STE14*, which was isolated in a screen designed to identify genes involved in mating (hence the <u>Ste</u>rile designation) (130, 131).

The gene product is predicted to contain multiple membrane-spanning regions, and it has no sequence resemblance to other known methyltransferases (132). The ste14 mutant is deficient in the ability to methylate prenylated peptide substrates such as a-factor and Ras proteins; it is the lack of a-factor methylation that results in the sterile phenotype (see below). The *STE14* gene has been expressed in bacteria and the protein product shown to possess a prenyl peptide methyltransferase activity (133). No molecular information is yet available on the mammalian counterpart of this enzyme.

FUNCTIONAL CONSEQUENCES OF PROTEIN PRENYLATION

Membrane Targeting and Cellular Localization

Most prenylated proteins are localized at cell membranes, at least for a portion of their lives, and the isoprenoid modification is generally essential for this membrane association. This property was first demonstrated with Ras proteins, for which either a mutation of the relevant Cys residue (i.e. the prenylation site) to Ser or a blocking of isoprenoid biosynthesis abolished both prenylation and membrane association of the protein (17–19). This finding has subsequently been confirmed with many other prenylated proteins (1, 3). Ras has also served as a model system for assessing the contribution of each of the processing steps of CaaX-type proteins—prenylation, proteolysis, and methylation—to stable membrane association. In rabbit reticulocyte lysates, K-Ras undergoes only the first prenylation step. Addition of microsomal membranes and the methylation inhibitor methylthioadenosine then leads to production of a cleaved farnesylated, but unmethylated, intermediate, whereas the addition of the microsomal membrane fraction without the methylation inhibitor leads to complete processing (105). Analysis of each intermediate for its ability to associate with membranes showed that 20% of that modified only by the farnesyl associated with the membrane, but after the proteolysis step this fraction increased to 40%. Membrane association of completely processed K-Ras was ~80% in this system (105). These results indicate that although prenylation is essential for the membrane binding of Ras, the proteolysis and methylation steps also contribute markedly to efficient membrane association. This conclusion is also supported by a characterization of the membrane association of the I187TY mutant of K-Ras. This mutant protein can be farnesylated but does not apparently undergo either proteolysis or methylation, and it displays about 50% membrane association (134).

The contribution of C-terminal methylation to membrane association of the βγ subunits of trimeric G proteins has also been explored. This subunit complex of the retinal G protein transducin contains a farnesylated γ polypeptide, and the processed but unmethylated complex can be obtained either by direct isolation from retina (135) or by treatment of the purified protein with a nonspecific esterase (136). In both cases, the unmethylated $\beta\gamma$ complex exhibited marginally reduced affinity for membranes, consistent with an expected role for methylation in this interaction. Unmethylated transducin $\beta\gamma$ also has a much reduced efficacy in activating a specific subtype of phospholipase C that is subject to regulation by this G protein complex, and the magnitude of the reduction correlates well with the membrane affinity differences between those molecules that contain just the farnesyl group and those that are also methylated (136) (and see below).

More precise quantitation of the contributions of the C-terminal modifications of prenylated proteins in membrane interactions has been obtained through biophysical studies in model systems. These model studies primarily involved assessment of the interactions of modified peptides with artificial liposomes (137, 138). The basic conclusions derived from these studies are that a farnesyl group alone provides only a modest degree of membrane association and that such species can rapidly dissociate from the bilayer. Methylation of the farnesylcysteine, though, markedly shifts the equilibrium toward the membrane-associated form. In contrast, geranylgeranyl-modified peptides exhibit quite avid membrane association even in the unmethylated state. The presence of two geranylgeranylcysteines on a peptide, such as is found on the C-terminus of proteins modified by GGTase-II, results in quite effective bilayer association, as would be expected (139). The results with the mono-prenylated peptides also fit well with those obtained by computational modeling of the hydrophobicity parameter, designated log(P), for prenylated cysteine compounds (140). Methylation of a prenylated cysteine, which removes the negative charge at the carboxyl terminus, increases hydrophobicity about two log(P) units. The basic message from these studies is that methylation would have a profound influence on the membrane association of farnesylated proteins in particular and would probably be less important in this regard for those modified by either one or two geranylgeranyl groups.

In addition to the C-terminal processing initiated by prenylation, many CaaX-containing proteins need a second signal for stable membrane association. This second signal can be provided by palmitoylation in some proteins (e.g. H-Ras and N-Ras) or by a stretch of basic residues just upstream of the CaaX motif (e.g. K-RasB) (141). In H-Ras, Cys181 and Cys184 serve as palmitoylation sites. Mutation of these residues to Ser abolishes the palmitoylation of H-Ras (17). In one study, only 8% of the nonpalmitoylated H-Ras associated with the membrane fraction compared with 90% wild-type protein. Moreover, the small amount of mutant protein which was membrane associated could be extracted by 0.25M NaCl, whereas the wild-type protein was resistant to even 1M NaCl (141). Similarly, altering the polybasic region of K-RasB can apparently affect its membrane association. This protein contains six Lys residues just upstream of the prenylation site, and in one study mutation of all six Lys residues to Gln produced a protein primarily found in the cytosol, whereas 90% of wild-type K-RasB was associated with membranes (141). However, another group performing a similar study found much less of an effect of removing the basic residues from this Ras (142). No second signal seems to be required for efficient membrane association of CC- and CxC-containing proteins, probably because of the increased hydrophobicity imparted when both Cys residues are geranylgeranylated.

Although prenylation is in general essential for directing membrane association, additional factors must be important in directing these proteins to specific cell membranes. For instance, farnesylated Ras proteins are localized to the plasma membrane and another farnesylated protein termed PxF is found on the cytoplasmic surface of peroxisomes (17, 43). Similarly, geranylgeranylated CaaX-type proteins can reside at such sites as the plasma membrane [e.g. trimeric G proteins containing a geranylgeranylated γ subunit] or the Golgi (e.g. some forms of the Ras-related protein Rap) (143, 144). Most striking in this regard are the Rab proteins, each form of which is localized to distinct intracellular membranes (31). Therefore, the presence of a farnesyl or geranylgeranyl per se does not determine the membrane to which the protein is targeted. Analysis of chimeras between Rab proteins has revealed that the highly variable C-terminal domain of these proteins determines their specific membrane localization (145). In the case of lamin B, a nuclear localization signal in the protein is required for its appropriate subcellular localization (146).

Since prenylated proteins contain diverse membrane targeting signals in their sequences, distinct receptor-type proteins that can recognize these signals may exist in membranes. However, no such proteins have yet been identified. A microsomal membrane protein that recognizes prenylated peptides and proteins has been detected by both binding studies and crosslinking analysis, but its selectivity for prenylated peptides retaining the C-terminal tripeptide suggests that this protein plays a role in compartmentalizing newly prenylated proteins for subsequent processing steps (i.e. proteolysis and methylation) rather than in specific membrane targeting (147, 148).

Membrane localization of some prenylated proteins can be modulated. Rhodopsin kinase, which phosphorylates light-activated rhodopsin, is farnesylated, and both the farnesylated and unfarnesylated forms are predominately cytosolic in dark-adapted retina (42). However upon exposure to light the farnesylated, but not the nonfarnesylated, enzyme translocates to the rod–outer segment membrane (149). Whether this process is an example of a "switch" mechanism for extrusion of the lipid upon activation—a process well-documented for some myristoylated proteins (150, 151)—or whether it merely reflects the increased affinity of the kinase for membranes containing activated rhodopsin is not yet clear.

A more defined case of reversible membrane association of prenylated proteins comes from studies of Rab proteins. These proteins need to cycle between membranes and cytosol as part of their mechanism of action (151a). In this case, it is now quite clear that this reversible membrane association is due to protein-protein interactions. A specific protein termed GDI can bind to the GDP-bound form of Rab and induce its dissociation from membranes (3, 151a). Similarly, a protein named SOS, which acts as a nucleotide exchange factor for Ras, can bind this prenylated protein and induce its dissociation from the membrane (152). In both these instances, binding of the factor to the prenylated protein is dependent on the presence of the isoprenoid, suggesting that the factor contains a binding site for the lipid which shields it from the polar environment of the cytosol. Such interactions are described in greater detail in the next section.

Protein-Protein Interactions

Most prenylated proteins play important roles in signal transduction. A major question in the field is which protein-protein interactions in those pathways involving prenylated proteins are dependent on prenylation. This is a difficult question to address because the effect of the modifications on protein-protein interactions are difficult to examine separately from their effects on membrane association. However, in a few cases, the evidence is quite good that the prenyl group is important for protein-protein interactions.

Farnesylation of yeast Ras2 increases its affinity for adenylyl cyclase, the enzyme it regulates, by about 100-fold (153). This effect is observed even with solubilized adenylyl cyclase, providing good evidence that the influence of the farnesyl modification of Ras2 is to increase its affinity for adenylyl cyclase rather than simply to affect its membrane association. As noted above, farnesylation of K-Ras affects its interaction with SOS (152, 154). SOS forms a complex with farnesylated, but not with unprocessed, K-Ras, and SOS catalyzes the nucleotide exchange much more efficiently on farnesylated K-Ras than the unprenylated form. Interestingly, geranylgeranylated K-Ras is as good a substrate of SOS as is the farnesylated form, indicating that the putative lipid binding site on SOS is not absolutely selective for a specific lipid (154). Nucleotide exchange factors for the Ras-related protein termed Rho are similarly active on prenylated Rho and inactive on the unprenylated species (3, 156).

Rab3A is a heavily studied Rab protein owing to its involvement in regulated secretion in neuronal cells. This Rab terminates with Cys-Ala-Cys and so contains two geranylgeranyl groups and a methylated C-terminus (38, 40). As noted above, a cellular protein termed GDI can form a complex with prenylated Rabs but not with unprenylated ones. The GDI can bring the GDP-bound form of prenylated Rab3A into the cytosol by inhibiting the GDP-bound form of prenylated Rab3A from binding to synaptic membranes and inducing the dissociation of the GDP-bound form of prenylated Rab3A from synaptic plasma membranes. However, the GDI is inactive on the GTP-bound form of Rab3A, and thus the GTP-bound form of Rab3A is found associated with membranes (157). Therefore, the GDI can direct the membrane localization of RAb3A in a GTP-dependent fashion; this type of direction appears to occur in the modulation of membrane association of other Rab subtypes as well (151a).

Processing of prenylated proteins can also influence their assembly into multi-subunit complexes. One example of this is the $\beta\gamma$ complex of trimeric G proteins. Assembly of $\beta\gamma$ occurs in the cytosol, and both prenylated and unprenylated γ polypeptides can form dimers with the β subunit (143, 158, 159). However, removal of the three C-terminal residues from a prenylated γ eliminates the ability of the polypeptide to assemble with the β subunit (160). This finding suggests that assembly of the $\beta\gamma$ dimer must occur prior to completion of γ subunit C-terminal processing. More recent findings indicate that unprenylated γ is more efficient in dimerizing with β than is the prenylated species and, furthermore, that the unprenylated γ in such a $\beta\gamma$ complex can still be modified by the appropriate protein prenyltransferase (161), suggesting that the preferred route in vivo is to assemble the $\beta\gamma$ dimer prior to C-terminal processing of γ. Prenylation of γ is required, however, for the $\beta\gamma$ subunit complex to interact with the α subunit (159, 161a, 162). Additionally, interaction of a trimeric protein with a receptor can be influenced by the specific isoprenoid (i.e. farnesyl or geranylgeranyl) attached to the γ subunit (162a). Another example of prenylation-dependent assembly of a protein complex occurs with the large antigen of hepatitis delta virus. This polypeptide is prenylated, and the modification is required for its assembly with the surface antigen of hepatitis B (163, 164). Although the viral protein was initially identified as a geranylgeranylated protein (163), recent work indicates that it is in fact farnesylated, as would be predicted from its CaaX sequence (207). This finding indicates that inhibitors of FTase, which are under active development as anti-cancer agents (see below), may also be useful in blocking replication of hepatitis delta virus.

Information from Inhibitor Studies

Development of protein prenylation inhibitors is a focus of many groups in this field in both academia and the pharmaceutical industry. The primary driving force for such efforts came from the finding noted earlier that oncogenic forms of Ras proteins required farnesylation for their ability to transform cells. Ras proteins are key players in signal transduction associated with cell growth and development. Oncogenic mutations in Ras lead to production of constitutively active forms. Such mutations in Ras have been found in >50% of colon cancers and >80% of pancreatic carcinomas, for example, and are thought to be a critical factor in proliferation of these cancers (165, 166). Since controlling farnesylation can control Ras function, FTase has become a very attractive target for anti-tumor drug design.

Numerous inhibitors of FTase have been synthesized or identified. Based on their design template, these inhibitors can be grouped into three classes: (a) FPP analogs such as (α-hydroxyfarnesyl) phosphonic acid and others (167, 168); (b) CaaX peptide analogs such as BZA-5B, L-731,734, B581, and Cys-AMBA-Met (169–171, 174); and (c) bisubstrate analogs such as the phosphonic acid analog 14–16 (172). In addition to chemically synthesized inhibitors, several natural compounds are also inhibitors. These include chaetomellic acid, zaragozic acid, pepticiinnamins, gliotoxin, and manumycin (173).

Among the inhibitors identified, a handful have been shown to be active in intact cells. The FPP analog (α-hydroxyfarnesyl)phosphonic acid can partially inhibit the processing of Ras proteins in H-Ras-transformed NIH3T3 fibroblasts (167). Better results have been obtained with peptidomimetic inhibitors. Several types of these inhibitors have been developed that can block Ras processing and inhibit the growth of Ras-transformed cells (169, 170, 174–176). Bisubstrate analogs (172) and tricyclic compounds (172a) have also been identified that can block Ras transformation of cells. Several inhibitors have been tested in mouse xenograft models and shown to inhibit Ras-dependent tumor growth at concentrations that are not toxic to the host (176, 177, 177a). The most dramatic demonstration to date of the potential for FTase inhibitors in cancer chemotherapy comes from a recent study in which an analog of the peptidomimetic L-739,749 was used against tumors arising in a transgenic mouse carrying the gene for an oncogenic form of H-Ras (178). Administration of the inhibitor to mice bearing tumors resulted in almost complete tumor regression without visible toxicity to the animal. Withdrawal of the inhibitor resulted in a reappearance of the tumors which, in most cases, regressed again upon retreatment (178).

One quite unexpected finding from these studies was that the inhibitors of FTase were not toxic to cells at dosages that completely blocked processing of H-Ras protein (169, 174) and, presumably, other crucial farnesylated proteins. Moreover, treatment of cells with inhibitor under conditions that inhibit the processing of Ras did not affect the stimulation of MAP kinase by EGF in untransformed cells, despite the fact that the EGF receptor requires functional Ras to signal through the MAP kinase pathway (179). This paradox has several possible explanations. One is that the processing of oncogenic Ras is much more sensitive to the presence of FTase inhibitors than are the normal Ras and other farnesylated proteins. As such, the small amount of processed Ras and other required farnesylated proteins left would be enough to support normal cell growth.

Another possibility is that GGTase-I may prenylate at some level proteins that would normally be processed by FTase and that the geranylgeranylated forms would be functional. This type of situation has been documented in *S. cerevisiae*, in which a *RAM1* deletion can be partially complemented by CDC43, but double mutants of *RAM1* and *CDC43* are lethal (82; see also p. 249). The discovery that K-RasB also serves as a substrate for GGTase-I, at least in vitro, also provides support for this notion (83). Alternatively, geranyl-geranylated relatives of Ras, such as the recently discovered R-RAS2/TC21 protein, which is a member of the Ras superfamily and shares 55% amino acid identity with Ras proteins, may be able to substitute for other Ras proteins in normal cell growth (180). Yet another possibility is that there are branches in signaling pathways that can bypass those that require normal Ras proteins. It is also still a formal possibility that unprenylated Ras proteins can support normal cell growth in vivo. This type of support has also been documented in yeast, as overexpression of Ras2 proteins lacking the prenylation site can support cell growth (181). An additional concern is that at least some of the ability of FTase inhibitors to reverse cell transformation may be the result of a block to the processing of farnesylated proteins other than Ras (182). Further evidence for this possibility comes from studies showing that many types of cancer cells not harboring oncogenic forms of Ras can respond to FTase inhibitors (183, 183a).

Another target in development of inhibitors against prenyl protein process-ing is the C-terminal methyltransferase. Much of this work has centered on the use of the acetylated prenylcysteine substrate of this enzyme, N-acetyl farnesylcysteine (AFC). This molecule, which acts as an inhibitor of the enzyme owing to competition for protein substrates, inhibits human platelet aggregation in response to many agonists that act through G protein–controlled pathways. AFC also inhibits the chemotactic response of mouse peritoneal macrophages and superoxide formation induced by the agonist f-Met-Leu-Phe in human neutrophils (120, 122, 123). In these studies, the physiologic effects of AFC treatment have been viewed as being primarily due to inhibition of prenyl protein, particularly G protein, methylation. However, recent reports suggest that these data should be interpreted with caution. Farnesylcysteine analogs that are inactive as inhibitors of the methyltransferase activity also inhibit the aggregation of human platelets, suggesting that the primary target of AFC is not the methyltransferase (184). Furthermore, AFC and related prenylcysteines can directly interfere with G protein signaling via interruption of protein-protein interactions (185). Nonetheless, these and related findings (186) (and see below) indicate that it may be possible to develop prenylcysteine analogs as pharmacologic agents designed to disrupt cellular processes dependent on interactions involving recognition of prenyl proteins.

Metabolism of Prenylated Proteins

The findings that prenylation is a stable modification (19) and that such proteins comprise up to 2% of total cellular protein (187) raise the question of how cells dispose of isoprenoid-modified cysteines which would be produced during prenyl-protein turnover. Accumulation of such metabolites in cells could result in critical problems, since, as noted above, many cellular processes can be disrupted by free prenylcysteines (120, 122). While at this point in time the ultimate disposition of these potential prenylcysteine-containing metabolites is unclear, several recent studies may bear on this problem.

One potential mechanism for the cell to rid itself of prenylcysteines is for the prenylcysteine to be transported out of the cell and into the circulatory system for clearance. Data in support of this model come from studies of a particular class of cell-surface transporters in both yeast and mammalian systems. The prototype transporter in this class is a mammalian protein termed P-glycoprotein; this transporter is also known as the multidrug resistance transporter because overexpression of the transporter in human cancers confers resistance to many cytotoxic drugs used in therapy (188, 189).

In yeast, a transporter similar to P-glycoprotein, termed Ste6, is responsible for the export of the prenylated peptide mating pheromone, a-factor (190, 191). In a Ste6 deletion mutant, expression of a mouse P-glycoprotein gene could complement the function of Ste6 and restore low, but detectable, mating activity (192). This experiment suggested that the homology between P-glycoprotein and Ste6 extends to function as well as structure. Additionally, these findings implied that structural domain(s) of the a-factor peptide that determines specific transport by Ste6 are also involved in its interaction with the mouse P-glycoprotein. In fact, key determinants for recognition of a-factor by Ste6 are the farnesylated cysteine and its methyl ester; the requirement for the methyl ester for recognition explains why loss of function of the methyltransferase gene (i.e. STE14) results in sterility (132). In a recent study, prenylcysteine methyl esters corresponding to the C-terminus of prenylated proteins were found to exhibit properties characteristic of P-glycoprotein substrates (193). Recognition of prenylcysteines by P-glycoprotein was dependent on both the isoprenoid moiety and carboxyl methylation. The specificity of this interaction supports the hypothesis that these prenylcysteines may be physiologic substrates of P-glycoprotein (Figure 2). Additionally, analogs of these prenylcysteines have been identified that act as inhibitors of drug transport by

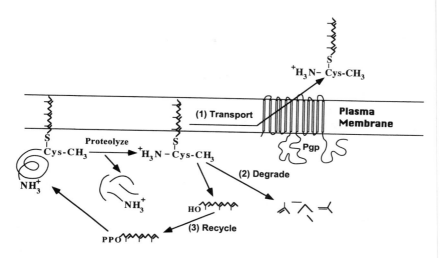

Figure 2 Potential fates of isoprenoid-containing compounds derived from the metabolism of prenylated proteins. Three potential fates of the prenylcysteines that could be produced upon the degradation of prenyl proteins (the "proteolyze" step) are depicted.

P-glycoprotein; such compounds can restore cytotoxic drug accumulation in cancer cells overexpressing P-glycoprotein (194).

Other possibilities for disposal of prenylcysteine-containing metabolites by cells include degradation of the isoprenoid or its recycling through some form of salvage pathway (Figure 2). While no data are yet available on potential degradative pathways for these compounds, several rather intriguing papers potentially related to the possibility that recycling can occur have recently appeared. These studies described the ability of cells to metabolize the prenyl alcohols farnesol and geranylgeraniol into forms capable of being incorporated into the prenyl protein pool and, in the case of farnesol, the sterol pool in cells (198–200). These data indicate the existence of cellular kinases capable of converting the prenyl alcohols to their corresponding di- phosphates for utilization by prenyltransferases. Thus, under certain conditions cells may be able to cleave prenylcysteines at the thioether bond to produce a prenyl alcohol which can be re-utilized for protein prenylation. Additionally, prenyl alcohols and their derivatives may serve regulatory functions in cells (195–197). S-oxidative cleavage of farnesylcysteine by a flavin-containing monooxygenase has in fact been detected in pig liver microsomes (201), and prenyl alcohols may also be produced by dephosphorylation of the corresponding diphosphates (202).

CONCLUSION

The past five years have yielded substantial progress on both the molecular mechanisms of protein prenylation and the functional properties bestowed on proteins by this processing. Efforts to design selective cell-active inhibitors of FTase in particular have been amazingly successful and have been greatly aided by the acquisition of mechanistic information on the protein prenyltransferases. Much remains to be learned about these enzymes, however, and they will likely yield many secrets to the enzymologist and structural biologist in the next five years. Crucial issues include the structural basis for specificity in substrate utilization, especially between FTase and GGTase-I, and the role of the Zn^{2+} atom in catalysis by the enzymes. The ever-increasing conviction of many that inhibitors of protein prenylation could be effective therapeutic agents in the treatment of many human cancers will continue to drive much of this effort.

The contribution of the attached lipid to the activities of prenylated proteins should also continue to be a fruitful area of investigation. The precise role of the prenylated Cys residue in protein-protein interactions needs to be more clearly defined. Here again, structural information such as that recently obtained for myristoyl binding sites on proteins would provide the beginning of a molecular model for these interactions (203, 204). Another important question that remains to be addressed for many prenylated proteins is exactly how they get sorted to specific cellular membranes. Many distinct routes will likely emerge, but all will almost assuredly involve interactions with additional proteins en route to or at the final destination (205). Additional post-translational modifications (e.g. phosphorylation) may also come into play in this process (3).

More information is sorely needed on the additional modifications of C-terminal proteolysis and methylation that occur on CaaX-type prenylated proteins. The findings that both processing steps apparently occur in a discrete subcellular compartment and that a single enzyme for each step modifies both farnesylated and geranylgeranylated forms suggest that both classes of CaaX-type prenylated proteins traffic through the same compartment in the cell on the way to their final destination. However, lack of molecular information on these processes in mammalian cells has seriously limited progress in ascribing functional consequences to the processing steps. However, these modifications can clearly contribute significantly to the biological activity of some prenylated proteins. Unraveling these pathways and the further molecular and functional details of protein prenylation should continue to provide new insight into both protein-membrane association and the regulation of the activities of many proteins involved in cellular signaling and membrane trafficking.

ACKNOWLEDGMENTS

We would like to thank the many colleagues who provided preprints of unpublished work. Special thanks are also due to the past and present members of the lab for their contributions to this work, to Julia Thissen, Lili Zhang, Tim Fields, Ruth Fu, Bill Tschantz, and Skip Waechter for their comments on the manuscript, John Moomaw for assistance in preparing figures, and Kristy Thompson for editorial assistance. Work from the authors' laboratory was supported by research grants from the NIH and American Cancer Society and an Established Investigator Award from the American Heart Association.

Any *Annual Review* chapter, as well as any article cited in an *Annual Review* chapter,
may be purchased from the Annual Reviews Preprints and Reprints service.
1-800-347-8007; 415-259-5017; email: arpr@class.org

Literature Cited

1. Clarke S. 1992. *Annu. Rev. Biochem.* 61:355–86
2. Schafer WR, Rine J. 1992. *Annu Rev. Genet.* 25:209–38
3. Glomset JA, Farnsworth CC. 1994. *Annu. Rev. Cell Biol.* 10:181–205
4. Caldwell GA, Naider F, Becker JM. 1995. *Microbiol. Rev.* 59:406–422
5. Goldstein JL, Brown MS. 1990. *Nature* 343:425–34
6. Kamiya Y, Sakurai A, Tamura S, Takahashi N. 1978. *Biochem. Biophys. Res. Commun.* 83:1077–83
7. Ishibashi Y, Sakagami Y, Isogai A, Suzuki A. 1984. *Biochemistry* 23:1399–404
8. Endo A. 1992. *J. Lipid Res.* 33:1569–82
9. Schmidt RA, Glomset JA, Wight TN, Habenicht AJ, Ross R. 1982. *J. Cell Biol.* 95:144–53
10. Fairbanks KP, Witte LD, Goodman DS. 1984. *J. Biol. Chem.* 259:1546–51
11. Schmidt RA, Schneider CJ, Glomset JA. 1984. *J. Biol. Chem.* 259:10175–80
12. Sinensky M, Logel J. 1985. *Proc. Natl. Acad. Sci. USA* 82:3257–61
13. Maltese WA, Sheridan KM. 1987. *J. Cell. Physiol.* 133:471–81
14. Beck LA, Hosick TJ, Sinensky M. 1988. *J. Cell. Biol.* 107:1307–16
15. Wolda SL, Glomset JA. 1988. *J. Biol. Chem.* 263:5997–6000
16. Farnsworth CC, Wolda SL, Gelb MH, Glomset JA. 1989. *J. Biol. Chem.* 264:20422–29
17. Hancock JF, Magee AI, Childs JE, Marshall CJ. 1989. *Cell* 57:1167–77
18. Schafer WR, Kim R, Sterne R, Thorner J, Kim S-H, Rine J. 1989. *Science* 245:379–85
19. Casey PJ, Solski PA, Der CJ, Buss JE. 1989. *Proc. Natl. Acad. Sci. USA* 86:8323–27
20. Burgering BM, Bos JL. 1995. *Trends in Biochem. Sci.* 20:18–22
21. Boguski MS, McCormick F. 1993. *Nature* 366:643–54
22. Powers S, Michaelis S, Broek D, Santa-Ana AS, Field J, et al. 1986. *Cell* 47:413–22
23. Anderegg RJ, Betz R, Carr SA, Crabb JW, Duntze W. 1988. *J. Biol. Chem.* 263:18236–40
24. Gibbs JB. 1991. *Cell* 65:1–4
25. Stimmel JB, Deschenes RJ, Volker C, Stock J, Clarke S. 1990. *Biochemistry* 29:9651–59
26. Cox AD, Der CJ. 1992. *Crit. Rev. Oncog.* 3:365–400
27. Farnsworth CC, Gelb MH, Glomset JA. 1990. *Science* 247:320–22
28. Rilling HC, Breunger E, Epstein WW, Crain PF. 1990. *Science* 247:318–20
29. Kinsella BT, Maltese WA. 1991. *J. Biol. Chem.* 266:8540–44
30. Khosravi-Far R, Clark GJ, Abe K, Cox AD, McLain T, et al. 1992. *J. Biol. Chem.* 267:24363–68
31. Novick P, Brennwald P. 1993. *Cell* 75:597–601
32. Reiss Y, Goldstein JL, Seabra MC, Casey PJ, Brown MS. 1990. *Cell* 62:81–88
33. Casey PJ, Thissen JA, Moomaw JF. 1991. *Proc. Natl. Acad. Sci. USA* 88:8631–35
34. Yokoyama K, Goodwin GW, Ghomash-

chi F, Glomset JA, Gelb MH. 1991. *Proc. Natl. Acad. Sci. USA* 88:5302–6

35. Moores SL, Schaber MD, Mosser SD, Rands E, O'Hara MB, et al. 1991. *J. Biol. Chem.* 266:14603–10
36. Seabra MC, Goldstein JL, Sudhof TC, Brown MS. 1992. *J. Biol. Chem.* 267:14497–503
37. Horiuchi H, Kawata M, Katayama M, Yoshida Y, Musha T, et al. 1991. *J. Biol. Chem.* 266:16981–84
38. Farnsworth CC, Seabra MC, Ericsson LH, Gelb MH, Glomset JA. 1994. *Proc. Natl. Acad. Sci. USA* 91:11963–67
39. Farnsworth CC, Kawata M, Yoshida Y, Takai Y, Gelb MH, Glomset JA. 1991. *Proc. Natl. Acad. Sci. USA* 88: 6196–200
40. Smeland TE, Seabra MC, Goldstein JL, Brown MS. 1994. *Proc. Natl. Acad Sci. USA* 91:10712–16
41. Deleted in proof
42. Inglese J, Glickman JF, Lorenz W, Caron M, Lefkowitz RJ. 1992. *J. Biol. Chem.* 267:1422–25
43. James GL, Goldstein JL, Pathak RL, Anderson RGW, Brown MS. 1994. *J. Biol. Chem.* 269:14182–90
44. Goldstein JL, Brown MS, Stradley SJ, Reiss Y, Gierasch LM. 1991. *J. Biol. Chem.* 266:15575–78
45. Brown MS, Goldstein JL, Paris KJ, Burnier JP, Marsters JC. 1992. *Proc. Natl. Acad. Sci. USA* 89:8313–16
46. Reiss Y, Stradley SJ, Gierasch LM, Brown MS, Goldstein JL. 1991. *Proc. Natl. Acad. Sci. USA* 88:732–36
47. Chen W-J, Andres DA, Goldstein JL, Russell DW, Brown MS. 1991. *Cell* 66:327–34
48. Chen W-J, Andres DA, Goldstein JL, Brown MS. 1991. *Proc. Natl. Acad. Sci. USA* 88:11368–72
49. Kohl NE, Diehl RE, Schaber MD, Rands E, Soderman DD, et al. 1991. *J. Biol. Chem.* 266:18884–88
50. Omer CA, Kral AM, Diehl RE, Prendergast GC, Powers S, et al. 1993. *Biochemistry* 32:5167–76
51. Andres DA, Goldstein JL, Ho YK, Brown MS. 1993. *J. Biol. Chem.* 268:1383–90
52. Fujiyama A, Matsumoto K, Tamanoi F. 1987. *EMBO J.* 6:223–28
53. Wilson KL, Herskowitz I. 1987. *Genetics* 115:441–49
54. Miyajima I, Nakayama N, Nakafuku M, Kaziro Y, Arai K, Matsumoto K. 1988. *Genetics* 119:797–804
55. He B, Chen P, Chen S, Vancura KL, Michaelis S, Powers S. 1991. *Proc. Natl. Acad. Sci. USA* 88:11373–77
56. Goodman LE, Judd SR, Farnsworth CC, Powers S, Gelb MH, et al. 1990. *Proc. Natl. Acad. Sci. USA* 87:9665–69
57. Schafer WR, Trueblood CE, Yang C-C, Mayer MP, Rosenberg S, et al. 1990. *Science* 249:1133–39
58. Chen W-J, Moomaw JF, Overton L, Kost TA, Casey PJ. 1993. *J. Biol. Chem.* 268:9675–80
59. Moomaw JF, Casey PJ. 1992. *J. Biol. Chem.* 267:17438–43
60. Reiss Y, Brown MS, Goldstein JL. 1992. *J. Biol. Chem.* 267:6403–8
61. Myers LC, Terranova MP, Ferentz AE, Wagner G, Verdine GL. 1993. *Science* 261:1164–67
62. Reiss Y, Seabra MC, Armstrong SA, Slaughter CA, Goldstein JL, Brown MS. 1991. *J. Biol. Chem.* 266:10672–77
63. Stradley SJ, Rizo J, Gierasch LM. 1993. *Biochemistry* 32:12586–90
64. Koblan KS, Culberson JC, deSolms SJ, Giuliani EA, Mosser SD, et al. 1995. *Protein Sci.* 4:681–88
65. Ying W, Sepp-Lorenzino L, Cai K, Aloise P, Coleman PS. 1994. *J. Biol. Chem.* 269:470–77
65a. Pompliano DL, Rands E, Schaber MD, Mosser SD, Anthony NJ, Gibbs JB. 1992. *Biochemistry* 31:3800–7
66. Pompliano DL, Schaber MD, Mosser SD, Omer CA, Shafer JA, Gibbs JB. 1993. *Biochemistry* 32:8341–47
67. Furfine ES, Leban JJ, Landavazo A, Moomaw JF, Casey PJ. 1995. *Biochemistry* 34:6857–62
68. Dolence JM, Poulter CD. 1995. *Proc. Natl. Acad. Sci. USA* 92:5008–11
69. Boguski MS, Murray AW, Powers S. 1992. *New Biol.* 4:408–11
70. Yokoyama K, Gelb MH. 1993. *J. Biol. Chem.* 268:4055–60
71. Zhang FL, Diehl RE, Kohl NE, Gibbs JB, Giros B, et al. 1994. *J. Biol. Chem.* 269:3175–80
72. Ohya Y, Goebl M, Goodman LE, Petersen-Bjorn S, Friesen JD, et al. 1991. *J. Biol. Chem.* 266:12356–60
73. Adams AE, Johnson DI, Longnecker RM, Sloat BF, Pringle JR. 1990. *J. Cell Biol.* 111:131–42
74. Finegold AA, Johnson DI, Farnsworth CC, Gelb MH, Judd SR, et al. 1991. *Proc. Natl. Acad. Sci. USA* 88:4448–52
75. Seabra MC, Reiss Y, Casey PJ, Brown MS, Goldstein JL. 1991. *Cell* 65:429–34
76. Mayer ML, Caplin BE, Marshall MS. 1992. *J. Biol. Chem.* 267:20589–93
77. Zhang FL, Moomaw JF, Casey PJ. 1994. *J. Biol. Chem.* 269:23465–70
78. Zhang FL. 1995. *Mammalian Protein Geranylgeranyltransferase Type I.* PhD thesis. Duke Univ., Durham, NC

79. Yokoyama K, McGeady P, Gelb MH. 1995. *Biochemistry* 34:1344–54
80. Bukhtiyarov YE, Omer CA, Allen CM. 1995. *J. Biol. Chem.* 270:19035–40
81. Mitsuzawa H, Esson K, Tamanoi F. 1995. *Proc. Natl. Acad. Sci. USA* 92:1704–8
82. Trueblood CE, Ohya Y, Rine J. 1993. *Mol. Cell. Biol.* 13:4260–75
82a. Ohya Y, Qadota H, Anraku Y, Pringle JR, Botstein D. 1995. *Mol. Biol. Cell* 4:1017–25
83. James GL, Goldstein JL, Brown MS. 1995. *J. Biol. Chem.* 270:6221–26
84. Adamson P, Marshall CJ, Hall A, Tilbrook PA. 1992. *J. Biol. Chem.* 267:20033–38
85. Armstrong SA, Hannah VC, Goldstein JL, Brown MS. 1995. *J. Biol. Chem.* 270:7864–68
86. Seabra MC, Brown MS, Slaughter CA, Sudhof TC, Goldstein JL. 1992. *Cell* 70:1049–57
87. Armstrong SA, Seabra MC, Sudhof TC, Goldstein JL, Brown M. 1993. *J. Biol. Chem.* 268:12221–29
88. Andres DA, Seabra MC, Brown MS, Armstrong SA, Smeland TE, et al. 1993. *Cell* 73:1091–99
89. Araki S, Kaibuchi K, Sasaki T, Hata Y, Takai Y. 1991. *Mol.Cell.Biol.* 11:1438–47
89a. Kinsella BT, Maltese WA. 1992. *J. Biol. Chem.* 267:3940–45
90. Wilson AL, Maltese WA. 1993. *J. Biol. Chem.* 268:14561–64
91. Beranger F, Cadwallader K, Porfiri E, Powers S, Evans T, et al. 1994. *J. Biol. Chem.* 269:13637–43
92. Seabra MC, Brown MS, Goldstein JL. 1993. *Science* 259:377–81
93. Cremers FPM, Armstrong SA, Seabra MC, Brown MS, Goldstein JL. 1994. *J. Biol. Chem.* 269:2111–17
94. Seabra MC, Ho YK, Anant JS. 1995. *J. Biol. Chem.* 270:24420–27
95. Li R, Havel C, Watson JA, Murray AW. 1993. *Nature* 366:82–84
96. Rossi G, Jiang Y, Newman AP, Ferro-Novick S. 1991. *Nature* 351:158–61
97. Li R, Havel C, Watson JA, Murray AW. 1994. *Nature* 371:438
98. Jiang Y, Rossi G, Ferro-Novick S. 1993. *Nature* 366:84–86
99. Fujimura K, Tanaka K, Nakano A, Toh-e A. 1994. *J. Biol. Chem.* 269:9205–12
100. Waldherr M, Ragnini A, Schweyen RJ, Boguski MS. 1993. *Nat. Genet.* 3:193–94
101. Wiesenberger G, Waldherr M, Schweyen RJ. 1992. *J. Biol. Chem.* 267:6963–69
102. Jiang Y, Ferro-Novick S. 1994. *Proc. Natl. Acad. Sci. USA* 91:4377–81
103. Vancura A, Sessler A, Leichus B, Kuret J. 1994. *J. Biol. Chem.* 269:19271–78
104. Gutierrez L, Magee AI, Marshall CJ, Hancock JF. 1989. *EMBO J.* 8:1093–98
105. Hancock JF, Cadwallader K, Marshall CJ. 1991. *EMBO J.* 10:641–46
106. Ma Y-T, Rando RR. 1992. *Proc. Natl. Acad. Sci. USA* 89:6275–79
107. Jang GF, Yokoyama K, Gelb MH. 1993. *Biochemistry* 32:9500–7
108. Ashby MN, King DS, Rine J. 1992. *Proc. Natl. Acad. Sci. USA* 89:4613–17
109. Ma Y-T, Chaudhuri A, Rando RR. 1992. *Biochemistry* 31:11772–77
110. Ma Y-T, Gilbert BA, Rando RR. 1993. *Biochemistry* 32:2386–93
111. Hrycyna CA, Clarke S. 1992. *J. Biol. Chem.* 267:10457–64
112. Ma Y-T, Rando RR. 1993. *FEBS Lett.* 332:105–10
113. Stephenson RC, Clarke S. 1990. *J. Biol. Chem.* 265:16248–54
114. Pillinger MH, Volker C, Stock JB, Weissmann G, Philips MR. 1994. *J. Biol. Chem.* 269:1486–92
115. Perez-Sala D, Tan EW, Canada FJ, Rando RR. 1991. *Proc. Natl. Acad. Sci. USA* 88:3043–46
116. Volker C, Lane P, Kwee C, Johnson M, Stock J. 1991. *FEBS Lett.* 295:189–94
117. Perez-Sala D, Gilbert BA, Tan EW, Rando RR. 1992. *Biochem. J.* 284:835–40
118. Tan EW, Perez-Sala D, Canada FJ, Rando RR. 1991. *J. Biol. Chem.* 266:10719–22
119. Shi YQ, Rando RR. 1992. *J. Biol. Chem.* 267:9547–51
120. Hazoor-Akbar, Wang WJ, Kornhauser R, Volker C, Stock JB. 1993. *Proc. Natl. Acad. Sci. USA* 90:868–72
121. Volker C, Miller RA, McCleary WR, Rao A, Poenie M, et al. 1991. *J. Biol. Chem.* 266:21515–22
122. Philips MR, Pillinger MH, Staud R, Volker C, Rosenfeld MG, et al. 1993. *Science* 259:977–80
123. Ding J, Lu DJ, Perez-Sala D, Ma Y-T, Maddox JF, et al. 1994. *J. Biol. Chem.* 269:16837–44
124. Backlund PS Jr, Simonds WF, Spiegel AM. 1990. *J. Biol. Chem.* 265:15572–76
125. Tan EW, Rando RR. 1992. *Biochemistry* 31:5572–78
126. Clarke S. 1985. *Annu. Rev. Biochem.* 54:479–506
127. Stock J, Simms S. 1988. *Adv. Exp. Med. Biol.* 231:201–12
128. Farnsworth CC, Kawata M, Yoshida Y, Takai Y, Gelb MH, Glomset JA. 1991. *Proc. Natl. Acad. Sci. USA* 88:6196–200

129. Giner JL, Rando RR. 1994. *Biochemistry* 33:15116–23
130. Marr RS, Blair LC, Thorner J. 1990. *J. Biol. Chem.* 265:20057–60
131. Hrycyna CA, Clarke S. 1990. *Mol. Cell.Biol.* 10:5071–76
132. Sapperstein S, Berkower C, Michaelis S. 1994. *Mol.Cell.Biol.* 14:1438–49
133. Hrycyna CA, Sapperstein SK, Clarke S, Michaelis S. 1991. *EMBO J.* 10:1699–709
134. Kato K, Cox AD, Hisaka MM, Graham SM, Buss JE, Der CJ. 1992. *Proc. Natl. Acad. Sci. USA* 89:6403–7
135. Fukada Y, Matsuda T, Kokame K, Takao T, Shimonishi Y, et al. 1994. *J. Biol. Chem.* 269:5163–70
136. Parish CA, Smrcka AV, Rando RR. 1995. *Biochemistry* 34:7722–27
137. Epand RF, Xue CB, Wang SH, Naider F, Becker JM, Epand RM. 1993. *Biochemistry* 32:8368–73
138. Silvius JR, l'Heureux F. 1994. *Biochemistry* 33:3014–22
139. Shahinian S, Silvius JR. 1995. *Biochemistry* 34:3813–22
140. Black SD. 1992. *Biochem. Biophys. Res. Commun.* 186:1437–42
141. Hancock JF, Paterson H, Marshall CJ. 1990. *Cell* 63:133–39
142. Jackson JH, Li JW, Buss JE, Der CJ, Cochrane CG. 1994. *Proc. Natl. Acad. Sci. USA* 91:12730–34
143. Muntz KH, Sternweis PC, Gilman AG, Mumby SM. 1992. *Mol.Biol.Cell* 3:49–61
144. Beranger F, Goud B, Tavitian A, De Gunzburg J. 1991. *Proc. Natl. Acad. Sci. USA* 88:1606–10
145. Chavrier P, Gorvel JP, Stelzer E, Simons K, Gruenberg J, Zerial M. 1991. *Nature* 353:769–72
146. Holtz D, Tanaka RA, Hartwig J, McKeon F. 1989. *Cell* 59:969–77
147. Thissen JA, Casey PJ. 1993. *J. Biol. Chem.* 268:13780–83
148. Thissen JA, Barrett MG, Casey PJ. 1995. *Methods Enzymol.* 250:158–68
149. Inglese J, Koch WJ, Caron MG, Lefkowitz RJ. 1992. *Nature* 359:147–50
150. McLaughlin S, Aderem A. 1995. *Trends Biochem. Sci.* 20:272–76
151. Dizhoor AM, Chen C-K, Olshevskaya E, Sinelnikova VV, Phillipov P, Hurley JB. 1993. *Science* 259:829–32
151a. Pfeffer SR, Dirac-Svejstrup B, Soldati S. 1995. *J. Biol. Chem.* 270:17057–59
152. Orita S, Kaibuchi K, Kuroda S, Shimizu K, Nakanishi H, Takai Y. 1993. *J. Biol. Chem.* 268:25542–46
153. Kuroda U, Suzuki N, Kataoka T. 1993. *Science* 259:683–88

154. Porfiri E, Evans T, Chardin P, Hancock JF. 1994. *J. Biol. Chem.* 269:22672–77
155. Deleted in proof
156. Hori Y, Kikuchi A, Isomura M, Katayama M, Miura Y, et al. 1991. *Oncogene* 6:515–22
157. Araki S, Kikuchi A, Hata Y, Isomura M, Takai Y. 1990. *J. Biol. Chem.* 265:13007–15
158. Simonds WF, Butrynski JE, Gautam N, Unson CG, Spiegel AM. 1991. *J. Biol. Chem.* 266:5363–66
159. Casey PJ. 1994. *Curr. Opin. Cell. Biol.* 6:219–25
160. Higgins JB, Casey PJ. 1994. *J. Biol. Chem.* 269:9067–73
161. Higgins JB. 1995. *Role of Prenylation in Assembly and Function of Trimeric G Proteins.* PhD thesis. Duke Univ., Durham, NC
161a. Iniguez-Lluhi Ja, Simon MI, Robishaw JD, Gilman AG. 1992. *J. Biol. Chem.* 267:23409–17
162. Fukada Y, Takao T, Ohguro H, Yoshizawa T, Akino T, Shimonishi Y. 1990. *Nature* 346:658–60
162a. Kisselev O, Ermolaeva M, Gautam N. 1995. *J. Biol. Chem.* 270:25356–58
163. Glenn JS, Watson JA, Havel CM, White JM. 1992. *Science* 256:1331–33
164. Lai MC. 1995. *Annu. Rev. Biochem.* 64:259–86
165. Barbacid M. 1987. *Annu. Rev. Biochem.* 56:779–827
166. Khosravi-Far R, Der CJ. 1994. *Cancer Metastasis Rev.* 13:67–89
167. Gibbs JB, Pompliano DL, Mosser SD, Rands E, Lingham RB, et al. 1993. *J. Biol. Chem.* 268:7617–20
168. Patel DV, Schmidt RJ, Biller SA, Gordon EM, Robinson SS, Manne V. 1995. *J. Med. Chem.* 38:2906–21
169. Kohl NE, Mosser SD, deSolms SJ, Giuliani EA, Pompliano DL, et al. 1993. *Science* 260:1934–37
170. Garcia AM, Rowell C, Ackermann K, Kowalczyk JJ, Lewis MD. 1993. *J. Biol. Chem.* 268:18415–18
171. Nigam M, Seong C-M, Qian Y, Hamilton AD, Sebti SM. 1993. *J. Biol. Chem.* 268:20695–98
172. Patel DV, Gordon EM, Schmidt RJ, Weller HN, Young MG, et al. 1995. *J. Med. Chem.* 38:435–42
172a. Bishop WR, Bond R, Petrin J, Wang L, Patton R, et al. 1995. *J. Biol. Chem.* 270:30611–18
173. Tamanoi F. 1993. *Trends Biochem. Sci.* 18:349–53
174. James GL, Goldstein JL, Brown MS, Rawson TE, Somers TC, et al. 1993. *Science* 260:1937–42
175. Vogt A, Qian Y, Blaskovich MA, Fos-

sum RD, Hamilton AD, Sebti SM. 1995. *J. Biol. Chem.* 270:660–64
176. Gibbs JB, Oliff A, Kohl NE. 1994. *Cell* 77:175–78
177. Hara M, Akasaka K, Akinaga S, Okabe M, Nakano H, et al. 1993. *Proc. Natl. Acad. Sci. USA* 90:2281–85
177a. Sun JZ, Qian YM, Hamilton AD, Sebti SM. 1995. *Cancer Res.* 55:4243–47
178. Kohl NE, Omer CA, Conner MW, Anthony NJ, Davide JP, et al. 1995. *Nat. Med.* 1:792–97
179. James GL, Brown MS, Cobb MH, Goldstein JL. 1994. *J. Biol. Chem.* 269:27705–14
180. Graham SM, Cox AD, Drivas G, Rush MG, D'Eustachio P, Der CJ. 1994. *Mol. Cell. Biol.* 14:4108–15
181. Deschenes RJ, Broach JR. 1987. *Mol. Cell. Biol.* 7:2344–51
182. Prendergast GC, Davide JP, deSolms SJ, Giuliani EA, Graham SL, et al. 1994. *Mol. Cell. Biol.* 14:4193–202
183. Sepp-Lorenzino L, Ma Z, Rands E, Kohl NE, Gibbs JB, et al. 1995. *Cancer Res.* 55:5302–09
183a. Nagasu T, Yoshimatsu K, Rowell C, Lewis MD, Garcia AM. 1995. *Cancer Res.* 55:5310–14
184. Ma Y-T, Shi YQ, Lim YH, McGrail SH, Ware JA, Rando RR. 1994. *Biochemistry* 33:5414–20
185. Scheer A, Gierschik P. 1995. *Biochemistry* 34:4952–61
186. Marciano D, Benbaruch G, Marom M, Egozi Y, Haklai R, Kloog Y. 1995. *J. Med. Chem.* 38:1267–72
187. Epstein WW, Lever D, Leining LM, Bruenger E, Rilling HC. 1991. *Proc. Natl. Acad. Sci. USA* 88:9668–70
188. Gottesman MM, Pastan I. 1993. *Annu. Rev. Biochem.* 62:385–427
189. Endicott JA, Ling V. 1989. *Annu. Rev. Biochem.* 58:137–71
190. Michaelis S. 1993. *Semin. Cell Biol.* 4:17–27
191. Kuchler K, Sterne RE, Thorner J. 1989. *EMBO J.* 8:3973–84
192. Raymond M, Gros P, Whiteway M, Thomas DY. 1992. *Science* 256:232–34
193. Zhang L, Sachs CW, Fine RL, Casey PJ. 1994. *J. Biol. Chem.* 269:15973–76
194. Zhang L, Sachs CW, Fu H, Fine RL, Casey PJ. 1995. *J. Biol. Chem.* 270:22859–65
195. Correll CC, Ng L, Edwards PA. 1994. *J. Biol. Chem.* 269:17390–93
196. Bradfute DL, Simoni RD. 1994. *J. Biol. Chem.* 269:6645–50
197. Forman BM, Goode E, Chen J, Oro AE, Bradley DJ, et al. 1995. *Cell* 81:687–93
198. Crick DC, Andres DA, Waechter CJ. 1995. *Biochem. Biophys. Res. Commun.* 211:590–99
199. Crick DC, Waechter CJ, Andres DA. 1994. *Biochem. Biophys. Res. Commun.* 205:955–61
200. Fliesler SJ, Keller RK. 1995. *Biochem. Biophys. Res. Commun.* 210:695–702
201. Park SB, Howald WN, Cashman JR. 1994. *Chem. Res. Toxicol.* 7:191–98
202. Bansal VS, Vaidya S. 1994. *J. Biol. Chem.* 315:393–99
203. Zheng J, Knighton DR, Xuong NG, Taylor SS, Sowadski JM, Ten Eyck LF. 1993. *Protein Sci.* 2:1559–73
204. Tanaka T, Ames JB, Harvey TS, Stryer L, Ikura M. 1995. *Nature* 376:444–47
205. Casey PJ. 1995. *Science* 268:221–25
206. Ma Y, Omer CA, Gibbs RA. 1996. *J. Am. Chem. Soc.* In press
207. Otto JC, Casey PJ. 1996. *J. Biol. Chem.* In press

Annu. Rev. Biochem. 1996. 65:271–303

PROTEIN TRANSPORT ACROSS THE EUKARYOTIC ENDOPLASMIC RETICULUM AND BACTERIAL INNER MEMBRANES

Tom A. Rapoport, Berit Jungnickel, and Ulrike Kutay

Department of Cell Biology, Harvard Medical School, Boston, Massachusetts 02115

KEY WORDS: endoplasmic reticulum, protein transport, signal hypothesis, membrane protein, translocation

ABSTRACT

Protein transport across the endoplasmic reticulum membrane can occur by two pathways, a co- and a post-translational one. In both cases, polypeptides are first targeted to translocation sites in the membrane by virtue of their signal sequences and then transported across or inserted into the phospholipid bilayer, most likely through a protein-conducting channel. Key components of the translocation apparatus have now been identified and the translocation pathways seem likely to be related to each other but mechanistically distinct. Protein transport across the bacterial inner membrane is both similar to and different from the process in eukaryotes. Other pathways of protein translocation exist that bypass the ones involving classical signal sequences.

CONTENTS

INTRODUCTION 272
PROTEIN TARGETING 272
COMPONENTS OF THE TRANSLOCATION APPARATUS 276
    The Sec61p/SecYEGp Complex 276
    The Signal Recognition Particle Receptor.................................... 278
    The Translocating–Chain Associating Membrane Protein...................... 278
    The Sec62/Sec63p Complex 279
    Kar2p (BiP).. 280
    The SecA Protein 280
    The SecD/SecFp Complex 280
    Enzymes in the Translocation Site... 281
    Other Components 282

271

0066-4154/96/0701-0271$08.00

MECHANISTIC ASPECTS OF PROTEIN TRANSLOCATION 282
    A Protein-Conducting Channel 282
    Cotranslational Translocation 283
    Post-Translational Translocation in Yeast 289
    Post-Translational Translocation in Escherichia coli. 292
INSERTION OF MEMBRANE PROTEINS 294
OTHER TRANSLOCATION PATHWAYS.................................... 295
CONCLUSIONS AND PERSPECTIVES...................................... 298

INTRODUCTION

In this review we attempt a survey of the processes of protein translocation into the endoplasmic reticulum (ER) of eukaryotes and across the cytoplasmic membrane of bacteria. The major focus is on the phase of translocation during which a polypeptide chain is actually transferred across the membrane. For many years this phase was a black box, but recently key components of the translocation apparatus have been identified, and the mechanistic details of the process are being unraveled. This article is written almost exactly 20 years after the proposal of the signal hypothesis by Blobel & Dobberstein (1). It is now clear that many of the predictions made at the time were correct, in particular the major one: Signal sequences direct nascent polypeptides to the ER membrane where they are further elongated by membrane-bound ribosomes and simultaneously transported across the membrane through an aqueous channel formed from membrane proteins. However, it is now also clear that there exist post-translational pathways of protein translocation that share important features with the cotranslational one, but differ from it in other ways. Perhaps one of the most gratifying discoveries during the past several years is that basic features of the translocation machineries are similar in organisms ranging from bacteria to man. Therefore we try to integrate the information on different systems into a coherent picture at the risk of an occasional over-simplification. We concentrate on the transport of proteins that are completely transferred across the membrane, such as secretory proteins, but also include a section on the integration of membrane proteins. In addition, we give a brief summary of other pathways that bypass the one involving classical signal sequences. This review is not comprehensive; rather, it focuses on unresolved questions and recent developments. Other reviews on the subject (2–9), in particular the excellent review by Walter & Johnson (3), deal with early events of the transport process, emphasizing the targeting pathway involving the signal recognition particle (SRP).

PROTEIN TARGETING

In all eukaryotes there are two pathways, a co-and a post-translational one, by which proteins are transported across the ER membrane or are integrated into it. In the cotranslational pathway, transport occurs while the polypeptide chain is being synthesized on a membrane-bound ribosome; in the post-translational pathway, the polypeptide chain is completed in the cytoplasm before being transported. In bacteria, ribosomes do not seem to be tightly bound to the membrane and most proteins may be transported post-translationally, or at least after much of the chain has been synthesized (10). Both translocation modes require that polypeptides destined for translocation or integration be specifically targeted to the membrane. This process is best understood for the cotranslational pathway in mammals (3); here we give only a short summary with special reference to recent developments.

The cotranslational targeting process is initiated when the signal sequence in a nascent polypeptide chain has emerged from the ribosome and SRP is bound (Figure 1, step 1). SRP binds to the signal sequence through the methionine-rich M-domain of its 54 kDa subunit (SRP54) which also contains a second domain (the G-domain) with a GTP-binding site (11–16). While bound to the signal sequence, SRP also interacts directly with the ribosome (17), and this interaction increases the affinity of SRP54 for GTP (step 2); the ribosome thus serves as a "GTP-loading factor" (G Bacher, H Luetcke, B Jungnickel, TA Rapoport & B Dobberstein, unpublished). In the next step of the targeting process, the entire complex consisting of the ribosome, the nascent chain, and SRP (containing bound GTP) binds to the ER membrane (step 3). This binding involves two distinct interactions, one between SRP and its membrane receptor (also called docking protein) (18, 19), and one between the ribosome and ER membrane proteins. The α subunit of the SRP receptor (SRα) interacts in its GTP-bound form with SRP. The latter is released from both the signal sequence and the ribosome (step 4) (20–23), and the nascent chain is transferred into the membrane. SRP remains bound to its receptor until, upon GTP-hydrolysis, it is released into the cytosol and can begin a new targeting cycle (step 5) (24, 25). The precise order of hydrolysis of the GTP-molecules bound to SRP54 and to SRα is unknown as yet, but it is likely that SRP54 and SRα act as "GTPase-activating proteins (GAPs)" for each other (24, 26, 27).

The fact that the ribosome shows a direct interaction with ER membrane proteins poses a conceptual problem: How can one explain that only SRP-containing complexes are targeted to the translocation site? Recent observations suggest that a protein complex, called "nascent polypeptide–associated complex" (NAC) (28), serves as an inhibitor for the SRP-independent interaction of the ribosome with the ER membrane (29–31). NAC can be removed from the ribosomes by salt washing; its readdition was found to restore SRP dependence and to prevent membrane binding of ribosome–nascent chain complexes that lack signal sequences (30). NAC has also been proposed to prevent the interaction of SRP with nascent chains lacking signal sequences; if NAC is removed from the ribosomes by a high-salt wash, SRP can interact with

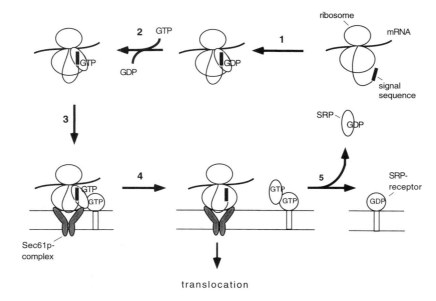

Figure 1 The SRP cycle. The scheme shows the first steps in cotranslational protein translocation across the ER membrane (for details and references, see text). When the signal sequence of a growing polypeptide chain has emerged from the ribosome, SRP is bound to both the nascent chain and the ribosome (step 1). In step 2, GDP bound to SRP54 is replaced by GTP. In step 3, the SRP-ribosome–nascent chain complex binds to the ER membrane by interactions with both the SRP receptor (in the GTP bound form) and the Sec61p complex. In step 4, SRP is released from both the signal sequence and the ribosome, and the membrane-bound ribosome–nascent chain complex can begin the actual translocation phase (see Figure 3). In step 5, GTP is hydrolyzed in both SRP and its receptor. SRP is released and can begin a new cycle. NAC may play a role in these processes by preventing ribosomes that synthesize nascent polypeptide chains lacking signal sequences from interacting with SRP and the Sec61p complex (not shown).

such nascent chains and targeting occurs (28). Thus, both proposed functions of NAC would prevent the indiscriminate targeting of polypeptide chains to the ER membrane.

An SRP-dependent targeting pathway is found in all eukaryotes but may be utilized to varying degrees. *S. cerevisiae* mutants lacking SRP or SRP receptor are viable, although they grow poorly (32, 33). These same deletions in other yeasts are lethal. Obviously, alternative targeting pathways must exist at least in *S. cerevisiae,* and they probably exist in other organisms as well. One possibility is that in the absence of the SRP pathway, the translating ribosome could bind directly to the ER membrane. However, some proteins can be targeted after completion of their synthesis, even in wild-type cells, and this post-translational targeting pathway is likely to be SRP independent. Indeed, most in vivo and in vitro experiments indicate that SRP does not promote translocation across the ER membrane of polypeptides that have been released from the ribosome (34–36; see however 37). This pathway may be more restrictive since the folding characteristics of a particular polypeptide may determine how effectively it can be kept in a translocation-competent conformation. Cytoplasmic chaperones, in particular Hsp70 and Ydj1p, a homolog of DnaJ, play a role (38–40), but the details of this process have not yet been elucidated.

Although the exact relationship among various targeting pathways is still unclear, SRP likely has the first "pick." SRP samples nascent chains for the presence of signal sequences by interacting with ribosomes at a discrete step of the elongation cycle of translation (36). Properties of the signal sequence, such as its hydrophobicity and charge distribution, may determine how efficiently SRP can interact with the growing nascent chain. Since the affinity of the signal sequence for SRP decreases with chain length (41), there is only a certain "time-window" during which SRP can bind and target polypeptides to the ER membrane (42, 43). As a result, some proteins might miss the chance to interact with SRP while they are being synthesized and therefore would require a post-translational pathway. Interestingly, recent results indicate that a homolog of SRP54 in the stroma of chloroplasts, 54CP, can function post-translationally to target some precursor proteins into the thylakoid membrane (44). Learning what sequence features 54CP recognizes and how it differs from the mammalian and yeast SRP will be very interesting.

Multiple targeting pathways also exist in *Escherichia coli* and other bacteria. An SRP-like particle and a homolog of the SRP receptor are involved in the translocation of a subset of proteins (45–48), but the majority of exported proteins in *E. coli* use a pathway mediated by the cytoplasmic chaperone SecBp (49–51). Although SecBp may interact with the signal sequence (52, 52a), other evidence indicates that it recognizes mostly internal regions of the precursor polypeptide chain (53, 54). Therefore, properties of the polypeptide other than the presence of a cleavable signal sequence may distinguish exported from nonexported proteins. Indeed, in an *E. coli* strain carrying a mutation that was selected to allow the export of proteins with defective signal sequences, alkaline phosphatase and several other proteins can be translocated even if their signal sequence is completely deleted, whereas cytosolic proteins remain intracellular (55). Secretion of alkaline phosphatase is normally independent of SecBp but becomes dependent on it in the mutant strain. One possible explanation for these surprising results is that exported proteins in *E. coli* generally fold more slowly than cytosolic ones, allowing them to interact with the chaperone SecBp (55). The presence of a signal sequence may further slow the folding process so that the interaction with SecBp becomes more efficient (56, 57).

COMPONENTS OF THE TRANSLOCATION APPARATUS

After targeting, polypeptides are transported through the membrane at specific translocation sites, also called translocons (58). These are complex structures consisting of several proteins with different functions. It is likely that only some of the components are directly involved in the translocation process, the others being required for chemical modification or folding of the translocating polypeptide chain. In this section, we list the known components of the translocation site and discuss their structural features, synthesizing our knowledge of analogous components in different organisms.

The Sec61p/SecYEGp Complex

All classes of organisms contain a heterotrimeric complex of membrane proteins, called Sec61p complex in eukaryotes and SecYEGp complex in prokaryotes, that appears to be a key component of the protein translocation apparatus. The α subunit of the eukaryotic complex was first discovered in *S. cerevisiae* in genetic screens for translocation defects and termed Sec61p (59). Sec61p is encoded by an essential gene, and the strongest temperature-sensitive alleles affect the translocation of all proteins tested at nonpermissive conditions (60). A mammalian homolog of Sec61p was found—Sec61α (61)—whose amino acid sequence is 56% identical to that of the yeast protein. In addition, both proteins have significant sequence similarity with SecYp of bacteria, the first member of this family to be identified (62). All 3 proteins have identical topologies; they each contain 10 membrane-spanning segments (Figure 2) (63, 61; B Wilkinson & C Stirling, personal communication). SecYp is likely to be a major component of the protein export apparatus in *E. coli* since it has appeared in several genetic selection schemes (7). Furthermore, the trimeric complexes from mammals and from *E. coli* have been demonstrated to be essential for the translocation of polypeptides into reconstituted proteoliposomes (64–66). However, it has also been reported that SecYp is not required for protein translocation in a reconstituted system (67, 68).

The smallest subunit of the trimeric complex is called Sec61γ in mammals, Sss1p in *S. cerevisiae* and SecEp in bacteria. The mammalian Sec61γ is highly related to Sss1p, a protein originally found as a suppressor of temperature-sensitive mutations in Sec61p (69), and it can replace Sss1p functionally in yeast cells (70). Both proteins span the membrane once via a segment close to the C terminus (Figure 2). SecEp is also a single-spanning protein in most bacteria (see 70); in *E. coli*, however, it has three membrane-spanning regions (Figure

SRP- receptor (mammals, S. cerevisiae)

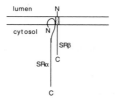

Sec61p/ SecYp/ Ssh1p - complex

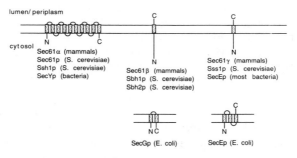

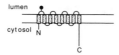

Sec61α (mammals)
Sec61p (S. cerevisiae)
Ssh1p (S. cerevisiae)
SecYp (bacteria)

Sec61β (mammals)
Sbh1p (S. cerevisiae)
Sbh2p (S. cerevisiae)

Sec61γ (mammals)
Sss1p (S. cerevisiae)
SecEp (most bacteria)

SecGp (E. coli)

SecEp (E. coli)

TRAM protein (mammals)

Sec62/ Sec63p- complex (S. cerevisiae)

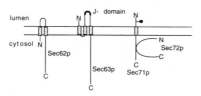

SecDp/SecFp- complex (E. coli)

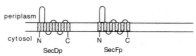

Figure 2 Membrane topologies of translocation components (for references, see text). The topologies are depicted in a schematic manner and the relative sizes of the proteins are not drawn to scale. The topology of some proteins is uncertain and only one possibility is indicated (see text). Black dots indicate N-linked carbohydrate chains. The thick line in Sec63p indicates the J-domain, a region of homology to DnaJ, that interacts with the lumenal chaperone Kar2p.

2), although only the C-terminal one is essential for its function (71), and all the sequence similarity to the eukaryotic γ subunits is contained in this region (70). In both yeast and *E. coli* the proteins are encoded by essential genes (69, 71). SecEp has also been found in several genetic selection schemes (7).

The intermediate-sized subunit of the trimeric complex is called Sec61β in mammals (64), Sbh1p in *S. cerevisiae* (72), and SecGp in *E. coli* (73, 74). Whereas Sec61β and Sbh1p are clearly related in sequence (72), SecGp does not show any obvious similarity to the eukaryotic proteins. Sec61β and Sbh1p span the membrane once via a C-terminal segment. The topology and orientation of SecGp are uncertain; a comparison of the sequences of SecGp from various bacteria suggests that it spans the membrane twice (Figure 2). All these proteins were originally identified by biochemical approaches and their roles in vivo are not yet entirely clear, particularly because neither Sbh1p (72a) nor SecGp (75) are essential for cell viability. In bacteria, translocation into reconstituted proteoliposomes in vitro can occur with only SecYp and SecEp (66), although SecGp has a large stimulatory effect (73).

Recent results indicate that ER membranes of *S. cerevisiae* contain a second trimeric complex related to the Sec61p complex, the Ssh1p complex (72a). The α subunit, Ssh1p, shares about 35% overall amino acid identity with Sec61p and has the same predicted membrane topology. The β subunit, Sbh2p, is a close homolog of Sbh1p, and the γ subunit, Sss1p, is common to both trimeric complexes. The Ssh1p complex is presumed to function exclusively in the cotranslational pathway of protein transport.

Mammalian cells also have a second transcribed gene coding for a highly homologous Sec61-protein (in this case the identity is higher than 95%) (61), but whether the protein is expressed is unknown.

The Signal Recognition Particle Receptor

The SRP receptor in mammals consists of two subunits (76) that are both GTPases (21, 26, 77). The α subunit was discovered by proteolytic dissection of the mammalian ER membrane (78, 79), and it is responsible for the interaction with SRP during the targeting process (18, 19). The α subunit is not a bona fide membrane protein, as it can be extracted under conditions that would not solubilize integral membrane proteins. It is most likely anchored to the membrane via the β subunit which is a classical membrane protein (Figure 2) (80). *S. cerevisiae* has homologs of both subunits (33, 77), whereas *E. coli* apparently has only a homolog of the α subunit, called FtsY (45, 46). The latter is, however, a soluble protein or peripheral membrane protein.

The Translocating–Chain Associating Membrane Protein

The translocating–chain associating membrane (TRAM) protein has been detected in chemical crosslinking experiments designed to identify membrane proteins that are in proximity to short nascent polypeptide chains after their transfer from SRP into the mammalian ER membrane (81). Upon purification, TRAM was shown to be required for the cotranslational translocation of most, but not all, polypeptides into reconstituted proteoliposomes (64, 81). So far, homologs of TRAM have only been discovered in organisms as divergent as *Caenorhabditis elegans*. TRAM is a glycoprotein, has 6 or 8 predicted membrane-spanning segments and a cytoplasmic, C-terminal tail of about 60 residues (Figure 2) (81).

The Sec62/Sec63p Complex

The Sec62/Sec63p complex contains four subunits: Sec62p, Sec63p, Sec71p, and Sec72p. In yeast microsomes, this complex is found associated with the trimeric Sec61p complex to form a heptameric Sec complex (72). However, it may also exist as a separate entity since the large Sec complex can be readily dissociated in vitro (72) and a sub-complex of the Sec62/Sec63p complex can be purified (82). Sec62p and Sec63p were detected in *Saccharomyces cerevisiae* in genetic screens for translocation components (59, 83–85). They are encoded by essential genes and span the ER membrane two and three times, respectively (Figure 2). Both proteins have extended C-terminal, cytoplasmic domains. Sec63p has a lumenal domain that is homologous to the J-domain of the *E. coli* chaperone DnaJ (85); it is via this domain that DnaJ interacts with its partner DnaK. Similarly, in the lumen of the ER, Sec63p interacts through its J-domain with the lumenal chaperone Kar2p (BiP) (86, 87), another member of the Hsp70 family to which DnaK belongs.

Sec71p and Sec72p were found in genetic screens for translocation components (88, 89), as well as by biochemical methods (originally called Sec66p and Sec67p, respectively), as associated proteins of Sec62p and Sec63p (90–92). Sec71p is a glycoprotein that spans the membrane once; Sec72p is a peripheral membrane protein located on the cytoplasmic side of the ER membrane (Figure 2). Deletion of the gene for Sec71p leads to the additional absence of Sec72p (93), suggesting that these two proteins interact with each other in the complex. In the deletion mutant, a translocation defect is seen but the growth of the cells is only affected at elevated temperatures. Deletion mutants of SEC72 do not show any growth defects (91, 93). A role for the constituents of the Sec62/63p complex in translocation has also been demonstrated in vitro by the use of reconstituted systems (72, 82, 94).

Homologs of Sec62p and probably of Sec63p exist in higher eukaryotes (95, 96). The SEC62 gene of *Drosophila* can functionally replace the corresponding gene in yeast (96). Interestingly, the level of expression of the gene in *Drosophila* varies among different tissues.

Some mutations in components of the Sec62/Sec63p complex of *S. cere-*

visiae show strong unexpected defects in nuclear protein localization kary-ogamy or DNA-replication [Sec63p = npl1(85); Sec71p, Sec72p (94); Sec72p = sim2 (232)]. The same is true for Kar2p (97). Although these mutations may affect the ER membrane translocation of one or more proteins involved in these nuclear processes and thus exert indirect effects, another possibility is that the complex has functions in addition to that in protein transport.

Kar2p (BiP)

A role for the ER-lumenal chaperone Kar2p in translocation is indicated by genetic data that show a rapid appearance of translocation defects in tempera-ture-sensitive mutants at nonpermissive temperatures (98), as well as by the stimulatory effect of Kar2p on in vitro translocation (72, 82, 99). Kar2p is a soluble ATPase that can interact with the J-domain of Sec63p (86, 87). Kar2p has been reported to copurify with Sec63p under certain conditions and to dissociate from it upon addition of the ATP-analog ATPγS (82).

The SecA Protein

SecAp is a key component of the translocation apparatus in *E. coli* and other bacteria (100). It has been detected in various genetic screens for translocation components and is an essential component for the translocation of proteins in bacterial in vitro assays (7, 65, 66). Homologs in eukaryotes have only been found in chloroplasts (101). SecAp is detected in *E. coli* as a homodimer (102) both in the cytoplasmic compartment and bound to the cytoplasmic membrane (103). It is an ATPase, and each monomer contains two ATP-binding sites (104, 105). SecAp has affinities for the cytoplasmic chaperone SecBp, for the signal sequence and mature region of precursor polypeptide chains, for acidic phospholipids in the membrane, and for the SecYEGp complex (103, 104, 106).

The SecD/SecFp Complex

SecD and SecF have been found in genetic screens in *E. coli* (107, 108) and are also present in other bacteria, but apparently not in *Mycoplasma geni-talium,* the free-living organism with the smallest known genome (108a). Deletion of the genes of SecD and SecF in *E. coli* leads to growth and translocation defects, but the cells are still viable (109). SecD and SecF are also not required for the translocation of proteins into reconstituted proteo-liposomes (65, 66). The proteins are related to each other, form a complex, and are coexpressed (108). Both are predicted to span the cytoplasmic mem-brane six times (Figure 2).

Enzymes in the Translocation Site

In eukaryotes, the translocation site appears to contain two enzymes that catalyze cotranslocational modifications of polypeptide chains: the signal peptidase and the oligosaccharyltransferase. Neither is essential for the translocation process per se (81). Both are unusual enzymes in that they are as abundant as their substrates, probably reflecting the fact that they are located in each translocation site so that they may act on nascent chains as they emerge into the lumen of the ER. In addition, both enzymes are surprisingly complex, each containing several polypeptide subunits.

The signal peptidase generally cleaves at a site that has small aliphatic residues at positions -1 and -3 (110). In mammals, the signal peptidase is a complex of five different membrane proteins (111). Two of the subunits, SP18 and SP21, span the membrane once and have sequence similarity to the yeast protein Sec11p and to the bacterial signal (leader) peptidase Lep1p (112, 113). They each contain an active site that includes an essential serine residue (114). The other subunits are SP22/23, a glycoprotein with the same membrane topology as SP18 and SP21 (113, 115), and SP25 and SP12, which span the membrane twice with both their N- and C-terminal domains in the cytoplasm and little of their sequence exposed to the lumen (116, 116a). These subunits are unlikely to play a role in the enzymatic reaction; they could perhaps be involved in interactions with components of the translocation site. The signal peptidase in *S. cerevisiae* is similarly complex (117), but is less well characterized. In bacteria, the signal (leader) peptidase consists of only one subunit that spans the membrane twice with closely spaced membrane anchors (118). The eukaryotic multisubunit signal peptidases may have aquired multiple catalytic subunits to broaden their range of substrate specificity. This possibility is suggested by the fact that the signal peptidase in the inner membrane of mitochondria of *S. cerevisiae* is a complex that contains two catalytic subunits with nonoverlapping substrate specificities (119).

The oligosaccharyltransferase protein transfers an oligosaccharyl moiety from the dolichol intermediate to Asn-residues located in the sequence context Asn-X-Ser or Thr (with X being any amino acid other than Pro). Glycosylation occurs when this site has reached a distance of at least 12 amino acids from the plane of the lumenal side of the membrane (120). The enzyme has been characterized both in mammals and yeast; it does not exist in bacteria. Three subunits (ribophorin I, ribophorin II, and Ost48) have been identified in the mammalian enzyme (121) but there are probably more. In *S. cerevisiae*, the enzyme can be purified as a complex of at least six subunits (122). Genetic experiments have revealed that several of the subunits have a role in N-glycosylation in vivo (123–127). The largest α subunit (Ost1) is a homolog of ribophorin I (125), the β subunit (Wbp1) of Ost48 (128), and the δ subunit

(Swp1) of ribophorin II (122). All these proteins span the membrane once. The γ- and ε-subunits (Ost3 and Ost2, respectively) span the membrane four and three times, respectively (126, 127). The sequence of an additional small subunit is unknown as yet. Genetic evidence suggests that Stt3p, a protein not copurified with the yeast enzyme complex, is also required in vivo for the activity of the oligosaccharyl transferase (129). So far, the precise function of any of these proteins is only poorly understood.

Other Components

Several other components have been implicated in the translocation process or are assumed to be present at the translocation site. The membrane protein p180 has been proposed as a ribosome receptor in the cotranslational translocation process (130–132), but arguments have been raised against its suggested role (133, 233). The data on whether it is essential for translocation in reconstituted systems conflict (64, 131). p180 may perhaps provide "storage sites" for ribosomes not engaged in translocation (132). Another membrane protein, p34, also has been proposed as a ribosome receptor (134, 135), but again counterarguments have been raised (133). Further components with unknown functions include the heterotetrameric translocon associated–protein (TRAP) com- plex, previously called signal-sequence receptor (SSR) (136), and RAMP4, a small protein with affinity for ribosomes (64). As judged from results with reconstituted systems (64, 72, 137), none of these components appears to be absolutely essential for translocation of the proteins tested thus far.

MECHANISTIC ASPECTS OF PROTEIN TRANSLOCATION

A Protein-Conducting Channel

It was postulated long ago that a protein-conducting channel in the membrane is transiently formed or opened so that a polypeptide chain can move across the phospholipid bilayer (1). Such a channel must exist for both co- and post-translational translocation pathways although the detailed mechanism by which the polypeptide chain is transferred through it may differ. Initial evidence for a channel came from the observation that translocating proteins can be extracted from the membrane by aqueous perturbants (138). More direct evidence for protein-conducting channels in the cotranslational mode of protein transport was provided by electrophysiological experiments. Simon & Blobel (139) fused rough microsomes into planar lipids and demonstrated that a large number of ion-conducting channels appear when the nascent chains are re- leased from the membrane-bound ribosomes by treatment with the drug puro- mycin. Presumably the channels were previously plugged by polypeptide chains in transit through the membrane. The channels close when the salt concentration is subsequently raised, causing the ribosomes to detach from the membrane. These conditions are thought to mimic the physiological termination of translation. Using the same methodology, evidence was provided for the existence of ion-conducting channels in the cytoplasmic membrane of *E. coli* that open upon addition of synthetic signal peptides (140). These conditions may imitate the post-translational pathway and suggest again the existence of a protein-conducting channel that is gated by the signal sequence.

Strong support for such hydrophilic protein-conducting channels during cotranslational translocation comes from experiments in which fluorescent probes are incorporated into translocating nascent chains of the secretory protein preprolactin (141, 142). Measurements of the fluorescent life-time of the probes indicate an aqueous environment inside the membrane. The probes cannot be quenched by iodide ions added to the cytoplasmic compartment, indicating that the ribosome makes a tight seal with the membrane. When the membrane is treated with streptolysin O to allow passage of the iodide ions through the bilayer into the lumenal compartment, the probes can be quenched even if they are located in the portion of the nascent chain expected to be inside the ribosome. These data suggest that a contiguous hydrophilic channel reaches from the lumen of the ER all the way up into the ribosome. If the nascent chains contain fewer than 64 amino acid residues, the iodide ions cannot enter the channel from the lumen of the ER (142), indicating that a certain polypeptide chain length is required to open the channel toward the lumen.

The protein-conducting channel must be different from a channel that transports small molecules because it must open in two dimensions: perpendicular to the membrane to let hydrophilic polypeptide chains across, and within the plane of the membrane to let hydrophobic anchors of membrane proteins into the phospholipid bilayer (139, 143). Recent data indeed show that the channel has an interface with the lipid: Photoreactive probes, incorporated into the hydrophobic portions of either a membrane anchor of a membrane protein or the signal sequence of a secretory protein, can react both with protein components of the translocation apparatus and with lipids (144).

Cotranslational Translocation

The translocation apparatus involved in cotranslational protein transport has been best characterized in the mammalian system. Following the key observation that translocation-active proteoliposomes can be produced from crude detergent extracts of dog pancreatic microsomes (145), a reconstituted system consisting only of highly purified membrane proteins was developed (64). Surprisingly, proteoliposomes containing only three components—the SRP receptor, the Sec61p complex, and the TRAM protein—are competent to translocate all proteins tested so far (the TRAM protein is dispensable for some proteins, see below). Although these three components constitute the minimum translocation apparatus in the membrane, additional factors may be required to stimulate or regulate the process. Also, because cotranslational translocation is studied with crude in vitro translation systems, cytosolic proteins in addition to SRP and NAC could play a role.

The molecular mechanism of cotranslational translocation and the specific functions of components of the minimum translocation apparatus are only now being unraveled. Most mechanistic studies make use of defined translocation intermediates which are produced by translating truncated mRNAs in vitro in the presence of SRP and microsomal membranes; this yields ribosome-bound nascent chains of varying length that are caught at different stages of their transfer through the membrane (146).

The first step following the targeting by SRP is the binding of the ribosome–nascent chain complex to the translocation site of the ER membrane (Figure 3, stage 1 to stage 2). An interaction of the nascent chain with the translocation site can already be observed with short polypeptides in which the signal sequence has just emerged from the ribosome and is thus able to interact with the SRP (in the case of the secretory protein preprolactin at a chain length of 51 to 64 residues). At this stage, the ribosome–nascent chain complex can be extracted from the membrane at high salt concentrations, indicative of an electrostatic interaction (147) (Figure 3, stage 2). The nascent chain is also susceptible to externally added protease (28). This initial weak membrane contact of the ribosome–nascent chain complex is most likely caused by a direct binding of the ribosome to Sec61p. When ribosome–nascent chain complexes are targeted by SRP, only the SRP-receptor and the Sec61p complex are required in the membrane to transfer the nascent chain from SRP into the vicinity of Sec61α (28). Ribosome–nascent chain complexes that are washed with high salt and can therefore be targeted in the absence of the SRP/SRP-receptor system (28–30) require only the Sec61p complex for membrane binding (28). This salt-sensitive interaction is probably also the basis for the specific binding to microsomes of ribosomes lacking nascent chains altogether (148). When the latter are incubated with proteoliposomes reconstituted from crude detergent extracts of microsomes, they bind at physiological salt concentrations (150 mM) almost exclusively to the Sec61p complex (133). At lower salt concentrations, however, nonspecific ribosome binding to other proteins may occur, explaining the conflicting reports on different ribosome receptors (130, 131, 133–135).

Clearly distinct from this initial weak membrane interaction of the ribosome–nascent chain complex is the much stronger membrane association that occurs when the polypeptide chain becomes longer (in the case of preprolactin,

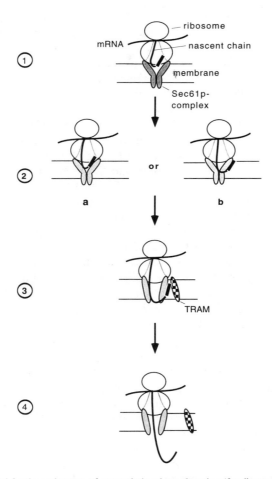

Figure 3   Model for the early steps of cotranslational translocation (for discussion and references, see text). Stage 1 represents a hypothetical state shortly after disengagement of SRP (following step 4 in Figure 1). At stage 2, the nascent chain is transferred into the membrane and the ribosome is bound loosely to the Sec61p complex. At this point, the signal sequence may contact either only the Sec61p complex (model a) or both the protein and lipids (model b). Following further chain elongation, the nascent polypeptide adopts a loop structure and its signal sequence is recognized in a process that involves the Sec61p complex (and the TRAM protein in the case of TRAM-dependent proteins). At this point (stage 3), the nascent chain is productively inserted into the translocation site, its signal sequence contacts the TRAM protein, the ribosome is firmly bound to the Sec61p complex, and the protein-conducting channel is open towards the lumen. At stage 4, the signal sequence has been cleaved off, the TRAM protein is not in immediate proximity of the nascent chain anymore, and the nascent chain has adopted a transmembrane orientation. The elongating polypeptide chain is transferred across a contiguous channel extending from the ribosome through the tightly linked Sec61p complex.

longer than 70 residues) (Figure 3, beginning with stage 3). This form of binding is not solely due to electrostatic interactions because even at high salt concentrations the nascent chains cannot be extracted from the membrane (28, 138, 147). The Sec61p complex must mediate this interaction as well since ribosome–nascent chain complexes targeted by SRP are bound in a salt-resistant manner to proteoliposomes containing only the SRP receptor and the Sec61p complex (28). Consistent with this assumption, the Sec61p complex remains bound to ribosomes at high salt concentrations after solubilization of rough microsomes in mild detergents, and this tight interaction is induced when a nascent chain is targeted by SRP to the ER membrane (61).

A tight seal between the ribosome and the membrane at this stage of translocation is also suggested by the finding that the nascent chain is now resistant to proteolysis (146) even if the membranes are solubilized before addition of the protease (146, 149). The membrane component responsible for shielding the polypeptide chain must be the Sec61p complex because in proteoliposomes it alone suffices to obtain full protection against proteolysis (28). Conversely, membrane-bound ribosomes in rough microsomes efficiently protect Sec61α from proteolytic attack (133). The tight membrane interaction is also characterized by the fact that the removal of ribosomes from the membrane requires both the release of the nascent chain from the ribosome by puromycin and high salt concentrations (150), conditions that lead to the dissociation of the ribosome into its two subunits and may therefore mimic physiological termination of translation. The same conditions are required to dissociate the solubilized ribosome-Sec61p complex (61) and to close the protein-conducting channel in the electrophysiological experiments (139). Taken together, the increase in the affinity of the ribosome for the Sec61p complex must be caused directly or indirectly by the ribosome-bound nascent polypeptide chain that is deeply inserted into the membrane.

The critical transition from the weak to the tight interaction of the ribosome–nascent chain complex with the membrane requires a functional signal sequence in the nascent chain (28). Strikingly, the chain length at which preprolactin attains resistance to treatment with high salt or protease (28) is the same as that required in the fluorescence quenching experiments for the opening of the protein-conducting channel towards the lumen of the ER (142). It is therefore likely that the signal sequence gates the protein-conducting channel. For the translocation substrate preprolactin, the only protein component required in the phospholipid bilayer to discriminate a functional from a nonfunctional signal sequence is the Sec61p complex (28). The fact that the polypeptide chain must be about 15 amino acids longer for signal recognition in the membrane than for its efficient interaction with the SRP suggests that the nascent chain inserts into the translocation site in a loop structure, in agreement with other data (151, 152); apparently the signal sequence must have emerged sufficiently from the ribosome in order to reach its binding site within the membrane. Whether the signal sequence enters the translocation site laterally through the lipid phase or directly from the cytoplasm through a proteinaceous environment (Figure 3, stages 2a or 2b) is still unknown. It is therefore not clear whether functional and nonfunctional signal sequences are discriminated by a protein-protein interaction with the Sec61p complex or by their ability to partition into the lipid phase, as demonstrated for synthetic signal peptides (153, 154). It is also possible that both membrane components are involved. After insertion into the translocation site, the hydrophobic core of the signal sequence is in contact with both Sec61α and lipid (stage 3) (144), whereas the N-terminal portion of the signal sequence is located in an aqueous environment (141).

Even though the entire translocation process and all individual steps analyzed are faithfully reproduced in the defined reconstituted system, it is possible that in this system the channel would be constantly open so that gating would not be required. In a physiological situation, the channel could be plugged by a "gating factor" (142) whose removal may be triggered by the interaction of the signal sequence with the Sec61p complex. Such a gating process could involve the function of BiP and of other lumenal proteins, providing an explanation as to why these proteins are not required for translocation in reconstituted systems (64) as they appear to be in native microsomes (155).

The insertion of the nascent chain of most secretory proteins into the translocation site requires, in addition to the Sec61p complex, the presence of the TRAM protein in proteoliposomes (S Voigt, B Jungnickel, E Hartmann & TA Rapoport, unpublished observations). Preprolactin is one of the few tested proteins that can be transported into reconstituted vesicles lacking TRAM (64). A signal sequence–exchange experiment carried out with preprolactin and prepro-α-factor, a TRAM-dependent secretory protein, indicates that the differing behavior of these proteins is due entirely to the structure of their signal sequences (S Voigt, B Jungnickel, E Hartmann & TA Rapoport, unpublished observations). The role TRAM plays in the process of signal sequence recognition of TRAM-dependent proteins is still unknown. Crosslinking experiments performed with preprolactin show that TRAM contacts the N-terminal domain of the signal sequence (152, 156) once the nascent chain is productively inserted into the translocation site (28).

The existence of two consecutive signal-sequence recognition events during cotranslational translocation, one in the cytosol by the SRP and one in the membrane, may increase the fidelity of the process. The hydrophobicity of the signal sequence is likely the decisive feature recognized in both steps, but more subtle differences in specificity may also exist. If the same features were recognized, however, an increased fidelity of signal sequence recognition would only be expected if a kinetic proofreading step existed that actively discarded mistargeted polypeptides, an effect for which there is no evidence. Another reason for a second signal sequence recognition step in the membrane may be to serve as the point at which different targeting pathways merge. The signal sequence–gated opening of the translocation channel could indeed be the common denominator of all co- and post-translational transport pathways.

Once the nascent polypeptide chain has been inserted into the translocation site, it may be committed to translocation but there appear to be further distinct steps in the process. One step has been identified as being sensitive to treatment with N-ethylmaleimide (157) and another by the return of protease sensitivity to the nascent chain when it exceeds a length of about 100 residues (146). The initiation phase is probably only completed when the signal sequence has been cleaved off by the signal peptidase and the polypeptide has adopted a trans-membrane orientation [which occurs for preprolactin when the chain reaches a length of about 140 residues (S Voigt & TA Rapoport, unpublished observations)] (Figure 3, stage 4). At this point, TRAM can no longer be crosslinked to the nascent polypeptide chain (152).

Although no direct evidence yet shows that the Sec61p complex forms a protein-conducting channel, numerous crosslinking studies with various proteins have demonstrated that Sec61α is in spatial proximity to translocating polypeptide chains throughout the translocation process (61, 152, 156, 158–160). For preprolactin chains that have adopted a transmembrane orientation after signal peptide cleavage, systematic, site-specific crosslinking experiments have shown that Sec61α contacts each of the approximately 40 consecutive amino acid residues of the nascent chain that directly precede the region of about 30 residues buried inside the ribosome (152). No major crosslinks are observed from amino acid residues in this region to other membrane proteins suggesting that these do not have easy lateral access to the polypeptide chain as it passes through the membrane. Taking into account the data from electro-physiological, fluorescence, and biochemical approaches, it seems likely that during the cotranslational translocation process the elongating nascent polypeptide chain is transferred directly from a channel in the ribosome into a Sec61p-channel in the membrane (see Figure 3, stage 4). No pushing or pulling machinery has to be invoked; the growing nascent chain has only one way out of the extended channel (3, 61, 64). In addition, the tight coupling of the ribosome channel with the membrane channel prevents both the folding of the polypeptide chain into a translocation-incompetent conformation and the passage of other molecules through the membrane.

The final step in the process of cotranslational translocation is likely to be coupled to the termination of translation. Upon arrival at the stop codon, the C-terminal 30 residues that are still in the ribosome must be transferred across the membrane, the channel must close, and the ribosome must detach from the membrane. The details of these events have not yet been analyzed.

Post-Translational Translocation in Yeast

The mechanism of post-translational protein transport across the ER membrane must differ in fundamental aspects from that of the cotranslational pathway. In the post-translational pathway the ribosome obviously does not play a role, and vectorial synthesis of the polypeptide chain cannot be the mechanism by which directionality of protein transport across the membrane is achieved. Whereas in the cotranslational pathway the ribosome makes a tight seal with the membrane and thus prevents the leakage of molecules through it (141), the membrane barrier must be maintained by alternative means during post-translational translocation. Also, as discussed above, the mechanisms of targeting to the ER membrane are likely to be different for the two pathways.

Post-translational protein translocation across the ER membrane has been best studied in *S. cerevisiae*. In contrast to microsomes from dog pancreas, those from yeast can transport the secretory protein prepro-α-factor in vitro in a post-translational manner (161–164), suggesting that they contain translocation components required for the post-translational mode which the canine microsomes seem to lack. The distinction is not absolute, however, since mammalian microsomes can transport some proteins, mostly small ones, in a post-translational manner (165–167), and yeast microsomes can transport some proteins only cotranslationally (34).

Post-translational protein translocation in yeast can be reproduced with reconstituted proteoliposomes containing a purified complex of seven proteins, the Sec complex, which is composed of the trimeric Sec61p complex (Sec61p, Sbh1p, and Sss1p) and the tetrameric Sec62/Sec63p complex (Sec62p, Sec63p, Sec71p, and Sec72p) (72). With in vitro–synthesized prepro-α-factor as a translocation substrate, about 10–20% of the molecules are translocated into the reconstituted vesicles as demonstrated by the fact that they become resistant to externally added protease. When dissociated, neither of the two subcomplexes shows significant translocation activity when reconstituted alone, but the activity is restored to the original level if they are recombined. Inclusion of Kar2p into the lumen of reconstituted vesicles and addition of ATP stimulates the rate and extent of translocation of prepro-α-factor by a factor of 3–5 (72), consistent with previous reports that showed an important role for this chaperone in protein translocation in vivo and in vitro (82, 98, 99). While the addition of Kar2p and ATP exerts only a stimulatory effect in the case of prepro-α-factor, these factors are essential for the translocation of proOmpA, a bacterial protein that can be post-translationally transported into yeast microsomes (168).

These and other results suggest a model for the mechanism of post-translational protein transport, but it is still largely hypothetical and based in some respects on analogy with the cotranslational pathway (Figure 4). Presumably, the completed polypeptide chain is presented to the ER membrane in a complex with cytosolic proteins, particularly with chaperones, and these must be released before translocation can occur. A component of the Sec complex or Ydj1p, a DnaJ-homolog bound to the ER membrane via a farnesyl chain (40), could interact with the chaperones and release them from the precursor polypeptide. The polypeptide chain would then bind to the Sec complex in the membrane and adopt a loop structure (Figure 4, step 1). The components that recognize the signal sequence are unknown. Sec72p could be one receptor (91), but Sec62p [which appears to function prior to Sec61p (169, 170)] and the constituents of the trimeric Sec61p complex (in analogy to the mammalian system) could also be involved in signal sequence recognition. Following the insertion of the protein into the translocation site, the signal sequence would be held in place by an interaction with its receptor while the C-terminal portion of the hairpin formed by the polypeptide chain could slide back and forth through the protein-conducting channel (Figure 4, step 2). When a portion of the chain arrives on the lumenal side of the membrane, Kar2p could bind to it. This binding requires the prior interaction of Kar2p with the DnaJ-domain of the Sec63p subunit (82, 85–87). Upon ATP-hydrolysis, Kar2p in its ADP-bound form would be associated with the polypeptide chain, thereby preventing its retrograde movement through the translocation channel (step 3). According to this model, Kar2p in conjunction with Sec63p would function as a molecular ratchet (171). In this way the model would be similar to one suggested for the import of proteins into the matrix of mitochondria (172–174) in which mt-Hsp70, a homolog of Kar2p in the mitochondrial matrix, cooperates with the inner membrane protein MIM44/Isp45 in an ATP-dependent manner. This model could also explain the basal level of transport seen for prepro-α-factor in the reconstituted system lacking Kar2p and ATP (72), i.e. the C-terminus of some prepro-α-factor molecules may reach the lumenal side of the membrane by passively sliding through the channel. However, in the case of proOmpA, folding on the cytoplasmic side may prevent its passive sliding; therefore binding to Kar2p on the lumenal side would be required either to pull the polypeptide chain across the membrane or merely to shift the equilibrium towards the lumen of the ER.

Kar2p may have an additional function earlier in the translocation process: In microsomes isolated from certain temperature-sensitive mutants, the insertion of prepro-α-factor into the translocation site, determined by its crosslinking to Sec61p, is prevented at nonpermissive temperatures (170). Perhaps Kar2p is involved in an early gating step during which the translocation channel would be opened in an ATP-dependent manner by the signal sequence. If one assumes that the channel is constantly open in reconstituted proteoliposomes, this would also explain why, in contrast to native microsomes, they can

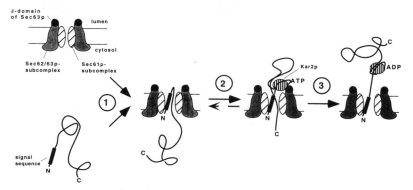

Figure 4   Hypothetical model for post-translational protein transport across the yeast ER membrane (for discussion and references, see text). In step 1, the completed polypeptide chain is inserted in a loop structure into the translocation site formed from the Sec-complex (which consists of the Sec61p- and Sec62/63p subcomplexes). In step 2, the C-terminal portion of the hairpin formed by the polypeptide chain can slide back and forth through the protein-conducting channel, but once on the lumenal side, Kar2p can bind to the incoming chain. This process involves an ATP-dependent interaction of Kar2p with the J-domain of the Sec63p subunit of the Sec complex. In step 3, upon ATP-hydrolysis, Kar2p binds to the polypeptide chain and prevents its retrograde movement through the protein-conducting channel.

transport prepro-α-factor in the absence of ATP (72). As discussed above, the function of Kar2p (BiP) in gating might also be required in cotranslational translocation, providing an explanation for its recently reported involvement in this pathway (170a).

Given the fact that all membrane proteins identified in *S. cerevisiae* in genetic screens for translocation components are contained in the heptameric Sec complex, and that the latter consists exclusively of them (plus the novel Sbh1-protein) (72), the Sec complex likely represents a functional unit that is involved in vivo in post-translational protein transport. In support of this assumption, the Sec complex shows little association with ribosomes upon solubilization of rough microsomes (72). In contrast, the trimeric Sec61p complex that also exists as a separate entity in yeast microsomes, has an affinity to ribosomes like its counterpart in mammals and is therefore presumably involved in cotranslational translocation. It seems possible that the translocation apparatus is a dynamic assembly of building blocks, with the trimeric Sec61p complex as the core of the machinery. In the cotranslational pathway it would cooperate with the SRP receptor and perhaps with a homolog of the TRAM protein, and in the post-translational pathway with the Sec62/Sec63p complex and Kar2p. Indeed, proteins whose translocation is blocked in certain mutants of Sec62p and Sec63p are not affected in mutants of a SRP subunit, and conversely, proteins impeded most severely in SRP mutants are insensitive to mutations in the Sec62 and Sec63 proteins (32, 83; P Walter, personal communication). These results suggest that the Sec62/Sec63p complex and SRP function in vivo in distinct pathways, although there may be some "cross-talk" between the two. By analogy to the role of the SRP pathway, one may hypothesize that the Sec62/Sec63p complex is involved in the targeting process during post-translational transport, although in this case the targeting complex would remain bound to the core complex and therefore be part of each trans-location site. Crosslinking experiments have indeed provided evidence that Sec62p functions prior to Sec61p during the post-translational translocation of prepro-α-factor (169, 170).

Post-Translational Translocation in Escherichia coli

Protein transport across the cytoplasmic membrane in *E. coli*, and probably other bacteria, not only is triggered by signal sequences that are similar to and exchangeable with those in eukaryotic proteins, but also involves the trimeric SecYEGp complex that is structurally related to the eukaryotic Sec61p complex (70). Like its homolog in eukaryotes, the SecYEGp complex is believed to form a protein-conducting channel (175). Despite these similarities, how-ever, the molecular mechanisms of protein transport must be significantly different. In bacteria, no homologs of the Sec62/Sec63p complex or of Kar2p (BiP) are known, and a molecular ratchet mechanism involving these compo-nents cannot be responsible for providing the driving force in the prevailing post-translational mode of protein transport. Instead, bacteria possess an es-sential ATPase—SecAp—in the cytoplasmic compartment and inner mem-brane. Obviously, the location of the two ATPases, SecAp and Kar2p, on opposite sides of the membrane in prokaryotic and eukaryotic organisms implies entirely different mechanisms for the coupling of ATP-hydrolysis with protein transport (2).

Much of our knowledge of the details of protein transport in *E. coli* comes from the fact that efficient translocation of the protein proOmpA can be reproduced in a completely reconstituted system consisting of purified precursor protein, the cytosolic chaperone SecBp, the "translocation ATPase" SecAp, and proteo-liposomes containing the SecYEGp complex (65, 66, 176). These and other in vitro experiments (103, 177–179) have led to a model depicted in Figure 5. The first step in the translocation process is the interaction of the polypeptide chain with SecBp in the cytoplasm. Next, the polypeptide is transferred to SecAp bound to the membrane (step 2). This high-affinity interaction of SecAp with the membrane involves both the SecYEGp complex and acidic phospholipids (180), and it is required to activate SecAp for binding the preprotein-SecBp complex (103). With its N-terminal domain SecAp recognizes the signal sequence but also interacts with the mature part of the preprotein (104, 106, 181). ATP-binding to

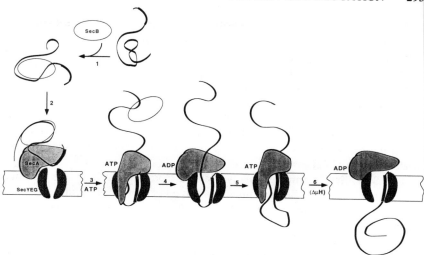

Figure 5   Model for post-translational protein transport in bacteria (for details, see text). In step 1, the completed polypeptide binds to the cytosolic chaperone SecBp. In step 2, the resulting complex associates with SecAp that is prebound to the SecYEGp complex in the cytoplasmic membrane. In step 3, SecAp inserts itself into the membrane, taking with it a bound stretch of the polypeptide chain of about 20 amino acids. Upon hydrolysis of ATP (step 4), SecAp releases the bound polypeptide and withdraws from the membrane. This cycle is repeated (steps 4 and 5) and the polypeptide chain is thereby transported one stretch at a time. The transport reaction can also be driven by the membrane potential once the polypeptide chain has been inserted into the translocation and is relased from SecAp (step 6).

the N-terminal nucleotide binding site of SecAp (182) then activates the protein to insert itself deeply into the membrane so that it actually reaches the periplasmic side even though it does not contain any long hydrophobic stretches of amino acids (179, 183, 183a) (step 3). During its membrane insertion, SecAp takes about 20 residues of the bound polypeptide chain with it into the membrane (177). Based on analogy with the mammalian Sec61p complex (28) and on the identification of signal sequence suppressor mutants in constituents of the SecYEGp complex (184, 185), it appears possible that the signal sequence is again recognized or proofread in the membrane. Upon ATP-hydrolysis, SecAp releases the polypeptide chain and comes out of the membrane (179, 183a) but remains bound to the SecYEGp complex (step 4). It is then able to start a new cycle of binding about 20 residues of the translocating polypeptide chain and bringing them into the membrane (steps 4 and 5). This process is repeated until the polypeptide has completely crossed the membrane. Such a "plunging" model would explain the translocation of the polypeptide chain in discrete portions (177), but the present evidence does not completely exclude an alternative model in which SecAp would remain in the membrane until the polypeptide chain is completely across.

The electrochemical potential across the cytoplasmic membrane is an alternative energy source that can provide the driving force for translocation (186) (Figure 5, step 6). It appears that the potential-dependent mechanism can be replaced by the SecAp-dependent one in vitro (177), but it is essential in vivo (187). The molecular basis for the effect of the membrane potential is still poorly understood; it does not seem to be a simple electrophoretic effect (188). The presence of a membrane potential decreases the number of ATP molecules hydrolyzed by SecAp per molecule of protein transported, and it may counteract the retrograde movement of the polypeptide chain when SecAp is released from it (177).

The SecD/SecFp complex likely plays a facilitating role in the outlined translocation process. Several functions have been proposed and need to be further explored: coupling the membrane potential across the cytoplasmic membrane with the translocation machinery (189), release of polypeptide chains from the translocation site into the periplasmic space (190), and regulation of the membrane insertion of SecAp (183, 183a). Genetic experiments indicate an interaction between the SecD/SecFp complex and SecYp (191, 192), suggesting that the components cooperate during the translocation process.

INSERTION OF MEMBRANE PROTEINS

The membrane insertion of many proteins occurs through the same targeting and translocation machinery used for the translocation of secretory proteins (64, 158, 193–199), but the molecular mechanism is still poorly understood. Reconstituted proteoliposomes containing the components of the minimum translocation apparatus of the mammalian ER membrane are able to insert at least single-spanning membrane proteins of all classes (64, 196). Thus, the Sec61p complex and the TRAM protein alone in the phospholipid bilayer must suffice to recognize hydrophobic segments of the nascent polypeptide chain as stop-transfer sequences and to achieve their correct orientation. Post-translational insertion of membrane proteins in eukaryotes may occur (200, 201) but has not been studied in any detail. In *E. coli*, some proteins of the cytoplasmic membrane use the SecYp-dependent pathway for their insertion (202, 203), but others, like the M13 phage coat protein, can insert post-translationally into membranes lacking the normal translocation machinery or even into protein-free liposomes (204).

A major question is at which point membrane anchors of a nascent membrane protein leave the translocation site to become embedded into the phospholipid bilayer. In vitro experiments, carried out with short N-terminal fragments of a eukaryotic membrane protein whose uncleaved signal sequence serves as the single membrane anchor (signal-anchor protein), demonstrate that the hydrophobic segment can be crosslinked to both Sec61α and lipids at early phases of translocation (144). Thus, the anchor seems to be located at the interface between the two. The precise location of the signal anchor in the translocation site may differ from that of a cleavable signal sequence (205). The complete release of the signal anchor from the channel into the lipid phase requires polypeptide-chain termination (144). A similar conclusion was reached in experiments with a polypeptide that has a cleavable signal sequence and a succeeding membrane anchor; crosslinks to proteins of the translocation site are seen even if the polypeptide segment following the anchor is as long as 100 residues, but they disappear upon chain termination (206). It would be of great interest to know whether the anchors of multi-spanning membrane proteins leave the translocation channel during the translation process. If so, they may either leave in a consecutive manner by being displaced from the interface site and pushed out into the lipid phase when the next anchor arrives, or two successive anchors could be transiently "stored" at the interphase site, or all anchors may be "stored" and only released upon polypeptide chain termination (207; S Simon, personal communication).

Another unresolved issue is how the correct orientation of a membrane protein is achieved (207a). In a cotranslational translocation process, the orientation of the first hydrophobic segment, be it a cleavable signal sequence or an uncleaved anchor, may determine the orientation of all subsequent membrane anchors. One predictive rule for the orientation of this first segment is that the flanking region with the more positive net charge is located in the cytoplasm (208). In some cases, however, the membrane anchors have an inherent preferred orientation regardless of their position in a polypeptide chain (209). In bacteria, the "positive inside rule" predicts that the loops between membrane anchors containing the highest number of positively charged residues (neglecting the negatively charged ones) will be cytoplasmic (210). It appears that cytoplasmic loops are the prime determinants for the orientation of membrane anchors in multi-spanning membrane proteins in *E. coli* and that, if these signals are conflicting with each other, the proteins may adopt topologies that leave out potential membrane anchors (211, 212). The correct topology with positive charges remaining in the cytosol requires an intact electrochemical potential across the cytoplasmic membrane (213). Such a mechanism is unlikely to play a role in eukaryotes because the ER membrane lacks a significant electrochemical potential.

OTHER TRANSLOCATION PATHWAYS

Both prokaryotes and eukaryotes contain several secretory and membrane proteins that do not have classical signal sequences and that bypass the general Sec-protein-dependent translocation machineries to reach the same final destinations, i.e. locations outside the cell or in membranes of the secretory pathway.

One example of a nonclassical translocation pathway involves ABC (ATP-binding cassette)-transporters (also called traffic ATPases) (for review, see 214–216). These transporters are a ubiquitous family of membrane proteins consisting of four domains: two membrane spanning domains (typically containing six membrane-spanning segments each) and two cytoplasmic ATP-binding domains (see 217, 218). The latter couple ATP-hydrolysis to transport of the specific substrate across the membrane. The mechanism is best understood for the secretion of the 110 kDa protein hemolysin (HlyA) from *E. coli*. In addition to the ABC-transporter HlyB, located in the inner membrane, at least two additional proteins are required to transport the protein through both membranes: HlyD in the inner membrane (which may, however, span the periplasmic space) and TolC in the outer membrane. The secretion signal for hemolysin is contained in its C-terminal 60 amino acids; this region is both necessary and sufficient for the export of hemolysin or of chimeric proteins.

In other bacteria, transport of members of the metalloprotease family by ABC-transporters also requires a signal at their C-terminus. In this case, a Dxxx-motif (with x being hydrophobic residues) is necessary, though not sufficient, for secretion (219). One possibility is that the secretion signal is formed by a (rather imperfect) amphipathic helix (215). The hydrophobic face of this helix may direct the protein molecule to the membrane where it could interact with the ABC-transporter. Such an interaction may be enhanced by the observed post-translational attachment of a fatty acid to hemolysin, although this modification is not essential for secretion (220). An initial interaction of substrates of ABC-transporters with lipid has been suggested in other cases as well (221), but specificity must also require substrate binding to the transporter. The hydrophobic substrates could laterally enter the translocation channel that is presumably formed from the membrane-spanning segments of the transporter (222). Such a model is reminiscent of one proposed for classical translocation pathways in which the signal sequence would enter the translocation channel laterally (see Figure 3, model b). More hydrophilic substrates, however, are probably bound directly from the aqueous phase to ABC-transporters. In any case, it is still unclear precisely how the substrate is recognized, how ATP-hydrolysis is coupled to protein transport, and how directionality of translocation is achieved. Hopefully, the study of other ABC-transporters that transport small peptides, such as the Ste6 protein from *S. cerevisiae* (see 223) or the TAP transporter in mammals (see 224), will help to simplify the analysis of these systems.

Another even less characterized nonclassical transport system is the secretion of certain proteins, such as interleukins IL-1a and -1b, basic and acidic fibroblast growth factors, and transglutaminase in eukaryotes (for review, see 218, 225). These proteins lack hydrophobic signal sequences and evidently do not follow the classical secretory pathway; they are not glycosylated even though they contain potential N-glycosylation sites, their transport cannot be blocked when vesicular transport is inhibited by the drug brefeldin A, and they are not found associated with membranes of the secretory pathway. In each case the secretory proteins are found inside the cell in rather high concentrations before export (for review, see 225). Quantitative export of the basic fibroblast growth factor has been observed in an energy-dependent reaction (234). The interleukin 1 polypeptides are synthesized as larger precursors which are then further processed and modified by attachment of a fatty acid molecule before secretion. These features are reminiscent of those of a-factor, the peptide substrate of the ABC-transporter Ste6 in yeast (223), and it has been speculated that interleukin 1 may be exported by a related system (225).

Another nonclassical pathway is used to insert a class of membrane proteins into phospholipid bilayers which have an N-terminal cytoplasmic domain and possess a C-terminal membrane anchor ("tail-anchored" membrane proteins) (226). These proteins do not contain hydrophobic signal sequences and their membrane anchor is so close to the C-terminus that it remains buried in the ribosome when termination of translation occurs; membrane insertion must therefore be post-translational. Tail-anchored proteins are found in various organelles of the secretory pathway. Examples include cytochrome b_5, protein tyrosine phosphatase 1B, heme oxygenase, resident in the ER, and a large number of SNARE proteins, located in various transport vesicles and target membranes (226). On the basis of in vitro results with cytochrome b_5, these proteins were initially believed to be able to insert indiscriminately into all membranes [for references, see (226)]. However, cytochrome b_5 specifically inserts into the ER in vivo (227, 228). Furthermore, several proteins have been shown to first insert into the ER membrane and subsequently be transported along the secretory pathway to their various destinations. These include synaptobrevin, a protein in synaptic vesicles, the yeast protein Sso2p, a homolog of syntaxin in the plasma membrane, and giantin, a Golgi protein (229–231). If vesicular transport is inhibited by brefeldin A, the proteins stay in the ER, indicating that only this membrane has the capacity to incorporate them.

Recent in vitro experiments carried out with synaptobrevin indicate that its insertion into the mammalian ER membrane is ATP-dependent and greatly stimulated by a trypsin-sensitive factor in the membrane (230). Proteoliposomes depleted of the Sec61p complex remain competent for membrane insertion, whereas proteoliposomes containing the purified SRP receptor, the Sec61p complex, and the TRAM protein are inactive. These results indicate that the Sec61p complex is not required and that at least one membrane protein is involved in the insertion process that differs from the known mammalian translocation components (230).

CONCLUSIONS AND PERSPECTIVES

With the progress of recent years, several important issues in the field of protein translocation have been clarified. The existence of a protein-conducting channel now seems likely, and most essential membrane components of the translocation machineries of the co- and post-translational pathways may have been identified. Both basic similarities and fundamental differences have emerged for the pathways in various classes of organisms. Major questions now on the agenda concern the molecular mechanism of the translocation process. How is the protein-conducting channel gated? How is loop-insertion of a polypeptide chain into the translocation site achieved? Is there a proofreading mechanism for signal sequences? How is the translocation process regulated? How precisely is ATP-hydrolysis coupled to translocation in the post-translational pathways?

The answers to these and other questions may well lead to the discovery of additional translocation factors and to the identification of a function for components that are known to exist but have been ignored in the current mechanistic models. Despite a general conservation of basic features of the translocation process, different cell types could vary in their content of translocation components or may contain different isoforms of them, and the process may be more subject to fine regulation than hitherto appreciated. Novel assays, biophysical methods, and their combination with established reconstituted systems, as well as structural studies on the purified translocation components will be required to address the mechanistic problems.

The biosynthesis of membrane proteins also clearly still presents a particular enigma. It is more hypothesis than fact that the protein-conducting channel opens laterally toward the lipid to let membrane anchors out, and neither the point at which this occurs nor the mechanism by which the orientation of a membrane protein is achieved are known. Finally, the analysis of nonclassical translocation pathways may well lead to the discovery of entirely novel mechanisms by which proteins insert into or cross membranes.

ACKNOWLEDGMENTS

We thank J Beckwith, R Gilmore, K Matlack, C Nicchitta, SM Rapoport, M Rolls, P Silver, P Stein, K Verhey, P Walter, and W Wickner for helpful remarks and critical reading of the manuscript. We also thank our colleagues who communicated to us their unpublished results. The work in the authors' laboratory was supported by the Deutsche Forschungsgemeinschaft and the National Institutes of Health.

Literature Cited

1. Blobel G, Dobberstein B. 1975. *J. Cell Biol.* 67:835–51
2. Wickner W. 1994. *Science* 266:1197–98
3. Walter P, Johnson AE. 1994. *Annu. Rev. Cell Biol.* 10:87–119
4. Johnson AE. 1993. *Trends Biochem. Sci.* 18:456–58
5. Gilmore R. 1993. *Cell* 75:589–92
6. Rapoport TA. 1992. *Science* 258:931–36
7. Schatz PJ, Beckwith J. 1990. *Annu. Rev. Genet.* 24:215–48
8. High S. 1995. *Prog. Biophys. Mol. Biol.* 63:233–50
9. Ito K. 1995. *Adv. Cell Mol. Biol. Membr. Organelles* 4:35–60
10. Randall LL. 1983. *Cell* 33:231–40
11. Walter P, Blobel G. 1981. *J. Cell Biol.* 91:557–61
12. Krieg UC, Walter P, Johnson AE. 1986. *Proc. Natl. Acad. Sci. USA* 83:8604–8
13. Kurzchalia TV, Wiedmann M, Girshovich AS, Bochkareva ES, Bielka H, Rapoport TA. 1986. *Nature* 320:634–36
14. Zopf D, Bernstein HD, Johnson AE, Walter P. 1990. *EMBO J.* 9:4511–17
15. High S, Dobberstein B. 1991. *J. Cell Biol.* 113:229–33
16. Lütcke H, High S, Römisch K, Ashford AJ, Dobberstein B. 1992. *EMBO J.* 11:1543–51
17. Walter P, Ibrahimi I, Blobel G. 1981. *J. Cell Biol.* 91:545–50
18. Gilmore R, Walter P, Blobel G. 1982. *J. Cell Biol.* 95:470–77
19. Meyer DI, Krause E, Dobberstein B. 1982. *Nature* 297:647–50
20. Gilmore R, Blobel G. 1983. *Cell* 35:677–85
21. Connolly T, Gilmore R. 1989. *Cell* 57:599–610
22. Rapiejko PJ, Gilmore R. 1992. *J. Cell Biol.* 117:493–503
23. Rapiejko PJ, Gilmore R. 1994. *Mol. Biol. Cell* 5:887–97
24. Connolly T, Gilmore R. 1993. *J. Cell Biol.* 123:799–807
25. Connolly T, Rapiejko PJ, Gilmore R. 1991. *Science* 252:1171–73
26. Miller JD, Wilhelm H, Gierasch L, Gilmore R, Walter P. 1993. *Nature* 366:351–54
27. Powers T, Walter P. 1995. *Science* 269:1422–24
28. Wiedmann B, Sakai H, Davis TA, Wiedmann M. 1994. *Nature* 370:434–40
29. Jungnickel B, Rapoport TA. 1995. *Cell* 82:261–70
30. Lauring B, Skai H, Kreibich G, Wiedmann M. 1995. *Proc. Natl. Acad. Sci. USA* 92:5411–15
31. Lauring B, Kreibich G, Wiedmann M. 1995. *Proc. Natl. Acad. Sci. USA* 92:9435–39
32. Hann BC, Walter P. 1991. *Cell* 67:131–44
33. Ogg SC, Poritz MA, Walter P. 1992. *Mol. Biol. Cell* 3:895–911
34. Hansen W, Walter P. 1988. *J. Cell Biol.* 106:1075–81
35. Wiedmann M, Kurzchalia TV, Bielka H, Rapoport TA. 1987. *J. Cell Biol.* 104:201–8
36. Ogg SC, Walter P. 1995. *Cell* 81:1075–84
37. Sanz P, Meyer DI. 1988. *EMBO J.* 7:3553–57
38. Chirico WJ, Waters MG, Blobel G. 1988. *Nature* 332:805–10
39. Deshaies RJ, Koch BD, Werner WM, Craig EA, Schekman R. 1988. *Nature* 332:800–5
40. Caplan AJ, Cyr DM, Douglas MG. 1992. *Cell* 71:1143–55
41. Siegel V, Walter P. 1988. *EMBO J.* 7:1769–75
42. Rapoport TA, Heinrich R, Walter P, Schulmeister T. 1987. *J. Mol. Biol.* 195:621–36
43. Johnsson N, Varshavsky A. 1994. *EMBO J.* 13:2686–98
44. Li X, Henry R, Yuan J, Cline K, Hoffman N. 1995. *Proc. Natl. Acad. Sci. USA* 92:3789–93
45. Bernstein H, Poritz M, Strub K, Hoben P, Brenner S, Walter P. 1989. *Nature* 340:482–86
46. Römisch K, Webb J, Herz J, Prehn S, Frank R, et al. 1989. *Nature* 340:478–82
47. Phillips GJ, Silhavy TJ. 1992. *Nature* 359:744–46
48. Luirink J, ten Hagen-Jongman CM, van der Weijden CC, Oudega B, High S, et al. 1994. *EMBO J.* 13:2289–96
49. Kumamoto CA. 1991. *Mol. Microbiol.* 5:19–22
50. Weiss J, Ray P, Bassford P. 1988. *Proc. Natl. Acad. Sci. USA* 85:8978–82
51. Collier D, Bankaitis V, Weiss J, Bassford P. 1988. *Cell* 53:273–83
52. Watanabe M, Blobel G. 1989. *Cell* 58:695–705
52a. Watanabe M, Blobel G. 1995. *Proc. Natl. Acad. Sci. USA* 92:10133–36
53. Randall LL, Topping TB, Hardy S. 1990. *Science* 248:860–63
54. Hardy SJS, Randall LL. 1991. *Science* 251:439–43

55. Derman AI, Puziss JW, Bassford PJ, Beckwith J. 1993. *EMBO J.* 12:879–88
56. Liu GP, Topping TB, Randall LL. 1989. *Proc. Natl. Acad. Sci. USA* 86:9213–17
57. Park S, Liu G, Topping TB, Cover WH, Randall LL. 1988. *Science* 239:1033–35
58. Walter P, Lingappa VR. 1986. *Annu. Rev. Cell Biol.* 2:499–516
59. Deshaies RJ, Schekman R. 1987. *J. Cell Biol.* 105:633–45
60. Stirling CJ, Rothblatt J, Hosobuchi M, Deshaies R, Schekman R. 1992. *Mol. Biol. Cell* 3:129–42
61. Görlich D, Prehn S, Hartmann E, Kalies KU, Rapoport TA. 1992. *Cell* 71:489–503
62. Ito K. 1984. *Mol. Gen. Genet.* 197:204–8
63. Akiyama Y, Ito K. 1987. *EMBO J.* 6:3465–70
64. Görlich D, Rapoport TA. 1993. *Cell* 75:615–30
65. Brundage L, Hendrick JP, Schiebel E, Driessen AJM, Wickner W. 1990. *Cell* 62:649–57
66. Akimaru J, Matsuyama SI, Tokuda H, Mizushima S. 1991. *Proc. Natl. Acad. Sci. USA* 88:6545–49
67. Watanabe M, Nicchitta C, Blobel G. 1990. *Proc. Natl. Acad. Sci. USA* 87:1960–64
68. Watanabe M, Blobel G. 1993. *Proc. Natl. Acad. Sci. USA* 90:9011–15
69. Esnault Y, Blondel MO, Deshaies RJ, Schekman R, Kepes F. 1993. *EMBO J.* 12:4083–93
70. Hartmann E, Sommer T, Prehn S, Görlich D, Jentsch S, Rapoport TA. 1994. *Nature* 367:654–57
71. Schatz PJ, Bieker KL, Ottemann KM, Silhavy TJ, Beckwith J. 1991. *EMBO J.* 10:1749–57
72. Panzner S, Dreier L, Hartmann E, Kostka S, Rapoport TA. 1995. *Cell* 81:561–70
72a. Finke K, Plath K, Panzner S, Prehn S, Rapoport TA, et al. 1995. *EMBO J.* 14:In press
73. Nishiyama K, Mizushima S, Tokuda H. 1993. *EMBO J.* 12:3409–15
74. Douville K, Leonard M, Brundage L, Nishiyama K, Tokuda H, et al. 1994. *J. Biol. Chem.* 269:18705–7
75. Nishiyama K, Hanada M, Tokuda H. 1994. *EMBO J.* 13:3272–7
76. Tajima S, Lauffer L, Rath VL, Walter P. 1986. *J. Cell Biol.* 103:1167–78
77. Miller JD, Tajima S, Lauffer L, Walter P. 1995. *J. Cell Biol.* 128:273–82
78. Walter P, Jackson RC, Marcus MM, Lingappa VR, Blobel G. 1979. *Proc. Natl. Acad. Sci. USA* 76:1795
79. Meyer DI, Dobberstein B. 1980. *J. Cell Biol.* 87:503
80. Andrews D, Lauffer L, Walter P, Lingappa V. 1989. *J. Cell Biol.* 108:797–810
81. Görlich D, Hartmann E, Prehn S, Rapoport TA. 1992. *Nature* 357:47–52
82. Brodsky JL, Schekman R. 1993. *J. Cell Biol.* 123:1355–63
83. Rothblatt JA, Deshaies RJ, Sanders SL, Daum G, Schekman R. 1989. *J. Cell Biol.* 109:2641–52
84. Deshaies RJ, Schekman R. 1989. *J. Cell Biol.* 109:2653–64
85. Sadler I, Chiang A, Kurihara T, Rothblatt J, Way J, Silver P. 1989. *J. Cell Biol.* 109:2665–75
86. Scidmore MA, Okamura HH, Rose MD. 1993. *Mol. Biol. Cell* 4:1145–59
87. Schlenstedt G, Harris S, Risse B, Lill R, Silver PA. 1995. *J. Cell Biol.* 129:979–88
88. Green N, Fang H, Walter P. 1992. *J. Cell Biol.* 116:597–604
89. Kurihara T, Silver P. 1993. *Mol. Biol. Cell* 4:919–30
90. Feldheim D, Yoshimura K, Admon A, Schekman R. 1993. *Mol. Biol. Cell* 4:931–39
91. Feldheim D, Schekman R. 1994. *J. Cell Biol.* 126:935–43
92. Deshaies RJ, Sanders SL, Feldheim DA, Schekman R. 1991. *Nature* 349:806–8
93. Fang H, Green N. 1994. *Mol. Biol. Cell* 5:933–42
94. Brodsky JL, Hamamoto S, Feldheim D, Schekman R. 1993. *J. Cell Biol.* 120:95–102
94a. Ng D, Walter P. 1996. *J. Cell Biol.* In press
95. Brightman SE, Blatch GL, Zetter BR. 1995. *Gene* 153:249–54
96. Noel PJ, Cartwright IL. 1994. *EMBO J.* 13:5253–61
97. Polaina J, Conde J. 1982. *Mol. Gen. Genet.* 186:253–58
98. Vogel JP, Misra LM, Rose MD. 1990. *J. Cell Biol.* 110:1885–95
99. Sanders SL, Schekman R. 1992. *J. Biol. Chem.* 267:13791–94
100. Oliver D, Beckwith J. 1982. *Cell* 30:311–19
101. Yuan JG, Henry R, McCaffery M, Cline K. 1994. *Science* 266:796–98
102. Akita M, Shinkai A, Matsuyama S, Mizushima S. 1991. *Biochem. Biophys. Res. Commun.* 174:211–16
103. Hartl FU, Lecker S, Schiebel E, Hendrick JP, Wickner W. 1990. *Cell* 63:269–79
104. Lill R, Dowhan W, Wickner W. 1990. *Cell* 60:271–80
105. Lill R, Cunningham K, Brundage LA,

Ito K, Oliver D, Wickner W. 1989. *EMBO J.* 8:961–66

106. Akita M, Sasaki S, Matsuyama S, Mizushima S. 1990. *J. Biol. Chem.* 265: 8164–69
107. Gardel C, Benson S, Hunt J, Michaelis S, Beckwith J. 1987. *J. Bacteriol.* 169: 1286–90
108. Gardel C, Johnson K, Jacq A, Beckwith J. 1990. *EMBO J.* 9:3209–16
108a. Fraser CM, Gocayne JD, White O, Adams MD, Clayton RA, et al. 1995. *Science* 270:397–403
109. Pogliano JA, Beckwith J. 1994. *EMBO J.* 13:554–61
110. von Heijne G. 1983. *Eur. J. Biochem.* 133:17–21
111. Evans EA, Gilmore R, Blobel G. 1986. *Proc. Natl. Acad. Sci. USA* 83:581–85
112. Shelness GS, Blobel G. 1990. *J. Biol. Chem.* 265:9512–19
113. Shelness GS, Lin LJ, Nicchitta CV. 1993. *J. Biol. Chem.* 268:5201–8
114. Dalbey RE, von Heijne G. 1992. *Trends Biochem. Sci.* 17:474–78
115. Shelness GS, Kanwar YS, Blobel G. 1988. *J. Biol. Chem.* 263:17063–70
116. Greenburg G, Blobel G. 1994. *J. Biol. Chem.* 269:25354–58
116a. Kalies K-U, Hartmann E. 1995. *J. Biol. Chem.* In press
117. Yadeau JT, Klein C, Blobel G. 1991. *Proc. Natl. Acad. Sci. USA* 88:517–21
118. Wolfe PB, Wickner W, Goodman JM. 1983. *J. Biol. Chem.* 258:12073–80
119. Nunnari J, Fox TD, Walter P. 1993. *Science* 262:1997–2004
120. Nilsson I, von Heijne G. 1993. *J. Biol. Chem.* 268:5798–801
121. Kelleher DJ, Kreibich G, Gilmore R. 1992. *Cell* 69:55–65
122. Kelleher DJ, Gilmore R. 1994. *J. Biol. Chem.* 269:12908–17
123. te Heesen S, Janetzky B, Lehle L, Aebi M. 1992. *EMBO J.* 11:2071–75
124. te Heesen S, Knauer R, Lehle L, Aebi M. 1993. *EMBO J.* 12:279–84
125. Silberstein S, Collins PG, Kelleher DJ, Rapiejko PJ, Gilmore R. 1995. *J. Cell Biol.* 128:525–36
126. Karaoglu D, Kelleher DJ, Gilmore R. 1995. *J. Cell Biol.* 130:567–77
127. Silberstein S, Collins P, Kelleher DJ, Gilmore R. 1995. *J. Cell Biol.* 131:371–83
128. Silberstein S, Kelleher DJ, Gilmore R. 1992. *J. Biol. Chem.* 267:23658–63
129. Zufferey R, Knauer R, Burda P, Stagljar I, te Heesen S, et al. 1995. *EMBO J.* 14:4949–60
130. Savitz AJ, Meyer DI. 1990. *Nature* 346: 540–44

131. Savitz AJ, Meyer DI. 1993. *J. Cell Biol.* 120:853–63
132. Wanker EE, Sun Y, Savitz AJ, Meyer DI. 1995. *J. Cell Biol.* 130:29–39
133. Kalies KU, Görlich D, Rapoport TA. 1994. *J. Cell Biol.* 126:925–34
134. Tazawa S, Unuma M, Tondokoro N, Asano Y, Ohsumi T, et al. 1991. *J. Biochem.* 109:89–98
135. Ichimura T, Shindo Y, Uda Y, Ohsumi T, Omata S, Sugano H. 1993. *FEBS Lett.* 326:241–45
136. Hartmann E, Görlich D, Kostka S, Otto A, Kraft R, et al. 1993. *Eur. J. Biochem.* 214:375–81
137. Migliaccio G, Nicchitta CV, Blobel G. 1992. *J. Cell Biol.* 117:15–25
138. Gilmore R, Blobel G. 1985. *Cell* 42: 497–505
139. Simon SM, Blobel G. 1991. *Cell* 65: 371–80
140. Simon SM, Blobel G. 1992. *Cell* 69: 677–84
141. Crowley KS, Reinhart GD, Johnson AE. 1993. *Cell* 73:1101–15
142. Crowley KS, Liao SR, Worrell VE, Reinhart GD, Johnson AE. 1994. *Cell* 78:461–71
143. Singer S, Maher P, Yaffe M. 1987. *Proc. Natl. Acad. Sci. USA* 84:1960–64
144. Martoglio B, Hofmann MW, Brunner J, Dobberstein B. 1995. *Cell* 81:207–14
145. Nicchitta CV, Blobel G. 1990. *Cell* 60: 259–69
146. Connolly T, Collins P, Gilmore R. 1989. *J. Cell Biol.* 108:299–307
147. Wolin SL, Walter P. 1993. *J. Cell Biol.* 121:1211–9
148. Borgese N, Mok W, Kreibich G, Sabatini DD. 1974. *J. Mol. Biol.* 88: 559–80
149. Matlack KE, Walter P. 1995. *J. Biol. Chem.* 270:6170–80
150. Adelman MR, Sabatini D, Blobel G. 1973. *J. Cell Biol.* 56:206–28
151. Shaw AS, Rottier PJ, Rose JK. 1988. *Proc. Natl. Acad. Sci. USA* 85:7592–6
152. Mothes W, Prehn S, Rapoport TA. 1994. *EMBO J.* 13:3937–82
153. Hoyt DW, Gierasch LM. 1991. *Biochemistry* 30:10155–63
154. Briggs MS, Gierasch LM, Zlotnick A, Lear JD, DeGrado WF. 1985. *Science* 228:1096–9
155. Nicchitta CV, Blobel G. 1993. *Cell* 73: 989–98
156. High S, Martoglio B, Görlich D, Andersen SSL, Ashford AJ, et al. 1993. *J. Biol. Chem.* 268:26745–51
157. Nicchitta C, Blobel G. 1989. *J. Cell Biol.* 108:789–95
158. High S, Andersen SSL, Görlich D, Hart-

mann E, Prehn S, et al. 1993. *J. Cell Biol.* 121:743–50
159. Kellaris KV, Bowen S, Gilmore R. 1991. *J. Cell Biol.* 114:21–33
160. Nicchitta CV, Murphy EC, Haynes R, Shelness GS. 1995. *J. Cell Biol.* 129:957-70
161. Waters M, Blobel G. 1986. *J. Cell Biol.* 102:1543–50
162. Hansen W, Garcia PD, Walter P. 1986. *Cell* 45:397–406
163. Rothblatt J, Meyer D. 1986. *EMBO J.* 5:1031–6
164. Garcia PD, Walter P. 1988. *J. Cell Biol.* 106:1043–8
165. Schlenstedt G, Gudmundsson GH, Boman HG, Zimmermann R. 1990. *J. Biol. Chem.* 265:13960–8
166. Schlenstedt G, Zimmermann R. 1987. *EMBO J.* 6:699–703
167. Müller G, Zimmermann R. 1988. *EMBO J.* 7:639–48
168. Sanz P, Meyer DI. 1989. *J. Cell Biol.* 108:2101–6
169. Müsch A, Wiedmann M, Rapoport TA. 1992. *Cell* 69:343–52
170. Sanders SL, Whitfield KM, Vogel JP, Rose MD, Schekman RW. 1992. *Cell* 69:353–65
170a. Brodsky JL, Goeckeler J, Schekman R. 1995. *Proc. Natl. Acad. Sci. USA* 92:9643–46
171. Simon SM, Peskin CS, Oster GF. 1992. *Proc. Natl. Acad. Sci. USA* 89:3770–74
172. Schneider HC, Berthold J, Bauer MF, Dietmeier K, Guiard B, et al. 1994. *Nature* 371:768–74
173. Rassow J, Maarse AC, Krainer E, Kubrich M, Muller H, et al. 1994. *J. Cell Biol.* 127:1547–56
174. Kronidou NG, Oppliger W, Bolliger L, Hannavy K, Glick BS, et al. 1994. *Proc. Natl. Acad. Sci. USA* 91:12818–22
175. Joly JC, Wickner W. 1993. *EMBO J.* 12:255–63
176. Hanada M, Nishiyama K, Mizushima S, Tokuda H. 1994. *J. Biol. Chem.* 269:23625–31
177. Schiebel E, Driessen AJM, Hartl F-U, Wickner W. 1991. *Cell* 64:927–39
178. Driessen AJM, Wickner W. 1991. *Proc. Natl. Acad. Sci. USA* 88:2471–75
179. Economou A, Wickner W. 1994. *Cell* 78:835–43
180. Hendrick JP, Wickner W. 1991. *J. Biol. Chem.* 266:24596–600
181. Kimura E, Akita M, Matsuyama S, Mizushima S. 1991. *J. Biol. Chem.* 266:6600–6
182. Shinkai A, Mei LH, Tokuda H, Mizushima S. 1991. *J. Biol. Chem.* 266:5827–33

183. Kim YJ, Rajapandi T, Oliver D. 1994. *Cell* 78:845–53
183a. Economou A, Pogliano JA, Beckwith J, Oliver DB, Wickner W. 1995. *Cell.* 83:1171–82
184. Emr SD, Hanley-Way S, Silhavy T. 1981. *Cell* 23:79–88
185. Osborne RS, Silhavy TJ. 1993. *EMBO J.* 12:3391–98
186. Geller BL, Movva NR, Wickner W. 1986. *Proc. Natl. Acad. Sci. USA* 83:4219–22
187. Zimmermann R, Watts C, Wickner W. 1982. *J. Biol. Chem.* 257:6529–36
188. Kato M, Tokuda H, Mizushima S. 1992. *J. Biol. Chem.* 267:413–18
189. Arkowitz RA, Wickner W. 1994. *EMBO J.* 13:954–63
190. Matsuyama S, Fujita Y, Mizushima S. 1993. *EMBO J.* 12:265–70
191. Bieker-Brady K, Silhavy TJ. 1992. *EMBO J.* 11:3165–74
192. Taura T, Akiyama Y, Ito K. 1994. *Mol. Gen. Genet.* 243:261–69
193. Lingappa VR, Katz FN, Lodish HF, Blobel G. 1978. *J. Biol. Chem.* 253:8667–70
194. McCune JM, Lingappa VR, Fu SM, Blobel G, Kunkel HG. 1980. *J. Exp. Med.* 152:463–68
195. Friedlander M, Blobel G. 1985. *Nature* 318:338–43
196. Oliver J, Jungnickel B, Goerlich D, Rapoport TA, High S. 1995. *FEBS Lett.* 362:126–30
197. High S, Flint N, Dobberstein B. 1991. *J. Cell Biol.* 113:25–34
198. High S, Görlich D, Wiedmann M, Rapoport TA, Dobberstein B. 1991. *J. Cell Biol.* 113:35–44
199. Wilson C, Connolly T, Morrison T, Gilmore R. 1988. *J. Cell Biol.* 1077:69–77
200. Mueckler M, Lodish HF. 1986. *Cell* 44:629–37
201. Wiedmann M, Wiedmann B, Voigt S, Wachter E, Muller HG, Rapoport TA. 1988. *EMBO J.* 7:1763–68
202. Wolfe PB, Rice M, Wickner W. 1985. *J. Biol. Chem.* 260:1836–41
203. Andersson H, von Heijne G. 1993. *EMBO J.* 12:683–91
204. Geller BL, Wickner W. 1985. *J. Biol. Chem.* 260:13281–85
205. Nilsson I, Whitley P, von Heijne G. 1994. *J. Cell Biol.* 126:1127–32
206. Thrift RN, Andrews DW, Walter P, Johnson AE. 1991. *J. Cell Biol.* 112:809–21
207. Rapoport TA. 1985. *FEBS Lett.* 187:1–10
207a. Spiess M. 1995. *FEBS Lett.* 369:76–79
208. Hartmann E, Rapoport TA, Lodish HF.

1989. *Proc. Natl. Acad. Sci. USA* 86:5786–90
209. Locker JK, Rose JK, Horzinek MC, Rottier PJM. 1992. *J. Biol. Chem.* 267:21911–18
210. von Heijne G. 1986. *EMBO J.* 5:3021–27
211. McGovern K, Ehrmann M, Beckwith J. 1991. *EMBO J.* 10:2773–82
212. Gafvelin G, von Heijne G. 1994. *Cell* 77:401–12
213. Andersson H, von Heijne G. 1994. *EMBO J.* 13:2267–72
214. Fath MJ, Kolter R. 1993. *Microbiol. Rev.* 57:995–1017
215. Koronakis V, Hughes C. 1993. *Semin. Cell Biol.* 4:7–15
216. Holland IB, Kenny B, Blight M. 1990. *Biochemistry* 72:131–41
217. Higgins CF. 1992. *Annu. Rev. Cell Biol.* 8:67–113
218. Kuchler K, Thorner J. 1992. *Endocr. Rev.* 13:499–514
219. Ghigo JM, Wandersman C. 1994. *J. Biol. Chem.* 269:8979–85
220. Issartel JP, Koronakis V, Hughes C. 1991. *Nature* 351:759–61
221. Higgins CF, Gottesman MM. 1992. *Trends Biochem. Sci.* 17:18–21
222. Higgins CF. 1994. *Cell* 79:393–95
223. Michaelis S. 1993. *Semin. Cell Biol.* 4:17–27
224. Roemisch K. 1994. *Trends Cell Biol.* 4:311–14
225. Muesch A, Hartmann E, Rohde K, Rubartelli A, Sitia R, Rapoport TA. 1990. *Trends Biochem. Sci.* 15:86–88
226. Kutay U, Hartmann E, Rapoport TA. 1993. *Trends Cell Biol.* 3:72–75
227. Darrigo A, Manera E, Longhi R, Borgese N. 1993. *J. Biol. Chem.* 268:2802–8
228. Mitoma JY, Ito A. 1992. *EMBO J.* 11:4197–203
229. Jantti J, Keranen S, Toikkanen J, Kuismanen E, Ehnholm C, et al. 1994. *J. Cell Sci.* 107:3623–33
230. Kutay U, Ahnert-Hilger G, Hartmann E, Wiedenmann B, Rapoport TA. 1995. *EMBO J.* 14:217–23
231. Linstedt AD, Foguet M, Renz M, Seelig HP, Glick BS, Hauri HP. 1995. *Proc. Natl. Acad. Sci. USA* 9295:5102–5
232. Dahmann C, Diffley JFX, Nasmyth KA. 1995. *Curr. Biol.* 5:1257–69
233. Nunnari JM, Zimmerman DL, Ogg SC, Walter P. 1991. *Nature* 352:638–40
234. Florkiewicz RZ, Majack RA, Buechler RD, Florkiewicz E. 1995. *J. Cell. Physiol.* 162:388–99

Annu. Rev. Biochem. 1996. 65:305–36

MOLECULAR BIOLOGY OF MAMMALIAN AMINO ACID TRANSPORTERS

Marc S. Malandro and Michael S. Kilberg

Department of Biochemistry and Molecular Biology, University of Florida College of Medicine, Gainesville, Florida 32610-0245

KEY WORDS: neurotransmitters, expression cloning, membrane proteins, nutrient transport, oocytes

ABSTRACT

Recently a number of α–amino acid transport proteins and corresponding cDNA clones have been isolated and categorized into gene families. The "CAT family" contains two members that mediate high-affinity Na^+-independent transport of cationic amino acids in many tissues, and a third member that encodes a liver-specific low-affinity activity. The "glutamate transporter family" contains at least four members that mediate Na^+-dependent glutamate/aspartate uptake and two members that are selective for neutral amino acids. The glutamate transporters are expressed at high levels in both glia and neurons of the central nervous system. The Na^+/Cl^--dependent proline transporter (PROT) belongs to a large superfamily of neurotransmitter transporters and is expressed in regions of the brain that contain glutamanergic neurons. All four glycine transporters of the "GLYT family" also belong to the neurotransmitter superfamily and exhibit the greatest expression in the central nervous system. The "rBAT/4F2hc family" of proteins induce both neutral and cationic amino acid uptake when expressed in Xenopus oocytes. Cystinuria is linked to specific mutations in the rBAT sequence.

CONTENTS

INTRODUCTION 306
CAT FAMILY FOR CATIONIC SUBSTRATES 308
  Cloning of the CAT Family 308
  Regulation of CAT Transporter Expression. 310
  Structure-Function Analysis 313
GLUTAMATE FAMILY FOR ANIONIC SUBSTRATES 314
  Cloning of the Glutamate Family 314

0066-4154/96/0701-0305$08.00

Distribution of Glutamate Transporter Expression................................ 316
Regulation of the Glutamate Transporters 318
Structure/Function Analysis.. 319
Additional Members of the Glutamate Transporter Family...................... 319
BRAIN-SPECIFIC PROLINE TRANSPORTER 321
Neurotransmitter Transporter Family.. 321
Cloning of the PROT Transporter... 321
PROT Expression in the Brain... 322
GLYCINE-SPECIFIC TRANSPORTERS... 322
Cloning of the GLYT Family.. 323
GLYT1 Isoforms 323
GLYT2 Transporter 325
rBAT/4F2hc FAMILY AND CYSTINURIA 327
Cloning of the rBAT/4F2hc Family.. 327
Structure/Function Analysis.. 328
Relation to 4F2hc Protein 330
Role of rBAT in Cystinuria... 330
SUMMARY 331

INTRODUCTION

The study of amino acid transport into animal cells actually began in 1914 with the observations of Van Slyke and Meyer who demonstrated tissue accumulation of amino acids against a concentration gradient. Description of individual transport activities began in the early 1960s with the pioneering work from the Christensen laboratory (1). It has been more than 20 years since the subject of mammalian amino acid transport was reviewed in this series (2). During that time, literally dozens of distinct amino acid transport activities have been described using intact cells, membrane vesicles, and reconstituted proteoliposomes. The paradigm that has emerged from the last three decades of investigation is that plasma membrane transport for each of the three classes of α-amino acids[1]—anionic, cationic, and zwitterionic—is mediated by several independent transport activities with overlapping substrate specificities. Some of these systems are secondary active transporters, energized by the Na^+-electrochemical gradient and result in net accumulation of amino acid against the concentration gradient. Others are Na^+-independent facilitated transporters[2] and permit flux in either direction depending on the chemical principle of mass action. The characterization of transport for any particular amino acid is made more complex because of cell-specific expression of a subset of the amino

---

[1]Throughout this review we will use the generic term "amino acids" to refer to those α-amino acids that are incorporated into protein. The transport of some synthetic analogs will be discussed, but most of the description will focus on transporters for the α-amino acids only.

[2]Although the term "facilitated diffusion" has been applied to Na^+-independent uptake systems, we see this term as an oxymoron. These activities exhibit saturable binding, contrary to what the term diffusion may imply. Of course, some diffusion may occur within a channel before or after the binding event, but "facilitated transport" distinguishes these activities from those showing channel characteristics.

Table 1 Amino acid transporters for which there are cDNA clones[a]

Clone	Alternate names	Deduced amino acid length	Substrate specificity	Ions coupled
CAT1 (12)	—	622–624	cationic	—
CAT2 (25, 26)	CAT2a (29)	658	cationic	—
CAT2a (29)	CAT2b (29)	659	cationic	—
GLAST (45)	GluT (48), EAAT1 (49)	543	D,L-aspartate L-glutamate	Na^+_{in}, K^+_{out}, $OH^-/HCO_3^-{}_{out}$
GLT (50)	EAAT2 (49), GLAST2 (55), GLTR (54)	573	D,L-aspartate L-glutamate	Na^+_{in}, K^+_{out}, $OH^-/HCO_3^-{}_{out}$
EAAC (57)	EAAT3 (49)	523 –525	D,L-aspartate L-glutamate	Na^+_{in}, K^+_{out}, $OH^-/HCO_3^-{}_{out}$
EAAT4 (62)	—	564	D,L-aspartate L-glutamate	Na^+_{in}, K^+_{out}, $OH^-/HCO_3^-{}_{out}$
ASCT1 (78)	SATT (77, 80)	532	neutral (pH 7.5) anionic (pH 5.5)	Na^+_{in}
AAAT (82)	—	553	neutral	Na^+_{in}
PROT (87, 91)	—	637	proline	Na^+_{in}, Cl^-_{in}
GLYT1a (101, 103)	GLYT2 (100)	633	glycine	Na^+_{in}, Cl^-_{in}
GLYT1b (102)	GLYT1 (100)	638	glycine	Na^+_{in}, Cl^-_{in}
GLYT1c (99)	—	692	glycine	Na^+_{in}, Cl^-_{in}
GLYT2 (104)	—	799	glycine	Na^+_{in}, Cl^-_{in}
rBAT (117)	Naa-Tr (115), D2 (119)	683	neutral, cationic, and cystine	—
4F2hc (138, 139)	—	529	neutral and cationic	Na^+_{in} (for neutral)

[a] Reference numbers appear in parentheses.

acid transporters known to exist in mammalian cells. The breadth of substrate specificity for individual transporters extends from the glycine transporters that accept only a single substrate to activities such as System L that accept nearly all naturally occurring zwitterionic amino acids.

Several excellent reviews (3–9) and two monologues (10, 11) have been published over the past five years that describe our knowledge of amino acid transporters and their associated regulation. To describe completely the characteristics, kinetics, regulation, and molecular biology of all of these transport activities is now well beyond the scope of any single review. Therefore, we have confined this presentation to the recent advances in molecular cloning, expression, and genomic characterization for the α-amino acid families. These cDNA clones are summarized in Table 1. This aspect of mammalian amino acid transport is relatively new, the first α–amino acid transporter cDNA having been identified in 1991. However, during that relatively short interval of time a significant amount of work has been published and several important themes have emerged. We have tried to be reasonably complete regarding the molecular biology of amino acid transport since 1991, but space constraints permit only limited discussion of how these recent observations fit into the context of the past 30 years of transport data. We apologize in advance to those investigators who contributed to that body of knowledge and who believe that their contributions have not been given sufficient credit. No doubt they are correct, but the scope of the many interesting potential discussions quickly becomes overwhelming. With this humbling realization in mind, we have taken this opportunity to review what has been accomplished over the past few years and, more importantly, to encourage additional experimentation by investigators already in the field and those who will join us in the future.

CAT FAMILY FOR CATIONIC SUBSTRATES

Three members of the CAT family of amino acid transporters have been cloned, each of which mediate the Na^+-independent transport of the cationic amino acids arginine, lysine, ornithine, and histidine when positively charged. Although there is significant similarity in the deduced amino acid sequence of these proteins and they share a common substrate specificity, the tissue-specific expression and apparent role of these proteins in metabolism differ.

Cloning of the CAT Family

The cloning of the first CAT family member by Cunningham and coworkers resulted from the search for the host cell protein responsible for infection by the murine ecotropic leukemia virus (MuLV) (12). The deduced amino acid sequence of the retroviral receptor cDNA clone revealed a 622–amino acid glycoprotein with a predicted molecular mass of 67 kDa. The hydropathy profile of the protein predicted 12–14 membrane spanning regions, similar to the previously cloned facilitated glucose transporters; however, the profile showed no significant amino acid sequence similarity to any known protein. The corresponding gene was then mapped to murine chromosome 5, the site of the previously defined REC-1 locus for ecotropic MuLV susceptibility, confirming its role in viral infectivity (13). Although the name EcoR has been applied to the murine protein and REC-1 applied to the gene to describe its role in viral infection, the cDNA has been termed murine CAT1 (cationic amino acid transporter 1) in the context of its amino acid transport properties.

The function of the CAT1 cDNA as a transporter was unknown until 1991 when two independent groups simultaneously described the cationic amino acid transport properties of the murine cDNA (14, 15). The sequence similarity between the CAT1 protein and the arginine and histidine permeases of *Saccharomyces cerevisiae* suggested the actual cellular function for the EcoR receptor. This function was confirmed when *Xenopus* oocytes injected with CAT1 cRNA were shown to exhibit a significantly higher rate of cationic amino acid transport. The substrate specificity and kinetics of the CAT1-mediated uptake were consistent with the previously defined transport activity, System y^+ (16, 17). However, some investigators (18) believe that the lack of inhibition by neutral amino acids in the presence of sodium, a known property of System y^+ (17), may be consistent with assignment to an activity termed System b^+ defined in mouse blastocysts.

The human gene homologous to murine CAT1 was cloned and mapped to chromosome 13q12-q13 (19, 20). The gene consists of 10 introns and 11 exons and codes for a mRNA of approximately 9 kb with an open reading frame of only 1.8 kb. The remainder of the gene consists of long untranslated regions. The human cDNA sequence predicts a 12–14–membrane spanning protein of 629 amino acids with a molecular mass of 68 kDa, and it shares 88% identity with the mouse sequence. A rat CAT1 cDNA encoding a 624–amino acid protein sharing 97% sequence homology with the murine transporter also has been isolated (21, 22). Puppi et al (23) sequenced 7.0 kb of the 7.9 kb rat mRNA, but the significance of the 5 kb 3′ untranslated region of the mRNA is unknown.

The cloning of the second member of the CAT transporter family also came about during the pursuit of unrelated genes. MacLeod and coworkers, investigating genes involved in T-cell function that have the ability to promote tumor formation in mice, isolated several cDNAs which were expressed at high levels in a tumorigenic murine T-lymphoma cell line (SL12.4) and at nearly undetectable levels in noninvasive control cells (24). One of these cDNAs coded for a protein with significant amino acid sequence identity to CAT1 (25). The 2.4 kb cDNA, termed Tea (T-cell early-activation gene), predicted a 50 kDa protein of 453 amino acids with 7 membrane-spanning domains, but sequence analysis and comparison to CAT1 suggested that the protein was truncated. The complete amino acid sequence, later described by the same group (26), coded for 658 amino acids, contained 12 putative transmembrane domains, and shared 61% amino acid identity with CAT1. The Tea gene has been localized to murine chromosome 8 (25).

The function of the Tea gene as a cationic amino acid transporter was suggested by its structural similarity to CAT1, as well as to a family of bacterial and yeast nutrient transporters (26). *Xenopus* oocytes injected with Tea cRNA exhibited increased uptake of cationic amino acids compared to water-injected controls (27, 28). The cationic amino acid transport elicited by the Tea cRNA was saturable and Na^+-independent, similar to that mediated by CAT1. Analysis of amino acid transport by electrophysiological measurements revealed a K_m of approximately 150–250 μM for both Tea and CAT1 (27). After a period of some variation in nomenclature due to the rapid isolation of these and

Amino acid sequence differences between murine CAT2 and CAT2a

```
CAT2      277 - IFPMPRVIYAMAEDGLLFKCLAQINSKTKTPVIATLSSGAVA - 318
                .**.**...***.*****..**..**..**.**...*..
CAT2a     277 - MFPLPRIFLAMARDGLLFRFLARV-SKRQSPVAATMTAGVIS - 317
```

Figure 1 Deduced amino acid sequence alignment of murine CAT2 and CAT2a from residues 277 to 318 corresponding to the fourth extracellular loop in the predicted structure. Asterisks indicate positions of identity and dots indicate conservative substitutions. The remainder of the amino acid sequence of the two proteins is identical.

homologous transporters by independent laboratories,[3] the Tea cDNA is now referred to as CAT2.

Using the original partial Tea sequence obtained by MacLeod et al (25), Cunningham and coworkers cloned a third member of the CAT family from murine liver, CAT2a (29). The CAT2 and CAT2a cDNAs arise from the same gene on mouse chromosome 8, and the two sequences are identical with the exception of an alternatively spliced region within the predicted fourth extracellular loop (Figure 1). Injection of oocytes with CAT2a cRNA induces cationic amino acid transport, but the apparent K_m for arginine is nearly tenfold higher than that for either CAT1 or CAT2 (29).

Regulation of CAT Transporter Expression

Consistent with its identification as System y$^+$, the tissue distribution of CAT1 expression appears to be nearly ubiquitous. In murine tissues, the primary CAT1 mRNA is 7.9 kb, but as a result of tissue-specific differential polyadenylation an additional 7.0 kb transcript is often observed (12, 14, 27). In rats, a primary CAT1 mRNA size of 7.9 kb as well as one of 3.4 kb has been observed in placenta (30), hepatoma cells (3), brain (21), and intestine (23). Human CAT1 mRNA species identified include one of 9.0 kb and a second of 7.9 kb (19, 31, 32).

The only tissue known to lack expression of high-affinity cationic amino acid transport is the liver (17). Consistent with the transport data, the lack of hepatic CAT1 mRNA has been reported by several groups (3, 14, 22, 27, 29), and Woodard et al (33) were unable to detect CAT1 protein in both rat and human liver by immunohistochemistry. Immunohistochemistry indicates that a CAT1 (or closely related protein) exists within discreet clusters randomly

---

[3]Closs et al (28) originally designated the lymphocyte clone as CAT2b and the liver-derived clone as CAT2a, but to reflect the chronological order of isolation the lymphocyte sequence will be referred to as CAT2 and the hepatic sequence as CAT2a, as suggested by MacLeod and Cunningham (personal communication).

distributed on the plasma membrane surface of human fibroblasts (33). Closs et al (29) prepared antibodies specific for CAT1 and CAT2a to demonstrate the lack of CAT1 expression in liver. Immunoblotting of oocyte membranes following injection of either CAT1 or CAT2a cRNA showed that both glycosylated proteins migrated at 70–80 kDa, but after N-glycosidase treatment the apparent molecular size was 60–65 kDa.

The CAT2 cDNA, encoding high-affinity (micromolar K_m values) arginine transport, was isolated from a murine T-lymphoma cell line (24). Subsequently, CAT2 expression also was detected in concanavalin A-activated spleen cells, thymic epithelial cells, liver cells, skeletal muscle, skin, lung, brain, uterus, and stomach (25, 27). Expression of CAT2a mRNA, low-affinity arginine transport (millimolar K_m values), is even more limited in that it is only detectable in normal liver tissue (29) and skeletal muscle (C MacLeod, personal communication).

Expression of the CAT1 transporter mRNA is enhanced when cells are shifted from a quiescent or differentiated state to an undifferentiated, rapidly growing state (149). For example, activation of murine T-cells by concanavalin A- or B-cells by lipopolysaccharide treatment increased steady-state CAT1 mRNA content. Splenomegaly, induced by mink cell focus-forming virus, also increased the expression of CAT1 mRNA. Conversely, down-regulation of CAT1 mRNA content was demonstrated when HL-60 promyelocytic leukemia cells progressed toward terminal differentiation into granulocytes or macrophages (149). Yoshimoto et al proposed (149) that regulation of CAT1 mRNA content is mediated by protein kinase C because: (a) CAT1 mRNA content increases in phorbol 12-myristate 13-acetate (PMA)-treated cells. (b) PMA analogs that do not induce the protein kinase C pathway do not alter CAT1 mRNA levels. (c) Inhibition of PMA induced CAT1 expression by protein kinase C-specific inhibitors. (d) CAT1 expression was induced by calcium ionophores. Direct evidence for transcriptional or post-transcriptional mechanisms will require more detailed analysis.

As mentioned above, CAT1 mRNA expression is not detectable in normal adult liver (3, 14, 22, 27, 28), a result consistent with the inability of MuLV to infect adult hepatocytes (34) and the lack of high-affinity arginine transport in freshly isolated hepatocytes (17). Infection of liver cells with retroviruses is possible after partial hepatectomy (34), therefore several groups have focused on the examination of CAT1 expression in liver regeneration. Closs et al (35) observed induction of CAT1 expression in murine hepatocytes after 48 h of primary culture, but no CAT1 expression in liver tissue 12–48 h after partial hepatectomy (27). Although the latter result seems to contradict the retroviral infection data, Wu et al (22) demonstrated an induction of CAT1 mRNA synthesis at 4–6 h after hepatectomy, paralleling the time-course of viral susceptibility, which then decays back to a level below detection by 12 h post-hepatectomy. CAT1 mRNA expression in normal liver tissue can also be induced by acute treatment of rats with insulin, dexamethasone, or excess arginine, or by adaptation to a low-protein diet (22).

Given that regulation of CAT2 paralleled that of CAT1 in activated T-cells (31), MacLeod and coworkers examined the expression of CAT2 in regenerating liver (27). No change in CAT2 steady-state mRNA level was seen in control, sham-operated, or regenerating liver 24 h after partial hepatectomy (27). Unfortunately, as noted above, the increased expression of CAT1 was detected only from 4–6 h after hepatectomy and returned to control values by 24 h (22). Therefore, whether or not the transient increase in CAT1 expression is paralleled by a transient increase in CAT2 is unknown. Interestingly, skeletal muscle CAT2 mRNA was increased following either hepatectomy, splenectomy, or fasting (36, 37). It is postulated that a corresponding increase in transporter may permit efflux of cationic amino acids from the muscle, a major source of catabolic fuel during surgical trauma or fasting (38, 39).

Developmental regulation of CAT1 has also been demonstrated. Induction of CAT1 steady state mRNA content in rat placenta occurs from 14–20 d gestation, equivalent to the final trimester of human pregnancy (30). The increase in CAT1 mRNA was paralleled by an increase in arginine uptake, and a corresponding lack of CAT2 expression supported the role of CAT1 as the transporter responsible. Campione and Van Winkle (39a) have described the regulation of both CAT1 and CAT2 in the developing mouse embryo, detecting expression as early as the one-cell stage (40, 41). The appearance of the CAT mRNAs paralleled the emergence of Na$^+$-independent, cationic amino acid transport. Puppi and Henning (23) observed a high level of CAT1 mRNA in rat fetal intestine that decreased upon birth and then returned to a high level in adulthood.

Regulated expression of the CAT2 gene isoforms involves specific mRNA splicing events in both the coding and noncoding regions of the gene. As mentioned above, the predicted protein products, CAT2 and CAT2a, differ in only a single stretch of 40/41 amino acids in the fourth extracellular loop of the transporter and are expressed in a tissue-specific manner (27, 35). Analysis by MacLeod and coworkers of many CAT2 cDNA clones from T-lymphoma cells led to the identification of four isoforms that contain the entire coding sequence, but differ in their 5′ untranslated regions (37). An independent laboratory isolated a fifth isoform altered in the 5′ untranslated region (42). Each of these mRNAs converge into a common sequence 16 base pairs upstream of the AUG initiation codon. The CAT2 gene 5′-untranslated exons span a distance of over 18 kb that may contain up to five distinct promoters (37). The sequence upstream of the first untranslated exon contains no TATA sequence, but it does contain DNA binding motifs common to TATA-less promoters including GC-rich sequences and SP1 binding sites. The product of this promoter appears to be present in all cell types that express CAT2. The sequence upstream of the second untranslated exon is a classic TATA promoter. Selective use of independent promoter regions is likely responsible for the tissue-specific expression of CAT2 isoforms.

Structure-Function Analysis

Wang et al (15) have demonstrated in voltage-clamped *Xenopus* oocytes injected with murine CAT1 cRNA an inwardly directed, saturable current in the presence of substrate. When the magnitude of the current is compared to transport of radiolabelled substrate in similarly injected oocytes, movement of one positive charge per molecule of substrate is observed. Kavanaugh (43) examined membrane current resulting from CAT1-dependent transport in oocytes and showed that at membrane potentials ranging from +20 mV to −120 mV arginine uptake followed values predicted by Michaelis-Menten kinetics. At arginine concentrations from 0.01 to 1.00 mM, influx increased exponentially with membrane hyperpolarization rather than exhibiting saturation kinetics. On the basis of this result, the author proposes that the rate-limiting step in transport, thought to be the reorientation of the transporter to the outward facing position, is voltage dependent. Thus, a charge translocation occurs during the reorientation of the empty transporter, and this charge movement significantly affects the voltage-dependent cationic amino acid flux mediated by CAT1.

In an effort to determine the structural features responsible for the differences in transport activity between CAT1 and CAT2/CAT2a, Closs et al (28) expressed chimeric proteins in *Xenopus* oocytes. CAT1 (14) and CAT2 (27, 28) exhibit high-affinity transport (micromolar K_m values) and are capable of trans-stimulation. CAT2a, the liver-specific isoform, has a tenfold greater apparent K_m for arginine than CAT1 or CAT2, but a higher V_{max} (29). Insertion of the CAT2a-specific 40–amino acid sequence into CAT1 resulted in transport with greater K_m and V_{max} values, consistent with the properties of CAT2a (28). Similarly, the reciprocal construct in which the 40–amino acid sequence from CAT2a was replaced resulted in high-affinity transport indicative of CAT1. With regard to specific amino acid residues important for CAT1 function, Wang et al (44) noted the conservation of a glutamate residue (position 107 in CAT1) residing in a putative membrane-spanning region of not only CAT1, CAT2, and CAT2a, but also in the homologous arginine and histidine permeases from yeast. The substitution of aspartate for glutamate 107 in CAT1 did not alter transporter trafficking or binding of the viral coat protein gp70, but completely abolished transport activity.

Recent reports on the function of CAT1 and CAT2 have led to some interesting findings that may lead to a re-evaluation of the "one protein-one transport activity" paradigm. Van Winkle and coworkers used *Xenopus* oocyte expression, as well as data by other investigators, to analyze the transport kinetics of CAT1 and CAT2 (18). When electrophysiological measurements were used to assay transport, the L-arginine uptake mediated by either CAT1 or CAT2 expressed individually yielded biphasic Eadie-Hoffstee plots, suggesting two kinetically distinct transport activities. Possible interpretations of these data are that both activities could be mediated by a single CAT protein, the CAT protein could associate with multiple accessory proteins to form distinct complexes, or the CAT protein itself represents a single activity and, in addition, up-regulates the activity of an endogenous oocyte transporter.

GLUTAMATE FAMILY FOR ANIONIC SUBSTRATES

Cloning of the Glutamate Family

Even before the cloning of the first glutamate transporter, the existence of several distinct activities was proposed because glutamate transport in selected brain regions was differentially sensitive to inhibitors such as dihydrokainate and L-α-aminoadipate (7). One of the earliest groups to identify a glutamate transporter cDNA, known as GLAST, was Storck et al (45). During the isolation of a galactosyltransferase from rat brain, a hydrophobic protein of about 66 kDa was copurified. Proteolytic fragments of this protein were sequenced, and oligonucleotide probes were generated to screen a rat brain cDNA library. A 3 kb clone was isolated that coded for a protein of 543 amino acids with a predicted molecular mass of 60 kDa. This GLAST cDNA showed considerable sequence similarity to the previously cloned glutamate and monocarboxylate transporters of bacteria (46). The predicted protein sequence also showed structural features typical of transporters such as multiple membrane-spanning domains and glycosylation sites. Computer analysis of the GLAST sequence predicts only six putative transmembrane domains within the N-terminal half of the protein. The C-terminal portion contains six short hydrophobic stretches, which also are conserved in the homologous prokaryotic transporters, but may be too short to span the membrane as α-helices (46). To determine the function of GLAST, *Xenopus* oocytes were injected with cRNA and transport assayed by uptake of radiolabelled substrate (45). Transport of both glutamate and aspartate was demonstrated to be Na+-dependent and inhibited by DL-threo-3-hydroxyaspartate, the strongest known inhibitor of Na+-glutamate transport in brain slices (47). The human homolog of the GLAST transporter was subsequently isolated by low-stringency screening of human brain cDNA libraries by Kawakami et al (48) and Arriza et al (49) who named their clones GluT and EAAT1, respectively. The human clone shares 97% amino acid sequence identity with the rat transporter. In this review, the original name of

GLAST will be used in reference to this transporter from either rat or human to eliminate ambiguity about the term GluT, a designation that is well established for facilitated glucose transporter clones.

The second member of the glutamate transporter family (GLT) was isolated by Pines et al (50). The cloning of GLT essentially began several years before the isolation of the cDNA with the purification to near-homogeneity of the GLT transport protein (51) and the subsequent generation of both polyclonal (52) and monoclonal antibodies (53). The polyclonal antibody was used to immunoscreen a cDNA expression library and a 4.6 kb clone was isolated (50). The GLT cDNA predicted an open reading frame of 1719 bp coding for 573 amino acids with a molecular mass of 63 kDa. The hydropathy profile of GLT is similar to that of GLAST in that it predicts 6–8 putative membrane spanning regions at the N-terminus and several shorter hydrophobic regions at the C-terminus. GLT shares homology with an *E. coli* glutamate transporter (46), as well as 44% amino acid identity and 68% homology with GLAST. To confirm the function of the GLT cDNA, transport of radiolabelled substrates was measured after expression of the cDNA in mammalian cells. GLAST mediated Na^+-dependent transport of L-glutamate that was dependent on internal K^+, but was not inhibited by the ionophore valinomycin, suggesting that K^+ served as a substrate rather than as a driving force for transport. Arriza et al (49) and Manfras et al (54) independently identified the human homolog of GLT. Although a few sequence differences exist between the two human clones, these cDNAs most likely arise from the same gene and share 90% amino acid sequence identity with rat GLT. Although alternative names were used for the clones isolated by Arriza et al (EAAT2) and Manfras et al (GLTR), once again the original name of GLT will be used in this review to eliminate ambiguity. A mouse homolog of GLT was isolated by Kirschner et al (55) and localized to mouse chromosome 2.

A second human clone highly homologous to GLT, unfortunately named GLAST2, has been reported by Shashidharan et al (56). Human GLAST2 has 89% identity to rat GLT (50) and 98% identity to the human homolog (49, 54). The major differences between GLT and GLAST2 are a deletion of the first 10 N-terminal amino acids and two single base deletions that introduce a frame-shift mutation corresponding to amino acids 253–281 of GLT. Shashidharan et al (56) isolated five additional clones from two independent cDNA libraries that confirmed the sequence differences between GLT and GLAST2, but isolation of genomic clones will be necessary to establish the basis for these differences.

A third member of the glutamate transport family was identified by Kanai and Hediger (57) who employed oocyte expression cloning using fractionated mRNA from rabbit intestine. A cDNA (EAAC) of 3.4 kb was isolated that contained an open reading frame coding for a protein of 524 amino acids and a predicted molecular mass of 57 kDa. EAAC shares 66% amino acid sequence similarity with GLAST and 61% with GLT. Expression of the EAAC cDNA resulted in a 1000-fold increase in Na^+-dependent L-glutamate transport over that of water-injected oocytes. Electrophysiological analysis of transport revealed high-affinity uptake for L-glutamate and L-aspartate (6–12 μM K_m values) and inhibition of this uptake by D-aspartate and DL-threo-3-hydroxy-aspartate, two strong inhibitors of synaptosomal glutamate uptake (58). Subsequently, both Arriza et al (49) and Kanai et al (59) identified human homologs of EAAC that share 96% amino acid sequence similarity with the rabbit clone. Hediger and coworkers have localized the human EAAC to chromosome 9p24 (60). A rat EAAC homolog that shares 95% similarity with the rabbit clone also has been reported (61). Antibodies against the C-terminal 120 amino acids of the rat EAAC recognize a broad protein band centered about 80 kDa in both brain and placenta (MS Malandro, DA Novak, MS Kilberg, unpublished results). The differential expression of the EAAC transporter and glutamate transport activity during development in rat placenta basal and apical plasma membrane subdomains is consistent with a role in fetal nutrition.

Recently, a fourth member of this family (EAAT4) has been isolated by Fairman et al (62) using degenerate oligonucleotide primers corresponding to conserved sequences within members of the glutamate family and low-stringency rtPCR from human cerebellum mRNA. By Northern analysis, the EAAT4 transporter mRNA was detected as a single mRNA species of 2.4 kb in the brain cerebellum and placenta, but rtPCR analysis revealed a low level of EAAT4 in brain stem, cortex, and hippocampus (62). The EAAT4 shares 65% (GLAST), 41% (GLT), and 48% (EAAC) amino acid sequence identity with the corresponding human glutamate transporter clones, named EAAT1-3, identified by the same laboratory. Expression of EAAT4 in oocytes selectively increased Na^+-dependent L-glutamate uptake. Substitution of extracellular chloride with gluconate completely inhibited the outwardly directed current induced by L-aspartate at depolarizing membrane potentials, suggesting a role for chloride in the transport mechanism. However, when membrane potentials were clamped at −60 mV, the substrate-induced current was not significantly different either in the presence or absence of chloride. The data are consistent with the authors' proposed model for EAAT4 as a chloride channel activated by substrate binding (glutamate/aspartate and sodium). This unique property of EAAT4 may be involved in not only the re-uptake of glutamate from the synapse, but also may aid in the re-establishment of membrane potential by influencing cellular chloride permeability.

Distribution of Glutamate Transporter Expression

In the initial identification of rat GLAST, a single mRNA species of 4.5 kb with expression exclusively in the brain was detected with Northern analysis

(45). In situ hybridization revealed a low-level uniform distribution across the entire cerebellum with high expression only in the Purkinje cells of the cerebellar cortex (45). Rat GLAST protein has been detected by immunoblotting, as a broad band of approximately 66 kDa (63). Expression of the protein correlates well with the localization of the mRNA, primarily in the cerebellum with expression restricted to astrocytes. No GLAST protein is detected in neurons. In situ hybridization of rat retina by Otori and coworkers demonstrated GLAST mRNA expression in retinal glial cells, specifically Müeller cells and astrocytes (64). For human tissue, Northern analysis revealed a single mRNA species of approximately 4.1 kb not only in brain, but also in heart, lung, placenta, and skeletal muscle (48, 49). Within the human brain, GLAST mRNA is abundant in motor cortex, frontal cortex, hippocampus, basal ganglia, and cerebellum (49). GLAST or related sequences may also be expressed in tissues other than the central nervous system. Van Winkle and colleagues have detected developmentally regulated GLAST mRNA expression in mouse blastocytes concurrent with the appearance of Na^+-dependent glutamate uptake (39a).

Initial detection of rat GLT transporter mRNA content revealed a single mRNA species of 10 kb that was restricted in expression to the brain (50). Human GLT, with a similar mRNA size, was also predominantly brain-specific, but there was a weak signal in placenta (49, 55). Human brain GLT mRNA content was greatest in the motor cortex, frontal cortex, hippocampus, and basal ganglia, with lower levels detected in the cerebellum (49). When the tissue distribution of GLT mRNA was measured in the mouse, a weak but detectable signal was observed in the liver in addition to the brain (55). The rat GLT protein has been localized to glial cells by immunocytochemical methods (52), and the highest content was in the hippocampus, lateral septum, cerebral cortex, and striatum (63). No GLT protein was seen in glutamanergic nerve terminals (52, 63). Although the GLAST and GLT transporters have a similar distribution in the brain, their distributions in the retina, a major site of glutamanergic nerve innervation, may differ. GLAST expression is present in retinal Müeller cells and astrocytes (64), whereas GLT is absent from these cell types as measured by immunocytochemistry (66).

EAAC is the most widely distributed of the glutamate transporters. Rabbit EAAC is expressed as 3.5 kb and 2.5 kb mRNA species in small intestine, kidney, brain, liver, and heart (57), with a similar distribution in humans (49, 59) and rat (61). The human brain EAAC mRNA is present in the motor cortex, frontal cortex, hippocampus, basal ganglia, and cerebellum (49). In situ hybridization of rat brain reveals highest expression of rat EAAC in the cerebellar granule cell layer, hippocampus, superior colliculus, and neocortex. In all cases, EAAC was localized to neuronal cell bodies, including some nonglutamanergic neurons (61). Collectively, the data are consistent with the suggestion that in the brain GLAST and GLT represent glial-specific glutamate transporters, whereas EAAC represents a neuron-specific activity (57).

Regulation of the Glutamate Transporters

Even before any of the glutamate transporter proteins or cDNA clones were identified, oocyte expression of mRNA isolated from developing rat brain permitted Blakely et al (67) to demonstrate a region-specific increase in glutamate transport activity throughout development from fetus to adult. Otori et al (64) examined the regulation of the GLAST glutamate transporter mRNA content by in situ hybridization following ischemia in the rat retina. Although retinal ischemia caused neuronal cell death of the non-GLAST expressing neurons, the glial cells containing GLAST mRNA survived. After reperfusion the glial GLAST transporter mRNA content was actually increased. It has been proposed that glutamate transport, possibly mediated by GLAST, may operate in a reverse mode under ischemic conditions, thus allowing glutamate efflux rather than uptake (68). Increased GLAST expression during reperfusion would permit removal of the high levels of extracellular glutamate resulting from the period of ischemia.

Regulation of GLT-mediated glutamate transport was investigated by transiently transfecting rat C6 glioma cells and HeLa cells with the GLT cDNA (69). Within 30 min after treating either cell line with phorbol ester there was a rapid increase in L-glutamate transport which paralleled the phosphorylation of the GLT protein. Site-directed mutation of serine 113 to asparagine, the single protein kinase C consensus phosphorylation site within GLT, did not affect the basal transport rate or transporter number as estimated by immunoblot analysis. However, in the mutant transporter the phorbol ester–mediated increase in glutamate transport was abolished (69). Smit et al (70) observed regulation of the GLAST and GLT transporters expressed in oocytes in response to arachidonic acid, a messenger molecule involved in ischemia and released upon activation of glutamate receptors (70). Arachidonic acid caused a 30% inhibition of glutamate transport via GLAST, but stimulated transport via GLT by increasing the substrate affinity twofold (71). These data suggest that specific members of the glutamate transporter family differentially respond to arachidonic acid, at least when it is expressed in oocytes.

McGivan and coworkers described the induction of a single kinetically defined glutamate transport activity by amino acid deprivation of NBL-1 bovine renal epithelial cells (72). However, the EAAC content, detected by low-stringency PCR, did not correlate with the regulation of glutamate transport activity in these cells. In fact, during the period of starvation-induced glutamate transport, the steady-state EAAC mRNA level actually decreased. The results suggest that these cells express a glutamate transporter gene other than EAAC that responds to the metabolic environment in an independent manner.

Structure/Function Analysis

The glutamate transporters share a similar structure of at least six transmembrane spanning domains in the N-terminal half of the protein and at least two conserved putative glycosylation sites within the extracellular loop between transmembrane segments three and four. Elimination of the two N-linked glycosylation sites by site-directed mutagenesis of GLAST revealed that both were glycosylated, but the nonglycosylated transporter was completely functional and properly targeted (73). To elucidate essential residues in GLT, Zhang et al (74) used site-directed mutagenesis to change two amino acids residues, lysine 298 and histidine 326, conserved in all glutamate transporters and proposed to be within a membrane-spanning region. Substitutions of arginine or histidine for lysine 298 resulted in functional proteins that retained nearly full activity. However, less conservative substitutions of glutamine and threonine at the same position resulted in targeting defects. In contrast to the lysine mutants, any substitution for histidine 326 resulted in a completely nonfunctional protein. Mutagenesis of highly conserved negatively charged residues in the hydrophobic portions of the C-terminal of the transporter (aspartate 398, aspartate 470, glutamate 404) resulted in decreased activity, even when these residues were substituted with amino acids of the same charge (75). Substitution of glutamate 404 reduced the glutamate transport by 80%, but resulted in only a 20% reduction in either D- or L-aspartate transport. Given that the binding of glutamate was unaffected in this mutant, the data suggest a possible selective impairment of glutamate translocation or release.

Kavanaugh and coworkers identified an inherent ion flux, in the absence of glutamate, associated with human GLAST, but not GLT [termed EAAT1 and EAAT2, respectively, in (76)]. This GLAST activity was attributed to an uncoupled flux of cations across the plasma membrane. A chimeric transporter was made by substitution of residues 366–441 from GLT into the GLAST sequence resulting in 18 amino acid changes. The chimeric transporter bound kainate with high-affinity, a property of the substituted domain of GLT not exhibited by GLAST, and, furthermore, kainate binding inhibited the constitutive ion flux of the modified GLAST transporter.

Additional Members of the Glutamate Transporter Family

Shafqat et al (77) and Arriza et al (78) independently isolated a cDNA from human brain that shares nearly 40% sequence identity with the glutamate transporters, but does not typically serve as a glutamate transporter. Although a few sequence discrepancies exist, even after publication of a revised sequence (77), we believe the two clones represent the same gene. Therefore, we have abandoned the term SATT and adopted the name ASCT1 as proposed by Arriza et al (78). The human gene for ASCT1 has been localized to chromosome

2p13-p15 (79). The cDNA identified an open reading frame of 532 amino acids with a predicted molecular size of about 56 kDa. With 50–60% similarity to GLAST, GLT, and EAAC, the ASCT1 sequence predicts six well-defined transmembrane sequences near the N-terminal of the protein and up to six additional hydrophobic stretches near the C-terminal. Arriza et al (78), utilizing *Xenopus* oocyte expression and electrophysiological assays, demonstrated Na^+-dependent uptake of neutral amino acids including alanine, serine, and cysteine, but not anionic amino acids. Transient expression of the transporter in HeLa cells gave similar results with the exception of cysteine (77), a discrepancy that was later resolved (80). The substrate specificity and ion selectivity were consistent with a previously described activity called System ASC (3, 81).

Although System ASC activity has been identified in nearly every cell type tested, Northern analysis revealed a family of three ASCT1 mRNAs, approximately 5.0, 3.5, and 2.2 kb, expressed at a high level in the brain, skeletal muscle, and pancreas only. These data suggest that, with circumstances similar to those for glutamate, glycine, and arginine transporter families, what historically has been called System ASC transport activity is actually mediated by different gene products in different tissues. Indeed, "ASCT2" may have been identified already by Liao and Lane (82). They isolated a cDNA, named AAAT, for which a mRNA was expressed in fully differentiated 3T3-L1 adipocytes, but not by the fibroblast-like preadipocytes. A full-length 2.7 kb cDNA contained an open reading frame of 1659 bp corresponding to a 553–amino acid protein of 65 kDa molecular size. The predicted amino acid sequence has 56% identity to ASCT1 and about 40% or less to the glutamate transporters (82). Northern analysis shows that tissues that express the mRNA in greatest abundance are lung and adipose; brain, kidney, liver, spleen, thymus, heart, and muscle give little or no detectable signal. Transient transfection of the preadipocyte resulted in a small but detectable increase in serine uptake, but only after acute insulin treatment.

A unique characteristic of System ASC activity is a pH-dependent change in substrate selectivity (83–85). When assayed at neutral pH, System ASC primarily mediates the uptake of neutral amino acids, whereas if the extracellular pH is decreased anionic amino acids become more effective inhibitors and substrates. Use of the anionic amino acid analogs cysteate or cysteinesulfinate, with sidechain pK values of less than 1.5, eliminates titration of the inhibitor to a zwitterion as a possible interpretation. Tamarappoo et al (86) have expressed the human brain ASCT1 transporter in HeLa cells to demonstrate that it is inhibited by cysteate in a pH-dependent manner. The observed acceptance of anionic amino acids in the presence of a decreased extracellular pH may help to explain the apparent evolutionary relationship between ASCT1 and the glutamate transporters (45, 50, 57, 62).

BRAIN-SPECIFIC PROLINE TRANSPORTER

Neurotransmitter Transporter Family

A cDNA sequence (PROT) that encodes a Na^+/Cl^--dependent proline transporter has been isolated from both rat and human brain libraries (87, 88). These cDNAs show significant sequence homology to a large family of Na^+/Cl^--dependent transporters that all exhibit significant conservation of amino acid sequence and transmembrane topology (89). Included in this family are transporters for glycine, GABA, betaine, taurine, creatine, norepinephrine, dopamine, and serotonin (89, 90). This transporter family does not show significant sequence similarity to other Na^+-dependent mammalian or prokaryotic transporters that do not require Cl^- cotransport, nor do they mediate counter-transport of K^+ and OH^-/HCO_3^- as do the glutamate transporters.

Cloning of the PROT Transporter

Oligonucleotides corresponding to conserved amino acid sequences within the first and sixth transmembrane domains of the GABA and norepinephrine transporters were utilized for PCR amplification of related sequences from rat brain. The PCR product hybridized to a single 4 kb mRNA present in multiple brain regions. Northern analysis of poly(A) mRNA revealed no signal in heart, spleen, lung, liver, skeletal muscle, kidney, or testes (87, 91). The PCR product was used to obtain a full-length cDNA (PROT) that contained a 1911 bp open reading frame that encoded a protein of 637 amino acids and had a predicted molecular mass of 71 kDa. The putative secondary structure contains 12 transmembrane domains typical of the Na^+/Cl^--dependent transporter family. The PROT cDNA had approximately 45% identity and 65–70% similarity to the other Na^+/Cl^--dependent transporters (87, 91). Anti-peptide antibodies against a region of the PROT C-terminal detected a primary protein at approximately 68 kDa in brain, but no immunoreactive protein in several peripheral tissues. Removal of the carbohydrate by treatment with glycosidase F resulted in a protein that migrated at 53 kDa rather than the predicted core size of 71 kDa (91). Anomalous migration in SDS gels is often observed for membrane proteins (92), and it has been reported for other members of this transporter family. Shafqat et al (88) used the rat cDNA to obtain the corresponding human PROT clone from a hippocampal cDNA library. The amino acid sequence of the human cDNA is 98% identical to the rodent sequence, and it is also specifically expressed in brain based on both Northern analysis for mRNA and immunoblotting (88). Interestingly, neither the rat nor the human PROT transporter shows significant similarity to a Na^+-dependent proline transporter cloned from E. coli (93).

PROT Expression in the Brain

The proposal has been made that proline plays a role in neurotransmission. As listed by Shafqat et al (88), the observations that support this proposal include: (a) Synaptosomes contain high-affinity Na^+-dependent proline transport. (b) L-proline is released in a Ca^{++}-dependent manner after K^+-induced depolarization. (c) Proline produces an electrophysiologic response in neurons. (d) Proline inhibits depolarization-dependent release of glutamate from brain slices. (e) Proline is neurotoxic when injected into the cerebrum. Despite these suggestive data, the exact role of proline in neurotransmission remains unknown, as does the physiologic function of the PROT transporter.

Although proline exhibits the highest affinity (K_m=10 µM) of all the amino acids tested as substrates for the PROT transporter, the K_i values for sarcosine (30 µM), phenylalanine (48 µM), histidine (83 µM), and cysteine (91 µM) suggest that they cannot be totally excluded as substrates in vivo. Isolated brain synaptosomes have been shown to have both high- and low-affinity proline transport (94), whereas nonneuronal tissues only express the low-affinity transporter, which is not dependent on cotransport with Cl^- (95). In situ hybridization of rat brain shows specific PROT hybridization to glutamanergic neurons (87). Based on a discordant distribution of the mRNA and protein, Fremeau and coworkers postulate that the PROT transporter is synthesized in the neuronal cell body and migrates to presynaptic termini (91). For example, the mRNA content is highest in the hippocampus and cerebral cortex, but it is not detectable in the caudate putamen (striatum). However, the caudate putamen contains significant levels of the PROT transporter protein based on immunohistochemistry (91). Further support for this proposal comes from immuno-electron microscopy with anti-PROT antibodies demonstrating punctate labelling in four brain regions that receive glutamanergic innervation. The majority of labelling was reported to be associated with organellar vesicles within the axon.

GLYCINE-SPECIFIC TRANSPORTERS

Glycine serves two important roles as a neurotransmitter in the central nervous system. First, it has been well documented to be an inhibitory neurotransmitter in the spinal cord, brain stem, and retina (96). This inhibitory action is mediated by a glycine receptor that functions as a ligand-gated Cl^- channel that is activated by glycine and competitively antagonized by strychnine (97). By blocking the action of glycine, strychnine causes severe seizures. The second role that glycine plays in neurotransmission is that of an obligatory modulator of glutamate-mediated excitatory neurotransmission. The N-methyl-D-aspartate (NMDA) glutamate receptors have a distinct binding site for glycine that must be occupied for receptor activation by glutamate to occur (98). Given these functions, glycine clearly represents an extremely important regulator of neurotransmission for both inhibitory and excitatory pathways.

Cloning of the GLYT Family

Two independent genes have been identified that encode Na^+/Cl^--dependent glycine transport in the brain. These two genes and the corresponding cDNA sequences, termed GLYT1 and GLYT2, have been identified as a result of their sequence homology to other members of the Na^+/Cl^--dependent family of neurotransmitter transporters. As a result of differential splicing and/or utilization of multiple promoters, transcription of the GLYT1 gene results in at least three different mRNA isoforms, and the nomenclature for these used within this review will follow that recommended by Kim et al (99) who assigned the names GLYT1a, GLYT1b, and GLYT1c based on the order of exon usage within the gene. The human GLYT1 gene has been localized to chromosome 1p3.1-3.3 and the mouse gene to chromosome 4 (99).

GLYT1 Isoforms

GLYT1a and 1b have been cloned from both rat and mouse (100–103), whereas the GLYT1b and GLYT1c homologs have been isolated from human (99). Sequence relationships between these three isoforms and a schematic illustrating the first three exons is shown in Figure 2 (99). GLYT1a and GLYT1b are identical in sequence except that the N-terminal GLYT1a has a 10–amino acid N-terminus that arises from exon 1, whereas these 10 amino acids are replaced by a 15–amino acid sequence in GLYT1b that arises from exon 2 (100, 104). The human GLYT1c isoform contains the same 15–amino acid N-terminus as GLYT1b, but differs from both GLYT1a and GLYT1b in that it contains a 54–amino acid insert following the initial 15 amino acids (Figure 2). Beginning with exon 3, the sequences for all three isoforms are identical. All three isoforms are detected as a single mRNA ranging in size from 3–4 kb depending on the species and tissue analyzed. The open reading frame for GLYT1a is 633 amino acids encoding a core protein of approximately 71 kDa; GLYT1b encodes a protein of 638 amino acids; and GLYT1c, containing the 54–amino acid insert, is the largest open reading frame of 692 amino acids encoding a core protein of approximately 88 kDa (99). All three of these proteins exhibit 35–45% identity and 50–60% similarity to the other transporters in the Na^+/Cl^--dependent neurotransmitter transporter family.

Antibody against the C-terminus of rat GLYT1b detects a broad smear of immunoreactivity at 80–100 kDa (105). Complete removal of the carbohydrate resulted in a protein that migrated at 47 kDa, significantly less than the expected size of about 72 kDa based on the amino acid composition. The GLYT1 sequence contains four potential asparagine-linked glycosylation sites

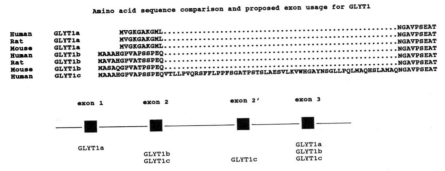

Figure 2 Amino acid sequence comparison and proposed genomic structure for GLYT1. The top panel shows the amino acid sequence alignments of the three GLYT1 isoforms from human, rat, and mouse species. The lower panel illustrates the proposed exon usage by each of the isoforms, based on composite data from Liu et al (104) and Kim et al (99) for the mouse and human genes, respectively. The intron and exon lengths are not drawn to scale. Adapted from the presentation of Kim et al (99).

that the transmembrane model would predict face the extracellular environment. Olivares et al (105) mutated each of those residues sequentially until all four were replaced with glutamine. There was a parallel loss of activity and reduction in apparent molecular size as each asparagine residue was substituted. The authors showed that the fully deglycosylated transporter still exhibits activity, but is not properly sorted to the plasma membrane.

Expression of the rat GLYT1a isoform in oocytes yields Na^+/Cl^--dependent high-affinity transport exhibiting a K_m for glycine of approximately 30 μM. GLYT1a-mediated transport is inhibited by sarcosine (N-methyl glycine), but not strongly by any of the naturally occurring zwitterionic amino acids (100, 101, 103). Likewise, Smith et al (102) demonstrated high-affinity Na^+/Cl^--dependent glycine transport following expression of the rat GLYT1b cDNA in COS cells. Once again, only glycine analogs showed any significant degree of inhibition. Expression of GLYT1c in COS cells also resulted in expression of a Na^+/Cl^--dependent glycine transport, exhibiting a K_m value of 90 μM (99). Whether or not any of these isoforms encode the activity previously described in nonneuronal cell types called System Gly (106) is unknown.

Expression of GLYT1 isoforms in tissues other than brain appears to be both isoform and species specific. For example, Smith et al (102), using Northern analysis, reported no detectable GLYT1b mRNA in kidney, spleen, and aorta of rats, but a low level of expression in liver. Guastella et al (103) observed a low level of GLYT1a mRNA in rat liver and no detectable signal in kidney, whereas Liu et al (101) reported just the opposite result for GLYT1a in mouse. Borowsky et al (100) used Northern analysis to detect GLYT1a (termed GLYT2 in their publication) in lung, liver, spleen, stomach, and uterus, but were unable to detect any significant hybridization to tissues other than brain for GLYT1b (called GLYT1 in their paper). Those authors used in situ hybridization on fixed rat tissues to show that the mRNA in the spleen, lung, and thymus was localized to a subset of cells identified by a macrophage-specific antibody. In the rodent, GLYT1a appears to be significantly expressed outside of the brain, and GLYT1b is expressed in these peripheral tissues to a lesser extent (100, 107).

Northern blotting by Kim et al (99) detected a strong signal for human GLYT1c at 3.6 kb in brain mRNA, but little or no signal in various other tissues. In contrast, analysis of these same human tissues with a cDNA probe against sequences common to all three isoforms resulted in readily detectable hybridization in most tissues, with the following intensity: brain = kidney > pancreas = lung > placenta = liver > heart > muscle (99). In comparison with the data from several laboratories that have investigated both rat and mouse tissues, the tissue distribution of the GLYT1 isoforms appears to be more widespread in humans. Further analysis with probes specific for isoforms GLYT1a and GLYT1b will be necessary to establish the identity of these mRNA species outside the brain.

Northern analysis, in situ hybridization, and immuno-detection of protein have all been used to document the distribution of the three GLYT1 isoforms in specific regions within the brain. Probes that did not distinguish among the GLYT1 isoforms revealed the greatest expression in the cerebellum, brain stem, olfactory bulb, and spinal cord (101–103, 107). Borowsky et al (100) used oligonucleotides specific for either GLYT1a (again, called GLYT2 in their paper) or GLYT1b (called GLYT1 in their paper) to demonstrate distinct differences in the distribution within the brain. Using in situ hybridization they detected GLYT1b mRNA in all areas containing white matter, but observed a more defined distribution of GLYT1a which was only present in gray matter corresponding to regions rich in strychnine-sensitive glycine receptors. However, neither Smith et al (102) nor Guastella et al (103) detected a significant amount of glycine transporter mRNA in white matter. Borowsky et al (100) propose that GLYT1a is present in both glia and neurons, whereas GLYT1b is glial specific, but the apparent discrepancies between laboratories remain to be resolved.

GLYT2 Transporter

Liu et al (104) screened at low stringency a mouse neonatal brain cDNA library with a GLYT1 probe. A partial clone was obtained and the sequence suggested that it encoded a novel transporter. This partial mouse clone was used to obtain a full-length cDNA of the corresponding sequence from a rat brain cDNA library. Sequencing of this clone, called GLYT2, indicated a 48% identity with the GLYT1 glycine transporter within the 622 overlapping amino acids, and injection of the corresponding cRNA into *Xenopus* oocytes resulted in expression of high-affinity (K_m of 17 μM) glycine transport. Although the GLYT2 transport activity was both Na^+ and Cl^--dependent, the glycine uptake mediated by this transporter was not inhibited by sarcosine (104). As mentioned, sarcosine inhibits the GLYT1 isoforms effectively with estimated K_i values below 100 μM, whereas 1 mM sarcosine inhibited GLYT2-mediated glycine uptake by less than 10%. This insensitivity to sarcosine inhibition was reported previously for the purified porcine brain stem glycine transporter (108, 109). The rat GLYT2 cDNA contained an open reading of 799 amino acids encoding a protein of 88 kDa; in vitro translation in the absence of glycosylation resulted in a polypeptide of approximately 90 kDa (104). An antibody prepared against the N-terminal sequence detected a single polypeptide of 90 kDa by immunoblotting of mouse brain proteins (110). The GLYT2 amino acid sequence showed significant similarity to other members of the neurotransmitter transporter superfamily, especially the PROT transporter (104).

The 799 amino acids of GLYT2 represent the largest protein of all the neurotransmitter transporters cloned to date, and the 90 kDa molecular mass is consistent with the size estimates (100 kDa) for a purified glycine transporter from pig brain stem (108, 109). Although the coding sequence for the GLYT2 protein is only 2397 bp in length, a single mRNA species of 8 kb in length is detected in brain tissue, and Liu et al (104) suggest that the 3508 bp cDNA obtained from the library is truncated at its 3' end. The cationic amino acid transporter CAT1 also is encoded by a mRNA containing a 4–5 kb 3' untranslated region (23). The physiologic significance of the long untranslated regions in these and other transporter mRNAs is unclear. The primary difference between GLYT2 and the GLYT1 isoforms is an extension of the N-terminal sequence by about 160 amino acids and a short extension of 13 amino acids at the C-terminus (104). The N-terminal extension of GLYT2 contains nine potential protein kinase phosphorylation sites and thus may be the basis for regulatory mechanisms not present in the GLYT1 isoforms.

Northern analysis reveals expression of the GLYT2 mRNA in mouse cerebellum, spinal cord, and brain stem, with little or no hybridization in other regions of the brain. No significant hybridization was detected against total RNA from heart, lung, spleen, testes, intestine, liver, and kidney (104). Immunohistochemistry with GLYT2-specific antiserum and in situ hybridization confirmed expression of the GLYT2 transporter in cerebellum, brain stem, and spinal cord (110, 111). The distribution within the brain corresponds to the distribution of strychnine-sensitive glycine receptors and suggests that the function of GLYT2 transporters may be to regulate the glycine concentration in regions where it represents an important inhibitory neurotransmitter. Liu et al (104) postulated that the distribution of the GLYT1 transporters suggests a role in regulating glycine concentration in those brain regions containing the glycine-dependent NMDA receptors. Further analysis of the cell-specific distribution and regulation of the GLYT1 and GLYT2 transporters should provide interesting insight into their role in modulating glycine-dependent neurotransmission.

rBAT/4F2hc FAMILY AND CYSTINURIA

A unique group of proteins associated with transport are those of the rBAT/4F2hc family. These proteins increase the transport of certain neutral and positively charged amino acids when their cRNAs are injected into *Xenopus* oocytes. However, the predicted transmembrane structure of these proteins is different from the majority of the solute transporters cloned to date in that the rBAT/4F2hc proteins are predicted to have only one to four membrane-spanning regions. This observation has led some investigators to suggest that these proteins do not function as transporters themselves, but as regulators or accessory subunits of a transporter complex.

Cloning of the rBAT/4F2hc Family

Induction of both cationic and neutral amino acid transport in oocytes following expression of size-fractionated kidney mRNA, by two independent laboratories (112–114), led to the identification of a new class of proteins identified in rat (115, 116), rabbit (117), and human (118, 119). We will use the nomenclature of Palacin (for review see 120) and refer to all of these clones as rBAT (related to $b^{o,+}$ amino acid transport). The proteins have 80–85% sequence identity with one another, but the human sequence reported by Lee et al (119) appears to lack 22 amino acids of the C-terminus present in the other clones. The gene for the human rBAT was localized to chromosome 2-p21 (119, 121), and the functional promoter of the rat rBAT gene also has been identified (121).

The rBAT cDNA contains an open reading frame of 2049 nucleotides coding for a protein of 683 amino acids with a molecular mass of about 78 kDa. The rBAT protein contains seven potential glycosylation sites, and a leucine-zipper motif is present in the rat protein from amino acids 548 to 569 (116). rBAT mRNA, as assayed by Northern analysis, is expressed primarily in the kidney and small intestine (115–117, 122). The rBAT protein has been detected by polyclonal antibodies as a 84–87 kDa species in both rat kidney and small intestine (123). Immunohistochemistry and immuno-electron microscopy have localized the rBAT protein to the epithelial membrane that lines the rat kidney

proximal tubule and to the rat jejunal microvilli (124, 125), and immunohistochemistry studies of chromaffin cells and autonomic neurons has localized
this protein to the rat adrenal gland, brain stem, and spinal cord (126).

Structure/Function Analysis

In situ hybridization has localized rBAT mRNA expression in rat kidney to
the S3 portion of the proximal tubule and, to a lesser extent, to the S1 and S2
segments (127). This localization is consistent with the role of rBAT in renal
amino acid resorption, as microperfusion experiments of the proximal tubule
have identified high-affinity transport in segments S2 and S3 (128, 129). There
are two primary rBAT mRNA species, 2.3 kb and 4.0 kb detected; both of
these mRNAs can induce transport in a similar fashion, and the difference in
size is the result of differential polyadenylation (130). A related mRNA,
slightly larger in size, appears in the brain and heart.

Hydrophobicity profiles of rBAT from rat, rabbit, or human have suggested
that it contains either one (116–119) or four (114) transmembrane domains.
Mosckovitz et al (123) used two experimental approaches to investigate the
topology of the rBAT protein. The first technique examined the expression of
rBAT in COS-7 cells by immunocytochemistry with fluorescently labelled
antibodies. Three antipeptide antibodies, for which the epitopes are separated
in the linear sequence but predicted to be oriented extracellularly in the four
membrane–spanning model, demonstrated fluorescent labelling of intact cells,
confirming the classification of these epitopes as extracellular. Three additional
antibodies, generated to sequences predicted to lie intracellularly (including
both the N- and C-termini) in the four membrane–spanning model, only demonstrated fluorescence after Triton X-100 permeabilization of the cells. Although these data are consistent with the four-transmembrane model proposed
by Tate and Udenfriend (115), they must be interpreted with caution because
expression of this protein in COS-7 cells does not increase amino acid transport
as it does in *Xenopus* oocytes. The protein expressed in COS-7 cells is either
nonfunctional, improperly targeted, or these cells lack additional components
necessary for proper rBAT function. The second approach used by Mosckovitz
et al (123) was limited proteolytic digestion of rat renal microvillous brush
border membrane vesicles by papain. The six antipeptide antibodies used in
the experiments described above were used to identify proteolytic fragments
of rBAT. The results were consistent with the four-transmembrane model
rather than a single transmembrane segment.

Expression of the rBAT cRNA in oocytes results in significantly increased
Na^+-independent uptake of both neutral and cationic amino acids compared to
water-injected controls. Originally, Tate et al (115) concluded that rBAT
(termed NAA-Tr in their report) was mediating a transport activity similar to

System L, an activity selective for large neutral amino acids (1). However, several features of the rBAT-induced amino acid uptake were later shown to be inconsistent with this interpretation (115). First, 2-aminobicyclo [2,2,1] heptane carboxylic acid (BCH), a model substrate for System L, does not inhibit the neutral amino acid uptake induced by rBAT. Second, there is no increase in the transport of tryptophan, a high-affinity substrate for System L in most cells. Third, the increased neutral amino acid uptake is significantly inhibited by cationic amino acids. The properties of the activity increased by rBAT expression are more consistent with uptake by System $b^{o,+}$, a transporter of both neutral and cationic amino acids (131), which serves as the primary endogenous uptake mechanism for these amino acids in oocytes (132). Induction of an endogenous activity, along with the unique structure of rBAT, has led to the suggestion that rBAT may be activating an endogenous oocyte transporter rather than functioning as a transport protein itself. Consistent with this proposal, rBAT expression in oocytes results in enhancement of two distinct Na^+-independent arginine transport activities, one high-affinity and one low-affinity. If the oocytes are pretreated with N-ethylmaleimide prior to rBAT expression, only the high-affinity activity is detected (P Taylor, personal communication). Short of purification to homogeneity and reconstitution into proteoliposomes, the precise function of rBAT may remain unknown.

Interaction of rBAT with other proteins may be necessary for function. Immunoblot analysis of rat kidney brush border membrane vesicles with rBAT specific antibodies revealed two separate polypeptides under nonreducing conditions: an 85 kDa species, presumably the glycosylated rBAT monomer, and an immunoreactive band at 135 kDa (133). When electrophoresis was performed under reducing conditions, only the 85 kDa rBAT monomer was detected. Chemical cross-linking of rat kidney membrane vesicles with dimethylsulfate also resulted in a 135 kDa band as the major cross-linked product. Similar immunoreactive species were also identified in *Xenopus* oocytes following injection of rBAT cRNA, suggesting recruitment of an oocyte protein similar in size to that in rat kidney membrane vesicles. Immunoblot analysis of rBAT-transfected COS-7 cells failed to reveal the 135 kDa species, possibly explaining the lack of increased transport in these rBAT-expressing cells (115).

Electrophysiological analysis of rBAT expression in oocytes has provided some interesting insight into the mechanism of the resulting transport. Unexpectedly, Na^+-independent neutral amino acid uptake induced by rBAT, a process anticipated to be electroneutral, resulted in the generation of outwardly directed currents (134, 135). Preloading of the oocytes with cationic amino acids increased the uptake of neutral amino acids and vice versa, but significant inhibition of uptake occurred when the preloaded amino acid had the same charge as the transported substrate. These data are consistent with a model of exchange transport in which uptake of a charged amino acid is enhanced by the outward flux of neutral amino acids and uptake of a neutral amino acid is stimulated by counter-transport of a cationic substrate. These observations led to the suggestion that the rBAT-mediated efflux of cationic amino acids from renal epithelial cells is the driving force for cystine and neutral amino acid clearance from urine (134).

Relation to the 4F2hc Protein

There are significant sequence similarities between the rBAT proteins and a previously reported membrane protein antigen called 4F2hc. The 4F2 protein complex was identified in a human T-cell line as a heterodimer composed of an 85 kDa glycosylated heavy chain, which spans the membrane a single time, in disulfide linkage with a 45 kDa nonglycosylated light chain (136, 137). The cDNA for the 4F2 heavy chain (4F2hc) was subsequently isolated from both mouse and human libraries (138–140). The sequence similarity with the rBAT proteins led Bertran et al (141) and Wells et al (142) to investigate amino acid uptake following injection of 4F2hc cRNA into oocytes. Interestingly, 4F2hc not only enhanced Na^+-independent cationic amino acid transport, but Na^+-dependent neutral amino acid transport as well. Independent isolation of a rat 4F2hc cDNA clone was obtained by expression cloning in *Xenopus* oocytes (S Bröer, B Hamprecht, personal communication). A fundamental difference between the activities induced by rBAT and 4F2hc was that 4F2hc did not increase cystine transport (141, 142). The activity enhanced by 4F2hc expression more closely resembled System y^+L, originally identified in human erythrocytes (143). System y^+L transport is characterized as high-affinity, Na^+-independent cationic and Na^+-dependent neutral amino acid uptake. Fei et al (144) showed that mRNA isolated from JAR human choriocarcinoma cells induced System y^+L activity when injected into oocytes. They confirmed the presence of 4F2hc and the absence of rBAT in these cells by Northern analysis. However, it cannot yet be ruled out that 4F2hc affects System y^+ activity (CAT1/CAT2) in a way that promotes its known acceptance of a neutral amino acid in the presence of Na^+ (17, 145). Whether or not the coincident expression of both 4F2hc and CAT2 in T-cells is of physiologic importance is unknown.

Role of rBAT in Cystinuria

The tissue distribution of rBAT and its role in renal uptake of cystine and dibasic amino acids made it a potential candidate for the defective gene in cystinuria. This autosomal recessive genetic disease is characterized by high levels of urinary cystine and cationic amino acids, ultimately leading to cystine stones and renal disease (146). It has been recognized that a defective transporter for these amino acids in the brush border of the proximal tubule epithelial cells results in elevated excretion and that the low solubility of cystine causes crystal formation farther along the urinary tract. Sequencing of rBAT mRNAs expressed from several cystinuric patients, and subsequent sequencing of the corresponding genomic DNA, revealed six different point mutations in the rBAT gene that represented 30% of the cystinuric patients analyzed (147). The most commonly occurring mutation, methionine 467 to threonine, resulted in an 80% decrease in rBAT-induced transport activity compared to wild-type after expression in oocytes. Additional support for the role of rBAT in cystinuria was provided by linkage analysis which localized the gene responsible for cystinuria to chromosome 2p (148), coincident with the location of rBAT to 2p21 (119, 121). Obviously, greater understanding of the exact molecular basis for cystinuria is tied to gaining additional knowledge about the mechanism by which rBAT affects transport. Regardless of their mechanism, the rBAT/4F2hc proteins appear to define a new class of solute transporter-associated proteins.

SUMMARY

After more than 30 years of describing amino acid transport activities at the cellular level, the identification of the first transporter proteins and the cloning of the corresponding cDNA sequences has been accomplished during the past five years. Even over this relatively short period, the body of published work is sufficient to illustrate several emerging themes. A number of gene "families" have been documented, the members of which share significant amino acid sequence similarity. Some of these families might have been predicted given the chemical differences between anionic, cationic, and zwitterionic amino acids and, therefore, the need for corresponding substrate discrimination by transporter binding sites. However, the observation that the ASCT1 zwitterionic amino acid transporter is a member of the glutamate transporter family demonstrates that predicting gene relatedness on substrate charge alone is not tenable. Likewise, it is interesting to note that the mammalian Na^+-dependent proline transporter, PROT, shows no sequence similarity to a bacterial Na^+-dependent proline transporter (88). Instead, the bacterial protein is related to the mammalian Na^+-dependent glucose transporter family (93).

Expression of transporter cDNA clones in heterologous systems has permitted us to establish that there are at least three distinct mechanisms for Na^+-linked cotransport. The neurotransmitter family represents the class of Na^+/Cl^--dependent transporters. In contrast, the Na^+-dependent glutamate transporters do not require extracellular Cl^- for cotransport, but rather involve countertransport of one K^+ and either one OH^- or HCO_3^-. Finally, several of the Na^+-dependent transport activities for which proteins or cDNA clones have yet to be identified do not appear to fall into either of these categories. Given these fundamental mechanistic differences and the lack of significant amino acid sequence similarity between these families, additional insight into the evolutionary development of these separate ion-linked energization mechanisms will be interesting.

The recent advances in the molecular biology of amino acid transporters confirmed many of the transporter properties described during the last three decades, but this new information has also been the source of much needed insight as well. We now appreciate that, in many cases, what was once thought to be a single activity is actually the result of cell-specific expression of closely related, but clearly distinct, proteins mediating transport activities sufficiently similar that kinetic characterization could not distinguish them. Indeed, as shown for GLYT1, CAT2, and several other examples, a single gene may code for different transporter isoforms based on tissue-specific promoter usage or differential splicing. We can easily imagine the identification of many additional members of the known transporter families as well as discovery of entirely new families. For example, although ASCT1 mRNA is only detectable in a few tissues, a System ASC–like activity has been described in nearly every cell and tissue tested. How many genes account for this activity? What is their relationship to each other? As for new families, there are some glaring examples of well-described transport activities for which neither proteins nor cDNA clones have been identified. Systems A and L, the first amino acid transport "systems" to be described (1), remain elusive despite many identification attempts by dozens of laboratories employing a variety of techniques. We believe that these and other amino acid transporter mysteries will be solved over the next few years; it is an exciting time to be working in this area of research. For those new to the field, we offer encouragement and the adage "Always expect the unexpected."

ACKNOWLEDGMENTS

The authors acknowledge support for their research by the National Institutes of Health, Institute for Diabetes, Digestive and Kidney Diseases (DK-28374). We thank Mrs. Jeri Stoner for the patient and careful secretarial assistance. The authors also want to gratefully thank the following colleagues who responded to a request for unpublished results or manuscripts in press: Drs. Stefan Broeer, Robert Fremeau, Jr., Gian Carlo Gazzola, Bernd Hamprect, Joseph Handler, Rose Johnstone, David Kabat, Michael Kavanaugh, Carol MacLeod, John McGivan, Heine Murer, Nathan Nelson, Marcal Pastor-Anglada, Wiley Souba, Wilhelm Stoffel, Suresh Tate, Peter Taylor, and Lon Van Winkle.

Literature Cited

1. Oxender DL, Christensen HN. 1963. *J. Biol. Chem.* 238:3686–99
2. Oxender DL. 1972. *Annu. Rev. Biochem.* 41:777–814
3. Kilberg MS, Stevens BR, Novak DA. 1993. *Annu. Rev. Nutr.* 13:137–65
4. McGivan JD, Pastor-Anglada M. 1994. *Biochem. J.* 299:321–34
5. Van Winkle LJ. 1995. *Biochim. Biophys. Acta* 1154:157–72
6. Reizer J, Reizer A, Saier MH Jr. 1994. *Biochim. Biophys. Acta* 1197:133–66
7. Kanai Y, Smith CP, Hediger MA. 1995. *FASEB J.* 7:1450–59
8. Hediger MA. 1995. *J. Exp. Biol.* 196:15–49
9. Mailliard ME, Stevens BR, Mann GE. 1995. *Gastroenterology* 108:888–910
10. Kilberg MS, Haussinger D. 1992. In *Mammalian Amino Acid Transport: Mechanisms and Control.* New York: Plenum
11. Harvey WR. 1994. In *Transporters.* Cambridge, UK: Co. Biol. Ltd.
12. Albritton LM, Tseng L, Scadden D, Cunningham JM. 1989. *Cell* 57:659–66
13. Kozak CA, Rowe WP. 1979. *Science* 204:69–71
14. Kim JW, Closs EI, Albritton LM, Cunningham JM. 1991. *Nature* 352:725–28
15. Wang H, Kavanaugh MP, North RA, Kabat D. 1991. *Nature* 352:729–31
16. White MF, Gazzola GC, Christensen HN. 1982. *J. Biol. Chem.* 257:4443–49
17. White MF. 1985. *Biochim. Biophys. Acta* 822:355–74
18. Van Winkle LJ, Kakuda DK, MacLeod CL. 1995. *Biochim. Biophys. Acta* 1233:213–16
19. Yoshimoto T, Yoshimoto E, Meruelo D. 1992. *J. Virol.* 66:4377–81
20. Albritton LM, Bowcock AM, Eddy RL, Morton CC, Tseng L, et al. 1992. *Genomics* 12:430–34
21. Stoll J, Wadhwani KC, Smith QR. 1993. *J. Neurochem.* 60:1956–59
22. Wu JY, Robinson D, Kung H, Hatzoglou M. 1994. *J. Virol.* 68:1615–23
23. Puppi M, Henning SJ. 1995. *Proc. Soc. Exp. Biol. Med.* 209:38–45
24. MacLeod CL, Fong AM, Seal BS, Walls L, Wilkinson MF. 1990. *Cell Growth Differ.* 1:271–79
25. MacLeod CL, Finley K, Kakuda D, Kozak CA, Wilkinson MF. 1990. *Mol. Cell. Biol.* 10:3663–74
26. Reizer J, Finley K, Kakuda D, MacLeod CL, Reizer A, Saier MH. 1993. *Protein Sci.* 2:20–30
27. Kakuda DK, Finley KD, Dionne VE, MacLeod CL. 1993. *Transgene* 1:91–101
28. Closs EI, Lyons CR, Kelly C, Cunningham JM. 1993. *J. Biol. Chem.* 268:20796–800
29. Closs EI, Albritton LM, Kim JW, Cunningham JM. 1993. *J. Biol. Chem.* 268:7538–44
30. Malandro MS, Beveridge MJ, Kilberg MS, Novak DA. 1994. *Am. J. Physiol.* 267:C804–11
31. Yoshimoto T, Yoshimoto E, Meruelo D. 1991. *Virology* 185:10–17
32. Pan M, Malandro MS, Stevens BR. 1995. *Am. J. Physiol.* 268:G578–85
33. Woodard MH, Dunn WA, Laine RO, Malandro M, McMahon R, et al. 1994. *Am. J. Physiol.* 266:E817–24
34. Jaenisch R, Hoffman E. 1979. *Virology* 98:289–97
35. Closs EI, Rinkes IHMB, Bader A, Yarmush ML, Cunningham JM. 1993. *J. Virol.* 67:2097–102
36. MacLeod CL, Finley KD, Kakuda DK. 1994. *J. Exp. Biol.* 196:109–21
37. Finley KD, Kakuda DK, Barrieux A, Kleeman J, Huynh PD, MacLeod CL. 1995. *Proc. Natl. Acad. Sci. USA* 92:9378–82
38. Souba WW. 1987. *J. Parenter. Enter. Nutr.* 11:569–79
39. Rennie MJ, Ahmed A, Low SY, Hundal HS, Watt PW, et al. 1990. *Biochem. Soc. Trans.* 18:1140–42
39a. Campione AL, Van Winkle LJ. 1994. *Biol. Reprod.* 50(Suppl.):71A
40. Van Winkle LJ, Iannaccone PM, Campione AL, Garton RL. 1990. *Dev. Biol.* 142:184–93
41. Van Winkle LJ, Campione AL. 1990. *Biochim. Biophys. Acta* 1028:165–73
42. Kavanaugh MP, Wang H, Zhang Z, Zhang WB, Wu Y-N, et al. 1994. *J. Biol. Chem.* 269:15445–50
43. Kavanaugh MP. 1993. *Biochemistry* 32:5781–85
44. Wang H, Kavanaugh MP, Kabat D. 1994. *Virology* 202:1058–60
45. Storck T, Schulte S, Hofmann K, Stoffel W. 1992. *Proc. Natl. Acad. Sci. USA* 89:10955–59
46. Tolner B, Poolman B, Wallace B, Konings WN. 1992. *J. Bacteriol.* 174:2391–93
47. Eisenberg D, Schwarz E, Komaromy M, Wall R. 1984. *J. Mol. Biol.* 179:125–42
48. Kawakami H, Tanaka K, Nakayama T, Inoue K, Nakamura S. 1994. *Biochem. Biophys. Res. Commun.* 199:171–76

334   MALANDRO & KILBERG

49. Arriza JL, Fairman WA, Wadiche JI, Murdoch GH, Kavanaugh MP, Amara SG. 1994. *J. Neurosci.* 14:5559–69
50. Pines G, Danbolt NC, Bjoras M, Zhang Y, Bendahan A, et al. 1992. *Nature* 360:464–67
51. Danbolt NC, Pines G, Kanner BI. 1990. *Biochemistry* 29:6734–40
52. Danbolt NC, Storm-Mathisen J, Kanner BI. 1992. *Neuroscience* 51:295–310
53. Hees B, Danbolt NC, Kanner BI, Haase W, Heitmann K, Koepsell H. 1992. *J. Biol. Chem.* 267:23275–81
54. Manfras BJ, Rudert WA, Trucco M, Boehm BO. 1994. *Biochim. Biophys. Acta* 1195:185–88
55. Kirschner MA, Copeland NG, Gilbert DJ, Jenkins NA, Amara SG. 1994. *Genomics* 24:218–24
56. Shashidharan P, Wittenberg I, Plaitakis A. 1994. *Biochim. Biophys. Acta* 1191:393–96
57. Kanai Y, Hediger MA. 1992. *Nature* 360:467–71
58. Rauen T, Jeserich G, Danbolt NC, Kanner BI. 1992. *FEBS Lett.* 312:15–20
59. Kanai Y, Nussberger S, Romero MF, Boron WF, Hebert SC, Hediger MA. 1995. *J. Biol. Chem.* 270:16561–68
60. Smith CP, Weremowicz S, Kanai Y, Stelzner M, Morton CC, Hediger MA. 1994. *Genomics* 20:335–36
61. Velaz-Faircloth M, McGraw TS, Malandro MS, Fremeau RT Jr, Kilberg MS, Anderson KJ. 1996. *Am. J. Physiol.* 270:C67–75
62. Fairman WA, Vandenberg RJ, Arriza JL, Kavanaugh MP, Amara SG. 1995. *Nature* 375:599–603
63. Lehre KP, Levy LM, Ottersen OP, Storm-Mathisen J, Danbolt NC. 1995. *J. Neurosci.* 15:1835–53
64. Otori Y, Shimada S, Tanaka K, Ishimoto I, Tano Y, Tohyama M. 1994. *Mol. Brain Res.* 27:310–14
65. Deleted in proof
66. Rauen T, Kanner BI. 1994. *Neurosci. Lett.* 169:137–40
67. Blakely RD, Clark JA, Pacholczyk T, Amara SG. 1991. *J. Neurochem.* 56:860–71
68. Atwell D, Barbour B, Szatkowski M. 1993. *Neuron* 11:401–7
69. Casado M, Bendahan A, Zafra F, Danbolt NC, Aragon C, et al. 1993. *J. Biol. Chem.* 268:27313–17
70. Smit NPM, van Roermund CWT, Aerts HMFG, Heikoop JC, Vandenberg M, et al. 1993. *Biochim. Biophys. Acta* 1181:1–6
71. Zerangue N, Arriza JL, Amara SG, Kavanaugh MP. 1995. *J. Biol. Chem.* 270:6433–35
72. Plakidou-Dymock S, McGivan JD. 1993. *Biochem. J.* 295:749–55
73. Conradt M, Storck T, Stoffel W. 1995. *Eur. J. Biochem.* 229:682–87
74. Zhang Y, Pines G, Kanner BI. 1994. *J. Biol. Chem.* 269:19573–77
75. Pines G, Zhang Y, Kanner BI. 1995. *J. Biol. Chem.* 270:17093–97
76. Vandenberg RJ, Arriza JL, Amara SG, Kavanaugh MP. 1995. *J. Biol. Chem.* 270:17668–71
77. Shafqat S, Tamarappoo BK, Kilberg MS, Puranam RS, McNamara JO, et al. 1993. *J. Biol. Chem.* 268:15351–55
78. Arriza JL, Kavanaugh MP, Fairman WA, Wu YN, Murdoch GH, et al. 1993. *J. Biol. Chem.* 268:15329–32
79. Hofmann K, Duker M, Fink T, Lichter P, Stoffel W. 1994. *Genomics* 24:20–26
80. Kilberg MS, Tamarappoo BK, Shafqat S, Fremeau RT, Jr. 1994. In *Transport in the Liver.* Lancaster, UK: Kluwer
81. Christensen HN, Liang M, Archer EG. 1967. *J. Biol. Chem.* 242:5237–46
82. Liao K, Lane MD. 1995. *Biochem. Biophys. Res. Commun.* 208:1008–15
83. Makowske M, Christensen HN. 1982. *J. Biol. Chem.* 257:14635–38
84. Makowske M, Christensen HN. 1982. *J. Biol. Chem.* 257:5663–70
85. Vadgama JV, Christensen HN. 1984. *J. Biol. Chem.* 259:3648–52
86. Tamarappoo BK, McDonald KK, Kilberg MS. 1996. *Biochim. Biophys. Acta.* 1279:131–36
87. Fremeau RT Jr, Caron MG, Blakely RD. 1992. *Neuron* 8:915–26
88. Shafqat S, Velaz-Faircloth M, Henzi VA, Whitney KD, Yang-Feng TL, et al. 1995. *Mol. Pharmacol.* 48:219–29
89. Shafqat S, Velaz-Faircloth M, Guadano-Ferraz A, Fremeau RT Jr. 1993. *Mol. Endocrinol.* 7:1517–29
90. Amara SG, Kuhar MJ. 1995. *Annu. Rev. Neurosci.* 16:73–93
91. Velaz-Faircloth M, Guadano-Ferraz A, Henzi VA, Fremeau RT Jr. 1995. *J. Biol. Chem.* 270:15755–61
92. Takagi T. 1995. *Adv. Electrophor.* 4:391–406
93. Hediger MA, Turk E, Wright EM. 1989. *Proc. Natl. Acad. Sci. USA* 86:5748–52
94. Nadler JV. 1995. *J. Neurochem.* 49:1155–60
95. Stevens BR, Wright EM. 1995. *J. Membr. Biol.* 87:27–34
96. Ottersen OP, Strom-Mathisen J. 1990.

Glycine Transmission. New York: Wiley

97. Grenningloh G, Rienitz A, Schmitt B, Methfessel C, Zensen M, et al. 1995. *Nature* 328:215–20

98. Johnson JW, Ascher P. 1987. *Nature* 325:529–31

99. Kim K-M, Kingsmore SF, Han H, Yang-Feng TL, Godinot N, et al. 1994. *Mol. Pharmacol.* 45:608–17

100. Borowsky B, Mezey E, Hoffman BJ. 1993. *Neuron* 10:851–63

101. Liu Q-R, Nelson H, Mandiyan S, Lopez-Corcuera B, Nelson N. 1992. *FEBS Lett.* 305:110–14

102. Smith KE, Borden LA, Hartig PR, Branchek T, Weinshank RL. 1992. *Neuron* 8:927–35

103. Guastella J, Brecha N, Weigmann C, Lester HA, Davidson N. 1992. *Proc. Natl. Acad. Sci. USA* 89:7189–93

104. Liu Q-R, Lopez-Corcuera B, Mandiyan S, Nelson H, Nelson N. 1993. *J. Biol. Chem.* 268:22802–8

105. Olivares L, Aragon C, Gimenez C, Zafra F. 1995. *J. Biol. Chem.* 270:9437–42

106. Christensen HN, Handlogten ME. 1981. *Biochem. Biophys. Res. Commun.* 98:102–7

107. Jursky F, Tamura S, Tamura A, Mandiyan S, Nelson H, Nelson N. 1994. *J. Exp. Biol.* 196:283–95

108. Lopez-Corcuera B, Vazquez J, Aragon C. 1991. *J. Biol. Chem.* 266:24809–14

109. Lopez-Corcuera B, Alcantara R, Vazquez J, Aragon C. 1993. *J. Biol. Chem.* 268:2239–43

110. Jursky F, Nelson N. 1995. *J. Neurochem.* 64:1026–33

111. Luque JM, Nelson N, Richards JG. 1995. *Neuroscience* 64:525–35

112. Tate SS, Urade R, Getchell TV, Udenfriend S. 1989. *Arch. Biochem. Biophys.* 275:591–96

113. Magagnin S, Bertran J, Werner A, Markovich D, Biber J, et al. 1992. *J. Biol. Chem.* 267:15384–90

114. Bertran J, Werner A, Stange G, Markovich D, Biber J, et al. 1992. *Biochem. J.* 281:717–23

115. Tate SS, Yan N, Udenfriend S. 1992. *Proc. Natl. Acad. Sci. USA* 89:1–5

116. Wells RG, Pajor AM, Kanai Y, Turk E, Wright EM, Hediger MA. 1992. *Am. J. Physiol.* 263:F459–65

117. Bertran J, Werner A, Moore ML, Stange G, Markovich D, et al. 1992. *Proc. Natl. Acad. Sci. USA* 89:5601–5

118. Bertran J, Werner A, Chillaron J, Nunes V, Biber J, et al. 1993. *J. Biol. Chem.* 268:14842–49

119. Lee W-S, Wells RG, Sabbag RV, Mohandas TK, Hediger MA. 1993. *J. Clin. Invest.* 91:1959–63

120. Palacin M. 1994. *J. Exp. Biol.* 196:123–37

121. Yan N, Mosckovitz R, Gerber LD, Mathew S, Murty VVVS, et al. 1994. *Proc. Natl. Acad. Sci. USA* 91:7548–52

122. Yan N, Mosckovitz R, Udenfriend S, Tate SS. 1992. *Proc. Natl. Acad. Sci. USA* 89:9982–85

123. Mosckovitz R, Heimer E, Felix A, Tate SS, Udenfriend S. 1993. *Proc. Natl. Acad. Sci. USA* 90:4022–26

124. Pickel VM, Nirenberg MJ, Chan J, Mosckovitz R, Udenfriend S, Tate SS. 1993. *Proc. Natl. Acad. Sci. USA* 90:7779–83

125. Furriols M, Chillaron J, Mora C, Castello A, Bertran J, et al. 1993. *J. Biol. Chem.* 268:27060–68

126. Nirenberg MJ, Tate SS, Mosckovitz R, Udenfriend, S, Pickel VM. 1995. *J. Comp. Neurol.* 356:505–22

127. Kanai Y, Stelzner MG, Lee W-S, Wells RG, Brown D, Hediger MA. 1992. *Am. J. Physiol.* 263:F1087–93

128. Voelkl H, Silbernagl S. 1982. *Pflügers Arch.* 359:190–95

129. Schafer JA, Watkins ML. 1984. *Pflügers Arch.* 401:143–51

130. Markovich D, Stange G, Bertran J, Palacin M, Werner A, et al. 1993. *J. Biol. Chem.* 268:1362–67

131. Van Winkle LJ. 1988. *Biochim. Biophys. Acta* 947:173–208

132. Campa MJ, Kilberg MS. 1991. *J. Cell. Physiol.* 141:645–52

133. Wang Y, Tate SS. 1995. *FEBS Lett.* 368:389–92

134. Busch AE, Herzer T, Waldegger S, Schmidt F, Palacin M, et al. 1994. *J. Biol. Chem.* 269:25581–86

135. Coady MJ, Jalal F, Chen X, Lemay G, Berteloot A, Lapointe JY. 1994. *FEBS Lett.* 356:174–78

136. Eisenbarth GS, Haynes BF, Schroer JA, Fauci AS. 1981. *J. Immunol.* 124:1237–44

137. Hemler ME, Strominger JL. 1982. *J. Immunol.* 129:623–28

138. Quackenbush E, Clabby M, Gottesdiener KM, Jones N, Strominger J, et al. 1995. *Proc. Natl. Acad. Sci. USA* 84:6526–30

139. Teixeira S, DiGrandi S, Kuhn LC. 1987. *J. Biol. Chem.* 262:9574–80

140. Parmacek MS, Karpinski BA, Gottesdiener KM, Thompson CB, Leiden JM. 1989. *Nucleic Acids Res.* 17:1915–31

141. Bertran J, Magagnin S, Werner A,

Markovich D, Biber J, et al. 1992. *Proc. Natl. Acad. Sci. USA* 89:5606–10

142. Wells RG, Lee W-S, Kanai Y, Leiden JM, Hediger MA. 1992. *J. Biol. Chem.* 267:15285–88

143. Deves R, Chavez P, Boyd CAR. 1992. *J. Physiol.* 454:491–501

144. Fei YJ, Prassad PD, Leibach FH, Ganapathy V. 1995. *Biochemistry* 34:8744–51

145. Christensen HN, Handlogten ME, Thomas EL. 1969. *Proc. Natl. Acad. Sci. USA* 63:948–55

146. Segal S, Thier SO. 1989. *The Metabolic Basis of Inherited Diseases,* pp. 2479–96. New York: McGraw-Hill

147. Calonge MJ, Gasparini P, Chillaron J, Chillon M, Gallucci M, et al. 1994. *Nat. Genet.* 6:420–25

148. Pras E, Arber N, Aksentijevich I, Katz G, Schapiro JM, et al. 1994. *Nat. Genet.* 6:415–19

149. Yoshimoto T, Yoshimoto E, Meruelo D. 1992. *J. Virol.* 66:4377–81

Annu. Rev. Biochem. 1996. 65:337–65

TELOMERE LENGTH REGULATION

Carol W. Greider

Cold Spring Harbor Laboratory, Cold Spring Harbor, New York 11724

KEY WORDS: telomeres, telomerase, chromosome, telomere binding proteins, cancer

---

ABSTRACT

Telomeres are the components of chromosome ends that provide stability and allow the complete replication of the ends. Telomere length is maintained by a balance between processes that lengthen and those that shorten telomeres. Telomerase is a ribonucleoprotein polymerase that specifically elongates telomeres. In human cells telomere length is not maintained and telomerase is not active in some tissues. In tumors, however, telomerase is active and may be required for the growth of cancer cells. Thus understanding telomerase and telomere length regulation may help us understand tumor progression. Evidence from various organisms suggests that several factors influence telomere length regulation, such as telomere binding proteins, telomere capping proteins, telomerase, and DNA replication enzymes. Understanding how these factors interact to coordinate the regulation of telomere length will allow a more complete understanding of telomere function in the cell.

CONTENTS

TELOMERE STRUCTURE AND FUNCTION................................. 338
TELOMERASE 339
   Telomerase Biochemistry...................................... 341
   Telomerase RNA Components 343
   Telomerase Protein Components.............................. 344
   In Vivo Effects of Telomerase Alteration..................... 345
TELOMERE REPLICATION 346
   Cytosine-Rich Strand Synthesis 347
TELOMERE BINDING PROTEINS....................................... 348
   Oxytricha Telomere Protein.................................. 348
   RAP1 349
TELOMERE PROCESSING 351
REGULATION OF TELOMERE LENGTH EQUILIBRIUM.................... 352
   Models for Length Regulation 353

337

0066-4154/96/0701-0337$08.00

TELOMERES IN CELLULAR SENESCENCE AND AGING.   355
TELOMERASE ACTIVATION IN CANCER CELLS   357
   Tissue Culture Models   357
   Telomeres and Telomerase in Human Cancer   358
   Mouse Models for Telomerase Regulation   360
PERSPECTIVES   361

TELOMERE STRUCTURE AND FUNCTION

Telomeres were defined in the 1930s as essential components that stabilize chromosome ends. In 1938 Muller found he could not recover terminal deletions at *Drosophila* chromosome ends. He suggested that chromosome ends are specialized structures and coined the term telomeres (1). Barbara McClintock was at that time correlating phenotypic changes in maize plants with cytogenetic alterations in the chromosomes. In crosses involving ring and normal linear chromosomes, chromosome breakage would occur and she could follow the fate of the broken ends. These studies led McClintock to propose that a natural chromosome must differ from broken chromosomes in possessing a structure that provides stability to the end (2–5). Since then many workers have used molecular techniques to confirm that telomeres are essential for chromosome stability, and they have begun to define their molecular components reviewed in (6, 7). In addition to maintaining chromosome stability, telomeres also participate in nuclear processes such as chromosome positioning in the nucleus, transcriptional repression, heterochromatin formation, and replication timing. These functions have been reviewed recently and will not be considered in detail here (8–12).

In eukaryotes from diverse taxa including protozoa, fungi, flagellates, plants, and animals telomeric DNA consists of tandem repeats of simple sequences (reviewed in 13, 14). These repeated sequences are usually short and rich in G residues on one strand. For example, the ciliate *Tetrahymena* contains tandem TTGGGG repeats, whereas humans and other mammals contain TTAGGG repeats. Ciliates have been featured prominently in telomere research because they have a unique developmental process in which chromosomes from one of two nuclear types are fragmented and telomeres are added onto the ends (15). Thus a single cell will have 40,000–1,000,000 telomeres depending on the species. Unlike ciliates and mammals, which have regular repeats, some organisms have an irregular repeat sequence. In yeast, for example, a T is followed by one, two, or three Gs (abbreviated TG_{1-3}). A different type of irregular repeat is exemplified by *Paramecium* in which the sequences TTGGGG and TTTGGG are interspersed along a given telomere. The orientation of the telomere sequences is also conserved; the G-rich strand runs 5′ to 3′ toward the end of the chromosome and thus makes up the molecular 3′ end of the chromosome.

In most organisms the number of repeats on any given end is not fixed, giving telomeres a typical heterogeneous or "fuzzy" appearance in Southern blots. The tract length of these repeated sequences ranges from 38 base pairs (bp) in the ciliates *Oxytricha* and *Euplotes* up to tens of kilobases (kb) in mammalian cells, each organism having a characteristic mean length (reviewed in 6). In addition to these simple repeated motifs, recent studies in budding yeasts have identified telomere repeat sequences with more complex structures (16, 17). The organism in which telomeres were first defined, *Drosophila*, appears to utilize a different kind of telomere repeat sequence than most other eukaryotes. The 6-kb repeated elements found on *Drosophila* chromosome ends are telomere-specific transposable elements that can be lost and re-added onto chromosome ends (18–21; reviewed in 7).

The length of the simple telomere repeat tracts found in most eukaryotic phyla is maintained by the enzyme telomerase. Telomerase is a specialized DNA polymerase that adds telomeric sequences onto chromosome ends (22, 23). This *de novo* addition of sequences balances the natural loss of repeats from chromosome ends (reviewed in 6). In the absence of an active elongation mechanism, telomeres shorten with each cell division. This shortening may be due to the inability of conventional polymerases to replicate the very end of a chromosome (24, 25; reviewed in 26). Shortening may also result from other mechanisms (see below). Thus telomeres, which are essential for chromosome stability, are maintained not by a stable static defense but rather by a dynamic equilibrium of shortening and lengthening. The fuzzy nature of telomere fragments on Southern blots stems in part from the variable number of repeats on each chromosome resulting from this dynamic equilibrium. The use of a dynamic equilibrium mechanism at chromosome ends in all eukaryotic phyla examined to date attests to the evolutionary success of this kind of mechanism. This review focuses on the mechanisms involved in the establishment and maintenance of simple sequence repeats and on the consequences of equilibrium disruption.

TELOMERASE

Telomerase activity was first identified in the ciliate *Tetrahymena* (22) and has since been identified in several other eukaryotes including the ciliates *Oxytricha* and *Euplotes*, humans, mice, *Xenopus*, and yeast (27–33). Telomerase is a ribonucleoprotein enzyme that contains both essential proteins and an essential RNA component. The RNA component provides the template for the telomeric repeats that are synthesized de novo onto chromosome ends in vitro and in vivo (23, 34). In *Tetrahymena* the telomerase RNA is 159 nucleotides long and contains nine potential template nucleotides 5′ CAACCCCAA 3′. The redundancy in the template region allows base pairing of the growing telomere with the RNA and still leaves a region that can serve as a template (Figure 1) (34). Although the length of the template region varies, all telom-

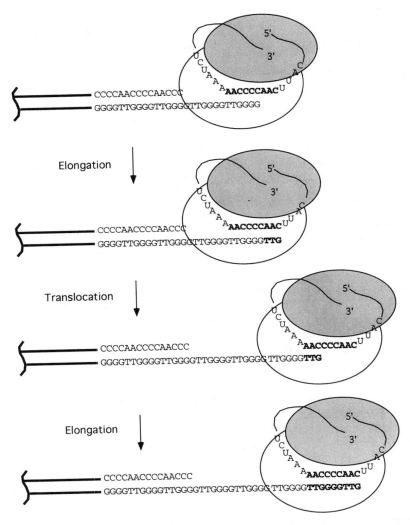

Figure 1   Telomerase elongation model. 1. Telomerase recognizes the telomere substrate and the terminal TTGGGG repeat is base paired with the 5′ CAACCCCAA 3′ template sequence in the RNA component. 2. The RNA is copied to the end of the template region. 3. Translocation repositions the terminal TTGGGGTTG sequence and exposes additional template sequences. 4. Another round of template copying produces additional TTGGGG repeats. This figure is adapted from (34).

Table 1 Telomerase RNA components[a]

Organism	Telomere sequence	RNA template sequence	RNA length	References
Tetrahymena	TTGGGG	CAACCCCAA	159	34
Euplotes	TTTTGGGG	CAAAACCCCAAAACC	190	35
Oxytricha	TTTTGGGG	CAAAACCCCAAAACC	190	36, 41
Human	TTAGGG	CUAACCCUAAC	450	40
Mouse	TTAGGG	CCUAACCCU	450	39
S. cerevisiae	TG(1-3)	CACCACACCCACACAC	~1300	37
K. lactis	TTTGATTAGGTATG-TGGTGTACGGA	UCAAAUCCGUACACCA AUACCUAAUCAAA	~1300	38

[a] Telomerase RNA has been isolated from several different eukaryotes. The sequence of the telomere repeat and the telomerase RNA template are shown for each organism, along with the length and the paper where this information is reported.

erase RNAs identified to date have a template region that is longer than a full telomeric repeat (Table 1) (35–41), suggesting that the elongation mechanism proposed for the *Tetrahymena* enzyme is conserved.

Telomerase Biochemistry

Telomerase biochemistry has been best studied in *Tetrahymena*, in which the enzyme is abundant and the purification has been well established (42, 43, 67). Many properties of the *Tetrahymena* enzyme are shared by enzymes isolated from other sources; exceptions are noted below. In a typical telomerase assay a telomeric sequence–DNA oligonucleotide primer is added to the enzyme in the presence of the appropriate dNTPs (for *Tetrahymena*, for which the repeat sequence is TTGGGG, dGTP and dTTP are used). Telomerase elongates the input primer, adding one nucleotide at a time to generate multiple tandem repeats of the telomeric sequence. Experimentally, the repeat synthesis is detected by a ^{32}P label on one of the dNTPs (22), an end-label on the primer (27, 36), or by subsequent amplification of the products by the polymerase chain reaction (PCR) (44). The reaction products are then resolved on an acrylamide gel. Typically a ladder of repeats with the periodicity of the telomere repeat sequence is generated.

Telomerase preferentially binds to and elongates telomeric sequence primers over nontelomeric sequences. However, telomere-like sequences other than the cognate telomere repeats will serve as efficient primers (29–31, 42, 45). Thus *Tetrahymena* telomerase will elongate primers consisting of the telomere repeat from *Oxytricha*, human, plants, and yeast. This property of telomerase parallels the ability of telomeres from different organisms to function in yeast

(46, 47). The ability of telomerase to preferentially elongate telomeric primers without specificity for a unique sequence suggests telomerase has a general affinity for G-rich oligonucleotide sequences (48–51). This affinity must be different from the high-affinity sequence-specific recognition of many DNA and RNA binding proteins.

Current models suggest that two independent sites on the telomerase enzyme interact with the primer (29, 48, 50, 51; reviewed in 52). The template site contains the complementary RNA and aligns the primer 3′ end for elongation at the catalytic center. The anchor site binds to the primer 5′ of the template site and provides a route for the growing chain to exit the enzyme. This model is based on the two-site models proposed for RNA polymerase interaction with the DNA template and the growing RNA transcript (53). For both RNA polymerase and for telomerase the presence of two sites for interaction with the growing polymer allows processive synthesis to occur: Many repeats are added without the enzyme dissociating from the growing polymer. During the translocation step that is necessary for processive synthesis one site can maintain contact with the growing chain while the active site is repositioned for another round of elongation (reviewed in 52, 53).

A second similarity between telomerase and RNA polymerase is the primer cleavage activity associated with telomerase. A telomerase-mediated primer cleavage activity that removes bases from the 3′ end of certain primers has been characterized in *Tetrahymena*, yeast, and *Euplotes* telomerases (32, 51; M Melek & D Shippen, personal communication). Template-mismatched bases and bases that extend past the 5′-most residue in the template are cleaved by telomerase. The cleavage activity appears to be specifically associated with the end of the template as altering the length of the template alters the position of cleavage (55). Experiments with *Euplotes* suggest that cleavage is an endonucleolytic, not an exonucleolytic, reaction (M Melek & D Shippen, personal communication). What role the cleavage activity plays in the normal telomerase reaction is not yet clear. Cleavage activity may be a proofreading function, or it may play a role in enzyme processivity.

RNA polymerases also have specific associated cleavage activities that cleave back the growing RNA chain when the polymerase stalls on the template. Cleavage appears to allow the polymerase to bypass the stall site and continue elongation (56). By analogy, cleavage by telomerase may be associated with processive synthesis, although many of the RNA mutants that alter cleavage do not affect processivity (55). Telomerase differs from all other polymerases in that it utilizes an internal template rather than an external one, which places specific steric constraints on primer elongation and catalysis. Current evidence suggests that a single catalytic site must move relative to the internal template in the RNA (reviewed in 52). Understanding the mechanism of this catalysis will be a primary task for future research.

Telomerase RNA Components

Telomerase RNA components have been isolated from a variety of species. These RNA range in size from ~160–190 nucleotides (nt) in the ciliates to ~1300 nt in the yeasts (see Table 1) (34–41). Each of the RNAs contain the appropriate template sequence to synthesize telomere repeats. Telomerase RNAs resemble RNaseP RNA (57), a well studied functional RNA in two important respects. First, telomerase RNAs share little primary sequence similarity. Second, in ciliate telomerase RNAs the secondary structure is conserved (58–62). Both the mammalian and yeast telomerase RNAs are significantly longer than the ciliate RNAs. In *Tetrahymena* only 159 nt are sufficient for full telomerase activity (63), so the role of the longer RNA regions is unclear. In yeast the U2 RNA is over 1000 nt while in mammals the RNA is only 250 nt (64). Much of the extra sequence in the yeast U2 is not needed for function (65). Determining if the larger telomerase RNAs in yeast and mammals share some components of the ciliate secondary structure will require comparative phylogenetic analysis.

To study the function of the *Tetrahymena* telomerase RNA, both in vitro and in vivo reconstitution assays have been developed. In vitro reconstitution uses wild-type enzyme isolated from *Tetrahymena*: The RNA component is then removed and activity is reconstituted by addition of wild-type or mutant RNA (63). In vivo reconstitution involves transforming *Tetrahymena* with a high copy–mutant telomerase RNA gene and isolating the telomerase enzyme from transformed cells (66). Mutational analysis of the *Tetrahymena* telomerase RNA has focused on the template region. Alterations in the sequence of the 5′-most six nucleotides (<u>CAACCCC</u>AA) results in altered telomere repeats in vivo and in vitro (23, 63, 66). The 3′-most nucleotides (CAACCC<u>CAA</u>) appear to be used to align the primer 3′ end and do not serve as template residues during processive elongation of telomeric primers. The C at position 49 can be but is not always used as a template (55, 63). Not all the effects of template mutations are directed by base pairing. In one mutant, 43 A (<u>A</u>AACCCCAA), premature dissociation of product occurs before nucleotide incorporation at the mutant position. This finding suggests that steric effects of altered nucleotides in the template can affect efficient polymerization at other sites, and it raises the possibility that the RNA may play a direct role in the catalysis of nucleotide addition (66).

The 5′ boundary of the template region appears to be specifically determined, perhaps by a steric mechanism. In all ciliate telomerase RNAs a conserved sequence is found two nucleotides upstream of the 5′ end of the template. The conserved position relative to the template 5′ end led to the proposal that this region defines the template boundary (36). When either the sequence or position of this region is altered and the enzyme reconstituted in vitro, the boundary of the telomerase template is also altered (55). These experiments suggest that the conserved sequence represents a protein binding site or a functional RNA element that blocks the progression of the active site 5' on the RNA template. This sequence upstream of the template in ciliate RNAs is not conserved in mammalian and yeast telomerase RNAs, suggesting that these enzymes may use a different sequence or mechanism to define the template boundary.

Telomerase Protein Components

Tetrahymena telomerase appears to consist of two polypeptides, p80 and p95, that copurifiy with the enzyme in many different fractionation schemes (43). Crosslinking experiments using partially or highly purified fractions have shown that p95 binds telomeric primers and p80 binds telomerase RNA (43, 67). In immunoprecipitation experiments using antibodies directed against p80, both telomerase activity and p95 are specifically precipitated. The apparent molecular mass of telomerase (250kD) is consistent with a stoichiometry of one of each protein and the RNA component in the active enzyme complex (43). Reconstitution of enzyme activity from the isolated proteins is necessary for formal proof that these two proteins and the RNA are the only essential components of the enzyme.

The sequence of p80 and p95 do not share any extensive sequence homology with other proteins in the database, indicating that these are novel proteins. Limited sequence similarity was found with p95 and RNA-dependent RNA polymerases and DNA polymerase α and β in a conserved polymerase motif region. Further alignment "by eye" allowed extension of this conserved domain (43). All four polymerase motifs can be found in the p95 protein, although the degree of conservation of this region even among polymerases is very low, and spurious similarity with nonpolymerase enzymes has been noted (68). Mutational analysis of the cloned proteins is needed to confirm the functional significance of the limited sequence homologies.

To date the only other candidate for a telomerase protein is the *EST1* gene from *S. cerevisiae*. Cells deleted for this gene gradually lose telomeric sequences and viability drops dramatically after a lag phase (69). The phenotype of these cells is the same as a deletion of the RNA component of telomerase (see below), suggesting that the *EST1* gene might encode a telomerase protein component. No homology is found between the protein encoded by *EST1* (Est1p) and either of the telomerase proteins cloned from *Tetrahymena*. Recently two different assays were reported for yeast telomerase activity. One study, in which crude extracts were assayed and PCR was used to detect telomerase products, showed that Est1p is required for telomerase activity (33). The second study, in which fractionated extracts were used, showed that Est1p was not required for yeast telomerase activity (32). The fact that fractions with telomerase activity can be generated in the absence of Est 1 indicates that even if the protein is associated with yeast telomerase it may not represent an essential catalytic component. Perhaps Est1p is in some manner peripherally associated with the yeast telomerase.

In Vivo Effects of Telomerase Alteration

The first evidence for the role of telomerase in telomere synthesis in vivo came from mutations introduced into the *Tetrahymena* RNA template region. Telomerase RNA genes with mutant templates transformed into *Tetrahymena* generated cells containing the mutant sequence on chromosome ends (23). The phenotype of cells with mutant telomere repeats is abnormal; very large "monster" cells are produced that appear to have problems in nuclear division. In addition, these cells are very sick, and they lose viability after several generations. In some mutants the telomeres are longer than wild type and in other cases they are shorter, suggesting that the normal equilibrium maintenance has been upset (23). A likely explanation of this effect is that telomere binding proteins (TBPs) play a role in equilibrium length establishment and that sequence alteration inhibits the binding of these essential proteins (see below) (23).

The *S. cerevisiae est 1* mutant also shows telomere length defects and cell senescence (69). Although the functional relationship of Est1p to telomerase is not yet clear, the phenotype of *est1* mutants looks like a telomerase mutant phenotype (see below). The levels of yeast telomerase RNA are unchanged in ΔEST1 cells indicating an indirect effect of Est1p on telomerase RNA is not responsible for the phenotype (33). Deletion of *est1* results in progressive telomere shortening with each cell division. Initially no effect on cell viability is seen. After about 50 cell doublings, however, the cells get very large and many cells die. In the presence of an active recombination system, however, "survivors" appear in the culture. Many of these survivors are cells that have lengthened their telomeres through a recombination pathway that increases both TG(1-3) repeats and telomere associated (Y′) elements in the cell. In mutant cells that are deficient in gene conversion "survivors" are not generated at a high rate, and the cells die (70). In addition to Y′ recombination other types of suppressors may be generated in ΔEST1 cells, although the nature of these revertants is not yet fully characterized (33, 70).

The phenotype of yeast cells deleted for the telomerase RNA component is very similar to that of ΔEST1. TLC1 (for Telomerase component) is a single-copy gene encoding the 1.3 kb RNA that contains the template for synthesis of yeast telomeres (37). Changing the template region of the RNA results in altered telomere repeats on yeast chromosomes. Disruption of TLC1 in vivo causes progressive telomere shortening. Initially cell growth and viability are identical to wildtype, indicating that telomerase does not play an essential function during most cell divisions. After 50 to 60 generations, when telomeres are very short, abnormal cell phenotypes are seen and the growth rate of the culture slows. The survivors generated are similar to those seen in the *est1* mutants. Yeast cells deleted for *TLC1* do not have detectable levels of telomerase activity in vitro (32).

The telomerase RNA component has also been isolated from the yeast *Kluveromyces lactis* (38). This budding yeast contains unusually long, 23 bp–telomere repeats. The long telomere repeat suggested there should be a long unique sequence template in the telomerase RNA, and probes against this sequence allowed cloning of the RNA gene. As with telomerase RNA template mutants in *Tetrahymena* and *S. cerevisiae*, changing the template sequence in vivo results in altered sequence telomere repeats on chromosome ends. When the gene for this RNA (*TEL1*) is deleted initially, telomeres shorten with no phenotype; later, cell viability is lost, and then survivors are generated as with Δ*est1* and Δ*tlc1*(38).

Using the cloned K. lactis telomerase RNA gene, the function of both telomerase and telomere repeat sequences can be studied. When mutations are present in the template region and the mutant gene replaces the wild-type copy, cells containing mutant telomere repeats are produced. Thus studying the effect of simultaneously altering the sequence of all telomeres in a cell is possible. In two mutants that introduce restriction sites in the template region, (AccI and BsiWI) the mean length changes rapidly; from 1 kilobase pair (kbp) up to over 10 kbp within 75 cell generations. In two other mutants (Bgl II and KpnI) telomere lengthening was delayed, and no change was seen until after 400 generations. The effects on telomere length are probably due to effects of the altered sequence telomere repeats in vivo and not the altered telomerase per se, since cells expressing both mutant and wild-type RNAs do not have these phenotypes. If the long telomeres were the result of increased telomerase processivity one might expect long telomeres to be dominant. Presumably in cells containing altered telomere repeats the interaction of TBPs is altered and the equilibrium that determines telomere length is upset (38).

TELOMERE REPLICATION

Telomeres consist predominantly of double-stranded telomere repeats with only the extreme terminus containing some single-stranded G-rich repeats. During each round of replication most of the double-stranded region of the telomere is replicated by conventional semiconservative polymerase mechanisms. In *Tetrahymena* cells expressing mutant telomerase RNA, only the terminal regions of the telomeres show altered sequences (23). And when mutant telomerase generates the first repeats on newly formed telomeres in *Tetrahymena*, after loss of the mutant enzyme, the internal mutant repeats are stable: They are not replaced by wild-type repeats (71). The invariance of internal telomere sequences in *S. cerevisiae* also suggests that internal repeats do not turn over at each cell division (72). Similarly the initial terminal location of mutant telomere repeats in *K. lactis* cells expressing telomerase RNA template mutations (38) indicates that in any given cell division, the centromere proximal telomere repeats are not synthesized by telomerase, but rather must be copied by conventional replication apparatus.

Telomere replication occurs late in the cell cycle in both yeast and mammalian cells. In yeast this late replication is due to the telomeric location of genes and not to an effect of the surrounding sequence (12). Yeast chromosomes also acquire single-stranded extensions during replication late in S-phase (73). These guanine-rich (G-rich) telomere repeat extensions are thought to be added by telomerase, although they could also be the result of incomplete lagging-strand replication, or nucleolytic activity acting on the cytosine-rich (C-rich) strand. These models can now be distinguished by looking for these G-rich extensions in cells that lack telomerase activity.

Cytosine-Rich Strand Synthesis

Although telomerase is clearly necessary to maintain telomere length (37, 38), it is only capable of elongating the G-rich strand. To obtain net telomere elongation, the C-rich strand must subsequently be filled in. Current models suggest that conventional DNA polymerases carry out this function, although there is as yet little experimental data on this process. The most likely candidate for filling in the C-rich strand is polα/primase because it has the ability to initiate replication de novo. Experiments in *Oxytricha* extracts suggest that a primase activity is able to initiate C-strand synthesis on single-stranded telomeric repeat sequences (74). The structure of newly formed telomeres during development in *Euplotes* indicates that synthesis of the C-rich strand is precise: The majority of cloned nascent telomeres are of defined length and begin with the sequence C4 (75). Thus copying of the single-stranded G-rich tail generated by telomerase appears to be a regulated, not random, process.

Several mutants in yeast implicate specific DNA polymerases in telomere length regulation. The *cdc17-1* mutant is a temperature-sensitive mutation in DNA polymerase α (76), and these cells have elongated telomere tracts when they are grown at the semipermissive temperature (77). Whether this telomere phenotype is related directly to polymerase α's role in filling in the C-strand after telomerase elongation, in conventional replication of the double-stranded region of the telomere, or in some other process is not clear. Mutations in the polymerase accessory factor RFC also show a telomere elongation phenotype

(A Adams & C Holm, personal communication). RFC is a primer binding protein that is required for leading-strand DNA synthesis by polymerase δ (78).

Several models have been proposed for how these replication proteins affect telomere length. The elongated single-stranded regions at yeast telomeres late in S-phase (73) may be stabilized by polymerases that synthesize the C-rich strand. In *cdc17* polymerase slippage during replication of double-stranded telomere repeats may lead to telomere elongation. However, why slippage would be biased in the direction of elongation but not shortening is unclear. To test whether telomere elongation is independent of telomerase as this model would predict, telomere elongation in *cdc17* or RFC mutants should be examined in Δ*TLC1* strains. Perturbation of the normal cell cycle may lead to telomere elongation by extending the cell cycle phase in which telomere replication occurs. Although not all yeast cell cycle mutants show telomere length effects. However, recent data showing that a telomere length regulation mutant *TEL1* is homologous to the cell cycle–checkpoint genes *MEC1* from yeast and Ataxia Telangictasia Mutated (ATM) from humans support a role for cell cycle effects on telomere length (79).

TELOMERE BINDING PROTEINS

Telomere length regulation is mediated at least in part by telomere binding proteins (TBPs). Studies of telomere chromatin structure gave the first indication that telomeres may be packaged into specialized nucleoprotein complexes (reviewed in 80). The terminal regions of telomeres appear to be bound by proteins different from the internal tracts (81, 82) (P Cohn & EH Blackburn, personal communication). Consistent with this data on chromatin structures, TBPs fall into two distinct classes: those that bind the single-stranded repeats at the extreme termini, and those that bind along the length of the double-stranded telomere repeats. Proteins that bind telomere sequences have been characterized in various organisms using biochemical assays. The properties of many of these proteins are described in several recent reviews (14, 80, 83). Since the biological role of many of these proteins has not yet been determined, here only the best-characterized example from each class is described: the *Oxytricha* single-stranded end binding complex and the yeast double-stranded binding protein *RAP 1*.

Oxytricha Telomere Protein

The prototype single-stranded TBP is the α/β heterodimer protein from Oxytricha that consists of a 41 (α) and a 56 (β) kDa subunit. The complex binds tenaciously but noncovalently to the *Oxytricha* G-rich single-stranded telomere overhang (84–86). The binding is very stable: Protein remains associated with telomeres even in 2.5 M NaCl or 6M CsCl. Cloning of the genes for the two subunits and expression of recombinant proteins showed that neither protein has regions homologous to other known DNA binding proteins (87, 88). Although both subunits have affinity for DNA, the α subunit alone will bind and protect telomeric DNA oligonucleotides (87). In the absence of DNA the two subunits are monomeric. When telomeric DNA is present, they interact with each other specifically and form a protein DNA complex with a 1:1:1 stoichiometry (89). The α protein uses hydrophobic interactions and base stacking for DNA binding. Two tyrosine residues and a histidine in the DNA binding region of the α protein specifically cross link to telomeric DNA (90). The hydrophobic component of binding probably accounts for the extreme salt stability of the complex. Both the α and β proteins are phosphorylated in vivo, and in vitro studies suggest that the phosphorylation of the β subunit is regulated in the cell cycle. This phosphorylation could play a role in regulating the availability of telomeres for elongation by telomerase (91). A telomere end-binding protein with similar properties is being characterized in Euplotes (92, 93). These end-binding proteins may represent a class of telomere proteins that stabilize telomeres by providing a cap at the extreme terminus.

RAP1

In addition to single-strand terminus-specific binding proteins, telomeric DNA is bound by specific proteins along the length of the double-stranded telomeric region. These binding proteins may be directly involved in maintaining telomere length equlibrium. The most well characterized double-stranded TBP is Repressor activator protein (Rap1p) from *S. cerevisiae*. Rap1p is a transcription factor needed for expression of a variety of genes, and it is also involved in establishing silent transcriptional domains (94, 95). Mutants in *rap1* also affect telomere length. The various functions of Rap1p have recently been reviewed (9, 80, 96). Only the role of Rap1p in telomere length regulation is considered here, although an overlap between functions is likely.

Rap1p is a 92 kDa protein with specific functional domains. The central region contains the DNA binding domain (97); the C terminus is involved in transcriptional activation, transcriptional silencing, and interaction with other proteins; and the amino terminal region is not essential for any function tested so far (98–104). The Rap1p consensus binding site (A/G)T(A/G) CACC CANNC(C/A)CC is present at yeast telomeres in the form ACACCCACA-CACC (105–107). On average there is a Rap1p binding site every 18 bp in yeast telomeres (108). Rap1 binds yeast telomeric sequences in vivo (104) and is found in meiotic cells at yeast chromosome termini (109).

Rap1 is an essential gene in yeast as evidenced by the fact that spores lacking the gene do not grow. This immediate lethality may reflect the protein's role in transcription, because other genes that affect only telomere length show delayed lethal phenotypes. Temperature-sensitive alleles of *rap1* show reversible telomere shortening. Telomeres shorten in cells grown at a semipermissive temperature, but telomeres return to normal when cells are grown at the permissive temperature (110). Mutations isolated in RAP1 are grouped into two classes: those that primarily affect mating type silencing (rap1^s) and those that primarily affect telomeres (rap1^t), although some telomere length alterations are seen in both classes. All the mutations in these two classes fall into the C terminal region of Rap1p and do not show transcriptional defects, which indicates that the telomere length, silencer, and transcriptional domains of Rap1p are separable (98–100, 104).

Overexpression of the C terminal half of Rap1p causes telomere lengthening, suggesting that this region normally interacts with other proteins that limit telomere growth (104). Interestingly, rap1^t mutants that lack the C-terminal domain also show a telomere-lengthening phenotype (100). In these strains telomere length can increase from less than1 kbp to up to more than 4 kbp. The length increase is due to tracts of telomeric TG_{1-3} repeats, not to subtelomeric recombination. The long telomeres in these strains are unstable and undergo rapid deletion events, probably through recombination that removes the extra telomeric repeats. These deleted telomere tracts can then re-elongate. This instability of long telomere repeats is also seen in trypanosomes and in *K. lactis* cells that have undergone unusual telomere elongation (38, 111, 112), suggesting that the cell can only maintain a certain number of extra telomeric repeats.

The behavior of the *rap1* mutants illustrates the complex role that TBPs play in telomere length regulation. Temperature-sensitive *rap1* mutants show telomere shortening, suggesting that Rap1p binding to telomeres is needed to stop telomere shortening. Yet telomere lengthening is seen in both deletions of the Rap1p C-terminal region and over expression of the C-terminal region, indicating that the interaction of Rap1p with other factors, not Rap1p alone, inhibits telomere elongation.

Several proteins are known to interact with the Rap1p C-terminus: Rifp1 and proteins involved in the establishment of transcriptionally silent chromatin in yeast, Sir3p and Sir4p (101, 113, 114). Rap1 interaction factor (*Rif1*), a likely mediator of some functions of Rap1p, was cloned based on its interactions with Rap1p (101). Mutants deleted for *rif1* are defective in silencing and show some telomere elongation, although the degree of tract lengthening is not as dramatic as in the *rap1^t* mutants. The effects of Rap1p on chromatin structure, silencing, and telomere length are likely interrelated. Recent evidence suggests that the increased telomere length of *rap1^s* mutants may mediate the silencing phenotype by titrating Rap1p away from the mating-type silencer elements (183).

TELOMERE PROCESSING

Mature telomeres probably result from both synthesis and processing activities. Evidence for telomere processing so far comes primarily from ciliates; however, the conservation of telomere structure and function suggest that processing activities are likely to be present in other eukaryotes as well. The best example of telomere processing is found during macronuclear development in *Euplotes*. During conjugation in ciliates, a macronucleus is formed from a copy of the zygotic nucleus. The chromosomes initially present in the zygotic nucleus are fragmented into many pieces and new telomeres are added to the ends of each piece (reviewed in 15). During the process of telomere addition in *Euplotes*, "oversized" telomeres are initially added onto maturing macronuclear chromosomes (115). These extra-long telomeres are subsequently trimmed from about 80 bp to the final mature size of 28 bp. The trimming occurs reproducibly at a discrete time during development. The G-rich strand of the oversized telomeres is longer than the C-rich strand and heterogeneous at the 3' end (75). Although many 3' terminal sequences are found, the majority of molecules end with -TT 3' or -TTT 3', whereas mature *Euplotes* telomeres end in -GG 3'. Trimming of the telomeres to the final size is not a result of under-replication, because efficient trimming occurs in cells where replication is blocked by aphidocholin (116). These data indicate that a telomere-specific endo-or exonuclease activity processes *Euplotes* telomeres to generate the final mature end. Telomere processing is likely to occur during each cell cycle and not just during macronuclear development since the unique -GG 3' end is maintained at all telomeres during vegetative growth (117). This position is not the position where *Euplotes* telomerase preferentially dissociates during telomere elongation (28, 35).

Telomere length regulation in *Euplotes* probably differs from that in other organisms as most telomeres are maintained at 28 bp and length heterogeneity is limited (75, 117). However, good evidence has also been found for telomere processing in *Tetrahymena,* which has considerable length heterogeneity. When *Tetrahymena* is kept in continuous exponential growth, the average telomere length increases (118), although the reason for this increase is not known. However when cells with long telomeres are returned to stationary phase in "stock" tubes, telomere length shortens dramatically. The rate of shortening is more consistent with active processing than with loss due to replication, although experiments with aphidicolin to inhibit replication have not been done. The structure at the very end of the telomere in vegetatively growing *Tetrahymena*, as in *Euplotes*, is consistent with active processing. In vitro, *Tetrahymena* telomerase preferentially dissociates from growing telomere repeats at the first G residue in the sequence TTGGGG (119). However, in vivo almost all telomeres apparently end with the sequence TTGGGG 3', since 3' end labeling of total DNA allows a unique sequence to be read (120).

Although the mechanism of telomere processing is not known, two enzymes have cleavage activities that suggest they could be involved in this process. The cleavage activity of telomerase itself could in principle participate in telomere processing. However, the predominant 3' ends created by telomerase cleavage are not what is found in vivo. In both *Tetrahymena* and *Euplotes*, cleavage removes a 3' dG from —TTG to generate TT 3', although removal of additional nucleotides can also occur (51) (M Melek & D Shippen personal communication). In yeast Kem1p binds to G-quartet structures often found at telomeres and cleaves adjacent to these structures (121). Although slight telomere shortening is seen in *kem1* deletion strains, the protein is not essential for telomere maintenance (122). Neither the cleavage activity of telomerase nor the Kem1p activity has the ability to cleave double-stranded telomere repeats to cause the dramatic shortening of repeat length seen in *Euplotes*. Thus additional telomere processing enzymes likely remain to be discovered in both ciliates and other eukaryotes.

REGULATION OF TELOMERE LENGTH EQUILIBRIUM

The overall maintenance of telomere length is likely to involve an intricate balance of competing factors. The major players in establishing the equlibrium have been introduced above. At minimum the forces involved include: telomere shortening (incomplete replication, telomere processing activities, recombination), telomere stabilization (TBPs, telomere chromatin structure), and telomere lengthening (telomerase, C-strand synthesis, recombination). Models for how these competing activities interact are just now beginning to come into focus.

Genetic experiments in yeast provide some insight into the dynamics of regulation. Different strains of yeast maintain different average telomere lengths (123). Although in each case the equilibrium length is tightly maintained, the "set point" around which the lengths vary can range from 100 bp up to 500 bp of telomeric TG_{1-3} sequences. When two haploid strains with different set points are crossed, the telomeres in the heterozygote are adjusted to a new average set point. Strain-specific differences in telomere length are also found in maize (124), *Arabidopsis* (125), and mice (126–128). In maize, quantitative trait analysis reveals that several distinct genes must contribute to the different telomere lengths in the various lines (124).

Experiments in yeast suggest that TBPs establish the telomere-length set point by negatively regulating telomere growth. When many extra telomeres are introduced into yeast, the average length of all telomeres increases (129). This finding suggests that some essential TBP is being sequestered by the added binding sites. Further evidence comes from the introduction of mutations into telomere repeats in vivo as described above. When certain mutations are introduced into telomere sequences in *K. lactis* by altering the sequence of the telomerase RNA template, dramatic telomere lengthening occurs (38). With some mutations the length change is immediate, whereas in other cases length changes are not seen until most of the telomere is replaced with mutant sequence. The difference in phenotype observed with mutations at different positions in the repeat could be due to effects on different binding proteins (double-stranded vs terminal-binding proteins) (38), or they could be due to differential effects on the affinity of a single protein for a given site. Some changes in the telomere repeats may hit a more critical residue in a protein binding site than others. Since little is known about *K. lactis* TBPs, it is not yet clear which positions in the repeat are essential.

Models for Length Regulation

Several models have been proposed for how proteins interact at telomeres to regulate telomere length (38, 75, 130, 131). Using components from these proposals, three models of increasing complexity are described. The simplest model (Figure 2a) proposes that TBPs limit telomere growth. The addition of repeats onto chromosome ends by telomerase is not regulated, but only those added repeats bound by TBPs are stabilized. If the newly made repeats are coated by cooperative binding of TBP's initiating on the previously bound internal TBPs, the number of telomere repeats protected will be determined by the amount of available TBP. The remaining repeats could be removed by processing or degradation and the end capped by an end-binding protein. If the equilibrium ratio of bound to free TBP is altered, the length of all telomeres in the cell will be altered. This simple model is consistent with the data on temperature-sensitive alleles of Rap1p that lead to telomere shortening. The presence of fewer active Rap1p molecules in the cell causes telomeres to shorten. In addition the model could explain the growth of telomeres in trypanosomes and *Tetrahymena* (112, 118); the increased availability of TBPs during rapid growth may allow telomere elongation. This limiting factor model, however, is inconsistent with some newer data.

A second generation model (Figure 2b) can account for yeast data that show that telomere growth must be actively inhibited. In this model the TBPs directly interact with telomerase to limit the number of repeats that are added at each cell division. This negative regulation of telomerase may act through TBPs or through TBP-associated factors (as drawn). This model can explain the telomere length increases in the presence of Rap1p C-terminal truncation (100), Rap1p C-terminal domain overexpression (104), and the addition of extra telomeres (129). Although telomerase can elongate telomeres that are bound

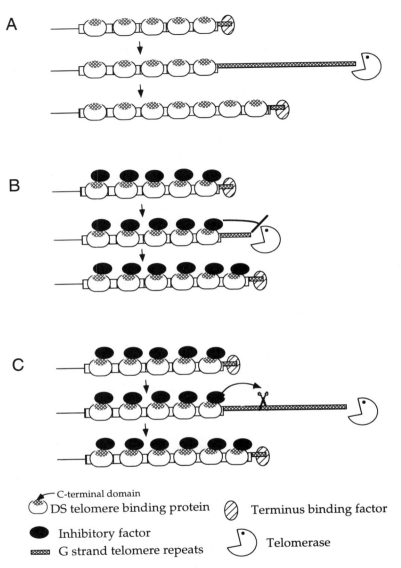

A

B

C

C-terminal domain
DS telomere binding protein   Terminus binding factor

Inhibitory factor
G strand telomere repeats   Telomerase

Figure 2 Models for telomere length regulation. (*a*) Limiting factor model. Telomeres are bound by double-stranded TBP. During replication telomerase elongates the GT-rich single-stranded region, and conventional polymerase fills in the CA-rich strand. Telomerase activity is unregulated; the final number of repeats added depends on the availability of telomere binding proteins to protect the ends from degradation. (*b*) Telomerase activity regulation model. Telomeres are bound by double-stranded telomere binding proteins that recruit factors to limit telomere elongation. A TBP or an accessory factor (shown in figure) directly limits the number of repeats that telomerase adds at each replication. (*c*) Telomere processing model. Telomerase activity is not regulated. Telomere binding proteins recruit a processing factor that limits telomere length. The processing factor might cleave the G strand as shown or might cleave the double-stranded telomere repeats.

by the Oxytricha terminal protein (130), no evidence yet shows that the number of repeats added by telomerase is regulated by interaction with other proteins.

A third model (Figure 2c) suggests that negative regulation of telomere length may occur via a telomere processing activity and not through telomerase regulation. The TBPs could recruit negative regulatory factors that limit telomere length by cleaving the newly generated repeats. This model is also consistent with the yeast data that show that TBPs negatively regulate telomere growth. Interference with telomere processing by sequestering the processing activity (by adding extra telomeres or by overexpression of the Rap1p C-terminus) or preventing its recruitment (by Rap1p truncation or mutant telomere repeats) would result in telomere elongation. The models presented in Figure 2b and c fit the data on telomere length regulation to date, although other models may also be consistent with these data.

Some aspects of telomere length regulation may not be shared by all eukaryotes. McEachern and Blackburn proposed a model to explain the delayed elongation of telomeres in *K. lactis* mutants (38). They suggest a direct protein-protein interaction between the terminus binding factor and the internal most–double-stranded TBP, with the telomere looping back on itself. The delayed elongation phenotype, however, could also be explained by variations of the models shown in Figure 2b and c. Cooperative binding of TBPs would allow internal sites to act as an anchor, stabilizing TBP binding to more-terminal sites when those sites are mutant. When the internal anchor is mutated, a domino effect of weakened protein binding would be propagated to the end of the chromosome, preventing recruitment of factors that limit telomere growth. Although this model is attractive, cooperative binding of TBPs has yet to be demonstrated. Regardless of whether cooperative protein binding, telomere processing, regulation of telomerase repeat addition, or some other model is substantiated by future results, telomerase is clearly necessary but not sufficient to maintain telomere length.

TELOMERES IN CELLULAR SENESCENCE AND AGING

Unlike the immortal single cell eukaryotes, in which telomere length must be continually maintained, human cells, which are not normally immortal, have telomeres that shorten in many somatic tissues. The first evidence that human telomeres were similar to those characterized in other eukaryotes came from analysis of the pseudoautosomal region of the X chromosome (132). The terminal restriction fragment of human chromosomes identified with this probe was heterogeneous in size, similar to telomeres in *Tetrahymena* and yeast. A curious finding is that the size of the terminal fragment is longer in sperm DNA than in blood DNA from the same individual (133). Subsequently, human telomeres were cloned and shown to contain tandem repeats of $(TTAGGG)n$

at their termini (134–138), and telomerase activity that synthesized this sequence was identified in HeLa cells (29).

Two observations set the stage for today's understanding of mammalian telomere dynamics. The first finding was that telomeres are shorter in many tumors than in normal adjacent tissue (135, 139). The second observation was that when primary human fibroblasts are grown in culture, telomeres shorten with each cell division (140), and in vivo telomeres are shorter in certain human tissues in older people than in younger people (139, 140). These observations, and the long telomere lengths in sperm cells, suggested that telomerase may be active in the germline and not in somatic tissue (133, 135, 139, 140).

In the 1960s, Hayflick and Moorehead demonstrated that normal human cells in culture have a finite doubling capacity. This limited ability of somatic cells to divide was proposed as a check against immortal cell growth in cancer and as an "in vitro" model for cellular aging (141, 142). Telomeres shorten as various primary human cell types divide in culture (140, 143–146; reviewed in 147). The number of divisions fibroblasts can undergo in culture is correlated with telomere length: Cells with long telomeres divide more times than cells with short telomeres (148). These observations suggested that telomere shortening may play a direct role in determining the number of divisions primary cells can undergo (140, 149, 150). Telomere shortening may thus be a checkpoint signaling entry into senescence (151, 152). Obtaining direct evidence in primary human cells for the role of telomere length in senescence is difficult because there are 92 chromosome ends, all of which can vary in length. Thus there is little direct experimental evidence yet to show that telomeres play a causal role in entry into senescence, although testing this theory is currently an active area of research.

Several models have been proposed for the mechanism by which telomere length might signal entry into senescence. A short telomere may look like damaged DNA to a cell, and the p53 mediated–checkpoint pathway (153, reviewed in 154) could be activated (149, 152, 155). Consistent with this idea, the cell cycle inhibitor p21 is elevated in senescent cells (156), and this gene is known to be induced by p53 (157). Transcriptional regulation of telomere-adjacent genes has also been proposed as a mechanism to regulate entry into senescence (158). In yeast, genes placed near telomeres are regulated by telomeric repression; however, despite considerable effort, telomere silencing in mammalian cells has yet to be demonstrated (159).

Other mechanisms are also consistent with current data such as the fact that the free TBPs released by telomere shortening send a signal for entry into senescence. Current data do not allow us to distinguish among the many possible models. If telomere shortening does play a causal role, we do not know what the signal might be. Is it the shortest telomere, the average length of all telomeres, or the total amount of (TTAGGG)n sequence? Evidence from several labs suggests that multiple pathways are involved in the entry into cellular senescence (reviewed in 160). Additional characterization of these pathways that limit cell division in normal cells is needed before a direct role of telomere length in senescence can be established.

Although the role of telomere length in signaling cellular senescence is not fully understood, telomere length is clearly not directly correlated with organismal aging. Cellular senescence as defined in vitro probably occurs in at least some cell types in vivo (reviewed in 161, 162). Models for the role of cellular senescence in vivo suggest that it may contribute to certain age-related diseases, although it is unlikely to be a direct determinant of organismal life span. Evidence from mice indicates that telomere length does not correlate with organismal life span. In the common laboratory mouse, *Mus musculus* telomere containing restriction fragments are around 50–150 kbp (126, 127), significantly longer than the ~15 kbp found in human germline cells. Some of the difference in telomere length can be attributed to the repetitive satellite sequences found near many telomeres in the mouse (163). Since these repeats are not frequently cut by restriction enzymes, the terminal fragment released from chromosome ends may seem unusually large. However in addition to large terminal restriction fragments, mouse telomeres contain quantitatively more (TTAGGG)n sequences (127). Since mice have a significantly shorter life span (~2 years) than humans (~90 years), telomere length is unlikely to play a direct role in life span determination. In addition, some strains of mice have very different telomere lengths. *Mus spretus* has terminal restriction fragments that are ~15 kbp on average (127, 128), yet these mice do not have a significantly shorter life span than the laboratory mouse *Mus musculus* which has 50–150 kbp telomere fragments. Thus, even if telomeres play a role in signaling cell senescence, short telomeres are unlikely to shorten the life span of the entire organism.

TELOMERASE ACTIVATION IN CANCER CELLS

Tissue Culture Models

Cellular senescence may be one mechanism cells have to limit their division capacity and decrease the likelihood of developing cancer (152). Normal human cells can be immortalized in culture if they are transfected or infected with certain viral onco-proteins. After expression of SV40 T-antigen, human cells have an extended life span, and yet most of the cells will die after about 100 generations when the culture undergoes crisis. Immortal cells can be derived from these cultures at a frequency of about 1×10^{-7} (164). This low rate of immortalization suggests that additional events are required in combination with viral protein expression. In preimmortal and extended–life span cells, telomere length shortens continuously as the cells divide and telomerase is not detected. After crisis, telomeres are maintained at a stable (often short) length and telomerase activity is present (144, 145, 165, 166). This finding suggests that at crisis, cells that express or have reactivated telomerase activity are selected. This selection further suggests that telomerase may be required for growth in cells where telomeres have shortened. The experiments in yeast that show loss of viability in cells deleted for telomerase RNA support the idea that telomerase is required in cells with shortened telomeres (37, 38).

Not all immortal cell lines express telomerase activity. Some cell lines established by viral transfection or by other means lack detectable telomerase activity (44, 167, 168). Most of these cell lines have very long telomeres and thus may not require telomerase for growth. The long telomeres in these lines likely result from transient telomerase activation or from telomere recombination as in yeast, although the presence of additional unknown pathways has not been excluded (167).

When normal and immortal human cells are fused the hybrid is usually mortal, indicating that senescence is dominant over the immortal phenotype (169, 170). When several telomerase positive immortal lines were fused with normal human cell lines all of the hybrids senesced as expected. However some of the senescing clones still expressed telomerase activity and had long telomeres (167). Thus the presence of telomerase is not sufficient for immortalization of cells grown in vitro. The presence of telomerase activity in some normal mouse and human tissues (see below) also argues that telomerase is not sufficient for immortalization.

Telomeres and Telomerase in Human Cancer

In vivo telomere shortening and telomerase activation appear to mimic the tissue culture model fairly well. Telomeres are shorter in tumor tissue than in normal adjacent tissue (135, 139, 171; reviewed in 172, 173). Since the tumor cells likely went through more cell divisions than the adjacent tissue, this shortening is probably analogous to the telomere shortening of primary cells in culture. In cells from human ovarian carcinoma, telomeres are short and telomerase activity is present; in the normal cells from the patients, telomeres are long and telomerase activity is not detected (174). Subsequently, with the development of a more sensitive telomerase assay (44), activity has been found in a wide variety of human tumors and is not detected in most adjacent normal tissues (reviewed in 172, 173). Preliminary evidence suggests that the presence or absence of telomerase in some cancers may be a good diagnostic for clinical outcome of the disease (178), although a thorough investigation of telomerase in many cancers is required before any clinically defined diagnostic parameters can be established.

The simple idea that telomere shortening in tumors mimics what is seen in cellular immortalization in tissue culture does present some problems. If telomeres shortened continually as the tumor initially grew, large tumors should have shorter telomeres than small tumors. This is clearly not the case in renal cell carcinoma, suggesting that telomerase activation can occur earlier in some tumors and tumor types than in others (171). In addition to cell growth, tumors also have a significant amount of cell deaths occurring through apoptosis and necrosis. Clonal evolution of cell populations within the tumor may lead to telomerase activation in some cells and not in others (173). Thus the tissue culture model of telomere shortening to a very short length followed by stabilization and telomerase activation may be only one of several possible mechanisms of telomerase activation in tumors (175).

Nevertheless, if telomerase is required for the growth of some tumor cells, as suggested by the apparent selection for telomerase positive cells in both immortalization in culture and in tumors, telomerase inhibition may be a good target for cancer therapy (44, 140, 166, 173). If telomerase inhibition is to be pursued as a cancer target, we need to know more about the normal regulation and expression of the enzyme. In human cells telomerase activity has not been detected in primary cell lines derived from various tissues. In addition, normal tissues such as liver, kidney, colon, lung, breast, and others have been shown to be telomerase negative (reviewed in 172). Although initial reports of telomerase-negative normal human tissues prompted speculation that all nontumor somatic tissues lack enzyme activity, this initial conclusion was somewhat premature. Recent work demonstrated telomerase activity in normal human white blood cells and in some noncancerous liver diseases (176–178). Ongoing experiments may identify additional human somatic tissues that express telomerase activity.

The presence of telomerase activity in normal human blood suggests that stem cells may require telomerase to sustain the many divisions they undergo. One curious fact about the blood cell data, however, is that the bulk of normal leukocytes show telomeres shortening with age (143). In addition, the candidate leukocyte stem cell, CD34$^+$ CD48(^{lo}), also shows telomere shortening with age in vivo (146). Telomerase activity is present in mature T and B cells, as well as in the progenitor CD34$^+$ CD48(^{lo}) stem cells (177, 179). If the telomerase activity were present in blood stem cells to allow telomere maintenance in this long-lived population, one would expect that telomere length in the blood would be maintained as it is in *Tetrahymena* and yeast. Thus telomerase activity may not be in normal blood to maintain stem cell telomeres, but to play some as yet unknown role in blood cell function, or it may be a remnant of embryonic telomerase expression in this cell lineage. The presence or absence of telomerase activity in the intestinal epithelial crypt cells and other stem cell types needs to be investigated.

Telomerase activity was reported in liver samples from patients with hepatitis and liver cirrhosis but no tumor pathology (178). This finding raises the question of whether telomerase can be reactivated during liver regeneration or other regenerative processes. Telomerase activity is present during development in various human fetal tissues such as muscle, lung, skin, and adrenal gland, suggesting that telomerase is initially active and then down-regulated during development (J Shay & MA Piatyszek, personal communication).

Mouse Models for Telomerase Regulation

To study the regulation of telomerase in normal and cancerous mammalian cells, mouse models are being developed to follow telomere length, telomerase activity, and RNA component expression. Since *Mus musculus* has terminal restriction fragments of 50–150 kbp, following changes of a few kbp on the large telomere fragments in this species is not practical. Studies of telomere shortening in mice have made use of the species *Mus spretus*, which has much shorter terminal fragments (128). Like human cells, primary skin fibroblasts from these mice show telomere shortening with increased divisions in vitro, and telomerase is not detected. Mouse fibroblasts grown in culture will spontaneously immortalize, unlike human cells (180). After prolonged growth and, presumably, immortalization in culture, the mouse fibroblasts have stabilized telomere lengths and telomerase activity is present (128). This finding suggests that the basic phenomenon of initial telomere shortening and subsequent stabilization and telomerase activation seen in human cells also occurs in mice.

Telomerase activity is present in several normal mouse tissues such as liver, kidney, and spleen, whereas it is reportedly absent in humans (30, 181). The presence of activity in these tissues suggests that telomerase may not be as tightly regulated in mouse cells as in human cells. The greater expression of telomerase in mice may predispose cells for immortalization and thus contribute to the higher rate of spontaneous immortalization of mouse vs human cells in culture (128).

Although regulation appears to differ somewhat in humans and mice, telomerase is clearly regulated during mouse development. Telomere length differs among specific adult mouse tissues; however in newborn mice, telomere length in all tissues is very similar, suggesting that differences in telomere length in various tissues arise postnatally as a result of different levels of telomerase activity (128). Consistent with the idea that telomerase activity is down regulated during development, the levels of the telomerase RNA component decrease after birth in liver, kidney, and brain (39). The adult liver has the highest level of telomerase RNA, consistent with the presence of telomerase activity in adult liver (30, 39, 181). Since telomerase activity is also present in some tissues in early human development, but not in the adult tissues (182), the pathways that regulate telomerase in mouse and human development may be similar.

PERSPECTIVES

Telomere stability and telomere length have been implicated in various biological processes at both the cellular and the organismal level. In the past 10 years, work on various organisms has led to the understanding that telomeres are maintained through a dynamic equilibrium of lengthening and shortening. With the recent cloning of telomerase components from yeast, ciliates, and mammals, rapid progress should soon be made on several fronts. Within the next 10 years we will undoubtedly understand how TBPs interact with telomerase to establish and maintain telomere length. In addition we may understand the answers to many other questions such as: Is there a direct effect of the TBPs on telomerase processivity, or is the role of TBPs separable from telomerase? Is the synthesis of the C-rich strand of the telomere regulated, and does this synthesis play a role in establishing telomere length? Does the developmental down regulation of telomerase play a functional role in mammalian development? Is the presence of telomerase in early development required for embryonic growth, or will telomerase deletion create a phenotypic lag, as seen with deletion of the telomerase RNA component gene in yeast?

The role that telomere length plays in cellular senescence will be an active area of research. Does telomere length directly affect entry into senescence? If the correlation does reflect an underlying cause and effect, what is the mechanism by which telomere shortening sends a signal? This question is extremely important, yet difficult to test because of the heterogeneous nature of telomeres and the difficulty in knowing precisely the length of each chromosome end at any given division.

The correlation of telomerase activation and cancer should receive much attention in the next few years. Evidence already suggests that the level and timing of telomerase activation differs among cancers. If current efforts at designing a highly specific inhibitor for telomerase are successful, then knowing which cancers might require telomerase activity and which might not will be essential. Potential inhibition therapies should initially be directed at those diseases associated with very short telomeres where telomerase is reproducibly activated in vivo.

Solving these puzzles of telomere length regulation will help complete the picture of how telomeres influence cell viability in normal and disease states. In the past, however, the unexpected findings led to the greatest insights, such as the discovery of repetitive DNA structure, telomerase, the unusual repeats at *Drosophila* and budding yeast telomeres, and mammalian telomere shortening and telomerase activation. In the future, undoubtedly the surprises will again catch our attention. Unexpected insights will lead telomere biologists to new questions that cannot yet be asked. The next generation of important issues is somewhere around the corner. The fun will be in rounding the corner, surveying the new terrain, and choosing the next path toward understanding the complexities of chromosome dynamics.

ACKNOWLEDGMENTS

I thank Titia de Lange, Chantal Autexier, and Nathaniel Comfort for critical reading of the manuscript. I would also like to thank my colleagues in the telomere field for many interesting discussions over the past years. Many of the models presented here are the cumulative result of ideas circulating and being continually refined within the telomere research community. Work in my lab described in this report was supported by the NIH grants GM43080 and AG09383 and by Geron Corporation.

Literature Cited

1. Muller HJ. 1938. *Collect. Net* 13:181–98
2. McClintock B. 1938. *Mo. Agric. Exp. Stn. Res. Bull.* 290:1–48
3. McClintock B. 1939. *Proc. Natl. Acad. Sci. USA* 25:405–16
4. McClintock B. 1941. *Genetics* 26:234–82
5. McClintock B. 1942. *Proc. Natl. Acad. Sci. USA* 28:458–63
6. Blackburn EH. 1991. *Nature* 350:569–73
7. Blackburn EH, Greider CW, eds. 1995. *Telomeres.* Cold Spring Harbor, NY: Cold Spring Harbor Lab. Press
8. Dernberg A, Sedat J, Cande Z, Bass H. 1995. See Ref. 7, pp. 295–338
9. Shore DA. 1995. See Ref. 7, pp. 139–99
10. Gilson E, Laroche T, Gasser SM. 1993. *Trends Cell Biol.* 3:128–34
11. Brewer BJ, Fangman WL. 1993. *Science* 262:1728–31
12. Ferguson BM, Fangman WL. 1992. *Cell* 68:333–39
13. Greider CW. 1991. *Curr. Opin. Cell Biol.* 3:444–51
14. Henderson ER. 1995. See Ref. 7, pp. 11–34
15. Prescott DM. 1994. *Microbiol. Rev.* 58:233–67
16. McEachern MJ, Hicks JB. 1993. *Mol. Cell Biol.* 13:551–60
17. McEachern MJ, Blackburn EH. 1994. *Proc. Natl. Acad. Sci. USA* 91:3453–57
18. Levis RW. 1989. *Cell* 58:791–801
19. Biessmann H, Mason JM. 1988. *EMBO J.* 7:1081–86
20. Biessmann H, Mason JM, Ferry K, d'Hulst M, Valgeirsdottir K, et al. 1990. *Cell* 61:663–73
21. Levis RW, Ganesan R, Houtchens K, Tolar LA, Sheen F. 1993. *Cell* 75:1083–93
22. Greider CW, Blackburn EH. 1985. *Cell* 43:405–13
23. Yu G-L, Bradley JD, Attardi LD, Blackburn EH. 1990. *Nature* 344:126–32
24. Olovnikov AM. 1973. *J. Theor. Biol.* 41:181–90
25. Watson JD. 1972. *Nature New Biol.* 239:197–201
26. Zakian VA. 1989. *Annu. Rev. Genet.* 23:579–604
27. Zahler AM, Prescott DM. 1988. *Nucleic Acids Res.* 16:6953–72
28. Shippen-Lentz D, Blackburn EH. 1989. *Mol. Cell. Biol.* 9:2761–64
29. Morin GB. 1989. *Cell* 59:521–29
30. Prowse KR, Avilion AA, Greider CW.

1993. *Proc. Natl. Acad. Sci. USA* 90: 1493–97

31. Mantell LL, Greider CW. 1994. *EMBO J.* 13:3211–17

32. Cohn M, Blackburn EH. 1995. *Science* 269:396–400

33. Lin J-J, Zakian VA. 1995. *Cell* 81:1127–35

34. Greider CW, Blackburn EH. 1989. *Nature* 337:331–37

35. Shippen-Lentz D, Blackburn EH. 1990. *Science* 247:546–52

36. Lingner J, Hendrick LL, Cech TR. 1994. *Genes Dev.* 8:1984–98

37. Singer MS, Gottschling DE. 1994. *Science* 266:404–9

38. McEachern MJ, Blackburn EH. 1995. *Nature* 376:403–9

39. Blasco M, Funk W, Villepounteau B, Greider CW. 1995. *Science* 269:1267–70

40. Feng J, Funk W, Wang S, Weinrich S, Avilion A, et al. 1995. *Science* 269: 1236–41

41. Melek M, Davis B, Shippen DE. 1994. *Mol. Cell Biol.* 14:7827–38

42. Greider CW, Blackburn EH. 1987. *Cell* 51:887–98

43. Collins K, Koybayashi R, Greider CW. 1995. *Cell* 81:677–86

44. Kim NW, Piatyszek MA, Prowse KR, Harley CB, West MD, et al. 1994. *Science* 266:2011–14

45. Blackburn EH, Greider CW, Henderson E, Lee M, Shampay J, Shippen-Lentz D. 1989. *Genome* 31:553–60

46. Szostak JW, Blackburn EH. 1982. *Cell* 29:245–55

47. Pluta AF, Dani GM, Spear BB, Zakian VA. 1984. *Proc. Natl. Acad. Sci. USA* 81:1475–79

48. Lee MS, Blackburn EH. 1993. *Mol. Cell Biol.* 13:6586–99

49. Morin GB. 1991. *Nature* 353:454–56

50. Harrington LA, Greider CW. 1991. *Nature* 353:451–54

51. Collins K, Greider CW. 1993. *Genes Dev.* 7:1364–76

52. Greider CW. 1995. See Ref. 7, pp. 35–68

53. Chamberlin MJ. 1993. *The Harvey Lectures* 88:1–21

54. Blackburn EH. 1993. In *The RNA World*, ed. RF Gestland, JF Atkins, pp. 557–76. Cold Spring Harbor, NY: Cold Spring Harbor Lab. Press

55. Autexier C, Greider CW. 1995. *Genes Dev.* 15:2227–39

56. Kassavetis GA, Geiduschek EP. 1993. *Science* 259:944–45

57. James BD, Olsen GJ, Liu JS, Pace NR. 1988. *Cell* 52:19–26

58. Romero DP, Blackburn EH. 1991. *Cell* 67:343–53

59. ten Dam E, Van Belkum A, Pleij K. 1991. *Nucleic Acids Res.* 19:6951

60. McCormick-Graham M, Romero DP. 1995. *Nucleic Acids Res.* 23:1091–97

61. Bhattacharyya A, Blackburn EH. 1994. *EMBO J.* 13:5521–31

62. Zaug AJ, Cech TR. 1995. *RNA* 1:363–74

63. Autexier C, Greider CW. 1994. *Genes Dev.* 8:563–75

64. Ares M. 1986. *Cell* 47:49–59

65. Igel AH, Ares M. 1988. *Nature* 334: 450–53

66. Gilley D, Lee M, Blackburn EH. 1995. *Genes Dev.* 9:2214–26

67. Harrington LA. 1993. *Purification and characterization of Tetrahymena telomerase.* PhD thesis. State Univ. NY at Stony Brook

68. Hennikoff S. 1991. *New Biol.* 3:1148–54

69. Lundblad V, Szostak JW. 1989. *Cell* 57:633–43

70. Lundblad V, Blackburn EH. 1993. *Cell* 73:347–60

71. Yu G-L, Blackburn EH. 1991. *Cell* 67: 823–32

72. Wang S-S. 1990. *Mol. Cell. Biol.* 10: 4415–19

73. Wellinger RJ, Wolf AJ, Zakian VA. 1993. *Cell* 72:51–60

74. Zahler AM, Prescott DM. 1989. *Nucleic Acids Res.* 17:6299–317

75. Vermeesch JR, Price CM. 1994. *Mol. Cell Biol.* 14:554–66

76. Campbell JL, Newlon CS. 1991. In *The Molecular and Cellular Biology of the Yeast Saccharomyces*, ed. JR Broach, JRPringle, EW Jones, pp. 41–146. Cold Spring Harbor, NY: Cold Spring Harbor Lab. Press

77. Carson M, Hartwell L. 1985. *Cell* 42: 249–57

78. Tsurimoto T, Stillman B. 1989. *EMBO J.* 8:3883–89

79. Greenwell P, Kronmal SL, Porter SE, Gassenhuber J, Obermaier B, Petes TD. 1995. *Cell.* 82:823–29

80. Fang G, Cech TR. 1995. See Ref. 7, pp. 69–105

81. Lejnine S, Makarov VL, Langmore JP. 1995. *Proc. Natl. Acad. Sci. USA* 92: 2393–97

82. Tommerup H, Dousmanis A, de Lange T. 1994. *Mol. Cell. Biol.* 14:5777–85

83. de Lange T. 1996. *Semin. Cell Biol.* 7:23–29

84. Gottschling DE, Cech TR. 1984. *Cell* 38:501–10

85. Gottschling DE, Zakian VA. 1986. *Cell* 47:195–205

86. Price CM, Cech TR. 1987. *Genes Dev.* 1:783–93

87. Gray JT, Celander DW, Price CM, Cech TR. 1991. *Cell* 67:807–14

88. Hicke BJ, Celander DW, MacDonald GH, Price CM, Cech TR. 1990. *Proc. Natl. Acad. Sci. USA* 87:1481–85
89. Fang GW, Cech TR. 1993. *Proc. Natl. Acad. Sci. USA* 90:6056–60
90. Hicke BJ, Willis MC, Koch TH, Cech TR. 1994. *Biochem.* 33:3364–73
91. Hicke BJ, Rempel R, Mahler J, Swank RA, Hamaguchi JR, et al. 1995. *Nucleic Acids Res.* 23:1887–93
92. Price CM. 1990. *Mol. Cell. Biol.* 10:3421–31
93. Wang W, Skopp R, Scofield M, Price C. 1992. *Nucleic Acids Res.* 20:6621–29
94. Huet J, Cottrelle P, Cool M, Vignais M-L, Thiele D, et al. 1985. *EMBO J.* 4:3648–49
95. Shore D, Naysmyth K. 1987. *Cell* 51:721–32
96. Shore D. 1994. *Trends Genet.* 10:408–12
97. Henry YA, Chambers A, Tsang JSH, Kingsman AJ, Kingsman SM. 1990. *Nucleic Acids Res.* 18:2617–23
98. Hardy CFJ, Balderes D, Shore D. 1992. *Mol. Cell. Biol.* 12:1209–17
99. Sussel L, Shore D. 1991. *Proc. Natl. Acad. Sci. USA* 88:7749–53
100. Kyrion G, Boakye K, Lustig A. 1993. *Mol. Cell. Biol.* 12:5159–73
101. Hardy CFJ, Sussel L, Shore D. 1992. *Genes Dev.* 6:801–14
102. Cockell M, Palladino F, Laroche T, Kyrion G, Liu C, et al. 1995. *J. Cell Biol.* 129:909–24
103. Liu C, Mao X, Lustig AJ. 1994. *Genetics* 138:1025–40
104. Conrad MN, Wright JH, Wolf AJ, Zakian VA. 1990. *Cell* 63:739–50
105. Buchman AR, Lue NF, Kornberg RD. 1988. *Mol. Cell. Biol.* 8:5086–99
106. Buchman AR, Kimmerly WJ, Rine J, Kornberg RD. 1988. *Mol. Cell. Biol.* 8:210–25
107. Graham IR, Chambers A. 1994. *Nucleic Acids Res.* 22:124–30
108. Gilson E, Roberge M, Giralado R, Rhodes D, Gasser SM. 1993. *J. Mol. Biol.* 231:293–310
109. Klein F, Laroche T, Cardenas M, Hofmann J, Schweizer D,Gasser S. 1992. *J. Cell Biol.* 117:935–48
110. Lustig AJ, Kurtz S, Shore D. 1990. *Science* 250:549–52
111. Van der Ploeg LTH, Liu AYC, Borst P. 1984. *Cell* 36:459–68
112. Bernards A, Michels PAM, Lincke CR, Borst P. 1983. *Nature* 303:592–97
113. Aparicio OM, Billington BL, Gottschling DE. 1991. *Cell* 66:1279–87
114. Moretti P, Freeman K, Coodly L, Shore D. 1994. *Genes Dev.* 2257–69
115. Roth M, Prescott DM. 1985. *Cell* 41:411–17
116. Vermeesch JR, Williams D, Price CM. 1993. *Nucleic Acids Res.* 21:5366–71
117. Klobutcher LA, Swanton MT, Donini P, Prescott DM. 1981. *Proc. Natl. Acad. Sci. USA* 78:3015–19
118. Larson DD, Spangler EA, Blackburn EH. 1987. *Cell* 50:477–83
119. Greider CW. 1991. *Mol. Cell. Biol.* 11:4572–80
120. Henderson E, Larson D, Melton W, Shampay J, Spangler E, et al. 1988. In *Cancer Cells,* ed. T Kelly, B Stillman, 6:453–61. Cold Spring Harbor, NY: Cold Spring Harbor Lab.
121. Liu Z, Gilbert W. 1994. *Cell* 77:1083–92
122. Liu Z, Lee A, Gilbert W. 1995. *Proc. Natl. Acad. Sci. USA* 92:6002–6
123. Walmsley RM, Petes TD. 1985. *Proc. Natl. Acad. Sci. USA* 82:506–10
124. Burr B, Burr F, Matz EC, Romero-Severson J. 1992. *Plant Cell* 4:953–60
125. Richards E. 1995. See Ref. 7, pp. 371–87
126. Kipling D, Cooke HJ. 1990. *Nature* 347:400–2
127. Starling JA, Maule J, Hastie ND, Allshire RC. 1990. *Nucleic Acids Res.* 18:6881–88
128. Prowse KR, Greider CW. 1995. *Proc. Natl. Acad. Sci. USA* 92:4818–22
129. Runge KW, Zakian VA. 1989. *Mol. Cell. Biol.* 9:1488–97
130. Shippen DE, Blackburn EH, Price CM. 1994. *Proc. Natl. Acad. Sci.* 91:405–9
131. Schulz VP, Zakian VA. 1994. *Cell* 76:145–55
132. Cooke HJ, Brown WRA, Rappold GA. 1985. *Nature* 317:687–92
133. Cooke HJ, Smith BA. 1986. *Cold Spring Harbor Symp. Quant. Biol.* 51:213–19
134. Moyzis RK, Buckingham JM, Cram LS, Dani M, Deaven LL, et al. 1988. *Proc. Natl. Acad. Sci. USA* 85:6622–26
135. de Lange T, Shiue L, Myers R, Cox DR, Naylor SL, et al. 1990. *Mol. Cell. Biol.* 10:518–27
136. Brown WRA. 1989. *Nature* 338:774–76
137. Cross SH, Allshire RC, McKay SJ, McGill NI, Cooke HJ. 1989. *Nature* 338:771–74
138. Cheng J-F, Smith CL, Cantor CR. 1989. *Nucleic Acids Res.* 17:6109–27
139. Hastie ND, Dempster M, Dunlop MG, Thompson AM, Green DK, Allshire RC. 1990. *Nature* 346:866–68
140. Harley CB, Futcher AB, Greider CW. 1990. *Nature* 345:458–60
141. Hayflick L. 1965. *Exp. Cell Res.* 37:614–36

142. Hayflick L, Moorhead PS. 1961. *Exp. Cell Res.* 25:585–621

143. Vaziri H, Schaechter F, Uchida I, Wei L, Zhu XM, et al. 1993. *Am. J. Hum. Genet.* 52:661–67

144. Shay JW, Wright WE, Brasiskyte D, Van der Hagen BA. 1993. *Oncogene* 8:1407–13

145. Klingelhutz AJ, Barber S, Smith PP, Dyer K, McDougall JK. 1994. *Mol. Cell Biol.* 14:961–69

146. Vaziri H, Dragowska W, Allsopp RC, Thomas TE, Harley CB, Lansdorp PM. 1994. *Proc. Natl. Acad. Sci. USA* 91:9857–60

147. Harley CB. 1995. See Ref. 7, pp. 247–63

148. Allsopp RC, Vaziri H, Patterson C, Goldstein S, Younglai EV, et al. 1992. *Proc. Natl. Acad. Sci. USA* 89:10114–18

149. Harley CB. 1991. *Mutat. Res.* 256:271–82

150. Harley CB, Vaziri H, Counter CM, Allsopp RC. 1992. *Exp Geront.* 27:375–82

151. Hartwell LH, Weinert TA. 1989. *Science* 246:629–34

152. Goldstein S. 1990. *Science* 249:1129–33

153. Kuerbitz SJ, Plunkett BS, Walsh WV, Kastan MB. 1992. *Proc. Natl. Acad. Sci. USA* 89:7491–95

154. Lane DP. 1992. *Nature* 358:15–16

155. Greider CW. 1990. *BioEssays* 12:363–69

156. Noda A, Ning Y, Venable SF, Pereira-Smith OM, Smith JR. 1994. *Exp. Cell Res.* 211:90–98

157. el-Deiry WS, Tokino T, Velculescu VE, Levy DB, Parsons R, et al. 1993. *Cell* 75:817–25

158. Wright W, Shay JW. 1992. *Trends Genet.* 8:193–97

159. Cooke HJ. 1995. See Ref. 7, pp. 219–45

160. Vojta PJ, Barrett JC. 1995. *Biochim. Biophys. Acta* 1242:29–41

161. Walton J. 1982. *Mech. Aging Dev.* 19:217–44

162. Stanulis-Praeger BM. 1987. *Mech. Aging Dev.* 38:1–48

163. Kipling D, Ackford HE, Taylor TA, Cooke HJ. 1991. *Genomics* 11:235–341

164. Shay JW, Wright WE. 1989. *Exp. Cell Res.* 184:109–18

165. Counter CM, Botelho FM, Wang P, Harley CB, Bacchetti S. 1994. *J. Virol.* 68:3410–14

166. Counter CM, Avilion AA, LeFeuvre CE, Stewart NG, Greider CW, et al. 1992. *EMBO J.* 11:1921–29

167. Bryan TR, Englezou A, Gupta J, Bacchetti S, Reddel R. 1995. *EMBO J.* 14:4240–48

168. Murnane JP, Sabatier L, Marder BA, Morgan WF. 1994. *EMBO J.* 13:4953–62

169. Norwood TH, Pendergrass WR, Sprague CA, Martin GM. 1974. *Proc. Natl. Acad. Sci. USA* 71:2231–35

170. Pereira-Smith OM, Smith JR. 1983. *Science* 221:965–66

171. Mehle C, Ljungberg B, Roos G. 1994. *Can. Res.* 54:236–41

172. Bacchetti S, Counter CM. 1995. *Int. J. Oncol.* 7:423–32

173. Harley CB, Kim NW, Prowse KR, Weinrich SL, Hirsch KS, et al. 1994. *Cold Spring Harbor Symp. Quant. Biol.* 59:307–15

174. Counter GM, Hirte HW, Bacchetti S, Harley CB. 1994. *Proc. Natl. Acad. Sci. USA* 91:2900–4

175. Greider CW. 1994. *Curr. Opin. Genet. Dev.* 4:203–11

176. Counter CM, Gupta J, Harley CB, Leber B, Bacchetti S. 1995. *Blood* 85:2315–20

177. Broccoli D, Young JW, de Lange T. 1995. *Proc. Natl. Acad. Sci. USA* 92:9082–86

178. Tahara H, Nakanishi T, Kitamoto M, Nakashio R, Shay JW, et al. 1995. *Cancer Res.* 55:2734–36

179. Hiyama K, Hirai Y, Kyoizumi S, Akiyama M, Hiyama E, et al. 1995. *J. Immunol.* 155:3711–15

180. Macieira-Coelho A, Azzarone B. 1988. *Anticancer Res.* 8:669–76

181. Chadeneau C, Siegel P, Harley CB, Muller W, Bacchetti S. 1995. *Oncogene* 11:893–98

182. Wright WE, Piatyszek MA, Rainey WE, Bryd W, Shay JW. 1996. *Dev. Genet.* 18:In press

183. Buck SW, Shore S. 1995. *Genes Dev.* 9:370–84

Annu. Rev. Biochem. 1996. 65:367–409

THE STRUCTURE AND FUNCTION OF PROTEINS INVOLVED IN MAMMALIAN PRE-mRNA SPLICING

Angela Krämer

Département de Biologie Cellulaire, Université de Genève, CH-1211 Genève 4, Switzerland

KEY WORDS: spliceosome, splicing factor, snRNA, snRNP

ABSTRACT

Intervening sequences are removed from nuclear pre-mRNAs in a well-defined multi-step pathway. Small nuclear ribonucleoprotein particles (snRNPs) and numerous protein factors are essential for the formation of the active spliceosome in which intron excision proceeds in two successive transesterification reactions. Important elements for catalysis are the RNA moieties of the snRNPs that align the pre-mRNA splice sites in the active center of the spliceosome. Although pre-mRNA splicing is almost certainly RNA-mediated, both snRNA-associated proteins and non-snRNP splicing factors participate in each step of the splicing reaction. Splicing proteins exert auxiliary functions in the recognition, selection, and juxtaposition of the splice sites and drive conformational changes during spliceosome assembly and catalysis. Many splicing factors have been isolated in recent years and corresponding cDNAs have been cloned. This review summarizes the structure and function of mammalian proteins which are essential components of the constitutive splicing machinery.

CONTENTS

INTRODUCTION 368
THE SPLICING REACTION 369
    Catalysis 369
    Spliceosome Assembly 371
THE SR FAMILY OF SPLICING PROTEINS 377
    Structural Organization 377
    Functions. 380
    Transcriptional and Post-Transcriptional Regulation. 384
    Regulation by Reversible Phosphorylation 385

367

0066-4154/96/0701-0367$08.00

POLYPYRIMIDINE TRACT–BINDING PROTEINS.   386
    U2 snRNP Auxiliary Factor (U2AF).   386
    Polypyrimidine Tract–Binding Protein (PTB).   388
    PTB-Associated Splicing Factor (PSF).   389
BRANCH SITE–BINDING PROTEINS.   390
hnRNP PROTEINS   392
snRNP-ASSOCIATED SPLICING ACTIVITIES   392
    The 17S U2 snRNP.   392
    The 20S U5 snRNP.   396
    The 25S U4/U6.U5 tri-snRNP   397
ATP-DEPENDENT RNA HELICASES   398
OTHER ACTIVITIES   399
    Protein Factors Required During Spliceosome Formation.   399
    Protein Factors Required for Catalysis.   400
    Cap-Binding Proteins.   401
    Nuclear Matrix–Associated Proteins.   402
CONCLUSION.   402

INTRODUCTION

The protein-coding regions in most primary transcripts of RNA polymerase II are interrupted by intervening sequences (introns) that are eliminated from the pre-mRNA in a process termed splicing (1). This reaction proceeds in the nucleus and can occur co- or post-transcriptionally (2–5). The coding sequences (exons) are joined to generate the mature mRNA that is exported to the cytoplasm and translated into protein. In higher eukaryotes the number of introns in a given pre-mRNA can vary from 0 to more than 50. Whereas exons are in general relatively short (10–400 nucleotides), intron sizes can extend up to more than 200,000 nucleotides. Despite their size, introns usually have little protein-coding potential, and their retention within a mRNA can introduce in-frame stop codons generating prematurely terminated proteins upon translation. The presence of more than one intron in a pre-mRNA, however, provides a means for the regulation of gene expression by alternative splicing events that can result in an on-off switch for a particular gene or translation of different protein isoforms from one and the same pre-mRNA. Alternative splicing is achieved in many ways and influenced by *cis*-acting elements as well as *trans*-acting factors (see 6, 7 for review).

Sequences that are essential for intron removal are limited to the exon/intron borders (8). The 5′ splice site in higher eukaryotes conforms to the consensus sequence AG|GURAGU (the splice site is denoted by a vertical bar and invariant nucleotides are underlined; R=purine, Y=pyrimidine, N=any nucleotide). The 3′ splice site is characterized by the sequence YAG| and is preceded by a stretch of pyrimidine residues in most vertebrate introns. Another sequence element, the branch site, is usually located at a distance of 18 to 40 nucleotides upstream of the 3′ splice site and displays the degenerate sequence

YNCUR<u>A</u>Y (the site of branch formation is shown in bold). In the yeast *Saccharomyces cerevisiae* these elements are more strictly conserved and, in particular, the branch site sequence UACUAAC is invariant (9). The higher conservation of splice sites in *S. cerevisiae* probably reflects the fact that only a few yeast pre-mRNAs contain introns and, with two exceptions, only a single intron is present. A minor class of introns in vertebrates and invertebrates has splice site sequences that differ from other introns (10, 11). The sequences |AUAUCUU and CAC| represent highly conserved elements at the 5′ and 3′ splice sites, respectively, and the sequence UCCUUAAC, located 16–19 nucleotides upstream of the 3′ splice site, probably serves as the branch site. Splice sites in plants resemble those of vertebrate pre-mRNAs (12); introns are usually rich in uridines and adenosines, and polypyrimidine tracts or branch sites are not conserved.

The contrast between the relatively short conserved sequences at the splice junctions and the enormous variation in intron numbers and sizes in pre-mRNAs of higher eukaryotes raises the issue of how accuracy and specificity in splicing are achieved. Furthermore, cryptic splice sites that resemble the bonafide splice sites are present in both exons and introns, and their use has to be prevented. As discussed below, splice sites are recognized by small nuclear ribonucleoprotein particles (snRNPs) and non-snRNP protein factors that engage in an intricate network of interactions with the pre-mRNA and one another to assemble the active splicing complex, known as a spliceosome, which is the site of the biochemical reactions that result in intron removal. The multitude and dynamic nature of the contacts between spliceosomal components provide a mechanism by which the accurate and specific selection of splice sites is controlled at more than one step in the reaction, and these interactions may also serve to ensure the directionality of events.

Many aspects of pre-mRNA splicing have been the subject of recent reviews (9, 13–16). This review will concentrate on questions concerning the functions of protein factors in constitutive splicing in mammalian cells.

THE SPLICING REACTION

Catalysis

The development of cell-free systems that accurately process model pre-mRNAs has been crucial for unraveling the process of intron removal and the processes that lead to the assembly of the active spliceosome (17–20). In vitro studies performed with synthetic radioactively labeled pre-mRNA substrates that were incubated in HeLa nuclear extracts demonstrated that introns are excised from a pre-mRNA in two distinct steps (21, 22; Figure 1). The reaction is initiated by a nucleophilic attack of the 2′ hydroxyl group of the adenosine at the branch site on the 3′,5′-phosphodiester bond at the 5′ splice site. Con-

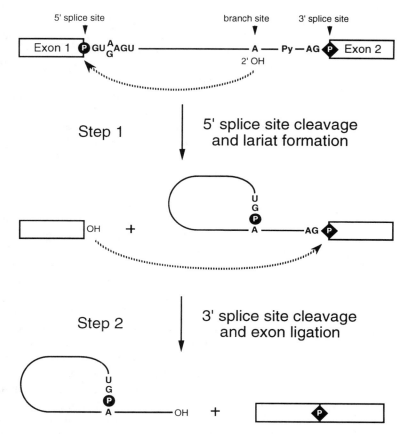

Figure 1   The catalytic steps of nuclear pre-mRNA splicing. Exons are shown as boxes, the intron as a line. Conserved nucleotides (in mammalian pre-mRNAs) and the phosphates at the splice sites are indicated. The dashed arrows represent the nucleophilic attack of the hydroxyl groups on the splice junctions.

comitant with cleavage, the 5' terminal guanosine of the intron is covalently attached to the branch site adenosine in a 2',5'-phosphodiester bond. The splicing intermediates thus formed are the cleaved-off 5' exon and the intron–3' exon intermediate in a branched circular, or lariat, configuration. The second step proceeds by attack of the 3' hydroxyl group of the 5' exon on the phosphodiester bond at the 3' splice site, resulting in the displacement of the lariat intron and the concurrent ligation of the exons. Incorporation of modified nucleotides at the splice junctions has shown that both steps of splicing represent in-line S_N2 nucleophilic displacement or transesterification reactions (23, 24).

The group II introns that are present in the genomes of organelles in lower eukaryotes, plants, and also bacteria are excised from the primary transcript by way of the same reaction intermediates and products (for review see 25). These introns can be spliced autocatalytically in vitro, i.e. the enzymatic activity is intrinsic to the intron itself, and the stereochemistry of the reaction is identical to that of nuclear pre-mRNA splicing (26). Given the fundamental similarities in the reaction parameters between nuclear pre-mRNA splicing and group II intron excision, an RNA-based mechanism for nuclear pre-mRNA splicing has been proposed, despite the finding that this process depends on *trans*-acting proteins as well as ATP (see 27, 28 for recent review and discussion). A requirement for *trans*-acting factors in nuclear pre-mRNA splicing can be explained by the paucity of conserved sequences that define the reactive sites in the substrate RNA. Whereas highly conserved sequence elements juxtapose splice sites and branch site by intramolecular base pairing interactions in group II introns, these functions are compensated by intermolecular interactions between the RNA substrate and the RNA moieties of snRNPs in nuclear pre-mRNA splicing. ATP appears to be required as an energy donor to drive conformational changes within the spliceosome.

Spliceosome Assembly

The catalytic steps are preceded by the assembly of the pre-mRNA into a splicing-competent structure. Fully assembled spliceosomes sediment with 50–60S, suggesting a complexity comparable to that of the ribosome (29–31). Major players in the alignment of the reactive sites and the formation of the catalytic core of the spliceosome are the U1, U2, U4/U6, and U5 small nuclear RNAs (snRNAs) which are associated with several proteins (reviewed in 32, 33). U4 and U6 snRNAs are extensively base-paired to one another and packaged into a single particle; other snRNPs contain only one RNA molecule. The primary sequences (and in some cases also the sizes) of the snRNAs vary between species, but secondary structures are highly conserved. U6 snRNA represents an exception; its primary sequence is very similar in human and yeast. The snRNAs are associated with eight proteins (Sm proteins) that are common to all snRNPs. These proteins bind to the conserved sequence RAU_{3-6} GR present in all snRNAs except for U6. The Sm proteins are recognized by autoantibodies from patients with autoimmune diseases. These autoantibodies were useful tools in the initial analysis of snRNP function. In addition, all spliceosomal snRNPs contain unique proteins, some of which play essential roles in splicing.

The first hint that snRNPs are involved in splicing was the finding that the

5' end of U1 snRNA is complementary to the conserved sequence at the 5' splice site (34, 35). With various techniques— antibody inhibition and depletion, oligonucleotide-targeting of snRNAs, immunoprecipitation of nuclease-treated spliceosomes, mutagenesis of snRNAs, genetic suppression experiments and specific photo cross-linking— functions for all the aforementioned snRNPs in splicing have been established (for review see 13, 14, 32, 36). The following section summarizes the interactions of snRNPs with the pre-mRNA and with one another during spliceosome assembly and catalysis. The function of protein factors in these reactions will be covered in the following chapters.

Upon incubation of a pre-mRNA substrate in a splicing extract, the RNA is packaged into the so-called H-complex which shows a heterogeneous distribution in native polyacrylamide gels (37). The formation of this complex is independent of ATP, temperature, and functional splice sites. Isolated H-complexes cannot be chased into spliced products when incubated in extracts in the presence of an excess of competitor RNA (38), suggesting that they are not specific to the splicing reaction.

The first specific stage in spliceosome assembly is the binding of U1 snRNP to the pre-mRNA, which occurs in an ATP- and temperature-independent fashion (39, 40) and involves base pairing of the 5' end of U1 snRNA to conserved sequences at the 5' splice site (41; Figures 2 and 3a). U1 snRNP binding is essential for the subsequent binding of U2 snRNP and for the commitment of the pre-mRNA to the splicing pathway. Complexes containing U1, but no other snRNPs, were first described in the yeast system, in which two commitment complexes (CC) can be resolved in native polyacrylamide gels (42–45). Formation of CC1 requires a functional 5' splice site and the 5' end of U1 snRNA, whereas CC2 formation also depends on the branch site (45). An ATP-independent complex (complex E) in the mammalian system that shares many of its characteristics with the yeast commitment complexes has been isolated by gel filtration (38) and is probably identical to a complex identified in a template commitment assay (46). In contrast to yeast, the binding of U1 snRNA to the 5' splice site is not absolutely necessary for the initial U1 snRNP/pre-mRNA interaction in the mammalian system (47), but the formation of complex E occurs more efficiently in the presence of intact 5' and 3' splice sites (46, 48).

Pre-splicing complex A is formed upon binding of U2 snRNP to the branch site (49–51; Figure 2). This step as well as all subsequent stages of the splicing reaction are ATP-dependent (19, 30, 52–53a). The assembly of complex A in mammalian cells requires the polypyrimidine tract and a functional 3' splice site, but the 5' splice site is dispensable (30, 54). In yeast, both the 5' splice site and branch site are essential for pre-spliceosome assembly, and nucleotides downstream of the branch site are not required until the second catalytic step (55). Base pairing between an internal region of U2 snRNA and the branch

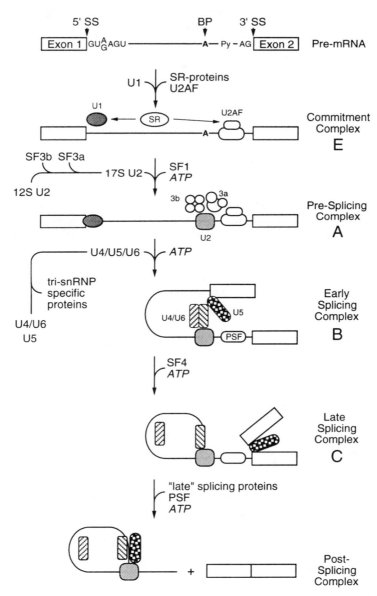

Figure 2   The spliceosome assembly pathway. SnRNPs are represented by filled symbols, splicing proteins by open symbols.

site results in the formation of a short helix from which the branch point adenosine is bulged out (56–59; Figure 3*a*). This interaction is thought to specify the adenosine as the nucleophile for 5′ splice site cleavage and lariat formation (59).

The U4/U6 and U5 snRNPs interact with complex A in the form of a U4/U6.U5 tri-snRNP complex to generate splicing complex B (49–51), which is converted into the catalytically active complex C after a conformational rearrangement (Figure 2; see below). One of the consequences of this structural change is the destabilization of the extensive base pairing interaction between U4 and U6 snRNAs (Figure 3). Although U4 snRNA appears to remain associated with the spliceosome (60), its presence is not essential for the subsequent catalytic steps (61).

The transesterification reactions are initiated during or subsequent to the structural reorganization of the spliceosome. The liberated intron remains associated with U2, U5, and U6 snRNAs, whereas the spliced mRNA is released from the spliceosome as a low molecular weight complex (50, 51). Spliceosome disassembly represents an active process that requires at least one protein factor and ATP (53, 62). Subsequent to the release of the snRNPs, the intron lariat is linearized by a debranching enzyme that specifically cleaves the 2′,5′-phosphodiester bond at the branch site followed by degradation of the intron (63).

In addition to the association of U1 and U2 snRNAs with the pre-mRNA, several new base-pairing interactions are established during spliceosome assembly and catalysis, whereas others are resolved (Figure 3). These interactions can be summarized as follows (see 14, 15, 63a for in-depth reviews): U2 snRNA, bound to the branch site, engages in multiple base-pairing interactions with U6 snRNA (64–68). The formation of U2-U6 helix I (Figure 3C, interaction 5) involves nucleotides just upstream of the U2/branch site helix and immediately downstream of the phylogenetically invariant sequence ACA-GAGA in U6 snRNA (67, 68) which is crucial for both catalytic steps (69–71). In addition, the penultimate guanosine in the invariant U6 sequence interacts with an adenosine residue of U2 snRNA that is located in the bulge of helix I (71a; Figure 3C). Nucleotides in the invariant sequence in U6 snRNA come into close contact with the conserved 5′ splice site sequence (72–74), and a uridine residue of U2 snRNA in the second part of helix I makes specific contacts with the first nucleotide of exon 2 (74a; Figure 3C). In combination, these interactions can serve to juxtapose the reactive sites of the pre-mRNA for the first and second catalytic steps. These results furthermore imply that U2 and U6 snRNAs (as well as U5 snRNA; see below) are located close to, or within, the catalytic core of the spliceosome.

A direct role of U6 snRNA in the catalysis of the splicing reaction is supported by additional findings. First, U6 snRNA is by far the most conserved

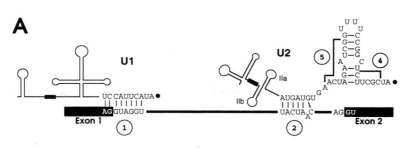

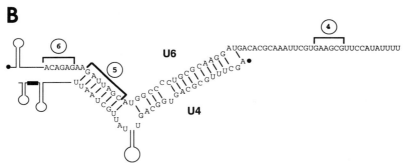

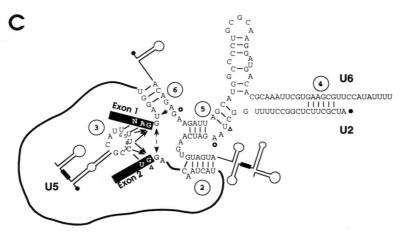

Figure 3 Interactions between spliceosomal RNAs. (*a*) Base pairing of U1 and U2 snRNAs with the pre-mRNA 5′ splice site and the branch site, respectively. (*b*) Base pairing between the U4 and U6 snRNAs. (*c*) Interactions between pre-mRNA, U2, U5, and U6 snRNAs. Relevant sequences of human pre-mRNA and snRNAs are given in letters; other regions are shown as line drawings. Brackets and circled numbers indicate regions of specific interactions. Base pairing interactions and contacts established by genetic suppression experiments are indicated by dashes or open circles; contacts demonstrated by site-specific cross-linking are shown by arrows or triangles. Please note that some of these interactions have been demonstrated in yeast. The black dots indicate the 5′ terminal cap structures of the snRNAs; the square boxes show the Sm-binding site. Modified from (13).

of all spliceosomal RNAs, both in primary sequence and in length (75). Second, several fungal U6 snRNA genes contain pre-mRNA-type introns in the vicinity of the ACAGAGA sequence, which are thought to have been inserted into the snRNA through errors during the splicing reaction (76, 77). Third, mutations of nematode U6 snRNA sequences upstream of the invariant sequence lead to an aberrant attack of the pre-mRNA branch site on U6 snRNA instead of the 5' splice site (78).

Other interactions crucial for the formation of the active spliceosome and accurate splice-site selection are contacts of a conserved loop in U5 snRNA and exon sequences at the 5' and 3' splice sites (74, 74a, 79, 80; Figure 3). On the one hand, this interaction could align the splice sites in the active site for the second catalytic step. On the other hand, the 5' exon must be held within the spliceosome after cleavage at the 5' splice site and lariat formation, and U5 snRNA could tether the liberated 5' exon to the intron–3' exon lariat. During 3'–splice site cleavage and exon ligation these interactions appear to be disrupted, and the U5 snRNA base-pairs with nucleotides downstream of the 5' splice site in the excised intron lariat (72).

An intramolecular non–Watson Crick interaction between the highly conserved guanosine residues at the 5' and 3' termini of the intron is essential for the second catalytic step and for the specification of the 3' cleavage site (see 81 and references therein; Figure 3). Reciprocal suppression experiments have shown that the guanosine residues can only be replaced by adenosine and cytidine at the 5' and 3' splice site, respectively. Interestingly, adenosine and cytidine are also the conserved terminal nucleotides at the splice junctions of the minor class of introns mentioned above (11).

In summary, contacts of U2, U5, and U6 snRNAs with the pre-mRNA and with one another are crucial elements of the catalytic core of the spliceosome. U1 snRNP does not appear to serve an essential function during catalysis (81a), but it is important for the initial recognition of the pre-mRNA and commitment to the splicing pathway. Its interaction with the 5' splice site is probably replaced by the binding of U6 snRNA (82). Similarly, U4 snRNA is not required for catalysis (61), and by base-pairing to U6 snRNA it could sequester the sequences that participate in the formation of the active site, thus acting as a regulatory element of U6 snRNA function (77). The base-pairing interactions within the spliceosome generate structures comparable to conserved and essential domains in group I and group II introns or hairpin ribozymes, supporting the proposal that the catalysis of nuclear pre-mRNA splicing is RNA-mediated (see 27).

Given the importance of RNA-RNA interactions during spliceosome assembly and intron removal, proteins that are essential for splicing are unlikely to participate directly in the catalysis of the reaction, but they serve important auxiliary functions. Protein factors have been most extensively studied in mammalian and yeast cells. By combination of genetic and biochemical approaches more than 30 proteins with functions at every step of the reaction have been identified in *S. cerevisiae* (see 9, 16 for review). The mammalian splicing apparatus has a comparable complexity, as documented by the work of Reed and colleagues who have analyzed components present in isolated spliceosomes (see 83 and references therein). The analysis of mammalian splicing proteins is based exclusively on biochemical methods, and most of the non-snRNP proteins identified to date function during the early phases of spliceosome assembly.

THE SR FAMILY OF SPLICING PROTEINS

The SR proteins represent a family of splicing factors that are remarkably conserved in vertebrates and invertebrates and have also been found in plants (84–85a). Members of the SR family in mammalian cells have been identified and isolated through several approaches. For example, alternative splicing factor/splicing factor 2 (ASF/SF2) has been purified as an essential component in constitutive splicing (86) as well as a factor that influences the use of different 5′ splice sites within the same pre-mRNA (87, 88). The spliceosome component of 35 kDa (SC35) and 9G8 have been identified by monoclonal antibodies raised against isolated spliceosomes or hnRNP complexes (89, 90), and PR264, which is identical to SC35, was found as a protein that is encoded on the antisense strand of the chicken and human *c-myb* ET exons (91). Moreover, proteins of approximately 20, 30, 40, 55, and 75 kDa were detected in cells and tissues of various invertebrate and vertebrate species by a monoclonal antibody (mAb104) that stains active sites of RNA polymerase II transcription (92). The mAb104 antigens were termed SR proteins because they contain a domain rich in arginine and serine residues (92). Individual proteins of this class are referred to as SRp followed by a number designating their apparent molecular mass (Table 1).

Structural Organization

Amino acid sequencing and cDNA cloning of the SR proteins have revealed a common structural organization (for references see Table 1). The N-terminal part of the SR proteins consists of an RNA recognition motif (RRM, also known as RNA-binding domain or RNP consensus sequence) that is characteristic of many RNA-binding proteins including snRNP and hnRNP proteins (84, 93). The RRM is organized into four antiparallel β sheets and two α helices which are arranged in the order $\beta1$-$\alpha1$-$\beta2$-$\beta3$-$\alpha2$-$\beta4$. The RNP-1 octamer and the RNP-2 hexamer, two highly conserved submotifs that are essential for RNA binding, are juxtaposed on the centrally positioned antiparallel

Table 1 Mammalian splicing activities

	Size (kDa)[a]	Structural motifs[b]	Function	Interaction with	Homologues/ related proteins[c]	References
SR proteins			E-complex, interactions across introns and exons, alternative splicing	U1-70K U2AF[35]		
SRp20	19.3	RRM, SR			X16 (Mm)	92, 109, 157
ASF/SF2	27.7	RRM, Ψ-RRM, SR	= SRp30a		RBP1 (Dm) ASF (Mm) –SR1 (At)	85, 155, 159, 327
SC35	25.6	RRM, SR	= SRp30b, PR264		PR264 (Gg)	91, 158
SRp30c	25.5	RRM, Ψ-RRM, SR				95
SRp40	31.2	RRM, Ψ-RRM, SR			HRS (Rn)	95, 160
SRp55	39.6	RRM, Ψ-RRM, SR			SRp55 (Dm)	95, 216, 328
SRp75	56.8	RRM, Ψ-RRM, SR			B52 (Dm)	94
9G8	27.4	RRM, CCHC, SR				90
Polypyrimidine-tract binding proteins						
U2AF[65]	53	SR, 2 RRM, Ψ-RRM	E-complex, 3′ splice site selection	U2AF[35]	U2AF[50] (Dm) U2AF[59] (Sp)	99, 182, 183
U2AF[35]	34	SR		U2AF[65]	U2AF[38] (Dm)	100[d]
PSF	76	RGG, pro/gln, 2 RRM	second catalytic step, = SAPs102/68			188
PTB	57–59	4Ψ-RRM	alternative splicing, = hnRNP I	SR proteins		191–193
SnRNP-associated splicing activities						
SF3a[60]	59	C2H2	A-complex, associated with 17S U2 snRNP, = SAP61	SF3a[120]	PRP9 (Sc)	241, 242, 245
SF3a[66]	49	C2H2, heptad repeats	= SAP62	SF3a[120]	PRP11 (Sc)	240, 244
SF3a[120]	89	2 SURP, pro, ubil	= SAP114	SF3a[60] SF3a[66]	PRP21 (Sc) PRP21 (Ce) SWAP proteins (Hs, Mm, Dm, Ce)	101, 243,
SF3b[50]	44	2RRM, pro/gly	A-complex, associated with 17S U2 snRNP, = SAP49	SAP145	HSH49 (Sc)	239, 251a

Table 1 (continued)

	Size (kDa)[a]	Structural motifs[b]	Function	Interaction with	Homologues/ related proteins[c]	References
SF3b[130] SF3b[145] SF3b[155]	97.6	pro, gn	= SAP130 = SAP145 = SAP155	SAP49	CUS1 (Sc)	239a
p220			Pre-mRNA binding, U5 snRNP protein			266, 267
HSLF			Functionally equivalent to tri-snRNP-specific proteins		Prp8 (Sc)	269
Other splicing activities						
CBP80	92.8		Cap-binding complex, E-complex assembly, nucleocytoplasmic export of snRNAs			321, 322
CBP20	18	RRM				
SF1	68	KH, CCHC, pro	A-complex, heat-stable, isoforms with different C-termini			313 e
SF4			C-complex/1st catalytic step			
88 kDa			A-complex			305
HRF late activity			A-complex, heat-resistant second catalytic step			307
HRH1	139.3	SR, DEAH	Putative ATP-dependent RNA helicase, possibly involved in release of mRNA from spliceosome	SR proteins	Prp22 (Sc)	311
SRPK1	74.3	S/T-kinase	Specifically phosphorylates serines in SR domains of splicing proteins			291 168

[a]Size predicted from conceptual translation.

[b]The structural motifs are listed according to their order in protein. Abbreviations used: RRM: consensus RNA recognition motif; Ψ-RRM: noncanonical RRM; RGG: arginine/glycine/glycine motif; C_2H_2: zinc finger; CCHC: zinc knuckle; DEAH: motif characteristic for ATP-dependent RNA helicases; SURP: motif characteristic for PRP21 and SWAP homologs; ubi: ubiquitin-like domain; KH: hnRNP K homology; pro: proline; gln: glutamine; gly: glycine; S/T-kinase: serine/threonine kinase catalytic domains.

[c]Abbreviations used: At: *Arabidopsis*; Ce: *C. elegans*; Dm: *Drosophila*; Gg: chicken; Mm: mouse; Rn: rat; Sc: *S. cerevisiae*; Sp: *S. pombe*.

[d]R Rudner, K Breger, R Kanaar & D Rio, personal communication.

[e]S Arning & A Krämer, in preparation.

β3 and β1 strands, respectively. Except for SRp20, SC35, and 9G8, all SR proteins contain a second domain with limited but significant homology to the RRM (Ψ-RRM) with the characteristic sequence SWQDLKD (84). RRM and Ψ-RRM are usually separated by a glycine-rich "hinge." This domain is lacking in 9G8 which contains a zinc knuckle of the CCHC type instead (90).

The C-terminal part of the SR proteins harbors a domain enriched in arginine (R) and serine (S) residues (SR domain), many of which are present as RS or SR dipeptides. The variable length of this domain contributes to the size differences observed among individual SR proteins (for examples see 94, 95). The SR domain is highly phosphorylated (94–96), and protein kinases that phosphorylate SR proteins have been identified (see below). Sequences enriched in alternating arginine and serine residues are also found in other proteins involved in splicing, such as the U1 70 kDa protein (97, 98) and both subunits of U2AF (99, 100; see below), and in the *Drosophila* Tra, Tra-2, and SWAP proteins that regulate alternative splicing events (101–104). These proteins differ in their overall structural organization from the SR proteins and have been classified as SR protein-related polypeptides (104a).

A subset of the SR proteins (SRp20, SRp40, SRp55, and SRp75, but not ASF/SF2 and SC35) share another structural feature that is recognized by a monoclonal antibody (105). The epitope appears to consist of arginines that alternate with glutamate or aspartate residues and may be defined by alternating positive and negative charges. The antibody reacts with about 15–20 additional nuclear proteins, including the 35- and 65-kDa subunits of U2AF and the U1 70-kDa protein. Consistent with its binding to splicing proteins, the antibody inhibits splicing in vitro; however, information about the function of the epitope is not available.

Functions

A simple procedure for the isolation of most SR proteins involves two successive precipitations, the first with ammonium sulfate at a saturation of 60–90%, the second with 20 mM $MgCl_2$ (92). The "Mg-pellet" is highly enriched in different SR proteins; they can be separated by SDS polyacrylamide gel electrophoresis and isolated from gels to be used for functional studies. The function of SR proteins in constitutive splicing can conveniently be tested in cytoplasmic S100 extracts (86), which contain all components necessary for splicing but only limited concentrations of SR proteins (94, 106) and are thus inactive in splicing. With this system an essential function for ASF/SF2 and other SR proteins in constitutive splicing has been demonstrated (86, 92, 107–109), and additional approaches have provided a detailed picture of their role at the onset of spliceosome assembly.

In the presence of an excess of competitor pre-mRNA, a splicing substrate that was preincubated with isolated SR proteins is spliced more efficiently than naked RNA, suggesting that SR proteins are among the first components that interact with the pre-mRNA and have the ability to commit the pre-mRNA to the splicing pathway (110). Consistent with this view, SR proteins are incorporated into spliceosomes at the time of E-complex assembly and stimulate the formation of this complex when added to a nuclear extract (111). SR proteins are thought to enhance the binding of U1 snRNP to 5′ splice sites (112–115) and, when present at high concentrations, they can abrogate the need for U1 snRNP in splicing (116–119), again suggesting an early and essential role in spliceosome assembly and splice-site selection. However, the functions of SR proteins are not limited to the formation of the commitment complex, but they may play a more general role in escorting snRNPs to the assembling spliceosome: SR proteins can recruit U2 snRNP to the branch site of a RNA substrate that lacks a 5′ splice site (119), and they are also required for the transition of complex A into complex B (53a), a step that involves the incorporation of the U4/U6.U5 snRNP into the spliceosome.

ASF/SF2 engages in a stable ternary complex with U1 snRNP and the pre-mRNA substrate, the formation of which is dependent on a functional 5′ splice site (113, 114). The recruitment of U1 snRNP to the 5′ splice site is most likely mediated by a direct interaction between SR proteins, which may be bound to the splicing substrate (111, 120), and the integral U1 snRNP 70-kDa protein (113, 121; Figure 4a). The interaction between ASF/SF2 and U1-70K can occur in the absence of pre-mRNA and requires the SR domains of both proteins (113), whereas either one of the two RRMs of ASF/SF2 is necessary in addition for the formation of the ternary complex with U1 snRNP and the pre-mRNA (114). With the yeast two-hybrid system (122), simultaneous contacts of ASF/SF2 or SC35 with U1-70K and U2AF[35], which is bound to the polypyrimidine tract via its intimate association with U2AF[65] (see below), have been demonstrated (121). These interactions are thought to bridge 5′ and 3′ splice sites (Figure 4a), consistent with observations that the first functional association between 5′ and 3′ splice sites occurs during the formation of complex E (48) and that SC35 is involved in these contacts (123).

SR proteins also engage in interactions across exons (124). Robberson et al (125) defined exons, but not introns, as the primary units within vertebrate pre-mRNAs that are initially recognized by trans-acting factors. According to the model, U1 snRNP binds to the 5′ splice site located at the downstream side of the exon and directs the binding of splicing factors to the 3′ splice site located upstream, followed by interactions across the upstream intron and intron removal (Figure 4b). In line with this proposal, a downstream 5′ splice site stimulates splicing of an upstream intron (125–128). Moreover, the binding of U1 snRNP is required for the function of the downstream 5′ splice site (125, 127, 129) and promotes the association of U2AF[65] with the upstream 3′ splice

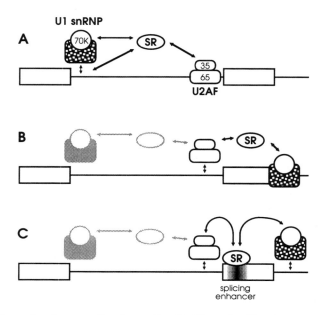

Figure 4 Interactions between splice sites mediated by SR proteins. The exons are represented by boxes, the intron as a line, and the splicing enhancer by shading. U1 snRNP is shown as a filled symbol; the SR proteins and U2AF are indicated by open symbols. (*a*) Interactions of SR proteins that bridge 5′ and 3′ splice sites across an intron. (*b*) Interactions of SR proteins across an exon. (*c*) Interactions of SR proteins with splicing enhancers and components that interact with downstream 3′ and upstream 5′ splice sites.

site (129). Furthermore, so-called splicing enhancers have been identified in several natural pre-mRNAs that are often subject to stage- or tissue-specific alternative splicing (124, 130–137, and references therein). The splicing enhancers are usually purine-rich and are found in exons downstream of a regulated 3′ splice site. In many cases splice sites surrounding these exons are poorly conserved, and the presence of a splicing enhancer potently stimulates splicing. Complexes similar to complex E assemble on splicing enhancers or on substrates containing only one exon with a downstream 5′ splice site (111, 138). SR proteins are present within these complexes; they bind directly to the splicing enhancer and are essential for enhancer function (111, 130, 131, 133–135, 137, 138). Significantly, a splicing enhancer or a downstream 5′ splice site and SR proteins are required for the splicing in *trans* between 5′ and 3′ splice sites located on separate substrates, emphasizing the role of SR proteins in mediating protein-protein contacts and the function of enhancer elements in the activation of splicing (139, 140). Taken together, these results support a model in which weak interactions between U2AF and a suboptimal 3' splice site (or U1 snRNP and a suboptimal downstream 5' splice site) are stabilized by interactions with SR proteins bound to a splicing enhancer (Figure 4c). These interactions could operate by the same principles shown for the communication across introns. In addition, Wang et al (137) have proposed an RNA chaperone effect as an alternative mechanism: The binding of proteins to the splicing enhancer could induce a conformational change within the pre-mRNA that allows U2AF to bind to the 3' splice site with higher affinity.

When initially tested for complementation of splicing activity in S100 extracts, the function of various SR proteins appeared to be redundant (92, 107–109), a finding that contrasts with the high structural conservation of individual SR proteins during evolution (84) and the result that the *Drosophila* SRp55 is encoded by an essential gene (141). Additional experiments have demonstrated that individual SR proteins do exhibit differential effects in both constitutive and alternative splicing. As mentioned above, ASF/SF2 was isolated as an activity that switches the use of duplicated 5' splice sites from the distal to the proximal site (87, 88; for review see 142). In these studies nuclear extracts were used that contain sufficiently high concentrations of SR proteins to allow for efficient splicing. Effects on splice-site selection were seen when ASF/SF2 was added to the nuclear extract in increasing concentrations. Moreover, ASF/SF2 and SC35 promote the use of proximal 3' splice sites at high concentrations (108). Other SR proteins also shift the use of alternative splice sites, and in vitro and in vivo studies have established qualitative and quantitative differences in the effect of individual SR proteins on alternative splice-site selection (95, 107, 109, 143–146a).

Individual SR proteins also differ in their ability to commit various pre-mRNAs to the splicing pathway (110) and in their interactions with competing 5' splices (115) and exonic splicing enhancers (111, 134, 135, 138, 147). These findings could be explained by differential recognition of specific RNA sequences by SR proteins. For example, specific target sequences are selected from random pools of oligonucleotides by human ASF/SF2 and SC35 (148) and *Drosophila* RBP1 (149). Diversity in interactions with various pre-mRNAs could also originate from an interplay between individual SR proteins. Both ASF/SF2 and SC35 bind to themselves and to one another and also interact with alternative splicing regulators (121, 150). Moreover, SR proteins vary in their ability to bind to U1-70K (113). Thus, combinations of SR proteins with specific RNA-binding properties are likely to affect the maturation of a wide range of pre-mRNA substrates in constitutive and alternative splicing.

The role of the structural elements in splicing has been determined in the case of ASF/SF2 (146, 151, 152). Both RRM and Ψ-RRM are essential for constitutive splicing and alternative splice-site switching. Each of these domains can independently interact with RNA; optimal binding however requires both RRMs. Furthermore, a derivative of ASF/SF2 that comprises the RRM and Ψ-RRM, but lacks the SR domain, exhibits a different RNA binding specificity than the N-terminal RRM alone, and it has been suggested that both RRMs of ASF/SF2 may cooperate to define the specific binding site(s) within a pre-mRNA (148). Mutations within the SR domain inhibit constitutive but not alternative splicing (151, 152). The side chains of both arginines and serines in this domain are important for splicing, and eight arginine-serine dipeptides appear to be sufficient for the function of ASF/SF2. However, at least one SR protein (SRp30c) has been isolated that contains only five arginine-serine dipeptides (95).

Transcriptional and Post-Transcriptional Regulation

The specific functions of individual SR proteins in constitutive and alternative splicing imply that their abundance and/or activity is regulated in different tissues or at defined stages of development. Several recent reports provide insight into possible mechanisms of regulation. First, the promoter elements of SC35 and 9G8 contain putative binding sites for transcriptional regulators, and the expression of SC35 in hematopoietic cells is modulated by c-myb proto-oncogene products, suggesting a first level of regulation during the synthesis of the primary transcript (153, 154). Second, the pre-mRNAs of several SR proteins are alternatively spliced (95, 109, 154–156) and mRNA species are expressed in a tissue-specific or developmentally regulated fashion (91, 95, 109, 154, 156–158). As a consequence, varying amounts of given SR proteins could be produced in different cells (144). On the other hand, translation of alternatively spliced mRNA species can generate proteins that lack certain domains or contain domain substitutions. Translation of a variant ASF/SF2 mRNA, for example, generates ASF-3 which lacks the SR domain (155). In vitro, ASF-3 inhibits splicing and acts as a dominant repressor (152). An in vivo function for the variant proteins has not been established, however, the expression of some of the SR protein isoforms is phylogenetically conserved (95, 155, 159, 160) which suggests a role in vivo. Alternative splicing and/or polyadenylation also give rise to mRNAs of both SC35 and 9G8 that differ from one another only in their 3′ untranslated regions (154, 156). The differential expression of these mRNAs observed in various cell types most likely reflects a modulation in mRNA stability, as shown for SC35 (156). It has been proposed that the variation in SC35 mRNA stability is affected by sequences in the 3′ untranslated regions of the different mRNAs (156). This proposal is consistent with studies of other systems that have established that 3′ untranslated sequences are involved in the control of mRNA degradation, translation, growth, and differentiation (161, 162).

Regulation by Reversible Phosphorylation

In addition to the regulation of the abundance of individual SR proteins that can occur at both the transcriptional and post-transcriptional level, the activity of SR proteins can be modulated by phosphorylation and dephosphorylation cycles. Addition to splicing extracts of either the catalytic subunits PP1 and PP2a of human protein phosphatases or specific phosphatase inhibitors affects the splicing reaction at distinct stages (163–165). Whereas at least some splicing components have to be phosphorylated at the onset of spliceosome formation, dephosphorylation is required for both catalytic steps. Endogenous SR proteins lose the phosphoepitope recognized by mAb104 during incubation of extracts with phosphatases, and the epitope is restored, concomitant with spliceosome formation, when phosphatase inhibitors are added (165). Moreover, purified SR proteins can rescue splicing in a phosphatase-inhibited extract, suggesting that SR proteins are one target for phosphorylation/ dephosphorylation events. U1-70K has been identified as a second target (166, 167). Splicing but not spliceosome assembly is inhibited in the presence of U1-70K that was phosphorylated in the presence of nonhydrolyzable ATPγS (166), an effect similar to that observed with phosphatase inhibitors. Consistent with these results, a protein kinase [SR protein kinase 1 (SRPK1)] that is essential for splicing has been isolated (168, 169). SRPK1 specifically phosphorylates the serine residues in the SR domains of several SR proteins and U2AF^{65}, but not proteins unrelated to splicing. Addition of purified SRPK1 to an in vitro splicing reaction inhibits splicing in a dose-dependent manner, suggesting that either hyperphosphorylation and/or a block in dephosphorylation interferes with activity. Recently, SR proteins have been identified as targets of CLK/ STY, a dual specificity kinase (169a). Phosphorylation of ASF/SF2 by CLK/ STY occurs mainly in the SR domain. Interestingly, this kinase contains a SR-like domain, which is apparently involved in specific substrate recognition. A protein kinase activity that phosphorylates U1-70K as well as ASF/SF2 has also been found in snRNP preparations (167). Moreover, an essential splicing protein of *S. pombe* (Prp4p) contains structural motifs characteristic of serine/threonine kinases (170). Mammalian homologs of Prp4p have been isolated, but the target is unknown (N Käufer, personal communication).

By altering the ratio of positive and negative charges, phosphorylation-dephosphorylation cycles are likely to affect protein folding, RNA binding activity, and protein-protein interactions of SR proteins and, consequently, spliceosome assembly and catalysis. SRPK1 also differentially phosphorylates SR proteins during the cell cycle and may itself be subject to regulation (168). In addition, SRPK1 affects the intranuclear localization of SR proteins (168, 171). Thus, reversible phosphorylation can modulate the activity of splicing proteins during the course of the reaction, during the cell cycle, and probably also in a tissue- or stage-specific fashion. Although comparable data concerning other splicing factors are not yet available, it is to be expected that regulation by kinases and phosphatases is not limited to the components mentioned above, but represents a more general means to modulate splicing activities, as is the case in numerous other cellular processes.

POLYPYRIMIDINE TRACT–BINDING PROTEINS

Most introns of higher eukaryotes contain a region of high pyrimidine content located between the branch site and the conserved AG dinucleotide at the 3′ splice site. This polypyrimidine tract plays an essential role in the definition of the 3′ splice site at the earliest stage of spliceosome assembly, and it is recognized by different proteins during the course of the reaction.

U2 snRNP Auxiliary Factor (U2AF)

The best-characterized splicing factor that binds to the polypyrimidine tract is U2AF (172). U2AF is essential for the association of U2 snRNP with the branch site (173), and it binds to the pre-mRNA rapidly at 4° C in the absence of ATP (172; Figure 2). These reaction parameters are similar to the conditions under which U1 snRNP interacts with the 5′ splice site (40), and they are consistent with the presence of U2AF in complex E (174, 175). U2AF dissociates from the spliceosome at later stages of the reaction (174, 175), possibly as early as during the transition from complex E to complex A in a process that requires ATP (175a).

Mammalian U2AF is composed of two subunits of 35 and 65 kDa (172). The large subunit (U2AF^{65}) binds to the polypyrimidine tract of various pre-mRNAs, and this interaction is detected in the absence of U2AF^{35} (176). The binding affinity of U2AF^{65} for its substrate depends on the pyrimidine content and the length of the polypyrimidine tract (99). By amplifying RNAs selected from pools of random oligonucleotides the consensus RNA binding site for U2AF^{65} has been determined as $U_6(U/C)CC(C/U)U_7CC$ (177). The pool of sequences selected by U2AF^{65} displayed a nucleotide distribution similar to that of a collection of metazoan polypyrimidine tracts, consistent with the role of U2AF^{65} as an essential splicing factor that must recognize a wide variety of natural polypyrimidine tracts. Recently, U2AF has also been shown to interact with the pre-mRNA immediately adjacent to the branch site adenosine (178). Binding requires the presence of the polypyrimdine tract and is apparently necessary for an initial ATP-independent interaction of U2 snRNA with the branch site.

The cDNA of human U2AF^{65} encodes a protein with a calculated molecular mass of 53 kDa (99). Its N-terminal portion is characterized by an arginine/serine-rich region, similar to the C-terminal SR domains of members of the SR protein family. Three RRMs are located in the C-terminal half. The first and second RRMs contain consensus RNP-1 and RNP-2 domains whereas the third RRM is less well conserved. Unique amino acid sequences that separate SR domain and RRMs harbor the binding site for U2AF^{35} (100). All three RRMs of U2AF^{65} contribute to the specific and high-affinity binding to the poly-pyrimidine tract and are essential for splicing activity (99). The SR domain is dispensable for RNA binding but required for spliceosome formation and splicing; thus, this domain has been referred to as an effector domain, and its function might be similar to that of the activation domains of DNA-binding transcription factors (179). This notion is supported by the fact that the SR domain of U2AF^{65}, when fused to the sex-lethal protein (which functions in *Drosophila* somatic sexual differentiation), is converted from a repressor to an activator of splicing (180).

The exact function of U2AF^{35} is less well understood. Despite the fact that it is tightly associated with U2AF^{65} in the purified protein, HPLC-purified or recombinant U2AF^{65} is sufficient to support splicing in U2AF-depleted extracts (99, 176). Moreover, a U2AF^{65} derivative that lacks the binding domain for U2AF^{35} is fully functional in splicing (M Green, personal communication). Thus, U2AF^{35} appears to be dispensable for the splicing of at least some pre-mRNA substrates. However, as discussed above, U2AF^{35} engages in protein-protein interactions that bridge 5' splice sites or exonic enhancers with 3' splice sites (121; Figure 4); hence, U2AF^{35} may be essential for the splicing of pre-mRNAs with weak splice sites or in alternative splice-site selection. A U2AF^{35} cDNA codes for a protein of 34 kDa, and amino acids in the central portion are essential for the interaction with U2AF^{65} (100). A SR domain that is interrupted by 12 consecutive glycine residues is located in the C-terminal part of the protein and no RRM or other RNA-binding motifs are present, a finding consistent with a lack of RNA-binding activity for U2AF^{35}.

Proteins antigenically related to both U2AF subunits have been detected in vertebrate and invertebrate cells (176), and homologs have been cloned from various organisms. Human and mouse U2AF^{65} (181) differ by only two amino acids, and the *Drosophila* (182) and *S. pombe* (183) cognates display overall identities of about 70% and 30%, respectively, with the human protein. The homology of these proteins to human U2AF^{65} is most pronounced within the RRM portion, whereas size differences between the proteins can be attributed mainly to the N-terminal region that also varies in the number of SR dipeptides. As may be expected for an essential splicing factor, the genes encoding the U2AF^{65} homologs of *Drosophila* and *S. pombe* provide essential functions in vivo (182, 183). Similarly, the *Drosophila* counterpart of human U2AF^{35} is encoded by an essential gene, and the amino acid conservation among the small U2AF subunits is higher than among the large subunits in these organisms (D Rudner, K Breger, R Kanaar & D Rio, personal communication). These findings are consistent with the notion that U2AF[35] plays an important role in splicing.

Typical polypyrimidine tracts are absent from *S. cerevisiae* introns, but shorter uridine-rich regions, which appear to be important for the second catalytic step (184), can be found upstream of the 3' splice site (15). Sequences between the branch site and the 3' splice site are dispensable for spliceosome assembly in yeast (55); instead, the highly conserved branch site is essential early during the reaction (42, 45). Therefore, whether a U2AF homolog is present in yeast, and, if so, whether the protein interacts with the branch site sequence is an open question (9). In a genetic screen, Abovich et al (186) have isolated the *MUD2* gene that encodes a protein of 58 kDa. Mud2p binds to the pre-mRNA in a branch site–dependent reaction and interacts with U1 snRNP. In line with the finding that the assembly of the CC2 commitment complex in yeast is dependent on an intact branch site (45), Mud2p is required for the formation of this complex and is found in association with it (186). A single C-terminal RRM in Mud2p, which is required for function, shows highest homology to the third (atypical) RRM of hU2AF[65], but Mud2p lacks a typical N-terminal SR domain. Moreover, Mud2p is not essential for yeast viability, and further experiments are necessary to confirm a functional homology between human U2AF[65] and yeast Mud2p. The finding that U2AF[65] also interacts with the mammalian branch site (178) could be taken as further evidence for a relationship between these proteins.

Polypyrimidine Tract–Binding Protein (PTB)

PTB was identified in a UV cross-linking assay as a protein of ~62 kDa that shows a preference for binding to polypyrimidine tract–containing RNAs (187). Because binding of PTB to wild-type and mutated pre-mRNAs correlated with the activity of these substrates in the formation of pre-splicing complex A, it was suggested that PTB represents an essential splicing factor. Further experimentation has provided evidence against participation of PTB in constitutive splicing (46, 188), however, PTB functions in alternative splice-site selection (177, 189, 190). PTB has been purified as a doublet or triplet of proteins of ~57 kDa (187, 191, 192), and cDNA cloning confirmed the existence of three isoforms, which most likely arise from alternative splicing to various 3' splice sites (191, 192). A fourth cDNA, which encoded a protein with an additional seven amino acids not present in the other forms, was isolated during the cloning of the hnRNP I protein, which also exists in several electrophoretic variants (193). Thus PTB and hnRNP I represent the same protein. The predicted amino acid sequence of PTB revealed the presence of four repeated domains of ~80 amino acids which are distantly related to the

RRM in their overall sequence, but lack the canonical RNP-1 and RNP-2 sequences (191–193). PTB not only shows reactivity to pyrimidine-rich RNA, but binds single-stranded pyrimidine-rich DNA sequences as well, which is a characteristic of several hnRNP proteins (194). Consistent with this property PTB has been isolated as a sequence-specific DNA-binding protein from mouse and rat cells (195–197). In addition, a protein that binds specifically to the internal ribosomal entry site of picornavirus mRNA has been identified as PTB (198). Clearly, further experiments are necessary to unambiguously determine a function of PTB in these diverse processes.

PTB-Associated Splicing Factor (PSF)

PSF, a protein of 100 kDa, was initially detected in fractions enriched in PTB, which exhibited splicing activity in a reconstituted system (192). Purification of the activity resulted in the isolation of a complex of ~320 kDa containing both PSF and PTB (188). In vitro reconstitution and antibody inhibition experiments revealed that PSF is required for constitutive splicing (188), and it functions during the second step of the reaction (83). PSF and a proteolytic breakdown product of 68 kDa are identical to the spliceosome-associated proteins SAP102 and SAP68 which bind to the spliceosome at the time of B-complex assembly (83; Figure 2). Higher levels of these proteins are detected in isolated C-complexes, consistent with the function of PSF during the second catalytic step. PSF binds to the polypyrimidine tract of introns (188) and could displace U2AF^{65} from its binding site at late stages of spliceosome assembly (83). The polypyrimidine tract is not only essential early during the splicing reaction, but may also aid in the recognition of the AG dinucleotide at the 3′ splice site during the second step, especially when the distance between the branch site and the AG exceeds 40 nucleotides (199). Thus, by binding to the pre-mRNA at late stages of spliceosome assembly PSF may trigger this recognition event (83).

Consistent with its RNA-binding activity, PSF contains two consensus RRMs in its central portion and three repeats of the sequence arginine-glycine-glycine (RGG) close to the N-terminus. RGG boxes were first found in the hnRNP U protein (200) and are important for RNA-protein interactions (see 84, 194). A domain of ~230 amino acids that separates the RGG repeats and RRMs is highly enriched in proline residues (33%), and a more limited stretch within this region consists of 88% proline (P) and glutamine (Q). Similar domains are present in other splicing factors, snRNP, and hnRNP proteins (201, 202; Table 1). P-rich regions in various proteins have been implicated in protein-protein interactions (203) and P- or P/Q-rich domains are landmarks of activation domains of several transcription factors (204, 205). Although no results on the function of these domains in splicing proteins are available to date, by analogy to transcription factors, they may be important for the recruitment of additional components into the spliceosome (188). In this respect, P- or P/Q-rich domains may be equivalent to SR domains, at least some of which mediate protein-protein contacts (113, 121, 150).

The apparent molecular mass of 100 kDa for PSF is very similar to the size of IBP (intron-binding protein), a protein of unknown function in splicing that interacts with the polypyrimidine tract (206). IBP has also been described as a polypeptide of 70 kDa (207, 208) which corresponds in size to the proteolytic product of PSF. IBP is recognized by anti-Sm antibodies and has been found associated with U5 snRNP. Similar observations have not been made for PSF (188), and whether these proteins are identical remains to be established.

A protein related in sequence to PSF is p54^{nrb} (nuclear RNA-binding protein of 54 kDa; 209). This protein was identified in an attempt to isolate the human homolog of the *S. cerevisiae* Prp18p, a U5 snRNP-associated protein that functions in the second catalytic reaction (210, 211). p54^{nrb} does not share any amino acid sequence homology with Prp18p, but both proteins nevertheless contain a common epitope (209). The amino acid sequence of p54^{nrb} is similar to an internal region of PSF including both RRMs and, in its somewhat shorter N-terminal region, p54^{nrb} contains stretches of glutamine and proline residues. The region of homology between p54^{nrb} and PSF is moreover conserved to a significant degree in the *Drosophila* NONA/BJ6 protein which is encoded by the *no-on-transient A* gene that is required for normal response to visual stimuli in *Drosophila* (212, 213). NONA/BJ6 is ubiquitously expressed in neural and non-neural nuclei throughout fly development (214), and the protein has been localized to transcriptionally active sites on polytene chromosomes (215), which is also the case for the *Drosophila* SR proteins B52 and RBP1 (109, 216). Whether p54^{nrb} has any activity in splicing is not currently known, but its homology to PSF may hint at a role in splicing (209).

BRANCH SITE–BINDING PROTEINS

In contrast to the branch site sequence for yeast, which is highly invariant, the branch point adenosine in mammalian introns is embedded in the degenerate sequence YNCURAY. The base pairing of these nucleotides with an internal sequence of U2 snRNA is well established (56–58). Given the degeneracy of the branch site and the short size of the duplex formed with U2 snRNA, whether additional components are required for the initial recognition event or for a stabilization of the duplex during the course of the splicing reaction is not clear. Moreover, the adenosine is bulged out from the branch site/U2 snRNA duplex (59), but whether this conformation is sufficient to activate the adenosine as the nucleophile in the first catalytic step is not known. It has also been shown that intron-3' exon intermediates with aberrant branch structures form in pre-mRNAs with certain mutations at the 5' splice site or branch site; however, the second catalytic step is blocked in the presence of these mutations (217–220). In yeast, mutations in Prp16p can suppress a branch site mutation and the associated inhibition of the second step (221). Thus, the branch site is important for both transesterification reactions, and correct lariat intermediates are selected before completion of the splicing reaction, but exactly how this selection is achieved is not understood.

By site-specific photochemical cross-linking several proteins that are in intimate contact with the branch site have been identified (222). An 80-kDa polypeptide binds to the pre-mRNA in an ATP- and temperature-independent fashion and requires the presence of U1 snRNP for binding. The protein is not associated with any of the spliceosomal complexes and may be required for an early recognition of the branch site. Polypeptides of 14, 35, and 150 kDa require the presence of U1 and U2 snRNP as well as ATP for binding and are first detected in complex A. Whereas binding of the 35- and 150-kDa proteins diminishes during the course of the reaction, the 14-kDa protein remains associated with splicing complexes B and C. Interaction of proteins of 70 and 220 kDa with the branch site depends on all spliceosomal snRNPs and, consistently, these proteins are only detected in complexes B and C. The 220-kDa protein also interacts directly with exon nucleotides adjacent to the 5' splice site and is probably identical to the U5 snRNP-associated protein p220 (see below). The identity of the other polypeptides is unknown, but it is possible that the 150-kDa protein corresponds to a U2 snRNP-specific protein that interacts with the 3' splice site region (175), whereas the 35-kDa polypeptide is similar in size to a protein that binds to the 3' splice site (222; see 178). With a different photo reagent incorporated immediately downstream of the branch site adenosine, specific cross-links of U2AF[65] and a 28-kDa protein have been detected (178). Whereas binding of U2AF[65] is rapid and ATP-independent, the 28-kDa protein interacts with the branch site only in the catalytically active spliceosomal complex C, and the contact is established after the completion of the first catalytic step.

Although specific functions of most of these proteins have not been described, the branch site is clearly recognized by different polypeptides at each step of spliceosome assembly and catalysis, another example of consecutive interactions with sequences essential for pre-mRNA splicing. The dynamic nature of these interactions parallels the conformational rearrangements in the spliceosome owing to changes in the base-pairing interactions of pre-mRNA and snRNAs. Future experiments will certainly provide important clues as to whether and how branch site–binding proteins affect these base-pairing interactions and whether these proteins participate in the selection of 3' splice sites.

hnRNP PROTEINS

Nascent RNA polymerase II transcripts associate in vivo with a set of more than 20 polypeptides, the hnRNP proteins, with sizes ranging from 34 to 120 kDa (reviewed in 194, 223). HnRNP proteins represent general RNA-binding proteins and bind readily to single-stranded nucleic acids. In splicing extracts they rapidly associate with the pre-mRNA to form complex H (30, 31, 37, 38, 224, 225). The binding is independent of temperature, ATP, and functional splice sites, suggesting that these interactions are not splicing specific. By UV cross-linking, binding of hnRNP proteins is observed only in H-complex, not in complex E (111). Thus, although hnRNP proteins are also found in later spliceosomal complexes (174, 225), their association with the RNA appears to be weakened during the course of the reaction. Individual hnRNP proteins exhibit distinct binding preferences for ribohomopolymers (226), and the binding of at least some hnRNP proteins is transcript dependent (225). Several hnRNP proteins have been shown to specifically interact with 5' or 3' splice sites (see 227 and references therein), and antibodies to hnRNP C inhibit splicing in vitro (228, 229). Despite these observations, direct roles for hnRNP proteins in constitutive splicing have remained elusive, but functions for hnRNP A/B, PTB/hnRNP I, and hnRNP F in alternative splice-site selection have been established (106, 145, 177, 190, 230–233).

snRNP-ASSOCIATED SPLICING ACTIVITIES

SnRNPs were isolated initially as particles of ~10S that contain the common Sm proteins and, in the case of U1 and U2 snRNP, polypeptides unique to these particles (reviewed in 33). Improved isolation techniques have led to the identification of larger snRNP assemblies that are characterized by the presence of additional particle-specific proteins. These proteins are less tightly bound than the other polypeptides, and they dissociate from the snRNPs at salt concentrations exceeding 150–200 mM, resulting in a concomitant decrease in sedimentation rate. Owing to these characteristics, some of the specific polypeptides have been isolated as non-snRNP splicing factors by chromatographic procedures.

The 17S U2 snRNP

The 17S U2 snRNP is associated with nine polypeptides in addition to the A', B", and Sm proteins that are present in the 12S U2 snRNP (234). The 17S U2 snRNP, but not the 12S form, is active in pre-spliceosome assembly, indicating that at least some of the 17S U2–specific proteins are essential for splicing (235). Two activities, SF3a and SF3b, that were purified as non-snRNP splicing factors are constituents of the 17S U2 snRNP and function in the assembly of the particle in a two-step reaction: SF3b interacts with isolated 12S U2 snRNP, and subsequent binding of SF3a converts the intermediate into the 17S form (236, 237; Figure 2). The function of SF3a and SF3b in this pathway is consistent with their essential role in splicing (236, 238).

SF3b consists of four polypeptides of 50, 130, 145, and 155 kDa (K Gröning, P Grüter & A Krämer, in preparation). The 50- and 145-kDa subunits can be UV cross-linked to U2 snRNA in the absence of the 12S U2 snRNP proteins, and competition experiments have located the binding site of the 50-kDa protein to the 5′ terminal stem-loop of U2 snRNA. SF3a comprises three subunits of 60, 66, and 120 kDa (236), none of which appear to bind to U2 snRNA directly. SF3a cannot bind to SF3b or the 12S U2 snRNP individually, suggesting that a certain conformational context, generated upon interaction of SF3b with the 12S U2 snRNP, is required for the integration of SF3a into the 17S U2 snRNP (K Gröning, P Grüter & A Krämer, in preparation). Mod- ification-interference, UV cross-linking, and nuclease-protection experiments, as well as structural analyses by electron microscopy are consistent with a model in which SF3b binds to the 5′ half of U2 snRNA (235; B Kastner, K Gröning & A Krämer, unpublished data). SF3a appears to be associated mainly with the 3′ half of U2 snRNA through contact with the core and/or the A′ and B″ proteins.

Yeast homologs of the SF3a subunits and SF3b[145] (see below) interact genetically with sequences immediately downstream of the branch site inter-action sequence in U2 snRNA and with stem-loops IIa and IIb (251–251b; see Figure 3a). Moreover, yeast SF3b[145] interacts with the SF3a[66] homolog, but not with the homologs of SF3a[60] and SF3a[120] (251a). With the assumption that similar interactions occur in the human system, SF3b[145], bound to the U2 snRNA downstream of the branch site interaction sequence, could mediate the binding of SF3a to the U2 snRNP by contacts with SF3a[66].

Seven spliceosome-associated proteins that are incorporated into the splice-osome during the assembly of pre-splicing complex A are part of the U2 snRNP (175). Co-electrophoresis in two-dimensional gels indicated that SAPs 61, 62, and 114 represent the three SF3a subunits (M Bennett, R Reed, R Brosi & A Krämer, unpublished results), whereas SAPs 49, 130, 145, and 155 almost certainly correspond to the SF3b subunits. The 17S U2 polypeptides of 35 and 92 kDa (234) have not been detected in other studies, and their identity remains to be defined. All U2 snRNP-associated SAPs, except for SAP130, are in direct contact with the pre-mRNA in the isolated spliceosomal complexes A, B, and C (83, 175) and can be cross-linked to the pre-mRNA upstream of the branch site in the order SF3b[50] - SF3a[120] - SF3a[60] - SF3b[145] - SF3a[66] - SF3b[155] (239, 239a). SF3b[155] apparently also binds to a site downstream of the branch site. Mutations within the sequences upstream of the branch site (termed anchoring site) abolish neither cross-linking of the polypeptides nor A-complex assembly;

however, an oligoribonucleotide complementary to the anchoring site inhibits both, protein binding and complex formation. Thus, interactions of SF3a and SF3b with the pre-mRNA appear to be sequence-independent and are required for pre-spliceosome formation.

These results are highly suggestive of a function of SF3a and SF3b in tethering the U2 snRNP to the pre-mRNA. Binding of the proteins to the pre-mRNA could aid in the initial formation of the helix or stabilize the interaction once it is established. Moreover, the 17S U2 snRNP could be guided to the branch site and/or anchoring site by interaction of one or more of its polypeptides with U2AF, which is bound to the adjacent polypyrimidine tract in complex E and is essential for U2 snRNP binding (172, 173, 175). In yeast, the putative U2AF[65] homolog Mud2p interacts with Prp11p, the counterpart of SF3a[66] (186). A similar interaction has so far not been reported in the human system. SF3a and SF3b may play an additional role in the formation of the intermolecular helix between the 5' end of U2 and the 3' end of U6 snRNAs (64, 65; Figure 3). Part of the interaction domain lies within the 5' terminal stem-loop of U2 RNA, whereas the 3' end of U6 RNA is unpaired. The structural change at the 5' end of U2 snRNA observed during the conversion of the 12S into the 17S particle (235) might serve to provide the suitable conformational context for this base pairing-interaction.

cDNAs for all three subunits of SF3a have been isolated (240–243). Analysis of the predicted amino acid sequences as well as previous immunological studies (235, 237, 240) demonstrated that SF3a[60] (SAP61), SF3a[66] (SAP62), and SF3a[120] (SAP114) represent human homologs of the essential yeast splicing proteins Prp9p, Prp11p, and Prp21p (244–247). In addition to structural features (see below), many functional aspects are conserved between the human and yeast counterparts. First, Prp9p, Prp11p, and Prp21p are essential for pre-spliceosome formation (247–250). Second, they interact genetically with U2 snRNA and respond to mutations in U2 snRNA in a very similar fashion (250, 251, 251b). In addition, Prp9p is loosely bound to U2 snRNA in yeast extracts (248). Third, complementation experiments with heat-inactivated extracts of temperature-sensitive *prp9* and *prp11* strains suggested that the proteins function in a complex (244, 250). This conclusion has been confirmed and extended by the demonstration of genetic and physical interactions among Prp9, Prp11, and Prp21 (246, 249–252). Most of these interactions are conserved in the human homologs (see below).

In addition, two putative SF3b subunits, SF3b[50] (SAP49) and SF3b[145] (SAP145), have been cloned (239, 239a). A yeast homolog of SF3b[145] has been isolated in a search for suppressors of a cold-sensitive mutation in the U2 snRNA stem-loop IIa, an essential element for the incorporation of U2 snRNP into the spliceosome (251a). The *CUS1* gene is essential, and Cus1p functions in the assembly of the pre-spliceosome.

The C-terminal domains of SF3a^{60} (SAP61) and Prp9p are relatively well conserved, and they contain a zinc finger motif of the C$_2$H$_2$-type that is essential for viability in yeast (241, 242, 245). A chimeric Prp9 protein in which amino acids encompassing the zinc finger domain of SF3a^{60} have been exchanged for the corresponding amino acids of Prp9p rescues the temperature-sensitive phenotype of the *prp9-1* strain. Thus, the function of this domain is evolutionarily conserved (242). Zinc finger domains are reminiscent of nucleic acid binding proteins (253). SF3a^{60} binds directly to the pre-mRNA, and this function could be fulfilled by the zinc finger (175). In Prp9p, amino acids encompassing the C-terminal zinc finger have been implicated in homodimer formation (249) which has so far not been observed for SF3a^{60} (241). However, this domain of Prp9p appears to interact with an additional unknown component (249) which, in correlation to SF3a^{60}, could represent the pre-mRNA. A second zinc finger, as well as a leucine-rich region present in Prp9p, is not conserved in the human protein (241, 242, 245). The interaction between SF3a^{60} and SF3a^{120} involves the N-terminal ~100 amino acids of SF3a^{60} (241; D Nesic & A Krämer, unpublished data). Similarly, temperature-sensitive mutations that abolish the interaction of Prp9p with Prp21p have been mapped to the N-terminal half of Prp9p (249).

The N-terminal two thirds of SF3a^{66} (SAP62) exhibits significant homology to Prp11p and contains a C$_2$H$_2$-type zinc finger domain (240, 244). The additional C-terminal third, which is lacking in the yeast counterpart, is organized into 22 proline-rich heptad repeats of the consensus sequence GVHPPAP. Similar heptad repeats, although of different sequence, comprise the C-terminus of the large subunit of RNA polymerase II and function in protein-protein interactions (254). The association between yeast Prp11p and Prp21p (252) is conserved between the human homologs (240, 243) and apparently does not involve the heptad repeats (see 239). Whether or not this domain engages in contacts with other proteins is currently unknown.

SF3a^{120} (SAP114) represents a new member of the SURP family of splicing factors (243). This protein family includes the *S. cerevisiae* Prp21p, the *Drosophila* protein *suppressor-of-white-apricot* (DmSWAP), and homologs in *C. elegans*, human, and mouse (101, 246, 255, 256). All proteins are characterized by a tandemly repeated domain of ~40 amino acids, termed the SURP module. The N-terminal half of this domain is highly conserved, whereas the C-terminal half is more variable in both amino acid sequence and length. Based on the spacing between the SURP modules and additional conserved structural domains, the SURP proteins can be grouped into two classes (256). Members of the SWAP subfamily share a conserved N-terminal domain of ~80 amino acids followed by the tandemly repeated SURP modules that are separated by 181–206 amino acids, and they contain a SR domain close to their C-termini. SF3a^{120}, Prp21p, and the *C. elegans* homolog CePrp21 lack the conserved

N-terminal and SR domains, and the spacing between the SURP modules is reduced to 45–71 amino acids. A proline-rich region and a ubiquitin-like domain present in SF3a[120] are not found in Prp21p. The ubiquitin-like domain is well conserved in CePrp21, but the homology within the proline-rich region is lower. The division of the SURP family members into two classes also holds true when functional aspects are considered. DmSWAP is involved in the alternative splicing of its own pre-mRNA in that it blocks the removal of the first and second introns (101). Alternatively spliced transcripts have also been identified in the human, mouse, and *C. elegans* SWAP proteins (255, 256), suggesting that they function in alternative splicing as well. In contrast, both SF3a[120] and Prp21p are essential for constitutive splicing events (236, 246, 247). Given that SF3a[60] and SF3a[66] both bind to SF3a[120], it will be interesting to see whether an interaction of these polypeptides with the human SWAP protein is required for its activity in alternative splicing.

The interaction of SF3a[120] with SF3a[60] requires sequences that include the second SURP module of SF3a[120] (243; C Wersig & A Krämer, unpublished data). A 60–amino acid domain located C-terminal to the SURP modules is required for binding to SF3a[66]. The interaction domains are in large part conserved during evolution (A Tartakoff, A Krämer & P Legrain, submitted). Roles of the proline-rich and ubiquitin-like domains in SF3a[120] have not been established, but functions in protein-protein interactions have been proposed (243).

Consistent with its direct association with pre-mRNA and U2 snRNA (175; K Gröning, P Grüter & A Krämer, in preparation), SF3b[50] (SAP49) contains two N-terminal RRMs (239). The C-terminal part is organized into proline/glycine-rich sequences, the function of which is unknown (239). A yeast protein sequence, retrieved during database searches, is 36% identical to human SF3b[50] and lacks the proline/glycine-rich domain (251a). A functional relationship between these proteins remains to be determined.

SF3b[145] (SAP145) contains proline-rich sequences in its N-terminal portion and a stretch of glutamine residues in the C-terminal part, but no known RNA-binding motifs (239a). Significant sequence homology between SF3b[145] and yeast Cus1p is limited to a ~200–amino acid region located in the central portions of both proteins (239a, 251a). SF3b[145] and SF3b[50] interact with one another in filter-binding assays (239). The regions of interactions have been mapped to the N-terminal region of SF3b[50] that contains the RRMs and a region in the C-terminal part of SF3b[145] (239, 239a).

The 20S U5 snRNP

As in the case of U2 snRNP, two distinct forms of U5 snRNP have been isolated (257). The 8S U5 snRNP is associated with the Sm proteins, whereas at least eight additional proteins are found in the 20S U5 snRNP. Information regarding the function of individual U5-specific proteins, except for one polypeptide of >200 kDa (p220), is not available. p220 is the human homolog of the essential yeast splicing factor Prp8p which is a component of yeast U5 snRNP (208, 258–260). Prp8p plays a role in the association of U5 with U4/U6 snRNP and the incorporation of these snRNPs into the spliceosome (258, 259, 261). In addition, Prp8p can be specifically cross-linked to exon nucleotides at the 5′ splice site before the initiation of the first transesterification reaction, and this interaction is most likely maintained through the second step (262). After completion of the first, but before the second, catalytic step Prp8p specifically contacts the pre-mRNA two nucleotides downstream of the branch site and the first nucleotide of the 3′ exon. Another study demonstrated interactions of Prp8p with the uridine-rich tract at the 3′ splice site and specific binding to the last nucleotide of the intron (263). Site-specific cross-links of human p220 with the 5′ exon and the branch site occur with apparently similar kinetics in HeLa cell extracts (222, 264), and they may explain the previous result that U5 snRNP participates in the second catalytic step (265, 265a). Thus, Prp8p and p220 play important roles in both the first and second transesterification reactions. Prp8p may stabilize interactions between the U5 snRNA loop and exon sequences at the 5′ and 3′ splice sites, thereby increasing the specificity of splice-site recognition. Vital functions of Prp8p and p220 at various stages of the splicing reaction are supported by a remarkable conservation of both primary structure and size of homologous proteins in other organisms (266, 267).

The 25S U4/U6.U5 tri-snRNP

U4/U6 and U5 snRNPs are incorporated into the spliceosome in the form of a tripartite snRNP particle (50, 51). This 25S U4/U6.U5 snRNP has been purified from HeLa cells and contains five polypeptides in addition to those associated with the 20S U5 and 10S U4/U6 snRNPs (268). Polypeptides released from the tri-snRNP by micrococcal nuclease treatment promote the assembly of the 25S particle from isolated 20S U5 and 10S U4/U6 snRNPs, which by themselves do not assemble into a larger structure. Moreover, splicing in extracts devoid of a functional 25S U4/U6.U5 snRNP is restored in the presence of the isolated tri-snRNP or the associated proteins, but not upon addition of U5 and U4/U6 snRNP alone (269). These results are consistent with a function of the tri-snRNP-specific proteins in the assembly of the 25S U4/U6.U5 particle, and they support the conclusion that U4/U6 and U5 join the spliceosome as one entity. Whether the tri-snRNP-specific proteins play additional roles in the incorporation of the particle into the spliceosome or in subsequent steps is not clear. Also, a comparison of these proteins with spliceosome-associated proteins is lacking; hence, whether the proteins remain bound to the spliceosome after the destabilization of the U4/U6 base-pairing interaction is unknown. Similar to SF3a and SF3b, an activity [heat shock-labile splicing factor (HSLF)] that is functionally equivalent to the tri-snRNP-specific proteins has been isolated from HeLa cell extracts as a snRNA-free chromatographic fraction (269).

ATP-DEPENDENT RNA HELICASES

As discussed in previous sections, pre-mRNA splicing requires energy input in the form of ATP at multiple steps, despite the fact that the phosphates of ATP are not incorporated into the newly formed phosphodiester bonds. One use for ATP at the onset of spliceosome assembly could be in the phosphory-lation of splicing factors. Moreover, the conformational rearrangements within the spliceosomal RNAs are likely to be energy-driven. Thus, the discovery of splicing proteins with putative ATP-dependent RNA helicase activities (270) may not be surprising. These activities, except for one, have only been de-scribed in *S. cerevisiae*; because they are likely to play key roles in splicing, their functions are briefly summarized here. Five yeast splicing factors [Prp2p, Prp5p, Prp16p, Prp22p, and Prp28p; (62, 271–274)] contain sequence motifs reminiscent of the DEAD/H (for amino acids asp-glu-ala-asp/his) family of RNA helicases, the prototype of which is translation initiation factor eIF4A (275, 276). Purified Prp2p and Prp16p hydrolyze ATP in the presence of RNA (277–279). Attempts to demonstrate RNA helicase activity have failed thus far, possibly owing to stringent sequence and/or structural requirements in RNA helices that are difficult to mimic in vitro in the absence of the assembling spliceosome. Moreover, accessory proteins may be required for full activity, as in the case of eIF4A which carries out many of its functions only in the presence of eIF4B (280). Nevertheless, the presently available information regarding the function of the DEAD/H splicing factors at consecutive stages of the reaction strongly suggests that these factors are directly involved in RNA conformational switches.

Prp5p associates with the commitment complex in the absence of ATP (250) and is required for A-complex assembly in yeast (272). *PRP5* interacts geneti-cally with U2 snRNA as well as *PRP9, PRP11, PRP21,* and *CUS1* (250–251b), suggesting a function in the preparation of U2 snRNP and/or of the pre-mRNA for the U2 snRNA/branch site interaction. Prp28p is essential for the first catalytic step, and genetic evidence has led to the proposal that it resolves the U4/U6 base pairing interaction just prior to catalysis (274, 282). Prp2p is likewise required for the first step of splicing (51, 283). It associates with the spliceosome in an ATP-independent fashion after the incorporation of the U4/U6.U5 snRNP and binds directly to the pre-mRNA, suggesting that the pre-mRNA is at least one of its substrates (284–286). Once bound, Prp2p hydrolyzes ATP and thus initiates catalysis (285, 286). In addition, an RNA conformational change and/or RNA displacement promoted by Prp2 appears to be essential for the progression to the first step and for release of the protein from the spliceosome (279). Prp16p functions in a manner similar to Prp2p in that it only transiently associates with the spliceosome, and it uses its ATPase activity to drive a conformational rearrangement that is required for the second catalytic step (277, 287–289). Because introns with a mutation in the branch site adenosine can be spliced in the presence of certain alleles of *Prp16*, the protein has been implicated in the proofreading of lariat intron intermediates (221, 273, 290). Prp22p acts after the splicing reaction itself is completed and is involved in the dissociation of the spliceosome and release of spliced mRNA (62). The requirement for a putative ATP-dependent RNA helicase at this stage suggests that spliceosome disassembly is an active process, not the result of mere diffusion of the individual components.

The only human RNA helicase implicated in splicing thus far is human RNA helicase 1 (HRH1) which is homologous to Prp22p (291). HRH1 can partially complement the temperature-sensitive phenotype of a *prp22* mutant when introduced into yeast cells. Sequence alignments demonstrate high homology to the helicase domain of Prp22p which spans about 300 amino acids and is also conserved in Prp2p and Prp16p (62, 271, 273). In addition, HRH1 contains a domain present in Prp22p which is similar to a putative RNA-binding domain in bacterial ribosomal protein S1 and polynucleotide phosphorylase (62). HRH1 is also characterized by a short arginine/serine-rich domain which is not found in Prp22p (291). This domain is involved in contacts with ASF/SF2 and possibly other SR proteins. This situation brings to mind the U1-70K protein which lacks a SR domain in yeast (292), whereas in mammalian cells the U1-70K SR domain interacts with SR proteins (113). The functional significance of the SR domain in HRH1 is not clear: It might serve to recruit the protein to the spliceosome, for example by interaction with exon-bound SR proteins.

OTHER ACTIVITIES

Protein Factors Required During Spliceosome Formation

SF1 is a 75-kDa protein that is required for pre-spliceosome assembly, but its exact role is unknown (238, 293). Several cDNAs encoding SF1 have been isolated, and conceptual translation generates at least four protein isoforms that vary in length and in the amino acid composition of their C-termini (S Arning & A Krämer, in preparation). The N-terminal half of SF1 is characterized by a KH domain and a zinc knuckle. The KH domain comprises ~40 amino acids and was first identified in hnRNP K, a major poly(C)-binding protein in HeLa cells (294, 295). This domain is highly conserved in several prokaryotic, archaebacterial, and eukaryotic proteins that are in physical and/or functional contact with RNA or DNA (296, 297), and mutations within this domain abolish RNA or DNA binding (298, 299).

KH domains are also found in the *Drosophila* splicing factor PSI which is involved in the soma-specific inhibition of P element transposition (300) and in the *S. cerevisiae* MER1 protein that activates the splicing of the MER2 pre-mRNA in meiotic cells (301). Both proteins affect specific splicing events through their interaction with sequences adjacent to the 5' splice sites of their target substrates. Zinc knuckles are found in retroviral nucleocapsid proteins and are implicated in specific RNA binding (302). In splicing proteins, zinc knuckles have been found in the human SR protein 9G8 and yeast Slu7p. In 9G8 the zinc knuckle plays a role in the determination of RNA-binding specificity (90; J Stévenin, personal communication). A function for the zinc knuckle of Slu7p has not been reported, but this protein interacts genetically with U5 snRNA and is involved in 3' splice site choice in *S. cerevisiae* (303, 304). Consistent with the presence of these domains, both purified and recombinant SF1 bind to RNA, but in a nonspecific fashion (S Arning & A Krämer, in preparation). Whether specific binding to the pre-mRNA substrate (or any of the snRNAs) occurs in the context of the spliceosome remains to be investigated. The C-terminal half of SF1 is composed of proline-rich sequences and could play roles in interactions with other splicing proteins. Given the finding that several SF1 mRNAs, which correspond in size to the different cDNAs isolated, are differentially expressed in several tissues, it will be interesting to see whether the individual protein isoforms have distinct functions, for example in the modulation of interactions with other splicing components or in alternative splicing events.

An 88-kDa protein has been identified by a monoclonal antibody raised against large ribonucleoprotein particles (305). These particles have a size of ~200S and contain unspliced transcripts, snRNPs, SR proteins, and other components (see 306). The 88-kDa protein functions in pre-splicing complex formation, but its exact function is unclear. A protein termed HRF (heat-resistant factor) has been purified from HeLa cells and shown to restore splicing in a heat-treated extract (307). HRF appears to act early during spliceosome formation, however, additional information is not available.

Protein Factors Required for Catalysis

In contrast to yeast, little is known about mammalian non-snRNP splicing factors that function after the formation of the pre-spliceosome. SF4 has been described as a protein that is dispensable for the assembly of complexes A and

B but required for the catalytic steps to occur (308). Pre-mRNA present in B-complexes that are formed in the absence of SF4 can be chased into spliced products upon addition of SF4, indicating that SF4 acts during the conversion of the functional but inactive complex B into the active spliceosome.

In *S. cerevisiae*, several proteins (Prp16p, Prp17p, Prp18p, Slu7p, and Ssf1p) have been identified that are required for the second transesterification reaction (15). Despite the early definition of two mammalian splicing activities that function at this stage (310), information regarding second-step factors in this system is limited. As mentioned above, PSF acts during the second catalytic step (83), but details concerning this function are unknown. An activity distinct from PSF has recently been described that appears to exist in a large complex with other components (311). When added to splicing reactions or isolated spliceosomes that have already undergone the first step, this activity promotes progression through the second step. The conversion of splicing intermediates into products mediated by this factor does not require ATP, suggesting that it acts after the ATP-dependent event in the second step. This characteristic resembles properties reported for both Prp18p and Slu7p (211, 312); a direct comparison between these activities, however, requires additional experiments.

Cap-Binding Proteins

All primary transcripts that are synthesized by RNA polymerase II carry a 5′ terminal monomethylated cap that is added co-transcriptionally (for references see 313). Cap structures play roles in different aspects of RNA metabolism including RNA stability, RNA export from the nucleus, and translation initiation. Additional roles in pre-mRNA splicing have been suggested by observations that splicing in HeLa cell extracts or Xenopus oocytes can be inhibited by addition of a large excess of cap analogs or capped RNAs that lack splice sites (19, 314–318). Cap-binding proteins involved in splicing have been identified in UV cross-linking assays using short, capped transcripts (319, 320), and cDNAs of two polypeptides (CBP80 and CBP20) have been cloned (313, 321, 322). CBP80 and CBP20 are tightly associated in a cap-binding complex in which only CBP20 appears to bind directly to the cap structure (321, 323). Antibodies raised against the recombinant proteins have a severe negative effect on splicing activity in vitro and in vivo and reduce the assembly of all ATP-dependent splicing complexes, and recent results indicate that the cap-binding complex is required for the formation of complex E (313, 321; J Lewis & I Mattaj, manuscript submitted). The amino acid sequence of CBP80 did not reveal any discernible structural motifs, but CBP20, consistent with its activity in cap binding, contains a consensus RRM (313, 321, 322). In addition to playing a role in splicing, the CBP20/CBP80 complex functions in the export of snRNAs containing trimethylated cap structures (313).

Nuclear Matrix–Associated Proteins

In recent years experimental evidence that pre-mRNA splicing may take place in association with the nuclear matrix has accumulated (for references see 324). The nuclear matrix is defined as an insoluble structure that remains after extraction of nuclei with non-ionic detergents, nucleases, and high salt, and it consists of a lattice of interconnected filaments composed of protein and RNA, including snRNPs and non-snRNP proteins. The extent of involvement in pre-mRNA splicing of components associated with the nuclear matrix has been addressed in studies with monoclonal antibodies directed against these components (324–326). Several of these antibodies inhibit splicing in vitro and precipitate splicing intermediates and products, suggesting that the antigens are incorporated into spliceosomes. At least some of these proteins are related to the SR family of splicing proteins (326a).

CONCLUSION

Since the discovery of spliced RNAs a large body of evidence has accumulated suggesting that nuclear pre-mRNA splicing is RNA-mediated. At the same time, many proteins that are essential to the splicing process have been identified. Whereas interactions between the spliceosomal RNAs provide the architectural frame for catalysis, protein factors aid in these interactions and play important roles in the recognition and selection of elements necessary for splicing, in the juxtaposition of splice sites by protein-protein contacts, and in conformational rearrangements in the assembling spliceosome. Splicing factors can be grouped into certain classes according to the presence of shared structural motifs, which also provide possible clues as to their function. In addition to their role in constitutive splicing, proteins fulfill tasks in alternative splice-site selection. The presence of variant cDNAs for splicing factors, the isolation of protein kinases that specifically phosphorylate splicing proteins, and the requirement for phosphatase action during splicing suggest that intron excision can be fine-tuned in a tissue- or developmental stage–specific fashion at both the posttranscriptional and posttranslational level. Moreover, as splicing can occur on nascent transcripts, links to the transcription machinery are to be expected. The powerful genetic approaches with yeast have provided information regarding interactions between known splicing proteins and have also led to the identification of novel splicing activities; the use of the yeast two-hybrid system should yield equivalent findings concerning the mammalian splicing apparatus. The isolation of homologs of splicing proteins from mammals, *Drosophila, C. elegans, S. cerevisiae, S. pombe,* and plants should help increase our understanding of the mechanistic similarities and differences in pre-mRNA splicing used in diverse organisms.

ACKNOWLEDGMENTS

I thank Graeme Bilbe and Ueli Schibler for comments on the manuscript, and I am grateful to many colleagues for sharing results prior to publication. Work in my group is supported by the Kanton of Geneva and the Schweizerischer Nationalfonds.

Literature Cited

1. Sharp PA. 1994. *Cell* 77:805–15
2. Beyer AL, Osheim YN. 1988. *Genes Dev.* 2:754–65
3. Baurén G, Wieslander L. 1994. *Cell* 76:183–92
4. Mattaj IW. 1994. *Nature* 372:727–28
5. Wuarin J, Schibler U. 1994. *Mol. Cell. Biol.* 14:7219–25
6. Rio DC. 1993. *Curr. Opin. Genet. Dev.* 3:574–84
7. Valcárcel J, Singh R, Green MR. 1995. See Ref. 329, pp. 97–112
8. Stephens RM, Schneider TD. 1992. *J. Mol. Biol.* 228:1124–36
9. Rymond BC, Rosbash M. 1992. In *The Molecular and Cellular Biology of the Yeast Saccharomyces*, ed. EW Jones, JR Pringle, JR Broach, 2:143–92. Cold Spring Harbor, NY: Cold Spring Harbor Lab.
10. Jackson IJ. 1991. *Nucleic Acids Res.* 19:3795–98
11. Hall SL, Padgett RA. 1994. *J. Mol. Biol.* 239:357–65
12. Filipowicz W, Gniadkowski M, Klahre U, Liu H-X. 1995. See Ref. 329, pp. 65–77
13. Moore MJ, Query CC, Sharp PA. 1993. In *The RNA World*, ed. RF Gesteland, JF Atkins, pp. 303–57. Cold Spring Harbor, NY: Cold Spring Harbor Lab.
14. Madhani HD, Guthrie C. 1994. *Annu. Rev. Genet.* 28:1–26
15. Umen JG, Guthrie C. 1995. *RNA* 1:869–85
16. Beggs JD. 1995. See Ref. 329, pp. 79–95
17. Hernandez N, Keller W. 1983. *Cell* 35:89–99
18. Padgett R, Hardy S, Sharp P. 1983. *Proc. Natl. Acad. Sci. USA* 80:5230–34
19. Krainer AR, Maniatis T, Ruskin B, Green MR. 1984. *Cell* 4:1158–71
20. Lin RJ, Newman AJ, Cheng SC, Abelson J. 1985. *J. Biol. Chem.* 260:14780–92
21. Padgett RA, Konarska MM, Grabowski PJ, Hardy SF, Sharp PA. 1984. *Science* 225:898–903
22. Ruskin B, Krainer AR, Maniatis T, Green MR. 1984. *Cell* 38:317–31
23. Maschhoff KL, Padgett RA. 1993. *Nucleic Acids Res.* 21:5456–62
24. Moore MJ, Sharp PA. 1993. *Nature* 365:364–68
25. Michel F, Ferat J-L. 1995. *Annu. Rev. Biochem.* 64:435–61
26. Padgett RA, Podar M, Boulanger SC, Perlman PS. 1994. *Science* 266:1685–88
27. Weiner AM. 1993. *Cell* 72:161–64
28. Wise JA. 1993. *Science* 262:1978–79
29. Brody E, Abelson J. 1985. *Science* 228:963–67
30. Frendewey D, Keller W. 1985. *Cell* 42:355–67
31. Grabowski PJ, Seiler SR, Sharp PA. 1985. *Cell* 42:345–53
32. Guthrie C, Patterson B. 1988. *Annu. Rev. Genet.* 22:387–419
33. Lührmann R, Kastner B, Bach M. 1990. *Biochim. Biophys. Acta* 1087:265–92
34. Lerner MR, Boyle JA, Mount SM, Wolin SL, Steitz JA. 1980. *Nature* 283:220–24
35. Rogers J, Wall R. 1980. *Proc. Natl. Acad. Sci. USA* 77:1877–79
36. Steitz JA, Black DL, Gerke V, Parker KA, Krämer A, et al. 1988. In *Structure and Function of Major and Minor Small Nuclear Ribonucleoprotein Particles*, ed. ML Birnstiel, pp. 115–54. Berlin: Springer-Verlag
37. Konarska MM, Sharp PA. 1986. *Cell* 46:845–55
38. Michaud S, Reed R. 1991. *Genes Dev.* 5:2534–46
39. Mount SM, Pettersson I, Hinterberger M, Karmas A, Steitz JA. 1983. *Cell* 33:509–18
40. Black DL, Chabot B, Steitz JA. 1985. *Cell* 42:737–50

41. Zhuang Y, Weiner AM. 1986. *Cell* 46: 827–35
42. Ruby SW, Abelson J. 1988. *Science* 242:1028–35
43. Legrain P, Séraphin B, Rosbash M. 1988. *Mol. Cell. Biol.* 8:3755–60
44. Séraphin B, Rosbash M. 1989. *Cell* 59: 349–58
45. Séraphin B, Rosbash M. 1991. *EMBO J.* 10:1209–16
46. Jamison SF, Crow A, García-Blanco MA. 1992. *Mol. Cell. Biol.* 12:4279–87
47. Barabino SM, Blencowe BJ, Ryder U, Sproat BS, Lamond AI. 1990. *Cell* 63: 293–302
48. Michaud S, Reed R. 1993. *Genes Dev.* 7:1008–20
49. Pikielny CW, Rymond BC, Rosbash M. 1986. *Nature* 324:341–45
50. Konarska MM, Sharp PA. 1987. *Cell* 49:763–74
51. Cheng S-C, Abelson J. 1987. *Genes Dev.* 1:1014–27
52. Sawa H, Ohno M, Sakamoto H, Shimura Y. 1988. *Nucleic Acids Res.* 16:3157–64
53. Sawa H, Shimura Y. 1991. *Nucleic Acids Res.* 19:6819–21
53a. Roscigno RF, García-Blanco MA. 1995. *RNA* 1:692–706
54. Roscigno RF, Weiner M, García-Blanco MA. 1993. *J. Biol. Chem.* 268: 11222–29
55. Rymond BC, Rosbash M. 1985. *Nature* 317:735–37
56. Parker RA, Siliciano PG, Guthrie C. 1987. *Cell* 49:229–39
57. Wu J, Manley J. 1989. *Genes Dev.* 3:1553–61
58. Zhuang Y, Weiner AM. 1989. *Genes Dev.* 3:1545–52
59. Query CC, Moore MJ, Sharp PA. 1994. *Genes Dev.* 8:587–97
60. Blencowe BJ, Sproat BS, Ryder U, Barabino S, Lamond AI. 1989. *Cell* 59:531–39
61. Yean SL, Lin RJ. 1991. *Mol. Cell. Biol.* 11:5571–77
62. Company M, Arenas J, Abelson J. 1991. *Nature* 349:487–93
63. Ruskin B, Green MR. 1985. *Science* 229:135–40
63a. Ares M Jr, Weiser B. 1995. *Prog. Nucl. Acid Res.* 50:131–59
64. Datta B, Weiner AM. 1991. *Nature* 352: 821–24
65. Wu JA, Manley JL. 1991. *Nature* 352: 818–21
66. Sun JS, Manley JL. 1995. *Genes Dev.* 9:843–54
67. Madhani HD, Guthrie C. 1992. *Cell* 71:803–17
68. McPheeters DS, Abelson J. 1992. *Cell* 71:819–31
69. Fabrizio P, Abelson J. 1990. *Science* 250:404–9
70. Madhani HD, Bordonné R, Guthrie C. 1990. *Genes Dev.* 4:2264–77
71. Wolff T, Menssen R, Hammel J, Bindereif A. 1994. *Proc. Natl. Acad. Sci. USA* 91:903–7
71a. Madhani HD, Guthrie C. 1994. *Genes Dev.* 8:1071–86
72. Wassarman DA, Steitz JA. 1992. *Science* 257:1918–25
73. Lesser CF, Guthrie C. 1993. *Science* 262:1982–88
74. Sontheimer EJ, Steitz JA. 1993. *Science* 262:1989–96
74a. Newman AJ, Teigelkamp S, Beggs JD. 1995. *RNA* 1:968–80
75. Brow DA, Guthrie C. 1988. *Nature* 334: 213–18
76. Tani T, Ohshima Y. 1991. *Genes Dev.* 5:1022–31
77. Brow DA, Guthrie C. 1989. *Nature* 337: 14–15
78. Yu YT, Maroney PA, Nilsen TW. 1993. *Cell* 75:1049–59
79. Newman AJ, Norman C. 1991. *Cell* 65:115–23
80. Newman AJ, Norman C. 1992. *Cell* 68:743–54
81. Deirdre A, Scadden J, Smith W. 1995. *EMBO J.* 14:3236–46
81a. Yean SL, Lin RJ. 1996. *Gene Expr.*:In press
82. Konforti BB, Koziolkiewicz MJ, Konarska MM. 1993. *Cell* 75:863–73
83. Gozani O, Patton JG, Reed R. 1994. *EMBO J.* 13:3356–67
84. Birney E, Kumar S, Krainer AR. 1993. *Nucleic Acids Res.* 21:5803–16
85. Lazar G, Schaal T, Maniatis T, Goodman H. 1995. *Proc. Natl. Acad. Sci. USA* 92:7672–76
85a. Lopato S, Mayeda A, Krainer A, Barta A. 1996. *Proc. Natl. Acad. Sci. USA*:In press
86. Krainer AR, Conway GC, Kozak D. 1990. *Genes Dev.* 4:1158–71
87. Ge H, Manley JL. 1990. *Cell* 62:25–34
88. Krainer AR, Conway GC, Kozak D. 1990. *Cell* 62:35–42
89. Fu X-D, Maniatis T. 1990. *Nature* 343: 437–41
90. Cavaloc Y, Popielarz M, Fuchs JP, Gattoni R, Stevenin J. 1994. *EMBO J.* 13:2639–49
91. Vellard M, Sureau A, Soret J, Martinerie C, Perbal B. 1992. *Proc. Natl. Acad. Sci. USA* 89:2511–15
92. Zahler AM, Lane WS, Stolk JA, Roth MB. 1992. *Genes Dev.* 6:837–47
93. Nagai K, Oubridge C, Ito N, Avis J, Evans P. 1995. *Trends Biochem. Sci.* 20:235–40

94. Zahler AM, Neugebauer KM, Stolk JA, Roth MB. 1993. *Mol. Cell. Biol.* 13:4023–28
95. Screaton G, Cáceres J, Mayeda A, Bell M, Plebanski M, et al. 1995. *EMBO J.* 14:4336–49
96. Roth MB, Murphy C, Gall JG. 1990. *J. Cell. Biol.* 111:2217–23
97. Theissen H, Etzerodt M, Reuter R, Schneider C, Lottspeich F, et al. 1986. *EMBO J.* 5:3209–17
98. Mancebo R, Lo PC, Mount SM. 1990. *Mol. Cell. Biol.* 10:2492–502
99. Zamore PD, Patton JG, Green MR. 1992. *Nature* 355:609–14
100. Zhang M, Zamore PD, Carmo-Fonseca M, Lamond AI, Green MR. 1992. *Proc. Natl. Acad. Sci. USA* 89:8769–73
101. Chou T-B, Zachar Z, Bingham PM. 1987. *EMBO J.* 6:4095–104
102. Boggs RT, Gregor P, Idriss S, Belote JM, McKeown M. 1987. *Cell* 50:739–47
103. Amrein H, Gorman M, Nöthiger R. 1988. *Cell* 55:1025–35
104. Goralski TJ, Edström J-E, Baker BS. 1989. *Cell* 56:1011–18
104a. Fu X-D. 1995. *RNA* 1:663–80
105. Neugebauer K, Stolk J, Roth M. 1995. *J. Cell Biol.* 129:899–908
106. Mayeda A, Helfman DM, Krainer AR. 1993. *Mol. Cell. Biol.* 13:2993–3001
107. Mayeda A, Zahler AM, Krainer AR, Roth MB. 1992. *Proc. Natl. Acad. Sci. USA* 89:1301–4
108. Fu X-D, Mayeda A, Maniatis T, Krainer AR. 1992. *Proc. Natl. Acad. Sci. USA* 89:11224–28
109. Kim Y-J, Zuo P, Manley JL, Baker BS. 1992. *Genes Dev.* 6:2569–79
110. Fu X-D. 1993. *Nature* 365:82–85
111. Staknis D, Reed R. 1994. *Mol. Cell. Biol.* 14:7670–82
112. Eperon IC, Ireland DC, Smith RA, Mayeda A, Krainer AR. 1993. *EMBO J.* 12:3607–17
113. Kohtz JD, Jamison SF, Will CL, Zuo P, Lührmann R, et al. 1994. *Nature* 368:119–24
114. Jamison S, Pasman Z, Wang J, Will C, Lührmann R, et al. 1995. *Nucleic Acids Res.* 23:3260–67
115. Zahler AM, Roth MB. 1995. *Proc. Natl. Acad. Sci. USA* 92:2642–46
116. Crispino JD, Blencowe BJ, Sharp PA. 1994. *Science* 265:1866–69
117. Tarn W-Y, Steitz JA. 1994. *Genes Dev.* 8:2704–17
118. Crispino J, Sharp P. 1995. *Genes Dev.* 8:2314–23
119. Tarn WY, Steitz JA. 1995. *Proc. Natl. Acad. Sci. USA* 92:2504–8
120. Zuo P, Manley JL. 1994. *Proc. Natl. Acad. Sci. USA* 91:3363–67
121. Wu JY, Maniatis T. 1993. *Cell* 75:1061–70
122. Fields S, Song OK. 1989. *Nature* 340:245–46
123. Fu X-D, Maniatis T. 1992. *Proc. Natl. Acad. Sci. USA* 89:1725–29
124. Black DL. 1995. *RNA* 1:763–71
125. Robberson BL, Cote GJ, Berget SM. 1990. *Mol. Cell. Biol.* 10:84–94
126. Talerico M, Berget SM. 1990. *Mol. Cell. Biol.* 10:6299–305
127. Kuo HC, Nasim FH, Grabowski PJ. 1991. *Science* 251:1045–50
128. Kreivi JP, Zerivitz K, Akusjärvi G. 1991. *Nucleic Acids Res.* 19:6956
129. Hoffman BE, Grabowski PJ. 1992. *Genes Dev.* 6:2554–68
130. Lavigueur A, La Branche H, Kornblihtt AR, Chabot B. 1993. *Genes Dev.* 7:2405–17
131. Sun Q, Mayeda A, Hampson RK, Krainer AR, Rottman FM. 1993. *Genes Dev.* 7:2598–608
132. Watakabe A, Tanaka K, Shimura Y. 1993. *Genes Dev.* 7:407–18
133. Tian M, Maniatis T. 1994. *Genes Dev.* 8:1703–12
134. Lynch KW, Maniatis T. 1995. *Genes Dev.* 9:284–93
135. Ramchatesingh J, Zahler A, Neugebauer K, Roth M, Cooper T. 1995. *Mol. Cell. Biol.* 15:4898–907
136. Tian H, Kole R. 1995. *Mol. Cell. Biol.* 15:6291–98
137. Wang Z, Hoffmann HM, Grabowski PJ. 1995. *RNA* 1:21–35
138. Tian M, Maniatis T. 1993. *Cell* 74:105–14
139. Bruzik J, Maniatis T. 1995. *Proc. Natl. Acad. Sci. USA* 92:7056–59
140. Chiara MD, Reed R. 1995. *Nature* 375:510–13
141. Ring HZ, Lis JT. 1994. *Mol. Cell. Biol.* 14:7499–506
142. Horowitz DS, Krainer AR. 1994. *Trends Genet.* 10:100–6
143. Harper JE, Manley JL. 1992. *Gene Expr.* 2:19–29
144. Zahler AM, Neugebauer KM, Lane WS, Roth MB. 1993. *Science* 260:219–22
145. Cáceres JF, Stamm S, Helfman DM, Krainer AR. 1994. *Science* 265:1706–9
146. Wang J, Manley J. 1995. *RNA* 1:335–46
146a. Himmelspach M, Cavaloc Y, Chebli K, Stévenin J, Gattoni R. 1995. *RNA* 1:794–806
147. Sun Q, Hampson RK, Rottman FM. 1993. *J. Biol. Chem.* 268:15659–66
148. Tacke R, Manley J. 1995. *EMBO J.* 14:3540–51
149. Heinrichs V, Baker B. 1995. *EMBO J.* 14:3987–4000

150. Amrein H, Hedley ML, Maniatis T. 1994. *Cell* 76:735–46
151. Cáceres JF, Krainer AR. 1993. *EMBO J.* 12:4715–26
152. Zuo P, Manley JL. 1993. *EMBO J.* 12:4727–37
153. Sureau A, Soret J, Vellard M, Crochet J, Perbal B. 1992. *Proc. Natl. Acad. Sci. USA* 89:11683–87
154. Popielarz M, Cavaloc Y, Mattei M-G, Gattoni R, Stévenin J. 1995. *J. Biol. Chem.* 270:17830–35
155. Ge H, Zuo P, Manley JL. 1991. *Cell* 66:373–82
156. Sureau A, Perbal B. 1994. *Proc. Natl. Acad. Sci. USA* 91:932–36
157. Ayane M, Preuss U, Köhler G, Nielsen PJ. 1991. *Nucleic Acids Res.* 19:1273–78
158. Fu X-D, Maniatis T. 1992. *Science* 256:535–38
159. Tacke R, Boned A, Goridis C. 1992. *Nucleic Acids Res.* 20:5482
160. Diamond RH, Du K, Lee VM, Mohn KL, Haber BA, et al. 1993. *J. Biol. Chem.* 268:15185–92
161. Sachs A. 1993. *Cell* 74:413–21
162. Wickens M. 1993. *Nature* 363:305–6
163. Mermoud JE, Cohen P, Lamond AI. 1992. *Nucleic Acids Res.* 20:5263–69
164. Tazi J, Daugeron MC, Cathala G, Brunel C, Jeanteur P. 1992. *J. Biol. Chem.* 267:4322–26
165. Mermoud JE, Cohen PTW, Lamond AI. 1994. *EMBO J.* 13:5678–88
166. Tazi J, Kornstädt U, Rossi F, Jeanteur P, Cathala G, et al. 1993. *Nature* 363:283–86
167. Woppmann A, Will CL, Kornstädt U, Zuo P, Manley JL, et al. 1993. *Nucleic Acids Res.* 21:2815–22
168. Gui JF, Lane WS, Fu X-D. 1994. *Nature* 369:678–82
169. Gui JF, Tronchere H, Chandler SD, Fu X-D. 1994. *Proc. Natl. Acad. Sci. USA* 91:10824–28
169a. Colwill K, Pawson T, Andrews B, Prasad J, Manley JL, Bell JC, Duncan PI. 1996. *EMBO J.* 15:265–75
170. Alahari SK, Schmidt H, Käufer NF. 1993. *Nucleic Acids Res.* 21:4079–83
171. Spector DL. 1994. *Nature* 369:604
172. Zamore PD, Green MR. 1989. *Proc. Natl. Acad. Sci. USA* 86:9243–47
173. Ruskin B, Zamore PD, Green MR. 1988. *Cell* 52:207–19
174. Bennett M, Michaud S, Kingston J, Reed R. 1992. *Genes Dev.* 6:1986–2000
175. Staknis D, Reed R. 1994. *Mol. Cell. Biol.* 14:2994–3005
175a. Champion-Arnaud P, Gozani O, Palandjian L, Reed R. 1995. *Mol. Cell. Biol.* 15:5750–56

176. Zamore PD, Green MR. 1991. *EMBO J.* 10:207–14
177. Singh R, Valcárcel J, Green MR. 1995. *Science* 268:1173–76
178. Gaur R, Valcárcel J, Green M. 1995. *RNA* 1:407–17
179. Ptashne M, Gann AAF. 1990. *Nature* 346:329–31
180. Valcárcel J, Singh R, Zamore PD, Green MR. 1993. *Nature* 362:171–75
181. Sailer A, MacDonald NJ, Weissmann C. 1992. *Nucleic Acids Res.* 20:2374
182. Kanaar R, Roche SE, Beall EL, Green MR, Rio DC. 1993. *Science* 262:569–73
183. Potashkin J, Naik K, Wentz-Hunter K. 1993. *Science* 262:573–75
184. Patterson B, Guthrie C. 1991. *Cell* 64:181–87
185. Deleted in proof
186. Abovich N, Liao XC, Rosbash M. 1994. *Genes Dev.* 8:843–54
187. García-Blanco MA, Jamison S, Sharp PA. 1989. *Genes Dev.* 3:1874–86
188. Patton JG, Porro EB, Galceran J, Tempst P, Nadal-Ginard B. 1993. *Genes Dev.* 7:393–406
189. Mullen MP, Smith CWJ, Patton JG, Nadal-Ginard B. 1991. *Genes Dev.* 5:642–55
190. Lin C-H, Patton JG. 1995. *RNA* 1:234–45
191. Gil A, Sharp PA, Jamison SF, García-Blanco MA. 1991. *Genes Dev.* 5:1224–36
192. Patton JG, Mayer SA, Tempst P, Nadal-Ginard B. 1991. *Genes Dev.* 5:1237–51
193. Ghetti A, Piñol-Roma S, Michael WM, Morandi C, Dreyfuss G. 1992. *Nucleic Acids Res.* 20:3671–78
194. Dreyfuss G, Matunis MJ, Piñol-Roma S, Burd CG. 1993. *Annu. Rev. Biochem.* 62:289–321
195. Bothwell ALM, Ballard DW, Philbrick WM, Lindwall G, Maher SE, et al. 1991. *J. Biol. Chem.* 266:24657–63
196. Brunel F, Alzari PM, Ferrara P, Zakin MM. 1991. *Nucleic Acids Res.* 19:5237–45
197. Jansen-Dürr P, Boshart M, Lupp B, Bosserhoff A, Frank RW, et al. 1992. *Nucleic Acids Res.* 20:1243–49
198. Hellen CU, Witherell GW, Schmid M, Shin SH, Pestova TV, et al. 1993. *Proc. Natl. Acad. Sci. USA* 90:7642–46
199. Reed R. 1989. *Genes Dev.* 3:2113–23
200. Kiledjian M, Dreyfuss G. 1992. *EMBO J.* 11:2655–64
201. Sillekens PTG, Beijer RP, Habets WJ, van Venrooij WJ. 1988. *Nucleic Acids Res.* 16:8307–28
202. Piñol-Roma S, Swanson M, Gall J, Dreyfuss G. 1989. *J. Cell Biol.* 109:2575–87

203. Williamson MP. 1994. *Biochem. J.* 297:249–60
204. Courey AJ, Tjian R. 1988. *Cell* 55:887–98
205. Tanese N, Pugh F, Tijan R. 1991. *Genes Dev.* 5:2212–24
206. Tazi J, Alibert C, Temsamani J, Reveillaud I, Cathala G, et al. 1986. *Cell* 47:755–66
207. Gerke V, Steitz JA. 1986. *Cell* 47:973–84
208. Pinto AL, Steitz JA. 1989. *Proc. Natl. Acad. Sci. USA* 86:8742–46
209. Dong BH, Horowitz DS, Kobayashi R, Krainer AR. 1993. *Nucleic Acids Res.* 21:4085–92
210. Horowitz DS, Abelson J. 1993. *Mol. Cell. Biol.* 13:2959–70
211. Horowitz DS, Abelson J. 1993. *Genes Dev.* 7:320–29
212. Besser H, Schnabel P, Wieland C, Fritz E, Stanewsky R, et al. 1990. *Chromosoma* 100:37–47
213. Jones KR, Rubin GM. 1990. *Neuron* 4:711–23
214. Rendahl KG, Jones KR, Kilkarni SJ, Bagully SH, Hall JC. 1992. *J. Neurosci.* 12:390–407
215. Frasch M, Saumweber H. 1989. *Chromosoma* 97:272–81
216. Champlin DT, Frasch M, Saumweber H, Lis JT. 1991. *Genes Dev.* 5:1611–21
217. Newman AJ, Lin RJ, Cheng SC, Abelson J. 1985. *Cell* 42:335–44
218. Fouser LA, Friesen JD. 1986. *Cell* 45:81–93
219. Parker R, Guthrie C. 1985. *Cell* 41:107–18
220. Aebi M, Hornig H, Padgett RA, Reiser J, Weissmann C. 1986. *Cell* 47:555–65
221. Burgess SM, Guthrie C. 1993. *Trends Biochem. Sci.* 18:381–84
222. MacMillan AM, Query CC, Allerson CR, Chen S, Verdine GL, et al. 1994. *Genes Dev.* 8:3008–20
223. Swanson MS. 1995. See Ref. 329, pp. 17–33
224. Mayrand SH, Pederson T. 1990. *Nucleic Acids Res.* 18:3307–18
225. Bennett M, Piñol-Roma S, Staknis D, Dreyfuss G, Reed R. 1992. *Mol. Cell. Biol.* 12:3165–75
226. Swanson MS, Dreyfuss G. 1988. *Mol. Cell. Biol.* 8:2237–41
227. Burd CG, Dreyfuss G. 1994. *EMBO J.* 13:1197–204
228. Choi YD, Grabowski PJ, Sharp PA, Dreyfuss G. 1986. *Science* 231:1534–39
229. Sierakowska H, Szer W, Furdon PJ, Kole R. 1986. *Nucleic Acids Res.* 14:5241–54
230. Mayeda A, Krainer AR. 1992. *Cell* 68:365–75
231. Mayeda A, Munroe SH, Cáceres JF, Krainer AR. 1994. *EMBO J.* 13:5483–95
232. Yang XM, Bani M-R, Lu S-J, Rowan S, Ben-David Y, et al. 1994. *Proc. Natl. Acad. Sci. USA* 91:6924–28
233. Min H, Chan R, Black D. 1995. *Genes Dev.* 9:2659–71
234. Behrens SE, Tyc K, Kastner B, Reichelt J, Lührmann R. 1993. *Mol. Cell. Biol.* 13:307–19
235. Behrens SE, Galisson F, Legrain P, Lührmann R. 1993. *Proc. Natl. Acad. Sci. USA* 90:8229–33
236. Brosi R, Hauri HP, Krämer A. 1993. *J. Biol. Chem.* 268:17640–46
237. Brosi R, Gröning K, Behrens SE, Lührmann R, Krämer A. 1993. *Science* 262:102–5
238. Krämer A, Utans U. 1991. *EMBO J.* 10:1503–9
239. Champion-Arnaud P, Reed R. 1994. *Genes Dev.* 8:1974–83
239a. Gozani O, Feld R, Reed R. 1996. *Genes Dev.* 10:233–43
240. Bennett M, Reed R. 1993. *Science* 262:105–8
241. Chiara MD, Champion-Arnaud P, Buvoli M, Nadal-Ginard B, Reed R. 1994. *Proc. Natl. Acad. Sci. USA* 91:6403–7
242. Krämer A, Legrain P, Mulhauser F, Gröning K, Brosi R, et al. 1994. *Nucleic Acids Res.* 22:5223–28
243. Krämer A, Mulhauser F, Wersig C, Gröning K, Bilbe G. 1995. *RNA* 1:260–72
244. Chang TH, Clark MW, Lustig AJ, Cusick ME, Abelson J. 1988. *Mol. Cell. Biol.* 8:2379–93
245. Legrain P, Choulika A. 1990. *EMBO J.* 9:2775–81
246. Chapon C, Legrain P. 1992. *EMBO J.* 11:3279–88
247. Arenas JE, Abelson JN. 1993. *Proc. Natl. Acad. Sci. USA* 90:6771–75
248. Abovich N, Legrain P, Rosbash M. 1990. *Mol. Cell. Biol.* 10:6417–25
249. Legrain P, Chapon C, Galisson F. 1993. *Genes Dev.* 7:1390–99
250. Ruby SW, Chang T-H, Abelson J. 1993. *Genes Dev.* 7:1909–25
251. Wells SE, Ares M Jr. 1994. *Mol. Cell. Biol.* 14:6337–49
251a. Wells SE, Neville M, Haynes M, Wang J, Igel H, Ares M Jr. 1996. *Genes Dev.* 10:220–32
251b. Yan D, Ares M Jr. 1996. *Mol. Cell. Biol.* 16:818–28
252. Legrain P, Chapon C. 1993. *Science* 262:108–10
253. Burd CG, Dreyfuss G. 1994. *Science* 265:615–21
254. Usheva A, Maldonado E, Goldring A,

Lu H, Houbavi C, et al. 1992. *Cell* 69:871–81

255. Denhez F, Lafyatis R. 1994. *J. Biol. Chem.* 269:16170–79

256. Spikes DA, Kramer J, Bingham PM, Van Doren K. 1994. *Nucleic Acids Res.* 22:4510–19

257. Bach M, Winkelmann G, Lührmann R. 1989. *Proc. Natl. Acad. Sci. USA* 86:6038–42

258. Lossky M, Anderson GJ, Jackson SP, Beggs J. 1987. *Cell* 51:1019–26

259. Anderson GJ, Bach M, Lührmann R, Beggs J. 1989. *Nature* 342:819–21

260. García-Blanco MA, Anderson GJ, Beggs J, Sharp PA. 1990. *Proc. Natl. Acad. Sci. USA* 87:3082–86

261. Brown JD, Beggs JD. 1992. *EMBO J.* 11:3721–29

262. Teigelkamp S, Newman AJ, Beggs JD. 1995. *EMBO J.* 14:2602–12

263. Umen JG, Guthrie C. 1995. *Genes Dev.* 9:855–68

264. Wyatt JR, Sontheimer EJ, Steitz JA. 1992. *Genes Dev.* 6:2542–53

265. Winkelmann G, Bach M, Lührmann R. 1989. *EMBO J.* 8:3105–12

265a. Patterson B, Guthrie C. 1987. *Cell* 49:613–24

266. Shea JE, Toyn JH, Johnston LH. 1994. *Nucleic Acids Res.* 22:5555–64

267. Hodges PE, Jackson SP, Brown JD, Beggs JD. 1995. *Yeast* 11:337–42

268. Behrens SE, Lührmann R. 1991. *Genes Dev.* 5:1439–52

269. Utans U, Behrens SE, Lührmann R, Kole R, Krämer A. 1992. *Genes Dev.* 6:631–41

270. Wassarman DA, Steitz JA. 1991. *Nature* 349:463–64

271. Chen JH, Lin RJ. 1990. *Nucleic Acids Res.* 18:6447

272. Dalbadie-McFarland G, Abelson J. 1990. *Proc. Natl. Acad. Sci. USA* 87:4236–40

273. Burgess S, Couto JR, Guthrie C. 1990. *Cell* 60:705–17

274. Strauss EJ, Guthrie C. 1991. *Genes Dev.* 5:629–41

275. Linder P, Lasko PF, Ashburner M, Leroy P, Nielsen PJ, et al. 1989. *Nature* 337:121–22

276. Fuller-Pace FV. 1994. *Trends Cell Biol.* 4:271–74

277. Schwer B, Guthrie C. 1991. *Nature* 349:494–99

278. Kim SH, Smith J, Claude A, Lin RJ. 1992. *EMBO J.* 11:2319–26

279. Plumpton M, McGarvey M, Beggs JD. 1994. *EMBO J.* 13:879–87

280. Rozen F, Edery I, Meerovitch K, Dever TE, Merrick WC, et al. 1990. *Mol. Cell. Biol.* 10:1134–44

281. Deleted in proof

282. Strauss EJ, Guthrie C. 1994. *Nucleic Acids Res.* 22:3187–93

283. Lin RJ, Lustig AJ, Abelson J. 1987. *Genes Dev.* 1:7–18

284. King DS, Beggs JD. 1990. *Nucleic Acids Res.* 18:6559–64

285. Kim SH, Lin RJ. 1993. *Proc. Natl. Acad. Sci. USA* 90:888–92

286. Teigelkamp S, McGarvey M, Plumpton M, Beggs JD. 1994. *EMBO J.* 13:888–97

287. Schwer B, Guthrie C. 1992. *Mol. Cell. Biol.* 12:3540–47

288. Schwer B, Guthrie C. 1992. *EMBO J.* 11:5033–39

289. Madhani HD, Guthrie C. 1994. *Genetics* 137:677–87

290. Burgess SM, Guthrie C. 1993. *Cell* 73:1377–91

291. Ono Y, Ohno M, Shimura Y. 1994. *Mol. Cell. Biol.* 14:7611–20

292. Smith V, Barrell BG. 1991. *EMBO J.* 10:2627–34

293. Krämer A. 1992. *Mol. Cell. Biol.* 12:4545–52

294. Matunis MJ, Michael WM, Dreyfuss G. 1992. *Mol. Cell. Biol.* 12:164–71

295. Siomi H, Matunis MJ, Michael WM, Dreyfuss G. 1993. *Nucleic Acids Res.* 21:1193–98

296. Siomi H, Siomi MC, Nussbaum RL, Dreyfuss G. 1993. *Cell* 74:291–98

297. Gibson TJ, Thompson JD, Heringa J. 1993. *FEBS Lett.* 324:361–66

298. Duncan R, Bazar L, Michelotti G, Tomonaga T, Krutzsch H, et al. 1994. *Genes Dev.* 8:465–80

299. Siomi H, Choi M, Siomi MC, Nussbaum RL, Dreyfuss G. 1994. *Cell* 77:33–39

300. Siebel CW, Admon A, Rio DC. 1995. *Genes Dev.* 9:269–83

301. Nandabalan K, Roeder GS. 1995. *Mol. Cell. Biol.* 15:1953–60

302. Katz RA, Jentoft JE. 1989. *BioEssays* 11:176–81

303. Frank D, Patterson B, Guthrie C. 1992. *Mol. Cell. Biol.* 12:5197–205

304. Frank D, Guthrie C. 1992. *Genes Dev.* 6:2112–24

305. Ast G, Goldblatt D, Offen D, Sperling J, Sperling R. 1991. *EMBO J.* 10:425–32

306. Miriami E, Angenitzki M, Sperling R, Sperling J. 1995. *J. Mol. Biol.* 246:254–63

307. Delannoy P, Caruthers MH. 1991. *Mol. Cell. Biol.* 11:3425–31

308. Utans U, Krämer A. 1990. *EMBO J.* 9:4119–26

309. Deleted in proof

310. Krainer AR, Maniatis T. 1985. *Cell* 42:725–36

311. Lindsey LA, Crow AJ, García-Blanco MA. 1995. *J. Biol. Chem.* 270:13415–21

312. Jones MH, Frank D, Guthrie C. 1995. *Proc. Natl. Acad. Sci. USA* 92:9687–91

313. Izaurralde E, Lewis J, Gamberi C, Jarmolowski A, McGuigan C, et al. 1995. *Nature* 376:709–12

314. Konarska M, Padgett R, Sharp P. 1984. *Cell* 38:731–36

315. Edery I, Sonenberg N. 1985. *Proc. Natl. Acad. Sci. USA* 82:7590–94

316. Ohno M, Sakamoto H, Shimura Y. 1987. *Proc. Natl. Acad. Sci. USA* 84: 5187–91

317. Patzelt E, Thalmann E, Hartmuth K, Blaas D, Kuechler E. 1987. *Nucleic Acids Res.* 18:6989–95

318. Inoue K, Ohno M, Sakamoto H, Shimura Y. 1989. *Genes Dev.* 3:1472–79

319. Ohno M, Kataoka N, Shimura Y. 1990. *Nucleic Acids Res.* 18:6989–95

320. Izaurralde E, Stepinski J, Darzynkiewicz E, Mattaj IW. 1992. *J. Cell Biol.* 118:1287–95

321. Izaurralde E, Lewis J, McGuigan C, Jankowska M, Darzynkiewicz E, et al. 1994. *Cell* 78:657–68

322. Kataoka N, Ohno M, Kangawa K, Tokoro Y, Shimura Y. 1994. *Nucleic Acids Res.* 22:3861–65

323. Kataoka N, Ohno M, Moda I, Shimura Y. 1995. *Nucleic Acids Res.* 23:3638–41

324. Blencowe BJ, Nickerson JA, Issner R, Penman S, Sharp PA. 1994. *J. Cell Biol.* 127:593–607

325. Zeng C, He D, Berget SM, Brinkley BR. 1994. *Proc. Natl. Acad. Sci. USA* 91:1505–9

326. Chabot B, Bisotto S, Vincent M. 1995. *Nucleic Acids Res.* 23:3206–13

326a. Blencowe BJ, Issner R, Kim J, McCaw P, Sharp P. 1995. *RNA* 1:852–65

327. Krainer AR, Mayeda A, Kozak D, Binns G. 1991. *Cell* 66:383–94

328. Roth MB, Zahler AM, Stolk JA. 1991. *J. Cell. Biol.* 115:587–96

329. Lamond AR, ed. 1995. *Pre-mRNA Processing.* Austin: Landes

Annu. Rev. Biochem. 1996. 65:411–440

MOLECULAR GENETICS OF SIGNAL TRANSDUCTION IN *DICTYOSTELIUM*

Carole A. Parent and Peter N. Devreotes

Department of Biological Chemistry, The Johns Hopkins School of Medicine, Baltimore, Maryland 21205

KEY WORDS: G protein–coupled receptors, aggregation, chemotaxis, gene expression

ABSTRACT

In conditions of starvation, the free living amoebae of *Dictyostelium* enter a developmental program: The cells aggregate by chemotaxis to form a multicellular structure that undergoes morphogenesis and cell-type differentiation. These processes are mediated by a family of cell surface cAMP receptors (cARs) that act on a specific heterotrimeric G protein to stimulate actin polymerization, activation of adenylyl and guanylyl cyclases, and a host of other responses. Most of the components in these pathways have mammalian counterparts. The accessible genetics of this unicellular organism facilitate structure-function analysis and enable the discovery of novel genes involved in the regulation of these important pathways.

CONTENTS

PERSPECTIVES 412
INTERCELLULAR SIGNALS IN GROWTH AND DEVELOPMENT 412
BIOCHEMICAL AND GENETIC TOOLS 415
SIGNAL TRANSDUCTION IN EARLY DEVELOPMENT 416
   Receptor Subtype Switching Programs Development 416
   Responses Mediated by Chemoattractant Stimulation 419
   Many G Protein α-Subunits: Functional Diversity or Redundancy? 420
   G Protein β-Subunit: A Single Gene Mediates All Responses 421
   Adenylyl Cyclase Activation Involves a Novel Cytosolic Regulator 422
   Ras and MAP Kinase Regulate Adenylyl Cyclase Activation 424
   Secreted and Extracellular Membrane-Bound Forms of Phosphodiesterase Regulate
         Ambient Levels of cAMP 425
   Is Intracellular cAMP Necessary for Gene Expression? 426
   Cyclic-GMP Is Essential for Chemotaxis 426
   PLC Activation Is Not Necessary for Chemotaxis or Gene Expression 428

411

0066-4154/96/0701-0411$08.00

DESENSITIZATION OF RECEPTOR-MEDIATED RESPONSES   428
   Receptor Phosphorylation Causes Loss-of-Ligand Binding   429
   Receptor Phosphorylation Is Not Necessary for Adaptation of Adenylyl Cyclase. . . .   429
RECEPTOR-MEDIATED G PROTEIN–INDEPENDENT PROCESSES   430
RAPID PHENOTYPIC SCREENS FOR RANDOM MUTAGENESIS   431
   Identification of Distinct Activation States of cAR1   432
   Isolation of Inactive, G Protein–Insensitive, and Constitutively Active Mutants of
      ACA.   435
CONCLUDING REMARKS   436

PERSPECTIVES

Signal transduction is involved in nearly all physiological events, and defects in signal transduction pathways often give rise to disease. These processes are difficult to study in complex multicellular organisms. Fortunately, studies of microorganisms show that certain signaling strategies have been conserved throughout eukaryotic evolution. In *Dictyostelium*, G protein–linked signal transduction events, in particular, are essential for chemotaxis, cell aggregation, morphogenesis, gene expression, and pattern formation. Many of the proteins involved in these events have mammalian counterparts (1). Thus, the sensing of extracellular stimuli is a fundamental cellular process. Just as the yeast genetic system is useful for studies of cell division and secretion, the genetics and biochemistry of *Dictyostelium* provide powerful tools for the study of signal transduction and chemotaxis (2). Null mutants are constructed to assess the roles of receptors, G-protein subunits, and various effectors. Phenotypic rescue of these mutants provides a convenient screen for random mutagenesis, a technique that can be used to study mammalian as well as endogenous proteins. Insertional mutagenesis is used to discover new genes that feed into these pathways.

This review focuses on signaling in the early stages of development: the transmembrane signal transduction events occurring about 4 h after the initiation of starvation. An in-depth examination of events occurring in late development, as well as cell-fate decision making and morphogenesis, is provided in several recently published reviews (3–7).

INTERCELLULAR SIGNALS IN GROWTH AND DEVELOPMENT

Dictyostelids are free-living protozoa that have developed strategies to survive during starvation. Their life cycle consists of two distinct phases: (*a*) a vegetative or growth stage in which individual amoebae use phagocytosis or pinocytosis to ingest bacteria or liquid media and (*b*) a starvation-induced developmental stage in which amoebae aggregate and, within 24 h, differentiate into a resistant form consisting of spores atop a stalk of vacuolated cells. When adequate environmental conditions recur, the spores germinate and the cycle is repeated. The most well-characterized species is *Dictyostelium discoideum*.

The transition from single cells to multicellularity is mediated by a variety of signaling molecules. One of them is the ubiquitous messenger adenosine 3'-5' cyclic monophosphate (cAMP). Its synthesis, detection, and degradation are exquisitely regulated. Within 4 h of starvation, when the necessary components are maximally expressed, cells begin to secrete the nucleotide. It binds to surface receptors leading to chemotaxis, the synthesis and secretion of additional cAMP (signal relay), and increased early gene expression. Cyclic AMP is produced at 6-min intervals— a specific form of extracellular phosphodiesterase serves to degrade it between pulses. In a cell monolayer these oscillations in the levels of cAMP generate propagating waves. Each passing wave provides a gradient that directs the cells further toward the aggregating center (Figure 1*a*). Cells are attracted to each other and form streams as they assemble. By 10 h, up to 10^5 cells have formed a loose aggregate. At this stage, high, constant levels of cAMP activate specific transcription factors, and the cells within the aggregate differentiate. As a migrating slug appears, the prestalk cells (15%) and the prespore cells (75%) sort to the front and back of the structure, respectively. "Anterior-like" cells (10%) form isolated islands in the prespore region. The cell-type decision is governed by position in the cell cycle at the time of starvation and through the action of specific morphogens including differentiation-inducing-factor (DIF), adenosine, and ammonia. These compounds direct a series of morphological rearrangements, and culmination into a mature fruiting body finally occurs (Figure 1*b*).

Even in the vegetative and preaggregatory stages, these free living amoebae display a high degree of interaction by secreting and sensing specific signaling molecules. Pterins bind to specific receptors on the cell surface, activate guanylyl and adenylyl cyclases and actin polymerization, and induce chemotactic responses similar to those induced by cAMP in early aggregation (8–12). The cells can move toward bacteria by sensing folic acid. Growing amoebae secrete a glycoprotein called prestarvation factor (PSF), which accumulates in proportion to cell density, serving as a sensor for the availability of nutrients (13–15). At high PSF/bacteria ratios, a prestarvation response is initiated. The expression of several genes involved in early aggregation is partially increased, thereby preparing cells to enter the developmental program. How PSF mediates these effects is not known; the components used for cell-cell signaling during aggregation are not required (16).

Once the cells have starved, other diffusible molecules are secreted, presumably also for sensing cell density (17). One of these, conditioned medium factor (CMF), is involved in regulating cAMP signaling (18–20). CMF has been cloned, and cells transformed with antisense genes are unable to undergo

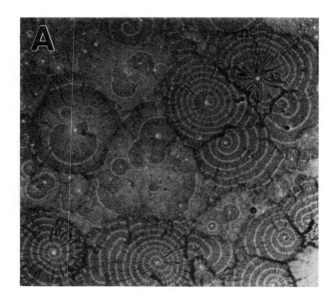

Figure 1   *Dictyostelium* chemotaxis and morphological development. (*a*) Dark field photography of aggregating cells (5 h after the onset of starvation). Each territory of 1–2 cm in diameter contains approximately 1×10^6 cells. (*b*) A composite picture depicting the various developmental stages of *Dictyostelium*. Clockwise from bottom right corner: cell aggregate, mound, tipped mound, first finger, migrating slug (off set; 2–3 mm in length), early-, mid-, late-culminant and fruiting body (2–3 mm tall). Photographs *a* and *b* courtesy of Drs. Peter C Newell and Larry Blanton, respectively.

development because they cannot respond to cAMP (21, 22). When recombinant CMF is given to the antisense cells, cAMP-mediated responses are restored within 30 s indicating that CMF is required, not for expression, but for function of one or more components of the signaling apparatus (23). Cells also respond to various other activators including platelet activating factor (PAF), lysophosphatidic acid (LPA), and yeast extract (24–26). In summary, throughout the life cycle, intercellular communication plays a central role in defining both growth and developmental outcome whether the cells are dispersed (vegetative stage) or assembled in a multicellular structure (developmental stage).

BIOCHEMICAL AND GENETIC TOOLS

GENOME *Dictyostelium's* genome is made of 40,000 kbp organized in six linkage groups (27). Through mutagenic and genetic studies, the total number of genes present in the genome is estimated to be around 7000 (28; WF Loomis, personal communication). Mutations in about 300 genes yield viable cells that display aberrations in development. A parasexual genetic system is used to assign mutants to linkage groups and to cross mutations into a single strain (29). Megabase-restriction and yeast artificial chromosome (YAC) contiguous maps spanning the entire genome have recently been completed. These can be used to rapidly localize the position of any gene and clone the surrounding locus. Completion of the maps has paved the way for sequencing of the genome (WF Loomis & A Kuspa, personal communication).

CELL CULTURE AND MUTANT HANDLING The availability of wild-type and axenic strains makes it possible to grow cells on bacterial lawns or in liquid cultures of defined media with doubling times of 4 and 12 h, respectively (30). Over 10^{11} clonal cells in 10 liters can be grown in a few days. The vegetative and developmental stages are completely independent, and switching between the two states is trivial. Since the process is readily reversible, developmental mutants can be easily selected and then propagated by returning them to liquid media (31). Mutant storage is accomplished by freezing amoebae or spores at $-70°$ C or by desiccating spores (30).

HOMOLOGOUS RECOMBINATION The disruption of endogenous genes by homologous recombination is efficient and predictable. Close to 100 genes involved in signal transduction, cell motility, and cell differentiation have been targeted (WF Loomis, personal communication). Double and triple knockouts are readily achieved in single or tandem transformation steps. Most signal transduction genes are not essential for growth. Since the cells are free living, gene deletions that might be lethal in an organism can often be studied.

HIGH-EFFICIENCY TRANSFORMATION   Establishment of permanent cell lines is achieved within two weeks of transformation. There are six different available selectable markers (neomycin, hygromycin, bleomycin, blasticidin, uracil, thymidine). Endogenous plasmids, present in *Dictyostelium*, have been used to construct shuttle vectors carrying specific promoters (32–35). These vectors segregate and have transformation efficiencies of 10^{-3}. They are used to carry out mutagenic analysis of selected genes by complementation of null mutants: Randomly mutagenized libraries are constructed in these extrachromosomal expression vectors, transfected into the mutants, and the resulting transformants are screened for phenotypic abnormalities (see below). Heterologous expression of mammalian genes with retention of function is also possible, allowing the use of *Dictyostelium* genetics to be applied to the study of mammalian proteins (36–39).

RESTRICTION ENZYME–MEDIATED INTEGRATION   Restriction enzyme–mediated integration (REMI), first described in yeast, has been successfully used to isolate greater than 20 new gene products involved in aggregation and late development (40, 41; WF Loomis, personal communication). REMI is performed by cotransforming a linearized plasmid with a restriction enzyme and generating random insertions of the plasmid within the genome at the corresponding restriction sites. The cloning of the DNA flanking the insertion is achieved by genomic DNA digestion, circularization, and transformation into *E. coli*. To verify that the recovered DNA sequence is responsible for the phenotype, the rescued plasmid is used to recreate the genotype by homologous recombination.

SIGNAL TRANSDUCTION IN EARLY DEVELOPMENT

A panoply of signaling components involved in cell-cell communication have been cloned and characterized, and the genes for most of these elements have been deleted. The resulting developmental and biochemical phenotypes have provided a detailed picture of how signaling leads to multicellular development. The expression pattern and properties of these genes will be discussed. For many of the signaling elements, multiple minor transcripts are expressed throughout development— the predominant ones are illustrated in Figure 2. Figure 3 depicts a model of the signaling pathways.

Receptor Subtype Switching Programs Development

Four cAMP receptors (cAR1–cAR4) are sequentially expressed throughout development (42–45). cAR1 is expressed early in development in all cells (Figure 2). A second cAR1 transcript, encoding the same protein, is expressed

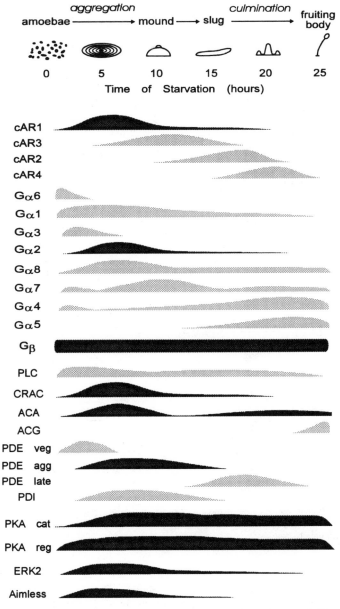

Figure 2 Time course of expression of major RNA transcripts. Top panel illustrates the developmental stages appearing after the onset of starvation. Receptors, G proteins, regulators, and several effectors are shown below. Components essential for early development are heavily shaded. See text for details.

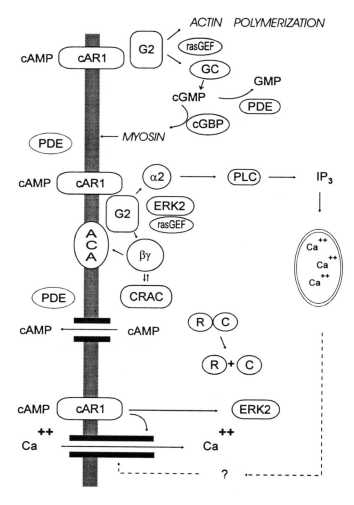

Figure 3 Signal transduction pathways. Model illustrating the proposed events following activation of surface receptors by cAMP leading to chemotaxis, signal relay, and gene expression. See text for details.

at low levels during the multicellular stages. The expression of the early transcript depends on cAMP pulses present in early development: The late transcript is induced by the presence of continuous levels of cAMP (46, 47). cAR3 appears next and is confined to prespore cells. cAR2 and cAR4 are expressed in the prestalk cells of mounds and slugs (45, 47a, 47b) (Figure 2).

The deduced amino acid sequence of each receptor conforms to the seven transmembrane topology common to members of the superfamily of G protein–coupled receptors. The proteins are 60% identical within their transmembrane domains and connecting loops (48). The C-terminal cytoplasmic domains differ extensively in sequence and twofold in length. cAR1 and cAR3 display high-affinity cAMP binding sites with Kds of about 200 and 700 nM, respectively. cAR2 binds cAMP with a Kd in the μM range (49). This range of affinities mirrors the cAMP concentrations thought to be present in the early and late developmental stages (50).

Deletions of individual and/or multiple receptors have been examined. Cells that lack cAR1 are aggregation deficient and, in response to low concentrations of cAMP, do not activate adenylyl cyclase and do not carry out chemotaxis (51, 52). However, *car1⁻* cells will respond to higher concentrations of cAMP and can be forced to differentiate into fruiting bodies (51). Deletion of both cAR1 and cAR3 renders the cell completely insensitive to cAMP, suggesting that cAR1 and cAR3 are functionally redundant (53). Recent studies show that cAR2 and a series of cAR1/cAR2 chimeras, when ectopically expressed in *car1⁻/car3⁻* cells, can trigger all the signal transduction events typically mediated by cAR1 and cAR3 (see below) (54; JY Kim & PN Devreotes, unpublished results). Cells that lack cAR2 or cAR4 aggregate normally but then arrest at multicellular stages of development. The *car2⁻* cells form tipless mounds; the *car4⁻* cells form abnormal slugs and fruiting bodies (43, 45). Taken together, these observations suggest that the cARs are linked to the same signaling pathways. The differential expression of four receptors with different affinities for the same ligand may allow particular cell types to respond differently to the same external cAMP signals. These differential responses might be used to drive differentiation and the morphological movements of the cells.

Responses Mediated by Chemoattractant Stimulation

The binding of cAMP to cells leads to multiple events having distinct time courses and patterns of regulation (reviewed in 55). Within 10 to 30 s after the addition of a stimulus, inositol triphosphate (IP_3) and guanosine 3′-5′ cyclic monophosphate (cGMP) levels increase and a dramatic rise in filamentous actin occurs. The light chains of myosin I and II are phosphorylated; myosin II is translocated to the cytoskeleton where its heavy chain is phosphorylated (56, 57; M Titus, personal communication). After a 5 s lag, a transient influx of calcium occurs and the level of cytosolic calcium rises (58–60). Within 30 s, a talin homologue is recruited to the tips of filopods (61). Translocation of a cytosolic regulator of adenylyl cyclase occurs, and accumulation of cAMP peaks within one minute (62). A MAP kinase is activated and a G protein α-subunit is phosphorylated with a similar time course (62a). As with other G protein–coupled receptors, cAMP binding to cAR1 results in persistent phosphorylation of the receptor and subsequent desensitization (see below). Some of these events presumably come together to mediate chemotaxis; others are involved in cell-cell signaling or in gene expression.

Many G Protein α-Subunits: Functional Diversity or Redundancy?

Eight G protein α-subunits have been isolated (63–67). The mRNA for each has been examined throughout development. Gα6 is present exclusively in growth stage cells; Gα1, Gα2, Gα3, Gα7, and Gα8 are expressed in early development; and Gα4 and Gα5 appear predominantly in the multicellular stages (Figure 2). All the α-subunits are proposed to couple to a unique constitutively expressed β-subunit (Figure 2). The γ-subunit has yet to be cloned. The eight α-subunits are ~35–50% identical to each other and their mammalian counterparts (68). They do not fall into distinct classes nor within any of the Gs, Gi, or Gq subfamilies. Most contain all of the highly conserved features of the guanine nucleotide binding domains found in mammalian α-subunits. Gα8 differs: It has a much longer and divergent C-terminal region, the putative site of receptor interaction.

The functions of seven of the α-subunits have been assessed by gene targeting (Gα6 remains to be analyzed) (67, 69–71; RA Firtel, personal communication). Only Gα2 and Gα4 show striking phenotypes, discussed in detail below. The deletions of the others have not displayed major growth or morphological aberrations. *gα1⁻* cells show slight abnormal morphogenesis and conditional defects in phospholipase C (PLC) regulation (69, 72). Cells lacking Gα5 do not display morphological abnormalities but seem defective in developmental timing (RA Firtel, personal communication). Moreover, in an experiment that might address redundancy, expression of constitutively active forms of Gα1 and Gα7 produced developmental abnormalities (67, 69). Additional experiments of this kind are needed to delineate the functions of the individual subunits.

Null mutants lacking functional Gα2 genes do not aggregate or differentiate (73). These mutants show no cAMP-induced stimulation of adenylyl cyclase, guanylyl cyclase, PLC, or actin polymerization. Moreover, membranes derived from *gα2⁻* cells do not exhibit guanosine triphosphate (GTP) regulation of cAMP binding affinity or guanosine-5′-0-(3-thiotriphosphate) (GTPγS) activation of PLC (74–76). Perhaps because of persistent desensitization, both dominantly active and inactive point mutations in Gα2 yield essentially loss-of-function phenotypes (77). Taken together, these data demonstrate that Gα2 is the G protein coupled to cAR1 and cAR3 and that this complex mediates the cAMP-dependent events in aggregating cells (Figure 3).

The $g\alpha2^-$ cells have also been used to test whether receptors for other agonists use G2. Folic acid–induced responses are normal in $g\alpha2^-$ cells (74, 78). PAF, on the other hand, seems to mediate some of its effects through the Gα2 subunit. The intracellular concentration of PAF increases following the addition of cAMP (24). Exogenously added PAF amplifies cAMP-dependent cellular oscillations as well as activation of adenylyl and guanylyl cyclases. In addition, PAF can independently stimulate a net calcium uptake, an effect which is abolished in $g\alpha2^-$ cells. Through the use of selective inhibitors, it has been proposed that PAF stimulates calcium influx via the IP_3 signaling pathway (79, 80).

As reported for several mammalian G protein α-subunits, Gα2 is phosphorylated in response to agonist (81). The phosphorylation occurs at physiological concentrations of cAMP, depends upon cAR1, and coincides with the time course of activation of adenylyl cyclase. Analysis of a series of Gα1/Gα2 chimeras and site-directed mutants of Gα2 demonstrated that serine 113 is the exclusive site of phosphorylation (82). The physiological significance of this receptor-mediated modification remains unknown— expression of the S113A mutant in $g\alpha2^-$ cells appears to restore the phenotype to that of wild-type cells (82).

Null mutants of Gα4 show both growth and developmental phenotypes (71). On bacteria, cells lacking Gα4 grow slower than wild-type cells. Upon starvation, they aggregate and differentiate normally to the mound stage but show abnormal late development and a low level of spore production. These phenotypes are consistent with the expression pattern of the Gα4 subunit: low in vegetative cells, absent in aggregation, and at high levels again at the mound stage (Figure 2). Biochemical analysis of wild-type and Gα4 null cells revealed that Gα4 couples to specific subtypes of folic acid receptors expressed in vegetative and differentiated cells (83). Consequently, the cells use folic acid not only as a chemoattractant, but also as a signal for multicellular differentiation in later development. Although $g\alpha4^-$ cells lose folic acid–stimulated responses, they retain normal cAMP-mediated adenylyl and guanylyl cyclase activation, as well as chemotaxis (71).

G Protein β-Subunit: A Single Gene Mediates All Responses

The unique G protein β-subunit in *Dictyostelium* is over 90% homologous to its *C. Elegans*, *Drosophila*, and mammalian counterparts (84). The homology extends over the seven characteristic conserved sequences designated WD repeats located in the C-terminal. Every WD repeat is more similar to those in other species than to neighboring repeats. The recent crystal structure of mammalian heterotrimeric G proteins shows that these repeats form the blades of a propeller structure (84a). Although the functional significance of these repeats is not understood, their high degree of conservation does imply functional equivalence (85).

The Gβ null mutants have been useful for evaluating the role of heterotrimeric G proteins in a variety of processes. G proteins are not essential for cell-cycle and growth-related processes (84). The $g\beta^-$ cells are normal in size and display typical growth in liquid culture. Although the cells show apparently normal motility on agar surfaces and glass, they do grow more slowly on bacterial lawns, suggesting a defect in phagocytosis. Upon starvation, $g\beta^-$ cells fail to aggregate. cAMP-mediated adenylyl cyclase, guanylyl cyclase, PLC activation, and actin polymerization are completely absent. The cells do not carry out chemotaxis to any chemoattractant tested and do not accumulate cGMP in response to folic acid (86; PN Devreotes, unpublished observations). Moreover, membranes of $g\beta^-$ cells display only low affinity, GTP-insensitive, cAMP binding sites. Taken together, these data show that all G protein–dependent pathways are mediated through this unique β-subunit (Figure 3). Whether this subunit dimerizes with a single γ-subunit or with a family of multiple γ-subunits, as observed in mammalian cells, remains to be seen.

Adenylyl Cyclase Activation Involves a Novel Cytosolic Regulator

Two distinct forms of adenylyl cyclase are expressed during development, adenylyl cyclase for aggregation (ACA) and adenylyl cyclase for germination (ACG) (Figure 2) (87). ACA is expressed during aggregation. It shares homology and a typical 12-transmembrane topology with the *Drosophila* and mammalian G protein–coupled adenylyl cyclases. This enzyme is responsible for the synthesis of cAMP required for cell-cell signaling. Cells lacking ACA are devoid of chemoattractant-induced adenylyl cyclase activity and, of course, will not spontaneously aggregate. These cells demonstrate normal cAMP-mediated guanylyl cyclase activation and show chemotaxis towards cAMP, indicating that the upstream signaling components are present and functional. Even though ACA topologically resembles a "transporter," aca^- cells expressing ACG are still capable of secreting cAMP, suggesting that there are independent cAMP transporters (87). ACG is a novel form of adenylyl cyclase, predicted to have a single transmembrane helix separating large intracellular and extracellular domains. ACG is normally expressed exclusively during germination (Figure 2). When expressed in aggregating cells the enzyme is constitutively active and insensitive to GTPγS. Nevertheless, ACG suppresses the aggregationless phenotype of the aca^- cells and produces miniature fruiting bodies (87). Recent evidence shows that ACG can be activated by changes in osmolarity and, for this reason, ACG is proposed to be involved in osmosensing in spores (P Schaap, personal communication).

Even though ACA resembles the mammalian enzymes, its mechanism of activation is novel. As described above, both $g\alpha2^-$ and $g\beta^-$ cells lack agonist-induced adenylyl cyclase activity. However, GTPγS will stimulate ACA in membranes of $g\alpha2^-$ cells, indicating that Gα2 does not directly activate the enzyme (73). The GTPγS activation appears to be mediated through the $\beta\gamma$-subunits since $g\beta^-$ cells are completely insensitive to GTPγS (86). Point mutations in Gβ specifically prevent activation of the enzyme without preventing coupling of G2 to cAR1 (L Wu, T Jin & PN Devreotes, unpublished observations). It has been proposed that in $g\alpha2^-$ cells, GTPγS releases $\beta\gamma$-subunits from G proteins other than G2 (Gα1, Gα3, Gα7, or Gα8), which are then able to activate ACA (86). In this respect, ACA is analogous to the mammalian type II and IV adenylyl cyclases which are conditionally activated by G protein $\beta\gamma$-subunits (88). Unlike other effectors/regulators that interact with G$\beta\gamma$-subunits, such as the β-adrenergic receptor kinase 1 and 2 (βARK1, βARK2), PLC, phosducin, and the atrial K^+ channel (GIRK1), adenylyl cyclases do not have pleckstrin homology (PH) domains (89, 90). A short sequence common to the mammalian type II and IV enzymes has been suggested as the $\beta\gamma$-subunit contact site (91). In *Dictyostelium*, however, a novel cytosolic protein containing a PH domain is essential for ACA regulation.

The discovery of this protein, named CRAC for cytosolic regulator of adenylyl cyclase, came from analysis of an aggregation-deficient mutant that lacked receptor- and G protein–stimulated adenylyl cyclase activity. GTPγS-induced activation of ACA in mutant lysates could be restored by the addition of cytosol derived from wild-type cells (92). This assay was used to purify the protein and obtain N-terminal sequence (93). The gene for CRAC was obtained from a REMI mutant displaying the same phenotype as the original mutant. The deduced amino acid sequence of 698 residues is hydrophilic and rich in threonine and serine residues— the PH domain is in the N-terminus. There are no other obvious sequence motifs (62). The original mutant bears an in-frame 53–amino acid deletion in the C-terminal region of the protein (P Lilly & PN Devreotes, unpublished observations). Like ACA and cAR1, CRAC is maximally expressed in early aggregation (Figure 2) (62).

The mechanism by which CRAC activates ACA is intriguing. The protein is translocated to membranes following chemoattractant stimulation of intact cells or during GTPγS activation of lysates. The GTPγS-induced relocalization does not require ACA, cAR1, or Gα2; it occurs in lysates of cells lacking each of these genes. However, this relocalization does not take place in mutants lacking the G protein β-subunit (94). These results suggest that CRAC binds directly to activated $\beta\gamma$-subunits or that the generation of its binding sites depends on $\beta\gamma$-subunits (Figure 3). Given what is known about PH domains, the N-terminal of CRAC likely mediates the $\beta\gamma$-subunits-CRAC association. Consequently, the mode of recruitment of CRAC to the membrane could be analogous to that of βARK's interaction with released G protein βγ-subunits (95). Since the G protein–coupled receptor cascade is highly conserved from mammals to *Dictyostelium*, we propose that a mammalian homologue of CRAC exists and that this novel mode of regulation is important for activation of adenylyl cyclases in higher eukaryotes.

Ras and MAP Kinase Regulate Adenylyl Cyclase Activation

The pathway leading from cAR1 to the activation of ACA displays further complexity (Figure 3). This pathway involves one of two *Dictyostelium* MAP kinase homologues, extracellular signal–regulated kinase 2 (ERK2) (96). ERK2 was isolated as an aggregation-deficient REMI mutant. Northern analysis reveals that ERK2 mRNA is present in vegetative cells and in early development, peaking at ~4 h after the onset of starvation (Figure 2) (96). Surprisingly, cells lacking ERK2 are specifically defective in cAMP-stimulated adenylyl cyclase activation. Receptor-mediated activation of guanylyl cyclase is unaffected in these cells (96). Recent experiments have demonstrated that ERK2 is transiently activated by chemoattractants. This is a receptor-mediated response: It is completely lost in cells lacking both cAR1 and cAR3 (62a). It is not clear whether ERK2 acts directly, for instance, by phosphorylation of CRAC, ACA, or G protein βγ-subunits, or more indirectly, by inducing the expression of yet another essential component required for the activation of ACA.

Another component required for activation of ACA was also originally identified as an aggregation-deficient REMI mutant. The gene is called *aimless* (*ale*) because the mutant is impaired in chemotaxis to cAMP, but shows normal random motility. Molecular cloning revealed that Aimless is a homologue of the yeast cdc25 gene, a ras guanine exchange factor (rasGEF) (173). Aimless displays strong homology to other known rasGEFs in its C-terminal region. Northern analysis shows that Aimless is present in growth and in early development (Figure 2). In lysates prepared from *ale*⁻ cells, GTPγS-mediated adenylyl cyclase activation is defective and cannot be corrected by the addition of exogenous CRAC. This suggests that the defect lies either in the generation of CRAC binding sites or in the capacity of ACA to respond to activated βγ-subunits and/or CRAC. The nature of the defect in chemotaxis is unknown.

The target of Aimless must be a ras-like protein. In *Dictyostelium*, at least six *ras* genes have been identified (*rasD, rasG, rasB, rap1, rasS, rasC*), each having a specific expression pattern (97–101). They share the four well-conserved GTP-binding domains as well as the C-terminal CAAX box, although rasC and rasS are considerably divergent from the mammalian consensus. Since specific null mutants are not yet available, little is actually known about the physiological functions of the ras family of proteins in *Dictyostelium*. The overexpression of an activated form of rasD results in the appearance of multi-tipped aggregates and a decrease in the level of cAR1 expression (102, 103). In mammals, rap1A is known to revert the transformed morphology of cells overexpressing activated ras (104). A similar effect is observed in *Dictyostelium*, where the co-expression of the rap1 protein partially reverses the multi-tipped phenotype (G Weeks, personal communication). Overexpressing rap1 in wild-type cells results in amoebae displaying an enlarged actin cortex, although little effect on development is measured (105). Finally, cells expressing an activated form of rasG, which is normally expressed in growth, show similar altered cytoskeletal function but fail to aggregate (G Weeks, personal communication).

Secreted and Extracellular Membrane-Bound Forms of Phosphodiesterase Regulate Ambient Levels of cAMP

The extracellular concentration of cAMP is tightly controlled through the expression of membrane bound and secreted forms of phosphodiesterase (PDE) and a specific phosphodiesterase inhibitor (PDI) (106). The PDE gene encodes an hydrophilic 452–amino acid protein (107). Three distinct PDE transcripts are expressed during growth, aggregation, and late development (Figure 2) (108). The transcripts contain the same protein coding sequence linked to three overlapping 5′-untranslated sequences. During aggregation, PDE is found in membrane bound (mPDE) and extracellular (ePDE) forms (109, 110). The mechanism by which mPDE is associated with the membrane is not understood; the protein contains a signal sequence but no transmembrane domain. The PDI is expressed in early aggregation (Figure 2) (111). This 26 kDa cysteine-rich soluble protein tightly binds ePDE, changing its K_m for cAMP from 5 µM to 2 mM (112).

Mutants lacking phosphodiesterase activity are unable to undergo aggregation and remain as a smooth monolayer indefinitely (113). Since the cells are unable to degrade cAMP, they cannot support the cAMP oscillations essential for proper development. These mutants can aggregate following the addition of exogenous PDE, suggesting that ePDE is sufficient for degradation of cAMP (114). Previous experiments have suggested that the mPDE alone is also sufficient (110). The physiological role of PDI is unclear because its deletion by homologous recombination does not cause an obvious developmental phenotype (115).

The expression of PDE and PDI is closely regulated by extracellular cAMP. Curiously, even though ePDE and mPDE are derived from the same transcript, they are differentially regulated. The secreted form is increased most effectively by continuous applications of cAMP; the membrane-bound form is induced by intermittent applications of cAMP and suppressed by the constant presence of cAMP (116). A post-translational modification or the presence of a membrane-associated PDE binding protein could explain these distinct expression patterns. As might be expected, the expression of the PDI gene is repressed by extracellular cAMP (111).

Is Intracellular cAMP Necessary for Gene Expression?

In *Dictyostelium* the cAMP-dependent protein kinase (PKA) is composed of a single regulatory (R) and catalytic (C) subunit (117). The 37 kDa R subunit shares extensive homology with that of mammals (118). However, the C subunit is almost twice the size of its mammalian, *Drosophila*, or yeast counterparts (73 vs 41 kDa) with similarity confined to the C-terminal domain (119, 120). The expression of the R and C subunits is developmentally regulated. Low levels are found in growing cells and expression increases throughout early aggregation (Figure 2) (119–122).

Mutants in either the C or R subunits revealed that PKA is essential for development and gene expression. Disruption of the C subunit yields viable, motile cells that are unable to aggregate and differentiate (119). A similar developmental phenotype is observed in cells constitutively overexpressing the R subunit or expressing a mutant R subunit, Rm, which cannot bind cAMP (123–125). Overexpression of Rm from either prespore or prestalk promoters essentially eliminates the corresponding cell types (126, 127). The former mutants form fruiting bodies with "glassy" heads filled with amoebae rather than spores; the latter form slugs that migrate indefinitely. Cells lacking the R subunit activity or overexpressing the C subunit progress more rapidly through the later stages of development (128, 129).

The expression of genes at all stages of development requires both cAMP and PKA. However, occupancy of PKA by cAMP is neither necessary nor sufficient to activate the genes (130, 131). Instead, most of the effects of cAMP are mediated by the cARs and do not require intracellular cAMP (132). The apparent cAMP-independent PKA activity may be due to mismatching in the levels of the R and C subunits or to stimulation by other compounds such as cGMP or calcium (130, 133).

Cyclic-GMP Is Essential for Chemotaxis

Several lines of evidence point to a fundamental role for cGMP in chemotactic orientation. Chemoattractant-stimulated guanylyl cyclase activation is very brief, peaking after ~10 s and returning to basal levels within 30 s (11, 134). The rapid turn-off may be mediated by an elevation in intracellular calcium (135). The time course of activation correlates with myosin II heavy chain phosphorylation (56). A mutant, *stm F*, defective in a specific cGMP PDE, displays an extended cGMP response, delayed myosin II phosphorylation, and prolonged chemotactic orientation, suggesting that cGMP is important in chemotaxis (reviewed in 136; 137). The properties of a series of nonchemotactic mutants, designated KI1-KI10, more directly address the potential role of cGMP in chemotaxis. The mutants were selected as aggregation-deficient and then screened for defects in both folic acid– and cAMP-induced chemotaxis. Nine of the ten mutants are deficient in signal transduction events involving cGMP, although they all display normal cGMP phosphodiesterase activity. Genetic analysis reveals that nine mutants are recessive and belong to distinct complementation groups. One mutant, KI10, is dominant (26).

Two mutants, KI-8 and KI-10, lack folic acid– and cAMP-mediated guanylyl cyclase activation. KI-8 is defective in guanylyl cyclase activity per se, while KI-10 is deficient in the activation of the enzyme (26). Additional experiments suggest a direct link between cGMP activation and myosin II heavy chain assembly and phosphorylation: Both responses are absent in KI-10 cells (138). Interestingly, cAMP-stimulated actin polymerization takes place in mutant KI-10, suggesting that actin polymerization is not sufficient to induce a chemotactic response (138). Moreover, cells lacking or overexpressing the myosin heavy chain kinase, a homologue of mammalian protein kinase C (PKC), show altered polarization, chemotaxis, and development (139; 139a). Cyclic GMP must be involved in more than regulation of myosin II, since cells in which the myosin heavy chain has been deleted are viable and display many forms of movement, including chemotaxis (140).

Further characterization of KI4/KI5 and KI2/KI7 also points to a central role for cGMP in chemotaxis. These mutants display normal guanylyl cyclase and phosphodiesterase activities although chemoattractant-induced cGMP accumulation is low in KI4/5 and high and delayed in KI2/7. Recent experiments indicate that these mutants have defects in a cytosolic cGMP-binding protein (cGBP) (141). It has been suggested that guanylyl cyclase is inhibited by phosphorylation. This inhibition is strongly promoted by 8-bromo-cGMP. Thus, cGBP is proposed to be a kinase involved in the regulation of guanylyl cyclase (H Kuwayama & PJM Van Haastert, personal communication).

In addition to being involved in chemotaxis, guanylyl cyclase activation plays a role in mediating responses to osmotic stress (141a). In wild-type cells modest changes in osmolarity induce the synthesis of cGMP which, in turn, stimulates phosphorylation of the myosin heavy chain and leads to a series of events which together bring resistance against osmotic stress. Cells lacking myosin heavy chain or altered in cGMP synthesis (KI4, KI5, KI7, and KI8 mutants) are more sensitive to osmotic shock than wild-type cells. The signal transduction mutants *car1⁻/car3⁻* and *gβ⁻* display normal osmotic shock responses, suggesting that the signal to the guanylyl cyclase does not use this pathway.

A series of mutants in the phototaxis loci were isolated and found to have altered cGMP responses to light or heat stimulation while displaying normal cAMP-stimulated cGMP synthesis (142). Additional experiments need to be performed in order to fully understand the signaling pathway involved in these responses since mutant *stm F* (which is defective in cGMP PDE) exhibited wild-type light and heat responses.

PLC Activation Is Not Necessary for Chemotaxis or Gene Expression

As noted above, cAMP stimuli elevate IP_3 levels. Moreover, an antagonist of chemotaxis, $3'-NH_2$-cAMP, causes a decrease in IP_3 levels (144). The only known PLC gene in *Dictyostelium* encodes a 91 kDa protein which shares homology with the mammalian PLCδ enzyme in its C-terminal domain (143). This gene is expressed throughout development and is responsible for the measurable cAMP-dependent PLC activity. Cells in which the PLC gene has been deleted by homologous recombination no longer elevate IP_3 levels in response to stimulation but otherwise show no developmental or biochemical phenotype (145). This result is surprising in view of the extensive roles that have been attributed to receptor-mediated PLC activation in mammalian cells (146).

Further studies have addressed the roles of lipids in signaling. Cells lacking PLC have tonic levels of intracellular IP_3 (145). Recent studies suggest that the IP_3 comes from a PLC-independent route of synthesis in which IP_5, synthesized through sequential phosphorylation of inositol, is degraded to IP_3 via two IP_4 isomers (147). The enzyme proposed to be involved in IP_5 metabolism displays similarity to the mammalian Multiple Inositol Polyphosphate Phosphatase (MIPP) (148; 148a). In addition, changes in the overall levels of diacylglycerol (DAG) during development have been measured (149).

DESENSITIZATION OF RECEPTOR-MEDIATED RESPONSES

Desensitization is a general term describing the waning of a response during persistent stimulation as well as diminished sensitivity to subsequent challenges. For G protein–coupled receptors, at least three processes are involved: (*a*) a loss of responsiveness without loss of cell surface binding sites, referred to as uncoupling or adaptation; (*b*) a loss of cell surface binding sites without loss of receptor molecules, referred to as sequestration or loss-of-ligand-binding; and (*c*) a loss of receptor molecules, referred to as down-regulation. Analysis of a series of mutants in *Dictyostelium* suggests that loss-of-ligand binding and adaptation occur in the absence of changes in known second messengers, whereas down-regulation requires intracellular cAMP (150).

A great deal of evidence suggests that agonist-induced receptor phosphorylation plays a role in desensitization. According to a well-established paradigm for rhodopsin and the β-adrenergic receptor, phosphorylation by rhodopsin kinase or βARK leads to the binding of arrestin or βarrestin, a competitive inhibitor of the receptor–G protein interaction. Thus, the phosphorylated receptor is essentially removed from the reaction, terminating all responses (151). Although the role for receptor modification in uncoupling from G protein is accepted, whether phosphorylation regulates sequestration or is needed for down-regulation is not generally known. Moreover, the sites of phosphorylation are complex and vary widely for different G protein–coupled receptors (151).

Receptor Phosphorylation Causes Loss-of-Ligand Binding

For cAR1, sites of basal and agonist-induced phosphorylation have been determined by analysis of an extensive series of serine substitution mutants in the C-terminal tail (152). Eighteen serine residues, grouped in four clusters, are found in this region. The results can be summarized as follows: (*a*) clusters 2 and 3 are phosphorylated in unstimulated cells; (*b*) occupancy of the receptor in intact cells triggers a rapid addition of phosphates to clusters 1 and 2; and (*c*) within cluster 1 the phosphorylation of two specific serines residues, S303 and S304, causes an increase in the apparent molecular weight of cAR1 from 40 to 43 kDa.

Receptor phosphorylation is responsible for loss-of-ligand binding. In studies of this process, the loss of binding induced by pretreatment of the cells with cAMP was shown to be substantially due to a reduction in the affinity of the low-affinity class of cAMP binding sites (153, 154). When phosphorylation is prevented by substitution of all the serines, the loss-of-ligand binding process is completely blocked. The inhibition is due primarily to the substitutions of S303 and S304. In summary, receptors phosphorylated on these positions display about a fivefold lower affinity for cAMP than unphosphorylated receptors.

Receptor Phosphorylation Is Not Necessary for Adaptation of Adenylyl Cyclase

The affinity reduction caused by phosphorylation of cAR1 cannot account for the adaptation of adenylyl cyclase. A low concentration of stimulus leaves many nonphosphorylated receptors. The persistent presence of agonist should continue to elicit a response, yet cells adapt completely when occupancy, however low, remains constant and only regain sensitivity when the stimulus is removed. These properties of adaptation have been modeled by assuming that all forms of the receptor contribute, to varying extents, to the overall state of activation (155–157). When the occupancy is initially increased, a preponderance of the occupied, unphosphorylated form triggers the rapid response. At steady-state, a negative contribution from the phosphorylated, unoccupied form offsets that from the more active forms and the response completely subsides. Unfortunately, in spite of the elegant models, the fully substituted or truncated forms of cAR1 in $car1^-$ cells support transient cAMP accumulation responses with wild-type time courses and show only subtle developmental phenotypes (D Hereld & PN Devreotes, unpublished observation).

Adaptation of adenylyl cyclase activation must therefore involve a mechanism other than receptor phosphorylation. In fact, pretreatment of cells with cAMP causes a decrease in the extent of subsequent GTPγS activation of adenylyl cyclase in cell lysates, suggesting that the receptor sends an inhibitory signal that persists in lysates (158). How is this signal transmitted and what is its target? In cells pretreated with cAMP, GTPγS can no longer generate CRAC binding sites (94). It follows that adenylyl cyclase cannot be activated. Since the generation of the CRAC binding sites requires the G protein βγ-subunits, these observations suggest that the signal from the receptor decreases the activity of the βγ-subunits. Mutant analyses show that the signal from cAR1 leading to this attenuation does not require Gα2 or CRAC (73; P Lilly & PN Devreotes, unpublished observations).

RECEPTOR-MEDIATED G PROTEIN–INDEPENDENT PROCESSES

Until recently, seven transmembrane receptors other than the sensory rhodopsin of *Halobacterium salinarium* were believed to mediate all of their physiological effects through the activation of heterotrimeric G proteins (159). However, analysis of several chemoattractant-mediated responses has revealed that these receptors are capable of activating effectors in a G protein–independent manner. The first of these responses to be identified was the receptor-operated calcium influx. Experiments performed in cells lacking Gα2 or Gβ subunits (as well as in $g\alpha1^-$, $g\alpha3^-$, $g\alpha4^-$, $g\alpha7^-$, and $g\alpha8^-$ cells) demonstrate that folic acid- and cAMP-mediated calcium influx still occurs in the absence of functional G proteins (160–162). The G protein–independent calcium influx leads to a rise in cytosolic calcium levels (60). The EC_{50} for the cAR1-mediated calcium influx response is close to the Kd for the lower affinity state of cAR1. However, the maximal responses are decreased by 50% in cells lacking Gα2 or Gβ, suggesting that calcium influx is partly dependent on the standard pathway. As in mammalian cells, half of the receptor-mediated calcium influx may depend on the depletion of the IP_3-sensitive calcium store (Figure 3) (163).

Agonist-induced cAR1 phosphorylation occurs in a G protein–independent fashion: It displays the same kinetics and concentration dependence in $g\beta^-$ and in wild-type cells (162). Similar results are observed with the phosphorylation of rhodopsin in mammalian cells. Light activation is known to be sufficient to allow receptor phosphorylation: The addition of purified $\beta\gamma$-subunits does not change the extent of rhodopsin phosphorylation (151). In the case of the β-adrenergic receptor, phosphorylation partly depends on the presence of G protein $\beta\gamma$-subunits. The $\beta\gamma$-subunits have been shown to form a complex with βARKs, thereby targeting the kinase to the membrane where it can interact with the receptor (164).

Recent evidence suggests that G protein–independent signaling may be quite common. Indeed, several other cAR1-mediated responses including $G\alpha2$ phosphorylation (82), loss-of-ligand binding (162), ERK2 activation (62a), and various gene expression events also take place in a G protein–independent manner. Continuous levels of cAMP accelerate expression of the secreted phosphodiesterase and suppress that of its inhibitor in the $g\beta^-$ cells (164a). The transcription factor GBF can be activated by extracellular cAMP to direct expression of the immediate early genes *rasD* and *CP2* in growth stage $g\beta^-$ cells (165). No quantification is available to assess whether these responses are partially dependent on G proteins.

RAPID PHENOTYPIC SCREENS FOR RANDOM MUTAGENESIS

The *car1⁻*, *gα2⁻*, *gβ⁻*, *crac⁻*, *aca⁻*, *pka⁻*, *erk2⁻*, *ale⁻*, and *pde⁻* cell lines are unable to aggregate. The *car2⁻* and *car4⁻* cells arrest at later stages of development. All of these mutants regain the ability to aggregate and differentiate when transformed with plasmids constitutively expressing the respective wild-type cDNAs. Consequently, these cell lines can be used in simple biological screens for loss- or gain-of-function mutations that affect the regulation and activity of these important signal transduction components (Figure 4).

These studies rely on the development of extrachromosomal expression vectors which provide high transformation efficiencies and complete segregation [each transformant contains a unique mutated plasmid (see above)]. Randomly mutagenized genes are subcloned into these vectors, the libraries electroporated into amoebae typically yielding 5000–10,000 transformants, and screened for phenotypic abnormalities. Screening is conveniently performed by clonally seeding cells on a lawn of *Klebsiella aerogenes* (30). As each clone expands, it depletes the bacteria, forming a 1 cm plaque, and undergoes development, making visual scoring easy and unequivocal. Wild-type cells aggregate and form normal fruiting bodies; loss-of-function mutants remain as smooth monolayers or as arrested developmental structures. Suppressors or gain-of-function mutants can be easily spotted as isolated positive plaques in a sea of aggregation-minus clones (Figure 4). Through biochemical analyses, the mutants can be classified and unique amino acid residues can be ascribed to specific functions. Here, the application of the technique is illustrated through the examples of cAR1 and ACA.

Identification of Distinct Activation States of cAR1

The structure-function relationships of seven transmembrane G protein–coupled receptors have been studied in various systems and several common structural features have been established (166). The existence of the four cARs as well as the description of multiple G protein–dependent and –independent responses (see above) makes *Dictyostelium* an excellent system to further study the mechanisms of agonist binding, activation/adaptation, and G protein and effector stimulation.

A region of cAR1 encompassing transmembrane span (TM) III to TMVII was randomly mutagenized. Although each sustained an average of two amino acid substitutions, 90% of the mutant receptor molecules supported wild-type development. The aggregation-deficient clones expressed mutant receptors that had defects in distinct biochemical parameters (JY Kim, MJ Caterina, KC Lin, JA Borleis & PN Devreotes, manuscript in preparation). The majority of the mutants displayed lower binding affinities but were fully functional at high agonist concentrations. These mutants clustered in distinct subclasses (classes IIIa, IIIb, and IIIc) that displayed affinities of either 5-fold, 100-fold, or >10,000-fold lower than wild type. The incremental decreases might suggest that these mutations disrupt individual ligand contact sites. The screen also gave rise to rare mutants which bound cAMP and were phosphorylated but showed neither G protein–coupling nor cAMP-mediated calcium influx (class I).

Random mutagenesis of the third intracellular loop was carried out. As noted for the larger region, the third intracellular loop could be extensively mutagenized without effect on receptor function. The few mutants that did appear displayed signaling or affinity defects (167; JLS Milne & PN Devreotes, manuscript in preparation). Most of the mutant receptors showed diminished (or absent) responses in both G protein–dependent and –independent pathways (classes IV and V).

These random mutagenesis studies provide valuable information on the structure-function relationship of cAR1. A model for receptor activation is depicted in Figure 5. In this scheme, ligand binding leads to activation of the receptor and the acquisition of an active conformation (cAR1*) which permits the interaction with the kinase and leads to receptor phosphorylation. A second

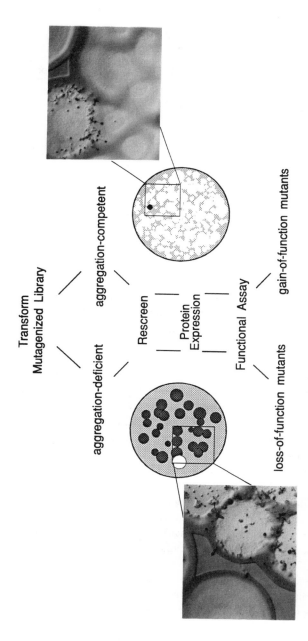

Figure 4 Isolation of loss- and gain-of-function mutants using biological screens. The mutagenized libraries are transformed into *Dictyostelium* cells and aggregation-deficient (white circles) as well as aggregation-competent (dark circles) clones are isolated from bacterial lawns. After rescreening for development and assessment of protein expression, the mutants are classified according to their biochemical characteristics.

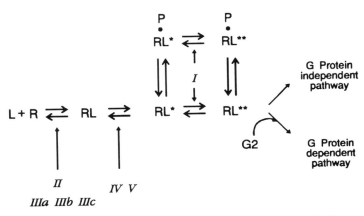

Figure 5 Model for receptor activation based on cAR1 random mutagenesis. Diagram shows interconversions between receptor states. R=cAR1, L=cAMP. The mutant classes are shown in italics. See text for details.

active state (cAR1**) enables the receptor to couple to G proteins as well as activate the G protein-independent pathway. The class III mutants lower affinity without altering activity or affecting the transitions between the activated states.

A procedure called random chimeragenesis was used to study the affinity difference between cAR1 and cAR2. As discussed earlier, under physiological conditions, cAR1 displays a much higher affinity (25 and 230 nM) than cAR2 (>2 μM). This difference disappears in the presence of ammonium sulfate; both receptors display Kds of about 5 nM (49). The residues responsible for the difference in affinity were localized to the C-terminal half of the second extracellular loop (54). In that region, there are only four nonconserved residues; two negatively charged amino acids in cAR2 appear to be the essential changes. cAR3, which has an intermediate affinity, has one negatively charged amino acid within that same region. It was proposed that the negatively charged amino acid(s) interfere with high-affinity binding by hindering the entry of cAMP into the binding pocket and that shielding by high salt eliminates this interference. Interestingly, a cAR1 mutant isolated by random mutagenesis displayed cAR2-like binding properties (class II): It carried a single amino acid substitution in the C-terminal half of the second extracellular loop. Similar results were obtained with the adenosine mammalian receptor: Residues in the extracellular loops were found to be determinants of affinity (168). These residues may comprise "gate keeper" regions modulating the accessibility of ligands to their binding pockets. The affinity differences displayed by the class

IIIa-IIIc mutants persist in ammonium sulfate and may represent differences in the binding pocket.

Isolation of Inactive, G Protein–Insensitive, and Constitutively Active Mutants of ACA

As noted above, G protein–coupled adenylyl cyclases, including ACA, form a family of transmembrane proteins predicted to span the membrane 12 times and contain large cytoplasmic domains between TMVI and TMVII and at the carboxy-terminus (C1 and C2; ~40 kDa each). Since these two cytoplasmic loops share amino acid homology with other adenylyl and guanylyl cyclases, it has been proposed that the catalytic sites are located within these intracellular domains (169). Site-directed mutagenesis and deletion analyses of the mammalian enzymes have demonstrated that catalytic activity and proper regulation require the interaction of the two cytoplasmic loops and suggested that the loops are not functionally equivalent, C1 being more important for regulation and C2 for catalysis (170, 170a).

The first five predicted transmembrane helices as well as the C1 domain of ACA were randomly mutagenized and loss-of-function mutants were isolated. These were found to be either catalytically inactive or resistant to chemoattractant and G protein activation. Of four catalytically inactive mutants analyzed, three harbored substitutions on highly conserved residues located halfway within the C1 loop. These results suggest that this region of the C1 loop is essential for substrate binding and/or catalytic activity. The G protein–insensitive mutants isolated displayed mutations that were grouped in two regions of the cytoplasmic domain: one close to the plane of the plasma membrane just after TMVI and the other halfway within the C1 loop. Mutations in the transmembrane domains led to missorting (35).

These results are interesting for several reasons. First, they suggest that, as is the case for receptors, regions of cytoplasmic loops abutting transmembrane domains are critical for G-protein interaction (171). Tang and Gilman have also demonstrated that the cytoplasmic domains of mammalian adenylyl cyclases are sufficient for G-protein activation (172). Second, these studies imply that the twelve transmembrane domains are needed for appropriate membrane association. Consistent with this possibility is the finding that three ACA mutants showed an abnormal cellular distribution following sucrose gradient fractionation. All three mutants harbored point mutations or deletions within the transmembrane domains, again suggesting that sorting is dependent on these regions.

Gain-of-function adenylyl cyclase mutants were isolated by transforming the same mutagenized libraries into *crac⁻* cells. Since CRAC is essential for G protein stimulation of ACA, suppression of the *crac⁻* phenotype screened for mutants with high unregulated activity or for those that retain regulated activity in the absence of CRAC (CA Parent & PN Devreotes, manuscript in preparation). The characterization of these clones along with the isolation of additional mutants will bring more insight into the structure-function relationship of adenylyl cyclases.

CONCLUDING REMARKS

The recent development of powerful genetic tools for *Dictyostelium* has made it an attractive model system, uniquely suited for studies of a variety of biological processes, including chemotaxis, cell aggregation, gene expression, differentiation, and pattern formation. These events are regulated by a family of cell surface cAMP receptors that are expressed at specific stages in a starvation-induced developmental program. The pathways activated by the cARs share many features with mammalian signaling systems. Analysis of null mutants created by gene disruption has shown that none of the familiar signal transduction components are required for growth or cell motility but many are needed for various aspects of development. The phenotypes of the mutants suggest several unexpected roles and interactions among these genes. Rescue of the phenotypes of these mutants is providing simple screens for loss- and gain-of-function mutations in these important signaling proteins. Screening of populations of gene-tagged cell lines for similar phenotypes is leading to the discovery of novel genes in these pathways.

The molecular genetic analysis of signal transduction in *Dictyostelium* has answered several questions but raised many others. How do seven transmembrane receptors cause calcium influx and mediate gene expression without coupling to heterotrimeric G proteins? What regions of the cARs control the affinity of agonist and antagonist binding? How many receptor-activated states are there and which regions control the activated states? What are the mechanism(s) of adaptation that are independent of receptor phosphorylation? What additional proteins are involved in the adenylyl cyclase activation pathway? Do mammalian adenylyl cyclases have such complex activation pathways in vivo? What regions of adenylyl cyclase are involved in activation by receptors and G proteins? What are the targets of PKA that control gene expression at the various stages of development? How can PKA be active in the absence of cAMP? What is the role of cGMP in chemotactic orientation? What are the links between large and small G proteins? Future studies will undoubtedly address these and other questions.

ACKNOWLEDGMENTS

Work performed in this laboratory is supported by grants from the National Institutes of Health and the American Cancer Society. CA Parent is the recipient of a fellowship from the Medical Research Council of Canada. The authors would like to thank the many groups who provided information in advance of publication, Dr. Pauline Schaap for helpful discussions, and Dr. Brenda Blacklock for reading the manuscript.

Literature Cited

1. Devreotes PN. 1994. *Neuron* 12:235–41
2. Devreotes P. 1989. *Science* 245:1054–58
3. Gross JD. 1994. *Microbiol. Rev.* 58:330–51
4. Kay RR. 1994. *Curr. Opin. Genet. Dev.* 4:637–41
5. Takeuchi I, Tasaka M, Okamoto K, Maeda Y. 1994. *Int. J. Dev. Biol.* 38:311–20
6. Williams J, Morrison A. 1994. *Prog. Nucleic Acid Res. Mol. Biol.* 47:1–27
7. Firtel RA. 1995. *Genes Dev.* 9:1427–44
8. de Wit RJW, Bulgakov R. 1986. *Biochim. Biophys. Acta* 886:76–87
9. de Wit RJW, Bulgakov R. 1986. *Biochim. Biophys. Acta* 886:88–95
10. van Haastert PJM, de Wit RJW, Konijn TM. 1982. *Exp. Cell Res.* 140:453–56
11. Wurster B, Schubiger K, Wick U, Gerisch G. 1977. *FEBS Lett.* 76:141–44
12. Tillinghast HS, Newell PC. 1987. *J. Cell Sci.* 87:45–53
13. Clarke M, Kayman SC, Riley K. 1987. *Differentiation* 34:79–87
14. Rathi A, Kayman SC, Clarke M. 1991. *Dev. Genet.* 12:82–87
15. Rathi A, Clarke M. 1992. *Mech. Dev.* 36:173–82
16. Burdine V, Clarke M. 1995. *Mol. Biol. Cell.* 6:311–25
17. Williams JG. 1988. *Development* 103:1–16
18. Mehdy MC, Firtel RA. 1985. *Mol. Cell. Biol.* 5:705–13
19. Gomer RH, Yuen IS, Firtel RA. 1991. *Development* 112:269–78
20. Yuen IS, Taphouse C, Halfant KA, Gomer RH. 1991. *Development* 113:1375–85
21. Jain R, Yuen IS, Taphouse CR, Gomer RH. 1992. *Genes Dev.* 6:390–400
22. Jain R, Gomer RH. 1994. *J. Biol. Chem.* 269:9128–36
23. Yuen IS, Jain R, Bishop JD, Lindsey DF, Deery WJ, et al.1995. *J. Cell Biol.* 129:1251–62
24. Bussolino F, Sordano C, Benfeneti E, Bozzaro S. 1991. *Eur. J. Biochem.* 196:609–16
25. Jalink K, Moolenaar WH, Van Duijn B. 1993. *Proc. Natl. Acad. Sci. USA* 90:1857–61
26. Kuwayama H, Ishida S, van Haastert PJM. 1993. *J. Cell Biol.* 123:1453–62
27. Loomis WF, Welker D, Hughes J, Maghakian D, Kuspa A. 1995. *Genetics* 141:147–57
28. Loomis WF. 1978. *Birth Defects: Orig. Artic. Ser.* 14:497–505
29. Newell PC. 1978. *Annu. Rev. Genet.* 12:69–93
30. Sussman M. 1987. *Methods Cell Biol.* 28:9–29
31. Soll DR. 1987. *Methods Cell Biol.* 28:413–31
32. Hughes JE, Ashktorab H, Welker DL. 1988. *Dev. Genet.*9:495–504
33. Leiting B, Lindner IJ, Noegel AA. 1990. *Mol. Cell. Biol.*10:3727–36
34. Hughes JE, Kiyosawa H, Welker DL. 1994. *Mol. Cell. Biol.* 14:6117–24
35. Parent CA, Devreotes PN. 1995. *J. Biol. Chem.*270:22693–96
36. Datta S, Gomer RH, Firtel RA. 1986. *Mol. Cell. Biol.* 6:811–20
37. Dingermann T, Troidl EM, Broker M, Nerke K. 1991. *Appl. Microbiol. Biotechnol.* 35:496–503
38. Dittrich W, Williams KL, Slade MB. 1994. *Bio-Technology* 12:614–18
39. Voith G, Dingermann T. 1995. *Biotechnology.* 13:1225–29
40. Schiestl RH, Petes TD. 1991. *Proc. Natl. Acad. Sci. USA* 88:7585–89
41. Kuspa A, Loomis WF. 1992. *Proc. Natl. Acad. Sci. USA* 89:8803–7
42. Klein PS, Sun TJ, Saxe CL III, Kimmel AR, Johnson RL, et al. 1988. *Science* 241:1467–72
43. Saxe CL III, Ginsburg GT, Louis JM, Johnson R, Devreotes PN, et al. 1993. *Genes Dev.* 7:262–72
44. Johnson RL, Saxe CL III, Gollop R, Kimmel AR, Devreotes PN. 1993. *Genes Dev.* 7:273–82

45. Louis JM, Ginsburg GT, Kimmel AR. 1994. *Genes Dev.* 8:2086–96

46. Saxe CL III, Johnson RL, Devreotes PN, Kimmel AR. 1991. *Genes Dev.* 5:1–8

47. Louis JM, Saxe CL III, Kimmel AR. 1993. *Proc. Natl. Acad. Sci. USA* 90: 5969–73

47a. Yu Y, Saxe CL III. 1996. *Dev. Biol.* In press

47b. Saxe CL III, Yu Y, Jones C, Bauman A, Haynes H. 1996. *Dev. Biol.* In press

48. Saxe CL III, Johnson R, Devreotes PN, Kimmel AR. 1991. *Dev. Genet.* 12:6–13

49. Johnson RL, van Haastert PJM, Kimmel AR, Saxe CL III, Jastorff B, et al. 1992. *J. Biol. Chem.* 267:4600–7

50. Schaap P, Wang M. 1993. *Experimental and Theoretical Advances in Biological Pattern Formation*, pp. 301–18. New York: Plenum

51. Sun TJ, Devreotes PN. 1991. *Genes Dev.* 5:572–82

52. Pupillo M, Insall R, Pitt GS, Devreotes PN. 1992. *Mol. Biol. Cell.* 3:1229–34

53. Insall RH, Soede RDM, Schaap P, Devreotes PN. 1994. *Mol. Biol. Cell.* 5:703–11

54. Kim JY, Devreotes PN. 1994. *J. Biol. Chem.* 269:28724–31

55. Caterina MJ, Devreotes PN. 1991. *FASEB J.* 5:3078–85

56. Berlot CH, Spudich JA, Devreotes PN. 1985. *Cell* 43:307–14

57. Berlot CH, Devreotes PN, Spudich JA. 1987. *J. Biol. Chem.* 262:3918–26

58. Wick U, Malchow D, Gerisch G. 1978. *Cell Biol. Int. Rep.* 2:71–79

59. Bumann J, Wurster B, Malchow D. 1983. *J. Cell Biol.* 98:173–78

60. Schlatterer C, Gollnick F, Schmidt E, Meyer R, Knoll G. 1994. *J. Cell Sci.* 107:2107–15

61. Kreitmeier M, Gerisch G, Heizer C, Muller-Taubenberger A. 1995. *J. Cell Biol.* 129:179–88

62. Insall R, Kuspa A, Lilly PJ, Shaulsky G, Levin LR, et al.1994. *J. Cell Biol.* 126:1537–45

62a. Maeda M, Aubry L, Insall R, Gaskins C, Devreotes PN, Firtel RA. 1996. *J. Biol. Chem.* In press

63. Pupillo M, Kumagai A, Pitt GS, Firtel RA, Devreotes PN.1989. *Proc. Natl. Acad. Sci. USA* 86:4892–96

64. Kumagai A, Pupillo M, Gundersen R, Miake-Lye R, Devreotes PN, et al. 1989. *Cell* 57:265–75

65. Hadwiger JA, Wilkie TM, Strathmann M, Firtel RA. 1991. *Proc. Natl. Acad. Sci. USA* 88:8213–17

66. Wu L, Devreotes PN. 1991. *Biochem. Biophys. Res. Commun.* 179:1141–47

67. Wu L, Gaskins C, Zhou KM, Firtel RA, Devreotes PN. 1994. *Mol. Biol. Cell.* 5:691–702

68. Wu L, Gaskins R, Gundersen R, Hadwiger JA, Johnson RL, et al. 1993. *Handbook of Experimental Pharmacology, Vol. 108/II, GTPases in Biology II*, ed. BF Dickey, L Birnbaumer, pp. 335–49. Heidelberg: Springer-Verlag

69. Dharmawardhane S, Cubitt AB, Clark AM, Firtel RA. 1994.*Development* 120: 3549–61

70. Kesbeke F, Snaar-Jagalska BE, van Haastert PJM. 1988. *J. Cell Biol.* 107: 521–28

71. Hadwiger JA, Firtel RA. 1992. *Genes Dev.* 6:38–49

72. Bominaar AA, van Haastert PJM. 1994. *Biochem. J.* 297:189–93

73. Kesbeke F, Snaar-Jagalska BE, van Haastert PJM. 1988. *J. Cell Biol.* 107: 521–28

74. Kumagai A, Hadwiger JA, Pupillo M, Firtel RA. 1991. *J. Biol. Chem.* 266: 1220–28

75. Hall A, Warren V, Condeelis J. 1989. *Dev. Biol.* 136:517–25

76. Bominaar AA, van der Kaay J, Kesbeke F, Snaar-Jagalska BE,van Haastert PJM. 1990. *Biochem. Soc. Symp.* 56:71–80

77. Okaichi K, Cubitt AB, Pitt GS, Firtel RA. 1992. *Mol. Biol. Cell.* 3:735–47

78. Kesbeke F, van Haastert PJM, de Wit RJW, Snaar-Jagalska BE. 1990. *J. Cell Sci.* 96:669–73

79. Sordano C, Cristino E, Bussolino F, Wurster B, Bozzaro S. 1993. *J. Cell Sci.* 104:197–202

80. Schaloske R, Sordano C, Bozzaro S, Malchow D. 1995. *J. Cell Sci.* 108: 1597–603

81. Gundersen RE, Devreotes PN. 1990. *Science* 248:591–93

82. Chen MY, Devreotes PN, Gundersen RE. 1994. *J. Biol. Chem.* 269:20925–30

83. Hadwiger JA, Lee S, Firtel RA. 1994. *Proc. Natl. Acad. Sci. USA* 91:10566–70

84. Lilly P, Wu L, Welker DL, Devreotes PN. 1993. *Genes Dev.* 7:986–95

84a. Wall MA, Coleman DE, Lee E, Iniguez-Lluhi JA, Posner BA, Gilman AG, Sprang SR. 1995. *Cell* 83:1047–58

85. Neer EJ. 1995. *Cell* 80:249–57

86. Wu L, Valkema R, van Haastert PJM, Devreotes PN. 1995. *J. Cell Biol.* 129: 1667–75

87. Pitt GS, Milona N, Borleis J, Lin KC, Reed RR, et al. 1992. *Cell* 69:305–15

88. Tang WJ, Gilman AG. 1991. *Science* 254:1500–3

89. Inglese J, Koch WJ, Touhara K, Lefkowitz RJ. 1995. *Trends Biochem. Sci.* 20:151–56

90. Musacchio A, Gibson T, Rice P, Thompson J, Saraste M. 1993. *Trends Biochem. Sci.* 18:343–48
91. Chen J, DeVivo M, Dingus J, Harry A, Li J, et al. 1995. *Science* 268:1166–69
92. Theibert A, Devreotes PN. 1986. *J. Biol. Chem.* 261:15121–25
93. Lilly PJ, Devreotes PN. 1994. *J. Biol. Chem.* 269:14123–29
94. Lilly PJ, Devreotes PN. 1995. *J. Cell Biol.* 129:1659–65
95. Touhara K, Inglese J, Pitcher JA, Shaw G, Lefkowitz RJ. 1994. *J. Biol. Chem.* 269:10217–20
96. Segall JE, Kuspa A, Shaulsky G, Ecke M, Maeda M, et al. 1995. *J. Cell Biol.* 128:405–13
97. Robbins SM, Williams JG, Jermyn KA, Spiegelman GB, Weeks G. 1989. *Proc. Natl. Acad. Sci. USA* 86:938–42
98. Reymond CD, Gomer RH, Mehdy MC, Firtel RA. 1984. *Cell* 39:141–48
99. Daniel J, Spiegelman GB, Weeks G. 1993. *Oncogene* 8:1041–47
100. Robbins SM, Suttorp VV, Weeks G, Spiegelman GB. 1990. *Nucleic Acids Res.* 18:5265–69
101. Daniel J, Bush J, Cardelli J, Spiegelman GB, Weeks G. 1994. *Oncogene* 9:501–8
102. Reymond CD, Gomer RH, Nellen W, Theibert A, Devreotes P, et al. 1986. *Nature* 323:340–43
103. Luderus MEE, Kesbeke F, Knetsch MLW, Van Driel R, Reymond CD, et al. 1992. *Eur. J. Biochem.* 208:235–40
104. Kitayama H, Sugimoto Y, Matsuzaki T, Noda M. 1989. *Cell* 56:77–84
105. Rebstein PJ, Weeks G, Spiegelman GB. 1993. *Dev. Genet.* 14:347–55
106. Franke J, Kessin RH. 1992. *Cell. Signal.* 4:471–78
107. Lacombe ML, Podgorski GJ, Franke J, Kessin RH. 1986. *J. Biol. Chem.* 261:16811–17
108. Faure M, Franke J, Hall AL, Podgorski GJ, Kessin RH. 1990. *Mol. Cell. Biol.* 10:1921–30
109. Malchow D, Nagele B, Schwartz H, Gerisch G. 1972. *Eur. J. Biochem.* 28:136–42
110. Gerisch G. 1976. *Cell Differ.* 5:21–25
111. Wu L, Franke J. 1990. *Gene* 91:51–56
112. Kessin RH, Orlow SJ, Shapiro RI, Franke J. 1979. *Proc. Natl. Acad. Sci. USA* 76:5450–54
113. Barra J, Barrand P, Blondelet MH, Brachet P. 1980. *Mol. Gen. Genet.* 177:607–13
114. Darmon M, Barra J, Brachet P. 1978. *J. Cell Sci.* 31:233–43
115. Wu L, Franke J, Blanton RL, Podgorski GJ, Kessin RH. 1995. *Dev. Biol.* 167:1–8
116. Yeh RP, Chan FK, Coukell MB. 1978. *Dev. Biol.* 66:361–74
117. De Gunzburg J, Part D, Guiso N, Veron M. 1984. *Biochemistry* 23:3805–12
118. Mutzel R, Lacombe ML, Simon MN, De Gunzburg J, Veron M. 1987. *Proc. Natl. Acad. Sci. USA* 84:6–10
119. Mann SKO, Firtel RA. 1991. *Mech. Dev.* 35:89–101
120. Burki E, Anjard C, Scholder JC, Reymond CD. 1991. *Gene* 102:57–65
121. Part D, De Gunzburg J, Veron M. 1985. *Cell Differ.* 17:221–27
122. De Gunzburg J, Franke J, Kessin RH, Veron M. 1986. *EMBO J.* 5:363–67
123. Simon MN, Driscoll D, Mutzel R, Part D, Williams J, et al. 1989. *EMBO J.* 8:2039–44
124. Firtel RA, Chapman AL. 1990. *Genes Dev.* 4:18–28
125. Harwood AJ, Hopper NA, Simon MN, Driscoll DM, Veron M, et al. 1992. *Cell* 69:615–24
126. Harwood AJ, Hopper NA, Simon MN, Bouzid S, Veron M, et al. 1992. *Dev. Biol.* 149:90–99
127. Hopper NA, Harwood AJ, Bouzid S, Veron M, Williams JG. 1993. *EMBO J.* 12:2459–66
128. Simon MN, Pelegrini O, Veron M, Kay RR. 1992. *Nature* 356:171–72
129. Anjard C, Pinaud S, Kay RR, Reymond CD. 1992. *Development* 115:785–90
130. Pitt GS, Brandt R, Lin KC, Devreotes PN, Schaap P. 1993. *Genes Dev.* 7:2172–80
131. Schulkes C, Schaap P. 1995. *FEBS Lett.* 368:381–84
132. Schaap P, Van Ments-Cohen M, Soede RDM, Brandt R, Firtel RA, et al. 1993. *J. Biol. Chem.* 268:6323–31
133. Mann SKO, Firtel RA. 1993. *Development* 119:135–46
134. Mato JM, van Haastert PJM, Krens FA, Rhijnsburger EH, Dobbe FCPM, et al. 1977. *FEBS Lett.* 79:331–36
135. Valkema R, van Haastert PJM. 1992. *Biochem. Biophys. Res. Commun.* 186:263–68
136. Newell PC, Liu G. 1992. *BioEssays* 14:473–79
137. Liu G, Newell PC. 1993. *Symp. Soc. Exp. Biol.* 47:283–95
138. Liu G, Kuwayama H, Ishida S, Newell PC. 1993. *J. Cell Sci.* 106:591–96
139. Ravid S, Spudich JA. 1992. *Proc. Natl. Acad. Sci. USA* 89:5877–81
139a. Abu-Elneel K, Karchi M, Ravid S. 1996. *J. Biol. Chem.* 271:977–84
140. De Lozanne A, Spudich JA. 1987. *Science* 236:1086–91

141. Kuwayama H, Viel GT, Ishida S, van Haastert PJM. 1995. *Biochim. Biophys. Acta* 1268:214–20
141a. Kuwayama H, Ecke M, Gerisch G, van Haastert PJM. 1996. *Science.* 271:207–09
142. Darcy PK, Wilczynska Z, Fisher PR. 1994. *Microbiol.* 140:1619–32
143. Drayer AL, van Haastert PJM. 1992. *J. Biol. Chem.* 267:18387–92
144. Bominaar AA, van Haastert PJM. 1993. *J. Cell Sci.* 104:181–85
145. Drayer AL, Van der Kaay J, Mayr GW, van Haastert PJM. 1994. *EMBO J.* 13:1601–9
146. Michell RH. 1992. *Trends Biochem. Sci.* 17:274–76
147. van Dijken P, Lammers AA, van Haastert PJM. 1995. *Biochem. J.* 308:127–30
148. Craxton A, Ali N, Shears SB. 1995. *Biochem. J.* 305:491–98
148a. van Dijken P, de Haas JR, Craxton A, Erneux C, Shears SB, van Haastert PJM. 1995. *J. Biol. Chem.* 270:29724–31
149. Cubitt AB, Dharmawardhane S, Firtel RA. 1993. *J. Biol.Chem.* 268:17431–39
150. van Haastert PJM, Wang M, Bominaar AA, Devreotes PN, Schaap P. 1992. *Mol. Biol. Cell.* 3:603–12
151. Lefkowitz RJ. 1993. *Cell* 74:409–12
152. Hereld D, Vaughan R, Kim JY, Borleis J, Devreotes P. 1994. *J. Biol. Chem.* 269:7036–44
153. Caterina MJ, Hereld D, Devreotes PN. 1995. *J. Biol. Chem.* 270:4418–23
154. Caterina MJ, Devreotes PN, Borleis J, Hereld D. 1995. *J. Biol. Chem.* 270:8667–72
155. Knox BE, Devreotes PN, Goldbeter A, Segel LA. 1986. *Proc. Natl. Acad. Sci. USA* 83:2345–49
156. Martiel JL, Goldbeter A. 1987. *Biophys. J.* 52:807–28
157. Goldbeter A. 1987. *Molecular Mecha-nisms of Desensitization to Signal Molecules.* NATO ASI Ser. H, Cell Biology, ed. TM Konijn, PJM van Haastert, H van der Starre, H van der Wel, MD Houslay, Vol. 6, pp. 43–62. Berlin: Springer-Verlag
158. Theibert A, Devreotes P. 1986. *J. Biol. Chem.* 261:15121–25
159. Spudich JL. 1994. *Cell* 79:747–50
160. Milne JL, Coukell MB. 1991. *J. Cell Biol.* 112:103–10
161. Milne JL, Devreotes PN. 1993. *Mol. Biol. Cell.* 4:283–92
162. Milne JLS, Wu L, Caterina MJ, Devreotes PN. 1995. *J. Biol. Chem.* 270:5926–31
163. Putney JW Jr, Bird GSJ. 1993. *Cell* 75:199–201
164. Pitcher JA, Inglese J, Higgins JB, Arriza JL, Casey PJ, et al. 1992. *Science* 257:1264–67
164a. Wu L, Hansen D, Franke J, Kessin RH, Podgorski GJ. 1995. *Dev. Biol.* 171:149–58
165. Schnitzler GR, Briscoe C, Brown JM, Firtel RA. 1995. *Cell* 81:737–45
166. Savarese TM, Fraser CM. 1992. *Biochem. J.* 283:1–19
167. Caterina MJ, Milne JLS, Devreotes PN. 1994. *J. Biol.Chem.* 269:1523–32
168. Olah ME, Jacobson KA, Stiles GL. 1994. *J. Biol. Chem.* 269:24692–98
169. Tang WJ, Gilman AG. 1992. *Cell* 70:869–72
170. Tang WJ, Krupinski J, Gilman AG. 1991. *J. Biol. Chem.* 266:8595–603
170a. Tang WJ, Stanzel M, Gilman AG. 1995. *Biochemistry* 34:14563–72
171. Baldwin JM. 1994. *Curr. Opin. Cell Biol.* 6:180–90
172. Tang WJ, Gilman AG. 1995. *Science* 268:1769–72
173. Insall RH, Borleis J, Devreotes PN. 1996. *Curr. Biol.*: In press

Annu. Rev. Biochem. 1996. 65:441–73

STRUCTURAL BASIS OF LECTIN-CARBOHYDRATE RECOGNITION

William I. Weis

Department of Structural Biology, Stanford University School of Medicine, Stanford, California 94305

Kurt Drickamer

Glycobiology Institute, Department of Biochemistry, University of Oxford, Oxford OX1 3QU, United Kingdom

KEY WORDS: binding, crystallography, mutagenesis, proteins, molecular recognition

---

ABSTRACT

Lectins are responsible for cell surface sugar recognition in bacteria, animals, and plants. Examples include bacterial toxins; animal receptors that mediate cell-cell interactions, uptake of glycoconjugates, and pathogen neutralization; and plant toxins and mitogens. The structural basis for selective sugar recognition by members of all of these groups has been investigated by x-ray crystallography. Mechanisms for sugar recognition have evolved independently in diverse protein structural frameworks, but share some key features. Relatively low affinity binding sites for monosaccharides are formed at shallow indentations on protein surfaces. Selectivity is achieved through a combination of hydrogen bonding to the sugar hydroxyl groups with van der Waals packing, often including packing of a hydrophobic sugar face against aromatic amino acid side chains. Higher selectivity of binding is achieved by extending binding sites through additional direct and water-mediated contacts between oligosaccharides and the protein surface. Dramatically increased affinity for oligosaccharides results from clustering of simple binding sites in oligomers of the lectin polypeptides. The geometry of such oligomers helps to establish the ability of the lectins to distinguish surface arrays of polysaccharides in some instances and to crosslink glycoconjugates in others.

441

0066-4154/96/0701-0441$08.00

CONTENTS

INTRODUCTION 442
STRUCTUCTURAL BASIS FOR SACCHARIDE-LECTIN INTERACTION AND
    RECOGNITION.. 443
    Hydrogen Bonding 443
    Recognition of Charged Groups on Sugars................................. 452
    Role of Divalent Cations 453
    Nonpolar Interactions.. 454
    Extended Sites... 456
    Secondary Sites.. 458
    Conformational Changes Upon Sugar Binding............................. 459
    Comparison with Other Carbohydrate-Binding Proteins 460
RECOGNITION OF DISTINCT SETS OF SUGARS BY CLOSELY RELATED
    LECTINS 461
    C-Type Animal Lectins... 462
    Legume Lectins 463
    Glycolipid-Binding Toxins 463
    Principles of Differential Sugar Recognition 464
MULTIVALENCY AND THE IMPORTANCE OF GEOMETRY IN
    OLIGOSACCHARIDE RECOGNITION 464
    Energetic Considerations.. 464
    Surface Recognition 466
    Lattice Formation by Bridging Lectins 468
    Other Geometrical Arrangements 469
SUMMARY AND PERSPECTIVES 469

INTRODUCTION

Many interactions at the cell surface require selective recognition of specific carbohydrate structures by cognate receptor proteins that are designated lectins. Although a large number of lectins from plant, animal, and bacterial sources have been characterized, our understanding of lectin functions is widely variable (1). Some relatively well characterized lectins are those utilized by pathogens as a means of attachment to eukaryotic cell surfaces. Examples of lectins involved in this process include the hemagglutinins of influenza and other viruses as well as the toxins produced by gram-negative bacteria. Several animal lectins serve as part of the innate host defense system by selectively binding to the surface of potential bacterial and viral pathogens and initiating steps toward their neutralization. Thus, selective recognition of sugars can provide a means of distinguishing self from nonself. Other animal lectins mediate adhesion between animal cells, sorting of newly synthesized glycosylated proteins within the luminal compartment of the endoplasmic reticulum, and endocytosis of selected subsets of circulating glycoproteins. In contrast, plant lectins remain somewhat enigmatic in function. The suggestion has been made that they may be involved in host-rhizobium interactions or in defense against pathogens (2), although recent evidence suggests that the lectins involved in such processes may be related to but distinct from the more commonly studied seed lectins (3).

In spite of the enormous diversity of lectins, two aspects of their organization aid us in understanding them. First, the sugar-binding activity can be ascribed to a limited portion of most lectin molecules, typically a globular carbohy-drate-recognition domain (CRD) of less than 200 amino acids. Second, com-parison of CRDs reveals that many are related in amino acid sequence to each other, so that most of the known lectins can be organized into a relatively small number of groups. In the plant kingdom, lectins isolated from the seeds of legumes are homologous to each other, whereas a second major category of lectins derives from the seeds of cereals (4). Although the number of animal lectins continues to increase, a recent classification indicates that most fall into one of five major groups: C-type or Ca^{2+}-dependent lectins, Gal-binding galect-ins, P-type Man 6-phosphate receptors, I-type lectins including sialoadhesins and other immunoglobulin-like sugar-binding proteins, and L-type lectins re-lated in sequence to the leguminous plant lectins (5). In addition, all of the structurally characterized bacterial toxins that use carbohydrates as cellular receptors display common structural features (6).

As anticipated from the similar amino acid sequences within each of these groups, the three-dimensional structures of lectins confirm that members of each category share fundamental structural features. Many of these overall features were discussed in a recent review (7). A summary of the lectin-sugar complexes that have been analyzed at the atomic level is provided in Table 1. Rather than attempt a systematic review of all the known lectin structures, the present review focuses on how lectins interact with sugar ligands. Examples of lectin-sugar complexes will be drawn from many of the lectin categories, but will be organized around basic principles that underlie the ability of different lectins to bind selected saccharide ligands with appropriate affinity. Abbreviations used throughout the review are given in Table 1.

STRUCTURAL BASIS FOR SACCHARIDE-LECTIN INTERACTION AND RECOGNITION

Hydrogen Bonding

DIRECT HYDROGEN BONDING WITH PROTEIN GROUPS   Cooperative hydrogen bonding, in which the hydroxyl group (OH) acts simultaneously as a hydro-gen-bond donor and acceptor, is characteristic of the interaction of lectins and other carbohydrate-binding proteins with sugar hydroxyls. The oxygen atom of the OH is sp^3 hybridized, giving an approximately tetrahedral arrangement of two lone pairs of electrons and a proton. The OH can therefore act as an acceptor of two hydrogen bonds and as a donor of a single hydrogen bond. The participation of different protein side-chain and main-chain groups in hydrogen bond formation with sugar ligands is summarized in Table 2 and

Table 1 Lectin-carbohydrate complexes of known three-dimensional structure

Lectin family	Lectin	Abbreviation	ligands	Resolution (Å)	References
Plant lectins					
Legume lectins	Concanavalin A	ConA	Manα1-Me	2.0	8, 9
	Pea lectin		Trisaccharide 3,4,4'[a]	2.6	10
	Lathyrus ochrus lectin	LOL I	Manα1-Me & Glcα1-Me	2.0/2.2	11
		LOL I	Trisaccharide 2,3,4[a]	2.1	12
		LOL I	Octasaccharide 2,3,4,4',5,5',6,6'[a]	2.3	13
		LOL I	Muramic acid & muramyl dipeptide	2.05	13a
		LOL II	Decasaccharide 1,1',2,3,4,4',5,5',6,6'[a]	2.8	14
	Griffonia simplicifolia lectin 4	GS4	Lewis b tetrasaccharide[b]	2.0	15
	Erythrina corallodendron lectin	EcorL	Galβ 1-4Glc	2.0	16
	Soybean agglutinin	SBA	Pentasaccharide[c]	2.6	17
Cereal lectins	Wheat germ agglutinin	WGA	NeuNAcα2,3Galβ 1-4Glc	2.2	18
			GlcNAcβ1,4GlcNAc	1.8	19
			Sialoglycopeptide[d]	2.0	20, 21
Plant toxin	Ricin		Galβ1-4Glc	2.5	22, 23
Bulb lectins	Snowdrop lectin	GNA	Manα 1-Me	2.3	24
Animal Lectins					
Galectins	Galectin-1		Galβ1-4Glc	1.9	25
	Galectin-1		Octasaccharide 2,3,4,4',5,5',6,6'[a]	2.15-2.45	26
	Galectin-2		Galβ1-4Glc	2.9	27
C-type	Mannose-binding protein-A	MBP-A	High Man octasaccharide[e]	1.7	28
	Gal-binding mutant of mannose-binding protein-A		Gal & GalNAc	1.9-2.0	29
	Mannose-binding protein-C	MBP-C	Manα1-Me, GlcNAcα1-Me,Fucα1-Me, Fucβ1-Me & Gal	1.7-1.9	30
Pentraxins	Serum amyloid P component	SAP	4,6-O-(1-carboxyethylidene)-Galβ1-Me	2.0	31
Viral lectins					
Influenza virus	Hemagglutinin—Leu^{226} (site 1)	HA	NeuNAcα2,6Galβ1-4Glc	3.0	32, 34
	Hemagglutinin—Gln^{226} (site 1)		NeuNAcα2,3Galβ1-4Glc	2.9	32
	Hemagglutinin (site 2)		NeuNAcα2,3Galβ1-4Glc	2.9	33
	Hemagglutinin—Leu^{226} (site 1)		4-O-Ac-NeuNAcα2-Me & NeuNAc derivatives	2.9	34
	Hemagglutinin—Leu^{226} (site 1)		NeuNAc derivatives	2.15-3.0	35
Polyoma virus	Viral protein 1	VP1	NeuNAcα2,3Galβ1-4Glc	3.65	36

Bacterial lectins

Toxins				
Enterotoxin	LT	Gal	2.2	37
Enterotoxin		Galβ1-4Glc	2.3	38
Cholerae toxin	CT	G$_M1$ pentasaccharide[f]	2.2	39
Pertussis toxin	PT	Undecasaccharide 1,2,3,4,4',5,5',6,6',7,7'[a]	3.5	40

[a]Oligosaccharides designated by numbers refer to the residues indicated on the following biantennary complex oligosaccharide:

```
  7        6       5         4
NeuNAcα2-6-Galβ1-4GlcNAcβ1-2Manα1-3
                                    3        2         1
                                     \ Manβ1-4-GlcNAcβ1-4GlcNAcβ1-Asn
NeuNAcα2-6-Galβ1-4GlcNAcβ1-2Manα1-6    Fucα1,6
  7'       6'      5'        4'            1'
```

b
```
Fucα1-2Galβ1-3GlcNAc
              |
           Fucα1-4
```

c
```
Galβ1-4GlcNAcβ1-2
                 \ GalOH
Galβ1-4GlcNAcβ1-2
```

d
```
NeuNAcα2-3Galβ1-3GalNAc
                  |
               NeuNAcα2-6
```

e
```
Manα1-2 Manα1-3
Manα1-3          \ Manβ1-4-GlcNAcβ1-4GlcNAcβ1-Asn
        Manα1-6 /
Manα1-6
```

f
```
Galβ1-3GalNAcβ1-4Galβ1-4Glc
              |
           NeuNAcα2-3
```

detailed examples of lectin-carbohydrate interactions are shown in Figures 1
to 6. Generally, one acidic side chain is used as a hydrogen bond acceptor
from one or two sugar OHs. Hydrogen-bond donors come primarily from
main-chain amide groups and the side-chain amide group of asparagine and,
less frequently, glutamine. Charged side-chain donors also occur with some
frequency. Protein OHs from tyrosine, serine, and threonine are much less
common as either donors or acceptors of hydrogen bonds with sugar OHs.
Thus, the most common hydrogen bonding scheme involving sugar OHs is:

$$(NH)_n \rightarrow OH \rightarrow O=C$$

where NH is a hydrogen bond donor group, O=C is a carbonyl or carboxylate
acceptor, OH is a nonanomeric sugar hydroxyl, and $n=1$ or 2. The hydrogen-
bond geometry of planar donors and acceptors is fixed, whereas an OH has
torsional freedom in the placement of the proton and lone pairs. Presumably,
rotational freedom permits optimization of hydrogen bonds between the OH
and neighboring groups, albeit with some entropic cost due to fixing the OH
rotamer. The infrequent use of protein OHs in hydrogen bonds with sugar OHs
suggests that the entropic cost of fixing the rotamers of both sugar and protein
OHs is large.

In some cases, a pair of vicinal sugar OHs or one OH and the ring oxygen
interact with two functional groups in a single amino acid side chain or with
consecutive main-chain amide groups. Excellent hydrogen-bond geometry can
be acheived between planar amino acid side chains and vicinal sugar OHs in
an equatorial/equatorial or equatorial/axial configuration, as both sugar con-
figurations give an OH-OH spacing of about 2.8 Å (43). For example, Nδ2 of
Asn[90] in CT donates to Gal 2-OH, and Oδ1 accepts a bond from Gal 3-OH
(Figure 2), and in EcorL, Asp[89] forms hydrogen bonds with 3-OH and 4-OH
of Gal (Figure 5). Likewise, the torsional freedom of the C6-O6 bond allows
formation of two hydrogen bonds between a single side chain and the 4- and
6-OHs of hexoses, with the 4-OH either equatorial or axial. An example is
found in the Man-binding legume lectins (Figure 1). In contrast, vicinal axial
OHs in hexoses are separated by about 3.7 Å (43) and thus do not make
multiple hydrogen bonds with a single amino acid side chain.

The sugar ring oxygen atom is also sp^3 hybridized and has two lone pairs
of electrons, in tetrahedral geometry, that can act as hydrogen-bond acceptors
but cannot participate in cooperative hydrogen bonds. When used in direct
protein interactions, this oxygen usually shares a hydrogen bond–donating
amino-acid side chain with one of the sugar OHs (Figures 1 and 3). Moreover,
this oxygen is common to all sugars and thus cannot be used to distinguish
among sugars. An interesting hydrogen bond that involves a ring oxygen of a
hexose occurs in the complex of a galectin with N-acetyl-lactosamine, where

Table 2 Lectin-carbohydrate hydrogen bonds[a]

Structure	Charged					Planar polar			OH-containing			Main-chain		H_2O[b]
	Asp	Glu	Arg	Lys	His	Asn	Gln	Trp	Tyr	Ser	Thr	N	O	
LOL, ConA, Pea lectin (Man)	1/O4,6					1/O4						3/O3,4; O3:O6		
GS4 (Gal)	1/O3,4					1/O3						1/O3	1/O3	1/O2
(Fuc α(1,4))			1/O3									2/O4;O4		1/O5
(Fuc α(1,2))								1/O2						2/O3;O5
EcoL,SBA (Gal)	1/O3,4					1/O2 1/O3	1/O6			1/O4		2/O3;O4		1/O6[c]
WGA 1° (GlcNAc/NeuNAc)		1/N5							1/O3(4)[d]	2/O10; O1A[e]				2/O4;O7[e]
2° (Gal)	1/N5								1/O3	1/O10				
Ricin B (Gal)	1/O3,4					1/O3 1/O2	1/O6 1/O3							1/O5
GNA (Man)	1/O2				1/O4	2/O4;O6			1/O4					1/O3,4 1/N5[f]
Galectin (Gal) (Glc/GlcNAc)		1/O6,3[f]	1/O4,5,3[f] 1/O3		1/O4									
MBP (Man) Gal-binding mutant (Gal)	1/O4	2/O3:O4 1/O3				2/O3:O4 1/O3	1/O4							
Influenza HA 1° (NeuNAc)		1/O9					1[g]/O1A,8		1/O8	2/O1B; O9		1/O1B	1/N5	
2° (NeuNAc)		1/O9								1/O4,10		2/O4;O4		
Polyoma VP1 (NeuNAc)			1/O1A,B		1/O4				1/N5					
CT/LT (Gal)		1/O4		1/O3,4		1/O2,3							1/O6	3/O2;O6; O6
CT (NeuNAc)												1/O1A		3/O1A; O8;O9
PT (NeuNAc)			1/O8,9							1/O1A		1/O1A	1/N5	

[a] Hydrogen bonds are listed only for interactions with the principal sugar specificity determinants. The number before the slash gives the number of side chains (or main chain NH, main-chain O, or water) that form contacts with the sugar. The sugar atoms with which hydrogen bonds are formed are given after the slash. If a single side chain forms hydrogen bonds with more than one sugar atom, then those atoms are separated by a comma (e.g. O3,4). In cases with more than one occurrence of a given side chain, the atoms interacting with different residues are separated with a semicolon. Primary (1°) and secondary (2°) binding sites are listed separately for WGA and HA.

[b] Hydrogen bonds mediated by a single water molecule. Note that the number of water molecules located depends in part on the resolution of the structure.

[c] EcoL only.

[d] O4 of NeuNAc is equivalent to O3 of GlcNAc.

[e] NeuNAc only.

[f] GlcNAc in N-acetyllactosamine.

[g] L226Q mutant only.

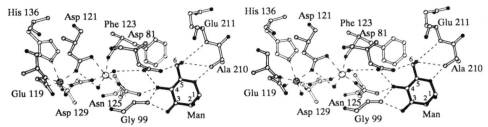

Figure 1   Stereo view of the terminal mannose of Manα1,3Manβ1,4GlcNAc bound to LOL I (PDB entry 1log) (12). The carbon atoms of Man have been numbered. Protein is shown with open bonds, carbohydrate with black bonds. White, grey, and black spheres represent carbon, nitrogen, and oxygen atoms, respectively. Isolated black spheres represent ordered water molecules. The Mn^{2+} and Ca^{2+} are shown as larger grey and white spheres, respectively. Short dashed lines represent hydrogen bonds, longer dashed lines coordination bonds. This and all other figures made with MOLSCRIPT (41).

one NH_2 group of a conserved arginine forms hydrogen bonds with the ring oxygen and the 4-OH of Gal, while the other NH_2 of the arginine forms a hydrogen bond with the 3-OH of GlcNAc (Figure 3) (25).

The acetamido moiety of GlcNAc, GalNAc, and NeuNAc is often a dominant or significant recognition determinant. Unlike OHs, the amide group and carbonyl oxygen of the acetamido substitutent have fixed, planar geometry.

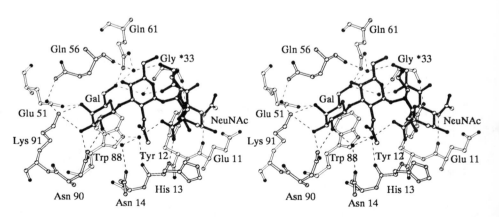

Figure 2   Stereo view of the pentasaccharide from ganglioside G_{M1} (Table 1) bound to CT (PDB entry 1chb) (39). The residue denoted with an asterisk is from a neighboring protomer in the pentamer. The Glc residue in G_{M1} has been omitted for clarity. The water molecule that connects Gal 2-OH with Asn^{14} is conserved in CT and LT. Symbols are as described in Figure 1.

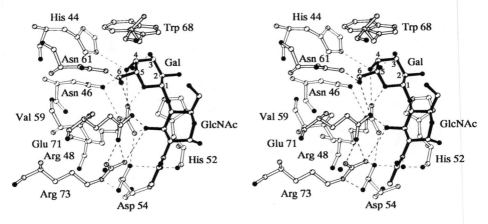

Figure 3 Stereo view of N-acetyllactosamine bound to galectin-1 (PDB entry 1slt) (25). Carbon atoms of the Gal residue are numbered. Symbols are as described in Figure 1.

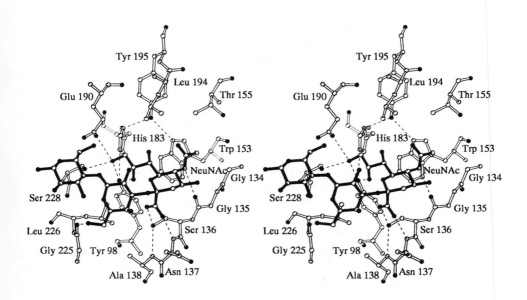

Figure 4 Stereo view of α(2,3) N-acetylneuramin-lactose complexed with influenza virus HA (PDB entry 1hgg) (42). Symbols are as described in Figure 1.

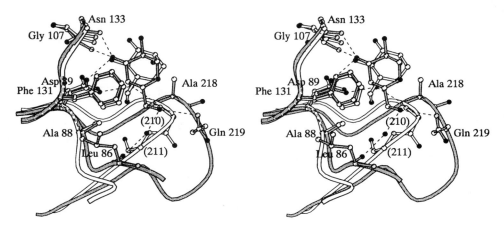

Figure 5 Stereo view of the binding site of LOL I with Man (PDB entry 1log; white bonds) (12) superimposed on EcorL bound to Gal (16) (PDB entry 1lte; grey bonds). The backbone is represented by a continuous tube in each case. Residue numbers are shown for EcorL. The main-chain atoms of LOL I residues 210 and 211, which correspond to EcorL residues 218 and 219, are marked in parentheses; the main-chain amide groups of these residues donate hydrogen bonds to the sugar ligand in both cases (also see Figure 1). Symbols are as described in Figure 1.

The amide group acts as a hydrogen bond donor to planar carbonyl or carboxylate oxygens in all cases observed to date (Table 2 and Figures 2 and 4), while the acetamido oxygen often accepts hydrogen bonds from serine.

The particular sugar functionalities that form hydrogen bonds with the lectin are those required for specific recognition and discrimination, whereas those positions that are not used as recognition elements tend to be solvent exposed and form no direct contacts with the protein. For example, ConA, pea lectin, and LOL all bind both Man and Glc and form specific hydrogen bonds with the 3-, 4- and 6-OHs of the sugars, but not the 2-OH, which differs between these two monosaccharides (9–11) (Figure 1). In contrast, GNA is specific for Man only and does form hydrogen bonds with the 2-OH in addition to other OHs (24). Similarly, the Gal-binding EcorL forms hydrogen bonds with the 3-, 4- and 6-OHs of Gal (Figure 5) (16); because Gal has an axial 4-OH, while that of Man and Glc is equatorial, specificity is derived from hydrogen bond geometry specific to a single epimer. Specificity for Gal in the galectins is achieved using the 4- and 6-OHs; an equatorial 4-OH would be sterically excluded by Trp[68] and would be unable to form the same hydrogen-bond interactions with arginine and histidine residues (Figure 3) (25). Recognition of the acetamido substituent of GlcNAc, GalNAc, and NeuNAc varies, again depending on the specificity of the sugar. WGA, which recognizes both

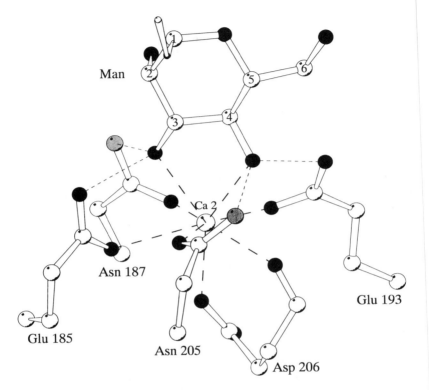

Figure 6 Mannose binding to rat mannose-binding protein A (PDB entry 2msb) (28). Symbols are as described in Figure 1.

GlcNAc and NeuNAc, interacts primarily with the acetamido group and the adjacent OH (3 for GlcNAc, 4 for NeuNAc) (18, 19).

In general, the hydrogen bonds between lectins and essential recognition determinants on sugars are shielded from bulk solvent, meaning that they exist in a lower dielectric environment and are probably enthalpically stronger than those formed with water (44). Moreover, the use of hydrogen bond donors and acceptors with fixed geometry may be important to specificity. A freely rotating group like the OH of serine has some plasticity in the formation of hydrogen bonds with the sugar and therefore may not be capable of discriminating absolutely between, for example, epimeric OHs.

WATER-MEDIATED HYDROGEN BONDS   Water molecules are observed to mediate hydrogen bonds between sugar and protein in many of the lectin-carbohydrate complex structures determined at high resolution. Studies of glycogen phosphorylase inhibitors have shown that water-mediated hydrogen bonds between sugar and protein can be as strong as direct protein-sugar hydrogen bonds (45). Comparison of a series of sugars bound to a given lectin, or a series of related lectins bound to a given sugar, sometimes reveals common water molecules which can be presumed to be important elements in recognition. For example, three water molecules are observed to mediate identical interactions between Gal and both LT and CT (37, 39), and in both the galactose and lactose complexes of LT (37, 38) (Figure 2). Also, comparison of the binding site in its liganded and unliganded states shows that some water molecules bound to the protein in the sugar-free state remain in the presence of sugar and form hydrogen bonds with the ligand. In this case the water molecules act as fixed structural elements, equivalent to hydrogen-bonding groups of the protein, and can therefore be considered part of the binding site architecture. The structure of CT observed at high resolution in several crystal forms reveals one such water molecule that interacts with the 2-OH of Gal (37) (Figure 2).

Recognition of Charged Groups on Sugars

The most extensive structural data available for recognition of charged sugars by lectins are for sialic-acid binding lectins, namely CT (39), influenza HA (32, 34, 35), and WGA (18, 19). Thus far only complexes with the most common sialic acid, N-acetylneuraminic acid (NeuNAc), and some of its derivatives (34, 35) have been studied. A complex between LOL and muramic acid indicates that the acid is relatively solvent exposed and interacts with the protein only through water molecules (13a).

Although salt-bridge formation between the NeuNAc carboxylate and positively charged amino acids might be expected to be important to recognition, the data show that, with the exception of polyoma virus (36), the carboxylate moiety interacts with main-chain amide groups, polar side chains (especially serine), and ordered water molecules rather than fully-charged side chains (Figures 2 and 4) in what are essentially hydrogen-bond interactions. In contrast, formation of salt bridges with the acid appears to be a common feature of neuraminidases (48–50). A curious symmetry exists between the common occurrence of acidic side chains as hydrogen bond acceptors from sugar OHs (see section on "Hydrogen Bonding" above) and the frequent use of the serine OH in the recognition of the NeuNAc carboxylate. The use of polar groups to provide electrostatic stabilization is well documented for the periplasmic sulfate- and leucine/isoleucine/valine-binding proteins, which bind to SO_4^{2-} or the NH_3^+ and COO^- moieties of zwitterionic amino acids primarily through main-chain amide and carbonyl groups; no charged amino acid side chains are used (51, 52).

Charge-charge interactions do appear to have roles in the interaction between polyoma virus and NeuNAc-containing ligands. A low-resolution (3.6 Å) structure of polyoma virus complexed with sialyllactose indicates that the NeuNAc carboxylate interacts with an arginine side chain (36). Based on the structure of the α(2,3) sialyllactose complex and modelling, it was proposed that the inability of one polyoma strain to bind branched ligands containing both α(2,3)- and α(2,6)-linked NeuNAc is due to charge repulsion by a glutamic acid residue in a putative second site specific for the α(2,6) branch (36). Precedent for charge repulsion comes from WGA, in which GlcNAc and NeuNAc are both observed to bind at the primary (P) site, whereas only GlcNAc binding is observed at the secondary (S) site. The difference has been ascribed to the presence of acidic residues in the lattice contact near the S site and is consistent with the observation that the neutral methyl ester of NeuNAc will bind at the S site (19).

Role of Divalent Cations

INDIRECT ROLES   Several classes of lectins require divalent cations for function. Of those with known three-dimensional structure, the legume lectins use Ca^{2+} and Mn^{2+} to stabilize the binding site and fix the positions of amino acids that interact with sugar ligands (Figure 1). The Ca^{2+} forms coordination bonds with the side-chain carbonyl oxygen of a conserved asparagine residue while the side-chain NH_2 of this asparagine donates a hydrogen bond to the sugar ligand. Likewise, one carboxylate oxygen of an acidic amino acid is bound to a water molecule that is in turn a Ca^{2+} ligand, and the other accepts a hydrogen bond from the sugar. The Ca^{2+} coordination shell thus fixes side chains for optimal binding to the sugar. The cations also stabilize the general architecture of the site by fixing the positions of second shell structural elements, i.e. those that interact with other protein groups that in turn contact the sugar ligand. Thus the Mn^{2+} does not coordinate any residues that interact directly with the protein, but instead fixes the Ca^{2+} position.

DIRECT INTERACTIONS   The C-type lectins are unique among the structurally characterized lectins in that a required Ca^{2+} forms direct coordination bonds with the sugar ligand. In this case, the full noncovalent bonding potential of two vicinal OHs is used (28–30) (Figure 6): one lone pair of electrons from each OH forms a coordination bond with the Ca^{2+}, the other lone pair accepts a hydrogen bond from a side-chain amide group, and the proton is donated to an acidic oxygen in a hydrogen bond. The hydrogen bonding in this scheme is reminiscent of that in the legume lectins in that the Ca^{2+} positions groups for hydrogen bonding with the ligand. However, the paucity of other interactions in this case relative to the other lectins, which display more extensive hydrogen bond and van der Waals interactions between protein and sugar, indicates that the direct coordination bonds are the primary determinants of affinity (30). Although thus far unique to the C-type lectins, direct coordination complexes have been observed in Ca^{2+} salts of sugars in small-molecule crystal structures (53). A Ca^{2+} also interacts directly with the carboxylate moiety of a 4,6-O-(1-carboxyethylidene) galactoside derivative in SAP (31). However, the Gal moiety does not interact with the Ca^{2+}, and whether binding of SAP to naturally-occuring sugars involves direct interactions with Ca^{2+} is not known. The only other direct sugar-metal interaction known in proteins is xylose isomerase, in which the sugar forms direct coordination bonds with two Mg^{2+} (54).

Nonpolar Interactions

Carbohydrates are highly polar and solvated molecules owing to the presence of the OHs and the ring oxygen. Nonetheless, sugars have significant nonpolar patches formed by the aliphatic protons and carbons at the various epimeric centers which extend out to the exocyclic 6 position of hexoses (Figure 7) and the glycerol moiety of neuraminic acids. With only two exceptions to date, this patch is observed to be packed against the face of one or more aromatic side chains of the protein. Removal of the apolar patch on the sugar, as well as the aromatic amino acid side chain, from bulk solvent is believed to provide significant binding energy.

In all lectin-Gal complex structures, the apolar patch of the Gal B face (Figure 7) packs against the face of tryptophan or phenylalanine (Figures 2, 3, 5, and 8b). This arrangement is frequently described as stacking, which implies that the ring of Gal is parallel to the plane of the aromatic ring. In fact, the sugar ring is always canted off somewhat from a parallel position. Superposition of the bound Gal seen in the various structures reveals that the aromatic rings form an angle between 17° and 52° with the least-squares plane through the pyranose ring and exocyclic carbon, with an average angle of 32° (29).

In addition to the B-face of the sugar ring, the methyl group of the acetamido moiety of GlcNAc, GalNAc, and NeuNAc often interacts with an aromatic ring in lectins that specifically recognize this group, including WGA (18, 19) and HA (32) (Figure 4). The carbon backbone of the glycerol moiety of NeuNAc can also present an apolar surface to the protein. For example, the glycerol moiety packs against the face of a tyrosine in WGA (18). NMR measurements of sugar binding to WGA (55), HA (42), and a Gal-binding mutant of MBP-A (56) reveal shifts in the resonances of specific sugar protons resulting from their interaction with the field produced by aromatic ring currents. As expected, these protons are attached to sugar carbons that are observed near the aromatic rings of amino acid side chains in the crystal structures of these lectin-sugar complexes.

The infrequent use of aliphatic side chains in nonpolar interactions with sugars suggests that interaction with the delocalized electron cloud of the aromatic ring is energetically significant beyond simply providing a geometrically complementary apolar surface. Most likely, the interaction is driven by the proximity of the aliphatic protons of the sugar ring, which carry a net positive partial charge, and the π-electron cloud of the aromatic ring. The same principle governs the edge-to-face stacking of aromatic residues in protein structures (57).

While Gal has thus far always been found packed against an aromatic side chain, the situation with Man is more variable. These observations may reflect the fact that the axial disposition of the Gal 4-OH, as opposed to the equatorial 4-OH of Man, creates a more extensive and continuous nonpolar surface (Figure 7). Man C5 and C6 pack against a conserved phenylalanine ring in the Man-binding legume lectins (Figure 1), whereas no aromatic interactions

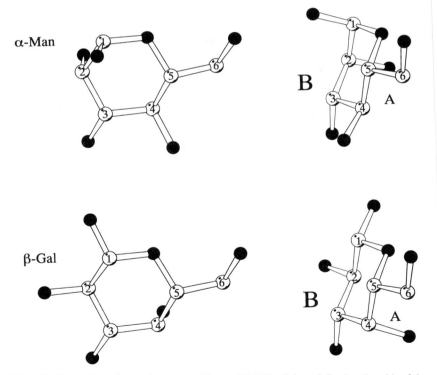

Figure 7 Comparison of nonpolar areas on Man and Gal. The B face, defined as that side of the ring that gives clockwise numbering along the ring (54a), is shown in two views for each sugar. The axial 4-OH of Gal gives a more continuous aliphatic patch formed by the 3, 4, 5 and 6-carbons relative to Man.

are found in MBPs or GNA. Analysis of a series of liver MBP (MBP-C)-monosaccharide complexes reveals that the less polar face is always turned in the direction of a solvent-exposed valine, whereas ordered water molecules are visible near the other face (30). In the case of serum MBP (MBP-A), a histidine residue is present at the position equivalent to this valine. In this case, however, it is the A face of Man that faces the histidine, and the rings are splayed apart so that the only van der Waals contact is between the 2-OH and the imidazole ring (Figure 8a) (28).

Extended Sites

Some lectins are specific for only a single monosaccharide unit and do not discriminate on the basis of what other sugars are linked to this recognition element. In other cases, substantial affinity enhancements are observed when additional sugars are linked to the primary determinant, even though the lectin does not show significant affinity for the second sugar as a monosaccharide. The term extended site is useful for describing binding sites that interact specifically with more than one sugar residue to provide increased affinity. This feature has also been termed subsite multivalency (7). For example, among the Man/Glc-binding legume lectins, LOL and pea lectin display higher affinity for N-linked oligosaccharides containing $\alpha(1,6)$-linked Fuc than for those lacking Fuc, whereas ConA does not (58–60). A complex between LOL II and a decasacccharide from human lactotransferrin has provided a structural explanation of these observations (14). In this case, Man 4 (see Table 1 for numbering) binds in the primary site in the manner previously observed in complexes of mono- and oligosaccharides with LOL I (Figure 1) and other Man/Glc-specific legume lectins. Fuc 1′ lies in a shallow crevice next to the Man 4 binding site. The Fuc 2- and 3-OHs form hydrogen bonds with a glutamic acid residue and this sugar also packs against the side chain of Phe[123], but on the face opposite that contacted by the 5- and 6-carbons of Man 4 (Figure 1). The region of ConA equivalent to this extended Fuc binding site in LOL II contains a histidine residue that would sterically clash with Fuc 1′.

Other lectins show no significant affinity for monosaccharides, but instead bind specifically to larger oligosaccharides. This class includes the selectins, for which the natural ligands are not known but which appear to require both Fuc and a charged group (61), and GS4, which binds to the Lewis[b] antigen (Le[b]) (Table 1) (15). The GS4-Le[b] complex shows that the binding site is similar to the Gal site in EcorL, but the loop at residues 220–222 is truncated. This region is occupied by one of the Fuc residues, and both Fuc residues interact directly with the protein (15). Another example is provided by the galectins, which display marginal affinity for Gal but bind lactose (Galβ1,4-

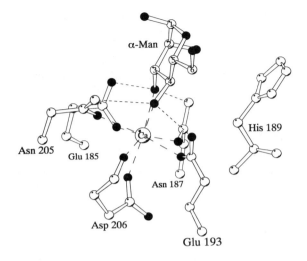

(a)

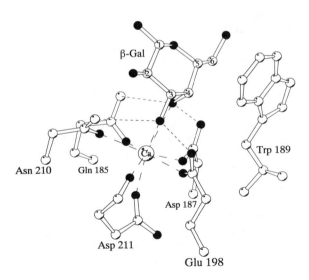

(b)

Figure 8 Comparison of wild-type MBP-A complexed with Man (a) and Gal-binding mutant QPDWG complexed with Gal (b) (29). In both cases the same number of coordination and hydrogen bonds are formed with the protein. Gal packs against Trp[189] in the mutant; in the orientation shown in (a), Man would be sterically excluded by Trp[189] in the mutant binding site. Symbols are as described in Figure 1.

Glc) and N-acetyllactosamine (Galβ1,4GlcNAc), with preference for the latter
(62). This specificity appears to be due to the formation of hydrogen bonds
by the 3-OH of Glc or GlcNAc with amino acids that are also ligands for the
Gal residue (Figure 3) (25, 27). Tighter binding of N-acetyllactosamine compared to lactose has been attributed to van der Waals contacts between the
acetamido group of GlcNAc and the protein (25) that are absent in the complex
with lactose (27).

Another kind of extended site is definable only from structures of lectins
complexed with oligosaccharides, which reveal a number of direct and water-
mediated interactions that cannot be inferred from binding data. Water molecules play a dual role in these cases. First, they stabilize the conformation of
the oligosaccharide by forming hydrogen bonds among OHs and ring oxygens
of different sugars. Second, they mediate interactions between protein and
ligand. There is a continuum ranging from numerous direct interactions such
as those observed in the LOL II-decasaccharide complex (i.e. in addition to
Man 4 and Fuc 1') (14), to those mediated almost exclusively by water molecules. Examples of the latter include LOL I-octasaccharide (13), WGA-glyco-
phorin glycopeptide (20), and MBP-A complexed with a Man_6 oligosaccharide
(28). Several lines of evidence make it difficult to argue that the lectin is
specific for the particular oligosaccharide in these cases. First, the specific,
direct interactions observed in the LOL II-decasaccharide complex suggest
that the extensive water-mediated contacts seen in the LOL I-octasaccharide
complex, which lacks Fuc 1' required for high affinity binding, represent a
nonspecific interface (14). Second, several high resolution structures of lectins
crosslinked by asymmetric oligosaccharides (20, 26, 28) have shown that
although recognition of the terminal monosaccharide by the primary binding
site is structurally identical, the water-mediated contacts are different on the
nonequivalent branches of the oligosaccharide. Thus water molecules can be
viewed as filler or mortar that allows more extensive contact with the protein
surface in a relatively nonspecific manner. As carbohydrates display a great
deal of conformational heterogeneity owing to the relatively small energetic
barriers to rotation about glycosidic linkages, the ability of water molecules
to mediate contacts with the protein surface is undoubtedly important in selecting a particular oligosaccharide conformation from the ensemble that exists
in solution. On the other hand, steric exclusion of certain oligosaccharide
conformations by features of the protein surface may be equally important to
the overall specificity and function of the lectin.

Secondary Sites

Lectins can have two clearly separated and thermodynamically independent
binding sites in a single protomer. The most well-characterized examples are

WGA (19) and GNA (24). In these cases, the three-dimensional protein structures as well as the amino acid sequences reveal internal duplications, so the primary and secondary binding sites are homologous. Secondary sites are most significant in terms of multivalent binding, since, as discussed below, multiple independent binding sites can give rise to enormous increases in affinity when the lectin is presented with multivalent ligands. In no case is there any evidence of molecular cooperativity, meaning changes in affinity at one site induced by binding to another, for either secondary sites or primary sites on a different protomer of an oligomer. Such cooperativity would be distinct from the simple sigmoidal curves obtained in inhibition assays in which monosaccharides compete with a highly valent ligand for binding to a multivalent lectin, as these curves simply reflect binding to a number of sites with identical affinities. As discussed in the following section, the observed lack of cooperativity is consistent with the conclusion that binding sites in lectins are preformed and that the proteins change little upon sugar binding.

In certain lectins, secondary sites with no sequence or structural homology to the primary sites have been identified. Examples include influenza HA crystals soaked in high concentrations of (2,3) sialyllactose (33) and MBP-C crystals soaked in 1.3 M α-methyl mannoside (30). In these cases, the secondary sites appear to have much lower affinity for the ligand than the originally characterized sites, and their significance is unclear.

Conformational Changes Upon Sugar Binding

Lectins undergo few, if any, changes in conformation upon binding to sugar. In no case have global changes in protein structure been observed; instead, small movements are restricted to the immediate vicinity of the sugar. For example, in LOL the loop containing residues 208–211 moves about 1 Å so that three hydrogen bonds are formed between main-chain amide groups and Man (Figure 1) (11, 13). Wright has pointed out that exposed regions with high thermal mobility exhibit larger conformational changes and lowering of mobility when contacting ligand (18). Ligand binding reduces the amount of conformational variability of certain residues by selecting from the ensemble in solution the conformation that optimizes the contacts. Thus a loop of LT that forms part of the binding site adopts several conformations in the five independent copies in the unliganded LT pentamer, but in all of the subunits it adopts a similar conformation and becomes better ordered upon ligand binding (37). An analogous observation has been made for WGA, although the analysis is complicated by the fact that the binding site is in a lattice contact and some of the decrease in mobility may be due to effects on the crystal lattice rather than the protein itself. On the other hand, comparison of the MBP-C structure in the presence and absence of ligand indicates no change in conformation upon ligand binding (30), perhaps reflecting the sparsity of contacts between the protein and ligand in this case.

The binding sites of lectins appear to be preformed, with ordered water molecules forming hydrogen bonds with the unliganded proteins in a pattern that closely mimics the hydrogen bonding by sugar OHs (10). This effect is a consequence of the use of planar hydrogen-bond donors and acceptors, which provide fixed geometry for interaction with ligands, and of interactions among amino acid residues and sometimes divalent cations in the binding site that maintain the conformation of the side chains in the absence of sugar ligands. Water molecules are found in the binding site of saccharide-free pea lectin near the positions occupied by the 3- and 4-OHs of mannose and form the same hydrogen bonds with the protein (10). Likewise, binding of sugars to MBP-C displaces two water molecules that form the same set of coordination and hydrogen bonds as the vicinal OHs of the sugar ligand, and one other water molecule that occupies a position that overlaps with a portion of the pyranose ring (30). Three other water positions are significantly different between the complexed and uncomplexed structures. Aside from these three, the network of water molecules that stabilizes the MBP-C conformation near the carbohydrate-binding site is unperturbed by sugar binding.

There is no evidence in any of the complexes cited in Table 1 for significant distortion of the sugar ring geometry upon binding to a lectin. This situation contrasts with hydrolytic enzymes that cleave the glycosidic bond between two sugars, such as lysozyme (63) and neuraminidase (64), in which strain introduced by distortion of the ring appears to be a necessary part of the enzymatic mechanism.

Comparison with Other Carbohydrate-Binding Proteins

In addition to lectins, many other types of proteins interact with sugars. Such proteins include enzymes that act on sugars as substrates, carbohydrate-specific antibodies, membrane transport proteins, and bacterial periplasmic binding proteins involved in sugar transport and chemotactic detection. The distinct functions served by these proteins are reflected in differences in their intrinsic affinities for monosaccharides, which are often much higher than the lectins, and in the relative importance of oligosaccharides as ligands, which form the primary high affinity ligands for the lectins.

Quiocho and colleagues have categorized carbohydrate-binding proteins into two groups (43, 65). The carbohydrate-binding sites of group I proteins are located in deep clefts and binding leaves little or no part of the ligand exposed to bulk solvent. Carbohydrate binding by this class of proteins is high-affinity, typically with sub-micromolar dissociation constants. Group I proteins include the bacterial periplasmic binding proteins and certain enzymes that act on sugar substrates. In contrast, the binding sites of group II proteins, which include the lectins, tend to be on the surface of the protein and are rather shallow clefts. Binding affinities are lower than for group I proteins; for example, lectins often have dissociation constants in the millimolar range.

Despite the differences between group I and group II proteins, many features of carbohydrate recognition are shared, with differences reflected primarily in the number of contacts formed and the degree of shielding from aqueous solution. The frequency of planar hydrogen bond donors and acceptors in protein-carbohydrate interactions was noted several years ago by Quiocho and coworkers, based largely on high-resolution structural studies of the bacterial periplasmic binding proteins (43, 44). In the periplasmic proteins, virtually all of the hydrogen bonding potential of the sugar is used, including that of the ring oxygen, whereas lectins form hydrogen bonds with only a subset of the OHs of the sugar. Moreover, in lectins a significant portion of the sugar ring is exposed to solvent, which is important in allowing recognition of different substituents at particular ring positions (see below). In contrast, the periplasmic binding proteins undergo a large conformational change that buries the entire sugar from bulk solvent. In this case it would be energetically costly to have unsatisfied hydrogen bonds. The larger number of hydrogen bonds formed in a low-dielectric environment, in addition to the favorable entropy gain of expelling large numbers of water molecules when the protein closes around the ligand, is believed to provide the tight affinity displayed by this class of proteins when compared to lectins (44).

RECOGNITION OF DISTINCT SETS OF SUGARS BY CLOSELY RELATED LECTINS

The classification of animal lectins discussed in the introduction reflects the related structures of the sugar-binding portions of these proteins. In some cases, lectins within a particular subgroup share the ability to bind similar sets of sugars. For example, all of the galectins bind β-galactosides and show related although not identical selectivity for particular ligands such as lactose and N-acetyllactosamine. In such cases, many of the conserved amino acids common to all members of the family are directly involved in forming the sugar-binding site. In other lectin families, various members display conserved protein structures yet bind to distinct sets of sugars. Comparison of lectins with relatively similar amino acid sequences that bind different saccharide ligands can be particularly informative about which interactions allow individual family members to discriminate among the different ligands. In two cases, the C-type animal lectins and the leguminous plant lectins, the discriminatory power of several different family members has been investigated at the structural level.

C-Type Animal Lectins

C-type lectins display many types of binding selectivity. Several of these lectins bind derivatives of Man and GlcNAc (Man-type ligands) although the relative affinities for sugars in this group vary. A second set of C-type lectins bind Gal and GalNAc (Gal-type ligands), again with varying degrees of discrimination. Although the only complexes of natural C-type CRDs with sugar that have been studied by x-ray diffraction are the MBPs (28, 30), mutants of MBP-A that have Gal-binding acitivity have been created by changing the amino acid sequence to make the protein resemble natural Gal-binding C-type CRDs in the region of the sugar-binding site (56). Structures of one such mutant in complex with Gal and GalNAc have recently been determined (Figure 8) (29), allowing a comparison of the mechanisms of Man and Gal binding.

Fundamental aspects of the sugar-binding mechanism are shared in the Man-type and Gal-type binding sites. In particular, the 3- and 4-OHs of both types of sugars, although differently disposed, are apical coordination ligands for a bound Ca^{2+}, and they form a network of hydrogen bonds with amino acid side chains that also serve as equatorial ligands for this same divalent cation. Discrimination results from the three effects described below.

AMIDE POSITIONS AT THE BINDING SITE   Exchanging the positions of a single amide and a carboxylate oxygen between two of the four amino acid side chains that bind both Ca^{2+} and sugar ligand is sufficient to invert the sugar-binding selectivity of the binding site (56). Structural analysis indicates that this change does not alter the positions of these amino acid side chains (29), so only a very minor modification of the local environment is necessary to change dramatically the way sugars are oriented in the binding site. Unfortunately, an explanation for this effect rooted in the fundamental physical properties of the sugar and binding site is not yet possible.

PACKING INTERACTIONS UNIQUE TO GALACTOSE   High affinity binding to Gal is achieved when tryptophan is inserted next to the binding site defined by the Ca^{2+} coordination complex. Evidence from both NMR and x-ray crystallography indicates that this tryptophan packs against the apolar B face of Gal (29, 56). The aromatic packing interaction is unique to the Gal-binding site, as no such interactions occur with Man bound to wild-type MBP. Since the absolute affinities of the wild-type CRD for Man and of the tryptophan-containing mutant for Gal are similar, it can be inferred that the energy derived from ligation of Man to Ca^{2+} may be greater than that resulting from Gal binding, since the packing interaction with tryptophan would compensate by strengthening binding to Gal.

STERIC EXCLUSION   Exclusion also plays an important role in determining selectivity. When the binding site of the MBP CRD is modified by the amide switch and addition of tryptophan, Man still binds only slightly less tightly than does Gal. However, insertion of an additional glycine-rich segment of peptide following the tryptophan creates an enlarged protein loop that locks the tryptophan into a position incompatible with Man entering the binding site in the orientation observed for the wild-type protein (Figure 8) (29).

Legume Lectins

Three of the leguminous plant lectins that have been investigated as complexes with saccharide at the structural level, including ConA, pea lectin, and LOL (9–11), bind primarily to Man-type ligands, whereas GS4, EcorL, and SBA (15–17) bind Gal-type ligands preferentially. Comparison of sugar-bound complexes of these proteins reveals that many of the features of the binding sites are conserved even when the specificity for sugars is dramatically different (Figure 5). Apolar portions of both Man and Gal are packed against an aromatic residue (tyrosine or phenylalanine), and sugar OHs form hydrogen bonds with many of the same side chains, paricularly a pair of asparagine and aspartic acid residues in the core of the binding sites (residues Asn[133] and Asp[89] in EcorL). However, the sugar is rotated so that a different set of OHs is involved in these interactions: the 4- and 6-OHs of Man versus the 4- and 3-OHs of Gal.

While it is not yet possible to provide a detailed molecular explanation for the striking differences in affinities observed for these two very closely related types of binding sites, both accommodation and exclusion appear to play a role. In EcorL, displacement of a portion of the backbone in a β-turn is required to make room for the 6-OH of Gal; the 6-OH is in an entirely different position, making an alternative set of hydrogen bonds, when a Man-type ligand binds to a lectin such as ConA (8, 16) (Figure 1). The conformation of this loop presumably results in exclusion of Gal-type ligands from the binding sites of ConA, pea lectin, and LOL. Steric exclusion of unfavored ligands may also be an important aspect of EcorL selectivity. However, although it has been proposed that the side chain of Ala[218] in EcorL would not allow Man-type ligands with equatorial 4-OHs to bind in the orientation observed in members of this family that bind Gal-type ligands, recent mutagenesis experiments do not bear out this suggestion (66). Additional reasons for preferential binding of different ligands, such as changes in the local electrostatic environment of the binding site, remain to be examined at both the theoretical and experimental levels.

Glycolipid-Binding Toxins

Both CT and LT bind to ganglioside GM_1 as their primary cell-surface receptor. The importance of the terminal Gal and NeuNAc residues (see Table 1) is revealed in the crystal structure of the CT-GM_1 pentasaccharide complex (39).

However, LT but not CT shows additional weak affinity for gangliosides GM_2 and asialo GM_1, in which either the terminal Gal or sialic acid residue has been removed. Only one amino acid residue in the binding site, at position 13, differs between the two toxins: In CT this position is occupied by a histidine (Figure 2) that is replaced by arginine in most strains of LT. On the basis of the structure of CT in complex with GM_1, it has been suggested that Arg^{13} in LT may make additional hydrogen bonds with nonterminal portions of the oligosaccharide which would partially compensate for loss of either of the terminal residues and thus account for the weak affinities for GM_2 and asialo GM_1 (39). It remains to be determined whether structural analysis of LT in complex with various oligosaccharide ligands will support this hypothesis.

Principles of Differential Sugar Recognition

These examples show clearly that differential binding of sugars to lectins depends on subtle changes in the disposition of amino acid side chains near the sugar-binding sites. The sites are broad and shallow, with selectivity resulting from three factors. First, the epimeric OH that distinguishes particular sugars, such as the 4-OH in Man-type and Gal-type ligands, is involved in hydrogen and sometimes coordination bonds to the lectins, so that the binding site directly reads out the differences among these sugars. A second factor in establishing selectivity is steric exclusion of the disfavored ligands, rather than just loss of certain contacts compared to lectins that bind the same ligands tightly. Conversely, accommodation of multiple ligands by one particular binding site is achieved by leaving exposed those portions of the ligands that would allow them to be distinguished, such as the 2-substituents of Man and GlcNAc in the binding site of MBP. As noted in the previous section, formation of alternative, compensating contacts with two different preferred ligands is not observed. Similarly, lectins can be insensitive to how a bound sugar is linked in a simple glycoside or an oligosaccharide as long as the attached substituents project away from the lectin surface.

MULTIVALENCY AND THE IMPORTANCE OF GEOMETRY IN OLIGOSACCHARIDE RECOGNITION

Energetic Considerations

Lectins often bind to natural polysaccharides with high affinities, reflected in dissociation constants in the nanomolar range, yet their interaction with simple monosaccharides is far weaker, with dissociation constants often in excess of 1 mM. As noted in the second section of this review, increases in affinities for oligosaccharides derive in many instances from extended binding sites in which more than just terminal sugar residues of an oligosaccharide make contact, either directly or through bridging water molecules, with the surface of the lectin. However, the affinities resulting from these additional contacts are still in the micromolar range at best. Many lectins achieve much higher affinity by clustering several similar or identical binding domains, often by formation of polypeptide oligomers. The importance of clustered binding sites derives from the fact that the free energy of binding to a multivalent ligand to multiple sites on an oligomeric lectin can be as large as the sum of the free energies of the individual binding interactions. In other words, the individual dissociation constants can potentially be multiplied to achieve the high affinity of the lectin for a complex ligand, without the need to involve any additional molecular interactions beyond those seen with individual sugars in relatively simple binding sites.

In reality, the conditions required for true additivity of binding energies are not often met, owing to additional geometrical considerations, but the affinity enhancements can still be quite substantial. Since the binding sites of many lectins can best accommodate sugars located in terminal positions, the importance of clustering is not always obvious from analysis of binding to linear polysaccharides. Several synthetic approaches to display of multiple terminal residues have been developed, including preparation of small cluster ligands as well as larger multivalent neoglycoproteins (67, 68). Cluster ligands consist of 2 to 3 sugar residues linked to a hub as small as tris(hydroxymethyl)aminomethane or at larger spacings created by insertion of linker arms of various types. Binding affinity measurements using such ligands clearly demonstrate the importance of multivalency independent of the specific structural constraints of natural oligosaccharides. For example, the affinity of Gal-terminated conjugates for the hepatic asialoglycoprotein receptor increases from 0.5 mM for monovalent, to 8 M for divalent, to 100 nM for trivalent versions of one cluster ligand (69). Similarly, small, bivalent sialic acid–containing ligands inhibit hemagglutination by influenza virus much more potently than do monovalent sialosides (70–72). Analysis of multiple synthetic ligands demonstrates that there is a critical spacing for optimal binding. However, the enhancement due to multivalency probably involves interaction with multiple copies of the HA trimer on the surface of the virus, since no enhancement in binding to individual trimers is observed (70).

In the case of the asialoglycoprotein receptor, the affinity of cluster ligands is diminished in detergent-solubilized receptor, but there is still a substantial enhancement due to multivalency for clusters with appropriate spacing between terminal Gal residues, so in this case the ligand probably binds to multiple subunits within a single protein oligomer (69). The importance of spacing is also reflected in the ability of the receptor to distinguish natural glycoprotein oligosaccharides with different linkages (73). The energy minima that define the preferred conformations of oligosaccharides are usually rela-

(a)

tively shallow, allowing latitude in the orientation of terminal residues (74); nevertheless, the conformational energy surface for the oligosaccharides places constraints on the possible arrangements of the terminal groups. Thus, the spacing of binding sites on a multivalent lectin must match a preferred oligosaccharide conformation or there will be an unacceptable energetic cost for rearrangement of the ligand to fit the lectin.

Surface Recognition

Although the clustering of identical sugar-binding sites can lead to substantial increases in affinity for complex ligands, the geometry of the ligands and the specific arrangement of binding sites can also have important effects on the selectivity with which multivalent ligands are recognized. As noted in the introduction, a number of lectins distinguish self from nonself, a property that is achieved by maximizing recognition of certain carbohydrate-coated surfaces while minimizing binding to sugars attached to soluble glycoproteins. Structural studies reveal that many lectins that bind surfaces achieve the required planar array of sugar-binding sites by arrangement of polypeptide subunits in oligomers that have cyclic symmetry, with all of the binding sites located near one end of the oligomer. In the case of mammalian serum MBPs, three subunits associate (Figure 9) through triple α-helical coiled-coil stalks (75, 76), while a similar trimeric arrangement in influenza virus HA involves more extensive interactions between the binding domains themselves (77). In a related manner, the disposition of viral protein 1 on the surface of polyoma virus presents an array of binding sites that can be viewed as clusters of five sites with fivefold cyclic symmetry on each face of the virion (36).

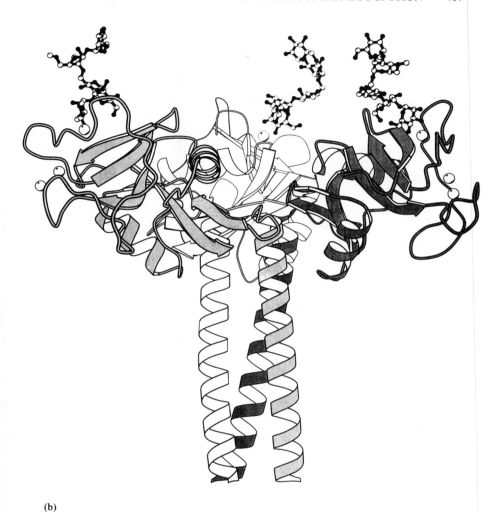

(b)

Figure 9 Arrangement of binding sites in oligomeric lectins. The CT B-chain pentamer (*a*) and MBP-A trimer (*b*) are shown with their molecular symmetry axes vertical; the target cell surfaces would be at the top. Models of oligosaccharides binding to each oligomer are meant to emphasize the large spacing between carbohydrate-binding sites. The G_{M1} pentasaccharide model was generated from that seen in the one protomer of CT where the entire pentasaccharide is visible (39). The oligosaccharide model in (b) was made by superposition of the Man_6 oligosaccharide structure complexed with the MBP-A (28) onto each protomer of trimeric MBP-A (75).

The glycolipid-binding bacterial toxins CT and LT display fivefold cyclic symmetry, forming toroid-shaped structures from which sugar-binding sites project on one face, near the base of the intact toxin (Figure 9) (38, 39). PT, which binds to glycoprotein receptors, is a heterooligomer with quasi fivefold cyclic symmetry (40). Although all of the subunits have three-dimensional folds similar to those of CT and LT, only two bind sugars, and these binding sites are located on a different surface of the subunits so that they are positioned on the sides rather than on one end of the toroidal pentamer (40). This difference may reflect the fact that oligosaccharides of glycoprotein receptors are projected further from the cell surface than are glycolipids.

In addition to facilitating the binding of these lectins to sugar-containing surfaces, the arrangement of binding sites in these proteins insures that single oligosaccharides will not bind with high affinity simply because typical glycoprotein- or glycolipid-linked oligosaccharides are not large enough to span the distances between binding sites, which are 45–53 Å for MBPs (75, 76), 45 Å for influenza virus HA, and 65 Å for PT (40). Thus, the intact lectins avoid binding to and being inhibited by soluble glycoconjugates.

Lattice Formation by Bridging Lectins

Since both lectins and oligosaccharides can be multivalent, they have the potential to form both linear and more complex arrays. In some cases, lattice formation results from lectins that have dihedral point symmetry, so that equivalent binding sites project from each end of the oligomer. Most natural oligosaccharides do not contain true symmetry elements, but the similarity in the termini of multiple terminal branches, combined with the fact that the branches often are directed away from each other, means they also display pseudodihedral symmetry and thus are able to form repeating networks with lectins. Linear assemblies of this kind have been studied crystallographically for mammalian galectin 1 with an octasaccharide (Table 1) (26). Additional numbers of binding sites on an oligomeric lectin, such as tetrameric SBA, allow for three-dimensional lattices with a synthetic pentasaccharide (17). More complex (i.e. triantennary) oligosaccharides can form other three-dimensional lattices that have not yet been characterized at the atomic level (78).

While the need for planar arrays of binding sites in the recognition of surfaces meshes nicely with the known biological functions of the lectins that display this subunit arrangement, a similar correlation remains to be established for the ability of lectins to form lattices. It is possible that interaction of lectins with soluble glycoconjugates could produce such arrays, but they have not been observed under biological conditions. Alternatively, the ability to form arrays may be an in vitro consequence of the natural role of lectins with this geometry in crosslinking or aggregating glycoconjugates such as those found on cell surfaces. Indeed, there is a longstanding correlation between the ability of lectins such as ConA to aggregate cell surface glycoproteins and their mitogenic activity (2). Additionally, one of the major proposed roles for galectins is in linking cell surfaces to extracellular matrix elements such as laminin (79).

Other Geometrical Arrangements

Several plant lectins display more complex arrangements of multiple binding sites. The repeating subdomains of WGA project a total of 8 binding sites in a pseudohelical arrangement (19, 80), while the recently described GNA structure reveals up to 12 potential binding sites arrayed over the surface of the tetrameric protein (24). A mechanism for binding of WGA to two branches of a tetrasaccharide linked to a glycopeptide from human erythrocyte glycophorin has been elucidated crystallographically (20, 21). The structural information accounts for the high affinity binding of this ligand, but in the absence of definitive information about the function of these lectins, it is not possible to say how this specific arrangement of binding sites is suited for a particular biological situation.

SUMMARY AND PERSPECTIVES

Comparison of the various lectin structures discussed in this review strongly indicates that the ability to bind carbohydrates has evolved independently on several separate occasions in the context of various protein structural frameworks. This conclusion is consistent with the lack of sequence similarity among the various groups of lectins (5). In spite of their independent origins, the binding sites that have been analyzed so far share a few key features, which are presumably related to the fact that the interaction of lectins with individual sugars is of modest affinity. Most of the binding sites are relatively shallow indentations on the protein surface. This arrangement differs from those in which monosaccharides are completely surrounded by the protein, as in the bacterial periplasmic sugar-binding proteins and hexokinase (81, 82), and the active sites of many other enzymes that utilize sugars or sugar conjugates as substrates, which are often deep clefts in the protein. An interesting comparison in the latter case is the distinction between the relatively shallow, low-affinity sialic acid binding site of influenza virus HA and the deeper active site of the neuraminidase found on the same viruses, which binds a transition state analog with roughly 1000-fold greater affinity than sialic acid for HA (64).

Not surprisingly, hydrogen bonding to sugar OHs is an essential part of lectin-carbohydrate interactions, and selective binding to epimeric OHs is an important aspect of differential binding of sugars. Two common features of hydrogen bonding between carbohydrates and lectins are evident. First, charged or polar planar groups form the majority of hydrogen-bond donors and acceptors with sugar OHs. Second, the sugar-binding sites are preformed, in the sense that few changes occur upon saccharide binding. High-resolution structures of unliganded lectins reveal that discrete water molecules form hydrogen bonds with those polar atoms that form hydrogen bonds with the sugars in the lectin-sugar complexes. These water molecules serve as mimics of the hydrogen-bonding pattern of sugar OHs.

Other features common to lectin binding sites are also evident. The van der Waals contacts between sugar and protein often include packing interactions with aromatic amino acid side chains. Such packing is particularly common in the Gal-specific sites, but it is also observed in legume lectins that bind Man and Glc. Both hydrophobic packing and hydrogen-bonding interactions with substituents on hexose rings, particularly the acetamido groups in GlcNAc, GalNAc, and NeuNAc, provide further selectivity for specific sets of saccharide ligands. The coordination bonds that link sugars to the Ca^{2+} in C-type animal lectins contribute substantially to the affinity of this binding site. These are the only known lectins that exploit this energetically favorable interaction, although the sugar binding to other groups of Ca^{2+}-dependent animal lectins has yet to be characterized at the structural level. In general, it will be interesting to see how many of the features of lectin-carbohydrate interactions are shared by lectin types still to be analyzed by crystallography or nuclear magnetic resonance.

A fundamental gap still exists between the structures observed and our thermodynamic understanding of the basis for selective binding of sugars (83). As with most protein-ligand interactions, hydrogen bonds and van der Waals contacts contribute to the binding energy. The combination of structural studies with mutagenesis of the protein and binding studies using various natural and modified sugar ligands provides clear evidence about which amino acid residues and which portions of the sugars are required for selective sugar binding. But these experiments alone do not tell us how each interaction contributes to the energy of binding. The difficulty is particularly apparent when comparing lectins of similar structure but distinct binding specificity, since the differences in specificity observed would not be predictable from the differences in the amino acids near the binding sites of these proteins. Establishing the connection between structure and thermodynamics for lectin-sugar interactions remains a major challenge.

The structural data make an important contribution to our understanding of how high affinity, highly selective binding is achieved for complex saccharides as biological targets for the lectins. The intrinsic low affinity of the lectins for monosaccharide ligands prevents interference from competing small sugar ligands. Although extending binding sites to include direct and water-mediated interactions with multiple sugars in larger oligosaccharide ligands results in some increase in affinity, high affinity is usually achieved only when multiple copies of simple sugar-binding sites are clustered in oligomers. To date, our best understanding of this phenomenon is in the area of cell surface recognition, where the lectin is confronted with large, nearly planar arrays of oligosaccharides. Selective binding to such surfaces is achieved through arrangement of binding sites on one side of an oligomer with sufficient spacing to insure that oligosaccharides alone, such as those found on soluble glycoproteins, will not be able to bridge the multiple sites and will thus not compete for binding to the surface. In contrast, analysis of the asialoglycoprotein receptor of hepatocytes suggests that appropriate arrangement of simple binding sites in multiple types of subunits may also be used to insure high affinity binding to selected oligosaccharide structures (84). The structural basis for this type of selective binding achieved through multivalency remains to be established.

ACKNOWLEDGMENTS

We thank Anand Kolatkar for help with preparation of the figures. WIW is supported by grant GM50565 from the National Institutes of Health and the Pew Scholars Program in the Biomedical Sciences. KD is a Wellcome Principal Research Fellow.

Literature Cited

1. Drickamer K, Taylor ME. 1993. *Annu. Rev. Cell Biol.* 9:237–64
2. Lis H, Sharon N. 1986. *Annu. Rev. Biochem.* 55:35–67
3. Etzler ME, Murphy JB, Ewing N. 1995. *Glycoconjugate J.* 12:443
4. Sharon N. 1993. *Trends Biochem. Sci.* 18:221–26
5. Drickamer K. 1995. *Curr. Opin. Struct. Biol.* 5:612–16
6. Burnette WN. 1994. *Structure* 2:151–58
7. Rini JM. 1995. *Annu. Rev. Biophys. Biomol. Struct.* 24:551–77
8. Derewenda Z, Yariv J, Helliwell JR, Kalb AJ, Dodson EJ, et al. 1989. *EMBO J.* 8:2189–93
9. Naismith JH, Emmerich C, Habash J, Harrop SJ, Helliwell JR, et al. 1994. *Acta Crystallogr.* D 50:847–58
10. Rini JM, Hardman KD, Einspahr H, Suddath FL, Carver JP. 1993. *J. Biol. Chem.* 268:10126–32
11. Bourne Y, Roussel A, Frey M, Rougé P, Fontecilla-Camps J-C, Cambillau C. 1990. *Proteins* 8:365–76
12. Bourne Y, Rougé P, Cambillau C. 1990. *J. Biol. Chem.* 265:18161–65
13. Bourne Y, Rougé P, Cambillau C. 1992. *J. Biol. Chem.* 267:197–203
13a. Bourne Y, Ayouba A, Rougé P, Cambillau C. 1994. *J. Biol. Chem.* 269:9429–35
14. Bourne Y, Mazurier J, Legrand D, Rougé P, Montreuil J, et al. 1994. *Structure* 2:209–19
15. Delbaere LTJ, Vandonselaar M, Prasad L, Quail JW, Wilson KS, Dauter Z. 1993. *J. Mol. Biol.* 230:950–65
16. Shaanan B, Lis H, Sharon N. 1991. *Science* 254:862–66
17. Dessen A, Gupta D, Sabesan S, Brewer CF, Sacchettini JC. 1995. *Biochemistry* 34:4933–42
18. Wright CS. 1990. *J. Mol. Biol.* 215:635–51

19. Wright CS. 1984. *J. Mol. Biol.* 178:91–104
20. Wright CS. 1992. *J. Biol. Chem.* 267:14345–52
21. Wright CS, Jaeger J. 1993. *J. Mol. Biol.* 232:620–38
22. Montfort W, Villafranca JE, Monzingo AF, Ernst SR, Katzin B, et al. 1987. *J. Biol. Chem.* 262:5398–403
23. Rutenber E, Robertus JD. 1991. *Proteins* 10:260–69
24. Hester G, Kaky H, Goldstein IJ, Wright CS. 1995. *Nat. Struct. Biol.* 2:472–79
25. Liao D-I, Kapadia G, Ahmed H, Vasta GR, Herzberg O. 1994. *Proc. Natl. Acad. Sci. USA* 91:1428–32
26. Bourne Y, Bolgiano B, Liao D-I, Strecker G, Cantau P, et al. 1994. *Nat. Struct. Biol.* 1:863–70
27. Lobsanov YD, Gitt MA, Leffler H, Barondes SH, Rini JM. 1993. *J. Biol. Chem.* 268:27034–38
28. Weis WI, Drickamer K, Hendrickson WA. 1992. *Nature* 360:127–34
29. Kolatkar A, Weis WI. 1996. *J. Biol. Chem.* 271:6679–85
30. Ng KK-S, Drickamer K, Weis WI. 1996. *J. Biol. Chem.* 271:663–74
31. Emsley J, White HE, O'Hara BP, Oliva G, Srinivasan N, et al. 1994. *Nature* 367:338–44
32. Weis WI, Brown JH, Cusack S, Paulson JC, Skehel DC, Wiley DC. 1988. *Nature* 333:426–31
33. Sauter NK, Glick GD, Crowther RL, Park SJ, Eisen MB, et al. 1992. *Proc. Natl. Acad. Sci. USA* 89:324–28
34. Sauter NK, Hanson JE, Glick GD, Brown JH, Crowther RL, et al. 1992. *Biochemistry* 31:9609–21
35. Watowich SJ, Skehel JJ, Wiley DC. 1994. *Structure* 2:719–31
36. Stehle T, Yan Y, Benjamin TL, Harrison SC. 1994. *Nature* 369:160–63
37. Merritt EA, Sixma TK, Kalk KH, van Zanten BAM, Hol WGJ. 1994. *Mol. Microbiol.* 13:745–53
38. Sixma TK, Pronk SE, Kalk KH, van Zanten BAM, Berghuis AM, Hol WGJ. 1992. *Nature* 355:561–64
39. Merritt EA, Sarfaty S, van den Akker F, L'Hoir C, Martial JA, Hol WGJ. 1994. *Protein Sci.* 3:166–75
40. Stein PE, Boodhoo A, Armstrong GD, Heerze LD, Cockle SA, et al. 1994. *Nat. Struct. Biol.* 1:591–96
41. Kraulis PJ. 1991. *J. Appl. Crystallogr.* 24:946–50
42. Sauter NK, Bednarski MD, Wurzburg BA, Hanson JE, Whitesides GM, et al. 1989. *Biochemistry* 28:8388–96
43. Vyas NK. 1991. *Curr. Opin. Struct. Biol.* 1:732–40
44. Quiocho FA. 1986. *Annu. Rev. Biochem.* 55:287–315
45. Watson KA, Mitchell EP, Johnson LN, Son JC, Bichard CJF, et al. 1994. *Biochemistry* 33:5745–58
46. Deleted in proof
47. Deleted in proof
48. Burmeister WP, Ruigrok RWH, Cusack S. 1992. *EMBO J.* 11:49–56
49. Varshese JN, Laver WG, Colman PM. 1983. *Nature* 303:35–44
50. Crennell S, Garman E, Laver G, Vimr E, Taylor G. 1994. *Structure* 2:535–44
51. Quiocho FA, Sack JS, Vyas NK. 1987. *Nature* 329:561–64
52. Pflugrath JW, Quiocho FA. 1985. *Nature* 314:257–60
53. Cook WJ, Bugg CE. 1977. In *Metal-ligand Interactions in Organic Chemistry and Biochemistry, Part 2*, ed. B Pullman, N Goldblum, pp. 231–56. Dordrecht, The Netherlands: Reidel
54. Lavie A, Allen KN, Petsko GA, Ringe D. 1994. *Biochemistry* 33:5469–80
54a. Rose IA, Hanson KR, Wilkinson KD, Wimmer MJ. 1980. *Proc. Natl. Acad. Sci. USA* 77:2439–41
55. Kronis KA, Carver JP. 1985. *Biochemistry* 24:826–33
56. Iobst ST, Drickamer K. 1994. *J. Biol. Chem.* 269:15512–19
57. Burley SK, Petsko GA. 1985. *Science* 229:23–28
58. Debray H, Decout D, Strecker G, Spik G, Montreuil J. 1981. *Eur. J. Biochem.* 117:41–55
59. Kornfeld K, Reitman ML, Kornfeld R. 1981. *J. Biol. Chem.* 256:6633–40
60. Debray H, Rougé P. 1984. *FEBS Lett.* 176:120–24
61. Varki A. 1994. *Proc. Natl. Acad. Sci. USA* 91:7390–97
62. Ramkumar R, Surolia A, Podder SK. 1995. *Biochem. J.* 308:237–41
63. Strynadka NC, James MN. 1991. *J. Mol. Biol.* 220:401–24
64. Janakiraman MN, White CL, Laver WG, Air GMA, Luo M. 1994. *Biochemistry* 33:8172–79
65. Quiocho FA, Vyas NK, Spulino JC. 1989. *Trans. Am. Crystallogr. Assoc.* 25:23–35
66. Adar R, Sharon N. 1995. *Glycoconjugate J.* 12:425
67. Wong SYC. 1995. *Curr. Opin. Struct. Biol.* 5:599–604
68. Lee YC, Lee RT. 1994. *Neoglyconjugates: Preparation and Applications.* New York: Academic

69. Lee RT, Lin P, Lee YC. 1984. *Biochemistry* 23:4255–61
70. Glick GD, Toogood PL, Wiley DC, Skehel JJ, Knowles JR. 1991. *J. Biol. Chem.* 266:23660–69
71. Glick GD, Knowles JR. 1991. *J. Am. Chem. Soc.* 113:4701–3
72. Sabesan S, Duus JO, Domaille P, Kelm S, Paulson JC. 1991. *J. Am. Chem. Soc.* 113:5865–66
73. Townsend RR, Hardy MR, Wong TC, Lee YC. 1986. *Biochemistry* 25:5716–25
74. Woods RJ. 1995. *Curr. Opin. Struct. Biol.* 5:591–98
75. Weis WI, Drickamer K. 1994. *Structure* 2:1227–40
76. Sheriff S, Chang CYY, Ezekowitz RAB. 1994. *Nat. Struct. Biol.* 1:789–94
77. Wilson IA, Skehel JJ, Wiley DC. 1981. *Nature* 289:366–73
78. Gupta D, Arango R, Sharon N, Brewer CF. 1994. *Biochemistry* 33:2503–8
79. Barondes SH, Cooper DNW, Gitt MA, Leffler H. 1994. *J. Biol. Chem.* 269:20807–10
80. Wright CS. 1980. *J. Mol. Biol.* 141:267–91
81. Quiocho FA. 1993. *Biochem. Soc. Trans.* 21:442–48
82. Fletterick RJ, Bates DJ, Steitz TA. 1975. *Proc. Natl. Acad. Sci. USA* 72:38–42
83. Toone EJ. 1994. *Curr. Opin. Struct. Biol.* 4:719–28
84. Rice KG, Weisz OA, Barthel T, Lee RT, Lee YC. 1990. *J. Biol. Chem.* 265:18429–34

Annu. Rev. Biochem. 1996. 65:475–502

CONNEXINS, CONNEXONS, AND INTERCELLULAR COMMUNICATION

Daniel A. Goodenough

Department of Cell Biology, Harvard Medical School, Boston, Massachusetts 02115

Jeffrey A. Goliger and David L. Paul

Department of Neurobiology, Harvard Medical School, Boston, Massachusetts 02115

KEY WORDS: gap junctions, connexin, intercellular communication

ABSTRACT

Cells in tissues share ions, second messengers, and small metabolites through clusters of intercellular channels called gap junctions. This type of intercellular communication permits coordinated cellular activity. Intercellular channels are formed from two oligomeric integral membrane protein assemblies, called connexons, which span two adjacent cells' plasma membranes and join in a narrow, extracellular "gap." Connexons are formed from connexins, a highly related multigene family consisting of at least 13 members. Since the cloning of the first connexin in 1986, considerable progress has been made in our understanding of the complex molecular switches that control the formation and permeability of the intercellular channels. Analysis of the mechanisms of channel assembly has revealed the selectivity of inter-connexin interactions and uncovered novel characteristics of the channel permeability and gating behavior. Structure-function studies provide a molecular understanding of the significance of connexin diversity and demonstrate the unique regulation of connexins by tyrosine kinases and oncogenes.

CONTENTS

Introduction.   476
Structure of the Connexins in the Intercellular Channels.   476
Assays for Function   484
Functional Anatomy.   485
Differences in Connexins Result in Selective Permeabilities   490
Lipids.   491
Phosphorylation and the Control of Connexin Function   492
Conclusion   497

475

0066-4154/96/0701-0475$08.00

Introduction

The connexins are a family of integral membrane proteins that oligomerize into clusters of intercellular channels called gap junctions, which join cells in virtually all metazoans. The channels permit rapid intercellular exchange of ions, and so they are used by neurons to avoid synaptic delays in escape responses when milliseconds dictate survival, and between neurons and other excitable cells (e.g. myocardiocytes) to entrain synchronous activity. Connexin-based intercellular channels are also selectively permeable to many small molecules, thus additionally providing a buffering system for transient intracellular perturbations of metabolites as well as ions. The ability of adjacent cells to exchange metabolites through gap-junctional channels also implies that connexins may act as suppressors of certain classes of somatic mutations involving enzymes in metabolic pathways.

Along with these roles, it has been hypothesized that intercellular communication through gap junctions may influence a wide variety of other cellular activities, including the regulation of growth, differentiation, and developmental signaling. However, experimental evidence in support of these ideas is extremely limited because no specific toxins or pharmacological agents permit selective blockage of the intercellular channel activity in vivo. Thus, many of the functions attributed to gap-junctional intercellular communication (GJIC) have been inferred from correlative data. Direct experimental support for these ideas has been recently provided by gene "knockouts" and other methods that specifically inhibit intercellular communication.

The connexin family of proteins is encoded by at least 13 genes in rodents, with many homologues cloned from other species (Figure 1). Connexins are highly related (50–80% identity), differing mainly in the sequences in the cytoplasmic portions of the molecules. The most commonly used system of nomenclature identifies each connexin by its species of origin and predicted molecular mass; thus, the most prominent connexin in the rat heart is a 43,036 D protein and is called rat connexin43, usually abbreviated rat Cx43 (1, 2). While Cx43 is highly conserved among many species, which makes interspecies comparisons straightforward, other connexins vary considerably in their molecular masses, making the assignment of some homologues ambiguous. For example, two connexins expressed by the differentiated lens fibers in the rat are Cx50 and Cx46. On the basis of sequence similarities and physiological properties as assayed in vitro, the two chicken counterparts are Cx45.6 and Cx56, respectively (3).

Structure of the Connexins in the Intercellular Channels

The structure of the connexin protein within the intercellular channel is inferred from topographic analysis, electron microscopy, and X-ray diffraction data.

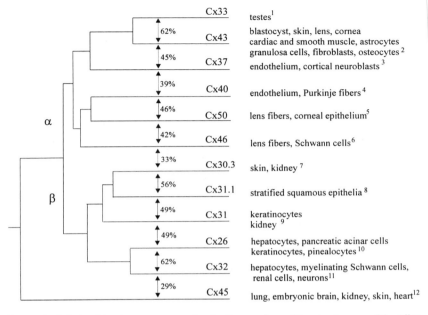

Figure 1 Relationships between connexin family members. The dendrogram (MegAlign, DNASTAR) illustrates primary sequence identity. At least one additional connexin is in the process of characterization, and likely there are more (K Willecke, personal communication). The connexin expression patterns are presented to illustrate their complex overlapping nature but are not comprehensive. These data in some cases represent only RNA expression in heterogeneous tissues; in other cases, immunohistochemical localization of protein in single cells. References: [1]: (200); [2]: (13, 201–204); [3]: (106, 200, 201); [4]: (200, 205, 206); [5]: (107); [6]: (207); [7]: (208, 209); [8]: (200, 201, 209); [9]: (208, 210); [10]: (201, 202, 204, 211); [11]: (199, 204, 212, 213); [12]: (202, 214–216).

Hydrophobicity plots of all members of the family reveal four hydrophobic domains. If each of the hydrophobic regions corresponds to a transmembrane segment, then a generic topology for the connexins can be deduced relative to the plasma membrane (Figure 2). The four transmembrane hydrophobic portions of the molecule are designated M1–M4, the two extracellular loops E1 and E2, and the cytoplasmic loop, CL. The N- and C-termini and the CL face the cytoplasm as demonstrated by biochemical and sequence-specific antibody binding data on Cx32, Cx26, and Cx43 (4–16). E1 and E2 are located at the extracellular surfaces of the junctional membranes. Antibodies recognizing these domains are denied access due to the 35 Å extracellular gap (17). However, separating the junctional membranes with an alkali-urea procedure (18) or a hypertonic disaccharide exposure (19–21) has permitted antibody binding and verified the extracellular location of these epitopes.

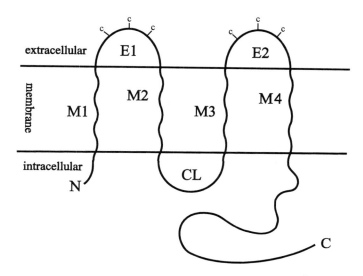

Figure 2 Topology of a generic connexin. All connexins display a similar transmembrane topology with well-conserved N-terminal, four transmembrane (M1–M4), and two extracellular domains (E1 and E2). The C-terminal and central cytoplasmic loop (CL) are very divergent between connexins. Each extracellular loop contains three conserved cysteines (C).

Connexins oligomerize to form half-intercellular channels, called connexons (22), which interact across a narrow extracellular space ("gap") to form the complete intercellular channel. In cells that express more than one connexin, different members of the connexin family likely co-oligomerize into heteromeric connexons, which has been shown to occur with recombinant Cx26 and Cx32 (23) and in vivo for Cx46 and Cx50 in the ovine lens (24). The co-oligomerization of different connexins into heteromeric connexons may be selective, as suggested by the demonstration that a dominant negative connexin construct shows a selective ability to block intercellular communication with Cx43 and Cx32, but not Cx37 (25). Connexons comprised of a single connexin have been shown to assemble into an intercellular channel with each connexon comprised of a different connexin, called a heterotypic intercellular channel (26–28). Intercellular channels cluster together into macular domains called gap junctions.

The tendency for the intercellular channels in some gap-junctional maculae to close-pack into hexagonal arrays is a structural feature that permits the morphological recognition of the junctions in electron micrographs of tissues and isolated plasma membrane fractions (17, 29–34). However, the density of channels in a junctional macula varies widely among cell types. For example, channels joining lens fibers are characteristically dispersed (35, 36) whereas those joining ovarian granulosa cells may close-pack into rectilinear arrays separated by "aisles" (37, 38). Rapid-freezing experiments have demonstrated that in selected cases a dispersed channel arrangement may be rapidly converted into a close-packed morphology following anoxia (39) or chemical fixation (40, 41), although the functional meaning of these changes in channel packing is obscure.

The characteristic channel packing visualized in the electron microscope, together with the relative insolubility of the gap-junctional maculae in chaotropic reagents, has been used to develop procedures for the isolation of gap junctions from liver and heart (31, 34, 42–47). These isolated maculae can be concentrated and partially oriented by centrifugation into specimens suitable for low-angle X-ray diffraction analysis (34), or they can be individually inspected in the electron microscope following stabilization by negative stains or low temperature. Examination of these isolated gap-junctional maculae with these methods reveals that the channels are not packed in crystalline arrays. While the channels may be in domains with long-range order, each channel may be displaced from its crystallographic lattice position by up to 1/6 of the lattice constant, resulting in short-range disorder (48, 49). This disorder limits the coherent X-ray and electron scattering to a resolution of 15–20 Å, denying any direct structural solution of connexin conformation within the intercellular channel (48). In addition, the innate curvature of isolated gap-junctional maculae (indicating different packing of connexons on either side of the gap junction), and the imperfect orientation of the isolated junctions by centrifugation or controlled drying, results in an angular spread of the X-ray diffraction data ± 15°, smearing high-resolution details.

Short-range disorder among close-packed elements suggests a combination of physical forces. Short-range disorder could be a direct consequence of mutual repulsive forces between the intercellular channels, whereas close-packing may result from the need to minimize membrane-membrane repulsive forces arising from close apposition of the two junctional membranes (50, 51). In addition to short-range disorder, channels may also exhibit rotational disorder in maculae with larger lattice constants presumed to exist in vivo (21, 39). In specimens with small lattice constants, the close-packed channels tend to skew ~8° (52–54), indicating that the close-packing has resulted in the interlocking of the intercellular channels and the removal of the rotational freedom. A similar skewing phenomenon has been reported for isolated cardiac gap junctions as well (55). Although channels skew in maculae with small lattice constants, their packing remains plagued by short-range disorder, possibly due to flexible interchannel interactions. Thus, currently available isolated gap-junctional maculae are not suitable for high-resolution structural analysis. For these reasons, efforts are in progress to crystallize purified connexons that are detergent-solubilized from bacculovirus-infected Sf9 cells (56).

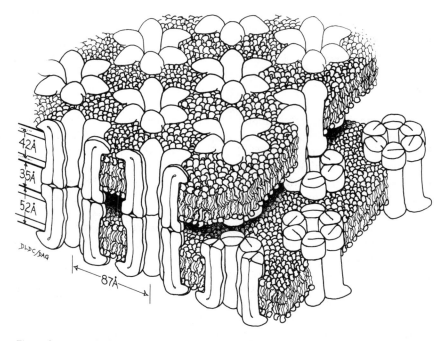

Figure 3 A model of gap-junction structure based on X-ray diffraction and electron microscopy. Each cell contributes a half-intercellular channel, or connexon, and these connexons interact in the intercellular space or gap to form the complete intercellular channel with an aqueous pore connecting the cytoplasms of adjacent cells. Each connexon is a multimer and can simultaneously contain more than one connexin type. Reproduced with permission (64).

Early X-ray and electronmicroscopic studies of isolated hepatocyte gap junctions revealed the structure shown in Figure 3 (57), in which each intercellular channel is composed of a pair of hexameric oligomers of connexins, termed a connexon. In X-ray patterns and Fourier-averaged electron micrographs, only those portions of the connexons that are ordered in the close-packed arrays will be imaged; disordered portions of the molecular assemblies will be invisible. Analysis of meridional diffraction data reveals that the cytoplasmic portions of the connexin molecules extend from the surface of the bilayer out to ~90 Å from the center of the gap (58). These cytoplasmic portions are usually sufficiently disordered so as not to be visible in three-dimensional reconstructions of electron micrographs of negatively stained or frozen-hydrated specimens (59, 60). In rare electron microscope images, portions of the cytoplasmic portions of the connexins appear to have become ordered on the cytoplasmic surfaces at the three-fold axes between the hexameric channel assemblies (61), as has been deduced from X-ray data (62). The cytoplasmic portions of the connexin molecules are also highly sensitive to beam damage and dehydration conditions associated with electron microscopy, which may also contribute to their invisibility in most reconstructed images (63). In most X-ray diffraction patterns and electron micrographs, therefore, the computed structures arise from transbilayer portions and the extracellular (gap) domains of the connexins. Direct support for this conclusion comes from the observation that proteolytic removal of the C-terminal cytoplasmic tails of the connexin molecules in liver and heart gap junctions produces no measurable differences in the Fourier-averaged channel structure (14, 58).

The diagram in Figure 3 shows the connexons each with six connexin molecules. This hexameric stoichiometry was originally suggested from an x-ray density–based estimate of the molecular weight of a connexon (between 140 and 170 kD), which was consistent with a six-subunit stoichiometry given the then-current estimate of 26–27 kD for the molecular weight of the connexin protein, based on measurements from SDS-PAGE (48, 58, 64). The assignment of hexagonal symmetry to the connexon was also supported by three-dimensional images of the gap-junction channels reconstructed from electron micrographs of negatively stained (59) and frozen-hydrated (60) gap junctions isolated from rat liver. The lattice was shown to have p6 symmetry, which is consistent with sixfold rotational symmetry of a connexon.

However, rotational and/or translational short-range disorder can influence Fourier averages of channel structure visualized by electron microscopy. For example, Fourier averaging of hexagonally packed channels with rotational disorder will result in smooth contoured toroids with no apparent substructure. Given the 20 Å resolution of most electron microscopic images of gap junctions, the p6 averaging and truncation of the data set in reciprocal space will result in a circular or slightly hexagonal image for the channel, regardless of its underlying symmetry. In addition, the packing of pentagons on a plane has several solutions, some of which result in a low-resolution hexagonal lattice (65). These arguments may be used to explain why cholera toxin B subunit pentamers (whose dimensions are similar to those of a connexon) appear to be hexameric at a resolution of 17 Å when packed into a hexagonal array (66), and they leave open the possibility that connexons may be pentamers.

Analyses of electron micrographs of frozen-hydrated and negatively stained rat liver gap junctions with lattice constants less than 86 Å have yielded asymmetric (2,1) and (1,2) lattice lines (52, 54, 67), providing strong data in support of a hexameric connexon structure. In this case, the close-packing of the connexons in the plane of the membrane presumably permits interlocking of the connexin molecules in adjacent connexons, minimizing rotational disorder. Stauffer et al (56) have calculated rotational power spectra of electron micrographs of selected individual detergent-solubilized rat liver connexons that are not part of a hexagonal lattice. These spectra are characterized by the dominance of a sixfold harmonic, which is strongly indicative of hexameric structure. Ghoshroy et al (68) have also investigated the power spectra of correlation-averaged connexons in hemi-gap-junctional maculae split with urea/EGTA, and they found positive reinforcement for a hexameric substructure. Finally, chemical cross-linking studies of gel filtration–purified rat liver connexons isolated using a high pH–detergent-dithiothreitol procedure clearly resolve a ladder of six subunits on SDS gels (69). Thus, while the individual connexin molecules cannot be directly inspected, owing to the inherent disorder in gap-junctional maculae, a hexameric stoichiometry is currently best supported by the available data.

By analogy with other membrane proteins, the M1–M4 domains of the connexins have been postulated to be α-helical (9, 10, 70). Evidence for α-helical secondary structure in the assembled intercellular channel comes from both circular dichroism and X-ray modeling data. In the X-ray patterns, the dominant features related to secondary structure are arcs on the equator centered at 11 Å, together with meridional arcs centered at 4.9 Å, which are sampled by an interference function due to two scattering centers separated by ~80 Å in each half of the gap-junction profile. Richardson and Richardson (71) have correlated the ratio of the peak intensities between 10 Å and 4.5 Å in protein diffraction data with the amount of α-helical content, and comparison of the gap-junction diffraction intensities with that from different proteins of known structure indicates that the ordered portions of the connexins are about 60% α-helical (72).

Cascio et al (73) have measured the α-helical content of detergent- and alkali-isolated liver gap junctions using circular dichroism. Detergent-isolated specimens are ~50% α-helical, whereas the less highly enriched alkali-treated specimens were ~40% α-helical. Removal of the cytoplasmic portions of the connexins with protease resulted in a slight increase in the α-helical content, suggesting that the majority of helical secondary structure resides in the junctional membranes or within the gap, which are both stearically inaccessible to proteolysis. Measurement of CD spectra of detergent-solubilized isolated connexons results in a higher (66% vs 52%) estimate for the α-helical content of the Cx32; the reasons for the difference between the membrane-bound and detergent-isolated secondary structure are not clear (69).

Tibbitts et al (72) have modeled gap-junction diffraction patterns using proteins with solved structures. The results of this modeling suggest that the intramembrane portions of the connexins are α-helical, that the helices are long enough to span the bilayer with varying degrees of tilt, and that the helices may extend beyond the hydrophobic portions of the phospholipids in the direction of the extracellular gap. They calculated spherically and cylindrically averaged diffraction patterns of assemblies of solved proteins that had been imperfectly oriented to mimic that of the isolated gap-junction specimens. Analysis of the fringe intensity in the gap-junction diffraction patterns in the meridional data suggests that the maximum length of the α-helices in the connexins is 40 Å, with a mean length of 35 Å. A similar analysis of the equatorial data indicates that the helices are tilted ~20° to the membrane normal, with a standard deviation of ~10°. Since the hydrocarbon cores of the bilayers in the gap junction are ~32 Å thick (64), the lengths of these helical segments are long enough to completely span the bilayer with varying degrees of tilt. Analysis of the width of the meridional fringes indicates that the distance from the center of the gap to the centers of the α-helical segments in the bilayers is 38 ± 1 Å, placing the center of the helices 4 Å closer to the gap than the cytoplasmic surfaces, raising the possibility that some of the α-helices may extend into the domain of the phospholipid headgroups or even into the gap itself.

Portions of the connexons in the extracellular gap have been imaged by deep-etching of hepatocyte gap junctions split with hypertonic disaccharides (20, 21). The junctions split centrosymmetrically, leaving the tips of the connexons protruding 10–20 Å from the membrane surfaces. For a brief time (1–2 minutes), an antigenic epitope on the E2 loop is accessible for antibody binding following this hypertonic splitting; however, after that time the site disappears for unknown reasons (8). Hoh et al (74) have studied the extracellular surfaces of gap junctions "dissected" free from the apposed junctional membrane by atomic force microscopy. The extracellular domains of the connexons protruded 14 Å above the extracellular surface, and the resistance of the structure to the scraping action of the tip of the cantilever suggests a highly rigid structure adopted by the connexin molecules assembled into a connexon.

Structural studies have revealed variability in the connexon structure, which has been postulated to arise from dynamic movements associated with channel gating. Makowski et al (64) observed that smaller gap widths closely correlated with decreased lattice constants, which suggested transitions between two structural states, but these differences were not ascribable to any understood manipulation of the gap junctions during isolation. Unwin and his associates (59, 60) measured changes in the twisting and tilting of the connexins in isolated junctions as a function of Ca^{++} concentration, and they postulated that these movements underlie a calcium-induced gate. Makowski et al (58) measured X-ray diffraction patterns of isolated gap junctions in varying concentrations of sucrose, placing the electron density calculations on an absolute scale. These studies reveal that while sucrose is able to diffuse up to the cytoplasmic surfaces of the gap junctions and into the 3 nm gap, the sucrose is only able to penetrate into the intercellular channel to the level of the phosphate headgroups of the cytoplasmic leaflets of the junctional membranes. Since sucrose is small enough to entirely traverse the intercellular channel, these data suggest that a gate near the mouth of the channel (61, 62) denies the sugar access to the full length of the channel at both cytoplasmic surfaces and that isolated gap junctions are in the closed state.

Caspar et al (61) correlated the gap-junction stacking with the packing of the connexons in the plane of the junction as a function of the state of dehydration. They observed dramatic reversible variability in connexon packing and lipid content as a function of water activity. Lowering the relative humidity to about 90% removes most of the bulk water from the specimens; further dehydration results in a decrease in the lattice constant of the intercellular channels in the gap-junction maculae without further changes in the stacking period. As the intercellular channels move closer together, there are reversible changes in the relative intensities of the equatorial and off-equatorial X-ray reflections, indicating that reversible changes in the connexon structure accompany lattice compaction. This plastic nature of the connexon/connexon interactions may provide some explanation for the variability of lattice constants seen between different gap-junction specimens. The invariance of the meridional diffraction fringes observed in the 4–5 Å region of the X-ray diffraction patterns with variation in the lipid, water content, or calcium concentration (61), or when the C-terminal tails of the connexins are proteolytically removed (62), makes it unlikely that connexon gating is associated with tilting of the transmembrane helices and that the transmembrane and extracellular portions of the connexin molecules form a highly rigid structure (72).

Assays For Function

Functional assays for GJIC depend on the cellular system being investigated. The assays fall into two categories: those designed to measure the transfer of molecules between cells and those that measure electrical currents carried by intracellular ions. In the first category, "metabolic cooperation" is measured by mixing donor cells metabolically labeled with molecules small enough to pass through junctional channels, such as ^{3}H-uridine, with nonlabeled acceptor cells in co-culture. Following autoradiography, these cultures are examined for evidence of contact-dependent exchange of the metabolites from the donors to the acceptors (75–77). Negative controls are accomplished by permitting the donor cells to incubate in cold medium following labeling, thus chasing the labeled metabolic pool into channel impermeant macromolecules.

Alternative methods to assay GJIC include direct iontophoresis of a membrane-impermeant fluorescent dye {the most popular is Lucifer Yellow [MW = 443 D; (78)]} into a cell using an intracellular microelectrode, followed by monitoring of the time-dependent diffusion of the dye to adjacent cells through the junctional channels. Typical controls include co-injection of a larger fluorescent molecule, such as rhodamine-dextran, which can permeate intercellular bridges but not gap-junctional channels, or the acidification of the cellular system with acetate or CO_2, which is known to close most junctional channels (79). Finally, capitalizing on the "scrape loading" technique of McNeil et al (80), cell cultures may be locally scratched in the presence of Lucifer yellow (LY) and rhodamine dextran, and the transfer of LY from rhodamine/LY-labeled cells to nonrhodamine-containing cells measured (81).

The measurement of currents passing through gap-junctional channels emerged directly from electrophysiological studies of action-potential conduction across synapses (82, 83), which was extended to nonexcitable tissues by measuring the cell-cell passage of evoked potentials with dual-cell recordings (84–86). This methodology was greatly improved by the development of a four-electrode dual–voltage clamp paradigm (87), which permitted a direct measure of the conductance through the junctional channels. With the advent of patch-clamp methodologies (88, 89), the use of the dual voltage clamp was extended to the study of single-channel conductances and their physiological properties (90, 91).

Channels have been reconstituted into artificial lipid bilayer systems using purified connexins (92–96) and their properties compared with those measured by patch-clamp analysis. Single-channel properties have also been measured by transfecting cell lines, which have very low endogenous gap-junction activity, with specific connexins and studying the properties of the superimposed exogenous channels using patch-clamp techniques (97, 98). In one case, the contribution of two different endogenous connexins to the electrophysiological record was dissected using an antisense approach (99).

Dahl, Werner, and their colleagues (100, 101) were successful in developing an in vitro expression system which has been used in many laboratories interested in structure-function correlation in connexins. In this assay, *Xenopus* oocytes are injected with connexin mRNA, devitellinized, and then manipulated into contact such that intercellular channels can form. The conductances developed between the oocytes can be quantitatively measured by the dual–voltage clamp method. The oocytes have at least one endogenous connexin, *Xenopus* Cx38 (*Xen*Cx38) (102, 103), which is capable of interacting with some rodent connexins (26, 104). The activity of *Xen*Cx38 can be ablated with prior injection of antisense mRNA, thus removing its contribution from the electrophysiological data (105–107).

Functional Anatomy

The E1 and E2 domains of the connexins are likely to participate in intercellular interactions across the 35 Å extracellular gap. Each domain contains three conserved cysteines. The three E1 Cys residues are separated by six and three amino acids, and the E2 Cys residues by four and five residues, with the exception of Cx31, in which the first and second cys in the E2 domain are separated by five amino acids. The cysteines between the E1 and E2 loops are joined by at least one disulfide bond, deduced from proteolysis studies of isolated hepatocyte junctions (11, 108–110). Dahl et al (111, 112) have shown that mutation of any of these six cysteine residues in Cx32 completely blocks the development of gap-junctional conductances between *Xenopus* oocyte pairs. In addition, they have shown that a 38-mer corresponding to the E1 domain and a 36-mer corresponding to the E2 domain both block the assembly of new channels between oocytes, but will not interfere with conductances already assembled (113).

In a novel approach, Rosinski and Nicholson (personal communication) used site-directed mutagenesis to map the three-dimensional interactions between the extracellular loops. These studies involve moving the first and third cysteines of each E-loop of Cx32 to various positions away from their wild-type positions. Following the Cys movements, each mutant is paired with wild-type Cx32, and the development of conductances is measured in *Xenopus* oocyte pairs. The movement of a single Cys one or two positions from its wild-type position results in a nonfunctional construct, consistent with the studies of Dahl et al (112). However, if the Cys residues are moved in pairs, such that their relative positions permit continued disulfide bonding, then functional channels result, with little observable change in the channel biophysical properties compared with the parent Cx32. The movement of the pairs of Cys is maximally effective when they are moved exactly two (i.e. not one or three) residues towards the amino- or carboxy-termini, a finding consistent with a β-sheet conformation. In addition, two-residue displacements of the first and third cysteines from each E-loop result in the highest conductances, indicating that the two β-loops are antiparallel. An unexpected finding is that two of the nonfunctional mutants, which would not assemble conductances with wild-type Cx32, would form intercellular channels with the endogenous *Xenopus* Cx38, which is normally noncompatible with Cx32. Thus a new structure is generated which has switched specificity for heterotypic interaction.

All these data are consistent with a model in which the E domains form stacked β-sheets, each with a central reverse turn, which are joined and held in a rigid conformation by three interloop disulfide bonds. This rigid structure of the extracellular domain is in agreement with the X-ray data reviewed above. The antiparallel β-structure confines the transmembrane α-helices to be arranged in sequential order from N- to C-termini, as proposed in modeling studies by Milks et al (9). Rosinski and Nicholson have developed a model in which the connexin extracellular loops form a β-barrel which interdigitates with its counterpart in the apposing connexon to fit in the 35 Å gap. Sequence differences between the connexins may then help to confer specificity to the intercellular interactions. Additionally, the need for the precise alignment of the cysteine residues explains why Cx31, which differs from all other known connexins in terms of the spacing of its E-loop cysteines, is only able to form functional channels with itself.

Antisera have been generated to synthetic peptides corresponding to the E1 and E2 domains of several connexins. In general, these antipeptide antisera will not bind their antigens in intact junctions, as predicted from tracer studies showing that the 35 Å gap is stearically inaccessible to protein molecules (114). The anti-E1 and anti-E2 reagents have been used to block gap-junction assembly and prevent intercellular communication. Meyer et al (115) demonstrated that Fab fragments from both anti-E1 and anti-E2 were capable of blocking the assembly of gap junctions between Novikoff Hepatoma cells, again supporting a direct role of the E1 and E2 domains in intercellular connexin interactions.

Blocking the assembly of gap junctions with antisera (115) also interfered with the assembly of adherens junctions. This finding places gap-junctional communication in a potential feedback loop with certain cell-adhesion systems. Cadherin adhesion between cells has been shown to be a prerequisite for the development of gap junctions in several cell systems (116–119). The data of Meyer et al (115) reveal that adherens junctions either cannot form or are unstable if gap-junctional communication is blocked. This blockade, however, did not interfere with all adhesion systems available to the Novikoff cells, since adhering cell pairs were available in this study for electrophysiological and morphological manipulations. A coordinate loss of adhesion with specific blockade of intercellular communication can also be inferred from experiments of Paul et al (25) who expressed a mutant connexin that dominantly inhibited intercellular communication in *Xenopus* embryos. Progeny of blastomeres injected with the mutant connexin mRNA either delaminated and were extruded from the embryos, or they were found in inappropriate locations within the developing embryos. The normal phenotype could be rescued by concomitant injection of Cx37, which was shown to be insensitive to the action of the connexin mutant.

The functional contributions to intercellular communication of the E1 and E2 domains have been studied by both mutational analysis and chimeric constructs. White et al (27) studied the ability of the three known lens connexins to form heterotypic interactions. While heterotypic channels Cx43/Cx46 and Cx46/Cx50 readily formed, Cx50 was unable to form functional channels with Cx43. Replacing the E2 domain of Cx46 with the E2 domain of Cx50 resulted in a chimeric connexin that was unable to form functional channels with Cx43. Conversely, Cx50 acquired the ability to form functional channels with Cx43 when supplied with the E2 domain of Cx46. These data indicate that the E2 domain is a determinant of heterotypic selectivity between these connexins.

Most connexins in intercellular channels are voltage sensitive (120). The connexins do not have an S4 domain (121), and experimental efforts have focused on defining those regions of the connexin molecules that participate in the voltage inactivation. Rubin et al (122) described a novel voltage dependence by studying heterotypic channels formed from Cx26 and Cx32 in the *Xenopus* oocyte-pair system. This dependence, not seen in either of the homotypic pairs, is an extremely fast sensitivity to membrane potential. Thus, the voltage sensors of the intercellular channel are not only determined by the individual connexins, but derive some of their properties from the interaction between the apposed connexins. These authors also showed that replacement of the two charged amino acids K-41 and E-42 in the E1 domain of Cx26 with residues E-41 and S-42 of Cx32's E1 domain affected the free energy difference between the open and closed channel states in the absence of a voltage field, suggesting that the E1 domain is part of the voltage-sensing mechanism underlying both the slow and fast responses.

Rubin et al (123) have also constructed chimeric molecules of Cx32 and Cx26 involving exchanges of E1, E2, CL, and portions of M3. These constructs displayed voltage inactivation properties different from either of the parent connexins. The changes were described as stabilization/destabilization of open and closed states of the hybrid channels. The data indicate that the voltage sensor is derived from an interaction between various portions of the connexin molecule and that the sensor may not be in the aqueous channel. Verselis et al (124) explained the novel rectification reported by Rubin et al (122) by postulating that the polarity of voltage sensitivity for Cx32 was opposite to that of Cx26, i.e. the Cx32 channels would close when their cytoplasmic sides were negative relative to their gap domains, while Cx26 channels would close when their cytoplasmic sides were relatively positive. Point mutations in either Cx32 or Cx26, which altered the charge of the N-terminal second amino acid, could reverse the gating polarity of these connexins. On the basis of these data, Verselis et al (124) concluded that gating polarity was an intrinsic property of the connexon. In addition, mutations at the M1/E2 boundary, similar to those of Rubin et al (122), were also sufficient to reverse the gating polarity of Cx32. Together, these data support the idea that an interaction between different regions of the connexin molecules forms a "charge complex" that is an integral part of the voltage sensor.

As with other physiological properties, the gating polarity of a connexin may change depending on its heterotypic interactions. White et al (125) showed that Cx46 hemichannels, which can spontaneously form in the *Xenopus* oolemma in low Ca^{++}, reverse their gating polarity upon assembly into intercellular channels. Thus, in contrast to the conclusions of Verselis et al (124), these data show that the polarity of the voltage sensor is not intrinsic to Cx46, but can be modulated by its interaction with opposing connexins.

Bruzzone et al (126) have studied the assembly and voltage properties of chimeric connexins formed from Cx32 and Cx43. These connexins are not able to form heterologous intercellular channels with each other (127), demonstrating the selectivity phenomenon found between other connexin pairs (27, 128, 129). When the chimeric connexins were expressed in paired *Xenopus* oocytes, all but one failed to show functional activity, underscoring the conclusion from the chimeric studies discussed above (122, 123) that the connexins are not composed of cassettes of functional domains, but derive their properties from the complex interactions among various portions of the molecules. The one chimera that was active consisted of Cx32 from the N-terminus through M2, fused to Cx43 from CL to the C-terminus. This construct was unable to form intercellular channels with itself or with Cx32, but was able to assemble heterotypic channels with Cx43 and with the endogenous *Xenopus* Cx38. Voltage inactivation of these heterotypic channels was asymmetric, not predicted from the properties of either parent connexin, and depended on the type of connexin in the apposed connexon, again demonstrating that voltage gating is not an intrinsic behavior for a given connexin, but can be modulated by the partner connexin.

In addition to the N-terminus and E1 domain, the M2 domain has been implicated in the regulation of voltage gating. Suchyna et al (130) have mutated an invariant Pro in M2 to Leu in Cx26 (P87L) and expressed this construct in *Xenopus* oocyte pairs. This construct was unable to form conductances between oocytes when paired with itself, but was able to make intercellular channels when paired with its wild-type parent, Cx26. The mutant connexons in these heterotypic channels acquired the property of being voltage activated, a reversal in the voltage-gating of their parent connexins. The authors suggest that this phenomenon reflects an increase in the single-channel open-probability time with increasing transjunctional voltage (cytoplasmic side positive), suggesting that M2 is active in the transduction event in voltage gating.

Deletion analysis suggests that the cytoplasmic portions of the connexin molecules have important regulatory roles. Liu et al (131) measured the pK_a for channel closure of both Cx43 and Cx32 in the oocyte-pair system. Initial acidification caused a significant increase in conductance with both connexins, although Cx32 increased conductance over a broader pH range. Further decreases in pH resulted in 100% closure of all channels. Deletion of the last 125 residues from the C-terminus of Cx43 caused a reduction in the pK_a to levels similar to Cx32 (6.1) and steepened the Hill coefficient to values similar to Cx32 (6.0), suggesting that the longer C-terminus of Cx43 facilitates acid-induced gating.

Morley et al (submitted) have proposed that pH gating may be modeled as a "particle-receptor" interaction between two separate domains of Cx43. These authors expressed a Cx43 construct truncated at met257 in *Xenopus* oocyte pairs and demonstrated a loss of pH sensitivity. This sensitivity could be rescued by coinjection of a soluble C-terminal peptide (residues 259–382), consistent with their model in which pH gating in Cx43 involves an interaction between the C-terminus and a second region, likely the CL domain including his95 (132). Interestingly, the soluble C-terminal peptide from Cx43 can also interact with channels formed from Cx32, suggesting that the receptor domain has been conserved among connexins. An involvement of his95 (133) in pH gating has also been shown by experiments in which this residue was mutated. Cx43 with acidic or uncharged residues at position 95 was less susceptible to acidification, whereas the opposite was true when basic residues were substituted (132).

Differences in Connexins Result in Selective Permeabilities

Individual connexins have been shown to have unique selective permeabilities with regard to molecular mass and/or charge, although the molecular basis is not clear. Steinberg et al (134) compared the Lucifer yellow (LY, MW = 443, charge = −2) permeability of Cx43 and Cx45 in two osteoblastic cell lines. They found that LY passed readily through Cx43 channels and poorly through Cx45 channels. Transfection experiments demonstrated that these differences in LY permeability were not likely due to differences in the numbers of channels joining the cells, or to the cellular background, but to intrinsic properties of each connexin. Veenstra et al (135) investigated the nature of Cx45 permeability further and found that it exhibited a marked cationic selectivity. These channels show selective permeability to 2',7'-dichlorofluorescein but not to 6-carboxyfluorescein, two tracer molecules of similar limiting dimensions but opposite charge.

In a comprehensive study of many different mouse connexins transfected into HeLa cells, Elfgang et al (127) also found clear evidence for selectivity in connexin channel permeabilities. In contrast with the study of Steinberg et al (134) reviewed above, Elfgang et al (127) found that Cx45 intercellular channels were permeable to LY, as were channels composed of Cxs 26, 31, 32, 37, 40, and 43. Although propidium iodide (MW 414, charge +2) and ethidium bromide (MW 314, charge +1) also passed through many connexin channels, these tracers failed to transfer through channels composed of Cx31 or Cx32. 4,6 diamidino-2-phenylindole (MW = 279, charge +1) showed poor transfer between intercellular channels composed of Cx31 and Cx43, and neurobiotin (MW 287, charge +1) was only weakly transferred between cells joined by Cx31 connexons. Although striking, the biological significance of differential permeability is not yet clear. Conceivably, differences in the permeability of biologically active molecules could result in limited spread of stimuli between subsets of cells in tissues and organs.

Lipids

The lipid composition of isolated gap junctions has been studied in several laboratories. These analyses report an enrichment of cholesterol in isolated gap-junctional maculae (34, 44; for a review, see 136). The relative insolubility of gap-junctional maculae in detergents may arise from both noncovalent protein-protein interactions between connexons and from specific interactions between the connexins and their "boundary" lipids. In this regard, X-ray diffraction patterns of isolated gap junctions have revealed an unusually high electron density in the bilayer profile at the level of the lipid phosphate headgroups in the cytoplasmic leaflet of the junctional membranes (58). These findings suggest interactions between the lipid polar headgroups and the connexins that result in a partial loss of the water of hydration. These interactions may not be characteristic of all members of the connexin family, however. Gap-junctional maculae composed of different connexins have very different solubilities in detergents, and protocols which have been classically used for liver gap–junction isolation have needed to be redesigned to isolate gap junctions from heart and lens (18, 46, 137–139).

Lipophilic molecules have been used to block gap-junctional channels in many cellular systems (91, 140, 141), although the mechanism of action of these compounds is unknown. Burt and her colleagues have shown that oleic acid shows differential effects on gap-junctional channels with differing single-channel conductances (γ_j), suggesting that these compounds interact directly with connexin proteins (142). Other studies using volatile anesthetics leave open the possibility that the junctional channels are closing in response to changes in the lipid environment (143). Takens-Kwak (144) found that heptanol reduced all nonjunctional membrane ionic currents, in addition to gap-junction conductances, suggesting a pan-membrane effect on channel function. These authors proposed that heptanol affects the membrane lipid structure, rather than interacting directly with gap-junctional channels.

Venance et al (145) have shown that anandamide, a naturally occurring arachidonic acid derivative, is a potent inhibitor of GJIC between striatal astrocytes. The mechanism of anandamide action on GJIC is not known. Anandamide activates cannabinoid receptors, but the effect on GJIC is neither mimicked by cannabinoid-receptor agonists nor prevented by antagonists. The functional significance of anandamide sensitivity may relate to the propagation of calcium waves by astrocytes. Glutamate released from neurons evokes calcium waves in astrocytes that propagate via gap junctions and may, in turn, activate neurons distant from their initiation sites in astrocytes (146–148). Thus, glia may be directly involved in CNS information processing in addition to providing trophic and support functions for neurons. GJIC seems likely to be an essential aspect of this behavior.

More recent studies (149) have combined measurements of fluorescence steady-state anisotropy (a measure of membrane lipid fluidity) with gap-junctional conductance, as measured with the dual whole-cell patch-clamp technique. Both heptanol and 2-(methoxy-ethoxy)ethyl 8-(cis-2-n-octylcyclopropyl)-octanoate (A2C) increased bulk membrane fluidity to the same extent, but only heptanol was able to block the junctional channels. Closer inspection of various lipid domains, however, revealed that while heptanol decreased the fluidity of cholesterol-rich domains, A2C had no measurable effects on cholesterol fluidity. This finding supports the biochemical findings, reported above, that gap-junctional maculae are enriched in cholesterol and the conclusion of Takens-Kwak et al (144) that heptanol indirectly affects channel activity by altering the fluidity of cholesterol-rich domains.

Phosphorylation and the Control of Connexin Function

Most of the connexins, except for Cx26, are phosphoproteins (150, 151). Phosphorylated connexins frequently display anomalous or multiple SDS-PAGE mobilities that are correctable upon phosphatase treatment (e.g. 3, 119), suggesting that phosphorylation is a common form of posttranslational modification. Amino acid analyses have demonstrated that under normal growth conditions, connexin phosphorylation occurs predominantly on serine residues; however, threonine phosphorylation also occurs (152, 153). The residues phosphorylated during normal growth have not been determined for any of the connexins, although preliminary biochemical analyses suggest that for Cx43, many of them are likely to reside in the COOH-tail between amino acids 271 and 382 (154), where several consensus kinase recognition sites have been identified (155). Treating cells with reagents such as cyclic AMP, phorbol esters, and growth factors can change the pattern or extent of connexin ser/thr phosphorylation, which often correlates with altered GJIC properties (*vide infra*). Saez et al (151) demonstrated that Ser233 in Cx32 was the major site phosphorylated in vitro and in vivo in response to cAMP, and it was also phosphorylated in conjunction with other serine residues in response to the phorbol ester 12-O-tetradecanoylphorbol 13-acetate (TPA). However, site-directed mutagenesis of ser233 together with ser240 caused no observable differences in permeability or pH-dependent gating properties of this connexin when expressed in paired *Xenopus* oocytes (156), suggesting that cAMP-induced phosphorylation of ser233 might affect other aspects of GJIC.

Several studies have addressed how phosphorylation affects connexin function. Experiments by Musil et al (119) have shown that the noncommunicating mouse cell lines L929 and S180 make Cx43 protein, but fail to phosphorylate it significantly and do not localize nonphosphorylated Cx43 to junctional maculae. However, S180 cells stably transfected with the chick cadherin gene

L-CAM display normal basal patterns of Cx43 phosphorylation and efficiently communicate via gap junctions (117). These results suggest that basal phosphorylation may regulate trafficking of Cx43. Other experiments by Musil and Goodenough (157) demonstrate that in NRK cells a phosphorylated form of Cx43 (Cx43-P2) is not solubilized by the detergent Triton-X-100 and is almost exclusively associated with junctional maculae. In contrast, most of the nonphosphorylated Cx43 (Cx43-NP) is detergent-soluble and is predominantly intracellular. Biotinylation experiments further demonstrated that conversion of Cx43-NP to Cx43-P2 occurs during or after assembly into maculae. Taken together, the data strongly suggest that under normal growth conditions, phosphorylation of Cx43 occurs subsequent to connexon assembly and surface localization and thus may regulate later steps of connexon incorporation into junctional maculae.

Ser/thr phosphorylation may also affect channel activity. Experiments by Moreno et al (152) measured single-channel conductance (γ_j) in SKHep-1 cells that were stably transfected with Cx43. The cells were cultured in the presence of okadaic acid (a phosphatase inhibitor) or staurosporine (an inhibitor of PKA and PKC), or injected with bacterial alkaline phosphatase. Untreated cells showed predominantly two populations of γ_j for Cx43 centered at 65 pS and 100 pS. Kinase inhibition or phosphatase treatment increased the relative frequency of large γ_j events, whereas phosphatase inhibition increased the relative frequency of small γ_j events. Western and phosphoamino-acid analyses correlated culture treatments with altered phosphorylation patterns of Cx43, suggesting that changes in phosphorylation affected the distribution frequencies of unitary conductance states.

As mentioned above, many reagents, growth factors, and viral oncogenes influence GJIC (reviewed in 158), and their effects are frequently associated with changes in the basal patterns of connexin phosphorylation. Perhaps the best understood example is the inhibition of GJIC by the $pp60^{v\text{-}src}$ tyrosine kinase. In cells or paired *Xenopus* oocytes co-expressing Cx43 and $pp60^{v\text{-}src}$, Cx43 is rapidly phosphorylated on tyrosine (159–164), and GJIC is inhibited. $pp60^{v\text{-}src}$ likely acts principally at the level of channel gating because (*a*) there is a tight temporal correspondence between $pp60^{v\text{-}src}$ activation and inhibition of GJIC, (*b*) gap junctions are not morphologically disrupted, and (*c*) the effect does not require protein synthesis (159). The effect of $pp60^{v\text{-}src}$ on channel gating is likely due to direct phosphorylation of a connexin tyrosine residue. For example, mutation of tyr265 in Cx43 to a phenylalanine residue completely abolished the ability of $pp60^{v\text{-}src}$ to phosphorylate Cx43 and inhibit GJIC (164), without changing any other measurable channel property. In addition, recent evidence shows that purified Cx43 can be directly phosphorylated by $pp60^{v\text{-}src}$ in vitro (165). Two other oncogenic viral tyrosine kinases, *v-fps* (166) and *v-yes* (167), may similarly inhibit GJIC by directly phosphorylating Cx43.

However, kinase activity alone is not sufficient to inhibit GJIC, since cells expressing a mutant pp60$^{v\text{-}src}$ (which retains kinase activity, but lacks myristic acid and fails to localize to the plasma membrane) communicate normally (168). In addition, tyrosine phosphorylation does not affect all connexins: Swenson et al (164) have shown that the *src* kinase does not affect communication induced by Cx32.

Activation of receptor tyrosine kinases can also inhibit GJIC, although these effects are not mediated by direct tyrosine phosphorylation of connexins (169–171). EGF stimulation induces specific phosphorylation of Cx43 only on serine residues, and this effect appears to be mediated by a mitogen-activated protein kinase signal cascade (155, 171). Similarly, the inhibition of communication by PDGF is also independent of tyrosine phosphorylation of Cx43 (172). PDGF stimulation activates several proteins involved in signal transduction, and inhibition of communication by PDGF may occur by multiple pathways. Hepatocyte growth factor/scatter factor, another ligand with a tyrosine kinase receptor, also inhibits dye coupling in a mouse keratinocyte cell line without tyrosine phosphorylation of Cx43 (173).

Several studies have also examined how TPA, a tumor promoting phorbol ester, affects GJIC. The effects of TPA can vary depending on the cell type (153). The most frequently reported effect is a rapid inhibition of 85–90% of GJIC followed by complete inhibition and loss of junctional maculae after longer exposures (174–178). For cells expressing Cx43, the TPA effects have been correlated with hyperphosphorylation. A similar TPA-dependent hyperphosphorylation has been reported for Cx32, but not for Cx26 (175, 179). Hyperphosphorylation is presumably mediated by protein kinase C (PKC) since this kinase is strongly activated by TPA (180) and has been previously reported to directly phosphorylate Cx43 (181). Although the amino acid residues that are phosphorylated in response to TPA treatment are not yet known, some of them are likely to reside within PKC consensus recognition sites located in the COOH-terminal tail of Cx43 (155). Matesic et al (177) examined the kinetics of TPA-mediated changes in phosphorylation, immunodetectable connexin levels, and intercellular communication in the rat liver epithelial cell line WB-F344. They showed that hyperphosphorylation of Cx43 was detectable as early as inhibition of GJIC-mediated dye transfer (5 min after TPA treatment), whereas loss of detectable immunostaining occurred somewhat more slowly (not detectable until 15 min). Similar results with primary mouse keratinocytes were reported earlier by Brisette et al (174). The results indicate that TPA acts at multiple levels, initially affecting channel permeability and subsequently affecting connexin trafficking and/or synthesis.

The initial effect of TPA on gating and/or permeability is complex. The inhibition of dye transfer by TPA between cultured cells has been shown to parallel a shift in the distribution of single-channel events toward the smallest

γ_j (20pS) of Cx43 (175, 182, 183). Surprisingly, TPA treatment increased total junctional conductance (g_j) between rat cardiomyocytes, suggesting that TPA may also affect open channel time and/or macula size (182). In other studies, TPA has the opposite effect on total junctional conductance (175, 183). Nevertheless, a consistent effect of increased phosphorylation of Cx43 is to increase the relative frequency of small γ_j events and to reduce dye transfer.

The presence of multiple connexins further complicates the analysis of TPA effects on intercellular communication. Both mouse primary keratinocytes (174) and WB-F344 cells (177) express Cx26 in addition to Cx43. As mentioned above, there is no evidence that Cx26 is phosphorylated. How then can we account for the rapid inhibition of GJIC by TPA? One possible explanation is that Cx26 is expressed at much lower levels than Cx43, which would be consistent with the lack of complete inhibition of GJIC (85–90%). Alternatively, Cx26 channels are abundant, but their activity may not have been detected because the dyes used to measure GJIC do not permeate these channels efficiently (127). The complete inhibition of GJIC at longer times of TPA exposure (15–30 min) correlates with the loss of all immunodetectable connexin expression. This finding suggests that the phosphorylation of Cx26 is not required for these long-term effects.

In some cell types, the principal effects of TPA appear to be on junction assembly. Lampe (184) showed that TPA inhibited dye transfer between Novikoff hepatoma cells that had been dissociated with EDTA and then reaggregated. However, TPA did not affect GJIC in undissociated cells. Moreover, TPA-induced inhibition of GJIC correlated with increased phosphorylation of Cx43 and did not appear to affect its levels. The most plausible explanation for these results is that TPA inhibited assembly of plasma membrane–associated Cx43 into functional gap junctions. Since TPA treatment rapidly stimulates translocation of PKC from cytosol to membrane (185), and since Cx43 reaches the plasma membrane after incorporation into connexons (186), PKC regulation in this system may affect the final steps of junction formation: assembly of intercellular channels from connexons and their aggregation into maculae. Why TPA inhibits GJIC in some undissociated cells, but has no effect on others, remains an intriguing and as-yet-unanswered question.

Another frequently studied reagent known to alter connexin phosphorylation is cyclic AMP (cAMP). Unlike TPA treatment, phosphorylation induced by cAMP is frequently associated with increased junctional permeability and increased total numbers of gap junctions (187–191). In some tumor cell lines this increase in permeability correlates with an increase in Cx43 transcription (192, 193). cAMP, however, may also regulate GJIC at the level of channel assembly. Experiments by Atkinson et al (191) demonstrated that cAMP treatment of the mouse mammary tumor cell line MMT22 resulted in increased levels of detergent-insoluble phosphorylated Cx43, increased numbers of gap-junctional maculae, and increased GJIC. No increases in total levels of mRNA or protein were observed, suggesting that cAMP modified the cellular distribution of Cx43 such that a greater proportion of the intracellular pool of this connexin was assembled into channels at the cell surface. These effects of cAMP were slow to develop and persisted with treatment.

The kinases regulating connexin phosphorylation under normal growth conditions are not known, and their identification will require mapping the phosphorylated residues. However, as discussed above, the proportion of the connexin pool in MMT22 cells assembled into intercellular channels varies directly with the intracellular concentrations of cAMP, indicating that PKA may regulate connexin phosphorylation. This finding is consistent with the observation that staurosporine, an inhibitor of PKA (and PKC) decreased the amounts of Cx43 phosphorylation in C9 cells (175). Other kinases may regulate hyperphosphorylation of connexins under altered growth conditions. For example, the receptor protein tyrosine kinases v-src, v-fps, and v-yes all may directly tyrosine phosphorylate Cx43, leading to the disruption of GJIC. In addition, the inhibitory effects of TPA and some other growth factors are likely to be mediated by PKC. Other effectors, including ras (174, 194), EGF (155), and the mitogenic phospholipid lysophosphatidic acid (195) may inhibit GJIC by upregulating MAP kinase, which has been shown to phosphorylate Cx43 on serine residues in vitro (155).

Why are some connexins sensitive to the actions of certain kinases while others are not? The reasons are not yet clear, although studies in the mammalian retina demonstrate a real function for differential regulation. While retinal cone bipolar cells synapse directly on ganglion cells, rod bipolar cells do not; rather, they synapse on AII amacrine cells, which in turn are connected by gap junctions both to each other and to the cone bipolar cells (196). Since both rod and cone inputs ultimately converge on cone bipolar cells, a mechanism must exist to distinguish between the rod and cone inputs that are active under different light levels. AII/AII intercellular transfer of neurobiotin is substantively reduced by dopamine or forskolin, indicating a cAMP-mediated regulation of these gap-junctional channels. Forskolin, however, has little effect on AII/cone bipolar communication. In contrast, cGMP agonists cause a reduction of neurobiotin transfer from AII cells to the cone bipolar cells (197, 198). Taken together, these data implicate cyclic nucleotides in the network switching and offer a functional rationale for why there may be different connexins with different regulatory properties within a single cell. The possibility that different connexins co-exist in AII cells is supported by the demonstration of size selectivity between the gap junctions joining these cells to each other and to cone bipolar cells (198). Neurobiotin (MW 286, charge +1) passes easily between both AII/AII and AII/cone bipolar junctions, whereas biotin-X cadaverine (MW 442, charge +1) passes readily only through AII/AII gap junctions.

Conclusion

While there has been substantive progress in the identification of connexin genes, and a remarkable new wealth of information is available about single channel physiology, many unanswered questions remain about the regulation and function of connexin intercellular channels. In the area of structure, the imperfect crystallization of channels in isolated gap-junctional maculae has precluded a molecular solution. Connexins have not been crystallized from solution, a difficulty shared with many integral membrane proteins, and the fact that a large portion of the molecule is buried in the hydrophobic membrane does not permit enzymatic cleavage and subsequent crystallization of functional domains. The most promising avenue of approach to this problem has been the expression of recombinant Cx32 in Sf9 cells, which is assembled into physiologically active connexons that can be purified in quantities sufficient to attempt various crystallization strategies (56, 96).

In the area of assembly, we know that cell-cell adhesion is a prerequisite, presumably involving a complex signaling cascade, and that connexins can be targeted to specific cell surfaces facing different neighbors. Phosphorylation seems to occur at very late stages of intercellular channel assembly, perhaps signaling the grouping of channels into maculae or their gating between various conductance states. We know that connexins turn over with surprisingly rapid half-lives, lending a dynamic quality to gap-junctional regulation, and that disassembly of connexons can be extremely rapid. But this knowledge offers few insights as to mechanisms or functional meanings and leaves unanswered many details of these complex processes.

Perhaps the most difficult questions to answer center on identifying which of the many molecules passing from cell to cell through gap junctions are the important ones subserving physiological functions. Which of the many molecules permitted intercellular passage are active in establishing informational gradients, in synchronizing cellular activities and hormonal responses, and in clustering groups of cells into functional compartments? While some clear and exciting answers have been found in some cellular systems, in most cases these questions remain enigmatic and experimentally daunting.

Finally, key questions remain about the significance of connexin diversity. This diversity results in differences in channel permeation and channel gating, in the ability to co-oligomerize homo- and heteromerically, and in the ability to assemble homo- and heterotypically, creating the potential for a bewildering array of channel possibilities. Whether the differences have physiological significance can now begin to be tested by studying which connexins can substitute for each other in knockout mice. What is clear about gap-junctional intercellular channels is that they are not receptors with specific ligands and specific downstream signaling pathways. Rather they are conduits which have likely evolved to subserve a host of diverse functions, from trans-myelin cytoplasmic buffering in an individual Schwann cell (199) to the coordination of the contraction of billions of myocardiocytes during a heartbeat. As nonselective conduits, gap-junctional channels permit diverse languages of communication among cells, rather than the simple phrase transmitted by a receptor tyrosine kinase. Thus, the functions of gap junctions will undoubtedly be as diverse as the myriad cell types they interconnect, each cell type having adapted this form of communication to meet its specialized physiological needs.

ACKNOWLEDGMENTS

This work was supported by GM18974, EY02430, and GM37751 from the NIH. JAG was supported by a grant from the Medical Foundation. We are grateful to Drs. Bruce Nicholson, Klaus Willecke, and Mario Delmar for access to unpublished data.

Literature Cited

1. Beyer EC, Paul DL, Goodenough DA. 1987. *J. Cell Biol.* 105:2621–29
2. Beyer EC, Goodenough DA, Paul DL. 1988. In *Gap Junctions*, ed. EL Hertzberg, RG Johnson, pp. 167–75. New York: Liss
3. Jiang JX, White TW, Goodenough DA, Paul DL. 1994. *Mol. Biol. Cell* 5:363–73
4. Nicholson BJ, Gros DB, Kent SBH, Hood LE, Revel JP. 1985. *J. Biol. Chem.* 260:6514–17
5. Zimmer DB, Green CR, Evans WH, Gilula NB. 1987. *J. Biol. Chem.* 262:7751–63
6. Manjunath CK, Nicholson BJ, Teplow D, Hood L, Page E, Revel JP. 1987. *Biochem. Biophys. Res. Commun.* 142:228–34
7. Evans WH, Rahman S. 1989. *Biochem. Soc. Trans.* 17:983–85
8. Goodenough DA, Paul DL, Jesaitis LA. 1988. *J. Cell Biol.* 107:1817–24
9. Milks LC, Kumar NM, Houghton R, Unwin N, Gilula NB. 1988. *EMBO J.* 7:2967–75
10. Hertzberg EL, Disher RM, Tiller AA, Zhou Y, Cook RG. 1988. *J. Biol. Chem.* 263:19105–11
11. Rahman S, Evans WH. 1991. *J. Cell Sci.* 100:567–78
12. Kistler J, Schaller J, Sigrist H. 1990. *J. Biol. Chem.* 265:13357–61
13. Beyer EC, Kistler J, Paul DL, Goodenough DA. 1989. *J. Cell Biol.* 108:595–605
14. Yeager M, Gilula NB. 1992. *J. Mol. Biol.* 223:929–48
15. Knudsen KA, Wheelock MJ. 1992. *J. Cell Biol.* 118:671–79
16. Zhang J-T, Nicholson BJ. 1994. *J. Membr. Biol.* 139:15–29
17. Goodenough DA, Revel JP. 1970. *J. Cell Biol.* 45:272–90
18. Manjunath CK, Goings GE, Page E. 1984. *J. Membr. Biol.* 78:147–55
19. Barr L, Dewey MM, Berger W. 1965. *J. Gen. Physiol.* 48:797–823
20. Goodenough DA, Gilula NB. 1974. *J. Cell Biol.* 61:575–90
21. Hirokawa N, Heuser J. 1982. *Cell* 30:395–406
22. Goodenough DA. 1975. In *Methods in*

Membrane Biology, ed. ED Korn, pp. 51–80. New York: Plenum

23. Stauffer KA. 1995. *J. Biol. Chem.* 270: 6768–72

24. Jiang JX, Goodenough DA. 1996. *Proc. Natl. Acad. Sci. USA.* 93:1287–91

25. Paul DL, Yu K, Bruzzone R, Gimlich RL, Goodenough DA. 1995. *Development* 121:371–81

26. Swenson Kl, Jordan JR, Beyer EC, Paul DL. 1989. *Cell* 57:145–55

27. White TW, Bruzzone R, Wolfram S, Paul DL, Goodenough DA. 1994. *J. Cell Biol.* 125:879–92

28. Sosinsky G. 1995. *Proc. Natl. Acad. Sci. USA* 92:9210–14

29. Robertson JD. 1963. *J. Cell Biol.* 19: 201–21

30. Benedetti EL, Emmelot P. 1965. *J. Cell Biol.* 26:299–305

31. Benedetti EL, Emmelot P. 1968. *J. Cell Biol.* 38:15–24

32. Revel JP, Karnovsky MJ. 1967. *J. Cell Biol.* 33:C7–C12

33. Kreutziger GO. 1968. *Proc. Electron Microsc. Soc. Am. Meet., 26th,* p. 234. Baton Rouge: Claitor's Publ.

34. Goodenough DA, Stoeckenius W. 1972. *J. Cell Biol.* 54:646–56

35. Benedetti EL, Dunia I, Bloemendahl H. 1974. *Proc. Natl. Acad. Sci. USA* 71: 5073–77

36. Goodenough DA. 1979. *Invest. Ophthalmol. Vis. Sci.* 18:1104–22

37. Albertini DF, Anderson E. 1974. *J. Cell Biol.* 63:234–50

38. Albertini DF, Fawcett DW, Olds PJ. 1975. *Tissue Cell* 7:389–405

39. Raviola E, Goodenough DA, Raviola G. 1980. *J. Cell Biol.* 87:273–79

40. Miller TM, Goodenough DA. 1985. *J. Cell Biol.* 101:1741–48

41. Hanna RB, Ornberg RL, Reese TS. 1994. In *Gap Junctions,* ed. MVL Bennett, DC Spray, pp. 23–32. Cold Spring Harbor, NY: Cold Spring Harbor Lab.

42. Evans WH, Gurd JW. 1972. *Biochem. J.* 128:691–700

43. Hertzberg EL, Gilula NB. 1979. *J. Biol. Chem.* 254:2138–47

44. Henderson D, Eibl H, Weber K. 1979. *J. Mol. Biol.* 132:193–218

45. Hertzberg EL. 1984. *J. Biol. Chem.* 259: 9936–43

46. Kensler RW, Goodenough DA. 1980. *J. Cell Biol.* 86:755–64

47. Manjunath CK, Goings GE, Page E. 1982. *Biochem. J.* 205:189–94

48. Caspar DLD, Goodenough DA, Makowski L, Phillips WC. 1977. *J. Cell Biol.* 74:605–28

49. Sosinsky GE, Baker TS, Caspar DLD,

Goodenough DA. 1990. *Biophys. J.* 58: 1213–26

50. Braun J, Abney JR, Owicki JC. 1984. *Nature* 310:316–18

51. Abney JR, Braun J, Owicki JC. 1987. *Biophys. J.* 52:441–54

52. Baker TS, Caspar DLD, Hollingshead CJ, Goodenough DA. 1983. *J. Cell Biol.* 96:204–16

53. Baker TS, Sosinsky GE, Caspar DLD, Gall C, Goodenough DA. 1985. *J. Mol. Biol.* 184:81–98

54. Gogol E, Unwin N. 1988. *Biophys. J.* 54:105–12

55. Yeager M. 1994. *Acta Crystallogr. D* 50:632–38

56. Stauffer KA, Kumar NM, Gilula NB, Unwin N. 1991. *J. Cell Biol.* 115:141–50

57. Vrensen GFJM, Graw J, DeWolf A. 1991. *Exp. Eye Res.* 52:647–59

58. Makowski L, Caspar DLD, Phillips WC, Goodenough DA. 1984. *J. Mol. Biol.* 174:449–81

59. Unwin PNT, Zampighi G. 1980. *Nature* 283:545–49

60. Unwin PN, Ennis PD. 1984. *Nature* 307:609–13

61. Caspar DLD, Sosinsky GE, Tibbitts TT, Phillips WC, Goodenough DA. 1988. In *Gap Junctions,* ed. EL Hertzberg, RG Johnson, pp. 117–33. New York: Liss

62. Makowski L. 1985. In *Gap Junctions,* ed. MVL Bennett, DC Spray, pp. 5–12. Cold Spring Harbor, NY: Cold Spring Harbor Lab.

63. Sosinsky GE, Jesior JC, Caspar DLD, Goodenough DA. 1988. *Biophys. J.* 53: 709–22

64. Makowski L, Caspar DLD, Phillips WC, Goodenough DA. 1977. *J. Cell Biol.* 74:629–45

65. Tarnai T, Gáspár G, Szalai L. 1995. *Biophys. J.* 69:612–18

66. Ludwig DS, Ribi HO, Schoolnik GK, Kornberg RD. 1986. *Proc. Natl. Acad. Sci. USA* 83:8585–88

67. Unwin PNT, Ennis PD. 1984. *Nature* 307:609–13

68. Ghoshroy S, Goodenough DA, Sosinsky GE. 1995. *J. Membr. Biol.* 146:15–28

69. Cascio M, Kumar NM, Safarik R, Gilula NB. 1995. *J. Biol. Chem.* 270: 18643–48

70. Unwin N. 1986. *Nature* 323:12–13

71. Richardson JS, Richardon DC. 1985. *Methods Enzymol.* 115:189–206

72. Tibbitts TT, Caspar DLD, Phillips WC, Goodenough DA. 1990. *Biophys. J.* 57: 1025–36

73. Cascio M, Gogol E, Wallace BA. 1990. *J. Biol. Chem.* 265:2358–64

74. Hoh JH, Sosinsky GE, Revel JP,

Hansma PK. 1993. *Biophys. J.* 65:149–63

75. Subak-Sharpe H, Burk RR, Pitts JD. 1969. *J. Cell Sci.* 4:353–67
76. Pitts JD, Burk RR. 1976. *Nature* 264:762–64
77. Gilula NB, Reeves OR, Steinbach A. 1972. *Nature* 235:262–65
78. Stewart WW. 1978. *Cell* 14:741–59
79. Turin L, Warner A. 1977. *Nature* 270:56–57
80. McNeil PL, Murphy RF, Lanni F, Taylor DL. 1984. *J. Cell Biol.* 98:1556–64
81. El-Fouly MH, Trosko JE, Chang CC. 1987. *Exp. Cell Res.* 168:422–30
82. Furshpan EJ, Potter DD. 1957. *Nature* 180:342–43
83. Furshpan EJ, Potter DD. 1959. *J. Physiol.* 145:289–325
84. Loewenstein WR. 1966. *Ann. NY Acad. Sci.* 137:441–72
85. Bennett MVL. 1966. *Ann. NY Acad. Sci.* 137:509–39
86. Potter DD, Furshpan EJ, Lennox EX. 1966. *Proc. Natl. Acad. Sci. USA* 55:328–35
87. Spray DC, Harris AL, Bennett MVL. 1981. *J. Gen. Physiol.* 77:75–94
88. Hamill OP, Marty A, Neher E, Sakmann B, Sigworth FJ. 1981. *Pfluegers Arch.* 391:85–100
89. Sakmann B, Neher E. 1984. *Annu. Rev. Physiol.* 46:455–72
90. Neyton J, Trautmann A. 1985. *Nature* 317:331–35
91. Burt JM, Spray DC. 1988. *Proc. Natl. Acad. Sci. USA* 85:3431–34
92. Young JD, Cohn ZA, Gilula NB. 1987. *Cell* 48:733–43
93. Mazet JL, Jarry T, Gros D, Mazet F. 1992. *Eur. J. Biochem.* 210:249–56
94. Donaldson P, Kistler J. 1992. *J. Membr. Biol.* 129:155–65
95. Harris AL, Walter A, Paul D, Goodenough DA, Zimmerberg J. 1992. *Mol. Brain Res.* 15:269–80
96. Buehler LK, Stauffer KA, Gilula NB, Kumar NM. 1995. *Biophys. J.* 68:1767–75
97. Veenstra RD, Wang HZ, Westphale EM, Beyer EC. 1992. *Circ. Res.* 71:1277–83
98. Eghbali B, Kessler JA, Spray DC. 1990. *Proc. Natl. Acad. Sci. USA* 87:1328–31
99. Moore LK, Burt JM. 1994. *Am. J. Physiol.* 267:C1371–80
100. Werner R, Miller T, Azarnia R, Dahl G. 1985. *J. Membr. Biol.* 87:253–68
101. Dahl G, Miller T, Paul D, Voellmy R, Werner R. 1987. *Science* 236:1290–93
102. Ebihara L, Beyer EC, Swenson KI, Paul DL, Goodenough DA. 1989. *Science* 243:1194–95

103. Gimlich RL, Kumar NM, Gilula NB. 1990. *J. Cell Biol.* 110:597–605
104. Werner R, Levine E, Rabadan-Diehl C, Dahl G. 1989. *Proc. Natl. Acad. Sci. USA* 86:5380–84
105. Barrio LC, Suchyna T, Bargiello T, Xu LX, Roginski RS, et al. 1991. *Proc. Natl. Acad. Sci. USA* 88:8410–14
106. Willecke K, Heynkes R, Dahl E, Stutenkemper R, Hennemann H, et al. 1991. *J. Cell Biol.* 114:1049–57
107. White TW, Bruzzone R, Goodenough DA, Paul DL. 1992. *Mol. Biol. Cell* 3:711–20
108. Goodenough DA. 1974. *J. Cell Biol.* 61:557–63
109. John SA, Revel JP. 1991. *Biochem. Biophys. Res. Commun.* 178:1312–18
110. Dupont E, el-Aoumari A, Briand JP, Fromaget C, Gros D. 1989. *J. Membr. Biol.* 108:247–52
111. Dahl G, Levine E, Rabadan-Diehl C, Werner R. 1991. *Eur. J. Biochem.* 197:141–44
112. Dahl G, Werner R, Levine E, Rabadan-Diehl C. 1992. *Biophys. J.* 62:187–95
113. Dahl G, Nonner W, Werner R. 1994. *Biophys. J.* 67:1816–22
114. Goodenough DA, Revel JP. 1971. *J. Cell Biol.* 50:81–91
115. Meyer RA, Laird DW, Revel JP, Johnson RG. 1992. *J. Cell Biol.* 119:179–89
116. Matsuzaki F, Miaege R-M, Jaffe SH, Friedlander DR, Gallin WJ, et al. 1990. *J. Cell Biol.* 110:1239–52
117. Mege RM, Matsuzaki F, Gallin WJ, Goldberg JI, Cunningham BA, Edelman GM. 1988. *Proc. Natl. Acad. Sci. USA* 85:7274–78
118. Jongen WMF, Fitzgerald DJ, Asamoto M, Piccoli C, Slaga TJ, et al. 1991. *J. Cell Biol.* 114:545–55
119. Musil LS, Cunningham BA, Edelman GM, Goodenough DA. 1990. *J. Cell Biol.* 111:2077–88
120. Bennett MVL, Barrio LC, Bargiello TA, Spray DC, Hertzberg E, Saez JC. 1991. *Neuron* 6:305–20
121. Noda M, Shimizu S, Tanabe T, Takai T, Kayano T, et al. 1984. *Nature* 312:121–27
122. Rubin JB, Verselis VK, Bennett MVL, Bargiello TA. 1992. *Biophys. J.* 62:183–93
123. Rubin JB, Verselis VK, Bennett MVL, Bargiello TA. 1992. *Proc. Natl. Acad. Sci. USA* 89:3820–24
124. Verselis VK, Ginter CS, Bargiello TA. 1994. *Nature* 368:348–51
125. White TW, Bruzzone R, Goodenough DA, Paul DL. 1994. *Nature* 371:208–9

126. Bruzzone R, White TW, Paul DL. 1994. *J. Cell Sci.* 107:955–67
127. Elfgang C, Eckert R, Lichtenberg-Frate H, Butterweck A, Traub O, et al. 1995. *J. Cell Biol.* 129:805–17
128. Bruzzone R, Haefliger J-A, Gimlich RL, Paul DL. 1993. *Mol. Biol. Cell* 4:7–20
129. White TW, Paul DL, Goodenough DA, Bruzzone R. 1995. *Mol. Biol. Cell* 6: 459–70
130. Suchyna TM, Xu LX, Gao F, Fourtner CR, Nicholson BJ. 1993. *Nature* 365: 847–49
131. Liu S, Taffet S, Stoner L, Delmar M, Vallano ML, Jalife J. 1993. *Biophys. J.* 64:1422–33
132. Ek JF, Delmar M, Perzova R, Taffet SM. 1994. *Circ. Res.* 74:1058–64
133. Spray DC, Burt JM. 1990. *Am. J. Physiol.* 258:C195–205
134. Steinberg TH, Civitelli R, Geist ST, Robertson AJ, Hick E, et al. 1994. *EMBO J.* 13:744–50
135. Veenstra RD, Wang HZ, Beyer EC, Brink PR. 1994. *Circ. Res.* 75:483–90
136. Malewicz B, Kumar VV, Johnson RG, Baumann WJ. 1990. *Lipids* 25:419–27
137. Manjunath CK, Going GE, Page E. 1985. *J. Membr. Biol.* 85:159–68
138. Paul DL, Goodenough DA. 1983. *J. Cell Biol.* 96:625–32
139. Kistler J, Goldie K, Donaldson P, Engel A. 1994. *J. Cell Biol.* 126:1047–58
140. Johnston MF, Simon SA, Ramon F. 1980. *Nature* 286:498–500
141. Bernardini G, Peracchia C, Peracchia LL. 1984. *Eur. J. Cell Biol.* 34:307–12
142. Hirschi KK, Minnich BN, Moore LK, Burt JM. 1993. *Am. J. Physiol.* 265: C1517–26
143. Burt JM, Spray DC. 1989. *Circ. Res.* 65:829–37
144. Takens-Kwak BR, Jongsma HJ, Rook MB, Van Ginneken AC. 1992. *Am. J. Physiol.* 262:C1531–38
145. Venance L, Piomelli D, Glowinski J, Giaume C. 1995. *Nature* 376:590–94
146. Charles AC, Dirksen ER, Sanderson MJ. 1991. *Neuron* 6:938–92
147. Cornell-Bell AH, Finkbeiner SM, Cooper MS, Smith SJ. 1990. *Science* 247: 470–74
148. Nedergaard M. 1994. *Science* 263: 1768–71
149. Bastiaanse EM, Jongsma HJ, van der Laarse A, Takens-Kwak BR. 1993. *J. Membr. Biol.* 136:135–45
150. Traub O, Look J, Dermietzel R, Brummer F, Hulser D, Willecke K. 1989. *J. Cell Biol.* 108:1039–51
151. Saez JC, Nairn AC, Czernik AJ, Spray DC, Hertzberg EL, et al. 1990. *Eur. J. Biochem.* 192:263–73
152. Moreno AP, Saez JC, Fishman GI, Spray DC. 1994. *Circ. Res.* 74:1050–57
153. Saez JC, Berthoud VM, Moreno AP, Spray DC. 1993. In *Advances in Second Messengers and Phosphoprotein Research,* ed. S Shenolikar, AC Nairn, pp. 163–98. New York: Raven
154. Laird DW, Revel JP. 1990. *J. Cell Sci.* 97:109–17
155. Kanemitsu MY, Lau AF. 1993. *Mol. Biol. Cell* 4:837–48
156. Werner R, Levine E, Rabadan-Diehl C, Dahl G. 1991. *Proc. R. Soc. London Ser. B* 243:5–11
157. Musil LS, Goodenough DA. 1991. *J. Cell Biol.* 115:1357–74
158. Hotz-Wagenblatt A, Shalloway D. 1993. *Crit. Rev. Oncog.* 4:541–58
159. Atkinson MM, Menko AS, Johnson RG, Sheppard JR, Sheridan JD. 1981. *J. Cell Biol.* 91:573–78
160. Azarnia R, Loewenstein WR. 1984. *J. Membr. Biol.* 82:191–205
161. Chang C-C, Trosko JE, Kung H-J, Bombick D, Matsumura F. 1985. *Proc. Natl. Acad. Sci. USA* 82:5360–64
162. Azarnia R, Reddy S, Kmiecik TC, Shalloway D, Loewenstein WR. 1988. *Science* 239:398–401
163. Crow DS, Beyer EC, Paul DL, Kobe SS, Lau AF. 1990. *Mol. Cell. Biol.* 10:1754–63
164. Swenson KI, Piwnica-Worms H, McNamee H, Paul DL. 1990. *Cell Regul.* 1:989–1002
165. Loo LWM, Berestecky JM, Kanemitsu MY, Lau AF. 1995. *J. Biol. Chem.* 270:12751–61
166. Kurata WE, Lau AF. 1994. *Oncogene* 9:329–35
167. Filson AJ, Azarnia R, Beyer EC, Loewenstein WR, Brugge JS. 1990. *Cell Growth Differ.* 1:661–68
168. Crow DS, Kurata WE, Lau AF. 1992. *Oncogene* 7:999–1003
169. Madhukar BV, Oh SY, Chang CC, Wade M, Trosko JE. 1989. *Carcinogenesis* 10:13–20
170. Maldonado PE, Rose B, Loewenstein WR. 1988. *J. Membr. Biol.* 106:203–10
171. Lau AF, Kanemitsu MY, Kurata WE, Danesh S, Boynton AL. 1992. *Mol. Biol. Cell* 3:865–74
172. Pelletier DB, Boynton AL. 1994. *J. Cell. Physiol.* 158:427–34
173. Moorby CD, Stoker M, Gherardi E. 1995. *Exp. Cell Res.* 219:657–63
174. Brissette JL, Kumar NM, Gilula NB, Dotto GP. 1991. *Mol. Cell. Biol.* 11: 5364–71
175. Berthoud VM, Rook MB, Traub O, Hertzberg EL, Saez JC. 1993. *Eur. J. Cell Biol.* 62:384–96

176. Asamoto M, Oyamada M, El Aoumari A, Gros D, Yamasaki H. 1991. *Mol. Carcinog.* 4:322–27
177. Matesic DF, Rupp HL, Bonney WJ, Ruch RJ, Trosko JE. 1994. *Mol. Carcinog.* 10:226–36
178. Murray AW, Gainer HSC. 1989. In *Cell Interactions and Gap Junctions,* ed. N Sperelakis, WC Cole, pp. 97–106. Boca Raton, FL: CRC Press
179. Brissette JL, Kumar NM, Gilula NB, Dotto GP. 1991. *Mol. Cell. Biol.* 11:5364–71
180. Kikkawa U, Kishimoto A, Nishizuka Y. 1989. *Annu. Rev. Biochem.* 58:31–44
181. Saez JC, Nairn AC, Czernik AJ, Spray DC, Hertzberg EL. 1993. *Prog. Cell Res.* 3:275–81
182. Kwak BR, Vanveen TAB, Analbers LJS, Jongsma HJ. 1995. *Exp. Cell Res.* 220:456–63
183. Munster PN, Weingart R. 1993. *Pfluegers Arch.* 423:181–88
184. Lampe PD. 1994. *J. Cell Biol.* 127:1895–905
185. Hu J, Engman L, Cotgreave IA. 1995. *Carcinogenesis* 16:1815–24
186. Musil LS, Goodenough DA. 1993. *Cell* 74:1065–77
187. Azarnia R, Dahl G, Loewenstein WR. 1981. *J. Membr. Biol.* 63:133–46
188. De Mello WC. 1984. *Fed. Proc.* 43:2692–96
189. Flagg-Newton JL, Dahl G, Loewenstein WR. 1981. *J. Membr. Biol.* 63:105–21
190. Saez JC, Spray DC, Nairn AC, Hertzberg E, Greengard P, Bennett MVL. 1986. *Proc. Natl. Acad. Sci. USA* 83:2473–77
191. Atkinson MM, Lampe PD, Lin HH, Kollander R, Li X, Kiang DT. 1995. *J. Cell Sci.* 108:3079–90
192. Mehta P, Yamamoto M, Rose B. 1992. *Mol. Biol. Cell* 3:839–50
193. Schiller PC, Mehta PP, Roos BA, Howard GA. 1992. *Mol. Endocrinol.* 6:1433–40
194. Wood KW, Sarnecki C, Roberts TM, Blenis J. 1992. *Cell* 68:1041–50
195. Hii CST, Oh SY, Schmidt SA, Clark KJ, Murray AW. 1994. *Biochem. J.* 303:475–79
196. Strettoi E, Dacheux RF, Raviola E. 1994. *J. Comp. Neurol.* 347:139–49
197. Hampson EC, Vaney DI, Weiler R. 1992. *J. Neurosci.* 12:4911–22
198. Mills SL, Massey SC. 1995. *Nature* 377:734–37
199. Bergoffen J, Scherer SS, Wang S, Scott MO, Bone LJ, et al. 1993. *Science* 262:2039–42
200. Haefliger J-A, Bruzzone R, Jenkins NA, Gilbert DJ, Copeland NG, Paul DL. 1992. *J. Biol. Chem.* 267:2057–64
201. Goliger JA, Paul DL. 1994. *Dev. Dyn.* 200:1–13
202. Butterweck A, Elfgang C, Willecke K, Traub O. 1994. *Eur. J. Cell Biol.* 65:152–63
203. Mikkelsen HB, Huizinga JD, Thuneberg L, Rumessen JJ. 1993. *Cell Tissue Res.* 274:249–56
204. Willecke K, Jungbluth S, Dahl E, Hennemann H, Heynkes R, Grzeschik K-H. 1990. *Eur. J. Cell Biol.* 53:275–80
205. Bastide B, Neyses L, Ganten D, Paul M, Willecke K, Traub O. 1993. *Circ. Res.* 73:1138–49
206. Hennemann H, Suchyna T, Lichtenberg-Frate H, Jungbluth S, Dahl E, et al. 1992. *J. Cell Biol.* 117:1299–310
207. Paul DL, Ebihara L, Takemoto LJ, Swenson KI, Goodenough DA. 1991. *J. Cell Biol.* 115:1077–89
208. Tucker MA, Barajas L. 1994. *Exp. Cell Res.* 213:224–30
209. Hennemann H, Dahl E, White JB, Schwarz HJ, Lalley PA, et al. 1992. *J. Biol. Chem.* 267:17225–33
210. Hoh JH, John SA, Revel JP. 1991. *J. Biol. Chem.* 266:6524–31
211. Zhang J-T, Nicholson BJ. 1989. *J. Cell Biol.* 109:3391–401
212. Paul DL. 1986. *J. Cell Biol.* 103:123–34
213. Naus CC, Belliveau DJ, Bechberger JF. 1990. *Neurosci. Lett.* 111:297–302
214. Kanter HL, Laing JG, Beyer EC, Green KG, Saffitz JE. 1993. *Circ. Res.* 73:344–50
215. Liang JG, Westphale EM, Engelmann GL, Beyer EC. 1994. *J. Membr. Biol.* 139:31–40
216. Butterweck A, Gergs U, Elfgang C, Willecke K, Traub O. 1994. *J. Membr. Biol.* 141:247–56

Annu. Rev. Biochem. 1996. 65:503–35

RHIZOBIUM LIPO-CHITOOLIGOSACCHARIDE NODULATION FACTORS:
Signaling Molecules Mediating Recognition and Morphogenesis

Jean Dénarié and Frédéric Debellé

Laboratoire de Biologie Moléculaire des Relations Plantes-Microorganismes, CNRS-INRA, B. P. 27, 31326 Castanet-Tolosan Cedex, France

Jean-Claude Promé

Laboratoire de Pharmacologie et de Toxicologie Fondamentales, CNRS, 205 Route de Narbonne, 31077 Toulouse Cedex, France

KEY WORDS: Nod factors, chitin oligomers, oligosaccharins, host specificity, plant organogenesis

---

ABSTRACT

Rhizobia elicit on their specific leguminous hosts the formation of new organs, called nodules, in which they fix nitrogen. The rhizobial nodulation genes specify the synthesis of lipo-chitooligosaccharide signals, the Nod factors (NFs). Each rhizobial species has a characteristic set of nodulation genes that specifies the length of the chitooligosaccharide backbone and the type of substitutions at both ends of the molecule, thus making the NFs specific for a given plant host. At extremely low concentrations, purified NFs are capable of eliciting on homologous legume hosts many of the plant developmental responses characteristic of the bacteria themselves, including cell divisions, and the triggering of a plant organogenic program. This review summarizes our current knowledge on the biosynthesis, structure, and function of this new class of signaling molecules. Finally we discuss the possibility that these signals could be part of a new family of plant lipo-chitooligosaccharide growth regulators.

503

0066-4154/96/0701-0503$08.00

CONTENTS

NOD GENES AND NOD FACTORS.. 504
   Rhizobial Infection and Nodulation of Legumes.......................... 504
   Rhizobial Nodulation Genes.. 506
   Nod Factor Identification and Characterization........................ 506
SYNTHESIS OF NOD FACTORS, SPECIFIC SIGNALING MOLECULES 508
   The Chitin Oligomer Backbone 508
   N-Substitutions at the Nonreducing End 510
   O-Substitutions at the Nonreducing End 513
   O-Substitutions at the Reducing End.................................. 514
   Nod Factor Transport... 516
   Nod Factor Structure and Host Specificity............................ 518
   Mechanisms Ensuring Signaling and Host-Range Variation 519
   Chemical and Enzymatic Synthesis 521
NOD FACTORS AS PLANT GROWTH REGULATORS 522
   Responses in Root Epidermis.. 522
   Cell Cycle Activation in Root Cortex................................. 525
   Responses in Stele... 526
   Nodule Primordia and Nodule Development.............................. 527
   Nod Factor Perception and Transduction............................... 527
   Lipo-Chitooligosaccharides: A New Class of Plant Growth Regulators? .. 529

NOD GENES AND NOD FACTORS

Rhizobial Infection and Nodulation of Legumes

All bacteria able to elicit nodule formation on legumes are called rhizobia. In fact these bacteria do not form a discrete clade. Some rhizobia are more closely related to nonsymbiotic bacteria than they are to other rhizobia. Rhizobia are now classified in four genera: *Rhizobium, Bradyrhizobium, Azorhizobium,* and *Sinorhizobium* (1, 2; see Table 1). The ability of a plant to establish a nitrogen-fixing symbiosis with rhizobia is restricted to legumes with one exception; the genus *Parasponia* of the Ulmaceae family. Leguminosae is a large family containing more than 15,000 species, most of which can be nodulated by rhizobia, and these nitrogen-fixing associations have considerable ecological and agronomical importance. Rhizobia-plant associations are specific, as a given bacterium can only nodulate a defined number of plants (3). However the degree of specificity varies. Some rhizobia, such as *R. meliloti* and *R. leguminosarum* bv. *trifolii* (*R. l.* bv. *trifolii*) exhibit a narrow host range of nodulation. In contrast, some strains have a broad host range; for example *R.* sp. NGR234 nodulates over 70 legume genera and even the nonlegume *Parasponia* (4).

    Rhizobia can infect their hosts by various mechanisms, such as intercellular penetration (crack-in) between epidermal and cortical cells, as in *Arachis* (peanut) and *Parasponia*, or by infection of root-hair cells as in most plant systems described in this review, such as alfalfa, pea, vetch, and soybean (5–7). Associated with infection is the induction of cell division in the cortex and the formation of nodule primordia. The development of primordia gives rise to nodules that are genuine organs, not mere tumors or deformed roots. Nodules

Table 1 Structure of Nod factors from various rhizobal species

Host plant[a]	Rhizobial species	Nod factor substituents[b]				n[c]	Ref.
		R1	R2[d]	R3	R4		
Medicago	*R. meliloti*	H	C16:2, C16:3 C18-C26(ω-1)OH	Ac(O-6),H	S	1,2,3	22,23,25 61
Vicia	*R. l. bv viciae*	H	C18:1,C18:4	Ac(O-6),H	H	2,3	38
Pisum cv. Afghanistan	*R. l. bv viciae* TOM	H	C18:1,C18:4	Ac(O-6)	Ac	2,3	92
Trifolium	*R. l. bv trifolii*	H	C18:1, C18:3 C20:3, C20:4	Ac(O-6)	H	1,2,3	62,74
Lotus	*R. loti*	Me	C18:1	Cb(O-4)	AcFuc	3	183
Phaseolus	*R. etli*	Me	C18:1	Cb(O-4),H	AcFuc	3	111,112
	R. tropici	Me	C18:1	H	S,H	3	31
Acacia	*R.* sp GRH2	Me,H	C18:1	H	S,H	2,3,4	162
	S. teranga	Me	C18:1	Cb	S,H	3	J. Lorquin
Lablab	*R.* sp NGR234	Me	C18:1	Cb(O-6 and O-3 or O-4), H	MeFuc AcMeFuc MeSFuc	3	27
Glycine	*R. fredii*	H	C18:1	H	MeFuc,Fuc	1,2,3	116
	B. japonicum	H	C18:1	H	MeFuc	3	114,80
	B. elkanii	H,Me	C18:1	Ac(O-6),H,Cb	MeFuc,Fuc	2,3	80
Sesbania	*A. caulinodans*	Me	C18:1	Cb(O-6)	D-Ara,H	2,3	28

[a] Plant genera from which rhizobial strains were isolated
[b] R5=H in all strains except *B. elkanii* where R5=H, glycerol
[c] The bold numbers indicate the number of glucosamine residues of the most abundant Nod factors
[d] Selected fatty acyl substituents

are always formed on the roots; however, some aquatic legumes, such as *Sesbania rostrata*, also exhibit stem nodulation. Nitrogen-fixing nodules of a given plant species have a characteristic morphology and anatomy. They also have a defined developmental pattern that is either indeterminate (with a persistent meristem) as in alfalfa, pea, and vetch, or determinate (with a transient meristem) as in soybean and beans (5, 6). The type of infection, as well as the nodule structural and developmental characteristics, are specified by the plant and not by the rhizobial strain, indicating that the host possesses the genetic information for symbiotic infection and nodulation and that the role of the bacteria is to turn on this program with specific signals (8, 9).

Rhizobial Nodulation Genes

Rhizobial molecular genetics, combined with the study of the symbiotic behavior of bacterial mutant strains, has led to the identification of nodulation genes (*nod*, *nol*, and *noe*) required for infection, nodule formation, and the control of host specificity (9–11). Both regulatory and structural nodulation genes are involved in the signal exchange that occurs during the early steps of symbiosis (4, 12–15). All rhizobia studied so far have at least one copy of *nodD*, whose gene-product is a transcriptional activator of the LysR family. In the presence of plant signals, usually flavonoids excreted in plant root exudates, NodD proteins activate the transcription of structural *nod* genes. In many species, multiple copies of *nodD* are present, coding for regulatory proteins that are activated by different plant or environmental signals (4, 9, 15, 16). Other regulatory *nod* genes can also be present (9, 15, 17). This activation of regulatory Nod proteins by diverse plant signals constitutes a first level of control of host specificity (9, 11, 18).

Structural *nod* genes can be divided into two sets. The *nodABC* genes, termed "common" because they are present in all rhizobial species, play a pivotal role as evidenced by the fact that their inactivation results in the complete loss of the ability to elicit any detectable plant response, regardless of the host. The second set comprises *nod* genes present in various combinations in different species or biovars (see Table 2). Mutations in these "species-specific" *nod* genes result in an alteration in the host range of nodulation. Since the introduction of the *nod* genes from a given rhizobial species into strains of the non-nodulating genus *Agrobacterium* results in the transfer of the ability to nodulate the appropriate legume, it is clear that *nod* genes are responsible for determining host recognition and the formation of a new plant organ.

Nod Factor Identification and Characterization

A link between the genetic determinants of nodulation and the production of excreted molecules was provided by the observation that the sterile supernatants of rhizobial cultures cause changes in root and root-hair morphology on homologous hosts and that this activity is dependent on the *nodABC* genes (19–21). Using an alfalfa root-hair deformation bioassay, extracellular Nod factors (NFs) (Nod because they are produced under the control of *nodABC* genes) were purified from cultures of *R. meliloti* genetically engineered to overexpress *nod* genes. NFs were found to be amphiphilic and were extracted

Table 2 Biochemical function of nodulation gene products involved in Nod factor synthesis and transport

nod gene	Species[a]	Gene product function	Selected references
Regulation of nod gene expression			
nodD	Common	LysR-type regulator	12,15,176
nodV	Bj	two-component family sensor	177
nodW	Bj	two-component family regulator	177
nolA	Bj	MerR-type regulator?	178
nolR	Rm	LysR-type regulator	179
syrM	Rm	LysR-type regulator	180
Synthesis of the chito oligosaccharide backbone			
nodM	Rm,Rlv,Rlt	D-glucosamine synthase	36,37
nodC	Common	UDP-GlcNAc transferase	32,33,34,45
nodB	Common	De-N-acetylase	33,51,52
N-substitutions at nonreducing end			
nodE	Rm,Rlv,Rlt	Beta-ketoacyl synthase	38,61,69
nodF	Rm,Rlv,Rl	Acyl carrier protein	61,63,64,65
nodA	Common	N-acyltransferase	51,73
nodS	Rn,Rt,Ac,Bj,Rf	S-adenosyl methionine methyl transferase	48,54,56,57
O-substitutions at nonreducing end			
nodL	Rm,Rlv,Rlt	6-O-acetyltransferase	75,77,78,79
nodU	Rn,Rt,Ac,Bj,Rf	6-O-carbamoyltransferase	54,81
O-substitutions at reducing end			
nodP	Rm,Rt	ATP sulfurylase	83,85,86
nodQ	Rm,Rt	ATP sulfurylase, APS kinase,	83,85,86
nodH	Rm,Rt	Sulfotransferase	24,88,89,90
nodZ	Bj,Rn	Fucosyl transferase	95,181
nodZ	Ac	Glycosyl transferase?	81
nolK	Ac	Sugar epimerase	96
nodX	Rlv TOM	Acetyl transferase	92
Secretion of Nod factors			
nodI	Common	ATP-binding protein	101,103,104,105
nodJ	Common	Membrane protein	101,103,104,105
nodT	Rlv,Rlt	Outer membrane protein	106,182
nolFGHI	Rm	Membrane proteins	35,108

[a]Abbreviations: Ac: *Azorhizobium caulinodans;* Be: *Bradyrhizobium elkanii;* Bj: *B. japonicum*; Re: *Rhizobium etli;* Rf: *R. fredii;* Rg: *R.* sp GRH2; Rl: *R. loti*; Rlt: *R. l.* bv *trifolii*; Rlv: *R. leguminosarum* bv *viciae*; Rm: *R. meliloti;* Rn: *R.* sp NGR234; Rt: *R. tropici.*

by liquid/liquid extraction (water/butanol) of culture filtrates and were purified by ion exchange chromatography and reverse phase HPLC. Purified NFs were analyzed by mass spectrometry, nuclear magnetic resonance, and chemical analysis and found to be lipo-chitooligosaccharides (LCOs). *R. meliloti* NFs consist of a chitin oligomer backbone of four or five β,1-4-linked glucosamine residues. The chitin oligomer backbone is mono-N-acylated at the nonreducing terminal end by unsaturated C16 fatty acids possessing one or two double bonds conjugated with the amide group, N-acetylated on the other residues, O-acetylated at the nonreducing end, and O-sulfated at the reducing end (22–26; see Table 1).

NFs have now been purified from various rhizobial species using similar methods. Extractions were performed from culture filtrates either with butanol (23) or by liquid/solid procedures using C18 reverse phase or styrene divinyl–benzene cross-polymer columns (27, 28). Analytical methods included mono- or multidimensional thin layer chromatography, with reverse or direct phases, after NF radiolabelling by growing rhizobia in the presence of the radioactive precursors ^{14}C acetate, ^{14}C glucosamine, or ^{35}S sulfate (27, 29–31). NFs from all rhizobial species belong to the same chemical family: They are mono-N-acylated chitooligosaccharides, mostly penta- and tetramers. The chitin oligomer backbone, however, is diversely substituted in the various species, on the two terminal glucosamine residues (Table 1).

Genetic loss- and gain-of-function experiments have clearly shown that *nod* genes control host-specificity, infection, and nodulation by specifying NF production. In addition, purified NFs are very active signal molecules and are capable of eliciting, at extremely low concentrations, many of the symbiotic plant responses induced by the bacteria themselves.

SYNTHESIS OF NOD FACTORS, SPECIFIC SIGNALING MOLECULES

Analysis of NF biosynthetic pathways has been facilitated by the use of bacterial molecular genetics. We describe current knowledge of the role of *nod* gene products in the synthesis of the chitin oligomer backbone and in the modifications at both ends of the molecules. We also give a few examples of relationships between NF and general metabolism.

The Chitin Oligomer Backbone

SYNTHESIS OF PRECURSORS   N-acetyl-D-glucosamine (GlcNAc) is an important constituent of lipopolysaccharide lipid A in the outer membrane and of peptidoglycan, both of which are important components of the bacterial envelope. Uridyl di-phosphate (UDP)-GlcNAc is a common precursor for the biosynthesis of these two components and can also be used as an activated sugar donor for the synthesis of the NF backbone in different rhizobial species (32–34). An important question is whether rhizobia draw from the lipid A-peptidoglycan UDP-GlcNAc pool for NF synthesis or whether they contribute to the synthesis of this substrate with the expression of specific nodulation genes.

The *nodM* gene, identified in *R. meliloti* and *R. l.* bv. *viciae* and *trifolii*, codes for a glucosamine synthase (35). *nodM* mutants exhibit a delay in nodulation and a lower level of NF production (35–37). In addition to NodM, *R. l.* bv. *viciae* possesses a "house-keeping" GlmS glucosamine synthase (37), and NodM and GlmS have interchangeable functions. A double *nodM/glmS* mutant is auxotrophic for glucosamine and unable to elicit nodulation (37), but the presence of either *nodM* or *glmS* restores prototrophy and symbiotic abilities (37). Thus the *nod* glucosamine synthase function appears to increase the GlcNAc concentration for NF synthesis, and both the *nod* and *glm* pathways appear to feed a common UDP-GlcNAc pool.

SYNTHESIS OF THE CHITOOLIGOSACCHARIDE BACKBONE   In all rhizobial species *nodABC* genes are required for the production of lipo-chitooligosaccharidic NFs. In a *R. l.* bv *viciae* strain cured of the pSym plasmid, the regulatory *nodD* gene and the common *nodABC* genes are sufficient to produce "core" NF molecules, chitooligosaccharides with four or five glucosamine residues that have no O-substitutions and are N-acylated with vaccenic acid (C18:1), the most abundant membrane fatty acid present in this species (38).

NodC exhibits significant homologies with several processing β-glycosyl transferases: chitin synthases (39–41), cellulose synthases (39), *Streptococcus* hyaluronan synthase (42), and the *Xenopus* developmental protein DG42 (40, 43). Sequence comparison by various methods including hydrophobic cluster analysis indicates that these β-glycosyl transferases constitute a single protein family that has a modular architecture with transmembrane domains attached to a globular region composed of two domains, presumably involved in the formation of β-linkage (44).

nodC-expressing cell extracts, in the presence of radioactive UDP-GlcNAc as a sugar donor, are able to incorporate GlcNAc into chitooligosaccharides of up to five glucosamine residues (32, 33, 45). NodC is localized in the cytoplasmic membrane (46), and membrane fractions prepared from *R. fredii* cultures grown in the presence of a *nod* gene inducer incorporate GlcNAc into chitin oligomers of up to five residues (34). For maximal activity the reaction requires free N-acetyl-D-glucosamine that cannot be substituted by glucosamine, galactosamine, or N-acetyl-D-galactosamine as acceptor molecules. Pulse-chase experiments indicate that the chain elongation of the oligosaccharide proceeds toward the nonreducing end and that the reaction follows a single-chain mechanism: Free chitobiose, chitotriose, or chitotetraose are not substrates for initiation and/or elongation (34).

Synthesis of oligosaccharides and polysaccharides often occurs by transfer of the activated glycosyl to a sugar residue that is glycosidically linked to an aglycone primer (see 47). However, no clear evidence shows the presence of lipid or protein intermediates in the NodC-dependent in vitro synthesis of chitooligosaccharides (34, 48).

NodC is the first example of a new class of $\beta(1,4)$ N-acetyl-D-glucosaminyl transferases. Unlike chitin synthases, NodC is not involved in the synthesis of polymers but of oligomers. NodC-containing cell extracts and membrane fractions specify the incorporation of GlcNAc into oligomers with a maximal size of five, and insoluble chitin-like products are not detected (32–34, 45). Rhizobial species produce NFs with a backbone constituted by a defined number of glucosamine residues. For example the majority of NFs excreted by *R. meliloti* have four glucosamine residues (22, 23, 25), whereas *R. tropici* produces pentamers (31). The introduction of a *R. tropici nodC* gene into a *R. meliloti nodC* mutant results in the production of a majority of pentamers, indicating that *nodC* genes are involved in the control of the length of the chitooligosaccharide backbone (49). *R. meliloti* NFs with five glucosamine residues are much less active than the tetramers on the host plant alfalfa (25), indicating that the *nodC* gene is likely to be a host-range determinant (49).

N-Substitutions at the Nonreducing End

All NFs are N-acylated at the nonreducing end, but N-substitutions can vary among species and can be involved in the control of host specificity. Two strategies seem to be used by rhizobia (see Figure 1). In most species the N-atom is acylated with a common fatty acid and can carry a methyl group as a second substitution. In a few *Rhizobium* species, NFs are N-acylated with specific polyunsaturated fatty acids that are synthesized under the control of *nod* genes.

N-DEACETYLATION AT THE NONREDUCING END   NodB is homologous to a fungal chitin deacetylase which requires a substrate of at least four GlcNAc residues for activity (50). Experiments carried out in vivo, by growing bacterial cells containing various combinations of *nodABC* genes in the presence of radioactive precursors, indicate that NodB is a specific deacetylase (33, 51). Purified *R. meliloti* NodB protein has been shown to deacetylate the terminal nonreducing N-acetyl glucosamine residue of chitooligosaccharides (52). The monosaccharide N-acetyl glucosamine is not deacetylated by NodB, but chitooligosaccharides with two to six N-acetylglucosamines are deacetylated (52, 53). The resulting free amino group at the nonreducing terminus can then be acylated and, in some species, methylated.

N-METHYLATION   The presence of an N-methyl group is reported in most rhizobial species studied so far (see Table 1). Genetic gain-of-function evidence indicates that *nodS* is responsible for N-methylation. The introduction of the *nodS* gene from *R*. sp. NGR234 into *R. fredii* USDA257 results in the production of N-methylated NFs (54, 55). NodS proteins have homologies to S-adenosylmethionine-utilizing methyltransferases (56) and in *A. caulinodans*

NodS can be cross-linked to radiolabelled S-adenosylmethionine (56). In cell extracts, in vitro methylation is observed with chitooligosaccharides that have polymerization degrees from 2 to 6, but not with the glucosamine monosaccharide (57). The study of various combinations of *A. caulinodans nodABCS* genes in *E. coli* indicates that NodS uses as substrates chitooligosaccharides deacetylated at the nonreducing end by NodB (48). The *A. caulinodans nodS* gene was fused to a glutathione gene, expressed in *E. coli,* and the fusion protein purified. This protein can methylate end deacetylated chitooligosaccharides, but cannot fully methylate acetylated and acylated chitooligosaccharides (57). These experiments indicate that NodS is a methyl transferase and suggest that N-methylation precedes acylation (48, 57).

FATTY ACID SUBSTITUENTS   NFs from most rhizobial species, including all those that are N-methylated, are acylated with common fatty acids (Table 3). The fatty acyl composition reflects that of membrane lipids: The major constituent is *cis*-vaccenic acid (Δ11-C18:1) and minor constituents include palmitic acid (C16:0), palmitoleic acid (C16:1), and stearic acid (C18:0) (58–60).

Particular polyunsaturated fatty acids, with *trans* double bonds conjugated to the carbonyl group, have been found in NFs from two *Rhizobium* species (see Table 3). The *nodFE* genes are involved in the synthesis of the polyunsaturated acyl groups in *R. leguminosarum* biovars *viciae* (38) and *trifolii* (62) and in *R. meliloti* (61). NodF is homologous to acyl carrier proteins (63) and carries a 4'-phosphopantetheine group bound to serine residue 45 (64, 65). This serine residue is essential for the biological function of NodF, indicating the crucial importance of the prosthetic group (65). NodE has homology to various β-ketoacyl synthases (66). In *R. meliloti* the *nodFE* genes are involved in alfalfa infection and specify the synthesis of unsaturated C16 acyl groups (61, 67, 68). The introduction of the *nodFE* genes from *R. l.* bv. *viciae* into a *R. meliloti nodFE* mutant leads to the formation of C18 unsaturated acyl groups (61).

The *nodE* gene is a major determinant of the difference in host specificity between the two *R. leguminosarum* biovars *viciae* and *trifolii* (38). A 198–amino acid central domain, containing only 44 nonconserved amino acid residues, determines the host specificity of the NodE protein (69). The biovar *viciae* NodE protein specifies the synthesis of the C18 polyunsaturated acid with the *cis* double bond in position 11 and allows nodulation of vetch and poor nodulation of clover. The biovar *trifolii* NodE determines the synthesis of C18/C20 polyunsaturated acids with no *cis* double bond and allows no nodulation of vetch and abundant nodulation of clover (69). Thus, the main difference between the *nodE*-determined lipo-chitooligosaccharides of biovars *viciae* and *trifolii* appears to be the length of the carbon chain and the presence or absence of the *cis* double bond, which result in different hydrophobicities.

Table 3 Fatty acyl substituents of Nod factors

Metabolic pathway	Fatty acids	*Rhizobium* species[a]	Selected references
Membrane lipids	C18:1 (Δ11Z), C18:0, C16:1 (Δ9Z), C16:0	All	38,62,74,80
Lipid A precursors?	(ω-1) OH C18 to C26	Rm	61
NodFE	C16:2 (Δ 2E,9Z), C16:3 (Δ 2E, 4E, 9Z)	Rm	22,25
	C18:4 (Δ 2E, 4E, 6E, 11Z)	Rlv	38
	C18:3 (Δ 2E, 4E, 6E)	Rlt	62
	C20:3 (Δ 2E, 4E, 6E), C20:4 (Δ 2E, 4E, 6E, 8E)		

[a]For abbreviations see Table 2.

However, how the central domain of NodE controls both the *trans* polyunsaturation conjugated to the carbonyl group and the absence of a *cis* double bond in positions 11 or 13 is not clear.

After induction of the *nodFE* genes in *R. l. viciae*, even in the absence of the *nodABC* genes, the C18:4 fatty acid is synthesized, indicating that the synthesis of the polyunsaturated fatty acids is completed before they are linked to the chitooligosaccharide backbone (70). C18:4 fatty acid is also found bound to the *sn*-2 position of all major phospholipids, suggesting a link between the Nod factor and phospholipid pathways and the intriguing possibility that such phospholipids, containing *nodFE*-derived fatty acids, might have a signal function of their own (70).

R. meliloti NFs can be acylated by a series of C18 to C26 (ω-1)-hydroxylated fatty acids (61) that are likely to be precursors of the C28 (ω-1)-hydroxylated acyl group present in lipid A (71, 72). The synthesis of NFs containing these fatty acids is under the control of the regulatory *nodD3* and *syrM* genes (72). C18 to C26 (ω-1)-hydroxylated fatty acids and *nodFE*-derived fatty acids are not found in lipid A, even in the absence of chitooligosaccharide acceptor molecules (in a *nodC* mutant), indicating that the N-acyl transferases involved in the amidification of the glucosamine residues of NFs and lipid A have different specificities (70, 72).

N-ACYL TRANSFER   Cell extracts of *R. meliloti* overproducing NodA convert a radiolabelled chitotetraose deacetylated at the nonreducing terminus by purified NodB into hydrophobic compounds that comigrate with NFs and exhibit NF biological activity (73). Similar results have been obtained with permeabilized *E. coli* cells expressing combinations of *R. meliloti nodA* and *nodB* genes (51). The acylation activity is sensitive to chitooligosaccharide chain length, with tetramers serving as a better acceptor than dimers and trimers (51). NodA also exhibits specificity for the acyl group. In *R. meliloti*, replacement of the *nodA* gene by the *R. tropici nodA* allele results in the production of NFs that are acylated by vaccenic acid (C18:1), the acyl substituent of *R. tropici* NFs, and not by C16 polyunsaturated and (ω-1)-hydroxylated acyl groups that normally substitute *R. meliloti* NFs (49). All these data show that NodA is required for the N-acylation of NFs and support the hypothesis that NodA is an N-acyl transferase.

O-Substitutions at the Nonreducing End

O-ACETYLATION   An O-acetyl group has been found at the C-6 position in the NFs of *R. meliloti* (23), *R. l. viciae* (38), and *R. l. trifolii* (62, 74). NodL protein has sequence similarity to bacterial acetyl transferases (75, 76), and mutations in the *nodL* gene result in the production of NFs that are not O-acetylated (38, 77). Purified NodL protein has transacetylating activity in vitro with acetyl-CoA as the acetyl donor (78). NodL specifically substitutes one O-acetyl group at C-6 of the terminal nonreducing sugar. Chitooligosaccharides can be acetylated, showing that the presence of the N-acyl chain is not required. Tetramers and pentamers of chitin and chitosan are more efficiently acylated than dimers and trimers (78). Quantitative substrate specificity studies have shown that chitin oligomers with a free amino group on the nonreducing terminal residue are the preferred acceptor molecules, indicating that de-N-acylated chitin oligosaccharides produced by the NodC and NodB enzymes might be the in vivo substrates for NodL (79). An O-acetyl group has also been found at C-6 in some strains of *B. japonicum* and *B. elkanii* (80).

O-CARBAMOYLATION   In contrast to O-acetylation, O-carbamoylation is observed not only at C-6 but also at C-4 and possibly C-3 positions. In *R.* sp. NGR234 most NFs are doubly carbamoylated with one substitution shown to be at C-6 and the other presumed to be at either C-3 or C-4 (27, 54). NodU has high sequence similarity with a *Nocardia* protein involved in the biosynthesis of carbamoylated β-lactam antibiotics (54, 81). The introduction of the *nodU* gene of *R.* sp. NGR234 into *R. fredii* USDA257 renders this bacterium able to produce 6-O-carbamoylated NFs (54). Mutation of *nodU* in *R.* sp. NGR234 decreases but does not abolish NF carbamoylation and suppresses the formation of bis-carbamoylated species (54). These results suggest that in NGR234, *nodU* is involved in 6-O-carbamoylation and that another gene controls O-carbamoylation at positions 3 or 4. An NGR234 *nodU* mutant has lost the ability to nodulate *Leucaena,* and the transfer of an active *nodU* gene into the soybean symbiont *R. fredii* USDA257 extends its host range to *Leucaena*, indicating that 6-O-carbamoylation of NFs is required for the specific nodulation of certain hosts (54). In *A. caulinodans,* the O-carbamoyl group is located at C-6 (28), and introduction into *E. coli* of the *nodU* gene, in addition to *nodABCS,* results in the production of carbamoylated NFs, further evidence that *nodU* controls 6-O-carbamoylation (81).

O-Substitutions at the Reducing End

Substitutions at the reducing end are mostly located on C-6, are varied, and are extremely important for host specificity.

6-O-SULFATION   To date, glucosamine substitution with sulfate, which causes a striking change in NF charge, has only been found in *R. meliloti, R. tropici, R.* sp. GRH2 and *S. teranga* (Table 1). In *R. meliloti* all NFs are sulfated, and three genes, *nodPQ* and *nodH,* have been shown to be involved in both NF sulfation and host specificity of nodulation (22, 24). *nodH* mutants produce nonsulfated factors and exhibit a shift in their host range (24). They have lost the ability to infect and nodulate alfalfa, a homologous host, and have acquired the ability to infect and nodulate vetch, a nonhost. *nodPQ* mutants produce a mixture of sulfated and nonsulfated NFs and infect both alfalfa and vetch (24). This leakiness in sulfation is due to the presence of a copy of *nodPQ*-homologs in the *R. meliloti* genome (82, 83), and mutants altered in both copies only produce nonsulfated factors (24). Purified sulfated factors are active on alfalfa, whereas nonsulfated factors are active on vetch (24).

Sulfation requires the synthesis of activated forms of sulfate. In *E. coli* two enzymes are involved in the formation of ATP-derived forms of sulfate, adenosine 5′-phosphosulfate (APS) and 3′-phosphoadenosine 5′-phosphosulfate (PAPS). CysN and CysD form an ATP-sulfurylase and CysC is an APS kinase. In the presence of ATP, these enzymes catalyse the formation of APS and PAPS, respectively (84). NodP is highly homologous to CysD, and NodQ to CysN and CysC (85). NodP and NodQ have been shown to exhibit both ATP sulfurylase and APS kinase activities and to form a multifunctional protein complex (85, 86). NodQ has a GTP-binding site (82, 87), and GTP is required for the activity of this sulfate activation complex (85). A possible function for the *nodPQ* homologs has recently been proposed (59). Lipopolysaccharides of *R. meliloti* have also been found to be sulfated (59). Their sulfation is *nodPQ*-dependent but flavonoid-independent and thus could be under the control of the *nodPQ* homologs, whose transcription is not flavonoid-dependent (83). *R. meliloti* double mutants in *nodPQ* and the *nodPQ* homologs are not auxotrophic for cysteine, indicating the existence of a third locus involved in the synthesis of activated sulfate for sulfur-containing intermediary metabolites (83). Indeed such a third locus, called *saa* (for sulfur amino acid) has been identified (83). Thus, in *R. meliloti* three pathways exist for synthesizing activated sulfate. Two can complement each other, at least partially; those involved in the sulfation of Nod factors and lipopolysaccharides. The third, involved in the "housekeeping" of sulfur metabolism, is independent of the other two.

NodH has homology with mammal sulfotransferases (24). In the presence of PAPS, partially purified NodH prepared from *R. meliloti* or *E. coli* transfers the sulfate group from PAPS to nonsulfated NFs or chitotetraose (51, 88–90). Kinetic analysis with NFs, chitooligosaccharides, and their deacetylated derivatives reveals that NFs are the preferred substrates and that tetrameric NFs are better substrates than trimers or pentamers (89, 90). These data suggest that the core lipo-chitooligosaccharide structure must be synthesized prior to its host-specific modification with a sulfate group.

R. tropici produces a mixture of 6-O-sulfated and nonsulfated NFs, and this ability likely contributes to its broad host range (31). The introduction of *R. meliloti nodPQ* genes into *R. tropici* results in the complete sulfation of NFs, whereas the introduction of *nodH* has no effect, suggesting that the partial sulfation is due to a limited synthesis of activated sulfate (91). The introduction of *R. meliloti nodPQ* genes also results in the production of a mixture of methylated and nonmethylated factors, suggesting a link between sulfation and S-adenosylmethionine-dependent methylation pathways, with competition between these two pathways for the limited pool of activated sulfate available for NF synthesis (91).

O-ACETYLATION   This rare modification has only been observed in a strain of *R. l. viciae* having a particular host range. Certain strains of *R. l. viciae* can efficiently nodulate varieties of pea such as cv. Afghanistan (92, 93). A representative strain (TOM) has been shown to have an additional gene (*nodX*) that is required to overcome the resistance to nodulation of cv. Afghanistan (92). TOM strain makes NFs similar to those produced by other *R. l. viciae* strains, but *nodX* specifies the synthesis of additional NFs O-acetylated on C-6 of the reducing glucosamine residue (92). Pentameric NFs have this extra-acetylation whereas tetrameric factors do not (92). TOM strain thus possesses two NF O-acetyltransferases, NodL and NodX. Although NodL and NodX both O-acetylate the same carbon (C-6) of glucosamine residues, there is remarkably little homology between these two enzymes which belong to two distinct acetyl transferase families (75, 94).

O-GLYCOSYLATION   Two types of glycosylation have been reported: fucosylation and arabinosylation. Many rhizobial species have a fucosyl substitution and this glycosylation allows further possibilities for modifications. Very little is known about the genetics and biochemistry of fucosylation. In *B. japonicum*, mutations in *nodZ* result in both the production of Nod factors lacking the 2-O-methyl fucosyl residue and defective nodulation of siratro (95).

O-arabinosylation of NFs has only been observed in rhizobia able to nodulate stems and roots of *Sesbania rostrata* (28) and in *S. saheli* (J Lorquin, personal communication). In *A. caulinodans,* a NodZ protein, highly homologous to NodZ of *B. japonicum,* is involved in arabinosylation, suggesting that proteins of the same family might be involved in the O-glycosylation of the reducing glucosamine residue with two different glycosyl residues (81). NolK may play a role in glycosylation as it is homologous to NAD/NADP-binding sugar epimerases/dehydrogenases (96). Recently, *S. saheli* bv. *sesbaniae* strains have been shown to make NFs that are both arabinosylated and fucosylated (J Lorquin, personal communication).

1-O-GLYCEROL SUBSTITUTION   There are rare reports of substitution on the anomeric carbon. In *B. elkanii,* minor NFs have the "reducing" glucosamine glycosidically linked to glycerol (80). In *B. japonicum, nolO* mutants produce some derivatives that are not 2-O-methyl fucosylated but are glycosidically linked to glycerol (97). The fact that glycerol-containing NFs are minor compounds in a wild-type bacterium and can be found in a mutant altered in NF synthesis suggests that they may be biosynthetic intermediates (47).

Nod Factor Transport

Studies comparing the affinity of Nod enzymes for various possible synthesis intermediates, in addition to molecular genetics and NF chemistry, have led to the proposal that the NF biosynthetic pathway involves a sequence of enzymatic steps (Figure 1). Protein sequence analysis and use of specific antibodies indicate that NodC and NodE are located in the inner membrane (46, 98), NodA is cytosolic but partly associated with the inner membrane (99), and NodF, NodB, and NodL are cytosolic (63, 78, 100). Nod enzymes are thus distributed between the cytosol and the inner membrane, and NF synthesis is likely to occur in enzyme complexes located at the interface between the cytoplasmic membrane and the cytosol.

nodIJ genes have been found in all genetically studied rhizobial species. NodI is related to a large family of proteins having an ATP-binding cassette (ABC transporters) (101) and is associated with the inner membrane (102). NodJ is a very hydrophobic membrane protein, possibly an integral membrane protein (101). NodI and NodJ proteins have strong similarities with pairs of bacterial proteins that are involved in polysaccharide secretion (103). In *R. l. trifolii,* NF excretion is dependent on both *nodI* and *nodJ* (104). *nodIJ* mutants do not accumulate NFs in the external medium, the outer membrane, or the periplasmic space, and they nodulate clover very poorly (104). However, in other species such as *R. l. viciae,* mutations in *nodIJ* have moderate effects on both NF excretion and the symbiotic phenotype, suggesting the existence of alternative transport systems (105).

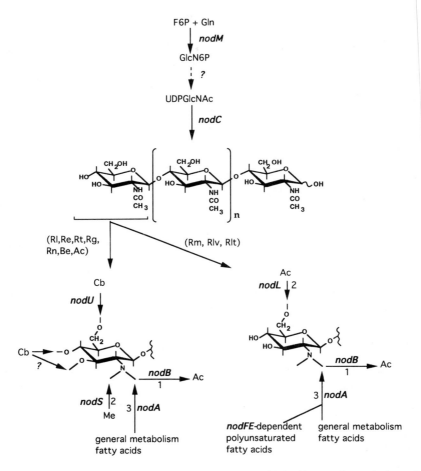

Figure 1   Biosynthesis of the Nod factors: chito-oligosaccharide backbone and nonreducing end. The two strategies used by rhizobia to substitute the nonreducing end are shown with the rhizobial species in which they occur. The numbers indicate the order of the reactions. The bacterial genes encoding the various biosynthetic enzymes are indicated in italics. Abbreviations: Gln: glutamine; F6P: fructose 6 phosphate; GlcN6P: glucosamine 6 phosphate; UDPGlcNAc: uridyl di-phosphate N-Acetyl glucosamine; Ac: acetate; Me: methyl; Cb: carbamoyl; for rhizobial species abbreviations see Table 2.

The *nodT* gene, which has been identified in *R. l. viciae* and *R. l. trifolii*, encodes an outer membrane protein similar to bacterial outer membrane proteins that are associated with ABC transport complexes (106). One hypothesis holds that NodT, together with NodI and NodJ, could constitute a secretion system that allows NFs to be secreted directly across the inner and outer membranes without periplasmic intermediates (107).

In *R. meliloti*, four nodulation genes, *nolFGHI* (35), code for proteins that show similarities with two families of integral membrane proteins that can function together (108). The proposal has been made that NolGHI are cytoplasmic membrane transport proteins that function together with the NolF membrane fusion protein to allow transport of NFs across both the inner and outer membranes (108). In rhizobia that produce NFs acylated with particular polyunsaturated fatty acids such specific transport systems could protect NFs from having their fatty acids shuffled and from translipidation with membrane lipid components (59).

After transport, the proportion of NFs that is excreted to the medium and that which remains bound to the bacterial membrane seems to vary among species. For example, in *R. l. trifolii*, NFs accumulate primarily in membranes (74) and are present at very low concentrations in the culture supernatant (62). In contrast, in *R. sp.* NGR234, labelling experiments indicate that more than 90% of the label is in excreted NFs (30). The amphiphilic nature of NFs could, in fact, allow two possible mechanisms of signaling to the plant: at a distance with excreted diffusible molecules and by surface interaction with molecules associated with the bacterial membrane (109).

Nod Factor Structure and Host Specificity

Several lines of evidence have shown that the way most structural nodulation genes control host-specificity of infection and nodulation is by determining the structure of NFs. Loss- and gain-of-function experiments have shown that in *R. meliloti* the three major NF substitutions are under the control of the major host-range genes and are involved in alfalfa infection (24, 49, 67, 77). Biological activity of purified NFs parallels the symbiotic behavior of the corresponding mutants. In *R. leguminosarum,* the role of *nodE* in the control of host specificity and in the synthesis of polyunsaturated fatty acids of various carbon-chain length and degree of unsaturation is well documented (62, 98), as is the role of *nodX* which specifies extra NF O-acetylation and the ability to nodulate new pea cultivars (92). In *B. japonicum, nodZ* mutants both make NFs that lack methyl-fucose and have defective nodulation of siratro (95). The introduction of either active *nodS* or *nodU* genes into *R. fredii* extends its host range to *Leucaena* (54).

Systematic studies of NF structure in various species have revealed a correlation between the type of NF produced and host range, but not with the rhizobial taxonomical position (110). For example, *B. japonicum* and *R. fredii* are soybean symbionts that are very distant genetically but produce similar fucosylated nonsulfated NFs. *A. caulinodans* and *S. saheli* strains that nodulate roots and stems of *Sesbania rostrata* are very distant genetically but both produce arabinosylated NFs. *R. loti* and *R. etli* are two clearly distinct species and have distinct host ranges, *Lotus* and *Phaseolus*, respectively (111). Surprisingly, NFs from representative strains of these two species were found to be similar (with O-4 acetylated methyl-fucose) (111, 112). *nodD* regulatory genes have been shown to be the major host-range determinants distinguishing these two species: Introduction of a flavonoid-independent *nodD* into *R. etli* and *R. loti* renders these species able to nodulate both *Lotus* and bean (111).

The stringency for NF structural requirements varies widely among plants. For example, efficient infection and nodulation of alfalfa requires rhizobia that produce NFs with three precise substitutions (24, 67, 77). All NFs from *R. meliloti* isolates, even those from varied geographical origins, have these same three types of substitutions (72). Similarly, *Sesbania rostrata* is only nodulated by rhizobia that produce arabinosylated NFs (28) (J Lorquin, personal communication). In contrast, *Phaseolus* can be nodulated by various *Rhizobium* species, including *R. etli* and *R. tropici*, that produce quite distinct NFs, acetyl-fucosylated and sulfated respectively, on their reducing moiety (31, 112).

Mechanisms Ensuring Signaling and Host-Range Variation

NF STRUCTURAL VARIATION BETWEEN RHIZOBIAL SPECIES   Regulatory *nod* genes play a key role in the control of NF synthesis and host range (see 9, 18). NodD proteins in various species are activated or inactivated by various sets of plant signals and control the level of NF synthesis and transport according to plant exudates. Quantitative variation in NF production seems to influence host range (15, 113). Most rhizobia have multiple copies of *nodD,* and this allelic variation permits better modulation of the level of *nod* gene expression in response to the diversity of plant and environmental signals.

Variation of NF structure among species involves diverse genetic mechanisms affecting mainly structural *nod* genes. Each species or biovar contains a combination of nonallelic genes that are present in certain species, absent in others (such as *nodH, nodL, nodX , nodSU,* and *nodZ*), and control specific substitutions on the chitooligosaccharide backbone. Another mechanism of nonallelic variation is observed in some strains in which certain genes can be present but are inactive because of genetic rearrangements that have not been counter-selected by "recent" symbiotic interactions. For example, in some *R. fredii, B. elkanii,* and *B. japonicum* strains NFs are not N-methylated and O-carbamoylated, in spite of the presence of *nodSU* genes. These substitutions are apparently not required to nodulate soybean (80, 114–116). Although *nodABC* genes are present in all species, allelic variation enables them to control NF structural variation and host specificity. *nodC* is involved in the control of the length of the chitooligosaccharide backbone, and the proportions of tetra- and pentamers vary, depending on the species (49). Similarly, *nodA* is a host-range determinant that can specify N-acylation by various fatty acids in a species-specific way (49). *nodFE* genes are responsible for both nonallelic variation, since they can be present or absent, and allelic variation, because in various species and biovars they specify the synthesis of particular unsaturated fatty acids that are involved in host specificity (38, 61, 62).

NOD-FACTOR DIVERSITY WITHIN A STRAIN   All strains produce a mixture of NFs, and whether this diversity is important for broadening the host range or to diversify signaling to the same host during various steps of the symbiotic interaction is not known. Frequently NFs from a given strain have diverse N-acyl substitutions (38, 61, 80). This variation may be caused by the relatively low specificity of NodA in some species. Some strains with a broad host range produce a mixture of sulfated and nonsulfated NFs. In *R.* sp. NGR234, the methylfucose moiety is either sulfated or acetylated, but not both, a mechanism that ensures the production of both charged and neutral NFs (27). In *R. tropici,* the partial sulfation seems to be due to a limitation in the synthesis of activated sulfate and not to a limitation in the sulfate transfer onto the lipo-chitooligosaccharide (31, 91).

Recent findings show that regulatory *nod* genes can control not only quantitative but also qualitative adaptation of NF production to plant signals. In *R. meliloti*, the *nodD3* gene, which mediates bacterial responses to plant signals that differ from those involving *nodD1* and *nodD2,* is required for the production of a particular class of NFs that are N-acylated by a series of (ω-1) hydroxylated fatty acids (72). Such a mechanism allows a bacterial strain to vary NF production in the presence of various hosts, or during different steps of the symbiotic interaction.

STRUCTURAL VARIATIONS AT BOTH ENDS   Numerous rhizobial species possess *nodSU* genes and can produce N-methylated Nod factors that are acylated with common fatty acids. N-substitutions, being the same in all these strains, are unlikely to be involved in host specificity. In such species a limited degree of variation can be due to the lack of N-methylation (80, 114, 116). Rhizobia such as *R. meliloti, R. l. viciae,* and *R. l. trifolii,* which nodulate modern legumes of the Trifolieae and Vicieae tribes, have *nodFE* genes and synthesize polyunsaturated fatty acids with conjugated double bonds. In these fatty acids, the number and location of double bonds and the carbon-chain length are species specific (38, 61, 62). Evolution from N-methyl to conjugated double bonds seems to have enabled the contribution of N-substitutions in structural and host-range variation.

Two simple groups (acetyl and carbamoyl) can be O-substituted at the nonreducing end, allowing only a limited number of combinations (Figure 1).

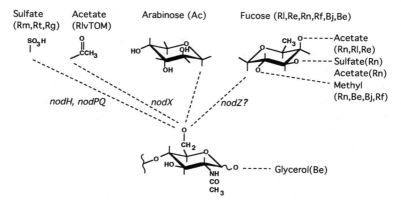

Figure 2 Biosynthesis of the Nod factors: reducing end. The various genes involved in the substitutions at the reducing terminal glucosamine residue and the rhizobial species in which they occur are indicated (for abbreviations see Table 2).

Steric constraints in the interaction between the plant NF-receptor(s) and the NF region that is carrying the fatty acid may not provide room for larger substituents such as glycosyl groups.

At the reducing end, the possibilities for structural variation are extensive because substituents can be sugars such as fucose and arabinose which can in turn be substituted (Figure 2). The case of *Sinorhizobium* strains nodulating stems of *Sesbania* is the most sophisticated in the sense that the reducing glucosamine is substituted by fucose and arabinose. Sulfation, a substitution that drastically changes the NF charge, has only been observed at the reducing end. If NF receptors of all legumes belong to the same family, the receptor domain interacting with the NF reducing end must have evolved to interact with a highly variable moiety of Nod signals.

Chemical and Enzymatic Synthesis

The first NFs to be chemically synthesized were the major *R. meliloti* factor with its three specific substitutions (sulfate, O-acetate and C16:2 (Δ2,9) fatty acid) and the nonsulfated or nonacetylated related LCO molecules (117). The sulfated compounds were found to be active in nodulation assays at a concentration range similar to that of the corresponding natural NFs (N Demont, F Maillet, JC Promé, J Dénarié, G Truchet, KC Nicolaou & JM Beau, in preparation). The major NF of *B. japonicum* has also been synthesized and exhibits biological activity (118, 119). These data confirm the validity of the proposed

NF structures and show that biological activity of natural NFs was not due to impurities.

R. meliloti NFs have also been synthesized using other strategies (120–122). The various strategies used have different advantages and limitations for the synthesis of NF derivatives and analogs (123). For example, methods introducing the N-acyl chain in the last steps facilitate the production of (*a*) a series of compounds with different acyl chains that can be used to analyze the role of the structure of the acyl chain in biological activity and host specificity (N Demont, M Maillet, JC Promé, J Dénarié, G Truchet, KC Nicolaou & JM Beau, in preparation), and (*b*) chain-labeled derivatives for studying NF-binding proteins in the search for putative NF receptors (124). Several *B. japonicum* NF derivatives that differ in the acyl chain, the presence or absence of 2-O-methylfucose, and the number of glucosamine residues have been synthesized for structure-function relationship studies (118, 125, 126).

Nod enzyme–enriched cell extracts and purified Nod enzymes have been used to synthesize chitooligosaccharides and lipo-chitooligosaccharides and modify them in specific ways (32, 51, 52, 79, 88, 89, 127). Several *nod* genes involved in the various steps of the NF synthesis pathways have been rapidly cloned and characterized, thanks to bacterial genetics. These genes could be used to facilitate the identification and cloning of homologous eukaryotic genes involved in the synthesis of biologically important glucosamine- or fucose-containing oligosaccharides. In the rapidly developing oligosaccharide biotechnology industry, bacteria containing *nod* genes could be used as sources of Nod enzymes to facilitate the controlled transfer of groups (acetyl, carbamyl, sulfate, methyl, fucosyl, arabinosyl, N-acetylglucosaminyl, etc) onto specific sites of oligosaccharidic substrates.

NOD FACTORS AS PLANT GROWTH REGULATORS

Purified NFs elicit in legume hosts "symbiotic" responses similar to those induced by bacteria themselves in the course of infection and nodulation. Responses are observed in different root tissues, epidermis, cortex, and stele and exhibit different structural requirements (Table 4).

Responses in Root Epidermis

The first contact between rhizospheric rhizobia and the host occurs on epidermal cells at the root surface. A number of legumes are infected via root hairs. The first visible sign of infection is root-hair deformation and curling, followed by the formation of an infection focus within the curl and the development of a tubular structure, the infection thread, which grows through the root-hair cell and into the root cortex where it ramifies (5, 7). All epidermal cells (trichoblasts and nontrichoblasts) in the root differentiation zone seem to be responsive to NFs.

DEVELOPMENTAL RESPONSES   Purified NFs, at nano- to picomolar concentrations, elicit differentiation of epidermal cells into root-hair cells in vetch (24, 38, 128). In the same range of concentrations, NFs elicit root-hair deformations in all tested host plants (22, 23, 25, 27, 28, 38, 80, 114, 116, 129). In contrast, plant hormones, hormone-like substances, and inhibitors of hormone action do not elicit root-hair deformations (129). This deformation starts by swelling, probably due to cytoskeletal reorganization, is detectable within 1 h, and results in clearly deformed root hairs within 3 h (130). Incubation with Nod factors for approximately 10 min is required to set the deformation process in motion (130).

Plant genes that are specifically expressed during early steps of symbiosis, infection, and nodule formation are called early nodulin (*ENOD*) genes (131, 132). *ENOD5* and *ENOD12* encode proline-rich proteins which are hypothesized to be cell-wall proteins involved in the infection process (131, 133). In the presence of specific rhizobia, *ENOD12* is rapidly expressed in epidermal cells prior to and during root hair emergence (134). Purified NFs elicit the same responses at concentrations as low as 10^{-12} or 10^{-13} M (135). NFs also induce the rapid and transient transcription of *rip1* (for *Rhizobium*-induced peroxidase) (136). Cytological studies have shown that NFs elicit the expression of these genes in all epidermal cells of the root differentiation region (135, 136) and that a direct contact between NFs and epidermal cells is required (135). Thus, it is likely that within the susceptible zone, all epidermal cells recognize NFs. The response of *ENOD12* and *rip1* to exogenously supplied NFs is restricted to the epidermis and occurs within a few hours after NF application (135, 136). NFs also induce transcription of *ENOD5* in the epidermis (137). Several plant genes are NF-activated in root-hair cells, such as *ENOD5* (138), *ENOD12* (135, 138), *rip1* (136), and a gene coding for a lipid transfer-like protein (139). Comparison of the epidermal *ENOD12* gene induction activity of purified NFs to that of *R. meliloti* mutants producing the corresponding modified NFs suggests that the physiological range of NF concentration may be from 10^{-10} to 10^{-12} M (135).

In this range of concentrations epidermal responses require the presence of an acyl chain, since chitooligosaccharides are inactive (38, 129, 130, 135). The presence of unsaturations in the acyl group, however, does not appear to be important (119, 128, 130, 135). Concerning O-substitutions, whereas the acetyl group at the nonreducing end has no significant effect (38, 128, 135), the presence or absence of sulfate or methyl fucose at the reducing end is extremely important (24, 25, 119, 135). The structural requirements are thus stringent at the reducing end and nonstringent at the nonreducing end (Table 4). The length of the chitooligosaccharide backbone is important with dimers and trimers being inactive (25, 130, 140). On alfalfa, tetrameric NFs are more active than pentamers, and the reverse is true for soybean (25, 119).

ELECTROPHYSIOLOGICAL RESPONSES   Epidermal cells, particularly root-hair cells, are directly in contact with the external medium and accessible for electrophysiological studies and direct microscopic observation. Addition of NFs causes a depolarization of the plasma membrane potential in root hairs and other epidermal cells (140–142). Proton efflux and changes in the flux of calcium have been observed in root hairs (143). Rearrangements of the actin filaments (143) and increased cytoplasmic streaming (130) indicate rapid cytoskeletal reorganization. These changes occur within 2 to 30 min after NF addition and may be part of a series of cellular events that lead to root-hair deformation, as NF structural requirements for both types of responses are similar (140, 142, 143). One difference however has been observed between electrophysiological and developmental root-hair responses: The presence of unsaturations in the acyl moiety is important for the elicitation of membrane depolarization (140). The secreted protein NodO of $R. l.$ bv. $viciae$ can complement the absence of polyunsaturations in NF acyl moiety (144, 145). NodO is a Ca^{2+}-binding protein that forms cation-selective channels allowing movement of monovalent cations across the membrane (145). One hypothesis holds that NodO either stimulates NF uptake or forms cation-specific channels that function synergistically with NFs for depolarization of the plasma membrane of leguminous plants (145).

INFECTION THREAD INITIATION   After being trapped in the root-hair curl, bacteria induce the formation of infection threads (ITs). NFs are also involved in IT initiation. In $R. meliloti$, $nodF$ and $nodL$ mutants, which produce NFs modified at the nonreducing end, show a strong decrease in the ability to induce IT initiation (67). Double $nodF/nodL$ mutants are completely unable to initiate IT formation and to penetrate into root hairs. However they induce multiple tip growth of the cell wall of both types of epidermal cells (67). These results indicate that NF structural requirements are more stringent for bacterial entry into the root hair and initiation of the infection process than for induction of developmental responses (Table 4). This finding has led to the hypothesis that at least two different NF receptors are present in the epidermis: (a) A "signaling receptor" involved in the induction of root-hair deformation and other developmental responses in all epidermal cells, and (b) an "entry receptor" which is activated only by molecules with very specific structures, located in developing root hairs and involved in bacterial entry and IT initiation (67). The involvement of two types of NF-recognition mechanisms is also suggested by studies of a pea variety carrying the $sym2$ gene. Strains of $R. l.$ bv. $viciae$ that lack the *nodX* gene and produce NFs without extra O-acetylation are able to induce root-hair deformation on this variety, but the ability to induce IT formation is strongly reduced (93, 146).

Evidence that specific NFs are keys for entry in legume roots via infection threads also comes from the following experiments. Purified NFs, prepared from the broad host range *R*. sp. NGR234 and supplied to a NF-production deficient *nodABC* mutant of this strain, restore the ability to penetrate, nodulate, and fix nitrogen on a host plant, *Vigna*. Moreover, these NFs allow a *R. fredii* strain to enter roots and fix nitrogen in *Calopogonium*, which is a nonhost of *R. fredii* (147). These complementation experiments require relatively high NF concentrations (10^{-6} to 10^{-7} M) and seem to work with plants that have determinate nodules such as *Vigna*, *Glycine*, and *Calopogonium* (147), but not with plants that have indeterminate nodules, such as alfalfa (67) and vetch (AAN van Brussel, personal communication).

PLANT PRODUCTS SECRETED IN THE RHIZOSPHERE   NFs are degraded by plant chitinases. Such enzymes are secreted in the rhizosphere and shorten the chitooligosaccharide backbone, generating acylated di- and trisaccharides that have drastically reduced biological activity (130, 148–150). Substitutions and length of the glucosamine backbone can influence the sensitivity of NF to chitinases (130, 148, 150). The preincubation of alfalfa roots with NFs at nanomolar concentrations induces their rapid degradation (150). These results indicate that the concentration of active NFs is likely to be determined by the balance between production and degradation.

NFs induce an increase in the level of specific flavonoids released in root exudates. NF structural requirements are stringent since, in vetch, substitutions with polyunsaturated C18:4 fatty acid and O-acetyl are required (38), and in soybean the absence of a sulfate group on the O-methylfucose (151, 152) is essential for flavonoid accumulation.

Cell Cycle Activation in Root Cortex

In the root cortex rhizobia realize two objectives: the induction of the formation of a niche in which to differentiate and fix nitrogen (the nodule) and the development of an infection apparatus to multiply and colonize the niche (the network of infection threads) (5, 7). In legumes that have indeterminate nodules, such as pea, vetch, and alfalfa, root-hair infection results in the activation of underlying outer cortical cells; nuclei move to a central position and radially aligned cytoplasmic bridges can be observed, paving the way for infection thread development. These structures are called pre-infection threads (128). In the inner cortex, at a distance from the infection thread tip, mitoses are induced, and cell division gives rise to a nodule primordium. Purified NFs elicit both the formation of pre-infection threads in the outer cortex and the induction of mitoses in the inner cortex (128, 153).

van Brussel et al (128) have proposed the following model. Rhizobia and NFs trigger activation of outer and inner cortical cells; in the outer cortex, reorganization of the cytoskeleton and cell wall deposits will not give rise to cell division but to the formation of pre-infection threads, whereas inner cortical cells will divide and reenter the cell cycle. Using molecular markers of the cell cycle in *in situ* hybridization experiments has provided direct evidence that Nod factors do activate the cell cycle in the outer and inner cortex and that the cell cycle seems to be G2-arrested in the outer cortex, whereas a cyclin characteristic of mitosis is expressed in the inner cortex only (154). NFs have been reported to activate the cell cycle machinery in cell suspension cultures also, at nanomolar concentrations (155). All these results suggest that a major role of NFs is to activate the cell cycle. Both cytokinin (156) and compounds that block polar auxin transport (157) also induce cortical cell divisions in roots, suggesting that NFs may act by modifying the auxin/cytokinin balance (146, 158). For induction of pre-infection threads and mitosis, substitutions at both reducing (sulfate) and nonreducing (O-acetate and polyunsaturated fatty acids) ends are required (38, 67, 128, 153).

Responses in Stele

The stele consists of root vascular bundles surrounded by pericycle. The stele is important for nodule formation physiologically, because phytohormones and stele factor(s) required for nodule primordium initiation are transported in vascular bundles (146, 158, 159). The stele is also important developmentally, because the new vascular bundles that are formed in the developing nodule will ultimately be connected with root vasculature to ensure metabolic integration between the nodule and the rest of the plant. *R. meliloti* NFs not only trigger cell division in the inner cortex but also the formation, in deeper cell layers, of new vascular structures perpendicular to the root vascular bundles (153). The *ENOD40* gene is expressed very early during nodule initiation, and *in situ* hybridization analysis has shown that *ENOD40* mRNA accumulates first in the root pericycle within 8 hours and then in dividing cortical cells after a few days (154, 160). Nod factors elicit *ENOD40* expression (126, 132, 161; Yang & Bisseling, personal communication), and NF structural requirements for induction of *ENOD40* transcription in the pericycle are less stringent than for induction of cortical cell divisions. In alfalfa, NFs doubly modified at the nonreducing end are unable to elicit the formation of nodule primordia, but trigger pericycle *ENOD40* expression within 6 hours (WC Yang, F de Billy, G Truchet & T Bisseling, personal communication). In soybean, a rapid and transient increase of *ENOD40* transcription, probably located in the pericycle, is induced by LCOs that are unable to elicit the formation of nodule primordia and even, remarkably, by unsubstituted (nonacylated) chitin pentamers (126) (Table 4).

Nodule Primordia and Nodule Development

NFs elicit nodule primordium formation on all tested host plants (28, 38, 115, 147, 153, 162). On some hosts, such as vetch and pea (38, 128) or *Medicago truncatula* (P Gamas, J Cock & J Cullimore, personal communication), nodule primordia do not develop into genuine nodules. On many hosts, however, NFs or synthetic LCOs induce the formation of differentiated nodules that have the origin and anatomy of bacteria-induced nodules. This is the case for plants that have indeterminate nodules, such as alfalfa (153), or determinate nodules, such as *Phaseolus* (110), *Sesbania rostrata* (28), *Vigna*, *Macroptilium* (147), and soybean (115).

NF concentrations required to trigger nodule primordium or nodule formation are higher than those required to elicit root-hair deformation by a factor of at least 1000 (24, 28, 38, 115, 129, 153). Structural requirements are often more stringent than for root-hair deformation. For example, with *R. meliloti* and *R. l.* bv. *viciae* that possess *nodFE* and *nodL* genes, the presence of the specific polyunsaturated fatty acyl chain and of the O-acetyl substitutions are important (38, 67, 153). Use of synthetic LCOs carrying various fatty acyl chains has revealed that the presence of the double bond conjugated to the carbonyl group is important for alfalfa nodule induction, as is the carbon length of the fatty acid (N Demont, F Maillet, JC Promé, J Dénarié, G Truchet, KC Nicolaou & JM Beau, in preparation). In alfalfa and soybean, C-6 substitutions at the reducing end are essential (24, 119). The length of the chitooligosaccharide backbone is also important, tetramers being more efficient for alfalfa (25) and pentamers for soybean (119), and dimers and trimers having no activity. An interesting interdependence has been observed between length of the chitin oligomer backbone and C-6 substitution at the reducing end for soybean nodulation. In the presence of O-methylfucose, the pentamer is active and the tetramer inactive, and in its absence the reverse is true (119).

Nod Factor Perception and Transduction

NFs are released in the vicinity of host epidermal cells. Substitutions on the chitin backbone can differentially protect NFs from degradation by plant chitinases (148, 149). The fact that structural differences can be recognized by plants in rapid responses, within a few minutes, suggests that this mechanism is unlikely to play a major role in the specificity of NF recognition by leguminous hosts (140, 142). The elicitation of epidermal responses at extremely low concentrations by factors with specific structures suggests a lock and key type of recognition mechanism involving high-affinity NF receptors.

The fatty acid moiety might be needed to allow NF insertion into the lipid bilayer of the host membrane, permitting both lateral diffusion towards putative membrane receptors and orientation of NF molecules to facilitate ligand-receptor interactions. In most rhizobia, namely those that nodulate tropical legumes, the N-acyl chain is vaccenic acid, and the structure of the acyl moiety does not seem to be important for receptor activation (115). However, the structure of the N-acyl moiety is important in *Rhizobium* species that nodulate legumes of the tribes Trifolieae and Vicieae. They produce NFs acylated with particular unsaturated fatty acids, and the acyl chain structure is important for eliciting some plant responses (38, 67). We can hypothesize that NF receptors involved in recognizing these particular NF types have structural properties that allow subtle discrimination between different acyl moieties. In these plant species root-hair cells have different structural requirements for the acyl chain structure, for root-hair deformation, and for the initiation of infection thread formation, suggesting the involvement of at least two types of NF receptors (67, 93, 146). Root-hair cells are directly in contact with exogenously added NFs, and the intracellular transduction pathway(s) triggered by NF-receptor activation should be directly amenable to electrophysiological and pharmacological approaches.

A major challenge is identifying the putative NF receptors. Chemical synthesis of NFs and NF derivatives and enzymatic NF modifications have opened the way to the synthesis of radio- and photoaffinity-labelled ligands which should facilitate the identification and characterization of NF-binding proteins (51, 88, 89, 117, 122, 124, 127). A NF-binding site has been identified in a particulate fraction prepared from *Medicago truncatula* roots (124). In tomato cell suspension cultures, chitin oligomers and NFs can induce a rapid and transient extracellular alkalinisation, suggesting that some receptors could be activated by both types of ligands (149, 163, 164). This idea is also supported by the observation that both NFs and chitin oligomers can activate transcription of *ENOD40* in the pericycle of soybean (126). Future studies will tell whether receptors for NFs and chitin oligomers belong to the same family.

In both superficial (epidermis) and deep tissues (inner cortex and pericycle) both stringent and nonstringent responses are observed (Table 4), indicating that differences in structural requirements are unlikely to be simply due to varied diffusion properties of NFs and chitin fragments through root tissues. A reasonable hypothesis is that different mechanisms of NF perception and transduction are operating. For the responses triggered in the underlying tissues, cortex and stele, future studies will address the question of whether NFs are diffusing either apoplastically or symplastically to reach the target cells directly, or if secondary messengers are required for the intercellular transduction of NF signals. The availability of various plant bioassays showing relatively rapid responses, and the availability of techniques permitting intracellular signal delivery (143, 165), should make possible the deciphering of the various NF signal transduction pathways.

Lipo-Chitooligosaccharides: A New Class of Plant Growth Regulators?

The discovery of the chemical nature of NFs was a great surprise. The *nod* genes were previously believed to control cell divisions and nodule organo-genesis induction directly by specifying phytohormone synthesis (100). Why do rhizobia use lipo-chitooligosaccharides to trigger these various plant responses? Certain oligosaccharides prepared by degradation of microbial and plant cell walls have been shown to be involved in the regulation of growth, development, and defense responses in plants, and the term oligosaccharins has been proposed to designate such molecules with regulatory activities (166, 167). In addition to classical plant hormones (cytokinins, auxins, giberellins, ethylene) that play a general role in growth and development, plants may use the structural complexity of oligosaccharides to regulate in a more refined way important physiological processes (166). NFs are clearly growth regulators in legumes and constitute the first class of oligosaccharins for which coherent genetic, biochemical, and physiological evidence demonstrates their biological significance (24, 38, 153).

Rhizobium, Bradyrhizobium, Sinorhizobium, and *Azorhizobium* species, which are very distant genetically, all use lipo-chitooligosaccharides to trigger nodule development. Are these signaling molecules unique to rhizobia or do rhizobia mimic plant endogenous growth regulators (13, 168)? Several lines of evidence indicate that lipo-chitooligosaccharides are also able to elicit developmental responses in nonlegume plants. First, some *Bradyrhizobium* and *Rhizobium* strains are able to infect and nodulate species of *Parasponia*, members of the Ulmaceae family distant to Leguminosae (169, 170). Mutations in the *nodABC* genes of such strains suppress both the ability to nodulate *Parasponia* (171) and to synthesize NFs (27). Thus NFs can act as growth regulators and trigger cell divisions and nodule organogenesis in these non-legumes also. Second, arrested embryo development in a mutant cell line of carrot (Umbelliferae) can be rescued by NF supplied exogenously (172, 173). Third, expression of *nodAB* genes in transgenic tobacco (Solanaceae) changes leaf morphology as well as whole plant development (174), suggesting that substrate(s) for NodB and NodA enzymes, probably chitin oligomers, exist in plants and that their modification influences morphogenesis (174). Fourth, synthetic LCOs alleviate the requirement for auxin and cytokinin to sustain growth of cultured tobacco protoplasts, in which they activate an auxin-responsive promoter and the expression of a gene implicated in auxin action (127). LCOs containing C18:1 *trans*-fatty acid substituents were found to be

Table 4 Nod factor structural requirements for plant responses

Plant responses	Nod factor structure					References
	Number of glucosamine	O-substitutions at reducing end	N-acyl chain		O-substitutions at non-reducing end	
			Presence	Structure		
Epidermis						
Hair induction	nt[a]	+[b]	+	−	−	24,38
Hair deformation	+	+	+	−	−	24,25,119,130,38
enod12 expression	nt	+	+	−	−	135
Membrane depolarization	+	+	+	+	−	140,142
IT initiation	nt	+	+	+	+	67
Cortex						
Nodule primordium	+	+	+	+	+	24,25,38,67,119,153
Pericycle						
enod40 expression	nt	−	−	−	−	126

[a] nt = not tested.
[b] +: important for triggering plant responses.

clearly more active than their *cis*-analogs, showing that the structure of the fatty acyl moiety is important for activity in this nonlegume system (127). Taken together these data indicate that LCOs are able to act as plant growth regulators in nonlegumes also and that they may be related to an as yet unidentified family of endogenous plant signals.

From a more general point of view, recent findings are of particular interest. The *Xenopus* DG42 gene is only expressed at precise embryonic developmental stages (175), and its gene product has striking sequence similarities with NodC proteins (40). The DG42 protein has been found to catalyze the synthesis of chitin oligosaccharides, suggesting that such chitin oligomer derivatives could act as signals in early vertebrate embryogenesis (43). Future research will tell whether the rhizobial NFs are just a subset of an important new class of widely distributed LCO signaling molecules.

ACKNOWLEDGMENTS

We thank our colleagues for sending us reprints and preprints and J Cullimore, C Gough and E Greene for reviewing this manuscript. Work in our laboratories is supported by grants from the Human Frontiers Science Programme and the European Union PTP Programme.

Literature Cited

1. Martinez-Romero E. 1994. *Plant Soil* 161:11–20
2. De Lajudie P, Willems A, Pot B, Dewettinck D, Maestrojuan G, et al. 1994. *Int. J. Syst. Bacteriol.* 44:715–33
3. Young JPW, Johnston AWB. 1989. *Trends Ecol. Evol.* 4:341–49
4. Fellay R, Rochepeau P, Relic B, Broughton WJ. 1995. In *Pathogenesis and Host Specificity in Plant Diseases. Histopathological, Biochemical, Genetic and Molecular Bases,* ed. US Singh, RP Singh, K Kohmoto, pp. 199–220. Oxford: Pergamon/Elsevier Sci.
5. Brewin NJ. 1991. *Annu. Rev. Cell Biol.* 7:191–226
6. Caetano-Anollés G, Gresshoff PM. 1991. *Annu. Rev. Microbiol.* 45:345–82
7. Kijne JW. 1992. In *Biological Nitrogen Fixation,* ed. G Stacey, RH Burris, HJ Evans, pp. 349–98. New York: Chapman & Hall
8. Dénarié J, Roche P. 1992. See Ref. 184, pp. 295–324
9. Dénarié J, Debellé F, Rosenberg C. 1992. *Annu. Rev. Microbiol.* 46:497–531
10. Göttfert M. 1993. FEMS *Microbiol. Rev.* 104:39–63
11. van Rhijn P, Vanderleyden J. 1995. *Microbiol. Rev.* 59:124–42
12. Fisher RF, Long SR. 1992. *Nature* 357:655–60
13. Dénarié J, Cullimore J. 1993. *Cell* 74:951–54
14. Spaink HP. 1992. *Plant Mol. Biol.* 20:977–86
15. Schultze M, Kondorosi E, Ratet P, Buiré M, Kondorosi A. 1994. *Int. Rev. Cytol.* 156:1–75
16. Kondorosi A. 1992. See Ref. 184, pp. 325–40
17. Stacey G. 1995. *FEMS Microbiol. Lett.* 127:1–9
18. Spaink HP. 1995. *Annu. Rev. Phytopathol.* 33:345–68

19. van Brussel AAN, Zaat SAJ, Canter-Cremers HCJ, Wijffelman CA, Pees E, et al. 1986. *J. Bacteriol.* 165:517–22
20. Faucher C, Maillet F, Vasse J, Rosenberg C, van Brussel AAN, et al. 1988. *J. Bacteriol.* 170:5489–99
21. Banfalvi Z, Kondorosi A. 1989. *Plant Mol. Biol.* 13:1–12
22. Lerouge P, Roche P, Faucher C, Maillet F, Truchet G, et al. 1990. *Nature* 344: 781–84
23. Roche P, Lerouge P, Ponthus C, Promé JC. 1991. *J. Biol. Chem.* 266:10933–40
24. Roche P, Debellé F, Maillet F, Lerouge P, Faucher C, et al. 1991. *Cell* 67:1131–43
25. Schultze M, Quiclet-Sire B, Kondorosi E, Virelizier H, Glushka JN, et al. 1992. *Proc. Natl. Acad. Sci. USA* 89:192–96
26. Lerouge P. 1994. *Glycobiology* 4:127–34
27. Price NPJ, Relic B, Talmont E, Lewin A, Promé D, et al. 1992. *Mol. Microbiol.* 6:3575–84
28. Mergaert P, Van Montagu M, Promé JC, Holsters M. 1993. *Proc. Natl. Acad. Sci. USA* 90:1551–55
29. Spaink HP, Aarts A, Stacey G, Bloemberg GV, Lugtenberg BJJ, Kennedy EP. 1992. *Mol. Plant-Microbe Interact.* 5: 72–80
30. Price NPJ, Carlson RW. 1995. *Glycobiology* 5:233–42
31. Poupot R, Martinez-Romero E, Promé JC. 1993. *Biochemistry* 32:10430–35
32. Geremia RA, Mergaert P, Geelen D, Van Montagu M, Holsters M. 1994. *Proc. Natl. Acad. Sci. USA* 91:2669–73
33. Spaink HP, Wijfjes AHM, van der Drift KMGM, Haverkamp J, Thomas-Oates JE, Lugtenberg BJJ. 1994. *Mol. Microbiol.* 13:821–31
34. Inon de Iannino N, Pueppke SG, Ugalde RA. 1995. *Mol. Plant-Microbe Interact.* 8:292–301
35. Baev N, Endre G, Petrovics G, Banfalvi Z, Kondorosi A. 1991. *Mol. Gen. Genet.* 228:113–24
36. Baev N, Schultze M, Barlier I, Ha DC, Virelizier H, et al. 1992. *J. Bacteriol.* 174:7555–65
37. Marie C, Barny MA, Downie JA. 1992. *Mol. Microbiol.* 6:843–51
38. Spaink HP, Sheeley DM, van Brussel AAN, Glushka J, York WS, et al. 1991. *Nature* 354:125–30
39. Atkinson EM, Long SR. 1992. *Mol. Plant-Microbe Interact.* 5:439–42
40. Bulawa CE. 1992. *Mol. Cell. Biol.* 12: 1764–76
41. Debellé F, Rosenberg C, Dénarié J. 1992. *Mol. Plant-Microbe Interact.* 5: 443–46
42. Dougherty BA, van de Rijn I. 1994. *J. Biol. Chem.* 269:169–75
43. Semino CE, Robbins PW. 1995. *Proc. Natl. Acad. Sci. USA* 92:3498–501
44. Saxena IM, Brown RM, Fevre M, Geremia RA, Henrissat B. 1995. *J. Bacteriol.* 177:1419–24
45. Kamst E, van der Drift KMGM, Thomas-Oates JE, Lugtenberg BJJ, Spaink HP. 1995. *J. Bacteriol.* 177: 6282–6285
46. Barny MA, Downie JA. 1993. *Mol. Plant-Microbe Interact.* 6:669–72
47. Carlson RW, Price NPJ, Stacey G. 1994. *Mol. Plant-Microbe Interact.* 7:684–95
48. Mergaert P, D'Haeze W, Geelen D, Promé D, Van Montagu M, et al. 1995. *J. Biol. Chem.* 270:29217–23
49. Debellé F, Roche P, Plazanet C, Maillet F, Pujol C, et al. 1995. In *Nitrogen Fixation: Fundamentals and Applications,* ed. IA Tikhonovich, VI Romonov, NA Provorov, WE Newton, pp. 275–80. Netherlands: Kluwer Academic.
50. Kafetzopoulos D, Thireos G, Vournakis JN, Bouriotis V. 1993. *Proc. Natl. Acad. Sci. USA* 90:8005–8
51. Atkinson EM, Palcic MM, Hindsgaul O, Long SR. 1994. *Proc. Natl. Acad. Sci. USA* 91:8418–22
52. John M, Röhrig H, Schmidt J, Wieneke U, Schell J. 1993. *Proc. Natl. Acad. Sci. USA* 90:625–29
53. Röhrig H, Schmidt J, Wieneke U, Kondorosi E, Barlier I, et al. 1994. See Ref. 185, pp. 76–80
54. Jabbouri S, Fellay R, Talmont F, Kamalaprija P, Burger U, et al. 1995. *J. Biol. Chem.* 270:22968–73
55. Lewin A, Cervantès E, Wong CH, Broughton WJ. 1990. *Mol. Plant-Microbe Interact.* 3:317–26
56. Geelen D, Mergaert P, Geremia RA, Goormachtig S, Van Montagu M, Holsters M. 1993. *Mol. Microbiol.* 9:145–54
57. Geelen D, Leyman B, Mergaert P, Klarskov K, Van Montagu M, et al. 1995. *Mol. Microbiol.* 17:387–97
58. Hubac C, Guerrier D, Ferran J, Tremolières A, Kondorosi A. 1992. *J. Gen. Microbiol.* 138:1973–83
59. Cedergren RA, Lee JG, Ross KL, Hollingsworth RI. 1995. *Biochemistry* 34:4467–77
60. Orgambide GG, Huang ZH, Gage DA, Dazzo FB. 1993. *Lipids* 28:975–79
61. Demont N, Debellé F, Aurelle H, Dénarié J, Promé JC. 1993. *J. Biol. Chem.* 268:20134–42
62. Spaink HP, Bloemberg GV, van Brussel AAN, Lugtenberg BJJ, van der Drift KMGM, et al. 1995. *Mol. Plant-Microbe Interact.* 8:155–64

63. Shearman CA, Rossen L, Johnston AWB, Downie JA. 1986. *EMBO J.* 5: 647–52
64. Geiger O, Spaink HP, Kennedy EP. 1991. *J. Bacteriol.* 173:2872–78
65. Ritsema T, Geiger O, van Dillewijn P, Lugtenberg BJJ, Spaink HP. 1994. *J. Bacteriol.* 176:7740–43
66. Bibb MJ, Biro S, Motamedi H, Collins JF, Hutchinson CR. 1989. *EMBO J.* 9:2727–36
67. Ardourel M, Demont N, Debellé F, Maillet F, De Billy F, et al. 1994. *Plant Cell* 6:1357–74
68. Debellé F, Rosenberg C, Vasse J, Maillet F, Martinez E, et al. 1986. *J. Bacteriol.* 168:1075–86
69. Bloemberg GV, Kamst E, Harteveld M, van der Drift KMGM, Haverkamp J, et al. 1995. *Mol. Microbiol.* 16:1123–36
70. Geiger O, Thomas-Oates JE, Glushka J, Spaink HP, Lugtenberg BJJ. 1994. *J. Biol. Chem.* 269:11090–97
71. Bhat UR, Mayer H, Yokota A, Hollingsworth RI, Carlson RW. 1991. *J. Bacteriol.* 173:2155–59
72. Demont N, Ardourel M, Maillet F, Promé D, Ferro M, et al. 1994. *EMBO J.* 13:2139–49
73. Röhrig H, Schmidt J, Wieneke U, Kondorosi E, Barlier I, et al. 1994. *Proc. Natl. Acad. Sci. USA* 91:3122–26
74. Orgambide GG, Lee J, Hollingsworth RI, Dazzo FB. 1995. *Biochemistry* 34: 3832–40
75. Downie JA. 1989. *Mol. Microbiol.* 3: 1649–51
76. Baev N, Kondorosi A. 1992. *Plant Mol. Biol.* 18:843–46
77. Ardourel M, Lortet G, Maillet F, Roche P, Truchet G, et al. 1995. *Mol. Microbiol.* 17:687–99
78. Bloemberg GV, Thomas-Oates JE, Lugtenberg BJJ, Spaink HP. 1994. *Mol. Microbiol.* 11:793–804
79. Bloemberg GV, Lagas RM, van Leeuwen S, Van der Marel GA, Van Boom JH, et al. 1995. *Biochemistry* 34:12712–20
80. Carlson RW, Sanjuan J, Bhat UR, Glushka J, Spaink HP, et al. 1993. *J. Biol. Chem.* 268:18372–81
81. Geelen D, Fernandez-Lopez M, D'Haeze W, Mergaert P, Van Montagu M, et al. 1995. *J. Bacteriol.* In press
82. Schwedock J, Long SR. 1989. *Mol. Plant-Microbe Interact.* 2:181–94
83. Schwedock JS, Long SR. 1992. *Genetics* 132:899–909
84. Leyh TS, Vogt TF, Suo Y. 1992. *J. Biol. Chem.* 267:10405–10
85. Schwedock JS, Liu CX, Leyh TS, Long SR. 1994. *J. Bacteriol.* 176:7055–64
86. Schwedock J, Long SR. 1990. *Nature* 348:644–47
87. Cervantes E, Sharma SB, Maillet F, Vasse J, Truchet G, Rosenberg C. 1989. *Mol. Microbiol.* 3:745–55
88. Bourdineaud JP, Bono JJ, Ranjeva R, Cullimore JV. 1995. *Biochem. J.* 306: 259–64
89. Schultze M, Staehelin C, Rohrig H, John M, Schmidt J, Kondorosi E, et al. 1995. *Proc. Natl. Acad. Sci. USA* 92:2706–9
90. Ehrhardt DW, Atkinson EM, Faull KF, Freedberg DI, Sutherlin DP, et al. 1995. *J. Bacteriol.* 177:6237–45
91. Poupot R, Martinez-Romero E, Maillet F, Promé JC. 1995. *FEBS Lett.* 368:536–40
92. Firmin JL, Wilson KE, Carlson RW, Davies AE, Downie JA. 1993. *Mol. Microbiol.* 10:351–60
93. Kozik A, Heidstra R, Horvath B, Kulikova O, Tikhonovich I, et al. 1995. *Plant Sci.* 108:41–49
94. Clark CA, Beltrame J, Manning PA. 1991. *Gene* 107:43–52
95. Stacey G, Luka S, Sanjuan J, Banfalvi Z, Nieuwkoop AJ, et al. 1994. *J. Bacteriol.* 176:620–33
96. Goethals K, Mergaert P, Gao M, Geelen D, Van Montagu M, Holsters M. 1992. *Mol. Plant-Microbe Interact.* 5:405–11
97. Luka S, Sanjuan J, Carlson RW, Stacey G. 1993. *J. Biol. Chem.* 268:27053–59
98. Spaink HP, Weinman J, Djordjevic MA, Wijffelman CA, Okker RJH, Lugtenberg BJJ. 1989. *EMBO J.* 8:2811–18
99. Schlaman HRM. 1992. *Regulation of nodulation gene expression in Rhizobium leguminosarum biovar viciae.* PhD thesis. Univ. Leiden, The Netherlands
100. Schmidt J, Wingender R, John M, Wieneke U, Schell J. 1988. *Proc. Natl. Acad. Sci. USA* 85:8578–82
101. Evans IJ, Downie JA. 1986. *Gene* 43: 95–101
102. Schlaman HRM, Okker RJH, Lugtenberg BJJ. 1990. *J. Bacteriol.* 172:5486–89
103. Vazquez M, Santana O, Quinto C. 1993. *Mol. Microbiol.* 8:369–77
104. McKay IA, Djordjevic MA. 1993. *Appl. Environ. Microbiol.* 59:3385–92
105. Spaink HP, Wijfjes AHM, Lugtenberg BJJ. 1995. *J. Bacteriol.* 177:6276–81
106. Rivilla R, Sutton JM, Downie JA. 1995. *Gene* 161:27–31
107. Downie JA. 1994. *Trends Microbiol.* 2:318–24
108. Saier MH, Tam R, Reizer A, Reizer J. 1994. *Mol. Microbiol.* 11:841–47
109. Hirsch AM. 1992. *New Phytol.* 122: 211–37
110. Martinez E, Laeremans T, Poupot R,

Rogel MA, Lopez L, et al. 1995. In *Nitrogen fixation: fundamentals and applications*. ed. IA Tikhonovich, VI Romonov, NA Provorov, WE Newton, pp. 281–86. Netherlands: Kluwer Academic.

111. Cardenas L, Dominguez J, Quinto C, Lopez-Lara IM, Lugtenberg BJJ, et al. 1995. *Plant Mol. Biol.* 29:453–64
112. Poupot R, Martinez-Romero E, Gautier N, Promé JC. 1995. *J. Biol. Chem.* 270:6050–55
113. Relic B, Staehelin C, Fellay R, Jabbouri S, Boller T, Broughton WJ. 1994. See Ref. 185, pp. 69–75
114. Sanjuan J, Carlson RW, Spaink HP, Bhat UR, Barbour WM, et al. 1992. *Proc. Natl. Acad. Sci. USA* 89:8789–93
115. Stokkermans TJW, Peters NK. 1994. *Planta* 193:413–20
116. Bec-Ferté MP, Krishnan HB, Promé D, Savagnac A, Pueppke SG, Promé JC. 1994. *Biochemistry* 33:11782–88
117. Nicolaou KC, Bockovich NJ, Carcanague DR, Hummel CW, Even LF. 1992. *J. Am. Chem. Soc.* 114:8701–2
118. Ikeshita S, Nakahara Y, Ogawa T. 1995. *Carbohydr. Res.* 266:C1–C6
119. Stokkermans TJW, Ikeshita S, Cohn J, Carlson RW, Stacey G, et al. 1995. *Plant Physiol.* 108:1587–95
120. Wang LX, Li C, Wang QW, Hui YZ. 1993. *Tetrahedron Lett.* 34:7763–66
121. Ikeshita S, Sakamoto A, Nakahara Y, Ogawa T. 1994. *Tetrahedron Lett.* 35:3123–26
122. Tailler D, Jacquinet JC, Beau JM. 1994. *J. Chem. Soc. Chem. Commun.*, pp. 1827–28
123. Promé JC, Demont N. 1996. In *Plant Microbe Interactions*, Vol. 1, ed. G Stacey, NT Keen. New York: Chapman Hall
124. Bono J-J, Riond J, Nicolaou KC, Bockovich NJ, Estevez VA, et al. 1995. *Plant J.* 7:253–60
125. Ikeshita S, Nakahara Y, Ogawa T. 1994. *Glycoconjugate J.* 11:257–61
126. Minami E, Kouchi H, Cohn JR, Ogawa T, Stacey G. 1995. *Plant J.* In press
127. Röhrig H, Schmidt J, Walden R, Czaja I, Miklasevics E, et al. 1995. *Science* 269:841–43
128. van Brussel AAN, Bakhuizen R, Van Spronsen PC, Spaink HP, Tak T, et al. 1992. *Science* 257:70–72
129. Relic B, Talmont F, Kopcinska J, Golinowski W, Promé JC, Broughton WJ. 1993. *Mol. Plant-Microbe Interact.* 6: 764–74
130. Heidstra R, Geurts R, Franssen H, Spaink HP, Van Kammen A, Bisseling T. 1994. *Plant Physiol.* 105:787–97

131. Scheres B, Van De Wiel C, Zalensky A, Horvath B, Spaink HP, et al. 1990. *Cell* 60:281–94
132. Vijn I, das Neves L, Van Kammen A, Franssen H, Bisseling T. 1993. *Science* 260:1764–65
133. Bauer P, Crespi MD, Szecsi J, Allison LA, Schultze M, et al. 1994. *Plant Physiol.* 105:585–92
134. Pichon M, Journet E-P, Dedieu A, De Billy F, Truchet G, Barker DG. 1992. *Plant Cell* 4:1199–211
135. Journet E-P, Pichon M, Dedieu A, De Billy F, Truchet G, Barker D. 1994. *Plant J.* 6:241–49
136. Cook D, Dreyer D, Bonnet D, Howell M, Nony E, Vandenbosch K. 1995. *Plant Cell* 7:43–55
137. Vijn I, Martinez-Abarca F, Yang WC, das Neves L, van Brussel A, et al. 1995. *Plant J.* 8:111–19
138. Horvath B, Heidstra R, Lados M, Moerman M, Spaink HP, et al. 1993. *Plant J.* 4:727–33
139. Krause A, Sigrist CJA, Dehning I, Sommer H, Broughton WJ. 1994. *Mol. Plant-Microbe Interact.* 7:411–18
140. Felle HH, Kondorosi E, Kondorosi A, Schultze M. 1995. *Plant J.* 7:939–47
141. Ehrhardt DW, Atkinson EM, Long SR. 1992. *Science* 256:998–1000
142. Kurkdjian AC. 1995. *Plant Physiol.* 107: 783–90
143. Allen NS, Bennett MN, Cox DN, Shipley A, Ehrhardt DW, Long SR. 1994. In *Advances in Molecular Genetics of Plant-Microbe Interactions*. ed. MJ Daniels, JA Downie, AE Osbourn, pp. 107–13. Dordrecht: Kluwer Academic
144. Economou A, Davies AE, Johnston AWB, Downie JA. 1994. *Microbiology* 140:2341–47
145. Sutton JM, Lea EJA, Downie JA. 1994. *Proc. Natl. Acad. Sci. USA* 91:9990–94
146. Mylona P, Pawlowski K, Bisseling T. 1995. *Plant Cell* 7:869–85
147. Relic B, Perret X, Estrada-Garcia MT, Kopcinska J, Golinowski W, et al. 1994. *Mol. Microbiol.* 13:171–78
148. Staehelin C, Schultze M, Kondorosi E, Mellor RB, Boller T, Kondorosi A. 1994. *Plant J.* 5:319–30
149. Staehelin C, Granado J, Muller J, Wiemken A, Mellor RB, et al. 1994. *Proc. Natl. Acad. Sci. USA* 91:2196–200
150. Staehelin C, Schultze M, Kondorosi E, Kondorosi A. 1995. *Plant Physiol.* 108: 1607–19
151. Schmidt PE, Broughton WJ, Werner D. 1994. *Mol. Plant-Microbe Interact.* 7: 384–90
152. Xie ZP, Staehelin C, Vierheilig H,

Wiemken A, Jabbouri S, et al. 1995. *Plant Physiol.* 108:1519–1525

153. Truchet G, Roche P, Lerouge P, Vasse J, Camut S, et al. 1991. *Nature* 351:670–73

154. Yang WC, de Blank C, Meskiene I, Hirt H, Bakker J, et al. 1994. *Plant Cell* 6:1415–26

155. Savoure A, Magyar Z, Pierre M, Brown S, Schultze M, et al. 1994. *EMBO J.* 13:1093–102

156. Cooper JB, Long SR. 1994. *Plant Cell* 6:215–25

157. Hirsch AM, Bhuvaneswari TV, Torrey JG, Bisseling T. 1989. *Proc. Natl. Acad. Sci. USA* 86:1244–48

158. Hirsch AM, Fang YW. 1994. *Plant Mol. Biol.* 26:5–9

159. Smit G, de Koster CC, Schripsema J, Spaink HP, van Brussel AAN, Kijne JW. 1995. *Plant Mol. Biol.* 29:869–73

160. Kouchi H, Hata S. 1993. *Mol. Gen. Genet.* 238:106–19

161. Crespi MD, Jurkevitch E, Poiret M, Daubenton-Carafa Y, Petrovics G, et al. 1994. *EMBO J.* 13:5099–112

162. Lopez-Lara IM, van der Drift KMGM, van Brussel AAN, Haverkamp J, Lugtenberg BJJ, et al. 1995. *Plant Mol. Biol.* 29:465–77

163. Baureithel K, Felix G, Boller T. 1994. *J. Biol. Chem.* 269:17931–38

164. Boller T. 1995. *Annu. Rev. Plant Physiol. Plant Mol. Biol.* 46:189–214

165. Spaink HP, Bloemberg GV, Wijfjes AHM, Ritsema T, Geiger O, et al. 1994. In *Advances in Molecular Genetics of Plant-Microbe Interactions,* ed. MJ Daniels, JA Downie, AE Osbourn, pp. 91–98. Dordrecht: Kluwer Academic

166. Darvill A, Augur C, Bergmann C, Carlson RW, Cheong JJ, et al. 1992. *Glycobiology* 2:181–98

167. Albersheim P, Darvill AG. 1985. *Sci. Am.* 253:58–64

168. Spaink HP, Wijfjes AHM, Van Vliet TB, Kijne JW, Lugtenberg BJJ. 1993. *Aust. J. Plant Physiol.* 20:381–92

169. Cocking EC, Webster G, Kothari SL, Pawlowski K, Jones J, et al. 1994. See Ref. 185, pp. 215–19

170. Trinick MJ, Hadobas PA. 1988. *Plant Soil* 110:177–85

171. Marvel DJ, Torrey JG, Ausubel FM. 1987. *Proc. Natl. Acad. Sci. USA* 84:1319–23

172. de Jong AJ, Heidstra R, Spaink HP, Hartog MV, Meijer EA, et al. 1993. *Plant Cell* 5:615–20

173. Schmidt EDL, de Jong AJ, de Vries SC. 1994. *Plant Mol. Biol.* 26:1305–13

174. Schmidt J, Röhrig H, John M, Wieneke U, Stacey G, et al. 1993. *Plant J.* 4:651–58

175. Rosa F, Sargent TD, Rebbert ML, Michaels GS, Jamrich M, et al. 1988. *Dev. Biol.* 129:114–23

176. Mulligan JT, Long SR. 1985. *Proc. Natl. Acad. Sci. USA* 82:6609–13

177. Sanjuan J, Grob P, Göttfert M, Hennecke H, Stacey G. 1994. *Mol. Plant-Microbe Interact.* 7:364–69

178. Sadowsky MJ, Cregan PB, Göttfert M, Sharma A, Gerhold D, et al. 1991. *Proc. Natl. Acad. Sci. USA* 88:637–41

179. Cren M, Kondorosi A, Kondorosi E. 1995. *Mol. Microbiol.* 15:733–47

180. Mulligan JT, Long SR. 1989. *Genetics* 122:7–18

181. Fellay R, Perret X, Viprey V, Broughton WJ, Brenner S. 1995. *Mol. Microbiol.* 16:657–67

182. Surin BP, Watson JM, Hamilton WDO, Economou A, Downie JA. 1990. *Mol. Microbiol.* 4:245–52

183. Lopez-Lara IM, van den Berg JDJ, Thomas-Oates JE, Glushka J, Lugtenberg BJJ, Spaink HP. 1995. *Mol. Microbiol.* 15:627–38

184. Verma DPS, ed. 1992. *Molecular Signals in Plant-Microbe Communications.* Boca Raton, FL: CRC Press

185. Kiss GB, Endre G, eds. 1994. *Proc. 1st European Nitrogen Fixation Conference.* Szeged: Officina Press

Annu. Rev. Biochem. 1996. 65:537–61

ELECTRON TRANSFER IN PROTEINS

Harry B. Gray and Jay R. Winkler

Beckman Institute, California Institute of Technology, Pasadena, California 91125

KEY WORDS: electron tunneling, electronic coupling, reorganization energy, cytochrome c, myoglobin, azurin, photosynthetic reaction center, cytochrome c oxidase

ABSTRACT

Electron-transfer (ET) reactions are key steps in a diverse array of biological transformations ranging from photosynthesis to aerobic respiration. A powerful theoretical formalism has been developed that describes ET rates in terms of two parameters: the nuclear reorganization energy (λ) and the electronic-coupling strength (H_{AB}). Studies of ET reactions in ruthenium-modified proteins have probed λ and H_{AB} in several metalloproteins (cytochrome c, myoglobin, azurin). This work has shown that protein reorganization energies are sensitive to the medium surrounding the redox sites and that an aqueous environment, in particular, leads to large reorganization energies. Analyses of electronic-coupling strengths suggest that the efficiency of long-range ET depends on the protein secondary structure: β sheets appear to mediate coupling more efficiently than α-helical structures, and hydrogen bonds play a critical role in both.

CONTENTS

INTRODUCTION 538
    Electron-Transfer Theory. 538
    Ru-Modified Proteins 540
THE NUCLEAR FACTOR 541
    Reorganization Energy. 541
    Self-Exchange Reactions 542
    Conformational Changes 543
    Cytochrome c 543
    Myoglobin. 545
ELECTRONIC COUPLING 547
    Theoretical Models. 547
    Rate vs Distance. 549
    Azurin, a β-Sheet Protein. 551
    α-Helical Proteins 554
    Coupling Zones. 556

0066-4154/96/0701-0537$08.00

INTRODUCTION

A combination of X-ray crystallographic experiments and biophysical investigations has produced a strikingly detailed picture of the initial events in photosynthesis (1, 2). The primary photochemical charge separation occurs with a time constant of 2 ps, creating a hole in the special pair (BCh$_2$) and placing an electron on the bacteriopheophytin (BPh) acceptor. Next there is a charge shift from the bacteriopheophytin to a menaquinone (BPh$^-$ → Q$_A$, τ ~ 100 ps), followed by hole-filling at the oxidized special pair by a reduced cytochrome (Fe^{2+}-cyt c → BCh$_2^+$, τ ~ 10 ns).

The final charge shift from the menaquinone radical to ubiquinone occurs in 100 ns. The overall result is charge separation across a membrane that stores roughly 0.3 eV of chemical potential. The efficiency of this charge-separation process is very high; most of the energy-wasting recombination reactions are several orders of magnitude slower than the competing charge-shift reactions. Understanding the relative rates of these processes, as well as the curious fact that only one arm of the nearly twofold symmetric reaction center is electron-transfer (ET) active (3, 4), represents a major challenge for both theoreticians and experimentalists.

Some of the most critical steps in the functioning of mitochondrial enzymes are long-range ET reactions. Cytochrome c oxidase, the terminal ET complex in aerobic respiration, catalyzes the four-electron reduction of O$_2$ to H$_2$O and pumps four protons across the inner mitochondrial membrane, creating a transmembrane potential that ultimately drives ATP synthesis (5–7). Cytochrome c oxidase contains four distinct redox centers: cytochrome a and binuclear Cu$_A$ are the primary electron acceptors; oxygen activation occurs at a binuclear cytochrome a_3/Cu$_B$ active site (8–10). Experimental studies of this enzyme have provided rate constants for many of its long-range ET reactions (8, 11), and recent time-resolved resonance Raman measurements have detailed the individual steps in the oxygen-activation reaction (12–15). In contrast to the primary events in photosynthesis, protein conformational dynamics as well as bond-breaking and bond-forming processes are intimately linked to the ET reactions of cytochrome c oxidase.

Electron-Transfer Theory

The photosynthetic reaction center and cytochrome c oxidase are just two of the many biological systems in which ET reactions play central roles. The unique simplicity of ET reactions has fostered the development of a powerful theoretical formalism that describes the rates of these processes in terms of a small number of parameters. The conceptual breakthrough that led to the development of ET theory was the recognition of the pivotal role played by

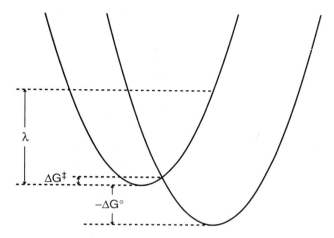

Figure 1 Schematic representation of reactant and product potential-energy surfaces in electron-transfer reactions indicating the driving force ($-\Delta G°$), the activation free energy ($\Delta G^{\ddagger}$), and the reorganization energy (λ).

the Franck-Condon principle (16). Owing to the much higher electron veloci-ties, nuclei will remain fixed during the actual transition from reactants to products. The transition state for this reaction must lie at a point in nuclear-configuration space where the reactant and product states are degenerate (Fig-ure 1). Hence, through fluctuations of the reacting molecules and their sur-roundings, the transition-state configuration will be reached and an electron can transfer.

Electron tunneling in proteins occurs in reactions where the electronic in-teraction between redox sites is relatively weak. Under these circumstances, the transition state for the electron-transfer reaction must be formed many times before reactants are successfully converted to products; the process is electronically nonadiabatic. Semiclassical theory (equation 1) (17) predicts that

$$k_{ET} = (4\pi^3/h^2\lambda k_B T)H_{AB}^2 \exp[-(\Delta G° + \lambda)^2/4\lambda k_B T], \qquad 1.$$

the reaction rate for ET from a donor (D) to an acceptor (A) at fixed separation and orientation depends on the reaction driving force ($-G°$), a nuclear reor-ganization parameter (λ), and the electronic-coupling strength between reac-tants and products at the transition state (H_{AB}) (17). This theory reduces a complex dynamical problem in multidimensional nuclear-configuration space to a simple expression comprised of just two parameters (λ, H_{AB}). Equation 1 naturally partitions into nuclear (exponential) and electronic (pre-exponential) terms: ET rates reach their maximum values (k_{ET}°) when the nuclear factor is optimized ($-\Delta G^{\circ} = \lambda$); these k_{ET}° values are limited only by the electronic-coupling strength (H_{AB}^2).

Ru-Modified Proteins

Investigations of the driving-force, temperature, and distance dependences of ET rates can be used to define the fundamental ET parameters λ and H_{AB}. Natural systems often are not amenable to the systematic studies that are required to explore the fundamental aspects of biological ET reactions. A successful alternative approach involves measurements of ET in metalloproteins that have been surface-labeled with redox-active molecules (18, 19). By varying the binding site and chemical composition of the probe molecule, it has been possible to elucidate the factors that control the rates of long-range ET reactions in proteins.

Ruthenium complexes are excellent reagents for protein modification and electron-transfer studies. Ru^{2+}-aquo complexes readily react with surface His residues on proteins to form stable derivatives (20, 21). Low-spin pseudo-octahedral Ru complexes exhibit small structural changes upon redox cycling between the Ru^{2+} and Ru^{3+} formal oxidation states. Hence, the inner-sphere barriers to electron transfer (λ_I) are small. With the appropriate choice of ligand, the $Ru^{3+/2+}$ reduction potential can be varied from < 0.0 to > 1.5 V vs NHE. ET in a Ru-modified protein was first measured in $Ru(NH_3)_5(His33)^{3+}$-ferricytochrome c (18). Photochemical methods were used to inject an electron into the Ru^{3+} site on the protein surface, and this kinetic product (Ru^{2+}-Fe^{3+}) subsequently converted to the thermodynamic product (Ru^{3+}-Fe^{2+}) by intramolecular electron transfer ($k_{ET} = 30$ s^{-1}, T = 295°C, $-\Delta G^{\circ} = 0.20$ eV). The early experimental measurements of long-range ET rates in Ru-ammine-modified Fe-cyt c (18, 22, 23) were followed by related studies of other Ru-ammine proteins [Zn-cyt c (24–26), myoglobin (Mb) (27–31), high-potential iron-sulfur protein (HiPIP) (32, 33), azurin (34), plastocyanin (35, 36), stellacyanin (37, 38), cyt b_5 (39), and cyt $c_{551}(40)$].

Recent work on Ru-modified proteins has involved a $Ru(bpy)_2(im)$ $(HisX)^{2+}$ (bpy = 2,2'-bipyridine; im = imidazole) label (41–45). In addition to the attractive ET properties of Ru-ammine systems, Ru-bpy complexes have an additional characteristic not found with the ammines: long-lived, luminescent metal-to-ligand charge-transfer excited states. These excited states enable a wider range of electron-transfer measurements than is possible with nonluminescent complexes (41). Furthermore, the bpy ligands raise the $Ru^{3+/2+}$ reduction potential to >1 V vs NHE, making observed ET rates closer to k_{ET}°, which leads to more reliable estimates of H_{AB} and λ. Systematic investigations of Ru bpy–modified proteins have provided a detailed picture of long-range protein electron transfer.

THE NUCLEAR FACTOR

Reorganization Energy

The nuclear factor in equation 1 results from a classical treatment of nuclear motions in which all reorganization is described by a single harmonic coordinate. The parameter λ is defined as the energy of the reactants at the equilibrium nuclear configuration of the products (Figure 1). The remarkable aspect of the nuclear factor is the predicted free-energy dependence (Figure 2). At low driving forces, rates increase with $-\Delta G°$, but as the driving force moves into the region where $-\Delta G° > \lambda$, ET rates are predicted to decrease (inverted effect). Experimental studies of electron-transfer rates in synthetic model complexes (46–50) and in biological systems (11, 51–53) have provided convincing evidence for inverted driving-force effects.

For ET reactions in polar solvents, the dominant contribution to λ arises from reorientation of solvent molecules in response to the change in charge distribution of the reactants (λ_s). Dielectric continuum models are commonly used in calculations of solvent reorganization. The earliest models treated the reactants as conducting spheres (17); later refinements dealt with charge shifts inside low dielectric cavities of regular (spherical, ellipsoidal) shape (54, 55). Embedding reactants in a low dielectric medium (e.g. a membrane) can dramatically reduce reorganization energies, but the effect on ET rates depends on the response of $\Delta G°$ to the nonpolar environment. Generally, low dielectric media will reduce the driving force for charge-separation reactions (D + A $\rightarrow$ D^+ + A^-), but will have a smaller effect on the energetics of charge-shift reactions (e.g. D^- + A $\rightarrow$ D + A^-).

The second component of the nuclear factor arises from changes in bond lengths and bond angles of the donor and acceptor following electron transfer. Classical descriptions of this inner-sphere reorganization (λ_i) usually are not adequate, and quantum-mechanical refinements to equation 1 have been developed (56). The most significant consequences of quantized nuclear motions are found in the inverted region. Owing to nuclear tunneling through the activation barrier, highly exergonic reactions will not be as slow as predicted by the classical model. Distortions along coordinates associated with high-frequency vibrations ($> 1000 \text{ cm}^{-1}$) can significantly attenuate the inverted effect (Figure 2).

The nuclear factor reflects the interplay between driving-force and reorganization energy that regulates ET rates. A reaction in the inverted region can be

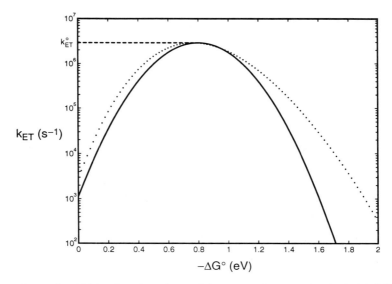

Figure 2 Predicted driving-force dependences of electron-transfer rates using classical (solid line) and quantum-mechanical (dotted line) descriptions of nuclear reorganization.

accelerated if a pathway is available that releases less free energy in the actual ET step. One such pathway leads to electronically excited products ($*D^+$, $*A^-$); formation of these products lowers the ET driving force by an amount equal to the energy of the excited electronic state. An ET process that forms excited products will be the preferred pathway if its driving force is closer to λ than that of a reaction forming ground-state products. Chemiluminescent ET reactions are a familiar example of such processes, and they are a clear demonstration of the inverted effect.

Self-Exchange Reactions

The simplest ET reactions are those in which the reactants are the same as the products. The driving force for these self-exchange reactions is zero, and the predicted activation free energy is just $\lambda/4$ (equation 1). In his formulation of ET theory, Marcus developed expressions describing the rates and reorganization energies for ET reactions between different reagents (cross reactions) in terms of the self-exchange rates and reorganization energies for each reactant (17). These simple expressions permit cross-reaction rates to be estimated from self-exchange data, or self-exchange rates to be estimated from cross-reaction data, and have proven to be a powerful predictive tool in ET research.

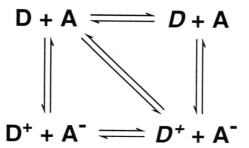

Figure 3 Redox reactions of a protein with two stable conformations in its reduced (D, *D*) and oxidized (D$^+$, *D*$^+$) states.

Conformational Changes

The multiple conformation, ionization, and ligand-binding states of biopolymers can severely complicate the electron-transfer problem. Oxidized and reduced proteins often have different conformations such that their redox reactions involve major nuclear rearrangements in addition to electron transfer. If a protein has multiple stable conformations, then new redox-reaction pathways become available. The redox reactions of a protein with two stable conformations in its reduced (D, *D*) and oxidized (D$^+$, *D*$^+$) states can be represented by the scheme in Figure 3 (57–59).

A concerted conversion of (D + A) to (*D*$^+$ + A$^-$) is accompanied by a large reorganization barrier (λ) due to the D $\rightarrow$ *D*$^+$ conformation change. If the reorganization energy is great enough, sequential pathways could be more favorable. In a sequential mechanism, the conformational change could either precede (D + A $\rightarrow$ *D* + A $\rightarrow$ *D*$^+$ + A$^-$) or succeed (D + A $\rightarrow$ D$^+$ + A$^-$ $\rightarrow$ *D*$^+$ + A$^-$) the ET step. Many, if not most, biological redox reactions will involve complex reaction schemes in which elementary ET steps are not the rate-limiting processes.

Cytochrome c

A study of the driving-force dependence of intramolecular ET in Ru(NH$_3$)$_4$L-(His33)-Zn-cyt *c* (L = NH$_3$, pyridine, isonicotinamide) found that rates could be described by the parameters $\lambda = 1.15$ eV and $H_{AB} = 0.1$ cm^{-1} (Figure 4) (25). Application of the Marcus cross relation, using a value of 1.2 eV for the electron self-exchange reorganization energy of Ru(NH$_3$)$_4$L(His33)$^{3+/2+}$, suggests that the self-exchange reorganization energy for Zn-cyt *c* is 1.2 eV. This value is in good agreement with the estimated self-exchange reorganization energy of native cyt *c* (1.0 eV) (17). This analysis shows that the hydrophilic

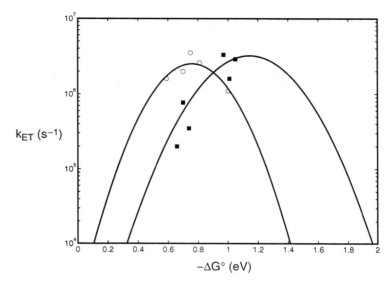

Figure 4 Driving-force dependence of electron-transfer rates in Ru-ammine-modified Zn-substituted cytochrome *c* (filled squares), and Ru-bpy-modified Fe-cytochrome *c* (open circles).

Ru-ammine complex is responsible for half of the total reorganization energy in Ru-ammine–modified cyt *c* ET reactions.

Studies of self-exchange reactions have demonstrated that replacing ammonia ligands with diimine ligands substantially reduces the reorganization energy associated with $Ru^{3+/2+}$ ET (60). The difference can be attributed to a decrease in solvent polarization by the larger Ru-diimine ions and to somewhat smaller inner-sphere barriers as well. The reorganization energy for ET in $Ru(bpy)_2(im)(His33)$-Fe-cyt *c* (bpy = 2,2'-bipyridine) was then expected to be less than 1.2 eV; a cross-relation calculation suggested a value of 0.8 eV. A recent study of the driving-force dependence of $Fe^{2+} \rightarrow Ru^{3+}$ ET rates in $Ru(LL)_2(im)(His33)$-Fe-cyt *c* (LL = bpy, 4,4'-$(CH_3)_2$-bpy, 4,4', 5,5'-$(CH_3)_4$-bpy, 4,4'-$(CONH(C_2H_5))_2$-bpy) is in excellent agreement with this estimate (λ = 0.74 eV, Figure 4) (44).

The significant difference in reorganization energy between Ru-ammine and Ru-bpy-modified cytochromes (Figure 4) highlights the important role of water in protein electron transfer. The bulky bpy ligands shield the charged metal center from the polar aqueous solution, reducing the solvent reorganization energy. In the same manner, the medium surrounding a metalloprotein active site will affect the reorganization energy associated with its ET reactions. A hydrophilic active site will lead to larger reorganization energies than a hydrophobic site. Consequently, the kinetics of protein ET reactions will be very sensitive to the active-site environment.

Some have suggested that the electron-transfer reactions of cyt c may be gated by protein conformational changes (61, 62). These ideas stem from observations that ferricytochrome c is not as tightly folded as ferrocytochrome c (63–65). Indeed, equilibrium titrations demonstrate that ferricytochrome c unfolds at lower concentrations of denaturants [e.g. urea, guanidine hydrochloride (GuHCl)] than ferrocytochrome c. Extrapolations to zero denaturant concentration indicate that the free energy of folding is ~8 kcal mol^{-1} greater for the reduced protein in aqueous solution (65). In addition, ^{1}H NMR spectra have been interpreted in terms of a more flexible structure for ferricytochrome c (63). A recent measurement of the $Fe^{2+} \rightarrow Ru^{3+}$ ET rate in Ru(NH$_3$)$_5$(His33)-Fe-cyt c demonstrates that, contrary to earlier suggestions, ET in cyt c is fully reversible and the ratio of rate constants for the $Ru^{2+} \rightarrow Fe^{3+}$ and $Fe^{2+} \rightarrow Ru^{3+}$ ET processes equals the equilibrium constant determined from electrochemical measurements (66). This observation provides compelling evidence that, at least on the timescales of these kinetics, conformational dynamics and energetics do not limit cyt c ET reactions.

Gated ET may be involved in the reactions of cytochrome c in the presence of denaturants (GuHCl). The reduction potential of folded cyt c ($E° = 0.26$ V vs NHE) is ~0.4 eV greater than that of unfolded cyt c (65). At intermediate GuHCl concentrations (3–4 M), ferrocytochrome c is folded and ferricytochrome c is unfolded, creating a situation similar to that represented in Figure 3. Owing to the scale of the nuclear rearrangement, redox cycling between oxidized and reduced proteins is unlikely to occur in concert with folding (or unfolding). Instead, the two processes should occur in sequence with the potential of the redox partner determining the course of the reaction. Recent work has shown that strong electron donors inject an electron into unfolded ferricytochrome to form a transient unfolded ferrocytochrome, which then rearranges to the folded structure in less than 100 ms (65). ET triggering methods have the potential to bridge the ns to ms measurement time gap for protein folding (65).

Myoglobin

Myoglobin (Mb) is an oxygen-binding heme protein that, like cyt c, can participate in electron-transfer reactions. Unlike cyt c, however, Mb undergoes a coordination change upon cycling between the Fe^{3+} and Fe^{2+} oxidation states. The heme in Fe^{3+}-Mb is a high-spin, six-coordinate complex with His93 and H$_2$O axial ligands; Fe^{2+}-Mb has a high-spin, five-coordinate heme bound only to His93 (67). This situation is analogous to that represented in Figure 3. The ET reactions of Mb are necessarily coupled to the dynamics and energetics of

H_2O binding to both the ferro- and ferri-hemes. This type of coordination change is not uncommon in heme enzymes, and it can have profound consequences for ET kinetics.

A cyclic voltammetric study suggests that water binding (k_b) and dissociation (k_f) in Fe^{2+}-Mb are quite slow ($k_f = 1.0(5)$, $k_b = 0.5(2)$ s^{-1}) (68). This study also led to the suggestion that five-coordinate, high-spin Fe^{2+}-Mb is electrochemically inactive and that water binding is the rate-limiting process for oxidation to met-Mb [$Fe(OH_2)^{3+}$-Mb]. Whether the same is true for homogeneous oxidations of Fe^{2+}-Mb depends very much upon the reaction driving force. Observed rates of Fe^{2+} oxidation in $Ru(NH_3)_4L$(His48)-modified Mb [L, k_{ET} s^{-1}(−ΔG°, eV): NH_3, 0.04 (0.02); py, 2.5 (0.28); isn, 3.0 (0.35) (30)] are within an order of magnitude of the proposed H_2O-dissociation rate, but since two of the rates are at least five times faster than k_b, the ET reaction is apparently not gated by the ligand-binding process. The slowest reaction (L = NH_3) could involve a ligand-binding pre-equilibrium step prior to electron transfer. In the two faster reactions (L = py, isn), water binding to the heme iron must occur during or after electron transfer. If this process is concerted, then λ will be substantially greater than the 1.2-eV value found for ET reactions in Ru-modified cyt c. In a sequential ET/ligand-binding mechanism, the driving force for the ET process will be lower by an amount equal to the free-energy change for water binding to Fe^{3+}-Mb. Although the dynamics of ligand binding do not appear to limit these Ru-modified Mb ET reactions, the energetics of water binding to both Fe^{2+}-Mb and Fe^{3+}-Mb play a critical role.

Much of the complexity of Mb redox reactions can be eliminated by replacing the heme with luminescent metalloporphyrins (e.g. ZnP, MgP, CdP, H_2P, PdP, PtP; P = porphyrin dianion). A previous analysis of ET reactions in Ru(ammine)(His48)-modified, metal-substituted Mb suggested a reorganization energy of 1.26 eV (30). The ET rates of Ru(ammine)(His48)-modified Fe-Mb could be described by this reorganization energy, but with a slightly smaller value of H_{AB}. These results are quite similar to those found for Ru(ammine)(His33)-cyt c, but they provide no information about the role of the coordination change in the ET reaction.

These investigations of ET in Ru-modified cyt c and Mb provide some insight into the factors affecting energies in heme proteins. The presence of water around one or both redox sites has a significant impact: More water leads to larger values of λ. Excluding water by burying redox centers inside hydrophobic pockets or in membranes can lead to unusually small reorganization energies. The redox cofactors of the photosynthetic reaction center are embedded in a membrane-spanning helical protein matrix, and extremely small reorganization energies have been estimated for the initial ET step in photosynthesis (69, 70). This small value of λ leads to a deeply inverted charge-recombination reaction and is likely to be responsible for the overall efficiency of photochemical energy storage. The recent x-ray crystal structures of cytochrome c oxidase will provide important information for the analysis of electron transfer in this enzyme (9, 10). Long-range ET reactions in cytochrome c oxidase proceed at remarkably fast rates. The rate constant for ET from cyt c to Cu_A is 10^5 s^{-1} ($-\Delta G° = 0.02$ eV) (71) and that for $Cu_A \rightarrow$ cyt a ET is 2 $\times$ 10^4 s^{-1} ($-\Delta G° = 0.1$ eV; 19 Å metal-metal distance) (9, 11, 72). Strong coupling (*vide infra*) and low reorganization energies are required to explain the observed ET rates.

ELECTRONIC COUPLING

Theoretical Models

Nonadiabatic ET reactions are characterized by weak electronic interaction between the reactants and products at the transition-state nuclear configuration ($H_{AB} \ll k_BT$). This coupling is directly related to the strength of the electronic interaction between the donor and acceptor (73). When donors and acceptors are separated by long distances (>10 Å), the D/A interaction will be quite small.

HOMOGENEOUS-BARRIER MODELS   In 1974 Hopfield described biological ET in terms of electron tunneling through a square potential barrier (74). In this model, H_{AB}^2 (and, hence, k_{ET}) drops off exponentially with increasing D-A separation. The height of the tunneling barrier relative to the energies of the D/A states determines the distance-decay constant (β). A decay constant in the range of 3.5–5 Å^{-1} has been estimated for donors and acceptors separated by a vacuum and, as a practical matter, ET is prohibitively slow at D-A separations (R) greater than 8 Å ($k_{ET}° < 10$ s^{-1}). An intervening medium between redox sites reduces the height of the tunneling barrier, leading to a smaller distance-decay constant. Hopfield estimated $\beta \sim 1.4$ Å^{-1} for biological ET reactions on the basis of measurements of the temperature dependence of ET from a cytochrome to the oxidized special pair in the photosynthetic reaction center of *Chromatium vinosum* (74). An 8-Å edge-edge separation was estimated on the basis of this decay constant; later structural studies revealed that the actual distance was somewhat greater (12.3 Å).

SUPEREXCHANGE-COUPLING MODELS   The square-barrier models assume that the distant couplings result from direct overlap of localized donor and acceptor wavefunctions. In long-range ET ($R > 10$ Å), the direct interaction between donors and acceptors is negligible; electronic states of the intervening bridge mediate the coupling via superexchange. If oxidized states of the bridge mediate the coupling, the process is referred to as hole transfer; mediation by reduced bridge states is known as electron transfer. In 1961 McConnell developed a superexchange coupling model to describe charge-transfer interactions between donors and acceptors separated by spacers comprised of m identical repeat units (equation 2) (75). The total coupling depends upon the interaction between adjacent hole

$$H_{AB} = (h_D/\Delta)(h_j/ \Delta)^m h_A \qquad\qquad 2.$$

or electron states in the bridge (h_j), the energy difference between the degenerate D/A states and the bridge states (Δ), and the interactions between the D and A states and the bridge (h_D, h_A).

The McConnell model assumes that only nearest-neighbor interactions mediate the coupling and consequently predicts that H_{AB} will vary exponentially with the number of repeat units in the bridge. Several studies of the distance dependence of ET in synthetic donor-acceptor complexes agree quite well with this prediction. *Ab initio* calculations of H_{AB} for bridges composed of saturated alkane spacers, however, suggest that the simple superexchange model is not quantitatively accurate (76–79). Nonnearest-neighbor interactions were found to dominate the couplings and, except in a few cases, nearest-neighbor interactions were relatively unimportant. A particularly significant finding in these studies is that nonnearest-neighbor interactions make the coupling along a saturated alkane bridge quite sensitive to its conformation.

The medium separating redox sites in proteins is comprised of a complex array of bonded and nonbonded contacts, and an *ab initio* calculation of coupling strengths is a formidable challenge. The homologous-bridge superexchange model (equation 2) is not suitable, because of the diverse interactions in proteins. Beratan, Onuchic, and their coworkers developed a generalization of the McConnell superexchange coupling model that accommodates the structural complexity of a protein matrix (80–84). In this tunneling-pathway model, the medium between D and A is decomposed into smaller subunits linked by covalent bonds, hydrogen bonds, or through-space jumps. Each link is assigned a coupling decay $(\varepsilon_C, \varepsilon_H, \varepsilon_S)$, and a structure-dependent searching algorithm is used to identify the optimum coupling pathway between the two redox sites. The total coupling of a single pathway is given as a repeated product of the couplings for the individual links:

$$H_{AB} \propto \Pi\varepsilon_C \, \Pi\varepsilon_H \, \Pi\varepsilon_S \qquad\qquad 3.$$

A tunneling pathway can be described in terms of an effective covalent tunneling path comprised of n (nonintegral) covalent bonds, with a total length equal to σ_l (equation 4). The relationship between σ_l and the direct D-A distance (R) reflects

$$H_{AB} \propto (\varepsilon_C)^n \qquad\qquad\qquad\qquad 4a.$$

$$\sigma_l = n \times 1.4 \text{ Å/bond} \qquad\qquad\qquad 4b.$$

the coupling efficiency of a pathway (45). The variation of ET rates with R depends upon the coupling decay for a single covalent bond (ε_C), and the magnitude of ε_C depends critically upon the energy of the tunneling electron relative to the energies of the bridge hole and electron states (85). In considering ET data from different protein systems, care must be taken to compare reactions in which oxidants (for hole tunneling) have similar reduction potentials.

Rate vs Distance

The D-A distance decay of protein ET rate constants depends on the capacity of the polypeptide matrix to mediate distant electronic couplings. In a seminal paper in 1992, Dutton and coworkers showed (86) that Hopfield's protein distance-decay constant (1.4 Å^{-1}) (74) could be used to estimate long-range ET rates in the photosynthetic reaction center (RC). Although Dutton's rate/distance correlation gives a rough indication of RC coupling strengths (86, 87), extensive theoretical work clearly shows that the intervening polypeptide structure must be taken into account in attempts to understand distant D-A couplings in other proteins (80–84, 88–97).

The tunneling-pathway model has proven to be one of the most useful methods for estimating long-range electronic couplings (80–82). Employing this model, Beratan, Betts, and Onuchic predicted in 1991 that proteins comprised largely of β-sheet structures would be more effective at mediating long-range couplings than those built from α helices (84). This analysis can be taken a step further by comparing the coupling efficiencies of individual protein secondary structural elements (β sheets, α helices). The coupling efficiency can be determined from the variation of σ_l as a function of R. A linear σ_l/R relationship implies that k_{ET}° will be an exponential function of R; the distance-decay constant is determined by the slope of the σ_l/R plot and the value of ε_C.

A β sheet is comprised of extended polypeptide chains interconnected by hydrogen bonds; the individual strands of β sheets define nearly linear coupling pathways along the peptide backbone spanning 3.4 Å per residue. The tunneling length for a β strand exhibits an excellent linear correlation with β-carbon separation (R_β, Figure 4); the best linear fit with zero intercept yields a slope of 1.37 σ_l/R_β (distance-decay constant = 1.0 Å^{-1}). Couplings across a β sheet depend upon the ability of hydrogen bonds to mediate the D/A interaction. The standard parameterization of the tunneling-pathway model defines the coupling decay across a hydrogen bond in terms of the heteroatom separation:

$$\varepsilon_H = \varepsilon_C^2 \exp[-1.7(R-2.8)] \qquad\qquad 5.$$

If the two heteroatoms are separated by twice the 1.4-Å covalent-bond distance, then the hydrogen-bond decay is assigned a value equal to that of a covalent bond (82). Longer heteroatom separations lead to weaker predicted couplings, but this relationship has not yet been confirmed experimentally.

In the coiled α-helix structure a linear distance of just 1.5 Å is spanned per residue. In the absence of mediation by hydrogen bonds, σ_1 is a very steep function of R_β, implying that an α helix is a poor conductor of electronic coupling (2.7 σ_1/R_β, distance-decay constant = 1.97 Å^{-1}, Figure 5) (45). If the hydrogen-bond networks in α helices mediate coupling, then the Beratan-Onuchic parameterization of hydrogen-bond couplings suggests a σ_1/R_β ratio of 1.72 (distance-decay constant = 1.26 Å^{-1}, Figure 5). Treating hydrogen bonds as covalent bonds further reduces this ratio (1.29 σ_1/R_β, distance-decay constant = 0.94 Å^{-1}, Figure 5). Hydrogen-bond interactions will determine whether α helices are vastly inferior to or slightly better than β sheets in mediating long-range electronic couplings.

The coiled helical structure leads to poorer σ_1/R_β correlations, especially for values of R_β under 10 Å. In this distance region, the tunneling pathway model predicts little variation in coupling efficiencies for the different secondary structures (Figure 5). The coupling in helical structures could be highly anisotropic. Electron transfer along a helix may have a very different distance dependence from ET across helices. In the latter cases, the coupling efficiency will depend on the nature of the interactions between helices. β sheets and α helices are described by quite different peptide bond angles (ϕ, φ). *Ab initio* calculations on saturated hydrocarbons have suggested that different conformations provide different couplings (76). Thus, different values of ε_C might be necessary to describe couplings in β sheets and α helices.

Analyses of ET rate/distance relationships require a consistent definition of the D-A distance. When comparing rates from systems with different donors and/or acceptors, identifying a proper distance measure can be difficult. All maximum ET rates should extrapolate to a common adiabatic rate as R approaches van der Waals contact. So-called edge-to-edge distances are often employed but introduce many ambiguities, not the least of which is defining the set of atoms that constitute the edges of D and A. For planar aromatic molecules (e.g. chlorophylls, pheophytins, quinones), edge-edge separations are usually defined on the basis of the shortest distance between aromatic carbon atoms of D and A. In transition-metal complexes (e.g. Fe-heme, Ru-ammine, Ru-bpy), however, atoms on the periphery are not always well coupled to the central metal, and empirical evidence suggests that metal-metal distances are more appropriate. This dichotomy is by no means rigorously supported by experimental data, but instead represents the best available compromise. In the following analyses, edge-edge distances will be used for ET

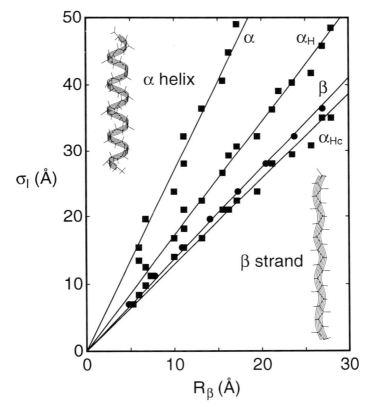

Figure 5   Plots of σ_1 vs R_β for an idealized β strand and α helix. The solid lines are the best linear fits with zero intercept. The slope of the β-strand line is 1.37 (circles). For the α helix (squares), three different treatments of the hydrogen-bond interaction were used: no mediation of coupling, α slope = 2.7; Beratan-Onuchic parameterization of hydrogen-bond couplings, α_H slope = 1.72; hydrogen bonds treated as covalent bonds, α_{Hc} slope = 1.22.

reactions between aromatic donors and acceptors, metal-metal separations will be used for reactions involving two transition-metal complexes, and edge-metal distances will be used for mixed metal/aromatic-molecule reactions.

Azurin, a β-Sheet Protein

A great deal of work has been done on blue copper proteins. In azurin, a prototypal blue protein with a β-barrel tertiary structure (Figure 6), the central Cu atom is coordinated to Cys112 (S), His117 (N), and His46 (N) donor atoms

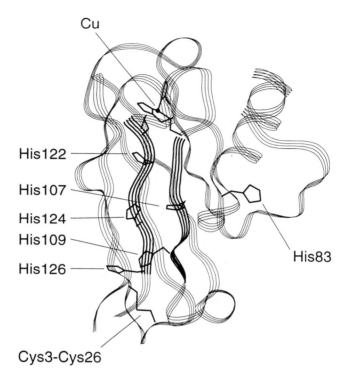

Figure 6 Structure of *Pseudomonas aeruginosa* azurin including the His residues that have been coordinated to Ru(bpy)$_2$(im)$^{2+}$; the Cys3-Cys26 disulfide group is also shown.

in a trigonal planar structure, with weakly interacting Met121 (S) and Gly45 (carbonyl O atom) ligands above and below the plane. Individual β strands that extend from these ligands form a β sheet. The structural similarity of oxidized and reduced azurin, as well as the large self-exchange rate constant [10^5 M^{-1} s^{-1} (17)], suggests relatively small reorganization barriers to electron transfer. A study of the temperature dependence of the redox potentials and ET rates in Ru(bpy)$_2$(im)(His83)-azurin is consistent with a reorganization energy of 0.8 eV, and it indicates that Cu$^+$ → Ru^{3+} ET rates measured in Ru(bpy)$_2$(im)(HisX)-azurins are close to k_{ET}° (85).

We have measured the coupling along β strands in Ru-modified derivatives of azurin (45). Five azurin mutants have been prepared with His residues at different sites on the strands extending from Met121 (His122, His124, His126) and Cys112 (His109, His107) (Figure 6); Ru(bpy)$_2$(im)$^{2+}$ has been coordinated to these surface His groups and intraprotein Cu$^+$ → Ru^{3+} ET rates have been

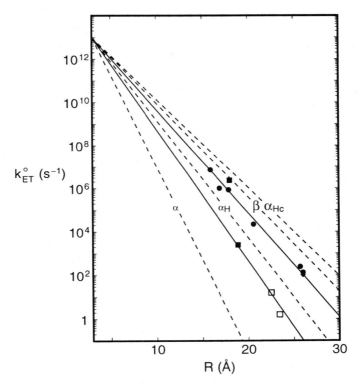

Figure 7 Plot of log k°_{ET} vs R: Ru-modified azurins (filled circles) (45, 85, 98); Cys3-Cys26(S$\overline{2}$) → Cu^{2+} ET in azurin (diamond) (103–105); Ru-modified Mb (filled squares) (106, 107) and the RC (open squares) (86). Solid lines are the best linear fits with an intercept at 10^{13} and correspond to distance decays of 1.1 Å^{-1} for azurin and 1.4 Å^{-1} for Mb and the RC.

measured using photochemical techniques (45, 98). The variation of k°_{ET} with direct metal-metal separation (R_M) is well described by an exponential function with a decay constant of 1.1 Å^{-1} (Figure 7). The result is in remarkably good agreement with the slope predicted for the coupling decay along a strand of an ideal β sheet.

Owing to the unusual Cu coordination, the couplings along different strands should show striking variations. Detailed electronic structure calculations indicate that the S atom of Cys112 has by far the strongest coupling to the Cu center; the His (imidazole) couplings are somewhat weaker than that of the Cys ligand, and the Met121 (S) and Gly45 (O) couplings are just a fraction of the Cys coupling (99, 100). These highly anisotropic ligand interactions strongly favor pathways that couple to the Cu through Cys112. Couplings along different β strands would be expected to have the same distance-decay constants, but different intercepts at close contact. In light of these findings, the fact that the distance dependence of ET in Ru-modified azurin can be described by a single straight line (Figure 7) is quite surprising.

One explanation for uniform distance dependence of couplings along the Met121 and Cys112 strands is that strong interstrand hydrogen bonds serve to direct all of the distant couplings through the Cys112 ligand. A hydrogen bond between Met121(O) and Cys112(NH) could mediate coupling from the Ru complex bound to His122. A second hydrogen bond [Gly123(O)-Phe110(NH)] would provide a coupling link for His124 and His126 ET reactions. The importance of the pathways that cross from the Met121 strand to the Cys112 strand depends upon the coupling efficiencies of the hydrogen bonds. Model-complex studies have demonstrated efficient electron transfer across hydrogen-bonded interfaces (101, 102). In the standard Beratan-Onuchic pathway model, hydrogen-bond couplings are distance-scaled and generally afford weaker couplings than covalent bonds (82). This procedure for calculating hydrogen-bond couplings cannot explain the similar distance dependences of ET along the Met121 and Cys112 strands in Ru-modified azurins. Treating the hydrogen bonds as covalent bonds in the tunneling-pathway model ($\varepsilon_H = \varepsilon_C^2$), however, does lead to better agreement with experiment (85).

Long-range ET from the Cys3-Cys26 disulfide radical anion to the copper in azurin has been studied extensively by Farver and Pecht (103–105). Estimates based on experimental rate data indicate that the S_2/Cu coupling is unusually strong for a donor/acceptor pair separated by 26 Å. Relatively strong Cu/Ru couplings also have been found for ET reactions involving Ru-modified His83 (85). Interestingly, both the Cys3-Cys26 and His83 couplings fit on the 1.1 Å^{-1} distance decay defined by the couplings along the Met121 and Cys112 strands (Figure 7). Strong interstrand hydrogen bonds may be responsible for the efficient couplings from the disulfide site and from His83. The tunneling-pathway model can only explain the electronic couplings to these two sites if hydrogen-bond couplings in this β-sheet protein are comparable to covalent-bond couplings. Thus, β sheets appear to be tightly knit structures that efficiently and isotropically mediate distant electronic couplings.

α-Helical Proteins

Donor-acceptor pairs separated by α helices include the heme-Ru redox sites in two Ru-modified myoglobins, Ru(bpy)$_2$(im)(HisX)-Mb (X = 83,95; Figure 8) (106–107) and the Q_A–BCh$_2$ and Q_B–BCh$_2$ (Q_A = menaquinone, Q_B = ubiquinone, BCh$_2$ = bacteriochlorophyll special pair; Figure 9) redox centers in the photosynthetic reaction center (1, 86, 108). The tunneling pathway from His95 to the Mb-heme is comprised of a short section of α helix terminating

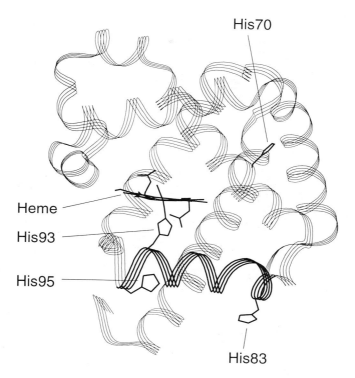

Figure 8   Structure of Mb including the His residues that have been coordinated to Ru(bpy)$_2$(im)$^{2+}$.

at His93, the heme axial ligand. The coupling for the [Fe^{2+} → Ru^{3+} (His95)]-Mb ET reaction (106) is of the same magnitude as that found in Ru-modified azurins with comparable D-A spacings. This result is consistent with the tunneling-pathway model, which predicts very little difference in the coupling efficiencies of α helices and β sheets at small D-A separations (Figure 5). The electronic couplings estimated from the [Fe^{2+} → Ru^{3+}(His83)]-Mb (107) and [Q$_{A,B}^-$ →BCh$_2^+$]-RC (86) ET rates, however, are substantially weaker than those found in β-sheet structures at similar separations, suggesting a larger distance-decay constant for α helices (Figure 7). Differences in hydrogen bonding in β sheets and α helices may be responsible for this behavior. Infrared spectra in the amide I (ν_{CO}, CO stretch) region show that hydrogen bonding in α helices (ν_{CO} = 1650–1660 cm^{-1}) is significant (nonhydrogen-bonded peptides, ν_{CO} = 1680–1700 cm^{-1}), but it is not as strong as that in β sheets (ν_{CO} ~ 1630 cm^{-1}) (109, 110). If spectroscopically derived hydrogen-bond strengths reflect

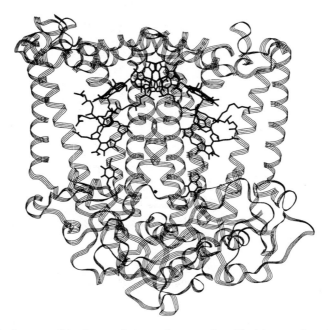

Figure 9   Structure of the photosynthetic reaction center from *Rhodobacter sphaeroides* (108).

electronic-coupling efficiencies, then long-range couplings at given distances along α helices will be weaker than those at corresponding distances along β strands.

Coupling Zones

The tunneling-pathway model suggests that different protein secondary structures mediate electronic coupling with different efficiencies, a notion supported by experimental evidence. We can define different ET coupling zones in a rate vs distance plot (Figure 10). The β-sheet zone, representing efficient mediation of electronic coupling, is bound by coupling-decay constants of 0.9 and 1.15 Å^{-1}. All of the ET rates measured with Ru-modified azurin fall in this zone. The α-helix zone describes systems with coupling-decay constants between 1.25 and 1.6 Å^{-1}. ET rates from Ru(His83)-modified myoglobin and the two RC Q-BCh$_2$ pairs lie in this zone. ET rate data are available for a Ru-modified myoglobin (His70) where the intervening medium is not a simple section of α helix; the His70-Mb ET rate lies in the β-sheet zone (106). In the photosynthetic reaction center, two BCh$_2^{+}$ hole-filling reactions occur over relatively

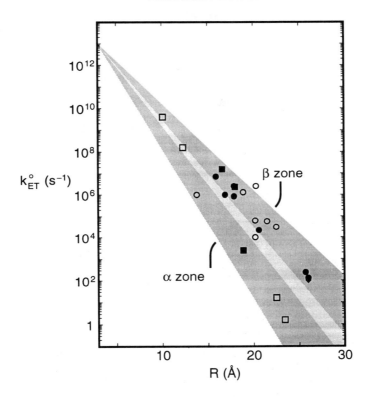

Figure 10 Plot of log k_{ET}° *vs R* illustrating the different ET coupling zones. Zones (shaded regions) are bounded by the following distance-decay lines: α zone, 1.25 and 1.6 Å^{-1}; β zone, 1.15 and 0.9 Å^{-1}. The lighter shaded region is the interface between the α and β zones. For Ru-bpy-modified proteins, metal-metal separation distances are used. Distances between redox sites in the RC are reported as edge-edge separations. Ru-modified azurin data (filled circles) (45, 85, 98): (Ru-label site, k_{ET}° s^{-1}, R Å] His122, 7.1×10^6, 15.9; His124, 2.2×10^4, 20.6; His126, 1.3×10^2, 26.0; His109, 8.5×10^5, 17.9; His107, 2.4×10^2, 25.7; His83 1.0×10^6, 16.9. Ru-modified myoglobin data (filled squares) (106, 107): His83 2.5×10^3, 18.9; His95 2.3×10^6, 18.0; His70 1.6 10^7, 16.6. Ru-modified cyt *c* data (open circles) (45): His39, 3.3×10^6, 20.3; His33, 2.7×10^6, 17.9; His66, 1.3×10^6, 18.9; His72, 1.0×10^6, 13.8; His58, 6.3×10^4, 20.2; His62, 1.0×10^4, 20.2; His54, 3.1×10^4, 22.5; His54(Ile52), 5.8×10^4, 21.5. Cys3-Cys26($S\bar{2}$) → Cu^{2+} ET in azurin (diamond) (103–105): 1.0×10^2, 26. RC data (open squares) (86): [donor to BCh_2^+, k_{ET}° s^{-1}, R Å] Q_A^-, 1.6×10^1, 22.5; Q_B^-, 1.6, 23.4; BPh^-, 4.0×10^9, 10.1; cyt c_{559}, 1.6×10^8, 12.3.

short distances where the differences between the β-sheet and α-helix zones are less distinct: The observed rates lie between the two zones (86).

The coupling-zone concept sets the stage for analyses of ET rates in a wide variety of proteins. Cyt *c* has a tightly packed structure but is not dominantly

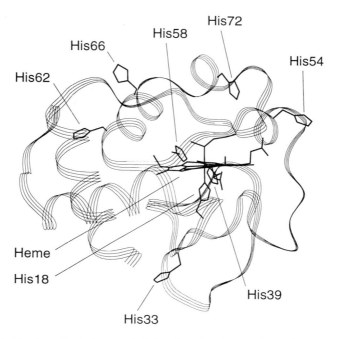

His72

His58

His66

His62   His54

Heme

His18

His39

His33

Figure 11   Structure of cytochrome c including the His residues that have been coordinated to Ru(bpy)$_2$(im)$^{2+}$.

α helical or β sheet (Figure 11). Analysis of the distance dependence of the Fe^{2+} → Ru^{3+} ET rates in eight Ru-modified derivatives of cyt c suggests that this protein mediates electronic coupling with an efficiency comparable to that found in azurin (45). Even though the structures separating redox sites are not simple β strands or sheets, most of the tunneling pathways are fairly direct and therefore provide effective coupling routes. An exception is His72-modified cyt c: The metal-metal separation distance in this protein is 13.8 Å, yet the ET rate is just 10^6 s^{-1} (a value in the α-helix zone; Figure 10). The poor coupling, however, is not due to an intervening α helix, but rather to a poor pathway that includes a through-space jump (42). The coupling estimated from a driving-force study of ET kinetics in Ru(bpy)$_3$$^{2+}$-modified cyt b_5 places this system in the region between the α-helix and β-sheet zones (53).

The variations in coupling efficiencies among different protein secondary structures could have important functional consequences. In a crude sense, β-sheet structures could act as conducting pathways through proteins while α helices might provide insulation against long-range electron transfer. In multisubunit redox enzymes such as cytochrome c oxidase, the structure between subunits may play a key role in directing and regulating electron flow. Inspection of the structure of the oxidase reveals that ET from Cu_A (subunit II) to cytochrome a (subunit I) occurs over a 19-Å Cu-Fe distance through a direct coupling pathway consisting of 14 covalent bonds and 2 hydrogen bonds (9, 10, 111). Based on the relative bond couplings extracted from work on Ru-modified proteins, the 19-Å Cu_A/cyt a ET rate falls in the efficient (β) coupling zone of Figure 9 (k_{ET}° between 4×10^4 and 8×10^5 s^{-1}). With these k_{ET}° values, the reorganization energy for Cu_A to cyt a ET must be between 0.15 and 0.5 eV (111). It apparently is the combination of a low reorganization energy and an efficient ET pathway that allows electrons to flow rapidly with only a small change in free energy from the Cu_A center of subunit II to cytochrome a in subunit I of the oxidase.

ACKNOWLEDGMENTS

We thank BG Malmström, JN Onuchic, JJ Regan, MJ Therien, and K Warncke for helpful discussions. Our work on protein electron transfer is supported by the National Institutes of Health, the National Science Foundation, and the Arnold and Mabel Beckman Foundation.

Literature Cited

1. Deisenhofer J, Epp O, Sinning I, Michel H. 1995. *J. Mol. Biol.* 246:429–57
2. Bixon M, Fajer J, Feher G, Freed JH, Gamliel D, et al. 1992. *Isr. J. Chem.* 32:369–518
3. Steffen MA, Lao KQ, Boxer SG. 1994. *Science* 264:810–16
4. Heller BA, Holten D, Kirmaier C. 1995. *Science* 269:940–45
5. Malmström BG. 1990. *Chem. Rev.* 90:1247–60
6. Malmström BG. 1993. *Acc. Chem. Res.* 26:332–38
7. Gennis R, Ferguson-Miller S. 1995. *Science* 269:1063–64
8. Einarsdóttir Ó. 1995. *Biochim. Biophys. Acta* 1229:129–47
9. Tsukihara T, Aoyama H, Yamashita E, Tomizaki T, Yamaguchi H, et al. 1995. *Science* 269:1071–74
10. Iwata S, Ostermeier C, Ludwig B, Michel H. 1995. *Nature* 376:660–69
11. Winkler JR, Malmström BG, Gray HB. 1995. *Biophys. Chem.* 54:199–209
12. Varotsis C, Zhang Y, Appelman EH, Babcock GT. 1993. *Proc. Natl. Acad. Sci. USA* 90:237–41
13. Babcock GT, Wikström M. 1992. *Nature* 356:301–9
14. Rousseau DL, Han SH, Song SH, Ching YC. 1992. *J. Raman Spectrosc.* 23:551–56
15. Kitagawa T, Mizutani Y. 1994. *Coord. Chem. Rev.* 135:685–735
16. Marcus RA. 1993. *Angew. Chem. Int. Ed. Engl.* 32:1111–21
17. Marcus RA, Sutin N. 1985. *Biochim. Biophys. Acta* 811:265–322
18. Winkler JR, Nocera DG, Yocom KM, Bordignon E, Gray HB. 1982. *J. Am. Chem. Soc.* 104:5798–800
19. Conrad DW, Zhang H, Stewart DE, Scott RA. 1992. *J. Am. Chem. Soc.* 114:9909–15
20. Yocom KM, Shelton JB, Shelton JR, Schroeder WE, Worosila G, et al. 1982. *Proc. Natl. Acad. Sci. USA* 79:7052–55
21. Matthews CR, Erickson PM, Van Vliet DL, Petersheim M. 1978. *J. Am. Chem. Soc.* 100:2260–62

22. Nocera DG, Winkler JR, Yocom KM, Bordignon E, Gray HB. 1984. *J. Am. Chem. Soc.* 106:5145–50
23. Isied SS, Kuehn C, Worosila G. 1984. *J. Am. Chem. Soc.* 106:1722–26
24. Elias H, Chou MH, Winkler JR. 1988. *J. Am. Chem. Soc.* 110:429–34
25. Meade TJ, Gray HB, Winkler JR. 1989. *J. Am. Chem. Soc.* 111:4353–56
26. Therien MJ, Selman MA, Gray HB, Chang I-J, Winkler JR. 1990. *J. Am. Chem. Soc.* 112:2420–22
27. Crutchley RJ, Ellis WR Jr, Gray HB. 1985. *J. Am. Chem. Soc.* 107:5002–4
28. Axup AW, Albin M, Mayo SL, Crutchley RJ, Gray HB. 1988. *J. Am. Chem. Soc.* 110:435–39
29. Karas JL, Lieber CM, Gray HB. 1988. *J. Am. Chem. Soc.* 110:599–600
30. Winkler JR, Gray HB. 1992. *Chem. Rev.* 92:369–79
31. Fenwick C, Marmor S, Govindaraju K, English AM, Wishart JF, Sun J. 1994. *J. Am. Chem. Soc.* 116:3169–70
32. Jackman MP, Lim MC, Salmon GA, Sykes AG. 1988. *J. Chem. Soc. Chem. Commun.*, pp. 179–80
33. Jackman MP, Lim MC, Sykes AG, Salmon GA. 1988. *J. Chem. Soc. Dalton Trans.*, pp. 2843–50
34. Kostic NM, Margalit R, Che C-M, Gray HB. 1983. *J. Am. Chem. Soc.* 105:7765–67
35. Jackman MP, Sykes AG. 1987. *J. Chem. Soc. Chem. Commun.*, pp. 65–66
36. Jackman MP, McGinnis J, Powls R, Salmon GA, Sykes AG. 1988. *J. Am. Chem. Soc.* 110:5880–87
37. Farver O, Pecht I. 1989. *FEBS Lett.* 244:379–82
38. Farver O, Pecht I. 1990. *Inorg. Chem.* 29:4855–58
39. Jacobs BA, Mauk MR, Funk WD, MacGillivray RTA, Mauk AG, Gray HB. 1991. *J. Am. Chem. Soc.* 113:4390–94
40. Osvath P, Salmon GA, Sykes AG. 1988. *J. Am. Chem. Soc.* 110:7114–18
41. Chang I-J, Gray HB, Winkler JR. 1991. *J. Am. Chem. Soc.* 113:7056–57
42. Wuttke DS, Bjerrum MJ, Winkler JR, Gray HB. 1992. *Science* 256:1007–9
43. Casimiro DR, Richards JH, Winkler JR, Gray HB. 1993. *J. Phys. Chem.* 97:13073–77
44. Mines GA, Bjerrum MJ, Hill MG, Casimiro DR, Chang I-J, et al. 1996. *J. Am. Chem. Soc.* 118:1961–65
45. Langen R, Chang I-J, Germanas JP, Richards JH, Winkler JR, Gray HB. 1995. *Science* 268:1733–35
46. Closs GL, Calcaterra LT, Green NJ, Penfield KW, Miller JR. 1986. *J. Phys. Chem.* 90:3673–83
47. Closs GL, Miller JR. 1988. *Science* 240: 440–47
48. Wasielewski MR, Niemczyk MP, Svec WA, Pewitt EB. 1985. *J. Am. Chem. Soc.* 107:1080–82
49. Fox LS, Kozik M, Winkler JR, Gray HB. 1990. *Science* 247:1069–71
50. Chen P, Duesing R, Graff DK, Meyer TJ. 1991. *J. Phys. Chem.* 95:5850–58
51. McLendon G, Miller JR. 1985. *J. Am. Chem. Soc.* 107:7811–16
52. McLendon G, Hake R. 1992. *Chem. Rev.* 92:481–90
53. Scott JR, Willie A, Mark M, Stayton PS, Sligar SG, et al. 1993. *J. Am. Chem. Soc.* 115:6820–24
54. Brunschwig BS, Ehrenson S, Sutin N. 1986. *J. Phys. Chem.* 90:3657–68
55. Brunschwig BS, Ehrenson S, Sutin N. 1987. *J. Phys. Chem.* 91:4714–23
56. Brunschwig BS, Sutin N. 1987. *Comments Inorg. Chem.* 6:209–35
57. Hoffman BM, Ratner MA. 1987. *J. Am. Chem. Soc.* 109:6237–43
58. Hoffman BM, Ratner MA, Wallin SA. 1990. In *Electron Transfer in Biology and the Solid State: Inorganic Compounds with Unusual Properties*, ed. M Johnson, pp. 125–46. Washington, DC: Am. Chem. Soc.
59. Brunschwig BS, Sutin N. 1989. *J. Am. Chem. Soc.* 111:7454–65
60. Brown GM, Sutin N. 1979. *J. Am. Chem. Soc.* 101:883–92
61. Williams RJP, Concar D. 1986. *Nature* 322:213–14
62. Bechtold R, Kuehn C, Lepre C, Isied SS. 1986. *Nature* 322:286–88
63. Moore GR, Williams RJP. 1980. *Eur. J. Biochem.* 103:523–32
64. Jeng M-F, Englander W, Elöve GA, Wand AJ, Roder H. 1990. *Biochemistry* 29:10433–36
65. Pascher T, Chesick JP, Winkler JR, Gray HB. 1996. *Science* 271:1558–60
66. Sun J, Wishart JF, Isied SS. 1995. *Inorg. Chem.* 34:3998–4000
67. Antonini E, Brunori M. 1971. *Hemoglobin and Myoglobin in Their Reactions with Ligands.* Amsterdam: North-Holland
68. King BC, Hawkridge FM, Hoffman BM. 1992. *J. Am. Chem. Soc.* 114: 10603–8
69. Wang ZY, Pearlstein RM, Jia YW, Fleming GR, Norris JR. 1993. *Chem. Phys.* 176:421–25
70. Jia YW, Dimagno TJ, Chan CK, Wang ZY, Du M, et al. 1993. *J. Phys. Chem.* 97:13180–91
71. Pan LP, Hibdon S, Liu R-Q, Durham B, Millett F. 1993. *Biochemistry* 32: 8492–98

72. Nilsson T. 1992. *Proc. Natl. Acad. Sci. USA* 89:6497–501
73. Newton MD. 1988. *J. Phys. Chem.* 92:3049–56
74. Hopfield JJ. 1974. *Proc. Natl. Acad. Sci. USA* 71:3640–44
75. McConnell HM. 1961. *J. Chem. Phys.* 35:508–15
76. Liang C, Newton MD. 1992. *J. Phys. Chem.* 96:2855–66
77. Liang C, Newton MD. 1993. *J. Phys. Chem.* 97:3199–211
78. Curtiss LA, Naleway CA, Miller JR. 1993. *Chem. Phys.* 176:387–405
79. Shephard MJ, Paddon-Row MN, Jordan KD. 1993. *Chem. Phys.* 176:289–304
80. Beratan DN, Onuchic JN, Hopfield JJ. 1987. *J. Chem. Phys.* 86:4488–98
81. Onuchic JN, Beratan DN. 1990. *J. Chem. Phys.* 92:722–33
82. Onuchic JN, Beratan DN, Winkler JR, Gray HB. 1992. *Annu. Rev. Biophys. Biomol. Struct.* 21:349–77
83. Beratan DN, Betts JN, Onuchic JN. 1992. *J. Phys. Chem.* 96:2852–55
84. Beratan DN, Betts JN, Onuchic JN. 1991. *Science* 252:1285–88
85. Regan JJ, Di Bilio AJ, Langen R, Skov LK, Winkler JR, et al. 1995. *Chem. Biol.* 2:489–96
86. Moser CC, Keske JM, Warncke K, Farid RS, Dutton PL. 1992. *Nature* 355:796–802
87. Farid RS, Moser CC, Dutton PL. 1993. *Curr. Opin. Struct. Biol.* 3:225–33
88. Skourtis SS, Regan JJ, Onuchic JN. 1994. *J. Phys. Chem.* 98:3379–88
89. Siddarth P, Marcus RA. 1990. *J. Phys. Chem.* 94:8430–34
90. Siddarth P, Marcus RA. 1990. *J. Phys. Chem.* 94:2985–89
91. Siddarth P, Marcus RA. 1992. *J. Phys. Chem.* 96:3213–17
92. Siddarth P, Marcus RA. 1993. *J. Phys. Chem.* 97:13078–82
93. Siddarth P, Marcus RA. 1993. *J. Phys. Chem.* 97:2400–5
94. Gruschus JM, Kuki A. 1992. *Chem. Phys. Lett.* 192:205–12
95. Gruschus JM, Kuki A. 1993. *J. Phys. Chem.* 97:5581–93
96. Friesner RA. 1994. *Structure* 2:339–43
97. Evenson JW, Karplus M. 1993. *Science* 262:1247–49
98. Langen R. 1995. *Electron Transfer in Proteins: Theory and Experiment.* PhD thesis. Calif. Inst. Technol., Pasadena. 208 pp.
99. Lowery MD, Solomon EI. 1992. *Inorg. Chim. Acta* 200:233–43
100. Guckert JA, Lowery MD, Solomon EI. 1995. *J. Am. Chem. Soc.* 117:2817–44
101. Turro C, Chang CK, Leroi GE, Cukier RI, Nocera DG. 1992. *J. Am. Chem. Soc.* 114:4013–15
102. de Rege PJF, Williams SA, Therien MJ. 1995. *Science* 269:1409–1413
103. Farver O, Skov LK, Vandekamp M, Canters GW, Pecht I. 1992. *Eur. J. Biochem.* 210:399–403
104. Farver O, Pecht I. 1994. *Biophys. Chem.* 50:203–16
105. Farver O, Skov LK, Pascher T, Karlsson BG, Nordling M, et al. 1993. *Biochemistry* 32:7317–22
106. Langen R, Colón JL, Casimiro DR, Karpishin TB, Winkler JR, Gray HB. 1996. *J. Biol. Inorg. Chem.* In press
107. Casimiro DR. 1994. *Electron Transfer in Ruthenium-Modified Recombinant Cytochromes and Myoglobins.* PhD thesis. Calif. Inst. Technol., Pasadena. 263 pp.
108. Ermler U, Fritzsch G, Buchanon SK, Michel H. 1994. *Structure* 2:925–36
109. Schellman JA, Schellman C. 1962. In *The Proteins*, ed. H Neurath. New York: Academic. 2nd ed.
110. Susi H. 1972. *Methods Enzymol.* 26:455–72
111. Ramirez BE, Malmström BG, Winkler JR, Gray HB. 1995. *Proc. Natl. Acad. Sci. USA* 92:11949–51

Annu. Rev. Biochem. 1996. 65:563–607

CROSSTALK BETWEEN NUCLEAR AND MITOCHONDRIAL GENOMES

Robert O. Poyton

Department of Molecular, Cellular, and Developmental Biology, University of Colorado, Boulder, Colorado 80309

Joan E. McEwen

Department of Medicine and Division on Aging, University of Arkansas for Medical Sciences, Little Rock, Arkansas 72205

KEY WORDS: respiratory proteins, mitochondrial biogenesis, intracellular signal transduction, gene expression, mitochondrial diseases

---

ABSTRACT

This review focuses on molecular mechanisms that underlie the communication between the nuclear and mitochondrial genomes in eukaryotic cells. These genomes interact in at least two ways. First, they contribute essential subunit polypeptides to important mitochondrial proteins; second, they collaborate in the synthesis and assembly of these proteins. The first type of interaction is important for the regulation of oxidative energy production. Isoforms of the nuclear-coded subunits of cytochrome c oxidase affect the catalytic functions of its mitochondrially coded subunits. These isoforms are differentially regulated by environmental and developmental signals and probably allow tissues to adjust their energy production to different energy demands. The second type of interaction requires the bidirectional flow of information between the nucleus and the mitochondrion. Communication from the nucleus to the mitochondrion makes use of proteins that are translated in the cytosol and imported by the mitochondrion. Communication from the mitochondrion to the nucleus involves metabolic signals and one or more signal transduction pathways that function across the inner mitochondrial membrane. An understanding of both types of interaction is important for an understanding of OXPHOS diseases and aging.

563

CONTENTS

PERSPECTIVES 564
NUCLEAR AND MITOCHONDRIAL CONTRIBUTIONS TO RESPIRATORY
    PROTEIN FUNCTION 566
NUCLEAR AND MITOCHONDRIAL CONTRIBUTIONS TO THE REGULATION
    OF ENERGY PRODUCTION..................................... 569
    Regulation of Cellular Energy Production by Cytochrome c Oxidase............ 569
    Subunit Function in Cytochrome c Oxidase 571
    Environmental Control of Cytochrome c Oxidase Levels..................... 574
    Nuclear-Coded Cytochrome c Oxidase Subunits and Regulation................. 575
    Yeast Subunit V Isoforms and Regulation 576
NUCLEAR AND MITOCHONDRIAL CONTRIBUTIONS TO MITOCHONDRIAL
    GENE EXPRESSION 577
    Transcription.. 578
    RNA Processing 580
    Translation 582
    Global Regulation... 585
    Gene Specific Control— Is There Differential Regulation of Mitochondrial Gene
        Expression?... 586
NUCLEAR CONTROL OF MITOCHONDRIAL ASSEMBLY 587
    Import, Export, and Assembly Pathways........................... 587
    Chaperonins 587
    Proteases.. 588
    Assembly of Multimeric Proteins and Specific Assembly Factors 588
    Mitochondrial Transmission.................................... 590
COORDINATION OF NUCLEAR AND MITOCHONDRIAL GENE EXPRESSION.. 591
SIGNALLING FROM THE MITOCHONDRION TO THE NUCLEUS 592
    Phenomenology.. 592
    Retrograde Regulation 594
    Intergenomic Signalling 596
    Reactive Oxygen Species as Second Messengers in Retrograde Regulation 597
CONCLUSIONS AND FUTURE PROSPECTS 599

PERSPECTIVES

The assembly and function of respiratory-competent mitochondria in eukaryotic cells result from a collaboration between gene products derived from both mitochondrial and nuclear genomes (1–4). Most of the proteins that reside in the mitochondrion are nuclear gene products. These proteins are translated in the cytoplasmic compartment of eukaryotic cells and subsequently transported into the mitochondrion post-translationally (5). These proteins play key roles in mitochondrial transcription and translation, mitochondrial lipid and heme synthesis, substrate oxidation by the TCA cycle, and mitochondrial electron transport and oxidative phosphorylation. In short, these proteins, which account for about 90% of the protein mass of the mitochondrion, play roles in all aspects of mitochondrial function, and they are distributed in all four mitochondrial compartments (inner and outer mitochondrial membranes, matrix, and intermembrane space). In contrast, the mitochondrial genome specifies only a few proteins, which reside mainly in the inner mitochondrial membrane.

The mitochondrial genome also encodes RNA molecules that co-assemble with proteins encoded in the nucleus.

The inventory of nuclear- and mitochondrially coded proteins required to assemble a functional mitochondrion shows clearly that nuclear and mitochondrial genomes interact in at least two ways. First, both nuclear and mitochondrial genes contribute to mitochondrial protein function. Second, both nuclear and mitochondrial genomes interact to affect the synthesis and assembly of mitochondrial proteins. The first level of interaction is obvious from the fact that mitochondrial gene products do not act alone; they are components of multimeric protein complexes that contain nuclear-encoded components as well. Four of these genetic chimeric complexes, common to all eukaryotes, are coenzyme Q cytochrome c reductase, cytochrome c oxidase, F_1F_0 ATPase, and the mitochondrial ribosome (6). Additional, but not universal, chimeric protein complexes include NADH dehydrogenases of vertebrates, plants, trypanosomes, and some fungal mitochondria, as well as small and large ribosomal subunits, RNase-P, and RNA splicing enzymes of fungal mitochondria (1).

The second level of interaction, biosynthesis of mitochondrial proteins, is obvious from the genetic circuitry involved in mitochondrial biogenesis. In addition to encoding proteins that are required for the energetic functions of the mitochondrion, the nuclear genome encodes many PET genes that function either in the expression of mitochondrial genes or in the assembly of respiratory proteins (7, 8). These trans-acting nuclear genes serve to modulate either the level of expression of the mitochondrial genome as a whole or the expression of individual genes on the mitochondrial genome. Similarly, some nuclear-coded "assembly factors" are required for assembly of many mitochondrial proteins, whereas others appear to act on individual respiratory complexes. Communication from the nuclear genome to the mitochondrion involves proteins that are translated in the cytosol and imported into the mitochondrion. Hence, the communication pathway by which the nucleus "talks to" the mitochondrion requires the machinery of mitochondrial protein import. Just as the nuclear genome affects the expression of mitochondrial genes, the mitochondrial genome can affect the expression of nuclear genes for mitochondrial (and other) proteins (9, 10; JD Trawick, LE Farrell, N Kraut, S Silve, D Raitt & RO Poyton, unpublished observations). The molecular details that underlie this recently discovered phenomenon are still being clarified, but this communication pathway is likely to be fundamentally different than the communication pathway from the nucleus to the mitochondrion. As discussed below, communication from the mitochondrion to the nucleus probably involves metabolic signals and one or more signal transduction pathways that function across the inner mitochondrial membrane.

Superimposed on the genetic circuitry that connects nuclear and mitochondrial genomes is environmental control of mitochondrial function and biogenesis. Several factors such as oxygen level, carbon source, and hormones are known to alter mitochondrial protein levels and/or function (12–17). This regulation is important for adjusting the energy-producing capacity of a cell to both short-term and long-term changes in energy demand. Progress is being made toward understanding the signal transduction pathways by which some of these environmental effectors alter the expression of genes encoding mitochondrial proteins. However, much remains to be learned about how the signal transduction pathways that are affected by environmental factors interface with the pathways that balance nuclear and mitochondrial gene expression and respiratory protein function. An understanding of this problem should impact a growing number of human diseases that are associated with defects in oxidative phosphorylation (18–20) and may enhance our understanding of human evolution (21).

In this review, we consider the two levels of crosstalk between nuclear and mitochondrial genomes. First, we focus on the contributions of nuclear- and mitochondrially coded proteins to the function of the respiratory chain and how the interplay between these two sets of proteins in cytochrome c oxidase plays a role in the regulation of cellular energy metabolism. Then we discuss the pathways that balance nuclear and mitochondrial gene expression and consider ways in which these pathways respond to environmental regulation.

NUCLEAR AND MITOCHONDRIAL CONTRIBUTIONS TO RESPIRATORY PROTEIN FUNCTION

The mitochondrial respiratory chain consists of a peripheral membrane protein, cytochrome c, and four multimeric membrane proteins designated Complex I (NADH: ubiquinone oxidoreductase), Complex II (succinate: ubiquinone oxidoreductase), Complex III (ubiquinol: ferricytochrome c oxidoreductase), and Complex IV (cytochrome c oxidase). These respiratory protein complexes act in sequence to accept reducing equivalents from NADH or $FADH_2$ and transfer them through a series of oxidation-reduction reactions to O_2, resulting in the formation of water and the generation of proton and ion gradients across the inner mitochondrial membrane. These latter gradients are used to drive the synthesis of ATP by another multimeric protein, Complex V (ATP synthase) (22). Combined, the respiratory proteins and ATP synthase consist of many electron-carrying prosthetic groups and many polypeptides. The total number of polypeptides, as well as the number of polypeptides encoded by nuclear and mitochondrial genes, varies from organism to organism (23). For example, in human (mammalian) cells, respiratory proteins and ATP synthase consist of about 74 polypeptides. Complex I contains 7 mitochondrial DNA (mtDNA) gene products and at least 25 nuclear DNA (nDNA) gene products (24, 25); Complex II contains no mtDNA gene products and 4 nDNA gene products (26); Complex III contains 1 mtDNA gene product and 10 nDNA gene products (27); Complex IV contains 3 mtDNA gene products and 10 nDNA gene products (28); and Complex V contains 2 mtDNA gene products and 11 nDNA gene products (29).

In the fungus *Saccharomyces cerevisiae*, the respiratory chain has a somewhat simpler composition. In this organism the respiratory chain and ATP synthase consist of 34 to 36 polypeptides. These differences are attributable to the absence of Complex I (30) in *S. cerevisiae* and to differences in the numbers of subunits contributed to complex III, complex IV, and ATP synthase by the nuclear and mitochondrial genomes. Complex II contains no mtDNA products and four nDNA products (31). Complex III contains one mtDNA gene product and eight nDNA gene products (32); Complex IV contains three mtDNA gene products and six–eight nDNA gene products (33–35); and Complex V contains three mtDNA gene products and eight nDNA gene products (36).

Are all of the polypeptides in preparations of Complexes I–V bonafide subunits or are some of them contaminating polypeptides that bind adventitiously to these complexes and copurify with them? This question is a general and persistent problem confronted by researchers who study multimeric membrane proteins. Because these proteins are removed from the membrane phospholipid bilayer by detergent solubilization, determining where the enzyme ends and the membrane begins is sometimes difficult. The assignment of a polypeptide as a subunit of a respiratory complex is important for an understanding of the structure and function of these proteins. Such an assignment is also important for understanding the evolution of these proteins, because the number of polypeptides associated with some complexes (i.e. Complexes III, IV, and V) differs among prokaryotes, lower eukaryotes, and higher eukaryotes (32–37). For Complexes III and IV these differences result from different numbers of polypeptides contributed by the nuclear genome. The number of polypeptides contributed to these two complexes by the mitochondrial genome is conserved in most eukaryotes. Moreover, the primary sequences of the mitochondrially coded polypeptides in these two complexes is conserved among eukaryotic species and between eukaryotes and prokaryotes (32, 38). This conservation in both number and primary sequence is one argument for the conclusion that the mitochondrially coded polypeptides are subunits of their respective complexes (32, 38). This argument cannot be made for the nuclear-coded polypeptides in Complex IV because they are absent from functional Complex IV preparations from prokaryotes. Similarly, except for cytochrome c_1 and the Fe-S protein, this argument cannot be made for the nuclear-coded subunits of Complex III. Moreover, these nuclear-coded polypeptides in Complexes III and IV have low primary sequence conservation among eukaryotic species (cf 39, 40).

The identification of a polypeptide as an integral part (i.e. bonafide subunit) of a respiratory complex has not been a problem for polypeptides that have easily demonstrable functions. For example, polypeptides that bind the electron-carrying prosthetic groups are obviously subunits because they are essential for the catalytic function of their respective complexes. These include three polypeptides (23 kDa, 24 kDa, and 75 kDa) that carry iron-sulfur clusters in Complex I (41, 42); CII-1, CII-2, and CII-3 in Complex II (43, 44); cytochrome b, cytochrome c_1, and the Rieske Fe-S proteins in Complex III (32); and subunits I and II in Complex IV (45). What about those subunits that lack prosthetic groups? In principle, these may also function in catalysis, they may play regulatory roles, or they may function in the structure or assembly of these complexes. Polypeptides that function in catalysis, assembly, or structure are expected to be essential; polypeptides that perform a regulatory function may or may not be essential. The essentiality of polypeptides in respiratory complexes has been addressed in three ways: through biochemical depletion experiments in which individual subunits are removed from the complex (46–48); through biochemical reconstitution studies, with purified polypeptides or groups of polypeptides (cf 49, 50); and through genetic studies, with strains that carry mutations (preferably deletions) in the genes for each polypeptide (12). The first two approaches have had limited usefulness. The third approach has been very useful, but only with genetically tractable organisms such as *S. cerevisiae* and *Neurospora crassa.*

In *S. cerevisiae,* the genetic approach has been successfully applied to the subunit polypeptides of Complexes III, IV, and V. For Complex III, cytochrome b, cytochrome c_1, the Reiske Fe-S protein, core 1, core 2, subunit 7, subunit 8, and subunit 9 are all essential in vivo (32). Yeast cells that lack these subunits have greatly reduced levels of Complex III activity and in most cases fail to grow on nonfermentable carbon sources (32). In contrast, yeast cells with a deletion in the gene for subunit 6 have half of the wild-type level of Complex III activity and grow on a nonfermentable carbon source (51). For Complex IV, the genetic approach demonstrated that subunits I, II, III, IV, Va or Vb, VI, VII, and VIIa are essential (12, 52, 53). Yeast cells that lack these subunits lack cytochrome c oxidase activity and do not grow on nonfermentable carbon sources. In contrast, a strain deleted for the gene for subunit VIII has 80% wild-type levels of cytochrome c oxidase activity and grows on nonfermentable carbon sources (54). Although subunit 6 of Complex III and subunit VIII of Complex IV are not essential, they are nonetheless part of their respective complexes, in which they play regulatory roles. Subunit 6 appears to facilitate the formation of a complex between cytochromes c_1 and c and is required for the reactivity of both catalytic sites in a Complex III dimer (51).

Subunit VIII of Complex IV is required for the production of stable cytochrome *c* oxidase dimers (55) and affects the low-affinity electron-transfer site from cytochrome *c* (56; TE Patterson, X-J Zhao, LA Allen & RO Poyton, unpublished observations) (see below). As discussed below, the subunit status of VIa and VIb (yeast nomenclature) in Complex IV remains uncertain.

Are the nuclear-coded polypeptides in respiratory complexes from other eukaryotes (e.g. mammals) also essential? With the exception of those polypeptides that contain prosthetic groups this question is largely unanswered. However, one or more of the following criteria have been used to argue that these polypeptides are integral components of their respective complexes: (*a*) They are present in stoichiometric amounts with other subunits and copurify with enzyme activity through several steps. (*b*) Antibodies to these polypeptides inhibit enzyme activity (47). (*c*) Antibodies to these polypeptides precipitate the enzyme (47). (*d*) They have primary sequence similarity to known polypeptide subunits in *S. cerevisiae*. New approaches to this question are currently being pursued. One approach is to use antisense RNA to inhibit the expression of genes for specific polypeptides in these complexes and ask if this results in the loss of enzyme activity (58). Another approach is to use heterologous complementation (59, 60) of yeast mutants with cDNA libraries or cloned genes encoding polypeptides from other species. The latter approach requires that the foreign polypeptide be able to function in yeast. This approach is especially promising because it can potentially show functional homology between polypeptides that may have only limited primary sequence conservation.

In summary, both nuclear and mitochondrially coded polypeptides clearly contribute to Complexes I, III, IV, and V. Genetic studies with *S. cerevisiae* have provided the most compelling evidence that the nuclear-coded polypeptides are essential components of Complexes III, IV, and V, and they have ruled out the assertion that these polypeptides in Complexes III and IV are supernumerary, as suggested previously (32). Moreover, as discussed below, the nuclear-coded polypeptide subunits of Complex IV play an important role in the regulation of energy production in eukaryotes.

NUCLEAR AND MITOCHONDRIAL CONTRIBUTIONS TO THE REGULATION OF ENERGY PRODUCTION

Regulation of Cellular Energy Production by Cytochrome c Oxidase

The major pathway by which eukaryotic cells produce energy in the presence of air is mitochondrial oxidative phosphorylation. Because of its high-energy yield, mitochondrial oxidative phosphorylation is capable of supplying most of the total ATP requirement for eukaryotic cells. How do proteins encoded by nuclear and mitochondrial genes contribute to the regulation of this aerobic energy production in eukaryotic cells? A consideration of the control points in oxidative phosphorylation is useful in addressing this question. Although several factors could interact to affect the rate of mitochondrial respiration and oxidative phosphorylation, several studies have led to the proposal that cytochrome c oxidase is a key enzyme in the overall regulation of cellular energy production in eukaryotes (61, 62). This enzyme, the terminal member of the respiratory chain, catalyzes the concerted transfer of four electrons from ferrocytochrome c to molecular oxygen, as protons are simultaneously pumped across the mitochondrial inner membrane from the matrix to the cytoplasmic side (45).

The hypothesis that cytochrome c oxidase plays a key role in the regulation of energy production is supported by the following findings: (*a*) The redox reactions between NADH and cytochrome c in the mitochondrial respiratory chain (i.e. those that include the first two sites of oxidative phosphorylation) are near equilibrium (63–65), whereas the redox reactions between cytochrome c and O_2, which are catalyzed by cytochrome c oxidase, are essentially irreversible (45). (*b*) Control theory (66, 67) revealed that cytochrome c oxidase is one of the two major steps that have "control strength" high enough to regulate oxidative phosphorylation in mitochondria from both lower eukaryotes (e.g. *S. cerevisiae*) (68) and higher eukaryotes (e.g. rat liver) (69). (*c*) Cytochromes aa$_3$ are limiting in amount with respect to other respiratory chain components in some organisms (e.g. *Candida parapsilosis, S. cerevisiae*) (70) and tissues (e.g. bovine liver) (71). (*d*) ATP (and other anions) have a marked effect on the kinetic properties of cytochrome c oxidase from various eukaryotes (72–75). Together, these findings suggest that one or more of the reactions catalyzed by cytochrome c oxidase is equivalent to the committed step in a metabolic pathway. Thus, these reactions are important control points that match the level of respiration and oxidative phosphorylation to cellular energy requirements.

How do eukaryotic cells alter their cytochrome c oxidase levels in response to changes in energy demand? In principle, two general types of regulation are possible: short-term and long-term. These two types of regulation are distinguishable by response time and by a requirement for protein synthesis and/or turnover. Short-term regulation is immediate, could be affected by allosteric regulation (via ATP or other metabolites) (76), and does not require protein synthesis or turnover. Long-term regulation would be affected by changing the number of functional holoenzyme molecules in the inner membrane or by the assembly of subunit isoforms (which affect the electron transfer or proton-pumping activities of cytochrome c oxidase) into holoenzyme molecules (see below). This type of control requires new protein synthesis and/or turnover. These two types of regulation are also distinguishable by their effect on cellular respiration. Short-term regulation through allosteric regulation would be expected to produce an effect on the holocytochrome c oxidase's Km for one or more of its substrates—O_2, ferrocytochrome c, and protons. In contrast, long-term regulation may modulate respiration though an overall change in respiration rate, effected by changes in the number of assembled holoenzyme molecules or by a change in the turnover numbers of individual holoenzyme molecules. Before discussing long-term and short-term regulation of cytochrome c oxidase levels, we consider the function of cytochrome c oxidase subunits.

Subunit Function in Cytochrome c Oxidase

Eukaryotic cytochrome c oxidases contain four redox-active metal centers (heme a, heme a_3, Cu_A, and Cu_B) embedded in a multisubunit protein matrix. Heme a_3 and Cu_B are bridged in the resting, fully oxidized form of the enzyme and constitute the binuclear reaction center (45). The three largest subunit polypeptides (I, II, and III) are encoded by mitochondrial genes and have primary sequence homology to the three polypeptides that constitute prokaryotic cytochrome c oxidases (38). The other subunit polypeptides are encoded by nuclear genes. The view that the mitochondrially coded subunits perform the catalytic functions of the holoenzyme is generally accepted. This theory is based on the primary sequence homologies between these polypeptides and the polypeptide subunits of prokaryotic cytochrome c oxidases (38), as well as experimental data from genetic, immunological, and biochemical studies (cf 38, 45–47). In both prokaryotic and eukaryotic cytochrome c oxidases, subunit I binds the heme a, heme a_3, and Cu_B redox centers (38, 45). Subunit II binds Cu_A (77) and also participates in cytochrome c binding, presumably at a site close to Cu_A (78). Subunit III appears to modulate, but is not required for, the proton-pumping activity of subunits I and II (46).

If the mitochondrially encoded subunits are sufficient for catalysis, what are the functions of the nuclear-coded subunits? This question has been most readily addressed through genetic studies with yeast cytochrome c oxidase. Active preparations of yeast cytochrome c oxidase contain a total of nine subunits (33, 79); three mitochondrially coded subunits (I, II, and III) and six subunits (IV, Va or Vb, VI, VII, VIIa, and VIII) encoded by nuclear genes (COX4, COX5a or COX5b, COX6, COX7, COX9, and COX8, respectively). COX5a and COX5b encode interchangeable isoforms, Va and Vb, of subunit V (80). The other subunits are specified by unique genes present in single copy on their respective genomes. This enzyme is composed of equimolar amounts of each subunit and, like mammalian cytochrome c oxidases, can exist in the inner mitochondrial membrane as either a monomer or dimer (55, 78).

Deletion mutations in *COX4, COX5a* and *COX5b, COX6, COX7,* and *COX9* lead to the complete loss of cellular respiration, cytochromes a and a$_3$, and cytochrome *c* oxidase activity (52, 53, 80–83). These findings reveal that subunits IV, Va or Vb, VI, VII, and VIIa are essential, but do not identify their role in the holoenzyme. In principle, these subunits could play a role in catalysis, the regulation of catalysis, or the folding or stability of a catalytic subunit (e.g. subunits I, II, and III). Based on the finding that a *COX4 null* mutant lacks detectable cytochromes a and a$_3$ and a fully assembled holoenzyme, the proposal has been made that subunit IV functions in holoenzyme assembly (82). However, an alternative interpretation of these findings is that subunit IV functions in catalysis or its regulation and that, because this polypeptide is part of a multimeric protein, its absence in the *null* mutant leads secondarily to an incompletely assembled, unassembled, or partially degraded holoenzyme (12). These opposing interpretations can also be applied to the role of subunits VI, VII, and VIIa, because *null* mutants in their structural genes have the same phenotype as a *cox4 null* mutant. Recently, subunit I levels have been shown to be greatly reduced in a *cox7 null* mutant, a decrease due to either a decreased rate of translation or rapid degradation of nascent subunit I (53). In addition, the mitochondrially coded subunits II and III (but not subunit I) have been shown to be less stable in a *cox4 null* mutant (84). Together, these results suggest that individual nuclear-coded subunits stabilize specific mitochondrial subunits in the holoenzyme.

In contrast to the complete loss of cytochrome *c* oxidase activity observed in *cox4, cox6, cox7, and cox9 null* mutants, a *null* mutant in *COX8* does not abolish cellular respiration or cytochrome *c* oxidase activity. Instead this mutation reduces them to 80% of wild-type levels (54, 56). Because this mutant has cytochrome *c* oxidase activity, subunit VIII obviously cannot play an indispensible role in holoenzyme assembly or catalysis. Recent studies with this mutant have revealed that subunit VIII affects the low-affinity electron-transfer reaction between cytochrome *c* and cytochrome *c* oxidase (56; TE Patterson, X-J Zhao, LA Allen & RO Poyton, unpublished observations) and is required for the formation of holoenzyme dimers (55). This finding suggests that subunit VIII functions to crosslink holoenzyme monomers and that the dimer participates in the low-affinity electron-transfer reaction. As discussed below (see section on Yeast Subunit V Isoforms and Regulation), the isoforms of subunit V, like subunit VIII, modulate the catalytic functions of the holoenzyme.

Together, the experimental results summarized above suggest that the mitochondrially coded subunits are responsible for catalysis and that the nuclear-coded subunits have at least two functions. Some of the subunits (V and VIII) modulate catalysis (see section on Nuclear-Coded Cytochrome *c* Oxidase Subunits and Regulation and section on Yeast Subunit V Isoforms and Regulation below) while others (subunits IV, VI, VII, and VIIa) are required for the stability of the catalytic subunits and/or the stable assembly of the holo-enzyme.

Recently, a milder purification protocol revealed what may be additional polypeptide subunits in yeast cytochrome c oxidase preparations (34). Lauryl maltoside, used in place of Triton X-100, revealed three additional polypeptides in purified cytochrome c oxidase preparations, bringing the total number of "subunits" to 12. At least two of these polypeptides have partial sequence homology to polypeptides [subunits VIa and VIb, see (78) for nomenclature] found in preparations of mammalian cytochrome c oxidases. Unfortunately, whether the catalytic properties of the 12-subunit enzyme are similar to those of the 9-subunit enzyme was not determined. Because the 9-subunit enzyme from yeast is fully active and has a CO-IR spectrum similar to that for bovine cytochrome c oxidase (85), it is appropriate to ask if these additional polypeptides are in fact bonafide subunits of the enzyme rather than contaminating polypeptides that copurify with it. This question is especially relevant because one of these polypeptides (subunit VIb) is also found in preparations of the F_1 ATPase inhibitor (B Trumpower, personal communication) and because none of them are coprecipitated with other subunits of the 9-subunit yeast enzyme by subunit antisera specific to either subunit VII or subunit Va (RO Poyton, unpublished observations). This question is also relevant to the evolution of cytochrome c oxidase, and it raises three possibilities. First, the mammalian enzyme may be more similar to the yeast enzyme than previously thought. Second, the yeast enzyme has fewer subunits than the bovine enzyme. Third, these polypeptides are not subunits of either the yeast or bovine enzyme.

To address the question of whether these polypeptides are subunits of the yeast enzyme, the genes (*COX12* and *COX13*) for two of them (subunit VIb and VIa, respectively) were disrupted (35, 88). A *cox12 null* mutant has a temperature-sensitive *petite* phenotype, severely diminished cytochrome c oxidase activity, but optically detectable cytochromes aa_3. The fact that this strain assembles cytochrome c oxidase with greatly diminished activity is subject to various interpretations. One possibility is that VIb, like subunit VIII, is required for activity but not assembly and that the enzyme assembled in the mutant has an altered turnover number. Another possibility is that an isoform of VIb is expressed in the *cox12 null* mutant and that the isoform decreases the turnover number of the holoenzyme, as is the case for the Va isoform of subunit V (89). A third possibility is that VIb is an assembly factor required for assembly of cytochrome c oxidase but not necessary for catalytic activity once the complex is assembled. This is a role similar to that ascribed to subunit III of cytochrome c oxidase from *Paracoccus denitrificans* (48). A *cox13 null* mutant has normal levels of assembled cytochrome c oxidase, but it produces a cytochrome c oxidase that is less active when assayed in high ionic strength buffers (88). In addition, the mutant enzyme appears to be differentially affected by adenine nucleotides when compared to the wild-type enzyme. These findings have been interpreted as an indication that the polypeptide is not required for assembly of the holoenzyme but plays a role in its regulation (88). Unfortunately, these later studies were done in whole mitochondria under conditions that were not well characterized with respect to high- and low-affinity electron-transfer reactions (see 85). Moreover, whether the deletion mutation has any differential effects on the expression of the isoforms of cytochrome c is not clear. These shortcomings, together with the finding that this polypeptide can be removed from bovine heart cytochrome c oxidase without affecting activity (90), raise serious doubts about its role as a subunit of holocytochrome c oxidase. With the recently published X-ray structure of the metal sites in bovine heart cytochrome c oxidase, it should soon be possible to determine if these polypeptides are subunits of the holoenzyme (90a).

Environmental Control of Cytochrome c Oxidase Levels

Long-term regulation of cytochrome c oxidase levels can be achieved in one of two ways: (*a*) by changing the number of functional holoenzyme molecules in the inner membrane, or (*b*) by assembly of subunit isoforms that have differential effects on the catalytic properties of the holoenzyme. Both types of regulation are exerted by environmental stimuli. In *S. cerevisiae*, oxygen concentration and carbon source are major regulators of cytochrome c oxidase levels (13–15). In mammals, oxygen concentration (91, 92), muscle contraction (93, 94), and hormone levels (95, 96) are known regulators of cytochrome c oxidase levels. Carbon source appears to have little, if any, effect.

In *S. cerevisiae*, fermentable and repressing carbon sources, such as glucose, reduce the level of respiration and cytochrome c oxidase about three-to fivefold relative to nonfermentable carbon sources like glycerol or lactate and to nonrepressing fermentable carbon sources like galactose (16; RO Poyton & PV Burke, unpublished observations). Much of this reduction in cytochrome c oxidase level appears to be brought about by transcriptional regulation of both nuclear and mitochondrial *COX* genes. However, the nuclear and mitochondrial *COX* genes are regulated by fundamentally different mechanisms. As discussed below, transcriptional regulation of the mitochondrial *COX1, COX2*, and *COX3* genes is coordinated, and is brought about in part by the modulation of levels of nuclear-coded mitochondrial RNA polymerase (98). In contrast, the nuclear *COX* genes are regulated independently of one another via specific transcription factors (99–106). The best understood of these genes is *COX6*. Carbon-source control of this gene requires the general transcription factor ABF1 (101, 102) and the HAP2/3/4 pathway (99–102). This gene regulation is mediated by a signal transduction pathway that results in the differential phosphorylation of ABF1 (102).

Oxygen concentration has more severe effects than carbon source on cytochrome c oxidase levels in *S. cerevisiae*. Cytochrome c oxidase activity is not expressed anaerobically and is maximal in cells grown in air (220 µM O_2) (107, 108). At oxygen concentrations between 0.1 µM and 1 µM, the level of enzyme is proportional to O_2 concentration, and above 1 µM the enzyme level becomes constant. The effects of oxygen on the nuclear-coded subunits of cytochrome c oxidase are exerted through the transcription of their genes (14; PV Burke & RO Poyton, unpublished observations). In contrast, oxygen does not appear to affect transcription of mitochondrial *COX* genes. Oxygen does, however, affect the levels of subunits I and II. These subunits are synthesized aerobically, but not anaerobically, both in vivo and in isolated mitochondria (110) under conditions in which transcription is inhibited (111). Hence, oxygen affects the expression of these two proteins posttranscriptionally. The level of cytochrome c oxidase activity that is produced at different oxygen tensions is determined in part by the number of holoenzyme molecules that are assembled and in part by the oxygen-regulated isoforms of subunit V (85, 89). The aerobic isoform, Va, is expressed at oxygen concentrations down to 0.5 µM (PV Burke & RO Poyton, unpublished observations). The anaerobic (hypoxic) isoform, Vb, is expressed in the absence of oxygen and at O_2 concentrations up to 1 µM (PV Burke & RO Poyton, unpublished observations). As discussed below, these isoforms alter enzyme level by altering the turnover number of holocytochrome c oxidase.

The mechanisms for regulating cytochrome c oxidase levels in mammals are less well understood. However, initial work suggests that the transcription of some of the mammalian *COX* genes is affected by oxygen concentration (112), muscle contraction (113), interferon (114), estrogen (115), vitamin D (116), cyclic nucleotides (117), and cocaine (KA Sevarino, personal communication). Moreover, the transcription of nuclear and mitochondrial genes appears to be coordinated under steady-state conditions (94, 119, 120) but discordant under non-steady-state conditions (95, 96, 121).

Nuclear-Coded Cytochrome c Oxidase Subunits and Regulation

At present, all of the short-term regulation exerted on cytochrome c oxidase appears to flow through one or more of the nuclear-coded subunits. This view is supported, in part, by the findings that ATP and reduced nucleotides, which affect the kinetics of cytochrome c oxidase, bind to one or more of the nuclear-coded subunits in mammalian cytochrome c oxidases (72–76). This theory is also supported by the discovery of isoforms for some of the nuclear-coded subunits in unicellular eukaryotes, *S. cerevisiae* and *D. discoideum* (12, 80,

85, 89, 122–124), and mammals (74). In unicellular eukaryotes, these isoforms are expressed in an oxygen-dependent manner (12, 124), and they affect the catalytic properties of holocytochrome c oxidase (85, 89) (see below). In mammals, these isoforms are tissue specific (125) and are developmentally regulated (126). However, their number varies from mammal to mammal. In bovine, there are two isoforms for subunits VIa (127, 128), VIIa (129), and VIII (130). One of these (the H type) is expressed in heart and skeletal muscle, the other (the L type) in liver (125). In rat, both types of isoform are found for subunits VIa (131) and VIII (132), but only the L isoform of subunit VIIa is expressed. And in humans, subunit VIII occurs as only the L isoform (133, 134), but both isoforms of subunits VIa (135, 136) and VIIa (137–139) are expressed. Recently, the suggestion was made that the VIa isoform can mediate the regulation of cytochrome c oxidase activity in response to allosteric effectors such as nucleotides and ADP (75, 76).

Yeast Subunit V Isoforms and Regulation

The mechanisms by which isoforms for the nuclear-coded subunits affect cytochrome c oxidase activity have been studied extensively in *S. cerevisiae* (85, 89). This organism is ideal for such studies for several reasons. First, in contrast to mammals, *S. cerevisiae* possesses only one nuclear-coded subunit (V) for which isoforms exist (12). Second, these isoforms, Va or Vb, have 66% primary sequence identity and are encoded by essential single-copy genes, *COX5a* and *COX5b* (80, 123). These genes are differentially regulated by oxygen and heme (12, 105, 140), and they allow cells to assemble different types of holoenzyme molecules in response to oxygen concentration. Third, yeast cells are easily manipulated genetically to synthesize cytochrome c oxidase molecules that contain one isoform or the other (80). Fourth, the catalytic functions of yeast cytochrome c oxidase can be assayed in vivo within whole cells (89), thereby avoiding purification related artifacts. Because the catalytic center (i.e. binuclear reaction center) lies within a conserved mitochondrially coded subunit (subunit I), and appears to be itself conserved between prokaryotes, yeast, and mammals (45, 85), findings with the subunit V isoforms in yeast will likely be applicable to other cytochrome c oxidases.

Waterland et al (89) found that the isoforms of subunit V affect the turnover number of holocytochrome c oxidase and that they do so by altering the rates of intramolecular electron transfer between heme a and the binuclear reaction center. In these studies the reaction velocity was enhanced three- to fourfold in the Vb isozyme, whereas the activation energy of the reaction remained unchanged. This finding was taken as evidence that the subunit V isoforms function allosterically to alter the conformation of the protein environment around the binuclear reaction center, within subunit I, so as to limit the accessibility of heme a_3 to electrons without altering the barrier height of the electron transfer reaction itself.

Recently, Allen et al (85) examined the effects of subunit V isoforms on the binuclear reaction center and on the interaction of holocytochrome c oxidase with cytochrome c. They found that the subunit V isoforms do not affect the Km for cytochrome c binding to cytochrome c oxidase but do alter the TNmax of the holoenzyme. Moreover, by using infrared spectroscopy of carbon monoxide–bound cytochrome c oxidase isozymes they found two conformers, CI and CII, for the binuclear reaction center in the Va isozyme; these conformers are similar to those observed in mammalian cytochrome c oxidases. In contrast, the Vb isozyme has only one conformer, CII. This finding supports the conclusion that subunit V affects ligand binding at the binuclear reaction center and alters the environment around heme a_3. To fit these findings with the observed differences in turnover numbers for Va and Vb isozymes, Allen et al (85) proposed that the CII conformer is productive for electron transfer between heme a and the binuclear reaction center and that the CI conformer is less productive. Because these conformers appear to be interconvertible, the overall rate of electron transfer from heme a to a_3 would be determined by the ratio of CII to CI. For the Vb isozyme, the binuclear reaction center has only one conformation; this conformation corresponds to the productive conformer, CII. Hence, the enhanced rate observed with the Vb isozyme results from the presence of a productive conformer that does not interconvert to a less productive one. So, according to this model, Vb locks the binuclear reaction center in the CII conformer.

Together, these studies clearly demonstrate that the isoforms of a nuclear-coded subunit, subunit V, can affect the catalytic functions of a mitochondrially coded subunit, subunit I. They also serve to illustrate the intricate interplay between environment and function. Under hypoxic growth conditions, the Vb isoform is expressed and the holoenzyme that is assembled has a high turnover number, whereas at high oxygen tensions the Va isoform is expressed and the enzyme has a low turnover number.

NUCLEAR AND MITOCHONDRIAL CONTRIBUTIONS TO MITOCHONDRIAL GENE EXPRESSION

In this section and the ones to follow, we focus heavily on work done with *S. cerevisiae* because current understanding of nuclear-mitochondrial cross talk is more complete for this organism than any other. When appropriate, we compare findings from yeast with those from other organisms. The mitochondrial genome of *S. cerevisiae* is between 75 and 80 KB in size, depending on the strain (4, 141–143). Most regions of the genome have been sequenced, although a few gaps in the sequence remain in intercistronic spacer regions. The protein-coding genes specify subunits of the respiratory enzyme complexes cytochrome *c* oxidase (subunits I, II and III), ATPase (subunits 6, 8, and 9) or ubiquinone-cytochrome *c* oxidoreductase (apocyto-chrome *b*), a ribosomal protein (Var1), several endonuclease subunits, and several maturases (intron-encoded proteins involved in splicing of specific introns from pre-mRNAs) (4, 141). In addition, the yeast mitochondrial genome encodes 25 tRNAs, two ribosomal RNAs (21S and 15S rRNAs of the large (60S) and small (30S) ribosomal subunits, respectively), and the 9S RNA component of mitochondrial RNase P (4, 141). Here, we consider three aspects of mitochondrial gene expression: transcription, RNA process-ing, and translation. Each of these steps in mitochondrial gene expression involves interactions between nuclear-coded gene products and the mitochon-drial genome or its gene products.

Transcription

Studies on mitochondrial transcription have focused largely on initiation. In comparison to other steps of mitochondrial gene expression, transcriptional initiation appears to require only a few nuclear-coded protein factors [see recent reviews by Jaehning (144) and Shadel & Clayton (145)]. Moreover, there is no evidence for differential regulation of mitochondrial transcription initiation by trans-acting factors. Instead, any differences in levels of transcrip-tion of specific genes are attributable to promoter strength (146, 147).

The yeast mitochondrial genome has at least 19 transcription units (2). Transcription initiates by binding of mitochondrial RNA polymerase to a nonanucleotide sequence (usually ATATAAGTA or TTATAAGTA) that pre-cedes each of these transcription units (1). (The underlined A residue in the promoter represents the +1 start site for transcription.) Yeast mitochondrial RNA polymerase consists of a core subunit, encoded by the *RPO41* gene, and a specificity factor encoded by the *MTF1* gene. The 145-kDa core subunit is homologous to the single subunit RNA polymerases of T7 and T3 bacterio-phages (148), and the 43-kDa specificity factor shows several blocks of amino acid–sequence similarity to prokaryotic sigma factors (149). Both subunits must be present in order to get specific initiation at a mitochondrial promoter element in vitro (144, 145). A temperature-sensitive *rpo41* mutant shows a temperature-sensitive defect in mitochondrial RNA synthesis (150). This block is suppressed by the *MTF1* gene on a high-copy plasmid, indicating functional interaction in vivo (151). Despite both in vitro and in vivo evidence for functional interaction, the two subunits are not tightly associated upon purifi-cation. Instead, the subunits appear to undergo cycles of association and dissociation in a manner that appears mechanistically similar to the association and dissociation of core polymerase and sigma factors in prokaryotes. Mtf1p is found in early initiation complexes and is released from complexes that are committed to elongation (152, 153).

A third nuclear-coded protein factor, sc-mtTFA (or MTF2p), encoded by the *ABF2* gene, may also be involved in yeast mitochondrial transcriptional initiation. Although this factor is not essential for yeast mitochondrial transcription per se (154), it is homologous to a mammalian factor (mammalian mtTFA) that is essential for mammalian mitochondrial transcription initiation in vitro (144, 145). These factors, which belong to the HMG-box family of DNA binding proteins, both bind DNA without much sequence specificity in vitro and cause bending, unwinding, and wrapping of the DNA (155, 156). The DNA binding activity is not entirely random, however. Short runs of dA appear to exclude binding by mtTFA and may be responsible for the observed preferential binding of the protein to critical control regions of DNA, such as nuclear ARS elements or mitochondrial *ori* sequences (155). In contrast to Rpo41p or Mtf1p, yeast Mtf2p is very abundant ($\sim$ 250,000 copies per cell) (154). Some estimates indicate that this is enough to bind every 15–30 bp of yeast mitochondrial DNA (154). The abundance and DNA binding characteristics of yeast Mtf2p suggest that it is involved in general packaging and compaction of mitochondrial DNA. This packaging and its observed phased binding to mitochondrial DNA may be important for making regulatory elements accessible to other factors, including Rpo41p and/or Mtf1p. In support of this theory are the findings that either sc-mtTFA or human mtTFA proteins produce a three- to fourfold stimulation of yeast mitochondrial transcription in vitro when added to a reaction containing Rpo41p, Mtf1p, and either linear or supercoiled template (157); the stimulatory activity is sensitive to the stoichiometry of mtTFA protein to template; and the optimal stoichiometry is different for linear or supercoiled templates.

Circumstantial evidence of additional nuclear-encoded factors that affect mitochondrial transcription has been found. Brohl et al (158) identified an extragenic suppressor mutation *sup1-798* of the temperature-sensitive *rpo41-ts798* mutation. A wild-type gene, *AZF1*, was also identified as a high-copy suppressor of both *rpo41-ts798* and *sup1-798*. In addition, the studies of Ulery et al (98) have demonstrated that global changes in mitochondrial transcription levels are not strictly correlated with levels of mitochondrial RNA polymerase or Mtf1p transcription factor, which suggests that additional factors are involved in global regulation.

The possibility that nuclear-coded factors affect mitochondrial transcriptional elongation has not been seriously addressed. One possible candidate for such a gene is *PET309* (159). The *COX1* intron-dependent defect in accumulation of *COX1* transcripts in *pet309* mutants may be due to either destabilization of intron-containing *COX1* transcripts or to attenuation of transcription of intron-containing alleles of *COX1* (159). A transcriptional elongation defect could also explain some of the phenotypes of *nam1* mutations (160, 161). Since mitochondrial RNA polymerase is related to bacteriophage T7 RNA polymerase, the fact that T7 RNA polymerase exhibits poor processivity when transcribing extended dA tracts in the template (162, 163), and is capable of terminating transcription when a run of uridine residues is present in the nascent RNA (164), may be relevant. Similar behavior by yeast mitochondrial RNA polymerase might be predicted to cause poor processivity or premature termination of transcription in many of the AT-rich intron or intercistronic regions of the genome, although this prediction is entirely speculative at this point. Perhaps the above-mentioned nuclear-coded trans-acting factors improve the processivity of RNA polymerase in specific intronic or intergenic regions of the genome.

Specific termination of mitochondrial transcription in mammals is due to interaction of a trans-acting protein with a specific 13 bp terminator element in the DNA (165, 166). Yeast shows no evidence of terminator elements or trans-acting termination factors in mitochondrial transcription. The existence of putative RNA processing signals at the ends of transcription units (167) may mean that specific termination signals are unnecessary for yeast mitochondrial DNA.

RNA Processing

Mitochondrial transcription units are usually multigenic, and primary transcripts derived from them are processed in one or more ways including tRNA excision, 5′ mRNA processing, 3′ mRNA processing, intron excision (in fungi and plants), and polyadenylation (in metazoans) (167, 168). In addition, specific mechanisms for stabilization of certain transcripts have been documented (167). Here, we focus on those types of processing that are universal (i.e. similar between genera). These types of processing are tRNA excision, 5′ mRNA processing, and 3′ mRNA processing. Some excellent recent reviews discuss mitochondrial mRNA splicing and polyadenylation (4, 141, 168).

tRNA EXCISION   In both fungal and mammalian mitochondria, ribosomal RNAs and mRNAs are derived from polycistronic primary transcripts that contain one or more tRNAs. Hence, excision of tRNAs is essential both for production of translationally functional tRNAs and for production of correctly matured rRNAs and mRNAs. This fact is illustrated by an RNA processing defect associated with a specific tRNA gene mutation in the human mitochondrial genome (169). This mutation results in decreased levels of the ND1 mRNA in skeletal muscle and has been implicated in the pathology of a mitochondrial myopathy in a human patient (169).

At least two nuclear-coded enzymatic activities are required for excision of yeast mitochondrial tRNAs from an RNA precursor (see 170 for review). Processing of the 5′ end of mitochondrial tRNAs is accomplished by mitochondrial RNase P. This enzyme consists of a catalytic RNA, which is encoded by the mitochondrial genome, and a protein, which is encoded by the nuclear genome. The RNA component, encoded by the mitochondrial *RPM1* locus, is homologous to RNase P RNA from bacteria. Processing of the 3′ end of yeast mitochondrial tRNAs is accomplished by an endonuclease activity. The gene(s) or gene product(s) responsible for this activity have not been identified yet. Additional posttranscriptional modifications of tRNAs include production of the 3′ CCA terminus by an ATP (CTP):tRNA nucleotidyltransferase and several base-modification reactions, carried out by proteins encoded in yeast by the nuclear *TRM1* and *MOD5* genes. The *TRM1* and *MOD5* genes produce gene products that are shared by the mitochondrial and nucleo/cytoplasmic systems and are therefore responsible for modification of tRNAs from both the mitochondrial and cytoplasmic translation systems (171).

5′ AND 3′ PROCESSING   For several mitochondrial genes the mature 5′ end of the mRNA is created directly by the 3′ processing of an upstream tRNA (167). An example is the *COX3* mRNA from yeast (172). For other genes, (e.g. the yeast *COB* gene) the mature 5′ end of the mRNA is generated by an additional endonucleolytic cleavage (167). After removal of $tRNA_{glu}$ from the $tRNA_{glu}$-*COB* primary transcript, a pre-*COB* intermediate is generated. Further processing of this pre-mRNA, to generate the mature 5′ end of *COB* mRNA, requires the product of the nuclear gene *CBP1*. Whether CBP1p is a component of the endonuclease itself, or whether it acts in another way, perhaps by putting the structure of the RNA substrate into a processing competent conformation (173), is not known. Other yeast mitochondrial mRNAs whose mature 5′ ends are generated by an endonucleolytic cleavage (other than the 3′ processing of a tRNA) are *VAR1* and the two polycistronic mRNAs for *AAP1-OLI2-ENS2* (167). The mature 3′ ends of yeast mitochondrial mRNAs correspond to either the end of the primary transcript or to an end created by an endonucleolytic cleavage (167). The endonucleolytic enzyme(s) responsible for 3′ cleavage have not been identified, although a protein complex that binds to a dodecamer motif may be involved (174).

The correct production of 5′ and 3′ ends of mitochondrial mRNAs is important for mRNA stability. Incorrectly processed mRNAs are often either unstable or untranslatable (167, 172). Little is known about the mechanisms of mitochondrial mRNA turnover or stabilization in either mammals or yeast. However, the steady-state levels of mitochondrial mRNAs, tRNAs, and rRNAs, are clearly greatly influenced by their differential stabilities, in both mammals and yeast (175, 176). A nucleoside triphosphate–regulated 3′ exonucleolytic activity from yeast may be involved in the generalized turnover of mitochondrial transcripts (177), and a protein complex that binds specifically to the dodecamer sequence found at the 3′ ends of yeast mitochondrial mRNAs may be involved in mRNA stabilization (174). Control of stability of specific transcripts also exists. CBP1p not only plays a role in generating the mature 5′ end of *COB* mRNA (noted above), but is required specifically for stabilization of the mature mRNA (173). Additional nuclear genes [*NCA1*, *NAM1*, and *PET309* (159, 167)] have also been implicated in stabilization of specific mitochondrial transcripts (see Table 1).

Translation

Although mitochondrial and eubacterial translation systems show many similarities, they also exhibit several differences (178). One of the most striking differences lies in the mechanism of translational initiation. In eubacteria, the initiation codon is selected primarily as a consequence of interactions between mRNA and 16S rRNA; the best understood of these interactions occurs at the Shine-Dalgarno sequence (179). In mitochondrial translation systems, no evidence convincingly shows interaction between mRNA and the small ribosomal 16S rRNA (172, 178). Hence, the mechanism of initiation codon selection may be quite different in mitochondria and eubacteria. Mammalian mitochondrial mRNAs lack, or have only very short, 5′ UTRs (untranslated leader regions), whereas yeast mitochondrial mRNAs have quite long 5′UTRs (ranging from 50 to over 900 bases, depending on the mRNA) (178). In mammalian mitochondrial mRNAs, the initiation codon is likely to be the first AUG from the 5′ end of the mRNA, whereas in yeast mitochondrial mRNAs, the initiation codon is rarely the first AUG in the mRNA. How the mitochondrial translation apparatus selects the correct initiation codon is not clear. However, yeast are thought to use mRNA-specific translational activator proteins for this purpose (180–183) (Table 1). Whether mammalian mitochondrial mRNAs use analogous activator proteins is not known.

The paradigm for function of the mRNA-specific translational activator proteins in yeast is provided by the *COX3* gene. Translation of this gene requires the products of three nuclear genes: *PET494*, *PET54*, and *PET122* (180–183). Currently, PET494p, PET122p, and PET54p are thought to work together to mediate the interactions between *COX3* mRNA, the small subunit of the mitochondrial ribosome, and the inner mitochondrial membrane (4, 167, 178, 184), thereby ensuring initiation of translation of *COX3* mRNA at the correct initiation codon by membrane-bound ribosomes. The following description summarizes what is known about the action of these proteins. First, a motif consisting of both structural and primary sequence elements exists within the 5′ untranslated leader region (5′ UTR) of *COX3* mRNA and spe-

Table 1 Nuclear PET genes that affect synthesis or assembly of yeast mitochondrial proteins

Gene Name	Location of Gene Product	Function	References[a]
RNA processing or stabilization			
NAM1	Matrix	Accumulation of transcripts from intron-containing transcription units containing *COB* and *COX1/AAP1/OLI2/ENS2*	290
PET309	Inner membrane	Accumulation of transcripts from intron-containing *COX1/AAP1/OLI2/ENS2* transcription unit and translation of *COX1* transcripts	159, 291
CBP1	Matrix	Processing and stabilization of *COB* mRNA 5′ end	292
NCA1	Not determined	Stabilization of *OLI1* mRNA	293
SUV3	Not determined	Processing or stabilization of mRNAs containing a dodeca-mer motif at the 3′ end	294
mRNA translation			
MSS51	Not determined	Translation of *COX1* mRNA	295
PET309	Inner membrane	Translation of *COX1* mRNA	159, 291
PET111	Inner membrane	Translation of *COX2* mRNA	190, 194
PET494	Inner membrane	Translation of *COX3* mRNA	190
PET122	Inner membrane	Translation of *COX3* mRNA	190
PET54	Dually localized: inner membrane and matrix	Translation of *COX3* mRNA and splicing of *COX1* intron aI5β	188, 190
CBP6	Mitochondrion	Translation of *COB* mRNA	296
CBS1	Inner membrane	Translation of *COB* mRNA	297
CBS2	Inner membrane	Translation of *COB* mRNA	297
AEP1	Not determined	Translation of *OLI1* mRNA	298
AEP2(ATP13)	Mitochondrion	Translation of *OLI1* mRNA	299, 300
Assembly factors			
COX10	Inner membrane	Heme a biosynthesis	223, 225
COX11	Inner membrane	Heme a biosynthesis	224, 225
PET117	Inner membrane	Cytochrome oxidase assembly	301; *
PET191	Inner membrane	Cytochrome oxidase assembly	301; *
PET100	Not determined	Cytochrome oxidase assembly	**
OXA1	Not determined	Cytochrome oxidase assembly	304
COX14	Mitochondrial membranes	Cytochrome oxidase assembly	318
SCO1	Mitochondrial membranes	Cytochrome oxidase assembly	305
ABC1	Not determined	Assembly or function of complexes II and III	306
CBP3	Mitochondrial membranes	Assembly of complex II	307
CBP4	Mitochondrial membranes	Assembly of complex II	308

Table 1 (*Continued*)

Gene Name	Location of Gene Product	Function	References
ATP10	Mitochondrial membranes	Assembly of ATPase	309
ATP11	Matrix	Assembly of ATPase	310
ATP12	Matrix	Assembly of ATPase	311
Proteases			
MAS1	Matrix	Subunit of mitochondrial processing peptidase	5
MAS2	Matrix	Subunit of mitochondrial processing peptidase	5
MIP1	Matrix	Mitochondrial intermediate peptidase	312
MSP1 (*YTA4*)	Outer membrane	Sorting of proteins to mitochondrial subcompartments	313
IMP1	Inner membrane	Processes exported proteins	212
IMP2	Inner membrane	Processes exported proteins	212
PIM1	Matrix	Protein degradation	314, 315, 317
YTA10	Inner membrane	Degrades unassembled mitochondrial gene products	210, 213, 218
YME1 (*YTA11*)	Inner membrane	Degrades unassembled cytochrome oxidase subunits	219
YTA12 (*RCA1*)	Inner membrane	Degrades unassembled mitochondrial gene products	210, 214
BCS1	Inner membrane	Assembly of complex III	316

[a]*JE McEwen, KH Hong, KM Calder, S Guha & D Culler, unpublished observations; ** C Church & RO Poyton, manuscript submitted.

cifically mediates translational control of *COX3* by *PET494*, *PET122*, and *PET54* (172). Based upon the overlap between the 5′UTRs of two different alleles, which conferred PET494/PET122/PET54 control on the downstream coding region (172, 185), the control region mediating PET494/PET122/PET54 function has been localized between nucleotides −480 and −183. A smaller region, located between −405 and −330 in the 5′ UTR, is necessary for translational control (172). Second, genetic evidence shows interaction between PET122p and appropriate regions of the *COX3* 5′UTR (172, 186). PET54p may also be important in RNA binding, as supported by the findings that it contains an amino acid sequence motif similar to the known RNA-binding domain of the bacteriophage T4 RegA protein and that a missense mutation of a residue within this region of PET54 that corresponds to a critical residue for the RegA RNA binding function abolishes PET54p function (187). Also, PET54p is necessary for a second function predicted to involve protein-RNA interaction, namely splicing of intron aI5β from *COX1* pre-mRNA (187, 188). Finally, PET54p produced in bacteria binds RNA but in a nonspecific fashion (C Johnson & JE McEwen, unpublished observations). Perhaps sequence specificity is conferred as a result of interaction of PET54p with PET122p and/or PET494p. Third, PET494p, PET54p, and PET122p interact with one another (187). PET494p and PET122p are membrane bound while the *PET54* protein is present in both membrane and soluble fractions of mitochondria (190). The distribution of PET54p in two mitochondrial subcompartments is consistent with its dual functions in *COX3* mRNA translation and *COX1* pre-mRNA splicing (188, 191). Fourth, at least one of the three proteins, PET122p, interacts functionally, and perhaps directly, with several proteins of the mitochondrial ribosomal small subunit (192, 193).

Although much less is known about the translation of other mRNAs in yeast mitochondria, the paradigm provided by *COX3* mRNA translation appears to be applicable. In addition to *COX3* mRNA, the mRNAs for *COX1*, *COX2*, *COB*, and *OLI1* are known to require specific translational activator proteins for their translation (see Table 1). In several cases, at least one of the translational activators for a specific mRNA is known to be membrane bound (see Table 1), and evidence suggests that the function of the activator is mediated by the 5'UTR of the mRNA (159, 194, 195; HB Lukins & P Nagley, personal communication). Whether any of the translational activators of these other mRNAs interact directly with the mitochondrial ribosome is not yet known.

In view of their membrane association and, in some cases, interaction with the mitochondrial ribosome, the suggestion has been made that these activator proteins function in membrane localization of specific mRNAs and in the cotranslational insertion of mitochondrial gene products into the inner membrane (183, 190). The precise mechanism by which these activators facilitate translation or insertion is unknown; elucidation of this mechanism will require the development of a mitochondrial mRNA–dependent translation system. So far, efforts to generate such a system have been unsuccessful (197).

Global Regulation

The respiratory capacity of *S. cerevisiae* changes in response to carbon source or oxygen availability. These changes are accompanied by global changes in mitochondrial gene expression (12–16, 167), which result from changes in mitochondrial DNA content, mitochondrial RNA levels, and/or mitochondrial protein synthetic rates. Many of the nuclear genes that encode proteins necessary for mitochondrial DNA metabolism, transcription, protein synthesis, and mitochondrial assembly are coordinately regulated in the nucleus, at the transcriptional level, and several transcriptional activator or repressor proteins that mediate the transcriptional regulation of these genes have been studied in detail (15). Some of these transcription factors and possible signal transduction mechanisms that serve to coordinate nuclear and mitochondrial gene expression are discussed further below (see section on Coordination of Nuclear and Mitochondrial Gene Expression and section on Signalling from the Mitochondrion to the Nucleus).

In mammals, tissue-specific differences in respiratory capacity are well documented (198). In addition, external stimuli, such as endurance training or thyroid hormone levels, lead to an increase in mitochondrial respiratory capacity in certain tissues (199). These observations provide the opportunity to study whether the differences are correlated with gene dosage, transcriptional or translational levels, or enzyme activity levels resulting from the expression of tissue-specific isoforms for nuclear-coded subunits of mitochondrial respiratory proteins (see above). Examples of each have been reported (126, 200–203). Unfortunately, these correlations do not prove cause and effect, and little is understood about the signal transduction pathways or molecular mechanisms that produce these global changes in mammalian mitochondrial biogenesis.

Gene Specific Control— Is There Differential Regulation of Mitochondrial Gene Expression?

A striking feature of mitochondrial gene expression in yeast is the fact that each gene for a respiratory enzyme component is controlled by multiple gene-specific trans-acting factors (see Table 1). The gene-specific factors identified so far are required for expression of mitochondrial genes at a posttranscriptional level, including processing of mitochondrial pre-mRNAs, stabilization of transcripts, translation of mRNAs, or assembly of mitochondrial translation products into functional holoenzymes (4, 15, 167, 204–206).

Despite the existence of distinct coteries of nuclear-encoded factors required for expression of each mitochondrial protein-coding gene, there is no convincing evidence that mitochondrial genes are differentially regulated under normal physiological conditions. Control of mitochondrial gene expression by carbon source or oxygen appears to be mediated globally, as described above. Other physiological conditions, such as nitrogen deprivation [leading to development of pseudohyphal morphology (207)], oxidative stress, sporulation, and germination have not been examined as yet for the presence or absence of differential expression of mitochondrial genes in yeast. However, a recent study of germinating N. crassa conidia suggests coordinated expression, at the level of translation, of the three mitochondrially coded subunits of cytochrome c oxidase (208). A circumstance in which discordant expression of subunits for the same complex (e.g. subunits I, II, and III of cytochrome c oxidase) would be advantageous to the cell is hard to imagine, unless individual subunits have other functions besides their roles as components of the complex.

NUCLEAR CONTROL OF MITOCHONDRIAL ASSEMBLY

Import, Export, and Assembly Pathways

The polypeptides encoded by nuclear and mitochondrial genes follow two fundamentally different pathways to their sites of residence within the mitochondrion. Those proteins encoded by nuclear genes are translated on cytosolic ribosomes (or, in some cases, ribosomes bound to the outer mitochondrial membrane) and are imported by mitochondria posttranslationally (5), whereas those proteins encoded by mitochondrial genes are translated on endogenous mitochondrial ribosomes that are bound to the matrix side of the inner membrane, and they are inserted into it cotranslationally (6, 183). These two pathways— the import pathway for proteins translated in the cytosol and the export pathway for proteins translated in the mitochondrion— must converge at some point into an assembly pathway for those multimeric proteins that are assembled from nuclear and mitochondrial gene products. Both the import (5) and export (6, 183) pathways have been reviewed recently and will not be covered here. However, one point to note is that the machinery for these pathways is composed entirely of nuclear-coded proteins. Some of these proteins act globally, whereas others have more specific effects (see Table 1). Those proteins with global effects include members of the hsp60 and hsp70 family of stress proteins (4), cyclophilin 20 (209), and proteases (5, 183, 210). Proteins with specific effects include assembly facilitators that work in a protein complex–specific manner.

Chaperonins

A member of the hsp70 family has been implicated in the proper folding of nascent polypeptides that are being translated on cytosolic ribosomes (5) and in the ATP-dependent unfolding of cytosolic precursors during, or prior to, their binding to mitochondrial surface receptors (5). Another member of the hsp70 family is present in the mitochondrial matrix where it functions in refolding newly imported nuclear-coded proteins (5). Mitochondrial Hsp70 also interacts with and assembles some mitochondrially coded proteins (211). Hsp60 is a chaperonin that resides in the mitochondrial matrix. This chaperonin works in conjunction with mitochondrial Hsp70 to refold newly imported unfolded proteins (5). Recently, cyclophilin 20, a peptidyl-prolyl cis-trans isomerase has been implicated in preprotein translocation across the inner membrane and in protein folding in the matrix (209).

Proteases

Recent studies have uncovered many mitochondrial proteases. Some of these proteases act globally, whereas others act on subsets of mitochondrial proteins (Table 1). For example, yeast MAS1p and MAS2p are required for processing nearly all proteins imported into the mitochondrial matrix (5). These two polypeptides are subunits of the matrix processing protease; this enzyme is analogous to the mitochondrial processing protease of mammals and *N. crassa*. The presequences of some proteins imported by mitochondria are processed in two steps (5). Two of these proteins, the Fe-S proteins of complex III and subunit IV of complex IV reside on the matrix side of the inner membrane; the other two, cytochrome c_1 of complex III and cytochrome b_2, reside on the cytoplasmic side of the inner membrane and in the intermembrane space, respectively. The first step in the processing of these proteins is carried out by the matrix protease. The intermediates of the Fe-S protein and cytochrome c oxidase subunit IV are processed by MIP, the mitochondrial intermediate processing protein; the intermediates for cytochrome c_1 and b_2 are processed by the inner membrane protease, a two-subunit protein composed of Imp1 and Imp2 (183, 212). This enzyme also processes a precursor to cytochrome c oxidase subunit II, a mitochondrial gene product (183, 212).

In addition to these enzymes, which are involved in mitochondrial import and export, many other proteases appear to be involved in either the bio-genesis or turnover of mitochondrial proteins. The YTA family of genes (210) encode global assembly factors; mutations in them produce pleiotropic deficiencies in mitochondrial functions (213, 214). This finding is somewhat paradoxical in view of the homology of the mitochondrial YTA proteins to subunits of the cytoplasmic proteosome complex (215), which is part of a major pathway for proteolytic degradation of cytoplasmic proteins (216), and the experimental evidence that the YTA proteins are involved in a mitochon-drial proteolytic pathway (217–219). A common function in protein assembly and proteosome-mediated proteolysis is polypeptide unfolding (216). Perhaps the YTA proteins have polypeptide unfoldase activity which is necessary for both assembly and degradation pathways in mitochondria. Even more in-triguing is the possibility that the YTA proteins are positioned at a decision point between the degradative and assembly pathways and thereby regulate the levels of unassembled subunits according to the availability of assembly partners in the mitochondrion.

Assembly of Multimeric Proteins and Specific Assembly Factors

How proteins that are encoded by nuclear and mitochondrial genomes assem-ble into multimeric proteins is not known. Although preliminary studies with cytochrome *c* oxidase (220), coenzyme Q cytochrome *c* oxidoreductase (221), and ATP synthase (31, 222) suggest that the subunits of these complexes assemble sequentially, the assembly pathways (i.e. the sequence in which subunits come together during the assembly of the multimer) for these and other multimeric mitochondrial proteins have not been elucidated. However, both genetic and biochemical strategies are being pursued to address this problem. These studies have begun to reveal the dynamics of assembly and the existence of nuclear-coded proteins that are required for the assembly of specific mitochondrial multimeric proteins, including cytochrome *c* oxidase, coenzyme Q cytochrome *c* oxidoreductase, and ATP synthase (Table 1).

A popular model for studies on multimeric protein assembly in mitochondria is yeast cytochrome *c* oxidase (Figure 1). Preliminary studies have suggested that the assembly of the nuclear- and mitochondrially coded subunits in this complex occurs in the inner membrane (220) and is either catalyzed or assisted by additional proteins (7) (designated here as assembly facilitators). These facilitators are unlike the chaperonins discussed above because they are specific for cytochrome *c* oxidase and do not affect the other complexes of the respiratory chain. The best understood proteins of this type are encoded by the *COX10* and *COX11* genes. These genes, which were first believed to encode cytochrome oxidase assembly factors (223, 224), are now believed to encode enzymes of the heme A biosynthetic pathway. The heme group, heme A, of cytochrome *c* oxidase is necessary for assembly or stability of protein subunits within the complex. *COX10* encodes a protein required for farnesylation of the vinyl group at carbon 2 of protoheme b, which is a direct precursor to heme A (225). Homologues of *COX10*, termed *cyoE* or *ctaB*, have been found in bacteria that contain either cytochrome o– or cytochrome a/a$_3$–type terminal oxidases (226–229). In addition, a human *COX10* homologue, which can functionally replace the yeast *COX10* gene in yeast, has been found (230). The evolutionary conservation of the *COX10* protein and the presence of a motif common to other farnesyl transferases suggest that *COX10* encodes the enzyme that farnesylates protoheme b. *COX11* encodes a protein that is proposed to be involved in the paired oxidation/oxygenation step that converts the methyl group at carbon 8 of the heme ring to a formyl group (225). Homologues of *COX11*, termed *ctaG*, have been found in some, but not all, bacteria that produce heme A. Interestingly, a different gene, *ctaA*, which is not homologous to *COX11*, is required for conversion of heme o to heme A in *B. subtilis*.

At least six additional nuclear genes (*SCO1, PET117, PET191, PET100, OXA1,* and *COX14*) (301, 304, 305, 318) are required for assembly of yeast cytochrome *c* oxidase (see Table 1). At least one of these genes is essential for normal mitochondrial morphology (C Church, personal communication). Because mutants carrying mutations in these genes have heme A (RO Poyton,

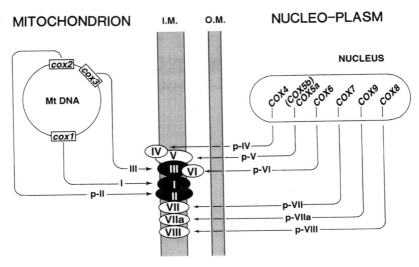

Figure 1   Nuclear and mitochondrial contributions to the assembly of yeast cytochrome *c* oxidase. The subunit polypeptides encoded by mitochondrial genes are shaded; those encoded by nuclear genes are clear. Subunit precursors are indicated by a p-prefix.

unpublished observations), these genes are unlikely to participate in heme A biosynthesis. However, they may encode proteins involved in inserting heme A into the holoenzyme, or they may function in subunit folding or assembly.

Mitochondrial Transmission

Genetic studies with *S. cerevisiae* have identified nuclear genes required for transmission of mitochondrial organelles to new cells (232). The *MDM1* gene is necessary for transfer of both mitochondria and nuclei, but not vacuoles, into newly formed daughter cells, i.e. buds (232). *MDM1* governs the formation of a type of cytoplasmic intermediate filament that is clearly distinct from the actin or tubulin cytoskeletal networks (233, 234). Mutations in *MDM2,* which is identical to the *OLE1* gene that encodes the Δ9 fatty acid desaturase (235), blocks inheritance of mitochondria but not other organelles. *MDM10* (236) and *MMM1* (237) are necessary for normal mitochondrial morphology, as well as inheritance of the organelle by buds. Each gene encodes a mitochondrial outer membrane protein partially exposed to the cytosol, and each mutation causes a nearly identical phenotype consisting of condensation of normally tubular mitochondria into giant spherical structures. These genes are clearly different, and determining if their gene products interact with one another will be extremely interesting. Both groups of investigators (236, 237) have speculated that the giant spherical mitochondria are produced as a result of a defect in attaching mitochondria to an internal framework, such as the cytoskeleton. Independent evidence from studies of singly substituted amino acid mutants of actin shows that mitochondria interact with cytoplasmic actin cables in wild-type yeast cells. A particularly interesting point is that the majority of actin mutants that perturbed mitochondrial morphology carried amino acid substitutions in a domain of actin thought to be important for actin-myosin interactions (238).

COORDINATION OF NUCLEAR AND MITOCHONDRIAL GENE EXPRESSION

As discussed above, the nuclear genome exerts a great deal of control over mitochondrial gene expression and the import, export, and assembly pathways required for biogenesis of a functional mitochondrion. Much less is known about feedback from the mitochondrial genome to the nucleus and the extent to which expression of nuclear genes for mitochondrial proteins is coordinated with expression of mitochondrial genes. Two types of models have been proposed. According to the first, one or both of the component subunits (the catalytic core and mtTFA transcription factor) of mitochondrial RNA polymerase, which are encoded by the nucleus, function in both the nucleus and mitochondrion to regulate the expression of genes on nuclear and mitochondrial genomes in parallel. In the second model, one or more transcription factors encoded by the nuclear genome coregulate the expression of the nuclear genes for mitochondrial RNA polymerase and nuclear structural genes for respiratory proteins. This parallel coregulation of mitochondrial RNA polymerase with other nuclear genes for mitochondrial proteins could coordinate the expression of mitochondrial and nuclear genes whose products coassemble into functional complexes, such as respiratory proteins.

The first model was suggested by the observation that the promoters of some nuclear genes in both yeast (239) and mammals (240, 241) contain *cis* elements that are part of mitochondrial promoters. This fact was initially observed for the promoter of the yeast *COX6* gene (239). Subsequently, similar *cis* sites were found in the promoters of several additional nuclear genes in yeast (242). Some of these encode mitochondrial proteins while others do not. In addition, some of these sites were found to be good promoters for yeast mitochondrial RNA polymerase in vitro (242). These findings led to the hypothesis that mitochondrial RNA polymerase or a mitochondrial transcription factor functions in both the nucleus and the mitochondrion. This hypothesis has been tested and ruled unlikely by two types of study. First, extensive mutational analysis has shown that the region of the *COX6* promoter that contains the mitochondrial promoter motif is not essential for either basal or regulated expression (100). Second, studies with *rpo41* mutant cells, which are deficient in the catalytic core of mitochondrial RNA polymerase, showed that mitochondrial RNA polymerase is not required for the expression of nuclear genes that contain the mitochondrial promoter sequence (242). Similar studies to test the validity of this model have not yet been done with mammalian cells.

The second model comes from the finding that transcription factors that regulate expression of nuclear genes for mitochondrial respiratory proteins also regulate expression of genes for the two components of mitochondrial RNA polymerase in both yeast and mammalian cells. Virbasius & Scarpulla (243, 244) found that transcription of human mtTFA is activated by NRF-1 and NRF-2, two general transcription factors that regulate expression of several cellular proteins including subunits of mitochondrial respiratory proteins (245, 246). Although these studies suggested the possibility that expression of nuclear and mitochondrial genes in mammals could be coregulated globally through NRF-1 or NRF-2, a regulatory role for NRF-1 and NRF-2 remains to be shown. In studies on the expression of the two components of yeast mitochondrial RNA polymerase, Ulery et al (98) found that the *RPO41* and *MTF1* genes (which encode the catalytic core and transcription factor components, respectively) are both glucose repressed, like many of the genes for subunits of respiratory proteins. In addition, at least two genes (*SNF1* and *REG1*) that are required for glucose derepression of genes for subunits of respiratory proteins are also required for derepression of *RPO41* and *MTF1*. These results are consistent with the hypothesis that common transcription factors that regulate nuclear genes for mitochondrial respiratory proteins also regulate mitochondrial transcription, but they do it by regulating the level of the components of mitochondrial RNA polymerase. However, this model appears to be incomplete because the level of mitochondrial transcription in yeast cells does not parallel the level of mitochondrial RNA polymerase (98). Hence, additional factors or pathways are likely involved. Some of these may be affected by mitochondrial genotype or phenotype, as discussed below.

SIGNALLING FROM THE MITOCHONDRION TO THE NUCLEUS

Phenomenology

Suggestive evidence for a reverse pathway from the mitochondrion to the nucleus initially came from studies with inhibitors of mitochondrial transcription or translation (247). These studies, with *N. crassa* and *Tetrahymena,* revealed that inhibition of mitochondrial protein synthesis leads to an increase in the levels of several nuclear-coded mitochondrial proteins, including mitochondrial RNA polymerase, DNA polymerase, and mitochondrial elongation factors (248). These studies led to the proposal that mitochondria produce a repressor that inhibits the expression of nuclear genes for some mitochondrial proteins (248). Unfortunately, these studies have not been followed up, and whether the increases observed are the result of changes in enzyme activity per se or changes in the level of these proteins is not clear. More recently, Parikh et al (9) demonstrated that transcription of several nuclear genes is up-regulated in yeast strains that either lack a mitochondrial genome (rho^0 strains) or carry mit$^-$ or rho$^-$ mutations on their mitochondrial genomes. These workers defined two classes of regulated transcripts: Class I consists of transcripts that are more abundant in respiration-deficient cells (either mit$^-$, rho^0, or rho$^-$) than in respiratory-competent cells (wild type), and Class II consists of transcripts that are more abundant in rho^0 or rho$^-$ (but not mit$^-$) cells. Because mit$^-$, rho^0, and rho$^-$ strains are all respiration deficient, Class II transcripts appear to respond to mitochondrial genotype and not respiration deficiency. In contrast, Class I transcripts respond to a deficiency in mitochondrial respiration, a conclusion supported by the fact that some of these transcripts are also up-regulated in rho$^+$ cells that are poisoned with antimycin A, an inhibitor of respiration (248, 249). Class I transcripts include those mRNAs encoded by *CIT2* (249, 250), the gene for peroxisomal citrate synthase, and the alternative oxidases in *N. crassa* and *Hansenula anomala* (252–254). Class II transcripts include those mRNAs encoded by the nontranscribed spacer region (NTS) of the nuclear ribosomal DNA repeat (255): *PUT1* (256); *MRP13* (257); and *ORF-D*, a gene that lies immediately downstream of *COX6* (9, JD Trawick, LE Farrell, N Kraut, S Silve, D Raitt & RO Poyton, unpublished observations). In mammals, Class I transcripts include GLUT1 mRNA and those mRNAs under the control of the transcription factor, NF-kB (see below).

Using a similar strategy to that of Parikh et al (9), Farrell et al (10) discovered a third class of transcripts in yeast cells. These transcripts are down-regulated in rho^0 or rho$^-$ cells. They are unaffected in respiration-deficient mit$^-$ or nuclear *pet* mutant cells (10; JD Trawick, LE Farrell, N Kraut, S Silve, D Raitt & RO Poyton, unpublished observations). In addition, these transcripts are not down-regulated in rho$^+$ cells poisoned with inhibitors of respiration. These transcripts have been designated Class III transcripts (10). Because these transcripts are not down-regulated in all respiratory-deficient cells, but only in those which contain large deletions of their mitochondrial genomes or have lost their mitochondrial genomes altogether, this down-regulation is clearly due to mitochondrial genotype and not to a respiration-deficient phenotype. Currently, it is not known what mitochondrial genes are involved. Class III transcripts include the transcripts from the nuclear *COX* genes *COX4*, *COX5a*, *COX6*, *COX8*, and *COX9* (JD Trawick, LE Farrell, N Kraut, S Silve, D Raitt & RO Poyton, unpublished observations). Other transcripts that are down-regulated in rho^0 (or rho$^-$) cells include some of the nuclear-coded subunits of complex III (258). Whether the down-regulation of these later transcripts is attributable to mitochondrial genotype or respiration-deficient phenotype is not yet known. The mechanisms for up-regulation of Class I transcripts (termed retrograde regulation) and down-regulation of Class III transcripts (termed intergenomic signalling) are currently under study and are discussed below.

Retrograde Regulation

The most extensive studies on retrograde regulation in yeast have been done with CIT2 (249, 250). Transcripts from this gene are up-regulated in rho^0 cells by 6- to 30-fold. In contrast, transcript levels from CIT1, the gene encoding the mitochondrial citrate synthase, are unaffected in rho^0 cells (249). Transcript levels from CIT2 are also increased in rho$^+$ cells whose respiration is inhibited by antimycin A or in rho$^+$ cells that carry a gene disruption in CIT1 or MDH1, the gene for mitochondrial malate dehydrogenase (259). By using promoter fusions that link the CIT2 promoter to a reporter gene, Liao et al (250) demonstrated that the up-regulation is transcriptional, and they identified a 76-bp region within the CIT2 promoter that is responsible. This region, des-ignated UASr, is essential for both basal and regulated transcription of CIT2. The protein products of at least three genes, RTG1, RTG2, and RTG3, are implicated in retrograde regulation of CIT2; these work through UASr. RTG1p has a helix-turn-helix motif and may be a transcription factor. RTG2p has an ATP binding domain that is similar to that of bacterial phosphatases that hydrolyze ppGpp and pppGpp (259). RTG3p is a component of a transcrip-tional complex that binds UASr, and it may form a heterodimer with RTG1p (RA Butow, personal communication). Recent studies have shown that RTG1p and RTG2p are required for expression of nuclear genes for other peroxisomal proteins [i.e. acetyl-CoA oxidase, catalase A, and the PMP27 protein of the peroxisomal membrane (259a)]. Moreover, yeast mutants (Δrtg1 and Δrtg2) that lack RTG1p and RTG2p do not grow well on media containing the long-chain fatty acid oleic acid as the sole carbon source. It is not yet clear if the growth defect in Δrtg1 and Δrtg2 is the result of defective transport of oleic acid into mitochondria, a defect in its metabolism via the β-oxidation pathway, or defective utilization of acetyl CoA that is produced by β-oxidation. Currently, it is not clear how a defect in mitochondrial respiration (in rho^0 cells, or on rho$^+$ cells inhibited with antimycin A) or a defect in the TCA cycle (in cit1$^-$ or mdh1$^-$ cells) leads to the up-regulation of yeast nuclear genes for peroxisomal proteins.

An interesting parallel to RTG1- and RTG2-mediated retrograde regulation in yeast can be found in the PPAR-activated gene expression pathway in mammals (259b). The PPAR superfamily of nuclear hormone receptors has at least four members (PPARα, PPARβ, PPARδ, and PPARγ) (259c). The first member of this superfamily was identified and cloned from the livers of rodents fed agents that lead to peroxisomal proliferation (e.g. long-chain fatty acids, xenobiotics, plasticizers) (259d). PPAR-responsive *cis* elements have been found in the promoters of several genes including cytochrome P450 4A6 (259e), three enzymes of the peroxisomal β-oxidation cycle (259f–h), and cytosolic malic enzyme (259i). Recently, PPAR has been found to activate expression of the nuclear gene for mitochondrial 3-hydroxyl-3-methylglutaryl CoA synthase (259j) as well as MCAD, the nuclear gene for the mitochondrial enzyme medium chain acyl CoA dehydrogenase (259k). This latter finding has led to the proposal that PPAR coordinates the expression of nuclear genes for mitochondrial and peroxisomal β-oxidative enzymes (259k). Interestingly, the activation of PPAR occurs when fatty acid uptake into mitochondria [via carnitine palmitoyl transferase, (CPTI)] is inhibited or in cells that use fatty acids (instead of carbohydrates) as their primary energy substrates. It has been proposed that one or more metabolites build up under these conditions and that these serve as ligands that activate PPAR. These ligands may be incompletely metabolized fatty acids or neutral acyl sterol esters (259k) and/or reactive oxygen species (ROS) resulting from an overproduction of H_2O_2 by the β-oxidation cycle (259c). It will be of interest to determine if lipid metabolites and/or ROS (see section on Reactive Oxygen Species as Second Messengers in Retrograde Regulation) are involved in the activation pathway of RTG1p or RTG2p in yeast cells with defective TCA cycles or respiratory chains.

Other nuclear-encoded proteins that provide insight concerning retrograde regulation are flavohemoglobin of *S. cerevisiae* and the alternative oxidase of *H. anomala*. The level of yeast flavohemoglobin increases in rho^0 cells, in cytochrome *c* oxidase–deficient nuclear pet mutants (89), and in rho$^+$ cells treated with antimycin A or cyanide (260). The gene *(YHB1)* for this protein is expressed in aerobic cells and induced slightly in cells treated with hyperbaric oxygen (251). In addition, transcript levels are elevated in cells treated with free radical generators (251). These observations led to the suggestion that this protein plays an antioxidant role (251). The alternative oxidase may have a similar role in *H. anomala*. The expression of this protein is induced by drugs (antimycin A) or mutations that block mitochondrial respiration. Interestingly, this induction is prevented by free radical scavengers (261). As discussed below (see section on Reactive Oxygen Species as Second Messengers in Retrograde Regulation), the induction of yeast *YHB1* and the alternative oxidase in *H. anomala* may result from the release of free reactive oxygen species by mitochondria in which the complete electron transport chain has been compromised by either mutation (i.e. rho^0 or rho$^-$) or respiration inhibitors (i.e. antimycin A).

In summary, retrograde regulation of nuclear genes appears likely to be the result of altered mitochondrial metabolism and not altered mitochondrial genotype. This type of regulation most likely results when mitochondrial respiratory functions are compromised. This type of regulation is intriguing, but not surprising, given the extensive metabolic crosstalk known to take place between mitochondria and the cytoplasm. At present, no evidence suggests that this type of regulation coordinates the expression on mitochondrial genes with the expression of nuclear genes for mitochondrial proteins.

Intergenomic Signalling

Three things distinguish intergenomic signalling from retrograde regulation. First, nuclear genes are down-regulated instead of up-regulated. Second, the down-regulation is attributable to mitochondrial genotype instead of respiration-deficient phenotype. In addition, intergenomic regulation affects the nuclear genes for mitochondrial proteins and hence may be instrumental in coordinating expression of mitochondrial genes with some nuclear genes. Currently, the best understood example of this type of regulation is the *COX6* gene, which encodes subunit VI of cytochrome *c* oxidase, from *S. cerevisiae.* Transcription of *COX6* is subject to combinatorial regulation by carbon source, oxygen, heme availability, and the mitochondrial genome. Transcription is reduced three- to fivefold in a repressing carbon source (99, 102), is turned off in heme-deficient (or anaerobic) cells (99; PV Burke & RO Poyton, unpublished observations), and is down-regulated five- to sixfold in rho^0 cells (10; JD Trawick, LE Farrell, N Kraut, S Silve, D Raitt & RO Poyton, unpublished observations). The down-regulation of *COX6* in rho^0 cells is independent of the down-regulation brought about by heme deficiency. However, this down-regulation in rho^0 cells suppresses glucose (catabolite) repression of *COX6.* Four *cis* elements are responsible for these three types of transcriptional regulation (100, 101). Two of these, HDS1 and HDS2, mediate heme- and oxygen-regulated transcription. Carbon-source control is mediated through two other elements— Domain 1 and Domain 2— but most of it goes through Domain 1 (101). Down-regulation in rho^0 cells is also mediated by Domain 1. This *cis* element spans 23 bp and contains consensus binding sites for two transcription factors, ABF1p and HAP2p. Both of these factors are involved in glucose regulation of *COX6,* but by different mechanisms. HAP2p acts through Domain1 but does not bind to it (101). ABF1p binds to Domain 1 (101). It is a multifunctional phosphoprotein (102) that is differentially phosphorylated under catabolite repressing and derepressing conditions (102).

These differences in phosphorylation are manifested as different numbers of retarded complexes that can be seen in gel retardation gels, suggesting that ABF1p combines with additional proteins in binding Domain 1 and that the phosphorylation of ABF1p may affect these protein-protein interactions (102). Mutational analysis has shown that the ABF1 site, but not the HAP2 site, is sufficient for mediating intergenomic signalling (JD Trawick, LE Farrell, N Kraut, S Silve, D Raitt & RO Poyton, unpublished observations). Moreover, ABF1p has been shown to be differentially phosphorylated in rho+ and rho0 cells and that the phosphorylation states in rho0 cells are different than those observed in rho+ cells in repressing or nonrepressing carbon sources (JD Trawick, LE Farrell, N Kraut, S Silve, D Raitt & RO Poyton, unpublished observations). These results suggest a complex regulatory pathway of ABF1p function. Both carbon source and mitochondrial genome affect its phosphorylation but in different ways. Although the way in which phosphorylation affects ABF1 function is not yet clear, its differential phosphorylation in response to mitochondrial genotype provides a useful clue to understanding the intergenomic signalling pathway from the mitochondrial genome to the nucleus. ABF1p is rich in phosphoacceptor amino acid residues, and at least 15 of these are consensus phosphorylation sites for known protein kinases (102, 261a). Thus, this protein may be under the control of more than one kinase cascade pathway; at least one of these responds to mitochondrial genotype.

Reactive Oxygen Species as Second Messengers in Retrograde Regulation

An estimated 1 to 2% of the oxygen consumed during respiration is not completely reduced to water but is instead only partially reduced to the reactive oxygen intermediates— superoxide (O_2 −) and hydrogen peroxide (H_2O_2), which can be converted to the highly reactive hydroxyl ion (OH•) (262–264). These partially reduced intermediates, referred to collectively as reactive oxygen species (ROS), are also produced during oxygen-consuming reactions in the cytosol. ROS are very unstable and highly reactive. They have been shown to mutate DNA (265), oxidize proteins (266), and damage membranes (267). Currently, the level of interest in oxidative damage is high because it has been implicated in several cellular processes [e.g. senescence and programmed cell death (268–272), in aging (273), and in the pathogenesis of several degenerative diseases (e.g. cancer, ischemic heart disease, Parkinson's disease, Alz-heimer's disease, and late onset diabetes) (274)].

Two sites along the respiratory chain have been implicated in the generation of ROS; NADH dehydrogenase (Complex I) and coenzyme Q (268). Under normoxic conditions the reduced flavins associated with Complex I and reduced forms of coenzyme Q (i.e. ubisemiquinone or ubiquinol) can be autoxidized to release superoxide or hydrogen peroxide. The levels of these that are produced and released by mitochondria are elevated when terminal steps in the respiratory chain are blocked, either by mutation or drugs. Both cyanide (which inhibits cytochrome c oxidase) and antimycin A (which inhibits the oxidation of cytochrome b) stimulate the production of superoxide and hydrogen peroxide (268).

Recently, it has become clear that ROS have additional functions besides those related to oxidative stress. These involve the use of ROS as second messengers in events required for cell growth and differentiation (275–277). These ROS are produced primarily by mitochondria, and their production is probably related to changes in electron flux through the respiratory chain, brought about by various physiological conditions. Some of these conditions are age-related accumulation of mitochondrial genome mutations (268, 278); heat shock (279–281); exposure to nitric oxide (282, 283); and natural variations in oxygen tension, either in the case of microorganisms exposed to changing natural environments, or in the case of cells located in relatively hypoxic tissues of a multicellular organism (284).

One of the best understood examples of the second messenger function of ROS is the role of ROS in the activation of the mammalian transcription factor NF-κB (285). When activated, NF-κB induces the expression of various genes involved in inflammatory responses and immune cell regulation and differentiation (286). A diverse set of conditions can cause NF-κB activation: These include viral infection, bacterial lipopolysaccharide, protein synthesis inhibitors, cytokines, T cell mitogens, and oxidative stress (286). The common factor in all of these conditions is the generation of ROS within the affected cells (275, 285). ROS activate NF-κB by causing the release of the active NF-κB heterodimer from an inactive cytoplasmic complex composed of an NF-κB dimer and an inhibitory subunit, IκB. Upon release from IκB, the activated NF-κB dimer migrates to the nucleus, binds to NF-κB enhancers located upstream of NF-κB regulated genes (286), and up-regulates their transcription. Although ROS can be generated by either mitochondrial or nonmitochondrial sources in the cell, a primary role for mitochondrially produced ROS in activation of NF-κB has been documented in the case of TNFα-induced cytotoxicity of tumor cells (287). Cell lines lacking mitochondrial DNA (rho^0 cell lines) are resistant to TNF cytotoxicity. The same rho^0 cell lines demonstrate a reduction in TNF-induced activation of NF-κB, and a parallel reduction in expression of an NF-κB regulated gene, for interleukin-6 (288). In cells containing mitochondrial DNA, electron transport chain inhibitors have varying effects, depending on the site of inhibition in the chain. Inhibitors acting at complexes I or II decrease the effects of TNF on both NF-κB activation and cell cytotoxicity, while antimycin A potentiates the effect of TNF on both NF-κB activation and cell cytotoxicity. These effects of electron transport chain inhibitors point to the respiratory complex III as the major site of ROS production in response to TNFα. In direct studies of the effect of TNFα on mitochondrial electron transport chain activity, inhibition of respiration was found to precede the onset of cell lysis, which is consistent with the hypothesis that inhibition of respiration is responsible for generation of ROS that then act as second messengers for inducing NF-κB–regulated genes and necrotic cell death (287).

Inhibitors of mitochondrial respiration also enhance the expression of other mammalian genes (16, 288a–d). One of these genes is *GLUT1*, the gene for one of the isoforms of the glucose transporter. Expression of this gene is enhanced by hypoxia and by exposing cells to inhibitors of mitochondrial respiration (16, 288a, 288d). Recently, the *GLUT1* promoter *cis* elements that are responsible for both hypoxic induction and induction by respiratory inhibitors have been identified and shown to be distinct from one another (288d). This indicates that any effects of hypoxia on *GLUT1* are independent of effects of inhibition of mitochondrial respiration and that there are two distinct signalling pathways that impact expression of *GLUT1*; one responds to hypoxia and the other to inhibition of mitochondrial electron transport. It is not yet known if ROS are involved in either of these pathways.

Signalling from the mitochondrion to the nucleus by ROS probably also occurs in fungi. As mentioned above, the *YHB*1 gene in *S. cerevisiae* is induced by both antimycin A and cyanide. This gene is also up-regulated in cells that lack superoxide dismutase, suggesting that superoxide serves as a signal for induction (251). Currently, the molecular players in this pathway are unknown. However, YAP1p, which belongs to the AP-1 family of eukaryotic transcription factors, may be involved, because the *YHB*1 promoter has YAP1p consensus binding sites (PV Burke & RO Poyton, unpublished observations), and YAP1-dependent transcription in yeast is induced by some of the same thiooxidants that induce expression of *YHB*1 (251, 289). As discussed above, the alternative oxidase in *Hansenula anomola* is also likely under the control of ROS that are released from mitochondria. How many other yeast nuclear genes are under the control of ROS is not known. However, a potentially useful experimental paradigm for the identification of genes regulated by mitochondrially produced ROS is to search for genes that are differentially expressed in response to antimycin A.

CONCLUSIONS AND FUTURE PROSPECTS

This review bears testimony to the impressive progress that has been made recently in understanding the interactions between nuclear and mitochondrial genomes. The fact that nuclear and mitochondrial genomes interact in at least two important ways is now clear. First, they contribute essential subunit polypeptides to important mitochondrial proteins; second, they collaborate in the synthesis and assembly of these proteins. The first type of interaction is important for the regulation of cellular energy production by cytochrome *c* oxidase. The isoforms of the nuclear-coded subunits of this multimeric protein

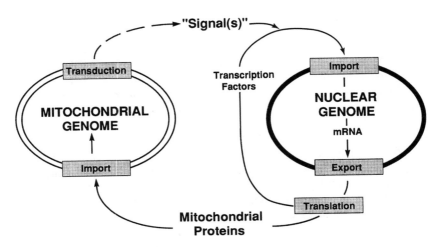

Figure 2   Communication pathways between nuclear and mitochondrial genomes. The vehicles of communication from the nucleus to the mitochondrion are nuclear-coded proteins that are translated in the cytosol and imported by mitochondria. The vehicles of communication from the mitochondrion to the nucleus are metabolic signals that are released from the mitochondrion.

affect the catalytic functions of its mitochondrially coded subunits. These isoforms are differentially regulated by environmental and developmental signals and probably allow tissues to adjust their energy production to different energy demands. They may hold a key to understanding degenerative OX-PHOS diseases, including aging.

The second type of interaction is also complex. It requires a bidirectional flow of information between the nucleus and the mitochondrion. Communication from the nucleus to the mitochondrion makes use of proteins that are translated in the cytosol and imported by the mitochondrion (Figure 2). Currently, the molecular details underlying the mode of action of these proteins are being studied. In contrast, communication from the mitochondrion to the nucleus involves metabolic signals (Figure 2) and one or more signal transduction pathways that function across the inner mitochondrial membrane. Some of these signals are reactive oxygen species. Others are metabolites that affect the phosphorylation or dephosphorylation of transcription factors that act in the nucleus.

An exciting area for future research is the signal transduction pathways that connect the mitochondrion with the nucleus. Especially intriguing will be the study of how they integrate input from the environment with input from the mitochondrion to mediate the expression of nuclear genes for mitochondrial proteins. This work should further our understanding of the regulation of cellular energy metabolism and, in addition, have important impact on our understanding of OXPHOS diseases.

ACKNOWLEDGMENTS

The preparation of this article was supported by grants GM30228 and GM39324 (to ROP) and GM36192 (to JEM) from the National Institutes of Health, grant 95-015720 (to JEM) from the American Heart Association, and a gift (to ROP) from Leslie F Bailey. We thank the many colleagues who provided reprints and preprints.

Literature Cited

1. Attardi G, Schatz G. 1988. *Annu. Rev. Cell. Biol.* 4:289–333
2. Grivell LA. 1989. *Eur. J. Biochem.* 182:477–93
3. Forsburg SL, Guarente L. 1989. *Annu. Rev. Cell. Biol.* 5:153–80
4. Pon L, Schatz G. 1991. In *The Molecular and Cellular Biology of the Yeast Saccharomyces Genome Dynamics, Protein Synthesis, and Energetics,* ed. EW Jones, JR Pringle, JR Broach, 1:333–406. Cold Spring Harbor, NY: Cold Spring Harbor Lab. Press
5. Glick B, Schatz G. 1991. *Annu. Rev. Genet.* 25:21–44
6. Poyton RO, Duhl DMJ, Clarkson GHD. 1992. *Trends Cell Biol.* 2:369–75
7. McEwen JE, Ko C, Kloeckener-Gruissem B, Poyton RO. 1986. *J. Biol. Chem.* 261:11872–79
8. Tzagoloff A, Dieckmann CL. 1990. *Microbiol. Rev.* 54:211–25
9. Parikh VS, Morgan MM, Scott R, Clements LS, Butow RA. 1987. *Science* 235:576–80
10. Farrell LE, Trawick JD, Poyton RO. 1990. *Structure, Function, and Biogenesis of Energy Transfer Systems,* ed. E Quagliariello, S Papa, F Palmieri, C Saccone, pp. 131–34. Amsterdam: Elsevier
11. Deleted in proof
12. Poyton RO, Trueblood CE, Wright RM, Farrell LE. 1988. *Ann. NY Acad. Sci.* 550:289–307
13. Zitomer RS, Lowry CV. 1992. *Microbiol. Rev.* 56:1–11
14. Poyton RO, Burke PV. 1992. *Biochim. Biophys. Acta* 1101:252–56
15. De Winde JH, Grivell LA. 1993. *Prog. Nucleic Acid Res. Mol. Biol.* 46:51–91
16. Bunn F, Poyton RO. 1996. *Physiol. Rev.* In press
17. Van Itallie CM, Dannies PS. 1988. *Mol. Endocrinol.* 2:332–37
18. Wallace DC. 1992. *Annu. Rev. Biochem.* 61:1175–212
19. Heddi A, Lestienne P, Wallace DC, Stepien G. 1993. *J. Biol. Chem.* 268:12156–63
20. Wallace DC. 1994. *J. Bioenerg. Biomembr.* 26:241–50
21. Stoneking M. 1994. *Evol. Anthropol.* 2:60–73
22. Futai M, Noumi T, Maeda M. 1989. *Annu. Rev. Biochem.* 58:111–36
23. Gillham NW. 1994. *Organelle Genes and Genomes.* New York: Oxford Univ. Press
24. Chomyn A, Mariottini P, Cleeter MWJ, Ragan CI, Matsuno-Yagi A, et al. 1985. *Nature* 314:592–97
25. Chomyn A, Cleeter MWJ, Ragan CI, Riley M, Doolittle RF, Attardi G. 1986. *Science* 234:614–18
26. Hatefi Y, Galante YM. 1980. *J. Biol. Chem.* 255:5530–37
27. Weiss H. 1987. *Curr. Top. Bioenerg.* 15:67–90
28. Kadenbach B, Jarausch J, Hartmann R, Merle P. 1983. *Anal. Biochem.* 129:517–21
29. Cox GB, Devenish RJ, Gibson F, Howitt SM, Nagley P. 1992. *Molecular Mechanisms of Bioenergetics,* ed. L Ernster, pp. 283–315. Amsterdam: Elsevier
30. Ohnishi T. 1970. *Biochem. Biophys. Res. Commun.* 41:344–50

31. Robinson KM, von Kieckebusch-Guck A, Lemire BD. 1991. *J. Biol. Chem.* 266:21347–50
32. Trumpower BL. 1990. *Microbiol. Rev.* 54:101–29
33. Power SD, Lochrie MA, Sevarino KA, Patterson TE, Poyton RO. 1984. *J. Biol. Chem.* 259:6564–70
34. Taanman JW, Capaldi RA. 1992. *J. Biol. Chem.* 267:22481–85
35. LaMarche AEP, Abate MI, Chan SHP, Trumpower BL. 1992. *J. Biol. Chem.* 262:22473–80
36. Nagley P. 1988. *Trends Genet.* 4:46–52
37. Bisson R, Schiavo G, Papini E. 1985. *Biochemistry* 24:7845–52
38. Saraste M. 1990. *Q. Rev. Biophys.* 23:331–66
39. Patterson TE, Trueblood CE, Wright RM, Poyton RO. 1988. *Cytochrome Systems: Molecular Biology and Bioenergetics,* ed. S Papa, B Chance, L Ernster, pp. 253–60. New York: Plenum
40. Trumpower BL, Gennis RB. 1994. *Annu. Rev. Biochem.* 63:675–716
41. Pilkington SJ, Skehel JM, Gennis RB, Walker JE. 1991. *Biochemistry* 30:2954–60
42. Dupuis A, Skehel JM, Walker JE. 1991. *Biochemistry* 30:2954–60
43. Hatefi Y. 1985. *Annu. Rev. Biochem.* 54:1015–69
44. Cole ST, Condon C, Lemire BD, Weiner JH. 1985. *Biochim. Biophys. Acta* 811:381–403
45. Babcock GT, Wikstrom M. 1992. *Nature* 356:301–9
46. Brunori M, Antonni G, Malatesta F, Sarti P, Ovilson MT. 1987. *Eur. J. Biochem.* 169:1–8
47. Cooper CE, Nicholls P, Freedman JA. 1991. *Biochem. Cell Biol.* 69:586–607
48. Haltia T, Finel M, Harms N, Nakari T, Raitio M, et al. 1989. *EMBO J.* 8:3571–79
49. Linke P, Weiss H. 1986. *Methods Enzymol.* 126:201–10
50. Hofhaus G, Weiss H, Leonard K. 1991. *J. Mol. Biol.* 221:1027–43
51. Schmitt ME, Trumpower BL. 1990. *J. Biol. Chem.* 265:17005–11
52. Aggeler R, Capaldi RA. 1990. *J. Biol. Chem.* 265:16389–93
53. Calder KM, McEwen JE. 1991. *Mol. Microbiol.* 5:1769–77
54. Patterson TE, Poyton RO. 1986. *J. Biol. Chem.* 261:17192–97
55. Droste M, Goehring B, Poyton RO. 1996. *FEBS Lett.* In press
56. Patterson TE. 1990. *Characterization of subunit VIII of cytochrome c oxidase in Saccharomyces cerevisiae.* PhD thesis. Univ. Colo., Boulder. 135 pp.
57. Deleted in proof
58. Sandona D, Bisson R. 1994. *Eur. J. Biochem.* 219:1053–61
59. Romanos MA, Scorer CA, Clare JJ. 1992. *Yeast* 8:423–88
60. Blachly-Dyson E, Zambronicz EB, Yu WH, Adams V, McCabe ER B., et al. 1993. *J. Biol. Chem.* 268:1835–41
61. Erecinska M, Wilson DF. 1982. *J. Membr. Biol.* 70:1–14
62. Wilson DF. 1982. *Membranes and Transport,* ed. A Martonosi, pp. 349–355. New York: Plenum
63. Wilson DF, Stubbs M, Veech RL, Erecinska M, Krebs HA. 1974. *Biochem. J.* 140:37–64
64. Erecinska M, Wilson DF. 1978. *Trends Biochem. Sci.* 3:219–22
65. Wilson DF, Stubbs M, Oshino N, Erecinska M. 1974. *Biochemistry* 13:5305–11
66. Kacser H, Burns JA. 1973. *Rate Control of Biological Processes,* pp. 65–104. London: Cambridge Univ. Press
67. Heinrich R, Rapoport TA. 1974. *Eur. J. Biochem.* 42:97–105
68. Mazat JP, Jean-Bart E, Rigoulet M, Guerin B. 1986. *Biochim. Biophys. Acta* 849:7–15
69. Tager JM, Wanders RJA, Groen AK, Kunz W, Bohnensack R, et al. 1983. *FEBS Lett.* 151:1–9
70. Briggs DR, Linnane AW. 1963. *Biochim. Biophys. Acta* 78:785–88
71. Kadenbach B. 1976. *Angew. Chem. Int. Ed. Engl.* 22:275–82
72. Ferguson-Miller S, Brautigan DL, Margoliash E. 1976. *J. Biol. Chem.* 251:1104–15
73. Bisson R, Schiavo G, Montecucco C. 1987. *J. Biol. Chem.* 262:5992–98
74. Kadenbach B, Kuhn-Nentwig L, Buge U. 1987. *Curr. Top. Bioenerg.* 15:113–61
75. Anthony G, Reinmann A, Kadenbach B. 1993. *Proc. Natl. Acad. Sci. USA* 90:289–307
76. Rohdich F, Kadenbach B. 1993. *Biochemistry* 32:8499–503
77. Hall J, Moubakak A, O'Brien P, Lian PP, Cho I, Millett F. 1988. *J. Biol. Chem.* 263:8142–49
78. Capaldi RA. 1990. *Annu. Rev. Biochem.* 59:569–96
79. Poyton RO, Goehring B, Droste M, Sevarino KA, Allen LA, Zhao X. 1995. *Methods Enzymol.* 260:97–116
80. Trueblood CE, Poyton RO. 1987. *Mol. Cell. Biol.* 7:3520–27
81. Wright RM, Dircks LK, Poyton RO. 1986. *J. Biol. Chem.* 261:17183–91

82. Dowhan W, Bibus CR, Schatz G. 1985. *EMBO J.* 4:179–84
83. Wright RM, Trawick JD, Trueblood CE, Patterson TE, Poyton RO. 1988. *Cytochrome systems: Molecular Biology and Bioenergetics,* ed. S Papa, B Chance, L Ernster, pp. 49–57. New York: Plenum
84. Nakai T, Mera Y, Yosuhara T, Ohasi A. 1994. *J. Biochem.* 116:752–58
85. Allen LA, Zhao X, Caughey W, Poyton RO. 1995. *J. Biol. Chem.* 270:110–18
86. Deleted in proof
87. Deleted in proof
88. Taanman JW, Capaldi RA. 1993. *J. Biol. Chem.* 268:18754–61
89. Waterland RA, Basu A, Chance B, Poyton RO. 1991. *J. Biol. Chem.* 266:4180–86
90. Capaldi RA, Malatesta F, Darley-Usmar VM. 1983. *Biochim. Biophys. Acta* 726:135–48
90a. Tsukihara T, Aoyama H, Yamashita E, Tomizaki T, Yamaguchi H, et al. 1995. *Science* 269:1069–74
91. Pious D. 1979. *Proc. Natl. Acad. Sci. USA* 65:1001–8
92. Kadowaki T, Kitagawa Y. 1991. *Exp. Cell Res.* 192:243–47
93. Williams RS, Garcia-Moll M, Mellor J, Salmons S, Harlan W. 1987. *J. Biol. Chem.* 262:2764–67
94. Hood DA, Zak R, Pette D. 1989. *Eur. J. Biochem.* 179:275–80
95. Luciakova K, Nelson BD. 1992. *Eur. J. Biochem.* 207:247–51
96. Wiesner RJ, Kurowski TT, Zak R. 1992. *Mol. Endocrinol.* 6:1458–67
97. Deleted in proof
98. Ulery TL, Jang SH, Jaehning JA. 1994. *Mol. Cell. Biol.* 14:1160–70
99. Trawick JD, Wright RM, Poyton RO. 1989. *J. Biol. Chem.* 264:7005–8
100. Trawick JD, Rogness C, Poyton RO. 1989. *Mol. Cell. Biol.* 9:5350–58
101. Trawick JD, Kraut ND, Simon FR, Poyton RO. 1992. *Mol. Cell. Biol.* 12:2302–12
102. Silve S, Rhode PR, Campbell JL, Poyton RO. 1992. *Mol. Cell. Biol.* 12:4197–208
103. Wright RM, Poyton RO. 1990. *Mol. Cell. Biol.* 10:1297–300
104. Trueblood CE, Poyton RO. 1988. *Genetics* 8:4537–40
105. Hodge MR, Singh K, Cumsky MG. 1990. *Mol. Cell. Biol.* 10:5510–20
106. Lambert JR, Bilanchone VW, Cumsky MG. 1994. *Proc. Natl. Acad. Sci. USA* 91:7345–49
107. Rogers PJ, Stewart PR. 1973. *J. Bacteriol.* 115:88–97
108. Rogers PJ, Stewart PR. 1973. *J. Gen. Microbiol.* 79:205–17
109. Deleted in proof
110. Groot GSP, Poyton RO. 1975. *Nature* 255:238–40
111. Bellus G. 1985. *Transcription and RNA processing in isolated yeast mitochondria.* PhD thesis., Univ. Conn., Storrs. 209 pp.
112. Kadowaki T, Kitagawa Y. 1991. *Exp. Cell Res.* 192:243–47
113. Williams RS, Garcia-Moll M, Mellor J, Salmons S, Harlan W. 1987. *J. Biol. Chem.* 262:2764–67
114. Shan B, Vazquez E, Lewis JA. 1990. *EMBO J.* 9:4307–14
115. Van Itallie CM, Dannies PS. 1988. *Mol. Endocrinol.* 2:332–37
116. Kessler M, Lamm L, Jarnagin K, DeLuca HF. 1986. *Arch. Biochem. Biophys.* 251:403–12
117. Ku CY, Lu Q, Ussuf KF, Weinstock GM, Sanborn BN. 1991. *Mol. Endocrinol.* 5:1669–76
118. Deleted in proof
119. Hood DA. 1990. *Biochem. J.* 269:503–6
120. Gagnon J, Kurowski TT, Wiesner RJ, Zak R. 1991. *Mol. Cell. Biochem.* 107:21–29
121. Luciakova K, Li R, Nelson BD. 1992. *Eur. J. Biochem.* 207:253–57
122. Cumsky MG, Trueblood CE, Ko C, Poyton RO. 1987. *Mol. Cell. Biol.* 7:3511–19
123. Cumsky MG, Trueblood CE, Ko C, Poyton RO. 1985. *Proc. Natl. Acad. Sci. USA* 82:2235–39
124. Schiavo G, Bisson R. 1989. *J. Biol. Chem.* 264:7129–34
125. Lomax MI, Grossman LI. 1989. *Trends Biochem. Sci.* 14:501–3
126. Bonne G, Siebel P, Possekel S, Marsac C, Kadenbach B. 1993. *Eur. J. Biochem.* 217:1099–107
127. Smith EO, Bement DM, Grossman LI, Lomax MI. 1991. *Biochim. Biophys. Acta* 1089:266–68
128. Ewart DD, Zhang Y-Z, Capaldi RA. 1991. *FEBS Lett.* 292:79–84
129. Seelan RS, Grossman LI. 1991. *J. Biol. Chem.* 266:19752–57
130. Lightowlers RN, Ewart G, Aggeler R, Zhang Y-Z, Calavetta L, Capaldi RA. 1990. *J. Biol. Chem.* 265:2677–81
131. Schlerf A, Droste M, Winter M, Kadenbach B. 1988. *EMBO J.* 7:2387–91
132. Scheja K, Kadenbach B. 1993. *Biochim. Biophys. Acta* 1132:91–93
133. Van Luilenburg ABP, Muijsers AO, Demol H, Dekker HL, Van Beeumen JJ. 1988. *FEBS Lett.* 240:127–32
134. Rizzuto R, Nakase H, Darras B, Francke U, Fabrizi GM, et al. 1989. *J. Biol. Chem.* 264:10595–600
135. Fabrizi GM, Rizzuto R, Nakase H, Mita

S, Kadenbach B, Schon EA. 1989. *Nucleic Acids Res.* 17:6409

136. Fabrizi GM, Sadlock J, Hirano M, Mita S, Koga Y, et al. 1992. *Gene* 119:307–12

137. Fabrizi GM, Rizzuto R, Nakase H, Mita S, Lomax MI, et al. 1989. *Nucleic Acids Res.* 17:7107

138. Van Beeumen JJ, Van Kuilenburg ABP, Van Bun S, Van den Bogert C, Tager JM, Muijsers AO. 1990. *FEBS Lett.* 263:213–16

139. Van Kuilenburg ABP, Van Beeumen JJ, Van der Meer NM, Muijsers AO. 1992. *Eur. J. Biochem.* 203:193–99

140. Trueblood CE, Wright RM, Poyton RO. 1988. *Mol. Cell Biol.* 8:4537–40

141. Pel HJ, Grivell LA. 1993. *Mol. Biol. Rep.* 18:1–13

142. de Zamaroczy M, Bernardi G. 1985. *Gene* 37:1–17

143. de Zamaroczy M, Bernardi G. 1986. *Gene* 47:155–77

144. Jaehning JA. 1993. *Mol. Microbiol.* 8:1–4

145. Shadel GS, Clayton DA. 1993. *J. Biol. Chem.* 268:16083–86

146. Biswas TK, Getz GS. 1990. *J. Biol. Chem.* 265:19053–59

147. Biswas TK. 1990. *Proc. Natl. Acad. Sci. USA* 87:9338–42

148. Masters BS, Stohl LL, Clayton DA. 1987. *Cell* 51:89–99

149. Jang SH, Jaehning JA. 1991. *J. Biol. Chem.* 266:22671–77

150. Lisowsky T, Schweizer E, Michaelis G. 1987. *Eur. J. Biochem.* 164:559–63

151. Lisowsky T, Michaelis G. 1988. *Mol. Gen. Genet.* 214:218–23

152. Mangus DA, Jang S-H, Jaehning JA. 1994. *J. Biol. Chem.* 269:26568–74

153. Biswas TK. 1992. *J. Mol. Biol.* 226:335–47

154. Diffley JFX, Stillman B. 1991. *Proc. Natl. Acad. Sci. USA* 88:7864–68

155. Diffley JFX, Stillman B. 1992. *J. Biol. Chem.* 267:3368–74

156. Fisher RP, Lisowsky T, Parisi MA, Clayton DA. 1992. *J. Biol. Chem* 267:3358–67

157. Parisi MA, Xu B, Clayton DA. 1993. *Mol. Cell. Biol.* 13:1951–61

158. Brohl S, Lisowsky T, Riemen G, Michaelis G. 1994. *Yeast* 10:719–31

159. Manthey GM, McEwen JE. 1995. *EMBO J.* 14:4031–43

160. Groudinsky O, Bousquet I, Wallis MG, Slonimski PP, Dujardin G. 1993. *Curr. Genet.* 240:419–27

161. Lisowsky T, Michaelis G. 1989. *Mol. Gen. Genet.* 219:125–28

162. Ling M-L, Risman SS, Klement JF, McGraw N, McAllister WT. 1989. *Nucleic Acids Res.* 17:1605–18

163. Martin CT, Muller DK, Coleman JE. 1988. *Biochemistry* 27:3966–74

164. Mead DA, Skorupa ES, Kemper B. 1986. *Protein Eng.* 1:67–74

165. Daga A, Micol V, Hess D, Aebersold R, Attardi G. 1993. *J. Biol. Chem.* 268:8123–30

166. Shang J, Clayton DA. 1994. *J. Biol. Chem.* 269:29112–20

167. Dieckmann CL, Staples RR. 1994. *Int. Rev. Cytol.* 152:145–81

168. Clayton DA. 1984. *Annu. Rev. Biochem.* 53:573–94

169. Bindoff LA, Howell N, Poulton J, McCullough DA, Morten KJ, et al. 1993. *J. Biol. Chem.* 268:19559–64

170. Hopper AK, Martin NC. 1992. In *The Molecular and Cellular Biology of the Yeast Saccharomyces Gene Expression,* ed. EW Jones, JR Pringle, JR Broach, pp. 99–141. Cold Spring Harbor, NY: Cold Spring Harbor Lab. Press

171. Hopper AK, Furukawa AH, Pham HD, Martin NC. 1982. *Cell* 28:543–50

172. Wiesenberger G, Costanzo MC, Fox TD. 1995. *Mol. Cell. Biol.* 15:3291–300

173. Chen W, Dieckmann CL. 1994. *J. Biol. Chem.* 269:16574–78

174. Min J, Zassenhaus HP. 1993. *Mol. Cell. Biol.* 13:4167–73

175. Mueller DM, Getz GS. 1986. *J. Biol. Chem.* 261:11816–22

176. King MP, Attardi G. 1993. *J. Biol. Chem.* 268:10228–37

177. Min J, Heuertz RM, Zassenhaus HP. 1993. *J. Biol. Chem.* 268:7350–57

178. Pel HJ, Grivell LA. 1994. *Mol. Biol. Rep.* 19:183–94

179. McCarthy JEG, Brimacombe R. 1994. *Trends Genet.* 10:402–7

180. Costanzo MC, Fox TD. 1986. *Mol. Cell. Biol.* 6:3694–703

181. Costanzo MC, Seaver EC, Fox TD. 1986. *EMBO J.* 5:3637–41

182. Kloeckener-Gruissem B, McEwen JE, Poyton RO. 1988. *J. Bacteriol.* 170:1399–402

183. Poyton RO, Sevarino KA, McKee EE, Duhl DMJ, Cameron V, Goehring B. 1996. *Adv. Mol. Cell Biol.* 18:In press

184. Gillham NW, Boynton JE, Hauser CR. 1994. *Annu. Rev. Genet.* 28:71–93

185. Costanzo MC, Fox TD. 1988. *Proc. Natl. Acad. Sci. USA* 85:2677–81

186. Costanzo MC, Fox TD. 1993. *Mol. Cell. Biol.* 13:4806–13

187. Brown NG, Costanzo MC, Fox TD. 1994. *Mol. Cell. Biol.* 14:1045–53

188. Valencik ML, Kloeckener-Gruissem B, Poyton RO, McEwen JE. 1989. *EMBO J.* 8:3899–904

189. Deleted in proof

190. McMullin TW, Fox TD. 1993. *J. Biol. Chem.* 268:11737–41
191. Valencik ML, McEwen JE. 1991. *Mol. Cell. Biol.* 11:2399–405
192. Haffter P, McMullin TW, Fox TD. 1991. *Genetics* 127:319–26
193. Haffter P, Fox TD. 1992. *Mol. Gen. Genet.* 235:64–73
194. Mulero JJ, Fox TD. 1993. *Mol. Biol. Cell.* 4:1327–35
195. Roedel G, Fox TD. 1987. *Mol. Gen. Genet.* 206:45–50
196. Deleted in proof
197. Dekker PJ, Papadopoulou B, Grivell LA. 1993. *Curr. Genet.* 23:22–27
198. Chretien D, Rustin P, Bourgeron T, Rotig A, Saudubray JM, Munnich A. 1994. *Clin. Chim. Acta* 228:53–70
199. Wiesner RJ. 1992. *Trends Genet.* 8:264–65
200. Williams RS. 1986. *J. Biol. Chem.* 261:12390–94
201. Wiesner RJ, Aschenbrenner V, Ruegg JC, Zak R. 1994. *Am. J. Physiol.* 267:C229–35
202. Town GP, Essig DA. 1993. *J. Appl. Physiol.* 74:192–96
203. Goglia F, Lanni A, Barth J, Kadenbach B. 1994. *FEBS Lett.* 346:295–98
204. Costanzo MC, Fox TD. 1990. *Annu. Rev. Genet.* 24:91–113
205. Pelissier PP, Camougrand NM, Manon ST, Velours GM, Guerin MG. 1992. *J. Biol. Chem.* 267:2467–73
206. Pel HJ, Tzagoloff A, Grivell LA. 1992. *Curr. Genet.* 21:139–46
207. Gimeno CJ, Ljungdahl PO, Styles CA, Fink GR. 1992. *Cell* 68:1077–90
208. Bittner-Eddy P, Monroy AF, Brambl R. 1994. *J. Mol. Biol.* 235:881–97
209. Rassow J, Mohrs K, Koidl S, Barthelmess IB, Pfanner N, Tropschug M. 1995. *Mol. Cell. Biol.* 15:2654–62
210. Schnall R, Mannhaupt G, Stucka R, Tauer R, Ehnle S, et al. 1994. *Yeast* 10:141–55
211. Herrmann JM, Stuart RA, Craig EA, Neupert W. 1994. *J. Cell Biol.* 127:893–902
212. Nunnari J, Fox T, Walter P. 1993. *Science* 262:1997–2004
213. Tauer R, Mannhaupt G, Schnall R, Pajic A, Langer T, Feldman H. 1994. *FEBS Lett.* 353:197–200
214. Tzagoloff A, Yue J, Jang J, Paul M-F. 1994. *J. Biol. Chem.* 269:26144–51
215. Fischer M, Hilt W, Richter-Ruoff B, Gonen H, Ciechanover A, Wolf DH. 1994. *FEBS Lett.* 355:69–75
216. Goldberg AL. 1995. *Science* 268:522–23
217. Nakai T, Yasuhara T, Fujiki Y, Ohashi A. 1995. *Mol. Cell. Biol.* 15:4441–52
218. Pajic A, Tauer R, Feldman H, Neupert W, Langer T. 1994. *FEBS Lett.* 353:201–6
219. Weber ER, Hanekamp T, Thorsness PE. 1996. *Mol. Biol. Cell.* In press
220. Poyton RO, McKemmie EE. 1976. In *The Genetics and Biogenesis of Mitochondria*, ed. T Bucher, W Neupert, W Sebald, S Werner, pp. 207–14. Amsterdam: North Holland
221. Crivellone MD, Wu M, Tzagoloff A. 1988. *J. Biol. Chem.* 263:14323–33
222. Hadikusumo RG, Meltzer S, Choo WM, Jean-Francois MJB, Linnane AW, Marzuki S. 1988. *Biochim. Biophys. Acta* 933:212–22
223. Nobrega MP, Nobrega FG, Tzagoloff A. 1990. *J. Biol. Chem.* 265:14220–26
224. Tzagoloff A, Capitanio N, Nobrega MP, Gatti D. 1990. *EMBO J.* 9:2759–64
225. Tzagoloff A, Nobrega M, Gorman N, Sinclair P. 1993. *Biochem. Mol. Biol. Int.* 31:593–98
226. Saiki K, Mogi T, Anraku Y. 1992. *Biochem. Biophys. Res. Commun.* 189:1491–97
227. Saiki K, Mogi T, Ogura K, Anraku Y. 1993. *J. Biol. Chem.* 268:26041–45
228. Saiki K, Mogi T, Hori H, Tsubaki M, Anraku Y. 1993. *J. Biol. Chem.* 268:26927–34
229. Svensson B, Lubben M, Hederstedt L. 1993. *Mol. Microbiol.* 10:193–201
230. Glerum DM, Tzagoloff A. 1994. *Proc. Natl. Acad. Sci. USA* 91:8452–56
231. Deleted in proof
232. McConnell SJ, Stewart LC, Talin A, Yaffe MP. 1990. *J. Cell Biol.* 111:967–76
233. McConnell SJ, Yaffe MP. 1992. *J. Cell Biol.* 118:385–95
234. McConnell SJ, Yaffe MP. 1993. *Science* 260:687–89
235. Stewart LC, Yaffe MP. 1991. *J. Cell Biol.* 115:1249–57
236. Sogo LF, Yaffe MP. 1994. *J. Cell Biol.* 126:1361–73
237. Burgess SM, Delannoy M, Jensen RE. 1994. *J. Cell Biol.* 126:1375–91
238. Drubin DG, Jones HD, Wertman KF. 1993. *Mol. Biol. Cell.* 4:1277–94
239. Wright RM, Ko C, Cumsky MG, Poyton RO. 1984. *J. Biol. Chem.* 259:15401–7
240. Suzuki H, Hosokawa Y, Nishikimi M, Osawa T. 1991. *J. Biol. Chem.* 266:2333–38
241. Nagley P. 1991. *Trends Genet.* 7:1–4
242. Marcznski GT, Schulz PW, Jaehning JA. 1989. *Mol. Cell. Biol.* 9:3193–202
243. Virbasius CA, Virbasius JV, Scarpulla RC. 1993. *Genes Dev.* 7:2431–45
244. Virbasius JV, Scarpulla RC. 1994. *Proc. Natl. Acad. Sci. USA* 91:1309–13

245. Chau CA, Evans MJ, Scarpulla RC. 1992. *J. Biol. Chem.* 267:6999–7006
246. Virbasius JV, Virbasius CA, Scarpulla RC. 1993. *Genes Dev.* 7:380–92
247. Poyton RO. 1980. *Curr. Top. Cell. Regul.* 17:231–95
248. Barath Z, Kuntzel H. 1972. *Proc. Natl. Acad. Sci. USA* 69:1371–74
249. Liao X, Small WC, Srere PA, Butow RA. 1991. *Mol. Cell. Biol.* 11:38–46
250. Liao X, Butow RA. 1993. *Cell* 72:61–71
251. Zhao X, Raitt D, Burke P, Clewell A, Pepperl S, Poyton RO. 1996. *J. Biol. Chem.* In press
252. Bertrand H, Argan CA, Szakacs NA. 1983. In *Mitochondria 1983. Nucleo-Mitochondrial Interactions,* ed. RJ Schweyen, K Wolf, F Kaudewitz, pp. 495–507. Berlin: Walter de Gruyter
253. Sakajo S, Minagawa N, Yoshimoto A. 1993. *FEBS Lett.* 3:310–12
254. Salcedo-Hernandez R, Escamilla E, Ruiz-Herrera J. 1994. *Microbiol. Rev.* 140:399–407
255. Parikh VS, Conrad-Webb H, Docherty R, Butow RA. 1989. *Mol. Cell. Biol.* 9:1897–907
256. Wang S-S, Brandriss MC. 1987. *Mol. Cell. Biol.* 7:4431–40
257. Partaledis JA, Mason TL. 1988. *Mol. Cell. Biol.* 8:3647–60
258. van Loon APGM, de Groot RJ, van Eyk E, van der horst GTJ, Grivell LA. 1982. *Gene* 20:323–37
259. Koonin EV. 1994. *Trends Biochem. Sci.* 19:156–57
259a. Chelstowska A, Butow R. 1995. *J. Biol. Chem.* 270:18141–46
259b. Kliewer SA, Forman BM, Blumberg B, Ong ES, Borgmeyer U, et al. 1994. *Proc. Natl. Acad. Sci. USA* 91:7355–59
259c. Zhu Y, Alvares K, Huang Q, Rao MS, Reddy JK. 1993. *J. Biol. Chem.* 268: 26817–20
259d. Issemann I, Green S. 1990. *Nature* 347: 645–50
259e. Muerhoff AS, Griffin KJ, Johnson EF. 1992. *J. Biol. Chem.* 267:19051–53
259f. Zhang B, Marcus SL, Sajjadi FG, Alvares K, Reddy JK, et al. 1992. *Proc. Natl. Acad. Sci. USA* 89:7541–45
259g. Kliewer SA, Umesono K, Noonan DJ, Heyman RA, Evans RM. 1992. *Nature* 358:771–74
259h. Tugwood JD, Issemann I, Anderson RG, Bundell KR, McPheat WL, Green S. 1992. *EMBO J.* 11:433–39
259i. Castelein H, Gulick T, Declerc PE, Mannaerts GP, Moore DD, Baes MI. 1994. *J. Biol. Chem.* 269:26754–58
259j. Rodriquez JC, Gil-Gomez G, Hegarat FG, Haro D. 1994. *J. Biol. Chem.* 269: 18767–72
259k. Gulick T, Cresci S, Caira T, Moore DD, Kelly DP. 1994. *Proc. Natl. Acad. Sci. USA* 91:11012–16
260. Ycas M. 1956. *Exp. Cell Res.* 11:1–6
261. Minagawa N, Koga S, Nakano M, Sakajo S, Yoshimoto A. 1992. *FEBS Lett.* 3:217–19
261a. Upton T, Wiltshire S, Francesconi S, Eisenberg S. 1995. *J. Biol. Chem.* 270: 16153–59
262. Boveris A, Chance B. 1973. *Biochem. J.* 134:707–16
263. Forman HJ, Boveris A. 1982. In *Free Radicals in Biology,* ed. WA Prior, 5: 65–90. Orlando: Academic
264. Cadenas E. 1989. *Annu. Rev. Biochem.* 58:79–110
265. Ames BN, Shigenaga MK. 1992. In *Molecular Biology of Free Radical Scavenging Systems,* ed. JG Scandalios, pp. 1–22. Cold Spring Harbor, NY: Cold Spring Harbor Lab. Press
266. Stadtman ER. 1992. *Science* 257:1220–24
267. Yu BP. 1993. *Free Radicals and Aging,* ed. BP Yu, pp. 57–88. Ann Arbor: CRC Press
268. Bandy B, Davison AJ. 1990. *Free Radic. Biol. Med.* 8:523–39
269. Hockenbery DM, Oltvai ZN, Yin XM, Milliman CL, Korsmeyer SJ. 1993. *Cell* 75:241–51
270. Kane DJ, Sarafian TA, Anton R, Hahn H, Gralla EB, et al. 1993. *Science* 262: 1274–77
271. Barinaga M. 1994. *Science* 263:754–56
272. Vayssiere J-L, Petit PX, Risler Y, Mignotte B. 1994. *Proc. Natl. Acad. Sci. USA* 91:11752–56
273. Shigenaga MK, Hagen TM, Ames BN. 1994. *Proc. Natl. Acad. Sci. USA* 91: 10771–78
274. Ames BN, Shigenaga MK, Hagen TM. 1993. *Proc. Natl. Acad. Sci. USA* 90: 7915–22
275. Schreck R, Rieber P, Baeuerle PA. 1991. *EMBO J.* 10:2247–58
276. Hansberg W, De Groot H, Sies H. 1993. *Free Radic. Biol. Med.* 14:287–93
277. Remacle J, Raes M, Toussaint O, Renard P, Rao G. 1995. *Mutat. Res.* 316:103–22
278. Miquel J. 1992. *Mutat. Res.* 275:209–16
279. Lambowitz AM, Kobayashi GS, Painter A, Medoff G. 1983. *Nature* 303:806–8
280. Maresca B, Kobayashi GS. 1989. *Microbiol. Rev.* 53:186–209
281. Patriarca EJ, Kobayashi GS, Maresca B. 1992. *Biochem. Cell. Biol.* 70:207–14
282. Henry Y, Lepoivre M, Drapier J-C, Ducrocq C, Boucher J-L, Guissani A. 1993. *FASEB J.* 7:1124–34
283. Kurose I, Kato S, Ishii H, Fukumura D,

Miura S, et al. 1993. *Hepatology* 18: 380–88

284. Park MK, Myers RAM, Marzella L. 1992. *Clin. Infect. Dis.* 14:720–40

285. Schmidt KN, Amstadt P, Cerutti P, Baeuerle PA. 1995. *Chem. Biol.* 2:13–22

286. Baeuerle PA, Henkel T. 1994. *Annu. Rev. Immunol.* 12:141–79

287. Schulze-Osthoff K, Beyaert R, Vandevoorde V, Haegeman G, Fiers W. 1993. *EMBO J.* 12:3095–104

288. Schulze-Osthoff K, Bakker AC, Vanhaesebroeck B, Beyaert R, Jacob WA, Fiers W. 1992. *J. Biol. Chem.* 267:5317–23

288a. Loike JD, Cao L, Brett J, Ogawa S, Silverstein SC, Stern D. 1992. *Am. J. Physiol.* 263:C326–33

288b. Sun MK, Reis DJ. 1994. *J. Physiol.* 476:101–16

288c. Gonzalez A, Almarez L, Obeso A, Rigual R. 1994. *Physiol. Rev.* 74:829–98

288d. Ebert BL, Firth JD, Ratcliffe PJ. 1995. *J. Biol. Chem.* 270:29083–89

289. Kuge S, Jones N. 1994. *EMBO J.* 13: 655–64

290. Wallis MG, Groudinsky O, Slonimski PP, Dujardin G. 1994. *Eur. J. Biochem.* 222:27–32

291. Manthey GM. 1995. *A study of the nuclear gene PET309 and its roles in regulating the expression of the mitochondrial COX1 gene in S. cerevisiae.* PhD thesis. Univ. Calif., Los Angeles. 160 pp.

292. Weber ER, Dieckmann CL. 1990. *J. Biol. Chem.* 265:1594–1600

293. Ziaja K, Michaelis G, Lisowsky T. 1993. *J. Mol. Biol.* 229:909–16

294. Stepien PP, Margossian SP, Landsman D, Butow RA. 1992. *Proc. Natl. Acad. Sci. USA* 89:6813–17

295. Decoster E, Simon M, Hatat D, Faye G. 1990. *Mol. Gen. Genet.* 224:111–18

296. Dieckmann CL, Tzagoloff A. 1985. *J. Biol. Chem.* 260:1513–20

297. Michaelis U, Korte A, Rodel G. 1991. *Mol. Gen. Genet.* 230:177–85

298. Payne MJ, Finnegan PM, Smooker PM, Lukins HB. 1993. *Curr. Genet.* 24:126–35

299. Finnegan PM, Payne MJ, Keramidaris E, Lukins HB. 1991. *Curr. Genet.* 20: 53–61

300. Ackerman SH, Gatti DL, Gellefors P, Douglas MG, Tzagoloff A. 1991. *FEBS Lett.* 2:234–38

301. McEwen JE, Hong KH, Park S, Preciado GT. 1993. *Curr. Genet.* 23:9–14

302. Deleted in proof

303. Deleted in proof

304. Bonnefoy N, Chalvet F, Hamel P, Slonimski PP, Dujardin G. 1994. *J. Mol. Biol.* 239:201–12

305. Schulze M, Rodel G. 1989. *Mol. Gen. Genet.* 216:37–43

306. Bousquet I, Dujardin G, Slonimski PP. 1991. *EMBO J.* 10:2023–2031

307. Wu M, Tzagoloff A. 1989. *J. Biol. Chem.* 264:11122–30

308. Crivellone MD. 1994. *J. Biol. Chem.* 269:21284–92

309. Ackerman SH, Tzagoloff A. 1990. *J. Biol. Chem.* 265:9952–59

310. Ackerman SH, Martin J, Tzagoloff A. 1992. *J. Biol. Chem.* 267:7386–94

311. Bowman S, Ackerman SH, Griffiths DE, Tzagoloff A. 1991. *J. Biol. Chem.* 266:7517–7523

312. Isaya G, Miklos D, Rollins RA. 1994. *Mol. Cell. Biol.* 14:5603–16

313. Nakai M, Endo T, Hase T, Mastubara H. 1993. *J. Biol. Chem.* 268:24262–69

314. Van Dyck L, Pearce DA, Sherman F. 1994. *J. Biol. Chem.* 269:238–42

315. Wagner I, Arlt H, Van Dyck L, Langer T, Neupert W. 1994. *EMBO J.* 13:5135–45

316. Nobrega FG, Nobrega MP, Tzagoloff A. 1992. *EMBO J.* 11:3821–29

317. Suzuki CK, Suda K, Wang N, Schatz G. 1994. *Science* 264:273–76

318. Glerum DM, Koerner TJ, Tzagoloff A. 1995. *J. Biol. Chem.* 270:15585-90

Annu. Rev. Biochem. 1996. 65:609–34

HEMATOPOIETIC RECEPTOR COMPLEXES

James A. Wells and Abraham M. de Vos

Department of Protein Engineering, Genentech, Inc., South San Francisco, California 94080

KEY WORDS: hormone receptor complexes, protein structure and function, protein-protein interactions, signal transduction

ABSTRACT

Hematopoietic hormones/cytokines and receptors regulate a wide variety of biological activities and are important in medicine. Through recent biochemical, biophysical, and structural studies we are beginning to understand how these molecules work at the molecular level. These extracellular hormones activate their transmembrane receptors by causing them to oligomerize. The receptor oligomers in turn activate intracellular tyrosine kinase molecules which then activate transcription factors (the JAK-STAT pathways). This review centers on the molecular basis for hormone-receptor binding, and how this information is useful in understanding protein-protein interactions and for the design of second generation molecules.

CONTENTS

PERSPECTIVE AND OVERVIEW 610

GENERAL STRUCTURAL ASPECTS 612
    Structure of the Ligands 612
    Structure of the Ligand Binding Domains 614

HEMATOPOIETIC RECEPTOR SIGNALING COMPLEXES 616
    The Homodimeric Group of Hematopoietic Receptor Complexes 616
    The Hetero-Oligomeric Group of Receptor Complexes 620

MOLECULAR RECOGNITION FOR hGH-RECEPTOR COMPLEXES 624
    Structure-Function of Site 1 624
    Structure-Function of Site 2 626
    The hGH-Prolactin Receptor Complex 627

GENERAL FEATURES AND CONCLUSIONS 629

FUTURE PERSPECTIVES 631

609

0066-4154/96/0701-0609$08.00

PERSPECTIVE AND OVERVIEW

Hormone receptors are sensory machines that allow external signals to be translated into internal events that direct cells to proliferate, differentiate, or even die. Once bound to their cognate hormone, the membrane receptor ignites a cascade of intracellular signaling events destined to carry out the hormonal message. Through high resolution structural and functional studies we are beginning to understand the fundamental mechanics and chemistry of how these signaling machines work. Such studies are of practical value, as hormone receptor complexes modulate important human biology and disease. Furthermore, protein-protein interactions like these are ubiquitous in biology; they are crucial elements of regulatory processes and cellular assemblies.

This review centers on aspects of molecular recognition for a class of hormones and receptors that govern a number of immune and growth functions, known as class 1 of the hematopoietic or cytokine superfamily (for general reviews see 1–5). This family of proteins consists of more than 20 known cytokines and growth factors and their corresponding receptors, linked by several common structural and functional features (Figure 1). Each hormone has a similar four-helix bundle motif. Each receptor has a single transmembrane segment with its amino-terminus outside and its carboxy-terminus inside. The extracellular portion contains at least one cytokine receptor homology region consisting of two fibronectin type III modules. The intracellular portions are quite diverse and structurally not well characterized. Unlike the tyrosine kinase family of receptors, the hematopoietic receptors do not contain an intrinsic tyrosine kinase domain but associate reversibly with a family of tyrosine kinases and cognate STAT proteins that activate transcription (for review see 6). These hormone-receptor complexes modulate a wide variety of immune functions [interleukin (IL)-2, IL-3, IL-4, IL-5, IL-6, IL-7, IL-9, IL-11, IL-12, IL-13, IL-15, granulocyte- and granulocyte macrophage–stimulating factors (G- and GM-CSF), thrombopoietin (TPO), erythropoietin (EPO), leukemia inhibitory factor (LIF), oncostatin M (OSM), and stem cell factor (SCF)], somato-lactogenic endocrinology [growth hormone (GH), prolactin (PRL), placental lactogen (PL)], and even survival of some neuronal cell types [ciliary neurotrophic factor (CNTF)] (for review see 7).

Here we discuss the common structural features that these hormones use to bind their receptors and activate them by receptor oligomerization. The intricacies of these binding and oligomerization reactions are perhaps best understood for the hGH-receptor complex and thus we review this system in most detail. Finally, these studies suggest ways of designing hormone variants and perhaps even small molecules that may be useful in expanding the therapeutic benefit for this important class of macromolecules.

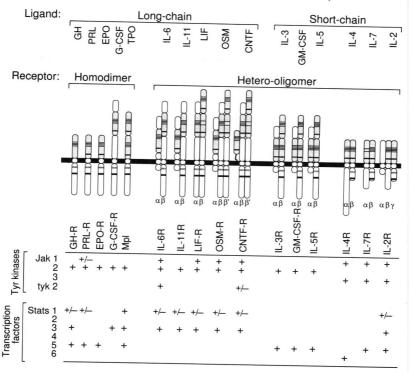

Figure 1 Class 1 of the hematopoietic ligand-receptor superfamily. The ligands are divided into two classes: long-chain and short-chain. The receptors are a single transmembrane type. Each receptor contains at least one cytokine receptor homology (CRH) region (shaded) that can bind the ligand. The thin bands within the CRH represent conserved cysteines and the thicker bands the WSXWS motif. Several of these receptors have additional domains, such as extra IgG-like domains on their extreme amino termini or three fibronectin-like repeats near the membrane. They are anchored to the membrane by a single transmembrane domain or by glycosyl phosphatidyl inositol linkage (in the case of the α receptor for CNTF). The intracellular domains vary considerably in sequence and length, but the signaling receptors contain two conserved regions known as box 1 and 2 (indicated by the banding). The receptors fall into two groups—the homodimeric and the hetero-oligomeric—depending on the number of nonidentical subunits necessary for binding and signaling. The hetero-oligomeric group can be further subdivided into three groups based on common β subunits used: gp130, gp140, or the IL-2γ chain. Signaling from these complexes is mediated by noncovalent binding to various tyrosine kinases (JAK1,2,3 or tyk 2) and transcription factors (STAT 1-6). The exact ones used in each case are marked, using a + or ± to designate whether they play a major or minor role in signaling, respectively. A blank space indicates that we are not aware of evidence implicating the particular JAK or STAT in signaling. The receptor diagram is expanded from (3) to include the IL-11 and homodimeric receptors. The JAK/STAT piece was adapted from (6). Recent evidence on TPO (8) is also included. The following is a list of abbreviations for the indicated receptors: GH-R, growth hormone receptor; PRL-R, prolactin receptor; EPO-R, erythropoietin receptor; G-CSF-R, granulocyte colony stimulating factor receptor; Mpl, myloproliferative leukemia receptor or the TPO receptor; IL-6R, interleukin-6 receptor; LIF-R, leukemia inhibiting factor receptor; LIFBP, LIF binding protein; OSM-R, oncostatin M receptor; CNTF-R, ciliary neurotropic factor receptor; GM-CSF-R, granulocyte colony stimulating factor receptor.

GENERAL STRUCTURAL ASPECTS

Structure of the Ligands

The first crystal structure reported for a hematopoietic hormone was that of porcine growth hormone (9). This 22-kDa hormone has an anti-parallel four-helix bundle core with a characteristic "up-up-down-down" topology (Figure 2a). The same basic fold on a smaller frame was then reported for GM-CSF (Figure 2b) (10, 11). As the structures of additional members of this family were solved, including hGH (12), G-CSF (13, 14), M-CSF (15), LIF (16), IL-2 (17, 18), IL-4 (19–22), IL-5 (23), and IL-6 (W Somers, personal communication), the hormones could be classified into two groups: "long-chain" and "short-chain" (24). The long-chain group (160 to 200 amino acids) has four long helices (each about 25 residues) and an angle between the AD and BC helix pairs of about 18°. Based on direct structural evidence this group consists of GH, G-CSF, LIF, and IL-6; based on sequence homology it is predicted that PRL, EPO, TPO, OSM, CNTF, IL-11 and IL-12 belong to this group as well (Figure 1). Members of the short-chain group (105 to 145 amino acids) have shorter helices (each about 15 residues) and a larger AD/BC packing angle (about 35°). Structural studies show this group includes GM-CSF, M-CSF, IL-2, IL-4, and IL-5, and sequence analysis suggests that the group also includes IL-3, IL-7, IL-9, IL-13, and SCF. Most of these molecules are monomers; however, M-CSF and IL-5 are disulfide-linked dimers with distinctly different dimerization interfaces.

The four-helix bundle of hGH consists of two long helices (A and D; 28 and 30 residues long, respectively) and two shorter helices (B and C; 19 and 23 residues, respectively) (Figure 2A). The internal core of the bundle is made up almost exclusively of hydrophobic residues with the exception of Asp169 (which hydrogen-bonds to the side chain of Trp86) and Ser79 (which hydrogen-bonds to the carbonyl oxygen of Leu75). Buried hydrophilic residues contribute to the four-helix bundle core in other hematopoietic hormones as well (25).

The first six residues are probably flexible in solution because they point away from the bundle core and have different conformations in five known crystal structures of wild-type and mutated hGH (12, 26, 27; AM de Vos & M Ultsch, unpublished results). The first crossover loop, linking helices A and B, is long and generally well defined. This loop contains two small segments of α helix, one near its beginning and one near its end. Another short helical segment is present in the short connection between helices B and C. In contrast to the first two loops, the long crossover connection between helices C and D is poorly ordered. The two long anti-parallel crossover loops are close together in space such that they cover the BC face of the molecule; this leaves the AC, AD, and BD faces exposed. The two short helical segments of the first cross-

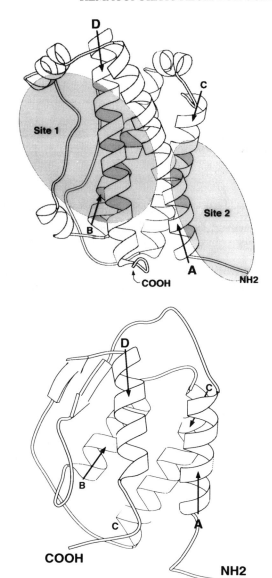

Figure 2 Basic structures for long-chain and short-chain hematopoietic ligands, exemplified by hGH and GM-CSF, respectively. These ligands share a common fold but differ in the length of their helices, the size of the connecting loops, and the positions of their disulfide bonds. (*a*) Ribbon diagram of a hGH taken from x-ray coordinates (12). Each of the four main helices (A–D) are labeled showing the up-up-down-down topology. The location of Sites 1 and 2 is indicated by shaded ovals. (*b*) Ribbon diagram of GM-CSF taken from the x-ray coordinates (10).

over connection hang over the top and bottom of the D helix, producing a concave shape to the AD side of the molecule. This is a site for binding one receptor and contrasts sharply to the flat AC face where a second receptor binds.

Structure of the Ligand Binding Domain

The signature of a hematopoietic receptor family member is the presence of a conserved cytokine receptor homology (CRH) region (28) within the extracellular portion, which is usually involved in hormone binding (Figure 1). Each CRH consists of 200 to 250 residues and was proposed to have two immunoglobin-like domains connected with a short linker. The N-terminal domain contains four strictly conserved cysteine residues, while a strongly-conserved Trp-Ser-X-Trp-Ser sequence, the so-called WSXWS box (X being any residue), is found near the C-terminus of the second domain. The structures of the extracellular portions of the hGH and hPRL receptors have been solved (12, 26) and confirm many of the structural predictions (28). However, instead of being immunoglobulin-like, the two domains of these receptors are better classified as fibronectin type III (FNIII) modules based on their overall fold and topology as well as characteristic "fingerprint" residues (see 29).

We focus on the three-dimensional structure of the hGH receptor because it is exemplary of the CRH region (Figure 3). Each domain contains seven β strands, labeled sequentially A, B, C, C', E, F, and G following the accepted FNIII convention (30); when necessary, a subscript N or C is used to distinguish the N-terminal and C-terminal domains, respectively. The seven β strands are divided into two sheets: The three-stranded sheet is made up of A, B, and E and the four-stranded sheet of C, C', F, and G. The linker between the domains is short and has helical conformation. This linker generates an acute domain-domain angle of about 80° and exposes the loops at the end of the domain barrels (AB_N, CC'_N, and EF_N as well as BC_C, $C'E_C$, and FG_C) so they can bind the hormone (Figure 2). Crystal structures are also known for tissue factor (31, 32) and the γ-interferon/receptor-α complex (33), which are members of class 2 of the cytokine superfamily. The domain architectures are virtually identical to the hGH and hPRL receptors. However, the angle between the domains for tissue factor and the γ-interferon/receptor-α complex is less acute, which means the AB_N loop is buried at the domain-domain interface. Thus, the domain-domain angle determines a great deal about how the receptor can bind the hormone.

The four conserved cysteines in the CRH are buried in the core of the N-terminal domain and form two disulfide bridges that link strands A to B and C' to E (Figure 3). The hGH receptor contains a third disulfide bond which is exposed and bridges strands F and G. The residues in the hGH receptor that

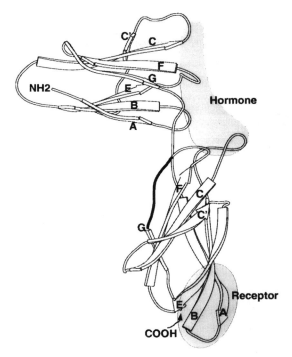

Figure 3 Basic structure for the hematopoietic receptor exemplified by a ribbon diagram of the hGH receptor taken from x-ray coordinates (12). The hormone-binding region is located at the hinge between the two fibronectin type III domains. The region involved in making receptor-receptor interactions is located on one side of the C-terminal domain and the WSXWS segment is located on the opposite side (filled).

correspond to the well-conserved WSXWS box are Tyr-Gly-Glu-Phe-Ser, which form part of an extended segment preceding β-strand G in the C-terminal domain. The main-chain conformation of this segment resembles two β-bulges (29) or two turns of poly-proline helix (34). The main-chain conformations are the same in this same region for the hPRL receptor, which contains Trp-Ser-Ala-Trp-Ser, as well as for tissue factor and even some nonreceptor FNIII domains, in which only the serine residues are conserved. Thus, the unusual main-chain conformation of the WSXWS box is neither absolutely dependent upon a WSXWS sequence nor unique to the hematopoietic receptor superfamily. The tryptophan side chains are packed between the aliphatic portions of several charged or hydrophilic residues from neighboring strands (12, 35, 26). A similar motif is conserved in all class 1 receptors (see for example 36, 37), supporting the supposition that this motif plays an important

Table 1   General biochemical and biophysical characteristics that may distinguish homodimeric and hetero-oligomeric receptor signaling complexes

Signaling class	Receptor homodimer	Receptor hetero-oligomer
Typical stoichiometry	1H: 2α	2H: 2α: 2β or H:α: β: β′ or H: α: 2β
Binding sites on hormone	2	2 or 3
Receptor-receptor sites	1	2
Hormone-binding receptor	α	α
Signaling receptor	α	β
Bell-shaped dose response	Possible	No
ECD-α	Antagonist	Sometimes agonist
Agonistic receptor MAbs	α	Possible for β
Agonistic receptor disulfides	α	Possible for β
Receptor dominant negative	α	β

structural and/or functional role. However, since the motif points away from both the ligand binding site and the receptor dimerization site its biological function remains unclear.

HEMATOPOIETIC RECEPTOR SIGNALING COMPLEXES

The expression of soluble hormone binding domains of receptors has made it possible to use biophysical and biochemical methods on highly purified components to begin to characterize the oligomeric structure of the hematopoietic receptor signaling complexes. These oligomers fall broadly into two basic categories, the homodimeric and hetero-oligomeric groups (Figure 1). For the homodimeric group, of which hGH is the paradigm, the α-receptor subunit that binds the hormone is responsible for signaling. The hetero-oligomeric group, of which IL-6 is a good example, has at least two receptors (α and β); the α receptor is generally responsible for binding and the β receptor for signaling. Here we survey the work on characterizing these complexes and summarize experimental criteria that may be helpful for categorizing them (Table 1).

The Homodimeric Group of Hematopoietic Receptor Complexes

A variety of biophysical methods, including x-ray crystallography, have been used to show that one molecule of hGH can bind to two molecules of the extracellular domain of the hGH receptor (also called the hGHbp). Gel filtration, titration calorimetry, and crystallization experiments showed that one equivalent of hGH can bind up to two equivalents of the hGHbp (38). Muta-

hGH

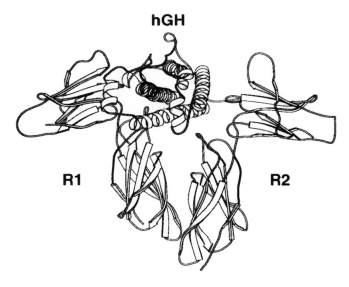

R1   **R2**

Figure 4 Ribbon diagram of the 1:2 hGH:hGH binding protein complex taken from x-ray coordinates (12).

tional studies (38, 39) and x-ray crystallography (12) showed two distinct sites on hGH for binding two equivalents of the hGHbp.

The crystal structure of the complex (Figure 4) shows how ligand-induced receptor dimerization is achieved. The two receptors bind in a roughly twofold symmetrical fashion to the AD and AC faces of hGH. Each receptor interacts with the hormone using the hinge region around their domain-domain interface. The complex positions the receptor stems in close proximity. A reasonable speculation is that the close joining of the extracellular portions immediately above the membrane brings the transmembrane and intracellular domains into close proximity below the membrane so as to initiate signaling.

A remarkable feature of the hGH receptor is that it has a single ligand binding site that can interact with the two sites on the hormone: Both receptors in the complex use their AB, CC', and EF loops of the N-terminal domain, as well as the linker and loops BC and FG of the C-terminal domain (Figure 3). Consequently, at very high concentrations hGH antagonizes the action of its receptor by saturating it in 1:1 complexes. This accounts for the bell-shaped dose-response curve (40). Furthermore, mutational studies show that hGH variants that block binding at Site 2 can form 1:1 complexes and act as potent antagonists, whereas mutations in Site 1 are not antagonists because they cannot form 1:1 complexes (38, 40). These results suggest that binding occurs

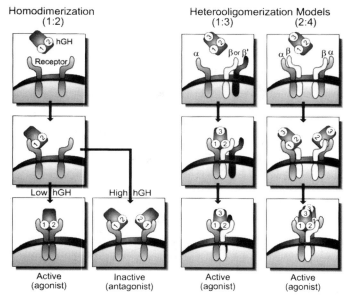

Figure 5 Mechanisms for receptor oligomerization for hGH (a member of the homodimeric group) and IL-6 (a member of the hetero-oligomeric group). (*a*) At physiologically relevant hormone concentrations (0.1 to 1 nM) hGH binds a first receptor using Site 1 and subsequently a second receptor using Site 2 to initiate signaling by receptor dimerization. At super physiologic hormone concentrations (> 1μM) the hormone receptors can become saturated as 1:1 complexes, thus preventing dimerization and signaling. (*b*) Two models have been suggested to form heterooligomeric complexes. One involves one hormone and three regular subunits ($\alpha\beta_2$ or $\alpha\beta\beta'$) (4; Grötzinger, Kurapkat, Wollmer, Kalai & Rose-John, unpublished results), the other two hormones and four receptors ($\alpha_2\beta_2$) (see 41, 42). In both models the hormone has three sites. The reactions begin by the α-receptor binding the hormone at Site 1 and then two β-subunits (or a β and β' subunit) are brought together to initiate signaling using Site 2 and Site 3. Three-dimensional structures of the final complex or receptor-bound intermediates have not yet been determined, so it is not clear what the order of formation is or what the symmetry of the final complex looks like.

sequentially: hGH first binds to a receptor using Site 1 and then to a second receptor using Site 2 (Figure 5).

The sequential dimerization mechanism is neatly supported by the structural results (Figure 4). The total area buried by the receptor at Site 1 of hGH is 1300 Å^2, whereas that buried at Site 2 is 850 Å^2. A third intermolecular contact surface located between the stems of the two receptors covers about 500 Å^2 per receptor. Therefore, once the first receptor is bound to Site 1, the total contact surface available for binding the second receptor is about 1350 Å^2. Thus, there are two asymmetric sites on the hormone for binding two receptors and one asymmetric site on each receptor for contacting the other receptor.

Additional molecular and cellular biology studies further support the view that the active signaling complex is a homodimer (Table 1). Many MAbs to the hGH receptor are potent agonists, which suggests that formation of a receptor dimer is all that is necessary for signaling (40). Deletions of the intracellular domain of the hGH receptor block signaling (43), showing that the intracellular domain is critical. In addition, the soluble extracellular domain acts as an antagonist in cell-based assays (G Fuh & J Wells, unpublished results) suggesting that it competes with the signaling receptor for binding.

The prolactin receptor is also very likely to be a member of the homodimeric group of receptors. Monoclonal antibodies can act as weak agonists for the PRL receptor (44, 45). Structural studies of the hGH-prolactin receptor complex (26), as well as mutational analysis of hGH and the prolactin receptor (46, 47), show that the structural and functional epitopes for forming the 1:1 complex with the hPRL receptor overlap those for forming the initial 1:1 complex with the hGH receptor. The dose-response curves for activating the hPRL receptor by hGH (or hPRL) are bell-shaped, as they are for activating the hGH receptor (48). Moreover, some of the same hGH variants that antagonize the hGH receptor are potent antagonists of the hPRL receptor (48, 49), suggesting that the receptor binding sites overlap. These studies indicate that the oligomerization mechanism is similar to that of the hGH receptor and that the Site 2 epitopes on hGH for the hGH and hPRL receptors overlap. Furthermore, mutations in the intracellular domain of the hPRL receptor affect signaling (50, 51), indicating that ligand-binding and signaling functions reside in the same polypeptide.

Other hematopoietic hormones, such as EPO, G-CSF, and TPO appear to be members of this homodimeric class of signaling complexes (Figure 1). For example, a constitutively activated variant of the EPO receptor has been isolated (R129C) that results in formation of a receptor homodimer (52). Moreover, a dominant negative receptor can be generated by truncating the intracellular domain of the EPO receptor (53), suggesting that oligomeric forms of the receptor involving the intracellular domain are important in signaling. Mutational studies have shown that residues in helix D as well as mutations in helices A and C of EPO affect signaling (54–56). A homology model of the 1:2 EPO-receptor complex, based on the hGH-receptor complex, neatly rationalizes these results (JA Caravella, PD Lyne & WG Richards, manuscript submitted). However, whether these sites bind sequentially and the order of binding are not known. In addition, cross-linking studies suggest there may be additional EPO receptor subunits (57). However, it is not clear if these are alternately spliced or post-translationally modified receptors.

The G-CSF and TPO receptors are apparently activated by homodimerization. Monomers and dimers of the G-CSF receptor have been isolated in the presence of G-CSF (58). A hybrid receptor containing the extracellular domain of the hGH receptor linked to the fibronectin, transmembrane, and intracellular domains of the G-CSF receptor induces a G-CSF response in the presence of hGH but not G-CSF (40). A similar result was found for a hybrid containing the extracellular domain and transmembrane domain of the hGH receptor linked to the intracellular domain of Mpl, the receptor for TPO (8). Furthermore, deletions in the intracellular domain of Mpl lead to inactivation, indicating that the intracellular domain of Mpl is required for signaling. Such hybrid and variant receptors provide evidence that these receptors require simple receptor homodimerization for signaling. However, more complicated mechanisms cannot be excluded, even for the structurally well-characterized receptors like hGH and hPRL. For example, additional membrane signaling proteins (endogenous to cell lines used to reconstitute signaling for hGH and hPRL receptors) could bind to the receptor once it has been dimerized.

Several other complexes between growth factors and tyrosine kinase receptors can fit the receptor homodimer model. Biophysical experiments, mutational analysis, and molecular modeling suggest that the complex between basic FGF and the flg-FGF receptor could bear resemblance to the 1:2 hGH-receptor complex (59, 60). The ability of agonistic monoclonal antibodies to activate EGF receptors has long been known (for review see 61). In fact, this observation first suggested that receptor clustering might be involved in signaling. Although the dimeric insulin receptor can bind two molecules of insulin, the first binds much tighter and appears to be all that is necessary for maximal signaling (for reviews see 62, 63). The negative cooperativity seen for insulin binding has been explained by analogy to the self-antagonism seen for hGH and its receptors (62).

The Hetero-Oligomeric Group of Receptor Complexes

A number of cytokines have more than one receptor subunit in the signaling complex (for reviews see 64, 3, 4). For this group (Figure 1), the α chain is generally responsible for initial hormone binding and one or more β chains are responsible for signaling. In some cases, just the extracellular domain of the α receptor along with the full-length β receptor is competent to signal in the presence of the hormone. Two different models have been proposed to describe the mechanism of receptor activation (Figure 5). Both of these suggest three sites on the hormone but differ in the stoichiometry of the final signaling complex.

Perhaps the best-characterized member of the heterooligomeric receptor family in biochemical and biophysical terms is IL-6, a cytokine that plays a major role in host defense. IL-6 binds to its α receptor (80 kDa) with a dissociation constant (K_d) of 10^{-9}M; this is a prerequisite for binding to the β receptor (gp130) to give a tight heterodimeric complex having an apparent K_d of 10^{-11}M (65).

Soluble forms of the receptor components of IL-6 have been expressed and the stoichiometry of the ligand-receptor complex(es) assessed by gel filtration and analytical ultracentrifugation (42). These studies indicate that formation of a hexameric complex consisting of two molecules of Il-6 (21 kDa), two soluble α subunits (~50 kDa), and two soluble β subunits (~80 kDa) is possible. There appears to be a sequential mechanism for formation of this dimer of heterotrimers (Figure 5). Isolated receptor components show no tendency to dimerize by themselves, and IL-6 will not bind the β subunit on its own. Various ratios of monomeric IL-6 and the α-receptor subunit only produce 1:1 complexes. Thus, one appears to form a 1:1 complex between IL-6 and the α subunit on the way to forming the hexameric complex. Intermediate complexes may form, but their exact nature has yet to be determined.

Further evidence for a hexameric stoichiometry has been provided by clever use of oligomerization assays (41). For example, an immobilized form of the β subunit can bind soluble ^{35}S-labeled β subunit only when it is mixed with IL-6 and the α subunit. Immobilizing the α subunit can trap ^{35}S-labeled α subunit only when IL-6 and the β subunit are added. Immobilized IL-6 can trap ^{35}S-labeled IL-6 only when the α and β subunits are present. Cross-linking studies show that IL-6 can contact the β subunit in the complex (66). Furthermore, in the presence of the α receptor, IL-6 induces formation of dimers of gp130 that are required for signaling (67).

A model of a possible 1:1:1 intermediate complex of IL-6 with the α and β subunits has been built based on the 1:2 hGH-receptor complex (68). Not surprisingly, the model predicts that IL-6 has two sites. Site 1 for the α receptor is structurally equivalent to Site 1 on hGH, and Site 2 for the β receptor is analogous to Site 2 on hGH. The structural validity of this model is supported by the recent determination of the x-ray structure of IL-6 (W Somers, personal communication), which shows it to be a four-helix bundle very similar to that of hGH. Furthermore, the α- and β-receptor subunits are homologous to each other and to the GH receptor, suggesting that their CRH domains are folded like the hGH receptor (2).

Various mutational studies lend functional support for the interfaces defined in the 1:1:1 model. For example, mutants largely confined to the C-terminal part of helix D and the loop connecting helices B and C affect binding to the α receptor (Site 1) and thereby affect signaling (68–72). Mutational studies guided by a homology model of IL-6 revealed determinants important for signaling and binding gp130, but not for binding to the α receptor (68, 71). This site, referred to as Site 2, comprises residues that cluster in helices A and C (see Figure 2A). Another set of experiments has identified residues in the loop following helix A and in the neighboring loop connecting helices C and D that do not affect α-receptor binding but do affect signaling (41, 73, 74). This third site has been proposed to be important for formation of the 2:2:2

complex by allowing IL-6 to either contact itself or a second molecule of gp130 in the hexameric complex (41). Mutations in either Site 2 or 3 can generate antagonists to the IL-6 receptor (41, 71, 73). In addition, mutations in the IL-6 α receptor that disrupt hormone binding map to the hinge between the two N- and C-terminal fibronectin-like domains on a model of the IL-6 α receptor (75). This hinge is located in a position similar to where hGH binds its receptor (Figure 3). A separate set of residues were identified that affect binding of the IL-6/α-receptor 1:1 complex to gp130. Thus, three sites appear to exist on IL-6 and perhaps at least two interfaces on the receptors allow for receptor-receptor interactions. These models and experiments suggest a 1:1:1 IL-6/α receptor/β receptor complex is structurally analogous to the 1:2 hGH/receptor complex.

Another structural model has been proposed (Grötzinger, Kurapkat, Wollmer, Kalai & Rose-John, unpublished results) involving one IL-6, one α receptor and two β receptors (Figure 5). This model accounts for the mutagenesis results but not the biochemical data showing that immobilized α-receptor can bind additional α-receptor in the presence of IL-6 and β-receptor (41). Nor does the 1:3 model explain the sedimentation results suggesting a 2:4 complex in solution (42). It is still possible that these in vitro studies are artifactual because of the high concentrations used to assess these complexes.

GM-CSF, IL-3, and IL-5 are structurally related cytokines; each has a specific α receptor, and they share a common β-receptor subunit, called gp140 (for a recent review see 3) (Figure 1). Although signaling is known to require the presence of the cytokine and both the α and β receptors, the exact stoichiometry of these complexes has not been reported. Nonetheless, extensive mutational analysis of GM-CSF has identified two sites important for receptor binding (76–79, 85) which are remarkably analogous to Sites 1 and 2 used by hGH (Figure 1). For example, some mutations at exposed residues between helix A and D affect binding to the α receptor (77, 80), whereas mutations at exposed residues between helix A and helix C can affect binding to the β receptor (76–79). On this basis mutants were made in helix A of GM-CSF (analogous to Site 2 in hGH) that were antagonists (76).

Chimeric mutations between IL-5 and GM-CSF suggest that IL-5 uses helix A determinants in the same region as GM-CSF to bind the common β receptor (80). Mutations in helix D of IL-5 affect binding to the α subunit (81) suggesting that its Site 1 is analogous to GM-CSF as well. IL-5 is structurally homologous to GM-CSF, but it is a disulfide-linked homodimer (23). Thus, it is tempting to speculate that one IL-5 dimer may bind two α receptors and two β receptors similar to the IL-6 receptor complex (see Figure 5). Gel filtration and cross-linking studies show a soluble form of the α receptor only forms 1:1 complexes with the IL-5 homodimer (82). Perhaps for two α subunits to bind to the IL-5 dimer at least one β receptor is required just as for IL-6. IL-3, which is a monomer like GM-CSF, also appears to have at least one site for binding its α receptor and one for binding the common β receptor. These determinants are located, at least in part, in the presumed helices D and A, respectively (83, 84). Some of these variants actually lead to increased binding affinity and biological activity. Models of IL-3 and GM-CSF in complex with one α and β receptor have been constructed that are analogous to the 1:2 hGH-receptor complex (85).

The oligomerization mechanism for IL-4 is very interesting in that it appears reversed from the other cytokines. The α receptor is again important for initial binding and the β receptor for signaling; the β receptor turns out to be the γ receptor of the IL-2 complex (for a review see 3). Models can be built of IL-4 bound with either two α receptors (36) or complexed as an α/β receptor heterodimer (37); these models are based upon the known structures of IL-4 and the hGH-receptor complex. As expected, these models suggest IL-4 has two sites for interacting with its receptor that are topologically similar to those used by hGH. Mutational and biophysical experiments support the model. However, unlike hGH, the first binding event involves residues analogous to Site 2 in hGH whereas the second event uses residues analogous to Site 1 (86–88). For example, native gel electrophoresis indicates that IL-4 forms a 1:1 complex with its α receptor (89). Mutations that disrupt binding to the α receptor are located in helices A and C (87); mutations in helix D disrupt signaling and are partial antagonists, but do not affect binding to the α receptor (86, 87). Neutralizing MAbs that block binding to the α receptor map to residues in helices A and C, whereas other MAbs that affect signaling but not α receptor binding map to helix D (88). Thus, the order of sites used in IL-4 to oligomerize its receptors is apparently reversed from the other members of the family.

IL-2 has three receptor subunits: α, β, and γ (for review see 90) (Figure 1). The hormone can form complexes of increasing apparent affinity: β < α < β/γ < α/β < α/β/γ. The exact stoichiometry of the signaling complex has not been reported, and the sites that interact with the cytokine have not been clearly elaborated. The fact that the apparent affinity for β/γ is higher in the presence of the α receptor does not prove the existence of the α/β/γ complex. It is possible that the α receptor in the α/β complex is displaced by the γ receptor to give the β/γ complex. There is no evidence for a α/γ complex with IL-2 and signaling can be achieved with only the β and γ subunits present. A mutation has been identified at Glu141 that is a weak antagonist and is believed to have reduced affinity for the γ receptor (91). Two other members of the class 1 family that utilize gp130 (OSM and CNTF) also have three receptor subunits (Figure 1); in these cases the α receptor may function in the same way the α receptor functions for IL-2. Although it is tempting to suggest these cytokines form a 1:3 complex (Figure 5), further experiments are necessary to show this directly.

MOLECULAR RECOGNITION FOR hGH-RECEPTOR COMPLEXES

It is now clear that the class 1 hematopoietins share similarities in their structures and mechanisms of receptor oligomerization. But how do these binding reactions work? What is the chemistry involved, and what are the important structural features that allow these proteins to associate with such high affinities? How is specificity achieved? These are fundamental questions in molecular recognition, and we can begin to address them anecdotally by mutational analysis in the context of the structure of the hGH-receptor complexes.

Structure/Function of Site 1

Alanine-scanning mutagenesis (for review see 92) was used to determine the importance of contact side chains to binding affinity at the Site 1 hormone-receptor interface (39, 93). Each of the 31 contact residues at Site 1 of hGH were mutated to alanine, and their effects on binding affinity and kinetics were measured using a biochip device developed by Pharmacia (BIAcore™) (93). By immobilizing a receptor variant (S201C) on the biochip that was incapable of dimerizing, it was possible to study the effects on formation of a 1:1 complex without the complication of receptor dimerization.

The largest decreases in binding free energy (1.6 to 2.4 kcal mol^{-1}) occur for alanine replacements at five residues: Arg64, Lys172, Thr175, Phe176, or Arg178. Three other residues (Lys41, Leu45, and Pro61) cause 1 to 1.5 kcal mol^{-1} reductions in binding free energy. Surprisingly, just these eight contact residues can account for >85% of the binding free energy. When these side chains are mapped upon the structure of hGH they form a patch or "hot spot" near the center of the contact epitope and are surrounded by residues of lesser or no importance (Figure 6b). In fact, some of the contact side chains surrounding the hot spot actually hinder binding because when they are mutated to alanine they enhance affinity for the receptor.

The primary role of these eight residues is to control the off-rate because the association rates for the corresponding alanine variants are close to wild-type, whereas the dissociation rates are much faster (93). Mutations at some of the basic residues cause two- to threefold reductions in on-rate but these effects are dwarfed by the 10 to 30-fold increases in off-rate caused by these mutations. Thus, these side chains function to keep the hormone there once it is bound and have little role in guiding the hormone to the receptor—diffusion is largely responsible for that (94).

The 30 residues from the receptor that contact Site 1 of hGH were alanine-scanned to test their role in binding affinity (95, 96). Alanine substitutions at

Trp104 or Trp169 cause the largest reductions in binding free energy (> 4.5 kcal mol^{-1}). Alanine replacements at Ile103, Ile105, or Pro106 cause substantial decreases in binding free energy (1.5 to 3.5 kcal mol^{-1}) and smaller but significant effects (1 to 2 kcal mol^{-1}) are observed for Arg43, Glu44, Asp126, Glu127, and Asp164. These 10 residues cluster in the structural epitope with the 2 crucial tryptophans and other hydrophobics near the center; these are surrounded by the hydrophilic residues that are of lesser importance (Figure 6b).

A remarkable complementarity exists when one compares the functional epitopes or hot spots from the hormone and receptor (Figure 6b). The critical residues from the receptor fit into grooves on hGH composed of the residues identified as most important for receptor binding. For example, Trp104 of the receptor is buried in a hole surrounded by the side chains of Asp171, Lys172, Thr175, Phe176, and Lys168 of hGH (Figure 7B). In addition, the entire receptor EF$_N$ loop 103-106 is covered by segment 60-63 of hGH containing Pro61. Trp169 on loop BC$_C$ from the receptor is buried between Arg43 of the receptor and Arg64, Thr175, and Ile179 of hGH. Pro106 from the receptor contacts Leu45 of hGH; Gly168 from the receptor touches Arg178 and Ile179 on hGH. Thus, important residues from the receptor are in contact with critical residues from the hormone.

Binding can also be affected by mutations in the residues underneath the functional epitope. For example, mutations at Phe10, Pro54, or Ile58 in hGH cause 4- to 15-fold reductions in affinity (39). Phe10 is likely to be important for maintaining overall packing of the four-helix bundle; Pro54 and Ile58 are buried between the first crossover connection and the four-helix bundle. Similarly, receptor residues Ile103 and Ile105 support the side chain of Trp104, and Asp126 of the receptor hydrogen bonds to the main-chain amide of Trp104. Thus, affinity can be affected through alanine mutations of residues that support the structure of the binding residues. These effects tend to be smaller than those that affect binding by directly deleting important contacts (see 92).

For the most part the electrostatic interactions appear to be less important for binding affinity than the hydrophobic interactions. For example, Lys172 and Thr175 on hGH contact Trp104 on the receptor, but they do so using the aliphatic portions of their side chains (Figure 7b). Arg43 of the receptor makes a salt bridge to Asp171 and a hydrogen bond to Thr175 of hGH, and the R43A receptor mutant causes a 2.2 kcal mol^{-1} reduction in affinity. However, a R43L mutant is reduced in affinity by only 0.6 kcal mol^{-1}, suggesting the guanido portion of Arg 43 is far less important than its aliphatic chain (T Clackson & J Wells, unpublished results).

Several other electrostatic interactions are seen that are of only moderate to mild importance to affinity: Residues Glu44 and Asp164 of the receptor salt bridge to Arg64 of hGH, and Glu127 of the receptor salt bridges to Lys41 and

Lys167 of hGH. Arg71 in the CC$'_N$ loop from the receptor forms a salt bridge to Glu56 on hGH; removal of this interaction costs only about 0.5 kcal mol^{-1} (93, 96). Gln46 of hGH forms a hydrogen bond with Glu120 of the receptor, but mutation of either of these residues does not affect binding. Finally, Arg217 and Asn218 on receptor loop FG$_C$ interact with His18, His21, and Glu174 on helix 1 of hGH, with the side chain of Asn218 sandwiched between the histidine side chains and Glu174 hydrogen bonding to both histidines. Alanine replacements at Asn218 on the receptor or His21 on hGH have no effect on affinity, whereas replacements at Glu174 and His18 in hGH increase affinity by 0.9 and 0.5 kcal mol^{-1}, respectively (93, 96).

Electrostatic interactions at the periphery of the structural epitope have little or no effect on binding affinity when they are deleted by alanine replacements. This peripheral–contact region has a number of trapped water molecules that make hydrogen bonds to these polar and charged residues (AM de Vos & M Ultsch, unpublished results; 96). Perhaps these trapped waters dampen the full energetic impact of the electrostatic interactions. To fully understand why some contact residues are more important than others would require an understanding of the structures of both the bound and free components as well as information on solvation/desolvation energies for the side chains involved.

Structure/Function of Site 2

Site 2 of hGH covers 24 side chains located in the N-terminus and N-terminal half of helix A and most of helix C (Figure 2a). The effect of alanine mutations at most of these residues in Site 2 of hGH on binding of the second receptor was tested using a fluorescence assay that measured receptor dimerization (38). Alanine substitutions at only five residues (Phe1, Ile4, Leu6, Arg8, and Asp116) resulted in greater than fivefold decreases in the EC$_{50}$ for dimerization. These residues form a small diffuse cluster toward the center of the structural epitope (Figure 6c).

The receptor uses virtually the same residues to bind to Site 2 as it does to bind to Site 1 (compare Figure 6b and 6c); however an assay has not been developed that would allow us to assess the energetic importance of these residues independent of the first binding reaction. Nonetheless, the structure shows that localized conformational changes in the two most important binding loops, EF$_N$ and BC$_C$, can accommodate the flat Site 2 surface on hGH without large overall structural changes such as adjustment of the domain-domain angle (compare Figure 7b and 7c).

Trp104 and Trp169 of the receptor are likely to be important for binding to Site 2 as they are for Site 1 (Figure 7c). The side chains of these tryptophan residues contact the residues that are important at Site 2 of hGH. In addition they are almost completely inaccessible to solvent in the complex, and together they

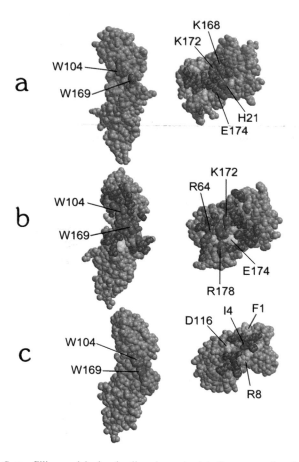

Figure 6 Space-filling models showing ligand-receptor interfaces mapped on the receptor (on the left) and on hGH (on the right). *(a)* The prolactin receptor and Site 1 on hGH, taken from the 1:1 hGH-prolactin receptor complex (87); *(b)* the growth hormone receptor and Site 1 on hGH, taken from the 1:2 hGH-hGH receptor complex (26); *(c)* the growth hormone receptor and Site 2 on hGH, taken from the 1:2 hGH-hGH receptor complex (26). For hGH and for the receptor in *(b)*, interface residues are color-coded according to the contribution made to binding affinity: red, > 1.5 kcal mol^{-1}; orange, 0.5 to 1.5 kcal mol^{-1}; green, –0.5 to 0.5 kcal mol^{-1}; yellow, < –0.5 kcal mol^{-1}; cyan, not tested. The contributions to affinity of Trp104 and Trp169 for the prolactin receptor *(a)* and for the Site 2 binding to hGH *(c)* have not been determined, but in analogy to *(b)*, both tryptophans are likely to be important for binding.

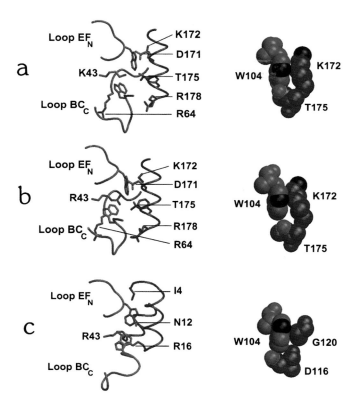

Figure 7 Close-up of the central region of hormone-receptor interfaces. On the left, receptor loops EF$_N$ and BC$_C$ (in magenta) are shown to interact with hGH (in green); on the right, close hydrophobic contacts between Trp104 and residues of hGH are displayed. *(a)* The prolactin receptor and Site 1 on hGH, taken from the 1:1 hGH-prolactin receptor complex (87); *(b)* the growth hormone receptor and Site 1 on hGH, taken from the 1:2 hGH-hGH receptor complex (26); *(c)* the growth hormone receptor and Site 2 on hGH, taken from the 1:2 hGH-hGH receptor complex (26). Note how the binding loops of the prolactin and growth hormone receptor have very similar structures for interacting with Site 1 (compare *a* to *b*), whereas when bound to Site 2 loop BC$_C$ has a very different conformation (compare *b* to *c*).

constitute over 33% of the total buried surface. The side chain of Trp104 is buried in a pocket on hGH, formed by the main chain of Gly120 and the side chains of Asp116, Glu119, and Thr123 in helix C. The importance of this interaction can be shown by substituting an arginine for Gly120 that sterically blocks receptor dimerization and antagonizes biological activity (97, 40). As in Site 1, Trp169 of the receptor is sandwiched between two arginines: the receptor Arg43 and hGH Arg16 plus hGH Asn12. Both tryptophan side chains are supported by contacts with Ile103 on the receptor. Pro106 of the receptor interacts with Phe1 and Pro2 of hGH, and Gly168 interacts with hGH Leu15. Several intermolecular hydrogen bonds and salt bridges are found immediately adjacent to the hydrophobic contacts; for example, between Arg43 of the receptor and Asn12 of hGH, between Glu44 of the receptor and Arg16 of hGH, and between receptor residue Asp126 and hGH Asn12. Comparing these interactions to those observed in Site 1, we see that the identical residues of the receptor interact with a completely different set of residues from the hormone, and that again only a small set of residues appear to be important for affinity.

The hGH/Prolactin Receptor Complex

Despite a relatively low sequence identity between hGH and prolactin (about 23%), as well as between their receptors (about 28%), hGH binds and activates the prolactin receptor. There are clear differences in the way hGH and prolactin bind the prolactin receptor. For example, hGH requires one equivalent of zinc for tight binding (Kd=33 pM), whereas prolactin does not require zinc and binds considerably more weakly (Kd=2.6 nM) (98). Furthermore, alanine-scanning of hGH shows that different, though overlapping, sets of residues are used by hGH to bind the prolactin and hGH receptors (46).

The structural basis for these effects is illuminated in the crystal structure of the 1:1 hGH–prolactin receptor complex (26). The hormone has essentially the same conformation when bound to either the hGH or the prolactin receptor (Figure 6a). In fact, only the orientation and conformation of the short helix near the beginning of the first crossover connection is different from the 1:1 hGH–hGH receptor complex (35). The structures of the hGH and hPRL receptors are very similar as well, and their domains can be superimposed to within 0.9 Å. In the complex, the prolactin and hGH receptors bury the same surface in Site 1 of hGH, and the ligand-binding determinants on the receptor are located on the same loops. However, a closer analysis of the interface reveals that while the conformation of the binding loops is very similar in the prolactin and hGH receptors, the domain positions are changed by a translation of 2–5Å and a rotation of 8°. The result is that although some of the interactions are conserved, many of the binding determinants are involved in different interactions than seen in the hGH-receptor complex. For example, of nine hydrogen bonds or salt bridges in Site 1 of the hGH-receptor complex, only one (between Lys 168 of hGH and the carbonyl oxygen of Trp104 of the receptor) is conserved in the hGH–prolactin receptor complex.

Alanine-scanning mutagenesis (46) of nearly all residues of hGH that contact the prolactin receptor in Site 1 has shown that only 7 of the 34 contact side chains dominate binding affinity (Figure 6a). These residues, which each contribute 2–4 kcal mol^{-1} of binding free energy, cluster near the center of the structural epitope and are surrounded by contact residues that are of lesser importance. One of the most surprising findings is that even though the structural epitopes are virtually the same for hGH to bind the hGH receptor or the prolactin receptor, the hot spots on hGH for binding them differ (compare Figure 6a and 6b).

The functional epitope on hGH (His18, His21, Arg167, Lys168, Lys172, Glu174, Phe176) for binding the prolactin receptor is composed predominantly of hydrophilic residues. Three of these residues (Arg167, Lys168, and Lys172) are involved in a series of salt bridges and van der Waals interactions with the receptor. Two others (His21 and Glu174) coordinate zinc in the complex (26), and a third (His18) could assist in coordinating zinc to hGH before the hPRL receptor binds (99). The hGH and hPRL receptors use residues from the FG_C loop to contact His18, His21, and Glu174 of hGH. The hPRL receptor has Asp217 and His218, which provide two ligands for coordinating zinc at the hormone-receptor interface. The hGH receptor does not require zinc to bind hGH, and it contains Arg217 and Asn218 at this point. However, binding in the hGH receptor has been made Zn dependent by mutating Asn218 to His (100). The zinc binding site is a dramatic example of how simple sequence changes have introduced new important interactions.

Trp104 and Trp169 of the hGH receptor are conserved in the prolactin receptor; these tryptophans are buried in the same pockets as in the hGH-receptor complex, and they touch the hot spot elaborated by alanine scanning (Figures 6a, 7a). In contrast to the small changes at the zinc site and in these tryptophans, big changes have occurred in other interactions with the hormone because of the repositioning of the receptor. The most important of these involves specificity determinant Arg167 of hGH. In the hGH-receptor complex, this arginine forms a salt bridge to Glu127 at the domain-domain linker of the receptor, but the R167A mutation has no effect on binding (93). However, in the complex with the prolactin receptor, the domain-domain linker has shifted such that Arg167 is in van der Waals contact with the Tyr127 of the receptor, and it forms a salt bridge with Asp124. In this complex, the R167A mutation decreases binding affinity 770-fold or 4 kcal mol^{-1} (46). Thus, subtle repositioning of binding determinants and the introduction of a metal site has allowed the same basic receptor scaffold to contact the same site on the hormone but bind it in a very different manner.

The differences in the functional epitopes on hGH for binding these receptors can be exploited to make receptor-selective analogs (46). Alanine mutations at the zinc-ligands overwhelmingly reduce binding to the hPRL receptor, yet preserve or even enhance binding to the hGH receptor. In fact, it is striking that the alanine mutations in hGH that increase binding to the hGH receptor are generally those crucial for binding to the hPRL receptor. Neither hPRL nor hPL bind detectably to the hGH receptor. However, by recruiting residues from hGH that reconstitute the hot spot these hormones can be coaxed into binding to the hGH receptor with virtually the same affinity (101, 102). These studies demonstrate that specificity can be modulated through a small set of substitutions.

GENERAL FEATURES AND CONCLUSIONS

Some common features of these hematopietic-receptor complexes emerge from the work described above. The hormones have a similar four-helix bundle with an up-up-down-down topology. They fall into two groups, a short-chain form and a long-chain form, which further distinguishes the groups of receptors with which they interact (Figure 1). All the ligands have at least two sites for binding two receptor molecules, and these sites are located generally in the same regions of these molecules (Figure 2). Binding occurs sequentially (Figure 5). The order in which they bind their receptors is usually similar to the order for hGH, although they can bind in the reverse order as seen in the case of IL-4.

The hormone binding region of the receptor (CRH) consists of two perpendicular fibronectin type III domains (Figure 3). This arrangement presents the loops from the end of the domain barrels and the linker region between the domains to bind the hormone. The complexes fall into two discrete groups: those that use a single receptor for binding and signaling (the homodimeric group) and those that use an α receptor for binding and one or more β receptor types for signaling (the hetero-oligomeric group). Signaling for both groups involves dimeric forms of the receptor: For the homodimeric group two α receptors need to come together, whereas for the hetero-oligomeric group two β receptors need to come together. In the best-characterized members of the hetero-oligomeric group, dimerization of the second receptor is all that is required for signaling.

Access to the α receptor is largely determined by Brownian diffusion, although long-range electrostatic interactions can play a minor role (93). The stoichiometry of 1:2, 1:3, or 2:2:2 reduces the number of these diffusion events needed to initiate signaling (94). In the case of hGH, only one three-dimensional diffusion step is necessary to bind its first receptor, because docking with the second receptor involves two-dimensional diffusion in the cell membrane (Figure 5). If the stoichiometry were 2:2, two three-dimensional diffusion events would be required to form the active complex. Similarly, a 2:2:2 stoichiometry is more economical than a 4:2:2 stoichiometry. Thus, the perplexing design problem nature faced of having the same receptor determinants bind two very different sites on the hormone may have been solved under the strong selective pressure to deliver the hormone most efficiently.

The oligomerization reactions then promote the intracellular activation of associated tyrosine kinases, leading to protein phosporylation of one or more STAT proteins (for review see 6). The exact kinase and STAT protein(s) used depends mostly on the intracellular domain of the α receptor or β receptor for the homo- or hetero-oligomeric receptors, respectively (Figure 1). Although the structures of the intracellular domains of the receptor have yet to be determined, sequence comparisons have identified two weakly conserved regions (called box 1 and 2). Deletions and mutations in the membrane proximal region, called box 1, generally knock out cell proliferative responses in these receptors, presumably by disrupting interactions with the tyrosine kinase.

The structures of the 1:1 and 1:2 hGH–hGH receptor and the 1:1 hGH–hPRL receptor complexes reveal the atomic contacts and suggest the types of conformational changes that occur upon formation of these protein-protein complexes. Large-scale movements do not appear to be involved, as the basic fold and conformation of each of the receptors and hormones in these complexes are virtually superimposable. The sequential dimerization of the 1:2 complex is not the result of large conformational changes, but rather the assembly into the 1:1 complex of two half-sites for binding the second receptor. Nevertheless, these molecules have sufficient flexibility to adapt to bind different molecular surfaces. For example, in the 1:2 complex the hGH receptor employs local conformational changes in the loops to bind two different sites on the hormone, whereas in the hGH–prolactin receptor complex a closely related receptor uses rigid-body adjustments of entire domains to interact with the same site on hGH. Similar structural adjustments may operate where the same receptor β-chain is shared among several different ligands.

Mutational analysis in the context of these high-resolution structures shows that only a small set of clustered contact residues are critical for binding. These side chains are generally important for keeping the complex together, not in driving its association. The "hot-spots" tend to be central to the structural epitope, hydrophobic in character, and surrounded by polar residues that are less important for conferring binding affinity (Figures 6 and 7). The reason these peripheral residues are less important for affinity may be that the energy to desolvate them is high relative to the energy gained in binding. Nevertheless, peripheral hydrophilic side chains could play an important role in solubilizing the free components as well as conferring specificity of the interaction. Specificity can be modulated by a few contact substitutions and/or by adjusting the registry of displayed side chains on the scaffold as evidenced in the hGH–hPRL receptor complex. Subtle changes can also be made in the binding interface by mutating residues buried just beneath it. These simple principles can account for the diversity of molecular interfaces produced in these structurally similar hormone and receptor complexes.

FUTURE PERSPECTIVES

The four helix–bundle architecture of the hematopoietin and its fibronectin-like receptor are robust and versatile scaffolds. Given the extent to which nature has utilized these structures for a vast array of general and specific functions, people will likely use them too. Generating antagonists to these molecules by simple mutations that allow the first binding event but block the subsequent oligomerization reaction(s) is clearly feasible. Such antagonists can be further optimized in two ways. First, one can build higher affinity Site 1 variants of the antagonist by removing hindrance determinants (40) or by optimizing determinants in and around the functional epitope using phage display (103) or structure-based methods. Second, one can build receptor-selective antagonist in cases in which the hormone can use Site 1 for binding two different receptor molecules.

Building new agonists to stimulate these receptors is also possible. The fact that simple dimerization is all that is necessary for activation of many of these receptors offers the possibility of using monoclonal antibodies to turn on specific receptors. These may have advantages over the natural hormones, such as longer half-lives and better specificity. One could also build pan-agonists (or pan-antagonists) that could react with more than one receptor by recruiting in common hot spots for different receptors on the same hormone scaffold. The observation that only a small set of residues are crucial for binding offers the hope that by mimicking these smaller contiguous structures one may produce orally active small molecules to do an equal or better job as the native hormone. Predicting whether or not such engineered molecules will become drugs is difficult. They can certainly be useful tools for pharmacologists to further penetrate the complex biology of this family of hormones and receptors. Finally, the principles of molecular recognition revealed through these and other structural and functional studies are useful in designing molecular interfaces for broad applications in both biology and chemistry.

ACKNOWLEDGMENTS

We thank our colleagues in the Protein Engineering Department at Genentech for their support and stimulating discussions, Kerri Andow, Wayne Anstine, and David Wood for graphics, and Brian Cunningham and Tony Kossiakoff for critical reading of the manuscript.

Literature Cited

1. Bazan JF. 1990. *Immunol. Today* 11: 350–54
2. Bazan JF. 1993. *Curr. Biol.* 3:603–6
3. Sato N, Miyajima A. 1994. *Curr. Opin. Cell Biol.* 6:174–79
4. Stahl NA, Yancopoulos GD. 1993. *Cell* 74:587–90
5. Wells JA, De Vos AM. 1993. *Annu. Rev. Biophys. Biomol. Struct.* 22:329–51
6. Schindler C, Darnell JE. 1995. *Annu. Rev. Biochem.* 64:621–51
7. Nicola NA. 1994. In *Guidebook to Cytokines and Their Receptors,* pp. 1–194. Oxford: Oxford Univ. Press
8. Gurney AL, Wong SC, Henzel WJ, DeSauvage FJ. 1995. *Proc. Natl. Acad. Sci. USA* 92:5292–96
9. Abdel-Meguid SS, Shieh H-S, Smith WW, Dayringer HE, Violand BN, Bentle LA. 1987. *Proc. Natl. Acad. Sci. USA* 84:6434–37
10. Diederichs K, Boone T, Karplus PA. 1991. *Science* 254:1779–82
11. Walter MR, Cook WJ, Ealick SE, Nagabhushan TL, Trotta PP, Bugg CE. 1992. *J. Mol. Biol.* 224:1075–85
12. De Vos AM, Ultsch M, Kossiakoff AA. 1992. *Science* 255:306–12
13. Hill CP, Osslund TD, Eisenberg D. 1993. *Proc. Natl. Acad. Sci. USA* 90: 5167–71
14. Lovejoy B, Cascio D, Eisenberg D. 1993. *J. Mol. Biol.* 234:640–53
15. Pandit J, Bohm A, Jancarik J, Halenbeck R, Koths K, Kim S-H. 1992. *Science* 258:1358–62
16. Robinson RC, Grey LM, Staunton D, Vankelecom H, Vernallis AB, et al. 1994. *Cell* 77:1101–16
17. Bazan JF, McKay DB. 1992. *Science* 257:410–13
18. Mott HR, Baines BS, Hall RM, Cooke RM, Driscoll PC, et al. 1995. *J. Mol. Biol.* 247:979–94
19. Powers R, Garrett DS, March CJ, Frieden EA, Gronenborn AM, Clore GM. 1992. *Science* 256:1673–77
20. Smith JL, Redfield C, Boyd J, Lawrence GMP, Edwards RG, et al. 1992. *J. Mol. Biol.* 224:899–904
21. Walter MR, Cook WJ, Zhao BG, Cameron RP Jr, Ealick SE, et al. 1992. *J. Biol. Chem.* 267:20371–76
22. Wlodawer A, Pavlovsky A, Gustchina A. 1992. *FEBS Lett.* 309:59–64
23. Milburn MV, Hassell AM, Lambert MH, Jordan SR, Proudfoot AEI, et al. 1993. *Nature* 363:172–76
24. Sprang SR, Bazan JF. 1993. *Curr. Opin. Struct. Biol.* 3:815–27
25. Rozwarski DA, Gronenborn AM, Clore GM, Bazan JF, Bohm A, et al. 1994. *Structure* 2:159–73
26. Somers W, Ultsch M, De Vos AM, Kossiakoff AA. 1994. *Nature* 372:478–81
27. Ultsch MH, Somers W, Kossiakoff AA, De Vos AM. 1994. *J. Mol. Biol.* 236: 286–99
28. Bazan JF. 1990. *Proc. Natl. Acad. Sci. USA* 87:6934–38
29. Muller YA, Ultsch MH, De Vos AM. 1996. *J. Mol. Biol.* 256:144–59
30. Leahy DJ, Hendrickson WA, Aukhil I, Erickson HP. 1992. *Science* 258: 802–10
31. Harlos K, Martin DMA, O'Brien DP, Jones EY, Stuart DI, et al. 1994. *Nature* 370:662–66
32. Muller YA, Ultsch MH, Kelley RF, De Vos AM. 1994. *Biochemistry* 33:10864–70
33. Walter MR, Windsor WT, Nagabhushan TL, Lundell DJ, Lunn CA, et al. 1995. *Nature* 376:230–35
34. Huber AH, Wang Y-M, Bieber AJ, Bjorkman PJ. 1994. *Neuron* 12:717–31
35. Kossiakoff AA, Somers W, Ultsch M, Andow K, Muller YA, De Vos AM. 1994. *Protein Sci.* 3:1697–705
36. Bamborough P, Hedgecock CJR, Richards WG. 1994. *Structure* 2:839–51
37. Gustchina A, Zdanov A, Schalk-Hihi C, Wlodawer A. 1995. *Proteins: Struct. Funct. Genet.* 21:140–48
38. Cunningham BC, Ultsch M, DeVos AM, Mulkerrin MG, Clauser KR, Wells JA. 1991. *Science* 254:821–25
39. Cunningham BC, Wells JA. 1989. *Science* 244:1081–85
40. Fuh G, Cunningham BC, Fukunaga R, Nagata S, Goeddel DV, Wells JA. 1992. *Science* 256:1677–80
41. Paonessa G, Graziani R, De Serio A, Savino R, Ciapponi L, et al. 1995. *EMBO J.* 14:1942–51
42. Ward LD, Howlett GJ, Discolo G, Yasu-

kawa K, Hammacher A, et al. 1994. *J. Biol. Chem.* 37:23286–89

43. Colosi P, Wong K, Leong SR, Wood WI. 1993. *J. Biol. Chem.* 268:12617–23

44. Dusanter-Fourt I, Dijiane J, Kelly PA, Houderine L-M, Teyssot B. 1984. *Endocrinology* 114:1021–27

45. Elberg G, Kelly PA, Djiane J, Binder L, Gertler A. 1990. *J. Biol. Chem.* 265:14770–76

46. Cunningham BC, Wells JA. 1991. *Proc. Natl. Acad. Sci. USA* 88:3407–11

47. Rozakis-Adock M, Kelly PA. 1992. *J. Biol. Chem.* 267:7428–33

48. Fuh G, Colosi P, Wood WI, Wells JA. 1993. *J. Biol. Chem.* 268:5376–81

49. Fuh G, Wells JA. 1995. *J. Biol. Chem.* 270:13133–37

50. Lebrun JJ, Ali S, Goffin V, Ullrich A, Kelly PA. 1995. *Proc. Natl. Acad. Sci. USA* 92:4031–35

51. Lebrun JJ, Ali S, Ullrich A, Kelly PA. 1995. *J. Biol. Chem.* 270:10664–70

52. Watowich SS, Yoshimura A, Longmore GD, Hilton DJ, Yoshimura Y, Lodish HF. 1992. *Proc. Natl. Acad. Sci. USA* 89:2140–44

53. Watowich SS, Hilton DJ, Lodish HF. 1994. *Mol. Cell. Biol.* 14:3535–49

54. Boissel J-P, Lee W-R, Presnell SR, Cohen FE, Bunn HF. 1993. *J. Biol. Chem.* 268:15983–93

55. Grodberg J, Davis KL, Sytkowski AJ. 1993. *Eur. J. Biochem.* 218:597–601

56. Wen D, Boissel JP, Showers M, Ruch BC, Bunn HF. 1994. *J. Biol. Chem.* 269:22839–46

57. Mayeux P, Lacombe C, Casadevall N, Chretien S, Dusanter I, Grisselbrecht S. 1991. *J. Biol. Chem.* 266:23380–85

58. Fukunaga R, Ishizaka-Ikeda E, Nagata S. 1990. *J. Biol. Chem.* 265:14008–15

59. Pantoliano MW, Horsick RA, Springer BA, Van Dyk DE, Tobery T, et al. 1994. *Biochemistry* 33:10229–48

60. Springer BA, Pantoliano MW, Barbera FA, Gunyuzlu PL, Thompson LD, et al. 1994. *J. Biol. Chem.* 269:26879–84

61. Ullrich A, Schlessinger J. 1990. *Cell* 61:203–12

62. DeMeyts P. 1994. *Diabetologia* 37 (Suppl. 2):S135–48

63. Schaffer LA. 1994. *Eur. J. Biochem.* 221:1127–32

64. Kishimoto T, Akira S, Taga T. 1992. *Science* 258:593–97

65. Hibi M, Murikami M, Saito M, Hirano T, Taga T, Kishimoto T. 1990. *Cell* 63:1148–57

66. D'Alessandro F, Colamonici OR, Nordan RP. 1993. *J. Biol. Chem.* 268:2149–53

67. Murikami M, Hibi M, Nakagawa N, Nakagawa T, Yasukawa K, et al. 1993. *Science* 260:1808–10

68. Savino R, Lahm A, Salvati AL, Ciapponi L, Sporeno E, et al. 1994. *EMBO J.* 13:1357–67

69. Hammacher A, Ward LD, Weinstock J, Treutlein H, Yasukawa K, Simpson RJ. 1994. *Protein Sci.* 3:2280–93

70. Leebeek FWG, Kariya K, Schwabe M, Fowlkes DM. 1992. *J. Biol. Chem.* 267:14832–38

71. Savino R, Ciapponi L, Lahm A, Demartis A, Cabibbo A, et al. 1994. *EMBO J.* 13:5863–70

72. Ward LD, Hammacher A, Zhang JG, Weinstock J, Yasukawa K, et al. 1993. *Protein Sci.* 2:1472–81

73. Brakenhoff JPJ, deHon FD, Fontaine V, Boekel ET, Schooltink H, et al. 1994. *J. Biol. Chem.* 269:86–93

74. Ehlers M, Grötzinger J, deHon FD, Mullberg J, Brakenhoff JPJ, et al. 1994. *J. Immunol.* 153:1744–53

75. Yawata H, Yasukawa K, Natsuka S, Murakami M, Yamasaki K, et al. 1993. *EMBO J.* 12:1705–12

76. Altmann SW, Kastelein RA. 1995. *J. Biol. Chem.* 270:2233–40

77. Hercus TR, Cambareri B, Dottore M, Woodcock J, Bagley CJ, et al. 1994. *Blood* 83:3500–8

78. Kastelein RA, Shanafelt AB. 1993. *Oncogene* 8:231–36

79. Lopez AF, Shannon MF, Hercus T, Nicola NA, Cambareri B, et al. 1992. *EMBO J.* 11:909–16

80. Shanafelt AB, Miyajima A, Kitamura T, Kastelein RA. 1991. *EMBO J.* 10:4105–12

81. McKenzie ANJ, Barry SC, Strath M, Sanderson CJ. 1991. *EMBO J.* 10:1193–99

82. Devos R, Guisez Y, Cornelis S, Verhee A, Van der Heyden J, et al. 1993. *J. Biol. Chem.* 268:6581–87

83. Lokker NA, Movva NR, Strittmatter U, Fagg B, Zenke G. 1991. *J. Biol. Chem.* 266:10624–31

84. Lopez AF, Shannon MF, Barry S, Phillips JA, Cambareri B, et al. 1992. *Proc. Natl. Acad. Sci. USA* 89:11842–46

85. Lyne PD, Bamborough P, Duncan D, Richards WG. 1995. *Protein Sci.* 4:2223–33

86. Kruse N, Tony H-P, Sebald W. 1992. *EMBO J.* 11:3237–44

87. Kruse N, Shen B-J, Arnold S, Muller T, Sebald W. 1993. *EMBO J.* 12:5121–29

88. Reusch P, Arnold S, Heusser C, Wagner

K, Weston B, Sebald W. 1994. *Eur. J. Biochem.* 222:491–99

89. Hoffman RJC, Schalk-Hihi C, Castner BJ, Gibson MG, Rasmussen BD, et al. 1994. *FEBS Lett.* 347:17–21

90. Minami Y, Kono T, Yamada K, Taniguchi T. 1992. *Biochim. Biophys. Acta* 1114:163–77

91. Zurawski SM, Imler J-L, Zurawski G. 1990. *EMBO J.* 9:3899–905

92. Wells JA. 1991. *Methods Enzymol.* 202: 390–410

93. Cunningham BC, Wells JA. 1993. *J. Mol. Biol.* 234:554–63

94. Wells JA. 1996. *Proc. Natl. Acad. Sci. USA.* In press

95. Bass SH, Mulkerrin MG, Wells JA. 1991. *Proc. Natl. Acad. Sci. USA* 88: 4498–502

96. Clackson T, Wells JA. 1995. *Science* 267:383–86

97. Chen WY, Wight DC, Wagner TE, Kopchick JJ. 1990. *Proc. Natl. Acad. Sci. USA* 87:5061–65

98. Cunningham BC, Bass S, Fuh G, Wells JA. 1990. *Science* 250:1709–12

99. Cunningham BC, Mulkerrin MG, Wells JA. 1991. *Science* 253:545–48

100. Matthews DJ, Wells JA. 1994. *Chem. Biol.* 1:25–30

101. Cunningham BC, Henner DJ, Wells JA. 1990. *Science* 247:1461–65

102. Lowman HB, Cunningham BC, Wells JA. 1991. *J. Biol. Chem.* 266:10982–88

103. Lowman HB, Wells JA. 1993. *J. Mol. Biol.* 234:564–78

Annu. Rev. Biochem. 1996. 65:635–92

DNA TOPOISOMERASES

James C. Wang

Department of Molecular and Cellular Biology, Harvard University, Cambridge, Massachusetts 02138

KEY WORDS: enzyme mechanism, DNA topology, DNA-protein interaction, DNA-protein covalent complexes, replication, transcription, gyrase, genome stability, chromosome condensation, chromosome segregation

ABSTRACT

The various problems of disentangling DNA strands or duplexes in a cell are all rooted in the double-helical structure of DNA. Three distinct subfamilies of enzymes, known as the DNA topoisomerases, have evolved to solve these problems. This review focuses on work in the past decade on the mechanisms and cellular functions of these enzymes. Newly discovered members and recent biochemical and structural results are reviewed, and mechanistic implications of these results are summarized. The primary cellular functions of these enzymes, including their roles in replication, transcription, chromosome condensation, and the maintenance of genome stability, are then discussed. The review ends with a summary of the regulation of the cellular levels of these enzymes and a discussion of their association with other cellular proteins.

CONTENTS

INTRODUCTION 636
PRIMARY STRUCTURES, NEW ENZYMES, AND CLASSIFICATION OF THE
    ENZYMES 637
MECHANISTIC STUDIES 640
    DNA Binding... 641
    Active-site Tyrosines and the Formation of Covalent Enzyme-DNA Complexes..... 647
    Nucleotide Binding to Type II DNA Topoisomerases 650
    Three-Dimensional Structures 652
    Mechanistic Implications of Biochemical and Structural Results................. 654
BIOLOGICAL FUNCTIONS OF DNA TOPOISOMERASES 656
    Removal of DNA Supercoils Generated by Various Cellular Processes 656
    Unlinking of Intertwined DNA Duplexes................................. 662
    Eukaryotic DNA Topoisomerase II and Chromosome Condensation 665
    DNA Topoisomerases and Genome Stability............................... 667
    Other Plausible Cellular Roles of DNA Topoisomerases...................... 672
    Regulation of the Cellular Levels of DNA Topoisomerases 673
    Association of DNA Topoisomerases with Other Proteins 677
CONCLUDING REMARKS 679

0066-4154/96/0701-0635$08.00

INTRODUCTION

DNA topoisomerases are enzymes that catalyze the passage of individual DNA strands or double helices through one another. These reactions are often manifested in interconversions between topological isomers of DNA rings; hence the name of the enzymes (1). The first known member of this class of enzymes, one which specifically relaxes negatively supercoiled DNAs, was partially purified from *Escherichia coli* cell extracts in 1971 (2). This finding was followed by the discovery of an activity in mouse cell extracts that relaxes both positively and negatively supercoiled DNAs (3) and of an ATP-dependent *E. coli* activity capable of introducing negative supercoils into a relaxed DNA (4). As it turned out, these three enzymes, now known as *E. coli* DNA topoisomerase I, mouse DNA topoisomerase I, and *E. coli* gyrase or DNA topoisomerase II, respectively, represent three distinct subfamilies of topoisomerases. Both *E. coli* and mouse DNA topoisomerase I are classified as type I enzymes, which effect topological changes of DNA by transiently cleaving one DNA strand at a time to allow the passage of another strand (reviewed in 5). Other than this common mechanistic feature, members of the *E. coli* DNA topoisomerase I subfamily (to be referred to as subfamily IA) and those of the mouse DNA topoisomerase I subfamily (to be referred to as subfamily IB) exhibit dissimilar structures and distinct reaction characteristics. *E. coli* gyrase or DNA topoisomerase II is a member of the type II DNA topoisomerase subfamily. The type II enzymes catalyze DNA topological transformations by transiently cleaving a pair of complementary DNA strands in one duplex segment to form an enzyme-operated gate in the segment. A second double-stranded DNA segment is then passed through the gate by the enzyme, after which the severed strands in the first DNA segment are rejoined (6–8). Type II enzymes from various sources, ranging from viruses to human, are structurally and evolutionarily related (9–15).

The DNA topoisomerases were last reviewed in these volumes over a decade ago (16). The topological problems of DNA are deeply rooted in its double-helix structure. As a consequence, these enzymes have been found to participate in nearly all cellular transactions of DNA. The study of the topoisomerases has also expanded into the realm of pharmacology and clinical medicine through the identification of bacterial gyrase as the target of antibiotics (17–19) and toxins (20–22), and eukaryotic DNA topoisomerases I and II as the targets of a large number of anticancer agents (23–27). The rapid expansion of the field in the past dedade is reflected by the publication of numerous reviews on particular members or specific aspects of this family of enzymes (28–54). Two monographs on DNA topology (55, 56) and several monographs and numerous reviews on the enzymes as targets of therapeutics have also appeared (57–67).

The purpose of the present review is to provide an overview of the DNA topoisomerase field. The emphasis is on key findings made in the past decade, and earlier developments are mentioned occasionally to provide a backdrop. Even with this set time interval, the scope of the field often forces a choice between being comprehensible and being comprehensive, and page limit alone precludes exhaustive literature coverage. The large body of literature on DNA topoisomerases as drug targets cannot be adequately covered in the context of this review and is left essentially untouched. These shortcomings are at least partially remedied, however, by the publication of the more specialized monographs and symposium volumes cited above.

PRIMARY STRUCTURES, NEW ENZYMES, AND CLASSIFICATION OF THE ENZYMES

Nucleotide sequencing of genes encoding the various DNA topoisomerases was initiated a decade ago, and by now the amino acid sequences of scores of enzymes from various organisms are known [for an annotated compilation and comparison of sequences, see (15); additional sequences can be found in the various data bases]. Cloning and sequence determination of the structural genes have facilitated genetic manipulations as well as the overexpression and purification of the enzymes, and these developments have in turn fostered rapid progress in biochemical and structural studies of the enzymes.

Primary structure information has provided strong support for grouping the enzymes into three subfamilies, which was previously done on the basis of their biochemical characteristics (15, 29, 68, 69). Since 1989, each of the three subfamilies has been enlarged by the discovery of new members. In the budding yeast, cloning and sequencing of a gene *EDR1*, which was known to increase the frequency of recombination between repetitive sequences when mutated, led to the conclusion that its product shares significant sequence homologies with *E. coli* DNA topoisomerase I. Henceforth the *EDR1* gene was renamed *TOP3*, and its product DNA topoisomerase III (70). Biochemical characterization of the protein overexpressed in yeast from cloned *TOP3* confirmed that the protein, similar to *E. coli* DNA topoisomerases I and III, has a single strand–specific DNA topoisomerase activity (71). Yeast DNA topoisomerase III appears to be more closely related to *E. coli* DNA topoisomerase III than *E. coli* DNA topoisomerase I in terms of the biochemical characteristics of the enzymes, especially the nucleotide sequences of their preferred DNA cleavage sites (71). The discovery of yeast DNA topoisomerase III and its classification as a type IA enzyme broke the prokaryote-eukaryote boundary which had previously separated the two type I subfamilies of enzymes. Recently, human cells were also found to possess a *TOP3* gene highly homologous to yeast *TOP3*, but with additional sequence elements in the C-terminal region of the enzyme resembling those found in the C-terminal region of *E. coli* DNA topoisomerase I (72). These features are also shared by another subfamily IA–like protein encoded by a *Bacillus anthracis* plasmid pXO1 (73).

The type IA subfamily was further expanded to include enzymes from the *archea* kingdom through recent work on "reverse gyrase," which was first discovered in the hyperthermophilic archeon *Sulfolobus acidocaldarius* and was found to catalyze ATP-dependent positive supercoiling of DNA (74, 75). Activities similar to the *S. acidocaldarius* enzyme were detected in a large number of thermophilic archea and bacteria (76–80). Sequencing of the structural gene of the *S. acidocaldarius* enzyme shows that the amino-terminal half of the single-polypeptide enzyme contains motifs commonly found in DNA helicases, and the carboxyl-terminal half of the enzyme shares sequence similarities with members of the type IA topoisomerase subfamily (81). A similar enzyme from a hyperthermophile *Pyrococcus furiosus*, which is phylogenetically rather removed from *S. acidocaldarius*, was also found to have the same bipartite structure (P Forterre, personal communication). Furthermore, *Desufurococcus amylolyticus* reverse gyrase was shown to cleave DNA with the same sequence specificity as bacterial DNA topoisomerase I (82). ATP-independent type I activities were also found in several hyperthermophilic archea including *D. amylolyticus* and *Thermoplasma acidophilum* (83–85), but it remains to be shown that these are distinct enzymes rather than derivatives of reverse gyrases.

The possibility that a prokaryotic hyperthermophilic methanogen might contain a topoisomerase of the IB subfamily was raised by biochemical studies of a *Methanopyrus kandleri* enzyme termed DNA topoisomerase V (86). Biochemically, this enzyme resembles IB subfamily members more than IA subfamily members: It relaxes both positively and negatively supercoiled DNA, and it becomes covalently linked to the 3′ end of the transiently cleaved DNA strand (86). Preliminary sequencing data of the structural gene revealed, however, that the protein shares little sequence similarity with the archetype DNA topoisomerases (A Slesarev, personal communication).

Within the type II subfamily, several new members have been identified. The *parC* and *parE* gene of *E. coli*, known to be involved in chromosomal partition, were shown to encode the two subunits of a type II enzyme termed DNA topoisomerase IV (87–89). The amino acid sequences of the ParC and ParE subunits are homologous to those of the GyrA and GyrB subunits, respectively. In the presence of ATP, purified *E. coli* DNA topoisomerase IV catalyzes the relaxation of supercoiled DNA and catenation/decatenation and knotting/unknotting of double-stranded DNA rings (similar to eukaryotic DNA topoisomerase II and phage T4 DNA topoisomerase); unlike bacterial gyrase, DNA topoisomerase IV does not catalyze DNA negative supercoiling. Thus bacterial gyrase remains the only type II topoisomerase capable of DNA negative supercoiling. Homologs of *E. coli* DNA topoisomerase IV have been found in *Salmonella typhimurium* (90, 91), in the Gram-positive pathogen *Staphylococcus aureus* (92), and in *Micoplasma genitalium* (94). A two-subunit type II DNA topoisomerase purified from the hyperthermophilic archeon *Solfolobus shibatae* was also found to catalyze ATP-dependent relaxation of positively or negatively supercoiled DNA, but not DNA supercoiling (95). Similar to type II enzymes other than bacterial gyrase, the *S. shibatae* enzyme is more efficient in decatenation than in supercoil removal (95). The presence of *gyrA*- and *gyrB*-like genes in a wide variety of other bacteria was also implicated from DNA sequence analysis (14, 96).

A second type II DNA topoisomerase was found in mammalian cells in the late 1980s. A 180-kDa murine enzyme IIβ with biochemical, pharmacological, and antigenic characteristics readily distinguishable from those of the more extensively studied 170-kDa IIα form was first reported in 1987 (97, 98). The coding sequences of human *TOP2α* and *TOP2β* genes have been determined (98–103); *TOP2α* was mapped to chromosome 17q21-22 (99), and *TOP2β* to chromosome 3p24 (104). Cross-species blot-hybridization experiments suggested that two closely related *TOP2* genes might exist in avian cells as well, but probably not in yeast or Drosophila (104). In *S. cerevisiae*, the absence of a second nuclear type II activity was strongly suggested by studies of catenation (105, 106) and supercoiling (71, 106, 107) of intracellular DNA in *top2ts* cells expressing a temperature-sensitive topoisomerase II.

The existence of a new viral topoisomerase was implicated by the nucleotide sequence of African swine fever virus DNA (108). The presence of an open reading frame encoding a protein highly homologous to other type II topoisomerases places this animal virus in the league of viruses encoding a topoisomerase, whose members were previously limited to the poxviruses and T-even phages. The possibility that an 11-kDa nucleocapsid protein of equine infectious anemia virus possesses a type I topoisomerase activity was also raised through biochemical and immunochemical experiments (109; E Priel, personal communication).

With the rapid accumulation of genomic sequencing data, one may well ask whether more topoisomerases are yet to be discovered. The nucleotide sequence of the *Haemophilus influenzae* genome (110) indicates that there is no DNA topoisomerase other than the four known bacterial enzymes. Similarly, the complete nucleotide sequence of *Microplasma genitalium* revealed no new topoisomerase genes (94). In the case of the budding yeast *S. cerevisiae* also, an answer should come soon as the yeast genome project progresses toward completion. Other than the well-characterized archetype DNA topoisomerases, however, many known or yet-to-be discovered enzymes likely possess DNA topoisomerase–like activities. Many DNA strand-transferases (111, 112) exhibit DNA topoisomerase activities under certain reaction conditions. More recently, it was reported that a *Spiroplasma* CpG methylase might possess a type I DNA topoisomerase activity (113). Studies of a restriction endonuclease *Nae I* also showed that a single mutation replacing a leucine by lysine converts the nuclease to an enzyme with a sequence-specific recombinase-topoisomerase activity (114). These examples illustrate that seemingly different classes of enzymes might be more closely related in the evolutionary as well as biochemical sense.

From the involvement of DNA topoisomerases in various cellular transactions of DNA, their presence in DNA-containing organelles such as chloroplasts and mitochondria is expected. Whether there are specialized organelle topoisomerases distinct from their nuclear counterparts, however, remains unclear. In the trypanosomatid *Crithidia fasciculata*, a type II enzyme was found to be enriched at two antipodal sites in its single mitochondrion termed the kinetoplast (115, 116). The kinetoplast contains a novel network of topologically linked DNA rings of two different sizes, termed the mini- and maxi-circles, and the replication machinery is likely to be located at the antipodal sites (117, 118). Trapping of covalent DNA-DNA topoisomerase complexes in *Trypanosoma equiperdum* (119), and in *C. fasiculata* (120), showed that both classes of rings are associated with a type II enzyme (120). Whether the enzyme enriched in kinetoplast is one distinct from the nuclear type II enzyme(s) is not known. Amino acid sequences deduced from cloned *TOP2* genes of *T. brucei* (121) and *C. fasiculata* (122) suggest that their products are similar to other well-characterized eukaryotic nuclear type II enzymes.

The chloroplast genome is a 135-kb negatively supercoiled DNA (123, 124). More recent results supported the earlier findings that chloroplasts contain both IA- and IB-type DNA relaxation activities, as well as one or more type II activities (125–133). Biochemical and immunochemical experiments suggested that a type II enzyme was associated with chloroplast DNA and reached a maximal level at the time of chloroplast DNA replication (129, 131). The presence of a bacterial gyrase-like activity in chloroplasts was also reported (127). Additional biochemical and genetic evidence is needed, however, to establish the identities of these activities and their relation to the nuclear enzymes.

MECHANISTIC STUDIES

Much progress was made in the past decade in structural and mechanistic studies of DNA topoisomerases. This section provides a review of interactions between various topoisomerases and DNA and between the type II enzymes and nucleoside triphosphates. Available data on the three-dimensional structures of these enzymes are then summarized and their mechanistic implications discussed. Together, these results are beginning to provide molecular sketches on how these intricate machines manipulate DNA.

DNA Binding

TYPE IA ENZYMES   Members of this subfamily appear to possess multiple DNA binding sites that interact with both double- and single-stranded DNA. *E. coli* DNA topoisomerase I was shown to preferentially bind to the junctions of duplex and single-stranded regions and catalyze transient breakage of DNA within the single-stranded region (134, 135). Nucleolytic and chemical probing of various DNA substrates in the presence and absence of the enzyme indicated that the enzyme interacts with both the 5′ and the 3′ side of the site of DNA cleavage and rejoining (135). The presence of a single-stranded DNA binding site in the catalytic pocket of the enzyme is also demonstrated by the ability of the enzyme to cleave oligodeoxynucleotides as small as heptamers: Cleavage occurs four nucleotides from the 5′ end and three nucleotides from the 3′ end (136). *E. coli* and yeast DNA topoisomerase III also appear to possess a site within the catalytic pocket for the cleavage of single-stranded DNA (71, 137) . In the case of the *E. coli* enzyme, RNA as well as DNA is cleaved (137). The minimal size of single-stranded nucleotides that are cleaved by *E. coli* DNA topoisomerase III is the same as that of deoxynucleotides cleaved by *E. coli* DNA topoisomerase I, but the site of cleavage is six nucleotides from the 5′ end and one from the 3′ end (138).

When *E. coli* DNA topoisomerase I binds to a negatively supercoiled DNA, the binding energy is sufficiently large to overcome the unfavorable free-energy of unpairing a short stretch of the DNA double helix. For relaxed or positively supercoiled substrates, however, the binding energy is insufficient, which accounts for the failure of the enzyme to form covalent enzyme-DNA intermediates with these substrates or to relax positively supercoiled DNA (2, 135). The catalytic actions of *E. coli* and yeast DNA topoisomerase III are even more strongly dependent on the presence of a single-stranded region in the DNA substrate, or on higher temperatures or a high degree of negative supercoiling to facilitate the formation of the enzyme-DNA complex containing a short stretch of single-stranded nucleotides (71, 139–141). Another IA subfamily member, the reverse gyrase of *Sulfolobus acidocaldarius*, was also found to bind single-stranded regions of DNA preferentially (79).

Sites outside the catalytic pockets of the type IA enzymes have also been implicated in DNA binding and/or DNA strand passage. The C-terminal 121 residues of *E. coli* DNA topoisomerase I, though dispensable in terms of the relaxation activity of the enzyme, can bind DNA and improve the stability of the enzyme-DNA complex (142). A region of *E. coli* DNA topoisomerase I spanning residues 598 to 737, which contains three tetracysteine motifs that are likely to bind three Zn(II) ions, may also participate in DNA binding and/or the manipulation of DNA strands (143, 144). Deletion of all three motifs leads to a loss of relaxation activity but not the ability of the enzyme to cleave single-stranded oligonucleotides (145–147). The first one, or maybe the first two of the three Zn(II)-bound motifs, is necessary for the DNA relaxation activity. C-terminal deletions extending into the first two motifs inactivate the enzyme (145). Stripping Zn(II) from the intact enzyme has a similar effect (142, 148), but depletion of Zn(II) does not eliminate the oligodeoxynucleotide cleavage activity of the enzyme (142). A lysine- and arginine-rich segment spanning residues 613 to 640 of the 653–amino acid *E. coli* DNA topoisomerase III also appears to participate in substrate binding (150). Gel electrophoretic mobility–shift measurements showed a progressive decrease in the affinity of the enzyme for single-stranded DNA as more residues in this region were removed from the C-terminus. Removal of this region was found to reduce processivity of the enzyme in its negative supercoil relaxation reaction, but it had no effect on cleavage-site selection (150).

TYPE IB ENZYMES   Eukaryotic DNA topoisomerase I binds double-stranded DNA (151) and covers a region of about 20 bp (152). These preferred DNA binding sites are nonrandom in their sequences (reviewed in 5; see also 152–154), and several attempts were made to deduce their structural basis (155, 156). Eukaryotic DNA topoisomerase I was shown to prefer supercoiled to relaxed DNA (157–160). In one interpretation, this preference was attributed to structural changes induced by DNA supercoiling (158–160). Because positive and negative supercoiling are expected to effect torsional distortions in the double helix in opposite ways, the ability of the enzyme to remove positive or negative supercoils with nearly equal efficiency (5) suggests that flexural rather than torsional strain might be responsible. The enzyme was reported to differentiate DNA topoisomers several kb in length that differ by as few as 1 in their linking numbers (158–160). This finding is surprising because such a small difference is associated with a free-energy difference no greater than the thermal energy (161, 162).

The preference of eukaryotic DNA topoisomerase I for supercoiled substrates could also be attributed to its preferential binding to a curved DNA segment (156, 163), or to a pair of DNA segments simultaneously (164, 165). The former interpretation stemmed primarily from observations that the enzyme appeared to recognize DNA sites that possess intrinsic curvature, the latter from observations that the enzyme preferred either positively or negatively supercoiled DNA to relaxed DNA and that in electron micrographs the enzyme was often seen at crossovers in linear or supercoiled DNA (164). Information on the stoichiometry of the enzyme-DNA complexes in the electron microscopy experiments is needed, however, to determine whether one enzyme molecule is simultaneously binding two DNA segments (one or both of which may be curved) or two enzyme molecules, each bound to a single DNA segment, are interacting to give the appearance of one enzyme molecule bound to a DNA crossover.

The identification of sequences to which eukaryotic DNA topoisomerase I binds tightly and catalyzes DNA relaxation efficiently (166, 167) led to extensive analysis of the sequence requirements of eukaryotic DNA topoisomerase I for enzyme-DNA complex formation (163, 168–170) and for catalysis of DNA breakage and rejoining (171, 172). Consistent with estimates of the size of a eukaryotic DNA topoisomerase I binding site (152, 170), nucleolytic and chemical footprinting data indicate that a bound enzyme protects both strands over a region of 15–19 nucleotides, within which the cleavage site is centrally located (169). On the scissle strand, the first base on the 5′ side of the break (position −1) can not be a purine (173, 174), but it can be absent (169); removal of bases at −2 to −7 eliminates DNA cleavage activity (169). Removal of other bases of the scissle or nonscissle strand leads to either partial inhibition, or, in some cases, stimulation of DNA cleavage activity. Interactions between the enzyme and a region spanning nucleotides +6 to +11 were also implicated (172, 175), but the footprint in this region appeared to be less prominent than that of the other side. DNA cleavage by eukaryotic topoisomerase I was shown to be strongly affected by methylation of the scissile strand cytosines near the cleavage site; cleavage is stimulated by methylation at position −4 but inhibited by methylation at position −3 (176). The inhibitory effect was attributed to the presence of a methyl group in the DNA major groove that interferes with enzyme-DNA major groove interactions in this region; the presence of thymine, but not uracil, at the −3 position was also found to inhibit DNA cleavage by the enzyme (176).

For vaccinia virus topoisomerase, the suggestion was made that the enzyme assumes a toroidal shape around duplex DNA (177). When this sequence-specific enzyme binds to its recognition site 5′-(C or T)CCTT↓ in a duplex DNA (178) (where the arrow indicates the position of DNA cleavage by the enzyme), it contacts four phosphates on the cleaved strand, including the scissle phosphate, and three phosphates on the noncleaved strand. All seven phosphates are located on one face of the pentameric motif. Photocrosslinking of bromopyrimidine-substituted DNA suggested that the enzyme makes base-specific contacts in the major groove of the pentameric motif, on the opposite side of the phosphate contacts. Unlike the findings for eukaryotic DNA topoisomerase I, no region in the DNA downstream of the scissle position was found to contact the enzyme (177).

Several studies were also carried out to map the various regions of type IB enzymes that are involved in DNA binding. The domain of yeast DNA topoisomerase I comprising residues 140–363 appears to participate in DNA binding, as evidenced by protein-footprinting as well as DNA-enzyme photo-crosslinking experiments (179, 180). In protein footprinting experiments probing the protection of lysines against citraconylation by DNA binding, a number of lysines within this domain, including Lys-245, 248, 249, 256, 257, 286, 301, and 304, were found to be protected (179, 180). In addition, one or more lysines in the regions of residues 356–407, 471–485, and 657–683 were also protected by DNA (179; NF Lue & JC Wang, unpublished observations). In vitro, a large N-terminal segment of eukaryotic DNA topoisomerase I is dispensable for enzymatic activity (5). The N-terminal 140 amino acids of yeast DNA topoisomerase I, for example, can be deleted with little effect on its DNA relaxation activity (181). In *E. coli* cells, however, expressing the yeast enzyme lacking the N-terminal 140 amino acids does not complement the lethal phenotype of inactivating *E. coli* DNA topoisomerase I, but expressing intact yeast DNA topoisomerase I does; thus the N-terminal domain might augment catalytic activity in vivo (181).

Partial proteolysis indicated that the small vaccinia virus topoisomerase has at least three domains separated by two protease-sensitive sites around Arg-80 and between residues 135 and 142 (182, 183). The 20-kDa carboxyl domain lacking the first 133 residues of the intact enzyme can bind DNA with lower affinity and sequence-specificity than the intact enzyme. The 9-kDa amino terminal domain was shown to participate in DNA binding. Deletion of the N-terminal 72 amino acids abolishes enzyme activity (183, 184), and DNA binding by the intact enzyme protects it from proteolytic cleavage at Arg-80 and in the Lys-135-to-Thr-142 region (183). Photocrosslinking experiments suggested that the 9-kDa N-terminal domain contacts the DNA major groove and is responsible for the DNA sequence-specificity of the viral enzyme (S Shuman, personal communication). In protein-footprinting of the lysines of vaccinia topoisomerase, Lys-35, Lys-83, two or three lysines in the Lys-133/135/138 cluster, Lys-213, Lys-249 and/or Lys-250, and Lys-271 were implicated in DNA binding (180; R Hanai, R Liu & JC Wang, unpublished data). Thus similar to eukaryotic DNA topoisomerase I, multiple regions of the viral enzyme participate in DNA binding.

TYPE II ENZYMES   In terms of interactions with the G- or the gate-segment of DNA, within which the enzyme-operated DNA gate resides, all type II enzymes interact with a segment of DNA duplex centered around the pair of 5′-staggered cleavage sites four base pairs apart. For *E. coli* DNA topoisomerase IV and phage and eukaryotic type II enzymes, a binding site spans 20–30 bp of DNA (185–187). Parts of the flanking regions within this stretch of protected DNA appear to be exposed to solvent. The major groove of all but 10 bp of the central region was shown to be available for triplex DNA formation (188).

For bacterial gyrases, a 140-bp DNA segment is wrapped around the enzyme in the enzyme-DNA complex, with the central 40 bp being the least accessible to DNase I (189). A degenerate consensus sequence within this central region was suggested by mapping the sites of cleavage in vitro and in intracellular plasmids upon treatment with nalidixic or oxolinic acid (190–192). Site-directed mutagenesis was also carried out to establish the importance of some of the highly conserved bases (192). The presence of prominent gyrase-mediated cleavage sites in the *E. coli* chromosome (193, 194), phage Mu (195), and pSC101 (196) was also implicated. Hydroxyl radical footprinting of DNA in the gyrase-DNA complex revealed a protected region of 128 bp (197). For the restriction fragment examined in the latter experiment, the protection pattern is asymmetric with respect to the dyad bisecting the gyrase-mediated site of cleavage: About 56 bp are protected on one side and 72 bp on the other; 13 bp adjacent to the cleavage site are the most strongly protected. The addition of AMPPNP or a quinolone drug that enhances enzyme-DNA covalent intermediate formation caused complex and asymmetric changes in this protection pattern, presumably as a result of conformational changes in the enzyme-DNA complexes (197; see also 198). Examination of additional sites is needed to establish the generality of these observations.

Studies of deletions in the GyrA subunit indicate that GyrA(1–572) is the smallest polypeptide that retains a low level of DNA supercoiling activity in the presence of GyrB, whereas GyrA(1–523) retains DNA relaxation but not supercoiling activity (199, 200). Substantial evidence shows that the C-terminal region of GyrA comprising residues 524–875 is involved in interaction with the flanking regions outside the central core of the 140-bp G-segment. Binding of the C-terminal fragment GyrA(572–875) to DNA was observed, and the fragment, when mixed with GyrA(1–571) and GyrB, restored DNA supercoiling activity (200).

Eukaryotic type II DNA topoisomerases were also reported to interact preferentially with curved DNA (201, 202) and with DNA crossovers (164, 201, 203–206). As noted above for the IB subfamily enzymes, these preferences could account for the preferential binding of a eukaryotic type II enzyme to positively or negatively supercoiled DNA (207–209). Uncertainties similar to those mentioned above (for the IB enzymes) remain, however, especially in view of the strong interaction observed between DNA-bound molecules of eukaryotic DNA topoisomerase II (210, 211).

Because the type II DNA topoisomerases catalyze the transport of a DNA double helix, termed the T-segment, through the DNA segment containing the enzyme-mediated DNA gate or the G-segment, the observation of protein located at a crossover in electron micrographs was often interpreted to be an enzyme molecule simultaneously interacting with a pair of G- and T-segments (164, 205). Biochemical analysis, however, raised strong doubt about this interpretation (206, 212, 213). In the decatenation of a supercoiled ring singly linked to a relaxed DNA by yeast DNA topoisomerase II, for example, the enzyme was found to catalyze decatenation efficiently while binding preferentially to a G-segment on the supercoiled ring (209, 212, 213). Thus even though the preferential binding of the enzyme to supercoiled DNA may involve its binding to DNA crossovers, the T-segment does not appear to be a part of the crossover.

In contrast to the preferential binding of eukaryotic DNA topoisomerase II to supercoiled DNAs of either sense, bacterial gyrase binds more strongly to relaxed than to negatively supercoiled DNA (214). A preference in the order of positively supercoiled > relaxed > negatively supercoiled DNA is expected from the right-handed wrapping of a DNA segment around the enzyme (215).

A special case of interest involves DNA binding by eukaryotic DNA topoisomerase II in the presence of AMPPNP. Upon addition of AMPPNP, Drosophila DNA topoisomerase II becomes resistant to dissociation by salt if it is bound to a circular DNA, but not if it is bound to a linear DNA (216). Studies of the binding of the DNA of various forms to yeast DNA topoisomerase II showed that in the absence of AMPPNP all forms can bind to the enzyme, with the positively or negatively supercoiled form binding several-fold more strongly than the nicked, relaxed, or linear form. Upon pre-incubation of the enzyme with AMPPNP, however, only the linear form can bind (209). These findings led to the proposal that the enzyme acts as an ATP-modulated protein clamp, and ATP or AMPPNP binding converts the enzyme from an open-clamp form to a closed-clamp form with a central hole large enough that a linear DNA can thread through it (209). Conversion of an enzyme bound to a DNA ring from the open-clamp form to the closed-clamp form would introduce a topological link between the enzyme and DNA, and thus the two would not come apart in concentrated salt solutions. Joining the ends of a linear DNA after it had threaded through the central hole in a closed enzyme clamp would similarly form a topological link between the enzyme and the DNA (209, 217). The formation of such topologically linked DNA-protein complexes was proposed earlier for the Ku-autoantigen (218), the replication complexes of phage T4 (219), E. coli (220), and yeast (221), as well as eukaryotic DNA topoisomerase II itself (208). In the earlier work on eukaryotic DNA topoisomerase II, a topological link between the enzyme and circular DNA in the absence rather than presence of AMPPNP was proposed, based on the retention of a DNA-protein complex on nitrocellulose that resisted elution by concentrated salt solutions (208). However, more recent results (209, 212, 217) indicate that the salt stability of the nucleotide-free complex retained on nitrocellulose filter is probably unrelated to the formation of a topological link between DNA and protein. Simultaneous binding of both N-terminal domains of the dimeric enzyme in its open-clamp form to the nitrocellulose filter, for example, may trap the enzyme-bound DNA.

Binding of DNA topoisomerases to two novel DNA substrates were reported. Eukaryotic DNA topoisomerase II was found to bind to the left-handed Z-helical form of DNA (222) and to parallel tetraplex structures composed of guanine-rich DNA (223, 224). Further experiments are needed to test the functional significance of these interactions. Novel DNA structures that readily form covalent complexes with DNA topoisomerases are reviewed in a later section.

Active-site Tyrosines and the Formation of Covalent Enzyme-DNA Complexes

IDENTIFICATION OF ACTIVE-SITE RESIDUES   Following the identification of the tyrosyl group as the nucleophile in transesterification between various DNA topoisomerases and DNA (225, 226), active-site tyrosines of *E. coli* DNA topoisomerase I, *E. coli* gyrase, and yeast DNA topoisomerases I and II were mapped by treating the covalent intermediates with trypsin, purifying the peptide-DNA covalent adducts, and sequencing the DNA-linked tryptic fragments (227–230). Because the entire amino acid sequences of the enzymes were known from nucleotide sequences of their structural genes, unambiguous identification of a tryptic fragment could be made from a few Edman cycles; the active-site tyrosine, being covalently attached to the DNA, showed as a blank in the Edman cycle corresponding to its position (227).

From sequence homology of enzymes in each of the three subfamilies, the active-site tyrosines of all DNA topoisomerases can be deduced. Site-directed mutagenesis with several topoisomerase structural genes confirmed assignments based on biochemical analysis and sequence comparisons (229, 231, 232). For *S. cerevisiae* and *Schizosaccharomyces pombe* DNA topoisomerase I, the active-site tyrosines were also inferred from a comparison of the cyanogen bromide or formic acid generated polypeptides in which the active-site tyrosines were linked to a radio-labeled phosphoryl group, transferred from the DNA by nucleolytic treatment of the enzyme-DNA covalent intermediates (233).

Aside from evidence of the involvement of the active-site tyrosines in the DNA cleavage-rejoining reactions, information on covalent catalysis by DNA topoisomerases is sparse. By analogy with other phosphoryl transfer reactions, DNA-protein transesterification by topoisomerases most likely falls into the general acid-base catalysis category. Some experimental evidence was provided by recent kinetic studies of vaccinia virus topoisomerase under single-turnover conditions. Base catalysis of the attack of the active-site tyrosine on the phosphodiester phosphorus, and acid catalysis of the departure of the 5′

hydroxyl group, were suggested based on the pH dependences of the cleavage and rejoining rates (234, 235). These measurements, as well as kinetic measurements of the effect of substitution of sulfur for a nonbridging oxygen of the scissle phosphorus, suggested that a conformation change of the DNA-enzyme complex follows DNA cleavage without a change of the protonated state of the complex. Furthermore, the rate of DNA cleavage appears to be limited by covalent chemistry and the rate of the rejoining step by reversing the conformation change (235). Catalytic residues other than the active-site tyrosine have not yet been established; histidine and cysteine residues were suggested based on the pH profiles of the rates, and additional groups including Arg-130 were thought to stabilize the transition state. Two mutations, K167D and G226N, were also reported to favor covalent intermediate formation (236), but the effects of these mutations are relatively small. The formation of covalent intermediates between DNA and various DNA topoisomerases is also affected by concentrations of divalent and monovalent salts and by temperature (reviewed in 153, 237, 238).

In transesterification catalyzed by DNA topoisomerases, the phosphotyrosyl intermediate in the catalytic pocket of an enzyme is presumably well-shielded from solvent hydroxyl groups, so rejoining of the DNA phosphoryl group to the hydroxyl group on the other end of the cleaved strand, rather than solvalysis of the phosphotyrosine link, would follow the cleavage and strand-passage steps. In the absence of a properly positioned deoxyribose hydroxyl group, hydrolysis and alcoholysis were observed with covalent complexes of *E. coli* and eukaryotic DNA topoisomerase I (239–242). The rate of solvalysis in each case is slow at neutral pH. Presumably, the solvent hydroxyl, unlike the deoxyribosyl hydroxyl group, is not optimally positioned in the catalytic pockets of the enzymes for efficient reaction with the tyrosine-linked phosphoryl group.

In the topoisomerase-mediated cleavage of one DNA for the passage of another, one interesting question is whether the passing-strand or duplex affects transesterification between the enzyme and the strand or duplex being cleaved. The effect of the passing-strand on the cleavage of the scissle-strand by a type I enzyme is unknown. However, it is conceptually helpful to view the DNA strands as active participants rather than passive "substrates" in the intricate reaction steps catalyzed by a topoisomerase; interactions between various parts of the DNA and the enzyme are most likely instrumental in the progression of the overall reaction (243). In DNA cleavage mediated by Drosophila DNA topoisomerase II in the presence of Ca(II) ions and in the absence of ATP, measurements of the dependence of the efficiency of DNA cleavage on DNA concentration led to the suggestion that the passing double helix must be present for efficient cleavage to occur (205). On the other hand, cleavage of the G-segment by yeast DNA topoisomerase II in its closed-clamp conformation, which excludes the T-segment from the interior of the enzyme, indicates that binding of the T-segment is not obligatory for the cleavage of the G-segment (209). Three-dimensional structural data suggest that the T-segment might have a more significant effect on the separation of the ends of the cleaved G-segment (see section on Mechanistic Implications of Biochemical and Structural Results).

SUBSTRATE SPECIFICITIES   Extensive studies have used various substrates to trap the covalent enzyme-DNA intermediates. In addition to providing information on substrate specificities in the topoisomerase-mediated DNA breakage and rejoining reactions, these studies firmly establish that the covalent complexes with phosphotyrosine links are the true reaction intermediates and provide the means of uncoupling the DNA breakage and rejoining steps (152, 171, 175, 186, 203, 239, 244–262). Studies of the topoisomerase-mediated DNA cleavage and rejoining reactions are particularly important in light of the findings that many antimicrobial and anticancer drugs act by stabilizing the topoisomerase-DNA covalent intermediates (reviewed in 59–61, 263).

Cleavage reactions were also carried out with oligodeoxynucleotides containing a phosphothioyl group at the site of cleavage. Substitution of a sulfur for a nonbridging oxygen led to the conclusion that E. coli DNA topoisomerase I can cleave an R_p but not an S_p phosphorothiodiester linkage, suggesting a stereospecific interaction between the enzyme and the phosphodiester group 3' to the cleavage site (240). Replacement of a phosphodiester linkage by a 5'-bridging phosphorothioate linkage was shown to give an efficient suicide substrate for eukaryotic DNA topoisomerase I, as the 5'-sulfhydryl group of the cleavage product can not undergo the DNA rejoining reaction (264).

CLEAVAGE OF NOVEL STRUCTURES   Apurinic sites in DNA were found to enhance double-stranded cleavage of DNA by DNA topoisomerase II in the absence of ATP (265). With a few apurinic sites in a 4-kb plasmid DNA, the extent of cleavage was found to be about the same as that in the presence of etoposide, a drug known to stabilize the covalent complex (59–61, 263). A comparison of these results is not straightforward, however, as DNA topoisomerase II–mediated cleavage of DNA in the presence of etoposide is strongly dependent on ATP. Drosophila DNA topoisomerase II was also shown to catalyze transesterification at a specific position in an oligodeoxynucleotide with a 19-nucleotide hairpin; DNA cleavage occurs near the 3' side of the base of the hairpin stem and requires a double-stranded/single-stranded junction at the 3' base of the hairpin (260). In mammalian cells, an activity that nicks duplex DNA with base mismatches was identified as DNA topoisomerase I (262). Nicking occurs at the first phosphodiester bond 5' to all eight possible mismatched bases, and is accompanied by DNA-enzyme covalent adduct formation. These reactions have important implications in vivo (see section on Topoisomerase-Mediated DNA Breakage and Its Effects on DNA Recombinations).

Nucleotide Binding to Type II DNA Topoisomerases

ATP BINDING   The mechanism by which ATP binding and hydrolysis and product release are coupled to the various steps in the transport of one DNA double helix through another is of key importance in understanding reactions catalyzed by the type II DNA topoisomerases. In the absence of ATP, *E. coli* gyrase can slowly relax negatively but not positively supercoiled DNA (266). Relaxation of positively or negatively supercoiled DNA in the absence or presence of ATP was also observed for a proteolytic product of *E. coli* gyrase lacking the N-terminal half of GyrB protein, within which the ATPase domain is located (266, 267). Similarly, phage T4 topoisomerase exhibits a low level of DNA transport activity in the absence of ATP, leading to knotting of supercoiled DNA rings at high enzyme concentrations (7, 268). For yeast and Drosophila DNA topoisomerase II, no topological transformation of DNA rings has been demonstrated in the absence of ATP (269, 270). In mutants with Gly-144 replaced by Pro, Val, or Ile, ATPase activity of the protein is greatly reduced and DNA relaxation activity is undetectable (271).

Early studies with nonhydrolyzable ATP analogs suggested that the binding of ATP is sufficient to effect DNA transport, and ATP hydrolysis and the release of ADP and orthophosphate allow the enzyme to turn over (216, 272, 273). The crystal structure of AMPPNP bound to dimers of the 43-kDa GyrB fragment shows that the nucleotide is bound to a subdomain comprised of residues 2–220 (274). Earlier studies using a reactive ATP analog pyridoxal 5′-diphospho-5′ adenosine had also implicated the same region for ATP binding, as the lysyl residues 103 and 110 were found to be the sites of adduct formation with the ATP analog (275). Site-directed mutagenesis of the nucleotide-binding subdomain implicated Glu-42 as a general base in the catalysis of ATP hydrolysis, and His-38 in aligning and polarizing the glutamate residue (276).

A key step in coupling ATP binding to DNA transport appears to be the dimerization at the GyrB N-terminus (274, 277, 278). The binding of AMPPNP to the 43-kDa N-terminal fragment of GyrB presumably causes a conformational change, which greatly favors dimerization. Dimerization of the 43-kDa fragment in the presence of ATP was also implicated in studies of the rate of ATP hydrolysis (277, 279) and by the appearance of a crosslinked dimer when the 43-kDa fragment was treated with dimethyl suberimidate in the presence of AMPPNP (278). In the case of yeast DNA topoisomerase II, interaction between the pair of ATPase domains in a homodimeric enzyme was implicated by the sigmoidal dependence of the DNA-dependent ATPase activity on ATP concentration (280). Pairing between the two N-terminal domains of the dimeric enzyme induced by the binding of AMPPNP or ATP to the enzyme was also assumed to be responsible for the conversion of eukaryotic DNA topoisomerase II from an open-clamp form to a closed-clamp form (209). Studies with heterodimeric yeast DNA topoisomerase comprised of one wild-type polypeptide and one G144I mutant polypeptide, in which Gly-144 is replaced by Ile, revealed that one bound AMPPNP is probably sufficient to trigger a global conformational change of the protein and close the N-terminal protein gate or N-gate (217). For a wild-type enzyme at a moderate ATP concentration, however, the positive cooperativity in ATP binding between the two enzyme halves (279–281) predicts that the predominant species with a closed N-gate has two bound ATP per enzyme.

Significantly, ADP does not promote dimerization of the 43-kDa GyrB fragment (277). The crystal structure of the AMPPNP-bound dimeric 43-kDa GyrB fragment indicates that residues from both polypeptides are involved in each ATPase pocket, and thus ATP turnover presumably occurs only if dimer contacts are made (274; see also 277, 279). A simplified picture is that the binding of one ATP to a type II DNA topoisomerase is followed rapidly by the formation of dimer contacts and the binding of a second ATP, closing the N-gate. ATP hydrolysis then follows, and the N-gate re-opens after ATP hydrolysis. The rate of a type II enzyme–catalyzed reaction is probably limited by the rate of the conformational change of the protein following ATP hydrolysis, or by the rate of product release itself (277, 278).

TRIPHOSPHATES OTHER THAN ATP   Early studies with Drosophila and yeast DNA topoisomerase II showed that dATP can substitute for ATP (282, 283). Conflicting results were reported for some of the other common triphosphates. In one study, relaxation of supercoiled DNA by Drosophila DNA topoisomerase II in the presence of 1 mM ATP was found to be strongly inhibited by CTP but not by GTP up to 3 mM (282). In a second study, however, pronounced inhibition of the same enzyme by GTP or dGTP at concentrations greater than 2 mM was seen, but CTP was not inhibitory up to 10 mM (222). Both studies found an inhibitory effect of >0.5 mM ITP, or ATP itself at concentrations >4 mM, and lack of an effect up to 10 mM of UTP. Others found no inhibition of the ATP-supported relaxation reaction by several mM GTP with either the Drosophila or the yeast enzyme (T Hsieh, personal communication; J Roca & JC Wang, unpublished results). The sources of these discrepancies are intriguing; plausible causes are the presence of other factors in the topoisomerase preparations, different states of modification of the enzyme, or impurities in the triphosphates.

Reports were also conflicting on the effects of triphosphates on type IB–

catalyzed relaxation of supercoiled DNA. Studies with DNA topoisomerase I from yeast (283), *Ustilago* (284), human (285, 286), and vaccinia virion (287, 288) showed no coherent pattern. More recently, the relaxation activity of vaccinia virus topoisomerase purified from the virion, or from *E. coli* cells expressing a cloned gene, was shown to be stimulated by nucleoside triphosphates in media containing 0.1 M NaCl (289). The authors suggested that NTPs may up-regulate the viral enzyme allosterically when the influx of the triphosphates into the virion and the transcription of the early viral genes are initiated.

Three-Dimensional Structures

Within the past several years the three-dimensional structures of six DNA topoisomerase fragments have been solved, five by X-ray crystallography and one by nuclear magnetic resonance (179, 182, 243, 274, 290, 291). A 67-kDa fragment of *E. coli* DNA topoisomerase I comprising the N-terminal 596 residues was crystallized and its structure solved to 2.2 Å resolution (243). In solution the 67-kDa polypeptide does not catalyze the relaxation of supercoiled DNA, but it can cleave single-stranded oligonucleotides to form a covalent enzyme-DNA complex (147). The polypeptide folds into four distinct domains: domains I and IV form the "base" of the fragment and domains III and II the "lid"; a pair of long strands connect the lid to the base, and the four domains and the connecting strands enclose a hole with an average side-chain to side-chain diameter of about 30 Å. The active-site nucleophile Tyr-319 is located in domain III at the junction between this domain and domains I and IV. This strategic location suggests that the DNA strand containing the scissle bond is bound to the enzyme across domain III and the base comprising domains I and IV. In order to provide room for the binding of a DNA strand, the lid comprising domains III and II and the base comprised of domains I and IV must move away relative to each other from their positions in the crystal structure, using the pair of connecting strands and other joints as hinges. Following covalent adduct formation between the enzyme and the DNA strand, the opening between the lid and the base is probably further increased to allow the passage of another DNA strand.

The structure of a C-terminal fragment of *E. coli* DNA topoisomerase I comprising residues 745–865, which is dispensable for enzymatic activity but appears to affect the processivity of the enzyme, has been solved by nuclear magnetic resonance (290). The structure of this domain resembles structures of protein domains that bind single-stranded nucleic acids (290). Three-dimensional structural information is lacking for the tetracysteine motifs between the 67-kDa N-terminal fragment and the 121-residue C-terminal domain.

For the type IB enzymes, the structures of two fragments, a 9-kDa N-terminal fragment of vaccinia virus topoisomerase comprising the first 77 amino acids, and a 26-kDa fragment of yeast DNA topoisomerase I comprising amino acids 135 to about 363, were determined by X-ray crystallography to 1.6 Å and 1.9 Å resolution, respectively (179, 182). The yeast fragment is composed of a 65 x 30 x 20 Å base and a tandem pair of α-helices that form a V-shaped arm across the base. The base of the 26-kDa yeast topoisomerase fragment shows a concave surface on one side. Mapping of lysines protected against citraconylation by DNA suggested that this surface and a region extending from the tip of the V-shaped pair of α-helices toward the concavity are in contact with DNA in the enzyme-DNA complex (179). Photocrosslinking and site-directed mutagenesis also led to a structural model for the participation of the 9-kDa N-terminal fragment of vaccinia virus topoisomerase in DNA binding (S Schuman, personal communication). Although the 9-kDa fragment of vaccinia virus topoisomerase and the 26-kDa fragment of yeast DNA topoisomerase I are both involved in DNA-binding and correspond to the N-terminal domain of a catalytically active enzyme, they are structurally distinct (179).

For the type II enzymes, a 392–amino acid N-terminal fragment containing the ATPase domain of E. coli GyrB subunit was crystallized in the presence of AMPPNP, and solved to 2.5 Å resolution (274). The 43-kDa fragment contains two subdomains. The N-terminal one (residues 2–220) comprises the ATP binding site; its N-terminal arm (residues 2–15) extends from the domain surface to wrap the other monomer. The C-terminal subdomain comprises a four-stranded β-sheet and four helices; the C-terminus proximal helix (residues 366–392) protrudes from the core of this subdomain to contact the corresponding helix in the other monomer. These N- and C-terminal contacts enclose a 20 Å hole in the dimer (274). Whereas the N-terminal dimer interface is substantial, the C-terminal interface probably represents crystallographic rather than native protein contacts.

The crystal structure of a 92-kDa fragment of yeast DNA topoisomerase II was solved recently to 2.7 Å resolution (291). This fragment comprises residues 410–1202 of the 1429–amino acid enzyme, the C-terminal region of which beyond residue 1166 is dispensable for catalytic activity (106). The 92-kDa fragment is dimeric in solution as well as in the crystal. In solution, the fragment does not relax supercoiled DNA, but retains the ability of the intact enzyme to cleave double-stranded DNA and to form a protein-DNA covalent complex. In the crystal, a pair of the crescent-shaped monomers contact each other to form a heart-shaped dimer with a large central hole 55 Å wide at its base, 25 Å wide at the top, and 60 Å long. The 92-kDa fragment can be thought of as two subfragments: The B′ subfragment (residues 410 to about 660) corresponds to the C-terminal half of the GyrB subunit, and the A′ subfragment corresponds to the GyrA subunit lacking a 250 residue C-terminal segment. The primary dimer contacts involve residues 1036–1128. The B′-B′ contacts are less substantial but appear significant. The dimer in the crystal can be viewed as a V-shaped A′-A′ dimer, with the opening of the V capped by a B′-B′ arch to form the large hole. The pair of active-site tyrosines are positioned near the tips of the V opening. A G-segment that has been cleaved and pulled apart by the topoisomerase was modeled into the protein crystal structure. In this model, each of a pair of 20–25 Å semicircular grooves near the opening of the V can accommodate 10–12 bp duplex DNA with a 4-nucleotide overhang; the 5′ phosphoryl end of this overhang is linked to Tyr-783 through a narrow tunnel formed by residues in both B′ fragments and the A′ fragment containing the active-site tyrosine.

Mechanistic Implications of Biochemical and Structural Results

Significant progress has been made in the past decade in understanding how the various DNA topoisomerases manipulate DNA strands. It is now well-established that the basic covalent chemistry in each case involves two successive transesterification reactions. In the first reaction, an enzyme tyrosyl group attacks an internucleotide phosphorous, forming a tyrosine-phosphate linkage at one end of the transiently broken DNA strand and leaving a hydroxyl group on the deoxyribose at the other end of the broken strand. The second reaction reverses the first, resulting in rejoining of the DNA strand and setting free the enzyme from covalent attachment to DNA. The two reactions are separated temporally by the transport of DNA strands through the DNA gate; one is likely to be the exact microscopic reversal of the other, but different groups of an enzyme could in principle participate in the two reactions. Catalysis of the transesterification reactions are most likely to fall in the general acid-base category by analogy to nucleolytic reactions, in which a water hydroxyl group is the nucleophile. Further advances in the covalent chemistry of DNA topoisomerase–catalyzed reactions require the identification of the catalytic groups other than the active-site tyrosines.

The most intriguing mechanistic aspects of DNA topoisomerases are the steps by which they move DNA strands or duplexes through one another. For the IA subfamily of enzymes, the combined biochemical and structural results strongly favor the enzyme-bridging model of DNA strand-passage. In this model, the enzyme holds onto both ends of the transiently severed DNA strand, and the enzyme-bound DNA ends are moved apart enough to allow the passage of another DNA strand. Although the crystal structure of the 67-kDa *E. coli* DNA topoisomerase I suggests that the scissle strand spans domain III and domains I and IV (243), a clear molecular picture of the strand-passage event has yet to emerge owing to the lack of detailed information on the binding sites for both the scissle and the passing strand, and on how these strands might be moved by the enzyme.

For the IB subfamily of enzymes, whether the enzyme-bridging model also applies is less certain. In the alternative DNA rotation model, the enzyme-DNA interaction is assumed to involve mainly DNA on the side of the scissle strand that becomes covalently linked to the enzyme in the first transesterification reaction; the enzyme has only a passive role in DNA strand passage, which occurs by rotation of the DNA segments on the two sides of the enzyme-linked DNA phosphoryl group relative to each other. Although there is some indication of interaction between eukaryotic DNA topoisomerase I and DNA downstream of the site of DNA breakage (169), a similar interaction between DNA and vaccinia virus topoisomerase has not been observed (177). In the limiting cases, the enzyme-bridging model requires the removal of only one supercoil per DNA breakage-rejoining cycle, whereas in the DNA-rotation model all supercoils can be removed by as few as one single DNA breakage-rejoining cycle. Other than the limiting cases, however, the distinction between the two models is not clear-cut, as interaction between the enzyme and DNA downstream of the scissle bond may range from very weak to very strong.

A molecular picture of how a type II enzyme catalyzes the transport of one DNA double helix through another has emerged from biochemical and structural studies of the enzymes. The crystal structure of the 92-kDa yeast enzyme fragment suggests that the two C-terminal or "GyrA" (A) halves of eukaryotic DNA topoisomerase II form a V-shaped dimer, with the G-segment of DNA held in a pair of semi-circular grooves near the top of the V. Binding of the G-segment to the intact enzyme involves mainly interactions between the DNA and residues in the A halves, but the N-terminal of the "GyrB" (B) halves are also involved. The pair of the B parts can either form a dimeric arch over the V, if the subfragments are in the ATP-bound conformation, or come apart upon ATP-hydrolysis. When the B-subfragments are apart, corresponding to the open-clamp form of the enzyme, a second DNA duplex, the T-segment, can enter the protein clamp through the N-terminal jaws or the N-gate. Closure of the N-gate, triggered by ATP-binding, captures the T-segment. A number of large conformational changes in the enzyme–G-segment complex are likely to occur following the capture of the T-segment. The DNA gate, which is formed by cleavage of a pair of DNA phosphodiester bonds and the simultaneous formation of a pair of DNA-protein phosphotyrosine links, is forced open. This opening involves movements as large as 35–40 Å along the helical axis of the G-segment and requires structural changes in both the A and B halves: The V-shaped A_2 dimer is splayed further apart, with coordinated changes in the B_2 cap. As the DNA gate opens, the T-segment is pushed into the large cavity below the G-segment. The enzyme-DNA complex, now temporarily free of steric repulsion, readjusts to close the DNA gate. This closure, however, reduces the size of the cavity containing the T-segment, and the resulting steric repulsion between the protein and the T-segment in turn forces the dimer interface at the bottom of the V-shaped A_2, termed the C-gate, to come apart. The exit of the T-segment completes a round of DNA transport, and ATP hydrolysis and product release allow the start of the next round. Two studies were carried out to determine the consequences of crosslinking the C-terminal domains of the two halves of yeast DNA topoisomerase II (292, 293). In the experiments of Roca et al (293), locking and unlocking of the C-gate without inactivating the enzyme were achieved through structure-based design of reversible disulfide links. Analysis of the products of AMPPNP-mediated decatenation of a pair of singly linked DNA rings by a crosslinked or uncrosslinked enzyme showed unequivocally that the DNA-ring containing the T-segment enters the dimeric enzyme through the N-gate, passes through the enzyme-bound G-segment and exits through the C-gate (293).

BIOLOGICAL FUNCTIONS OF DNA TOPOISOMERASES

The participation of topoisomerases in nearly all cellular processes involving DNA makes it impractical to review all aspects of their functions. Because the enzymes affect the topology and organization of intracellular DNA, the primary effects of inactivating a topoisomerase are also likely to generate far-reaching ripples. In the sections below, the emphasis is on the primary functions of the enzymes. The regulation of the cellular levels of the enzymes themselves and the association of the enzymes with other cellular proteins are also reviewed because these aspects are closely tied to the cellular functions of the enzymes.

Removal of DNA Supercoils Generated by Various Cellular Processes

Following the discovery of DNA gyrase in 1976, it was generally thought that, in bacteria, gyrase was responsible for DNA negative supercoiling and that DNA topoisomerase I served a check-and-balance role of removing negative supercoils. This view has been modified in the past decade. Supercoils, both negative and positive ones, appear to be generated by a number of processes involving the movement of macromolecular assemblies along DNA. One major cellular function of the topoisomerases is to prevent excessive supercoiling of intracellular DNA. However, this more general role of DNA topoisomerases does not mean that supercoiling is not utilized in vivo to drive a particular region of intracellular DNA into a conformation suitable for a particular process. Initiation of DNA replication, for example, often requires that the DNA be in a negatively supercoiled state [for a comprehensive treatise on DNA replication, see (294)].

REPLICATION   Replication is the best known process that generates supercoils in intracellular DNA. During semiconservative replication, positive supercoils are generated ahead of the replication fork as the parental strands come apart behind it. The separation of parental strands behind the fork can be viewed as a special case of negative supercoiling: The maximal extent of negative supercoiling of a DNA ring occurs when the two complementary DNA strands are not linked at all (specific linking difference = −1.0).

The involvement of various topoisomerases in the removal of positive supercoils generated by replication is generally in accordance with their known in vitro specificities. Namely, eukaryotic DNA topoisomerases I and II, and bacterial DNA topoisomerase IV, can efficiently remove supercoils of either sense; bacterial DNA topoisomerases I and III, and eukaryotic DNA topoisomerase III, can remove negative supercoils, but not positive supercoils, unless a single-stranded region is present in the DNA. Bacterial gyrase is unique in its ability to convert positive to negative supercoils; depending on how fast the positive supercoils are generated and how fast they are converted to negative supercoils, gyrase can either prevent accumulation of positive supercoils in an intracellular DNA segment or keep the segment in a negatively supercoiled state.

In *E. coli* and other bacteria, gyrase is probably the major activity responsible for the removal of positive supercoils accompanying nascent DNA chain elongation. The right-handed wrapping of DNA around gyrase allows preferential binding of the enzyme to a positively supercoiled rather than to a relaxed or negatively supercoiled region (215). DNA synthesis in *E. coli* gyrase temperature-sensitive mutants is much reduced immediately after subjecting the cells to a nonpermissive temperature (295, 296). For a strain of *E. coli* permeable to the gyrase inhibitor novobiocin, addition of the drug results in the accumulation of positive supercoils in a plasmid (297). The sensitivity of *E. coli* DNA topoisomerase IV to novobiocin is yet to be examined, however, and thus its participation in positive supercoil removal in vivo can not be discounted. The properties of purified bacterial DNA topoisomerase IV suggest that, in addition to decatenation of DNA rings, it might also be involved in the removal of positive as well as negative supercoils in vivo (reviewed in 298). Overexpression of the enzyme in *E. coli topA* cells was found to suppress the lethal phenotype of DNA topoisomerase I deficiency, suggesting that topoisomerase IV is capable of preventing excessive supercoiling when present at a higher than normal cellular concentration (87).

Recent studies in reconstituted in vitro replication systems for pBR322 and a plasmid bearing the *oriC* replication origin of *E. coli* indicated that three of the four known bacterial DNA topoisomerases could support nascent DNA chain elongation, with bacterial DNA topoisomerase I being the only exception (299, 300). The ability of purified *E. coli* DNA topoisomerase III to support the unwinding of parental strands during replication in vitro is surprising in view of its inability to remove positive supercoils (139–141). The enzyme could perhaps catalyze the passage of one parental strand through another by acting in a single-stranded region near the replication fork (299, 300), but then it is puzzling why *E. coli* DNA topoisomerase I cannot act in the same way. Perhaps *E. coli* DNA topoisomerase III is associated with other macromolecular components in these in vitro replication systems to place itself at the required position. Whether the enzyme has a significant role in the removal of positive supercoils in vivo remains to be shown.

In yeast, inactivation of DNA topoisomerase II alone does not alter significantly the rate of nascent DNA chain lengthening in cells released from cell cycle arrest. A temporal delay in the extension of these chains was observed, however, when cells lacking DNA topoisomerase I were similarly treated (301). Elongation of newly synthesized DNA chains longer than 5 kilonucleotides, but not the synthesis of shorter chains, was found to be stopped by inactivating both yeast DNA topoisomerases I and II (301). In both *S. Pombe* and *S. cerevisiae*, inactivation of DNA topoisomerases I and II was found to reduce DNA synthesis (105, 301, 302). These experiments suggest that in the yeasts DNA topoisomerase I probably serves as the major replication swivel, but its role can be substituted by DNA topoisomerase II. A similar conclusion was reached for simian virus 40 DNA replication in vitro (303). The requirement of either DNA topoisomerase I or II in nascent DNA chain elongation in yeast also shows that yeast DNA topoisomerase III is insufficient to fulfill the role of a replication swivel.

In principle, in the removal of positive supercoils generated by replication, a DNA topoisomerase can act anywhere within the topological domain containing the supercoils. Positive supercoils in a θ-shaped replicative intermediate, for example, can be removed by a type I or type II DNA topoisomerase acting within the unreplicated portion of θ, by a type II enzyme passing one duplex sister arm through another in the replicated portion of the θ-shaped DNA, or by a type I enzyme passing one parental strand through another within a single-stranded region immediately behind the replication fork (304). Avemann et al (305) suggested that eukaryotic DNA topoisomerase I might be present in the vicinity of the replication fork in SV40 DNA. The possibility that bacterial gyrase acts preferentially ahead of the fork, and DNA topoisomerase IV acts preferentially behind it, has been suggested by several authors to account for the predominant role of DNA topoisomerase IV in unlinking catenated pairs of newly replicated DNA molecules, and the sensitivity of the enzymes to certain drugs in vitro and in vivo (51, 298, 306, 307; K Marians, personal communication).

TRANSCRIPTION   A second example of cellular processes that generate supercoils is transcription. It was suggested that under certain conditions transcription drives positive supercoiling of the DNA template ahead of the polymerase and negative supercoiling of the DNA template behind it (308). This twin-su-percoiled-domain model, and the involvement of one or more DNA topoisom-erases in the removal of the resulting supercoils, has received substantial support (107, 309–326).

Two conditions are necessary for transcriptional supercoiling. One involves hindrance to the rotation of a transcription ensemble around the DNA template, and the other involves the presence of a barrier to the rotation of the DNA segment connecting two oppositely supercoiled domains (308, 327–332). Both conditions can be met by anchoring directly a transcription ensemble on the DNA template, as illustrated by fusing an extra DNA-binding domain, derived from the yeast GAL4 protein, onto the N-terminus of phage T7 RNA polymerase (333). Initiation of transcription by the fusion protein was shown to move and turn the DNA relative to the polymerase, resulting in lengthening and supercoil-ing of the DNA loop between the sites for RNA synthesis and DNA-binding in the chimeric protein (333). Removal of supercoils in such a loop would require either a topoisomerase or the dissociation of one of the DNA-bound domains of the chimeric protein from the DNA. Transcription by an RNA polymerase tightly associated with one or more DNA-bound regulatory factors is likely to create a situation similar to the one described above (334, 335).

Several other possibilities that can prevent the circling of the transcription ensemble around its template were suggested (308). Of these, insertion of mRNA-attached nascent polypeptides into cytoplasmic membranes appears to be the most significant in bacteria (323, 332, 336). Because transcription and translation are coupled in bacteria, anchoring of a transcription ensemble can occur if the nascent N-terminal chains of a membrane protein, or a protein for export (323), are inserted into the cytoplasmic membrane before the transcript encoding the protein is released from the RNA polymerase–DNA complex. This anchoring would in turn prevent the rotation of the transcription ensemble around the DNA template.

Little is known about barriers hindering the rotation of the DNA segment connecting two oppositely supercoiled domains. The frictional resistance against rotating a DNA segment around its helical axis can be greatly increased through association of a protein bound to it with a large cellular structure, or by the presence of a DNA loop or replication bubble in the segment (327, 328, 330–332, 337). In bacteria, the barrier could also be a membrane-anchored transcript in the connecting segment. If a pair of adjacent transcription ensem-bles are both prevented from circling around the template, supercoiling of the DNA in between would occur if the ensembles are moving away from each other (divergent transcription) or in tandem but asynchronously (308, 326, 327).

OTHER PROCESSES   Several processes other than replication and transcription are also known to generate supercoiled domains by the type of mechanism described above for the GAL4-T7 RNA polymerase chimera. Examples of proteins that supercoil DNA as they move along the double helix are bacterial type I restriction endonucleases (338), and helicases such as the SV40 T-antigen and the *uvrABC* complex in excision repair (339, 340). "Reverse gyrase" also appears to comprise a domain with an ATP-dependent DNA tracking activity and a domain with a type IA topoisomerase activity; the formation of two oppositely supercoiled domains is expected as the enzyme moves along the DNA in the presence of ATP, and relaxation of the negatively supercoiled domain by the topoisomerase component would lead to accumulation of positive supercoils (81; P Forterre, personal communication). The generation of supercoiled domains by triphosphate-dependent enzymes that can track along DNA and the involvement of one or more DNA topoisomerases in such a process are probably more common than generally appreciated.

CONSEQUENCES OF HYPERSUPERCOILING   DNA supercoiling is known to have strong effects on DNA structure and on its interactions with other molecules. Thus in addition to impeding directly processes that generate supercoils, the failure to prevent excessive supercoiling may affect many other cellular processes. This pleiotropic nature makes it difficult to pinpoint the various consequences of excessive supercoiling.

E. *coli* cells cannot survive inactivation of DNA topoisomerase I without compensatory mutations (341–344). The *topA*-mediated inviability is most likely a direct consequence of excessive negative supercoiling, perhaps in certain critical regions of the chromosome. Down-mutation in the *gyrA* or *gyrB* genes (341, 342), a low level expression of eukaryotic DNA topoisomerase I (181), and overexpression of E. *coli* DNA topoisomerase IV (87), have been shown to rescue the *topA* lethality.

Transcription of individual genes in prokaryotes is known to be affected differentially by supercoiling, and several recent reviews have appeared (329, 345). A S. *typhimurium* mutant promoter *leu500*, for example, is known to be activated in cells lacking DNA topoisomerase I (346 and references therein). Extensive studies show that transcription and translation of adjacent genes are necessary for this activation, suggesting that negative supercoiling of the region of DNA containing the promoter, driven by transcription and translation of adjacent genes, is responsible for the *topA*-dependent activation of *leu500* (319–321, 347–349). Experiments with various plasmid clones of S. *typhimurium* chromosomal DNA segments containing the *leu500* promoter showed that the mutant promoter was not activated in a *topA* strain unless the cloned segment also contained the *ilvIH* promoter, which is 1.9 kb upstream of *leu500* and transcribes away from it (349).

In most of the cases examined, template supercoiling probably affects mostly the initiation event. It may, however, also interfere directly with polymerase movement or indirectly with the elongation step. For example, in the synthesis of *E. coli* rRNA in vitro in the presence or absence of the transcription elongation factor NusA, more prominent pausing sites were seen in negative supercoiled than in relaxed template (350). More experiments specifically designed to examine the effects of template supercoiling on elongation and termination of transcripts are needed.

Recent studies indicate that DNA hypernegative supercoiling in *topA⁻* cells can result in R-loop formation between nascent RNA and the DNA template, which may in turn lead to aberrant replication and/or other cellular events [(326); see (327) for a review of earlier work on R-loop formation between nascent RNA and DNA templates of varying degrees of negative supercoiling]. In support of this notion, overexpression of RNase H was found to partially complement the growth defect of an *E. coli* Δ*topA* mutant (326, 351).

In eukaryotes, spatial and temporal association of DNA topoisomerase I and chromatin regions undergoing transcription is well-documented (352–358). Reduction of total RNA synthesis was seen in *S. pombe top1 top2* double mutants under restrictive conditions (302), and inactivation of topoisomerases I and II of *S. cerevisiae* similarly led to a drastic reduction in rRNA synthesis as well as a reduction of the rate of poly(A)-terminated mRNA synthesis by about threefold (105, 301). Both initiation and elongation of rRNA synthesis in *S. cerevisiae* were found to depend on a topoisomerase activity (359). Topoisomerase-mediated changes in template topology were reported to affect transcription of the yeast *ADH2* gene in vivo (360), and positive supercoiling of intracellular DNA in yeast was shown to reduce transcription of the GAL1 promoter to an undetectable level (337). The repression of cellular genes when cells enter the stationary phase was also reported to depend on DNA topoisomerase I (361).

In addition to studies using various topoisomerase mutants, microinjection of IgG against DNA topoisomerase I into nuclei of living salivary gland cells of *Chironomus tentans* was found to inhibit the RNA polymerase I–mediated synthesis of nucleolar preribosomal 38S rRNA by about 80%, and RNA polymerase II–mediated synthesis of chromosomal RNA by about 50% (362). Analysis of the size distribution of transcripts of the highly extended chromatin fibers in the Balbiani rings, which are large tissue-specific puffs, suggested that chain elongation was affected by inhibition of DNA topoisomerase I (362).

Not all RNA synthesis is affected by inactivation of topoisomerases, however. In the budding yeast, although transcription of chromosomal copies of rDNA is sensitive to simultaneous inactivation of DNA topoisomerases I and II, transcription of a plasmid-borne rDNA gene is not (105, 310). Similarly, neither the rate nor the extent of induction from the galactose-inducible *GAL1* gene promoter is affected by inactivation of both DNA topoisomerases I and

II (310). These cases probably reflect the existence of diffusional pathways for the cancellation of oppositely supercoiled domains in a plasmid.

Unlinking of Intertwined DNA Duplexes

Major advances have been made in the past decade in understanding the roles of DNA topoisomerases in chromosome segregation. As the length of the unreplicated DNA segment separating two converging replication forks is reduced to about 200 base pairs, untwining of the strands in this spatially confined segment is often incomplete at the time of complete unpairing of the parental strands. Thus multiply intertwined pairs of duplex DNA molecules may form, and, depending on how fast single-stranded gaps or nicks are repaired, these catenated replicative products may be covalently closed or may contain single-stranded gaps or nicks (29, 51, 298, 363–367). Although the formation of multiply intertwined progeny pairs may not be obligatory (see 368, 369 for examples), there is strong support for the formation of catenated replicative products in yeasts, bacteria, mammalian cells, and reconstituted replication systems (reviewed in 43, 51, 298, 365, 366; see also 370–374). Decatenation by a type II topoisomerase is also obligatory during meiosis in yeast (375, 376). The finding that DNA topoisomerase II is localized to miotic chromosomes of pachytene spermatocytes and to nuclei of elongating spermatotids adds further support to the involvement of this enzyme in meiosis (377).

In *E. coli*, earlier results through inhibition of gyrase by the use of a quinolone drug implied that gyrase was the major decatenation activity, at least in unlinking DNA catenanes formed by site-specific recombination (378). The discovery of DNA topoisomerase IV, and the finding that it is also sensitive to the quinolone drugs (88, 92, 307, 379, 380; S Nakamura, personal communication), led to the question of whether bacterial DNA topoisomerase IV, instead of or in conjunction with gyrase, was the quinolone-sensitive decatenation activity observed in the earlier experiments. Strong genetic and biochemical evidence now shows that DNA topoisomerase IV is the main enzyme in unlinking multiply intertwined chromosomal pairs. Upon inactivation of DNA topoisomerase IV in a *parC* or *parE* mutant, catenated plasmid replicons were found to accumulate to a steady-state level of about 10% (91). Pulse-labeling showed that essentially all newly replicated plasmids were in the catenated form in the absence of a functional DNA topoisomerase IV, but these catenanes were slowly unlinked by gyrase (306). The relative rates of decatenation by DNA topoisomerase IV and by gyrase in wild-type *E. coli* cells were estimated to be about 100 to 1 (306).

The weak decatenation activity of gyrase in vivo is probably significant. Simultaneous introduction of plasmid-borne *gyrA*[+] and *gyrB*[+] genes showed that the partition defect of *parC* and *parE* mutants could be compensated by a higher than normal cellular level of gyrase (88). In purified systems, both type II enzymes are capable of decatenation, but DNA topoisomerase IV is much more efficient, a finding consistent with the in vivo results (89, 299, 300, 380). The question remains, however, of why mutations in gyrase genes should lead to a partition defect (381; reviewed in 382, 383). Cells of one gyrase mutant were shown, for example, to accumulate dumbbell-shaped nucleoids under restrictive conditions, as expected for a defect in unlinking catenated pairs of chromosomes (384). Gyrase mutants that exhibit a partition defect are probably rather leaky, so progression of the replication forks is not completely impeded. It is plausible that incomplete removal of positive supercoils in such mutants might hinder a stage of replication before the formation of mutiply catenated progeny molecules, and the dumbbell-shaped nucleoids (384) might represent late-stage replication intermediates rather than catenated chromosomal pairs.

The finding that *E. coli* DNA topoisomerase III is effective in the segregation of replicating pairs of pBR322 and *oriC* plasmids in vitro suggests that chromosome segregation in these systems can also proceed by an alternative pathway. In this pathway, the unlinking of parental strands may precede the completion of nascent chain synthesis, so that multiply intertwined catenanes are rarely formed, or multiply linked gapped DNA rings may form first, but are then unlinked by DNA topoisomerase III (reviewed in 31; see also 300, 385).

In *S. cerevisiae*, *S. pombe*, and Drosophila, only one type II DNA topoisomerase is known. Catenated pairs of plasmid DNA rings are formed in *S. cerevisiae* and *S. pombe top2ts* cells at nonpermissive temperatures (106, 302, 342, 386), and lethality resulting from inactivation of DNA topoisomerase II occurs during mitosis, when chromosomal pairs are separated (387, 388). In an elegant experiment, a *top2ts-ndacs* double mutant was used so that spindle dynamics in the cells could be arrested by inactivation of the mutant *nda* gene encoding β-tubulin (389). At a low temperature that inactivated the mutant β-tubulin, spindle dynamics was shown to be inhibited but chromosome condensation proceeded to completion in the presence of an active DNA topoisomerase II. The fully condensed chromosome pairs remained catenated, however, in the absence of a driving force for the complete separation of the chromosomes. By upshifting the temperature to simultaneously inactivate DNA topoisomerase II and activate β-tubulin, it was shown that progression through anaphase with the entangled chromatids resulted in chromosome loss and/or breakage. A similar experiment was carried out in mammalian cells by administering successively a microtubule inhibitor and a topoisomerase II inhibitor, and the same conclusion was reached (373). These and other experiments cited earlier show that DNA topoisomerase II is directly involved in decatenation of linked chromosomes in vivo. The requirement of yeast DNA topoisomerase II in chromosomal segregation was shown to be particularly important for the larger chromosomes (372). In the absence of DNA topoisomerase II, large chromosomes were found to break in a 200-kbp region around the centromere, but intertwined chromosomes with arms shorter than 200 kbp appeared to resolve without breakage, presumably through movements of their ends (372).

The crucial role of DNA topoisomerase II in chromosome segregation was also observed in several other systems. Injection of antibodies that inhibit DNA topoisomerase II into Drosophila blastoderm embryos was found to hinder or prevent chromosomal segregation during anaphase (390). Similarly, treatment of mammalian cells with ICRF-193, which inhibits eukaryotic DNA topoisomerase II by locking the enzyme in its inactive closed-clamp conformation (391), leads to incomplete chromosome condensation and inhibition of anaphase chromosome segregation (374, 392–394). Studies of mammalian cell lines using etoposide, teniposide, or amsacrine, inhibitors of eukaryotic DNA topoisomerase II that trap the DNA-enzyme covalent intermediate, led to basically the same conclusion (390, 395). Interpretation of in vivo results using the etoposide class of drugs is more complicated, however, because of the complex cellular response to the trapping of the enzyme-DNA covalent intermediate. Complications by multiple cellular events are largely absent in experiments in vitro. Studies of simian virus 40 DNA replication in vitro and in vivo in the presence of specific inhibitors of eukaryotic DNA topoisomerase II showed that a type II enzyme is essential in the segregation of the catenated pairs of progeny molecules (367, 374, 396–400). Replication of SV40 in extracts of mammalian cells which had been immunologically depleted of DNA topoisomerase II (303, 401), and replication of chromosomes in a frog-egg extract system lacking the enzyme (402), provided strong additional evidence that DNA topoisomerase II is required in decatenation of DNA rings and in chromosome separation.

In mammalian cells, the division of labor between the two known type II enzymes IIα and IIβ is largely unknown. DNA topoisomerase IIα is probably the major activity in the decatenation of replicative products, based on the cell-cycle dependence of its cellular concentration (see section on Transcriptional Regulation). Expression of an antisense RNA targeting specifically the IIα message indicated that the essential role of IIα during mitosis can not be substituted by endogenous IIβ (T Andoh, personal communicastion).

In some of the in vitro studies on SV40 DNA replication, the requirement of a type II DNA topoisomerase in decatenation, as well as in processing the late Cairns-type structures, was suggested (397–400). More recent in vivo studies suggest, however, that in the absence of a functional type II DNA topoisomerase replication can proceed to catenane formation (374).

Eukaryotic DNA Topoisomerase II and Chromosome Condensation

CHROMOSOME CONDENSATION AND DECONDENSATION   Genetic and bio-chemical evidence shows that eukaryotic DNA topoisomerase II is involved in the final stage of chromosome condensation, and probably in certain phases of decondensation. The final stage of chromosome condensation in *S. pombe* *top2ts* cells was found to be blocked when grown at a restrictive temperature for DNA topoisomerase II (302, 389, 403). Studies of other eukaryotes, in-cluding surf clam (404), Drosophila (390), and mammalian cells (370, 373, 392, 395, 405, 406), led to essentially the same conclusion. Some of the earlier discrepancies between results obtained with thermal sensitive *top2* mutants of *S. pombe* and with mammalian cell lines can be attributed to complications in the mammalian experiments due to the use of inhibitors that trap the covalent topoisomerase II-DNA intermediates. When used at low concentrations (407), this class of inhibitors probably does not block decatenation completely; at high concentrations, they induce DNA damage-mediated G2 delay (408–410). Use of inhibitors that do not trap the covalent intermediates, such as merbarone and the bis-(2, 6-dioxopiperazine) derivative ICRF-193, indicates that mam-malian cells are similar to *S. pombe* in their response to DNA topoisomerase inactivation. Cells were found to traverse the cell cycle with incomplete chro-mosomal condensation and segregation, resulting in polyploidy and cell death (373, 392, 406). Several minor disagreements remain, however. In one study with tsBN$_2$ baby hamster kidney cells, which carry a temperature-sensitive mutation and can be induced to undergo premature chromosome condensation at a nonpermissive temperature, the presence of ICRF-193 was found to inhibit the compaction of 300-nm diameter chromatin fibers to 600-nm diameter chromatids (373). In another study with a rat kangaroo cell line PtK$_2$, however, interference with the late stages of chromosome condensation by the same drug was not observed (393). Similarly, the question of what role eukaryotic DNA topoisomerase II could play in chromosome decondensation is yet to be resolved (404, 411).

The development of in vitro systems has provided a way of biochemically dissecting the various components involved in chromosome assembly (402, 411–414). Reconstitution of nuclear structure around purified DNA in frog-egg extracts showed that nuclei assembly progressed through various stages of condensation to a highly compact state, after which the nuclei became less condensed; DNA topoisomerase II was found to be required in the final phase of compaction and during subsequent decondensation (411). When deconden-sation of demembraned sperm head chromatin in frog-egg extracts was in-hibited by inactivation of DNA topoisomerase II, assembly of the nuclear envelope-lamina around the chromatin continued, and initiation of DNA replication within the condensed chromatin was retarded, but not completely blocked (415). These results also confirm the findings in yeasts that inactivation of DNA topoisomerase II does not inhibit DNA synthesis.

The mechanism by which DNA topoisomerase II affects chromosome condensation and decondensation is unclear. The enzyme is probably required for its catalytic function of transporting one DNA double helix through another, so as to alleviate any topological problem associated with coiling and uncoiling of a long duplex DNA. Alternatively, or in addition, the enzyme could affect directly the condensation-decondensation process. The possibility of a stoichiometric rather than a catalytic role of eukaryotic DNA topoisomerase II in chromosome condensation-decondensation is closely related to the question of whether the enzyme has a structural role in chromatin organization (see below).

CHROMATIN ORGANIZATION   Eukaryotic DNA topoisomerase II was shown to multimerize under certain conditions, and it was reported that this multimerization was stimulated by phosphorylation of the enzyme (416). Purified eukaryotic DNA topoisomerase II was also found to bind cooperatively to DNA fragments containing AT-rich sequences, covering the entire DNA if a sufficient amount of the protein was present (211). These results led to the suggestion that the enzyme might have a stoichiometric role in DNA condensation. Eukaryotic DNA topoisomerase II was identified as a major chromosome scaffold/nuclear matrix protein (417–427). The term "chromosome scaffold" originally referred to an insoluble fraction obtained by extraction of nuclease-treated metaphase chromosomes with 2 M NaCl, dextran sulfate-heparin, or lithium diiodosalicylate. However, because the material so isolated retains the overall chromosome morphology, this term has also assumed a structural implication. Application of the same preparation procedure to isolated nuclei yields material with protein components similar to those of the chromosome scaffolds (see, however, 428, 429), and this material has been termed the nuclear matrix. In expanded metaphase chromosomes, DNA topoisomerase II was found to be concentrated in the axial region (419, 427). These and other findings led to the suggestion that DNA topoisomerase II might serve the role of fastening large chromosomal loops to the chromosome scaffold (420, 430–433). Mapping of the sites of DNA cleavage by topoisomerase II in the presence of appropriate drugs in intact cells, or in salt-extracted nuclear matrix, showed that strong sites, or perhaps clusters of sites, were present with a spacing of 50–300 kb (434–436; reviewed in 52).

On the other hand, in chromosomes assembled in frog-egg extracts DNA topoisomerase II was found to be evenly distributed rather than concentrated in the axial region (437). A large fraction of the metaphase chromosome-associated enzyme was also found to be easily removed without changing the shape of the chromosomes (437). Optical sectioning of living Drosophila embryos injected with rhodamine-labeled DNA topoisomerase II also showed that the enzyme was distributed uniformly throughout the chromosome and not restricted to a central axis (438). Thus either DNA topoisomerase II has no structural role in chromosome organization, or only a subpopulation of the enzyme is involved in this respect. One potentially significant finding was that the central axis of metaphase chromosomes assembled in frog-egg extracts was found to be stained by a monoclonal antibody MPM-2 (437), which had subsequently been shown to recognize a phosphorylated epitope in DNA topoisomerase IIα and several other phosphoproteins including DNA topoisomerase IIβ (439). The presence of AMPPNP, which could be related to the conversion of chromosome-associated DNA topoisomerase II to the closed-clamp form (209), was also found to stabilize metaphase chromosomes (440). Further experiments are needed to clarify the structural role of DNA topoisomerase II in chromosome organization. In addition to biochemical and cytological approaches, uncoupling of the enzymatic and structural function of the enzyme might be feasible through the use of genetic screens for different classes of mutants.

DNA Topoisomerases and Genome Stability

GENOME STABILITY   Genetic and biochemical studies of various DNA topoisomerases suggest that these enzymes are involved in genome stabilization (reviewed in 30, 31). In the budding yeast, inactivation of any one of the three known DNA topoisomerases was found to effect a hyper-recombination (hyper-rec) phenotype. A null $top1$ mutant or a $top2ts$ mutant grown at a semi-permissive temperature was shown to have a much higher frequency of mitotic recombination in the rDNA cluster than its TOP^+ parent (441). In cells of a $\Delta top1$-$top2ts$ yeast double mutant grown at a permissive temperature, destabilization of the rDNA gene cluster was revealed by the presence of over half of the rDNA genes as extrachromosomal rings containing one or more copies of the 9-kb rDNA unit (442). These excised rings were found to integrate back into the rDNA locus, presumably through homologous recombination, when a plasmid-borne $TOP1$ or $TOP2$ gene was expressed (442). Recombination in tandem arrays of repetitive nucleotide sequences other than the rDNA cluster appears to be unaffected by reducing the cellular level of DNA topoisomerase I or II (441). On the other hand, null mutations in $TOP3$ greatly increase recombination between a variety of sequence repeats (70, 443), including subtelomeric repeats (444). In $E.\ coli$, inactivation of DNA topoisomerase III was also found to increase the rates of deletion formation (445).

Several possibilities have been suggested to account for the roles of DNA topoisomerases as suppressors of mitotic recombination. One frequently invoked mechanism is the removal of supercoils, especially negative supercoils. This mechanism is particularly attractive in the case of rDNA excision in the Δtop1-top2ts double mutant. In a heavily transcribed region, the supercoils generated may persist when the combined DNA relaxation activity of the topoisomerases is low, and supercoiling in general and negative supercoiling in particular may stimulate recombination. Explaining the stimulation of recombination in top1, top2 , or top3 single mutants in terms of DNA supercoiling is more difficult. In yeast cells expressing E. coli DNA topoisomerase I, which specifically removes negative supercoils, accumulation of positive supercoils in plasmids was detected only when both DNA topoisomerases I and II were inactivated (107, 442). These results suggest that the presence of either DNA topoisomerase I or II is sufficient to relieve excessive supercoiling. However, bursts of supercoils may not be removed fast enough to suppress processes such as the initiation of recombination (330, 331).

A second explanation of the suppression of mitotic recombination by DNA topoisomerases invokes the existence of processes that eliminate recombinational synapsis in mitotic cells (30, 31, 446–448). Reversing a plectonemic recombinational junction is similar to the separation of interwound parental strands during replication, and the DNA topoisomerases might play a role in such a process (30, 31).

So far these nonexclusive possibilities have not been tested experimentally. In yeast, mutations in the SGS1 gene were found to suppress the top3 hyper-rec phenotype (443). The primary sequence of SGS1 protein suggests that it is a member of a DNA helicase family (443). This protein also appears to be physically associated with DNA topoisomerase III, and perhaps with DNA topoisomerase II as well (see section on Association of DNA Topoisomerases with Other Proteins). This association is reminiscent of reverse gyrase, an enzyme comprised of a helicase-like domain and a subfamily IA-type topoisomerase domain (81). The presence of a DNA topoisomerase III-SGS1 positive supercoiling activity in yeast cells is yet to be confirmed, however. The average linking numbers of intracellular rDNA rings in yeast top1 top2ts cells, under conditions in which both DNA topisomerases I and II were inactivated, appeared to be unaffected by the presence or absence of the SGS1 protein (R Rothstein, personal communication).

SGS1 protein by itself, or in association with a topoisomerase, was also suggested to participate in chromosome segregation. Deletion of the gene in a TOP+ genetic background was reported to increase the levels of chromosome missegregation by more than an order of magnitude during both mitotic and meiotic division (449). Whereas inactivation of SGS1 protein suppresses top3-mediated hyper-recombination, paradoxically sgs1 TOP+ strains themselves were reported to exhibit mitotic hyper-recombination (449). In this connection, it is probably significant that a human gene tied to Bloom syndrome, which is characterized by genome instability and a predisposition to many cancers, was recently found to encode a protein homologous to yeast SGS1 protein (450). To place these findings about DNA topoisomerases and genome instability within a common mechanistic and molecular framework is a challenging task. However, the available information indicates generally that the resolution of replicative intermediates, including both the late-Cairns' structures and catenanes with and without single-stranded gaps, is likely to be closely tied to mitotic recombination (30, 31, 70, 443, 449).

In addition to the hyper-rec phenotypes of yeast DNA topoisomerase mutants, mutations in *SIR2*, a gene known to be involved in yeast mating type switch, were found to elevate the frequency of mitotic recombination in the rDNA cluster (451). Mutations in *HPR1* gene were also shown to greatly increase mitotic recombination of nonribosomal contiguous repeats (452). The *HPR1* gene encodes a protein with limited sequence similarity to DNA topoisomerase I (30, 452), but the protein does not appear to be a topoisomerase (H Klein, personal communication). In combination with a *top1*, *top2*, or *top3* mutation, *hpr1* mutants show a synthetic lethal phenotype (452, 453). Mutations in a yeast gene *TRF4* also exhibit synthetic lethality in combination with inactivation of topoisomerase I (454). The molecular mechanisms underpinning such synthetic lethality is unknown.

TOPOISOMERASE-MEDIATED DNA BREAKAGE AND ITS EFFECT ON DNA RECOMBINATION   Whereas the DNA topoisomerases are important in the normal maintenance of genome stability, they are paradoxically also the cause of genome instability under certain conditions. Illegitimate recombination mediated by type II DNA topoisomerases was detected in vitro (455, 456), and both type I and type II topoisomerases were found to promote recombination in vivo (457–467). Expression of vaccinia virus topoisomerase in *E. coli* was also shown to promote illegitimate recombination (468).

For several yeast *top1* mutants encoding enzymes that appear to form covalent enzyme-DNA complexes more stable than that of the wild-type enzyme, recombination frequency in the rDNA locus is greatly increased (469). Furthermore, the level of a mutant enzyme that can be tolerated by cells appears to correlate inversely with the stability of its covalent complex with DNA (469). Both observations are consistent with an increase in double-stranded DNA breaks in the mutant cells. Similarly, expression of vaccinia virus topoisomerase with a K167D or G226N mutation in *E. coli* was shown to induce genes that are known to be induced in response to DNA damage, presumably owing to the formation of a more stable covalent intermediate between DNA and the mutant enzymes (236). DNA topoisomerase II–mediated recombination was also implicated in chromosomal translocation in the mouse gene encoding the κ-chain of immunoglobulin (462).

Treatment of cells with drugs that stabilize the covalent intermediates between DNA and DNA topoisomerases was found to stimulate chromosomal deletions and rearrangements, including sister-chromatid exchanges (reviewed in 66). For example, treatment of drug-permeable yeast strains with either the topoisomerase I drug camptothecin or the topoisomerase II drug amsacrine was found to induce high levels of both gene conversion and reciprocal exchange (470). Treatment of Chinese hamster cells with the topoisomerase II drug teniposide or amsacrine, but not with the topoisomerase I drug camptothecin, was found to increase greatly the frequency of quadriradial chromosomes, which are formed by reciprocal exchange of duplex DNA between single chromatids of different chromosomes (370). The formation of the quadriradial structures was found to be independent of replication. The topoisomerase II drug teniposide, but not the topoisomerase I drug camptothecin, was found to promote the integration of SV40 DNA into the host genome (471). Integration of adeno-associated virus vectors in either S-phase or non-S-phase human fibroblasts was shown to increase, however, by a low concentration of drugs that stabilize covalent intermediates between DNA and either DNA topoisomerases I or II (472). An understanding of the extents of genetic rearrangements induced by drugs targeting the DNA topoisomerases is of key importance in their medical applications. A high cumulative dose of topoisomerase-II targeting anticancer drugs that stabilize the DNA-enzyme covalent intermediates, for example, appears to increase the risk of secondary malignancies, especially acute myeloid leukemia (reviewed in 66).

The molecular steps from the formation of a covalent DNA-DNA topoisomerase intermediate to recombination are largely unknown. Plausible mechanisms can be divided into two categories, and both may contribute significantly. In the first mechanism, a functional type I topoisomerase or type II topoisomerase protomer is retained at a DNA end. This DNA-linked polypeptide then acts as a transferase to join the attached DNA strand to an acceptor DNA strand with an appropriate end group (reviewed in 5, 153, 260). Illegitimate recombination in *E. coli* expressing vaccinia virus topoisomerase provides an example of recombination mediated directly by a topoisomerase; nucleotide sequences of the recombination junctions suggest that recombination by the viral enzyme involves the binding of enzyme molecules to both recombination partners (473).

For eukaryotic DNA topoisomerase I, cleavage within a single stranded region (reviewed in 5), opposing a nick (251, 259), or at a mismatch in a duplex DNA (262) can yield a broken DNA strand with a 5'-hydroxyl group on one end and a covalently attached topoisomerase on the other end. In addition, when eukaryotic DNA topoisomerase I is trapped on a duplex DNA as a covalent complex, by the presence of camptothecin for example, collision with a replication fork may also generate a recombinogenic structure with a functional enzyme attached to a 3′-phosphoryl terminus (258, 259, 474–477). DNA topoisomerase I–mediated cleavage of a single-stranded region of intracellular DNA undergoing repair may contribute significantly to genome instability. A mammalian cell line deficient in poly(ADP-ribose) polymerase, an enzyme normally induced in cells by DNA damage, was found to be hypersensitive to camptothecin (478; see however, 479). Poly(ADP-ribosylation) is known to inactivate eukaryotic DNA topoisomerase I (see section on Post-translational Modifications). When DNA damage increases the cellular concentration of single-stranded DNA without polyribosylation of topoisomerase I, more recombinogenic DNA ends might be generated by the topoisomerase.

For a type II DNA topoisomerase, the dimeric nature of the enzyme suggests that it could mediate recombination through subunit exchange between two DNA-linked holoenzymes brought into close proximity, or between two DNA-linked halves of two enzyme molecules, each of which had separated from its original partner (reviewed in 480). Although dissociation of dimeric yeast DNA topoisomerase II in solution and reassociation of the monomers were reported to occur freely (481, 482), these findings could not be confirmed in several laboratories (T-S Hsieh & O Westergaard, personal communications; J Roca, W Li, & JC Wang, unpublished data). Abortive DNA-DNA topoisomerase II covalent adducts are more likely to form through interactions of the enzyme with specific DNA structures such as hairpins (260), or between a drug-trapped covalent complex and a DNA helicase tracking along the DNA (483). Replication and transcription might also convert a covalent adduct to a DNA break (475, 484).

The second category of models postulates that a DNA end generated by a topoisomerase is recombinogenic whether a topoisomerase is linked to the end or not. Both free and protein-linked DNA ends may initiate recombination, and the presence of a covalently linked topoisomerase may not be important. The protein moiety might, for example, be removed by a cellular repair system to give a recombinogenic DNA end or a duplex DNA with a single-stranded gap. Protein-free double-stranded DNA breaks were detected in mammalian cells treated with drugs that trap covalent adducts between DNA and DNA topoisomerase II (485). Marking of topoisomerases I and II by ubiquitination for proteolytic action was also implicated in the repair of the covalent intermediates (486).

In yeast, sensitivity to both camptothecin and amsacrine was found to be greatly increased in *rad52* null mutants (470), indicating that recombinational repair, rather than direct strand-transfer by topoisomerase molecules linked to the broken DNA ends, is the major path of salvaging DNA breaks generated by topoisomerases. A similar observation was reported with a mammalian cell line deficient in the repair of DNA double-stranded breaks (485, 487). Cell killing owing to the expression of a mutant yeast or human DNA topoisomerase

I that forms a more stable covalent enzyme-DNA intermediate than the wild-type enzyme was also found to be much more pronouced in yeast *rad52* null mutants than their *RAD52*[+] controls (469, 489). The yeast RAD2 and RAD6 pathways involved in excision repair and in DNA damage-induced mutagenesis, respectively, appear to be unimportant in processing protein-DNA covalent adducts (490).

Other Plausible Cellular Roles of DNA Topoisomerases

In *E. coli*, the effects of DNA topoisomerases on plasmid partition were reported (196, 491–493). Efficient partition of pSC101, for example, appears to depend on negative supercoiling of a region of the plasmid that has been implicated in partition as well as in the initiation of replication (491, 493). Binding of gyrase to a particular site in the 37-kb bacteriophage Mu was reported to have a major effect on its replicative transposition, perhaps by affecting locally the state of supercoiling, or by serving as an organizer for the formation of a larger nucleoprotein complex (195).

Some earlier studies implicating the involvement of DNA topoisomerases in a number of cellular processes require reexamination. A role of topoisomerase II in excision repair in mammalian cells, for example, was extrapolated mainly from experiments in which novobiocin was used as an inhibitor of mammalian DNA topoisomerase II. Novobiocin is not specific, however, and thus the role of topoisomerase II in excision repair is questionable (reviewed in 33). An earlier report on the involvement of mammalian topoisomerase I in DNA repair (494) is also unsupported by more recent work (495). Caution is needed even in experiments employing better characterized and more specific inhibitors such as teniposide. The drug was found, for example, to affect nucleoside uptake into HeLa cells (496) in addition to its well-established function of trapping the covalent complexes between DNA and eukaryotic type II topoisomerases. The possibility that secondary targets of various topoisomerase-targeting drugs are responsible for an observed effect should be entertained whenever independent confirmation is lacking. The suggestion has been made that secondary drug targets might be responsible for the effect of teniposide on the nuclear-to-cytoplasmic ratio of specific RNAs (496) and for the high sensitivity of Tat-mediated HIV-1 transcription to camptothecin (67).

Autoantibodies against mammalian DNA topoisomerase I have been found in sera of scleroderma patients (497–499). Autoantibodies against mammalian DNA topoisomerase II have been found in sera of patients suffering systemic lumpus erythematosus (500, 501), liver cancer (502), and insulin-dependent diabetes mellitus (J Hwang, personal communication). The immune response pathways leading to the presence of these autoantibodies are unknown.

Regulation of the Cellular Levels of DNA Topoisomerases

The involvement of DNA topoisomerases in multiple cellular processes and the importance of the cellular levels of DNA topoisomerases in cancer chemotherapy have led to several studies on the regulation of the cellular levels of these enzymes. Nuclear localization and posttranslational modifications also affect the cellular levels of functional enzymes. These aspects are reviewed below.

TRANSCRIPTIONAL REGULATION   Earlier studies suggest that the cellular levels of DNA topoisomerase I and gyrase in *E. coli* are homeostatically regulated to control the average degree of supercoiling of intracellular DNA: Relaxation of intracellular DNA stimulates *gyrA* and *gyrB* transcription (503–505), and a higher than normal level of negative supercoiling increases *topA* expression (143). Expression of the genes also appears to be modulated by environmental stimuli. Cold shock increases the synthesis of GyrA protein, probably mediated by the binding of a cold shock positive regulator CS7.4 to the promoter (506). The level of GyrB protein, unlike the levels of many other proteins, does not decrease upon heat shock (506). For the *E. coli topA* gene, one of the cluster of four promoters utilizes the σ^{32} factor of heat shock genes (507); transcription from this promoter is elevated upon heat shock to balance the reduction of transcription from the other *topA* promoters utilizing σ^{70}.

In the yeasts *S. cerevisiae* and *S. pombe*, the regulation of intracellular topoisomerase activities appears to be less stringent than in bacteria. Yeast cells can survive without DNA topoisomerase I, III, or both I and III (70, 508). Furthermore, although DNA topoisomerase II is indispensable, variation of the cellular level of this enzyme by several fold has no major effect on growth. The sequence of 5′-flanking region of yeast *TOP1* suggests that it is a constitutive gene (509). Nucleolytic probing of the chromatin structure of the TOP1 promoter revealed two nucleosomes, each with a set of rotationally phased positions, and a nucleosome-free region between them. The putative TATA sequence and the RNA initiation sites are located within the internucleosomal region. This chromatin pattern persists throughout logarithmic and linear growth phases, but disappears when cells enter the advanced stationary phase, coincident with the arrest of *TOP1* transcription (E DiMauro, personal communication).

Analysis of potential regulatory regions of human and mouse *TOP1* gene by nucleotide sequencing and nucleolytic footprinting indicates that most if not all of the regulatory elements are located within a 1-kb region upstream of a cluster of transcriptional start sites (511–516). Mammalian *TOP1* is expressed in all nucleated cells, and a region as short as 250 bp upstream of the start of transcription can drive the expression of a linked reporter gene in HeLa cells (511). This 250-bp stretch exhibits sequence features found in a number of housekeeping genes, such as a high G+C content, a high frequency of CpG dinucleotide, and absence of the TATA and CCAAT box. Nucleolytic foot-printing data support the presence of a variety of regulatory factors within the 250-bp region and beyond, and the majority of putative regulatory elements are conserved in human and mouse (515). Cellular transcription factors that were implicated in the regulation of mammalian *TOP1* included Sp1, OTF, CRE/ATF (511), AP1 (517), MF-IL6 (518), NF-κB, and a Myc-related DNA binding factor (513). The roles of many of these transcription factors in *TOP1* regulation remain to be established. In the case of CRE/ATF, for example, studies of deletion derivatives of the promoter suggested a role of the CRE/ATF site (513), but treatment of human skin fibroblasts with cAMP showed no change in TOP1 transcription level (516). Transcription of *TOP1* in human cells was also found to be elevated several hours after adenovirus-5 infection, although the protein and activity level appeared unchanged (519; see however, 520). This stimulation of transcription appeared to be mediated by the viral E1A protein (519).

For mammalian DNA topoisomerase II, studies in the past decade showed that the level of a type II activity correlates with cell proliferation (421, 521–531) and progressively decreases in cells induced to differentiate (421, 532–538). Following the discovery of two isoforms IIα and IIβ of the mammalian enzyme, the cellular concentration of the former but not the latter was generally found to correlate with cell proliferation (98, 540–546). During rat brain development, for example, IIα mRNA synthesis correlates well with mitotic activity in rapidly growing brain regions, and IIβ but not IIα mRNA was found in differentiated neronal cells (546). During a cell cycle, mammalian IIα enzyme or mRNA increases when cells progress from the G_0/G_1 phase, reaching a maximum in G_2 (98, 428, 516, 547–549). In HeLa cells, the increase in mRNA level at late S/G_2 was reported to result primarily from increased mRNA stability (549). In contrast, neither the protein level nor the cellular activity of the IIβ form in cultured cells is significantly altered during the cell cycle (428, 545, 548).

Nucleotide sequences of the 5'-flanking regions of human (550, 551) and Chinese hamster (552) *TOP2α* genes have been reported, and primer extension experiments showed several transcriptional starts about 90 bp upstream of the initiation codon. Potential regulatory elements of human *TOP2α* were examined by transient expression in HeLa cells of a reporter gene linked to 5' truncated fragments derived from a 2.5-kb DNA segment upstream of the *TOP2α* initiation codon (550). Maximal promoter activity was seen with a fragment extending to nucleotide −562. The nucleotide sequence of this region, similar to the promoter region of *TOP1*, also lacks a TATA box and is rich in CpG dinucleotides. Potential regulatory sites include Sp1 sites at −562 and −51, inverted CCAAT sites at −385, −259, −175, −108, and −68, and an ATF site at −226. It remains to be shown, however, whether sequences downstream of −562, or within the entire 2.5 kb 5′-upstream fragment, are sufficient to respond to various physiological signals, such as proliferation (516), transformation by *ras* (554), and infection by cytomeglovirus (555), in the same way as chromosomal *TOP2α*.

DNA topoisomerase IIβ in a variety of human cells and tissues was found to undergo differential splicing (556). The longer and less abundant IIβ-2 mRNA encodes an extra five amino acids, Thr-Leu-Phe-Asp-Gln after Val-23, which are spliced out in the shorter and more abundant IIβ-1. The deleted segment does not appear to affect enzymatic activity in vitro (557), but whether the presence or absence of the pentapeptide affects subcellular localization, interaction with other cellular entities, or posttranslation modification of the enzyme is unknown.

DECATENATION AND CELL-CYCLE CONTROL   In yeasts, the essentiality of DNA topoisomerase II during a particular stage in the cell-cycle, namely at the time of unlinking newly replicated pairs of chromosomes, qualifies *TOP2* as a cell-cycle gene. Without the action of a type II DNA topoisomerase at this juncture, cells cannot progress further without loss of viability (302, 387, 389). In *top2ts* strains of yeast, progression of the cell-cycle is not stopped by thermal inactivation of the enzyme; cells with improperly decatenated chromosomes nevertheless traverse mitosis.

For mammalian cell lines, several DNA topoisomerase II inhibitors were shown to trigger G2 arrest. In most of the cases, the inhibitors were of the class that trap the covalent topoisomerase-DNA intermediate, and the observed cell-cycle arrest can be attributed to activation of the checkpoint monitoring DNA damage (for examples, see 66, 405, 408–410, 558, 559). When the drug ICRF193 was used, which apparently does not cause DNA damage, progression of mammalian cells from G2 to mitosis was nevertheless found to be delayed by a few hours (374, 392, 394). Furthermore, this delay was not observed if caffeine, a compound known to promote cell-cycle checkpoint evasion, was also present (394). These observations led to the suggestion that there might be a cell-cycle checkpoint for completion of decatenation (394). However, aside from this delay, other mitotic events in mammalian cells, such as activation of cdc2 kinase, spindle apparatus reorganization, and nuclear envelope reassembly, are apparently not blocked, and cells containing catenated chromatids traversed the M-phase without their proper segregation (373, 392, 394). Thus in general the effects of inactivation of DNA topoisomerase II in mammalian cells by ICRF-193 resemble those in yeast *top2ts* strains by thermal inactivation of the enzyme. Whether the G2 to M delay observed in mammalian cells treated with ICRF-193 is solely due to inactivation of DNA topoisomerase II remains to be determined.

NUCLEAR LOCALIZATION   Eukaryotic DNA topoisomerase I possesses a large N-terminal section that can be removed without affecting enzymatic activity in vitro (560, 561). This region is about 140 amino acids in length for *S. cerevisiae* DNA topoisomerase I (181). A nuclear localization signal was identified within this dispensable N-terminal region of eukaryotic DNA topoisomerase I, but not in the cytoplasmically located vaccinia virus topoisomerase (562). In the case of eukaryotic DNA topoisomerase II, one or more nuclear localization signals were found in the C-terminal region dispensable for catalytic activity (106, 563–566), and an additional signal was identified in the N-terminal region of the fission yeast enzyme (564).

The preferential localization of mammalian DNA topoisomerase IIβ in nucleolus was suggested in one study (567), but other reports indicated that both IIα and IIβ were found at the boundary regions between heterochromatin and decondensed chromatin in the nucleoplasm and in the fibrillar zones of the nucleolus (568).

POSTTRANSLATIONAL MODIFICATIONS   Several types of modifications of eukaryotic DNA topoisomerases, including phosphorylation, poly(ADP-ribosylation) and ubiquitination, are known. Glycosylation of mammalian DNA topoisomerase II, probably IIα, was also observed (R Hancock, personal communication).

Mammalian DNA topoisomerase I was shown to be phosphorylated in vitro, predominantly at serines, by casein kinase II and protein kinase C (432, 569–574). Phosphorylation of eukaryotic topoisomerase I was found to increase its relaxation activity and its ability to form enzyme-DNA covalent adducts in the presence of camptothecin (432, 569, 570, 575). Difference in camptothecin sensitivity of two murine lymphoma sublines was attributed to a difference in DNA topoisomerase I phosphorylation (576). Phosphorylation of the enzyme in mammalian cells appears to increase in response to mitogenic stimuli (577–579), and protein kinase C–mediated phosphorylation of mammalian DNA topoisomerase I was suggested to be a link in the propagation of several signal transduction pathways (580, 581).

The level of phosphorylation of eukaryotic DNA topoisomerase II in avian and mammalian cells increases during G_2 and M (547, 582, 583). Analysis of phosphopeptides of the enzyme from interphase and mitotic cells revealed alterations in phosphorylation sites between G_2 and M (584). Phosphorylation of cell-cycle dependent sites was also reported for synchronized HeLa cells (585).

Several kinases were suggested to phosphorylate DNA topoisomerase II. In a *S. cerevisiae* temperature-sensitive casein kinase mutant *cka1 cka2*, no significant phosphorylation of DNA topoisomerase II was seen at a nonpermissive temperature (586). Analysis of tryptic phosphopeptides of the yeast enzyme

^{32}P-labeled in wild-type cells showed, however, that two of the seven major phosphopeptides were different from those derived from the enzyme ^{32}P-labeled in vitro by casein kinase II. The authors suggested that either phosphorylation by casein kinase II was a prerequisite for phosphorylation by a second kinase, or activation of the second kinase by casein kinase II was necessary for its phosphorylation of yeast DNA topoisomerase II (586). Recent studies in HeLa cells implicated the involvement of p34^{cdc2} and the mitogen-activated MAP kinase (585). Five sites of human DNA topoisomerase II phosphorylation in vivo, or by p34^{cdc2} or MAP kinase in vitro, were mapped to Ser-1212, Ser-1246, Ser-1353, Ser-1360, and Ser-1392 (585). Ser-1376 and Ser-1524 were also identified as the phosphorylation sites by casein kinase II in vitro.

The effects of phosphorylation on the activity and biological function of eukaryotic DNA topoisomerase II remain controversial. Numerous reports suggested that phosphorylation of purified eukaryotic DNA topoisomerase II enhanced its catalytic activity by several fold. Phosphorylation of purified Drosophila DNA topoisomerase II by casein kinase II or protein kinase C, for example, was reported to enhance catalytic activity by increasing the rate of ATP hydrolysis (587, 588). Several issues are raised, however, by these results. First, all known sites of phosphorylation are confined to the C-terminal portion of the enzyme dispensable for its catalysis of DNA transport, and how phosphorylation in this region can affect enzyme activity is not apparent. Second, in the case of S. pombe DNA topoisomerase II, dephosphorylation or phosphorylation appears to have no significant effect on its activity in vitro (564). More recent studies of mouse DNA topoisomerase IIα also revealed that treatment of the enzyme with casein kinase or potato acid phosphatase (the latter dephosphorylates the enzyme to near completion) had a negligible effect on its enzymatic activity when precautions were taken to avoid artifacts of enzyme dilution and incomplete removal of phosphatase (T Enomoto, personal communication).

In contrast to the elevation of eukaryotic DNA topoisomerase I activity upon phosphorylation, poly(ADP-ribosylation) of the enzyme inactivates it (reviewed in 589; see also 590–592). Cells deficient in poly(ADP-ribose) polymerase are known to be hypersensitive to camptothecin (478). The down-regulation of the enzyme in mammalian cells following ionized radiation has been postulated to be a consequence of its poly(ADP-ribosylation), as inhibitors of poly(ADP-ribose) transferase prevent this down-regulation (495).

Association of DNA Topoisomerases with Other Proteins

TYPE I DNA TOPOISOMERASES   As mentioned above, genetic and biochemical evidence indicates that yeast DNA topoisomerase III is associated with a protein SGS1 with helicase-like motifs (443). A reverse gyrase purified from the hyperthermophile *Methanopyrus kandleri* was reported to contain two separate subunits, a 50-kDa subunit containing the nucleophile for DNA cleavage and rejoining, and a 150-kDa subunit containing presumably the ATPase activity (593).

Several recent studies indicate that eukaryotic DNA topoisomerase I is probably a part of a nucleoprotein structure involved in the regulation of a number of genes. Purification of factors necessary for activation of RNA-polymerase II–mediated transcription in a reconstituted system led to the identification of human DNA topoisomerase I as a part of the TFIID complex (594, 595). The topoisomerase appeared to augment the response of TATA-contaning promoters to transcription activators and repress basal level transcription in the absence of an activator. A mutant human DNA topoisomerase I, Y723F, in which the active-site tyrosine is replaced by a phenylalanine, was also found to repress basal level transcription and enhance activation of transcription, and thus the relaxation activity of the topoisomerase is apparently dispensable in these effects (594). Recent work has also implicated a direct interaction between eukaryotic DNA topoisomerase I and nucleolin, a major nucleolar protein (596). Association of eukaryotic DNA topoisomerases with casein kinase II–like activity was also reported (575, 597).

TYPE II DNA TOPOISOMERASES   Gyrase is known to interact with two natural toxins. Microcin B17 is a peptide antibiotic produced in strains of *E. coli* bearing one of a group of naturally occuring plasmids. This peptide is first synthesized as a 69–amino acid precursor, which is then extensively processed by plasmid-encoded enzymes to give a 43-residue product with four thiazole and four oxazole rings (22). The *ccdB* (*letD*) gene product of *E. coli* F plasmid is an 11.7-kDa protein that interacts with the GyrA subunit (20, 21). This protein appears to induce covalent complex formation between gyrase and DNA in the presence of ATP (598). Among gyrase mutants that are tolerant to growth inhibition by CcdB protein, the most common mutation was found to map to Arg-462 of GyrA, corresponding to Arg-1121 of yeast DNA topoisomerase II. In the crystal structure of the dimeric yeast enzyme (291), the pair of Arg-1121 residues are located at the "bottom" of the large central hole and project into it, raising the possibility that the CcdB (LetD) toxin might interact with the interior of the hole or the GyrA-GyrA dimer interface in gyrase.

The possibility that bacterial DNA topoisomerase IV might be associated with an essential inner membrane protein ParF was suggested (90). The *parF* gene encoding the protein appears to be a part of the *parC* operon, and two *S. typhimurium parF* mutants were reported to exhibit partition defects similar to those displayed by *parC* and *parE* mutants (90). In *E. coli*, however, no partition defect was seen in *parF* mutants, and the ParF protein is believed to be an acyltransferase (599). The finding that mutations in the *E. coli minB* gene affect both chromosomal partitioning and plasmid supercoiling also raised the possibility that MinB protein might interact with a DNA topoisomerase (600).

Association of eukaryotic DNA topoisomerase II with other proteins was reported in several studies. Association between silkworm DNA topoisomerase II and a 50-kDa Ca(II)-binding protein was proposed to give a complex with limited negative supercoiling activity if present in molar excess over DNA (601, 602). Plausible association between yeast DNA topoisomerase II and SGS1 protein (451) was mentioned earlier. The identification of an extragenic suppressor *tos1* that suppresses a particular yeast *top2* mutation also raised the possibility that yeast DNA topoisomerase II might interact with the product of *TOS1* (603). Eukaryotic DNA topoisomerase II from several sources was reported to be associated with a protein kinase (604–606). Association between DNA topoisomerase II and casein kinase II, with both enzymes retaining their cataytic activities, was found to require the regulatory β subunit but not the catalytic α subunit of the kinase (606).

DNA topoisomerase II, probably in association with other proteins, was found to form a tight complex with a sequence element in the promoter of an immunoglobulin μ heavy chain gene. This element appears to be required for the induction of the promoter by interleukin-5 and antigen (607). DNA topoisomerase II and a protein termed chromosome scaffold protein 2 (ScII) were also identified in a nuclear protein factor UB2, which was detected in undifferentiated mouse erythroleukemia and other proliferating cell lines (608). UB2 was found to recognize a specific sequence element present in a number of vertebrate genes, although it does not appear to be involved in the regulation of these genes (609). ScII belongs to a family of proteins including the SMC1 protein and the *cut3* and *cut14* gene products of the budding yeast (429). Although the distribution of ScII and DNA topoisomerase II is nearly identical in cytological spreads of mitotic chromosomes, only the latter is present in interphase nuclear matrix (429).

CONCLUDING REMARKS

The DNA topoisomerases presumably co-evolved with the formation of very long and/or ring-shaped DNA molecules. To solve a variety of problems that are rooted in the double-helix structure of DNA, nature has created not one but three distinct enzymes. In eukaryotes, members of all three subfamilies of DNA topoisomerases have been found in the same cells; in bacteria, four members from two subfamilies participate in nearly all cellular transactions of DNA. The past decade saw much progress in the study of the DNA topoisomerases, but many questions remain.

In terms of the mechanisms of their actions, a molecular sketch has emerged on how a single type II enzyme molecule manages the transport of one DNA double helix through another. The large conformational changes in the enzyme-DNA complex during the various reaction steps are, however, yet to be described in atomic detail. For the type I enzymes, how they maneuver DNA strands remains mysterious despite progress in biochemical dissections of the enzymes and in the solution of the three-dimensional structures of several fragments. With the establishment of the topoisomerases as targets of many antimicrobial and anticancer drugs, structural and mechanistic studies of the topoisomerases have assumed a new urgency: Results of what used to be intellectual exercises are now eagerly awaited in the pharmacological laboratories and clinical wards. Each of these remarkable enzymes is also clearly several proteins in one, owing to their large conformation changes during catalysis; each topoisomerase therefore offers not one but several potential targets in drug design.

In terms of their biological roles, how these enzymes coordinate their actions in some cellular processes, and divide up the chores in others, is not completely understood. Eukaryotic DNA topoisomerase III, for example, is an enigma in terms of its strong effects on growth and sporulation in the presence of two other topoisomerases. The molecular processes underlying the hyper-recombination phenotype of any of the topoisomerases need to be unveiled.

The key to answering many of the questions may lie in the elucidation of interactions between the DNA topoisomerases and other cellular proteins. Complexes between these enzymes and transcription factors and chromosomal proteins illustrate new avenues yet to be fully explored. Furthermore, whereas the information available on topoisomerase-DNA interactions is substantial, that on interactions in the context of chromatin is scarce; whether eukaryotic DNA topoisomerase II has a structural role in the organization of interphase and/or metaphase chromosomes, for example, is yet to be settled.

Studies of DNA-DNA topoisomerase covalent complexes trapped by various drugs show that such a complex can lead to double-stranded DNA breakage. A DNA-bound topoisomerase is thus the Achilles' heel of intracellular DNA, and it is plausible that in the absence of such drugs a topoisomerase might become trapped as a covalent complex under certain conditions, for example when the enzyme is damaged. Studies with partially single-stranded DNA substrates have also shown that such a DNA might be broken by a topoisomerase when the enzyme-linked and the enzyme-free end of the severed DNA come apart. There is presumably a cellular repair system for the removal of covalently linked proteins, and the identification of such systems should be of much interest both in terms of the biological fuctions of the DNA topoisomerases and in terms of drug development. These and many other questions are sure to keep the topoisomerase field an active and productive one in future years.

ACKNOWLEDGMENTS

I thank those who kindly sent me manuscripts before publication. Without their help it would have been even more impossible to give an up-to-date account of an active field. I thank all my present and former associates and collaborators for making the study of DNA topoisomerases a most stimulating and enjoyable one. The three decades of continued support by NIH in the study of DNA topology and topoisomerases in my laboratory is gratefully acknowledged.

Any *Annual Review* chapter, as well as any article cited in an *Annual Review* chapter, may be purchased from the Annual Reviews Preprints and Reprints service.
1-800-347-8007; 415-259-5017; email: arpr@class.org

Literature Cited

1. Wang JC, Liu LF. 1979. In *Molecular Genetics,* ed. JH Taylor, pp. 65–88. New York: Academic
2. Wang JC. 1971. *J. Mol. Biol.* 55:523–33
3. Champoux JJ, Dulbecco R. 1972. *Proc. Natl. Acad. Sci. USA* 69:143–46
4. Gellert M, Mizuuchi K, O'Dea MH, Nash HA. 1976. *Proc. Natl. Acad. Sci. USA* 73:3872–76
5. Champoux JJ. 1990. See Ref. 55, pp. 217–42
6. Brown PO, Cozzarelli NR. 1979. *Science* 206:1081–83
7. Liu LF, Liu CC, Alberts BM. 1980. *Cell* 19:697–707
8. Kreuzer KN, Cozzarelli NR. 1980. *Cell* 20:245–54
9. Lynn R, Giaever G, Swanberg SL, Wang JC. 1986. *Science* 233:647–49
10. Uemura T, Morikawa K, Yanagida M. 1986. *EMBO J.* 5:2355–61
11. Huang WM. 1986. *Nucleic Acids Res.* 14:7379–90
12. Huang WM. 1986. *Nucleic Acids Res.* 14:7751–65
13. Wyckoff E, Natalie D, Nolan JM, Lee M, Hsieh T-S. 1989. *J. Mol. Biol.* 205:1–14
14. Huang YM. 1994. See Ref. 59, pp. 201–25
15. Caron PR, Wang JC. 1994. See Ref. 60, pp. 271–97
16. Wang JC. 1985. *Annu. Rev. Biochem.* 54:655–97
17. Gellert M, O'Dea MH, Itoh T, Tomi-

zawa JI. 1976. *Proc. Natl. Acad. Sci. USA* 73:4474–78
18. Gellert M, Mizuuchi K, O'Dea MH, Itoh T, Tomizawa JI. 1977. *Proc. Natl. Acad. Sci. USA* 74:4772–76
19. Sugino A, Peebles CL, Kreuzer KN, Cozzarelli NR. 1977. *Proc. Natl. Acad. Sci. USA* 74:4767–71
20. Maki S, Takiguchi S, Miki T, Horiuchi T. 1992. *J. Biol. Chem.* 267:12244–51
21. Bernard P, Couturier M. 1992. *J. Mol. Biol.* 226:735–45
22. Yorgey P, Lee J, Kordel J, Vivas E, Warner P, et al. 1994. *Proc. Natl. Acad. Sci. USA* 91:4519–23
23. Nelson EM, Tewey KM, Liu LF. 1984. *Proc. Natl. Acad. Sci. USA* 81:1361–65
24. Tewey KM, Chen GL, Nelson EM, Liu LF. 1984. *J. Biol. Chem.* 259:9182–87
25. Tewey KM, Rowe TC, Yang L, Halligan BD, Liu LF. 1984. *Science* 226:466–68
26. Chen GL, Yang L, Rowe TC, Halligan BD, Tewey KM, Liu LF. 1984. *J. Biol. Chem.* 259:13560–66
27. Hsiang Y, Hertzberg R, Hecht S, Liu LF. 1985. *J. Biol. Chem.* 260:14873–78
28. Maxwell A, Gellert M. 1986. *Adv. Protein Chem.* 38:69–107
29. Wang JC. 1987. *Biochim. Biophys. Acta* 909:1–9
30. Wang JC, Caron PR, Kim RA. 1990. *Cell* 62:403–6
31. Wang JC. 1991. *J. Biol. Chem.* 266:6659–62

32. Yanagida M, Wang JC. 1987. In *Nucleic Acids and Molecular Biology,* ed. F Eckstein, DMJ Lilley, 1:196–209. Berlin: Springer-Verlag

33. Downes CS, Johnson RT. 1988. *BioEssays* 8:179–84

34. Sternglanz R. 1989. *Curr. Opin. Cell Biol.* 1:533–35

35. Reese RJ, Maxwell A. 1991. *Crit. Rev. Biochem. Mol. Biol.* 26:335–75

36. Fink GR. 1990. *Cell* 58:225–26

37. Austin CA, Fisher LM. 1990. *FEBS Lett.* 266:115–17

38. Hsieh T. 1990. *Curr. Opin. Cell Biol.* 2:461–63

39. Osheroff N, Zechiedrich EL, Gale KC. 1991. *BioEssays* 13:269–75

40. Drlica K. 1992. *Mol. Microbiol.* 6:425–33

41. Bjornsti M-A. 1991. *Curr. Opin. Struct. Biol.* 1:99–103

42. Schmid MB, Sawitzke JA. 1993. *BioEssays* 15:445–49

43. Holm C. 1994. *Cell* 77:955–57

44. Watt PM, Hickson ID. 1994. *Biochem. J.* 303:681–89

45. Orphanides G, Maxwell A. 1994. *Curr. Biol.* 4:1006–9

46. Luttinger A. 1994. *Mol. Microbiol.* 15:601–6

47. Lima C, Mondragon A. 1994. *Structure* 2:559–60

48. Roca J. 1995. *Trends Biochem. Sci.* 20:156–60

49. Duguet M. 1995. See Ref. 610, pp. 84–114

50. Gupta M, Fujimori A, Pommier Y. 1995. *Biochim. Biophys. Acta* 1262:1–14

51. Ullsperger CJ, Vologodskii AV, Cozzarelli NR. 1995. See Ref. 610, pp. 115–42

52. Poljak L, Kas E. 1995. *Trends Cell Biol.* 5:348–54

53. Froelich-Ammon SJ, Osheroff N. 1995. *J. Biol. Chem.* 21429–32

54. Wigley D. 1995. *Annu. Rev. Biophys. Biomol. Struct.* 24:185–208

55. Cozzarelli NR, Wang JC. 1990. *DNA Topology and its Biological Effects.* Cold Spring Harbor, NY: Cold Spring Harbor Lab.

56. Bates AD, Maxwell A. 1993. *DNA Topology.* Oxford: IRL Press

57. Potmesil M, Ross WE. 1987. *1st Conf. on DNA Topoisomerases in Cancer Chemotherapy. NCI Monogr. 4.* Bethesda: Natl. Cancer Inst.

58. Andoh T, Ikeda H, Oguro M, eds. 1992. *Molecular Biology of DNA Topoisomerases and Its Application to Chemotherapy.* Boca Raton: CRC Press

59. Liu LF. 1994. *DNA Topoisomerases: Biochemistry and Molecular Biology.* Boca Raton: Academic

60. Liu LF. 1994. *DNA Topoisomerases and Their Applications in Pharmacology.* Boca Raton: Academic

61. Liu LF. 1989. *Annu. Rev. Biochem.* 58:351–75

62. Potmesil M, Kohn KW, eds. 1991. *DNA Topoisomerases in Cancer.* New York: Oxford Univ.

63. Capranico G, Zunino F. 1992. *Eur. J. Cancer* 28A:2055–60

64. Corbett AH, Osheroff N. 1993. *Chem. Res. Toxicol.* 6:585–97

65. Chen A, Liu LF. 1994. *Annu. Rev. Pharmacol. Toxicol.* 34:191–218

66. Anderson RD, Berger NA. 1994. *Mutat. Res.* 309:109–42

67. Li CJ, Dezube BJ, Biswas DK, Ahlers CM, Pardee AB. 1994. *Trends Microbiol.* 2:164–69

68. Wang JC. 1987. See Ref. 57, pp. 3–6

69. Slesarev AI, Lake JA, Stetter KO, Gellert M, Kozyavkin SA. 1994. *J. Biol. Chem.* 269:3295–303

70. Wallis JW, Chrebet G, Brodsky G, Rolfe M, Rothstein R. 1989. *Cell* 58:409–19

71. Kim RA, Wang JC. 1992. *J. Biol. Chem.* 267:17178–85

72. Hanai R, Caron PR, Wang JC. 1996. *Proc. Natl. Acad. Sci. USA.* In press

73. Fouet A, Sirard J-C, Mock M. 1994. *Mol. Microbiol.* 11:471–79

74. Kikuchi A, Asai K. 1984. *Nature* 309:677–81

75. Forterre P, Mirambeau G, Jaxel C, Nadal M, Duguet M. 1985. *EMBO J.* 4:2123–28

76. Collin RG, Morgan HW, Musgrave DR, Daniel RM. 1988. *FEMS Microbiol. Lett.* 55:235–39

77. Bouthier de la Tour C, Portemer C, Nadal M, Stetter KO, Forterre P, et al. 1990. *J. Bacteriol.* 172:6803–8

78. Bouthier de la Tour C, Portemer C, Huber R, Forterre P, Duguet M. 1991. *J. Bacteriol.* 173:3921–23

79. Slesarev AI, Kozyavkin SA. 1990. *J. Biomol. Struct. Dyn.* 7:935–42

80. Forterre P, Confalonieri F, Charbonnier F, Duguet M. 1995. *Orig. Life Evol. Biosph.* 25:235–49

81. Confalonieri F, Elie C, Nadal M, Bouthier de la Tour C, Forterre P, et al. 1993. *Proc. Natl. Acad. Sci. USA* 90:4753–57

82. Kovalsky OI, Kozyavkin SA, Slesarev AI. 1990. *Nucleic Acids Res.* 18:2801–5

83. Slesarev AI. 1988. *Eur. J. Biochem.* 173:395–99

84. Forterre P, Elie C, Sioud M, Hamal A. 1989. *Can. J. Microbiol.* 35:228–33

85. Slesarev AI, Zaitzev DA, Kopylov VM,

Stetter KO, Kozyavkin SA. 1991. *J. Biol. Chem.* 266:12321–28

86. Slesarev AI, Stetter KO, Lake JA, Gellert M, Krah R, et al. 1993. *Nature* 364:735–37

87. Kato J, Nishimura Y, Imamura R, Niki H, Hiraga S, et al. 1990. *Cell* 63:393–404

88. Kato J, Suzuki H, Ikeda H. 1992. *J. Biol. Chem.* 267:25676–84

89. Peng H, Marians KJ. 1993. *Proc. Natl. Acad. Sci. USA* 90:8571–75

90. Luttinger AL, Springer AL, Schmid MB. 1991. *New Biol.* 3:687–97

91. Adams DE, Shekhtman EM, Zechiedrich EL, Schmid MB, Cozzarelli NR. 1992. *Cell* 71:277–88

92. Ferrero L, Cameron B, Manse B, Langneaux D, Crouzet J, et al. 1994. *Mol. Microbiol.* 13:641–53

93. Deleted in proof

94. Fraser CM, Gocayne JD, White O, Adams MD, Clayton RA, et al. 1995. *Science* 270:397–403

95. Bergerat A, Gadelle D, Forterre P. 1994. *J. Biol. Chem.* 269:27663–69

96. Huang WM. 1992. See Ref. 58, pp. 39–48

97. Drake FH, Zimmerman JP, McCabe FL, Bartus HF, Per SR, et al. 1987. *J. Biol. Chem.* 262:16739–47

98. Drake FH, Hofmann GA, Bartus HF, Mattern MR, Crooke ST, et al. 1989. *Biochemistry* 28:8154–60

99. Tsai-Pflugfelder M, Liu LF, Liu AA, Tewey KM, Whang-Pheng J, et al. 1988. *Proc. Natl. Acad. Sci. USA* 85: 7177–81

100. Chung TDY, Drake FH, Tan KB, Per SR, Crooke ST, et al. 1989. *Proc. Natl. Acad. Sci. USA* 86:9431–35

101. Austin CA, Fisher LM. 1990. *FEBS Lett.* 266:115–17

102. Jenkins JR, Ayton P, Jones T, Davies SL, Simmons DL, et al. 1992. *Nucleic Acids Res.* 20:5587–92

103. Austin CA, Sng J-H, Patel S, Fisher LM. 1993. *Biochim. Biophys. Acta* 1172:283–91

104. Tan KB, Dorman TE, Falls KM, Chung TDY, Mirabelli CK, et al. 1992. *Cancer Res.* 52:231–34

105. Brill SJ, DiNardo S, Voelkel-Meiman K, Sternglanz R. 1987. *Nature* 326:414–16

106. Caron PR, Watt P, Wang JC. 1994. *Mol. Cell. Biol.* 14:3197–207

107. Giaever GN, Wang JC. 1988. *Cell* 55:849–56

108. Garcia-Beato R, Freje JM, Lopez-Otin C, Blasco R, Vinuela E, et al. 1992. *Virology* 188:938–47

109. Priel E, Showater SD, Roberts M,

Oroszlan S, Segal S, et al. 1990. *EMBO J.* 9:4167–72

110. Fleischmann RD, Adams MD, White O, Clayton RA, Kirkness EF, et al. 1994. *Science* 269:496–512

111. Craig NL. 1988. *Annu. Rev. Genet.* 22: 77–105

112. Cox M. 1989. In *Mobile DNA*, ed. DE Berg, MM Howe, pp. 661–70. Washington, DC: Am. Soc. Microbiol.

113. Matsuo K, Silke J, Gramatikoff K, Schaffner W. 1994. *Nucleic Acids Res.* 22:5354–59

114. Kiwon J, Topal MD. 1995. *Science* 267:1817–21

115. Simpson L. 1986. *Int. Rev. Cytol.* 99:119–79

116. Englund PT, Hajduk SL, Marini JC. 1982. *Annu. Rev. Biochem.* 51:695–726

117. Melendy T, Sheline C, Ray DS. 1988. *Cell* 55:1083–88

118. Ferguson M, Torri AF, Ward D, Englund P. 1992. *Cell* 70:621–29

119. Shapiro TA, Klein VA, Englund PT. 1989. *J. Biol. Chem.* 264:4173–78

120. Ray DS, Hines JC, Anderson M. 1992. *Nucleic Acids Res.* 20:3353–56

121. Strauss PR, Wang JC. 1990. *Mol. Biochem. Parasitol.* 38:141–50

122. Pasion SG, Hines JC, Aebersold R, Ray DS. 1992. *Mol. Biochem. Parasitol.* 50:57–69

123. Dyer TA. 1984. In *Chloroplast Biogenesis*, ed. NR Baker, J Barber, pp. 23–69. Amsterdam: Elsevier

124. Siedlecki J, Zimmermann W, Weissbach A. 1983. *Nucleic Acids Res.* 11:1523–26

125. Castora FJ, Lazarus M, Kunes D. 1985. *Biochem. Biophys. Res. Commun.* 130:854–66

126. Lam E, Chua N-H. 1987. *Plant Mol. Biol.* 8:415–24

127. Nielson BL, Tewari KK. 1988. *Plant Mol. Biol.* 11:3–14

128. Pyke KA, Marrison J, Leech RM. 1989. *FEBS Lett.* 242:305–8

129. Fukata H, Mochida A, Maruyama N, Fukasawa H. 1991. *J. Biochem.* 109:127–31

130. Marrison JL, Leech RM. 1992. *Plant J.* 2:783–90

131. Obernauerova M, Subik J, Ebringer L. 1992. *Curr. Genet.* 21:443–46

132. Ebringer L, Polanyi J, Krajcovic J. 1993. *Arzneim-Forsch./Drug Res.* 43:777–81

133. Mukherjee SK, Reddy MK, Kumar D, Tewari KK. 1994. *J. Biol. Chem.* 269:3793–801

134. Kirkegaard K, Pflugfelder G, Wang JC. 1984. *Cold Spring Harbor Symp. Quan. Biol.* 49:411–19

135. Kirkegaard K, Wang JC. 1985. *J. Mol. Biol.* 185:625–37
136. Tse-Dinh Y-C, McCarron BGH, Arentzen R, Chowdhry V. 1983. *Nucleic Acids Res.* 11:8691–701
137. DiGate RJ, Marians KJ. 1992. *J. Biol. Chem.* 267:20532–35
138. Zhang HL, Malpure S, Digate RJ. 1995. *J. Biol. Chem.* 270:23700–5
139. Dean FB, Krasnow MA, Otter R, Matzuk MM, Spengler SJ, et al. 1983. *Cold Spring Harbor Symp. Quant. Biol.* 47:769–78
140. Srivenugopal KS, Lockshon D, Morris DR. 1984. *Biochemistry* 23:1899–906
141. DiGate RJ, Marians KJ. 1988. *J. Biol. Chem.* 263:13366–73
142. Zhu C-X, Samuel M, Pound A, Ahumada A, Tse-Dinh Y-C. 1995. *Biochem. Mol. Biol. Int.* 35:375–85
143. Tse-Dinh Y-C, Beran RK. 1988. *J. Mol. Biol.* 202:735–42
144. Zhu C-X, Tse-Dinh Y-C. 1994. *Biochem. Mol. Biol. Int.* 33:195–204
145. Zumstein L, Wang JC. 1986. *J. Mol. Biol.* 191:333–40
146. Tse-Dinh Y-C. 1991. *J. Biol. Chem.* 266:14317–20
147. Lima CD, Wang JC, Mondragon A. 1993. *J. Mol. Biol.* 232:1213–16
148. Samuel M, Zhu C-X, Villanueva GB, Tse-Dinh Y-C. 1993. *Arch. Biochem. Biophys.* 300:302–8
149. Deleted in proof
150. Zhang HL, DiGate RJ. 1994. *J. Biol. Chem.* 269:9052–59
151. Been MD, Champoux JJ. 1984. *J. Mol. Biol.* 180:515–31
152. Trask DK, Muller MT. 1983. *Nucleic Acids Res.* 11:2779–800
153. Andersen AH, Svejstrup JQ, Westergaard O. 1994. See Ref. 59, pp. 83–101
154. Tanizawa A, Kohn KW, Pommier Y. 1993. *Nucleic Acids Res.* 21:5157–66
155. Shen C, Shen CKJ. 1990. *J. Mol. Biol.* 212:67–78
156. Camilloni G, Caserta M, Amadei A, Di Mauro E. 1991. *Biochim. Biophys. Acta* 1129:73–82
157. Muller MT. 1985. *Biochim. Biophys. Acta* 824:263–67
158. Camilloni G, Di Martino E, Caserta M, Di Mauro E. 1988. *Nucleic Acids Res.* 16:7071–85
159. Camilloni G, Di Martino E, Di Mauro E, Caserta M. 1989. *Proc. Natl. Acad. Sci. USA* 86:3080–84
160. Caserta M, Amadei A, Camilloni G, Di Mauro E. 1990. *Biochemistry* 29:8152–57
161. Depew RE, Wang JC. 1975. *Proc. Natl. Acad. Sci. USA* 72:4275–79
162. Pulleyblank DE, Shure M, Tang D, Vinograd J, Vosberg H-P. 1975. *Proc. Natl. Acad. Sci. USA* 72:4280–84
163. Krogh S, Mortensen UH, Westergaard O, Bonven BJ. 1991. *Nucleic Acids Res.* 19:1235–41
164. Zechiedrich EL, Osheroff N. 1990. *EMBO J.* 9:4555–62
165. Madden KR, Stewart L, Champoux JJ. 1995. *EMBO J.* 14:5391–409
166. Thomsen B, Mollerup S, Bonven BJ, Frank R, Blocker H, et al. 1987. *EMBO J.* 6:1817–23
167. Busk H, Thomsen B, Bonven BJ, Kjeldsen E, Nielsen OF, Westergaard O. 1987. *Nature* 327:638–40
168. Andersen AH, Gocke E, Bonven BJ, Nielsen OF, Westergaard O. 1985. *Nucleic Acids Res.* 13:1543–57
169. Stevnsner T, Mortensen UH, Westergaard O, Bonven BJ. 1989. *J. Biol. Chem.* 264:10110–13
170. Jaxel C, Capranico G, Kerrigan D, Kohn KW, Pommier Y. 1991. *J. Biol. Chem.* 266:20418–23
171. Svejstrup JQ, Christiansen K, Andersen AH, Lund K, Westergaard O. 1990. *J. Biol. Chem.* 265:12529–35
172. Christiansen K, Svejstrup ABD, Andersen AH, Westergaard O. 1993. *J. Biol. Chem.* 268:9690–701
173. Edwards KA, Halligan BD, Davis JL, Nivera NL, Liu LF. 1982. *Nucleic Acids Res.* 10:2565–76
174. Been MD, Burgess RR, Champoux JJ. 1984. *Nucleic Acids Res.* 12:3097–114
175. Christiansen K, Westergaard O. 1994. *J. Biol. Chem.* 269:721–29
176. Leteurtre F, Kohlhagen G, Fesen MR, Tanizawa A, Kohn KW, Pommier Y. 1994. *J. Biol. Chem.* 269:7893–900
177. Sekiguchi J, Shuman S. 1994. *J. Biol. Chem.* 269:29760–64
178. Shuman S, Prescott J. 1990. *J. Biol. Chem.* 265:17826–36
179. Lue NF, Sharma A, Mondragon A, Wang JC. 1995. *Structure* 3:1315–22
180. Hanai R, Wang JC. 1994. *Proc. Natl. Acad. Sci. USA* 91:11904–8
181. Bjornsti M-A, Wang JC. 1987. *Proc. Natl. Acad. Sci. USA* 84:8971–75
182. Sharma A, Hanai R, Mondragon A. 1994. *Structure* 2:767–77
183. Sekiguchi J, Shuman S. 1995. *J. Biol. Chem.* 270:11636–45
184. Morham SG, Shuman S. 1992. *J. Biol. Chem.* 267:15984–92
185. Lee MP, Sander M, Hsieh T-S. 1989. *J. Biol. Chem.* 264:21779–87
186. Freudenreich CH, Kreuzer KN. 1993. *EMBO J.* 12:2085–97
187. Peng H, Marians KJ. 1995. *J. Biol. Chem.* 270:25286–90

188. Spitzner JR, Chung IK, Muller MT. 1990. *Nucleic Acids Res.* 18:1–11
189. Liu LF, Wang JC. 1978. *Cell* 15:697–707
190. Morrison A, Cozzarelli NR. 1979. *Cell* 17:175–84
191. Lockshon D, Morris DR. 1985. *J. Mol. Biol.* 181:63–74
192. Fisher LM, Barot HA, Cullen ME. 1986. *EMBO J.* 5:1411–18
193. Franco R, Drlica K. 1988. *J. Mol. Biol.* 201:229–33
194. Condemine G, Smith CL. 1990. *Nucleic Acids Res.* 18:7389–97
195. Pato ML, Howe MM, Higgins NP. 1990. *Proc. Natl. Acad. Sci. USA* 87:8716–20
196. Miller CA, Beaucage SL, Cohen SN. 1990. *Cell* 62:127–33
197. Orphanides G, Maxwell A. 1994. *Nucleic Acids Res.* 22:1567–75
198. Krueger S, Zaccai G, Wlodawer A, Langowski J, O'Dea M, et al. 1990. *J. Mol. Biol.* 211:211–20
199. Reece RJ, Maxwell A. 1989. *J. Biol. Chem.* 264:19648–53
200. Reece RJ, Maxwell A. 1991. *J. Biol. Chem.* 266:3540–46
201. Howard MT, Lee MP, Hsieh T-S, Griffith JD. 1991. *J. Mol. Biol.* 217:53–62
202. Bechert T, Diekmann S, Arndt-Jovin DJ. 1994. *J. Biomol. Struct. Dyn.* 123:605–23
203. Zechiedrich EL, Christiansen K, Andersen AH, Westergaard O, Osheroff N. 1989. *Biochemistry* 28:6229–36
204. Howard MT, Griffith JD. 1991. *J. Mol. Biol.* 232:1060–68
205. Corbett AH, Zechiedrich EL, Osheroff N. 1992. *J. Biol. Chem.* 267:683–86
206. Roca J, Berger JM, Wang JC. 1993. *J. Biol. Chem.* 268:14250–55
207. Osheroff N, Brutlag D. 1983. In *Mechanism of DNA Replication and Recombination,* ed. NR Cozzarelli, pp. 55–64. New York: Liss
208. Pommier Y, Kerrigan D, Kohn K. 1989. *Biochemistry* 28:995–1002
209. Roca J, Wang JC. 1992. *Cell* 71:833–40
210. Chen Y-D, Maxwell A, Westerhoff HV. 1986. *J. Mol. Biol.* 190:201–14
211. Adachi T, Kas E, Laemmli UK. 1989. *EMBO J.* 8:3997–4006
212. Roca J, Wang JC. 1994. *Cell* 77:609–16
213. Roca J, Wang JC. 1996. *Genes Cells* 1:In press
214. Higgins NP, Cozzarelli NR. 1982. *Nucleic Acids Res.* 10:6833–47
215. Wang JC. 1979. *Cold Spring Harbor Symp. Quant. Biol.* 43:29–33
216. Osheroff N. 1986. *J. Biol. Chem.* 261:9944–50
217. Lindsley JE, Wang JC. 1993. *Nature* 361:749–50
218. Paillard S, Strauss F. 1991. *Nucleic Acids Res.* 19:5619–24
219. Alberts B. 1987. *Philos. Trans. R. Soc. London Ser. B* 317:395–420
220. Kong X-P, Onrust R, O'Donnell M, Kuriyan J. 1992. *Cell* 69:425–38
221. Kristna TSR, Kong X-P, Gary S, Bergers P, Kuriyan J. 1994. *Cell* 79:1233–44
222. Arndt-Jovin DJ, Udvardy A, Garner MM, Ritter S, Jovin TM. 1993. *Biochemistry* 32:4862–72
223. Chung IK, Muller MT. 1991. *J. Biol. Chem.* 266:9508–14
224. Chung IK, Mehta VB, Spitzner JR, Muller MT. 1992. *Nucleic Acids Res.* 20:1973–77
225. Tse Y-C, Kirkegaard K, Wang JC. 1980. *J. Biol. Chem.* 255:5560–65
226. Champoux JJ. 1981. *J. Biol. Chem.* 256:4805–9
227. Horowitz D, Wang JC. 1987. *J. Biol. Chem.* 262:5339–44
228. Lynn RM, Bjornsti M-A, Caron PR, Wang JC. 1989. *Proc. Natl. Acad. Sci. USA* 86:3359–63
229. Lynn R, Wang JC. 1989. *Proteins* 6:231–39
230. Worland S, Wang JC. 1989. *J. Biol. Chem.* 264:4412–16
231. Shuman S, Kane EM, Morham SG. 1989. *Proc. Natl. Acad. Sci. USA* 86:9793–97
232. Wilkinson A, Wang JC. 1990. In *Structure and Function of Nucleic Acids and Proteins,* ed. FY-H Wu, C-W Wu, pp. 61–75. New York: Raven
233. Eng W-K, Pandit SD, Sternglanz R. 1989. *J. Biol. Chem.* 264:13373–76
234. Stivers JT, Shuman S, Mildvan AS. 1994. *Biochemistry* 33:327–39
235. Stivers JT, Shuman S, Mildvan AS. 1994. *Biochemistry* 33:15449–58
236. Gupta M, Zhu C-X, Tse-Dinh Y-C. 1994. *J. Biol. Chem.* 269:573–78
237. Hsieh T-S. 1990. See Ref. 55, pp. 243–63
238. Tse-Dinh Y-C. 1994. See Ref. 59, pp. 21–37
239. Tse-Dinh Y-C, Wang JC. 1986. *J. Mol. Biol.* 191:321–31
240. Domanico PL, Tse-Dinh Y-C. 1988. *Biochemistry* 27:6365–71
241. Domanico PL, Tse-Dinh Y-C. 1988. *J. Inorg. Biochem.* 42:87–96
242. Christiansen K, Knudsen BR, Westergaard O. 1994. *J. Biol. Chem.* 269:11367–73
243. Lima CD, Wang JC, Mondragon A. 1994. *Nature* 367:138–46
244. Been MD, Champoux JJ. 1980. *Nucleic Acids Res.* 8:6129–42

245. Been MD, Champoux JJ. 1981. *Proc. Natl. Acad. Sci. USA* 78:2883–87

246. Prell B, Vosberg H-P. 1980. *Eur. J. Biochem.* 108:389–98

247. Halligan BD, Davis JL, Edwards KA, Liu LF. 1982. *J. Biol. Chem.* 257:3995–4000

248. Been MD, Burgess RR, Champoux JJ. 1984. *Nucleic Acids Res.* 12:3097–14

249. Been MD, Burgess RR, Champoux JJ. 1984. *Biochim. Biophys. Acta* 782:304–12

250. Champoux JJ, McCoubrey WK, Been MD. 1984. *Cold Spring Harbor Symp. Quant. Biol.* 49:435–42

251. McCoubrey WK Jr, Champoux JJ. 1986. *J. Biol. Chem.* 261:5130–37

252. Osheroff N, Zechiedrich EL. 1987. *Biochemistry* 26:4303–9

253. Andersen AH, Christiansen K, Zechiedrich EL, Jensen PS, Osheroff N, Westergaard O. 1989. *Biochemistry* 28:6237–44

254. Svejstrup JQ, Christiansen K, Gromova II, Andersen AH, Westergaard O. 1991. *J. Mol. Biol.* 222:669–78

255. Gale KC, Osheroff N. 1990. *Biochemistry* 29:9538–45

256. Gale KC, Osheroff N. 1992. *J. Biol. Chem.* 267:12090–97

257. Andersen AH, Sorensen BS, Christiansen K, Svejstrup JQ, Lund K, et al. 1991. *J. Biol. Chem.* 266:9203–10

258. Shuman S. 1992. *J. Biol. Chem.* 267:16755–58

259. Shuman S. 1992. *J. Biol. Chem.* 267:8620–27

260. Froelich-Ammon SJ, Gale KC, Osheroff N. 1994. *J. Biol. Chem.* 269:7719–25

261. Schmidt VK, Sorensen BS, Sorensen HB, Alsner J, Westergaard O. 1994. *J. Mol. Biol.* 241:18–25

262. Yeh Y-C, Liu H-F, Ellis CA, Lu A-L. 1994. *J. Biol. Chem.* 269:15498–504

263. Pommier Y, Bertrand R. 1993. In *The Causes and Consequences of Chromosomal Aberrations,* ed. IR Kirsch, pp. 277–309. Boca Raton: CRC Press

264. Burgin AB, Huizenga BN, Nash HA. 1995. *Nucleic Acids Res.* 23:2973–79

265. Kingma PS, Corbett AH, Burcham PC, Marnett LJ, Osheroff N. 1995. *J. Biol. Chem.* 270:21441–44

266. Brown PO, Peebles CL, Cozzarelli NR. 1979. *Proc. Natl. Acad. Sci. USA* 76:6110–14

267. Gellert M, Fisher LM, O'Dea MH. 1979. *Proc. Natl. Acad. Sci. USA* 83:7152–56

268. Wasserman SA, Cozzarelli NR. 1991. *J. Biol. Chem.* 266:20567–73

269. Goto T, Wang JC. 1982. *J. Biol. Chem.* 257:5866–72

270. Hsieh T-S. 1983. *J. Biol. Chem.* 258:8413–20

271. Lindsley JE, Wang JC. 1991. *Proc. Natl. Acad. Sci. USA* 88:10485–89

272. Sugino A, Higgins NP, Brown PO, Peebles CL, Cozzarelli NR. 1978. *Proc. Natl. Acad. Sci. USA* 75:4838–42

273. Wang JC, Gumport RI, Javaherian K, Kirkegaard K, Klevan L, et al. 1980. In *Mechanistic Studies of DNA Replication and Genetic Recombination,* ed. BM Alberts, CF Fox, pp. 769–84. New York: Academic

274. Wigley DB, Davies GJ, Dodson EJ, Maxwell A, Dodson G. 1991. *Nature* 351:624–29

275. Tamura JK, Gellert M. 1990. *J. Biol. Chem.* 265:21342–49

276. Jackson AP, Maxwell A. 1993. *Proc. Natl. Acad. Sci. USA* 90:11232–36

277. Ali JA, Jackson AP, Howell AJ, Maxwell A. 1993. *Biochemistry* 32:2717–24

278. Ali JA, Orphanides G, Maxwell A. 1995. *Biochemistry* 34:9801–8

279. Tamura JK, Bates AD, Gellert M. 1992. *J. Biol. Chem.* 267:9214–22

280. Lindsley JE, Wang JC. 1993. *J. Biol. Chem.* 268:8096–104

281. Maxwell A, Rau DC, Gellert M. 1986. In *Biomolecular Stereodynamics. Proc. 4th Conversat. Discipl. Biomol. Stereodyn.,* ed. RH Sarma, MH Sarma, 3:137–46. Albany: Adenine

282. Osheroff N, Shelton ER, Brutlag DL. 1983. *J. Biol. Chem.* 258:9536–43

283. Goto T, Laipis P, Wang JC. 1984. *J. Biol. Chem.* 259:10422–29

284. Rowe TC, Rusche JR, Brougham MJ, Holloman WK. 1981. *J. Biol. Chem.* 256:10354–61

285. Low RL, Holden JA. 1985. *Nucleic Acids Res.* 13:6999–14

286. Castora FJ, Kelly WG. 1986. *Proc. Natl. Acad. Sci. USA* 83:1680–84

287. Fogelsong PD, Bauer WR. 1984. *J. Virol.* 49:1–8

288. Shaffer R, Traktman P. 1987. *J. Biol. Chem.* 262:9309–15

289. Sekiguchi J, Shuman S. 1994. 269:29760–64

290. Yu L, Zhu CX, Tse-Dinh Y-C, Fesik SW. 1995. *Biochemistry* 34:7622–28

291. Berger JM, Gamblin S, Harrison SC, Wang JC. 1996. *Nature* 379:225–32

292. Lindsley J. 1996. *Proc. Natl. Acad. Sci. USA* 93:In press

293. Roca J, Berger JM, Harrison SC, Wang JC. 1996. *Proc. Natl. Acad. Sci. USA* 93:In press

294. Kornberg A, Baker TA. 1992. *DNA Replication.* New York: Freeman. 2nd ed.

295. Kreuzer KN, Cozzarelli NR. 1979. *J. Bacteriol.* 140:424–35
296. Filutowicz M, Jonczyk P. 1983. *Mol. Gen. Genet.* 191:282–87
297. Lockshon D, Morris DR. 1983. *Nucleic Acids Res.* 11:2999–3017
298. Nitiss JL. 1994. See Ref. 59, pp. 103–34
299. Hiasa H, Marians KJ. 1994. *J. Biol. Chem.* 269:16371–75
300. Hiasa H, Marians KJ. 1994. *J. Biol. Chem.* 269:32655–59
301. Kim RA, Wang JC. 1989. *J. Mol. Biol.* 208:257–67
302. Uemura T, Yanagida M. 1986. *EMBO J.* 5:1003–10
303. Yang L, Wold MS, Li JJ, Kelly TJ, Liu LF. 1987. *Proc. Natl. Acad. Sci. USA* 84:950–54
304. Champoux JJ, Been MD. 1980. In *Mechanistic Studies of DNA Replication and Genetic Recombination,* ed. B Alberts, pp. 809–15. New York: Academic
305. Aveman K, Knippers R, Koller T, Sogo JM. 1988. *Mol. Cell. Biol.* 8:3026–34
306. Zechiedrich EL, Cozzarelli NR. 1995. *Genes Dev.* 9:2859–69
307. Khodursky AB, Zechiedrich EL, Cozzarelli NR. 1995. *Proc. Natl. Acad. Sci. USA* 92:11801–5
308. Liu LF, Wang JC. 1987. *Proc. Natl. Acad. Sci. USA* 84:7024–27
309. Wu H-Y, Shyy S, Wang JC, Liu LF. 1988. *Cell* 53:433–40
310. Brill SJ, Sternglanz R. 1988. *Cell* 54:403–11
311. Tsao YP, Wu H-Y, Liu LF. 1989. *Cell* 56:111–18
312. Rahmouni AR, Wells RD. 1989. *Science* 246:358–63
313. Rahmouni AR, Wells RD. 1992. *J. Mol. Biol.* 223:131–44
314. Koo H-S, Wu H-Y, Liu LF. 1990. *J. Biol. Chem.* 265:12300–5
315. Wu H-Y, Liu LF. 1991. *J. Mol. Biol.* 219:615–22
316. Droge P, Nodheim A. 1991. *Nucleic Acids Res.* 19:2941–46
317. Cook DN, Ma D, Pon NG, Hearst JE. 1992. *Proc. Natl. Acad. Sci. USA* 89:10603–7
318. Dayn A, Malkhosyan S, Mirkin SM. 1992. *Nucleic Acids Res.* 20:5991–97
319. Chen D, Bowater RP, Dorman C, Lilley DMJ. 1992. *Proc. Natl. Acad. Sci. USA* 89:8784–88
320. Chen D, Bowater RP, Lilley DMJ. 1993. *Biochemistry* 32:13162–70
321. Chen D, Bowater RP, Lilley DMJ. 1994. *J. Bacteriol.* 176:3757–64
322. Wittig B, Wolff S, Dorbic T, Vahrson W, Rich A. 1992. *EMBO J.* 11:4653–63
323. Lynch AS, Wang JC. 1993. *J. Bacteriol.* 175:1645–55
324. Bowater RP, Chen D, Lilley DMJ. 1994. *Biochemistry* 33:9266–75
325. Bowater RP, Chen D, Lilley DMJ. 1994. *EMBO J.* 13:5647–55
326. Drolet M, Wu H-Y, Liu LF. 1994. See Ref. 59, pp. 135–46
327. Wang JC. 1992. In *Transcriptional Regulation,* ed. S McKnight, K Yamamoto, pp. 1253–69. Cold Spring Harbor, NY: Cold Spring Harbor Lab.
328. Wang JC, Lynch AS. 1993. *Curr. Opin. Genet. Dev.* 3:764–68
329. Wang JC, Lynch AS. 1995. In *Regulation of Gene Expression in Escherichia coli,* ed. ECC Lin, AS Lynch, pp. 127–47. Boca Raton: CRC Press
330. Droge P. 1993. *Proc. Natl. Acad. Sci. USA* 90:2759–63
331. Droge P. 1994. *BioEssays* 16:91–99
332. Cook DN, Ma D, Hearst JE. 1994. In *Nucleic Acids and Molecular Biology,* Vol. 8, ed. F Eckstein, DMJ Lilley, pp. 133–46. Berlin: Springer-Verlag
333. Ostrander EO, Benedetti P, Wang JC. 1990. *Science* 249:1261–65
334. Wang JC. 1985. In *Interrelationship Among Aging, Cancer and Differentiation. Jerusalem Symp. Quantum Chem. Biochem.,* ed. B Pullman, POP T'so, EL Schneider, 10:173–81. Holland: Reider
335. Wang JC. 1987. *Harvey Lect.* 81:93–110
336. Lodge JK, Kasic T, Berg DE. 1989. *J. Bacteriol.* 171:2181–87
337. Gartenberg MR, Wang JC. 1993. *Proc. Natl. Acad. Sci. USA* 90:10514–18
338. Yuan R, Hamilton DL, Burckhardt J. 1980. *Cell* 20:237–44
339. Yang L, Jessee CB, Lau K, Zhang H, Liu LF. 1989. *Proc. Natl. Acad. Sci. USA* 86:6121–25
340. Koo H-S, Claassen L, Grossman L, Liu LF. 1991. *Proc. Natl. Acad. Sci. USA* 88:1212–16
341. Pruss GJ, Manes SH, Drlica K. 1982. *Cell* 31:35–42
342. DiNardo S, Voelkel K, Sternglanz R, Reynolds AE, Wright A. 1982. *Cell* 31:43–45
343. Richardson SMH, Higgins CF, Lilley DMJ. 1984. *EMBO J.* 3:1745–52
344. Raji A, Zabel DJ, Laufer CS, Depew RE. 1985. *J. Bacteriol.* 162:1173–79
345. Menzel R, Gellert M. 1994. See Ref. 59, pp. 39–69
346. Margolin P, Zumstein L, Sternglanz R, Wang JC. 1985. *Proc. Natl. Acad. Sci. USA* 82:7178–82
347. Lilley DMJ, Higgins CF. 1991. *Mol. Microbiol.* 5:779–85
348. Tan J, Shu L, Wu H-Y. 1994. *J. Bacteriol.* 176:1077–86
349. Wu H-Y, Tan J, Fang M. 1995. *Cell* 82:445–51

350. Krohn M, Pardon B, Wagner R. 1992. *Mol. Microbiol.* 6:581–89
351. Drolet M, Bi X, Liu LF. 1994. *J. Biol. Chem.* 269:2068–74
352. Fleischmann G, Pflugfelder G, Steiner EK, Javaherian K, Howard GC, et al. 1984. *Proc. Natl. Acad. Sci. USA* 81: 6958–62
353. Muller MT, Pfund WP, Mehta VB, Trask DK. 1985. *EMBO J.* 4:1237–43
354. Gilmour DS, Pflugfelder G, Wang JC, Lis JT. 1986. *Cell* 44:401–7
355. Zhang H, Wang JC, Liu LF. 1988. *Proc. Natl. Acad. Sci. USA* 85:1060–64
356. Rose KM, Szopa J, Han FS, Cheng YC, Richter A, et al. 1988. *Chromosoma* 96:411–16
357. Stewart AF, Herrera RE, Nordheim A. 1990. *Cell* 60:141–49
358. Kroeger PE, Rowe TC. 1992. *Biochemistry* 31:2492–501
359. Schultz MC, Brill SJ, Ju QD, Sternglanz R, Reeder RH. 1992. *Genes Dev.* 6: 1332–41
360. Di Mauro E, Camilloni G, Verdone L, Caserta M. 1993. *Mol. Cell. Biol.* 13: 6702–10
361. Choder M. 1991. *Genes Dev.* 5:2315–26
362. Egyhazi E, Durban E. 1987. *Mol. Cell. Biol.* 7:4308–16
363. Sundin O, Varshavsky A. 1980. *Cell* 21:103–14
364. Sundin O, Varshavsky A. 1980. *Cell* 25:659–69
365. Varshavsky A, Levinger L, Sundin O, Barsoum J, Ozkaynak E, et al. 1983. *Cold Spring Harbor Symp. Quant. Biol.* 47:511–28
366. Wang JC, Liu LF. 1990. See Ref. 55, pp. 321–40
367. Snapka RM, Permana PA. 1993. *BioEssays* 15:121–27
368. Weaver DT, Fields-Berry SC, DePamphilis ML. 1985. *Cell* 41:565–75
369. Fields-Berry SC, DePamphilis ML. 1989. *Nucleic Acids Res.* 17:3261–73
370. Charron M, Hancock R. 1990. *Biochemistry* 29:9531–37
371. Funabiki H, Hagan I, Uzawa S, Yanagida M. 1993. *J. Cell Biol.* 121: 961–76
372. Spell RM, Holm C. 1994. *Mol. Cell. Biol.* 14:1465–76
373. Ishida R, Sato M, Narita T, Utsumi KR, Nishimoto T, et al. 1994. *J. Cell. Biol.* 126:1341–51
374. Ishimi Y, Ishida R, Andoh T. 1995. *J. Mol. Biol.* 247:835–39
375. Rose D, Thomas W, Holm C. 1990. *Cell* 60:1009–17
376. Rose D, Holm C. 1993. *Mol. Cell. Biol.* 13:3445–55
377. McPherson SM, Longo FJ. 1993. *Dev. Biol.* 158:122–30
378. Bliska JB, Cozzarelli NR. 1987. *J. Mol. Biol.* 194:205–18
379. Belland RJ, Morrison SG, Ison C, Huang WM. 1994. *Mol. Microbiol.* 14: 371–80
380. Hoshino K, Kitamura A, Morrissey I, Sato K, Kato J-I, et al. 1994. *Antimicrob. Agents Chemother.* 38:2623–27
381. Filutowicz M. 1980. *Mol. Gen. Genet.* 177:301–9
382. Hiraga S. 1992. *Annu. Rev. Biochem.* 61:283–306
383. Donachie WD. 1993. *Annu. Rev. Microbiol.* 47:199–230
384. Steck TR, Drlica K. 1984. *Cell* 36:1081–88
385. Hiasa H, DiGate RJ, Marians KJ. 1994. *J. Biol. Chem.* 269:2093–99
386. Shekhtman EM, Wasserman SA, Cozzarelli NR, Solomon MJ. 1993. *New J. Chem.* 8:757–63
387. Holm C, Goto T, Wang JC, Botstein D. 1985. *Cell* 41:553–63
388. Holm C, Stearns T, Botstein D. 1989. *Mol. Cell. Biol.* 9:159–68
389. Uemura T, Ohkura H, Adachi Y, Morino K, Shiozaki K, et al. 1987. *Cell* 50:917–25
390. Buchenau PH, Saumweber H, Arndt-Jovin DJ. 1993. *J. Cell Sci.* 104:1175–85
391. Roca J, Ishida R, Berger JM, Andoh T, Wang JC. 1994. *Proc. Natl. Acad. Sci. USA* 91:1781–85
392. Andoh T, Sato M, Narita T, Ishida R. 1993. *Biotech. Appl. Biochem.* 18:165–74
393. Clarke DJ, Johnson RT, Downes CS. 1993. *J. Cell Sci.* 105:563–69
394. Downes CS, Clarke DJ, Mullinger AM, Gimenez-Abian JF, Creighton AM, et al. 1994. *Nature* 372:467–70
395. Downes CS, Mullinger AM, Johnson RT. 1991. *Proc. Natl. Acad. Sci. USA* 88:8895–99
396. Snapka RM. 1986. *Mol. Cell. Biol.* 6: 4221–27
397. Richter A, Strausfeld U, Knippers R. 1987. *Nucleic Acids Res.* 15: 3455–68
398. Snapka RM, Powelson MA, Strayer JM. 1988. *Mol. Cell. Biol.* 8:515–21
399. Ishimi Y, Sugasawa K, Hanaoka F, Eki T, Hurwitz J. 1995. *J. Biol. Chem.* 267: 462–86
400. Ishimi Y, Ishida R, Andoh T. 1992. *Mol. Cell. Biol.* 12:4007–14
401. Kelly T. 1991. In *The Harvey Lectures,* ed. L Altieri, pp. 173–88. New York: Wiley-Liss
402. Shamu CE, Murray AW. 1992. *J. Cell Biol.* 117:921–34
403. Uemura T, Yanagida M. 1984. *EMBO J.* 3:1737–44

404. Wright SJ, Shatten G. 1990. *Dev. Biol.* 142:224–32

405. Roberge M, Th'ng J, Hamaguchi J, Bradbury EM. 1990. *J. Cell Biol.* 111: 1753–62

406. Chen M, Beck WT. 1993. *Cancer Res.* 53:5946–53

407. Sumner AT. 1992. *J. Cell Sci.* 103:105–15

408. Tobey RA. 1972. *Cancer Res.* 52:2720–25

409. Lock RB. 1992. *Cancer Res.* 52:1817–22

410. Lock RB, Ross WE. 1990. *Cancer Res.* 50:3761–66

411. Newport J. 1987. *Cell* 48:205–7

412. Newport J, Spann T. 1987. *Cell* 48:219–30

413. Wood ER, Earnshaw WC. 1990. *J. Cell. Biol.* 111:2839–50

414. Adachi Y, Luke M, Laemmli UK. 1991. *Cell* 64:137–48

415. Takasuga Y, Andoh T, Yamashita J, Yagura T. 1995. *Exp. Cell Res.* 217: 378–84

416. Vessetzsky YS, Dang Q, Benedetti P, Gasser SM. 1994. *Mol. Cell. Biol.* 14: 6962–74

417. Earnshaw WC, Halligan B, Cooke CA, Heck MM, Liu LF. 1985. *J. Cell Biol.* 100:1706–15

418. Berriors M, Osheroff N, Fisher PA. 1985. *Proc. Natl. Acad. Sci. USA* 82: 4142–46

419. Gasser SM, Laroche T, Falquet J, Boy de la Tour E, Laemmli UK. 1986. *J. Mol. Biol.* 188:613–29

420. Gasser SM, Amati BB, Cardenas ME, Hofmann JF. 1989. *Int. Rev. Cytol.* 119: 57–96

421. Heck MMS, Earnshaw WC. 1986. *J. Cell. Biol.* 103:2569–81

422. Tsutsui Ken, Tsutsui Kimiko, Oda T. 1989. *J. Biol. Chem.* 264:7644–52

423. Earnshaw WC. 1991. *Curr. Opin. Struct. Biol.* 1:237–44

424. Compton DA, Yen TJ, Cleveland DW. 1991. *J. Cell Biol.* 112:1083–97

425. Compton DA, Szilak I, Cleveland DW. 1992. *J. Cell Biol.* 116:1395–408

426. Laemmli UK, Kas E, Poljak L, Adachi Y. 1992. *Curr. Opin. Genet. Dev.* 2:275–85

427. Earnshaw WC, MacKay AM. 1994. *FASEB J.* 8:947–56

428. Woessner RD, Mattern MR, Mirabelli CK, Johnson RK, Drake FH. 1991. *Cell Growth Differ.* 2:209–14

429. Saitoh N, Goldberg IG, Wood ER, Earnshaw WC. 1994. *J. Cell Biol.* 127:303–18

430. Gasser SM, Laemmli UK. 1987. *Trends Genet.* 3:16–21

431. Blasquez VC, Sperry AO, Cockerill PN, Garrard WT. 1989. *Genome* 31:503–9

432. Pommier Y, Cockerill PN, Kohn KW, Garrard WT. 1990. *J. Virol.* 64:419–23

433. Saitoh N, Laemmli UK. *Cold Spring Harbor Symp. Quant. Biol.* 58: 755–65

434. Filipski J, Leblanc J, Youdale T, Sikorska M, Walker PR. 1990. *EMBO J.* 9:1319–27

435. Razin SV, Hancock R, Tarovaia O, Westergaard O, Gromova I, Georgiev GP. 1994. *Cold Spring Harbor Symp. Quant. Biol.* 58:25–35

436. Gromova II, Thomsen B, Razin SV. 1995. *Proc. Natl. Acad. Sci. USA* 92: 102–6

437. Hirano T, Mitchison TJ. 1993. *J. Cell Biol.* 120:601–12

438. Swedlow JR, Sedat JW, Agard DA. 1993. *Cell* 73:97–108

439. Taagepera S, Rao PN, Drake FH, Gorbsky GJ. 1993. *Proc. Natl. Acad. Sci. USA* 90:8407–11

440. Saitoh Y, Laemmli UK. 1994. *Cell* 76: 609–22

441. Christman MF, Dietrich FS, Fink GR. 1989. *Cell* 55:413–25

442. Kim RA, Wang JC. 1989. *Cell* 57:975–85

443. Gangloff S, McDonald JP, Bendixen C, Arthur L, Rothstein R. 1994. *Mol. Cell. Biol.* 14:8391–98

444. Kim RA, Caron PR, Wang JC. 1995. *Proc. Natl. Acad. Sci. USA* 92:2667–71

445. Schofield MA, Agbunag R, Michaels ML, Miller JH. 1992. *J. Bacteriol.* 174: 5168–70

446. Thaler DS, Sampson E, Seddiqi I, Rosenberg SM, Thomason LC, et al. 1989. *Genome* 31:53–67

447. Kodadek T. 1991. *J. Biol. Chem.* 266: 9712–18

448. Morel P, Hejna JA, Ehrlich SD, Cassuto E. 1993. *Nucleic Acids Res.* 21:3205–9

449. Watt PM, Louis EJ, Borts RH, Hickson ID. 1995. *Cell* 81:253–60

450. Ellis NA, Groden J, Ye T-Z, Straughen J, Lennon DJ, et al. 1995. *Cell* 83:655–66

451. Gottlieb S, Esposito R. 1989. *Cell* 56: 771–76

452. Aguilera A, Klein HL. 1990. *Mol. Cell. Biol.* 10:1439–51

453. Fan H, Klein H. 1994. *Genetics* 137:1–12

454. Sadoff BU, Heath-Pagliuso S, Castano IB, Yingfang Z, Kieff FS, et al. 1995. *Genetics* 141:465–79

455. Ikeda H. 1986. *Proc. Natl. Acad. Sci. USA* 83:922–26

456. Bae Y-S, Kawasaki I, Ikeda H, Liu LF.

1988. *Proc. Natl. Acad. Sci. USA* 85: 2076–80

457. Marvo SL, King SR, Jaskunas SR. 1983. *Proc. Natl. Acad. Sci. USA* 80:2542–46

458. Bullock P, Champoux JJ, Botchan M. 1985. *Science* 230:954–58

459. Pommier Y, Zwelling LA, Kao-Shen C-S, Whang-Pheng J, Bradley MO. 1985. *Cancer Res.* 45:3143–49

460. Saing K, Orii H, Tanaka Y, Yanagisawa K, Miura A, et al. 1988. *Mol. Gen. Genet.* 214:1–5

461. Dillehay LE, Jacobsen-Kram D, Williams JR. 1989. *Mutat. Res.* 215:15–23

462. Sperry AO, Blasquez VC, Garrard WT. 1989. *Proc. Natl. Acad. Sci. USA* 86: 5497–501

463. Miura-Masuda A, Ikeda H. 1990. *Mol. Gen. Genet.* 220:345–52

464. Chartrand P. 1991. See Ref. 62, pp. 240–45

465. Negrini M, Felix CA, Martin C, Lange BJ, Nakamura T, et al. 1993. *Cancer Res.* 53:4489–92

466. Han Y-H, Austin MJ, Pommier Y, Po-virk LF. 1993. *J. Mol. Biol.* 229: 52–66

467. Kumagai M, Yamashita T, Honda M. 1993. *Genetics* 135:255–64

468. Shuman S. 1989. *Proc. Natl. Acad. Sci. USA* 86:3489–93

469. Levin NA, Bjornsti M-A, Fink GR. 1993. *Genetics* 133:799–14

470. Nitiss J, Wang JC. 1988. *Proc. Natl. Acad. Sci. USA* 85:7501–5

471. Bodley AL, Huang H-C, Yu C, Liu LF. 1993. *Mol. Cell. Biol.* 13:6190–200

472. Russell DW, Alexander IE, Miller AD. 1995. *Proc. Natl. Acad. Sci. USA* 92: 5719–23

473. Shuman S. 1991. *Proc. Natl. Acad. Sci. USA* 88:10104–8

474. D'Arpa P, Liu LF. 1989. *Biochim. Biophys. Acta* 989:163–77

475. D'Arpa P, Beardmore C, Liu LF. 1990. *Cancer Res.* 50:6919–24

476. Ryan A, Squires S, Strutt HL, Johnson RT. 1991. *Nucleic Acids Res.* 19:3295–300

477. Tsao Y-P, Russo A, Nyamuswa G, Silber R, Liu LF. 1993. *Cancer Res.* 53: 5908–14

478. Chatterjee S, Trivedi D, Petzold SJ, Berger NA. 1990. *Cancer Res.* 50:2713–18

479. Bertrand R, Solary E, Jenkins J, Pommier Y. 1993. *Exp. Cell Res.* 207:388–97

480. Ikeda H. 1994. See Ref. 59, pp. 147–65

481. Lamhasni S, Larsen AK, Barray M, Monnot M, Delain E, et al. 1995. *Biochemistry* 34:3632–39

482. Frere V, Sourgen F, Monnot M, Troalen

F, Fermandjian S. 1995. *J. Biol. Chem.* 270:17502–7

483. Howard M, Neece S, Matson S, Kreuzer K. 1994. *Proc. Natl. Acad. Sci. USA* 91:12031–35

484. Kaufmann SH. 1991. *Cancer Res.* 51: 1129–36

485. Caldecott K, Jeggo P. 1991. *Mutat. Res.* 255:111–21

486. D'Arpa P. 1994. See Ref. 60, pp. 127–43

487. Jeggo PA, Caldecott K, Pidsley S, Banks GR. 1989. *Cancer Res.* 49:7057–63

488. Deleted in proof

489. Knab AM, Fertala J, Bjornsti M-A. 1993. *J. Biol. Chem.* 268:22322–30

490. Nitiss J, Wang JC. 1991. See Ref. 62, pp. 77–90

491. Beaucage SL, Miller CA, Cohen SN. 1991. *EMBO J.* 10:2583–88

492. Austin SJ, Eichorn BG. 1992. *J. Bacteriol.* 174:5190–95

493. Conley DL, Cohen SN. 1995. *Nucleic Acids Res.* 23:701–7

494. Boothman DA, Trask DK, Pardee AB. 1987. *Cancer Res.* 49:605–12

495. Boothman DA, Fukunaga N, Wang M. 1994. *Cancer Res.* 54:4618–26

496. Schaak J, Schedl P, Shenk T. 1990. *Nucleic Acids Res.* 18:1499–508

497. Shero JH, Bordwell B, Rothfield NF, Earnshaw WC. 1986. *Science* 231:737–40

498. Maul GG, French BT, van Venrooij WJ, Jimenez SA. 1986. *Proc. Natl. Acad. Sci. USA* 83:5145–49

499. Samuels DS, Tojo T, Homma M, Shimizu N. 1986. *FEBS Lett.* 209:231–34

500. Hoffmann A, Heck MMS, Bordwell BJ, Rothfield NF, Earnshaw WC. 1989. *Exp. Cell Res.* 180:409–18

501. Meliconi R, Bestagno M, Sturani C, Negri C, Galavotti V, et al. 1989. *Clin. Exp. Immunol.* 76:184–89

502. Imai H, Furuta K, Landberg G, Kiyosawa K, Liu LF, Tan EM. 1995. *Clin. Cancer Res.* 1:417–24

503. Menzel R, Gellert M. 1983. *Cell* 34: 105–13

504. Menzel R, Gellert M. 1987. *Proc. Natl. Acad. Sci. USA* 84:4185–89

505. Adachi T, Mizuuchi K, Menzel R, Gellert M. 1984. *Nucleic Acids Res.* 12: 6389–95

506. Jones PG, Krah R, Tafuri SR, Wolffe AP. 1992. *J. Bacteriol.* 174: 5798–802

507. Lesley SA, Jovanovich SB, Tse-Dinh Y-C, Burgess RR. 1990. *J. Bacteriol.* 172:6871–74

508. Yanagida M, Sternglanz R. 1990. See Ref. 55, pp. 299–320

509. Trash C, Bankier AT, Barrell BG,

510. Deleted in proof
511. Kunze N, Klein M, Richter A, Knippers R. 1990. *Eur. J. Biochem.* 194:323–30
512. Kunze N, Yang G, Dolberg M, Sundarp R, Knippers R, et al. 1991. *J. Biol. Chem.* 266:9610–16
513. Heiland S, Knippers R, Kunze N. 1993. *Eur. J. Biochem.* 217:813–22
514. Baumgartner B, Heiland S, Kunze N, Richter A, Knippers R. 1994. *Biochim. Biophys. Acta* 1218:123–27
515. Baumgartner B, Klett C, Hameister H, Richter A, Knippers R. 1994. *Mamm. Genome* 5:19–25
516. Hwang J, Hwong C-L. 1994. See Ref. 59, pp. 167–89
517. Hwong CL, Chen MS, Hwang JL. 1989. *J. Biol. Chem.* 264:14923–26
518. Heiland S, Knippers R. 1995. *Mol. Cell. Biol.* 15:6623–31
519. Romig H, Richter A. 1990. *Nucleic Acids Res.* 18:801–8
520. Chow K-C, Pearson GD. 1985. *Proc. Natl. Acad. Sci. USA* 82:2247–51
521. Duguet M, Lavenot C, Harper F, Mirambeau G, DeRecondo A-M. 1983. *Nucleic Acids Res.* 11:1059–75
522. Miskimins R, Miskimins WK, Bernstein H, Shimizu N. 1983. *Exp. Cell Res.* 146:53–62
523. Taudou G, Mirambeau G, Lavenot C, der Garabedian A, Vermeersch J, et al. 1984. *FEBS Lett.* 176:431–35
524. Tricoli JV, Sahai BM, McCormick PJ, Jarlinski SJ, Bertram JS, et al. 1985. *Exp. Cell Res.* 158:1–14
525. Sullivan DM, Glisson BS, Hodges PK, Smallwood-Kentro S, Ross WE. 1986. *Biochemistry* 25:2248–56
526. Deleted in proof
527. Markovits J, Pommier Y, Kerrigan D, Covey JM, Tilchen EJ, et al. 1987. *Cancer Res.* 47:2050–55
528. Nelson WG, Cho KR, Hsiang Y-H, Liu LF, Coffey DS. 1987. *Cancer Res.* 47: 3246–50
529. Fairman R, Brutlag DL. 1988. *Biochemistry* 27:560–65
530. Heck MMS, Hittelman WN, Earnshaw WC. 1988. *Proc. Natl. Acad. Sci. USA* 85:1086–90
531. Hsiang Y-H, Wu H-Y, Liu LF. 1988. *Cancer Res.* 48:3230–35
532. Bodley AL, Wu H-Y, Liu LF. 1987. *Natl. Cancer Inst. Monogr.* 4:31–35
533. Zwelling LA, Chan D, Hinds M, Silberman L, Mayes J. 1988. *Biochem. Biophys. Res. Commun.* 152:808–17
534. Constantinou A, Henning-Chubb C, Huberman E. 1989. *Cancer Res.* 49: 1110–17
535. Zwelling LA, Hinds M, Chan D, Altschuler E, Mayes J, et al. 1990. *Cancer Res.* 50:7116–22
536. Kaufmann SH, McLaughlin SL, Kastan MB, Liu LF, Karp JE, et al. 1991. *Cancer Res.* 51:3534–43
537. Ellis AL, Zwelling LA. 1994. *Biochem. Pharmacol.* 48:1842–45
538. Ellis AL, Altshuler E, Bales E, Hinds M, Mayes J, et al. 1994. *Biochem. Pharmacol.* 47:387–96
539. Deleted in proof
540. Tsutsui Kimiko, Tsutsui Ken, Sakurai H, Shohmori T, Oda T. 1986. *Biochim. Biophys. Res. Commun.* 138:1116–22
541. Hadlaczky G, Praznovszky T, Sofi J, Udvardy A. 1988. *Nucleic Acids Res.* 16:10013–23
542. Roca J, Mezquita C. 1989. *EMBO J.* 8:1855–60
543. Holden JA, Rolfson DH, Wittwer CT. 1990. *Biochemistry* 29:2127–34
544. Capranico G, Tinelli S, Austin CA, Fisher LM, Zunino F. 1992. *Biochim. Biophys. Acta* 1132:43–48
545. Negri C, Chiesa R, Cerino A, Bestagno M, Sala C, et al. 1992. *Exp. Cell Res.* 200:452–59
546. Tsutsui Kimiko, Tsutsui Ken, Okada S, Watanabe M, Shohmori T, et al. 1993. *J. Biol. Chem.* 268:19076–83
547. Saijo M, Ui M, Enomoto T. 1992. *Biochemistry* 31:359–63
548. Kimura K, Nozaki N, Saijo M, Kikuchi A, Ui M, Enomoto T. 1994. *J. Biol. Chem.* 269:24523–26
549. Goswami PC, Roti R, Hunt CR. 1996. *Mol. Cell. Biol.* In press
550. Hochhauser D, Stanway CA, Harris AL, Hickson ID. 1992. *J. Biol. Chem.* 267: 18961–65
551. Loflin PT, Hochhauser D, Hickson ID, Morales F, Zwelling LA. 1994. *Biochem. Biophys. Res. Commun.* 200:489–96
552. Ng S-W, Eder JP, Schnipper LE, Chan VTW. 1995. *J. Biol. Chem.* 270:25850–58
553. Deleted in proof
554. Woessner RD, Chung TD, Hofmann GA, Mattern MR, Mirabelli CK, et al. 1990. *Cancer Res.* 50:2901–8
555. Benson JD, Huang E-S. 1990. *J. Virol.* 64:9–15
556. Davies S, Jenkins JR, Hickson ID. 1993. *Nucleic Acids Res.* 21:3719–23
557. Austin CA, Marsh KL, Wasserman RA, Willmore E, Sayer PJ, et al. 1995. *J. Biol. Chem.* 270:15739–46
558. Krishan A, Paika K, Frei E III. 1975. *J. Cell Biol.* 66:521–30
559. Murray AW. 1993. *Curr. Opin. Genet. Dev.* 5:5–11

560. Liu LF, Miller KG. 1981. *Proc. Natl. Acad. Sci. USA* 78:3487–91
561. D'Arpa P, Machlin PS, Ratrie HR III, Rothfield NF, Cleveland DW, et al. 1988. *Proc. Natl. Acad. Sci. USA* 85:2543–47
562. Alsner J, Svejstrup JQ, Kjeldsen E, Sorensen BS, Westergaard O. 1992. *J. Biol. Chem.* 267:12408–11
563. Dingwall C, Laskey RA. 1991. *Trends Biochem. Sci.* 16:478–81
564. Shiozaki K, Yanagida M. 1992. *J. Cell Biol.* 119:1023–26
565. Crenshaw DG, Hsieh T. 1993. *J. Biol. Chem.* 268:21328–34
566. Crenshaw DG, Hsieh T. 1993. *J. Biol. Chem.* 268:21335–43
567. Zini N, Martelli AM, Sabatelli P, Santi S, Negri C, et al. 1992. *Exp. Cell Res.* 200:460–66
568. Petrov P, Drake FH, Loranger A, Huang W, Hancock R. 1993. *Exp. Cell Res.* 204:73–81
569. Durban E, Mills JS, Roll D, Busch H. 1983. *Biochim. Biophys. Res. Commun.* 111:897–905
570. Kaiserman HB, Ingebritsen TS, Benbow RM. 1988. *Biochemistry* 27:3216–22
571. Samuels DS, Shimizu Y, Shimizu N. 1989. *FEBS Lett.* 259:57–60
572. Coderoni S, Paparelli M, Gianfranceschi GL. 1990. *Int. J. Biochem.* 22:737–46
573. Samuels DS, Shimizu N. 1992. *J. Biol. Chem.* 267:11156–62
574. Cardellini E, Bramucci M, Gianfranceschi GL, Durban E. 1994. *Biol. Chem. Hoppe-Seyler* 375:255–59
575. Turman MA, Douvas A. 1993. *Biochem. Med. Metab. Biol.* 50:210–25
576. Staron K, Kowalska-Loth B, Szumiel I. 1995. *Biophys. Biochim. Acta* 1260:35–42
577. Samuels DS, Shimizu Y. 1992. *J. Biol. Chem.* 267:11156–62
578. Samuels DS, Shimizu Y, Nakabayashi T, Shimizu N. 1994. *Biochim. Biophys. Acta* 1223:77–83
579. Cardellini E, Durban E. 1993. *Biochem. J.* 291:303–7
580. Nambi P, Mattern M, Bartus JO, Aiyar N, Crooke ST. 1989. *Biochem. J.* 262:485–89
581. Mattern MR, Mong S, Mong S-M, Bartus JO, Sarau HM et al. 1990. *Biochem. J.* 265:101–7
582. Heck MMS, Hittelman WN, Earnshaw WC. 1989. *J. Biol. Chem.* 264:15161–64
583. Burden DA, Goldsmith LJ, Sullivan DM. 1993. *Biochem. J.* 293:297–304
584. Kimura K, Saijo M, Ui M, Enomoto T. 1994. *J. Biol. Chem.* 269:1173–76
585. Wells NJ, Hickson ID. 1995. *Eur. J. Biochem.* 231:491–97
586. Cardenas ME, Dang Q, Glover VC, Gasser SM. 1992. *EMBO J.* 11:1785–96
587. Corbett AH, DeVore RF, Osheroff N. 1992. *J. Biol. Chem.* 267:20513–18
588. Corbett AH, Fernald AW, Osheroff N. 1993. *Biochemistry* 32:2090–97
589. Higgins NP, Ferro AM, Olivera BM. 1990. See Ref. 55, pp. 361–70
590. Ferro AM, Olivera BM. 1984. *J. Biol. Chem.* 259:547–59
591. Darby MK, Schmitt B, Jongstra-Bilen J, Vosberg HP. 1985. *EMBO J.* 4:2129–34
592. Kasid UN, Halligan B, Liu LF, Dritschilo A, Smulson M. 1989. *J. Biol. Chem.* 264:18687–92
593. Kozyavkin SA, Krah R, Gellert M, Stetter KO, Lake JA, et al. 1994. *J. Biol. Chem.* 269:11081–89
594. Merino A, Madden KR, Lane WS, Champoux JJ, Reinberg D. 1993. *Nature* 365:227–32
595. Kretzschmar M, Meisterernst M, Roeder RG. 1993. *Proc. Natl. Acad. Sci. USA* 90:11508–12
596. Bharti AK, Olson MOJ, Kufe DW, Rubin EH. 1996. *J. Biol. Chem.* 271:1993–97
597. Kordiyak GJ, Jakes S, Ingebritsen TS, Benbow RM. 1994. *Biochemistry* 33:13484–91
598. Bernard P, Kezdy KE, van Melderen L, Steyaert J, Wyns L, et al. 1993. *J. Mol. Biol.* 234:534–41
599. Coleman J. 1992. *Mol. Gen. Genet.* 232:295–303
600. Mulder E, El'Bouhali M, Pas E, Woldringh CL. 1990. *Mol. Gen. Genet.* 221:87–93
601. Ohta T, Hirose S. 1990. *Proc. Natl. Acad. Sci. USA* 87:5307–11
602. Ohta T, Kobayashi M, Hirose S. 1995. *J. Biol. Chem.* 270:15571–75
603. Thomas W, Spell RM, Ming ME, Holm C. 1991. *Genetics* 128:703–14
604. Sander M, Nolan JM, Hsieh T-S. 1984. *Proc. Natl. Acad. Sci. USA* 81:6938–42
605. Saijo M, Enomoto T, Hanaoka F, Ui M. 1990. *Biochemistry* 29:583–90
606. Bojanowski K, Filhol O, Cochet C, Chambaz EM, Larsen AK. 1993. *J. Biol. Chem.* 268:22920–26
607. Webb CF, Eneff KL, Drake FH. 1993. *Nucleic Acids Res.* 21:4363–68
608. Ma X, Saitoh N, Curtis PJ. 1993. *J. Biol. Chem.* 268:6182–88
609. Ma X, Frazer P, Curtis PJ. 1991. *Differentiation* 47:135–41
610. Eckstein F, Lilley DMJ, eds. 1995. *Nucleic Acids and Molecular Biology,* Vol. 9. Berlin: Springer-Verlag

Annu. Rev. Biochem. 1996. 65:693–739

INTERRELATIONSHIPS OF THE PATHWAYS OF mRNA DECAY AND TRANSLATION IN EUKARYOTIC CELLS

Allan Jacobson

Department of Molecular Genetics and Microbiology, University of Massachusetts Medical School, Worcester, Massachusetts 01655-0122

Stuart W. Peltz

Department of Molecular Genetics and Microbiology, Robert Wood Johnson Medical School, University of Medicine and Dentistry of New Jersey, Rutgers University, Piscataway, New Jersey 08854

KEY WORDS   mRNA stability, mRNA degradation, mRNA decay, translation, RNA structure, poly(A), cap, ribosome, RNases, mRNP

ABSTRACT

While the potential importance of mRNA stability to the regulation of gene expression has been recognized, the structures and mechanisms involved in the determination of individual mRNA decay rates have just begun to be elucidated, particularly in mammalian systems and yeast. It is now well established that mRNA decay is not a default process, in which an array of nonspecific nucleases degrades indiscriminately based on target size or ribosome protection of the substrate. Rather, like transcription, RNA processing, and translation, mRNA decay is a precise process dependent on a variety of specific *cis*-acting sequences and *trans*-acting factors. Entry into the pathways of mRNA decay is triggered by at least three types of initiating event: poly(A) shortening, arrest of translation at a premature nonsense codon, and endonucleolytic cleavage. Steps subsequent to poly(A) shortening or premature translational termination converge in a pathway that progresses from removal of the 5′ cap to exonucleolytic digestion of the body of the mRNA. mRNA fragments generated by endonucleolytic cleavage are most likely removed by exonucleolytic decay as well, but these events have not been characterized in detail. Nucleases and other factors (including mRNA sequence elements and autoregulatory proteins) required for the promotion or inhibition of these pathways have been identified by both biochemical and

693

0066-4154/96/0701-0693$08.00

genetic methods and systematic attempts to understand their respective roles have begun. mRNA sequences whose presence or absence has marked effects on mRNA decay rates include the ubiquitous cap and poly(A) tail, sequences that comprise endonuclease cleavage sites, and sequences that promote poly(A) shortening. The latter are found in the 3'-UTR (untranslated region) and in coding regions. Evidence that poly(A) stimulates translation initiation, that some destabilization sequences must be translated in order to function, and that premature translation termination promotes rapid mRNA decay indicates a close linkage between the elements regulating mRNA decay and components of the protein synthesis apparatus. This linkage, and other data, leads us to propose a model for a functional mRNP. In this model, interactions between factors associated with opposite ends of an mRNA stimulate translation initiation and minimize the rate of entry into the pathways of mRNA decay. Events that initiate mRNA decay are postulated to be those that can disrupt this functional complex and create substrates for exonucleolytic digestion.

CONTENTS

INTRODUCTION   694
mRNA DECAY INITIATED BY POLY(A) SHORTENING   696
   Poly(A) Shortening is the Rate-Limiting Step in the Turnover of Many mRNAs   697
   After Poly(A) Shortening: A Pathway for mRNA Decay   697
   Enzymology and Genetics of Poly(A) Shortening, Decapping, and Exonucleolytic
               Decay in Yeast and Mammals   699
   Sequence Elements That Regulate mRNA Decay Dependent on Prior Poly(A)
               Shortening   702
   Poly(A) Shortening and Translation   706
   A Possible Mimic of the Poly(A) Shortening-Dependent Pathway: Histone mRNA
               Decay   712
mRNA DECAY INITIATED BY IMPAIRED TRANSLATION   713
   Nonsense-Mediated mRNA Decay   714
   Decay of mRNAs in Which Translation Initiation is Inhibited   726
   Decay of mRNAs With Aberrant Translational Termination   727
mRNA DECAY INITIATED BY ENDONUCLEOLYTIC CLEAVAGE   728
CONCLUSION   730

INTRODUCTION

The unit of mRNA decay is half-life ($t_{1/2}$), the time required for the disappearance of 50% of a given mRNA, and its derivation has been summarized in previous reviews (5, 11–17). In a given cell, the half-lives of individual mRNAs may differ by as much as 10- to 100-fold (5, 14, 15, 17). Unstable mRNAs in mammalian cells and yeast have half-lives as short as 15–30 min and 1–5 min, respectively, whereas their stable counterparts have respective half-lives as long as 10 or more h or 30–60 min. Individual mRNA decay rates are generally invariant, but can fluctutate in response to numerous environmental stimuli (5, 10, 15, 26, 27). Collectively, such differences in mRNA decay rates have significant effects on the expression of specific genes and provide the cell with flexibility in effecting change. Most obvious is the recognition that the net yield of polypeptide product is directly related to mRNA decay rate. More subtle, but recognized as early as the original theoretical framework of the mRNA hypothesis (1–3), is the contribution of mRNA decay rate to the rate of adaptation to a new steady-state. For example, upon a change in transcription rate, the time required to reach a new steady-state will be fastest with an unstable mRNA (4).

Sequences, factors, and environmental signals that regulate mRNA decay rates have been studied in many experimental organisms. Here, we address the mRNA turnover problem in mammals and in the yeast *Saccharomyces cerevisiae,* a bias that is both a direct reflection of current research activity and the attributes of these two systems. Other recent reviews discuss systems not addressed in depth here (5–10, 15, 26, 27). Within this context our emphasis will be on mRNA decay pathways, i.e. on multiple dependent steps that have been identified or inferred. Of particular interest is the identification and characterization of the rate-determining event(s) that lead to the subsequent rapid decay of the transcript. Understanding these events should help to explain both the range in decay rates and the mechanisms by which they can be modulated. Sequences that function as determinants to promote mRNA degradation have been identified in all regions of the transcript, including the 5′-UTR, the protein coding region, and the 3′-UTR. On the surface, this diversity in locations does not readily suggest global pathways for mRNA decay. Nevertheless, we believe that cells have evolved a small number of general pathways to degrade most mRNAs and that regulation of mRNA decay rates should be targeted to the initial events that dictate flow into one of these decay pathways.

As a consequence of transcription, processing, transport, and translation, mRNA structure and the set of mRNA-associated proteins is in constant flux (41) and might be expected to influence the rate of mRNA decay and the time of its onset. While nuclear events may well influence the cytoplasmic fate of mRNA (33, 76–78), they have yet to be described in detail. In contrast, a large set of observations point to an important role for translation in mRNA decay, including: (*a*) Inhibition of translational elongation, by the use of inhibitors or as a consequence of mutation, can reduce mRNA decay rates (37, 80–87). (*b*) Instability elements can be localized to mRNA coding regions (20, 21, 72, 73, 88–91). (*c*) The activity of some instability elements depends on ribosome translocation up to, or near, the element (28, 40, 87–89, 91–96). (*d*) Some 3′-UTR instability elements can influence mRNA translational activity (97–101). (*e*) Factors involved in decay can be polysome associated (53, 60, 102, 103). (*f*) Premature translational termination can enhance mRNA decay rates (32, 33). (g) Metabolism of the poly(A) tail is a rate-limiting step in the decay of several mRNAs, yet this structure has a role in translational initiation (44).

Until recently, the most well understood role of the relationship between translation and decay was autoregulation of the ß-tubulin mRNA. Numerous experiments demonstrated that this mRNA is destabilized by a co-translational protein:protein interaction in which free β-tubulin subunits activate or bind to a cellular factor that then binds the nascent tubulin polypeptide as it emerges a sufficient distance from the ribosome (28, 29). This event then stimulates mRNA decay, possibly by activating a nuclease or promoting an interaction with a nuclease. Translation may also be linked to decay since factors essential for the degradation of some mRNAs need to be synthesized de novo, possibly because they are unstable (71, 104). The bulk of the available data, however, indicates that the processes and components required for mRNA turnover and translation are intimately linked. Events that optimize translation may limit mRNA availability to factors that trigger the onset of rapid decay. Conversely, factors that promote rapid mRNA decay may, in part, compromise mRNA translational efficiency. We see an understanding of this interrelationship as one of the keys to understanding how mRNA stability is controlled and have thus focused this review on experimental results that begin to explain it. We address primarily three events that appear to trigger the initiation of mRNA decay—poly(A) shortening, premature translation termination, and endonucleolytic cleavage—and consider their involvement with translation.

mRNA DECAY INITIATED BY POLY(A) SHORTENING

With rare exceptions, mRNAs have a 3′ poly(A) tract, an appendage for which roles in nuclear RNA processing and transport, mRNA stability, and protein synthesis have been proposed (15, 42–44, 105). A cytoplasmic function for the poly(A) tract is implied by the observation that some mammalian viruses replicate exclusively in the cytoplasm yet synthesize polyadenylated mRNAs (106–110). There has been a long and contentious debate, however, as to whether that role in the cytoplasm is as a determinant of mRNA stability. The possibility that the polyadenylation status of an mRNA influences its decay rate is supported by several experiments (19–25, 69, 71, 83, 111–131) and refuted by others (15, 36, 37, 39, 132–135). This discrepancy most likely reflects the fact that the poly(A) tail is not, in itself, a determinant of mRNA stability. After mRNA enters the cytoplasm, poly(A) tracts are, in an mRNA- and organism-specific manner, gradually shortened from lengths of 80 to 250 adenylate residues to lengths of 10 to 60 adenylate residues and, in some instances, may be completely removed (42–44). The way that different mRNAs respond to such shortening may account for the apparently conflicting results. In some instances, shortening of the tail to an oligo(A) form is the rate-determining event that triggers turnover of the body of the transcript. In other cases, deadenylation occurs and may be an obligate event in a decay pathway, but it is not the rate-determining step (22, 24, 136; see below).

Poly(A) Shortening is the Rate-Limiting Step in the Turnover of Many mRNAs

Four types of experiments support a role for poly(A) shortening or removal in mRNA decay: (a) In oocytes and in vitro systems, most polyadenylated substrates are degraded more slowly than their deadenylated counterparts, implying a protective function for the poly(A) tail (69, 71, 111, 114, 125, 126). (b) In intact cells, poly(A) removal precedes the degradation of many mRNAs (19, 21, 22, 24, 25, 71, 115, 116, 136–138). (c) The rate of poly(A) removal from several yeast and mammalian mRNAs has been shown to be attributable to specific sequence elements (see below), and mutations in these elements have been shown to simultaneously reduce the rates of mRNA decay and deadenylation (21, 22, 136, 139). (d) Inhibition of the final 5′→3′ exonucleolytic step of a yeast decay pathway (see below) led to the accumulation of decay intermediates that all had shortened poly(A) tails (23, 64).

The precedence of poly(A) shortening to decay is best illustrated by transcriptional pulse-chase experiments monitoring the decay of mammalian and yeast mRNAs (19–24). In addition to establishing such precedence, these experiments have demonstrated that poly(A) shortening rates correlate with mRNA decay rates and that shortening of a specific poly(A) tail occurs in at least two kinetically distinct steps (21, 22). The latter include an initial slow mRNA-independent phase that may occur during cytoplasmic mRNP biogenesis followed by more rapid shortening at mRNA-specific rates (21, 22, 136). Shortening of the poly(A) tails of unstable yeast and mammalian mRNAs occurs at rates of approximately 2–8 and 0.2–1.0 adenylate residues per minute, respectively; shortening of the poly(A) tails of stable mRNAs in the same organisms is 2–100-fold slower (22, 25, 37, 139). For most mRNAs, poly(A) shortening slows, or reaches a steady-state with re-addition, prior to complete deadenylation. For example, poly(A) shortening of yeast mRNAs leaves them with an oligo(A) tract length of approximately 15 adenylate residues (22, 37, 44, 65, 113, 140–143).

After Poly(A) Shortening: A Pathway for mRNA Decay

Decay events that follow poly(A) shortening to an oligo(A) length have been characterized extensively in yeast and have led to the first elaboration of a eukaryotic mRNA decay pathway. Recent results indicate that, following poly(A) shortening, mRNA is cleaved one or two nucleotides from its 5′-end, removing the 5′ cap structure. The uncapped and deadenylated mRNA is subsequently degraded by a 5′→3′ exoribonuclease. Initial support for this decay pathway was the obervation that in cells harboring a deletion of the *XRN1* gene, encoding the major 5'→3' exoribonuclease, uncapped mRNAs with shortened poly(A) tails accumulated (64). Subsequent transcriptional pulse-chase experiments, utilizing *PGK1* and *MFA2* mRNAs harboring oligo(G) tracts in various positions, identified decay intermediates that were trimmed to the 5' side of the oligo(G) insertions (23). The presence of the oligo(G) insertions was key to these experiments because they are thought to form strong RNA secondary structures that inhibit progression of the Xrn1p 5'→3' exoribonuclease (23, 144). Inclusion of multiple oligo(G) tracts within the *PGK1* and *MFA2* transcripts identified a predominant oligoadenylated decay intermediate that was trimmed to the oligo(G) insertion most proximal to the mRNA 5' end, indicating that degradation was occurring in a 5'→3' direction (23). In an *xrn1Δ* strain the level of oligo(G)-terminated decay intermediates was greatly reduced (23). Furthermore, transcriptional pulse-chase experiments in *xrn1Δ* strains demonstrated that, following deadenylation to the oligoadenylate form, the *MFA2* transcript is decapped, shortened by one or two nucleotides from its 5' end, and substantially stabilized relative to its decay rate in wild-type cells (23).

Comparisons of the rates at which the *PGK1* and *MFA2* mRNAs progress through the poly(A) shortening, decapping, and 5'→3' decay pathway illustrate that poly(A) shortening can be an obligate or a rate-determining step in mRNA decay. The yeast *PGK1* gene encodes a stable mRNA (22, 37, 38, 40, 88) whose decay also progresses through this pathway (22, 24). Rates of poly(A) shortening, mRNA decapping, and mRNA decay are rapid for the *MFA2* mRNA, but slow for the *PGK1* mRNA (22, 24, 136). Increasing the deadenylation rate of the *PGK1* mRNA, by fusing the *MFA2* 3'UTR to the *PGK1* coding region, decreased the time of onset of *PGK1* mRNA decay, but not its rate (22). This result indicates that the stability of the oligoadenylated form of the *PGK1* mRNA did not change and is consistent with previous results demonstrating that a poly(A)-deficient form of this transcript is stable (37, 144). Thus, for certain mRNAs, such as *MFA2,* decapping of the transcript follows rapidly after poly(A) shortening and is not a rate-determining event while, for other mRNAs, such as *PGK1,* decapping following poly(A) shortening is not rapid and is, therefore, a rate-determining event.

When 5'→3' decay of wild-type *PGK1* mRNA, or *PGK1* mRNA with an early nonsense codon, is blocked by insertion of an oligo(G) tract into its 5'-UTR, mutation of the *XRN1* gene, or treatment of cells with cycloheximide, decay progressing in a 3'→5' direction can be detected (24, 145). This conclusion follows from experiments in which these mRNAs were shortened at their 3'-ends up to the position of a 3' oligo(G) insertion. The enzyme responsible for this shortening has not been identified. *PGK1* transcripts shortened at their 3' ends can also be detected in cells in which 5'→3' decay occurs, but at a substantially diminished level. Decay of the *MFA2* mRNA, however, shows no evidence of 3'→5' decay, whether or not the 5'→3' pathway is active (24). These data suggest that, after deadenylation, some yeast mRNAs can also be degraded 3'→5', but this mode of decay is a minor activity compared to the 5'→3' pathway. Evidence that some mammalian mRNAs may be degraded 3'→5' has been obtained in studies of decay in cell-free extracts (60, 67, 68, 71).

Enzymology and Genetics of Poly(A) Shortening, Decapping, and Exonucleolytic Decay in Yeast and Mammals

POLY(A) SHORTENING: IS IT DEPENDENT ON POLY(A)-BINDING PROTEIN?   Insight into, and controversy about, the mechanism of poly(A) shortening has followed from an analysis of poly(A)-binding proteins (PABPs), proteins that interact with the poly(A) tail. By far the most abundant of the numerous mRNA-binding proteins, the ubiquitous PABPs associate with the poly(A) tracts of essentially all polyadenylated mRNAs, spaced approximately 25 nt apart and requiring a minimal binding site of 12 adenosines (146–150). Cloning of the yeast PABP gene (*PAB1*; 148, 151) led to the demonstration that it is essential; depletion of PABP by promoter repression, or inactivation of its function with a temperature-sensitive (ts) mutation, inhibited growth, translation (see below), and poly(A) shortening (65). The latter observation led to the purification of a yeast poly(A) nuclease (PAN) activity that is dependent on poly(A)-binding protein for activity (66, 152). In vitro, PAN is a 3' to 5' exonuclease that rapidly shortens poly(A) tracts to a length of 15 to 25 nt, liberating 5' mononucleotides in the process (66). In the appropriate buffer, substrate recognition depends not only on PABP, but also on the proximity of the poly(A) tract to the 3' end of the mRNA (66). Additional activities associated with PAN include a slow removal of adenosines from poly(A) tails of less than 25 nt and a spermidine-dependent poly(A) shortening activity that functions independent of PABP (66). At present, whether these are PAN activities or whether they are due to other co-purifying enzymes is not clear. Yeast PAN is a multimeric enzyme, and genes thought to encode possible subunits have been cloned and sequenced (152). Recent experiments, however, indicate that the 161-kDa *PAN1* polypeptide thought to be such a subunit (152) is not required for poly(A) shortening activity in vitro or in vivo (A Sachs, personal communication).

The PABP dependence of the poly(A) shortening activity of yeast PAN differs from results observed in mammalian cells. Mammalian ribonucleases that degrade poly(A) 3'→5' in the absence of PABP have been identified (60–62). Moreover, experiments utilizing an in vitro mRNA turnover system indicate that purified PABP protects the poly(A) tail from shortening by a 3'→5' exoribonuclease, because removal of the PABP from the poly(A) tail made it susceptible to this activity (111). These results suggest several possibilities, including differences between mammalian cells and yeast, inaccuracy of the in vitro decay system, and heterogeneity among poly(A) nucleases. Current evidence suggests remarkable similarities between yeast and mammalian cells in most aspects of the biochemistry of gene expression, and the in vitro decay system certainly reflects several aspects of mRNA turnover observed in whole cells (5, 15).

Recent experiments indicate that the PABP/PAN proteins cannot be the only enzymes degrading poly(A) tails in yeast cells. Cells harboring a disruption of the *PAB1* gene are normally inviable (65), but growth can be restored by mutations in either the *SPB2* gene, encoding ribosomal protein L46 (65), or the *XRN1* gene, encoding the aforementioned $5' \rightarrow 3'$ exoribonuclease (14, 154). In *pab1Δ* strains carrying either of these two suppressor mutations, poly(A) shortening, albeit at a reduced rate, was observed (14, 154). This indicates that poly(A) shortening in yeast cells can occur in the absence of PABP.

Complications inherent in resolving the PABP-dependency of PAN activity include the following: (*a*) In the absence of PABP, other RNA binding proteins likely interact with the poly(A) tail. (*b*) The poly(A)-PABP complex found on virtually all RNAs also appears to function in modulating translation (44; see below). Thus, the effects of PABP depletion or inactivation on poly(A) tail lengths may be indirectly caused by changes in other functions of the poly(A)-PABP complex. For example, it is conceivable that, in the absence of PABP, mRNAs are inappropriately localized and translated inefficiently, making them inaccessible for poly(A) shortening. Consistent with this view, in *pab1Δ* strains overall translation rates are reduced, levels of polysomes are decreased, and levels of 80S monosomes are increased (65; see below).

In spite of the theoretical problems inherent in defining PABP-independent poly(A) degrading enzymes, at least one such enzyme has already been identified. Originally characterized as a $3' \rightarrow 5'$ exonuclease activity responsible for histone mRNA decay in a cell-free system, this enzyme has now been purified and shown to also have poly(A)-degrading activity (60). The 33-kDa Mg^{+2}-dependent protein, purified from cytoplasmic extracts of K562 erythro-leukemia cells, can degrade polysome-bound histone mRNA and other single-stranded RNAs, including transcripts containing poly(A) tracts that are not bound with PABP. Changing the 3'-hydroxyl group of histone H4 mRNA to a 3' PO_4 reduces its decay rate 10- to 20-fold, suggesting that the exonuclease recognizes 3'-hydroxyl groups. Consistent with this theory is the fact that observed products of digestion are nucleoside 5' monophosphates.

DECAPPING ENZYMES   Nuclease activities that remove the 5' cap structures from mRNAs have been identified. Pyrophosphatases that can hydrolyze 5' caps have been purified from potato, tobacco, HeLa cells, and yeast cells (63,

155–157). A decapping activity from yeast cells was initially observed in the course of purification of 5′→3′ exoribonucleases and shown to catalyze the hydrolysis of a capped transcript, releasing [³H]m7GDP and 5′-pRNA (58). Hydrolysis of m7GpppA(G) and UDP glucose was not observed, suggesting that the enzyme is a specific pyrophosphatase. Additional experiments have demonstrated that: (a) Decapping rate decreases significantly with decreases in RNA length. (b) The decapping enzyme hydrolyzes the 5′ cap structure but does not degrade the body of the transcript. (c) G5′ppp5′G-RNA is hydrolyzed at the same rate as m7G5′ppp5′G-RNA, indicating that the enzyme functions independently of methylation. (d) RNAs with a 5′-triphosphate end are not substrates for this activity. A 30-kDa decapping enzyme has been purified and characterized, and a gene encoding this protein has just been cloned (63; R Parker & A Stevens, personal communication). The gene is nonessential although cells in which it has been inactivated grow very slowly. Cells with disruptions of this gene fail to remove caps from both nonsense-containing mRNAs and wild-type mRNAs, supporting the notion of a convergent pathway (see below).

5′→3′ EXONUCLEASES: WHAT A LONG STRANGE TRIP IT'S BEEN   The best characterized ribonuclease known to be involved in mRNA turnover is the 5′→3′ exoribonuclease expressed from the *XRN1* gene in the yeast *S. cerevisiae* (57, 58). The Xrn1 protein (Xrn1p) has been purified to near homogeneity and has a mass of 160 kDa (59). Xrn1p degrades RNAs with a 5′-terminal phosphate, while capped RNAs are resistant to degradation (57). RNAs harboring a 5′ triphosphate are approximately 10-fold poorer substrates than RNAs with a monophosphate at the 5′ end (57, 59). Three other observations are consistent with a role of Xrn1p in mRNA turnover: (a) It is a very abundant protein [approximately 29,000 molecules/cell; (159)]. (b) Immunofluorescence microscopy suggests that Xrn1p is present only in the cytoplasm of cells (159). (c) Cells that are deleted for *XRN1* and either *SKI2* or *SKI3*, whose products are thought to be negative regulators of translation of uncapped and deadenylated viral mRNAs (160), are not viable. The latter result indicates that uncapped and deadenylated mRNAs, which are normally degraded by Xrn1p, may now allow unregulated translation of a subset of mRNAs, resulting in cell death.

The *XRN1* gene has been cloned and sequenced (161) and shown to be identical to several other independently isolated genes, i.e. *DST2*, *SEP1*, *RAR5*, and *KEM1* (162–166). The latter genes were isolated in screens thought to identify factors in nuclear fusion, plasmid stability, and DNA strand exchange during genetic recombination. The *XRN1* gene is nonessential, but its deletion causes a slow growth phenotype (161). Cells that are deleted for *XRN1* have altered ribosomal RNA processing and accumulate an internal transcribed spacer (ITS1) fragment of the pre-rRNA (167), as well as the aforementioned oligoadenylated decapped mRNAs (64). The *XRN1* gene has homology with the *HKE1/RAT1/TAP1* gene in the yeast *S. cerevisiae* and the *DHP1* gene in *Schizosaccharomyces pombe* (168–170). The *HKE1* gene is essential, and its protein, called Xrn2p, has been purified and demonstrated to be a $5' \rightarrow 3'$ exoribonuclease with activites similar to those of Xrn1p (171). The role of this nuclease in mRNA metabolism is unclear, however, since a mutation in this gene inhibited nucleocytoplasmic mRNA transport, suggesting a nuclear localization of its protein product (168). Yeast cells likely do have more than one $5' \rightarrow 3'$ exonuclease capable of degrading mRNA, however, because low levels of decay intermediates trimmed to the site of a $5'$ oligo(G) insertion can still be detected in *xrn1Δ* strains (24).

The most perplexing features of Xrn1p are its biochemical activities outside the presumed sphere of mRNA metabolism. Purified Xrn1p has a modest $5' \rightarrow 3'$ DNase activity (172), a potent DNA strand exchange activity (165), and the ability to promote microtubule bundling and polymerization (173). Interestingly, *xrn1Δ* strains are sensitive to benomyl, a drug that promotes microtubule disassociation (164). Xrn1p has also been suggested to cleave a G4 DNA tetraplex structure located in telomeres (174). Compared with wild-type strains, telomeres are shorter in *xrn1Δ* strains, and the suggestion has been made that the slow growth phenotype associated with *xrn1Δ* strains is a consequence of reduced telomere size (175). While these observations indicate that Xrn1p is a very interesting protein involved in a diverse set of biochemical activities, they may be difficult to reconcile with phenotypes in *XRN1*-deficient cells. This concern arises from the observation that the uncapped oligoadenylated mRNAs that accumulate in *xrn1Δ* cells (64) have significant reductions in specific translational activity (R Ganesan & A Jacobson, unpublished experiments) and may thus promote numerous pleiotropic effects.

Cytoplasmic exoribonuclease activities have been identified and characterized in other systems. A $5' \rightarrow 3'$ exoribonuclease activity was observed and partially purified in cytoplasmic extracts from mouse sarcoma 180 ascites cells (56, 177). Characterization of this activity demonstrated that it degrades RNAs in the $5' \rightarrow 3'$ direction, releasing $5'$ monoribonucleotides that were cleaved at the first, second, or third phosphodiester bond. RNAs with either capped or triphosphate $5'$ ends are degraded equally well. Further experiments are required to determine the role of this exoribonuclease in mRNA turnover.

Sequence Elements That Regulate mRNA Decay Dependent on Prior Poly(A) Shortening

The differences in decay rates in stable and unstable mRNAs are attributable to the presence or absence of specific sequence elements. With the exception of the stabilizer sequences that appear to be associated with the yeast *PGK1* mRNA (38, 40) and the human α-globin mRNA (34, 35), all identified elements promote mRNA destabilization, a function most probably accomplished by regulating the rates of poly(A) shortening or decapping. These sequences include the well known 50–100-nt AU-rich elements (AREs) located in the 3′-UTRs of mammalian mRNAs, as well as other less–well characterized sequences in mRNA coding regions and 3′-UTRs. Their properties are summarized below.

AU-RICH ELEMENTS (AREs)   The importance of 3′-UTR AREs to the regulation of eukaryotic gene expression was initially recognized as a consequence of three experimental observations: (*a*) Oncogenic forms of the *c-fos* gene were discovered to have lost a 67-nt AT-rich 3′-UTR sequence present in non-oncogenic forms of the gene (178). (*b*) AU-rich sequences, in particular the sequence AUUUA, were found in the 3-UTRs of highly regulated proto-oncogene, transcription factor, cytokine, and lymphokine mRNAs (179, 180). (*c*) Prototypical chimeric mRNA experiments demonstrated that AU-rich segments of the GM-CSF and *c-fos* 3′-UTRs could destabilize otherwise stable β-globin reporter mRNAs (119, 180). A large number of unstable mRNAs have since been shown to have A+U-rich sequences within their 3′-UTRs (20, 74, 83, 90, 119, 138, 139, 180–200), but not all such sequences have comparable destabilizing effects on reporter mRNAs. Although the presence of one or more AUUUA motifs was once considered prima facie evidence for the existence of a destabilizing element, it is now recognized that there is considerable sequence and functional heterogeneity among the AREs (200). UTRs with destabilization activity include those with (e.g. *c-fos* mRNA) and without (e.g. *c-jun* mRNA) AUUUA pentamers, and the number of such pentamers, when present, can vary from 1–7 (200). Moreover, the functional context of the pentamer appears to be UUAUUUAU/AU/A (190, 192), and the general activity of the A+U-rich sequences appears to be regulated by flanking U-rich domains (139, 198).

ApRES promote rapid poly(A) shortening and subsequent decay of the mRNA "body" (19, 21, 71, 122, 137). Several experimental approaches form the basis for this conclusion, one of which was the insertion of ARE sequences into the 3′-UTRs of stable reporter mRNAs. Experiments with chimeric β-globin mRNAs containing 3′-UTR insertions of the *c-fos* ARE demonstrated that, while poly(A) shortening still preceded decay of the remainder of the mRNA, the rate and extent of poly(A) shortening were greatly enhanced as was the rate of decay of the mRNA body (21). Subsequent experiments identified mutations in the ARE that generally either slow both the rate of poly(A) shortening and the overall rate of decay, or have little or no effect on poly(A) shortening rates but markedly reduce decay rates for the mRNA body (21,

139). The former class, but not the latter, obviously has a longer than normal lag before the onset of decay. The U-rich segment of the c-fos ARE (domain II) appears to enhance the rate of poly(A) shortening promoted by the segment containing the AUUUA pentamers (domain I), but has no autonomous desta-bilizing activity. In wild-type mRNAs domain II sequences do not appear to have a role in subsequent decay of the mRNA body; however, domain II can enhance both poly(A) shortening rates and mRNA decay rates in mRNAs with mutations in domain I (139, 200).

The destabilizing and poly(A) shortening functions of the AREs are thought to be mediated by one or more factors with specific binding activities. ARE-containing mRNAs form larger than average mRNPs (201), and numerous proteins that bind ARE-containing 3'-UTRs have been identified (186, 202–219) but their functions are largely unknown. Their abundance, binding activity, and cellular localization does not necessarily correlate with decay rates of mRNAs that are their presumed targets. Some (e.g. see 213) appear to shuttle between the nucleus and the cytoplasm, suggesting several possibilities. For example, factors involved in cytoplasmic mRNA decay may form part of an mRNP complex assembled in the nucleus and may also be involved in mRNA transport. Independent evidence for nuclear determination of cytoplasmic fates of mRNAs has been reported (77–79). Alternatively, proteins bound to the newly transported mRNP may exchange with specific destabilization factors in the cytoplasm. The position-independence of the activity of domain II of the *c-fos* ARE (139) suggests that it may enhance factor binding at a nearby specific site, e.g. the AUUUA-containing domain. These results, the multiplic-ity of the factors, and sedimentation analysis of ARE-containing mRNPs suggest that the poly(A) shortening and mRNA decay activities of the 3'-UTR are dictated by a complex of bound proteins.

Several of the proteins with apparent ARE-binding activity correspond to proteins known to have other functions. These include the somewhat under-standable detection of proteins with known RNA-binding activity, such as hnRNP A1, A0, and C (210, 212, 213), and Hel-N1 (215). These proteins may participate in the formation of mRNP complexes essential for mRNA splicing, transport, translation, or decay. Quite unexpected, however, was the discovery of ARE-binding proteins known to have enzymatic functions in intermediary metabolism. The latter group includes a thiolase (211), an enoyl-CoA hydratase (218), and glyceraldehyde-3-phosphate dehydrogenase (GAPDH) (217). The significance of such dual function proteins is unclear, but may be related to the observations that GAPDH also has tRNA-binding activity (220) and that other dehydrogenases also have specific RNA-binding activities (221, 222). The RNA binding activity of GAPDH has been localized to the NAD^+-binding domain (217), suggesting that small metabolites may regulate mRNA decay. In this regard, it is interesting to note that the IRE-BP, a protein that regulates accessibility of an endonucleolytic cleavage site in the transferrin receptor mRNA, is an aconitase whose RNA-binding activity is regulated by cellular levels of a Fe-S complex (see below).

OTHER 3'-UTR INSTABILITY ELEMENTS THAT REGULATE POLY(A) SHORTENING
The most well characterized of the non-ARE 3'-UTR instability elements is that associated with the yeast *MFA2* mRNA (136). Like the ARE-containing elements of mammalian mRNAs, this element promotes both rapid deadenylation and rapid onset of decay, and mutations that block one of these functions generally block the other (22, 136). The *MFA2* 3'-UTR appears to have two functional domains that are somewhat U-rich. Proteins that bind to these regions have yet to be identified, but the claim has been made that these sequences regulate the activity of poly(A) nuclease (PAN; see above) in vitro. Poly(A) tails of synthetic mRNA substrates lacking the *MFA2* 3'-UTR were shortened by purified PAN to 10–25 adenylate residues before the reaction slowed markedly (66). In mRNAs containing the *MFA2* 3'-UTR, however, deadenylation rates were stimulated and poly(A) removal was complete, a result attributed to a switch from a distributive to a processive activity (66). These results are of interest in light of recent experiments evaluating the deadenylation kinetics and poly(A) shortening intermediates observed for mRNAs containing either the *c-fos* or GM-CSF ARE (223). mRNAs with the *c-fos* ARE showed synchronous poly(A) shortening at a uniform rate whereas mRNAs with the GM-CSF ARE shortened their poly(A) tails asynchronously, accumulating deadenylated intermediates. These results, like those of mRNAs with and without the *MFA2* 3'-UTR, have been interpreted as an indication of processive and distributive deadenylation promoted by the GM-CSF and *c-fos* AREs, respectively (223).

CODING REGION INSTABILITY ELEMENTS   A role in poly(A) shortening has also been demonstrated for coding region instability elements in the mammalian *c-fos* mRNA and the yeast *MATα1* mRNA. Instability elements in both of these mRNAs were originally identified by their ability to destabilize chimeric reporter mRNAs (20, 21, 88, 90, 91). The coding region element of the *MATα1* mRNA is the only detectable instability determinant in that mRNA (88, 89) whereas the two coding region determinants of the *c-fos* mRNA complement a potent 3'-ARE (20, 21, 91). Surprisingly, the activity of the *c-fos* elements is limited to serum-stimulated cells and is not detectable when the mRNA is expressed constitutively (21). The function of these elements has been shown to be analogous to that of the 3'-UTR elements in that they enhance mRNA decay rates by promoting rapid poly(A) shortening (21, 91; G. Caponigro & R Parker, submitted). Theoretically, such a function in a coding region element could depend on the actual mRNA sequence or on the sequence of the nascent polypeptide derived from the mRNA. For the *c-fos* elements, the latter possibility has been excluded by demonstrating that a +1 frameshift had no effect on destabilizing activity (225). This experiment also excluded the possibility that the destabilization activity depended on unusual codon usage within the elements, an issue considered because the 5′ half of the 65-nt *MATα1* element is enriched for rare codons that appear to ensure its activity (89; see below). The coding region elements in both mRNAs require ongoing mRNA translation for their activity (88, 91; A Hennigan & A Jacobson, submitted), suggesting a role for the ribosome in delivering or removing site-specific stability or instability factors (discussed below). Two polypeptides that bind one of the *c-fos* determinants have been identified (227), but their role in promoting destabilization has not been determined.

Poly(A) Shortening and Translation

POLY(A) TAILS INFLUENCE mRNA TRANSLATABILITY   A variety of different experimental approaches have indicated that, in addition to being a modulator of mRNA turnover, the poly(A) status of an mRNA can be a determinant of mRNA translational efficiency (reviewed in 44). Electroporated tobacco, carrot, maize, rice, CHO, and yeast cells, and microinjected *Xenopus* oocytes all show substantially higher translation activity with adenylated, as opposed to deadenylated, mRNAs (127, 128, 229, 230). In developing frogs, mice, and flies translational activation or inactivation of many mRNAs is paralleled by the respective lengthening or shortening of their poly(A) tails (231–239). In extracts from reticulocytes, Ehrlich ascites tumor cells, and yeast cells poly(A)$^+$ mRNAs have approximately two to three times the translational activity of poly(A)$^-$ mRNA (240–243). The differences in translational activity in reticulocyte extracts were shown to be attributable to a reduced ability of poly(A)$^-$ mRNAs to join 60S ribosomal subunits to 48S preinitiation complexes (243). Consistent with this view are observations that, in the same in vitro system, poly(A)$^-$ histone mRNA formed a significantly higher percentage of "half-mers" (polysomes containing bound 40S subunits, but lacking 60S subunits) than poly(A)$^+$ globin mRNA (244).

Further evidence for a poly(A) role in translational initiation comes from the observation that the stimulatory effect of a poly(A) tail in electroporated CHO, yeast, and tobacco cells depends on the cap status of the respective mRNAs (230) and from recent studies of the translational repression of the uncapped and unadenylated mRNAs expressed by yeast cytoplasmic dsRNA viruses. In the latter study, the yeast *SKI2, SKI3,* and *SKI8* gene products were shown to be negative regulators of translation of these mRNAs as well as other nonviral uncapped unadenylated mRNAs (245–247). The effect was shown to be specific for poly(A)$^-$ mRNAs (as opposed to uncapped mRNAs) and has been interpreted as a reflection of the existence of a normal component of the protein synthesis apparatus that represses translation unless antagonized by the presence of a poly(A) tail on an mRNA (247). Expression of poly(A)⁻ viral mRNAs could also be reduced by mutations in *MAK* genes, most of which appear to reduce the levels of 60S ribosomal subunits (248). The simultaneous reduction of both 60S subunits and the translation of poly(A)⁻ mRNA implies, as did the translation results in vitro (see above), that the poly(A) tail stimulates 60S joining. These and other results (see 44) suggested that factors bound to the 5′-cap and the 3′-poly(A) tail interact.

THE PABP, COMPLEXED TO POLY(A), HAS A ROLE IN REGULATING TRANSLATION
Experiments characterizing the effects of PABP depletion or inactivation in yeast provided evidence that the PABP is the likely mediator of the translational effects of a poly(A) tail. In PABP-depleted cells, the level of polysomes was shown to decrease and was paralleled by an increase in the quantity of monosomes (65). Although this was originally interpreted as direct evidence for an effect on translation initiation, the fact that the size distribution of the remaining polysomes did not change is inconsistent with classic effects of initiation inhibition (249) and suggests a more complex effect.

An effect on translation was also indicated by an analysis of the extragenic suppressors of a temperature-sensitive mutation in the yeast *PAB1* gene. The suppressors comprised seven complementation groups, and representatives of each group showed marked decreases in their relative levels of 60S ribosomal subunits (65). Two suppressors, *spb2-1* and *spb4-1,* were shown to have mutations, respectively, in the gene for the 60S ribosomal protein L46 and in a gene encoding a putative rRNA helicase thought to be involved in the maturation of 25S rRNA (65, 250). These data indicate that translation in the absence of functional PABP may require an alteration in the structure of the 60S ribosomal subunit, a conclusion consistent with the function ascribed to poly(A) in reticulocyte extracts and in yeast viral systems (see above).

Other data supporting a role for the PABP in translation are indirect. They include the demonstration that, under conditions of translational discrimination between poly(A)⁺ and poly(A)⁻ mRNAs in vitro, the mRNP proteins cross-linked to otherwise identical representatives of these two classes of mRNA are the same except for PABP (243, 251). In these experiments the appearance of PABP multimers was dependent on poly(A) tail length which, in turn, correlated with increases in translational efficiency. Other indirect experiments showed that exogenous poly(A) is a potent and specific inhibitor of the translation of capped poly(A)⁺, but not poly(A)⁻ mRNAs in rabbit reticulocyte, wheat germ, L-cell, and pea seed extracts (243, 252–256), as well as in *Xenopus* oocytes (127). Inhibition activity appeared to be targeted to initiation of protein synthesis and was inversely related to poly(A) size, with the smallest effective size being similar to that protected by a monomer of PABP (252). These results have been interpreted to indicate that exogenous poly(A) inhibits translation by limiting the availability of unbound PABP, a conclusion supported by the observation that addition of purified PABP to reticulocyte lysates overcomes the inhibitory activity (256). Other experiments, however, have shown that poly(A) inhibition of translation can be reversed by simultaneous addition of translation initiation factors eIF-4F, eIF-4B, and eIF-4A (257), a result suggesting that exogenous poly(A) has a high affinity for another component(s) of the translation apparatus and that addition of purified PABP simply competes for this interaction. Since the poly(A)-PABP complex is more of an enhancer of translation than an essential requirement, an alternative interpretation of this result is that increased concentration of translation factors may be able to stimulate translation initiation independent of the poly(A) tail.

3′-UTR INSTABILITY ELEMENTS MAY HAVE A ROLE IN REGULATING TRANSLATION
As noted above, sequence elements within the 3′-UTRs of many mRNAs promote poly(A) shortening and rapid mRNA decay. Prototypical of such elements are the mammalian AREs, such as those present in the *c-fos*, GM-CSF, and β-interferon mRNAs (200). How such elements promote decay and deadenylation remains to be established, as does the role of the large number of factors that appear to interact with the elements. Described below (see section on Simultaneous Control of mRNA Translation and Turnover by Poly-(A) Length: A Model) is a model in which factors associated with mRNA 5′ and 3′ ends interact to form a functional mRNP, i.e. an mRNP with optimal translation activity. An impetus to the formulation of this model was the set of observations indicating a translational role for poly(A) and the PABP. Since poly(A) and PABP have apparent roles in mRNA translation and turnover it would follow that other 3′-UTR elements and factors with roles in mRNA turnover may also function in translation. This appears to be the case for the 3′-ARE associated with the *c-fos*, GM-CSF, and β-interferon mRNAs. Several independent experiments have shown that the 3′-UTRs of these mRNAs inhibit their translation, or the translation of chimeric mRNAs, in reticulocyte extracts and in *Xenopus* oocytes (97–101). For β-interferon mRNA, the inhibitory effect could be localized to a region of the UTR containing several UUAUUUAU repeats (99). Moreover, translational inhibition by this segment in reticulocyte extracts correlated with poly(A) tail length, i.e. maximal inhibition was observed when mRNAs had an ARE and long poly(A) tails (100). The latter experiments, and others assessing accessibility of the ARE, suggested that the ARE and the poly(A) tail interact, possibly via ARE-binding proteins and the PABP (100).

SOME INSTABILITY ELEMENTS, KNOWN TO PROMOTE POLY(A) SHORTENING, HAVE TO BE TRANSLATED TO FUNCTION   Numerous experiments have shown that inhibitors of translational elongation will stabilize mRNAs, including those for which poly(A) shortening is a prerequisite to decay (5, 15, 19, 37, 81, 180, 181, 258–261). The global effects of translation inhibitors on mRNA decay can be mimicked by mutational alteration of translation elongation rates (80), but such experiments still do not distinguish between direct and indirect effects of inhibition. Ongoing translation has been shown to be required for the de novo synthesis of decay factors (72, 104), for the decapping of yeast mRNAs (81), for interaction between a nascent polypeptide and the decay machinery (262), and, possibly, for ribosome involvement in decay (see below). To distinguish among these possibilities, study of the effects of translational inhibition in *cis* has been useful.

Several coding region and 3′-UTR instability elements known to promote poly(A) shortening and mRNA decay have been shown to require ongoing translation for their activity. For the *c-fos* coding region elements, this conclusion was drawn from experiments in which a strong stem-loop was inserted into the 5′-UTR of an element-containing mRNA. Sucrose gradient analysis demonstrated that the stem-loop had blocked translation of the mRNA, and standard analyses of mRNA levels and poly(A) shortening showed that translational inhibition markedly reduced the rates of mRNA deadenylation and decay (91). Experiments in yeast have shown that the deadenylation-promoting element from the *MATα1* mRNA is also dependent on ribosome progression for activity (88). In two chimeric mRNAs normally destabilized by the *MATα1* element, insertion of an upstream nonsense codon led to mRNA stabilization. Subsequent experiments have shown that ribosome progression midway through the 65-nt element is sufficient to promote decay. This transition point is preceded by a sequence highly enriched for rare codons and followed by an AU-rich region that contains a 19-nt sequence reiterated immediately downstream of the element (88, 89; A Hennigan & A Jacobson, submitted). Earlier experiments showed that the rare codon region, although essential for rapid decay, could be replaced by a completely different rare codon segment (89). Detailed mutagenesis of the transition region showed that deletion of three nucleotides 3′ to this site shifted the stability boundary one codon 5′ to its wild-type location. Conversely, constructs containing an additional three nucleotides at this same location shifted the transition downstream by an equivalent sequence distance (A Hennigan & A Jacobson, submitted). These results suggest that *MATα1* mRNA destabilization, and hence the initiation of poly(A) shortening, is triggered by an interaction among translating ribosomes and a downstream sequence element. The interaction is likely to be facilitated by reduced translational elongation rates that occur in response to the contiguous rare codons and may represent either the delivery or displacement of a factor by the ribosome.

Utilizing translational inhibition in *cis,* the role of ongoing protein synthesis for the activity of 3'-UTR instability elements from the yeast *MFA2* mRNA and the mammalian *c-fos* and GM-CSF AREs has been tested. Studies with the two mammalian AREs have had conflicting results (92, 201, 223, 261, 263). Three approaches have been used to inhibit ribosome loading onto mRNA, including insertion of a strong stem-loop into the mRNA 5'-UTR (92, 223), insertion of an iron-response element (IRE) into the mRNA 5'-UTR followed by cell culture in the presence or absence of iron (261, 263), and mutation of the mRNA initiation codon (201). Sedimentation analysis of cell extracts or immunoprecipitation of pulse-labeled protein was used to demonstrate that ribosome association with mRNA correlated with the availability of an initiation site. Using each of the aforementioned modes of inhibition, several studies showed that mRNAs containing the GM-CSF ARE were dependent on translation for destabilization (92, 201, 263). This ARE did not function when placed within coding sequences (201), and the inhibitory effect of a strong 5' stem-loop was shown to specifically target translation (as opposed to processive exonucleases) since insertion of a poliovirus internal ribosome entry site downstream of the stem-loop restored rapid mRNA decay (92). Experiments utilizing an IRE and iron deprivation to inhibit translation also showed that the *c-fos* ARE only functioned on translated mRNAs (263).

In contrast, two other studies have shown that decay of mRNAs containing the *c-fos* or GM-CSF ARE is unaffected by translational status (223, 261). The basis for these inconsistencies is, at present, unclear, but may reflect differences in decay pathways in different cells, differences that occur in a given cell type grown under different conditions, or differences attributable to the respective reporter mRNAs (200). For example, the experiments of Chen et al (223) and Winstall et al (263) examined the effects of the two AREs in NIH3T3 cells, but used different reporters and different cell-growth conditions. In the former study, the two AREs were incorporated into chimeric β-globin mRNAs, with or without strong 5' stem-loops, and mRNA synthesis was controlled by the inducible *c-fos* promoter (223). These cells were thus in G_0 to G_1 transition during decay rate measurements. In the latter study, the two AREs were inserted into α-globin mRNAs, translation was regulated by a 5' IRE and iron avaliability, and cells were cultured under normal growth conditions (263).

An alternative explanation for the discrepancies follows from experiments in yeast that show that insertion of a strong stem-loop into the 5'-UTR of the *MFA2* mRNA markedly reduces its translatability but does not alter the rates of mRNA deadenylation and decay or the nature of decay intermediates (81). This result at first appears to indicate that decay of the *MFA2* mRNA occurs independently of its translation. However, insertion of the same stem-loop into the 5'-UTR of the *PGK1* mRNA accelerated the turnover of that mRNA to such a large extent that its kinetics of deadenylation and decay were almost indistinguishable from those of the *MFA2* mRNA (24). These results suggest that, in the proper context, secondary structures inserted into mRNA 5'-UTRs may promote rapid mRNA decay, and that decay may have all the hallmarks of normal decay. Proper context may well include not only the location of the secondary structure within the 5'-UTR, but the presence or absence of atypical sequence features elsewhere in the mRNA.

An additional test of the importance of translation to the function of 3'-UTR elements evaluated the effects of positioning a strong stem-loop just 5' or just 3' to the GM-CSF ARE (264). These experiments showed that a very stable secondary structure placed 5' to the ARE, either before or after the normal translational terminator, enhanced mRNA stability approximately 20-fold. The same stem-loop placed 3' to the ARE, however, had only a minor effect. In light of earlier experiments that showed that the GM-CSF ARE was not functional when placed within coding sequences (201), these experiments suggest a mechanism similar to that occurring in the *MATα1* mRNA (see above). Translation upstream of the ARE appears to be important for the delivery of a ribosome-associated factor or for ribosome-mediated changes in downstream mRNP structure. Since destabilization does not occur when the ARE is present in coding sequences, translational termination may be required to release one ribosome subunit or to shed or add a specific ribosome-associated factor.

SIMULTANEOUS CONTROL OF mRNA TRANSLATION AND TURNOVER BY POLY(A) LENGTH: A MODEL   For many mRNAs, the extent of poly(A) shortening concurrently determines the time of onset of mRNA decay and the efficiency of translational initiation (see above). We believe that this coincidence reflects the existence of a transient form of an mRNA that is optimized for translation and resistance to degradation. Complexed with proteins, this mRNP may acquire this state by interactions between factors bound to its 5' and 3' ends (the closed loop; 44, 105, 238, 243, 252, 265–267) or by localization to a particular subcellular site (268–270). Crucial to the establishment of this state is the presence of the PABP or other specialized mRNA binding protein. These factors, in turn, require the presence of their respective binding sites which, for PABP, is a poly(A) tract of sufficient length (44). When in this state, the efficiency of translational initiation is optimized, possibly by enhancement of the 60S joining step. Enhancement may occur because factors interacting with the poly(A)-PABP complex, or proteins bound to other 3'-UTR elements, interact directly with the 60S subunit, an initiation factor, or a repressor of 60S joining (44, 65, 105, 243, 247, 257). The enhancement effect is thought to maximize initiation efficiency, not to be essential for it, and to be most pronounced with inherently inefficient initiators or when some component of the translational apparatus is limiting. Establishment of this state may occur as the mRNA is being transported, after it has completely entered the cytoplasm, or as it is being traversed by the first ribosome. The delay between mRNA processing and formation of the functional mRNP may be an opportunity for regulation of aberrant mRNAs, particularly those that are translation-impaired (see below).

A consequence of promoting translational efficiency is postulated to be the negative regulation of decapping. mRNAs with poly(A) tails shortened beyond a critical minimum length would thus be unable to bind PABP efficiently, stimulate initiation efficiently, or protect the cap efficiently. While such mRNAs should, in general, be substrates for rapid decay, that fate might be avoided if translation were enhanced by other non-poly(A)-dependent mechanisms or if decapping and exonucleolytic digestion were also regulated (271). Such alternative mechanisms of translation enhancement may explain why some mRNAs fail to be destabilized by substantial poly(A) shortening.

This model implies that the regulation of the poly(A) shortening rate is an important determinant of the translatibility and stability of a given mRNA. mRNA-specific *cis*-acting elements may influence this rate by stabilizing or destabilizing the poly(A)-PABP complex (15), recruiting a poly(A) nuclease, or associating with factors that interact with the poly(A) nuclease and increase its activity.

A Possible Mimic of the Poly(A) Shortening-Dependent Pathway: Histone mRNA Decay

Replication-dependent histone mRNAs differ from most other metazoan mRNAs in that their 3' ends lack poly(A) tails and their accumulation is tightly coupled to DNA replication (15, 31, 272–275). Although their structure and metabolism have often been considered to be substantially different from the rest of the cellular mRNA population, it may actually be quite similar.

Processing of histone pre-mRNAs leaves a 6-nt double-stranded stem and a 4-nt loop close to their 3'-termini (274, 276). This stem-loop is essential for the regulation of histone mRNA stability as well as histone mRNA translation (276–279), and its effects may be mediated by specific proteins that interact with the stem-loop structure [stem-loop binding proteins (SLBPs); 280–285]. A 50-kDa SLBP has been characterized and shown to bind to the stem-loop with high specificity. Mutations in the loop that abolish its in vivo function reduce binding of the 50-kDa protein (280). Stem-loop-dependent destabilization of histone mRNA requires ongoing translation, and it is deregulated if a ribosome terminates translation 300 or more nucleotides from the stem-loop structure or if a ribosome is able to translate into the mRNA 3'-UTR (94, 276–278, 286). These experiments suggest that the cell cycle regulation of histone mRNA stability depends on ribosomes positioned at a specific location relative to the stem-loop structure.

The decay pathway for histone mRNA was initially studied in an in vitro mRNA decay system (67–69). Histone H4 mRNA was shown to be degraded in a 3′ to 5′ direction, leading to the sequential accumulation of decay intermediates lacking 5 or 12 nucleotides from the mRNA 3′ terminus (67). During this step of decay, the 5′ region of the mRNA remains intact. The significance of the decay intermediates was demonstrated in experiments showing that identical intermediates were detectable in cells treated with drugs that inhibit DNA replication (68). In the in vitro system, subsequent 3′→5′ decay degrades the body of the mRNA (67), whereas, in cells, the absence of additional decay intermediates precluded definitive determination of a mode of decay. The suggestion has been made that histone mRNA is degraded in a 3′→5′ direction in vivo (68), although other nucleolytic events cannot be ruled out. The nuclease most likely to be responsible for at least the initial exonucleolytic decay has been purified (60; see above).

Additional factors that may be involved in the degradation of histone mRNAs include the histones themselves. Several lines of evidence indicate that histone mRNA turnover is autoregulated by histones that accumulate in late S-phase or early G2-phase soon after DNA synthesis stops (15, 70, 287, 288, 290–292). This possibility has been tested in a cell-free mRNA decay system, and the inclusion of the core histones and a post-ribosomal supernatant fraction (S130) was shown to accelerate the decay of the H4 histone mRNA four- to sixfold (70, 289, 292). Histone mRNA stability did not change in reaction mixtures lacking either the core histones or the S130 fraction, suggesting that the histones are modified or activated by the S130 fraction or that the S130 may provide additional factors that are required for accelerated decay (70).

An intriguing possiblity is that factors involved in histone gene expression may target the stem-loop/SLBP complex for either normal or autoregulated decay or translational enhancement. This complex could be an analog of the poly(A)/PABP complex (31) such that exonucleolytic digestion of the stem-loop eliminates its ability to bind the SLBP. Alternatively, modification of the SLBP may impair its ability to bind the stem-loop (31). Either event could disrupt a functional mRNP complex and promote further exonucleolytic digestion.

mRNA DECAY INITIATED BY IMPAIRED TRANSLATION

mRNAs that are normally stable can be targeted for rapid decay if their translation is aberrant. Examples of such mRNAs include those in which translation never initiates because of strong secondary structures in the 5′-UTR, mRNAs in which termination occurs 3′ to its normal site, and mRNAs in which termination occurs prematurely as a result of a nonsense mutation. Of these, the latter events have been characterized in the most detail.

Nonsense-Mediated mRNA Decay

A second pathway for the initiation of mRNA decay that also exemplifies the link between mRNA translation and turnover is the nonsense-mediated mRNA decay pathway. In both prokaryotes and eukaryotes, nonsense mutations in a gene can reduce the abundance of the mRNA transcribed from that gene (15, 17, 40, 48, 49, 102, 293–311). This reduced abundance is attributable to mRNA destabilization, and the cis-acting sequences and trans-acting factors responsible for this phenomenon are currently being investigated in mammalian cells, nematodes, and yeast (32, 33, 312). Presently at issue is the question of the site of mRNA decay. Experiments in yeast indicate that the accelerated turnover of nonsense-containing mRNAs is attributable to cytoplasmic events that are concurrent with mRNA translation (reviewed in 32). Some experiments in mammalian cells support this interpretation, but others indicate that certain nonsense-containing mRNAs are degraded while still in association with the nucleus, presumably targeted for turnover by either the mRNA transport pathway or by a signal transduction pathway (33, 302, 303, 308, 313–318). mRNA degradation in the nucleus would, at first, appear to be a phenomenon likely to be independent of translation and distinct from a cytoplasmic decay pathway. As will be discussed below, these apparently discordant results can be accomodated by a single model in which translation plays a prominent role.

SUBSTRATES AND FUNCTIONS OF THE NONSENSE-MEDIATED mRNA DECAY PATHWAY   The existence of a cellular mechanism for degrading nonsense-containing mRNAs raises the question of whether such mRNAs are the sole substrates of this decay pathway. Experiments in yeast have demonstrated that, in addition to mRNAs containing nonsense and frameshift mutations (40, 48, 49, 294, 304), substrates include transcripts harboring upstream open reading frames (32, 50, 319, 320) and inefficiently spliced intron-containing RNAs that escape from the nucleus to the cytoplasm (310). The latter RNAs were recognized as prominent substrates in experiments showing that the cytoplasmic abundance and stability of the CYH2, RP51B, and MER2 pre-mRNAs increased two- to fivefold in a yeast strain in which nonsense-mediated mRNA decay was inactivated (310). The prevalence of introns at the 5' ends of yeast genes (321) ensures rapid cytoplasmic degradation of unspliced pre-mRNAs. Degradation of these transcripts may minimize the generation of potentially deleterious polypeptide fragments. Support for such an mRNA surveillance function comes from the observation that recessive myosin heavy chain mutants in C.

elegans are dominant in mutant strains in which this mRNA decay pathway is inactivated (312).

An alternative function of the nonsense-mediated mRNA pathway may be to regulate the decay rates of transcripts with upstream open reading frames (uORFs). This hypothesis was initially suggested by the observation that a mutation that inserts an ATG codon in the leader region of the *CYC1* gene in yeast, creating a short uORF 5' of the *CYC1* protein coding region, results in a 100-fold reduction in the steady-state level of the mutant transcript (319). Mutations in two trans-acting factors that suppress the latter effect were identified and shown to be genes involved in nonsense-mediated mRNA decay (50, 319). Similarly, Oliveira & McCarthy (320) demonstrated that a transcript expressing the chloramphenicol acetyltransferase (CAT) gene in yeast was rapidly degraded if a uORF is inserted in the 5'UTR of the mRNA. These results suggest that a termination codon within a short uORF is recognized as a nonsense codon and promotes nonsense-mediated accelerated mRNA turnover. A small, but significant, percentage of mRNAs have uORFs (322) and, unless protected by other sequence features (see below), they are thus likely to be degraded by this pathway.

NONSENSE-MEDIATED mRNA DECAY CAN BE A CYTOPLASMIC TRANSLATION-DE-PENDENT EVENT   Numerous independent approaches demonstrate that decay of many nonsense-containing mRNAs occurs in the cytoplasm and depends on components of the translational apparatus. Evidence in support of these conclusions includes the following: (*a*) Nonsense-containing mRNAs are polysome-associated (48, 310). (*b*) Nonsense-suppressing tRNAs, mutants that inhibit translation, and inhibitors of translational elongation stabilize nonsense-containing mRNAs (80, 294, 304, 315). (*c*) A significant fraction of the protein encoded by the yeast *UPF1* gene, whose product is involved in nonsense-mediated mRNA decay, co-localizes with polysomes (102, 103). (*d*) A dominant-negative form of another yeast protein required for nonsense-mediated mRNA decay, Upf2p/Nmd2p, is only inhibitory when localized to the cytoplasm (51). (*e*) Microinjection of synthetic nonsense-containing, but not wild-type, transcripts into *Xenopus* oocyte cytoplasm is sufficient to trigger their rapid decay (323). (*f*) Nonsense-containing transcripts of the yeast *URA1, URA3, HIS4, LEU2,* and *PGK1* genes, the Rous sarcoma virus gag gene, the mouse and rabbit β-globin genes, as well as the intron-containing *CYH2, RP51B,* and *MER2* pre-mRNAs in yeast that escape nuclear retention systems and enter the cytoplasm, all disappear with kinetics that are consistent with turnover occurring in the cytoplasm (40, 48, 294, 295, 297, 305, 306, 310). (*g*) Yeast nonsense-containing mRNAs, stabilized by the translation elongation inhibitor cycloheximide, accumulate on polysomes yet resume decay upon removal of the drug and its translation block (S Zhang, E Welch, A Brown, A Jacobson & SW Peltz, in preparation).

RAPID mRNA DECAY DEPENDS ON A PREMATURE NONSENSE CODON AND SPECIFIC DOWNSTREAM SEQUENCES   All functional mRNAs contain a translation termination codon, yet they vary markedly in decay rates. It follows, then, that nonsense codon position must play a role in triggering rapid decay. This issue, and others pertaining to the sequence requirements for nonsense-dependent acceleration of mRNA decay, has been investigated most extensively in yeast. Early work showed that 5'-proximal nonsense mutations destabilized the *URA3* and *URA1* transcripts to a greater degree than those that were 3'-proximal (294, 297). More recent experiments, in which nonsense mutations have been systematically inserted into the protein coding regions of the *PGK1, HIS4, CYC1, MATα1*, and *HIS3* genes demonstrated that: (*a*) Nonsense codons, independent of the type, located within the first two-thirds to three-quarters of the respective protein coding regions accelerated the decay rate of the encoded transcripts up to 20-fold, whereas nonsense mutations located within the remaining portions of the coding regions had no effect on mRNA decay (40, 311, 325, 326). (*b*) Specific sequences 3' of the nonsense codon, which have been defined as "downstream elements" or "sensitive sites," are required for nonsense-mediated mRNA decay (40, 311, 325, 326). (*c*) The *PGK1* and *HIS4* transcripts contain a stabilizer element that, when translated, inactivates nonsense-mediated mRNA decay in *cis* (40, 311).

A requirement for sequences downstream of the stop codon was first suggested by experiments demonstrating that deletion of most of the *PGK1* protein coding region downstream of an early nonsense mutation resulted in a 10-fold stabilization of this "mini-*PGK1*" mRNA (40). Re-insertion of short regions of the deleted sequence back into the mini-*PGK1* nonsense allele demonstrated that a 106-nt sequence, when positioned 3' of the nonsense codon, could promote rapid mRNA (40). Comparable experiments showed that other sequences within the *PGK1* coding region exhibited this activity and that nonsense-mediated decay of the *HIS4* mRNA also relied on a downstream sequence element (40, 311, 325). Deletion analysis of the 106-nt *PGK1* element indicated that approximately 80 nt of 5'-proximal sequence is necessary for its function (40). Within this region are two identical ATG-containing sequences that are complementary to 18S rRNA (5' - TGCTGATGC TTTCTC TGCTGATGC - 3'). This complementary region has been defined as the sequence motif characteristic of a downstream element (40, 325).

Sequences essential to the function of a downstream element have been examined further by data base searching and by analysis of the activity of synthetic elements. Using a less stringent version of the downstream element motif (5'-TGYYGATGYYYYY-3'), a computer search of yeast DNA sequences showed that more than 75% of yeast genes harbor at least one copy of an imperfect motif (i.e. containing a single base change), and several genes contain two copies of the motif (325). Five different DNA fragments from the *PGK1*, *HIS4,* and *ADE3* genes harboring two copies of the sequence motif were able to function as downstream elements when inserted 3' of the nonsense mutation in the mini-*PGK1* gene (325). Studies with synthetic elements suggest that the region harboring the sequence motifs is the most critical component of the downstream element and that the flanking sequences serve as modulators of element activity (325). Such modulatory activity may be comparable to that observed for sequences that regulate mRNA splicing (327–335) or the promotion of mRNA decay by the *c-fos* ARE (139).

WHAT IS THE ROLE OF THE DOWNSTREAM ELEMENT?   Although downstream elements have been identified in protein coding regions of numerous genes, they do not promote rapid mRNA decay unless they are preceded by an upstream translational termination codon (40, 311, 325, 326). This finding indicates that these instability elements are not functional when translocated by ribosomes in the course of normal in-frame translation. Further support for this conclusion comes from recent experiments showing that downstream elements are functional only after at least one translational initiation/termination cycle has been completed (336). A downstream element inserted into the 5'-UTR of the *GCN4* transcript, upstream of uORF1, did not promote nonsense-mediated mRNA decay, whereas the same sequence inserted 3' of uORF1 caused accelerated mRNA turnover. These results suggest that a downstream element may, in a sense, function as a 3'-UTR instability element in promoting decay of a nonsense-containing transcript.

Several hypotheses can explain the function of the downstream element in nonsense-mediated mRNA decay. One is that failure to translate the downstream element unmasks it and promotes binding of a factor that leads to the degradation of the transcript (325, 336). Another is that reinitiation competes with normal initiation for a factor that indirectly prevents mRNA decapping. This competition leads to decapping and rapid mRNA decay (44). Another possibility is that the downstream element causes a scanning ribosome to pause in a manner that is not productive for reinitiation, but nevertheless leads to subsequent degradation of the RNA. Consistent with the view that a scanning 40S subunit searches and interacts with the downstream element is the observation that ts mutations in the *PRT1* gene, which is a subunit of the eIF3 complex, inactivate nonsense-mediated mRNA decay (E Welch & A Jacobson, in preparation).

IDENTIFICATION OF GENES ENCODING *TRANS*-ACTING FACTORS INVOLVED IN NONSENSE-MEDIATED mRNA DECAY   Studies on the nonsense-mediated mRNA decay pathways of *S. cerevisiae* and *Caenorhabditis elegans* have been particularly fruitful in the identification of genes whose products are involved in mRNA turnover. Yeast mutants that affect nonsense-mediated mRNA decay were initially isolated in a genetic screen designed to identify allosuppressors of the *his4-38* frameshift mutation (338). Subsequent screens that identified additional mutants in this pathway included those seeking omnipotent suppressors (340), mutations that increased frameshifting efficiencies (341, 342), suppressors of upstream initiation codons (319, 343, 344), or strains temperature-sensitive for nonsense-mediated mRNA decay (D Zuk & A Jacobson, in preparation). The respective wild-type genes have been isolated by complementation of these mutants or by the use of two-hybrid screens (51). To date, analyses of the alleles identified in these studies have demonstrated that mutations in the non-essential *UPF1* (also called *IFS2/SAL2*), *UPF2* (also called *NMD2/SUA1/IFS1*) and *UPF3* (also called *SUA6*) genes result in stabilization and increased accumulation of nonsense-containing mRNAs while having no effect on the abundance or stability of most wild-type transcripts (32, 40, 48–51, 365; D Zuk, AH Brown S Liebman & A Jacobson, in preparation). Mutations similar to these alleles have been identified in the nematode *Caenorhabditis elegans*. The nematode *smg* mutants comprise seven complementation groups and were identified as extragenic suppressors of myosin heavy chain B mutations that increase the abundance of nonsense-containing myosin transcripts while not affecting the abundance of wild-type mRNAs (312, 346). Interestingly, in *smg* strains, some recessive myosin alleles have a dominant-negative phenotype, presumably because of the accumulation of toxic myosin fragments (312). Genes encoding *SMG1, SMG2, SMG5,* and *SMG7* have been identified, and *SMG2* has been shown to be a homolog of the yeast *UPF1* gene (P Anderson, personal communication).

UPF1 and Upf1p   The yeast *UPF1* gene has been cloned and sequenced and shown to be: (*a*) nonessential for viability on rich media (49), (*b*) capable of encoding a 109-kDa protein with zinc finger, nucleotide (GTP) binding site, and RNA helicase (superfamily I) motifs (49, 348, 349), (*c*) identical to *NAM7,* a nuclear gene that was isolated as a high copy suppressor of mitochondrial RNA splicing mutations (348), (*d*) identical to *SAL2*, a gene identified in a screen for omnipotent suppressors, (*e*) identical to *IFS2,* a gene identified in a screen for frameshifting mutants (342, Y Cui, J Dinman, Y Weng & SW Peltz, submitted), and (*f*) partially homologous to the yeast *SEN1* gene (49, 348, 349).

   The subcellular localization of Upf1p has been investigated by insertion of a FLAG epitope tag (102) at the amino terminal end of the *UPF1* coding region or a triple HA tag from the infleunza virus at the carboxyl terminus (103). Molecular and genetic assays demonstrated that the activity of Upf1p produced by these alleles was largely unaffected by the epitope tags (32, 40, 103). Polysome-association of Upf1p was assessed by fractionation of post-mito-chondrial extracts on sucrose gradients and demonstrated that most Upf1p co-sediments with polysomes, although a small amount of the protein is also contained in the monosome and mRNP fractions (102, 103). These observations are supported by fluorescence and confocal microscopy studies that indicate a predominantly cytoplasmic localization for Upf1p (32, 103).

Upf1p has been purified from cells by exploiting the FLAG epitope tag and ion exchange and immunoaffinity chromatography (351). Purified Upf1p has a nucleic acid–dependent ATPase activity with a Km for ATP of 37 μM, indicating that it binds ATP quite well compared to other helicases in this superfamily. The Upf1p ATPase activity is stimulated by either RNA or DNA, and hydrolysis of ATP or dATP occurs at a much greater rate than that of any of the other NTPs or dNTPs. As estimated by immunoblot comparisons between whole cell extracts and purified recombinant Upf1p, there are approximately 1600 Upf1p molecules per cell (D Mangus & A Jacobson, unpublished observations). This estimate is consistent with measurements of the abundance of the *UPF1* mRNA that indicate that this mRNA is at least 100-fold less abundant than some ribosomal protein mRNAs (103) and may be present in as few as three copies per cell (D Zuk & A Jacobson, unpublished observations).

Genetic analyses indicate that Upf1p is a multifunctional protein with activities in mRNA turnover and translation termination (50; Y Weng, K Czaplinski & SW Peltz, submitted). Deletion of *UPF1* inactivates nonsense-mediated mRNA decay (40, 48, 49) and leads to marked reduction in translation reinitiation monitored in *PGK1-luc* fusions (R Ganesan & A Jacobson, in preparation), but not in *GCN4/PGK1* fusions (336). Failure to reinitiate in the former assay may be due to defects in reinitiation per se or in upstream termination events. Certain mutations in the helicase region of the *UPF1* gene inactivate its decay activity while maintaining a functional translation termination activity (Y Weng, K Czaplinski & SW Peltz, submitted). A mutant in which aspartic and glutamic acid residues of the ATP binding and hydrolysis motifs were changed to alanines lacked ATPase and helicase activity, and it formed a Upf1p:RNA complex in the absence of ATP, but had reduced complex formation in the presence of ATP. This indicates that ATP binding, but not its hydrolysis, modulates RNA binding in this mutant. The altered RNA binding activity of this mutant, as well as other Upf1p mutants, may account for their altered nonsense suppression phenotypes (Y Weng, K Czaplinski & SW Peltz, submitted).

Further evidence for a role of Upf1p in translation and mRNA turnover was uncovered in an analysis of mutations that affected programmed frameshifting (341; Y Cui, J Dinman, Y Weng & SW Peltz, submitted). One mechanism for regulating gene expression that is used predominantly by viruses is to induce elongating ribosomes to shift reading frame in response to specific mRNA signals (355–358). Frameshifting events produce fusion proteins, in which the N- and C-terminal domains are encoded by two distinct, overlapping open reading frames. In the yeast dsRNA virus L-A, a −1 ribosomal frameshift event is responsible for the production of a Gag-Pol fusion protein. M_1, a satellite dsRNA virus of L-A that encodes a secreted killer toxin, is encapsidated and replicated using the gene products synthesized by the L-A virus (355). A screen for mutations that increased the programmed −1 ribosomal frameshift efficiencies in cells identified eight chromosomal mutants that promoted the loss of the satellite killer M_1 and were called *mof* mutants (for Maintenance of Frame; 341). Molecular and genetic analysis of the *mof4-1* allele demonstrated that it led to inactivation of the nonsense-mediated mRNA decay pathway and was allelic to *UPF1* (Y Cui, J Dinman, Y Weng & SW Peltz, submitted). Two results rule out the possiblity that the loss of M_1 in *mof4-1* strains is a consequence of simply stabilizing the frameshift-containing L-A mRNA: (*a*) Overexpression of the L-A mRNA from a cDNA clone does not promote loss of the M_1 virus (359). (*b*) Strains containing other *upf1* alleles or harboring chromosomal deletions of the *UPF1* gene, which also inactivate the nonsense-mediated mRNA decay, do not promote loss of M_1 (Y Cui, J Dinman, Y Weng & SW Peltz, submitted). The fact that the *mof4-1* allele of the *UPF1* gene cannot maintain M_1 suggests that this mutation, in addition to inactivating nonsense-mediated decay and sensitizing cells to paromomycin, specifically affects programmed −1 ribosomal frameshifting efficiency (Y Cui, J Dinman, Y Weng & SW Peltz, submitted).

UPF2/NMD2 and Upf2p/Nmd2p   The *UPF2/NMD2* gene was recently isolated by two different approaches (50, 51). One strategy utilized a genetic screen to complement a *upf2* allele (50) while the second approach utilized the two-hybrid system to detect protein-protein interactions in vivo (51, 360). Screening of yeast genomic libraries identified six genes encoding potential Upf1p-interacting proteins. These included four previously uncharacterized genes, *NMD1-4* (Nonsense-Mediated mRNA Decay); *DBP2*, a gene encoding a putative RNA helicase with homology to mammalian p68 RNA helicase; and *SNP1*, a gene encoding a U1 snRNP 70-kDa protein homolog. Comparison of the DNA sequences of the *UPF2* and *NMD2* genes revealed that they are the same gene.

Characterization of the *UPF2/NMD2* gene indicates that it encodes a protein with a predicted mass of 127 kDa (50, 51). Comparisons of the Upf2p/Nmd2p sequence with those in the available data bases did not reveal any extensive identity with known protein sequences, but did identify a putative bipartite nuclear localization signal (spanning residues 26 to 46) and a putative helical transmembrane domain (spanning residues 470 to 490). These regions of the protein have been shown to be essential for function, but their actual role remains to be determined (361). The Upf1p-interacting domain of Upf2p/Nmd2p was originally localized to a 286–amino acid segment of its C-terminus that includes a sizeable hyperacidic region (51). Higher resolution mapping of the Upf1p-interacting domain of Upf2p/Nmd2p shows that it is confined to a 157–amino acid domain at the C-terminus of the molecule that does not include this highly acidic region (361). An analysis of the mRNA decay activity of missense and nonsense alleles of NMD2 indicates that the C-terminal regions of Upf2p/Nmd2p required for decay function generally overlap with those required for two-hybrid interaction with Upf1p (361). These results imply that Upf1p:Nmd2p interaction is required for nonsense-mediated mRNA decay.

High level expression of a 764–amino acid C-terminal fragment of Upf2p/Nmd2p has a dominant-negative effect on nonsense-mediated mRNA decay when the protein is localized to the cytoplasm, but not when it is localized to the nucleus, indicating that this decay pathway has a cytoplasmic component (51). Since this region of Upf2p/Nmd2p includes the Upf1p-interacting domain, a likely explanation for this effect was that cytoplasmic Upf1p was required for activity of the pathway and that the overexpressed fragment simply reduced the pool of functional Upf1p by direct and nonproductive interaction. However, recent data show that dominant-negative inhibition can also occur in the absence of a functional Upf1p-interacting domain, that a Upf1p-interacting domain is not sufficient for dominant-negative inhibition, and that overexpression of Upf1p cannot reverse the inhibition (361). The dominant-negative effects thus reflect the titration of yet another component required for decay, a conclusion supported by two-hybrid analyses that map a Upf3p-interacting domain to this fragment (F He & A Jacobson, in preparation; Y Weng, Y Kwan & SW Peltz, in preparation; see below).

Strains containing deletions of the UPF2/NMD2 gene (upf2Δ/nmd2Δ strains) are phenotypically identical to strains harboring deletions of the UPF1 gene (upf1Δ strains; 50, 51). Mutants with deletions in either gene are viable, show no apparent growth defects, and restore wild-type decay rates to nonsense-containing transcripts without affecting the decay of most wild-type transcripts. Mutations in either gene can function as omnipotent suppressors and promote slight sensitivity to the translation elongation inhibitor cycloheximide (49, 50). Strains harboring deletions of both genes (as well as strains doubly mutant for the UPF2/NMD2 and UPF3 genes; see below) are also fully viable and show no additivity of effects on mRNA decay. Taken together, these results indicate that UPF1 and UPF2/NMD2 are involved in the same pathway and, as implied by the two-hybrid results, may even be part of a functional complex (50, 51).

Mutations in the UPF2/NMD2 gene were also identified in a screen to identify factors that suppress the inhibitory effects of translation start sites located upstream of a gene's normal ATG initiator. The *sua* alleles were identified previously by selecting for suppressors of the *cyc1-362* allele, a mutation that results in an ATG codon 5' to the normal *CYC1* translation start site and out of frame with the *CYC1* protein coding sequence (296). Characterization of a subset of the *sua* alleles demonstrated that *sua1* and *sua6* alleles suppress the *cyc1-362* mutation by stabilizing the *CYC1* mRNA (50, 319, 343, 344). Further analysis demonstrated that *UPF2/NMD2* and *SUA1* encode the same gene while *SUA6* is the same gene as *UPF3* (50; see below). These results provide further evidence that factors involved in regulating nonsense-mediated mRNA decay are also involved in the regulation of mRNAs with uORFs (see above).

Any postulated role for Upf2p/Nmd2p must take into account its extremely low abundance. Immunoblot comparisons of whole cell extracts and purified recombinant Upf2p/Nmd2p indicate that there are fewer than 200 molecules of this protein per cell (D Mangus & A Jacobson, unpublished observations). Estimates of *NMD2* cellular mRNA concentration are consistent with the very low abundance of the protein (D Zuk & A Jacobson, unpublished observations).

UPF3 and Upf3p   Mutations in the *UPF3* gene, like those in the *UPF1* and *UPF2/NMD2* genes, lead to stabilization of nonsense-containing mRNAs while not affecting the decay of most wild-type transcripts (49, 325). mRNA stabilization in *UPF3* mutants is comparable to that observed in *upf1Δ* or *upf2Δ/nmd2Δ* strains, and *upf3Δ* strains have no apparent growth defects (49, 365). Strains doubly mutant for the *upf2Δ/nmd2Δ* and *upf3-1* alleles do not show a further stabilization of nonsense-containing mRNAs (50). The *UPF3* gene has recently been cloned and has no apparent homologies with other genes whose products have been characterized (365). Recent results from studies utilizing the two-hybrid system indicate that Upf3p interacts with Upf2p/Nmd2p and Upf1p (F He & A Jacobson, in preparation; Y Weng, Y Kwan & SW Peltz, in preparation). These results, and the nonadditive phenotypes of the double mutants (see above), suggest that Upf1p, Upf2p/NMd2p, and Upf3p may be part of a complex that regulates nonsense-mediated mRNA decay and, possibly, certain aspects of translation (50, 51).

THE POLY(A) SHORTENING-DEPENDENT AND NONSENSE-MEDIATED DECAY PATHWAYS INTERSECT   In a transcriptional pulse-chase experiment, a nonsense-containing β-globin transcript was shown to be degraded in mammalian cells prior to loss of its poly(A) tail (21). Similar results were obtained in yeast, using a transcriptional pulse-chase to monitor the decay of a nonsense-containing *PGK1* transcript (145). Although these results are suggestive of independent pathways for mRNAs whose decay is promoted by poly(A) shortening or early nonsnese codons, respectively, other data suggest that the two pathways converge. These data include the observations that: (*a*) The abundance of nonsense-containing *PGK1*, mini-*PGK1* containing a downstream element, *HIS4* mRNAs, and the intron-containing *CYH2* pre-mRNA, increases dramatically in an *xrn1*Δ strain when compared to an *XRN1*+ strain (145, 311). (*b*) The nonsense-containing mRNAs that accumulate in an *xrn1*Δ mutant lack one or two nucleotides from their 5′-termini (145, 311). These results suggest that nonsense-containing mRNAs are decapped by endonucleolytic cleavage near their 5′ ends and are subsequently degraded by the Xrn1p exonuclease. The simplest interpretation of these results is that either poly(A) shortening or a premature nonsense codon can trigger entry into a single pathway that involves decapping and 5′→3′ degradation (44, 145, 311). Confirmation of a common pathway must await further identification and characterization of the factors and ribonucleases involved in the degradation of these two types of substrates.

In an *xrn1*⁻ strain, the poly(A) tails of the stabilized mRNAs are eventually shortened (145, 311). This suggests that these mRNAs are not immune to poly(A) shortening, but are degraded so early in their "life cycles" that poly(A) shortening has not yet occurred. The possibility that degradation of these mRNAs occurs shortly after their synthesis and processing provides a possible explanation for differences between results observed in mammalian cells and yeast (see below).

WHY DOES PREMATURE TRANSLATION TERMINATION PROMOTE mRNA DECAPPING AND DECAY?   If we start with the assumption that premature translation termination and poly(A) shortening are independent means by which to initiate mRNA decapping and exonucleolytic decay, then the obvious question is whether both events stimulate a common causative event. As described earlier (see above) we envision a fully functional mRNA as an mRNP structure in which factors bound to opposite ends of the mRNA have a role in promoting translation initiation and minimizing the rate of decapping. Two ways in which this may occur are by formation of a "closed loop" (44) or by targeting the mRNA to a specific subcellular locale. Disruption of such an mRNP structure is proposed to be an event that renders the 5′ cap more accessible to removal, and decapped mRNAs are considered to be excellent substrates for the Xrn1p exonuclease or other 5′→3′ exonucleases. Poly(A) shortening might thus trigger decay by eliminating the binding site for the poly(A)-binding protein which, in turn, could be an essential contributor to maintenance of mRNP structure.

We propose that the events triggered by an early nonsense codon also lead to the disruption of a functional mRNP structure. In one scenario a fraction of the terminating ribosomes (or subunits) resume scanning and reinitiate at an AUG codon within a downstream element. Internal reinitiation is visualized as an event that competes with normal initiation for a 3'-UTR-binding protein, possibly the PABP or a protein with which it interacts. Such competition, in turn, is proposed to reduce the effectiveness with which this protein normally promotes formation of the functional mRNP. The cap's rate of removal is thus accelerated and 5'→3' digestion ensues (44). An alternative scheme with the same end result suggests that the mRNP-disrupting event is not reinitiation, but some other form of interaction between the downstream element and a factor or ribosome subunit that had scanned downstream after termination (325). Both scenarios assume that a functional mRNP forms before translation begins. If a functional mRNP is not completely assembled before attachment of the first ribosome, then the events subsequent to premature termination may even prevent the 5'/3' interactions necessary to create that mRNP.

Since poly(A) shortening is not required for the degradation of nonsense-containing mRNAs, and since poly(A) tail length is, in essence, a clock of mRNA age (44, 134, 366, 367), the decay-initiating event must occur very early in the functional lifetime of the mRNA. We speculate that, in some instances, this may occur before the mRNA has completely entered the cytoplasm. If, for, example, the first ribosome could have access to the mRNA before it had completely severed its association with the nuclear pore, then nonsense-mediated decay of some mRNAs may appear to be associated with the nucleus (33; see below) and may even involve components of the mRNA trafficking machinery.

The *UPF/NMD* gene products are integrated into these models by postulating that they modulate translation termination at the nonsense codon. For example, they may facilitate ribosomal recognition of the translation termination codon and/or unwinding of the RNA downstream of the nonsense codon, allowing a ribosome subunit or factor(s) to scan 3' and interact with the downstream element. Their role in termination must be regulatory, however, as opposed to direct, since *upf*Δ alleles do not affect growth rates and at least two of the factors, Upf1p and Upf2p/Nmd2p, are 5- to 50-fold less abundant per cell than factors thought to play a major role in translation termination (103, 368; D Mangus & A Jacobson, unpublished observations). It is possible that the Upfp complex is a general modulator of translation, since the *mof4* allele of the *UPF1* gene also affects −1 programmed frameshifting (Y Cui, J Dinman, Y Weng & SW Peltz, submitted).

DEGRADATION OF A SUBSET OF MAMMALIAN NONSENSE-CONTAINING mRNAS APPEARS TO BE ASSOCIATED WITH THE NUCLEUS   Decay of nonsense-containing transcripts in yeasts (see above) and frogs (323) appears to take place in the cytoplasm. The cytoplasm also appears to be the site of decay for some nonsense transcripts in mammalian cells (295, 305, 306, 309). However, the degradation of many mammalian nonsense-containing transcripts, including those of the human triosephosphate isomerase (TPI) gene, the T cell receptor-β gene, the hamster dihydrofolate reductase gene, the mouse major urinary protein (*MUP*), and the human β-globin gene transcribed from an SV40-based vector expressed in Syrian hamster cells, appears to be associated with the nucleus (33, 301–303, 308, 314–318). The principal evidence for this conclusion is that, after cell fractionation, decreased abundance of these mRNAs is observed to comparable extents in both nuclear and cytoplasmic fractions and cannot be accounted for by reductions in transcription or efficiency of RNA processing.

The effect of nonsense mutations on decay of the TPI transcript has been analyzed in greater detail than for any other mammalian mRNA (300, 302, 303, 313, 315, 316). Nonsense alleles of the TPI gene, expressed from either the metallothionein promoter or the inducible *c-fos* promoter, cause approximately four- to fivefold reductions in the abundance of nuclear and cytoplasmic TPI mRNA. They also have the following characteristics: (*a*) Nonsense transcripts are degraded in a position-dependent manner, i.e. TPI mRNA is rapidly degraded if the translation termination codon is situated at or before codon 189 of the TPI protein coding sequence. (*b*) The substrate for accelerated decay is the spliced transcript, but the decrease in nonsense-containing TPI mRNA is linked to the splicing of the last intron. (*c*) Nonsense-containing TPI mRNA that enters the cytoplasm is stable and is not susceptible to nonsense-mediated mRNA decay. (*d*) In spite of evidence for nuclear-association of decay, TPI nonsense-containing mRNAs are stabilized by phenomena that imply participation of cytoplasmic translation, namely secondary structures near the 5' end of the transcript or overexpression of a nonsense-suppressing tRNA.

When comparing these observations to those in the yeast system, several questions arise, including: (*a*) Is there a relationship between the mechanisms of decay of nonsense-containing transcripts in mammals and yeast? (*b*) How can decay of a nonsense transcript be associated with the nucleus, yet be sensitive to events that imply participation by cytoplasmic translation components? (*c*) Why are some mammalian nonsense-containing mRNAs degraded in the cytoplasm? (*d*) How does splicing affect the subsequent decay phenotype of the mRNA? While several models that offer answers to these questions have been proposed (33), few have as an objective the reconciliation of the mammalian and yeast data. Our bias is that pathways of gene expression in different eukaryotes should be similar, and we can therefore envisage a limited number of explanations for the discrepancies. Moreover, we consider ribosome involvement and early onset of decay to be essential hallmarks of any interpretation of the data. One model that accomodates both systems is that mRNA association with ribosomes can occur during (or immediately after) transport from the nucleus to the cytoplasm and that degradation can ensue shortly thereafter (33). If the overall kinetics of mRNA transport, mRNP assembly, and translation initiation were significantly faster in yeast cells than in mammalian cells, and faster than the onset of decay, then decay of nonsense transcripts in mammalian cells could commence while they were still nucleus-associated, but those in yeast would already have entered the cytoplasm and formed full-fledged polysomes. An alternate view is that the mammalian transcripts are degraded by the same mechanisms used by yeast, but, in mammalian cells, newly formed mRNPs are closely associated with the outside of the nucleus until the initial rounds of translation have transpired.

The apparent cytoplasmic stability of some mammalian nonsense-containing transcripts (301, 302, 303, 308) is particularly surprising. Failure to degrade these mRNAs rapidly may be an indirect consequence of poor translation or may indicate that a pretranslational step is the only one at which their capacity for aberrant translation can be detected. The requirement for intron sequences, but not the actual process of splicing, may indicate that the mRNA is "marked" in some manner by the spliceosome, possibly by deposition of a specific protein required at some stage of the decay pathway (33, 302, 303, 313). Pre-mRNAs lacking introns, or appropriate introns, might be resistant to nonsense-mediated mRNA decay because they lack such "tags." Since most genes in yeast do not contain introns, the intron requirement for activation of nonsense-mediated mRNA decay in this organism may have been lost. In this regard, however, the detection of the *SNP1* protein as a putative Upf1p-interacting protein is tantalizing (51).

Decay of mRNAs in Which Translation Initiation is Inhibited

The nonsense-mediated mRNA decay pathways described above demonstrate that cells have evolved a means to rid themselves of mRNAs that do not synthesize proteins. Results with the yeast *PGK1* transcript suggest the existence of a second pathway for the degradation of untranslated mRNAs (24). Normally, the stable *PGK1* mRNA has a half-life on the order of 40–60 min (24, 37, 40). However, inhibiting translation of this mRNA, by insertion of a strong secondary structure upstream of its translation initiation site, reduced its half-life to 7 min (24). Decay of this aberrant mRNA occurred by the deadenylation-dependent decapping pathway common to many yeast mRNAs (24; see above), and accelerated decay was attributable to enhanced rates of both deadenylation and decapping. The accelerated decay of this mRNA suggests that there may be a subset of mRNAs within the cell whose expression could be regulated by 5'-UTR secondary structures. In such mRNAs, formation of the secondary structures would inhibit translation and promote rapid decay of the transcript. Conversely, melting of the secondary structures would allow translation to occur and would stabilize the mRNA.

Transcript destabilization caused by inhibiting translation initiation with secondary structures has not been universally observed. The half-life of a CAT mRNA expressed in yeast is not changed by stem-loop insertions (369), and stem-loop insertions into the 5'-UTRs of either the GM-CSF or nonsense-containing TPI mRNAs stabilized these normally labile transcripts (92, 315). Moreover, inhibiting translation initiation by an alternative means, namely inactivation of eIF3, has no effect on the decay rates of normal mRNAs (E Welch & A Jacobson, in preparation).

The accelerated degradation of a *PGK1* mRNA harboring a stem-loop in its 5'-UTR also demonstrates the potential danger of utilizing these structures to determine the directionality of decay or the role of translation in the decay of a given mRNA. If inhibiting translation of a given mRNA shunts it into this default pathway, then evidence for translation-independent mechanisms of decay may be erroneous. This may be especially likely for an unstable mRNA that is degraded by the deadenylation-dependent decapping pathway.

Decay of mRNAs With Aberrant Translational Termination

Another step in translation at which interference with normal events affects mRNA decay rates is termination. Studies of α2 globin mRNA, which is ordinarily very stable, have shown that failure to terminate translation at the normal termination codon can destabilize the mRNA substantially (34, 35, 370).

The human globins, and their naturally occurring mutant forms, have served as a useful paradigm for gene regulation in eukaryotes and, in particular, for the identification of nucleotides essential to different aspects of gene regulation. This is, in part, attributable to the fact that quantitative changes in α or β globin production cause the autosomal recessive diseases known as thalassemias (371–373). Studies of one group of α-thalassemias have shown that the affected α2 alleles have point mutations in the normal UAA terminator that convert it from a nonsense to a sense codon (34, 374–377). The α2[Constant Spring] (CS) allele, for example, has a U-to-C substitution in the terminator (UAA→CAA). A consequence of this mutation is that the globin reading frame is extended 31 codons, ending at the UAA embedded within the normal AAUAAA polyadenylation signal (34). The mRNA encoded by the α2 CS allele is very unstable in cells of erythroid lineage, a result shown to be attributable to ribosome progression into the 3'-UTR by experiments demonstrating that this instability could be reverted by insertion of a new terminator two codons upstream of the mutated terminator (34). Analyses of other α2 globin mutants that allowed translation into the 3′-UTR showed that destabilization occurred when the ribosome had translated between one and four additional codons (34).

These results suggested that the 3′-UTR of this mRNA was bound by factors that promoted its stabilization and that such factors were displaced by ribosomal translocation. This interpretation has been supported by the identification of three noncontiguous C-rich sites in the α2 globin 3′-UTR that, when mutated, stabilize the α2 globin mRNA irrespective of whether the upstream terminator was mutated or whether the mRNA had a functional initiation codon (35). These sites fall within the region traversed by ribosomes using the "new" downstream terminator, and they appear to form a complex (the α-complex) containing at least three different proteins whose binding is disrupted by mutations that also destabilize the mRNA (370, 378). Two of these proteins (αCP1 and 2), purified by their poly(C)-binding activity, are 80% identical and contain three KH domains known to be present in other RNA-binding proteins (378). αCP1 and 2 are not able to bind the α2 globin 3′-UTR directly, but can restore such binding activity to extracts previously depleted of poly(C)-binding activity (378). Interestingly, the α-complex, and its binding sites, are specific for α-globin mRNAs (34, 35, 370). Thus, although the α- and β-globin mRNAs are comparably stable and coordinated regulated, they have evolved independent mechanisms to maintain their stability.

We have already alluded to another instance in which a termination defect alters mRNA stability. The experiments of Curatola et al (264; see above) showed that ARE-dependent decay of the GM-CSF mRNA could be blocked by a strong stem-loop inserted just upstream of the normal site of translation termination. This result implied that ribosome progression into the 3′-UTR is essential for the rapid decay of some mRNAs and that termination events which inhibited such progression could block ribosome-mediated changes in 3′-UTR structure or delivery of specific destabilizing factors.

mRNA DECAY INITIATED BY ENDONUCLEOLYTIC CLEAVAGE

Endonuclease cleavage sites have been identified in the 3′-UTRs and coding regions of a small number of mRNAs. Those with 3′-UTR sites include the mammalian insulin-like growth factor II (IGF-II; 379–384), transferrin receptor (TfR; 385–388), apolipoprotein II (apoII; 27, 389), and *groα* (390) mRNAs, the chicken 9E3 mRNA (391), and the *Xenopus Xlhbox2B* and *Xoo1* mRNAs (54, 392). Those with coding region sites include the mammalian *c-myc* mRNA (73) and the *Xenopus* albumin mRNA (55, 393) and possibly mammalian IL2 mRNA (394) and the yeast *PGK1* mRNA (24, 144). Endonucleolytic cleavage of many of these mRNAs was initially suspected because they showed precursor/product relationships in vivo or in vitro with 5' and 3' fragments. Subsequent experiments characterizing the cleavage sites have shown that they range in complexity from short linear sequences to much longer sequences with substantial secondary structure and long range RNA:RNA interactions. For example, the endonucleolytic cleavage site in *Xlhbox2B* mRNA is encompassed within the 17-nt linear sequence ACCUACCUACCCACCUA (54), and cleavage of *Xenopus* estrogen-regulated mRNAs occurs 3' to the sequence APyrUGA (55). In contrast, cleavage within the 4kb 3'-UTR of IGF-II mRNA requires the presence of two sequence elements (I and II) separated by approximately 2.1 kb (382). Elements I and II interact to form a very stable ds-RNA stem essential for mRNA decay. Element I is contained within a region of approximately 100 nucleotides and has no noteworthy sequence features. The 400-odd nucleotides of element II, however, include two highly conserved stem-loops, the cleavage site, and a highly conserved G-rich region that appears to form a guanosine quadruplex (379, 383, 384). The complex structure of the IGF-II 3'-UTR is likely required at a minimum for nuclease recognition and exposure of the site of cleavage as a single-strand (379, 384). The suggestion has been made that the guanosine quadruplex may also protect the 3' cleavage fragment from further exonucleolytic digestion, although it is recognized that a prospective biological function for this mRNA fragment is, at present, unclear (379, 384).

Endonucleolytic cleavage of at least some mRNAs appears to be regulated by proteins capable of binding in the vicinity of the respective cleavage sites, rendering them inaccessible to nucleolytic attack (30, 54, 73, 261, 385–388). The most well characterized example of such regulated cleavage is that described for the TfR mRNA, whose stability is contingent on cellular iron concentration (30, 261, 385–388). Iron regulation of TfR mRNA stability is dependent on the presence in the 3'-UTR of at least three highly conserved stem-loop structures termed iron responsive elements (IREs), a functional cleavage site embedded between the IREs, and the IRE-binding protein (IRE-BP; also known as IRP, i.e. iron regulatory protein). When cellular iron levels are depressed, the IRE-BP has a relatively high affinity for the IREs, and binding apparently occludes nuclease access to the cleavage site. When iron levels are elevated, IRE:IRE-BP interaction is diminished and endonucleolytic cleavage occurs (30, 385, 388). An understanding of a likely mechanism for this iron-dependent switch in binding affinities followed from the recognition that the IRE-BP was identical to the cytoplasmic form of the enzyme aconitase (395) and approximately 30% identical to its mitochondrial counterparts (396, 397). Aconitase was known to be dependent on a bound [4Fe-4S] iron-sulfur cluster for activity so the IRE-BP, too, was tested for the presence of such a cluster relative to its RNA-binding activity. IRE-BP isolated from cells with high iron levels was shown to have aconitase activity, a [4Fe-4S] cluster, and low RNA binding activity, whereas IRE-BP from cells containing little iron was shown to be capable of RNA-binding but devoid of aconitase activity (398, 399). Binding of an iron-sulfur cluster is thought to maintain the IRE-BP in a closed, non-RNA-binding conformation that is reversed by cluster dissociation (385). The notion that the open conformation exposes an otherwise sequestered RNA-binding domain is supported by the demonstration that the RNA-binding site is localized to amino acids known to be part of the aconitase active site (400).

Other conditions and factors that affect endonucleolytic cleavage are suggestive of a simple interaction between a nuclease and a cleavage site. Endonucleolytic cleavage of some mRNAs is dependent on ongoing translation (391), whereas cleavage of others is not (54, 261). A translation dependence of endonucleolytic cleavage may reflect the need for ribosome displacement of a bound protein from a coding region cleavage site (73, 87), ribosome melting of secondary structure that exposes the cleavage site as a single strand, or the de novo accumulation of a specific endonuclease (55). Like the decay of nonsense-containing transcripts, endonucleolytic cleavage of at least the *Xlhbox2B*, TfR, and IGF-II mRNAs does not require prior poly(A) shortening (54, 384, 388, 392). The end result of both events (poly(A) shortening and endonucleolytic cleavage) may be the same, however, in that deadenylation or endonucleolytic cleavage may be two independent means to disrupt a functional mRNP structure and create substrates for exonucleases resonsible for the bulk of mRNA degradation.

CONCLUSION

At least three different types of events, poly(A) shortening, translational impairment, and endonucleolytic cleavage, appear to initiate mRNA decay. Seeking a common thread, we suggest that the consequence of any of these events is the disruption of an mRNP structure otherwise optimized for translation. Such disruption is considered to render the mRNP a more favorable substrate for the nucleases that degrade the mRNA to completion. *cis*-acting instability elements are perceived as those that promote mRNP disruption while *cis*-acting translational elements are perceived as those that foster assembly or maintenance of the mRNP. Sometimes, as in the case of the poly(A) tail and the 5' cap, these functions overlap, suggesting that these appendages are not just "protectors" of mRNA stability, but part of an integrated system that regulates gene expression post-transcriptionally.

In classical models, mRNA decay is a purely cytoplasmic phenomenon.

While this view is true to a large extent, nuclear structures and factors are beginning to emerge as players in this process. The involvement of the nucleus may be as simple as the association of the first ribosome with an mRNA before it has completely exited the nuclear pore (33), or it may be significantly more complicated and include the deposition of specific mRNA binding proteins during RNA processing (33, 77) or transcription (78, 79). Such mRNA binding proteins may dictate mRNA stability or translatability in the cytoplasm. The cytoplasm, too, has been underrated in its potential to complicate matters. In particular, the likelihood that many mRNAs are translated or stored at particular subcellular cites (269) suggests another mode for regulation of mRNA decay.

Numerous *trans*-acting factors that play a role in mRNA decay have now been identified and have prompted several questions about their function. In what pathways do these factors participate, and what are their substrates? Are specific RNA sequences or structures sufficient for decay, or must some of these factors bind to mRNA regions where decay is initiated? Are the factors soluble, ribosome-associated, or localized to specific sites within the cell, and what fraction of them are nucleolytic? Although the characterization of these factors is still in its earliest phases, several basic conclusions are apparent: (*a*) *Trans*-acting factors involved in mRNA decay include those that either promote or hinder mRNA turnover. (*b*) These factors include those whose substrates are global, as well as factors whose function appears to be limited to only a subset of mRNAs. (*c*) The functions of these factors range from nucleolytic activity to potentially more complex roles such as targeting mRNAs to specific degradative sites or pathways. (*d*) Although most factors discussed here are polypeptides, other types of molecules may be involved (210, 219).

ACKNOWLEDGMENTS

Research in the authors' laboratories is supported by grants from the National Institutes of Health (AJ: GM27757; SWP: GM48631), the American Cancer Society (AJ: NP-843), and the National Science Foundation (SWP: MCB9303898), and a grant in aid from the American Heart Association to SWP. SWP is also supported by an Established Investigator Award from the American Heart Association. We thank Andrew Bond, Fran Delaney, Feng He, Aidan Hennigan, Alan Maderazo, David Mangus, Ellen Welch, Shuang Zhang, and Dorit Zuk for helpful comments and other assistance with the preparation of this review.

Literature Cited

1. Jacob F, Monod J. 1961. *J. Mol. Biol.* 3:318–56
2. Brenner S, Jacob F, Meselson M. 1961. *Nature* 190:576–81
3. Gros F, Hiatt H, Gilbert W, Kurland CG, Risebrough RW, Watson JD. 1961. *Nature* 190:581–85
4. Hargrove JL, Schmidt FH. 1989. *FASEB J.* 3:2360–70
5. Ross J. 1995. *Microbiol. Rev.* 59:423–50
6. Green PJ. 1994. *Annu. Rev. Plant Physiol. Plant Mol. Biol.* 45:421–45
7. Rochaix J-D. 1992. *Annu. Rev. Cell Biol.* 8:1–28
8. Sullivan ML, Green PJ. 1993. *Plant Mol. Biol.* 23:1091–104
9. Belasco JG, Brawerman G, eds. 1993. *Control of mRNA Stability.* San Diego: Academic. 517 pp.
10. Surdej P, Riedl A, Jacobs-Lorena M. 1994. *Annu. Rev. Genet.* 28:263–82
11. Kafatos F. 1972. *Acta Endocrinol.* 168:319–45
12. Parker R, Herrick D, Peltz SW, Jacobson A. 1991. In *Methods in Enzymology: Molecular Biology of Saccharomyces cerevisiae,* ed. C Guthrie, G Fink, pp. 415–23. New York: Academic
13. Belasco JG, Brawerman G. 1993. See Ref. 9, pp. 475–93
14. Caponigro G, Parker R. 1996. *Microbiol. Rev.* 60:233–49
15. Peltz SW, Brewer G, Bernstein P, Hart PA, Ross J. 1991. *Crit. Rev. Eukaryot. Gene Exp.* 1:99–126
16. Hentze M. 1991. *Biochim. Biophys.* Acta 1090:281–92
17. Peltz SW, Jacobson A. 1993. See Ref. 9, pp. 291–328
18. Ross J, Pizarro A. 1983. *J. Mol. Biol.* 167:607–17
19. Wilson T, Treisman, R. 1988. *Nature* 336:396–99
20. Shyu A-B, Greenberg ME, Belasco JG. 1989. *Genes Dev.* 3:60–72
21. Shyu A-B, Belasco JG, Greenberg ME. 1991. *Genes Dev.* 5:221–31
22. Decker CJ, Parker R. 1993. *Genes Dev.* 7:1632–43
23. Muhlrad D, Decker CJ, Parker R. 1994. *Genes Dev.* 8:855–66
24. Muhlrad D, Decker CJ, Parker R. 1995. *Mol. Cell. Biol.* 15:2145–56
25. Mercer JFB, Wake SA. 1985. *Nucleic Acids Res.* 13:7929–43
26. Atwater JA, Wisdom R, Verma IM. 1990. *Annu. Rev. Genet.* 24:519–41
27. Williams D, Sensel M, McTigue M, Binder R. 1993. See Ref. 9, pp. 161–97
28. Cleveland DW. 1988. *Trends Biochem. Sci.* 13:339–43
29. Theodorakis NG, Cleveland DW. 1993. See Ref. 9, pp. 219–38
30. Harford JB. 1993. See Ref. 9, pp. 239–66
31. Marzluff WF, Hanson RJ. 1993. See Ref. 9, pp. 267–90
32. Peltz SW, He F, Welch E, Jacobson A. 1994. *Prog. Nucleic Acid Res. Mol. Biol.* 47:271–97
33. Maquat LE. 1995. *RNA* 1:453–65
34. Weiss IM, Liebhaber SA. 1994. *Mol. Cell. Biol.* 14:8123–32
35. Weiss IM, Liebhaber SA. 1995. *Mol. Cell. Biol.* 15:2457–65
36. Shapiro RA, Herrick D, Manrow RE, Blinder D, Jacobson A. 1988. *Mol. Cell. Biol.* 8:1957–69
37. Herrick D, Parker R, Jacobson A. 1990. *Mol. Cell. Biol.* 10:2269–84
38. Heaton B, Decker C, Muhlrad D, Donahue J, Jacobson A, Parker R. 1992. *Nucleic Acids Res.* 20:5365–73
39. Santiago TC, Bettany AJE, Purvis IJ, Brown AJP. 1987. *Nucleic Acids Res.* 15:2417–29
40. Peltz SW, Brown AH, Jacobson A. 1993. *Genes Dev.* 7:1737–54
41. Dreyfuss G, Matunis MJ, Pinol-Roma S, Burd CG. 1993. *Annu. Rev. Biochem.* 62:289–321
42. Brawerman G. 1981. *Crit. Rev. Biochem.* 10:1–38
43. Baker EJ. 1993. See Ref. 9, pp. 367–415
44. Jacobson A. 1996. In *Translational Control,* ed. J Hershey, M Mathews, N Sonenberg, pp. 451–80. Cold Spring Harbor, NY: Cold Spring Harbor Press
45. Stevens A. 1993. See Ref. 9, pp. 449–71
46. Peltz SW, Jacobson A. 1992. *Curr. Opin. Cell Biol.* 4:979–83
47. Ross J. 1993. See Ref. 9, pp. 417–48
48. Leeds P, Peltz SW, Jacobson A, Culbertson MR. 1991. *Genes Dev.* 5:2303–14
49. Leeds P, Wood JM, Lee BS, Culbertson MR. 1992. *Mol. Cell. Biol.* 12:2165–77
50. Cui Y, Hagan KW, Zhang S, Peltz SW. 1995. *Genes Dev.* 9:423–36
51. He F, Jacobson A. 1995. *Genes Dev.* 9:437–54
52. Deleted in proof
53. Bandyopadhyay R, Coutts M, Krowczynska A, Brawerman G. 1990. *Mol. Cell. Biol.* 10:2060–69
54. Brown BD, Zipkin ID, Harland RM. 1993. *Genes Dev.* 7:1620–31
55. Dompenciel RE, Garnepudi VR,

Schoenberg DR. 1995. *J. Biol. Chem.* 270:6108–18

56. Coutts M, Brawerman G. 1993. *Biochim. Biophys. Acta* 1173:57–62
57. Stevens A. 1978. *Biochem. Biophys. Res. Commun.* 81:656–61
58. Stevens A. 1980. *J. Biol. Chem.* 255: 3080–85
59. Stevens A, Maupin MK. 1987. *Arch. Biochem. Biophys.* 252:339–47
60. Caruccio N, Ross J. 1994. *J. Biol. Chem.* 269:31814–21
61. Astrom J, Astrom A, Virtanen A. 1991. *EMBO J.* 10:3067–71
62. Astrom J, Astrom A, Virtanen A. 1992. *J. Biol. Chem.* 267:18154–59
63. Stevens A. 1988. *Mol. Cell. Biol.* 8: 2005–10
64. Hsu CL, Stevens A. 1993. *Mol. Cell. Biol.* 13:4826–35
65. Sachs AB, Davis RW. 1989. *Cell* 58: 857–67
66. Lowell JE, Rudner DZ, Sachs AB. 1992. *Genes Dev.* 6:2088–99
67. Ross J, Kobs G. 1986. *J. Mol. Biol.* 188:579–93
68. Ross J, Kobs G, Brewer G, Peltz SW. 1987. *J. Biol. Chem.* 262:9374–81
69. Peltz SW, Brewer G, Kobs G, Ross J. 1987. *J. Biol. Chem.* 262:9382
70. Peltz SW, Ross J. 1987. *Mol. Cell. Biol.* 7:4345–56
71. Brewer G, Ross J. 1988. *Mol. Cell. Biol.* 8:1697–708
72. Brewer G, Ross J. 1989. *Mol. Cell. Biol.* 9:1996–2006
73. Bernstein P, Herrick D, Prokipcak RD, Ross J. 1992. *Genes Dev.* 6:642–54
74. Herrick DJ, Ross J. 1994. *Mol. Cell. Biol.* 14:2119–28
75. Krikorian CR, Read GS. 1990. *J. Virol.* 65:112–22
76. Sorenson CM, Hart PA, Ross J. 1991. *Nucleic Acids Res.* 19:4459–65
77. Schneiter R, Kadowaki T, Tartakoff AM. 1995. *Mol. Biol. Cell.* 6:357–70
78. Enssle J, Kugler W, Hentze MW, Kulozik AE. 1993. *Proc. Natl. Acad. Sci. USA* 90:10091–95
79. Gunkel N, Braddock M, Thorburn AM, Muckenthaler M, Kingsman AJ, Kingsman SM. 1995. *Nucleic Acids Res.* 23: 405–12
80. Peltz SW, Donahue JL, Jacobson A. 1992. *Mol. Cell. Biol.* 12:5778–84
81. Beelman CA, Parker R. 1994. *J. Biol. Chem.* 269:9687–92
82. Baim SB, Pietras DF, Eustice DC, Sherman F. 1985. *Mol. Cell. Biol.* 5: 1839–46
83. Fort P, Rech J, Vie A, Piechaczyk M, Bonnieu A, et al. 1987. *Nucleic Acids Res.* 15:5657–67

84. Kelly K, Cochran BA, Stiles CD, Leder P. 1983. *Cell* 35:603–10
85. Stimac E, Groppi VE Jr, Coffino P. 1984. *Mol. Cell. Biol.* 4:2082–90
86. Wisdom R, Lee W. 1990. *J. Biol. Chem.* 265:19015–21
87. Wisdom R, Lee W. 1991. *Genes Dev.* 5:232–43
88. Parker R, Jacobson A. 1990. *Proc. Natl. Acad. Sci. USA* 87:2780–84
89. Caponigro G, Muhlrad D, Parker R. 1993. *Mol. Cell. Biol.* 13:5141–48
90. Kabnick KS, Housman DE. 1988. *Mol. Cell. Biol.* 8:3244–50
91. Schiavi SC, Wellington CL, Shyu A-B, Chen C-YA, Greenberg ME, Belasco JG. 1994. *J. Biol. Chem.* 269:3441–48
92. Aharon T, Schneider RJ. 1993. *Mol. Cell. Biol.* 13:1971–80
93. Yen TJ, Gay DA, Pachter JS, Cleveland DW. 1988. *Mol. Cell. Biol.* 8:1224–35
94. Graves RA, Pandey NB, Chodchoy N, Marzluff WF. 1987. *Cell* 48:615–26
95. Laird-Offringa IA. 1992. *BioEssays* 14: 119–24
96. Pierrat B, Lacroute F, Losson R. 1993. *Gene* 131:43–51
97. Kruys V, Wathelet M, Poupart P, Contreras R, Fiers W, et al. 1987. *Proc. Natl. Acad. Sci. USA* 84:6030–34
98. Kruys V, Marinx O, Shaw G, Deschamps J, Huez G. 1989. *Science* 245: 852–55
99. Kruys VI, Wathelet MG, Huez GA. 1988. *Gene* 72:191–200
100. Grafi G, Sela I, Galili G. 1993. *Mol. Cell. Biol.* 13:3487–93
101. Marinx O, Bertrand S, Karsenti E, Huez G, Kruys V. 1994. *FEBS Lett.* 345:107–12
102. Peltz SW, Trotta C, He F, Brown A, Donahue J, et al. 1993. In *Protein Synthesis and Targetting in Yeast,* ed. M Tuite, J McCarthy, A Brown, F Sherman, H71:1–10. Berlin: Springer-Verlag
103. Atkin AL, Altamura N, Leeds P, Culbertson MR. 1995. *Mol. Biol. Cell* 6: 611–25
104. Bouvet P, Paris J, Philippe M, Osborne HB. 1991. *Mol. Cell. Biol.* 11:3115–24
105. Munroe D, Jacobson A. 1990. *Gene* 91:151–58
106. Armstrong JA, Edmonds M, Nakazato H, Phillips BA, Vaughan MH. 1972. *Science* 176:526–28
107. Ehrenfeld E, Summers DF. 1972. *J. Virol.* 10:683–88
108. Pridgen C, Kingsbury DW. 1972. *J. Virol.* 10:314–17
109. Soria M, Huang AS. 1973. *J. Mol. Biol.* 77:440–55
110. Weiss SR, Bratt MA. 1974. *J. Virol.* 13:1220–30

111. Bernstein P, Peltz SW, Ross J. 1989. *Mol. Cell. Biol.* 9:659–70

112. Marbaix G, Huez G, Burny A, Cleuter Y, Hubert E, et al. 1975. *Proc. Natl. Acad. Sci. USA* 72:3065–67

113. Minvielle-Sebastia L, Winsor B, Bonneaud N, Lacroute F, 1991. *Mol. Cell. Biol.* 11:3075–87

114. Nudel U, Soreq H, Littauer UZ, Marbaix G, Huez G, et al. 1976. *Eur. J. Biochem.* 64:115–21

115. Wilson MC, Sawicki SG, White PA, Darnell JE. 1978. *J. Mol. Biol.* 126:23–36

116. Wilson MC, Sawicki SG, Salditt-Georgieff M, Darnell JE. 1978. *J. Virol.* 25:97–103

117. Avadhani NG. 1979. *Biochemistry* 13:2673–78

118. Colot HV, Rosbash M. 1982. *Dev. Biol.* 94:79–86

119. Treisman R. 1985. *Cell* 42:889–902

120. Restifo LL, Guild GM. 1986. *Dev. Biol.* 115:507–10

121. Swartwout SG, Preisler H, Guan W, Kinniburgh AJ. 1987. *Mol. Cell. Biol.* 7:2052–58

122. Swartwout SG, Kinniburgh AJ. 1989. *Mol. Cell. Biol.* 9:288–95

123. Green LL, Dove WF. 1988. *J. Mol. Biol.* 200:321–28

124. Harland R, Misher L. 1988. *Development* 102:837–52

125. Huez G, Marbaix G, Hubert E, Leclercq M, Nudel U, et al. 1974. *Proc. Natl. Acad. Sci. USA* 71:3143–46

126. Huez G, Marbaix G, Hubert E, Cleuter Y, Leclercq M, et al. 1975. *Eur. J. Biochem.* 59:589–92

127. Drummond DR, Armstrong J, Colman A. 1985. *Nucleic Acids Res.* 13:7375–94

128. Deshpande AK, Chatterjee B, Roy AK. 1979. *J. Biol. Chem.* 254:8937–42

129. Huez G, Bruck C, Cleuter Y. 1981. *Proc. Natl. Acad. Sci. USA* 78:908–11

130. Paek I, Axel R. 1987. *Mol. Cell. Biol.* 7:1496–507

131. Cochrane AW, Deeley RG. 1988. *J. Mol. Biol.* 203:555–67

132. Krowczynska A, Yenofsky R, Brawerman G. 1985. *J. Mol. Biol.* 181:231–39

133. Medford RM, Wydro RM, Nguyen HT, Nadal-Ginard B. 1980. *Proc. Natl. Acad. Sci. USA* 77:5749–53

134. Palatnik CM, Storti RV, Capone AK, Jacobson A. 1980. *J. Mol. Biol.* 141:99–118

135. Sehgal PB, Soreq H, Tamm I. 1978. *Proc. Natl. Acad. Sci. USA* 75:5030–33

136. Muhlrad D, Parker R. 1992. *Genes Dev.* 6:2100–11

137. Laird-Offringa IA, De Wit CL, Elfferich P, van der Eb AJ. 1990. *Mol. Cell. Biol.* 10:6132–40

138. Peppel K, Baglioni C. 1991. *J. Biol. Chem.* 266:6663–66

139. Chen C-YA, Chen TM, Shyu A-B. 1994. *Mol. Cell. Biol.* 14:416–26

140. Groner B, Hynes N, Phillips S. 1974. *Biochemistry* 13:5378–83

141. Hynes NE, Phillips SL. 1976. *J. Bacteriol.* 125:595–600

142. Phillips SL, Tse C, Serventi I, Hynes N. 1979. *J. Bacteriol.* 138:542–51

143. Saunders CA, Bostian KA, Halvorson HO. 1980. *Nucleic Acids Res.* 8:3841–49

144. Vrecken P, Raue HA. 1992. *Mol. Cell. Biol.* 12:2986–96

145. Muhlrad D, Parker R. 1994. *Nature* 370:578–81

146. Baer BW, Kornberg RD. 1980. *Proc. Natl. Acad. Sci. USA* 77:1890–92

147. Baer BW, Kornberg RD. 1983. *J. Cell Biol.* 96:717–21

148. Sachs AB, Bond MW, Kornberg RD. 1986. *Cell* 45:827–35

149. Sachs AB, Davis RW, Kornberg RD. 1987. *Mol. Cell. Biol.* 7:3268–76

150. Sachs A. 1990. *Curr. Opin. Cell. Biol.* 2:1092–98

151. Adam SA, Nakagawa T, Swanson MS, Woodruff TK, Dreyfuss G. 1986. *Mol. Cell. Biol.* 6:2932–43

152. Sachs AB, Deardorff JA. 1992. *Cell* 70:961–73

153. Deleted in proof

154. Caponigro G, Parker R. 1995. *Genes Dev.* 9:2421–32

155. Kole R, Sierakowska H, Shugar D. 1976. *Biochim. Biophys. Acta* 438:540–50

156. Shinshi H, Miwa M, Sugimura T. 1976. *FEBS Lett.* 65:254–57

157. Nuss DL, Furuichi Y, Koch G, Shatkin AJ. 1975. *Cell* 6:21–27

158. Deleted in proof

159. Heyer WD, Johnson AW, Reinhart U, Kolodner RD. 1995. *Mol. Cell. Biol.* 15:2728–36

160. Johnson AW, Kolodner RD. 1995. *Mol. Cell. Biol.* 15:2719–27

161. Larimer FW, Stevens A. 1990. *Gene* 95:85–90

162. Dykstra CC, Hamatake RK, Sugino A. 1990. *J. Biol. Chem.* 265:10968–73

163. Dykstra CC, Kitada K, Clark AB, Hamatake RK, Sugino A. 1991. *Mol. Cell. Biol.* 11:2583–92

164. Kim J, Ljungdahl PO, Fink GR. 1990. *Genetics* 126:799–812

165. Tishkoff D, Johnson AW, Kolodner R. 1991. *Mol. Cell. Biol.* 11:2593–608

166. Kipling D, Tambini C, Kearsey SE. 1991. *Nucleic Acids Res.* 19:1385–91

167. Stevens A, Hsu CL, Isham KR, Larimer FW. 1991. *J. Bacteriol.* 173:7024–28
168. Amberg DC, Goldstein AL, Cole CN. 1992. *Genes Dev.* 6:1173–89
169. Kenna M, Stevens A, McCammon M, Douglas MG. 1993. *Mol. Cell. Biol.* 13:341–50
170. Sugano S, Shobuike T, Takeda T, Sugino A, Ikeda H. 1994. *Mol. Gen. Genet.* 243:1–8
171. Stevens A, Poole TL. 1995. *J. Biol. Chem.* 270:16063–69
172. Johnson AW, Kolodner RD. 1991. *J. Biol. Chem.* 266:14046–54
173. Interthal H, Bellocq C, Bahler J, Bashkirov VI, Edelstein S, Heyer WD. 1995. *EMBO J.* 14:1057–66
174. Liu Z, Gilbert W. 1994. *Cell* 77:1083–92
175. Liu Z, Lee A, Gilbert W. 1995. *Proc. Natl. Acad. Sci. USA* 92:6002–6
176. Deleted in proof
177. Coutts M, Krowczynska A, Brawerman G. 1993. *Biochim. Biophys. Acta* 1173:49–56
178. Meijlink F, Curran T, Miller AD, Verma IM. 1985. *Proc. Natl. Acad. Sci. USA* 82:4987–91
179. Caput D, Beutler B, Hartog K, Thayer R, Brown-Shimer S, Cerami A. 1986. *Proc. Natl. Acad. Sci. USA* 83:1670–74
180. Shaw G, Kamen R. 1986. *Cell* 46:659–67
181. Rahmsdorf HJ, Schonthal A, Angel P, Litfin M, Ruther U, Herrlich P. 1987. *Nucleic Acids Res.* 15:1643–59
182. Jones TR, Cole MD. 1987. *Mol. Cell. Biol.* 7:4513–21
183. Whittemore L-A, Maniatis T. 1990. *Mol. Cell. Biol.* 64:1329–37
184. Wodnar-Filipowicz A, Moroni C. 1990. *Proc. Natl. Acad. Sci. USA* 87:777–81
185. Walz G, Stevens C, Zanker B, Melton LB, Clark SC, et al. 1991. *Cell. Immunol.* 134:511–19
186. Brewer G. 1991. *Mol. Cell. Biol.* 11:2460–66
187. Altus MS, Nagamine Y. 1991. *J. Biol. Chem.* 266:21190–96
188. Deleted in proof
189. Zaidi SHE, Malter JS. 1994. *J. Biol. Chem.* 269:24007–13
190. Zubiaga AM, Belasco JG, Greenberg ME. 1995. *Mol. Cell. Biol.* 15:2219–30
191. Nanbu R, Menoud P-A, Nagamine Y. 1994. *Mol. Cell. Biol.* 14:4920–28
192. Lagnado CA, Brown CY, Goodall GJ. 1994. *Mol. Cell. Biol.* 14:7984–95
193. Nair APK, Hahn S, Banholzer R, Hirsch HH, Moroni C. 1994. *Nature* 369:239–42
194. Alberta JA, Rundell K, Stiles CD. 1994. *J. Biol. Chem.* 269:4532–38
195. Stoecklin G, Hahn S, Moroni C. 1994. *J. Biol. Chem.* 269:28591–97
196. Henics T, Sanfridson A, Hamilton BJ, Nagy E, Rigby WFC. 1994. *J. Biol. Chem.* 269:5377–83
197. Gorospe M, Baglioni C. 1994. *J. Biol. Chem.* 269:11845–51
198. Chen C-YA, Shyu A-B. 1994. *Mol. Cell. Biol.* 14:8471–82
199. Chen FY, Amara FM, Wright JA. 1994. *Biochem. J.* 302:125–32
200. Chen C-YA, Shyu A-B. 1995. *Trends Biochem. Sci.* 20:465–70
201. Savant-Bhonsale S, Cleveland DW. 1992. *Genes Dev.* 6:1927–39
202. Malter JS. 1989. *Science* 246:664–66
203. Gillis P, Malter JS. 1991. *J. Biol. Chem.* 266:3172–77
204. Bohjanen PR, Petryniak B, June CH, Thompson CB, Lindsten T. 1991. *Mol. Cell. Biol.* 11:3288–95
205. Bohjanen PR, Petryniak B, June CH, Thompson CB, Lindsten T. 1992. *J. Biol. Chem.* 267:6302–9
206. Vakalopoulou E, Schaack J, Schenk T. 1991. *Mol. Cell. Biol.* 11:3355–64
207. Bickel M, Iwai Y, Pluznik DH, Cohen RB. 1992. *Proc. Natl. Acad. Sci. USA* 89:10001–5
208. Port JD, Huang L-Y, Malbon CC. 1992. *J. Biol. Chem.* 267:24103–8
209. You Y, Chen C-YA, Shyu A-B. 1992. *Mol. Cell. Biol.* 12:2931–40
210. Myer VE, Lee SI, Steitz JA. 1992. *Proc. Natl. Acad. Sci. USA* 89:1296–300
211. Nanbu R, Kubo T, Hashimoto T, Natori S. 1993. *J. Biochem.* 114:432–37
212. Hamilton BJ, Nagy E, Malter JS, Arrick BA, Rigby WFC. 1993. *J. Biol. Chem.* 268:8881–87
213. Katz DA, Theodorakis NG, Cleveland DW, Lindsten T, Thompson CB. 1994. *Nucleic Acids Res.* 22:238–46
214. Zaidi SH, Denman R, Malter JS. 1994. *J. Biol. Chem.* 269:24000–6
215. Levine TD, Gao F, King PH, Andrews LG, Keene JD. 1993. *Mol. Cell. Biol.* 13:3494–504
216. Zhang W, Wagner BJ, Ehrenman K, Schaefer AW, DeMaria CT, et al. 1993. *Mol. Cell. Biol.* 13:7652–65
217. Nagy E, Rigby WFC. 1995. *J. Biol. Chem.* 270:2755–63
218. Nakagawa J, Waldner H, Meyer-Monard S, Hofsteenge J, Jen P, Moroni C. 1995. *Proc. Natl. Acad. Sci. USA* 92:2051–55
219. Myer VE, Steitz JA. 1995. *RNA* 1:171–82
220. Singh R, Green MR. 1993. *Science* 259:365–68
221. Hentze MW. 1994. *Trends Biochem. Sci.* 19:101–3

222. Elzinga SDJ, Bednarz AL, von Osterum K, Dekker PJT, Grivell LA. 1993. *Nucleic Acids Res.* 21:5328–31
223. Chen C-YA, Xu N, Shyu A-B. 1995. *Mol. Cell. Biol.* 15:5777–88
224. Deleted in proof
225. Wellington CL, Greenberg ME, Belasco JG. 1993. *Mol. Cell. Biol.* 13:5034–42
226. Deleted in proof
227. Chen C-YA, You Y, Shyu A-B. 1992. *Mol. Cell. Biol.* 12:5748–57
228. Galili G, Kawata E, Smith LD, Larkins BA. 1988. *J. Biol. Chem.* 263:5764–70
229. Gallie DR, Lucas W, Walbot V. 1989. *The Plant Cell* 1:301–11
230. Gallie DR. 1991. *Genes Dev.* 5:2108–16
231. Bachvarova RF. 1992. *Cell* 69:895–97
232. Wickens M. 1992. *Semin. Dev. Biol.* 3:399–412
233. Wormington M. 1993. *Curr. Opin. Cell. Biol.* 5:950–54
234. Paris J, Richter JD. 1990. *Mol. Cell Biol.* 10:5634–45
235. Vassalli J-D, Huarte J, Belin D, Gubler P, Vassalli A, et al. 1989. *Genes Dev.* 3:2163–71
236. Salles FJ, Darrow AL, O'Connell ML, Strickland S. 1992. *Genes Dev.* 6:1202–12
237. Salles FJ, Lieberfarb ME, Wreden C, Gergen JP, Strickland S. 1994. *Science* 266:1996–99
238. Richter JD. 1996. In *Translational Control*, ed. J Hershey, M Mathews, N Sonenberg, pp. 481–503. Cold Spring Harbor, NY: Cold Spring Harbor Press
239. Wickens M, Kimble J, Strickland S. 1996. In *Translational Control*, ed. J Hershey, M Mathews, N Sonenberg, pp. 411–50. Cold Spring Harbor, NY: Cold Spring Harbor Press
240. Doel MT, Carey NH. 1976. *Cell* 8:51–58
241. Hruby DE, Roberts WK. 1977. *J. Virol.* 23:338–44
242. Iizuka N, Najita L, Franzusoff A, Sarnow P. 1994. *Mol. Cell. Biol.* 14:7322–30
243. Munroe D, Jacobson A. 1990. *Mol. Cell. Biol.* 10:3441–55
244. Nelson EM, Winkler MM. 1987. *J. Biol. Chem.* 262:11501–6
245. Wickner RB. 1992. *Annu. Rev. Microbiol.* 46:347–75
246. Widner WR, Wickner RB. 1993. *Mol. Cell. Biol.* 13:4331–41
247. Masison DC, Blanc A, Ribas JC, Carroll K, Sonenberg N, Wickner RB. 1995. *Mol. Cell. Biol.* 15:2763–71
248. Ohtake Y, Wickner RB. 1995. *Mol. Cell. Biol.* 15:2772–81
249. Hinnebusch AG, Liebman SW. 1991 In
250. Sachs AB, Davis RW. 1990. *Science* 247:1077–79
251. Munroe D. 1989. PhD thesis. Univ. Mass. Med. Sch., Worcester
252. Jacobson A, Favreau M. 1983. *Nucleic Acids Res.* 11:6353–68
253. Bablanian R, Banerjee A. 1986. *Proc. Natl. Acad. Sci. USA* 83:1290–94
254. Lemay G, Millward S. 1986. *Arch. Biochem. Biophys.* 249:191–98
255. Sieliwanowicz B. 1987. *Biochim. Biophy. Acta* 908:54–59
256. Grossi MF, Standart N, Martins C, Akhayat O, Huesca M, Scherrer K. 1988. *Eur. J. Biochem.* 176:521–26
257. Gallie DR, Tanguay R. 1994. *J. Biol. Chem.* 269:17166–73
258. Laird-Offringa IA, Elfferich P, van der Eb AJ. 1991. *Nucleic Acids Res.* 19:2387–94
259. Thompson CB, Challoner PB, Nieman PE, Groudine M. 1986. *Nature* 319:374–80
260. Mullner EW, Kuhn LC. 1988. *Cell* 53:815–25
261. Koeller DM, Casey JL, Hentze MW, Gerhardt EM, Chan L-NL, et al. 1989. *Proc. Natl. Acad. Sci. USA* 86:3574–78
262. Yen TJ, Machlin PS, Cleveland DW. 1988. *Nature* 334:580–85
263. Winstall E, Gamache M, Raymond V. 1995. *Mol. Cell. Biol.* 15:3796–804
264. Curatola AM, Nadal MS, Schneider RJ. 1995. *Mol. Cell. Biol.* 15:6331–340
265. Munroe D, Jacobson A. 1989. In *Structure, Function, and Evolution of Ribosomes*, ed. W Hill, pp. 299–305. Washington, DC: ASM Press
266. Brown AJP. 1993. *Trends Cell Biol.* 3:180–83
267. Brawerman G. 1993. See Ref. 9, pp. 149–59
268. Jackson RJ, Standart N. 1990. *Cell* 62:15–24
269. Kislauskis EH, Singer RH. 1992. *Curr. Opin. Cell Biol.* 4:975–78
270. Decker CJ, Parker R. 1995. *Curr. Opin. Cell Biol.* 7:386–92
271. Decker CJ, Parker R. 1994. *Trends Biol. Sci.* 19:336–40
272. Adesnik M, Darnell JE. 1972. *J. Mol. Biol.* 67:397–406
273. Greenberg JR, Perry RP 1972. *J. Mol. Biol.* 72:91–98
274. Schumperli D. 1988. *Trends Genet.* 4:187–91
275. Marzluff WF, Pandey NB. 1988. *Trends Biochem. Sci.* 13:49–52

The Molecular and Cellular Biology of the Yeast Saccharomyces, ed. JR Broach, JR Pringle, EW Jones, 1:627–735. Cold Spring Harbor, NY: Cold Spring Harbor Lab. Press

276. Pandey NB, Marzluff WF. 1987. *Mol. Cell. Biol.* 7:4557–59
277. Levine BJ, Chodchoy N, Marzluff WF, Skoultchi AI. 1987. *Proc. Natl. Acad. Sci. USA* 84:6189–93
278. Capasso O, Bleecker GC, Heintz N. 1987. *EMBO J.* 6:1825–31
279. Sun J-H, Pilch DR, Marzluff WF. 1992. *Nucleic Acids Res.* 20:6057–66
280. Pandey NB, Sun J-H, Marzluff WF. 1991. *Nucleic Acids Res.* 19:5653–59
281. Pandey NB, Williams AS, Sun J-H, Brown VD, Bond U, Marzluff WF. 1994. *Mol. Cell. Biol.* 14:1709–20
282. Eckner R, Birnstiel ML. 1992. *Nucleic Acids Res.* 20:1023–30
283. Melin L, Soldati D, Mital R, Streit A, Schumperli D. 1992. *EMBO J.* 11:691–97
284. Williams AS, Ingledue TC, Kay BK, Marzluff WF. 1994. *Nucleic Acids Res.* 22:4660–66
285. Williams AS, Marzluff WF. 1995. *Nucleic Acids Res.* In press
286. Luscher B, Stauber C, Schindler R, Schumperli D. 1985. *Proc. Natl. Acad. Sci. USA* 82:4389–93
287. Stein GS, Stein JL. 1984. *Mol. Cell. Biochem.* 64:105–10
288. Butler WB, Mueller GC. 1973. *Biochim. Biophys. Acta* 294:481–96
289. McLaren RS, Ross J. 1993. *J. Biol. Chem.* 268:14637–44
290. Senshu T, Yamada F. 1980. *J. Biochem.* 87:1659–68
291. Bonner WM, Wu RS, Panusz HT, Muneses C. 1988. *Biochemistry* 27:6542–50
292. Peltz SW, Brewer G, Groppi V, Ross J. 1989. *Mol. Biol. Med.* 6:227–38
293. Morse DE, Yanofsky C. 1969. *Nature* 224:329–31
294. Losson R, Lacroute F. 1979. *Proc. Natl. Acad. Sci. USA* 76:5134–37
295. Maquat LE, Kinniburgh AJ, Rachmilewitz EA, Ross J. 1981. *Cell* 27:543–53
296. Stiles JI, Szostak JW, Young AT, Wu R, Consaul S, Sherman F. 1981. *Cell* 25:277–84
297. Pelsy F, Lacroute F. 1984. *Curr. Genet.* 8:277–82
298. Baumann B, Potash MJ, Kohler G. 1985. *EMBO J.* 4:351–59
299. Nilsson G, Belasco JG, Cohen SN, Von Gabain A. 1987. *Proc. Natl. Acad. Sci. USA* 84:4890–94
300. Daar IO, Maquat LE. 1988. *Mol. Cell. Biol.* 8:802–13
301. Urlaub G, Mitchell PJ, Ciudad CJ, Chasin LA. 1989. *Mol. Cell. Biol.* 9:2868–80

302. Cheng J, Fogel-Petrovic M, Maquat LE. 1990. *Mol. Cell. Biol.* 10:5215–25
303. Cheng J, Maquat LE. 1993. *Mol. Cell. Biol.* 13:1892–902
304. Gozalbo D, Hohmann S. 1990. *Curr. Genet.* 17:77–79
305. Barker GF, Beemon K. 1991. *Mol. Cell. Biol.* 11:2760–68
306. Barker GF, Beemon K. 1994. *Mol. Cell. Biol.* 14:1986–96
307. Gaspar ML, Meo T, Bourgarel P, Guenet JL, Tosi M. 1991. *Proc. Natl. Acad. Sci. USA* 88:8606–10
308. Baserga SJ, Benz EJ Jr. 1992. *Proc. Natl. Acad. Sci. USA* 89:2935–39
309. Lim SK, Sigmund CD, Gross KW, Maquat LE. 1992. *Mol. Cell. Biol.* 12:1149–61
310. He F, Peltz SW, Donahue JL, Rosbash M, Jacobson A. 1993. *Proc. Natl. Acad. Sci. USA* 90:7034–38
311. Hagan KW, Ruiz-Echevarria MJ, Quan Y, Peltz SW. 1995. *Mol. Cell. Biol.* 15:809–23
312. Pulak R, Anderson P. 1993. *Genes Dev.* 7:1885–97
313. Cheng J, Belgrader P, Zhou X, Maquat LE. 1994. *Mol. Cell. Biol.* 14:6317–25
314. Naeger LK, Schoborg RV, Zhao Q, Tullis GE, Pintel DJ. 1992. *Genes Dev.* 6:1107–19
315. Belgrader P, Cheng J, Maquat LE. 1993. *Proc. Natl. Acad. Sci. USA* 90:482–86
316. Belgrader P, Cheng J, Zhou X, Stephenson LS, Maquat LE. 1994. *Mol. Cell. Biol.* 14:8219–28
317. Kessler O, Jiang Y, Chasin LA. 1993. *Mol. Cell. Biol.* 13:6211–22
318. Qian L, Theodor L, Carter M, Vu MN, Sasaki AW, Wilkinson MF. 1993. *Mol. Cell. Biol.* 13:1686–96
319. Pinto I, Na JG, Sherman F, Hampsey M. 1992. *Genetics* 132:97–112
320. Oliveira CC, McCarthy JEG. 1995. *J. Biol. Chem.* 270:8936–43
321. Fink GR. 1987. *Cell* 49:5–6
322. Kozak M. 1987. *Nucleic Acids Res.* 15:8125–48
323. Whitfield TT, Sharpe CR, Wylie CC. 1994. *Dev. Biol.* 165:731–34
324. Deleted in proof
325. Zhang S, Ruiz-Echevarria MJ, Quan Y, Peltz SW. 1995. *Mol. Cell. Biol.* 15:2231–44
326. Yun DF, Sherman F. 1995. *Mol. Cell. Biol.* 15:1021–33
327. Hedley ML, Maniatis T. 1991. *Cell* 65:579–86
328. Ryner LC, Baker BS. 1991. *Genes Dev.* 5:2071–85
329. Hoshijima K, Inoue K, Higuchi I,

Sakamoto H, Shimura Y. 1991. *Science* 252:833–36

330. Inoue K, Hoshijima K, Higuchi I, Sakamoto H, Shimura Y. 1992. *Proc. Natl. Acad. Sci. USA* 89:8092–96
331. Tian M, Maniatis T. 1993. *Cell* 74:105–14
332. Tian M, Maniatis T. 1992. *Science* 256:237–40
333. Horabin JI, Schedl P. 1993. *Mol. Cell. Biol.* 13:7734–46
334. Huh GS, Hynes RO. 1994. *Genes Dev.* 8:1561–74
335. Yeakley JM, Hedjran F, Morfin JP, Merillat N, Rosenfeld MG, Emeson RB. 1993. *Mol. Cell. Biol.* 13:5999–6011
336. Ruiz-Echevarria MJ, Peltz SW. 1996. *EMBO J.* In press
337. Deleted in proof
338. Culbertson MR, Underbrink KM, Fink GR. 1980. *Genetics* 95:833–53
339. Deleted in proof
340. Cox BS. 1977. *Genet. Res.* 30:187–205
341. Dinman JD, Wickner RB. 1994. *Genetics* 136:75–86
342. Lee SI, Umen JG, Varmus HE. 1995. *Proc. Natl. Acad. Sci. USA* 92:6587–91
343. Hampsey M, Na JG, Pinto I, Ware DE, Berroteran RW. 1991. *Biochimie* 73:1445–55
344. Pinto I, Ware DE, Hampsey M. 1992. *Cell* 68:977–88
345. Deleted in proof
346. Hodgkin J, Papp A, Pulak R, Ambros V, Anderson P. 1989. *Genetics* 123:301–13
347. Deleted in proof
348. Altamura N, Groudinsky O, Dujardin G, Slonimski PP. 1992. *J. Mol. Biol.* 224:575–87
349. Koonin EV. 1992. *Trends Biochem. Sci.* 17:495–97
350. Deleted in proof
351. Czaplinski K, Weng Y, Hagan KW, Peltz SW. 1995. *RNA* 1:610–23
352. Deleted in proof
353. Deleted in proof
354. Deleted in proof
355. Dinman JD. 1995. *Yeast* 11:1115–27
356. Chandler M, Fayet O. 1993. *Mol. Microbiol.* 7:497–503
357. Farabaugh PJ. 1995. *J. Biol. Chem.* 270:10361–64
358. Hayashi SI, Murakami Y. 1995. *Biochem. J.* 306:1–10
359. Wickner RB, Icho T, Fujimura T, Widner WR. 1991. *J. Virol.* 65:151–61
360. Fields S, Song OK. 1989. *Nature* 340:245–46
361. He F, Jacobson A. 1996. *RNA*. In press
362. Deleted in proof
363. Deleted in proof
364. Deleted in proof

365. Lee B-S, Culbertson MR. 1995. *Proc. Natl. Acad. Sci. USA* 92:10354–58
366. Palatnik CM, Storti RV, Jacobson A. 1979. *J. Mol. Biol.* 128:371–97
367. Palatnik CM, Storti RV, Jacobson A. 1981. *J. Mol. Biol.* 150:389–98
368. Stansfield I, Tuite MF. 1994. *Curr. Gene.* 25:385–95
369. Sagliocco FA, Zhu DL, Laso MRV, McCarthy JE, Tuite MF, Brown AJP. 1994. *J. Biol. Chem.* 269:18630–37
370. Wang X, Kiledjian M, Weiss IM, Liebhaber SA. 1995. *Mol. Cell. Biol.* 15:1769–77
371. Orkin SH, Kazazian HH Jr. 1984. *Annu. Rev. Genet.* 18:131–71
372. Karlsson S, Nienhuis AW. 1985. *Annu. Rev. Biochem.* 54:1071–108
373. Kazazian HH Jr. 1990. *Semin. Hematol.* 27:209–28
374. Clegg JB, Weatherall DJ, Milner PF. 1971. *Nature* 234:337–40
375. Clegg JB, Weatherall DJ, Conto-polou-Griva I, Caroutsos K, Poungouras P, Tsevrenis H. 1974. *Nature* 251:245–47
376. Milner PF, Clegg JB, Weatherall DJ. 1971. *Lancet* 1:729–32
377. DeJong WW, Meera Khan P, Bernini LF. 1975. *Am. J. Hum. Genet.* 27:81–90
378. Kiledjian M, Wang X, Liebhaber SA. 1995. *EMBO J.* In press
379. Scheper W, Meinsma D, Holthuizen PE, Sussenbach JS. 1995. *Mol. Cell. Biol.* 15:235–45
380. De Pagter-Holthuizen P, Jansen M, Van der Kammen A, Van Schaik FMA, Sussenbach JS. 1988. *Biochim. Biophys. Acta* 950:282–95
381. Meinsma D, Holthuizen PE, Van den Brande JL, Sussenbach JS. 1991. *Biochem. Biophys. Res. Commun.* 179:1509–16
382. Meinsma D, Scheper W, Holthuizen PE, Sussenbach JS. 1992. *Nucleic Acids Res.* 20:5003–9
383. Nielsen FC, Christiansen J. 1992. *J. Biol. Chem.* 267:19404–11
384. Christiansen J, Kofod M, Nielsen FC. 1994. *Nucleic Acids Res.* 22:5709–16
385. Klausner RD, Rouault T, Harford JB. 1993. *Cell* 72:19–28
386. Mullner EW, Neupert B, Kuhn L. 1989. *Cell* 58:373–82
387. Casey JL, Koeller DM, Ramin VC, Klausner RD, Harford JB. 1989. *EMBO J.* 8:3693–99
388. Binder R, Horowitz JA, Basilion JP, Koeller DM, Klausner RD, Harford JB. 1994. *EMBO J.* 13:1969–80
389. Binder R, Hwang S-PL, Ratnasabapathy R, Williams DL. 1989. *J. Biol. Chem.* 264:16910–18

390. Stoeckle MY. 1992. *Nucleic Acids Res.* 20:1123–27
391. Stoeckle MY, Hanafusa H. 1989. *Mol. Cell. Biol.* 9:4738–45
392. Brown BD, Harland RM. 1990. *Genes Dev.* 4:1925–35
393. Pastori RL, Moskaitis JE, Schoenberg DR. 1991. Biochemistry 30: 10490–98
394. Hua J, Garner R, Paetkau V. 1993. *Nucleic Acids Res.* 21:155–62
395. Kennedy MC, Mende-Mueller L, Blondin GA, Beinert H. 1992. *Proc. Natl. Acad. Sci. USA* 89:11730–34

396. Rouault TA, Sout CD, Kaptain S, Harford JB, Klausner RD. 1991. *Cell* 64: 881–83
397. Hentze MW, Argos P. 1991. *Nucleic Acids Res.* 19:1739–40
398. Haile DJ, Rouault TA, Tang CK, Chin J, Harford JB, Klausner RD. 1992. *Proc. Natl. Acad. Sci. USA* 89:7536–40
399. Constable A, Quick S, Gray NK, Hentze MW. 1992. *Proc. Natl. Acad. Sci. USA* 89:4554–58
400. Basilion JP, Rouault TA, Massinople CM, Klausner RD. 1994. *Proc. Natl. Acad. Sci. USA* 91:574–78

Annu. Rev. Biochem. 1996. 65:741–68

RECODING: Dynamic Reprogramming of Translation

Raymond F. Gesteland and John F. Atkins

Howard Hughes Medical Institute at the Department of Human Genetics, University of Utah, Salt Lake City, Utah 84112

KEY WORDS: recoding, frameshifting, bypassing, genetic code, ribosome

---

ABSTRACT

A minority of genes in probably all organisms rely on "recoding" for translation of their mRNAs. In these cases, the rules for decoding are temporarily altered through the action of specific signals built into the mRNA sequences. Three classes are described. 1. Frameshifting at a particular site allows expression of a protein from an mRNA with overlapping open reading frames, often giving two protein products from one mRNA. 2. The meanings of code words are altered: specific stop codons can be redirected to encode selenocysteine, tryptophan, or glutamine. 3. Ribosomes can translate over coding gaps in mRNA. These novel mechanisms expand the repertoire of the genetic code and are at the heart of several regulatory schemes.

CONTENTS

INTRODUCTION 742
HISTORY OF RECODING... 743
RECODING CLASSES 744
REDIRECTION OF LINEAR READOUT..................................... 745
  +1 Frameshifting 745
  −1 Frameshifting.. 749
REDEFINITION: PROGRAMMED ALTERATION OF CODON MEANING......... 753
  Murine Leukemia Virus 754
  Selenocysteine... 755
SUBVERSION OF CONTIGUITY.. 758
CONSEQUENCES OF NATURAL "NOISE" 760
EVOLUTIONARY IMPLICATIONS.. 762
NONCANNONICAL PAIRING VS NONPAIRING............................. 763
CONCLUSION... 763

741

0066-4154/96/0701-0741$08.00

INTRODUCTION

By now we should be used to the wiliness of evolution. For just as each ecological niche seems to have a resident who makes it home, each molecular opportunity for turning some biological process to some advantage seems to be exercised. However, we have been startled again by discoveries revealing that the process of reading-out the genetic code is itself subject to programmed changes that allow for another level of diversity of gene expression. The essential universality of the code and the machinery for its readout somehow lulled us into the comfortable view that at least some processes are sacrosanct. However, even the genetic code is not sacrosanct. Some specialized niches, especially mitochondria, show exceptions to the universal code, but in those cases all mRNAs are decoded in accordance with reassigned specificities of the tRNAs involved in their "hard-wired" decoding. The phenomenon of "recoding" is different. Rule changes are programmed in mRNA sequences. In some cases specialized translational components facilitate recoding, but instructions for the rule changes are carried in mRNA itself; rule changes are under genetic control.

One of the lessons learned from studying recoding is that mRNA contains more than linear information in its nucleotide sequence. This is not a new lesson; mRNA is rich with other signals. mRNA in prokaryotes contains sequences to pair with rRNA during initiation and to allow internal ribosome entry. mRNAs also have structures: 5' stem loops to regulate accessibility of ribosome start sites, 5' stem loops to bind protein repressors, and 3' elements that play a much more important role in gene expression than previously appreciated.

In recoding, the mechanism of readout itself is under dynamic control of mRNA information through the combination of specific codons and other signals (1). These other signals are stimulators that are often manifested through RNA:RNA interactions either between mRNA and rRNA or within the mRNA in the form of folded structures. Recoding is dynamic. Decoding rules are changed temporarily, one ribosome at a time. Recoding sites have two meanings: the hardwired and the dynamically altered one. Recoding events play various roles. They all reprogram the ribosome to do something different from what is specified by strict translation of triplets according to the genetic code, changing either the linearity of readout or the definition of code words. So far redefinition of code words is only known to occur for stop codons which can be redefined to mean glutamine, tryptophan, or selenocysteine. There is no reason in principle why sense codons could not also be redefined.

"Suppression" and "readthrough" are terms used to describe insertion of an amino acid for a stop codon. "Suppression" is a genetic term that is best used in its genetic context. "Readthrough" captures the idea of getting through a barrier but does not suggest the fundamental phenomenon, a programmed change that yields a protein product with a function. In the case of selenocysteine insertion, the new amino acid itself often plays a key role in the activity of the new protein. "Redefinition" comes closer to describing the phenomenon.

DNA sequence information from many organisms is accumulating rapidly, and one of the goals is clearly to be able to predict amino acid sequences of gene products from exon sequences or cDNAs. RNA splicing and RNA editing both contribute to a lack of a one-to-one correspondence between DNA sequence and mRNA sequence, so cDNA sequence comparisons are crucial. However, recoding poses a special concern for sequence data. The open reading frame (ORF) is all important. An ORF-like sequence that shows a stop codon or information in two overlapping reading frames is not held in much esteem by the community. The temptation is great to find the "problem" in the sequencing gel that made a nice clean ORF turn ugly. We must resist this type of "cleansing" of the data bases and be mindful of the more interesting alternative of recoding. However, even with a clean sequence the frequency of recoding will not be ascertained merely by inspection of the number of long open reading frames whose translation needs to be joined together. Recoding may give a proportion of a shortened product, when some ribosomes shift frame and quickly terminate in the new frame. In other cases the carboxy terminal extension may be short. Alternatively, the important feature may be that ribosomes reach an otherwise closed region of mRNA—in order, for instance, to alter mRNA structure and provide a link to other processes, perhaps message degradation. In this case the extra protein product may be unimportant and rapidly degraded, a fate akin to that of the product of the leader peptide regions of amino acid biosynthetic operons.

This review aims to give a broad perspective on recoding. Detailed reviews have recently been published (2–6).

HISTORY OF RECODING

Hints of "funny business" in translation have been around for sometime. Early experiments showed that at least in viruses, the readout of the genetic code could be more flexible than previously appreciated: A crucial phage Qβ protein (7) required decoding UGA as Trp with a 3 to 5% efficiency (8). The synthesis of several plant virus proteins (9) and Murine Leukemia virus gag pol polyprotein was stimulated by tRNAs that decoded UAG or UGA stop codons (10). A shift in reading frame was required for synthesis of a phage MS2 coat lysis hybrid protein (11, 12), and a phage T7 protein (13). Both of these products are inessential at least in common lab host strains (11, 14, 15), but it wasn't long before the synthesis of an essential protein for transposition of the yeast

TY1 element (16–18) and of the gag pol polyprotein of Rous Sarcoma Virus (19) were also shown to require a ribosomal frameshift.

The culmination of the initial studies on recoding was twofold. First came the findings of an essential chromosomal gene, *E. coli* release factor 2, whose DNA sequence did not predict the known protein without a switch in reading frame (20). Next came the finding that there are 21 encoded amino acids, not 20 (21, 22), and that the codon (UGA) for the 21st amino acid (selenocysteine) has multiple meanings determined by signals in mRNA.

Many aspects of decoding are subject to redirection. The meaning of codons can be redefined either as standard amino acids or as the selenocysteine. The reading frame can be reset. The linearity of mRNA readout can be changed so that internal nucleotides in mRNA are skipped. All these examples share the features of dynamic reprogramming: (*a*) rule changes that only temporarily alter the "hard-wired" rules; (*b*) competition between recoding and standard decoding events; (*c*) a site of action on the mRNA; and (*d*) stimulatory signals in the mRNA.

RECODING CLASSES

To illustrate frameshifting, redefining, and bypassing, we describe several recoding events in detail, each of which exemplifies a different style and provides a framework for the many related examples (6).

Recoding phenomena are often first noticed as mRNA sequences that do not make sense, having two overlapping reading frames which together would account for the protein product or a stop codon that interrupts a logical ORF. The obvious constructs can be made, usually with reporter sequences downstream, to understand how the mRNA is translated. These constructs can be assayed either in vivo by transformation (or transfection) or in vitro by transcription and translation in a cell-free system. In most cases where in vitro and in vivo constructs have been compared, similar conclusions are reached. However, using just a small recoding window in reporter constructs is always cause for concern (4). Distant stimulators may be missed, and reporter sequences in constructs may influence recoding. This problem is of particular concern because many stimulators involve folded mRNA structures that can be influenced by unexpected, competing interactions. Testing defined cassettes in different contexts is a good precaution against these possibilities.

Evidence does show that recoding signals respond differently, at least quantitatively, if they are placed at a short distance into the mRNA. A site that is still within the region covered by an initiating ribosome can have diminished frameshifting (23). Hungry codons awaiting aminoacyl tRNAs are somehow handled differently if they are within the first 10 or so codons (24, 25).

Although analysis of many constructs permits very strong inferences about the site and mechanism of recoding, knowing the amino acid sequence of the protein product is invaluable.

REDIRECTION OF LINEAR READOUT

Both +1 and −1 examples of natural programmed ribosomal frameshifting have been described. Plus 2 or −2 have not been found except in special test systems.

+1 Frameshifting

E. coli RELEASE FACTOR 2   The DNA sequence of the release factor 2 (RF2) gene revealed a stop codon at position 26 where the open reading frame switches to the +1 frame (Figure 1). Comparison of amino acid and mRNA sequences showed that ribosomes can shift reading frame at codon 25, avoid the stop codon, and decode the main portion of the message (20). Further work showed that this site-specific shift in reading frame is not just "noise" in translation but has an efficiency of up to 30% (26–28), that is 30% of the ribosomes make complete RF2 and 70% terminate at the zero frame UGA after 25 codons. The RF2 frameshift site is a "slippery" codon where the mRNA slides within the ribosome complex one nucleotide by breaking codon:anticodon pairing with peptidyl tRNA in the ribosome P site and re-establishing pairing with an overlapping codon in the new frame. In this case tRNA^{Leu} with anticodon $^{3'}$GAG$^{5'}$ pairs with CUU in the first frame and then with UUU in the +1 frame (CUU UGA), inserting one leucine for four nucleotides in the mRNA.

Competition between termination and frameshifting is over which process captures the U of the UGA sequence. Release factor 2, the recoding product, promotes ribosome termination at UGA (and UAA) and so competes with recoding by enhancing termination at UGA, the 26th codon in the original frame. Competition is tilted in favor of recoding in two ways. The presence of a C 3' of the UGA makes a poor termination context (29), and CUU is a particularly shift-prone codon (30). If extra RF2 is present, frameshifting decreases so that less RF2 is made. When RF2 is in short supply termination loses, frameshifting wins, and RF2 concentration is restored.

The stimulatory signal is a short sequence three nucleotides 5' of the shift site that pairs with 16S rRNA of the translating ribosome (28, 31, 32, 33), just like the Shine-Dalgarno pairing that occurs 5' of the AUG start codon during ribosome initiation. Thus, RF2 mRNA has two Shine-Dalgarno sequences, one for initiation and one in the coding sequence for promoting frameshifting. [It is striking that the gram positive bacterium *Bacillus subtilis* that is evolutionarily distant from *E. coli* has virtually the same frameshifting cassette (34).] How the rRNA:mRNA interaction promotes frameshifting is unclear. One

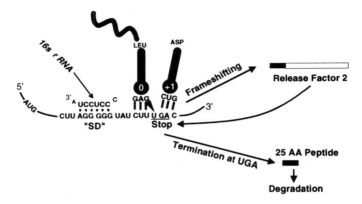

Figure 1 Model of *E. coli RF2* regulation through frameshifting.

possibility is that this interaction may create a pause that expands the window of time for frameshifting, but it may also play a more direct realigning role in the mechanism.

ANTIZYME   The only known example of a mammalian chromosomal gene that requires recoding is Ornithine Decarboxylase Antizyme which plays a key role in polyamine biosynthesis. (See 35 for a review.) Expression is autoregulatory because the required frameshift event is modulated by polyamines (see 36–38) (Figure 2). The antizyme protein binds to ornithine decarboxylase (ODC), the key enzyme in polyamine synthesis, and targets it for degradation by 26S proteasomes (see 35, 39). [Antizyme plays a second role in polyamine regulation by modulating the polyamine transporter (40, 41).] The nucleotide sequence of a rat cDNA clone (42) [and later human (43) and Xenopus (44)] gave two overlapping reading frames with the crucial region for interaction with ODC encoded in the second, +1 frame (45). A frameshift site was localized by assaying various fusion constructs in the reticulocyte lysate system (37, 38). The shift was shown definitively to occur at UCC UGA by protein sequencing of radioactive product made in vitro (38). Mutational analyses showed that the +1 shift did not occur by re-pairing of the peptidyl tRNA^{Ser} with CCU but rather by occlusion of the next nucleotide, U, of the UGA terminator by this tRNA, a process that competes with termination (38) (much like previously known occlusion with yeast TY3, described below). How this tRNA causes occlusion is not clear, especially since several different tRNAs in this position have the same capability, indicating that other signals may be participating (38). The efficiency of frameshifting at this site is dependent on

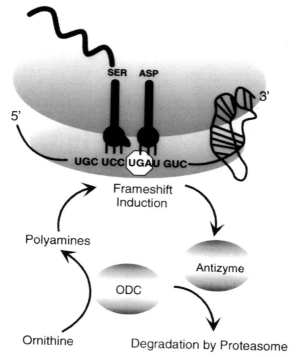

Figure 2   Model of the mechanism of frameshifting of mammalian antizyme and its regulatory role in regulation of polyamine biosynthesis.

the concentration of polyamines in cells or in the in vitro reaction mixtures (35, 37, 38). High concentrations of polyamines lead to greater frameshifting, giving more antizyme. Antizyme causes faster turnover of ODC and synthesis of fewer polyamines, completing an autoregulatory circuit.

Other nearby sequences in antizyme mRNA are needed to get the frameshift efficiency into the range useful for regulation. A 3' pseudoknot, spaced 3 nucleotides from the site provides a 2.5-fold stimulation (38, 46). A 5' sequence that stimulates about twofold is currently being characterized (Matsufuji, unpublished data).

How polyamines stimulate antizyme frameshifting is not clear. The regulatory pathway implies that polyamine stimulation should be specific for antizyme frameshifting. The very limited data available suggest that this prediction is true [MMTV gag-pro frameshifting is not stimulated (38)]. Since polyamines bind to RNA, the idea that they might alter some crucial part of the antizyme mRNA recoding signals is tempting. Polyamines still stimulate the remaining frameshifting if either the pseudoknot or the stop codon is deleted, so these at least are unlikely candidates for the site of action. Poly-amines could have their effect through alteration of the frameshift tRNA, since they are known to bind to tRNAs in specific places (47), not just through electrostatic interactions (48). However, the most likely mode of polyamine influence is through ribosomal interactions, but this effect will be very difficult to tease out.

To try to use the power of yeast genetics to identify components that interact with antizyme mRNA recoding signals, rat antizyme was expressed in *S. cerevisiae*. Frameshifting was found to be at the expected site, and the stimulatory response to the pseudoknot was even greater than in reticulocyte lysates. However, protein sequencing showed that the shift was −2 instead of +1, making this model less than perfect (46).

YEAST TY ELEMENTS   As with stop codons, reduced occupancy of a ribosomal A-site sense codon can provide the opportunity for frameshifting. "Hungry" codons and the specificity of their frameshifting effects were previously investigated in *E. coli* (49–51), but their ultilization in programmed frameshifting was discovered in the yeast element Ty1 (52) and later in Ty3 (53).

The *S. cerevisiae* transposable element Ty3 is a retrotransposon that has overlapping genes termed *GAG3* and *POL3*. *GAG3* encodes the protein constituents of the Ty3 virus-like particle, and *POL3* encodes the enzyme activities required for transposition of Ty3 (54). *POL3* overlaps the last 38 bp of *GAG3* in the +1 frame. A programmed +1 frameshifting event within the overlap permits 11% of ribosomes to enter the *POL3* gene which, like its retroviral counterparts, does not have independent ribosome entry. The sequence GCG AGU U (shown as codons of *GAG3*) yields the shift sequence of alanine-valine (GCG A GUU) (53) (Figure 3). The zero frame AGU is a rare serine codon, and the low availability of cognate tRNA induces a pause that is important for frameshifting (53). The effect of the hungry AGU codon is augmented by the sequence of 12 nucleotides 3′ of it (53). How this important sequence acts to stimulate frameshifting is unknown.

The striking feature of Ty3 frameshifting, which was discovered by Farabaugh and colleagues, is that the mRNA does not slip along the anticodon. Messenger RNA-tRNA re-pairing is not involved (53, 55). Rather the peptidyl tRNA in the P site somehow limits access to the 3′ adjacent base so that the A site tRNA must read the +1 codon. This result could be a direct effect of the tRNA occluding the next base or an indirect effect involving the tRNA-EF1α complex (55, 56) or the ribosome. However, the identity of the shift-site tRNA, and consequently its corresponding codon, are important (55). All possible shift-site codons were tested, and only a small minority caused frameshifting, corresponding to a total of eight tRNAs (55). The greater the

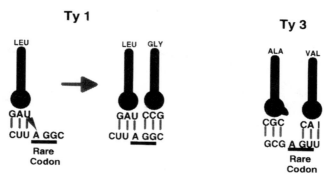

Figure 3 Models of frameshifting in yeast TY1 and TY3 elements. TY1 re-pairs and TY3 occludes.

abundance of the tRNA for the first codon in the +1 frame the higher the level of frameshifting (57), but evidence that the level of this tRNA varies in cells is lacking.

The related element TY1 also uses +1 frameshifting (18, 23) to make a TYA- TYB protein needed for transposition (Figure 3). The complete site is only seven nucleotides, CUU AGG C (52), and the efficiency of the shift is greatly enhanced by the hungry zero-frame AGG codon in the A site (58). However, here the peptidyl tRNA^{Leu} does not occlude the next base but rather re-pairs with the slipped mRNA (52). This tRNA^{Leu} is unusual in being able to recognize the four CUN leucine codons (59, 60), and it probably gives weak pairing with CUU, which likely facilitates dissociation and therefore slippage to UUA so that the +1 frame GGC codon is read by tRNA^{Gly}.

In both TY1 and TY3 the rare codon that stimulates frameshifting calls for a rare tRNA. The level of the aminoacylated form of this tRNA reflects at least one aspect of cellular physiology. In hard times the severe shortage of this charged tRNA will lead to increased transposition through expression of the TY1 or TY3 frameshift product. This movement of transposable elements may be advantageous in creating genetic diversity (61, 62).

−1 Frameshifting

MOUSE MAMMARY TUMOR VIRUS   MMTV typifies many retroviruses in having overlapping *gag, pro,* and *pol* genes such that one −1 frameshift event gives the gag-pro product and a second gives the gag-pro-pol product (63, 64). The sites of frameshifting are heptanucleotide sequences of the general form X XXY YYZ where the mRNA slips one base with respect to the two tRNAs on the ribosome (in A and P sites) (Figure 4). This process is called a tandem

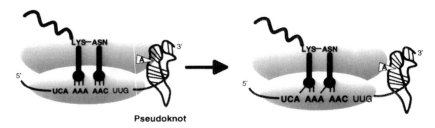

Pseudoknot

Figure 4   Model of tandem shift at gag-pro junction in MMTV.

shift and was first described for frameshifting in Rous Sarcoma Virus (RSV) decoding (65). Ribosomes initiate at the beginning of the MMTV *gag* gene. The majority read in a standard manner through the heptanucleotide sequence near the end of the gene and terminate at a downstream stop codon making the gag protein. In MMTV, a fraction (23%) hesitate at the heptanucleotide site where the mRNA slips backwards by one nucleotide, fueled by new pairing with the two tRNAs in the −1 frame. Most of these ribosomes terminate to make gag-pro but 8% shift frame again at the pro-pol junction to make the full gag-pro-pol product which is important for the virus life cycle (66, 67). The heptanucleotide sites—x_{-1}, $x_1x_2y_3$, $y_4y_5z_6$—allow the two tRNAs to pair with nucleotides 1 to 6 and then to realign with −1 to +5 in the new frame. The amino acid sequence of the transframe product shows that the actual shift occurs at the A site codon (YYZ to YYY). The tandem shift mechanism was revealed by experiments that showed the importance of the P site codon (XXY) and especially the next nucleotide upstream (X), because it must be able to pair with the P site tRNA in the new frame. This nucleotide is read twice, once in each frame. Sequences of the many examples of tandem shift sites, coupled with mutational analyses, show that the codon:anticodon pairing strength of the tRNA in the A site (reading YYZ) influences the efficiency. Weak pairing in the original frame and strong pairing in the new frame favors frameshifting.

Parallel studies of Infectious Bronchitis Virus, a coronavirus, revealed a heptanucleotide −1 shift that needed downstream sequences to reach full efficiency (68). This stimulator was proven to be a pseudoknot structure in the mRNA (68, 69). Pseudoknots—stem loops with their loop paired to downstream sequences (see Figure, p. 768)—proved to be a frequent theme for recoding events (70). The MMTV *gag-pro* frameshift also has a pseudoknot stimulator which is located seven nucleotides downstream of the shift site (71). This pseudoknot has been dissected both by mutagenic and structural analyses. Probing this region of the RNA using chemical modifications or nuclease digestion gave general support to the pseudoknot structure and showed that the junction region between stems 1 and 2 had unusual reactivity (72). Shen and Tinoco have derived an atomic model for the MMTV pseudoknot by NMR (Figure 5) (73). This pseudoknot has a slightly altered sequence in which some of the G-C pairs were reversed to enable the study, but it is still active as a stimulator. The structure is very revealing: It does not have coaxially stacked helices. A wedge base A on one side forces an angle between the helices, and this residue is important for stimulatory activity. The generality of a wedge base in pseudoknot-dependent recoding events is unclear (46).

Many examples provide information on what is important about pseudoknots and how they stimulate recoding. Ribosomes need to melt out secondary structure. Perhaps pseudoknots make this particularly difficult (68, 70), resulting in a ribosome pause that expands the window of opportunity for alternative decoding. The pseudoknots of yeast LA virus (74) and IBV (75) have been shown to cause ribosome pausing. However, substitution of a simple stem loop with equivalent base pairs did not promote frameshifting (69, 75) but did cause pausing. Thus, pausing may be necessary for frameshifting, but it is not sufficient. Pseudoknots must be playing a more complicated role. Attempts to find specific interactions with cellular components have not yet been fruitful.

More complicated stimulatory "knot" structures with three stems and at least one large loop have been described in human coronavirus (76) and transmissible gastroenteritis virus (77). Detailed analysis of these structures is guaranteed to be interesting.

Not all retroviruses of this class use a pseudoknot as a stimulator. HIV1 has just a weak 3' stem loop (78), but it has a shift site U UUU UUA that is very shifty on its own (79, 80).

dnaX   Another *E. coli* chromosomal gene, *dnaX,* requires frameshifting for its expression (Figure 6). The single message encodes both the tau and gamma subunits of DNA polymerase III (81–83). Only the tau subunit is essential for growth (84), but the holoenzyme without gamma has slightly different kinetic parameters (85). Standard translation makes the longer protein (tau) (71 kDa). A −1 frameshift two thirds of the way through the tau coding region leads quickly to a −1 stop codon to make the smaller protein (gamma) (47 kDa). The shift site is a particularly slippery tandem site, A AAA AAG. The efficiency of frameshifting approximates 50% (81, 82, 86, 87). Two stimulatory signals are required: a 5' Shine-Dalgarno sequence which pairs with 16S ribosomal RNA (86) and a 3' stem loop that plays a role in pausing the ribosome at the shift site (87). Deletion of the Shine Dalgarno sequence decreases the efficiency twofold and deletion of the stem loop decreases it tenfold (86, 87).

The Shine-Dalgarno interaction plays roles in initiation, −1 frameshifting (*dnaX*), and +1 frameshifting (RF2). Spacing between the Shine-Dalgarno sequence and the site of action is crucial and varies for the three cases. With

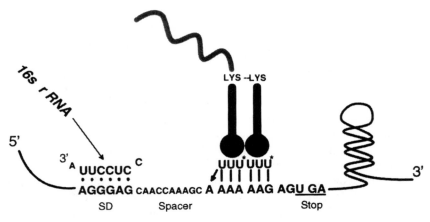

Figure 6   Model of *E. coli dnaX* −1 frameshifting showing 5′ and 3′ stimulators.

initiation the optimal spacing to the AUG is 5 (88), although there is considerable latitude (3 to 12). With the *dnaX* −1 shift, optimal spacing is 10–13 (86). With RF2 required spacing is 3 (86, 28). The implication is that pairing between mRNA and rRNA at these sites distorts the complex to either put the ribosome in an "initiation mode," a "minus mode" to force the complex back on the mRNA, or a "plus mode" to force it forward.

BACTERIAL IS ELEMENTS   Following the initial discovery of programmed frameshifting for expression of transposase in insertion sequence (IS) elements (89) it soon became apparent that IS sequences are a rich source of cases of −1 frameshifting (90). In IS911 the efficiency of a −1 tandem shift is regulated by translation initiation factor 3 (91) (MF Prère, P Polard, M Chandler, JF Atkins & O Fayet, unpublished observations). Further investigations are likely to reveal new mechanistic complexity and new regulatory schemes.

OTHER VIRUSES   Most, if not all, plant RNA viruses use frameshifting and/or stop codon redefinition to express polyproteins. These known frameshifts are −1 except for Closteroviruses which contain very long genomes of up to 20 kb (91a, b). Many of the −1 frameshifts are tandem shifts stimulated by downstream structures as in retroviruses (4, 92).

One unusual case is potato virus M which expresses a zinc finger protein by frameshifting from the capsid protein gene (93). The shift is −1 but does not require a tandem site. The peptidyl tRNA in the ribosome P site shifts back by one with re-pairing at the sequence A AAA UGA. This re-pairing is stimulated by the zero frame stop codon immediately after the shift codon.

Such "shifty stops" have been described before in bacteria in test constructs where both + and − frameshifting were detected (31, 28). However, the only previous example of a programmed "shifty stop" is *E. coli* RF2 which is +1.

Cocksfoot Mottle Sobemovirus has a UUU AAAC −1 shift site with a 3′ stem loop that has an efficiency of frameshifting much higher— 26%— than other plant viruses (93a). Barley Yellow Dwarf Virus employs a tandem −1 shift that is stimulated by RNA sequences that are four kilobases downstream, in the 3′ untranslated region (4, 94). One other case of a distant stimulator of frameshifting is in gene 10 of bacteriophage T7 which has a tandem shift-like site stimulated by sequences 200 nucleotides downstream in the 3′ UTR (95). How distant sequences might stimulate is not clear, but long-range RNA-RNA interactions or protein factors can be imagined.

Studies on yeast LA-virus have shown that the efficiency of frameshifting at its −1 tandem site is crucial for viability. An increase or decrease of more than twofold prevented viral replication (96). A particularly powerful approach to finding cellular components involved in frameshifting takes advantage of the power of yeast genetics. Mutants that increase frameshifting in LA-virus have revealed involvement of 5S ribosomal RNA (97) (long a mysterious molecule) and genes (5) previously shown to be involved in mRNA stability (98). Genetic approaches of this sort, as exemplified dramatically in studies on bacterial selenocysteine insertion, are key to understanding the many interactions in recoding. Parallel studies of frameshift mutant suppressors are providing insights into the mechanism of normal frame maintenance (99–102).

REDEFINITION: PROGRAMMED ALTERATION OF CODON MEANING

Recoding can reprogram the meaning of stop codons either as glutamine at UAG or as tryptophan or selenocysteine at UGA. The competing reaction to redefinition is termination which itself is context dependent. The three stop codons have inherently different efficiencies. In *E. coli* UAA is the tightest and is most frequently represented in highly expressed genes. UAG is less efficient, and UGA is naturally leaky at a level 1–3%. The first base 3′ to the stop codon has a large effect on the efficiency of termination with U > A > G > C in *E. coli* (29) and A = G >> C = U in mammals (103). As expected, the efficiency of the competing redefinition reaction has the inverse dependence on the next base (104).

Two cellular genes are known to use stop codon redefinition. In entertoxigenic *E. coli* the expression of genes for pili assembly involves decoding of a UAG stop codon (105). The *Kelch* gene in *Drosophila* codon gives two proteins, one an elongated form of the other, owing to redefinition of a UGA codon (106).

Many RNA viruses, especially of plants, use programmed redefinition of a stop codon to make elongated proteins. In some cases the signals are simple. Sindbis, an animal virus, redefines a UGA with 10% efficiency, and the stimulatory activity is essentially all in the 3'-adjacent C (107, 108), which merely provides a poor context for the competing termination.

With Tobacco Mosaic Virus (TMV), a UAG CAR YYA sequence is sufficient to promote 5% redefinition to make a form of a 126-kDa protein elongated to 183 kDa (109–111). The mechanism of stimulation of readthrough by 3' codons is not clear: It does not seem to depend on the abundance of the particular tRNAs or the tRNAs themselves. A structural role cannot be discerned. Perhaps there is an unrecognized interaction with rRNA (109).

Murine Leukemia Virus

Moloney Murine Leukemia Virus (MuLV) is an example of the class of retroviruses that use stop codon redefinition to express their gag-pol precursor protein. The amino acid sequence of the gag-pol junction showed glutamine insertion for the UAG stop codon that separates the two reading frames (112) (Figure 7). The 5% efficiency of redefinition is important for viral replication. If the stop codon is replaced by a GAG sense codon no virions are formed (113). However, either of the other two stop codons can replace UAG, in vivo (114, 115) or in vitro (115). The sequence downstream of the stop codon was necessary for redefinition (70, 116) and suggested a possible pseudoknot sequence that could be a stimulator. It is now clear that the 3' element is a pseudoknot (117, 118) that bears resemblance to the well characterized pseudoknot involved in MMTV frameshifting, but it has specific and interesting differences. The spacer region between the UAG and the first stem of the pseudoknot has crucial nucleotides, although they do not obviously fit into a structure (117, 119). Whether the two stems stack on each other is unclear. If the last base pair in stem 1 as drawn (117, Figure 1B) does not form there could be a "wedge" base that would prevent coaxial stacking as in MMTV. The identities of some of the bases in loop 2 are crucial, although other parts of the loop can be deleted or substituted (117). Again the important loop 2 bases cannot be incorporated into a structure in any obvious way. Probing experiments using nucleases and chemicals support the predicted structure and show that the crucial bases in loop 2 are unusually accessible, as if they are in a special configuration. The implication is that recognition of this pseudoknot by the ribosome complex or other factors must involve not only the folded structure but also base interactions.

The MMTV pseudoknot cannot substitute for its MuLV counterpart. The two knots are different, and determination of the structure of the MuLV pseudoknot knot would be most interesting for comparison.

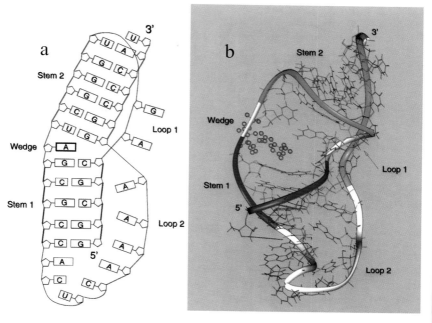

Figure 5 *(a)* Graphical representation of the MMTV pseudoknot. *(b)* The atomic model of the three-dimensional structure from NMR analyses. From Shen & Tinoco (73).

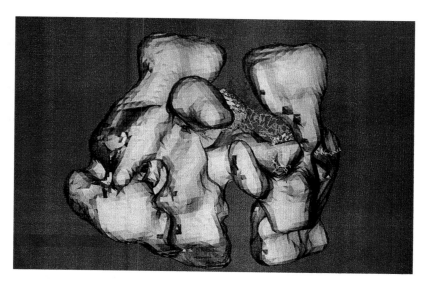

Figure 11 Ribosome model based on image reconstruction from cryo-electron micrographs. The A and P site tRNAs are shown in blue and green, and the mRNA is light blue. The 30S subunit is on the right and the 50S subunit is on the left. Photograph provided by H Stark, M van Heel, F Mueller & R Brimacombe.

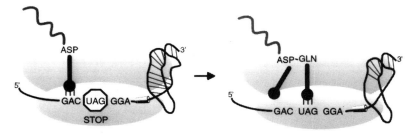

Figure 7 Model of redefinition of the UAG stop codon at gag-pol junction in MuLV. The advancing ribosome melts out the pseudoknot.

Selenocysteine

In 1976 the amino acid selenocysteine (Sec) was first found in a protein (120). At that time it was taken for granted that only 20 amino acids are directly encoded, and the 64 triplet codons were assigned. The subsequent discovery that selenocysteine is the 21st encoded amino acid came as a surprise (see 2 for a review). In 1986 the genes encoding two selenocysteine-containing proteins were sequenced, and each was found to have an in-frame TGA (UGA) codon, a stop codon in the "universal" code. One of the two genes encoded *E. coli* formate dehydrogenase H (22), the other mouse glutathione peroxidase (21). Later it was shown that UGA can directly specify selenocysteine (121). Selenocysteine-containing proteins are rare (see 2, 122 for a review): *E. coli* may have only three (all formate dehydrogenases, but *Clostridium sticklandi* has a glycine reductase and *Methanobacterium thermoautotrophicum* a hydrogenase) and mammals maybe only a dozen [so far four glutathione peroxidases, three iodothyronine deiodinases (123a), selenoprotein P, and selenoprotein W (139b)]. However, even these limited cases show that in organisms as diverse as *E. coli* and humans, UGA has three different meanings. It specifies translation termination, and it can be redefined by appropriate context to tryptophan or selenocysteine. [UGA also specifies selenocysteine in an archaebacteria (124)]. In *E. coli* the insertion of selenocysteine is highly efficient (exactly how termination is precluded is unknown). Why one of the redundant codons, as in the AGN box, has not evolved as the codon for selenocysteine does seem curious.

PROKARYOTIC SELENOCYSTEINE INSERTION   The mRNA "enabling element" that specifies a particular UGA as a selenocysteine codon rather than as a termination codon is a stem loop structure immediately 3′ of the UGA (125–127) that directs selenocysteyl tRNASec to the UGA (see 2 for a review) (Figure 8). Elongation factor–SELB complexes with selenocysteyl tRNASec

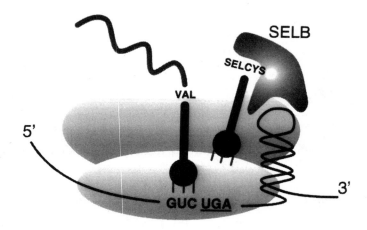

Figure 8 Model of redefinition of UGA as selenocysteine as understood for *E. coli* formate dehydrogenase.

and binds to the loop, positioning the tRNA near the UGA codon (128, 129). This positioning increases the local concentration of the crucial tRNA to provide effective competition with termination.

The specificity of this process for selenocysteine insertion is multifaceted. The tRNA^{Sec} is first charged with serine, which is then modified to selenocysteine, which can then be recognized by SELB (130). SELB is homologous to elongation factor Tu but has an "extra" domain at its carboxy terminal end that is a good candidate for interaction with the loop of the stem loop (2). The binding of SELB to selenocysteyl tRNA^{Sec} and of EF-Tu to tRNAs for the other 20 amino acids are mutually exclusive (128, 131).

The stem loop sequence appropriately positioned will promote insertion of selenocysteine at any UGA (2). This sequence will also redefine UAA, UAG (127), or even a UGC sense codon (132) if the anticodon of the selenocysteyl tRNA is altered to recognize these codons. Thus, even a sense codon can be redefined, at least under these conditions.

The notion of a "tethered" aminoacyl tRNA poised close to its UGA cognate codon contrasts sharply with the view of the entry of aminoacyl tRNA to the ribosomal A site by diffusion from a cytoplasmic soup. However, evidence is accumulating that in *E. coli* (133, 134), and especially in mammalian cells (135, 136), tRNAs for canonical amino acids are channeled and not free to diffuse. A model is emerging in which elongating ribosomes are the core of an elaborate complex with aminoacyl tRNA synthetases, tRNAs, and elongation factors that cycle tRNAs directly in a closed environment to the ribosome and pick up deacylated tRNAs for recycling (136). The bacterial selenocysteine insertion mechanism might be viewed as an extreme example of channeling for which the channeling mechanism is built into the mRNA sequence.

EUKARYOTIC SELENOCYSTEINE INSERTION   In mammals the essential enabling element, designated SECIS, is not adjacent to the UGA selenocysteine codon but rather is in the 3' untranslated region (UTR) of mRNA (137–139). The linear separation of UGA and the enabling RNA structure permits flexibility in UGA codon position in this mRNA (138). However, if the UGA is close enough to the SECIS element, it is read as a terminator (139a), which explains how two genes that encode selenocysteine can have UGA terminators (139b, 139c).

As in prokaryotes the selenocysteine tRNA is not acylated directly with selenocysteine: Serine is first conjugated to the tRNA and then converted to selenocysteine (see 2 for a review). The acceptor stem of mammalian seleno-cysteine tRNA is, like its prokaryotic counterpart, extra long (140), and the acylated tRNA is not recognized by the elongation factor, eEF-1, that forms a ternary complex with the other acylated tRNA species (141). So far a SELB-like protein that would form a complex with the selenocysteneyl tRNA has not been found. However, autoantibodies present in a subgroup of patients with a severe form of autoimmune chronic active hepatitis precipitate acylated selenocysteineyl tRNA, and some sera also precipitate an associated protein that is unlikely to be the aminoacyl tRNA synthetase (142). Whether the precipitated protein is the SELB equivalent remains to be seen.

The most interesting issue is at what point in translation the SECIS element "communicates" with the ribosome to redefine UGA in its mRNA to encode selenocysteine and not cause termination. Is it like *E. coli* in which the element "delivers" the selenocysteneyl tRNA to the waiting UGA? This long-range interaction could be accomplished through mRNA folding that would bring the 3' UTR near the UGA, although this folding might become difficult as multiple ribosomes move along the mRNA. Alternatively a protein tRNA complex in the 3' UTR could have an affinity for the translating ribosome and track with it, but this would allow it to service only one ribosome at a time. The complex would need to dissociate after one delivery and reform with the next advancing ribosome. A third alternative, which is perhaps the simplest to imagine, is that the 3' element does not deliver the tRNA but rather alters the initiating ribosome so that it is primed to redefine UGAs. Perhaps the 3' element triggers some conformational change in the ribosome or, more likely, delivers some component that tracks with the initiating ribosome.

These models are severely challenged by the fact that some mammalian mRNAs have multiple UGA selenocysteine codons. Selenoprotein P (SelP) has 10 UGA codons at least 7 of which (and probably all) encode selenocysteine (143, 144). One region of the human SelP mRNA has the sequence ...UGA—one codon—UGA—six codons—UGA—one codon—UGA... (144). How the 3′ UTR element can deliver tRNAs to each of the UGAs and to every ribosome that comes by is difficult to imagine. Imagining how the element could form a complex with each passing ribosome is equally hard unless a separate mechanism ensures that only one ribosome is on each mRNA at one time. Modification of the ribosome at initiation seems most compatible with the multiple UGAs in Selenoprotein P. If this model is correct, this type of recoding differs from other types of recoding in that the reprogramming of the ribosome is not transitory, but persists for the reading of that mRNA. None of the models is satisfactory without invoking additional elements. If ribosomes are modified at initiation how are they reset near the SECIS for termination? If the SECIS acts directly by delivery of the selenocysteine complex or by tracking with the ribosome, how can more than one ribosome be serviced? Elucidating the mechanism presents an exciting challenge.

SUBVERSION OF CONTIGUITY

TRANSLATIONAL BYPASSING   Noncoding information that disrupts genes is commonly dealt with by RNA splicing. However, in at least one case, disruptive sequences remain in the mRNA and are avoided by translational bypassing (146). Bacteriophage T4 gene *60* encodes a topoisomerase subunit that is dispensable, at least in conventional laboratory strains. The mRNA for this protein has a UAG terminator after codon 46, followed by 47 nucleotides (a 50-nucleotide coding gap) that the ribosome skips (Figure 9). Note that formally this is a frameshift since 50 is not divisible by 3, but whether the new frame is −1, 0, or +1 is not relevant for the bypass mechanism. Synthesis picks up again at a CUU leucine codon for amino acid 47 and continues to make the complete protein. Ribosome bypassing on this mRNA is remarkably efficient (146). Shorter bypasses found earlier in test systems were much less efficient (1–3%) (28, 147, 148), but set the stage for an understanding of the basic mechanism. These bypasses require matched codons where the peptidyl tRNA pairs first with one codon and then with the other by slipping of the mRNA. As reflected by the amino acid sequence, one amino acid is inserted for the two matched codons and any intervening nucleotides. The 50-nucleotide coding gap in gene *60* is preceded by a GGA glycine codon and terminated by a GGA glycine codon, and only one glycine is inserted for the whole region as if the mRNA moves from pairing with the first to the second GGA with no intermediate protein synthesis. The matched GGA codons are crucial. The mRNA could scan across the coding gap or fold into a structure that moves through the ribosome, allowing easy access to the matched codons (146, 149). Recent experiments support a scanning mechanism (FM Adamski, in preparation).

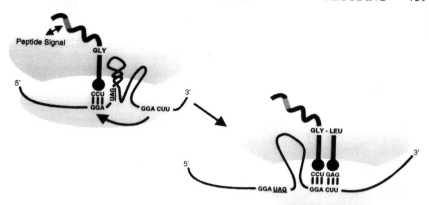

Figure 9   Model of the bypassing of a 50-nucleotide gap in bacteriophage T4 *gene 60.*

The most surprising requirement for the *gene 60* translational bypass is a specific sequence of 16 amino acids in the growing peptide chain (149). The sequence is basic and hydrophobic. It acts in *cis* and probably within an exit channel of the ribosome. When the ribosome is poised at the take-off site the nascent chain will be 46 amino acids long. The important peptide sequence is amino acids 17–32 which makes it just short of the 30 amino acids that are thought to be needed to emerge from the ribosome. Presumably the basic and hydrophobic peptide sequence interacts with some ribosome component in the exit channel to enable the bypass. The bacterial *cat86* gene also uses nascent chain information: A pentapeptide sequence in a leader peptide inhibits peptidyl transferase to regulate expression of the protein encoded by the downstream structural gene (150).

Little is known about the path that nascent chains take in the ribosome. Cross-linking (151) flourescent energy transfer (152) and cryo-electron microscopy (153, 154) studies are beginning to reveal complexity of the exit channel. Multiple channels with multiple functions are possible. The *gene 60* nascent peptide signal provides one tool to functionally probe the exit channel(s).

Although little is known about the mechanism of bypassing, a genetic approach has identified the involvement of ribosomal protein L9. A mutant *gene 60* mRNA with a much longer stem loop in the coding gap, and consequently a greatly reduced bypass efficiency, was used to look for host mutants that would restore bypassing. A mutant of ribosomal protein L9 was found (155). The crystal structure of L9 (156) reveals an intriguing dumbbell-shaped molecule with a long alpha helix connecting two globular domains predicted to bind RNA. On the ribosome L9 is located in the vicinity of the peptidyl transfer site (157), in a location where it could directly influence detachment of peptidyl tRNA.

Few other examples of bypassing have been found. Expression of *E. coli* repressor for tryptophan biosynthesis (trpR) when fused to lacZ gives a protein sequence suggesting that the ribosome hops from trpR sequences to lacZ sequences (158). This finding has led this group to strongly imply that bypassing is an important part of trpR regulation (159). However, the key experiment to show that a hop occurs within trpR itself remains to be done. The rules for bypassing here must be quite different from *gene 60*. A lectin-like protein in *Prevotella loescheii* (160) and an endoglucanase in *Bacteroides ruminicola* (161) may also involve ribosome bypassing.

Understanding the mechanism of bypassing will be interesting in its own right, but since it is not commonly found the more pertinent question is what limits it. If peptidyl tRNA on ribosomes could unpair with mRNA and find the next matching codon with great ease, these errors would quickly lead to chaos. Ribosomes must have evolved so as to keep bypassing to the optimal minimum.

TRANS-TRANSLATION   A dramatic example of subversion of contiguity was recently described by Keiler et al through investigation of a COOH peptide tagging mechanism for targeting abnormal proteins for degradation in *E. coli* (161a). It was know previously that expression of interleukin 6 in *E. coli* resulted in a protein truncated at its COOH end, but with an 11–amino acid extension (161b), the last 10 of which are encoded by the 10Sa RNA. It was noted (161b) that the 10Sa RNA can be aminoacylated by alanine (161c, 161d), which is the first amino acid of the extension. Keiler et al found that proteins translated from mRNAs that lack a stop codon because of nuclease cleavage or premature transcription termination, have this 11–amino acid sequence, and they proposed a translational model. In this model (Figure 10), ribosomes at the 3′ truncated end use the charged alanine on 10Sa RNA in the A site, without a corresponding codon. Then translation picks up on the 10Sa RNA itself decoding 10 codons to a stop signal. This unusual mechanism provides a way to recycle ribosomes stuck at the end of aberrant mRNAs and to target abnormal proteins for degradation.

This contrasts with the case of *gene 60* translation where noncontiguous reading was from a take-off site to a landing site on one mRNA. Here the striking fact is that ribosomes can read first off one mRNA and transfer to another— trans-translation.

CONSEQUENCES OF NATURAL "NOISE"

Despite the sophistication of the translation apparatus as manifested by intricate recoding, a low level of framing errors can be readily detected with frameshift mutants (162–164, 49). This error rate is not uniform: Runs of like-bases are

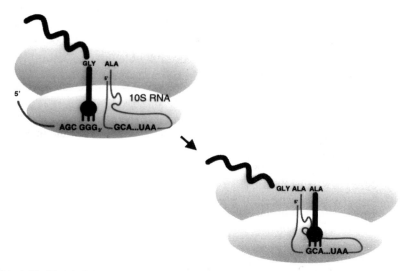

Figure 10   Model of trans-translation from the 3' terminal codon of the message to the *E. coli* 10Sa RNA with noncoded insertion of alanine from the 10S RNA (161a).

more prone to frameshifting. The most interesting case to date of a framing error at a run of repeat bases is in Herpes virus thymidine kinase gene (165; Horsburgh & Coen, personal communication). Viral thymidine kinase activates the antiviral compound acyclovir and is largely responsible for making the virus sensitive to the drug. Many viral mutants that are resistant to acyclovir have an extra base in a run of Gs in their thymidine kinase gene (runs of repeat bases are hot spots for frameshift mutations). At the site of the insertion 99% of the ribosomes continue triplet reading and terminate at a stop codon that was previously out of frame to yield a truncated product that lacks residues critical for enzyme activity. One percent of the ribosomes change reading frame at or near the mutation site, thereby remaining in the original frame, giving 1% residual enzyme activity. This level is below the threshold for sensitivity to acyclovir. However, it is enough to provide recovery from latency for reactivation of the virus (null mutants fail to reactivate). Thus, this low-level error in framing has medical consequences that allow the virus to survive and avoid the antiviral therapy of choice.

Error reading of mutant mRNA can also serve to alleviate the symptoms of some inherited diseases that are due to single base mutations. For instance, progression of a disease due to a premature stop codon can be slowed by low-level readthrough to produce a small amount of almost normal protein. Mutants of human cystic fibrosis transmembrane conductance regulator with

UGA instead of the glycine codon$_{542}$ or arginine$_{553}$, which are about one third of the way through the gene, cause less severe pulmonary problems than some missense mutants (166–168). A study with UGA premature stop codons at the equivalent positions of a member of the same protein family in yeast revealed that one was read through with an efficiency of 10% (169). Similarly, frame-shift mutant leakiness has been described for patients from several families with defects in their carbonic anhydrase II (170). Loss of this enzyme results in osteopetrosis. Patients in each of five families have the same single base deletion. Compensatory ribosomal frameshifting 12 codons 3′ of the mutation site yields a low level of active product. These patients have less severe symptoms than would have been expected. Even though the lesion in each of the families is the same they manifest different levels of symptoms (171, 170). More such examples will likely turn up.

Another manifestion of translational errors is of concern for overproduction of proteins in a heterologous system. Mammalian genes overexpressed in *E. coli* encounter codon usage, which on occasion leads to no product (172), but more commonly to spurious products. Frameshifting, bypassing, and read-through have all been encountered (148, 173, 174) and there are likely some unreported cases. The solution is to reengineer the construct to avoid rare codons, but a deeper understanding of the problem would also be helpful.

EVOLUTIONARY IMPLICATIONS

Upon first consideration, it is tempting to regard the recoding mechanism for selenocysteine as a useful appendage on a preformed code. However, Böck and colleagues (2, 175–177) propose that the encoding of selenocysteine by UGA predates the use of UGA as a stop codon. Selenocysteine is highly susceptible to oxidation (and heavy metal ions), and if it was commonly encoded before oxygen became part of the earth's atmosphere, it may then have become restricted to the very few proteins for which its high reactivity at the active site is particularly advantageous (the function of selenocysteine in some proteins such as selenoprotein P is unknown). In this scenario the "latter day" additional use of UGA as a termination codon predated the divergence of archae, eubacteria, and eukaryotes. However, nothing is intrinsically special in UGA being the codon for selenocysteine, because other codons will substitute provided that the anticodon of the appropriate tRNA is altered (see above).

Sequences in the 3′ untranslated regions of an mRNA, far removed from the recoding site, are essential for selenocysteine insertion in mammals and important for frameshifting in Barley Yellow Dwarf Virus and phage T7 gene 10. (Whether the aminoacylatable tRNA-like pseudoknot at the 3′ end of many plant viral genomic RNAs plays a role in recoding is unknown). It is not a big step from distant cis-acting RNA stimulators to the possibility of stimulatory elements on various RNA molecules, i.e. trans RNA stimulators. In some sense, the stimulatory effect of the anti–Shine Dalgarno sequence in *E. coli* 16S rRNA in pairing with mRNA just before the shift site in the RF2 and *dnaX* mRNAs is such a trans stimulator.

NONCANNONICAL PAIRING VS NONPAIRING

A general issue in decoding is the possibility of non–Watson-Crick, non-wobble pairing which is increasingly being found in folded RNAs (178). The nearly universal genetic code has none of this. However, in mitochondria a more limited set of tRNAs is sufficient for decoding, and some decode all four codons of one box (179–181). Either no pairing occurs in the third position and two out of three is enough, or noncannonical pairs form. If noncannonical pairs are allowed in codon:anticodon interactions then caution must be taken in interpreting the various re-pairings observed in cases of recoding.

CONCLUSION

Recoding phenomena provide entrees into further understanding of ribosome structure and function. The details of how the tRNAs align in A and P sites, the constraints of the mRNA path, and the nature of peptide exit channels are of intimate importance both to recoding and a general understanding of decoding. The most recent images of ribosome structure (153, 154) coupled to the arrangement of folded rRNA within the structure are beginning to reveal a much clearer picture of the whole ribosome complex. As seen in Figure 10, a model based on images reconstructed from cryo electron–microscopic pictures gives a strong feeling for how the A and P site tRNAs are nestled in a cavity between the subunits and for how the mRNA winds through the complex. More refined structural models, which can be coupled with the emerging structures of individual ribosomal proteins, promise a very fruitful time ahead.

Recoding has been described in a relatively small number of genes but these have been discovered because of their special efficiency or because of particularly detailed knowledge. It seems very likely that recoding tricks are used for many genes, perhaps particularly in cases in which additional forms of a gene product might be desired in small amounts for special functions.

Prediction of recoding sites in raw sequence is still difficult for all but a few classes. However, the fruits of the genome project will provide the exhilarating opportunity to examine and compare thousands of genes in detail. This opportunity will bring an appreciation of the extent of the phenomenon but will also undoubtedly enrich the spectrum of recoding possibilities.

ACKNOWLEDGMENTS

Research supported by the Howard Hughes Medical Institute and the National Institute of Health (R01-GM48152) to JFA.

Literature Cited

1. Gesteland RF, Weiss RB, Atkins JF. 1992. *Science* 257:1640–41
2. Baron C, Bock A. 1995. In *tRNA Structure, Biosynthesis and Function*, ed. D Soll, UL RajBhandary, pp. 529–44. Washington, DC: Am. Soc. Microbiol.
3. Brierley I. 1995. *J. Gen. Virol.* 76:1885–92
4. Miller WA, Dinesh-Kumar SP, Paul CP. 1995. *Crit. Rev. Plant Sci.* 14:179–211
5. Dinman JD. 1995. *Yeast* 11:1115–27
6. Atkins JF, Gesteland RF. 1995. In *tRNA Structure, Biosynthesis and Function*, ed. D Soll, UL RajBhandary, pp. 471–90. Washington, DC: Am. Soc. Microbiol.
7. Hofstetter H, Monstein HJ, Weissmann C. 1974. *Biochim. Biophys. Acta* 374:238–51
8. Weiner AM, Weber K. 1973. *J. Mol. Biol.* 80:837–55
9. Pelham HR. 1978. *Nature* 272:469–71
10. Philipson L, Andersson P, Olshevsky U, Weinberg R, Baltimore D, et al. 1978. *Cell* 13:189–99
11. Beremand MN, Blumenthal T. 1979. *Cell* 18:257–66
12. Atkins JF, Gesteland RF, Reid BR, Anderson CW. 1979. *Cell* 18:1119–31
13. Dunn JJ, Studier FW. 1983. *J. Mol. Biol.* 166:477–535
14. Condron BG, Atkins JF, Gesteland RF. 1991. *J. Bacteriol.* 173:6998–7003
15. Coleman J, Inouye M, Atkins JF. 1983. *J. Bacteriol.* 153:1098–100
16. Mellor J, Fulton SM, Dobson MJ, Wilson W, Kingsman SM, et al. 1985. *Nature* 313:243–46
17. Clare J, Farabaugh P. 1985. *Proc. Natl. Acad. Sci. USA* 82:2829–33
18. Wilson W, Malim MH, Mellor J, Kingsman AJ, Kingsman SM. 1986. *Nucleic Acids Res.* 14:7001–16
19. Jacks T, Varmus HE. 1985. *Science* 230:1237–42
20. Craigen WJ, Cook RG, Tate WP,

Caskey CT. 1985. *Proc. Natl. Acad. Sci. USA* 82:3616–20
21. Chambers I, Frampton J, Goldfarb P, Affara N, McBain W, et al. 1986. *EMBO J.* 5:1221–27
22. Zinoni F, Birkmann A, Stadtman TC, Bock A. 1986. *Proc. Natl. Acad. Sci. USA* 83:4650–54
23. Clare JJ, Belcourt M, Farabaugh PJ. 1988. *Proc. Natl. Acad. Sci. USA* 85:6816–20
24. Goldman E, Rosenberg AH, Zubay G, Studier FW. 1995. *J. Mol. Biol.* 245:467–73
25. Chen G-FT, Inouye M. 1990. *Nucleic Acids Res.* 18:1465–73
26. Adamski FM, Donly BC, Tate WP. 1993. *Nucleic Acids Res.* 21:5074–78
27. Craigen WJ, Caskey CT. 1986. *Nature* 322:273–75
28. Weiss RB, Dunn DM, Atkins JF, Gesteland RF. 1987. *Cold Spring Harbor Symp. Quant. Biol.* 52:687–93
29. Poole ES, Brown CM, Tate WP. 1995. *EMBO J.* 14:151–58
30. Curran JF. 1993. *Nucleic Acids Res.* 21:1837–43
31. Weiss RB, Dunn DM, Atkins JF, Gesteland RF. 1990. *Prog. Nucleic Acid Res. Mol. Biol.* 39:159–83
32. Weiss RB, Dunn DM, Dahlberg AE, Atkins JF, Gesteland RF. 1988. *EMBO J.* 7:1503–7
33. Curran JF, Yarus M. 1988. *J. Mol. Biol.* 203:75–83
34. Pel HJ, Rep M, Grivell LA. 1992. *Nucleic Acids Res.* 20:4423–28
35. Hayashi S, Murakami Y. 1995. *Biochem. J.* 306:1–10
36. Atkins JF, Lewis JB, Anderson CW, Gesteland RF. 1975. *J. Biol. Chem.* 250:5688–95
37. Rom E, Kahana C. 1994. *Proc. Natl. Acad. Sci. USA* 91:3959–63
38. Matsufuji S, Matsufuji T, Miyazaki Y, Murakami Y, Atkins JF, et al. 1995. *Cell* 80:51–60

39. Li X, Coffino P. 1993. *Mol. Cell. Biol.* 13:2377–83
40. Mitchell JLA, Judd GG, Bareyal-Leyser A, Ling SY. 1994. *Biochem. J.* 299:19–22
41. Suzuki T, He Y, Kashiwagi K, Murakami Y, Hayashi S, et al. 1994. *Proc. Natl. Acad. Sci. USA* 91:8930–34
42. Matsufuji S, Miyazaki Y, Kanamoto R, Kameji T, Murakami Y, et al. 1990. *J. Biochem.* 108:365–71
43. Tewari DS, Qian Y, Thornton RD, Pieringer J, Taub R, et al. 1994. *Biochim. Biophys. Acta* 1209:293–95
44. Ichiba T, Matsufuji S, Miyazaki Y, Hayashi S. 1995. *Biochim. Biophys. Acta* 1262:83–86
45. Li X, Coffino P. 1994. *Mol. Cell. Biol.* 14:87–92
46. Matsufuji S, Matsufuji T, Wills NM, Gesteland RF, Atkins JF. 1996. *EMBO J.* 15:In press
47. Garcia A, Giege R, Behr J-P. 1990. *Nucleic Acids Res.* 18:89–95
48. Frydman L, Rossomando PC, Frydman V, Fernandez CO, Frydman B, et al. 1992. *Proc. Natl. Acad. Sci. USA* 89: 9186–90
49. Weiss R, Gallant J. 1983. *Nature* 302: 389–93
50. Spanjaard RA, Chen K, Walker JR, van Duin J. 1990. *Nucleic Acids Res.* 18: 5031–36
51. Gallant J, Lindsley D. 1993. *Biochem. Soc. Trans.* 21:817–21
52. Belcourt MF, Farabaugh PJ. 1990. *Cell* 62:339–52
53. Farabaugh PJ, Zhao H, Vimaladithan A. 1993. *Cell* 74:93–103
54. Hansen L, Chalker D, Orlinsky K, Sandmeyer S. 1992. *J. Virol.* 66:1414–24
55. Vimaladithan A, Farabaugh PJ. 1994. *Mol. Cell. Biol.* 14:8107–16
56. van Noort JM, Kraal B, Bosch L. 1986. *Proc. Natl. Acad. Sci. USA* 83:4617–21
57. Pande S, Vimaladithan A, Zhao H, Farabaugh PJ. 1995. *Mol. Cell. Biol.* 15:298–304
58. Kawakami K, Pande S, Faiola B, Moore DP, Boeke JD, et al. 1993. *Genetics* 135:309–20
59. Weissenbach J, Dirheimer G. 1977. *FEBS Lett.* 82:71–76
60. Randerath E, Gupta RC, Chia L-LSY, Chang SH, Randerath K. 1979. *Eur. J. Biochem.* 93:79–94
61. Voytas DF, Boeke JD. 1993. *Trends Genet.* 9:421–27
62. Naas T, Blot M, Fitch WM, Arber W. 1994. *Genetics* 136:721–30
63. Moore R, Dixon M, Smith R, Peters G, Dickson C. 1987. *J. Virol.* 61:480–90
64. Jacks T, Townsley K, Varmus HE, Majors J. 1987. *Proc. Natl. Acad. Sci. USA* 84:4298–302
65. Jacks T, Madhani HD, Masiarz FR, Varmus HE. 1988. *Cell* 55:447–58
66. Wickner RB. 1989. *FASEB J.* 3:2257–65
67. Atkins JF, Weiss RB, Gesteland RF. 1990. *Cell* 62:413–23
68. Brierley I, Digard P, Inglis SC. 1989. *Cell* 57:537–47
69. Brierley I, Rolley NJ, Jenner AJ, Inglis SC. 1991. *J. Mol. Biol.* 220:889–902
70. ten Dam EB, Pleij CW, Bosch L. 1990. *Virus Genes* 4:121–36
71. Chamorro M, Parkin N, Varmus HE. 1992. *Proc. Natl. Acad. Sci. USA* 89: 713–17
72. Chen X, Chamorro M, Lee SI, Shen LX, Hines JV, et al. 1995. *EMBO J.* 14:842–52
73. Shen LX, Tinoco I. 1995. *J. Mol. Biol.* 247:963–78
74. Tu C, Tzeng TH, Bruenn JA. 1992. *Proc. Natl. Acad. Sci. USA* 89:8636–40
75. Somogyi P, Jenner AJ, Brierley I, Inglis SC. 1993. *Mol. Cell. Biol.* 13:6931–40
76. Herold J, Siddell SG. 1993. *Nucleic Acids Res.* 21:5838–42
77. Eleouet J-F, Rasschaert D, Lambert P, Levy L, Vende P, et al. 1995. *Virology* 206:817–22
78. Parkin NT, Chamorro M, Varmus HE. 1992. *J. Virol.* 66:5147–51
79. Jacks T, Power MD, Masiarz FR, Luciw PA, Barr PJ, et al. 1988. *Nature* 331: 280–83
80. Wilson W, Braddock M, Adams SE, Rathjen PD, Kingsman SM, et al. 1988. *Cell* 55:1159–69
81. Tsuchihashi Z, Kornberg A. 1990. *Proc. Natl. Acad. Sci. USA* 87:2516–20
82. Flower AM, McHenry CS. 1990. *Proc. Natl. Acad. Sci. USA* 87:3713–17
83. Blinkowa AL, Walker JR. 1990. *Nucleic Acids Res.* 18:1725–29
84. Blinkova A, Hervas C, Stukenberg PT, Onrust R, O'Donnell ME, et al. 1993. *J. Bacteriol.* 175:6018–27
85. Kim S, Marians KJ. 1995. *Nucleic Acids Res.* 23:1374–79
86. Larsen B, Wills NM, Gesteland RF, Atkins JF. 1994. *J. Bacteriol.* 176:6842–51
87. Tsuchihashi Z. 1991. *Nucleic Acids Res.* 19:2457–62
88. Chen HY, Bjerknes M, Kumar R, Jay E. 1994. *Nucleic Acids Res.* 22:4953–57
89. Sekine Y, Ohtsubo E. 1989. *Proc. Natl. Acad. Sci. USA* 86:4609–13
90. Chandler M, Fayet O. 1993. *Mol. Microbiol.* 7:497–503
91. Polard P, Prere MF, Chandler M, Fayet O. 1991. *J. Mol. Biol.* 222:465–77

91a. Agranovsky AA, Koonin EV, Boyko VP, Maiss E, Frötschl R, Lunin NA, Atabekov JG. 1994. *Virology* 198:311–24

91b. Karase AV, Boyko VP, Gowda S, Nikolaeva OV, Hilf ME, Koonin EV, Niblett CL, Cline K, Gumpf DJ, Lee RF, Garnsey SM, Lewandowski DJ, Dawson WO. 1995. *Virology* 208:511–20

92. Rohde W, Gramstat A, Schmitz J, Tacke E, Prufer D. 1994. *J. Gen. Virol.* 75:2141–49

93. Gramstat A, Prufer D, Rohde W. 1994. *Nucleic Acids Res.* 22:3911–17

93a. Mäkinen K, Naess V, Tamm T, Truve E, Aaspõllu A, Saarma M. 1995. *Virology* 207:566–71

94. Wang S, Miller WA. 1995. *J. Biol. Chem.* 270:13446–52

95. Condron BG, Gesteland RF, Atkins JF. 1991. *Nucleic Acids Res.* 19:5607–12

96. Dinman JD, Wickner RB. 1992. *J. Virol.* 66:3669–76

97. Dinman JD, Wickner RB. 1995. *Genetics* 140:95–105

98. Leeds P, Wood JM, Lee B-S, Culbertson MR. 1992. *Mol. Cell. Biol.* 12:2165–77

99. Atkins JF, Weiss RB, Thompson S, Gesteland RF. 1991. *Annu. Rev. Genet.* 25:201–28

100. Culbertson MR, Leeds P, Sandbaken MG, Wilson PG. 1990. In *The Ribosome, Structure, Function and Evolution*, ed. WE Hill, AE Dahlberg, RA Garrett, PB Moore, D Schlessinger, JR Warner, pp. 559–70. Washington, DC: Am. Soc. Microbiol. Press

101. O'Connor M, Willis NM, Bossi L, Gesteland RF, Atkins JF. 1993. *EMBO J.* 12:2559–66

102. O'Connor M, Dahlberg AE. 1995. *J. Mol. Biol.* 254:838–847

103. McCaughan KK, Brown CM, Dalphin ME, Berry MJ, Tate WP. 1995. *Proc. Natl. Acad. Sci. USA* 92:5431–35

104. Tate WP, Brown CM. 1992. *Biochemistry* 31:2443–50

105. Jalajakumari MB, Thomas CJ, Halter R, Manning PA. 1989. *Mol. Microbiol.* 3:1685–95

106. Xue F, Cooley L. 1993. *Cell* 72:681–93

107. Li G, Rice CM. 1993. *J. Virol.* 67:5062–67

108. Strauss JH, Strauss EG. 1994. *Microbiol. Rev.* 58:491–562

109. Skuzeski JM, Nichols LM, Gesteland RF, Atkins JF. 1991. *J. Mol. Biol.* 218:365–73

110. Zerfass K, Beier H. 1992. *EMBO J.* 11:4167–73

111. Stahl G, Bidou L, Rousset J-P, Cassan M. 1995. *Nucleic Acids Res.* 23:1557–60

112. Yoshinaka Y, Katoh I, Copeland TD, Oroszlan S. 1985. *Proc. Natl. Acad. Sci. USA* 82:1618–22

113. Felsenstein KM, Goff SP. 1988. *J. Virol.* 62:2179–82

114. Jones DS, Nemoto F, Kuchino Y, Masuda M, Yoshikura H, et al. 1989. *Nucleic Acids Res.* 15:5933–45

115. Feng YX, Levin JG, Hatfield DL, Schaefer TS, Gorelick RJ, et al. 1989. *J. Virol.* 63:2870–73

116. Panganiban AT. 1988. *J. Virol.* 62:3574–80

117. Wills NM, Gesteland RF, Atkins JF. 1991. *Proc. Natl. Acad. Sci. USA* 88:6991–95

118. Felsenstein KM, Goff SP. 1992. *J. Virol.* 66:6601–8

119. Feng YX, Yuan H, Rein A, Levin JG. 1992. *J. Virol.* 66:5127–32

120. Cone JE, Martin del Rio R, Davis JN, Stadtman TC. 1976. *Proc. Natl. Acad. Sci. USA* 73:2659–63

121. Zinoni F, Birkmann A, Leinfelder W, Bock A. 1987. *Proc. Natl. Acad. Sci. USA* 84:3156–60

122. Burk RF, ed. 1994. *Selenium in Biology and Human Health*. New York: Springer-Verlag. 221 pp.

123. Croteau W, Whittemore SL, Schneider MJ, St. Germain DL. 1995. *J. Biol. Chem.* 270:16569–75

123a. Davey JC, Becker KB, Schneider MJ, St. Germain DL, Galton VA. 1995. *J. Biol. Chem.* 270:26786–89

124. Halboth S, Klein A. 1992. *Mol. Gen. Genet.* 233:217–24

125. Zinoni F, Heider J, Bock A. 1990. *Proc. Natl. Acad. Sci. USA* 87:4660–64

126. Berg BL, Baron C, Stewart V. 1991. *J. Biol. Chem.* 266:22386–91

127. Heider J, Baron C, Bock A. 1992. *EMBO J.* 11:3759–66

128. Forchhammer K, Leinfelder W, Bock A. 1989. *Nature* 342:453–56

129. Baron C, Heider J, Bock A. 1993. *Proc. Natl. Acad. Sci. USA* 90:4181–85

130. Forchhammer K, Bock A. 1991. *J. Biol. Chem.* 266:6324–28

131. Forster C, Ott G, Forchhammer K, Sprinzl M. 1990. *Nucleic Acids Res.* 18:487–91

132. Baron C, Heider J, Bock A. 1990. *Nucleic Acids Res.* 18:6761–66

133. Spirin AS, Baranov VI, Ryabova LA, Ovodov SY, Alakhov YB. 1988. *Science* 242:1162–64

134. Kigawa T, Yokoyama S. 1991. *J. Biochem.* 110:166–68

135. Ryabova LA, Ortlepp SA, Baranov VI. 1989. *Nucleic Acids Res.* 17:4412

136. Stapulionis R, Deutscher MP. 1995. *Proc. Natl. Acad. Sci. USA* 92:7158–61
137. Berry MJ, Banu L, Chen Y, Mandel SJ, Kieffer JD, et al. 1991. *Nature* 353:273–76
138. Berry MJ, Banu L, Harney JW, Larsen PR. 1993. *EMBO J.* 12:3315–22
139. Shen Q, Chu F-F, Newburger PE. 1993. *J. Biol. Chem.* 268:11463–69
139a. Martin GW, Harney JW, Berry MJ. 1996. *RNA* 2: In press
139b. Williams DL, Pierce RJ, Cookson E, Capron A. 1991. *Mol. Biochem. Parasitol.* 52:127–30
139c. Vendeland SC, Beilstein MA, Yeh J-Y, Ream W, Whanger PD. 1995. *Proc. Natl. Acad. Sci. USA* 92:8749–53
140. Sturchler-Pierrat C, Hubert N, Totsuka T, Mizutani T, Carbon P, et al. 1995. *J. Biol. Chem.* 270:18570–74
141. Jung JE, Karoor V, Sandbaken MG, Lee BJ, Ohama T, et al. 1994. *J. Biol. Chem.* 269:29739–45
142. Gelpi C, Sontheimer EJ, Rodriguez-Sanchez JL. 1992. *Proc. Natl. Acad. Sci. USA* 89:9739–43
143. Hill KE, Lloyd RS, Yang J-G, Read R, Burk RF. 1991. *J. Biol. Chem.* 266:10050–53
144. Hill KE, Lloyd RS, Burk RF. 1993. *Proc. Natl. Acad. Sci. USA* 90:537–41
145. Deleted in proof
146. Huang WM, Ao SZ, Casjens S, Orlandi R, Zeikus R, et al. 1988. *Science* 239:1005–12
147. O'Connor M, Gesteland RF, Atkins JF. 1989. *EMBO J.* 8:4315–23
148. Kane JF, Violand BN, Curran DF, Staten NR, Duffin KL, et al. 1992. *Nucleic Acids Res.* 20:6707–12
149. Weiss RB, Huang WM, Dunn DM. 1990. *Cell* 62:117–26
150. Rogers EJ, Lovett PS. 1994. *Mol. Microbiol.* 12:181–86
151. Stade K, Junke N, Brimacombe R. 1995. *Nucleic Acids Res.* 23:2371–80
152. Odom OW, Picking WD, Hardesty B. 1990. *Biochemistry* 29:10734–44
153. Frank J, Zhu J, Penczek P, Li Y, Srivastava S, et al. 1995. *Nature* 376:441–44
154. Stark H, Mueller F, Orlova EV, Schatz M, Dube P, et al. 1995. *Structure* 8:815–21
155. Herbst KL, Nichols LM, Gesteland RF, Weiss RB. 1994. *Proc. Natl. Acad. Sci. USA* 91:12525–29
156. Hoffman DW, Davies C, Gerchman SE, Kycia JH, Porter SJ, et al. 1994. *EMBO J.* 13:205–12
157. Nag B, Akella SS, Cann PA, Tewari DS, Glitz DG, et al. 1991. *J. Biol. Chem.* 266:22129–35
158. Benhar I, Engelberg-Kulka H. 1993. *Cell* 72:121–30
159. Benhar I, Miller C, Engelberg-Kulka H. 1992. *Mol. Microbiol.* 6:2777–84
160. Manch-Citron JN, London J. 1994. *J. Bacteriol.* 176:1944–48
161. Matsushita O, Russell JB, Wilson DB. 1991. *J. Bacteriol.* 173:6919–26
161a. Keiler KC, Waller PRH, Sauer RT. 1996. *Science.* 271:990–93
161b. Tu G-F, Reid GE, Shang J-G, Moritz RL, Simpson RJ. 1995. *J. Biol. Chem.* 270:9322–26
161c. Komine Y, Kitabatake M, Yokogawa T, Nishikawa K. 1994. *Proc. Natl. Acad. Sci. USA* 91:9223–27
161d. Ushida C, Himeno H, Watanabe T, Muto A. 1994. *Nucl. Acids Res.* 22:3392–96
162. Atkins JF, Elseviers D, Gorini L. 1972. *Proc. Natl. Acad. Sci. USA* 69:1192–95
163. Atkins JF, Nichols BP, Thompson S. 1983. *EMBO J.* 2:1345–50
164. Fox TD, Weiss-Brummer B. 1980. *Nature* 288:60–63
165. Hwang CB, Horsburgh B, Pelosi E, Roberts S, Digard P, et al. 1994. *Proc. Natl. Acad. Sci. USA* 91:5461–65
166. Cutting GR, Kasch LM, Rosenstein BJ, Tsui L-C, Kazazian HH, et al. 1990. N. *Engl. J. Med.* 323:1685–89
167. Kerem B, Zielenski J, Markiewicz D, Bozon D, Gazit E, et al. 1990. *Proc. Natl. Acad. Sci. USA* 87:8447–51
168. Cuppens H, Marynen P, de Boeck C, de Baets F, Eggermont E, et al. 1990. *J. Med. Genet.* 27:717–19
169. Fearon K, McClendon V, Bonetti B, Bedwell DM. 1994. *J. Biol. Chem.* 269:17802–8
170. Hu PY, Waheed A, Sly WS. 1995. *Proc. Natl. Acad. Sci. USA* 92:2136–40
171. Hu PY, Ernst AR, Sly WS, Venta PJ, Skaggs LA, et al. 1994. *Am. J. Hum. Genet.* 54:602–8
172. Gerchman SE, Graziano V, Ramakrishnan V. 1994. *Protein Expr. Purif.* 5:242–51
173. de Smit MH, van Duin J, van Knippenberg PH, van Eijk HG. 1994. *Gene* 143:43–47
174. Meng S-Y, Hui JO, Haniu M, Tsai LB. 1995. *Biochem. Biophys. Res. Commun.* 211:40–48
175. Leinfelder W, Zehelein E, Mandrand-Berthelot M-A, Bock A. 1988. *Nature* 331:723–26
176. Bock A, Forchhammer K, Heider J, Leinfelder W, Sawers G, et al. 1991. *Mol. Microbiol.* 5:515–20

177. Tormay P, Wilting R, Heider J, Bock A. 1994. *J. Bacteriol.* 176: 1268–74
178. Tinoco I. 1993. In *The RNA World*, ed. RF Gesteland, JF Atkins, pp. 603–7. Cold Spring Harbor, NY: Cold Spring Harbor Lab. Press
179. Barrell BG, Anderson S, Bankier AT, de Bruijn MHL, Chen E, et al. 1980. *Proc. Natl. Acad. Sci. USA* 77:3164–66
180. Heckman JE, Sarnoff J, Alzner-DeWeerd B, Yin S, RajBhandary UL. 1980. *Proc. Natl. Acad. Sci. USA* 77: 3159–63
181. Claesson C, Lustig F, Boren T, Simonsson C, Barciszewska M, et al. 1995. *J. Mol. Biol.* 247:191–96

FIGURE ADDED IN PROOF:

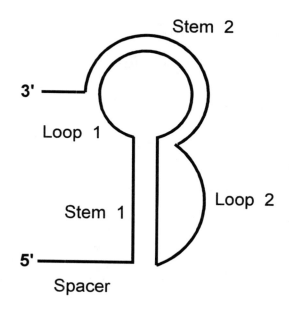

Annu. Rev. Biochem. 1996. 65:769–799

BIOCHEMISTRY AND STRUCTURAL BIOLOGY OF TRANSCRIPTION FACTOR IID (TFIID)

S. K. Burley

Laboratories of Molecular Biophysics and Howard Hughes Medical Institute, The Rockefeller University, New York, New York 10021

R. G. Roeder

Laboratory of Biochemistry and Molecular Biology, The Rockefeller University, New York, New York 10021

KEY WORDS: eukaryotic transcription, RNA polymerase II, transcriptional activators, TATA box binding protein, TBP, TBP-associated factors, TAF, TFIIA, TFIIB, TFIID, TFIIE, TFIIH, histone proteins

ABSTRACT

Eukaryotes have three distinct RNA polymerases that catalyze transcription of nuclear genes. RNA polymerase II is responsible for transcribing nuclear genes encoding the messenger RNAs and several small nuclear RNAs. Like RNA polymerases I and III, pol II cannot recognize its target promoter directly and initiate transcription without accessory factors. Instead, this large multisubunit enzyme relies on both general transcription factors and transcriptional activators and coactivators to regulate transcription from class II promoters. At the center of this process is TFIID, a 700-kD complex composed of the TATA box binding protein (TBP) and a set of phylogenetically conserved, polymerase-specific TBP-associated factors or TAF$_{II}$s. Together, TBP and the TAF$_{II}$s direct assembly of the transcription machinery and play critical regulatory roles in eukaryotic gene expression.

CONTENTS

INTRODUCTION 770

MOLECULAR ARCHITECTURE OF TFIID 771

769

TATA BOX BINDING PROTEIN AND TFIIB 776
   Apo-TBP: A Quasisymmetric Molecular Saddle 777
   TBP-DNA: Minor Groove Recognition/Core-Promoter Bending 778
   TFIIB-TBP-DNA: Recognition of the TBP-DNA Complex 782

HISTONE-LIKE TAF$_{II}$S 784
   dTAF$_{II}$42/dTAF$_{II}$62 Heterodimer 786
   dTAF$_{II}$42/dTAF$_{II}$62 Heterotetramer 788
   Histone-Like Octamer in TFIID 788

REGULATED RNA POLYMERASE II TRANSCRIPTION INITIATION 789
   Activators—TFIID 791
   Activators—Other General Transcription Factors 791
   Activators—Non-TAF$_{II}$ Coactivators—General Transcription Factors 792
   TAF$_{II}$ Coactivators—General Transcription Factors 792
   Mechanistic Considerations in Transcriptional Activation 793

TFIID AND TATA-LESS PROMOTERS 794

CONCLUSIONS AND PERSPECTIVES 795

INTRODUCTION

Eukaryotes have three distinct RNA polymerases (forms I, II, and III) that catalyze transcription of nuclear genes (1). Despite their structural complexity, these multisubunit enzymes require sets of auxiliary proteins known as general transcription initiation factors to catalyze transcription from corresponding class I, II, and III nuclear gene promoters (2–5). TATA box binding protein (TBP), first identified as a component of a class II initiation factor [transcription factor (TF) IID], participates in transcription by all three nuclear RNA polymerases (reviewed in 6). TBP is, therefore, a universal transcription initiation factor component (a situation formally analogous to that of essential subunits common to the three RNA polymerases).

RNA polymerase II (pol II) is responsible for transcribing nuclear genes encoding the messenger RNAs and several small nuclear RNAs. Like RNA polymerases I and III, pol II cannot recognize its target promoter directly. Class II nuclear gene promoters contain combinations of DNA sequences, which include core or basal promoter elements, promoter proximal elements, and distal enhancer elements. Transcription initiation by pol II is precisely regulated by transcription factors (proteins) that interact with these three classes of DNA targets and also with each other (reviewed in 3, 7). The best characterized core-promoter elements, which can function independently or synergistically, are the TATA element, located 25 base pairs (bp) upstream of the transcription start site, and a pyrimidine-rich initiator element (Inr), located at the start site. Individual core promoters may contain both, one, or neither of these elements and are respectively denoted as TATA$^+$Inr$^+$, TATA$^+$Inr$^-$, TATA$^-$Inr$^+$, or TATA$^-$Inr$^-$. Promoter proximal elements occur anywhere between 50 and 200 bp upstream of the cap site, and transcriptional activators binding to these sequences modulate transcription. Finally, distal enhancer elements, which can be found far from the transcription initiation site in either direction and orientation, constitute another group of DNA targets for factors modulating pol II activity.

Several distinct groups of proteins interact with each segment of a typical class II nuclear gene promoter during pol II–catalyzed transcription. The general initiation factors TFIIA, -B, -D, -E, -F, and -H must assemble on the core promoter with pol II before transcription begins. TFIID is the only one of these components with site-specific DNA binding ability, and in the most general case the orderly process of transcription initiation begins with TFIID recognizing and binding tightly to the TATA box. Thereafter, the TFIID-TATA box complex directs accretion of the remaining general initiation factors and pol II to form the preinitiation complex or PIC, a large multiprotein-DNA assembly that supports accurate initiation of transcription at one or at most a few start sites (Figure 1; reviewed in 7). Thus, the PIC is functionally equivalent to the much simpler *E. coli* holoenzyme, which is composed of the core RNA polymerase subunits and one of a modest number of σ-factors (reviewed in 8).

Consistent with its central role in transcription initiation, the earliest studies of transcriptional regulatory mechanisms also identified TFIID as a target for gene-specific activators. As a result, considerable effort has been focused on TFIID in recent years, significantly expanding our knowledge of its structure and mechanisms of action in the initiation and regulation of pol II transcription. Genes encoding virtually all of the stoichiometrically associated subunits of TFIID from human, Drosophila, and yeast have been cloned, sequenced, and expressed. TFIID has been both wholly and partially reconstituted, and these defined multiprotein complexes have revealed much about how TFIID functions. Considerable progress has also been made on the structural biology of TFIID. Three-dimensional structures include: two TBPs, two TBPs recognizing distinct TATA elements, a TFIIB-TBP-TATA element ternary complex, and a complex of two other TFIID subunits that resembles the histone H3-H4 heterotetramer.

MOLECULAR ARCHITECTURE OF TFIID

TFIID's critical role has made it the focus of considerable biochemical and genetic study since its discovery in human cells in 1980 (9). DNA binding by human TFIID was first demonstrated with the Adenovirus major late promoter (AdMLP) (10). DNase I footprinting studies of the AdMLP and selected human gene promoters revealed sequence-specific interactions between human TFIID and the TATA element, which are primarily mediated by the universal transcription factor TBP (see below). In contrast, protection both upstream and downstream of the TATA box is largely sequence independent, displays a nucleosome-like pattern of DNase I hypersensitivity, varies radically between

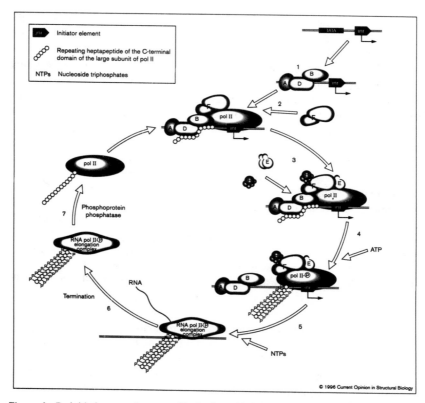

Figure 1 Preinitiation-complex assembly begins with TFIID recognizing the TATA element, followed by coordinated accretion of TFIIA, TFIIB, the nonphosphorylated form of pol II and TFIIF (RAP30/RAP74), TFIIE, and TFIIH. Prior to elongation, pol II is phosphorylated by TFIIH. Following termination, a phosphatase recycles pol II to its nonphosphorylated form, allowing the enzyme to reinitiate transcription. TBP (and TFIID) binding to the TATA box is an intrinsically slow step, yielding a long-lived protein-DNA complex. Efficient reinitiation of transcription can be achieved if recycled pol II reenters the preinitiation complex before TFIID dissociates from the core promoter. Reproduced with permission (145).

promoters (e.g. the AdMLP and the human hsp70 promoter show protection over sequences −47 to +35 and −35 to −19, respectively) (10–12), and can be induced by some activators (see below). A remarkable fact is that TATA-box binding by either TFIID (13) or TBP (14) precludes packaging of the core promoter with the histone proteins (H2A, H2B, H3, and H4). Conversely, core-promoter packaging by histone octamers into nucleosomes prevents TFIID or TBP from binding to the TATA element, effectively repressing transcription (reviewed in 15).

Publication of the sequence of yeast TBP by five laboratories was followed rapidly by the sequences of homologous genes from various eukaryotes and an archaebacterium (amino acid identities within the conserved 180-residue C-terminal portion range between 38% and 100%; reviewed in 6). Recombinant TBP alone can bind both general and regulatory factors and direct PIC assembly and basal (core-promoter) transcription in vitro (reviewed in 3). Activator-dependent transcription, however, requires TBP and the remaining subunits of TFIID, the TBP-associated factors or TAF_{II}s (see 16 and 17 for two early examples), and several non-TAF_{II} coactivators (reviewed in 18). Gene disruption studies of four yeast TAF_{II}s revealed that they are essential for viability (19, 20).

Affinity purification of TFIID using anti-epitope or anti-TBP antibodies permitted identification of a large set of TAF_{II}s, denoted by their origins and apparent molecular weight (19–26). In an attempt to simplify the discussion, the TAF_{II} nomenclatures of the Roeder, Tjian, and Weil laboratories have been adopted for human, Drosophila, and yeast TFIID, respectively. The main exception is the use of the Nakatani nomenclature for $dTAF_{II}62$ and $dTAF_{II}42$, respectively denoted $dTAF_{II}60$ and $dTAF_{II}40$ by Tjian, the structures of which have been determined by X-ray crystallography (27).

Figure 2 illustrates our current census of cloned TAF_{II}s from human (h), Drosophila (d), and yeast (y), ranging in molecular weight from 250 to 15 kD. The majority of these TFIID subunits display significant conservation among the three organisms, implying a common ancestral TFIID. Although no descriptions of archaebacterial TAFs have been published, these organisms likely also possess at least one multiprotein complex containing TBP. Primary sequence analyses of the TAF_{II}s also revealed the presence of various well-known motifs, including bromo- and HMG-domains in $hTAF_{II}250$ and $dTAF_{II}250$, SP1-like glutamine-rich regions in $hTAF_{II}135$ and $dTAF_{II}110$, WD40 repeats in $hTAF_{II}95$ and $dTAF_{II}80$, and a number of histone-like TAF_{II}s (see below). The $hTAF_{II}250/dTAF_{II}250$ bromo-domain is not present in the yeast homolog $yTAF_{II}130$ (19, 20). Instead, it may be provided in yeast by a non-TAF_{II} coactivator such as GCN5, which is required by some transcriptional activators (28, 29). This interpretation argues that protein-protein interaction domains responsible for the regulation of pol II activity are modular and can be distributed among transcription factors during evolution.

The architecture of TFIID is also phylogenetically conserved, and it seems likely that all TFIID complexes are stabilized by similar if not identical patterns of interactions between subunits. A large TAF_{II} ($hTAF_{II}250$, $dTAF_{II}250$, $yTAF_{II}130$) is thought to be a primary anchor to TBP in human, Drosophila, and yeast TFIID (19, 25, 30–32). Nakatani and coworkers have detected a secondary TBP interaction involving the N-terminus of $dTAF_{II}250$, which interferes with TATA-element binding and may be of some regulatory significance (32, 33).

Biochemical tools have also demonstrated that the following smaller $TAF_{II}s$ interact with TBP: $dTAF_{II}150$ (34), $dTAF_{II}80$ (35) and its homolog $hTAF_{II}95$ (Y Tao & RG Roeder, unpublished observations), $dTAF_{II}62$ (36, 37) and its homolog $hTAF_{II}80$ (36, 38), $dTAF_{II}42$ (37), $dTAF_{II}30\alpha/22$ (37, 39) and its

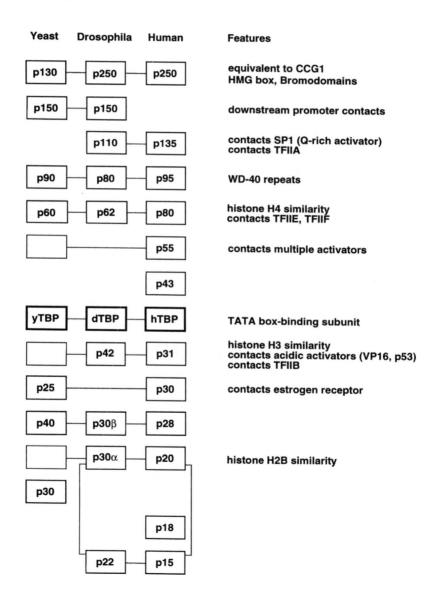

homolog $hTAF_{II}20/15$ (40, 41), $hTAF_{II}30$ (42), and $hTAF_{II}28$ (41). Although the TBP binding sites for these TAF_{II}s have not been extensively characterized, the conserved 180-residue C-terminal region of human TBP is known to be sufficient for TFIID assembly (43), and the corresponding portion of yeast TBP (81% identical to human) can bind the full complement of human TAF_{II}s to give a functional mixed-species TFIID (44).

Similar methods have been used to map TAF_{II}-TAF_{II} interactions, including $dTAF_{II}250$-$dTAF_{II}110$ (31, 45) and the homologous $hTAF_{II}250$-$hTAF_{II}135$ (41) (S Stevens & RG Roeder, unpublished observations), $dTAF_{II}250$-$dTAF_{II}150$ (34), $dTAF_{II}250$-$dTAF_{II}62$ (36) and the homologous $hTAF_{II}250$-$hTAF_{II}80$ (36, 38), $dTAF_{II}250$-$dTAF_{II}30\beta$ (39), $hTAF_{II}250$-$hTAF_{II}30$ (42), $dTAF_{II}150$-$dTAF_{II}30\beta$ (39), $dTAF_{II}110$-$dTAF_{II}80$ (35), $dTAF_{II}110$-$dTAF_{II}30\alpha$ (37, 39) and the homologous $hTAF_{II}135$-$hTAF_{II}20/15$ (40), $dTAF_{II}62$-$dTAF_{II}42$ (37, 36) and the homologous $hTAF_{II}80$-$hTAF_{II}31$ (38, 46), $hTAF_{II}80$-$hTAF_{II}20/15$ (40), $hTAF_{II}55$-$hTAF_{II}20/15$ (40), $hTAF_{II}30$-$hTAF_{II}20$ (41), $hTAF_{II}30$-$hTAF_{II}18$ (41), and $hTAF_{II}28$-$hTAF_{II}18$ (41).

The literature contains three useful schematic representations of TBP-TAF_{II} and TAF_{II}-TAF_{II} interactions (37, 39, 41). We have not included a comparable drawing in this review because most of these interactions have been demonstrated using biochemical methods and must await genetic proof before they can be considered definitive. Nonetheless, the enormous variety of interactions and their evolutionary conservation suggests that most TFIID subunits engage in multiple intermolecular interactions. Thus, TAF_{II}s may be recruited to TBP-containing complexes by both direct and indirect interactions, and there may be multiple pathways leading to assembly of a functional TFIID complex. The current data do not exclude the possibility of heterogeneous forms of TFIID, distinguished by the presence or absence of certain (perhaps loosely associated) TAF_{II}s (41, 42) or by alternatively spliced forms of individual TAF_{II}s (36, 37).

Figure 2 Evolutionary conservation and properties of TFIID subunits from human, Drosophila, and yeast. Homologous sets of TAF_{II}s include: $hTAF_{II}250$ (30, 146), $dTAF_{II}250$ (31, 32), and $yTAF_{II}130$ (19, 20); $dTAF_{II}150$ (34) and $yTAF_{II}150$ (20, 34, 147); $hTAF_{II}135$ (M Horikoshi & RG Roeder, unpublished observations) and $dTAF_{II}110$ (45, 98); $hTAF_{II}95$ (M Horikoshi & RG Roeder, unpublished observations), $dTAF_{II}85$ (35, 148), and $yTAF_{II}90$ (19, 20); $hTAF_{II}80$ (36, 38), $dTAF_{II}62$ (36, 37), and $yTAF_{II}60$ (20); $hTAF_{II}55$ (100); $hTAF_{II}43$ (M Guermah, C-M Chiang & RG Roeder, unpublished observations); hTBP, dTBP, and yTBP (papers reporting the cloning of various TBPs are reviewed in 6); $hTAF_{II}31$ (38, 46, 102) and $dTAF_{II}42$ (37, 103); $hTAF_{II}30$ (42) and $yTAF_{II}25$ (PA Weil, personal communication); $hTAF_{II}28$ (41), $dTAF_{II}30\beta$ (39), and $yTAF_{II}40$ (PA Weil, unpublished observations); $hTAF_{II}20/15$ (40, 41) and $dTAF_{II}30\alpha/22$ (37, 39); $yTAF_{II}30$ (118), which is also known as ANC-1 (149); and $hTAF_{II}18$ (41). A distinct $hTAF_{II}30$ (24), as well as $hTAF_{II}19$ (24) and $dTAF_{II}21$ (26), have been omitted from this figure because gene cloning and characterization are not complete. Yeast polypeptides that have only been tentatively identified as homologs of the indicated Drosophila and human TAF_{II}s are denoted with open boxes (Y Nakatani; PW Weil, personal communications).

Loss of loosely-associated TAF$_{II}$s during TFIID purification may also explain minor discrepancies in polypeptide compositions among various yeast, Drosophila, and human TFIID preparations (Figure 2).

TATA BOX BINDING PROTEIN AND TFIIB

TBP's role in eukaryotic transcription initiation and its regulation is best understood for genes transcribed by RNA polymerase II. TBP engages in physical and functional interactions with the general initiation factors TFIIA and TFIIB, the C-terminus of the large subunit of pol II, some negative cofactors (NC1, NC2/DR1) that inhibit TFIIB binding during PIC formation, and some transcriptional activators (reviewed in 6). Following TFIID or TBP recognition of the TATA element in a typical class II nuclear gene promoter (which can be blocked by the ATP-dependent inhibitor ADI; 47), TFIIB is the next general transcription factor to enter the PIC. The resulting TFIIB-TFIID (TBP)-DNA platform is in turn recognized by a complex of pol II and TFIIF (pol/F). In vitro studies with negatively supercoiled templates demonstrated that transcription initiation can be reconstituted with TBP, TFIIB, and pol II, suggesting that together TBP and TFIIB position pol II (48). Mutations in TFIIB alter pol II start sites in yeast, as do mutations in the large subunit of pol II, providing compelling evidence for its function as a precise spacer/bridge between TBP and pol II on the core promoter that determines the transcription start site (49, 50). In vivo and under different conditions in vitro, pol II transcription initiation depends on TFIIE, TFIIF, TFIIH, and possibly TFIIA. The latter may stabilize DNA binding by TBP or TFIID, and it appears to be instrumental in preventing or reversing various negative cofactor interactions that block either TATA-binding (ADI) or TFIIB-recognition of a preformed TBP-DNA complex (NC1, NC2/DR1) (reviewed in 47).

Once PIC assembly is complete, and in the presence of nucleoside triphosphates, strand separation at the transcription start site occurs to give an open complex, the C-terminal domain of the large subunit of pol II is phosphorylated, and pol II initiates transcription and is released from the promoter. During elongation in vitro, TFIID can remain bound to the core promoter to support rapid reinitiation of transcription by pol II and the other general factors (Figure 1; reviewed in 51). Core-promoter binding by the TBP subunit of TFIID is an intrinsically slow step because of the dramatic DNA deformation induced in the TATA element (reviewed in 52). An abbreviated PIC assembly mechanism has also been proposed, following recent discoveries of various pol II holoenzymes containing many if not all of the general initiation factors other than TFIID (reviewed in 53).

Studies of the mechanisms of action of TBP in nuclear gene transcription by RNA polymerases I (pol I) and III (pol III) are less well advanced. A defined

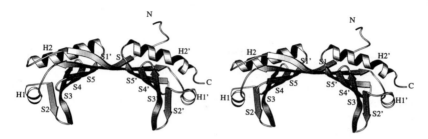

Figure 3 MOLSCRIPT (150) stereodrawing of the structure of TBP2 viewed perpendicular to the internal pseudodyad axis. The N- and C-termini of the protein are indicated. The α-helices are shown as ribbon spirals and labelled H. The β-strands are shown as ribbon arrows (S). Loops and turns are drawn as double lines. The symbol (') refers to the second structural domain or repeat. Reproduced with permission (58).

TBP-TAF complex (selectivity factor 1 or SL1) acts in pol I transcription, and its three TAF_Is are distinct from the pol II–specific subunits of TFIID (reviewed in 54). TFIIIB is the pol III–specific TBP-TAF complex consisting of at least two TAF_{III}s, one of which is similar to TFIIB (reviewed in 55). Another TBP-containing complex known as B-TFIID has been described (56). Although B-TFIID is active in basal or core-promoter transcription, it cannot substitute for TFIID in activator-dependent transcription, and its precise biological function is unknown (57). Analyses of the subunit structure and biochemical activities of B-TFIID do, however, provide some clues as to its identity. This TBP-TAF complex has considerable ATPase activity and appears to have only one large TAF that may be the mammalian homolog of ADI, the ATP-dependent inhibitor of TBP characterized in yeast (47).

Apo-TBP: A Quasisymmetric Molecular Saddle

In 1992, we reported the structure of TBP isoform 2 (TBP2) from *Arabidopsis thaliana* at 2.6 Å resolution (58). More recent progress on crystallographic studies of uncomplexed TBPs includes further refinement of TBP2 at 2.1 Å resolution (6) and a molecular replacement structure of the C-terminal 180 residues of yeast TBP (59). The structure of TBP2 determined at 2.1 Å resolution is illustrated in Figure 3. The two apo-TBP structures are very similar, with two α/β structural domains of 89–90 amino acids related by approximate intramolecular twofold symmetry. TBP2 also has a relatively flexible 18–amino acid N-terminal segment. The C-terminal or core region of TBP binds to the TATA consensus sequence (TATAa/tAa/t) with high affinity and slow off-rate, recognizing minor groove determinants and promoting DNA bending. The N-terminal portion of TBP varies in length, shows little or no conservation among different organisms, and is largely unnecessary for transcription in certain yeast strains.

TBP2 resembles a molecular saddle with approximate maximal dimensions 32 Å × 45 Å × 60 Å. DNA binding is supported by the concave underside of the saddle, which is lined by the central 8 strands of the 10-stranded anti-parallel β-sheet. The convex upper surface of TBP2 is composed of the four α-helices, the basic peptide linking the two domains, parts of strands S1 and S1′, and the nonconserved 18 N-terminal residues. This extensive upper surface binds various components of the transcription machinery (reviewed in 6). Each domain or structural repeat comprises approximately half of the phylogenetically conserved C-terminus of TBP, consisting of a five-stranded, curved antiparallel β-sheet and two α-helices. The two helices, lying approximately perpendicular to each other, abut the convex side of the sheet forming the hydrophobic core of each domain. The two structural domains of TBP2 are topologically identical with root-mean-square (r.m.s.) deviation between equivalent α-carbon atomic positions = 1.1 Å, corresponding to the two imperfect repeats in amino acid sequence (30% identical at the amino acid level and 50% identical at the nucleotide level). TBP's ancestor may, therefore, have functioned as a dimer, with gene duplication and fusion giving rise to a monomeric, quasisymmetric TBP.

The two crystal forms of apo-TBP each have two copies of TBP in the asymmetric unit. For TBP2 this appears to result from weak molecular self-association (buried surface area = 1700 Å^2 and measured K_d=1 μM; DB Nikolov & SK Burley, unpublished observations), which can be disrupted by dilution or addition of duplex oligonucleotides bearing a TATA element (58). Human TBP and TFIID have also been reported to form dimers at physiologic intranuclear concentration (60).

TBP-DNA: Minor Groove Recognition/Core-Promoter Bending

Structure of TBP2 complexed with the AdMLP TATA element (TATAAAAG) (52, 61) and of the C-terminus of yeast TBP complexed with the yeast CYC1 -52 TATA element (TATATAAA) (62) have been reported at 1.9 Å and 2.5 Å resolution, respectively (Figure 4). Although the two cocrystal structures differ slightly in detail, both demonstrate an induced-fit mechanism of protein-DNA recognition. DNA-binding is mediated by the protein's curved, eight-stranded, antiparallel β-sheet, which provides a large concave surface for minor groove and phosphate-ribose contacts with the 8 bp TATA element. The 5′ end of standard B-form DNA enters the underside of the molecular saddle, where the C-terminal portion of TBP produces an abrupt transition to an unprecedented, partially unwound form of the right-handed double helix induced by insertion of two phenylalanine residues into the first T:A base step.

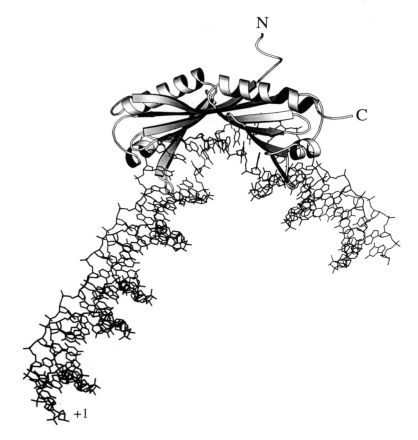

Figure 4   Three-dimensional structure of TBP2 from *Arabidopsis thaliana* complexed with the AdMLP TATA element. The molecular saddle (N- and C-termini labelled) is depicted with a ribbon drawing, and the DNA is shown as a stick figure with the transcription start site labelled with +1. When TBP recognizes the minor groove of the TATA element, the DNA is kinked and unwound to present the minor groove edges of the bases to the underside of the molecular saddle. The coding strand is denoted with solid bonds. Reproduced with permission (61).

Thereafter, the widened minor groove face of the unwound, smoothly bent DNA is approximated to the underside of the molecular saddle, burying a total surface area of about 3100 Å^2 and permitting direct interactions between protein side chains and the minor groove edges of the central 6 bp. A second large kink is induced by insertion of two phenylalanine residues in the base step between the last 2 bp of the TATA element, and there is a corresponding abrupt return to B-form DNA.

Despite this massive distortion, Watson-Crick base-pairing is preserved throughout, and no apparent strain is induced in the DNA, because partial unwinding has been compensated for by right-handed supercoiling of the double helix. Side-chain base contacts are restricted to the minor groove, including the four phenylalanines described above, plus five hydrogen bonds and many van der Waals contacts (Figure 5). There are no water molecules mediating side chain base interactions, and the majority of the hydrogen-bond donors and acceptors on the minor groove edges of the bases remain unsatisfied (13/17 in the AdMLP TATA box). Detailed analysis of the TBP2-DNA cocrystal structure at 1.9 Å resolution demonstrates that the protein also undergoes a modest conformational change upon DNA binding, involving a twisting motion of one domain with respect to the other (52).

DNA packaging into nucleosomes involves wrapping a double helix around the histone octamer (radius of curvature = 45 Å). TA-containing sequences and certain trinucleotide sequences (AAA and AAT) show strong preferences for approximating the minor groove to the histone octamer (63, 64). Sequence-dependent nucleosome positioning correlates with bending A+T-rich sequences toward the minor groove, and packaging of TATA elements into nucleosomes probably results in minor groove compression, precluding TBP binding. Conversely, a preformed PIC remains transcriptionally active after nucleosome assembly (13), and recombinant yeast TBP alone prevents nucleosome-mediated repression of transcription (14). Thus minor groove recognition and bending of DNA by TBP may provide a mechanism for stabilizing the PIC through multiple rounds of transcription initiation in the presence of chromatin (65).

TBP's unusual mode of DNA binding and recognition may also explain the profound toxicity of minor groove-binding anti-tumor agents, such as netropsin and distamycin A. Both of these drugs interact as monomers in the minor groove of B-form DNA, displacing a spine of water molecules, donating hydrogen bonds to A+T-rich sequences, and narrowing the minor groove (66, 67). In at least one case, two molecules of distamycin A have been shown to bind side by side in the minor groove of B-form DNA (68). These two similar modes of drug binding appear opposite to the way TBP recognizes the TATA element, and we previously suggested that TBP and netropsin or distamycin A might compete with one another for the same binding sites in vivo (61). Support for this prediction comes from a recently published study by Beerman and coworkers (69), who demonstrated that distamycin A is a potent inhibitor of TBP-DNA complex formation in vitro.

DNA deformation by TBP may also be important for coordinating and/or stabilizing PIC assembly and activator-PIC interactions. PIC assembly around a bend could produce a more compact complex on the promoter. In addition, DNA bending by TBP could aid in the looping of DNA to bring remotely

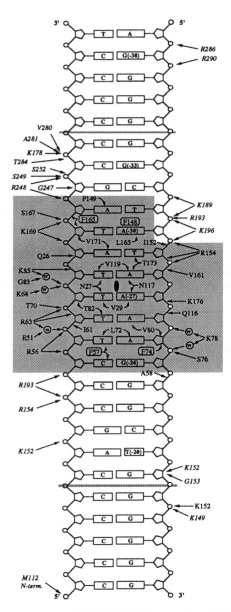

Figure 5 Schematic view of the cTFIIB-TBP2-DNA interaction. TBP2-DNA interactions are restricted to the shaded area and are printed in regular typeface (52). cTFIIB-DNA interactions are printed in italics. Reproduced with permission (82).

bound transcription factors closer to the core promoter for interactions with components of the PIC. This function would be similar to those of the HMG proteins (70, 71) and to the bacterial protein CAP (72), which are thought to activate transcription through bending of DNA upstream of the initiation site (reviewed in 73).

Other biophysical methods have been used to study interactions between TBP and DNA. Site-selection experiments with *Acanthamoeba* TBP showed a marked preference for a site very similar to those studied crystallographically (TATATAAG) (74). DNA bending by TBP in solution was confirmed using circular permutation assays (75). TBP binding was also shown to be enhanced by prebending of DNA toward the major groove and inhibited by prebending toward the minor groove (76). TBP-DNA association kinetics have been studied by various techniques (77–80), and three of the four studies gave results consistent with simultaneous binding and bending with a single second-order rate constant of about 10^5 M^{-1} s^{-1}. Coleman and Pugh opted for a completely different model, involving dissociation of a tight human TBP dimer, tight nonspecific DNA binding by TBP, and sliding of TBP on DNA (80). Finally, a novel chemical probe was used to demonstrate that core-promoter distortion transiently extends beyond the confines of the TATA box during TBP binding (81).

TFIIB-TBP-DNA: Recognition of the TBP-DNA Complex

The crystal structure of a TFIIB-TBP-TATA element ternary complex has been determined at 2.7 Å resolution (82). Core TFIIB (cTFIIB) is a two-domain α-helical protein that resembles cyclin A (83, 84) (Figure 6). The ternary complex is formed by cTFIIB clamping the acidic C-terminal stirrup of TBP2 (S2′-S3′) in its cleft, interacting with H1′, the C-terminus, and the phosphoribose backbone up- and downstream of the center of the TATA element (Figures 5 and 7). Although the two domains of cTFIIB have the same fold, they do not have chemically identical surfaces and cannot make equivalent interactions with TBP2. Contacts between cTFIIB and the C-terminal stirrup of TBP2 are made by BH3, BH4, and BH5. The interdomain peptide interacts with H1′ and the C-terminal stirrup of TBP2. cTFIIB's BH2′-BH3′ loop interacts with the same stirrup and the C-terminus of TBP2.

Despite the very extensive intermolecular contacts visualized in the ternary complex structure (total buried surface area ~5600 Å^2), the structure of the TBP2-TATA element complex itself is essentially unchanged. cTFIIB recognizes the preformed TBP-DNA complex, including the path of the phosphoribose backbone created by the unprecedented DNA deformation induced by binding of TBP. In addition to stabilizing the TBP-DNA complex, TFIIB binding contributes to the polarity of TATA element recognition. If TBP were to bind to the quasisymmetric TATA box in the wrong orientation (i.e. if the

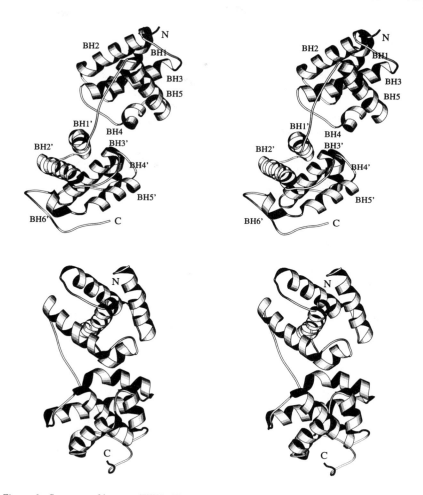

Figure 6 Structure of human cTFIIB. The N- and C-termini of the protein are labelled, and the α-helices of each domain are numbered (BH). The symbol (') refers to the second structural domain. The BH6' helix is absent from the first domain. The top portion of the figure is a stereodrawing of a ribbon representation of the three-dimensional structure viewed down the cleft between the two quasi-identical domains. The lower half of the figure is a stereodrawing viewed perpendicular to the cleft. Reproduced with permission (82).

N-terminal half of the molecular saddle were to interact with the 5' end of the TATA element), the basic/hydrophobic surface of the N-terminal stirrup (S2-S3) would make unfavorable electrostatic interactions with the basic cleft of TFIIB. The solution NMR structure of cTFIIB displays a slightly different arrangement of the two domains (85), suggesting that cTFIIB undergoes a

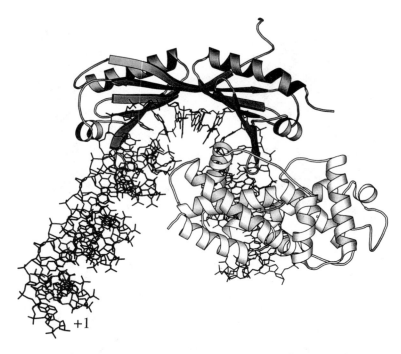

Figure 7 cTFIIB and TBP interacting with the AdMLP. Three-dimensional structure of the ternary complex of human cTFIIB recognizing TBP2 from *Arabidopsis thaliana* complexed with the AdMLP TATA element. cTFIIB (light grey) and TBP (dark grey) are depicted as shaded ribbons, and the DNA is shown as a stick figure with the transcription start site labelled with +1. The coding strand is denoted with solid bonds. The view is identical to that shown in Figure 4. When cTFIIB recognizes the TBP-DNA complex the trajectory of the negatively charged phosphoribose backbone remains essentially unchanged. Reproduced with permission (82).

modest conformational change on recognizing the TBP-DNA complex. The first domain of cTFIIB forms the downstream surface of the cTFIIB-TBP-DNA ternary complex, where, together with the putative Zn^{++}-binding, N-terminal domain of full-length TFIIB, it could readily act as a bridge between TBP and pol II to fix the transcription start site. The remaining solvent-accessible surfaces of TBP (~7900 $Å^2$) and the TFIIB (~8300 $Å^2$) are very extensive, providing ample recognition sites for binding of TAF_{II}s, other class II initiation factors, and transcriptional activators and coactivators.

HISTONE-LIKE TAF_{II}S

Primary structure analyses of some of the TAF_{II}s have revealed considerable amino acid sequence identity with nonlinker histone proteins (Figure 8) (37,

```
                    1  11          20        30        40        50        60        70        80        90       100
                                  ++  ++ ++ + +++ ++    ++    +++ +++ ++   +       +                   *  **   *    *  **
                                         Helix |  Loop |        Helix                                      Loop |  Helix
dTAF42  M--K ISAQIKHVPKDAQVIMSILKELNVQEYEPRVNQLLEFTFRVVTCILDDAKYANHARKKTIDLDDVRLATEVTLDKSFTGPLER H--K
                                   :     :| |::   :|  :::|    : |/:|  ||  :|| |:|:  :|  | ::  :
H3      67      FQRLVREIAQDFKTDLRFQSSAVMALQEASEAYLVGLFEDTNLCAIHAKRVTIMPKDIQLARRIR                                      131
                                         Helix |  Loop |        Helix

                    1                   10        20        30        40        50        60        70        80        91
                                +++++   ++  ++ + +++ ++    ++         ++         +++ ++ +
                                        Helix |  Loop |        Helix                      Loop |  Helix
dTAF62  1       MLYGSSISAESMKVIAESIGVGSLSDDAAKELAEDVSIKLKRIVQDAAKFMNHAKRQKLSVRDIDMSLKVRNVEPQYGFVAK D--R
                                :: :|||   || | :  ::  ::::   : ||:    |||   :::|  : |:||  ::|:
H4      30               TKPAIRRLARRGGVKRISGLIYEETRGVLKVFLENVIRDAVTYTEHAKRKTVTAMDVVYALKRQ                               93
                                        Helix |  Loop |        Helix

                   87 90         100       110       120       130       140       150       160       170       180   190   196
                                        Helix |  Loop |        Helix                                         Loop |  Helix
dTAF28  87 90   ENTPMLTKPRLTELVREVDTTTQLDEDVEEILLQIIDDFVEDTVKSTSAFAKHRKSNKIEVRDVQLHFERKYNMWIPGFGTDELRPYKRAAVTEAHKQRLALIRKTIKKY
                   : :::|    ::  :: ::|: |  :::| |   |:| :: :|  ::|::     |: | :|::|: :::: :         |: :|  | ||:  | ||
H2B     34      RKESYSIYVYKVLKQVHPDTGISSKAMGIMNSFVNDIFERIAGEASRLAHYNKRSTITSREIQT-AVRLLL---PGELA-------KHA-VSEGTK---AVTKYTSSK    126
```

Figure 8 Sequence alignments of histone-like portions of dTAF$_{II}$28/22, dTAF$_{II}$42, and dTAF$_{II}$62s with chicken H2B, H3, and H4. Amino acid identities are denoted with (|), and similarities with (:). dTAF$_{II}$42 is an H3 homolog with 14/67 identical plus 14/67 similar residues. dTAF$_{II}$62 is an H4 homolog, with 15/63 identical plus 17/63 similar residues. dTAF$_{II}$28/22 appears to be an H2B homolog, with 21/102 identical plus 21/102 similar residues. The α-helical regions were assigned from the X-ray structures of the TAF$_{II}$42/dTAF$_{II}$62 complex and the octameric histone core (90). In addition, residues are labelled with (+) and (*), indicating involvement in heterodimer and heterotetramer contacts, respectively. Solvent-accessible amino acids are denoted with (=). Reproduced with permission (27). Similar sequence relationships to core histones exist for hTAF$_{II}$80, hTAF$_{II}$31, and hTAF$_{II}$20/15 (38, 40).

38, 40, 86). In Drosophila, dTAF$_{II}$42 and dTAF$_{II}$62 appear to be H3 and H4 homologs, respectively, corresponding to hTAF$_{II}$31 and hTAF$_{II}$80 in human. Both Drosophila and human TFIID also contain putative histone H2B homologs (dTAF$_{II}$30α/22 and hTAF$_{II}$20/15), but appear to lack histone H2A homologs. A direct connection between components of the eukaryotic transcription apparatus and of the machinery of DNA packaging has already been demonstrated for the linker histones. The cocrystal structure of the DNA-binding domain of the liver-specific transcription factor HNF3-γ (87) is virtually identical to the structure of the chicken erythrocyte linker histone H5 obtained without DNA (88). Moreover, HNF-3α, a related factor, stabilizes a precisely positioned nucleosomal array in the liver-specific enhancer of the mouse albumin gene, where it may function as a sequence-specific linker histone (89).

dTAF$_{II}$42/dTAF$_{II}$62 Heterodimer

The cocrystal structure of a complex of two *Drosophila melanogaster* TAF$_{II}$s (dTAF$_{II}$42/dTAF$_{II}$62) has been determined at 2.0 Å resolution (27). dTAF$_{II}$42(17-86) and dTAF$_{II}$62(1-70) are illustrated in Figure 9 with their respective histone homologs. Both dTAF42$_{II}$(17-86) and dTAF62$_{II}$(1-70) are folded into a classical histone core protein motif, consisting of a long central α-helix flanked on each side by a random coil segment and a short α-helix (Figure 9). Truncation of dTAF$_{II}$42 for crystallization removed the portion corresponding to H3's additional N-terminal α-helix, which is present in the histone octamer core structure (90). The r.m.s. deviations between α-carbon atomic positions for dTAF$_{II}$42(17-84) and H3(67-131) and for dTAF$_{II}$62(9-70)

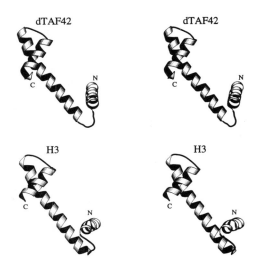

a.

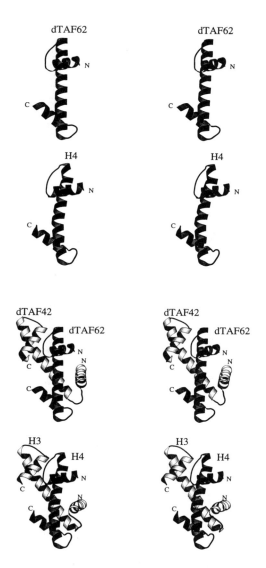

b.

c.

Figure 9 Stereodrawings of the three-dimensional structures of dTAF$_{II}$42(17-86) and dTAF$_{II}$62 (1-70), and their binary complex dTAF$_{II}$42(17-86)/dTAF$_{II}$62(1-70). The corresponding views of histones H3 and H4 have been included for comparison (90). (*a*) dTAF$_{II}$42(17-86) (left) and H3 (right). The additional N-terminal helix of H3 visualized in the structure of the histone octamer core has been omitted for clarity. (*b*) dTAF$_{II}$62(1-70) (left) and H4 (right). (*c*) dTAF$_{II}$42(17-86)/ dTAF$_{II}$62(1-70) (left) and H3/H4 (right). Reproduced with permission (27).

and H4(30-93) are 2.4 Å and 1.6 Å, respectively. These values compare favorably with those obtained by comparing individual histone proteins with one another (91), reflecting differences in the trajectory of the long α-helix. In H3 this helix is nearly straight, whereas it is somewhat kinked in $dTAF_{II}42(17-86)$ near its C-terminus. The converse is true for H4 and $dTAF_{II}62(1-70)$. H3 and H4 demonstrate a single cooperative unfolding transition (92), and the ternary structures of $dTAF_{II}42(17-86)$ and $dTAF_{II}62(1-70)$ are almost certainly not folded in the absence of one another (the number of intramolecular polar and nonpolar contacts between segments of each polypeptide chain is small).

Figure 9 illustrates the structure of the $dTAF_{II}42(17-86)/dTAF_{II}62(1-70)$ heterodimer. As in the H3/H4 heterodimer, also depicted in Figure 9, the two polypeptide chains adopt the histone fold and interact with one another in a head-to-tail fashion (91). Stabilizing contacts between $dTAF_{II}42(17-86)$ and $dTAF_{II}62(1-70)$ are largely hydrophobic, span the entire length of both molecules, and are conserved with H3 and H4 (Figure 8). Binary complex formation buries about 3390 Å^2 of solvent-accessible surface area (48% of the buried surface is hydrophobic, with the remainder either polar or charged).

$dTAF_{II}42/dTAF_{II}62$ Heterotetramer

The structure of the $dTAF_{II}42(17-86)/dTAF_{II}62(1-70)$ heterotetramer is depicted in Figure 10. Like the histone core octamer structure (90), the symmetry axis within the TAF_{II} tetramer coincides with a crystallographic twofold. Interactions between α-helices of the H3 homolog, $dTAF_{II}42(17-86)$, stabilize the tetramer, burying about 670 Å^2 of solvent-accessible surface area (48% of the buried surface is hydrophobic, with the remainder either polar or charged). These values are typical for biologically productive protein-protein molecular recognition events (reviewed in 93), and they are entirely consistent with the measured equilibrium dissociation constant of 10^{-6} M (27). Analysis of the $dTAF_{II}42(17-86)/dTAF_{II}62(1-70)$ heterotetramer reveals a configuration of surface-accessible residues similar to that found in the H3/H4 heterotetramer (reviewed in 94–96), suggesting that it may be capable of interacting with DNA.

Histone-Like Octamer in TFIID

Our crystallographic study of the $dTAF_{II}42(17-86)/dTAF_{II}62(1-70)$ complex suggests that TFIID contains a $(dTAF_{II}42/dTAF_{II}62)_2$ heterotetramer. Compelling, albeit indirect, support for this assertion comes from the results of recent studies of the human TAF_{II} homolog of histone H2B. The measured $hTAF_{II}20$:TBP ratio in TFIID is 4:1, and a histone-like pattern of protein-protein interactions has been demonstrated for $hTAF_{II}31$, $hTAF_{II}80$, and $hTAF_{II}20$ (40). Thus, TFIID may contain a TAF_{II} substructure that resembles the histone

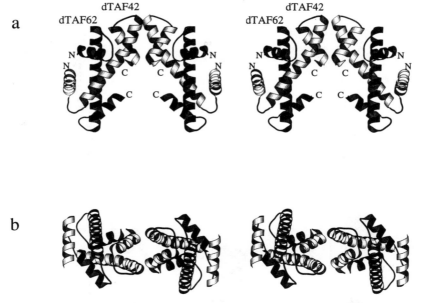

Figure 10 Stereodrawing ribbon representations of the dTAF$_{II}$42(17-86)/dTAF$_{II}$62(1-70) heterotetramer, generated by twofold crystallographic symmetry. (*a*) View perpendicular to the twofold symmetry axis. (*b*) View along the twofold symmetry axis. Reproduced with permission (27).

octamer and mediates some of TFIID's nonspecific interactions with DNA (see below).

REGULATED RNA POLYMERASE II TRANSCRIPTION INITIATION

Transcription initiation of a class II nuclear gene in response to a developmental/environmental signal is controlled by regulating the assembly of a large multiprotein complex on the gene's promoter. Figure 11 illustrates a diverse array of molecular recognition events that represent distinct control points for regulating pol II activity. In their simplest form, these interactions involve direct protein-protein contacts between components of the general transcription machinery (TBP, TAF$_{II}$s, TFIIA, TFIIB, pol II, TFIIF, TFIIE, and TFIIH) and transcriptional activators (bound either to promoter proximal or distal enhancer elements). Indirect interactions between the basal machinery and transcriptional activators mediated by non-TAF$_{II}$ coactivators have also been observed.

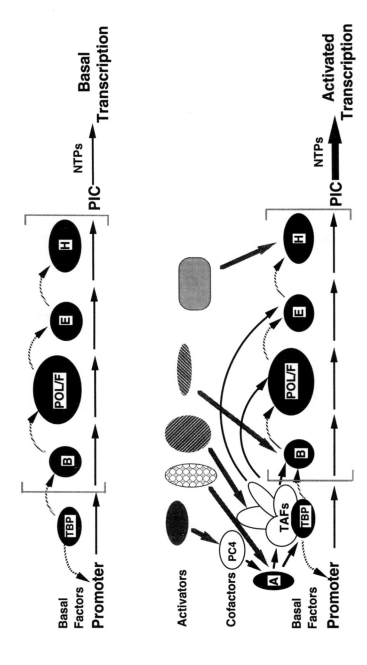

Figure 11 Schematic representation of functional interactions that modulate basal (upper) and activator-dependent transcription (lower). The basal transcription factors (TBP, TFIIA, TFIIB, TFIIF, TFIIE, TFIIH, and pol II are denoted by solid symbols, with the general factor contents of a possible holoenzyme enclosed by square brackets. TAF$_{II}$ and non-TAF$_{II}$ coactivators (open symbols) and transcriptional activators (shaded symbols) are shown interacting with their targets in the PIC.

Before discussing some mechanistic considerations in transcriptional activation, we review the evidence for each class of functional interaction depicted in Figure 11.

Activators—TFIID

Although TBP is a potential target for both viral and cellular transcriptional activators, the subject has already been reviewed extensively (6, 97), and we have elected to restrict our discussion to activator-TAF_{II} interactions. Our current picture of activator-TFIID interactions suggests that the TAF_{II}s can be regarded as a large multiprotein complex that sits atop TBP and integrates signals from many activators and non-TAF_{II} coactivators. The TAF_{II}- and activator-specific interactions that have been observed include: interactions of the glutamine-rich activation domain of Sp1 with $dTAF_{II}110$ (31, 98, 99); interactions of the Sp1 DNA-binding domain, four other activators (USF, CTF, Tat, E1A), and the Inr-binding factor YY1 with $hTAF_{II}55$ (100); interactions of the isoleucine-rich activation domain of NTF-1 with $dTAF_{II}150$ and $dTAF_{II}62$ (99); interactions of p53 and NFκB/p65 activation domains with two of the three histone-like TAF_{II}s, $dTAF_{II}42$ or its homolog $hTAF_{II}31$, and $dTAF_{II}62$ or its homolog $hTAF_{II}80$ (101, 102; M Guermah & RG Roeder, unpublished observations); interactions of the VP16 activation domain with $dTAF_{II}42$ and $hTAF_{II}31$ (46, 103); interaction of the estrogen receptor with $hTAF_{II}30$ (42); interaction of the retinoblastoma-susceptibility gene product with $hTAF_{II}250$ (104); and interactions of the CR3 activation domain of adenovirus E1A with internal and C-terminal regions of $hTAF_{II}250$ and $dTAF_{II}110$, respectively (105). In some cases, the significance of these interactions is supported by correlations between effects of activation-domain mutations on these interactions and on the levels of activator-dependent transcription and, most importantly, by correlations between activator-specific function and the presence of specific TAF_{II}s in partially reconstituted TFIID species (99).

Activators—Other General Transcription Factors

The remaining general transcription factors TFIIA, TFIIB, TFIIF, TFIIE, and TFIIH represent distinct targets within the PIC for interactions with transcriptional activators. Early in PIC assembly, TFIIA and TFIIB stabilize the TFIID-promoter complex (106–108), and their independent effects can be potentiated by direct interactions with various activators, stimulating transcription initiation (109, 110). Different activators have also been found to interact with TFIIF (111) and with pol II (112), which subsequently join the PIC via primary interactions with TFIIB (reviewed in 82). TFIIH, the last factor to enter the PIC, has also been shown to interact with transcriptional activators that modulate pol II activity (113). Indeed, it seems likely that every component of the

PIC is the target of at least one transcriptional activator during transcription from one or more of the estimated 100,000 class II nuclear gene promoters. Support for the relevance of the various interactions comes from the results of studies of activation-domain mutations, which block both physical interactions and transcriptional activation. It has also been possible to isolate TFIIB (109) and TFIIF (111) mutations that selectively affect the function of interacting activators, leaving basal (core-promoter) transcription unchanged.

Activators—Non-TAF$_{II}$ Coactivators—General Transcription Factors

Transcriptional activators need not bind directly to general transcription factors in the PIC (Figure 11). Non-TAF$_{II}$ coactivators, such as human PC4 (18, 114), human OCA-B (115), and the yeast Ada2-Ada3-GCN5 complex (29), can serve as adaptors between activators and basal factors. In some cases, the DNA-binding activator may function solely as a tether for the coactivator. Oct-1, on the other hand, both presents its own activation domain to the PIC and works synergistically with the activation domain of the tethered OCA-B (115). Another population of non-TAF$_{II}$ coactivators, such as yeast Gal11 and the yeast SRBs, is stably associated with pol II, and possibly a subset of other general factors, in a so-called holoenzyme complex (reviewed in 53). Finally, under some conditions TFIIA can be regarded as another non-TAF$_{II}$ coactivator, because it is required for activator-dependent transcription (116) but not for basal (core-promoter) transcription in a minimal system containing TBP instead of TFIID (117).

TAF$_{II}$ Coactivators—General Transcription Factors

The cocrystal structure of core TFIIB recognizing the preformed TBP-DNA complex provided a direct look at the second step of PIC assembly, explaining stabilization of the TBP-TATA element complex by TFIIB. With TFIID instead of TBP, however, the situation is more complicated. Affinity chromatography demonstrated that dTAF$_{II}$42 and its homolog hTAF$_{II}$31 interact with TFIIB (46, 103). Similar methods have been used to identify interactions between TAF$_{II}$s and other general transcription factors, including yTAF$_{II}$30-TFIIF (118), hTAF$_{II}$80-TFIIF and hTAF$_{II}$80-TFIIE (38), and dTAF$_{II}$110-TFIIA (119) and the homologous hTAF$_{II}$135-TFIIA (S Stevens & RG Roeder, unpublished observations). Although the functional significance of these interactions has not been established, it seems likely that they will increase PIC stability and, consequently, facilitate recruitment of the corresponding general transcription factors. They may even do so via conformational changes induced in the TFIID complex by interactions with transcriptional activators or non-TAF$_{II}$ coactivators.

Mechanistic Considerations in Transcriptional Activation

Figure 11 (upper panel) depicts a minimal set of purified polypeptides (basal factors) capable of initiating pol II transcription from many promoters. In vivo or under different conditions in vitro, however, these components effect only low levels of transcription, which may reflect limiting concentrations of basal factors, effects of inhibitors (chromatin, negative cofactors NC1, NC2/DR1), or suboptimal basal factor conformations. Transcriptional activators are generally assumed to overcome such limitations by engaging in direct or indirect protein-protein or protein-DNA interactions (Figure 11, lower panel). These specific molecular recognition events could stabilize and/or complement basal factor interactions, thereby enhancing the rate of formation or the stability of the entire PIC or its catalytic functions during initiation, elongation, or termination.

In the simplest case, direct contact between an activator and a given basal factor could stabilize an intermediate in PIC assembly, ultimately leading to an increased rate of PIC formation or increased PIC stability. Conformational changes important for stronger binding at this or a later stage in the assembly, disassembly, or catalytic processes may also be involved, as suggested for TFIIB interactions with Gal4-VP16 (120). Freely diffusible non-TAF$_{II}$ coactivators appear to modulate transcription using molecular mechanisms similar to those used by DNA-bound transcriptional activators (18, 114). In contrast, recognition of Gal11 by an upstream activator targets the pol II holoenzyme to the core promoter and activates transcription (121). An abbreviated PIC assembly pathway does not, at least in our view, preclude interactions between activators and basal factors that have been demonstrated during step-wise assembly of the PIC in vitro. These interactions may simply represent other molecular mechanisms to recruit the holoenzyme to the core promoter.

When transcriptional activators or non-TAF$_{II}$ coactivators recognize TAF$_{II}$s, increased recruitment and or stabilization of TFIID on the core promoter is observed (122–125). The results of studies with hybrid proteins consisting of TBP fused with heterologous DNA-binding domains suggest that TBP recruitment can be a limiting step (126–128), which otherwise can be overcome by activator-TAF$_{II}$ interactions. If so, what is the functional relevance of the myriad of interactions (both direct and indirect) observed between upstream activators and diverse components of the PIC? In vivo footprinting of the promoter proximal regions of some liver-specific genes has demonstrated that many transcriptional activators appear to be bound simultaneously (129), which is consistent with the view that two or more activators can exert synergistic effects on transcription through concerted interactions with multiple components of the PIC (reviewed in 18, 121, 125, 130). In direct support of this hypothesis, Sauer et al (131, 132) have recently demonstrated that synergy between two different activators (Bicoid and Hunchback) bound to the same promoter results, at least in part, from specific interactions with two distinct $dTAF_{II}s$ that enhance TFIID recruitment.

Another significant aspect of activator-TAF_{II} interactions concerns induced conformational changes in TFIID and TFIID-promoter complexes. Early studies by Horikoshi et al (133, 134) showed activator-induced changes in TFIID-promoter complexes, manifested by downstream extension of the TFIID footprint well beyond the initiation site, that correlated with increased recruitment of other general factors. Later studies with highly purified TFIID have confirmed qualitative and quantitative effects of activators on TFIID binding, although these effects sometimes required TFIIA (124, 125). Photoaffinity DNA-protein crosslinking studies have also revealed large changes in the disposition of $TAF_{II}s$ along the DNA in response to interactions of TFIIA with promoter-bound TFIID (T Oelgeschlager & RG Roeder, unpublished observations). Together, these data suggest that binding of transcriptional activators or coactivators can cause substantial rearrangements in the relative disposition of TFIID subunits and DNA. In this context, one might imagine that activator-induced changes in TAF_{II}-DNA interactions permit the presumptive histone octamer-like substructure within TFIID to engage DNA. We speculate that such a structure might be involved in stabilizing an activator-TFIID-promoter complex, yielding a stereospecific nucleoprotein complex that can support transcriptional activation (73).

TFIID AND TATA-LESS PROMOTERS

Class II nuclear genes that have Inr elements (135) but lack a functional TATA element in their core promoters ($TATA^-Inr^+$) have recently become the focus of intensive study. Surprisingly, these bona fide TATA-less promoters require TBP and $TAF_{II}s$ for both basal and activator-dependent transcription (136). Unlike the $TATA^+Inr^+$ promoters that depend on both high-affinity TBP-DNA interactions and $TAF_{II}s$ for synergy between TATA and Inr (135–137), transcription from $TATA^-Inr^+$ promoters proceeds normally with $TAF_{II}s$ and mutant forms of TBP unable to recognize the TATA element (138). This behavior precisely reflects the roles played by TBP in pol I and pol III transcription. SL1 and TFIIIB function are both unaffected by TBP mutations (139) that interfere with the canonical mechanism of TATA-box binding revealed by X-ray crystallography (reviewed in 52).

While the specific $TAF_{II}s$ and mechanisms involved in transcription from $TATA^-Inr^+$ promoters are unknown, studies of Inr elements and adjacent downstream sequences in composite $TATA^+Inr^+$ promoters have revealed a role for $dTAF_{II}150$ (140, 141). Unlike all other $TAF_{II}s$, $dTAF_{II}150$ is capable of promoter-specific DNA binding (34). Although $dTAF_{II}150$ is important for transcription from Inr-dependent TATA⁺ promoters, whether $dTAF_{II}150$ directly recognizes the Inr is not clear. In fact, crosslinking studies have documented close contacts only in downstream regions of $dTAF_{II}150$-dependent TATA⁺Inr⁺ promoters (140, 142). This uncertainty is compounded by the suggestion that other TFIID-interacting factors are required for $dTAF_{II}150$ function (141), and by reports of Inr-binding factors (TFII-I, YY1) active on certain composite TATA⁺Inr⁺ promoters (143, 144). We still do not know if $dTAF_{II}150$ or its presumptive human homolog are required for initiator function on TATA⁻Inr⁺ promoters. In reconstituted human transcription systems, these promoters require at least two other cofactor activities that are not found in purified human TFIID, and one is required for Inr synergy with TATA elements whereas the other is not (E Martinez & RG Roeder, unpublished observations).

These data suggest at least two alternate pathways for PIC assembly (in addition to the one involving TFIID recognition of a functional TATA element in either TATA⁺Inr⁺ or TATA⁺Inr⁻ promoters). Thus, $TAF_{II}s$ may facilitate TFIID recruitment to TATA⁻Inr⁺ promoters either through direct contact with the Inr or through indirect interactions with a preformed Inr-protein complex, yielding one or at most a few transcription start sites. For TATA⁻Inr⁻ promoters, activators bound to either promoter proximal or distal enhancer elements could recruit TFIID to the promoter, supporting transcription initiation from a multitude of start sites.

CONCLUSIONS AND PERSPECTIVES

Early appreciation of TFIID's importance in promoter recognition and PIC assembly encouraged biologists to think that this large macromolecular complex would play a central regulatory role in pol II transcription. The enormous body of work summarized above clearly documents that these expectations have been realized. Three-dimensional structures of the primary subunit (TBP) and its complex with the core promoter, and of the TFIIB-TBP-DNA ternary complex, revealed novel protein-DNA interactions and an appreciation of how these polypeptides support both initiation and activation. Purification, cloning, and functional characterization of the TFIID "regulatory" subunits ($TAF_{II}s$) revealed a remarkable structural complexity that is phylogenetically conserved between yeast, Drosophila, and human. Within this framework, individual $TAF_{II}s$ serve as activator-selective receptors that modulate, in a synergistic fashion, TFIID binding and structural changes important for PIC assembly and action. Structures of the histone-like $TAF_{II}s$ have suggested an additional evolutionary relationship between the histone octamer and a TFIID substructure, which may be functionally significant.

Although the complexity of TFIID and eukaryotic transcription poses a formidable barrier to our hopes of developing a detailed understanding of transcriptional regulatory mechanisms, the progress reviewed here is decidedly encouraging. Our high-resolution structural data provide a basis for designing a program of rational mutagenesis to confirm the importance of intermolecular interactions identified by biochemical and genetic analyses. In the immediate future, structural work can be combined with biophysical and biochemical studies of the kinetic behavior of TFIID (and interacting components) during PIC assembly and function. Finally, and perhaps most important, the discovery of yeast $TAF_{II}s$ and the power of yeast genetics hold the promise of providing critical insights into the function of TFIID and the remainder of the transcription machinery in vivo.

ACKNOWLEDGMENTS

We are grateful to the many transcription factor biologists who provided us with reprints and access to unpublished material. We apologize for failing to cite work that may have been relevant, but was excluded because of space limitations. We thank Drs. JL Kim, EM Martinez, DB Nikolov, S Stevens, and X Xie for help with figure and manuscript preparation. This work was supported by the Howard Hughes Medical Institute (S.K.B.) and the NIH and Pew Trust (R.G.R.).

Literature Cited

1. Sentenac A. 1985. *CRC Crit. Rev. Biochem.* 18:31–90
2. Reeder RH. 1992. In *Transcription Regulation,* ed. S McKnight, K Yamamoto, pp. 315–48. Cold Spring Harbor, NY: Cold Spring Harbor Lab. Press
3. Roeder RG. 1991. *Trends Biochem. Sci.* 16:402–8
4. Maldonado E, Reinberg D. 1995. *Curr. Opin. Cell Biol.* 7:352–61
5. Gabrielsen OS, Sentenac A. 1991. *Trends Biochem. Sci.* 16:412–16
6. Nikolov D, Burley SK. 1994. *Nat. Struct. Biol.* 1:621–37
7. Zawel L, Reinberg D. 1993. *Prog. Nucleic Acid Res. Mol. Biol.* 44:67–108
8. Helmann JD, Chamberlin MJ. 1988. *Annu. Rev. Biochem.* 57:839–72
9. Matsui T, Segall J, Weil PA, Roeder RG. 1980. *J. Biol. Chem.* 255:11992–96
10. Sawadogo M, Roeder RG. 1985. *Cell* 43:165–75
11. Van Dyke MW, Roeder RG, Sawadogo M. 1988. *Science* 241:1335–38
12. Nakajima N, Horikoshi M, Roeder RG. 1988. *Mol. Cell. Biol.* 8:4028–40
13. Workman JL, Roeder RG. 1987. *Cell* 51:613–22
14. Meisterernst M, Horikoshi M, Roeder RG. 1990. *Proc. Natl. Acad. Sci. USA* 87:9153–57
15. Owen-Hughes T, Workman JL. 1994. *Crit. Rev. Eukaryot. Gene Expr.* 4:403–41
16. Pugh B, Tjian R. 1990. *Cell* 61:1187–97
17. Hoffmann A, Sinn E, Yamamoto T, Wang J, Roy A, et al. 1990. *Nature* 346:387–90
18. Ge H, Roeder RG. 1994. *Cell* 78:513–23
19. Reese JC, Apone L, Walker SS, Griffin LA, Green MR. 1994. *Nature* 371:523–27
20. Poon D, Bai Y, Campbell AM, Bjorklund S, Kim Y-J, et al. 1995. *Proc. Natl. Acad. Sci. USA* 92:8224–28
21. Tanese N, Pugh B, Tjian R. 1991. *Genes Dev.* 5:2212–24
22. Dynlacht BD, Hoey T, Tjian R. 1991. *Cell* 66:563–76

23. Zhou Q, Lieberman PM, Boyer TG, Berk AJ. 1992. *Genes Dev.* 6:1964–74
24. Chiang C-M, Ge H, Wang ZX, Hoffmann A, Roeder RG. 1993. *EMBO J.* 12:2749–62
25. Takada R, Nakatani Y, Hoffmann A, Kokubo T, Hasegawa S, et al. 1992. *Proc. Natl. Acad. Sci. USA* 89:11809–13
26. Kokubo T, Takada R, Yamashita S, Gong D-W, Roeder RG, et al. 1993. *J. Biol. Chem.* 268:17554–58
27. Xie X, Kokubo T, Cohen SL, Mirza UA, Hoffmann A, et al. 1996. *Nature.* 380:316–22
28. Georgakopoulos T, Thireos G. 1992. *EMBO J.* 11:4145–52
29. Horiuchi J, Silverman N, Marcus G, Guarente L. 1995. *Mol. Cell. Biol.* 15:1203–9
30. Ruppert S, Wang EH, Tjian R. 1993. *Nature* 362:175–79
31. Weinzierl ROJ, Dynlacht BD, Tjian R. 1993. *Nature* 362:511–17
32. Kokubo T, Gong D-W, Yamashita S, Horikoshi M, Roeder RG, Nakatani Y. 1993. *Genes Dev.* 7:1033–46
33. Kokubo T, Yamashita S, Horikoshi M, Roeder RG, Nakatani Y. 1994. *Proc. Natl. Acad. Sci. USA* 91:3520–24
34. Verrijzer CP, Yokomori K, Chen J-L, Tjian R. 1994. *Science* 264:933–41
35. Kokubo T, Gong D-W, Yamashita S, Takada R, Roeder RG, et al. 1993. *Mol. Cell. Biol.* 13:7859–63
36. Weinzierl ROJ, Ruppert S, Dynlacht BD, Tanese N, Tjian R. 1993. *EMBO J.* 12:5303–9
37. Kokubo T, Gong D-W, Wootton JC, Horikoshi M, Roeder RG, Nakatani Y. 1994. *Nature* 367:484–87
38. Hisatake K, Ohta T, Takada R, Guermah M, Horikoshi M, et al. 1995. *Proc. Natl. Acad. Sci. USA* 92:8195–99
39. Yokomori K, Chen J-L, Admon A, Zhou S, Tjian R. 1993. *Genes Dev.* 7:2587–97
40. Hoffmann A, Chiang C-M, Xie X, Burley SK, Nakatani Y, et al. 1996. *Nature* 380:356–59
41. Mengus G, May M, Jacq X, Staub A, Tora L, et al. 1995. *EMBO J.* 14:1520–31
42. Jacq X, Brou C, Lutz Y, Davidson I, Chambon P, Tora L. 1994. *Cell* 79:107–17
43. Zhou Q, Boyer TG, Berk AJ. 1993. *Genes Dev.* 7:180–87
44. Zhou Q, Berk AJ. 1995. *Mol. Cell. Biol.* 15:534–39
45. Kokubo T, Gong D-W, Roeder RG, Horikoshi M, Nakatani Y. 1993. *Proc. Natl. Acad. Sci. USA* 90:5896–900
46. Klemm R, Goodrich J, Zhou S, Tjian R. 1995. *Proc. Natl. Acad. Sci. USA* 92:5788–92
47. Auble DT, Hansen KE, Mueller CGF, Lane WS, Thorner J, Hahn S. 1994. *Genes Dev.* 8:1920–34
48. Parvin JD, Sharp PA. 1993. *Cell* 73:533–40
49. Li Y, Flanagan PM, Tschochner H, Kornberg RD. 1994. *Science* 263:805–7
50. Pinto I, Wu W-H, Na JG, Hampsey M. 1994. *J. Biol. Chem.* 269:30569–73
51. Zawel L, Kumar K, Reinberg D. 1995. *Genes Dev.* 9:1479–90
52. Kim JL, Burley SK. 1994. *Nat. Struct. Biol.* 1:638–53
53. Koleske A, Young R. 1995. *Trends Biochem. Sci.* 20:113–16
54. Zomerdijk J, Beckmann H, Comai L, Tjian R. 1994. *Science* 266:2015–18
55. Kassavetis GA, Nguyen ST, Kobayashi R, Kumar A, Geiduschek EP, et al. 1995. *Proc. Natl. Acad. Sci. USA* 92:9786–90
56. Timmers HTM, Sharp PA. 1991. *Genes Dev.* 5:1946–56
57. Timmers HTM, Meyers R, Sharp PA. 1992. *Proc. Natl. Acad. Sci. USA* 89:8140–44
58. Nikolov DB, Hu SH, Lin J, Gasch A, Hoffmann A, et al. 1992. *Nature* 360:40–46
59. Chasman DI, Flaherty KM, Sharp PA, Kornberg RD. 1993. *Proc. Natl. Acad. Sci. USA* 90:8174–78
60. Coleman R, Taggart A, Benjamin L, Pugh B. 1995. *J. Biol. Chem.* 270:13842–49
61. Kim JL, Nikolov DB, Burley SK. 1993. *Nature* 365:520–27
62. Kim YC, Geiger JH, Hahn S, Sigler PB. 1993. *Nature* 365:512–20
63. Drew HR, Travers AA. 1985. *J. Mol. Biol.* 186:773–90
64. Satchwell SS, Drew HR, Travers AA. 1986. *J. Mol. Biol.* 191:659–79
65. Prioleau M-N, Huet J, Sentenac A, Mechali M. 1994. *Cell* 77:439–49
66. Kopka ML, Yoon C, Goodsell D, Pjura P, Dickerson RE. 1985. *J. Mol. Biol.* 183:553–63
67. Coll M, Aymami J, vander Marel GA, van Boom JH, Rich A, Wang A-HJ. 1989. *Biochemistry* 28:310–20
68. Chen X, Ramakrishnan B, Rao ST, Sundaralingam M. 1994. *Nat. Struct. Biol.* 1:169–75
69. Chiang S-Y, Welch J, Rauscher F, Beerman TA. 1994. *Biochemistry* 33:7033–40
70. Werner M, Huth J, Gronenborn A, Clore M. 1995. *Cell* 81:705–14
71. Love JJ, Li XA, Case DA, Giese K,

Grosschedl R, et al. 1995. *Nature* 376: 791–95

72. Schultz SC, Shields GC, Steitz TA. 1991. *Science* 253:1001–7
73. Tjian R, Maniatis T. 1994. *Cell* 77:5–8
74. Wong JM, Bateman E. 1994. *Nucleic Acids Res.* 22:1890–96
75. Starr DB, Hoopes BC, Hawley DK. 1995. *J. Mol. Biol.* 250:434–46
76. Parvin JD, McCormick RJ, Sharp PA, Fisher DE. 1995. *Nature* 373:724–27
77. Hoopes BC, LeBlanc JF, Hawley DK. 1992. *J. Biol. Chem.* 267:11539–46
78. Perez-Howard GM, Weil PA, Beechem JM. 1995. *Biochemistry* 34:8005–17
79. Parkhurst K, Brenowitz M, Parkhurst L. 1996. *Biochemistry* 35:In press
80. Coleman R, Pugh B. 1995. *J. Biol. Chem.* 270:13850–59
81. Sun D, Hurley L. 1995. *Chem. Biol.* 2:457–69
82. Nikolov DB, Chen H, Halay ED, Hisatake K, Lee DK, et al. 1995. *Nature* 377:119–28
83. Jeffrey PD, Russo AA, Polyak K, Gibbs E, Hurwitz J, et al. 1995. *Nature* 376: 313–20
84. Brown NR, Noble MEM, Endicott JA, Garman EF, Wakatsuki S, et al. 1995. *Structure* 3:1235–47
85. Bagby S, Kim S, Maldonado E, Tong KI, Reinberg D, Ikura M. 1995. *Cell* 82:857–67
86. Baxevanis A, Arents G, Moudrianakis E, Landsman D. 1995. *Nucleic Acids Res.* 23:2685–91
87. Clark KL, Halay ED, Lai E, Burley SK. 1993. *Nature* 364:412–20
88. Ramakrishnan V, Finch JT, Graziano V, Lee PL, Sweet RM. 1993. *Nature* 362:219–23
89. McPherson CE, Shim E-Y, Friedman DS, Zaret KS. 1993. *Cell* 75:387–98
90. Arents G, Burlingame RW, Wang B-C, Love WE, Moudrianakis EN. 1991. *Proc. Natl. Acad. Sci. USA* 88:10148–52
91. Arents G, Moudrianakis E. 1995. *Proc. Natl. Acad. Sci. USA* 92:11170–74
92. Karantza V, Friere E, Moudrianakis E. 1996. *Biochemistry* 35:2037–46
93. Janin J. 1995. *Proteins* 21:30–39
94. Klug A, Rhodes D, Smith J, Finch JT, Thomas JO. 1980. *Nature* 287:509–16
95. Arents G, Moudrianakis EN. 1993. *Proc. Natl. Acad. Sci. USA* 90:10489–93
96. Pruss D, Hayes J, Wolffe A. 1995. *Bio-Essays* 17:161–70
97. Hernandez N. 1993. *Genes Dev.* 7: 1291–308
98. Hoey T, Weinzierl ROJ, Gill G, Chen J-L, Dynlacht BD, Tjian R. 1993. *Cell* 72:247–60
99. Chen J-L, Attardi LD, Verrijzer CP, Yokomori K, Tjian R. 1994. *Cell* 79:93–105
100. Chiang C-M, Roeder RG. 1995. *Science* 267:531–36
101. Thut C, Chen J-L, Klemm R, Tjian R. 1995. *Science* 267:100–4
102. Lu H, Levine A. 1995. *Proc. Natl. Acad. Sci. USA* 92:5154–58
103. Goodrich JA, Hoey T, Thut CJ, Admon A, Tjian R. 1993. *Cell* 75:519–30
104. Shao Z, Ruppert S, Robbins P. 1995. *Proc. Natl. Acad. Sci. USA* 92:3115–19
105. Geisberg J, Chen J-L, Ricciardi R. 1995. *Mol. Cell. Biol.* 15:6283–90
106. Buratowski S, Hahn S, Guarente L, Sharp PA. 1989. *Cell* 56:549–61
107. Lee DK, DeJong J, Hashimoto S, Horikoshi M, Roeder RG. 1992. *Mol. Cell. Biol.* 12:5189–96
108. Imbalzano AN, Zaret KS, Kingston RE. 1994. *J. Biol. Chem.* 269:8280–86
109. Roberts SGE, Ha I, Maldonado E, Reinberg D, Green MR. 1993. *Nature* 363: 741–44
110. Kobayashi N, Boyer T, Berk A. 1995. *Mol. Cell. Biol.* 15:6465–73
111. Joliot V, Demma M, Prywes R. 1995. *Nature* 373:632–34
112. Cheong J, Yi M, Lin Y, Murakami S. 1995. *EMBO J.* 14:143–50
113. Xiao H, Pearson A, Coulombe B, Truant R, Zhang S, et al. 1994. *Mol. Cell. Biol.* 14:7013–24
114. Kretzschmar M, Kaiser K, Lottspeich F, Meisterernst M. 1994. *Cell* 78:525–34
115. Luo Y, Roeder RG. 1995. *Mol. Cell. Biol.* 15:4115–24
116. Meisterernst M, Roy AL, Lieu HM, Roeder RG. 1991. *Cell* 66:981–93
117. Cortes P, Flores O, Reinberg D. 1992. *Mol. Cell. Biol.* 12:413–21
118. Henry NL, Campbell AM, Feaver WJ, Poon D, Weil PA, Kornberg RD. 1994. *Genes Dev.* 8:2868–78
119. Yokomori K, Zeidler MP, Chen J-L, Verrijzer CP, Mlodzik M, Tjian R. 1994. *Genes Dev.* 8:2313–23
120. Roberts SGE, Green MR. 1994. *Nature* 371:717–20
121. Barberis A, Pearlberg J, Simkovich N, Farrell S, Reinagle P, et al. 1995. *Cell* 81:359–68
122. Abmayr SM, Workman JL, Roeder RG. 1988. *Genes Dev.* 2:542–53
123. Workman JL, Abmayr SM, Cromlish WA, Roeder RG. 1988. *Cell* 55:211–19
124. Lieberman PM, Berk AJ. 1994. *Genes Dev.* 8:995–1006
125. Chi T, Lieberman P, Ellwood K, Carey M. 1995. *Nature* 377:254–57
126. Chatterjee S, Struhl K. 1995. *Nature* 374:820–22

127. Klages N, Strubin M. 1995. *Nature* 374:822–24
128. Xiao H, Friesen J, Lis J. 1995. *Mol. Cell. Biol.* 15:5757–61
129. Rigaud G, Roux J, Pictet R, Grange T. 1991. *Cell* 67:977–86
130. Choy B, Green MR. 1993. *Nature* 366:531–36
131. Sauer F, Hansen S, Tjian R. 1995. *Science* 270:1783–88
132. Sauer F, Hansen S, Tjian R. 1995. *Science* 270:1825–28
133. Horikoshi M, Carey MF, Kakidani H, Roeder RG. 1988. *Cell* 54:665–69
134. Horikoshi M, Hai T, Lin Y-S, Green MR, Roeder RG. 1988. *Cell* 54:1033–42
135. Smale S, Schmidt M, Berk A, Baltimore D. 1990. *Proc. Natl. Acad. Sci. USA* 87:4509–13
136. Martinez E, Chiang C-M, Ge H, Roeder RG. 1994. *EMBO J.* 13:3115–26
137. Kaufmann J, Smale ST. 1994. *Genes Dev.* 8:821–29
138. Martinez E, Zhou Q, L'Etoile ND, Oelgeschlager T, Berk AJ, Roeder RG. 1995. *Proc. Natl. Acad. Sci. USA* 92:11864–68
139. Schultz MC, Reeder RH, Hahn S. 1992. *Cell* 69:697–702
140. Verrijzer CP, Chen J-L, Yokomori K, Tjian R. 1995. *Cell* 81:1115–25
141. Hansen S, Tjian R. 1995. *Cell* 82:565–75
142. Sypes MA, Gilmour DS. 1994. *Nucleic Acids Res.* 22:807–14
143. Roy AL, Malik S, Meisterernst M, Roeder RG. 1993. *Nature* 365:355–59
144. Usheva A, Shenk T. 1994. *Cell* 76:1115–21
145. Burley SK. 1996. *Curr. Opin. Struct. Biol.* 6:69–75
146. Hisatake K, Hasegawa S, Takada R, Nakatani Y, Horikoshi M, Roeder RG. 1993. *Nature* 362:179–81
147. Ray BL, White CI, Haber JE. 1991. *Curr. Genet.* 20:25–31
148. Dynlacht BD, Weinzierl ROJ, Admon A, Tjian R. 1993. *Nature* 363:176–79
149. Welch MD, Drubin DG. 1994. *Mol. Biol. Cell* 5:617–32
150. Kraulis PJ. 1991. *J. Appl. Crystallogr.* 24:946–50

Annu. Rev. Biochem. 1996. 65:801–47

STRUCTURE AND FUNCTIONS OF THE 20S AND 26S PROTEASOMES

Olivier Coux
Department of Cell Biology, Harvard Medical School, Boston, Massachusetts 02115

Keiji Tanaka
The Tokyo Metropolitan Institute of Medical Science 18-22, Honkomagome 3-chome, Bunkyo-ku, Tokyo 113, Japan

Alfred L. Goldberg
Department of Cell Biology, Harvard Medical School, Boston, Massachusetts 02115

KEY WORDS: protein degradation, ubiquitin, protease, ATPase, multicatalytic proteinase

ABSTRACT

The proteasome is an essential component of the ATP-dependent proteolytic pathway in eukaryotic cells and is responsible for the degradation of most cellular proteins. The 20S (700-kDa) proteasome contains multiple peptidase activities that function through a new type of proteolytic mechanism involving a threonine active site. The 26S (2000-kDa) complex, which degrades ubiquitinated proteins, contains in addition to the 20S proteasome a 19S regulatory complex composed of multiple ATPases and components necessary for binding protein substrates. The proteasome has been highly conserved during eukaryotic evolution, and simpler forms are even found in archaebacteria and eubacteria. Major advances have been achieved recently in our knowledge about the molecular organization of the 20S and 19S particles, their subunits, the proteasome's role in MHC-class 1 antigen presentation, and regulators of its activities. This article focuses on recent progress concerning the biochemical mechanisms and intracellular functions of the 20S and 26S proteasomes.

CONTENTS

INTRODUCTION   802
THE 20S PROTEASOME.   805

801

0066-4154/96/0701-0801$08.00

Components of the 20S Proteasome.. 806
Assembly of the Proteasome.. 808
Discovery of Proteasomes in Eubacteria.................................. 810
X-Ray Analysis of the 20S Proteasome 811
The Proteasome's Novel Catalytic Mechanism............................. 812
Peptidase Activities of the Eukaryotic Particle........................ 814
Regulators of the 20S Proteasome's Activity............................. 816
Selective Inhibitors of Proteasome Function............................. 819

THE 26S PROTEASOME 821
The PA700 or 19/22S Regulatory Complex.................................. 824
Subunits of the 26S Proteasome.. 826
Other Functions and Cofactors of the 26S Complex........................ 831

ADDITIONAL FUNCTIONS OF THE PROTEASOME 834
Proteolytic Processing of Precursor Proteins (NF-κB) 834
Proteasomes and Antigen Presentation.................................... 835
Ubiquitin-Independent Proteolysis....................................... 838

REGULATION OF PROTEASOME CONTENT AND COMPOSITION 839

INTRODUCTION

The past few years have brought dramatic advances in knowledge about the mechanisms for intracellular protein degradation (1). In eukaryotic cells, the proteasome is the site for degradation of most cell proteins (2) and is necessary for viability (3, 4). This particle is the major neutral proteolytic activity in mammalian cells and constitutes up to 1% of the cell protein (5). However, its concentration varies considerably among cell types and is greater in organs (e.g. liver) in which average rates of protein breakdown are higher than in other tissues (e.g. muscle) (5). Proteasomes are present in the nucleus and cytosol of all eukaryotic cells examined, and some particles are also found associated with the endoplasmic reticulum (6) and with the cytoskeleton (7). Recent studies of the architecture of the 20S particle and its role in antigen presentation have uncovered novel catalytic and regulatory mechanisms, and these exciting findings have stimulated wide interest. A number of informative reviews concerning the proteasome have appeared (8–11), including two recent volumes that contain articles by many different authors (12, 13) and several minireviews that focus on the structure of the proteasome (14, 15) or its role in the ubiquitin pathway (16–18) and antigen presentation (19). The present article provides a more systematic review of the proteasome's structure, biochemical mechanisms, pharmacology, and physiological functions.

The proteasome is an essential component of the ATP-dependent proteolytic pathway, which catalyzes the rapid degradation of many rate-limiting enzymes (e.g. ornithine decarboxylase), transcriptional regulators (e.g. IκB), and critical regulatory proteins (e.g. cyclins). The proteasome is also essential for the rapid elimination of highly abnormal proteins, which may arise by mutation or postsynthetic damage. However, recent studies have shown that the proteasome also plays a primary role in the slower degradation of the bulk of proteins in mammalian cells, in the acceleration of this process in muscle in pathological states, and in the turnover of membrane proteins (see below). Proteasomes generally degrade proteins to short peptides, most of which are rapidly hydro-lyzed by cytoplasmic exopeptidases. However, in higher vertebrates, some of the peptides generated are delivered to the cell surface for MHC-class 1 antigen presentation.

The involvement of the proteasome in a degradative process is now relatively easy to test. For example, in yeast, many mutants are available in the 20S proteasome (20), in the associated proteins in the 26S complex (4), and in the enzymes for ubiquitin conjugation (17, 18). In mammalian cells, the role of the proteasome can be studied by the use of recently described inhibitors of the proteasome (see below). Definitive evidence for a direct role for proteasomes requires the use of cell-free extracts, in which the degradative process is blocked when these particles are inhibited or removed [by ultracentrifugation (21, 22) or immuno-depletion (23)] and restored by their readdition.

Despite the abundance, characteristic appearance, and unique properties of the proteasome, progress in understanding its function has, until recently, been slow. The 20S proteasome was originally discovered independently by several laboratories working in different areas. The existence of a 700-kDa ATP-ac-tivated, neutral protease in rat liver (24) and reticulocytes (25) was reported in 1979 and suggested to play a critical role in the ATP-dependent degradative pathway, based on similarities in their pH optimum and inhibitor sensitivity. In 1983, a large "multicatalytic protease" complex containing chymotryptic- and tryptic-like activities was first isolated from the pituitary by Wilk and Orlowski (26), who proposed that it functioned in neuropeptide metabolism. These particles were independently isolated from mRNP preparations as ri-bonucleoprotein particles by Scherrer and coworkers, who proposed the name "pro-somes" in the belief that they programmed mRNA translation (27). In 1984, Monaco and McDevitt, using anti-MHC antibodies, precipitated from cells complexes of "low-molecular-weight proteins" or LMPs (28). These particles were then shown to be identical by functional, morphological, and immunological criteria (29, 30, 31).

In the literature, over 21 different names had been used for this structure. We, therefore, proposed the name "proteasome" to indicate its proteolytic and particulate nature (29), and this term has now come into wide use. However, to prevent ambiguity, in many contexts, a distinction must be made between the two forms of the particle, the 20S (or 700-kDa) and the 26S (2000-kDa) proteasome complexes. Because the 20S particles are abundant and contami-nate preparations of many large enzymes, they have been assigned incorrectly many additional functions (e.g. tRNA processing) (see discussion in 29). At present, their only firmly established role is in protein breakdown or processing by the ATP-dependent pathway.

Twenty years ago, a number of findings [first compiled in the *Annual Review of Biochemistry* (32)] led us to conclude that most protein breakdown in eukaryotic and prokaryotic cells requires ATP. This energy requirement was initially viewed with skepticism, since no known protease exhibited an ATP dependence. However, further studies led to the establishment of soluble cell-free systems in which protein breakdown was ATP-dependent (33) and to the important demonstration by Hershko, Ciechanover, and coworkers of the role of ubiquitin conjugation in marking proteins for degradation in eukaryotic cells (34). ATP plays multiple roles in this process. Ubiquitin conjugation to substrates requires ATP hydrolysis for the activation of ubiquitin, and this step was initially proposed to account for the energy requirement for proteolysis. However, studies in *Escherichia coli,* which lacks the ubiquitin pathway, led to the discovery of large multimeric proteases, in which protein and ATP hydrolysis are linked (9). Subsequent work demonstrated that, in mammalian cells, after ubiquitination of a substrate, ATP is still necessary for its proteolysis and also for the breakdown of proteins whose degradation is independent of ubiquitin (9). This process involves the 20S proteasome plus associated proteins that make it ATP-dependent and allow the selective recognition of ubiquitinated proteins. Thus, ATP-dependent proteases appear to be responsible for intracellular (but not extracellular) proteolysis in all cells and to be even more ubiquitous than ubiquitin.

Although the form of the proteasome most often isolated and studied is the 20S particle, the species that functions in the ubiquitin-dependent pathway is a much larger complex. Rechsteiner's (35) and Goldberg's (36) laboratories demonstrated the existence in mammalian cells of two large proteolytic complexes. The smaller one is the 20S proteasome, which by itself does not degrade ubiquitinated proteins. The larger structure, which was initially believed to be a distinct enzyme, is the 26S (2000-kDa) complex that selectively degrades ubiquitinated proteins by an ATP-dependent process. A variety of studies involving immunoprecipitation (23), reconstitution experiments (37, 38), genetic studies in yeast (39), and electron microscopic studies (40) clearly demonstrated that this larger complex contains the full 20S particle as its proteolytic core. This particle is, therefore, now generally called the 26S proteasome or the 26S proteasome complex. Other terms used [26S protease, Ubiquitin-Conjugate-Degrading Enzyme (UCDEN)] appear misleading in light of its multiple activities. In addition to the 20S particle, the 26S complex contains approximately 20 additional polypeptides found in a distinct complex called the "PA700 proteasome activator," the "19S complex," or the "22S Regulator," which determines substrate specificity and provides multiple enzymatic functions necessary for proteolysis and viability.

THE 20S PROTEASOME

Early electron microscopic (EM) studies of 20S proteasomes purified from different tissues and species revealed a cylinder-shaped particle composed of four stacked rings (41, 42). This complex, containing multiple subunits ranging in size between 20 and 35 kDa (Table 1), has a molecular weight of 700–750 kDa, an apparent diameter of ≈12 nm, and a length of ≈ 17 nm, and it possesses a large internal cavity (43–45, see below). Our present understanding of this structure has benefited greatly from studies of the 20S proteasome from the archaebacterium *Thermoplasma acidophilum*, which contains only two distinct types of subunits, α and β (46). Electron microscopic analyses and decoration with antibodies have shown that its two outer rings contain only α subunits, and its two inner ones only β subunits (47). These findings, together with the demonstration of the sevenfold symmetry of the complex and the determination of its subunits' molecular weights, indicated that this particle possesses 14 α- and 14 β subunits, with 7 subunits per ring ($\alpha_7\ \beta_7\ \beta_7\ \alpha_7$) (48), as confirmed now by X-ray crystallography (49).

The composition of the eukaryotic 20S proteasome is more complicated than that of the archaebacterial complex because the number of distinct sub-units has increased during evolution: 14 different subunits are present in yeast (39), and more in higher eukaryotes (50, 51), although probably no more than 14 distinct subunits can exist in a given 20S complex. Nevertheless, the quaternary structure of the eukaryotic complex is very similar to that of the *Thermoplasma* proteasome (45, 52). EM observations of the eukaryotic proteasome (53) and analysis of subunit sequences, which can be classified into two families, α and β (referring to the archaebacterial subunits) (54, 55, see below), indicate that each ring contains seven different subunits. As in *Thermoplasma*, the α sub-units form each of the two outer rings, and the β subunits form each of the two inner rings. Therefore, each type of subunit is present twice within the complex, once in each α or β ring, as confirmed by decoration with antibodies (56, 57). Thus, the eukaryotic proteasome has a pseudo-seven-fold symmetry, and is composed of two identical halves (56, 58).

In higher eukaryotes, the subunit composition of the 20S proteasome can vary in a given species and is subject to precise regulation. For example, γ-interferon treatment of cells induces the replacement of 3 β subunits by distinct, newly synthesized ones, the LMP subunits (59, 60), which alter the activities of the proteasome so as to favor antigen presentation (61, 62; see below). Subunit expression also varies among tissues and at different stages of development, and these modifications can have major physiological consequences (63–65; see below). In the mouse, for example, the normal subunit composition of the 20S proteasome differs in liver and muscle (66), and, in

Table 1   The subunits of the 20S proteasome

Human gene	MW	pI	Chromosomal locus	GDB symbol[b]	Rat gene	Yeast gene	Essential
α-Type subunits							
HC2[a]	29555	6.16		PSMA1	RC2	PRE5	+
Pros30[a]	30239	6.56	11q15.1	PSMA1			
HC3	25898	7.29	6q27	PSMA2	RC3	Y7	−
HC8	28433	5.06	14q23	PSMA3	RC8	PRS1	+
HC9	29483	7.69		PSMA4	RC9	Y13	+
Zeta	26425	4.59		PSMA5	rZeta	PUP2	+
Pros27 (Iota)	27374	5.97	14q13	PSMA6	rIota	PRS2	+
XAPC7-S[a]	27900	8.69		PSMA7	RC6-I-S	PRE6	+
XAPC7-L[a]				PSMA7	RC6-I-L		
β-Type subunits							
HC5	26489	8.20	7p12-13	PSMB1	RC5	PRS3	+
HC7-I	22836	6.61		PSMB2	RC7-I	PRE1	+
HC10-II	22931	6.15		PSMB3	RC10-II	PUP3	+
HN3	29192	5.63		PSMB4	RN3	PRE4	+
X (MB1, ε)	22897	8.67	14q11.2	PSMB5	rX	PRE2	+
Y (Delta)	25315	4.65	17p13	PSMB6	rDelta	PRE3	+
Z	29965	7.61	9q34.11-34.12	PSMB7		PUP1	+
LMP7-E1[a]	29769	5.46	6p21	PSMB8			−
LMP7-E2[a]	30354	7.18	6p21	PSMB8	RC1		−
LMP2	23245	4.75	6p21	PSMB9	rLMP2		−
MECL1	28936	7.73	16q22.1	PSMB10			

[a]These three pairs of subunits (HC2 and Pros30, XAPC7-S and XAPC7-L, LMP7-E1 and LMP7-E2) are almost identical. Therefore their mRNAs may arise by alternative splicing or use of different transcription initiation sites.

[b]GDB: human Genome Database.

Drosophila, isoforms of one subunit are expressed in different tissues at specific stages of development (J Belote, personal communication).

Components of the 20S Proteasome

To date, the sequences of more than 70 proteasomal subunits from a variety of species have been reported. In yeast (67), rat, and human (4), the sequences of all 14 major subunits are known (Table 1), although it is possible that minor isoforms are still to be discovered in the higher eukaryotes. In yeast, 13 of the 14 subunits are essential for viability (20). The proteasomal subunits constitute a large family of proteins with no obvious similarity to any other family. There are several glycine residues and a GxxxD motif common to nearly all α- and β subunits, but their functional significance is still unclear. Although these subunits share only limited sequence similarities, they are all clearly related, and most probably originated from a common ancestor (55, 58, 68).

Based on their similarities, all proteasomal sequences can be classified into two groups, α and β, referring to the homologous archaebacterial subunits. These two groups have distinct structural and functional roles. The α subunits comprise the outer rings, and the β subunits the inner rings of the 20S proteasome (47, 56, 69). Within each subunit group in eukaryotes, seven subfamilies can be distinguished (55, 68), and a member of each subfamily seems to be present in each particle. EM localization of antigens (56, 69) and chemical cross-linking experiments (K Hendil, personal communication) suggest that each subunit is located in a unique position within the α- or β rings. It is therefore likely that two closely homologous subunits from the same subfamily (e.g. LMPs and their homologs) are not normally present within the same particle. However, overproduction of a subunit by gene transfection can cause it to be incorporated in supranormal amounts, presumably by replacement of some members of other subfamilies (70).

The sequences of the α and β families differ in several respects. The α subunits are more conserved than the β, both within a given species and between species (55). Their N-termini possess a highly conserved motif, which is absent in β subunits and essential for proteasome assembly (see below). All α subunits also contain a RPxG motif of unknown function. These two α-specific motifs are present in the contact region between the α subunits (49). Several of the α subunits contain a nuclear localization signal (NLS), as well as a putative motif "complementary" to the NLS (cNLS) that may regulate nuclear localization (71). The NLS signal appears to function in vivo, since the proteasome is found in both cytoplasmic and nuclear compartments and can shuttle between these compartments during oocyte maturation or cell cycle, for example (72–74). Recently, it has been shown that the NLS of several proteasome subunits can function in nuclear import in permeabilized cells (74a). Tanaka et al suggested that phosphorylation(s) may regulate proteasome translocation across the nuclear membrane by inducing conformational changes that favor or impair its binding to the NLS receptor on the nuclear pore (71, 75) but evidence for this proposal is lacking.

The α subunits have several clear functions: 1. By themselves, they have the capacity to form rings, unlike the β chains (76). 2. The assembly of the α rings is necessary for the formation of the β rings (76). 3. They constitute a physical barrier that limits access of cytosolic proteins into the inner proteolytic chamber (49, see below). 4. They are the sites for binding of the 19S (PA700) and 11S (PA28) regulatory complexes, which modulate the proteolytic activities of the proteasome (40, 77).

The β subunits show a higher degree of sequence diversity than α subunits. Almost all βs initially possess an N-terminal prosequence, which is cleaved off during particle formation. This step exposes on most β subunits a terminal threonine residue that is necessary for activity (see below). Several observations indicate that the β subunits are responsible for the proteolytic activities: 1. All the amino acid substitutions that affect peptide hydrolysis are located on β subunits (3). 2. Unlike the α subunit, the mature β subunit of *Thermoplasma*, when expressed alone in *E. coli,* exhibits some peptidase activity (even though it forms only undefined aggregates) (76). 3. Transfection of individual mammalian β-genes can alter specific peptidase activities (70, 78, 79; see below). The presence of proteolytic sites on the β subunits has now been confirmed by X-ray analysis (49) and by the finding that certain inhibitors block activity by covalently modifying these sites (80; see below).

Early studies had reported that proteasomes are ribonucleoprotein particles and contain small RNAs of 80 to 120 nucleotides (27, 50), which are protected against RNase digestion in the complex (82). These RNAs are mainly tRNAs (83, 84), are not necessary for activity, and are present at much less than one RNA molecule per particle (ranging from one per 2400 to one per 22 particles, depending on their source) (84). Since tRNAs are also minor contaminants of other complexes of similar size (e.g. GroEL), it seems most likely that their presence in the proteasome is due to accidental binding or entrapment (84). It has also been recently reported that highly purified 20S particles have associated RNase activity (85). However, the biological significance of this observation remains to be established.

Assembly of the Proteasome

Expression of the archaebacterial subunits in *E. coli* has shown that both α- and β subunits are required for the formation of the 20S complex (76). When expressed alone, the α subunits (unlike βs) form rings of seven subunits. Interestingly, these α rings can associate in pairs, apparently mimicking the interaction of the α and β rings in the 20S complex (76). Thus, the particle's sevenfold symmetry is inherent in the α subunits, but the four-ring structure requires β subunits.

Genetic analyses of recombinant subunits have identified sequence motifs necessary for assembly of the complex. Deletion or mutation of the highly conserved motif present in the N-terminal part of α subunits prevents formation of the α ring in *Thermoplasma* (76). Similarly, when the *Drosophila* Dm25 subunit is expressed in mouse cells, it is incorporated into mouse proteasomes, but not if its N-terminal domain is deleted (86).

The β subunits possess a N-terminal "prosequence" which is removed upon their incorporation into the complex (54, 55, 58). These prosequences vary more than the processed β subunits: For example, the processed rat and human C5 subunits are 97% identical, but their prosequences are only 54% identical (87,

88). Presumably, the composition of the cleaved peptide was less conserved because these regions do not have to interact with other proteins. In fact, deletion of this prosequence in the archaebacterial proteasome does not prevent the synthesis of functional complexes (76). However, the prosequence is important for the incorporation of the subunits and the normal assembly of the 20S proteasome in eukaryotes. Yeast expressing a truncated proteasomal subunit (Doa3/Pre2) that lacks its prosequence are nonviable (89), and the failure of one of the human LMP7 precursors to generate a subunit that is incorporated into the proteasome appears to be due to its N-terminal extension (90). N-terminal extensions are common in proteolytic enzymes, such as ClpP in *E. coli,* and function to prevent premature activation (91) or to promote proper folding as internal chaperones (92). Similarly, the prosequences of the β subunits seem to be important for their proper folding, and their removal is essential for enzyme activity, since it leads to the exposure at the N-terminus of the threonine residue required for catalytic activity (49, 93). Indeed, when truncated archaebacterial β subunits that lack the prosequence are expressed alone (i.e. without the α subunits) in *E. coli,* they form large aggregates that possess some peptidase activity. In contrast, when the full length β subunits are expressed alone, they are not processed and accumulate as inactive monomers (76).

Recent data have shown that before processing, the β-type subunits accumulate in 13-16S complexes (90, 94). These precursor complexes contain the α and the unprocessed β subunits, and perhaps other unidentified polypeptides, and are converted into the active, fully assembled 20S proteasome after the cleavage of the β subunits (90, 94). Interestingly, treatment of cells with phosphatase inhibitors allows the formation of the precursor complexes but slows the processing of the β subunits and the formation of the 20S proteasome (90). Whether or not the proteolytic processing of the β subunits is autocatalytic is still unclear. Their processing does require the α subunits (76). However, the folding of the β subunits resembles that of other processed enzymes, and this new type of fold may favor an autocatalytic processing (94a).

In several physiological conditions, certain subunits are differentially expressed, and the composition of the 20S proteasome changes in important ways. Upon γ-interferon treatment, the β subunits ε (X), δ (Y), and Z are replaced by subunits LMP7, LMP2, and MECL1, respectively (51, 95, 96), which facilitate antigen presentation (see below). The mechanisms of such substitutions are not understood. Upon transfection of LMP2 and LMP7 genes, the extent of their incorporation into the complex depends on their levels of expression (97), and the excluded homologous subunits continue to be expressed but are not processed (97). Thus subunit exclusion occurs at the assembly stage.

Discovery of Proteasomes in Eubacteria

Eubacteria had been widely assumed to lack proteasomes, but recently, proteasomal genes and novel forms of proteasomes have been discovered in these organisms. The *E. coli* genome project uncovered the presence of a gene related to proteasome β subunits (98). Similar genes were also discovered in *Pasteurella haemolytica, Bacillus subtilis, Mycobacterium leprae* (99), and *Haemophilus influenzae* (100). These observations led to the isolation of active 20S proteasomes from the actinomycete *Rhodococcus* (101) and of related complexes in *E. coli* (102; SJ Yoo, JH Seol, DH Shin, M Rohrwild & MS Kang, submitted; D Missiakas, personal communication). The study of these simple eubacterial complexes has just begun, but should provide valuable new insights into the proteasome's origin, mechanisms of action, and physiological roles. Surprisingly, these recent results indicate the existence of two distinct types of proteasome in different eubacteria.

The *Rhodococcus* particles have a quaternary structure similar to that of other 20S proteasomes. Four different subunits are present; two of them are clearly β subunits, and the two others are distantly related to the α family (101). How these subunits are arranged, or whether all four subunits are present in the same particles, is unclear. The complex exhibits a chymotrypsin-like activity, but no trypsin-like or postglutamyl activities (101). The four genes are distributed in two operons, each containing one α and one β gene. At least one similar operon exists in the related actinomycete, *M. leprae* (101).

E. coli contains a proteasome-like complex that differs considerably from that of *Rhodococcus*. The product of the *hslV* gene is clearly related to the β subunits (99) and is transcribed together with the adjacent *hslU* gene, which codes for an ATPase related to ClpX (98). These proteins are found in a large complex (102) and can associate in vitro in the presence of ATP to form an active complex (SJ Yoo, JH Seol, DH Shin, M Rohrwild, MS Kang, et al, submitted). This complex degrades some hydrophobic peptides and proteins, only in the presence of ATP (102; CH Chung, personal communication). Like other ATP-dependent proteases in *E. coli,* HslV and HslU are heat-shock proteins that are induced in stressful conditions (104) in response to the accumulation of abnormal proteins in cells. Similar *hslV* and *HslU* genes are present in *P. haemolytica* and *H. influenzae.* Interestingly, evidence for an α subunit in the *E. coli* complex is lacking. In the *E. coli* operon, the *hslU* gene is located where the α gene is found in the *Rhodococcus* and *M. leprae* operons. Moreover, no α-related sequence could be detected in the genome of *H. influenzae* (100). The *E. coli* HslV protein can form a complex by itself, but the HslU ATPase is required for peptidase activity. In addition, association of the HslU ATPase with HslV activates ATP hydrolysis (SJ Yoo, JH Seol, DH Shin, M Rohrwild, MS Kang, et al, submitted). In these bacteria, the proteasome thus appears to be a novel structure resembling the ClpAP protease complex, which contains a central proteolytic ring-like structure surrounded by ATPase sub-units (105).

These findings raise many interesting questions, such as whether the *Rhodococcus* and *Thermoplasma* proteasomes associate with an ATPase similar to HslU or to an ATPase resembling those of the 26S complex, whether similar complexes exist in eukaryotic cells or organelles, and what is the functional significance of the various organizations of the proteasome in different bacteria. Possibly, the HslU ATPase serves similar functions as the ATPases in the 26S proteasome of eukaryotes, although they are not closely related in sequence (see below). Clearly, further study of these newly discovered particles should expand our understanding of Ub-independent proteolysis and the role of ATP in proteasome function.

X-Ray Analysis of the 20S Proteasome

Dramatic advances in our understanding of proteasome function have emerged recently through the X-ray diffraction analysis of the archaebacterial particle in the laboratories of R Huber and W Baumeister (49). A 3.4 Å resolution of this structure has been achieved, and unusually detailed insights into its architecture have been obtained owing to its important symmetry. Recently, two- and three-dimensional crystals of the more complex mammalian proteasome have been prepared (106, 107). Therefore, a detailed understanding of its architecture should also emerge in coming years. Most likely, the eukaryotic particle will resemble closely the archaebacterial one, because their subunit sequences and their quaternary organizations are very similar.

The X-ray analysis (49) has confirmed that the *Thermoplasma* 20S particle is a barrel-shaped cylinder (148 Å by 113 Å in diameter) with a penetrating central channel and three large internal chambers, as suggested by prior electron tomography (44, 48). The two outer chambers are enclosed by the α- and β rings. The much larger central compartment, where proteolysis occurs, is enclosed by the inner walls of the β rings. Access to these chambers is through narrow openings in the center of each ring. The entrance through the outer α rings is only 13 Å in diameter, which is only slightly larger than the diameter of an α-helix and is clearly insufficient to allow entry of folded proteins. Accordingly, only fully unfolded and reduced lactalbumin and only the linearized form of a cyclic peptide can be digested by this particle (108). Also, when bulky nanogold moieties were linked to a polypeptide substrate, the nanogold accumulated at the opening of the α rings (108), apparently because it was too large to enter. Interestingly, cross-linked proteins are less susceptible to degradation by the eukaryotic proteasome than their native forms (108a). This opening, which may be an important site for regulation of proteolysis (see below), is bordered by seven hydrophobic loops that emerge from each α subunit. The opening in the β rings that provides access to the central proteolytic chamber is much larger. These narrow orifices have probably evolved both to limit the entry of proteins and to insure a processive degradative mechanism by preventing the ready dissociation of partially digested polypeptides.

Even though the primary sequences of the α- and β subunits have only limited homologies to one another, their tertiary structures are remarkably similar (49). In fact, their C-α chains can be superimposed with only minor deviations. Each subunit is a sandwich of five anti-parallel β sheets, flanked by α helices on the top and bottom. These outer α helices are involved in extensive, tight contacts between the neighboring α subunits. Antiparallel helices at the periphery of the α- and β rings also appear to form wedges that are inserted into the adjacent subunit. Thus, the walls of the proteasome are tightly sealed and unlikely to separate readily. Since no gaps are evident in the walls, peptides generated during protein breakdown probably exit from the large chamber through the opening in the outer α rings, either during or following protein digestion.

By analyzing crystals containing the competitive inhibitor, Acetyl-leu-leu-norleucinal, Löwe et al (49) were able to localize the enzyme's active sites within the central chamber, one on each of the 14 β subunits. This tripeptide inhibitor was bound in an extended conformation in a hydrophobic cleft formed by a loop between two β sheets. The aldehyde moiety was adjacent to the threonine residue on the N-terminus of the β chains, which appears to function as the critical nucleophile in a novel proteolytic mechanism (see below). Interestingly, the distance between these active sites on neighboring β chains is about 28 Å, which corresponds to the length of a hepta- or octapeptide in an extended conformation. If substrates are maintained in an extended conformation within the central chamber, then this distance between active sites is likely a primary determinant of the mean size of peptides produced (see below). The proteasome has often been assumed to cut peptides of a standard mean length (generally believed to be eight amino acids in discussions of antigen presentation), according to a "molecular ruler" (109–111a). Preliminary studies with archaebacterial proteasome, in conditions preventing secondary cleavages of the substrate, indicate a mean size of the products ranging from 8 to 11 residues, with diverse protein substrates (A Kisselev, TN Akopian & AL Goldberg, unpublished data).

The Proteasome's Novel Catalytic Mechanism

The hydrolytic mechanism of the proteasome has long been a mystery, in large part because its subunits show no sequence homology to those of other proteases (54, 55, 58, 112) and because its pattern of sensitivity to standard inhibitors differs from those of the classical families of proteases. Several investigators initially identified the proteasome as a type of cysteine protease (113), and, by analogy to the papain family, names such as "macropain" (114) or "megapain" (35) were proposed. However, the 20S particle is not inhibited by thiol-reacting groups or other inhibitors of cysteine proteases, such as E-64 (5); also, the active β subunit of the archaebacterial proteasome lacks cysteine residues (54). Several investigators had suggested that it is an unusual type of serine protease (5) because of its sensitivity to dichloroisocoumarin and to peptide aldehydes. However, unlike typical serine proteases, it is quite insensitive to covalent modification by diisopropylfluorophosphate (115) or peptide chloromethyl ketones. Moreover, systematic mutagenesis studies showed that no serine residue (nor any histidine) is essential for the activity of the *Thermoplasma* enzyme (116).

Several findings demonstrate that the active-site nucleophile is the hydroxyl group of the N-terminal threonine on the β subunit (49, 80, 93). In the crystal structure, this hydroxyl group appears to form a hemiacetal intermediate with the aldehyde residue of the inhibitor, Ac-leu-leu-norleucinal, thus mimicking the catalytic transition state of serine proteases. Accordingly, mutation of this threonine to an alanine prevented activity, but its replacement with a serine allowed peptide hydrolysis (93). Moreover, a terminal threonine is conserved in most eukaryotic β subunits and in the β subunits of the recently discovered bacterial proteasomes (93). During biosynthesis of the particle, this terminal threonine is exposed by an endoproteolytic cleavage which releases the N-terminal propeptide, and, as expected, this cleavage step or deletion of the propeptide is essential for proteolytic activity (76). Interestingly, the fold of the N-terminus of the β subunits, that is shared by other hydrolytic enzymes having an N-terminal active site (e.g. glutamine PRPP amidotransferase or penicillin acylase), may allow the nucleophilic attack on peptide bonds by the N-terminal residue (threonine) of the subunit (94a).

Strong biochemical evidence supporting this mechanism and its applicability to eukaryotic proteasomes has come from the unexpected discovery that the natural product lactacystin and related lactones are covalent inhibitors of the mammalian proteasome (80; see below). These agents became linked to the terminal threonine on a single β subunit, (X), of the eukaryotic proteasome, apparently through nucleophilic attack by the hydroxyl group. Recently, dichloroisocoumarin has been shown to also inhibit irreversibly the proteasome by modification of this hydroxyl group (117), just as it modifies active sites of serine proteases. Interestingly, replacement of the terminal threonine by serine greatly enhances sensitivity to this inhibitor (93).

The proteasome clearly lacks the catalytic triad characteristic of serine or cysteine proteases. Nevertheless, to catalyze peptide bond hydrolysis, the active site must contain a basic group, in place of the histidine, to accept a proton from the threonine hydroxyl in the transition state. The most likely candidate is the α-amino group of the N-terminal threonine, which is appropriately located to bind the hydroxyl proton. An analogous mechanism has been proposed for penicillin acylase, in which catalysis involves a terminal serine whose amino group acts as a base (118). In both cases, a lysine residue is found near the active site, and replacement of this residue by mutagenesis in the proteasome abrogates activity (93). It remains possible that this lysine serves as the catalytic base, as lysine does in active sites of certain serine proteases (119).

Many features of this intriguing mechanism remain uncertain and have not been subjected to rigorous enzymological study. The rate-limiting step in this reaction scheme is unclear. The proposal has been made that the acyl-enzyme intermediate is relatively stable (49), which could reduce the chances of polypeptide release after only single cleavage and thus favor processive degradation to small products. The *Thermoplasma* proteasome does function in such a processive manner and at V_{max} takes approximately one to two minutes to digest a protein (casein) before attacking the next molecule (117). The average size of peptides generated should depend on the residence time for polypeptides in the central chamber and on the distance between the catalytic sites. Another critical requirement must be whether an easily cleaved peptide-bond is situated in a favorable position in the polypeptide. In addition, Löwe et al proposed that proteins are digested by an oligopeptidyl carboxypeptidase activity, i.e. from C to N termini (49). These important predictions are now open to experimental test.

Peptidase Activities of the Eukaryotic Particle

The 20S proteasome was initially described as a complex with multiple peptidase activities by Wilk and Orlowski (26). Subsequent work with fluorogenic peptide substrates and inhibitors defined three easily assayed activities: a "chymotrypsin-like" activity, which cleaves after large hydrophobic residues; a "trypsin-like" activity, which cleaves after basic residues; and a "postglutamyl" hydrolase, which cleaves after acidic residues (8, 120). These activities appear to be catalyzed mainly by distinct active sites, since inhibitors, point mutations in β subunits, or changes in subunit composition (e.g. after γ-interferon treatment) alter these activities differentially (19, 20). When additional substrates or inhibitors were used with mammalian proteasomes, evidence was obtained for additional peptidase activities. For example, although these three activities are inactivated with 3,4-dichloroisocoumarin (DCI), the degradation of certain proteins is accelerated (121, 122). Under these conditions, two additional endopeptidase activities appear to be stimulated: one cleaving preferentially after the branched-chain amino acids (BrAAP activity), and the other after small neutral amino acids (SNAAP activity) (123).

However, the sharp distinction among these various activities, based simply on the nature of the residue in the P1 position of the substrate, appears to be oversimplified. For example, incorporation of LMP7 can enhance the rate of cleavage after both hydrophobic and basic residues (78). Thus, the individual active sites have overlapping specificities, and a single peptide may be cleaved at multiple active sites, although at different rates and with different affinities (124, 125). In addition, the *Thermoplasma* proteasome, which contains only one type of active site and exhibits primarily chymotryptic-like activity, can also cleave standard basic and acidic substrates, though at 10-fold lower rates (117), and, upon prolonged incubation, can cleave almost all bonds in insulin (109). Presumably, during evolution of eukaryotic proteasomes, the various active sites developed increasing (but not absolute) specificities. Moreover, the residues contributing to specificity can lie at some distance from the P1 position (126). For example, attachment of a hydrophobic residue at the P4 position of a peptide aldehyde can increase its binding up to 1000-fold and transform an inhibitor of the tryptic activity, leupeptin, into a potent blocker of the chymotryptic activity (TN Akopian, B Gilbert, AL Goldberg, R Rando, submitted). Thus, the P4 position can be as important a determinant of specificity as P1.

The simplest model to explain these findings is that the individual peptidase activities are catalyzed by active sites on different β subunits. Recently, Wolf and coworkers, by mutagenizing the active site threonines on individual β subunits, have assigned the chymotryptic and postglutamyl activities to individual subunits, while the tryptic activity appears to be due to multiple active sites (DH Wolf, personal communication). Similarly, the replacement of mammalian β subunits has shown that the sequences of the β subunits determine the specificity of peptide cleavages. For example, transfection of the LMP7 gene enhances the chymotryptic activity (70), and LMP7 is quite close in sequence to the yeast PRE2 subunit, which catalyzes this same activity. Transfection of the Y (δ) gene enhances cleavage after acidic residues (78), and this subunit is highly homologous to the yeast PRE3 subunit required for postglutamyl activity.

How these multiple peptidases function together to degrade proteins is still unclear. Since treatment with DCI can enhance the BrAAP activity and the degradation of certain proteins (while inhibiting the other peptidase activities) (121), and since rates of protein breakdown are not altered by various agents, such as PA28, which markedly stimulate the hydrolysis of some small peptides, the argument has been made that the BrAAP activity is primarily responsible for the degradation of proteins and that the other peptidases catalyze primarily the degradation of protein fragments to small peptides (127). However, such results may simply indicate that the rate-limiting steps for protein and peptide hydrolysis differ. Distinct proteinase and peptidase sites appear very unlikely because yeast mutants lacking one peptidase activity or cells treated with inhibitors of individual peptidase sites show reduced rates of protein degradation (2, 128).

Proteins therefore are most likely to be degraded in a processive fashion all the way to small peptides through the combined actions of multiple peptidase sites. As discussed above, active site threonine residues are present in most β subunits in eukaryotes. However, at least three mammalian β subunits lack a N-terminal-threonine. The argument has therefore been made that these subunits play only a structural role (93). However, the possibility of other catalytic mechanisms for these subunits should not yet be ruled out (especially since the particles contain proteolytic activity insensitive to DCI). Also, the evolutionary pressures that would have favored the loss of degradative activity in multiple subunits of the proteasome are hard to imagine.

Regulators of the 20S Proteasome's Activity

For a protease to exist in the cytosol and nucleus, precise regulatory mechanisms must have evolved to allow selective protein breakdown while preventing generalized proteolytic damage to the cell (9). Two structural features of the proteasome, the isolation of its proteolytic sites within the internal chamber and the small opening in the α rings, clearly reduce the chances of nonspecific digestion of cell constituents. In eukaryotes, the requirement for ubiquitination of substrates and the machinery of the 19S complex probably evolved to insure that polypeptide entry is highly selective. In addition, certain inherent regulatory properties and cellular inhibitory factors appear to help maintain the proteasome in an inactive form.

The 20S proteasome in freshly prepared extracts, or when isolated by gentle methods, is in a latent form; it is inactive against protein substrates and degrades peptide substrates only slowly. Glycerol helps maintain the particle in this latent state (5, 129) and presumably preserves the 20 and 26S complexes in their most physiological forms. A dramatic activation can be induced by various treatments, including purification in the absence of glycerol; incubation at 37°C; heating to 55°C; incubation with basic polypeptides, SDS, guanidine HCl, or fatty acids; or dialysis against water (5, 124, 130). The 20S proteasome is remarkably resistant to heat and denaturing agents (131), although its unusual activation by low concentrations of SDS requires the presence of a substrate [otherwise, the enzyme is irreversibly inactivated (132)]. This activating effect has been very useful to distinguish the 20S from the 26S particles, which are inhibited under these conditions (133, 134).

Most of the treatments that activate the 20S particle are known to disturb protein conformations, so activation presumably involves either a conformational change in the β subunits that enhances catalytic activity or an alteration in the α ring that facilitates substrate entry. The structural basis for this activation has not been elucidated, although considerable changes in the proteasome's sedimentation rate occur with certain of these treatments (135). The stimulation of peptide hydrolysis has been correlated with a proteolytic cleavage of the C2 subunit (129, 136). However, many of the treatments that activate the proteasome are reversible and lead to distinct, interconvertible states (137). These treatments are likely to mimic conformational changes induced by physiological regulators, e.g. PA28 or PA700, which dramatically stimulate peptide hydrolysis (10–50x) by the 20S particle (see below). Presumably, this activation of proteolytic activity of the 20S proteasome within the 26S complex is important in the ATP-dependent degradation of proteins.

Spontaneous activation of the mammalian 20S proteasome occurs also upon incubation at 37°C, which is the normal temperature in mammals. So, maintenance of the latent state in vivo probably involves special mechanisms. Cytosolic proteins have been isolated that inhibit the 20S proteasome and presumably serve to maintain it in an inactive form in vivo. Etlinger and colleagues first reported a 240-kDa inhibitor (a hexamer of a 40-kDa subunit) that reduces the particle's multiple peptidase activities (138). This 240-kDa inhibitor has been reported to be identical to δ-aminolevulinic acid dehydratase, a critical enzyme in heme biosynthesis (139), and to correspond to the ATP-stabilized complex, CF2, necessary for reconstitution of the 26S proteasome in reticulocyte extracts (140). Its 40-kDa subunit has also been reported to be present in the 26S proteasome in a ubiquitinated form of 50 kDa (141). Etlinger and coworkers also described a 200-kDa inhibitor composed of a single 50-kDa subunit (142), which suppresses some of the proteasome's activities. Another interesting factor, PI31 (a dimer of a 31-kDa subunit), inhibits its peptidase activities, but not the hydrolysis of large proteins (143). Its carboxyl terminus is unusual in being very rich in proline residues (GN DeMartino, personal communication). Hsp90, which often copurifies with the 20S proteasome, inhibits one peptidase activity of the 20S proteasome (144) and has been shown to protect its peptidase activities against oxidative damage (144a). Whether these various factors function as inhibitors of the proteasome in vivo either in all cells or in specific cell types remains unclear.

THE PA28 ACTIVATOR   Two complexes that associate with the 20S proteasome and dramatically enhance its activity in eukaryotic cells have been characterized. One is PA700, the 19S complex that combines with the 20S proteasome in an ATP-dependent reaction to form the 26S complex (see below). The other activator, PA28 (145) or 11S regulator (146), can associate with the 20S proteasome in the absence of ATP. This activator causes a large increase in the V_{max} and a decrease in the K_m for multiple peptides, without affecting the rate of hydrolysis of proteins. Presumably, therefore, it does not function to

818   COUX ET AL

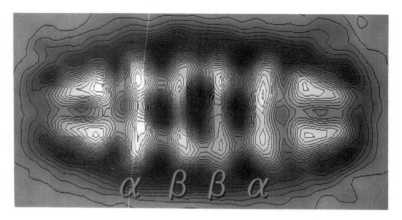

Figure 1 Averaged image, based on electron micrographs, of the complex of the PA28 (11S) activator and 20S proteasome from rat. The α- and β rings of the 20S proteasome are indicated. Photograph kindly provided by W Baumeister.

regulate the initial cleavages of proteins. In crude extracts, PA28 is found either free, or associated with the 20S proteasome and larger complexes (147), although it has not been found in purified 26S particles.

The pure activator is a complex of two subunits, PA28α and PA28β, which are approximately 50% identical (148, 149). Recently, DeMartino and coworkers showed that PA28 is composed of alternating α- and β subunits (GN DeMartino, personal communication). These subunits are 30–40% identical to a nuclear protein of unknown function, Ki antigen, which reacts with sera of patients with lupus erythematosus (150). PA28α alone can stimulate proteasome activity (151), while PA28β alone has no such effect. However, these two subunits together have kinetic effects distinct from either alone (CH Chung, personal communication). PA28β appears to be phosphorylated (GN DeMartino, personal communication), and, recently, Etlinger and coworkers found that treatment with a phosphatase greatly increased PA28 activity (J Etlinger, personal communication). Thus, in vivo, PA28 appears to be negatively regulated by phosphorylation.

Electron microscopy demonstrates that PA28 is a ring-shaped particle, which, like PA700 (see below), caps the 20S proteasome at both or either end (77; Figure 1). This binding of PA28 to the α rings requires its intact C-terminus (152). The PA28-proteasome complex is easily dissociated with low concentrations of salt. Thus, the association is much weaker than that of the proteasome with PA700. In fact, the binding of PA28 can be blocked by antibodies against a single α subunit of the proteasome, and addition of these antibodies will even dissociate the preformed PA28-proteasome complex (153).

PA28α contains in its central part a unique sequence, named by Rechsteiner and coworkers the "KEKE motif" (154). This unusual sequence consists of a very hydrophilic domain rich in alternating positively charged (lysine) residues and negatively charged (glutamate) residues. A similar sequence is present in several subunits of the 20S and 26S proteasomes (see below), as well as in certain heat-shock proteins and Ca^{2+}-binding ER proteins (154). Interestingly, PA28α can bind calcium, and calcium, although at concentrations higher than found in the cytosol, can reversibly inhibit the activation of the proteasome's peptidase activity by PA28 (154a). The proposal has been made that this motif promotes protein-protein interactions and may play a role in antigen presentation. When sequenced, PA28α was found to be identical to IGUPI-5111, which is a major gene product induced by γ-interferon (155). In fact, γ-interferon induces markedly the levels of mRNA for PA28α and PA28β, leading to a large increase in the PA28α protein. There is presently controversy on whether (149) or not (151) PA28β also rises. Together with PA28, γ-interferon induces three novel proteasome subunits; presumably, these adaptations function synergically to favor production of peptides suitable for class I presentation (see below).

The activation of peptide hydrolysis by PA28 is of unclear physiological significance (e.g. free peptides are quite rare in cells). This activity may even be misleading. For example, if PA28 actually functions to promote the release of peptides produced during protein degradation by enlarging the opening in the α ring, it would also stimulate peptide hydrolysis in vitro. Such a mechanism could alter the number, size, and nature of the peptides produced, and could be important in favoring antigen presentation.

Selective Inhibitors of Proteasome Function

The recent identification of inhibitors that can selectively block proteasome function in intact cells has already allowed dramatic progress in understanding its physiological roles (2, 22, 156). In addition to proving useful as investigational tools, these inhibitors also have therapeutic potential as anti-inflammatory agents owing to their ability to block NF-κB production (22, 157, 158). Early studies indicated that certain activities of the proteasome are inhibitable by specific peptide aldehydes that are substrate analogs. For example, chymostatin and Acetyl-leu-leu-norleucinal (MG101, also called calpain inhibitor 1) reversibly inhibit the chymotryptic-like activity, whereas leupeptin inhibits the tryptic-like activity (125). A systematic effort to synthesize more potent proteasome inhibitors was undertaken at ProScript, Inc. (formerly named Myogenics, Inc.), Cambridge, Massachusetts. One such agent, CBZ-leu-leu-leucinal (MG132), is a potent inhibitor of the chymotryptic, postglutamyl and BrAAP activities, and it readily enters cells (unpublished data). This agent blocks protein breakdown in mammalian cells (158) and in permeable strains of yeast (DH Lee & AL Goldberg, unpublished data) and has been made available for investigative purposes. A similar but less potent inhibitor is CBZ-leu-leu-norvalinal (MG115) (2). Both aldehydes reduce degradation of Ub-conjugated proteins by the 26S complex without affecting its ATPase or isopeptidase activities (AL Goldberg, unpublished data). Interestingly, their potency against the activated 20S particle is greater than against the 26S proteasome (2).

Most importantly, these agents block up to 90% of the proteolysis in cultured mammalian cells, including the degradation of highly abnormal proteins, and short- and long-lived normal proteins, which represent the bulk of cell proteins (2). Thus, the proteasome is the primary site for breakdown of most proteins in growing cells, although lysosomal proteolysis can be important in nongrowing cultures and in specialized cells (159, 160). These inhibitors or other peptide aldehydes have been used to implicate the proteasome in the rapid degradation of many short-lived proteins, such as IκB-α (22, 157), in MHC-class I antigen production (2), in the limited processing of NF-κB (22), and in the breakdown of membrane proteins including CFTR (161, 161a) on the endoplasmic reticulum and transmembrane tyrosine kinase receptors (162).

Several features of MG132 and 115 are of particular advantage for research purposes: (a) Their effects are reversible, and, after removal, normal rates of protein degradation are restored within 30–60 min. (b) Their efficacy in reducing proteolysis in intact cells appears greater than in crude extracts or with purified 26S particles (2). (c) Cells are fully viable for at least 10–20 h in their presence, and during this time rates of protein synthesis and ATP content are unaltered. With time, ubiquitinated proteins accumulate and cells express heat-shock genes, a classic stress response (K Bush, S Nigam, DH Lee & AL Goldberg, unpublished data) and, as a consequence, such inhibitors can induce thermotolerance (DH Lee & AL Goldberg, unpublished data). Because these peptide aldehydes also affect lysosomal and Ca^{2+}-activated proteases, it is important in such studies to also show that selective inhibitors of lysosomal proteolysis (e.g. weak bases) or of calpains (e.g. E64) do not have similar effects, and/or that the sensitivity of a response to different aldehyde inhibitors correlates with their potency against the proteasome (2).

Several other proteasome inhibitors have been reported, but have not been studied as extensively (125). Wilk and coworkers introduced the peptide aldehyde CBZ-ile-glu(O-t-Bu)-ala-leucinal, which can also enter cells and reduce the breakdown of IκB (157) and causes accumulation of Ub-conjugates (156). Siman et al have synthesized a number of potent dipeptide inhibitors that contain a hydrophobic chain on the α-amino terminus; they inhibit the chymotryptic activity, can reduce antigen presentation on MHC-class I molecules (163, 164), and can stabilize short-lived mutant proteins (164a). A novel group of potent peptide aldehydes has been synthesized recently (TN Akopian, B Gilbert, AL Goldberg & R Rando, submitted) that contain at the P4 position large hydrophobic groups which cause a dramatic increase in potency. Finally, Adams and coworkers have synthesized very potent, new nonaldehyde inhibitors that block reversibly multiple activities of the 20S and 26S particles and inhibit ubiquitin-dependent proteolysis in vivo (165), and other new nonaldehyde inhibitors were reported recently (165a, 165b).

A very different, and much more specific, proteasome inhibitor is the natural product lactacystin, originally isolated by Omura and colleagues from actinomycetes because of its ability to promote neurite outgrowth on cultured neurons and to block cell division (166). Elegant studies by Schreiber and coworkers (80) showed that lactacystin and related lactones are covalently linked to one polypeptide in the neuronal cell, which they identified as the X-subunit of the 20S proteasome. However, these agents also irreversibly associate with and modify other subunits including the LMP subunits induced by γ-interferon treatment (M Gaczynska & AL Goldberg, unpublished data). Lactacystin modifies the terminal threonine residue involved in catalysis (80), which accounts for the blockage of the chymotryptic- and tryptic-like activities. Like the peptide aldehydes, lactacystin reduces overall proteolysis and class I antigen presentation (KL Rock, personal communication), and degradation of specific proteins (e.g. CFTR and ODC) (161, 166a), without affecting lysosomal proteolysis. Because lactacystin does not inhibit other known proteases, it will certainly prove to be a very useful reagent for research purposes. Recent studies indicate that the lactacystin is converted in aqueous solution into the lactone, which is actually the molecule reacting with the proteasome (166b). Finally, 3,4 dichloroisocoumarin can block multiple activities of proteasomes (121) by reacting with the N-terminal threonine residues of the β subunits (TN Akopian, B Gilbert, AL Goldberg & R Rando, submitted). However, because it is nonspecific and, under certain conditions, activates the proteasome (121), it is not useful for studies in intact cells.

THE 26S PROTEASOME

In eukaryotic cells, degradation of many proteins involves their initial modification by conjugation of ubiquitin (Ub). Ubiquitinated proteins are rapidly degraded by the 26S proteasome (34; Figure 2), although they may also be deubiquitinated by isopeptidases or degraded in lysosomes or yeast vacuoles (166c). The 26S proteasome complex appears to be present in the cytosol and nucleus of all eukaryotic cells, and a variety of yeast mutants have clearly established that this particle is essential for growth and viability. The complex is unfortunately a labile structure, which explains why it was not discovered earlier and why many aspects of its biochemistry are difficult to study. Since the discovery that glycerol and ATP markedly stabilize the ATP-dependent

UBIQUITIN CONJUGATION   PROTEIN DEGRADATION

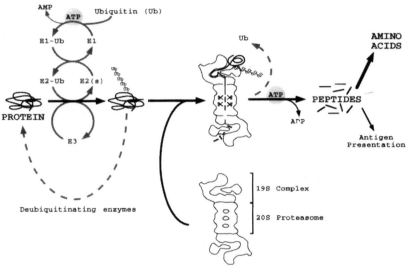

Figure 2 Ubiquitin-dependent proteolytic pathway. The protein substrate is first poly-ubiquitinated in a reaction involving three types of ubiquitinating enzymes: the ubiquitin activating protein E1, an ubiquitin carrier protein E2, and an ubiquitin-protein ligase E3. Deubiquitinating enzymes can reverse this modification. After its poly-ubiquitination, the substrate is then rapidly degraded into small peptides by the 26S proteasome, an ATP-dependent enzyme formed by the association of the 20S proteasome (the proteolytic core) and the 19S regulatory complex (see text).

proteolytic system (5, 167) and the 26S proteasome in particular (35, 36), this enzyme has been purified to apparent homogeneity from a variety of sources, including rabbit reticulocytes (35, 36), human kidney (168), rabbit skeletal muscle (169), *Drosophila* embryo (170), *Xenopus* oocytes (171), Spinach leaves (172), and recently *Saccharomyces cerevisiae* (173).

Although generally called the "26S proteasome" and initially assigned molecular weights of 1000 (35) or 1500 kDa (36), more accurate measurements indicate a sedimentation coefficient of 30S and a molecular weight of 2000 kDa (53). Detailed images of 26S proteasomes have been obtained by electron microscopy coupled with digital image analysis in Baumeister's laboratory (Figure 3) (40, 53, 172). The complexes from amphibia (40), mammals (53), higher plants (172), and yeast (173) appear indistinguishable (58). The 26S

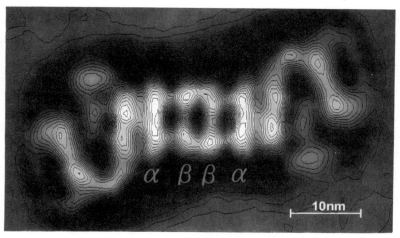

Figure 3 Averaged image, based on electron micrographs, of the rat 26S proteasome. The α- and β rings of the 20S proteasome are indicated. Photograph kindly provided by W Baumeister.

proteasome is a "dumbbell-shaped" symmetric structure that contains two large, rounded complexes on either end of the barrel-shaped 20S particle. These caps, which correspond to the PA700 activator, contain a large roughly V-shaped invagination that may be the site for substrate entry. They are bound to the 20S core in opposite orientations, which clearly reflect a twofold rotational (C2) symmetry of the 26S proteasome.

However, most preparations also contain a large fraction of asymmetric "mushroom-like" complexes, in which the 20S proteasome is associated with only one cap complex (40, 174). Whether these asymmetric forms (which may comprise up to 50% of the total particles in some preparations) differ in any functional properties from the symmetric "dumbbell"-shaped particles is unclear. Another unresolved question is whether the "mushroom"-shaped complexes correspond to distinct particles found in vivo or are in vitro breakdown products. Alternatively, the doubly-capped particles may be more stable in vitro or may be generated preferentially by the conditions used for isolation. Sawada et al (169) suggested that the mushroom- and dumbbell-shaped structures correspond to the two forms of the 26S complex which can be resolved by native electrophoresis and differ in stability (172, 175).

In its general organization, the 26S proteasome thus resembles the *E. coli* ClpAP (Ti) protease (105). This enzyme contains a central cylindrical proteolytic particle, ClpP, which is composed of two seven-membered rings containing a single subunit, and a large ATPase complex, ClpA, at either end. However, the ClpP protease and the proteasome subunits have no sequence similarity. Another complex with organization similar to ClpAP and the 26S

Table 2 The subunits of the 19/22S regulator

Human gene	MW	pI	Chromosomal locus	GDB[b] symbol	S. cerevisiae gene (S. pombe)	Essential
ATPase subunits						
S4 (p56)	49240	5.71	19p13.3	PSMC1	YTA5, YHS4 (MTS2)	+
MSS1 (S7)	48633	5.62	7q22.1-22.3	PSMC2	CIM5, YTA3	+
TBP1	45164	5.18	11p11.2	PSMC3	YTA1	+
TBP7 (S6/p48)	51550	5.42	19q13.11-13.12	PSMC4	YTA2	+
P45 (TRIP1, S8)	45770	8.35	17q23.1-23.3	PSMC5	CIM3, SUG1	+
p42	43990	7.26	5q14.3-15	PSMC6		
Non-ATPase subunits						
P112-L (S1)[a]	105865	5.12	2q37.1-37.2	PSMD1	SEN3	−
P112-S[a]	90512	5.47		PSMD1		
P97 (S2, TRAP2, 55.11)	100184	4.94		PSMD2	NAS1	−
P58 (S3)	60977	8.71		PSMD3	SUN2	+
S5a				PSMD4		
S5b (p50.5)	56195	5.23		PSMD5		
P44 (S10)	45531	5.34		PSMD6		
P40 (S12, Mov-34)	37020	6.03	16q23-24	PSMD7		
P31 (S14)	30004	7.15		PSMD8	NIN1 (MTS3)	+
p27	24682	6.50	12q24.31-24.32	PSMD9		
P28	24428	5.68		PSMD10		
P44.5				PSMD11		
P55				PSMD12		
P40.5				PSMD13		

[a]Subunits p112-L and p112-S are probably generated by alternative splicing.
[b]GDB: human Genome Database.

complex is the *E. coli* proteasome homolog HslV/HslU, which may represent the evolutionary ancestor of the eukaryotic 26S particle.

The PA700 or 19/22S Regulatory Complex

The 26S proteasome complex can be formed in vitro by the ATP-dependent association of the 20S proteasome with another large complex composed of at least 18 proteins that range in molecular weight from 25 to 110 kDa (Table 2). This complex seems to function as the "mouth" for the 20S's digestive machinery, and it provides components necessary for selective degradation of ubiquitinated proteins, including a binding site for Ub-chains (176, 176a), isopeptidase activity, and many ATPases, as well as the capacity to activate the peptidase activities of the 20S particle. Variants of this complex, having different properties, have been isolated by several laboratories and given diverse names, such as the "ball" (10, 175), "19S cap complex" (134), "μ particle" (170), "PA700" (147), "22S complex" (N Tanahasi, personal communication) and "multipain" (9, 177). These complexes are roughly similar in size and protein composition, but vary somewhat in subunit content and functional properties. In the presence of ATP, the purified ball was shown to associate with the 20S proteasome to form a larger proteolytically active complex, likely to be the 26S proteasome (175). PA700 was originally isolated as an ATP-dependent activator of the 20S proteasome. It is clearly distinct from the PA28 activator (147), and, in the presence of ATP, the pure PA700 associates with the 20S proteasome to yield a complex of about 26S that is very active in peptide hydrolysis (147) and can degrade Ub-conjugated proteins, unlike the 20S proteasome (GN DeMartino & AL Goldberg, unpublished observations). In crude extracts containing ATP, the 19S cap and the μ particle also are able to form a complex with the 20S proteasome, but not when they are purified. Multipain is a complex of similar size that has inherent proteolytic activity (9); it also undergoes ATP-dependent association with the 20S proteasome to form a larger complex that degrades Ub-conjugates. Most likely, these differing properties reflect differences in modes of purification of a single structure found in all eukaryotic cells. Alternatively, they may represent isoforms of a single complex present in different cells. However, which of these forms are normally found in vivo and which are artifacts of purification is unclear.

The 26S proteasome is not detectable in lysates of ATP-depleted cells, but can be reconstituted in vitro by addition of ATP (133, 178). Thus the 26S proteasome can be readily disassembled and reassembled. Hershko and his colleagues (178) showed that the 26S proteasome can be reconstituted, by a process that requires ATP, from three different subcomplexes, termed CF1 (600 kDa), CF2 (250 kDa) and CF3 (650 kDa). Subsequent studies demonstrated that CF3 corresponds to the 20S proteasome (37, 38) and that purified 20S particles associate with the other two components to form a large complex capable of degrading Ub-conjugates and casein by an ATP-dependent reaction (38). The relationship of CF1 and CF2 to the 19S complex present in normal cells, and the question of whether the reassembly of 26S particles from CF-1, CF-2, and CF-3 in such ATP-depleted extracts mimics the normal assembly process, are unresolved. CF2 was reported to correspond to the 240/250-kDa inhibitor of the 20S proteasome (140) whose 40-kDa subunit appears to be incorporated into the 26S proteasome in a ubiquitinated form (141). Although CF1 has not yet been completely purified, its polypeptide composition resembles that of the 19S cap complex of the μ particle, and of multipain (J Estrella

& AL Goldberg, unpublished data), despite their differing properties. Also, preparations of CF1, like PA700, contain proteasome-stimulating activity (J Frydman, J Driscoll & AL Goldberg, unpublished data).

Although its components and mode of assembly are still poorly defined, most likely the 26S proteasome is formed in normal cells by the interaction of at least two subcomplexes, the 20S proteasome and the PA700/ μ particle. Their association can probably be stimulated by other factors which may be transiently associated with them or the resulting complex. For example, De-Martino and colleagues recently isolated from normal cells a 300-kDa complex, termed the modulator, that contains three distinct polypeptides and apparently promotes the association of PA700 to the proteasome and enhances the degradation of Ub-conjugates (178a). In fact, the 26S complex may not be a single homogenous entity; instead, like the ribosome, its functional capacity may depend on transient associations with enzymatic or regulatory factors that are removed during purification. For example, in muscle, a protein kinase appears to be associated with the 26S proteasome (179), and, as discussed below, the degradation of certain ubiquitinated proteins is stimulated by iso-peptidase(s) (180), EF-1α (181), or chaperones which may be loosely associated with the 26S proteasome in vivo. Thus, the subunit composition and the functional capacity of the 26S complex may not be uniform and may vary with physiological status. In fact, in *Manduca sexta* during the atrophy and programmed death of the intersegmental muscles, large changes occur in the expression of certain ATPase subunits of the 26S proteasome (182), which presumably contribute to the acceleration of overall protein breakdown.

Subunits of the 26S Proteasome

The recent cloning and determination of the sequences of most of the subunits of the 19S cap complex have provided valuable clues about their functioning. The discovery of yeast mutants in many of these subunits has indicated that the 26S proteasome and therefore proteolysis are of crucial importance for the progression through the cell cycle. The important finding that these mutants have distinct phenotypes suggests that different 26S subunits are necessary for degradation of different protein substrates.

ATPase SUBUNITS   Several groups have undertaken a systematic analysis of the individual polypeptides comprising the PA700 component (Table 2) in order to understand their specific roles. Sequence analysis has revealed that at least six of these subunits are putative ATPases, and high ATPase activity has been found in 26S particles purified from various sources including mammals (168, 183), plants (172), and yeast (DH Hwang & AL Goldberg, unpublished results). Although studies of ATP-depleted reticulocyte extracts had suggested that assembly of the 26S particle is required for ATPase activity (183), purified PA700 and 22S regulator complex display intrinsic ATPase activity (184, 185). Most likely, the functioning of its ATPase subunits accounts for the ATP-requirement for proteolysis and for assembly of the 26S proteasome. Although CTP, GTP, and UTP can partially replace ATP in degradation of ubiquitinated proteins, only ATP and CTP can support the formation of the 26S proteasome from its subcomponents (147, 178, 183). Thus, distinct ATPase subunits probably catalyze these two processes.

The ATPase subunits of the 26S proteasome belong to a new family of ATP-binding proteins that is part of a larger group of ATPases involved in a wide variety of processes (for review, see 186). The hallmark of these ATPases is the presence of one or two copies of a highly conserved module of about 230 amino acids. The ATPases of the 26S proteasome possess only one copy of this module. So far six distinct, but related, ATPases of this family have been found in 26S proteasomes purified from various eukaryotes (187; GN DeMartino, personal communication) (Table 2). Strong sequence homologies exist between the human and yeast subunits, and the human S4 protein can even function in place of its yeast homolog within the 26S complex (188).

Why so many ATPases are present in the 26S proteasome is presently unclear, and whether they are all present within the same complex remains to be established. When expressed in *E. coli*, these proteins have intrinsic ATPase activity (189). Interestingly, human TBP1 and MSS1 expressed in *E. coli* display an ATPase activity equivalent to or higher than that of the 26S proteasome (CH Chung, personal communication). Thus these activities are probably inhibited in the complex, and peptide substrates can activate ATP-hydrolysis by the particle (NE Tawa & AL Goldberg, unpublished data). Mutations in specific ATPases can markedly reduce the degradation of some ubiquitinated proteins, while not affecting others. For example, ts-mutations in CIM3 or in CIM5 stabilize certain short-lived proteins, such as the CLB2 and CLB3 cyclins, and the fusion protein Ub-Pro-β-galactosidase, but not N-end rule substrates (e.g. Leu-β-galactosidase) (190). Thus certain ATPases of the 26S are necessary only for the degradation of specific types of substrates.

Surprisingly, most of the cDNAs coding for the proteasomal ATPases were cloned earlier, and these proteins were initially identified as regulators of gene transcription: (*a*) Human TBP1 was isolated as a protein that binds HIV Tat protein, and its over-expression was able to suppress Tat-mediated transactivation (191), even though it did not bind to DNA by itself (192). (*b*) The same study identified TBP7, a TBP1-homolog, which however did not influence transcription (192). (*c*) The human MSS1 protein was isolated as a suppressor of a defect in the cell-cycle in *S. cerevisiae*, and its expression was found also to enhance Tat-mediated transactivation of HIV genes (193). (*d*) The yeast SUG1 was originally identified because a SUG1 mutant suppressed the requirement for the activation domain of the GAL4 transcription factor (194). SUG1 was also reported to be a component of the mediator complex of the RNA polymerase II (195), and its presence within the 26S proteasome had been questionned (196). However, all these ATPases have now been clearly shown to be subunits of the 26S proteasome (173, 187, 197). Therefore, the changes in gene transcription in cells carrying mutations in these ATPases or overexpressing one such ATPase may all be indirect, owing to the reduced degradation of a short-lived normal or mutant transcriptional factor.

In yeast, the genes of five of these ATPase subunits have been cloned and their roles analyzed by several laboratories. Deletion of each is lethal (188, 190, 198; DM Rubin & D Finley, personal communication), because they prevent either assembly of the 26S complex or its function in proteolysis. Cells carrying point mutations in these ATPases display an abnormal accumulation of ubiquitinated proteins (188) and reduced rates of degradation of mitotic cyclins (190). Moreover, in yeast, site-directed mutations that should inactivate ATP-binding by the S4 and TBP1 homologs stabilize short-lived proteins and cause growth defects (DM Rubin & D Finley, personal communication). Inactivation of one of these ATPases prevents normal development of *Dictostelium discoideum* (199). Interestingly, temperature-sensitive mutations in the yeast homologs of S4, MSS1, and p45 all result in defects in the metaphase/anaphase transition of the cell cycle, causing abnormal chromosome segregation (188, 190, 198). Thus the Ub-proteasome system plays an essential role in exit from the M phase of the cell cycle.

POSSIBLE ROLES OF THE ATPases IN 26S FUNCTION   At least one of these subunits must be responsible for the ATP hydrolysis necessary for the association of the 20S proteasome and PA700 to form the 26S complex. An important issue is whether this assembly reaction occurs only during the biosynthesis of the 26S complex or whether dissociation and reassembly of the particle occur frequently under physiological conditions. In vivo, these association-dissociation processes might be regulated to control overall rates of proteolysis in cells; another possibility is that complex formation is also promoted by binding of ubiquitinated substrates to the 19S regulatory complex. Presumably, this ATP-requiring assembly reaction involves mechanisms similar to the ATP-dependent reactions in *E. coli* that allow complex formation between the ClpA ATPase and ClpP protease complex and between the HslV proteasome homolog and the HslU ATPase.

Within the 26S, the degradation of Ub-conjugated and certain nonubiquitinated proteins (see below) also requires hydrolysis of ATP, but proteolysis, unlike complex formation, can also utilize CTP, GTP, or UTP (178, 183). The hydrolysis of small hydrophobic peptides is also activated by ATP. Nonhydrolyzable ATP analogs or ATPase inhibitors (e.g. vanadate) can block degradation of Ub-conjugates and casein. Thus, cleavage of peptide bonds and ATP hydrolysis seem to be linked, perhaps through a stoichiometric relationship, as found for protease La (lon) (200).

The energy requirement for the functioning of the 26S proteasome and of the large multimeric proteases found in *E. coli* and mitochondria (9) is unexpected on thermodynamic grounds, and peptide bond hydrolysis is catalyzed by numerous proteases independently of ATP. Although investigations of the mechanism of the 26S complex are difficult because of the limited amounts and lability of this complex and of ubiquitin-conjugated substrates, studies of the *E. coli* ATP-dependent proteases La (lon) (201) and Clp AP (Ti) (202) have indicated multiple functions that ATP hydrolysis may serve. One clear role is to regulate proteolytic activity so as to prevent excessive digestion of cell constituents. The activity of protease La is normally inhibited allosterically by bound ADP, but the binding of an unfolded polypeptide to a regulatory domain triggers release of the inhibitory ADP and the binding of ATP, which leads to exposure of its proteolytic sites. After proteolysis, the ADP remains bound and returns the enzyme to an inactive state (9). In the ClpAP protease, ATP binding to the ClpA component enlarges the peptidase site on ClpP, allowing degradation of larger peptides (203). Both enzymes exhibit a substrate-activated ATPase, and such a property has been found associated with muscle 26S proteasome (NE Tawa, E Folco & AL Goldberg, unpublished data). Furthermore, in the *E. coli* proteasome homolog, ATP hydrolysis by HslU is activated by HslV (SJ Yoo, JH Seol, DH Shin, M Rohrwild, MS Kang, et al, submitted) and is absolutely necessary for the cleavage of peptides by HslV (102).

One other function of the ATPases could be to help inject the substrate into the 20S proteasome and thus to promote its processive degradation. In protease La, ATP hydrolysis is necessary for processive degradation, and nonhydrolyzable analogs allow only initial cleavages of the substrate. By itself, the 20S archaebacterial proteasome degrades fully unfolded proteins in a processive manner (TN Akopian & AL Goldberg, unpublished data). However, the processive degradation of ubiquitinated substrates, which may be globular or partially unfolded, may require additional unfolding reactions. Thus a likely role for the 26S ATPases could be to promote the unfolding of substrates in a chaperone-like function (14–16). As discussed above, entry of a polypeptide into the 20S particle requires that it is fully unfolded, and mutants in the various ATPases can affect the breakdown of different polypeptides in distinct fashions. Thus, the components required for guiding different substrates into the 20S particle seem to differ, perhaps depending on the substrate's conformation. It is also noteworthy that SiS1, a molecular chaperone of the DnaJ family, is required for degradation of certain abnormal proteins by the 26S complex (204; DH Lee, M Sherman & AL Goldberg, submitted). Interestingly, in *E. coli*, the

ClpA and ClpX ATPases, when not associated with the ClpP protease complex, can function as molecular chaperones in promoting dissassembly of multimeric substrates (205, 206). Since the pure PA700 regulatory complex can be isolated, these various possible functions can now be tested.

NON-ATPase SUBUNITS   The 19S regulatory complex (PA700) also contains approximately 15 polypeptides that seem to lack the capacity to bind ATP (184, 185, 187) (Table 2). Efforts to clone and characterize these non-ATPase subunits are progressing rapidly (see below), but our understanding of their functions is still quite limited. These subunits may serve diverse functions, such as allowing association with or activation of the 20S particle, capture of substrates, release and disassembly of the polyUb chains, maintenance of the particle's structure, and delivery of the substrate into the 20S particle. Interestingly, mutants in these various components can have very different phenotypic defects, presumably because they have distinct effects on the degradation of different cell proteins.

The "ubiquitin receptor," S5a (MBP1)   Deveraux et al demonstrated that a 50-kDa component of the human 26S proteasome, S5a, binds with high affinity multi-Ub chains that contain at least four Ub moieties (176). A homologous *Arabidopsis* cDNA, MBP1, encodes a 41-kDa acidic protein with a similar high affinity for multiubiquitin chains, and homologous genes exist in many organisms (176a). Interestingly, purified recombinant MBP1 can competitively inhibit the degradation of ubiquitinated proteins by the purified 26S proteasome, without affecting the peptidase activities of the complex (206a). Moreover, in cell extracts, MBP1 inhibits Ub-dependent proteolysis, including the Ca^{2+}-induced degradation of cyclin B. Thus, MBP1 appears to function as the receptor for ubiquitinated polypeptides in the 26S proteasome. However, surprisingly, the yeast *Mbp1* gene is not essential (Van Nocker, personal communication). Presumably, the 26S proteasome, in cells lacking MBP1, can still efficiently bind most substrates.

P31   P31 is a human homolog of the yeast protein Nin1p (207), which is required for both G1/S and G2/M transitions of the cell cycle. In these steps, Nin1p is necessary for the activation of the Cdc28p kinase. Most likely, *nin1-1* mutants cannot degrade the $p40^{sic1}$ Cdc28p kinase inhibitor, whose degradation by the Ub pathway is required for the G1/S transition (208). The inactivation of Nin1p has different physiologic effects than mutations in the ATPase subunits Cim3p (Sug1p) and Cim5p (MSS1), which block exit from M-phase by causing stabilization of the mitotic cyclins, Clb2p and Clb3p (190).

P58   P58 is the product of the human gene homolog to *sun2* (suppressor of *nin1-1*), which, when overexpressed, restores the ability of the *nin1-1* ts-mutants to activate the Cdc28p kinase (A Toh-e, personal communication). Deletion of *sun2* arrests yeast at the G2/M boundary, and the expression of its human homolog, p58 (184), can suppress this defect. P58 is highly homologous to the mouse P91A protein, a tumor transplantation antigen (209), and to the *Drosophila* diphenol oxidase (210), but neither p58 nor the 26S proteasome has any diphenol oxidase activity (184).

p112   p112 is the largest subunit of the 26S proteasome. Human cells contain two species of mRNA for p112 (p112-L and p112-S), which differ only by a block of 170 nucleotides in their 5′ regions. Both sequences contain two KEKE motifs in their C-terminal part and resemble closely that of the yeast protein, Sen3p, which appears to interact functionally with Nin1p, as suggested by a synthetic lethal screen (210a). Sen3 is essential for viability in some strains (210b). When not essential, disruption of this gene prevents growth at high temperature (210a), when protein breakdown is accelerated (204). In various stressful conditions, Sen3⁻ cells show defects in Ub-dependent proteolysis, cell cycle progression, and nuclear protein translocation. Expression of the human p112 cDNA in the Sen3⁻ yeast suppresses the temperature-sensitive defect.

p97   p97 has a sequence very similar to those of the proteins TRAP-2 (211) and 55.11 (212), which bind to the cytoplasmic portion of the type-1 tumor necrosis factor receptor and have been implicated in cell death induced by TNF-α. A KEKE motif is present in the C-terminus of p97. Disruption of the homologous yeast gene, *nas1* (non-ATPase subunit 1), is not lethal, but causes a growth defect and an accumulation of ubiquitinated proteins at high temperature (C Tsurumi, personal communication).

p40 (S12)   p40 (S12) is homologous to the mouse or *Drosophila* Mov-34 gene product (213, 214), whose disruption results in a recessive embryonic lethality in mouse (215). A KEKE motif is present in the C-terminus of p40.

Other subunits   p27, the smallest subunit of the modulator complex (see above), p28, p44.5, and p55 have no significant homology to any known protein (GN DeMartino, personal communication). p44 (S10) is identical to a gene product of unknown function (187). S5b (p50.5) cDNA was cloned on the assumption that it encoded the Ub-chain-binding protein, now identified as S5a (216). Its function is unknown. Several other proteins, S9 (45 kDa), S11 (43 kDa), S14 (32 kDa), and S15 (25 kDa), have been identified by SDS-PAGE as 26S subunits (187), but no information is presently available about their primary structures or possible functions.

Other Functions and Cofactors of the 26S Complex

The degradation of ubiquitinated proteins must involve several integrated steps, whose precise order is uncertain: (*a*) substrate binding, (*b*) removal and breakdown of the polyUb chain, releasing free Ub moieties, (*c*) unfolding of the substrate and its injection into the 20S particle, (d) release of oligopeptide products. Additional steps may also be important (e.g. activation of the 20S particle), and ATP hydrolysis probably occurs at several points to allow these steps. As discussed in detail below, specific components of the 26S proteasome or associated enzymes have been implicated in certain of these processes.

SUBSTRATE RECOGNITION   The MBP1 (S5a) subunit has a high affinity for poly-Ub chains containing four or more Ub moieties (176). Possibly, MBP1 serves as an anchor that holds the polypeptide on the 26S complex during degradation. However, other recognition mechanisms must also exist, since *Mbp1* is not essential in yeast (see above), for binding the ODC/AZ complex and other protein substrates, which are not ubiquitinated (217, 218; see below). Mutations in individual ATPases of the 26S proteasome can reduce the break-down of some substrates of the Ub-pathway, without stabilizing others (190). Thus, these ATPases may also be involved somehow in substrate recognition, in a similar fashion to the ClpA and ClpX ATPases, which direct the substrate specificity of the Clp (Ti) protease in *E. coli* (202).

UBIQUITIN RECYCLING   Prior to degradation, multiple Ub molecules are con-jugated to the substrates by isopeptide bonds linking ubiquitin's C-terminal residue to ε-amino groups of lysines on the substrate or another Ub molecule (34). Degradation of ubiquitinated proteins involves the removal of the polyUb chain and its depolymerization by isopeptidase(s), releasing free Ub molecules. A large, diverse family of Ub-isopeptidases (34) called Ub-specific processing proteases (Ubp) or Ub-carboxy-terminal hydrolases are found in cells and have the capacity to degrade isopeptide bonds. At least 15 distinct Ubp enzymes exist in yeast (17), and a similar number has been found in chick skeletal muscle (219). Their precise functions are unclear: some of them appear to compete with the proteasome for degradation of conjugates (219a), but several appear to function in the release of Ubs during protein breakdown (34).

The best-characterized such enzyme is isopeptidase T, which was originally isolated by Hershko and colleagues owing to its high affinity for Ub. This enzyme converts large polyUb-chains to smaller species with the release of free Ub molecules (180) and attacks preferentially polyUb chains having a free Ub-carboxyl group, i.e. chains that have been released from the polypep-tide substrate by the 26S proteasome (220). This enzyme stimulates the deg-radation of polyubiquitinated proteins by the 26S proteasome (180), apparently by destroying the polyUb remnants produced during substrate degradation. These remnants presumably inhibit the 26S proteasome by binding to the MBP1 subunit (see above) and competing with polyubiquitinated proteins for degradation. Accordingly, in yeast mutants lacking Ubp14, a deubiquitinating enzyme homologous to isopeptidase T, overall proteolysis is reduced and ubiquitinated proteins accumulate (17).

Another isopeptidase important in proteolysis is Doa4, a 105K protein required in yeast for the rapid degradation of MATα2 repressor (221). In *doa4* null mutants, small Ub-containing molecules accumulate, slightly larger in size than free Ub, diUB, or triUb (221). Most likely, these chains correspond to Ub molecules linked to small peptides and are generated during proteolysis by the 26S proteasome. Thus, in *doa4* mutants, one (or a few) Ub molecules appear to remain linked to the susbtrate during its degradation into small peptides. The inhibition of protein breakdown in these cells is probably due to the accumulation of these ubiquitinated peptides, which may remain associated with the 26S proteasome and thus limit its ability to degrade new substrates. The mechanism generating these Ub-peptides is unclear, since it is unlikely that a protein linked to Ub could enter into the 20S core particle. Recombinant Doa4 has isopeptidase activity (221), but it is not known whether Doa4 functions in vivo before, during, or after the substrate is digested. This enzyme was not found in highly purified mammalian 26S proteasomes (184, 187), presumably because it is only loosely associated and is lost during purification. Interestingly, Doa4 contains several domains similar to the human oncoprotein Tre-2, which also has isopeptidase activity (221). Moreover, the ability of Tre-2 to generate tumors correlates with the loss of isopeptidase activity and may result from a failure to degrade some short-lived, regulatory protein (221).

An additional isopeptidase activity associated with the 26S proteasome that may remove Ub from the substrate has been reported (222). This activity seems to require ATP hydrolysis, but its relationship to the other isopeptidases implicated in proteolysis is uncertain. Interestingly, this activity was not found in the 26S components (CF1, CF2, and CF3 complexes) isolated from ATP-depleted reticulocytes, but was detected only after they associated to form the 26S complex. However, isopeptidase activity is found in the PA700 activator complex, and this activity is augmented upon association with the modulator (AL Goldberg, DH Hwang & GN DeMartino, unpublished data).

Another intriguing cofactor of the 26S proteasome is Factor H (FH), originally isolated by Ciechanover and coworkers (223), because it is required for the degradation of some proteins whose amino terminals are blocked by acetylation. Such proteins constitute the majority of polypeptides in eukaryotic cells, but whether FH is required for degradation of all these proteins is unclear. Upon purification, FH was found to correspond to the elongation factor EF-1α, a major cell protein which catalyzes protein translation at the ribosome (181). Recombinant EF-1α, as well as its bacterial homolog, EF-Tu, can substitute for FH in proteolysis. Interestingly, the stimulation of proteolysis by EF-1α is inhibited by Ub-aldehyde, which specifically inhibits isopeptidases. Presumably, therefore, EF-1α acts as (or together with) a substrate-specific isopeptidase, although why such an activity would be necessary for the degradation of only acetylated proteins is unclear. One intriguing possibility is that the activities of EF-1α in translation and 26S function [and perhaps in microtubule stability (224)] are mechanistically related. Alternatively, these effects of EF-1α may represent a mechanism for linking the regulation of protein synthesis and breakdown, since EF-Tu and EF-1α bind charged tRNA, and, in bacteria, levels of uncharged tRNA regulate overall protein degradation (225).

PROTEIN UNFOLDING: INVOLVEMENT OF CHAPERONES   A polypeptide has to be in an extended, unfolded form to be hydrolyzed by the 20S proteasome (49, 108, 108a), and it is therefore likely that the 19S-cap particle helps unfold globular or partially unfolded proteins. As discussed above, the ATPases may serve as molecular chaperones that unfold ubiquitinated substrates, by analogy to ClpA (205) and ClpX ATPases (206), which have some chaperone-like activities and are necessary for ATP-dependent proteolysis by ClpP in *E. coli*. In prokaryotes, molecular chaperones have been shown to function in the ATP-dependent pathway (226, 227). Recently, DH Lee and colleagues found in yeast that the chaperones of the DnaJ and Hsp70 families also function at multiple sites in Ub-dependent degradation of certain abnormal proteins. The DnaJ homolog, Sis1, appears necessary for degradation of certain specific short-lived proteins by the 26S complex, and Sis1 mutants accumulate large amounts of Ub-conjugated proteins (DH Lee, M Sherman, A Merün & AL Goldberg, submitted). Presumably, the chaperones are required to maintain certain ubiquitinated proteins in a conformation that allows their degradation by the 26S complex.

ADDITIONAL FUNCTIONS OF THE PROTEASOME

Proteolytic Processing of Precursor Proteins (NF-κB)

The primary function of the proteasome is to catalyze the complete breakdown of proteins to small peptides, most of which are then quickly hydrolyzed to amino acids. Recently, it has been discovered that the Ub-proteasome pathway can also catalyze the regulated proteolytic processing of a large, inactive precursor into an active protein (22). The transcription factor NF-κB is necessary for the expression of various proteins critical in immune and inflammatory responses, including many cytokines and cell adhesion molecules (158). One of its subunits, p50, is a 50-kDa polypeptide generated by proteolytic processing of an inactive 105-kDa precursor. p50 is then maintained in the cytosol together with the p65 subunit in an inactive complex bound to the inhibitory protein, IκBα. Inflammatory signals (e.g. TNF-α) activate NF-κB by signalling the complete degradation of the inhibitor, IκB-α, and stimulating the proteolytic processing of the p105 precursor. This maturation of p105 is an ATP-dependent process in which the C-terminus is destroyed, leaving the N-terminal active p50 fragment (228). The processing of p105 and the complete degradation of IκBα require ubiquitination and the 26S proteasome (22, 229), and both processes could be prevented with proteasome inhibitors (22, 157). Similar processing of p105 occurs if it is expressed in yeast, and this process, like the complete degradation of many short-lived proteins, requires the Pre1, Pre2, and Cim5 proteasome subunits, as well as Doa4 (C Sears & T Maniatis, personal communication) and the Sis1 chaperone (DH Lee & AL Goldberg, unpublished data).

Why hydrolysis of the ubiquitinated p105 ceases after generation of the 50-kDa fragment is unclear. Perhaps the conformation of p50 may simply make it resistant to further degradation and promotes its release from the 26S complex. Release of partially degraded polypeptides from the proteasome may occur often; in fact, during breakdown of β-galactosidase, a relatively stable 90-kDa fragment is released, perhaps by similar mechanisms. Alternatively, limited proteolysis of p105 by the 26S proteasome involves additional factors that prevent complete digestion. Although evidence thus far has only been obtained for NF-κB, the proteasome may play a more general role in the proteolytic maturation of other soluble or membrane-associated proteins or in the partial cleavage of mature cell proteins. A related transcription factor, NF-κB2, undergoes a similar proteolytic processing from an inactive precursor (230), and, presumably, this process also involves ubiquitination and the 26S proteasome.

Proteasomes and Antigen Presentation

One important function of the proteasome in higher vertebrates is to generate the peptides presented on MHC-class 1 molecules to circulating lymphocytes (19, 177, 231). In this process, peptides generated during protein breakdown are transported into the endoplasmic reticulum, where they become bound to MHC-class 1 molecules and then delivered to the cell surface. The presentation of these peptides enables the immune system to screen for and destroy cells expressing viral or other unusual polypeptides. Clear evidence that the proteasome generates most of these antigenic peptides comes from studies in lymphoblasts microinjected with ovalbumin (2). Concentrations of inhibitors that block the ATP-dependent degradation of cell proteins prevent MHC-class 1 presentation of the ovalbumin-derived antigenic peptide (2). These treatments block reversibly the production of antigenic peptides, but do not reduce the transport of peptides into the ER or their delivery to the surface. In addition, these studies indicate that the great majority of presented peptides are generated by proteasomes (2). More selective proteasome inhibitors, e.g. lactacystin (KL Rock, personal communication), reduce antigen presentation similarly. These results are consistent with the finding that ubiquitin-conjugation is required for class I presentation of ovalbumin (232) and the observation that gene fusions that promote Ub-dependent degradation of a protein stimulate antigen presentation (233, 234).

Of particular interest was the discovery that the cytokine γ-interferon (γ-IFN), which s⁺imulates antigen presentation, as well as β-IFN and TNFα (51), alters proteasome subunit composition and functional activity. In 1984, Monaco and McDevitt described large complexes of "low–molecular weight proteins" (LMPs), which are encoded in the MHC-region (28). Eventually, the sequencing of the LMP2 and LMP7 genes showed that these proteins resemble closely the β subunits of the proteasome (235–237). Subsequent work established that LMP2 and LMP7 are proteasome subunits and are induced by γ-IFN (31, 238). This cytokine also induces one other β subunit, MECL1 (96). Upon γ-IFN treatment, all three new subunits are incorporated into newly synthesized particles in place of the subunits found normally: LMP2 for X, LMP7 for Y, and MECL1 for Z (95, 96, 239). The name immunoproteasome has been proposed for the particles containing these subunits to emphasize their special function in antigen presentation (240).

These changes in subunit composition lead to alterations in the particle's activity against standard peptide substrates (51, 61, 62). Upon γ-IFN treatment, the maximal capacity (V_{max}) of purified 20S and 26S particles to cleave after hydrophobic and basic residues increases substantially (51, 61, 62), while the postglutamyl activity decreases (61). However, γ-IFN does not affect the rate of degradation of casein or ubiquitinated lysozyme or the cellular content of proteasomes (61). These findings have now been confirmed by multiple laboratories in multiple cell types, although one group reported distinct functional changes (79, 241) and another failed to obtain similar results, probably owing to the use of unfavorable experimental conditions (242).

A variety of findings demonstrate that these changes in peptidase activity are due to the incorporation of the LMP subunits (78, 243). Proteasomes from lymphoblastoid lines that carry a homozygous MHC-deletion covering the LMP genes exhibit enhanced postglutamyl activity and reduced chymotryptic- and tryptic-like activities. Also, these latter activities did not rise upon γ-IFN treatment.

Furthermore, when the LMP7 gene was transfected into cells (78), LMP7 content increased and, in parallel, the chymotryptic and tryptic activities rose. Transfection of the LMP2 gene increased LMP2 content and caused a proportional suppression of postglutamyl activity, apparently because LMP2 replaces the homologous β subunit Y (51) [also called 2 (97), or δ (244)], which is important for such cleavages. When expression of Y was increased by transfection, the postglutamyl activity increased. Thus subunit Y appears responsible for cleavage after acidic residues, and its programmed elimination upon γ-IFN treatment can account for the loss of this activity. Although the incorporation of the LMPs and elimination of X and Y can account for the changes seen after γ-IFN treatment, the functional consequences of incorporation of MECL1 have not been studied, nor have possible changes in BrAAP or SNAAP activities. Most likely, these β subunits alter peptidase activity, because they contain distinct active sites (see above). The three γ-IFN-induced subunits, like the subunits they replace, contain N-terminal threonine residues and are therefore likely to be catalytically active.

These modifications of peptidase activity predict that, during protein breakdown, the proteasomes from the γ-IFN-treated cells should generate more peptides that have hydrophobic or basic carboxyl termini and fewer peptides with acidic carboxyl termini. In fact, the vast majority of peptides presented on MHC class I molecules are of this type. A hydrophobic or basic carboxy-terminal residue is required for strong binding of a peptide to MHC-class I molecules (245) and for selective uptake into the ER by the TAP transporter (246, 247). Proteasomes have been shown to generate some antigenic peptides from model substrates (oligopeptides or proteins), but these results were obtained in lengthy experiments under highly nonphysiological conditions (241, 248). Quantitative comparisons of the rates at which normal particles and immunoproteasomes generate antigenic peptides from ubiquitinated or other proteins are needed to test the actual importance of the γ-IFN-induced subunits.

The demonstration that LMP2 and LMP7 are not essential for class I presentation (249–251) was initially interpreted to indicate that these subunits (and proteasomes in general) were not important for antigen presentation. However, proteasomes lacking these subunits should still generate some antigenic peptides (although less efficiently than LMP-containing particles), since they still have significant chymotryptic- and tryptic-like activities. Recent studies have demonstrated that cells lacking LMPs have a decreased efficiency, rather than an absolute defect, in antigen processing (66, 252, 253). In mice carrying a deletion in LMP2, macrophages and spleen cells exhibit a clear defect in presentation of a viral antigen, and their proteasomes have altered peptidase activities similar to those seen in LMP-deficient lymphoblasts (66). Mice lacking LMP7 have reduced surface levels of MHC-class I molecules and impaired presentation of viral antigens (252). Clearly, the γ-IFN-induced modification of proteasome composition can increase the efficiency of immune surveillance. Although the proteasome appears responsible for the cleavages that determine the C-termini of antigenic peptides (78, 243, 254), there is no evidence to suggest that this particle also generates the N-termini, and some additional trimming of the N-terminus appears to occur in the ER (255–257).

LMP-containing proteasomes are found in liver and spleen of normal mice, but not in muscle and brain, whose content of MHC-class I molecules is also very low (66). Thus, proteasomes in different cells normally differ in subunit composition and functional activities in a way that correlates with the cell's capacity for antigen presentation. In addition to the three proteasome subunits and the peptide transporters (TAP1 & TAP2), as well as the class 1 molecules, γ-IFN treatment also induces a 180-kDa complex (PA28) which greatly stimulates peptide hydrolysis by the 20S particle (145, 146). As discussed above, this activator probably also helps increase the fraction of peptides generated by the proteasome that are appropriate for MHC-class 1 presentation (151, 257a).

Ubiquitin-Independent Proteolysis

The 26S proteasome was originally isolated by its ability to degrade Ub-conjugated proteins (35, 36). However, this complex can also rapidly hydrolyze certain nonubiquitinated proteins in an ATP-dependent manner, including casein (38), a polypeptide with little tertiary structure and the oncoprotein c-jun (218). [Curiously, in vivo, c-jun appears to be rapidly degraded by a Ub-dependent process (258)]. Which structural features eliminate the need for ubiquitination in these cases and whether many other polypeptides can also be degraded in a Ub-independent fashion in vitro or in vivo are unclear. When the amino groups on casein or albumin are modified so as to prevent ubiquitination, their degradation is reduced several-fold but is still ATP-activated (167).

The best-studied example of Ub-independent proteolysis by the 26S complex is the short-lived enzyme ornithine decarboxylase (ODC), which is rate-limiting in polyamine biosynthesis. Because polyamines are toxic in high levels, cells have evolved negative feedback mechanisms, which cause the inactivation and rapid degradation of ODC. When polyamines accumulate, they induce the specific inhibitory protein, Antizyme (AZ) (259), which binds to ODC, inactivates it, and triggers its degradation. When associated with AZ, ODC is rapidly hydrolyzed by the 26S complex, provided ATP is present. The regulatory protein, AZ, however, is not degraded in this process (217, 260, 261). Apparently, the binding of AZ to the N-terminal part of the ODC exposes a cryptic degradation signal on the enzyme's C-terminus (262). Which components of the 26S are necessary for the binding of the ODC/AZ complex and for ODC degradation, and whether AZ triggers the degradation of other cellular proteins, are unclear. Possibly, for other highly regulated proteins, analogous systems exist that cause their Ub-independent degradation.

Free 20S proteasomes are present in cells (90) and may not simply represent inactive precursors for the 26S complexes. Purified 20S particles, especially when activated with SDS, can also degrade many polypeptides without ubiquitin conjugation or even ATP (37, 126). However, whether 20S proteasomes ever function in such a fashion either in normal cells or even in pathological states is unclear. Although most intracellular proteolysis requires ATP, intact mammalian cells do show significant ATP-independent proteolysis (263). In erythrocytes and reticulocytes, oxidatively damaged hemoglobin is rapidly degraded after ATP depletion (264, 265), which prevents hydrolysis of most cell proteins. The 20S particle has been proposed to be responsible for this degradation of damaged polypeptides (266–268), since oxidative damage of proteins can enhance their susceptibility to the 20S proteasome (269, 270). However, removal of 20S proteasomes from extracts does not block this ATP-independent process (23), which may involve multiple proteolytic systems (266).

REGULATION OF PROTEASOME CONTENT AND COMPOSITION

Most studies have assumed that the rate-limiting step in the Ub-proteasome pathway is Ub-conjugation and that the subsequent degradation of the substrate is a rapid, unregulated process. However, growing evidence shows that proteasome content, composition, and functional properties can be regulated. As discussed above, the induction of three proteasome subunits by γ-interferon represents an elegant mechanism to regulate the products of protein digestion, without influencing the rates of protein disappearance. Substitution of one 20S subunit by another also occurs in certain cells during *Drosophila* development (J Belote, personal communication). Interestingly, in some cases, monoclonal antibodies against different proteasome subunits can stain tissues in distinct fashions (7), suggesting the presence of heterogeneous proteasome populations with distinct cellular distribution patterns and presumably specialized functions. For example, one proteasomal antigen was found primarily in the pericanalicular zone of rat hepatocytes, whereas others were distributed more uniformly in these cells (271).

Rates of proteolysis in a specific region of the cell may, in principle, be regulated by changes in the cellular distribution of the proteasomes. In fact, during oocyte maturation, dramatic shifts occur in the distribution of these particles from the cytoplasm into the nucleus (72), and during the normal cell cycle, programmed changes in their distribution occur (73, 74). Another possible mode of regulation of proteolysis may be by conversion of 20S particles into the 26S complex. The ratio of 20S to 26S particles has been reported to differ in the cytoplasm and nucleus of *Xenopus* oocytes, and in these cells the nucleus, unlike the cytoplasm, seems to be devoid of free 20S particles (134). Although prior studies had concluded that the proteasome activity is unchanged during the cell cycle (271a), in the ascidian meiotic cycle, the activity of the 26S proteasome increases transiently during metaphase-anaphase transition, owing to interconversion between 20S and 26S forms (271b). This change could be blocked by Ca^{2+} chelator and induced by Ca^{2+} ionophores (271b).

Moreover, changes in cytosolic Ca^{2+} also promote assembly of the 26S complex during *Xenopus* egg activation (271c).

Dramatic differences in the proteasome content of different cell populations are evident in *Drosophila* embryos, and these patterns change during development (272). Presumably, these differences reflect large differences in rates of proteolysis during development, as seem to occur in the abdominal intersegmental muscles of the moth *Manduca sexta* (182, 273). When these muscles undergo a dramatic atrophy and programmed cell death, ubiquitin and ubiquitin-protein conjugates increase, as does the level of 20S proteasomes (273). In addition, the subunits in both the 20S and 26S proteasomes change (182, 273), apparently through changes in the levels of corresponding mRNAs (182). In mammals, skeletal muscles also undergo rapid weight loss in various pathological states, including fasting (274), denervation atrophy (275), acidosis (276), cancer cachexia (277, 277a), or sepsis (278), primarily owing to accelerated proteolysis by the Ub-proteasome pathway. In these atrophying muscles, as in *Manduca*, Ub-protein conjugates also rise in the cell. Thus, their degradation by the 26S proteasome becomes the rate-limiting step in the pathway under these conditions. On the other hand, in hypothyroid animals, or animals on protein-deficient diets, overall ATP-dependent proteolysis in the muscles falls within a few days (279), as does their total content of 20S proteasomes (279a).

The increase in muscle protein breakdown in various forms of atrophy is accompanied by the coordinated induction of mRNA for at least four 20S subunits (276, 277, 280). Simultaneous induction of several mRNAs for proteasome subunits also occurs in cells stimulated by mitogens (281, 282). Therefore, similar organizations of the genes encoding the major proteasome subunits would seem likely. However, the few proteasomal genes described are located on different chromosomes (20, 283, 284) and appear to be very different. No consensus sequences have been found in their 5'-flanking regions, which display significantly different promoter activities (284–287). Within the same species, the exon/intron organization of the different proteasome genes is not uniform (287). Thus, the mechanisms coordinating their expression is unclear. In addition, in several cases, the intracellular levels of proteasome mRNAs rise, but the levels of corresponding subunits do not (97, 282). Therefore, the production of new proteasomes may be regulated at the translational level or during particle assembly, and the excess subunits not incorporated may be rapidly degraded.

Presently, our understanding of the mechanisms regulating the expression, cellular content, and intracellular distribution of the proteasome is clearly very limited. The observations reviewed here indicate that these areas are important topics deserving systematic study. The dramatic recent progress in elucidating the structure and function of the proteasome should facilitate the elucidation of these regulatory mechanisms in coming years.

ACKNOWLEDGMENTS

We are grateful to Mrs. Aurora Scott for helpful assistance in preparing this manuscript, and to our colleagues, especially Dr. D Finley, for their critical advice. This work has been supported by grants from the National Institute of Health (NIGMS), the Muscular Dystrophy Association (MDA) and the Human Frontier Science Program (HFSP) to ALG, from the Monbusho to KT, and by a fellowship from the HFSP to OC.

Literature Cited

1. Ciechanover A. 1994. *Cell* 79:13–21
2. Rock KL, Gramm C, Rothstein L, Clark K, Stein R, et al. 1994. *Cell* 78:761–71
3. Hilt W, Heinemeyer W, Wolf DH. 1993. *Enzym. Protein* 47:189–201
4. Tanaka K. 1995. *Mol. Biol. Rep.* 21:21–26
5. Tanaka K, Li K, Ichihara A, Waxman L, Goldberg AL. 1986. *J. Biol. Chem.* 261:15197–203
6. Rivett AJ, Palmer A, Knecht E. 1992. *J. Histochem. Cytochem.* 40:1165–72
7. Scherrer K, Bey F. 1994. *Prog. Nucleic Acid Res. Mol. Biol.* 49:1–64
8. Orlowski M. 1990. *Biochemistry* 29: 10289–97
9. Goldberg AL. 1992. *Eur. J. Biochem.* 203:9–23
10. Rechsteiner M, Hoffman L, Dubiel W. 1993. *J. Biol. Chem.* 268:6065–68
11. Rivett AJ, Savory PJ, Djaballah H. 1994. *Methods Enzymol.* 244:331–50
12. Wilk S, ed. 1994. *Enzyme Protein* 47(4–6)
13. Schmid HP, Briand Y, eds. 1995. *Mol. Biol. Rep.* 21(1)
14. Weissman JS, Sigler PB, Horwich AL. 1995. *Science* 268:523–24
15. Goldberg AL. 1995. *Science* 268:522–23
16. Rubin DM, Finley D. 1995. *Curr. Biol.* 5:854–58
17. Hochstrasser M. 1995. *Curr. Opin. Cell Biol.* 7:215–23
18. Jentsch S, Schlenker S. 1995. *Cell* 82: 881–84
19. Goldberg AL, Gaczynska M, Grant E, Michalek M, Rock KL. 1995. *Cold Spring Harbor Symp.* LX:In press
20. Hilt W, Wolf DH. 1995. *Mol. Biol. Rep.* 21:3–10
21. Hegde AN, Goldberg AL, Schwartz JH. 1993. *Proc. Natl. Acad. Sci. USA* 90: 7436–40
22. Palombella VJ, Rando OJ, Goldberg AL, Maniatis T. 1994. *Cell* 78:773–85
23. Matthews W, Driscoll J, Tanaka K, Ichihara A, Goldberg AL. 1989. *Proc. Natl. Acad. Sci. USA* 86:2597–601
24. DeMartino GN, Goldberg AL. 1979. *J. Biol. Chem.* 254:3712–15
25. Goldberg AL, Strnad N, Swamy KHS. 1980. In *Protein Degradation in Health and Disease*, pp. 227–51. Amsterdam/ Oxford/New York: Excerpta Medica
26. Wilk S, Orlowski M. 1983. *J. Neurochem.* 40:842–49
27. Schmid HP, Akhayat O, de Sa CM, Puvion F, Koehler K, Scherrer K. 1984. *EMBO J.* 3:29–34
28. Monaco JJ, McDevitt HO. 1984. *Nature* 309:797–99
29. Arrigo AP, Tanaka K, Goldberg AL, Welch WJ. 1988. *Nature* 331:192–94
30. Falkenburg PE, Haass C, Kloetzel PM, Niedel B, Kopp F, et al. 1988. *Nature* 331:190–92
31. Brown MG, Driscoll J, Monaco JJ. 1991. *Nature* 353:355–57
32. Goldberg AL, St. John AC. 1976. *Annu. Rev. Biochem.* 45:747–803
33. Etlinger J, Goldberg AL. 1977. *Proc. Natl. Acad. Sci. USA* 74:54–58
34. Hershko A, Ciechanover A. 1992. *Annu. Rev. Biochem.* 61:761–807
35. Hough R, Pratt G, Rechsteiner M. 1987. *J. Biol. Chem.* 262:8303–13
36. Waxman L, Fagan JM, Goldberg AL. 1987. *J. Biol. Chem.* 262:2451–57
37. Eytan E, Ganoth D, Armon T, Hershko A. 1989. *Proc. Natl. Acad. Sci. USA* 86:7751–55

38. Driscoll J, Goldberg AL. 1990. *J. Biol. Chem.* 265:4789–92
39. Heinemeyer W, Kleinschmidt JA, Saidowsky J, Escher C, Wolf DH. 1991. *EMBO J.* 10:555–62
40. Peters JM, Cejka Z, Harris JR, Kleinschmidt JA, Baumeister W. 1993. *J. Mol. Biol.* 234:932–37
41. Kleinschmidt JA, Hugle B, Grund C, Franke WW. 1983. *Eur. J. Cell Biol.* 32:143–56
42. Baumeister W, Dahlmann B, Hegerl R, Kopp F, Kuehn L, Pfeifer G. 1988. *FEBS Lett.* 241:239–45
43. Kopp F, Steiner R, Dahlmann B, Kuehn L, Reinauer H. 1986. *Biochim. Biophys. Acta* 872:253–60
44. Hegerl R, Pfeifer G, Puhler G, Dahlmann B, Baumeister W. 1991. *FEBS Lett.* 283:117–21
45. Coux O, Nothwang HG, Scherrer K, Bergsma-Schutter W, Arnberg AC, et al. 1992. *FEBS Lett.* 300:49–55
46. Dahlmann B, Kopp F, Kuehn L, Niedel B, Pfeifer G, et al. 1989. *FEBS Lett.* 251:125–31
47. Grziwa A, Baumeister W, Dahlmann B, Kopp F. 1991. *FEBS Lett.* 290:186–90
48. Pühler G, Weinkauf S, Bachmann L, Muller S, Engel A, et al. 1992. *EMBO J.* 11:1607–16
49. Löwe J, Stock D, Jap B, Zwickl P, Baumeister W, Huber R. 1995. *Science* 268:533–39
50. Martins de Sa C, Grossi de Sa MF, Akhayat O, Broders F, Scherrer K, et al. 1986. *J. Mol. Biol.* 187:479–93
51. Aki M, Shimbara N, Takashina M, Akiyama K, Kagawa S, et al. 1994. *J. Biochem.* 115:257–69
52. Dahlmann B, Kopp F, Kuehn L, Hegerl R, Pfeifer G, Baumeister W. 1991. *Biomed. Biochim. Acta* 50:465–69
53. Yoshimura T, Kameyama K, Takagi T, Ikai A, Tokunaga F, et al. 1993. *J. Struct. Biol.* 111:200–11
54. Zwickl P, Grziwa A, Puhler G, Dahlmann B, Lottspeich F, Baumeister W. 1992. *Biochemistry* 31:964–72
55. Coux O, Nothwang HG, Silva-Pereira I, Recillas-Targa F, Bey F, Scherrer K. 1994. *Mol. Gen. Genet.* 245:769–80
56. Kopp F, Dahlmann B, Hendil KB. 1993. *J. Mol. Biol.* 229:14–19
57. Schauer TM, Nesper M, Kehl M, Lottspeich F, Muller-Taubenberger A, et al. 1993. *J. Struct. Biol.* 111:135–47
58. Lupas A, Koster AJ, Baumeister W. 1993. *Enzyme Protein* 47:252–73
59. Brown MG, Driscoll J, Monaco JJ. 1993. *J. Immunol.* 151:1193–204
60. Akiyama K, Kagawa S, Tamura T, Shimbara N, Takashina M, et al. 1994. *FEBS Lett.* 343:85–88
61. Gaczynska M, Rock KL, Goldberg AL. 1993. *Nature* 365:264–67
62. Driscoll J, Brown MG, Finley D, Monaco JJ. 1993. *Nature* 365:262–64
63. Haass C, Kloetzel PM. 1989. *Exp. Cell Res.* 180:243–52
64. Ahn JY, Hong SO, Kwak KB, Kang SS, Tanaka K, et al. 1991. *J. Biol. Chem.* 266:15746–49
65. Hong SO, Ahn JY, Lee CS, Kang MS, Ha DB, et al. 1994. *Biochem. Mol. Biol. Int.* 32:723–29
66. Van Kaer L, Ashtonrickardt PG, Eichelberger M, Gaczynska M, Nagashima K, et al. 1994. *Immunity* 1:533–41
67. Heinemeyer W, Trondle N, Albrecht G, Wolf DH. 1994. *Biochemistry* 33:12229–37
68. Pühler G, Pitzer F, Zwickl P, Baumeister W. 1994. *Syst. Appl. Microbiol.* 16:734–41
69. Kopp F, Kristensen P, Hendil KB, Johnsen A, Sobek A, Dahlmann B. 1995. *J. Mol. Biol.* 248:264–72
70. Gaczynska M, Goldberg AL, Tanaka K, Hendil K, Rock KL. 1996. *J. Biol. Chem.* In press
71. Tanaka K, Yoshimura T, Tamura T, Fujiwara T, Kumatori A, Ichihara A. 1990. *FEBS Lett.* 271:41–46
72. Gautier J, Pal JK, Grossi de Sa MF, Beetschen JC, Scherrer K. 1988. *J. Cell Sci.* 90:543–53
73. Kawahara H, Yokosawa H. 1992. *Dev. Biol.* 151:27–33
74. Amsterdam A, Pitzer F, Baumeister W. 1993. *Proc. Natl. Acad. Sci. USA* 90:99–103
74a. Nederlof PM, Wang HR, Baumeister W. 1995. *Proc. Natl. Acad. Sci. USA* 92:12060–64
75. Tanaka K, Tamura T, Yoshimura T, Ichihara A. 1992. *New Biol.* 4:173–87
76. Zwickl P, Kleinz J, Baumeister W. 1994. *Nat. Struct. Biol.* 1:765–70
77. Gray CW, Slaughter CA, DeMartino GN. 1994. *J. Mol. Biol.* 236:7–15
78. Gaczynska M, Rock KL, Spies T, Goldberg AL. 1994. *Proc. Natl. Acad. Sci. USA* 91:9213–17
79. Kuckelkorn U, Frentzel S, Kraft R, Kostka S, Groettrup M, Kloetzel PM. 1995. *Eur. J. Immunol.* 25:2605–11
80. Fenteany G, Standaert RF, Lane WS, Choi S, Corey EJ, Schreiber SL. 1995. *Science* 268:726–31
81. Deleted in proof
82. Dineva B, Tomek W, Kohler K, Schmid HP. 1989. *Mol. Biol. Rep.* 13:207–11
83. Nothwang HG, Coux O, Keith G, Silva-

Pereira I, Scherrer K. 1992. *Nucleic Acids Res.* 20:1959–65
84. Pamnani V, Haas B, Puhler G, Sanger HL, Baumeister W. 1994. *Eur. J. Biochem.* 225:511–19
85. Pouch MN, Petit F, Buri J, Briand Y, Schmid HP. 1995. *J. Biol. Chem.* 270: 22023–28
86. Seelig A, Multhaup G, Pesold-Hurt B, Beyreuther K, Kloetzel PM. 1993. *J. Biol. Chem.* 268:25561–67
87. Tamura T, Tanaka K, Kumatori A, Yamada F, Tsurumi C, et al. 1990. *FEBS Lett.* 264:91–94
88. Tamura T, Lee DH, Osaka F, Fujiwara T, Shin S, et al. 1991. *Biochim. Biophys. Acta* 1089:95–102
89. Chen P, Hochstrasser M. 1995. *EMBO J.* 14:2620–30
90. Yang Y, Fruh K, Ahn K, Peterson PA. 1995. *J. Biol. Chem.* 270:27687–94
91. Maurizi MR, Clark WP, Katayama Y, Rudikoff S, Pumphrey J, et al. 1990. *J. Biol. Chem.* 265:12536–45
92. Baker D, Agard DA. 1994. *Biochemistry* 33:7505–9
93. Seemuller E, Lupas A, Stock D, Lowe J, Huber R, Baumeister W. 1995. *Science* 268:579–82
94. Frentzel S, Pesold-Hurt B, Seelig A, Kloetzel PM. 1994. *J. Mol. Biol.* 236: 975–81
94a. Brannigan JA, Dodson G, Duggleby HJ, Moody PCE, Smith JL, et al. 1995. *Nature* 378:416–19
95. Akiyama K, Yokota K, Kagawa S, Shimbara N, Tamura T, et al. 1994. *Science* 265:1231–34
96. Hisamatsu H, Shimbara N, Saito Y, Kristensen P, Hendil KB, et al. 1996. *J. Exp. Med.* In press
97. Fruh K, Gossen M, Wang K, Bujard H, Peterson PA, Yang Y. 1994. *EMBO J.* 13:3236–44
98. Plunkett GD, Burland V, Daniels DL, Blattner FR. 1993. *Nucleic Acids Res.* 21:3391–98
99. Lupas A, Zwickl P, Baumeister W. 1994. *Trends Biochem. Sci.* 19:533–34
100. Fleischmann RD, Adams MD, White O, Clayton RA, Kirkness EF, et al. 1995. *Science* 269:496–512
101. Tamura T, Nagy I, Lupas A, Lottspeich F, Cejka Z, et al. 1995. *Curr. Biol.* 5:766–74
102. Rohrwild M, Coux O, Huang HC, Moerschell RP, Yoo SJ, et al. 1996. *Proc. Natl. Acad. Sci. USA.* In press
103. Deleted in proof
104. Chuang SE, Burland V, Plunkett G, Daniels DL, Blattner FR. 1993. *Gene* 134:1–6
105. Kessel M, Maurizi MR, Kim B, Kocsis

E, Trus BL, et al. 1995. *J. Mol. Biol.* 250:587–94
106. Perkins GA, Bergsma-Schutter W, Keegstra W, Arnberg AC, Coux O, Scherrer K. 1994. *J. Struct. Biol.* 113: 124–34
107. Morimoto Y, Mizushima T, Yagi A, Tanahashi N, Tanaka K, et al. 1995. *J. Biochem.* 117:471–74
108. Wenzel T, Baumeister W. 1995. *Nat. Struct. Biol.* 2:199–204
108a. Friguet B, Stadtman ER, Szweda LI. 1994. *J. Biol. Chem.* 269:21639–43
109. Wenzel T, Eckerskorn C, Lottspeich F, Baumeister W. 1994. *FEBS Lett.* 349: 205–9
110. Dick LR, Moomaw CR, DeMartino GN, Slaughter CA. 1991. *Biochemistry* 30: 2725–34
111. Tokunaga F, Goto T, Koide T, Murakami Y, Hayashi S, et al. 1994. *J. Biol. Chem.* 269:17382–85
111a. Ehring B, Meyer TH, Eckerskorn C, Lottspeich F, Tampe R. 1996. *Eur. J. Biochem.* 235:404–15
112. Zwickl P, Lottspeich F, Dahlmann B, Baumeister W. 1991. *FEBS Lett.* 278: 217–21
113. Rivett AJ. 1985. *J. Biol. Chem.* 260: 12600–6
114. McGuire MJ, DeMartino GN. 1986. *Biochim. Biophys. Acta* 873:279–89
115. Djaballah H, Harness JA, Savory PJ, Rivett AJ. 1992. *Eur. J. Biochem.* 209: 629–34
116. Seemuller E, Lupas A, Zuhl F, Zwickl P, Baumeister W. 1995. *FEBS Lett.* 359: 173–78
117. Akopian TN, Kisselev A, Goldberg AL. 1996. *J. Biol. Chem.* In preparation
118. Duggleby HJ, Tolley SP, Hill CP, Dodson EJ, Dodson G, Moody PC. 1995. *Nature* 373:264–68
119. Rawlings ND, Barrett AJ. 1993. *Biochem. J.* 290:205–18
120. Rivett AJ. 1989. *J. Biol. Chem.* 264: 12215–19
121. Cardozo C, Vinitsky A, Hidalgo MC, Michaud C, Orlowski M. 1992. *Biochemistry* 31:7373–80
122. Pereira ME, Nguyen T, Wagner BJ, Margolis JW, Yu B, Wilk S. 1992. *J. Biol. Chem.* 267:7949–55
123. Orlowski M, Cardozo C, Michaud C. 1993. *Biochemistry* 32:1563–72
124. Rivett AJ. 1993. *Biochem. J.* 291:1–10
125. Wilk S, Figueiredo-Pereira ME. 1993. *Enzyme Protein* 47:306–13
126. Cardozo C, Vinitsky A, Michaud C, Orlowski M. 1994. *Biochemistry* 33: 6483–89
127. Mykles DL, Haire MF. 1995. *Biochem. J.* 306:285–91

128. Richter-Ruoff B, Wolf DH, Hochstrasser M. 1994. *FEBS Lett.* 354:50–52
129. McGuire MJ, McCullough ML, Croall DE, DeMartino GN. 1989. *Biochim. Biophys. Acta* 995:181–86
130. Dahlmann B, Becher B, Sobek A, Ehlers C, Kopp F, Kuehn L. 1993. *Enzyme Protein* 47:274–84
131. Tanaka K, Ichihara A. 1989. *Biochem. Biophys. Res. Commun.* 158:549–54
132. Tanaka K, Yoshimura T, Ichihara A. 1989. *J. Biochem.* 106:495–500
133. Orino E, Tanaka K, Tamura T, Sone S, Ogura T, Ichihara A. 1991. *FEBS Lett.* 284:206–10
134. Peters JM, Franke WW, Kleinschmidt JA. 1994. *J. Biol. Chem.* 269:7709–18
135. Djaballah H, Rowe AJ, Harding SE, Rivett AJ. 1993. *Biochem. J.* 292:857–62
136. Weitman D, Etlinger JD. 1992. *J. Biol. Chem.* 267:6977–82
137. Mykles DL. 1993. *Enzyme Protein* 47:220–31
138. Murakami K, Etlinger JD. 1986. *Proc. Natl. Acad. Sci. USA* 83:7588–92
139. Guo GG, Gu M, Etlinger JD. 1994. *J. Biol. Chem.* 269:12399–402
140. Driscoll J, Frydman J, Goldberg AL. 1992. *Proc. Natl. Acad. Sci. USA* 89:4986–90
141. Li XCS, Etlinger JD. 1992. *Biochemistry* 31:11964–67
142. Li XC, Gu MZ, Etlinger JD. 1991. *Biochemistry* 30:9709–15
143. Chu-Ping M, Slaughter CA, DeMartino GN. 1992. *Biochim. Biophys. Acta* 1119:303–11
144. Tsubuki S, Saito Y, Kawashima S. 1994. *FEBS Lett.* 344:229–33
144a. Conconi M, Szweda LI, Levine AL, Stadtman AR, Friguet B. 1996. *Arch. Biochem. Biophys.* In press
145. Chu-Ping M, Slaughter CA, DeMartino GN. 1992. *J. Biol. Chem.* 267:10515–23
146. Dubiel W, Pratt G, Ferrell K, Rechsteiner M. 1992. *J. Biol. Chem.* 267:22369–77
147. Chu-Ping M, Vu JH, Proske RJ, Slaughter CA, DeMartino GN. 1994. *J. Biol. Chem.* 269:3539–47
148. Mott JD, Pramanik BC, Moomaw CR, Afendis SJ, DeMartino GN, Slaughter CA. 1994. *J. Biol. Chem.* 269:31466–71
149. Ahn JY, Tanahashi N, Akiyama KY, Hisamatsu H, Noda C, et al. 1995. *FEBS Lett.* 366:37–42
150. Nikaido T, Shimada K, Shibata M, Hata M, Sakamoto M, et al. 1990. *Clin. Exp. Immunol.* 79:209–14
151. Realini C, Dubiel W, Pratt G, Ferrell K, Rechsteiner M. 1994. *J. Biol. Chem.* 269:20727–32
152. Chu-Ping M, Willy PJ, Slaughter CA, DeMartino GN. 1993. *J. Biol. Chem.* 268:22514–19
153. Kania M, DeMartino GN, Baumeister W, Goldberg AL. 1996. *Eur. J. Biochem.* 236:510–16
154. Realini C, Rogers SW, Rechsteiner M. 1994. *FEBS Lett.* 348:109–13
154a. Realini C, Rechsteiner M. 1995. *J. Biol. Chem.* 270:29664–67
155. Honore B, Leffers H, Madsen P, Celis JE. 1993. *Eur. J. Biochem.* 218:421–30
156. Figueiredo-Pereira ME, Berg KA, Wilk S. 1994. *J. Neurochem.* 63:1578–81
157. Traenckner EB, Wilk S, Baeuerle PA. 1994. *EMBO J.* 13:5433–41
158. Read MA, Neish AS, Luscinskas FW, Palombella VJ, Maniatis T, Collins T. 1995. *Immunity* 2:493–506
159. Gronostajski RM, Goldberg AL, Pardee AB. 1984. *J. Cell. Physiol.* 121:189–98
160. Dice JF. 1987. *FASEB J.* 1:349–57
161. Jensen TJ, Loo MA, Pind S, Williams DB, Goldberg AL, Riordan JR. 1995. *Cell* 83:129–35
161a. Ward CL, Omura S, Kopito RR. 1995. *Cell* 83:121–27
162. Sepplorenzino L, Ma ZP, Lebwohl DE, Vinitsky A, Rosen N. 1995. *J. Biol. Chem.* 270:16580–87
163. Iqbal M, Chatterjee S, Kauer JC, Das M, Messina P, et al. 1995. *J. Med. Chem.* 38:2276–77
164. Harding CV, France J, Song R, Farah JM, Chatterjee S, et al. 1995. *J. Immunol.* 155:1767–75
164a. Hoffman EK, Wilcox HM, Scott RW, Siman R. 1996. *J. Neurol. Res.* In press
165. Adam J, Ma YT, Stein R, Baevsky M, Grenier L, Planondon L. 1996. *Eur. Patent.* In press
165a. Igbal M, Chatterjee S, Kauer JC, Mallamo JP, Messina PA, et al. 1996. *Bioorg. Med. Chem. Lett.* 6:287–90
165b. Spaltenstein A, Leban JJ, Huang JJ, Reinhardt KR, Viveros OH, et al. 1996. *Tetrahedron Lett.* 37:1343–46
166. Omura S, Matsuzaki K, Fujimoto T, Kosuge K, Furuya T, et al. 1991. *J. Antibiot.* 44:117–18
166a. Murakami Y, Tanhashi N, Tanaka K, Onura S, Hayashi S. 1996. *Biochem. J.* In press
166b. Dick LR, Cruikshank AA, Grenier L, Melandri FD, Nunes SL, Stein RL. 1996. *J. Biol. Chem.* 271:7273–76
166c. Hicke L, Riezman H. 1996. *Cell* 84:277–87
167. Tanaka K, Waxman L, Goldberg AL. 1983. *J. Cell Biol.* 96:1580–85
168. Kanayama HO, Tamura T, Ugai S, Kagawa S, Tanahashi N, et al. 1992. *Eur. J. Biochem.* 206:567–78

169. Sawada H, Muto K, Fujimuro M, Akaishi T, Sawada MT, et al. 1993. *FEBS Lett.* 335:207–12
170. Udvardy A. 1993. *J. Biol. Chem.* 268: 9055–62
171. Peters JM, Harris JR, Kleinschmidt JA. 1991. *Eur. J. Cell Biol.* 56:422–32
172. Fujinami K, Tanahashi N, Tanaka K, Ichihara A, Cejka Z, et al. 1994. *J. Biol. Chem.* 269:25905–10
173. Rubin DM, Coux O, Wefes I, Hengartner C, Young RA, et al. 1996. *Nature* 379:655–57
174. Ikai A, Nishigai M, Tanaka K, Ichihara A. 1991. *FEBS Lett.* 292:21–24
175. Hoffman L, Pratt G, Rechsteiner M. 1992. *J. Biol. Chem.* 267:22362–68
176. Deveraux Q, Ustrell V, Pickart C, Rechsteiner M. 1994. *J. Biol. Chem.* 269: 7059–61
176a. Van Nocker S, Deveraux Q, Rechsteiner M, Vierstra RD. 1996. *Proc. Natl. Acad. Sci. USA* 93:856–60
177. Goldberg AL, Rock KL. 1992. *Nature* 357:375–79
178. Ganoth D, Leshinsky E, Eytan E, Hershko A. 1988. *J. Biol. Chem.* 263: 12412–19
178a. DeMartino GN, Proske RJ, Moomaw CR, Strong AA, Song X, et al. 1996. *J. Biol. Chem.* 271:3112–18
179. Satoh K, Nishikawa T, Yokosawa H, Sawada H. 1995. *Biochem. Biophys. Res. Commun.* 213:7–14
180. Hadari T, Warms JV, Rose IA, Hershko A. 1992. *J. Biol. Chem.* 267:719–27
181. Gonen H, Smith CE, Siegel NR, Kahana C, Merrick WC, et al. 1994. *Proc. Natl. Acad. Sci. USA* 91:7648–52
182. Dawson SP, Arnold JE, Mayer NJ, Reynolds SE, Billett MA, et al. 1995. *J. Biol. Chem.* 270:1850–58
183. Armon T, Ganoth D, Hershko A. 1990. *J. Biol. Chem.* 265:20723–26
184. DeMartino GN, Moomaw CR, Zagnitko OP, Proske RJ, Chu-Ping M, et al. 1994. *J. Biol. Chem.* 269:20878–84
185. Tanahashi N, Tsurumi C, Tamura T, Tanaka K. 1993. *Enzyme Protein* 47: 241–51
186. Confalonieri F, Duguet M. 1995. *BioEssays* 17:639–50
187. Dubiel W, Ferrell K, Rechsteiner M. 1995. *Mol. Biol. Rep.* 21:27–34
188. Gordon C, McGurk G, Dillon P, Rosen C, Hastie ND. 1993. *Nature* 366:355–57
189. Lucero HA, Chojnicki EW, Mandiyan S, Nelson H, Nelson N. 1995. *J. Biol. Chem.* 270:9178–84
190. Ghislain M, Udvardy A, Mann C. 1993. *Nature* 366:358–62
191. Nelbock P, Dillon PJ, Perkins A, Rosen CA. 1990. *Science* 248:1650–53
192. Ohana B, Moore PA, Ruben SM, Southgate CD, Green MR, Rosen CA. 1993. *Proc. Natl. Acad. Sci. USA* 90: 138–42
193. Shibuya H, Irie K, Ninomiya TJ, Goebl M, Taniguchi T, Matsumoto K. 1992. *Nature* 357:700–2
194. Swaffield JC, Bromberg JF, Johnston SA. 1992. *Nature* 357:698–700
195. Kim YJ, Bjorklund S, Li Y, Sayre MH, Kornberg RD. 1994. *Cell* 77:599–608
196. Swaffield JC, Melcher K, Johnston SA. 1995. *Nature* 374:88–91
197. Akiyama K, Yokota K, Kagawa S, Shimbara N, DeMartino GN, et al. 1995. *FEBS Lett.* 363:151–56
198. Schnall R, Mannhaupt G, Stucka R, Tauer R, Ehnle S, et al. 1994. *Yeast* 10:1141–55
199. Cao JG, Firtel RA. 1995. *Mol. Cell. Biol.* 15:1725–36
200. Menon AS, Goldberg AL. 1987. *J. Biol. Chem.* 262:14929–34
201. Goldberg AL, Moerschell RP, Chung CH, Maurizi MR. 1994. *Methods Enzymol.* 244:350–75
202. Maurizi MR, Thompson MW, Singh SK, Kim SH. 1994. *Methods Enzymol.* 244:314–31
203. Woo KM, Kim KI, Goldberg AL, Ha DB, Chung CH. 1992. *J. Biol. Chem.* 267:20429–34
204. Sherman M, Goldberg AL. 1996. In *Stress-Inducible Cellular Responses,* ed. U Feige, R Morimoto, I Yahara, BS Polla. Basel: Birkhäuser/Springer-Verlag. In press
205. Wickner S, Gottesman S, Skowyra D, Hoskins J, McKenney K, Maurizi MR. 1994. *Proc. Natl. Acad. Sci. USA* 91: 12218–22
206. Wawrzynow A, Wojtkowiak D, Marszalek J, Banecki B, Jonsen M, et al. 1995. *EMBO J.* 14:1867–77
206a. Deveraux Q, Van Nocker S, Mahaffey D, Vierstra RD, Rechsteiner M. 1995. *J. Biol. Chem.* 270:29960–63
207. Kominami K, Demartino GN, Moomaw CR, Slaughter CA, Shimbara N, et al. 1995. *EMBO J.* 14:3105–15
208. Schwob E, Bohm T, Mendenhall MD, Nasmyth K. 1994. *Cell* 79:233–44
209. Lurquin C, Van Pel A, Mariame B, De Plaen E, Szikora JP, et al. 1989. *Cell* 58:293–303
210. Pentz ES, Black BC, Wright TR. 1986. *Genetics* 112:823–41
210a. Yokota K, Kagawa S, Shimizu Y, Akioka H, Tsurumi C, et al. 1996. *Mol. Biol. Cell.* In press

210b. DeMarini DJ, Papa FR, Swaminathan S, Ursic D, Rasmussen TP, et al. 1995. *Mol Cell. Biol.* 15:6311–21

211. Song HY, Dunbar JD, Zhang YX, Guo D, Donner DB. 1995. *J. Biol. Chem.* 270:3574–81

212. Boldin MP, Mett IL, Wallach D. 1995. *FEBS Lett.* 367:39–44

213. Dubiel W, Ferrell K, Dumdey R, Standera S, Prehn S, Rechsteiner M. 1995. *FEBS Lett.* 363:97–100

214. Tsurumi C, DeMartino GN, Slaughter CA, Shimbara N, Tanaka K. 1995. *Biochem. Biophys. Res. Commun.* 210:600–8

215. Gridley T, Gray DA, Orr-Weaver T, Soriano P, Barton DE, et al. 1990. *Development* 109:235–42

216. Deveraux Q, Jensen C, Rechsteiner M. 1995. *J. Biol. Chem.* 270:23726–29

217. Murakami Y, Matsufuji S, Kameji T, Hayashi S, Igarashi K, et al. 1992. *Nature* 360:597–99

218. Jariel-Encontre I, Pariat M, Martin F, Carillo S, Salvat C, Piechaczyk M. 1995. *J. Biol. Chem.* 270:11623–27

219. Woo SK, Lee JI, Park IK, Yoo YJ, Cho CM, et al. 1995. *J. Biol. Chem.* 270:18766–73

219a. Huang Y, Baker RT, Fischer-Vize JA. 1995. *Science* 270:1828–31

220. Wilkinson KD, Tashayev VL, O'Connor LB, Larsen CN, Kasperek E, Pickart CM. 1995. *Biochemistry* 34:14535–46

221. Papa FR, Hochstrasser M. 1993. *Nature* 366:313–19

222. Eytan E, Armon T, Heller H, Beck S, Hershko A. 1993. *J. Biol. Chem.* 268:4668–74

223. Gonen H, Schwartz AL, Ciechanover A. 1991. *J. Biol. Chem.* 266:19221–31

224. Shiina N, Gotoh Y, Kubomura N, Iwamatsu A, Nishida E. 1994. *Science* 266:282–85

225. St. John AC, Conklin K, Rosenthal E, Goldberg AL. 1978. *J. Biol. Chem.* 253:3945–51

226. Sherman M, Goldberg AL. 1994. *J. Biol. Chem.* 269:31479–83

227. Kandror O, Busconi L, Sherman M, Goldberg AL. 1994. *J. Biol. Chem.* 269:23575–82

228. Fan CM, Maniatis T. 1991. *Nature* 354:395–98

229. Chen ZJ, Hagler J, Palombella VJ, Melandri F, Scherer D, et al. 1995. *Genes Dev.* 9:1586–97

230. Thanos D, Maniatis T. 1995. *Cell* 80:529–32

231. Heemels MT, Ploegh H. 1995. *Annu. Rev. Biochem.* 64:463–91

232. Michalek MT, Grant EP, Gramm C, Goldberg AL, Rock KL. 1993. *Nature* 363:552–54

233. Grant EP, Michalek MT, Goldberg AL, Rock KL. 1995. *J. Immunol.* 155:3750–58

234. Townsend A, Bastin J, Gould K, Brownlee G, Andrew M, et al. 1988. *J. Exp. Med.* 168:1211–24

235. Glynne R, Powis SH, Beck S, Kelly A, Kerr LA, Trowsdale J. 1991. *Nature* 353:357–60

236. Kelly A, Powis SH, Glynne R, Radley E, Beck S, Trowsdale J. 1991. *Nature* 353:667–68

237. Martinez CK, Monaco JJ. 1991. *Nature* 353:664–67

238. Ortiz-Navarrete V, Seelig A, Gernold M, Frentzel S, Kloetzel PM, Hammerling GJ. 1991. *Nature* 353:662–64

239. Belich MP, Glynne RJ, Senger G, Sheer D, Trowsdale J. 1994. *Curr. Biol.* 4:769–76

240. Tanaka K. 1994. *J. Leukocyte Biol.* 56:571–75

241. Boes B, Hengel H, Ruppert T, Multhaup G, Koszinowski UH, Kloetzel PM. 1994. *J. Exp. Med.* 179:901–9

242. Ustrell V, Pratt G, Rechsteiner M. 1995. *Proc. Natl. Acad. Sci. USA* 92:584–88

243. Gaczynska M, Rock KL, Goldberg AL. 1993. *Nature* 365:264–67

244. DeMartino GN, Orth K, McCullough ML, Lee LW, Munn TZ, et al. 1991. *Biochim. Biophys. Acta* 1079:29–38

245. Rammensee HG, Falk K, Rotzschke O. 1993. *Annu. Rev. Immunol.* 11:213–44

246. Heemels MT, Schumacher TNM, Wonigeit K, Ploegh HL. 1993. *Science* 262:2059–63

247. Momburg F, Neefjes JJ, Hammerling GJ. 1994. *Curr. Opin. Immunol.* 6:32–37

248. Dick LR, Aldrich C, Jameson SC, Moomaw CR, Pramanik BC, et al. 1994. *J. Immunol.* 152:3884–94

249. Arnold D, Driscoll J, Androlewicz M, Hughes E, Cresswell P, Spies T. 1992. *Nature* 360:171–14

250. Momburg F, Ortiz NV, Neefjes J, Goulmy E, van de Wal Y, et al. 1992. *Nature* 360:174–77

251. Yewdell J, Lapham C, Bacik I, Spies T, Bennink J. 1994. *J. Immunol.* 152:1163–70

252. Fehling HJ, Swat W, Laplace C, Kuhn R, Rajewsky K, et al. 1994. *Science* 265:1234–37

253. Sibille C, Gould KG, Willardgallo K, Thomson S, Rivett AJ, et al. 1995. *Curr. Biol.* 5:923–30

254. Gabathuler R, Reid G, Kolaitis G, Driscoll J, Jefferies WA. 1994. *J. Exp. Med.* 180:1415–25

255. Eisenlohr LC, Bacik I, Bennink JR,

Bernstein K, Yewdell JW. 1992. *Cell* 71:963–72
256. Elliott T, Willis A, Cerundolo V, Townsend A. 1995. *J. Exp. Med.* 181:1481–91
257. Snyder HL, Yewdell JW, Bennink JR. 1994. *J. Exp. Med.* 180:2389–94
257a. Groettrup M, Rupper T, Kuehn L, Seeger M, Standera S, et al. 1995. *J. Biol. Chem.* 270:23808–15
258. Treier M, Staszewski LM, Bohmann D. 1994. *Cell* 78:787–98
259. Hayashi S, Murakami Y. 1995. *Biochem. J.* 306:1–10
260. Mamroud-Kidron E, Kahana C. 1994. *FEBS Lett.* 356:162–64
261. Elias S, Bercovich B, Kahana C, Coffino P, Fischer M, et al. 1995. *Eur. J. Biochem.* 229:276–83
262. Li XQ, Coffino P. 1993. *Mol. Cell. Biol.* 13:2377–83
263. Gronostajski RM, Pardee AB, Goldberg AL. 1985. *J. Biol. Chem.* 260:3344–49
264. Fagan JM, Waxman L, Goldberg AL. 1986. *J. Biol. Chem.* 261:5705–13
265. Davies KJ, Goldberg AL. 1987. *J. Biol. Chem.* 262:8227–34
266. Fagan JM, Waxman L. 1991. *Biochem. J.* 277:779–86
267. Friguet B, Szweda LI, Stadtman ER. 1994. *Arch. Biochem. Biophys.* 311:168–73
268. Sahakian JA, Szweda LI, Friguet B, Kitani K, Levine RL. 1995. *Arch. Biochem. Biophys.* 318:411–17
269. Davies KJ. 1993. *Biochem. Soc. Trans.* 21:346–53
270. Grune T, Reinheckel T, Joshi M, Davies K. 1995. *J. Biol. Chem.* 270:2344–51
271. Briane D, Olink-Coux M, Vassy J, Oudar O, Huesca M, et al. 1992. *Eur. J. Cell Biol.* 57:30–39
271a. Mahaffey D, Yoo Y, Rechsteiner M. 1993. *J. Biol. Chem.* 268:21205–11
271b. Kawahara H, Yokosawa H. 1994. *Dev. Biol.* 166:623–33
271c. Aizawa H, Kawahara H, Tanaka K, Yokosawa H. 1996. *Biochem. Biophys. Res. Commun.* 218:224–28

272. Klein U, Gernold M, Kloetzel PM. 1990. *J. Cell Biol.* 111:2275–82
273. Jones ME, Haire MF, Kloetzel PM, Mykles DL, Schwartz LM. 1995. *Dev. Biol.* 169:436–47
274. Medina R, Wing SS, Haas A, Goldberg AL. 1991. *Biomed. Biochim. Acta* 50:347–56
275. Furuno K, Goodman MN, Goldberg AL. 1990. *J. Biol. Chem.* 265:8550–57
276. Mitch WE, Medina R, Grieber S, May RC, England BK, et al. 1994. *J. Clin. Invest.* 93:2127–33
277. Baracos VE, DeVivo C, Hoyle DH, Goldberg AL. 1995. *Am. J. Physiol.* 268:E996–1006
277a. Temparis S, Asensi M, Taillandier D, Aurousseau E, Larbaud D, et al. 1994. *Cancer Res.* 54:5568–73
278. Tiao G, Fagan JM, Samuels N, James JH, Hudson K, et al. 1994. *J. Clin. Invest.* 94:2255–64
279. Tawa NE Jr, Goldberg AL. 1992. *Am. J. Physiol.* 263:E317–25
279a. Tawa NE, Goldberg AL. 1991. *Surg. Forum* 42:25–8
280. Medina R, Wing SS, Goldberg AL. 1995. *Biochem. J.* 307:631–37
281. Kanayama H, Tanaka K, Aki M, Kagawa S, Miyaji H, et al. 1991. *Cancer Res.* 51:6677–85
282. Shimbara N, Orino E, Sone S, Ogura T, Takashina M, et al. 1992. *J. Biol. Chem.* 267:18100–9
283. Bey F, Silva-Pereira I, Coux O, Viegas-Pequignot E, Recillas-Targa F, et al. 1993. *Mol. Gen. Genet.* 237:193–205
284. Akioka H, Forsberg NE, Ishida N, Okumura K, Nogami M, et al. 1995. *Biochem. Biophys. Res. Commun.* 207:318–23
285. Frentzel S, Troxell M, Haass C, Pesold-Hurt B, Glatzer KH, Kloetzel PM. 1992. *Eur. J. Biochem.* 205:1043–51
286. Seelig A, Troxell M, Kloetzel PM. 1993. *Biochim. Biophys. Acta* 1174:215–17
287. Tamura T, Osaka F, Kawamura Y, Higuti T, Ishida N, et al. 1994. *J. Mol. Biol.* 244:117–24

AUTHOR INDEX

A

Aaspollu A, 753
Aaltonen LA, 118, 119
Aarts A, 508
Abate MI, 567, 573
Abdel-Meguid SS, 612
Abdel-Monem M, 171, 200
Abe K, 244, 250
Abeles RH, 96
Abelson J, 369, 371, 372, 374,
    378, 388, 390, 391, 394,
    395, 397–99, 401
Abelson JN, 394, 396
Aber VR, 228
Abmayr SM, 793
Abney JR, 479
Aboussekhra A, 16, 33, 57, 147,
    149–52
Abovich N, 388, 394
Abramic M, 61, 150, 153
Abu-Elneel K, 427
Abuin A, 161
Ackerman SH, 583, 584
Ackermann K, 260
Ackford HE, 357
Adachi T, 645, 666, 673
Adachi Y, 663, 665, 666, 675
Adam J, 821
Adam SA, 698
Adamczewski JP, 19, 20, 31, 33,
    58, 70
Adami GR, 72, 73
Adams AE, 248
Adams DE, 639, 662
Adams LB, 236
Adams MD, 280, 639, 810
Adams SE, 751
Adams V, 569
Adamski FM, 745
Adamson P, 249
Adar R, 463
Adelman MR, 286
Aderem A, 257
Adesnik M, 712
Admon A, 61, 150, 279, 400,
    774, 775, 791, 792
Aebersold R, 580, 640
Aebi M, 281, 391
Aerts HMFG, 318
Afendis SJ, 818
Affara N, 85, 744, 755
Agard DA, 667, 809
Agbunag R, 667
Aggeler R, 568, 572, 576
Agranovsky AA, 752
Aguilera A, 669
Aharon T, 695, 710, 727
Ahlers CM, 636, 672
Ahmed A, 312

Ahmed H, 444, 448, 458, 468
Ahn JY, 805, 818, 819
Ahn K, 809, 838
Ahnert-Hilger G, 297
Ahumada A, 641, 642
Air GMA, 460, 469
Aiyar N, 676
Aizawa H, 840
Akaishi T, 822, 823
Akasaka K, 260
Akella SS, 759
Akhayat O, 707, 708, 803, 805,
    808
Aki M, 805, 809, 836, 840
Akimaru J, 276, 278, 280, 292
Akinaga S, 260
Akino T, 259
Akioka H, 831, 840
Akira S, 620
Akita M, 280, 293
Akiyama K, 805, 809, 828, 836
Akiyama KY, 818, 819
Akiyama M, 359
Akiyama Y, 276, 294
Aksentijevich I, 331
Akusjärvi G, 381
Alahari SK, 385
Alakhov YB, 756
Alani E, 102, 111–13, 121, 122,
    125, 126, 128, 160
Albersheim P, 529
Alberta JA, 703
Albertini DF, 479
Alberts BM, 71, 176, 196, 197,
    636, 646, 650
Alberty RA, 200
Albin M, 540
Albrecht G, 806
Albritton LM, 307–11, 313
Alcantara R, 326
Aldrich C, 837
Alexander IE, 670
Al-Harithy R, 144, 146, 155
Ali JA, 650, 651
Ali N, 428
Ali S, 619
Alibert C, 390
Allay E, 46
Allen B, 227
Allen CM, 249
Allen E, 102
Allen KN, 454
Allen LA, 571, 573
Allen NS, 524, 529
Allen SA, 158
Allerson CR, 391, 397
Allison LA, 523
Allshire RC, 352, 356–58
Allsopp RC, 356, 359
Almarez L, 599

Almouzni G, 107
Aloise P, 246
Aloni Y, 22
Alseth I, 140
Alsner J, 649, 676
Altamirano M, 228
Altamura N, 695, 715, 718, 719,
    724
Altmann SW, 622
Altschuler E, 674
Altus MS, 703
Alvares K, 594, 595
Alzari PM, 227, 389
Alzner-DeWeerd B, 763
Amadei A, 642
Amara FM, 703
Amara SG, 307, 314–21
Amaratunga M, 174–76, 182,
    190, 194, 195, 198, 203,
    205, 206, 209
Amati BB, 666
Amberg DC, 702
Ambros V, 718
Ames BN, 75, 89, 120, 140, 597
Ames JB, 264
Amicosante G, 218
Amin AA, 57, 63, 74
Amrein H, 380, 383, 390
Amstadt P, 598
Amsterdam A, 807, 839
Analbers LJS, 495
Anant JS, 251
Anderegg RJ, 242
Andersen AH, 642, 643, 645,
    648, 649, 670
Andersen J, 106, 107
Andersen SSL, 287, 288, 294
Anderson CF, 172, 173
Anderson CW, 74, 159, 743, 746
Anderson E, 479
Anderson GJ, 397
Anderson KJ, 316, 317
Anderson M, 640
Anderson P, 714, 718
Anderson RD, 636, 670, 675
Anderson RGW, 245, 257, 595
Anderson S, 763
Andersson H, 294, 295
Andersson J, 743
Andoh T, 636, 662, 664–66, 675
Andow K, 615, 627
Andres DA, 245–47, 250, 251,
    263
Andrew M, 836
Andrews AD, 145
Andrews B, 385
Andrews CL, 185
Andrews D, 278
Andrews DW, 295
Andrews JS, 92

Andrews LG, 704
Androlewicz M, 837
Angel P, 703, 709
Angenitzki M, 400
Anjard C, 426
Anraku Y, 249, 589
Ansari A, 19, 31, 33, 35, 54, 58, 64, 150
Anthony G, 570, 575, 576
Anthony NJ, 247, 260
Anton R, 597
Antonini E, 545
Antonni G, 568, 571
Ao SZ, 758
Aoyama H, 538, 547, 559, 574
Aoyama T, 93
Aparicio OM, 350
Apone L, 773, 775
Appeldoorn E, 31, 36, 54, 55, 147, 150
Appelman EH, 538
Aquilina G, 116, 117, 120, 137
Aragon C, 318, 323, 324, 326
Arai KI, 174, 181–83, 185, 186, 188–90, 246
Arai N, 174, 181, 183, 185, 186, 188, 190, 201, 205
Araki S, 250, 259
Arango R, 468
Arber N, 331
Arber W, 749
Archer EG, 320
Ardourel M, 507, 511–13, 518–20, 524, 530
Arenas J, 374, 398, 399
Arenas JE, 394, 396
Arents G, 785, 786
Arentzen R, 641
Ares M, 343
Ares M Jr, 374, 378, 393, 394, 396, 398
Argan CA, 593
Argos P, 729
Ariza RR, 153
Arkowitz RA, 96, 294
Arlett CF, 69
Arlt H, 584
Armon T, 804, 825, 826, 833, 838
Armstrong EL, 226, 231
Armstrong GD, 445, 468
Armstrong JA, 696, 706, 707
Armstrong SA, 246, 249, 250, 251
Arnberg AC, 805, 811
Arndt-Jovin DJ, 645, 647, 651, 664, 665
Arnold D, 837
Arnold JE, 826, 840
Arnold S, 623
Arraj JA, 102, 148
Arrick BA, 704
Arrigo AP, 803

Arriza JL, 307, 314, 316, 318–20, 431
Arthur L, 667, 677
Arvai AS, 140
Asahina H, 148
Asai K, 638
Asamoto M, 487, 494
Asano Y, 282, 284
Aschenbrenner V, 586
Ascher P, 322
Asensi M, 840
Ashburner M, 398
Ashby MN, 252, 253
Ashford AJ, 273, 287, 288
Ashktorab H, 416
Ashtekar DR, 217, 236
Ashtonrickardt PG, 805, 837
Ast G, 379, 400
Aster JC, 72, 73
Astrom A, 698
Astrom J, 698
Atabekov JG, 752
Atkin AL, 695, 715, 718, 719, 724
ATKINS JF, 741–68; 742–48, 750–54, 758, 760
Atkinson EM, 507, 509, 510, 512, 513, 515, 522, 524, 528
Atkinson MM, 493, 495
Attardi G, 564, 565, 567, 578, 580, 581
Attardi LD, 339, 343, 345, 346, 791
Atwater JA, 694, 695
Atwell D, 318
Au KG, 103–5, 110, 112, 113
Auble DT, 776, 777
Aubry L, 420, 424, 431
Auer B, 144
Augur C, 529
Aukhil I, 614
Aumann KD, 95, 98
Aurelle H, 505, 507, 511, 512, 520
Aurousseau E, 840
Austin CA, 636, 639, 674, 675
Austin MJF, 669
Austin SJ, 672
Ausubel FM, 529
Autexier C, 342, 343
Avadhani NG, 696
Aveman K, 658
Avilion AA, 339, 341, 343, 358–60
Avis J, 377
Axel R, 696
Axley MJ, 97
Axup AW, 540
Ayaki H, 45, 154
Ayane M, 378, 384
Aymami J, 780
Ayouba A, 444, 452
Ayton P, 639

Ayyagari R, 67
Azarnia R, 485, 493, 495
Aziz A, 221
Azzarone B, 360

B

Babcock GT, 538, 568, 570, 571, 576
Bablanian R, 707
Bacchetti S, 358–60
Bacerdo M, 222
Bach FL, 216, 232
Bach M, 371, 392, 396, 397
Bachmann L, 805, 811
Bachvarova RF, 706
Bacik I, 837
Backlund PS Jr, 254
Bader A, 311, 312
Bae I, 73, 74
Bae Y-S, 669
Baer BW, 698
Baes MI, 595
Baeuerle PA, 598, 819, 820
Baev N, 507, 509, 513, 518
Baevsky M, 821
Bagby S, 783
Bagley CJ, 622
Baglioni C, 697, 703
Bagully SH, 390
Bahler J, 702
Bai G-H, 218, 219, 230
Bai Y, 773, 775
Bailis AM, 125, 126, 128
Bailly V, 34, 57, 59, 146, 153, 171
Baim SB, 695
Baines BS, 612
Bairoch A, 59, 146
Bajaksouzian S, 223
Bakalkin G, 72
Baker BS, 378, 380, 383, 384, 390, 717
Baker D, 809
Baker EJ, 696
Baker RT, 832
Baker S, 105, 112, 113, 119
Baker SM, 105, 112, 113, 121, 123, 124
Baker TA, 198, 199, 656
Baker TS, 479, 481
Bakhuizen R, 523, 525
Bakker AC, 598
Bakker J, 526
Balaban RS, 85, 89
Balasubramanian V, 227, 230, 235
Balderes D, 349, 350
Baldwin JM, 435
Bales ES, 33, 55, 150, 153, 674
Ballard DW, 389
Baltimore D, 743, 794
Bambara RA, 66, 67, 115, 143

Bamborough P, 615, 622, 623
Banaszak LJ, 228
Bandy B, 597, 598
Bandyopadhyay R, 695
Banecki B, 830, 834
Banerjee A, 216, 227, 230, 235, 707
Banfalvi Z, 506, 507, 509, 515, 518
Bang DD, 24, 26, 59, 150
Bang Y-J, 121
Banga SS, 67
Bange FC, 219
Banholzer R, 703
Bani M-R, 392
Bankaitis V, 275
Bankier AT, 673, 763
Banks GR, 671
Bansal VS, 263
Banu L, 88, 92, 93, 98, 757
Bapat B, 121, 123
Barabino SM, 372, 374
Baracos VE, 840
Barajas L, 477
Baranov VI, 756
Barath Z, 592, 593
Barbacid M, 260
Barber S, 356, 358
Barbera FA, 620
Barberis A, 793
Barbour B, 318
Barbour WM, 505, 519, 520, 523
Barciszewska M, 763
Barclay WR, 226
Bardwell AJ, 16, 18, 20, 30–32, 34, 36, 58, 60, 146, 148, 150, 153
Bardwell L, 16, 18, 20, 30–32, 34–36, 58, 60, 146, 148, 150, 153, 171, 197
Bareyal-Leyser A, 746
Bargiello T, 485
Bargiello TA, 488, 489
Barinaga M, 597
Barker DG, 523, 530
Barker GF, 714, 715, 725
Barlier I, 507, 509, 510, 512
Barnes D, 16
Barnes DE, 158
Barnhart K, 181
Barny MA, 507, 509, 516
Barofsky E, 92
Baron C, 84, 85, 87, 88, 92, 95, 743, 755, 756, 762
Barondes SH, 444, 458, 469
Barot HA, 645
Barr L, 477
Barr PJ, 751
Barra J, 425
Barrand P, 425
Barray M, 671
Barrell BG, 399, 673, 763
Barrett AJ, 814

Barrett JC, 108, 109, 120, 357
Barrett MG, 257
Barrieux A, 312
Barrio LC, 485, 488
Barrone C, 35
Barrow WW, 218
Barry SC, 622, 623
Bar-Shira A, 160
Barsoum J, 662
Barta A, 377
Bartek J, 20
Bartfeld N, 141
Barth J, 586
Barthel T, 471
Barthelmess IB, 587
Bartlett JD, 26
Barton DE, 831
Bartus HF, 639, 674
Bartus JO, 676
Barzilay G, 142
Baserga SJ, 714, 725, 726
Bashkirov VI, 702
Basilion JP, 728, 730
Bass A, 338
Bass SH, 624, 627
Bassford P, 275
Bassford PJ, 275
Bastiaanse EM, 492
Bastide B, 477
Bastin J, 836
Basu A, 573, 575, 576, 595
Basu J, 236
Bateman E, 782
Bates AD, 636, 650, 651
Bates DJ, 469
Batschauer A, 155
Battula N, 118, 119
Bauer G, 143
Bauer MF, 290
Bauer P, 523
Bauer WR, 652
Baughn CO, 230
Bauman A, 418
Baumann B, 714
Baumann WJ, 491
Baumeister W, 804–14, 816, 822, 823, 834, 839
Baumgartner B, 673, 674
Baureithel K, 528
Baurén G, 368
Baxevanis A, 786
Bayle JH, 72
Bazan F, 87, 91, 93
Bazan JF, 610–12, 614, 621
Bazar L, 400
Beach D, 46, 73, 148, 155
Beall EL, 378, 387
Bean DW, 170, 172
Bear DG, 175, 179, 185, 192
Beardmore C, 671
Beau JM, 522, 528
Beaucage SL, 645, 672
Bec-Ferté MP, 505, 519, 520, 523

Bechberger JF, 477
Becher B, 816
Bechert T, 645
Bechtold R, 545
Beck LA, 242
Beck S, 833, 836
Beck WT, 665
Becker JM, 241, 253, 256
Becker KB, 755
Beckmann H, 777
Beckwith J, 272, 275, 276, 278, 280, 293–95
Bednarski MD, 449, 454
Bednarz AL, 704
Bedwell DM, 762
Beechem JM, 782
Beelman CA, 695, 709, 710
Beemon K, 714, 715, 725
Been MD, 642, 643, 649, 658
Beerman TA, 780
Beetschen JC, 807, 839
Beggs JD, 369, 374, 376, 377, 379, 397–99
Behr J-P, 748
Beier H, 754
Beijer RP, 389
Beilstein MA, 92, 757
Beinert H, 729
Belasco JG, 694, 695, 703, 705, 706, 709, 714, 722
Belcourt M, 744, 749
Belcourt MF, 748, 749
Belgrader P, 714, 715, 725–27
Belich MP, 836
Belin D, 706
Belisle JT, 235
Bell IM, 95
Bell JC, 385
Bell M, 378, 380, 383, 384
Belland RJ, 662
Belliveau DJ, 477
Bellocq C, 702
Bellus G, 575
Belmaaza A, 125, 127
Belote JM, 380
Beltrame J, 515
Bement DM, 576
Benbaruch G, 262
Benbow RM, 676, 678
Bendahan A, 307, 315, 317, 318, 320
Ben-David Y, 392
Bendixen C, 667, 677
Benedetti EL, 478, 479
Benedetti P, 659, 666
Benfeneti E, 415, 421
Bengal E, 22
Benhar I, 760
Benjamin L, 778
Benjamin TL, 444, 452, 453, 466
Benneche T, 140

Bennett M, 87, 91, 378, 386, 392, 394, 395
Bennett MN, 524, 529
Bennett MVL, 485, 488, 489, 495
Bennett SE, 140
Bennink J, 837
Bennink JR, 835, 837
Bensaude O, 21
Benson CJB, 174, 175, 177, 207
Benson FE, 160
Benson JD, 675
Benson S, 280
Bentle LA, 612
Benveniste R, 218, 219
Benz E, 200
Benz EJ Jr, 714, 725, 726
Beran RK, 642, 673
Beranger F, 251, 257
Beratan DN, 548, 549, 554
Bercovich B, 838
Beremand MN, 743
Berestecky JM, 493
Berg BL, 755
Berg DE, 659
Berg KA, 819, 820
Berg P, 149, 158, 160
Berg RJW, 161
Berger JM, 645, 646, 652, 653, 656, 664, 678
Berger NA, 636, 670, 671, 675, 677
Berger W, 477
Bergerat A, 639
Bergers P, 646
Berget SM, 381, 402
Berghammer H, 144
Berghuis AM, 445, 452
Bergmann C, 529
Bergoffen J, 477, 498
Bergsma-Schutter W, 805, 811
Berk AJ, 773, 775, 791, 793, 794
Berkower C, 255, 262
Berlot CH, 419, 426
Bernard P, 636, 678
Bernardi G, 577
Bernardini G, 491
Bernards A, 350, 353
Bernini LF, 727
Berns A, 117, 121, 123, 127, 137
Bernstein A, 123
Bernstein HD, 273, 275, 278, 674
Bernstein JA, 178, 225
Bernstein K, 837
Bernstein P, 694–97, 700, 709, 712
Bernstein PL, 728–30
Berriors M, 666
Berroteran RW, 718, 722
Berry MJ, 84, 87, 88, 91–93, 98, 753, 757
Berry SA, 36, 54
Berteloot A, 329
Berthold J, 290

Berthoud VM, 492, 494
Bertram JS, 674
Bertran J, 307, 327, 328, 330
Bertrand H, 593
Bertrand R, 649, 671
Bertrand S, 695, 708
Bertrand-Burggraf E, 51, 65, 145
Bertsch LL, 201, 205
Besser H, 390
Bessho T, 46, 60, 141
Bestagno M, 672, 674
Bettany AJE, 696
Betts JN, 548, 549
Betz R, 242
Beutler B, 703
Beveridge MJ, 310, 312
Bey F, 802, 805, 812, 839, 840
Beyaert R, 598, 599
Beyer AL, 368
Beyer EC, 476–78, 485, 490, 493
Beyreuther K, 808
Bezanilla M, 174, 177
Bharati S, 139
Bharti AK, 678
Bhat UR, 505, 512, 513, 516, 519, 520, 523
Bhatt AD, 217, 236
Bhattacharyya A, 343
Bhattacharyya NP, 119, 120
Bhuvaneswari TV, 526
Bi X, 661
Bianchi L, 151
Bibb MJ, 511
Biber J, 327, 328, 330
Bibus CR, 572
Bichard CJF, 452
Bickel M, 704
Bidou L, 754
Bieber AJ, 615
Biedermann KA, 159
Bieker KL, 278
Bieker-Brady K, 294
Bielka H, 273, 275
Biessmann H, 339
Biggerstaff M, 33, 57, 147–50, 152
Bignami M, 116, 117, 137
Bigsby BM, 115
Bilanchone VW, 574
Bilbe G, 378, 394
Biller SA, 260
Billett MA, 826, 840
Billings PC, 61
Billington BL, 350
Binder L, 619
Binder R, 694, 695, 728
Bindereif A, 374
Bindoff LA, 580
Bingham PM, 378, 380, 395, 396
Binns G, 378
Bird GSJ, 430
Birkeland N, 140
Birkmann A, 85, 744, 755

Birnboim HC, 47, 155
Birney E, 377, 380, 383, 389
Birnstiel ML, 712
Biro S, 511
Bishop DK, 106, 107
Bishop JD, 415
Bishop JE, 189
Bishop JM, 44, 118
Bishop T, 119, 137
Bishop WR, 260
Bisotto S, 402
Bisseling T, 523–26, 528, 530
Bisson R, 567, 569, 570, 575, 576
Biswas DK, 636, 672
Biswas EE, 189
Biswas SB, 189
Biswas TK, 578, 579
Bittner-Eddy P, 586
Bixon M, 538
Bjelland S, 140
Bjerknes M, 751
Bjerrum MJ, 540, 544, 558
Bjoras M, 140, 307, 315, 317, 320
Bjorklund S, 773, 775, 828
Bjorkman PJ, 615
BJORNSON KP, 169–214; 179, 194, 195, 198, 203, 205, 206
Bjornsti M-A, 636, 644, 647, 660, 669, 672, 676
Blaas D, 401
Blachly-Dyson E, 569
Black BC, 831
Black DL, 372, 381, 382, 386, 392
Black E, 142
Black SD, 256
Black WA, 228
Blackburn EH, 338, 339, 341–43, 345–47, 350, 351, 353, 355, 358
Blackwell LJ, 175
Blair LC, 254
Blais FX, 217, 235
Blake RD, 173
Blakely RD, 307, 318, 321, 322
Blanc A, 706, 707, 711
BLANCHARD JS, 215–39; 227, 228, 230
Blank A, 143
Blanton RL, 425
Blasco M, 341, 343, 360
Blasco R, 639
Blaskovich MA, 260
Blasquez VC, 666, 669
Blatch GL, 279
Blattner FR, 810
Bleecker GC, 712
Blencowe BJ, 372, 374, 381, 402
Blenis J, 496
Blight M, 296
Blinder D, 696
Blinkova A, 751
Blinkowa AL, 751

Bliska JB, 662
Blobel G, 272, 273, 275, 276, 278, 281–83, 286–89, 294
Bloch AB, 216, 221, 233, 234
Blocker H, 643
Bloemberg GV, 505, 507, 508, 511, 513, 516, 518, 520, 522, 529
Bloemendahl H, 478
Blondel MO, 276, 278
Blondelet MH, 425
Blondin GA, 729
Bloom BR, 216
Bloomfield VA, 172
Blot M, 749
Blow JJ, 73
Blumberg B, 594
Blumenthal T, 743
Blunt T, 159
Boakye K, 349, 350, 353
Bochkareva ES, 273
Böck A, 84–90, 92, 95, 97, 743, 744, 755, 756, 762
Bockovich NJ, 521, 522, 528
Bockrath RC, 68
Bodley AL, 670, 674
Bodmer WF, 119, 137
Boehm BO, 307, 315
Boehmer PE, 175
Boeke JD, 749
Boekel ET, 621, 622
Boes B, 836, 837
Boesmiller K, 85
Bogenhagen DF, 23, 143
Boggs RT, 380
Boguski MS, 242, 247, 252
Bohjanen PR, 704
Bohm A, 612
Bohm T, 830
Bohmann D, 838
Bohnensack R, 570
Bohr VA, 24, 26, 67, 70, 150
Boissel J-P, 619
Boiteux S, 141
Bojanowski K, 679
Boldin MP, 831
Bolgiano B, 444, 458, 468
Bollag RJ, 123
Boller T, 519, 525, 527, 528
Bolliger L, 290
Boman HG, 289
Bombick D, 493
Bominaar AA, 420, 428, 429
Bond MW, 698
Bond R, 260
Bond U, 712
Bone LJ, 477, 498
Boned A, 378, 384
Bonetti B, 762
Bonk RT, 141
Bonk S, 222
Bonne G, 576, 586
Bonneaud N, 696, 697

Bonnefoy N, 583, 589
Bonner CA, 52, 71
Bonner WM, 713
Bonnet D, 523
Bonney WJ, 494, 495
Bonnieu A, 695, 696, 703
Bono J-J, 507, 515, 522, 528
Bonven BJ, 642, 643, 655
Boodhoo A, 445, 468
Boone T, 612, 613
Boorstein RJ, 140
Boosalis M, 140
Boothman DA, 672, 677
Bootsma D, 18, 29, 30, 36, 43, 44, 53, 54, 58, 61, 68, 69, 136, 145, 149, 151
Borden LA, 307, 323
Bordignon E, 540
Bordonné R, 374
Bordwell B, 672
Bordwell BJ, 672
Boren T, 763
Borgese N, 284, 297
Borgmeyer U, 594
Borleis J, 422, 424, 429
Bornigia S, 223
Boron WF, 316, 317
Borowiec JA, 175–77, 187, 207
Borowsky B, 307, 323
Borst P, 350, 353
Borts RH, 125, 127, 668, 669
Bos JL, 242
Bosch L, 748, 750, 751, 754
Boshart M, 389
Bosserhoff A, 389
Bossi L, 753
Bostian KA, 697
Botchan M, 669
Botelho FM, 358
Bothwell ALM, 389
Botstein D, 249, 663, 675
Bottger EC, 219, 221
Boucher J-L, 598
Boulanger SC, 371
Boulikas T, 70
Bourdineaud JP, 507, 515, 522, 528
Bourgarel P, 714
Bourgeron T, 586
Bouriotis V, 510
Bourne Y, 444, 448, 450, 452, 456, 458, 459, 463, 468
Bousquet I, 580, 583
Bouthier de la Tour C, 638, 660, 668
Bouvet P, 696, 709
Bouzid S, 426
Boveris A, 597
Bowater RP, 659, 660
Bowcock A, 120
Bowcock AM, 309
Bowen S, 288
Bowerfind GK, 119

Bowman BU, 227
Bowman KK, 18, 46, 155
Bowman S, 584
Boxer SG, 538
Boyd CAR, 330
Boyd JB, 67, 612
Boy de la Tour E, 666
Boyer JC, 54, 108, 109, 117, 120
Boyer TG, 773, 775, 791
Boyko VP, 752
Boyle JA, 372
Boyle RW, 226
Boynton AL, 494
Boynton JE, 582
Bozon D, 762
Bozzaro S, 415, 421
Brachet P, 425
Bradbury EM, 159, 665, 675
Braddock M, 704, 731, 751
Bradfute DL, 263
Bradley A, 161
Bradley DJ, 263
Bradley JD, 339, 343, 345, 346
Bradley MO, 669
Bradsher JN, 22
Brakenhoff JPJ, 621, 622
Brambl R, 586
Bramucci M, 676
Branch P, 116, 117, 120, 137
Branchek T, 307, 323
Branden C-I, 228
Brandriss MC, 593
Brandt R, 426
Brannigan JA, 809, 813
Brash DE, 69
Brasiskyte D, 356, 358
Bratt MA, 696
Braun J, 479
Brautigan DL, 570, 575
Bravo R, 66, 151
Brawerman G, 694–96, 702, 711
Brecha N, 307, 323
Breimer LH, 141
Bren G, 118, 119
Brennan CA, 171, 178, 208
Brennan PJ, 218, 224, 226, 230, 231
Brenner S, 275, 278, 507, 695
Brennwald P, 244, 257
Brenowitz M, 782
Brett J, 599
Breunger E, 244
Brewer BJ, 338
Brewer CF, 444, 463, 468
Brewer G, 694–96, 700, 703, 704, 709, 712, 713
Brewin NJ, 504, 505, 522, 525
Briand JP, 486
Briand Y, 802, 808
Briane D, 839
Brierley I, 743, 750, 751
Brigelius-Flohé R, 84, 92, 95, 98
Briggs DR, 570

Briggs MS, 287
Brightman SE, 279
Brill SJ, 55, 639, 658, 659, 661, 662
Brimacombe R, 582, 759
Brink PR, 490
Brinkley BR, 402
Briscoe C, 431
Brissette JL, 494–96
Broach JR, 261
Broccoli D, 359
Broders F, 805, 808
Brodsky G, 637, 667, 669, 673
Brodsky JL, 279, 280, 289–91
Brody E, 371
Brody T, 61, 149, 150
Broek D, 242, 243, 245, 249
Brohl S, 579
Broker M, 416
Bromberg JF, 828
Bronner CE, 105, 112, 113, 119, 121, 123, 124
Brookman KW, 54, 147, 158
Brooks P, 107, 140
Brosh RM Jr, 176
Brosi R, 378, 393–96
Brou C, 775, 791
Brougham MJ, 652
Broughton BC, 35
Broughton WJ, 504, 506, 507, 510, 519, 523, 525, 527
Brouwer J, 24, 26, 27
Brow DA, 376
Brow MAD, 143
Brown AH, 695, 698, 703, 714, 715, 718, 719, 726
Brown AJP, 696, 711, 727
Brown BD, 728, 730
Brown CM, 745, 753
Brown CY, 703
Brown D, 328
Brown GM, 544
Brown JD, 379, 397
Brown JH, 444, 452, 454
Brown JM, 159, 431
Brown MG, 803, 805, 836
Brown MS, 242, 244–51, 254, 257, 258, 260, 261
Brown NG, 584, 585
Brown NR, 782
Brown PO, 636, 650
Brown R, 63, 149
Brown RM, 509
Brown S, 526
Brown T, 51, 140
Brown TC, 106, 108
Brown VD, 712
Brown WC, 197
Brown WRA, 355, 356
Brownlee G, 836
Brown-Shimer S, 703
Bruck C, 696
Bruckner RC, 175

Bruenger E, 262
Bruenn JA, 751
Brugge JS, 493
Bruhn SL, 61
Brummer F, 492
Brundage L, 276, 278, 280, 292
Brundage LA, 280
Bründl K, 102
Brune M, 210
Brunel C, 385
Brunel F, 389
Bruni R, 103, 104
Brunner J, 283, 287, 295
Brunori M, 545, 568, 571
Brunschwig BS, 541, 543
Brush GS, 74
Brutlag D, 645
Brutlag DL, 651, 674
Bruzik J, 382
Bruzzone R, 477, 478, 485, 487–89
Bryan TR, 358
Bryd W, 360
Buchanon SK, 554, 556
Buchenau PH, 664, 665
Buchman AR, 349
Buck SW, 350
Buckingham JM, 356
Budd ME, 67, 152
Buehler LK, 485, 497
Buge U, 570, 575, 576
Bugg CE, 454, 612
Buiré M, 506, 507, 519
Bujalowski W, 173–75, 179, 181, 184–86, 188, 189, 203, 207
Bujard H, 809, 836, 840
Bukhtiyarov YE, 249
Bulawa CE, 509, 531
Bulgakov R, 412
Bullock P, 115, 143, 669
Bumann J, 419
Bundell KR, 595
Bunn F, 566, 574, 585, 599
Bunn HF, 619
Buratowski S, 19, 20, 30, 31, 54, 58, 150, 791
Burcham PC, 649
Burckhardt J, 660
Burd CG, 389, 392, 395, 695
Burda P, 282
Burden DA, 676
Burdine V, 412
Burdsall A, 98
Burger U, 507, 510, 513, 514, 518
Burgering BM, 242
Burgers P, 67
Burgers PM, 151
Burgers PMJ, 171
Burgess RR, 643, 649, 673
Burgess SM, 391, 398, 399, 590
Burgin AB, 649
Buri J, 808
Burk PG, 53

Burk RF, 92, 755, 758
Burk RR, 484
Burke PV, 566, 574, 575, 585, 595, 599
Burkhart WA, 67, 158
Burki E, 426
Burkle A, 144
Burland V, 810
BURLEY SK, 769–99; 455, 770, 772, 773, 775, 776, 778, 780, 781, 785, 786, 788, 791, 794
Burlingame RW, 785–88
Burmeister WP, 452
Burnette WN, 443
Burnier JP, 245
Burns JA, 570
Burny A, 696
Burr B, 352
Burr F, 352
Burt JM, 485, 490, 491
Busch AE, 329, 330
Busch DB, 54
Busch H, 676
Busconi L, 834
Bush J, 424
Bushnell DA, 19, 32, 37, 151, 153
Busk H, 643
Buss JE, 242, 255, 257, 262
Bussolino F, 415, 421
Butel JS, 61
Butler WB, 713
Butler WR, 232
Butow RA, 565, 583, 593, 594
Butrynski JE, 259
Butterweck A, 477, 489, 490, 495
Buvoli M, 378, 394, 395

C

Cabibbo A, 621, 622
Cáceres J, 378, 380, 383, 384
Cáceres JF, 383, 384, 392
Cadenas E, 597
Cadwallader K, 251, 252, 255
Caetano-Anollés G, 504, 505
Cai K, 246
Caira T, 595
Calavetta L, 576
Calcaterra LT, 541
Caldecott K, 67, 158, 671
Caldecott KW, 158
Calder KM, 568, 572
Caldwell GA, 241, 253
Call K, 140
Calmettes P, 179, 185
Calonge MJ, 331
Calvin HI, 94
Cambareri B, 622, 623
Cambau E, 224
Cambillau C, 444, 448, 450, 452, 458, 459, 463
Cameron B, 639, 662

Cameron RP Jr, 612
Cameron V, 582, 585, 587, 588
Camilloni G, 642, 661
Camougrand NM, 586
Campa MJ, 329
Campbell AM, 773, 775, 792
Campbell JL, 67, 152, 347, 574, 575, 596, 597
Campione AL, 312, 317
Camut S, 526, 527, 529, 530
Canada FJ, 254
Cande Z, 338
Candia AF, 121
Cann PA, 759
Cannon-Carlson S, 140
Cantau P, 444, 458, 468
Canter-Cremers HCJ, 506
Canters GW, 553, 554, 557
Cantor CR, 356
Cao JG, 828
Cao L, 122, 599
Capaldi RA, 567, 568, 571–74, 576
Capasso O, 712
Capecchi MR, 106
Capitanio N, 583, 589
Caplan AJ, 275, 290
Caplin BE, 248
Capone AK, 696, 724
Caponigro G, 694, 695, 700, 705, 706, 709
Capony J-P, 20
Capranico G, 636, 643, 674
Capron A, 755, 757
Caput D, 703
Carbon P, 757
Carcanague DR, 521, 528
Cardelli J, 424
Cardellini E, 676
Cardenas L, 505, 519
Cardenas ME, 349, 666, 676, 677
Cardozo C, 814, 815, 821, 838
Carethers JM, 117
Carey M, 19, 793, 794
Carey MF, 794
Carey NH, 706
Carillo S, 832, 838
Carlson M, 36
Carlson RW, 505, 507–9, 512, 513, 515, 516, 518–21, 523, 524, 527, 529, 530
Carmo-Fonseca M, 378, 380, 387
Caron M, 245, 257
Caron MG, 257, 307, 321, 322
Caron PR, 636–39, 647, 653, 663, 667–69, 676
Carothers AM, 27
Caroutsos K, 727
Carpenter ATC, 122, 123
Carr AM, 16, 47, 144, 155, 160
Carr SA, 242
Carrano AV, 158
Carraway M, 103, 123

Carrier F, 72
Carroll K, 706, 707, 711
Carson MJ, 347
Carswell-Crumpton C, 70
Carter K, 67, 158
Carter KC, 105, 108, 112, 113, 117, 119, 120
Carter M, 714, 725
Cartwright IL, 279
Carty MP, 57, 74
Caruccio N, 695, 698, 700, 713
Caruthers MH, 379, 400
Carver JP, 444, 450, 454, 460, 463
Casadevall N, 619
Casado M, 318
Casciano I, 117, 120
Cascio D, 612
Cascio M, 482
Case DA, 782
Caserta M, 642, 661
Casey JL, 709, 710, 728–30
CASEY PJ, 241–69; 242, 244–49, 255, 257, 259, 262–64, 431
Cashman JR, 263
Casimiro DR, 540, 544, 553, 557
Casjens S, 758
Caskey CT, 744, 745
Caspar DLD, 479–84, 491
Cassan M, 754
Cassuto E, 668
Castano IB, 669
Castelein H, 595
Castello A, 328
Castner BJ, 623
Castora FJ, 640, 652
Caterina MJ, 419, 429–32
Cathala G, 385, 390
Caughey W, 573–77
Cavaloc Y, 377, 378, 380, 383, 384, 400
Ceccotti S, 116
Cech TR, 341, 343, 348, 349
Cedergren RA, 511, 514, 518
Cejka Z, 804, 807, 810, 822, 823, 826
Celander DW, 349
Celis JE, 66, 151, 819
Cerami A, 703
Cerino A, 674
Cerundolo V, 837
Cerutti P, 598
Cervantés E, 510, 514
Chabot B, 372, 382, 386, 402
Chadeneau C, 360
Chafin DR, 22
Chakrabati P, 236
Chakravarti D, 140
Chalberg MD, 55
Chalker D, 748
Challoner PB, 709

Chalut C, 19, 150
Chalvet F, 583, 589
Chambaz EM, 679
Chamberlin MJ, 342, 771
Chambers A, 349
Chambers I, 85, 744, 755
Chambon P, 775, 791
Chamorro M, 750, 751
Champion-Arnaud P, 378, 386, 393–95
Champlin DT, 378, 390
Champoux JJ, 636, 642–44, 647, 649, 658, 669, 670, 678
Chan CK, 546
Chan DW, 159, 674
Chan FK, 426
Chan J, 328
Chan L-NL, 709, 710, 729, 730
Chan R, 392
Chan SHP, 567, 573
Chan VTW, 674
Chance B, 573, 575, 576, 595, 597
Chandler M, 720, 752
Chandler SD, 385
Chang C-C, 485, 493, 494
Chang CK, 554
Chang CYY, 466, 468
Chang E, 61, 149
Chang GJ, 61, 149
Chang I-J, 540, 544, 549, 550, 552, 553, 557, 558
Chang KY, 223
Chang SH, 749
Chang T-H, 378, 394, 395, 398
Chao K, 174, 176, 180, 182, 183, 188, 201, 203, 205
Chao KL, 174, 181, 186, 203
Chapman AL, 426
Chapon C, 394, 395
Charbonnier F, 638
Chardin P, 258
Charles AC, 491
Charles WC, 67, 146
Charron M, 662, 665, 670
Chartrand P, 125, 127, 669
Chase JW, 104
Chasin LA, 27, 714, 725, 726
Chasman DI, 777
Chatterjee B, 696, 706
Chatterjee S, 671, 677, 793, 820, 821
Chattopadhyay R, 236
Chau CA, 592
Chaudhuri A, 253
Chauhan DP, 117
Chavez P, 330
Chavrier P, 257
Che C-M, 540
Chebli K, 383
Chelstowska A, 594
Chen A, 636
Chen C-K, 257

Chen CL, 92
Chen CY, 73, 74
Chen C-YA, 695, 697, 703–6, 708–10, 717
Chen DJ, 159, 659, 660
Chen E, 763
Chen FY, 703
Chen G-FT, 88, 744
Chen GL, 636
Chen HY, 751, 781, 791
Chen I-T, 38, 73, 74
Chen J, 140, 263
Chen JH, 398, 399
Chen J-L, 774, 775, 791, 792, 794, 795
Chen JR, 423
Chen JW, 67, 158
Chen K, 748
Chen M, 63, 665
Chen MS, 674
Chen MY, 421, 431
Chen P, 246, 541, 809
Chen R-H, 27, 69
Chen SL, 148, 246, 391, 397
Chen TM, 697, 703, 704, 717
Chen W-J, 245, 246, 581, 582
Chen WY, 627
Chen XY, 329
Chen XZ, 750, 780
Chen Y-D, 645, 757
Chen Y-H, 23, 88
Chen Z, 105, 110
Chen ZH, 149
Chen ZJ, 835
Cheng J-F, 356, 714, 715, 725–27
Cheng S-C, 369, 372, 374, 391, 397, 398
Cheng YC, 661
Chenna A, 140
Cheong JH, 791
Cheong JJ, 529
Cherian SP, 117
Chesick JP, 545
Chi NW, 110–12, 115, 116
Chi T, 793, 794
Chia L-LSY, 749
Chiang A, 279, 280, 290
Chiang C-M, 773, 775, 785, 786, 788, 791, 794
Chiang S-Y, 780
Chiara MD, 378, 382, 394, 395
Chiesa R, 674
Childs JE, 242, 255
Chillaron J, 327, 328, 331
Chillon M, 331
Chin J, 729
Ching W-M, 97
Ching YC, 538
Chinsky JM, 105, 110
Chipoulet AJ, 36
Chipoulet M, 54, 55
Chirico WJ, 275
Chisolm G, 98

Chiu LN, 140
Cho CM, 832
Cho I, 571
Cho KR, 674
Chodchoy N, 695, 712
Choder M, 661
Choi B, 145
Choi D-J, 23
Choi M, 400
Choi S, 808, 813, 821
Choi YD, 392
Chojnacki T, 224, 230, 231
Chojnicki EW, 827
Chomyn A, 567
Choo WM, 589
Chou MH, 540
Chou T-B, 378, 380, 395, 396
Choulika A, 378, 394, 395
Chovnick A, 125, 127
Chow K-C, 674
Chow Y, 199
Chowdhry V, 641
Choy B, 793
Chrebet G, 637, 667, 669, 673
Chretien D, 586
Chretien S, 619
Christensen HN, 306, 309, 320, 324, 329, 330, 332
Christians FC, 27, 37
Christiansen J, 728–30
Christiansen K, 643, 645, 648, 649
Christie DM, 105, 112, 114, 115, 122, 123
Christman MF, 667
Chu F-F, 88, 757
Chu G, 45, 61, 149, 159
Chua N-H, 640
Chuang SE, 810
Chubatsu LS, 61
Chui GSJ, 66, 152
Chung CH, 829
Chung IK, 644, 647
Chung TDY, 639, 675
Chu-Ping M, 817, 818, 825, 827, 830, 831, 833, 838
Ciapponi L, 618, 621, 622
Ciechanover A, 588, 802, 804, 821, 832, 833
Cismowski MJ, 21
Ciudad CJ, 714, 725, 726
Civitelli R, 490
Claassen L, 660
Clabby M, 307, 330
Clackson T, 624, 626
Cladaras MH, 171, 180
Claesson C, 763
Clare JJ, 569, 743, 744, 749
Clark AB, 701
Clark AM, 420
Clark CA, 515
Clark GJ, 244, 250
Clark JA, 318

Clark KJ, 496, 802, 816, 819, 820, 835
Clark KL, 786
Clark MW, 378, 394, 395
Clark S, 102–4, 107, 126, 128
Clark SC, 703
Clark SH, 125, 127
Clark WP, 809
Clarke DD, 53
Clarke DJ, 664, 665, 675
Clarke M, 412
Clarke S, 241, 243, 253–55
Clarkson GHD, 565, 587
Clarkson SG, 29, 30, 37, 59, 145, 146
Claude A, 115, 143, 180, 398
Clauser KR, 616, 617, 626
Claussen TJ, 22
Claverys J-P, 102–4, 106, 108, 113, 122
Clayton DA, 578, 579, 580
Clayton RA, 280, 639, 810
Cleaver JE, 29, 30, 36, 53, 60, 67, 145, 146
Cleeter MWJ, 567
Clegg JB, 727
Clements LS, 565, 593
Cleuter Y, 696, 697
Cleveland DW, 666, 676, 695, 696, 704, 709–11
Clewell A, 595, 599
Cline K, 275, 280, 752
Clore GM, 612
Clore M, 782
Closs EI, 307–13
Closs GL, 541
Clugston CK, 63, 149
Coady MJ, 329
Cobb MH, 260
Cochet C, 679
Cochran BA, 695
Cochrane AW, 696
Cochrane CG, 257
Cockell M, 349
Cockerill FR, 227
Cockerill PN, 666, 676
Cocking EC, 529
Cockle SA, 445, 468
Cocks BG, 87, 91, 93
Coderoni S, 676
Coffey DS, 674
Coffino P, 695, 746, 838
Cohen FE, 619
Cohen PR, 118
Cohen PTW, 385
Cohen RB, 704
Cohen SL, 773, 785
Cohen SN, 645, 672, 714
Cohn J, 521, 523, 524, 527, 530
Cohn JR, 522, 526, 530
Cohn ML, 226, 339, 342, 345, 346
Cohn ZA, 485

Colamonici OR, 621
Cole CN, 702
Cole MD, 703
Cole ST, 218, 219, 221, 222, 227, 568
Coleman DE, 421
Coleman JE, 580, 679, 743
Coleman PS, 246
Coleman R, 778, 782
Coll M, 780
Collier D, 275
Collin RG, 638
Collins ARS, 70
Collins JF, 511
Collins K, 341, 342, 344, 352
Collins PB, 226, 281, 282, 284, 286, 288
Collins PG, 281
Collins T, 819, 820, 834
Colman A, 696, 706, 707
Colman PM, 452
Colón JL, 553–57
Colosi P, 619
Colot HV, 696
Colwill K, 385
Comai L, 777
Company M, 374, 398, 399
Compton DA, 666
Conaway JC, 19, 20
Conaway JW, 19, 22, 57, 58, 64, 69, 70, 150, 151
Conaway RC, 19, 20, 22, 57, 58, 64, 69, 70, 150, 151
Concar D, 545
Conconi M, 817
Conde J, 280
Condeelis J, 420
Condemine G, 645
Condon C, 568
Condron BG, 743, 753
Cone JE, 96, 755
Confalonieri F, 638, 660, 668, 827
Conklin K, 834
Conley DL, 672
Conner MW, 260
Connolly T, 273, 278, 284, 286, 288, 294
Conrad DW, 540
Conrad MN, 349, 350
Conradt M, 319
Conrad-Webb H, 593
Consaul S, 714, 722
Constable A, 729
Constantinou A, 674
Conti CJ, 161
Contopolou-Griva I, 727
Contreras R, 695, 708
Conway GC, 377, 380, 383
Coodly L, 350
Cook D, 523
Cook DN, 659
Cook PR, 73, 151

Cook RG, 477, 482, 744, 745
Cook WJ, 454, 612
Cooke CA, 666
Cooke HJ, 352, 355–57
Cooke RM, 612
Cookson E, 755, 757
Cool M, 349
Cooley L, 753
Cooper AJ, 146
Cooper CE, 568, 569, 571
Cooper DL, 104, 105, 115
Cooper DNW, 469
Cooper GW, 94
Cooper JB, 526
Cooper MJ, 121
Cooper MS, 491
Cooper PK, 27, 37
Cooper T, 382, 383
Copeland NG, 105, 110, 119, 307, 315, 317, 477
Copeland TD, 754
Corbett AH, 636, 645, 648, 649, 677
Cordeiro-Stone M, 54
Corey EJ, 808, 813, 821
Corlet J, 59, 146
Cornelis S, 622
Cornell-Bell AH, 491
Correll CC, 263
Cortes P, 792
Costa-Perira R, 217, 236
Costanzo MC, 581, 582, 584–86
Cote GJ, 381
Cotgreave IA, 495
Cotton FA, 89
Cottrelle P, 349
Coukell MB, 426, 430
Coulombe B, 72, 73, 791
Counter CM, 356, 358, 359
Counter GM, 358
Courey AJ, 389
Couto JR, 398, 399
Coutts M, 695, 702
Couturier M, 636, 678
COUX O, 801–47; 805, 808, 810–12, 822, 828, 829, 840
Cover WH, 276
Coverley D, 57, 149, 151
Covey JM, 674
Cox AD, 243, 244, 250, 255, 261
Cox BS, 718
Cox DN, 524, 529
Cox DR, 356, 358
Cox EC, 102
Cox GB, 567
Cox MM, 125, 639
Cozzarelli NR, 636, 639, 645, 646, 650, 657, 658, 662, 663
Crabb JW, 242
Craig EA, 275, 587
Craig NL, 639
Craig RJ, 148
Craig RW, 72

Craigen WJ, 744, 745
Crain PF, 244
Cram LS, 356
Crawford JT, 222
Craxton A, 428
Cregan PB, 507
Creighton AM, 664, 675
Cremers FPM, 251
Cren M, 507
Crennell S, 452
Crenshaw DG, 676
Cresci S, 595
Crespi MD, 523, 526
Cresswell P, 837
Crick DC, 263
Crispino JD, 381
Cristino E, 421
Crivellone MD, 583, 589
Croall DE, 816, 817
Crochet J, 384
Cromlish WA, 793
Crooke ST, 639, 674, 676
Cross FR, 144
Cross SH, 356
Croteau W, 92, 93, 98, 755
Crothers D, 172
Crouse GF, 105, 110, 114, 115, 123, 125, 126
Crouzet J, 639, 662
Crovato F, 35
Crow A, 372, 388
Crow AJ, 379, 401
Crow DS, 493, 494
Crowley KS, 283, 286, 287, 289
Crowther RL, 444, 452, 459
Cruikshank AA, 821
Crutchley RJ, 540
Crute JJ, 175
Cruz C, 102
Cubitt AB, 420, 428
Cui Y, 714, 715, 718
Cukier RI, 554
Culberson JC, 246
Culbertson MR, 695, 714, 715, 718, 719, 721, 722, 724, 753
Cullen ME, 645
Cullimore J, 506, 529
Cullimore JV, 507, 515, 522, 528
Cullmann G, 151
Cumsky MG, 574, 576, 591
Cunningham BA, 487, 492, 493
Cunningham BC, 616, 617, 619, 620, 624, 626–29, 631
Cunningham JM, 307–13
Cunningham K, 280
Cunningham RP, 51, 141, 142
Cuomo CA, 159
Cuppens H, 762
Curatola AM, 711, 728
Curran DF, 758, 762
Curran JF, 745
Curran T, 142, 703
Curtis PJ, 679

Curtiss LA, 548
Cusack S, 444, 452, 454
Cusick ME, 378, 394, 395
Cutler M, 94
Cutting GR, 762
Cuzick RA, 116–18
Cyr DM, 275, 290
Czaja I, 522, 528, 529, 531
Czaplinski K, 719
Czernik AJ, 492, 494
Czyzyk L, 72

D

Daar IO, 714, 725
Dacheux RF, 496
Daffe M, 218, 226, 230
Daga A, 580
Dahl E, 477, 485
Dahl G, 485, 486, 492, 495
Dahlberg AE, 745, 753
Dahlberg JE, 143
Dahlmann B, 805, 807, 808, 811–13, 816
Dahmann C, 280
Dahmus ME, 19
Dalalian H, 216, 232
Dalbadie-McFarland G, 398
Dalbey RE, 281
D'Alessandro F, 621
Dalphin ME, 753
Daly G, 158
Danbolt NC, 307, 315–18, 320
Danehower S, 67, 158
Danesh S, 494
Dang Q, 666, 676, 677
Dani GM, 342
Dani M, 356
Daniel J, 424
Daniel RM, 638
Daniels DL, 810
Dannies PS, 566, 575
Dansbury KG, 216, 221, 233, 234
Darby MK, 677
Darcy PK, 428
Darley-Usmar VM, 574
Darmon M, 425
Darnell JE, 610, 611, 630, 696, 697, 712
D'Arpa P, 671, 676
Darras B, 576
Darrigo A, 297
Darrow AL, 706
Darvill A, 529
Darvill AG, 529
Darzynkiewicz E, 379, 401
Das M, 820
Das RH, 175, 182, 185, 186
DasGupta C, 125
Dasgupta UB, 72
Dash PK, 175
das Neves L, 523, 526
Datta B, 374, 394

Datta S, 416
Daubenton-Carafa Y, 526
Daugeron MC, 385
Daum G, 279, 292
Dauter Z, 444, 456, 463
Davey JC, 755
Davey S, 46, 148, 155
David HL, 218, 221, 226, 230
Davide JP, 260, 261
Davidson I, 775, 791
Davidson LA, 226
Davidson N, 173, 307, 323
Davidson PT, 221
Davie JR, 114
Davies AA, 18, 37, 59, 60, 65, 146–48
Davies AE, 505, 507, 515, 518, 524
Davies C, 759
Davies GJ, 650–53
Davies J, 218, 219
Davies KJ, 839
Davies SL, 639, 675
Davis B, 341, 343
Davis JL, 643, 649
Davis JN, 84, 86, 89, 96, 97, 755
Davis KL, 619
Davis RW, 697, 698, 700, 707, 711
Davis TA, 273, 274, 284, 286, 287, 293
Davison AJ, 597, 598
Dawson SP, 826, 840
Dawson WO, 752
Dawut L, 155
Day RS, 116
Day RSI, 159
Dayn A, 659
Dayringer HE, 612
Dazzo FB, 505, 511–13, 518
Dean FB, 55, 63, 73, 149, 174, 176, 177, 179, 207, 641, 657
Deardorff JA, 698
Deaven LL, 356
de Baets F, 762
DEBELLÉ F, 503–35; 505–15, 518–20, 523, 524, 527, 529, 530
De Billy F, 511, 518, 519, 523, 524, 530
de Blank C, 526
de Boeck C, 762
Debray H, 456
de Bruijn MHL, 763
Decker CJ, 696, 698, 702, 703, 705, 710–12, 726, 729
Declerc PE, 595
Decoster E, 583
Decout D, 456
Dedieu A, 523, 530
Deeley RG, 696
Deery WJ, 415
DeGrado WF, 287

De Groot H, 598
deGroot N, 59, 150
de Groot RJ, 593
De Gunzburg J, 257, 426
de Haas JR, 428
de Haseth PL, 173
Dehning I, 523
de Hoffmann E, 224, 230, 231
deHon FD, 621, 622
Deirdre A, 376
Deisenhofer J, 538, 554
de Jong AJ, 529
DeJong J, 791
DeJong WW, 727
Dekker HL, 576
Dekker M, 117, 121, 123, 127, 137
Dekker PJT, 585, 704
de Koster CC, 526
Delain E, 671
De Lajudie P, 504
de Lange T, 348, 356, 358, 359
Delannoy M, 590
Delannoy P, 379, 400
Delbaere LTJ, 444, 456, 463
Delcourt SG, 173
Delmar M, 489, 490
De Lozanne A, 427
del Rio RM, 96
DeLuca HF, 575
DeMaria CT, 704
DeMarini DJ, 831
DeMartino GN, 803, 807, 812, 813, 816–18, 825–28, 830, 831, 833, 836, 838
Demartis A, 621, 622
De Mello WC, 495
Demengeot J, 159
DeMeyts P, 620
Demma M, 791, 792
Demol H, 576
Demont N, 505, 507, 511, 512, 518–20, 522, 524, 530
Demple B, 46, 47, 67, 141, 142
Dempster M, 356, 358
DÉNARIÉ J, 503–35; 505–7, 509, 511, 512, 519, 520, 529
Deng WP, 123
Denhez F, 395, 396
Denman R, 704
de Oliveira R, 141
De Pagter-Holthuizen P, 728
DePamphilis ML, 662
DePaulo J, 54
Depew RE, 642, 660
De Plaen E, 831
Der CJ, 242, 243, 255, 257, 260–62
Derbyshire MK, 159
DeRecondo A-M, 674
de Rege PJF, 554
Derencourt J, 20
Derewenda Z, 444, 463

Derfler B, 140
der Garabedian A, 674
Derman AI, 275
Dermietzel R, 492
Dernberg A, 338
de Ruijter M, 51
de Sa CM, 803, 808
DeSauvage FJ, 611, 620
Deschamps J, 695, 708
Deschenes RJ, 243, 253, 261
De Serio A, 618, 621, 622
Deshaies RJ, 275, 276, 278, 279, 292
Deshpande AK, 696, 706
DeSio G, 223
de Smit MH, 762
deSolms SJ, 246, 260, 261
Dessen A, 227, 228, 230, 444, 463, 468
Detloff P, 123
D'Eustachio P, 261
Deutscher MP, 756
Devary Y, 72
Devenish RJ, 567
Dever TE, 398
Deveraux Q, 824, 830–32
Deves R, 330
DeVivo C, 840
DeVivo M, 423
DeVore RF, 677
DE VOS AM, 609–34; 610–12, 614–17, 619, 626–28
Devos R, 622
DEVREOTES PN, 411–40; 412, 416, 418–26, 429–31, 434, 435
de Vries A, 161
de Vries SC, 529
De Weerd-Kastelein EA, 53
Dewettinck D, 504
Dewey MM, 477
de Wind N, 105, 110, 117, 121, 123, 127, 137
De Winde JH, 566, 574, 585, 586
De Wit CL, 697, 703
de Wit J, 29, 36, 53, 54, 68, 69, 146
de Wit RJW, 412, 421
DeWolf A, 480
DeYoung DR, 223
de Zamaroczy M, 577
Dezube BJ, 636, 672
D'Haeze W, 507, 509, 511, 513, 514, 516
Dharmawardhane S, 420, 428
d'Hulst M, 339
Diamond RH, 378, 384
Dianov G, 16, 36, 139, 143
Di Bilio AJ, 549, 552–54, 557
Dice JF, 820
Dick LR, 812, 821, 837
Dickerson RE, 780
Dickson C, 749

Dickson JA, 59
Dieckmann CL, 565, 580–83, 585, 586
Diederichs K, 612, 613
Diehl RE, 245, 248, 249
Diekmann S, 645
Dietmeier K, 290
Dietrich FS, 667
Diffley JFX, 280, 579
Digard P, 750, 751, 761
DiGate RJ, 641, 642, 657, 663
DiGrandi S, 307, 330
Dijiane J, 619
Dillehay LE, 669
Dillon P, 827, 828
Dillon PJ, 827
Dimagno TJ, 546
Di Martino E, 642
Di Mauro E, 642, 661
Dimpfl J, 126
Din SU, 55
DiNardo S, 639, 658, 660, 661, 663
Dinesh-Kumar SP, 743, 744, 752, 753
Dineva B, 808
Ding J, 254, 261
Ding R, 144
Dingermann T, 416
Dingus J, 423
Dingwall C, 676
Dinman JD, 718, 720, 743, 753
Dionne VE, 309–13
Dirac-Svejstrup B, 258, 259
Dircks LK, 572
Dirheimer G, 749
Dirksen ER, 491
Discolo G, 618, 621, 622
Disher RM, 477, 482
Dittrich W, 416
Divecha N, 159, 160
Dixon K, 57, 74
Dixon M, 749
Dizhoor AM, 257
Djaballah H, 802, 813, 817
Djiane J, 619
Djordjevic MA, 507, 516, 518
Dobbe FCPM, 426
Dobberstein B, 272, 273, 278, 282, 283, 287, 294, 295
Dobson MJ, 743
Docherty R, 593
Dodson EJ, 444, 463, 650, 814
Dodson G, 650, 809, 813, 814
Dodson M, 174, 175, 179
Dodson ML, 46
Dodson MS, 175
Doel MT, 706
Doetsch PW, 18, 23, 46, 141, 155
Dohet C, 103, 107, 128
Dolberg M, 673
Dolence JM, 247
Domaille P, 465

Domanico PL, 648, 649
Dombroski AJ, 171, 178, 208
Dominguez J, 505, 519
Domon M, 151
Dompenciel RE, 728–30
Donachie WD, 663
Donahue BA, 24, 30, 68
Donahue J, 695, 698, 703, 714, 715, 718, 719
Donahue JL, 695, 709, 714, 715
Donahue SA, 235
Donahue TF, 21, 54
Donaldson P, 485, 491
Dong BH, 390
Dong F, 175–79
Dong Q, 116
Donini P, 351
Donly BC, 745
Donnabella MV, 222
Donner DB, 831
Doolittle RF, 69, 567
Dorbic T, 659
Dorman C, 659, 660
Dorman TE, 639
Dortant PM, 161
Dosanjh MK, 140
Dotto GP, 494
Dottore M, 622
Dougherty BA, 509
Douglas MG, 275, 290, 583, 702
Douglass J, 219
Dousmanis A, 348
Douvas A, 676, 678
Douville K, 278
Dove WF, 696
Dover R, 151
Dowhan W, 280, 293, 572
Downes CS, 24, 26, 70, 636, 664, 665, 672, 675
Downie JA, 505, 507, 509, 511, 513, 515–18, 524
Dragowska W, 356, 359
Drake FH, 639, 660, 666, 667, 674, 676, 679
Drapier J-C, 598
Drapkin R, 19, 20, 30, 31, 33, 35, 54, 57, 58, 64, 69, 70, 150
Drayer AL, 428
Dreier L, 278, 282, 289
Dresler SL, 151
Drew HR, 780
Dreyer D, 523
Dreyfuss G, 378, 388, 389, 392, 395, 400, 695, 698
DRICKAMER K, 441–73; 442–44, 451, 453, 454, 456, 458–60, 462, 466, 467, 469
Driessen AJM, 276, 280, 292
Driscoll D, 98, 426
Driscoll DM, 426
Driscoll J, 803–5, 817, 825, 836–39
Driscoll PC, 612

Drissi R, 153
Dritschilo A, 160, 677
Drivas G, 261
Drlica K, 148, 636, 645, 660, 663
Drobetsky EA, 27
Droge P, 659, 668
Drolet M, 659, 661
Droste M, 569, 571, 572, 576
Drouhard T, 119
Drubin DG, 591, 775
Drummond DR, 696, 706, 707
Drummond JT, 105, 108, 109,
    111, 112, 117, 120
Du K, 378, 384
Du M, 546
Dualan R, 61, 150
Dube P, 759, 763
Dubiel W, 802, 817–19, 825,
    827, 828, 830, 831, 833, 838
Dubnau E, 227, 230, 235
Dubois M-F, 21
Ducrocq C, 598
Duesing R, 541
Duffin KL, 758, 762
Duggleby HJ, 809, 813, 814
Duguet M, 636, 638, 674, 827
Duhl DMJ, 565, 582, 585, 587,
    588
Dujardin G, 580, 583, 589, 718
Dujon B, 115, 144
Duker M, 320
Dulbecco R, 636
Dumdey R, 831
Dumenco LL, 46
Dunbar JD, 831
Duncan D, 622, 623
Duncan PI, 385
Duncan R, 400
Dunia I, 478
Dunlap RB, 94, 95
Dunlop MG, 356, 358
Dunn DM, 745, 752, 753, 758
Dunn JJ, 743
Dunn RL, 68, 69, 103
Dunn WA, 310, 311
Duntze W, 242
Dupont E, 486
DuPont M-A, 230
Dupuis A, 568
Durban E, 661, 676
Durham B, 547
Durwald H, 171, 200
Dusanter I, 619
Dusanter-Fourt I, 619
Dutta A, 72, 73
Dutton PL, 549, 553, 557
Duus JO, 465
Dyer K, 356, 358
Dyer TA, 640
Dykstra CC, 175, 206, 701
Dynan WS, 159
Dynlacht B, 19, 20, 58, 70
Dynlacht BD, 773–75, 791

E

Ealick SE, 612
Earley MC, 114, 115
Earnshaw WC, 665, 666, 672,
    674, 676, 679
Ebert BL, 599
Ebert RH, 226
Ebihara L, 477, 485
Ebner R, 121
Ebringer L, 640
Echols H, 126
Ecke M, 424, 427
Eckerskorn C, 812, 815
Eckert R, 489, 490, 495
Eckner R, 712
Economou A, 292–94, 507, 524
Ecudero KW, 185
Eddy RL, 309
Edelman GM, 143, 487, 492, 493
Edelstein S, 702
Edenberg H, 67
Eder JP, 674
Edery I, 398, 401
Edmonds K, 223
Edmonds M, 696
Edström J-E, 380
Edwards KA, 643, 649
Edwards PA, 263
Edwards RG, 612
Eftedal I, 139
Egawa S, 119
Egelman EH, 160, 174, 175, 177,
    181, 184, 185, 207
Eggermont E, 762
Eggleston AK, 199
Eghbali B, 485
Egli M, 148
Egly J-M, 19, 31, 150
Egozi Y, 262
Egyhazi E, 661
Ehlers C, 816
Ehlers M, 621
Ehnholm C, 297
Ehnle S, 584, 587, 588, 828
Ehrenfeld E, 696
Ehrenman K, 704
Ehrenreich A, 85, 90
Ehrenson S, 541
Ehrhardt DW, 507, 515, 524, 529
Ehring B, 812
Ehrlich SD, 668
Ehrmann M, 295
Eibl H, 479, 491
Eichelberger M, 805, 837
Eichorn BG, 672
Eick D, 19
Einarsdóttir Ó, 538
Einspahr H, 444, 450, 460, 463
Eisen MB, 444, 459
Eisenach K, 222
Eisenbarth GS, 330
Eisenberg D, 314, 612

Eisenberg S, 200, 597
Eisenlohr LC, 837
Ejima Y, 73
Ek JF, 490
Eker APM, 61, 149, 150, 154
Eki T, 176, 177, 207, 664
El Aoumari A, 486, 494
Elberg G, 619
El'Bouhali M, 679
El-Deiry WS, 72, 73, 356
Elenbaas B, 38, 72
Eleouet J-F, 751
Elfferich P, 697, 703, 709
Elfgang C, 477, 489, 490, 495
El-Fouly MH, 485
Elias H, 540
Elias S, 838
Elias-Arnanz M, 149, 160
Elie C, 638, 660, 668
Elledge SJ, 33, 55, 61, 72, 73, 153
Elliott T, 837
Ellis AL, 674
Ellis CA, 649, 670
Ellis J, 123
Ellis NA, 158, 669
Ellis RL, 194, 208
Ellis TM, 116
Ellis WR Jr, 540
Ellner JJ, 223
Ellwood K, 793, 794
Elöve GA, 545
Elseviers D, 760
Elwood DR, 123
Elzinga SDJ, 704
Emeson RB, 717
Emmelot P, 478, 479
Emmerich C, 444, 450, 463
Emmerson PT, 175
Emmerson C, 444, 454
Endicott JA, 262, 782
Endo A, 242
Endo T, 584
Endre G, 507, 509, 518
Eneff KL, 679
Eng W-K, 647
Engel A, 491, 805, 811
Engelberg-Kulka H, 760
Engelmann GL, 477
Engels W, 125, 127
Engelsberg BN, 61
England BK, 840
Englander W, 545
Englezou A, 358
English AM, 540
Englund PT, 640
Engman L, 495
Ennis PD, 480, 481, 483
Enomoto T, 674, 676, 679
Enssle J, 695, 704, 731
Epand RF, 256
Epand RM, 256
Eperon IC, 381

Epp O, 538, 554
Epstein LH, 159
Epstein WW, 244, 262
Erecinska M, 570
Erickson HP, 614
Erickson L, 116
Erickson PM, 540
Ericsson LH, 245, 251, 254, 258
Ermler U, 554, 556
Ermolaeva M, 259
Erneux C, 428
Ernst AR, 762
Ernst JF, 123
Ernst SR, 444
Errington RJ, 73, 151
Esaki N, 94
Escamilla E, 593
Escher C, 804, 805
Eshleman JR, 118, 119
Esnault Y, 276, 278
Esposito R, 669, 679
Essig DA, 586
Essigmann JM, 61, 142, 146, 149
Esson K, 249
Estevez VA, 522, 528
Estrada-Garcia MT, 525, 527
Etlinger JD, 804, 817, 825
Etzerodt M, 380
Etzler ME, 442
Eum HM, 218, 236
Eustice DC, 695
Evans EA, 281
Evans IJ, 507, 516
Evans JW, 159
Evans MJ, 592
Evans MK, 150
Evans P, 377
Evans RM, 595
Evans T, 251, 258
Evans WH, 477, 479, 486
Even LF, 521, 528
Evenson JK, 86
Evenson JW, 549
Ewart DD, 576
Ewart G, 576
Ewel A, 111, 121, 123
Ewing N, 442
Eytan E, 804, 825, 827, 828, 833, 838
Ezekowitz RAB, 466, 468

F

Fabrizi GM, 576
Fabrizio P, 374
Facca C, 21
Fagan JM, 804, 822, 838–40
Fagg B, 623
Faiola B, 749
Fairbanks KP, 242
Fairchok M, 218, 230
Fairman MP, 55, 159
Fairman R, 674

Fairman WA, 307, 314, 316, 319, 320
Fajardo-Cavazos P, 45
Fajer J, 538
Falk K, 837
Falkenburg PE, 803
Falls KM, 639
Falquet J, 666
Fan CM, 835
Fan H, 669
Fang GW, 348, 349
Fang H, 279
Fang L, 88
Fang M, 660
Fang W-H, 107–9, 117–20, 137
Fang YW, 526
Fangman WL, 338, 347
Farabaugh PJ, 720, 743, 744, 748, 749
Farah JM, 820
Farid RS, 549, 553, 557
Farnsworth CC, 241–46, 248, 251, 254, 255, 258, 264
Farrell LE, 565, 566, 568, 572, 575, 576, 585, 593, 596
Farrell S, 793
Farver O, 540, 553, 554, 557
Fath MJ, 296
Fattorini L, 218
Faucher C, 505–8, 510, 512, 514, 515, 518, 519, 523, 527, 529, 530
Fauci AS, 330
Faull KF, 507, 515
Faure M, 425
Favreau M, 707, 708, 711
Fawcett DW, 479
Faye G, 21, 583
Fayet O, 720, 752
Fearon ER, 118
Fearon K, 762
Feaver WJ, 19–21, 30–32, 34, 37, 58, 64, 150, 151, 153, 775, 792
Feher G, 538
Fehling HJ, 837
Fehrenbach FJ, 234
Fei YJ, 330
Feinstein SI, 122, 125, 128
Feitelson MA, 37, 72
Feld R, 379, 393, 394, 396
Feldberg RS, 149
Feldheim D, 279, 280, 290
Feldheim DA, 279
Feldman BJ, 45
Feldman H, 584, 588
Feldmann FM, 232
Felix A, 327, 328
Felix CA, 669
Felix G, 528
Fellay R, 504, 506, 507, 510, 513, 514, 518, 519
Felle HH, 524, 527, 530

Fellows R, 61
Felsenstein KM, 754
Feng JL, 341, 343
Feng W-Y, 118, 122
Feng YX, 754
Fenteany G, 808, 813, 821
Fenwick C, 540
Ferat J-L, 371
Ferentz AE, 246
Ferguson BM, 338, 347
Ferguson M, 640
Ferguson-Miller S, 538, 570, 575
Fermandjian S, 671
Fernald AW, 677
Fernandez CO, 748
Fernandez-Lopez M, 507, 513, 514, 516
Ferran J, 511
Ferrara P, 389
Ferrari ME, 201
Ferrell K, 817–19, 827, 828, 830, 831, 833, 838
Ferrer JV, 46, 155
Ferrero L, 639, 662
Ferro AM, 677
Ferro M, 512, 519, 520
Ferro-Novick S, 251, 252
Ferry K, 339
Fertala J, 672
Fesen MR, 643
Fesik SW, 652
Fesquest D, 20
Fevre M, 509
Field J, 242, 243, 245, 249
Fields S, 381, 720
Fields-Berry SC, 662
Fien K, 151
Fiers W, 598, 599, 695, 708
Figueiredo-Pereira ME, 815, 819, 820
Fijalskowska IJ, 68, 69
Filhol O, 679
Filipowicz W, 369
Filipski J, 666
Filson AJ, 493
Filutowicz M, 657, 663
Finan PJ, 119
Finch JT, 786, 788
Fine RL, 262, 263
Finegold AA, 248
Finel M, 568, 573
Finger LR, 175, 179
Fink GR, 586, 636, 667, 669, 672, 701, 702, 714, 718
Fink T, 320
Finkbeiner SM, 491
Finke K, 278
Finken M, 219
Finlay CA, 72
Finley D, 802, 805, 829, 836
Finley K, 307, 309
Finley KD, 309, 312
Finnegan PM, 583

Finnie NJ, 159
Firmenich AA, 149, 160
Firmin JL, 505, 507, 515, 518
Firtel RA, 412, 416, 420, 421,
    424, 426, 428, 431, 828
Firth JD, 599
Fischer M, 588, 838
Fischer-Vize JA, 832
Fishel R, 105, 110, 111, 119, 159
Fisher DE, 782
Fisher LM, 636, 639, 645, 650,
    674
Fisher MA, 173
Fisher PA, 666
Fisher PR, 428
Fisher RF, 506, 507
Fisher RP, 19, 20, 58, 70, 579
Fishman GI, 492, 493
Fitch WM, 749
Fitzgerald DJ, 487
FitzGerald M, 228
Fiumicino S, 116
Flagg-Newton JL, 495
Flaherty KM, 777
Flanagan PM, 776
Fleischmann G, 661
Fleischmann RD, 639, 810
Fleming GR, 546
Fletterick RJ, 469
Fliesler SJ, 263
Flint N, 294
Flohé L, 95
Flores O, 22, 792
Flores-Rozas H, 73, 171
Flory MA, 223, 233
Floth C, 151
Flower AM, 751
Fogel S, 105, 106, 112, 114, 122,
    123, 126
Fogel-Petrovic M, 714, 725, 726
Fogelsong PD, 652
Foguet M, 297
Folger KR, 106
Fong AM, 309, 311
Fontaine V, 621, 622
Fontecilla-Camps J-C, 444, 450,
    459, 463
Forchhammer K, 84, 85, 87, 89,
    90, 92, 756, 762
Ford JM, 30, 38
Ford R, 54
Fores-Rozas H, 73
Forester K, 72
Forman BM, 263, 594
Forman HJ, 597
Fornace AJ Jr, 38, 72
Forrester K, 37
Forsberg NE, 840
Forsburg SL, 564
Forster C, 756
Fort P, 695, 696, 703
Forterre P, 638, 639, 660, 668
Fossum RD, 260

Fotedar R, 55
Fouet A, 638
Fourtner CR, 489
Fouser LA, 391
Fowlkes DM, 621
Fox LS, 541
Fox ME, 45
Fox MS, 102
Fox TD, 281, 581–86, 588, 760
Fraenkel CH, 140
Frampton J, 85, 744, 755
France J, 820
Francesconi S, 597
Francke U, 576
Franco R, 645
Frank D, 400, 401
Frank J, 759, 763
Frank R, 275, 278, 643
Frank RW, 389
Franke J, 425, 426, 431
Franke WW, 805, 816, 825, 839
Franssen H, 523, 524, 526, 530
Franzusoff A, 706
Frasch M, 378, 390
Fraser CM, 280, 432, 639
Frattini MK, 151
Frazer P, 679
Freed JH, 538
Freedberg DI, 507, 515
Freedman JA, 568, 569, 571
Freeman K, 350
Frei E III, 675
Freje JM, 639
Fremeau RT Jr, 307, 316, 317,
    320–22
French BT, 672
Frendewey D, 371, 372, 392
Frentzel S, 808, 809, 836, 840
Frere V, 671
Freudenreich CH, 644, 649
Frey M, 444, 450, 459, 463
Freyer GA, 46, 155
Fried LM, 159
FRIEDBERG EC, 15–42; 16–18,
    21, 27, 30–32, 34–36, 44,
    46, 52, 57, 58, 60, 64, 67,
    102, 136, 137, 143, 144,
    146, 148, 151, 153, 154,
    159, 160, 171, 197
Frieden EA, 612
Frieden TR, 216
Friedlander DR, 487
Friedlander M, 294
Friedman DS, 786
Friere E, 788
Friesen JD, 248, 391, 793
Friesner RA, 549
Friguet B, 811, 817, 834, 839
Fritz E, 390
Fritz H-J, 102
Fritzsch G, 554, 556
Froelich-Ammon SJ, 636, 649,
    670, 671

Fromaget C, 486
Frötschl R, 752
Früh K, 809, 836, 838, 840
Frutiger S, 20
Frydman B, 748
Frydman J, 817, 825
Frydman L, 748
Frydman V, 748
Fryer GA, 18
Fu H, 263
Fu SM, 294
Fu X-D, 377–81, 383–85
Fuchs BPP, 145
Fuchs JP, 377, 378, 380, 400
Fuchs RPP, 50, 65
Fuh G, 617, 619, 620, 627, 631
Fujii H, 105, 110
Fujiki Y, 588
Fujimori A, 636
Fujimoto T, 821
Fujimura K, 252
Fujimura T, 720
Fujimuro M, 822, 823
Fujinami K, 822, 823, 826
Fujita Y, 294
Fujiwara T, 807, 809
Fujiwara Y, 61, 149
Fujiyama A, 245, 252
Fukada Y, 256, 259
Fukasawa H, 640
Fukata H, 640
Fukumura D, 598
Fukunaga N, 672, 677
Fukunaga R, 617, 619, 620, 627,
    631
Fuller-Pace FV, 398
Fulton SM, 743
Funabiki H, 662
Funk WD, 341, 343, 360, 540
Funnell BE, 198, 199
Furdon PJ, 392
Furfine ES, 247
Furriols M, 328
Furshpan EJ, 485
Furuichi M, 141
Furuichi Y, 701
Furukawa AH, 581
Furuno K, 840
Furuta K, 672
Furuya T, 821
Futai M, 566
Futcher AB, 356, 359
Futscher BW, 116

G

Gabathuler R, 837
Gabrielsen OS, 770
Gaczynska M, 802, 805, 807,
    808, 814, 815, 835–37
Gadelle D, 639
Gafvelin G, 295
Gage DA, 511

Gagnon J, 575
Gainer HSC, 494
Galante YM, 567
Galas S, 20
Galavotti V, 672
Galceran J, 378, 388
Gale KC, 636, 649, 670, 671
Galili G, 695, 708
Galisson F, 392, 394, 395
Gall C, 479
Gall JG, 380, 389
Gallant J, 748, 760
Gallie DR, 706, 708, 711
Gallin WJ, 487, 493
Gallinari P, 105, 109, 111, 112, 120
Gallucci M, 331
Gally JA, 143
Galton VA, 755
Gamache M, 710
Gamberi C, 379, 401
Gamblin S, 652, 653, 678
Game JC, 112, 114, 122, 123, 159
Gamliel D, 538
Gamper H, 23
Ganapathy V, 330
Ganesan R, 339
Ganesan S, 187, 188, 206
Ganesh A, 119, 120, 159
Ganges MB, 150
Gangloff S, 667, 677
Gann AAF, 387
Ganoth D, 804, 825–28, 838
Ganten D, 477
Gao F, 489, 704
Gao M, 507, 516
Gao Q, 148
Garbe T, 227
Garcia AM, 260, 261, 748
Garcia GE, 88, 96, 98
Garcia PD, 289
Garcia-Beato R, 639
García-Blanco MA, 372, 378, 379, 381, 388, 389, 397, 401
Garcia-Moll M, 574, 575
Gardel C, 280
Garman E, 452
Garman EF, 782
Garnepudi VR, 728–30
Garner MM, 647, 651
Garner R, 729
Garnsey SM, 752
Garrard WT, 666, 669, 676
Garrett DS, 612
Gartenberg MR, 659, 661
Garton RL, 312
Gary SL, 67, 151, 646
Gasc AM, 106, 108
Gasch A, 777, 778
Gaskins C, 420, 424, 431
Gaskins R, 420
Gáspár G, 481
Gaspar ML, 714

Gasparini P, 331
Gassenhuber J, 348
Gasser SM, 338, 349, 666, 676, 677
Gatti D, 583, 589
Gatti DL, 583
Gattoni R, 377, 378, 380, 383, 384, 400
Gaur R, 386, 388, 391
Gautam N, 259
Gautier J, 807, 839
Gautier N, 505, 519
Gay DA, 695
Gay JD, 223
Gay NJ, 180
Gazdar A, 120
Gazit E, 762
Gazzola GC, 309
Ge H, 377, 378, 383, 384, 773, 775, 792–94
Geacintov NE, 23
Geelen D, 507, 509–11, 513, 514, 516, 522
Gefter ML, 175, 182, 185, 186, 193, 200, 201, 203
Geider K, 172, 182
Geiduschek EP, 23, 342, 777
Geiger A, 141
Geiger JH, 778
Geiger O, 507, 511, 512, 529
Geisberg J, 791
Geiselmann J, 174, 175, 177–79, 185, 188, 201, 202, 207
Geist ST, 490
Gelb MH, 242–46, 248, 249, 251–54, 258
Gell D, 159, 160
Gellefors P, 583
Geller BL, 294
Gellert M, 159, 636–38, 650, 651, 660, 673, 678
Gelpi C, 757
Gennis R, 538
Gennis RB, 568
Georgakopoulos T, 773
George JW, 170, 172, 176, 196, 197
Georgiev GP, 666
Gerber LD, 327, 331
Gerchman SE, 759, 762
Geremia RA, 507, 509–11, 522
Gergen JP, 706
Gergs U, 477
Gerhardt EM, 709, 710, 729, 730
Gerhold D, 507
Gerisch G, 412, 419, 425–27
Gerke V, 372, 390
Germanas JP, 540, 549, 550, 552, 553, 557, 558
Gernold M, 836, 840
Gerson SL, 46
Gertler A, 619

GESTELAND RF, 741–68; 742–48, 750–54, 758, 759
Getchell TV, 327
Getz GS, 578, 581
Geurts R, 523, 524, 530
Gherardi E, 494
Ghetti A, 378, 388, 389
Ghigo JM, 296
Ghislain M, 827, 828, 830, 832
Ghomashchi F, 244, 245, 248, 249
Ghoshroy S, 482
Giaccia AJ, 159
Giaever G, 636
Giaever GN, 639, 659, 668
Gianfranceschi GL, 676
Giaume C, 491
Gibbs E, 57, 782
Gibbs JB, 242, 247–49, 260, 261
Gibson F, 567
Gibson MG, 623
Gibson TJ, 176, 400, 423
Giedroc DP, 181
Giege R, 748
Gierasch LM, 245, 246, 273, 278, 287
Gierschik P, 262
Giese K, 782
Gil A, 378, 388, 389
Gilad S, 160
Gilbert BA, 253, 254
Gilbert DJ, 307, 315, 317, 477
Gilbert S, 210
Gilbert W, 352, 695, 702
Gileadi O, 20, 21
Gil-Gomez G, 595
Giliani S, 35, 54
Gill G, 775, 791
Gill SC, 179, 185
Gillespie D, 142
Gillette T, 175
Gilley D, 343
Gillham NW, 566, 582
Gillis P, 704
Gilman AG, 257, 259, 421, 423, 435
Gilmore R, 272, 273, 278, 281, 282, 284, 286, 288, 294
Gilmour DS, 661, 795
Gilson E, 338, 349
Gilula NB, 477, 479, 481–86, 494, 497
Gimenez C, 323, 324
Gimenez-Abian JF, 664, 675
Gimeno CJ, 586
Gimlich RL, 478, 485, 487, 489
Giner JL, 254
Ginsburg GT, 416, 418, 419
Ginter CS, 488
Giralado R, 349
Giros B, 248, 249
Girshovich AS, 273
Gitt MA, 444, 458, 469

Giuliani EA, 246, 260, 261
Gladyshev VN, 97
Glaser DA, 54
Glass RS, 85, 89
Glatzer KH, 840
Gleig F, 59, 150
Glenn JS, 259
Glerum DM, 583, 589
Glick BS, 290, 297, 564, 584, 587, 588
Glick GD, 444, 452, 459, 465
Glickman B, 102
Glickman BW, 102
Glickman JF, 245, 257
Glisson BS, 674
Glitz DG, 759
Glomset JA, 241–45, 248, 249, 251, 254, 255, 258, 264
Glover VC, 676, 677
Glowinski J, 491
Glushka JN, 505, 507–13, 516, 519, 520, 523–25, 527, 530
Glynne R, 836
Glynne RJ, 836
Gniadkowski M, 369
Gocayne JD, 280, 639
Gocke E, 643
Godbout R, 159
Godinot N, 307, 323
Godwin AR, 123
Goebl M, 248, 827
Goeckeler J, 291
Goeddel DV, 617, 619, 620, 627, 631
Goehring B, 569, 571, 572, 582, 585, 587, 588
Goerlich D, 294
Goethals K, 507, 516
Goff SP, 754
Goffin V, 619
Goglia F, 586
Gogol EP, 174, 175, 177, 179, 479, 481, 482
Goh KS, 230, 234
Goings GE, 477, 479, 491
Gokhale H, 140
Gold LS, 75, 120
GOLDBERG AL, 801–47; 588, 802–5, 807, 808, 816, 819, 820, 822, 825, 829, 834–36, 838, 839
Goldberg IG, 666, 679
Goldberg JI, 487, 493
Goldbeter A, 430
Goldblatt D, 379, 400
Goldfarb P, 85, 744, 755
Goldie K, 491
Goldmacher VS, 116–18
Goldman E, 744
Goldring A, 395
Goldsmith JS, 66, 152
Goldsmith LJ, 676

Goldstein AL, 702
Goldstein IJ, 444, 450, 459, 469
Goldstein JL, 242, 244–51, 254, 257, 258, 260, 261
Goldstein S, 356, 357
GOLIGER JA, 475–502; 477
Golinowski W, 523, 525, 527
Gollnick F, 419, 430
Gollop R, 416
Gomer RH, 412, 415, 416, 424, 425
Gomes XV, 55, 63
Gonen H, 588, 826, 833
Gong D-W, 773–75, 784
Gonzalez A, 599
Goodall GJ, 703
Goode E, 263
GOODENOUGH DA, 475–502; 476–81, 483, 484, 486, 487, 489, 491, 492
Goodman DS, 242
Goodman H, 377, 378
Goodman JM, 281
Goodman LE, 246, 248
Goodman MF, 52, 71
Goodman MN, 840
Goodrich JA, 21, 22, 26, 33, 35, 57, 58, 64, 775, 791, 792
Goodsell D, 780
Goodwin GW, 244, 245, 248, 249
Goormachtig S, 507, 510, 511
Goosen N, 50, 51
Gopinathan KP, 228
Goralski TJ, 380
Gorbalenya AE, 179, 180, 188
Gorbsky GJ, 660, 667
Gordon C, 827, 828
Gordon EM, 260
Gorelick RJ, 754
Goridis C, 378, 384
Gorini L, 760
Görlich D, 276, 278, 279, 281–84, 286–88, 292, 294
Gorman M, 380
Gorman N, 583, 589
Gorospe M, 703
Gorvel JP, 257
Gossen M, 809, 836, 840
Gossett J, 141
Goswami PC, 674
Goto T, 650–52, 663, 675, 812
Gotoh Y, 834
Gottesdiener KM, 307, 330
Gottesman MM, 262, 296
Gottesman S, 830, 834
Göttfert M, 506, 507
Gottlieb RA, 72
Gottlieb S, 669, 679
Gottlieb TM, 159
Gottschling DE, 341, 343, 345, 347, 349, 350, 358
Goud B, 257

Gould K, 836
Gould KG, 837
Goulian M, 115
Goulmy E, 837
Govindaraju K, 540
Gowans BJ, 151
Gowda S, 752
Goyon C, 122, 123
Gozalbo D, 714, 715
Gozani O, 377, 379, 386, 389, 393, 394, 396, 401
Grabowski PJ, 369, 371, 381–83, 392
Graff DK, 541
Grafi G, 695, 708
Grafstrom RH, 103, 104
Graham IR, 349
Graham SL, 261
Graham SM, 255, 261
Gralla EB, 142, 597
Gramatikoff K, 640
Gramm C, 802, 816, 819, 820, 835, 836
Gramstat A, 752
Granado J, 525, 527, 528
Grange T, 793
Grant E, 802, 814, 835
Grant EP, 836
Grant RA, 185
Graves RA, 695, 712
Graw J, 480
Gray CW, 807, 818
Gray DA, 831
GRAY HB, 537–61; 538, 540, 541, 543, 545–50, 552–54, 557–59
Gray JT, 349
Gray NK, 729
Gray PJ, 23
Graziani R, 618, 621, 622
Graziano V, 762, 786
Green CR, 477
Green DK, 356, 358
Green JM, 176
Green KG, 477
Green LL, 696
Green MR, 368, 369, 372, 374, 378, 380, 386–88, 391, 392, 394, 401, 704, 773, 775, 791–94, 827
Green N, 279
Green NJ, 541
Green PJ, 695
Green S, 595
Greenberg JR, 712
Greenberg ME, 695, 703, 705, 706, 709, 722
Greenberg RB, 148
Greenburg G, 281
Greengard P, 495
Greenwell P, 348
Gregor P, 380
Gregory PE, 92

GREIDER CW, 337–65; 338, 339, 341–44, 351, 352, 356–60
Greiner C, 54
Grenier L, 821
Grenningloh G, 322
Gresshoff PM, 504, 505
Grey LM, 612
Grey SJ, 73
Gridley T, 831
Grieber S, 840
Griffin KJ, 595
Griffin LA, 773, 775
Griffith J, 38, 50, 72, 103, 104, 108, 111
Griffith JD, 160, 200, 645
Griffiths DE, 584
Grifo JA, 171, 180
Grigliatti TA, 67
Grilley M, 103, 104, 108, 110, 112, 113
Grilley MM, 110
Grisselbrecht S, 619
Grivell LA, 564, 566, 574, 577, 578, 580, 582, 585, 586, 593, 704, 745
Grob P, 507
Grodberg J, 619
Groden J, 119, 120, 158, 669
Groen AK, 570
Groen NA, 27
Groettrup M, 808, 836, 838
Grollman AP, 102
Gromova II, 649, 666
Gronenborn AM, 612, 782
Groner B, 697
Gröning K, 378, 393–95
Gronostajski RM, 820, 839
Groot GSP, 575
Groppi VE Jr, 695, 713
Gros DB, 477, 485, 486, 494
Gros F, 695
Gros P, 262
Gross JD, 412
Gross KW, 714, 725
Grosschedl R, 782
Grosset J, 223, 225
Grossi MF, 707, 708
Grossi de Sa MF, 805, 807, 808, 839
Grossman L, 18, 43, 44, 47, 50, 149, 175, 660
Grossman LI, 576
Grötzinger J, 621
Groudine M, 709
Groudinsky O, 580, 583, 718
Gruenbaum Y, 114
Gruenberg J, 257
Grunberger D, 27
Grund C, 805
Grune T, 839
Gruschus JM, 549
Grzeschik K-H, 477

Grziwa A, 805, 807, 808, 812, 813
Gsell B, 95
Gu H, 143
Gu J-R, 37
Gu M, 817
Gu MZ, 817
Guadano-Ferraz A, 307, 321, 322
Guan L, 228
Guan W, 696
Guarente L, 564, 773, 791, 792
Guastella J, 307, 323
Guckert JA, 553
Gudmundsson GH, 289
Guenet JL, 714
Guerin B, 570
Guerin MG, 586
Guermah M, 774, 775, 785, 786, 792
Guerrero C, 224
Guerrier D, 511
Gui JF, 379, 385
Guiard B, 290
Guild GM, 696
Guimaraes MJ, 87, 91, 93
Guisez Y, 622
Guiso N, 426
Guissani A, 598
Gulick T, 595
Gulyas KD, 21
Gumpf DJ, 752
Gumport RI, 650
Gundersen R, 420
Gundersen RE, 421, 431
Gunkel N, 704, 731
Gunyuzlu PL, 620
Guo D, 831
Guo GG, 817
Gupta D, 444, 463, 468
Gupta J, 358, 359
Gupta M, 636, 648, 669
Gupta RC, 749
Gurd JW, 479
Gurney AL, 611, 620
Gustchina A, 612, 615, 623
Guthrie C, 369, 371, 372, 374, 376, 388, 390, 391, 397–401
Gutierrez L, 252
Gutman GA, 103
Gutman L, 236
Guzder S, 18
Guzder SN, 21, 34, 47, 55, 57, 59, 64, 66, 148, 150, 152, 153

H

Ha DB, 805, 829
Ha DC, 507, 509
Ha I, 791, 792
Ha J-H, 173
Haas A, 840

Haas B, 808
Haase W, 315
Haass C, 803, 805, 840
Habash J, 444, 450, 463
Habenicht AJ, 242
Haber BA, 378, 384
Haber JE, 121–28, 148, 775
Haber LT, 103, 110, 112
Habets WJ, 389
Habraken Y, 18, 47, 55, 57, 60, 66, 146, 148, 150, 152
Hackney DD, 175, 210
Hadari T, 826, 832
Hadikusumo RG, 589
Hadlaczky G, 674
Hadobas PA, 529
Hadwiger JA, 420, 421
Haefliger J-A, 477, 489
Haegeman G, 598, 599
Haemers A, 223
Haffter P, 585
Hagan I, 662
Hagan KW, 714–19, 723
Hagen TM, 597
Hagerman PJ, 172
Hagler J, 835
Hahn H, 597
Hahn S, 139, 703, 776–78, 791
Hai T, 794
Haidacher D, 144
Haile DJ, 730
Haire MF, 815, 840
Hajduk SL, 640
Hajibagheri MAN, 146
Hake R, 541
Haklai R, 262
Halay ED, 781, 786, 791
Halboth S, 755
Halenbeck R, 612
Halfant KA, 412
Hall A, 249, 420
Hall AL, 425
Hall JC, 390, 571
Hall NR, 118–20
Hall PA, 151
Hall RM, 612
Hall SL, 369, 376
Halligan BD, 636, 643, 649, 666, 677
Halsey C, 105, 123, 124
Halter R, 753
Haltia T, 568, 573
Halvorson HO, 697
Hamaguchi JR, 349, 665, 675
Hamal A, 638
Hamamoto S, 279, 280
Hamatake RK, 701
Hameister H, 673, 674
Hamel P, 583, 589
Hamill OP, 485
Hamilton AD, 260
Hamilton BJ, 703, 704
Hamilton DL, 660

Hamilton SR, 119, 120
Hamilton WDO, 507
Hammacher A, 618, 621, 622
Hammel J, 374
Hammerling GJ, 836, 837
Hampsey DM, 123
Hampsey M, 714, 715, 718, 722, 776
Hampson EC, 496
Hampson RK, 382, 383
Hamrick MR, 222
Han FS, 661
Han HJ, 105, 112, 113, 307, 323
Han SH, 538
Han Y-H, 669
Hanada M, 278, 292
Hanafusa H, 728, 730
Hanai R, 638, 644, 652, 653
Hanaoka F, 664, 679
Hanawalt PC, 16, 24, 26, 27, 29, 30, 32, 37, 38, 67–70, 145, 146, 171
Hancock JF, 242, 252, 255–58
Hancock R, 662, 665, 666, 670, 676
Handlogten ME, 324, 330
Hanekamp T, 584, 588
Hang B, 140
Haniu M, 762
Hanley-Way S, 293
Hann BC, 274, 292
Hanna RB, 479
Hannah VC, 249
Hannavy K, 290
Hannon GJ, 73
Hansberg W, 598
Hansen D, 431
Hansen KE, 776, 777
Hansen L, 748
Hansen S, 133, 794, 795
Hansen W, 275, 289
Hansma HG, 174, 177
Hansma PK, 174, 177, 483
Hanson JE, 444, 449, 452, 454
Hanson KR, 455
Hanson RJ, 712, 713
Hansson J, 145, 146
Hao HL, 66, 152
Hara M, 260
Hara R, 73
Harada Y, 59
Hardesty B, 759
Harding CV, 820
Harding SE, 817
Hardman KD, 444, 450, 460, 463
Hardy CFJ, 349, 350
Hardy MR, 465
Hardy SF, 369
Hardy SJS, 275
Hare JT, 106, 114
Harford JB, 728, 729
Hargrove JL, 695
Harlan W, 574, 575

Harland RM, 696, 728, 730
Harley CB, 341, 356, 358–60
Harlos K, 614
Harms N, 568, 573
Harness JA, 813
Harney JW, 87, 88, 91, 92, 757
Haro D, 595
Harosh I, 34, 171
Harper F, 674
Harper JE, 383
Harper JW, 72, 73
Harrington JJ, 59, 65, 143, 146
Harrington LA, 341, 342, 344
Harris AL, 485, 674
Harris CC, 37, 72
Harris EB, 236
Harris JR, 804, 807, 822, 823
Harris S, 126, 128, 279, 280, 290
Harrison DJ, 148
Harrison L, 46, 47, 67, 141
Harrison PP, 85
Harrison SC, 444, 452, 453, 466, 652, 653, 656, 678
Harrop SJ, 444, 450, 463
Harry A, 423
Hart PA, 694, 695, 700, 709, 712
Harteveld M, 507, 511
Hartig PR, 307, 323
Hartl FU, 280, 292–94
Hartley KO, 159, 160
Hartmann E, 276, 278, 279, 281, 282, 286, 288, 289, 292, 294, 295, 297
Hartmann R, 567
Hartmuth K, 401
Hartog K, 703
Hartog MV, 529
Hartwell L, 347
Hartwell LH, 356
Hartwig J, 257
Harvey TS, 264
Harvey WR, 307
Harwood AJ, 426
Hase T, 584
Hasegawa S, 773, 775
Hashimoto S, 791
Hashimoto T, 704
Hassan AB, 73, 151
Hassell AM, 612, 622
Hastie ND, 352, 356–58, 827, 828
Hastings RC, 236
Hata M, 818
Hata S, 526
Hata Y, 250, 259
Hatahet Z, 141
Hatat D, 583
Hatefi Y, 567, 568
Hatfield D, 84, 86
Hatfield DL, 84, 86, 754
Hatfull GF, 222, 235
Hattman S, 114
Hatzoglou M, 309–12
Hauri HP, 297, 393, 396

Hauser CR, 582
Haussinger D, 307
Havel A, 233
Havel C, 251, 252
Havel CM, 259
Haverkamp J, 505, 507, 511, 527
Hawkridge FM, 546
Hawley DK, 782
Hawn MT, 117
Hayashi SI, 720, 746, 747, 812, 821, 832, 838
Hayes J, 788
Hayflick L, 356
Haynes BF, 330
Haynes H, 418
Haynes M, 378, 393, 394, 396, 398
Haynes R, 288
Hays JB, 118, 122
Hays S, 52, 71
Hays SL, 160
Hazoor-Akbar, 254, 261, 262
He B, 246
He D, 402
He F, 695, 714, 715, 718–21, 726
He SH, 97
He Y, 746
He ZG, 55, 57, 63, 65, 149, 153
Heard CJ, 115
Hearst JE, 23, 47, 50–52, 67, 113, 145, 146, 659
Heath-Pagliuso S, 669
Heaton B, 698, 703
Hebert SC, 316, 317
Hecht S, 636
Heck MMS, 666, 672, 674, 676
Heckman JE, 763
Heddi A, 566
Hederstedt L, 589
Hedgecock CJR, 615, 623
Hediger MA, 307, 314–18, 320, 321, 327, 328, 330, 331
Hedjran F, 717
Hedley ML, 383, 390, 717
Heemels M-T, 835, 837
Heerze LD, 445, 468
Hees B, 315
Hegarat FG, 595
Hegde AN, 803
Hegerl R, 805, 811
Heider J, 84, 85, 88, 92, 755, 756, 762
Heidstra R, 515, 523–25, 528–30
Heifets LB, 217, 221, 223, 225, 232, 233
Heikoop JC, 318
Heiland S, 673, 674
Heimer E, 327, 328
Heinemeyer W, 802, 804–6, 808
Heinrich R, 275, 570
Heinrichs V, 383
Heintz N, 712
Heitmann K, 315

Heizer C, 419
Hejna JA, 668
Helfman DM, 380, 383, 392
Helland DE, 141
Helle NM, 139
Hellen CU, 389
Heller BA, 538
Heller H, 833
Helliwell JR, 444, 450, 463
Helmann JD, 771
Hemler ME, 330
Hemminki A, 119, 120
Henderson DS, 67, 479, 491
Henderson ER, 338, 341, 348, 352
Hendil KB, 805, 807–9, 815, 836
Hendrick JP, 276, 280, 292
Hendrick LL, 341, 343
Hendrickson WA, 444, 451, 453, 456, 458, 462, 467, 614
Hengartner C, 822, 828
Hengel H, 836, 837
Henics T, 703
Henkel T, 598
Hennecke F, 102
Hennecke H, 507
Hennemann H, 477, 485
Henner DJ, 629
Henner WD, 141
Hennikoff S, 344
Henning KA, 34, 36, 68, 69
Henning SJ, 309, 310, 312, 326
Henning-Chubb C, 674
Henricksen LA, 55, 57, 63, 65, 74, 149, 153
Henrissat B, 509
Henry LH, 19, 21
Henry NL, 775, 792
Henry R, 275, 280
Henry YA, 349, 598
Hensel I, 230
Hentze MW, 694, 695, 704, 709, 710, 729, 731
Henzel WJ, 611, 620
Henzi VA, 307, 321, 322, 331
Herbst KL, 759
Hercus TR, 622
Hereld D, 429
Heringa J, 400
Herman G, 102, 103
Herman GE, 102
Hernandez N, 369, 791
Hernandez RR, 59
Herold J, 751
Herrera RE, 661
Herrick D, 694–96, 709, 726
Herrick DJ, 703, 728
Herrlich P, 703, 709
Herrmann JM, 587
Hershko A, 804, 821, 825–28, 832, 833, 838
Herskowitz I, 246

Hertzberg EL, 477, 479, 482, 488, 492, 494, 495
Hertzberg R, 636
Hervas C, 751
Herz J, 275, 278
Herzberg O, 444, 448, 458, 468
Herzer T, 329, 330
Hess D, 580
Hess P, 116, 117, 120
Hesse J, 159
Hester G, 444, 450, 459, 469
Heuertz RM, 582
Heumann H, 19
Heuser J, 477, 479, 483
Heusser C, 623
Heyer WD, 159, 701, 702
Heym B, 216, 227
Heyman RA, 595
Heynkes R, 477, 485
Hiasa H, 657, 658, 663
Hiatt H, 695
Hibdon S, 547
Hibi M, 620, 621
Hick E, 490
Hicke BJ, 349
Hicke L, 821
Hicks JB, 339
Hicks PS, 185
Hickson ID, 142, 636, 668, 669, 674–77
Hidalgo MC, 814, 815, 821
Hieter P, 126
Higgins CF, 296, 660
Higgins D, 180, 192
Higgins JB, 259, 431
Higgins NP, 645, 646, 650, 672, 677
High S, 272, 273, 275, 287, 288, 294
Higuchi I, 717
Higuti T, 840
Hii CST, 496
Hilf ME, 752
Hill CP, 612, 814
Hill KE, 92, 758
Hill MG, 540, 544
Hill TL, 171, 185, 200, 201, 203
Hilliker AJ, 125, 127
Hilt W, 588, 802, 803, 806, 808, 814, 840
Hilton DJ, 619
Hilvert D, 94, 95
Himeno H, 760
Himmelspach M, 383
Hinds M, 674
Hindsgaul O, 507, 510, 512, 513, 515, 522, 528
Hines JC, 640
Hines JV, 750
Hingorani MM, 174–76, 180, 181, 184, 185, 188, 189, 192, 207
Hinnebusch AG, 707

Hinterberger M, 372
Hiraga S, 638, 657, 660, 663
Hirai Y, 359
Hirano M, 576
Hirano T, 620, 666, 667
Hiriyanna KT, 217
Hirling H, 171
Hirokawa N, 477, 479, 483
Hirose S, 679
Hirsch AM, 518, 526
Hirsch HH, 703
Hirsch KS, 358, 359
Hirschfeld S, 149
Hirschi KK, 491
Hirt H, 526
Hirte HW, 358
Hisaka MM, 255
Hisamatsu H, 809, 818, 819, 836
Hisatake K, 774, 775, 781, 785, 786, 791, 792
Hittelman WN, 674, 676
Hiyama E, 359
Hiyama K, 359
Ho YK, 245, 247, 251
Hoben P, 275, 278
Hochhauser D, 674
Hochstrasser M, 802, 803, 809, 816, 832, 833
Hockenbery DM, 597
Hodge MR, 574, 576
Hodges PE, 379, 397
Hodges PK, 674
Hodgkin J, 718
Hodgman TC, 179
Hoeijmakers JHJ, 16, 18, 19, 29–31, 36, 43, 44, 54, 58, 68, 69
Hoess RH, 104
Hoey T, 773, 775, 791, 792
Hoffman BE, 381, 382
Hoffman BJ, 307, 323
Hoffman BM, 543, 546
Hoffman DW, 759
Hoffman E, 311
Hoffman EK, 820
Hoffman L, 802, 823, 825
Hoffman NE, 275
Hoffman RJC, 623
Hoffmann A, 672, 773, 775, 777, 778, 785, 786, 788
Hoffmann HM, 382, 383
Hoffmann-Berling H, 171, 172, 182, 200
Hoffner SE, 221
Hofhaus G, 568
Hofhuis FMA, 161
Hofmann GA, 639, 674, 675
Hofmann JFX, 349, 666
Hofmann K, 307, 314, 317, 320
Hofmann MW, 283, 287, 295
Hofsteenge J, 704
Hofstetter H, 743
Hogle JM, 94
Hoh JH, 477, 483

Hohmann S, 714, 715
Hol WGJ, 445, 448, 452, 459, 463, 464, 467, 468
Holden JA, 652, 674
Holland IB, 296
Holliday RA, 122
Hollingshead CJ, 479, 481
Hollingsworth NM, 105, 123, 124
Hollingsworth RI, 505, 511–14, 518
Holloman WK, 652
Hollstein MC, 140
Holm C, 636, 662–64, 675, 679
Holmes J, 107, 108
Holsters M, 505, 507, 508, 510, 511, 514, 516, 519, 522, 523, 527
Holten D, 538
Holthuizen PE, 728, 729
Holtz D, 257
Homma M, 672
Honda M, 669
Hong KH, 583, 589
Hong SO, 805
Honore B, 819
Honore N, 216, 219, 221, 222, 227
Hood DA, 574, 575
Hood LE, 477
Hoopes BC, 782
Hopfield JJ, 547, 548, 549
Hopper AK, 581
Hopper NA, 426
Horabin JI, 717
Hori H, 589
Hori Y, 258
Horii A, 105, 112, 113
Horikoshi M, 772–75, 780, 784–86, 791, 792, 794
Horiuchi H, 245, 250
Horiuchi J, 773, 792
Horiuchi T, 636, 678
Hornig H, 391
Horowitz DS, 383, 390, 401, 647
Horowitz JA, 728–30
Horsburgh B, 761
Horsick RA, 620
Horvath B, 515, 523, 525, 528
Horwich AL, 802, 829
Horzinek MC, 295
Hoshijima K, 717
Hoshino K, 662, 663
Hosick TJ, 242
Hoskins J, 830, 834
Hosobuchi M, 276
Hosokawa Y, 591
Hotz-Wagenblatt A, 493
Houbavi C, 395
Houderine L-M, 619
Hough PVC, 174, 177, 179
Hough R, 804, 813, 822, 838
Houghton R, 477, 482, 486
House KL, 94, 95

Housman DE, 61, 695, 703, 705
Houtchens K, 339
Howald WN, 263
Howard GA, 495
Howard GC, 661
Howard MT, 645, 671
Howe MM, 645, 672
Howell AJ, 650, 651
Howell M, 523
Howell N, 580
Howitt SM, 567
Howlett GJ, 618, 621, 622
Hoy CA, 60, 148
Hoyle DH, 840
Hoyt DW, 287
Hruby DE, 706
Hrycyna CA, 253–55
Hsiang Y-H, 636, 674
Hsieh T-S, 636, 644, 645, 648, 650, 676, 679
Hsu CL, 697, 698, 702
Hsu DS, 33, 45, 47, 49–51, 54, 55, 57, 58, 60, 63, 65, 66, 145, 149, 152
Hsuan J, 111
Hu J, 495
Hu PY, 762
Hu SH, 777, 778
Hua J, 729
Huang AS, 696
Huang E-S, 675
Huang HC, 670, 810, 829
Huang HV, 125, 126, 128
Huang JC, 19, 31, 33, 35, 47, 49, 54, 58, 61, 64, 65, 67, 145, 146, 150
Huang JJ, 821
Huang L, 143
Huang L-Y, 704
Huang Q, 594, 595
Huang WM, 224, 636, 639, 662, 676, 758
Huang Y, 832
Huang YM, 636, 639
Huang ZH, 511
Huarte J, 706
Hubac C, 511
Huber AH, 615
Huber R, 638, 805, 807, 809, 811, 813, 814, 816, 834
Huberman E, 674
Hubert E, 696, 697
Hubert N, 757
Hubscher U, 66, 149, 151, 158, 170, 172
Hudson K, 840
Huesca M, 707, 708, 839
Huet J, 349, 780
Huez G, 695–97, 708
Huez GA, 695, 708
Hughes C, 296
Hughes EN, 61, 837
Hughes GJ, 20

Hughes J, 415
Hughes JE, 416
Hughes MJ, 105, 109, 111, 112, 120
Hügle B, 805
Huh GS, 717
Hui JO, 762
Hui YZ, 522
Huizenga BN, 649
Huizinga JD, 477
Hulser D, 492
Hultner ML, 36
Humbert S, 19, 30–32, 54, 147, 150
Hummel CW, 521, 528
Hundal HS, 312
Hung J, 120
Hunt CR, 674
Hunt J, 280
Hunt T, 20
Hunting DJ, 151
Hupp TR, 72
Hurley JB, 257
Hurley L, 782
Hurwitz J, 55, 57, 63, 74, 115, 143, 149, 171, 174–77, 179, 207, 664, 782
Husain I, 67, 158
Hutchinson CR, 511
Hutchinson F, 102
Huth J, 782
Huynh PD, 312
Hwang BJ, 61, 149
Hwang CB, 761
Hwang J, 673–75
Hwang JL, 674
Hwang S-PL, 728
Hwong C-L, 673, 674
Hynes NE, 697
Hynes RO, 717

I

Iaccarino I, 105, 109, 111, 112, 120
Iannaccone PM, 312
Ibeanu GC, 140
Ibrahimi I, 273
Ichiba T, 746
Ichihara A, 802–4, 807, 813, 816, 822, 823, 825, 826, 839
Ichimura T, 282, 284
Icho T, 720
Ide H, 141
Idriss S, 380
Igarashi K, 832, 838
Igel AH, 343
Igel H, 378, 393, 394, 396, 398
Ihara M, 45, 154, 155
I'Heureux F, 256
Iizuka N, 706
Ikai A, 805, 822, 823

Ikeda H, 636, 638, 662, 663, 669, 671, 702
Ikeda JE, 200
Ikeda M, 669
Ikeshita S, 521–24, 527, 530
Ikura M, 264, 783
Imai H, 672
Imamura R, 638, 657, 660
Imbalzano AN, 791
Imboden P, 222
Imler J-L, 623
Impellizzeri KJ, 67
Inaba T, 93
Inamine JM, 218
Ingebristen TS, 676, 678
Ingledue TC, 712
Ingles CJ, 55, 57, 63, 65, 149, 153
Inglese J, 245, 257, 423, 424, 431
Inglis SC, 750, 751
Iniguez-Lluhi Ja, 259, 421
Inman RB, 125
Inon de Iannino N, 507–10
Inoue H, 154, 155, 157
Inoue K, 307, 314, 317, 401, 717
Inouye M, 88, 743, 744
Insall R, 419, 420, 423, 424, 431
Insall RH, 419, 424
Interthal H, 702
Iobst ST, 454, 462
Ionov Y, 118–20
Iordanescu S, 222
Iqbal M, 820, 821
Ireland DC, 381
Irie K, 827
Isaya G, 584
Isham KR, 702
Ishibashi Y, 242
Ishida H, 119
Ishida N, 840
Ishida R, 662–65, 675
Ishida S, 415, 427
Ishii C, 18, 46, 155, 157
Ishii H, 598
Ishimi Y, 115, 143, 662, 664, 675
Ishimoto I, 317, 318
Ishizaka-Ikeda E, 619
Ishizaki K, 45, 73, 154
Isied SS, 540, 545
Isogai A, 242
Isomura M, 258, 259
Ison C, 662
Issartel JP, 296
Issemann I, 595
Issner R, 402
Ito A, 297
Ito K, 272, 276, 280, 294
Ito N, 377
Ito T, 54
Itoh T, 636
Ivanov EL, 148
Iwai Y, 704
Iwamatsu A, 834
Iwata S, 538, 547, 559

Iyer N, 32, 34, 36, 68, 69, 153
Iype LE, 125
Izaurralde E, 379, 401
Izban MG, 23

J

Jabbouri S, 507, 510, 513, 514, 518, 519, 525
Jacken J, 54
Jackett PS, 228
Jackman MP, 540
Jacks T, 72, 744, 749–51
Jackson AP, 650, 651
Jackson DA, 73, 151
Jackson IJ, 369
Jackson JH, 257
Jackson KW, 22
Jackson RC, 278
Jackson RJ, 711
Jackson SP, 379, 397
Jacob F, 695
Jacob WA, 598
Jacobs BA, 540
Jacobs MR, 223
Jacobs WR Jr, 227, 228, 230, 235
Jacobsen-Kram D, 669
Jacobs-Lorena M, 694, 695
JACOBSON A, 693–739; 694–96, 698, 700, 703, 706, 707, 709, 711, 714, 715, 717–20, 723, 724, 726
Jacobson KA, 434
Jacq A, 280
Jacq X, 775, 791
Jacquier A, 144
Jacquinet JC, 522, 528
Jaeger J, 444, 469
Jaehning JA, 574, 578, 579, 591, 592
Jaenisch R, 311
Jaffe SH, 487
Jain R, 415
Jakes S, 678
Jalajakumari MB, 753
Jalal F, 329
Jalife J, 489
Jalink K, 415
James BD, 343
James GL, 245, 249, 257, 260, 261
James JH, 840
James MN, 460
Jameson SC, 837
Jamison SF, 372, 378, 381, 383, 388–90, 399
Jamrich M, 531
Janakiraman MN, 460, 469
Jancarik J, 612
Janetzky B, 281
Jang GF, 252, 253
Jang J, 584, 588
Jang SH, 574, 578, 579, 592

Janin J, 788
Jankowska M, 379, 401
Jansen M, 728
Jansen-Dürr P, 389
Jantti J, 297
Jap B, 805, 807, 811, 834
Jarausch J, 567
Jariel-Encontre I, 832, 838
Jarlier V, 224
Jarlier VL, 236
Jarlinski SJ, 674
Jarmolowski A, 379, 401
Jarnagin K, 575
Jarry T, 485
Järvinen H, 119, 120
Jaskunas SR, 669
Jaspers NGJ, 54, 147, 150, 159, 160
Jaspers-Dekker I, 21
Jastorff B, 419, 434
Jautzke G, 234
Javaherian K, 650, 661
Jaxel C, 638, 643
Jay E, 751
Jayaraman L, 38, 72
Jean-Bart E, 570
Jean-Francois MJB, 589
Jeanteur P, 385
Jefferies WA, 837
Jeffrey PD, 782
Jeggo P, 671
Jeggo PA, 16, 159, 671
Jen J, 105, 110, 119, 120, 146
Jen P, 704
Jencks WP, 171, 200, 201
Jeng M-F, 545
Jenkins JR, 639, 671, 675
Jenkins NA, 105, 110, 119, 307, 315, 317, 477
Jenner AJ, 750, 751
Jensen C, 831
Jensen ON, 92
Jensen PS, 649
Jensen RE, 590
Jensen TJ, 820, 821
Jentoft JE, 400
Jentsch S, 276, 278, 292, 802, 803
Jermyn KA, 424
Jesaitis LA, 477, 483
Jeserich G, 316
Jesior JC, 481
Jessberger R, 158
Jessee CB, 66, 67, 660
Jezewska MJ, 175, 179, 184, 185, 189
Ji B, 223, 225
Jia YW, 546
Jiang JX, 476, 478, 492
Jiang Y, 251, 252
Jiang YJ, 714, 725
Jiang YQ, 66, 152
Jimenez SA, 672
Jin J, 67

Jin SH, 218
Jiricny J, 102–6, 108, 111, 112, 120, 141–43
Johansen RF, 140
John M, 507, 510, 515, 516, 522, 528, 529
John SA, 477, 486
Johnsen A, 807
Johnson AE, 272, 273, 283, 286–89, 295
Johnson AP, 159
Johnson AW, 142, 701, 702
Johnson DI, 248
Johnson DK, 146
Johnson EF, 595
Johnson JW, 322
Johnson KA, 191, 210, 280
Johnson LN, 452
Johnson M, 254
Johnson PF, 62
Johnson RE, 115, 144, 416, 419
Johnson RG, 487, 491, 493
Johnson RK, 666, 674
Johnson RL, 416, 418–20, 434
Johnson RT, 24, 26, 70, 636, 664, 665, 671, 672
Johnsson K, 225, 227, 228, 230
Johnsson N, 275
Johnston AWB, 504, 507, 510, 511, 516, 524
Johnston LH, 379, 397
Johnston MF, 491
Johnston SA, 828
Johzuka K, 160
Joliot V, 791, 792
Joly JC, 292
Jonczyk P, 657
Jones CJ, 55, 63, 73, 145, 148, 418
Jones DS, 754
Jones EY, 614
Jones HD, 591
Jones J, 529
Jones KR, 390
Jones M, 116
Jones ME, 840
Jones MH, 401
Jones MM, 148
Jones NJ, 153, 158, 307, 330, 599
Jones PG, 673
Jones T, 639
Jones TR, 703
Jongen WMF, 487
Jongsma HJ, 491, 492, 495
Jongstra-Bilen J, 677
Jonsen M, 830, 834
Jordan JR, 478, 485
Jordan KD, 548
Jordan SR, 612, 622
Joshi M, 839
Journet E-P, 523, 530
Jovanovich SB, 673
Jovin TM, 647, 651

Ju QD, 661
Judd GG, 746
Judd SR, 246, 248
Julin DA, 180, 192, 201
June CH, 704
Jung J-E, 98, 757
Jung M, 160
Jung W, 85, 89
Jungbluth S, 477
JUNGNICKEL B, 271–303; 273, 284, 294
Junke N, 759
Jurkevitch E, 526
Jursky F, 325, 326

K

Kabat D, 308, 313
Kabnick KS, 695, 703, 705
Kacser H, 570
Kadenbach B, 567, 570, 575, 576, 586
Kadonaga J, 19, 23
Kadowaki T, 574, 575, 695, 704, 731
Kafatos F, 694
Kafetzopoulos D, 510
Kagawa S, 805, 809, 822, 826, 828, 831, 836, 840
Kahana C, 746, 747, 826, 833, 838
Kahn R, 145, 146
Kaibuchi K, 250, 258
Kaina B, 45, 137
Kaiser II, 86
Kaiser K, 792, 793
Kaiserman HB, 676
Kaiser-Rogers KA, 52, 170, 172, 182, 190
Kakidani H, 794
Kakuda DK, 307, 309, 312, 314
Kakuma T, 141
Kaky H, 444, 450, 459, 469
Kalb AJ, 444, 463
Kalies KU, 276, 278, 281, 282, 284, 286, 288
Kalk KH, 445, 452, 459
Kallenius G, 221
Kamakaka RT, 19, 23
Kamalaprija P, 507, 510, 513, 514, 518
Kameji T, 746, 832, 838
Kamen R, 703, 709
Kameyama K, 805, 822
Kamiya Y, 242
Kamst E, 507, 509–11
Kanaar R, 378, 387
Kanai Y, 307, 314–18, 320, 327, 328, 330
Kanamori A, 92, 93, 98
Kanamoto R, 746
Kanayama HO, 822, 826, 840
Kandror O, 834

Kane CM, 22, 69
Kane DJ, 597
Kane EM, 647
Kane JF, 758, 762
Kane MF, 118–20
Kanemitsu MY, 492–94, 496
Kanes W, 221
Kang MS, 805
Kang SS, 805
Kangawa K, 379, 401
Kania M, 818
Kanner BI, 315–17, 319
Kanter HL, 477
Kanwar YS, 281
Kao-Shen C-S, 669
Kapadia G, 444, 448, 458, 468
Kapp LN, 160
Kaptain S, 729
Kapur V, 222–24
Karantza V, 788
Karaoglu D, 281, 282
Karas JL, 540
Karas M, 97
Karase AV, 752
Karcagi V, 150
Karchi M, 427
Karimpour I, 94
Karin M, 72
Kariya K, 621
Karlsson BG, 553, 554, 557
Karlsson S, 727
Karmas A, 372
Karnovsky MJ, 478
Karoor V, 98, 757
Karp JE, 674
Karpinski BA, 330
Karpishin TB, 553–57
Karplus M, 549
Karplus PA, 612, 613
Karran P, 116–18, 136, 137
Karsenti E, 695, 708
Kartenbeck J, 200
Kas E, 636, 645, 666
Kasai H, 46, 141
Kasaka T, 226
Kasch LM, 762
Kashiwagi K, 746
Kashuba E, 72
Kasic T, 659
Kasid UN, 677
Kasperek E, 832
Kassavetis GA, 23, 342, 777
Kastan MB, 72, 356, 674
Kastelein RA, 622
Kastner B, 371, 392, 393
Kat A, 109, 117, 118, 137
Kataoka H, 61, 149
Kataoka N, 379, 401
Kataoka T, 258
Katayama M, 245, 250, 258
Katayama Y, 809
Kato J, 638, 657, 660, 662, 663
Kato J-I, 662, 663

Kato K, 255
Kato M, 294
Kato S, 598
Kato T Jr, 45, 154
Katoh I, 754
Katz DA, 704
Katz FN, 294
Katz G, 331
Katz RA, 400
Katzin B, 444
Kauer JC, 820, 821
Käufer NF, 385
Kaufmann J, 794
Kaufmann SH, 671, 674
Kaufmann WK, 54, 70
Kavanaugh MP, 307, 308, 312–14, 316, 318–20
Kavli B, 139, 140
Kawabata S, 141
Kawahara H, 807, 839, 840
Kawakami H, 307, 314, 317
Kawakami K, 749
Kawamura Y, 840
Kawasaki I, 669
Kawashima S, 817
Kawata M, 245, 250, 254
Kay BK, 712
Kay RR, 412, 426
Kayano T, 488
Kayman SC, 412
Kazantsev A, 47, 49, 61, 65, 74, 145
Kazazian HH Jr, 727, 762
Kaziro Y, 185, 246
Kearsey SE, 701
Keegstra W, 811
Keene JD, 704
Keeney S, 61, 149, 150
Kehl M, 805
Keiler KC, 760, 761
Keith G, 808
Kellaris KV, 288
Kelleher DJ, 281, 282
Keller RK, 263
Keller W, 369, 371, 372, 392
Kelley RF, 614
Kelly A, 836
Kelly C, 309, 313
Kelly DP, 595
Kelly GD, 216, 221, 233, 234
Kelly K, 695
Kelly PA, 619
Kelly TJ, 55, 74, 192, 658, 664
Kelly WG, 652
Kelm S, 465
Kelman Z, 52, 73, 151
Kemper B, 580
Kenna M, 702
Kennedy EP, 507, 508, 511
Kennedy MC, 729
Kenny B, 296
Kenny MKR, 57, 63, 66, 146, 149, 151

Kensler RW, 479, 491
Kent SBH, 477
Kepes F, 276, 278
Keramidaris E, 583
Keranen S, 297
Kerem B, 762
Kerppola TK, 22, 69
Kerr LA, 836
Kerrigan D, 643, 645, 646, 674
Kesbeke F, 420, 421, 423, 425, 430
Keske JM, 549, 553, 557
Kessel M, 811, 823
Kessin RH, 425, 426, 431
Kessler JA, 485
Kessler M, 575
Kessler O, 714, 725
Keyomarsi K, 72, 73
Keyse SM, 145, 153
Kezdy KE, 678
Khan R, 181
Khangulov SV, 97
Khodursky AB, 658, 662
Khosravi-Far R, 244, 250, 260
Khuller GK, 231
Kiang DT, 495
Kieff FS, 669
Kieffer JD, 88, 757
Kieser T, 235
Kigawa T, 756
Kijne JW, 504, 522, 525, 526, 529
Kikkawa U, 494
Kikuchi A, 258, 259, 638, 674
Kikuchi S, 226
KILBERG MS, 305–36; 307, 310–12, 316, 317, 319, 320, 329
Kilburn JO, 216, 221, 231, 232
Kiledjian M, 389, 727, 728
Kilkarni SJ, 390
Kim B, 143, 811, 823
Kim C, 67
Kim DK, 55, 153
Kim E, 140
Kim IY, 85, 87, 90, 91
Kim J, 61, 141, 402
Kim J-K, 145
Kim JL, 776, 778, 780, 781, 794
Kim JM, 701, 702
Kim JW, 307, 308, 310, 311, 313
Kim JY, 419, 429, 434
Kim K, 143
Kim KI, 829
Kim K-M, 307, 323
Kim NK, 121
Kim NW, 341, 358, 359
Kim R, 242, 255
Kim RA, 636, 637, 639, 641, 658, 661, 667–69
Kim S-H, 180, 242, 255, 398, 399, 612, 783, 829, 832
Kim S-J, 121

Kim SS, 751
Kim ST, 45, 50, 51, 154, 155
Kim Y, 143
Kim YC, 778
Kim Y-J, 293, 294, 378, 380, 383, 384, 390, 773, 775, 828
Kimble J, 706
Kimmel AR, 416, 418, 419, 434
Kimmerly WJ, 349
Kimura E, 293
Kimura K, 674, 676
Kimura Y, 96
King A, 50, 51
King B, 217, 235
King BC, 546
King CH, 235
King DS, 225, 228, 230, 252, 253, 399
King MP, 581
King PH, 704
King SR, 669
Kingma PS, 649
Kingsbury DW, 696
Kingsman AJ, 349, 704, 731, 743, 749
Kingsman SM, 349, 704, 731, 743, 749, 751
Kingsmore SF, 307, 323
Kingston J, 386, 392
Kingston RE, 791
Kinney D, 222
Kinniburgh AJ, 696, 703, 714, 715, 725
Kinsella BT, 244, 250
Kinzler KW, 118
Kipling D, 352, 357, 701
Kirchgessner CU, 159
Kirchhoff S, 155
Kirk-Bell S, 69
Kirkegaard K, 641, 647, 650
Kirkness EF, 639, 810
Kirmaier C, 538
Kirschner MA, 307, 315, 317
Kirschner P, 219
Kishimoto A, 494
Kishimoto T, 620
Kislauskis EH, 711, 731
Kisselev A, 813–15
Kisselev O, 259
Kistler J, 477, 485, 491
Kitabatake M, 760
Kitada K, 701
Kitagawa T, 538
Kitagawa Y, 574, 575
Kitamoto M, 358–60
Kitamura A, 662, 663
Kitamura T, 622
Kitani K, 839
Kitayama H, 425
Kiwon J, 640
Kiyosawa H, 416
Kiyosawa K, 672
Kjeldsen E, 643, 676

Klages N, 793
Klahre U, 369
Klarskov K, 507, 511
Klausner RD, 728, 729, 730
Kleckner N, 122
Kleeman J, 312
Kleene KC, 94
Kleijzer W, 53
Klein A, 97, 755
Klein C, 281
Klein F, 349
Klein HL, 669
Klein M, 673, 674
Klein PS, 416
Klein U, 840
Klein VA, 640
Kleinschmidt JA, 804, 805, 807,
    816, 822, 823, 825, 839
Kleinz J, 807, 813
Klement JF, 580
Klemm R, 775, 791, 792
Klett C, 673, 674
Klevan L, 650
Kliewer SA, 594, 595
Klingelhutz AJ, 356, 358
Klobutcher LA, 351
Kloeckener-Gruissem B, 565,
    582, 583, 585, 589
Kloetzel PM, 803, 805, 808, 809,
    836, 837, 840
Klonowska MM, 175, 179, 188,
    189, 207
Kloog Y, 262
Klopman G, 223
Klug A, 788
Klungland A, 140, 144
Kmiecik TC, 493
Knab AM, 672
Knauer R, 281, 282
Knecht E, 802
Knetsch MLW, 425
Knighton DR, 264
Knippers R, 171, 175, 658, 664,
    673, 674
Knoll G, 419, 430
Knowles JR, 465
Knox BE, 430
Knudsen BR, 648
Knudsen KA, 477
Ko C, 565, 576, 589, 591
Kobayashi GS, 598
Kobayashi M, 679
Kobayashi N, 791
Kobayashi R, 151, 390, 777
Kobayashi T, 55, 57, 63, 149,
    153, 154
Kobe SS, 493
Koblan KS, 246
Kobs G, 696, 713
Koch BD, 275
Koch G, 701
Koch TH, 349
Koch WJ, 257, 423

Kochera M, 121
Koch-Weser D, 226
Kocsis E, 811, 823
Kodadek T, 197, 198, 668
Kodo N, 148, 153
Koehler K, 803, 808
Koeller DM, 709, 710, 728–30
Koepsell H, 315
Koerner TJ, 583, 589
Kofod M, 728–30
Koga S, 595
Koga Y, 576
Kohl NE, 245, 248, 249, 260, 261
Köhler G, 378, 384, 714
Kohler K, 808
Kohlhagen G, 643
Kohn KW, 636, 642, 643, 645,
    646, 666, 676
Kohn SR, 118
Kohtz JD, 381, 383, 390, 399
Koi M, 117
Koide T, 812
Koidl S, 587
Kokame K, 256
Koken MHM, 21
Kokubo T, 773
Kolaitukudy PE, 226
Kolatkar A, 444, 453, 454, 457,
    462, 463
Kolb JM, 122, 124
Kole R, 379, 382, 392, 397, 398,
    701
Koleske AJ, 19, 153, 776, 792
Kollander R, 495
Koller T, 658
Kolmar H, 102
Kolodner RD, 102, 104–8, 110–
    16, 118–26, 128, 159, 160,
    701, 702
Kolter R, 296
Komaromy M, 314
Kominami K, 830
Komine Y, 760
Konarska MM, 369, 372, 374,
    376, 392, 396, 401
Kondo S, 151
Kondorosi A, 506, 507, 509, 511,
    513, 518, 519, 524, 525,
    527, 530
Kondorosi E, 505–8, 510, 512,
    515, 519, 522–25, 527, 528,
    530
Konforti BB, 376
Kong X-P, 151, 646
Konijn TM, 412
Konings WN, 314, 315
Konno K, 232
Kono T, 623
Koo H-S, 659, 660
Koonin EV, 179, 180, 188, 594,
    718, 752
Kopchick JJ, 627
Kopcinska J, 523, 525, 527

Kopito RR, 820
Kopka ML, 780
Kopp F, 803, 805, 807, 816
Kopylov VM, 638
Korangy F, 180, 192, 201
Kordel J, 636, 678
Kordiyak GJ, 678
Kornberg A, 174, 181–83, 185,
    186, 188–90, 198–201, 205,
    656, 751
Kornberg RD, 19–21, 30–32, 34,
    58, 64, 349, 481, 698, 775–
    77, 792, 828
Kornblihtt AR, 382
Kornfeld K, 456
Kornfeld R, 456
Kornhauser R, 254, 261, 262
Kornstädt U, 385
Koronakis V, 296
Korsmeyer SJ, 597
Korte A, 583
Koseleva E, 72
KOSHLAND DE JR, 1–13
Kossiakoff AA, 612, 614, 615,
    617, 619, 627, 628
Kost TA, 246
Koster AJ, 805, 806, 808, 812,
    822
Kostic NM, 540
Kostka S, 278, 282, 289, 808, 836
Kosuge K, 821
Koszinowski UH, 836, 837
Kothari SL, 529
Koths K, 612
Kouchi H, 522, 526, 530
Kovalsky OI, 638
Kovvali GK, 115, 144
Kow YW, 49, 141
Kowalczyk JJ, 260
Kowalczykowski SC, 197–99,
    201
Kowalska-Loth B, 676
Koybayashi R, 341, 344
Kozak CA, 307–9
Kozak D, 377, 378, 380, 383
Kozak M, 715
Kozik A, 515, 525, 528
Kozik M, 541
Koziolkiewicz MJ, 376
Kozyavkin SA, 637, 638, 641,
    678
Kraal B, 748
Kraemer KH, 53, 60, 145
Kraft R, 282, 808, 836
Krah R, 638, 673, 678
Krainer AR, 369, 372, 377, 378,
    380–84, 389, 390, 392,
    401
Krainer E, 290
Krajcovic J, 640
Kral AM, 245–48
KRÄMER A, 367–409; 372,
    378, 379, 393–99, 401

Kramer B, 105, 106, 112, 114, 122, 123, 125, 127
Kramer GF, 89
Kramer JC, 171, 180, 395, 396
Kramer W, 105, 106, 112, 114, 122, 123, 125, 127
Krasnow MA, 641, 657
Kraulis PJ, 448, 777
Krause A, 523
Krause E, 273, 278
Krauskopf A, 22
Kraut ND, 574, 596
Krebs HA, 570
Kreibich G, 273, 281, 284
Kreiswirth BN, 223, 224, 228
Kreitmeier M, 419
Kreivi JP, 381
Krell H, 200
Krens FA, 426
Kretzschmar M, 678, 792, 793
Kreutziger GO, 478
Kreuzer KN, 636, 644, 649, 657, 671
Krieg UC, 273
Krikorian CR
Krishan A, 675
Krishna TSR, 151
Krishnan HB, 505, 519, 520, 523
Kristensen P, 807, 809, 836
Kristna TSR, 646
Kroeger PE, 661
Krogh S, 642, 643
Krohn M, 661
Kronidou NG, 290
Kronis KA, 454
Kronmal SL, 348
Krowczynska A, 695, 696, 702
Krueger S, 645
Krupinski J, 435
Kruse N, 623
Krutzsch H, 400
Kruys VI, 695, 708
Ku CY, 575
Kubo T, 704
Kubomura N, 834
Kubrich M, 290
Kuchino Y, 754
Kuchler K, 262, 296, 297
Kuckelkorn U, 808, 836
Kuechler E, 401
Kuehn C, 540, 545
Kuehn L, 805, 816, 838
Kuerbitz SJ, 72, 356
Kufe DW, 678
Kuge S, 599
Kugler W, 695, 704, 731
Kuhar MJ, 321
Kuhl SB, 196, 197
Kuhn B, 200
Kuhn LC, 307, 330, 709, 728, 729
Kuhn R, 837
Kuhn-Nentwig L, 570, 575, 576
Kuismanen E, 297

Kuki A, 549
Kulikova O, 515, 525, 528
Kulozik AE, 695, 704, 731
Kumagai A, 420, 421
Kumagai M, 669
Kumamoto CA, 275
Kumar A, 777
Kumar D, 640
Kumar KP, 22, 26, 33, 35, 57, 64, 776
Kumar NM, 477, 479, 481, 482, 485, 486, 494, 497
Kumar R, 751
Kumar S, 377, 380, 383, 389
Kumar VV, 491
Kumatori A, 807, 808
Kundu M, 236
Kunes D, 640
Kung H, 309–12
Kung H-J, 493
Kunkel HG, 294
Kunkel TA, 107–9, 115, 117, 120
Kuntzel H, 592, 593
Kunugi KA, 231
Kunz W, 570
Kunze N, 673, 674
Kuo CF, 51, 142
Kuo HC, 381
Kupper JH, 144
Kuraoka I, 55, 57, 63, 148, 149, 153
Kurata WE, 493, 494
Kuret J, 252
Kurihara T, 279, 280, 290
Kurimasa A, 159
Kuriyan J, 151, 646
Kurkdjian AC, 524, 527, 530
Kurland CG, 695
Kuroda S, 258
Kuroda U, 258
Kurose I, 598
Kurowski TT, 574, 575
Kurtz S, 350
Kurzchalia TV, 273, 275
Kurzrock R, 118
Kushner SR, 175, 176, 192, 206, 216, 232
Kuska J, 233
Kuspa A, 415, 416, 419, 423, 424
KUTAY U, 271–303; 297
Kuwayama H, 415, 427
Kwak BR, 495
Kwak KB, 805
Kwee C, 254
Kwong AD, 55, 63, 149
Kycia JH, 759
Kyoizumi S, 359
Kyrion G, 349, 350, 353

L

Labbé J-C, 20
La Branche H, 382

Labrousse V, 230, 234
Lacave C, 226
Lacks SA, 102, 113, 122
Lacombe C, 619
Lacombe ML, 425, 426
Lacroute F, 695–97, 714
Lados M, 523
Laemmli UK, 645, 665–67
Laeremans T, 518, 527
Laff GM, 21
Lafyatis R, 395, 396
Lagas RM, 507, 513, 522
Lagnado CA, 703
Lahm A, 621, 622
Lahue EE, 175, 196, 197, 200
LAHUE R, 101–33
Lahue RS, 103–5, 111, 113, 115, 125, 126
Lai E, 786
Lai MC, 259
Laine RO, 310, 311
Laing JG, 477
Laipis P, 651, 652
Laird DW, 487, 492
Laird-Offringa IA, 695, 697, 703, 709
Lake JA, 637, 638, 678
Lalande V, 225
Lalley PA, 477
Lam E, 640
LaMarche AEP, 567, 573
Lambert JR, 574
Lambert MH, 612, 622
Lambert P, 751
Lambert WC, 145
Lambowitz AM, 598
Lamhasni S, 671
Lamm L, 575
Lammers AA, 428
Lamond AI, 372, 374, 378, 380, 385, 387
Lampe PD, 495
Landavazo A, 247
Landberg G, 672
Landsman D, 583, 786
Lane DP, 57, 72, 149, 151, 179, 356
Lane MD, 307, 320
Lane P, 254
Lane WS, 377–80, 383–85, 678, 776, 777, 808, 813, 821
Laneelle G, 226, 230
Lang EZ, 119
Lange BJ, 669
Langen R, 540, 549, 550, 552–54, 557, 558
Langer T, 584, 588
Längle-Rouault F, 104, 113
Langmore JP, 348
Langneaux D, 639, 662
Langowski J, 645
Lanni A, 586
Lanni F, 485

Lansdorp PM, 356, 359
Lanspa SJ, 118, 119
Lao KQ, 538
Lapham C, 837
Laplace C, 837
Lapoint J, 19, 32
Lapointe JY, 329
Larbaud D, 840
Larimer FW, 701, 702
Laroche T, 338, 349, 666
Larsen AK, 671, 679
Larsen B, 751, 752
Larsen CN, 832
Larsen PR, 84, 88, 92, 93, 98, 757
Larson DD, 351–53
Laskey RA, 676
Lasko PF, 398
Laso MRV, 727
Lau AF, 492–94, 496
Lau K, 660
Laufer CS, 660
Lauffer L, 278
Lauppe HF, 200
Laurent BC, 36
Lauring B, 273, 284
Laval F, 140
Laval J, 141
Lavenot C, 674
Laver G, 452
Laver WG, 452, 460, 469
Lavie A, 454
Lavigueur A, 382
Lawler S, 121
Lawrence CW, 145
Lawrence GMP, 612
Lawson TG, 171, 180
Lazar G, 377, 378
Lazarus M, 640
Lea EJA, 524
Leach FS, 105, 110, 118–20
Leadon SA, 27, 37
Leahy DJ, 614
Lear JD, 287
Learmonth D, 59
Learn BA, 103
Leban JJ, 247, 821
Leber B, 359
Leblanc J, 666
LeBlanc JF, 782
Lebowitz JH, 194, 198, 199, 208
Lebrun JJ, 619
Lebwohl DE, 820
Lecker S, 280, 292
Leclercq M, 696, 697
Leder P, 695
LeDoux SP, 26
Lee A, 352, 702
Lee BJ, 84, 86, 98, 757
Lee B-S, 714, 718, 719, 721, 722, 753
Lee C-G, 171
Lee CS, 805
Lee DH, 809

Lee DK, 781, 791
Lee EH, 118, 122, 421
Lee F, 87, 91, 93
Lee JG, 505, 511–14, 518, 636, 678
Lee JI, 832
Lee K, 141
Lee LW, 836
Lee MM, 145, 341, 343, 636
Lee MP, 644, 645
Lee MS, 196, 342
Lee MYWT, 66, 152
Lee PL, 786
Lee RF, 752
Lee RT, 465, 471
Lee S, 38, 72, 111, 160, 421
Lee SH, 55, 63, 149, 153, 175
Lee SI, 704, 718, 731, 750
Lee TH, 61
Lee VM, 378, 384
Lee VS, 226
Lee W-R, 619, 695, 730
Lee W-S, 307, 327, 328, 330, 331
Lee YC, 465, 471
Leebeek FWG, 621
Leech RM, 640
Leeds P, 695, 714, 715, 718, 719, 721, 722, 724, 753
Leem SH, 143
Lees-Miller SP, 159
LeFeuvre CE, 358, 359
Leffers H, 819
Leffler H, 444, 458, 469
Lefkowitz RJ, 245, 257, 423, 424, 429, 431
Legakis NJ, 223
LeGall J, 97
Legerski R, 36, 58, 73
Legerski RJ, 33, 34, 55, 57, 63, 149, 150, 153
Legrain P, 144, 372, 378, 392, 394, 395
Legrand D, 444, 456, 458
Lehle L, 281
Lehman IR, 104, 175
Lehmann AR, 16, 29, 30, 35, 47, 69, 136, 144–46, 151, 155, 158, 160
Lehre KP, 317
Leibach FH, 330
Leichus B, 252
Leiden JM, 330
Leinfelder W, 84, 85, 87, 89, 90, 92, 755, 756, 762
Leining LM, 262
Leiting B, 416
Lejnine S, 348
Lemay G, 329, 707
Lemire BD, 567, 568, 589
Lengauer C, 109, 117, 120
Lennon DJ, 158, 669
Lennox EX, 485
Leonard JL, 93, 96

Leonard K, 568
Leonard M, 278
Leong SR, 619
Lepoivre M, 598
Lepre C, 545
Lerner MR, 372
Leroi GE, 554
Lerouge P, 505, 507, 508, 510, 512–15, 518, 519, 523, 526, 527, 529, 530
Leroy P, 398
Lescoe MK, 105, 110, 111, 119
Leshinsky E, 825, 827, 828
Lesley SA, 673
Lesser CF, 374
Lester HA, 307, 323
Lestienne P, 566
Leteurtre F, 643
L'Etoile ND, 794
Lettieri T, 105, 109, 111, 112, 120
Leung WY, 125, 127
Lever D, 262
Levin JG, 754
Levin LR, 419, 423
Levin ME, 222, 235
Levin NA, 669, 672
Levine A, 38, 72
Levine AJ, 72, 775, 791
Levine AL, 817
Levine AS, 61, 149, 153
Levine BJ, 712
Levine E, 485, 486, 492
Levine RL, 839
Levine TD, 704
Levinger L, 662
Levinson G, 103
Levis RW, 339
Levy DB, 72, 73, 356
Levy LM, 317, 751
Lewandowski DJ, 752
Lewin A, 505, 508, 510, 513, 520, 523, 529
Lewin B, 23
Lewis J, 379, 401
Lewis JA, 575
Lewis JB, 746
Lewis MD, 260, 261
Ley RD, 45
Leyh TS, 507, 514
Leyman B, 507, 511
Leysen DC, 223
L'Hoir C, 445, 448, 452, 463, 464, 467, 468
Li C, 522
Li CJ, 636, 672
Li G-M, 105, 107–9, 111–13, 117–20, 137, 754
Li JJ, 192, 658, 664
Li JR, 423
Li JW, 257
Li JY, 223
Li K, 802, 813, 816, 822

Li L, 33, 34, 36, 55, 57, 63, 68, 69, 73, 149, 150, 153
Li L-L, 222
Li R, 73, 251, 252, 575
Li XA, 782
Li XCS, 817, 825
Li XQ, 746, 838
Li XR, 495
Li XX, 275
Li XY, 171
Li YF, 45, 154, 759, 763, 776, 828
Lian PP, 571
Liang C, 178, 179, 548, 550
Liang JG, 477
Liang M, 320
Liao D-I, 444, 448, 458, 468
Liao K, 307, 320
Liao SR, 283, 286, 287
Liao XC, 388, 394, 593, 594
Lichten M, 122, 123
Lichtenberg-Frate H, 477, 489, 490, 495
Lichter P, 320
Lieb M, 102
Lieber CM, 540
Lieber MR, 59, 65, 143, 146, 159
Lieberfarb ME, 706
Lieberman HB, 155
Lieberman PM, 773, 793, 794
Liebhaber SA, 703, 727, 728
Liebman SW, 707
Lieu HM, 792
Lightowlers RN, 576
Liljas A, 228
Lill R, 279, 280, 290, 293
Lilley DMJ, 659, 660
Lilly PJ, 419, 421–23, 430
Lim MC, 540
Lim SK, 714, 725
Lim YH, 261
Lima CD, 636, 642, 648, 652, 654
Lin A, 69
Lin C-H, 388, 392
Lin GY, 59
Lin HH, 495
Lin J-J, 18, 47, 50, 51, 339, 344, 345, 777, 778
Lin KC, 422, 426
Lin LJ, 281
Lin P, 465
Lin RJ, 180, 369, 374, 376, 391, 398, 399
Lin Y, 791
Lin Y-S, 794
Lincke CR, 350, 353
Lindahl T, 16, 45, 137–40, 142–45, 148, 158
Linder D, 97
Linder P, 171, 172, 180, 398
Lindholm-Levy PJ, 221, 223, 232, 233
Lindner IJ, 416

Lindsey DF, 415
Lindsey LA, 379, 401
Lindsley D, 748
Lindsley JE, 646, 650, 651, 656
Lindsten T, 704
Lindwall G, 389
Ling M-L, 580
Ling SY, 746
Ling V, 262
Lingappa VR, 276, 278, 294
Lingham RB, 260
Lingner J, 341, 343
Linke P, 568
Linn S, 61, 66, 140, 141, 149–52
Linnane AW, 570, 589
Linstedt AD, 297
Linton JP, 105, 110
Lipford J, 118–20
Lipford JR, 119
Lippard SJ, 47, 61, 65, 145
Lis H, 442, 444, 450, 463, 469
Lis JT, 378, 383, 390, 661, 793
Liskay RM, 105, 112–15, 122, 123, 125, 127
Lisowsky T, 578–80, 583
Litfin M, 703, 709
Littauer UZ, 696, 697
Little JW, 71
Liu AA, 639
Liu AYC, 350
Liu B, 71, 105, 109, 110, 112, 113, 117, 119, 120
Liu CC, 176, 196, 197, 349, 636, 650
Liu CX, 507, 514
Liu GP, 276, 427
Liu H-F, 649, 670
Liu H-X, 369
Liu JS, 343
Liu K, 105, 110, 114, 115, 123
Liu LF, 636, 639, 643, 645, 649, 650, 658–62, 664, 666, 669–72, 674, 676, 677
Liu Q-R, 307, 323
Liu R-Q, 547
Liu SG, 489
Liu VF, 55, 74
Liu Z, 352, 702
Ljungberg B, 358, 359
Ljungdahl PO, 586, 701, 702
Ljungquist S, 158
Lloyd RS, 46, 92, 142, 758
Lo PC, 380
Lobsanov YD, 444, 458
Lochrie MA, 567, 571
Lock RB, 665, 675
Locker JK, 295
Lockshon D, 641, 645, 657
Lodge JK, 659
Lodish HF, 294, 295, 619
Loeb LA, 118, 119, 143, 148
Loewenstein WR, 485, 493–95
Loflin PT, 674

Logel J, 242
Lohman PHM, 69
LOHMAN TM, 169–214; 52, 170–74, 180–83, 186, 188, 190, 195–97, 199, 201, 203, 205, 209
Loike JD, 599
Lokker NA, 623
Lomax MI, 576
Lommel L, 70
Lonberg N, 197, 199
London J, 760
Long SR, 506, 507, 509, 510, 512–15, 522, 524, 526, 528, 529
Longhese MP, 149
Longhi R, 297
Longley MJ, 105, 107–9, 111, 112, 117–20, 137
Longmore GD, 619
Longnecker RM, 248
Longo FJ, 662
Loo LWM, 493
Loo MA, 820, 821
Look J, 492
Loomis WF, 415, 416
Lopato S, 377
Lopez AF, 622, 623
Lopez AR, 121
Lopez L, 518, 527
Lopez-Corcuera B, 307, 323, 326
Lopez-Lara IM, 505, 519, 527
Lopez-Otin C, 639
Loranger A, 676
Lorenz W, 245, 257
Lortet G, 507, 513, 518, 519
Lossky M, 397
Losson R, 695, 714
Lott WA, 225
Lottspeich F, 380, 792, 793, 805, 808, 810, 812, 813, 815
Louis EJ, 668, 669
Louis JM, 416, 418, 419
Love JJ, 782
Love WE, 785–88
Lovejoy B, 612
Lovett PS, 759
Lovett ST, 104
Low KB, 122, 125, 128
Low RL, 652
Low SC, 87, 91
Low SY, 312
Löwe J, 805, 807, 809, 811, 813, 814, 816, 834
Lowell JE, 698, 705
Lowery MD, 553
Lowman HB, 629, 631
Lowrie D, 222
Lowrie DB, 228
Lowry CV, 566, 574, 585
Lu A-L, 102–4, 142, 649, 670
Lu DJ, 254, 261
Lu H, 395, 772, 775, 791

Lu Q, 575
Lu S-J, 392
Lu XY, 55, 57, 63, 72, 149
Lubben M, 589
Lubiniecki AS, 116
Lucas W, 706
Lucchini G, 149
Lucero HA, 827
Luciakova K, 574, 575
Luciw PA, 751
Luderus MEE, 425
Ludwig B, 538, 547, 559
Ludwig DS, 481
Lue NF, 349, 644, 652, 653
Lugtenberg BJJ, 505, 507–13,
    516, 518–20, 527, 529
Lührmann R, 371, 379, 381, 383,
    390, 392–94, 396–99
Luirink J, 275
Luka S, 507, 515, 516, 518
Lukash L, 27
Luke M, 665
Lukins HB, 583
Lund K, 643, 649
Lundblad V, 344, 345
Lundell DJ, 614
Lunin NA, 752
Lunn CA, 614
Luo M, 460, 469
Luo Y, 792
Lupas A, 805, 806, 808–10, 812–
    14, 816, 822
Lupp B, 389
Luque JM, 326
Lurquin C, 831
Luscher B, 712
Luscinskas FW, 819, 820, 834
Luse DS, 23
Lusnak K, 122, 123
Lustig A, 349, 350, 353
Lustig AJ, 349, 350, 378, 394,
    395, 398
Lustig F, 763
Lütcke H, 273
Luttinger A, 636
Luttinger AL, 639, 678
Lutz Y, 19, 31, 775, 791
Lutzner MA, 53
Lyamichev V, 143
Lynch AS, 659, 660
Lynch HT, 118–20
Lynch JF, 118
Lynch KW, 382, 383
Lynch P, 119
Lyne PD, 622, 623
Lynn R, 636
Lynn RM, 647
Lyons CR, 309, 313

M

Ma D, 659
Ma L, 31

Ma X, 679
Ma Y-T, 252–54, 261, 821
Ma Z, 261
Ma ZP, 820
Maandag ER, 127
Maarse AC, 290
MacDonald GH, 349
MacDonald NJ, 387
MacGeoch C, 117, 120
MacGillivray RTA, 540
Machaness GB, 232
Machlin PS, 676, 709
Macieira-Coelho A, 360
MacInnes MA, 59
MacKay AM, 666
Mackey ZB, 67, 158
MacLeod CL, 307, 309, 311,
    312, 314
MacMillan AM, 391, 397
Madden KR, 642, 678
Maddox JF, 254, 261
Madhani HD, 369, 372, 374, 399,
    750
Madhukar BV, 494
Madsen P, 66, 151, 819
Maeda M, 420, 424, 431, 566
Maeda Y, 412
Maenhaut MG, 104, 113
Maestrojuan G, 504
Magagnin S, 327, 330
Magee AI, 242, 252, 255
Maghakian D, 415
Magliozzo RS, 228
Magyar Z, 526
Mahaffey D, 830, 839
Maher PA, 283
Maher SE, 389
Maher VM, 27, 69
Mahler J, 349
Maillet F, 505–8, 510–15, 518–
    20, 523, 524, 527, 529, 530
Mailliard ME, 307
Maine IP, 197, 198
Maiorino M, 84, 92, 95, 98
Maiss E, 752
Majors J, 749
Makarov VL, 348
Mäkelä TP, 19, 20, 58, 70
Maki S, 636, 678
Mäkinen K, 753
Makowske M, 320
Makowski L, 479–81, 483, 484,
    491
MALANDRO MS, 305–36; 310–
    12, 316, 317
Malatesta F, 568, 571, 574
Malbon CC, 704
Malchow D, 419, 421, 425
Maldonado E, 22, 395, 770, 783,
    791, 792
Maldonado PE, 494
Malewicz B, 491
Malhotra K, 45, 155

Malik S, 795
Malim MH, 743, 749
Malkhosyan S, 118–20, 659
Mallamo J, 821
Malmström BG, 538, 541, 547,
    559
Malone RE, 121–23
Malpure S, 641
Malter JS, 703, 704
Maltese WA, 242, 244, 250, 251
Maltzman W, 72
Mamet-Bratley MD, 23
Mamroud-Kidron E, 838
Mancebo R, 380
Manch-Citron JN, 760
Mandel SJ, 88, 757
Mandiyan S, 307, 323, 325, 827
Mandrand-Berthelot M-A, 85,
    89, 762
Manera E, 297
Manes SH, 660
Manfras BJ, 307, 315
Mangus DA, 579
Maniatis T, 19, 369, 372, 377,
    378, 380–84, 387, 390, 401,
    703, 717, 782, 794, 803,
    819, 820, 834, 835
Manjunath CK, 477, 479, 491
Mankovich JA, 103, 110
Manley JL, 374, 377, 378, 380,
    381, 383–85, 390, 394
Mann C, 827, 828, 830, 832
Mann GE, 307
Mann SKO, 426
Mannaerts GP, 595
Manne V, 260
Mannhaupt G, 584, 587, 588, 828
Manning GS, 173
Manning PA, 515, 753
Manon ST, 586
Manrow RE, 696
Manse B, 639, 662
Mantell LL, 339, 341
Manthey GM, 579, 582, 583, 585
Mao X, 349
Maquat LE, 695, 714, 715, 725–
    27, 731
Marbaix G, 696, 697
March CJ, 612
Marchesi F, 222
Marciano D, 262
Marcinkeviciene J, 228
Marcus G, 773, 792
Marcus MM, 278
Marcus RA, 539, 541, 549, 552
Marcus SL, 595
Marcznski GT, 591, 592
Marder BA, 358
Maresca B, 598
Margalit R, 540
Margoliash E, 570, 575
Margolin P, 660
Margolis JW, 814

Margossian SP, 583
Mariame B, 831
Marians KJ, 196, 198–200, 638, 641, 644, 657, 658, 663, 751
Marie C, 507, 509
Marini JC, 640
Marinis E, 223
Marino-Alessandri DJ, 23
Marinus MG, 102, 103, 111, 116, 118, 123, 148
Marinx O, 695, 708
Mariottini P, 567
Mark M, 541, 558
Markiewicz D, 762
Markovich D, 307, 327, 328, 330
Markovits J, 674
Markowitz SD, 118–21
Marmor S, 540
Marnett LJ, 649
Marom M, 262
Maroney PA, 376
Marostenmaki J, 228
Marr RS, 254
Marra G, 117
Marrison JL, 640
Marsac C, 576, 586
Marsh KL, 675
Marshall CJ, 242, 249, 252, 255–57
Marshall MS, 248
Marsters JC, 245
Marszalek J, 830, 834
Martelli AM, 676
Marth JD, 143
Martial JA, 445, 448, 452, 463, 464, 467, 468
Martiel JL, 430
Martin AM, 46, 155
Martin CT, 580, 669
Martin D, 103, 104
Martin DMA, 614
Martin F, 832, 838
Martin GM, 358
Martin GW, 757
Martin J, 584
Martin NC, 581
Martin del Rio R, 755
Martinerie C, 377, 378, 384
Martinez CK, 836
Martinez E, 511, 518, 527, 794
Martinez R, 180
Martinez-Abarca F, 523
Martinez-Romero E, 504, 505, 508, 510, 515, 519, 520
Martiniuk F, 222
Martins C, 707, 708
Martins de Sa C, 805, 808
Martoglio B, 283, 287, 288, 295
Marty A, 485
Maruyama N, 640
Marvel DJ, 529
Marvo SL, 669
Mary JL, 119

Marynen P, 762
Marzella L, 598
Marzluff WF, 695, 712, 713
Marzuki S, 589
Maschhoff KL, 371
Mascotti DP, 173
Masiarz FR, 750, 751
Masison DC, 706, 707, 711
Mason JM, 339
Mason TL, 593
Massey SC, 496
Massinople CM, 730
Masta A, 23
Masters BS, 578
Mastrangelo IA, 174, 177, 179
Mastubara H, 584
Masuda M, 754
Masurel R, 30
Masutani C, 58, 59, 61, 150
Matesic DF, 494, 495
Mathew S, 327, 331
Mathis JM, 120
Matic I, 125, 128
Matlack KE, 286
Mato JM, 426
Matson SW, 34, 52, 170–72, 175, 176, 182, 190, 193, 194, 196, 197, 200, 208, 671
Matsuda T, 55, 57, 63, 148, 149, 153, 256
Matsufuji S, 746–48, 751, 832, 838
Matsufuji T, 746–48, 751
Matsui P, 54
Matsui T, 771
Matsukuma S, 46
Matsumoto K, 245, 246, 252, 827
Matsumoto Y, 143
Matsumura F, 493
Matsunaga T, 33, 45, 54, 55, 57, 58, 60, 65, 66, 73, 145, 149, 152, 154, 155
Matsuno-Yagi A, 567
Matsuo K, 640
Matsushita O, 760
Matsuyama SI, 276, 278, 280, 292–94
Matsuzaki F, 487, 493
Matsuzaki K, 821
Matsuzaki T, 425
Mattaj IW, 368, 401
Mattei M-G, 384
Mattern MR, 639, 666, 674–76
Mattes WB, 47
Matthews CR, 540
Matthews DJ, 628
Matthews W, 803, 804, 839
Matunis MJ, 389, 392, 400, 695
Matz EC, 352
Matzuk MM, 641, 657
Mauch H, 234
Mauk AG, 540
Mauk MR, 540

Maul GG, 672
Maule J, 352, 357
Maupin MK, 701
Maurizi MR, 809, 811, 823, 829, 830, 832, 834
Maxon ME, 57, 58, 64
Maxwell A, 636, 645, 650, 651
May M, 775
May RC, 840
Mayeda A, 377, 378, 380–84, 392
Mayer H, 512
Mayer ML, 248
Mayer MP, 246
Mayer NJ, 826, 840
Mayer SA, 378, 388, 389
Mayes J, 674
Mayeux P, 619
Mayne LV, 27, 37, 54, 68, 69
Mayo SL, 540
Mayr GW, 428
Mayrand SH, 392
Mazat JP, 570
Mazet F, 485
Mazet JL, 485
Mazur SJ, 172, 200
Mazurier J, 444, 456, 458
McAllister WT, 580
McAlpine PJ, 29, 30, 145
McAuley-Hecht K, 51, 140
McBain W, 85, 744, 755
McBlane J, 159
McBride TJ, 148
McCabe ERB, 569
McCabe FL, 639
McCaffery M, 280
McCammon M, 702
McCarron BGH, 641
McCarthy JEG, 582, 714, 715, 727
McCaughan KK, 753
McCaw P, 402
McClanahan T, 87, 91, 93
McClatchy JK, 221
McCleary WR, 254
McClendon V, 762
McClintock B, 338
McClure WR, 199
McConnell HM, 548
McConnell SJ, 590
McCormick F, 242
McCormick JJ, 27, 69
McCormick PJ, 674
McCormick RJ, 782
McCormick-Graham M, 343
McCoubrey WK Jr, 649, 670
McCready S, 47, 155
McCready SJ, 155
McCullough DA, 580
McCullough ML, 816, 817, 836
McCune JM, 294
McDaniel LD, 34, 36, 68, 69
McDermott W, 232
McDevitt HO, 803, 836

McDonald JP, 122, 123, 667, 677
McDonald KK, 320
McDougall JK, 356, 358
McEachern MJ, 339, 341, 343,
    346, 347, 350, 353, 355, 358
McEntee K, 52, 71
MCEWEN JE, 563–607; 565,
    568, 572, 579, 582, 583,
    585, 589
McGarvey M, 398, 399
McGeady P, 248, 249
McGhee JD, 197
McGill NI, 356
McGinnis J, 540
McGivan JD, 307, 318
McGoldrick JP, 142
McGovern K, 295
McGrail SH, 261
McGrath S, 57, 74
McGraw N, 580
McGraw TS, 316, 317
McGregor WG, 27
McGuigan C, 379, 401
McGuire MJ, 813, 816, 817
McGurk G, 827, 828
McHenry CS, 751
McIntyre CA, 103
McKay DB, 612
McKay IA, 507, 516
McKay SJ, 356
McKee EE, 582, 585, 587, 588
McKee PH, 151
McKemmie EE, 589
McKenna A, 69
McKenney K, 830, 834
McKenzie ANJ, 622
McKeon F, 257
McKeown CK, 158
McKeown M, 380
McKnight SL, 62
McLain T, 244, 250
McLaren RS, 713
McLaughlin K, 63, 149
McLaughlin S, 257
McLaughlin SL, 674
McLendon G, 541
McMacken RM, 194, 198, 199,
    208
McMahon R, 310, 311
McMullin TW, 583, 585
McNamara JO, 307, 319, 320
McNamee H, 493, 494
McNeil M, 218, 226
McNeil MR, 224, 230, 231
McNeil PL, 485
McPheat WL, 595
McPheeters DS, 374
McPherson CE, 786
McPherson SM, 662
McSwiggen JA, 179, 185, 197
McTigue M, 694, 695, 728
McWhir J, 148
Mead DA, 580

Meade TJ, 540, 543
Mechali M, 107, 780
Mecklin J-P, 119, 120
Medford RM, 696
Medina R, 840
Medoff G, 598
Meera Khan P, 727
Meerovitch K, 398
Mege RM, 487, 493
Mehdy MC, 412, 424
Mehle C, 358, 359
Mehta PP, 495
Mehta VB, 647, 661
Mei LH, 293
Meier A, 219
Meijer EA, 529
Meijlink F, 703
Meinsma D, 728, 729
Meisterernst M, 678, 772, 780,
    792, 793, 795
Mejean V, 103, 104, 113
Melamede RJ, 141
Melancon P, 172, 200
Melandri FD, 821, 835
Melcher K, 828
Melek M, 341, 343
Melendy T, 640
Meliconi R, 672
Melin L, 712
Mellon I, 24, 26, 29, 67
Mellor J, 574, 575, 743, 749
Mellor RB, 525, 527, 528
Melton DW, 148
Melton LB, 703
Melton RE, 235
Melton W, 352
Meltzer S, 589
Menage H, 151
Mendelman LV, 194
Mende-Mueller L, 729
Mendenhall MD, 830
Meng S-Y, 762
Mengus G, 775
Menissier-De Murcia J, 144
Menko AS, 493
Menon AS, 829
Menon NK, 97
Menoud P-A, 703
Menssen R, 374
Menzel R, 660, 673
Meo T, 714
Mera Y, 572
Mercer JFB, 696, 697
Mergaert P, 505, 507–11, 513,
    514, 516, 519, 522, 523,
    527
Merillat N, 717
Merino A, 678
Merle P, 567
Mermelstein F, 19, 20, 58, 70
Mermoud JE, 385
Merola AJ, 227
Merrick WC, 398, 826, 833

Merritt EA, 445, 448, 452, 459,
    463, 464, 467, 468
Meruelo D, 309–12
Meschini R, 59
Meselson M, 102, 103, 106, 113,
    114, 695
Meselson MS, 122
Meskiene I, 526
Messina P, 820, 821
Methfessel C, 322
Mett IL, 831
Meuth M, 119, 120
Meyer DI, 273, 275, 278, 282,
    284, 289
Meyer R, 419, 430
Meyer RA, 487
Meyer SA, 116
Meyer TH, 812
Meyer TJ, 541
Meyer UA, 93
Meyer-Monard S, 704
Meyers R, 777
Mezard C, 125, 126, 128
Mezey E, 307, 323
Mezquita C, 674
Mezzina M, 19, 31, 54, 58
Miaege R-M, 487
Miake-Lye R, 420
Miao GG, 142
Michael WM, 378, 388, 389, 400
Michaelis G, 578–80, 583
Michaelis S, 242, 243, 245, 246,
    249, 255, 262, 280, 296, 297
Michaelis U, 583
Michaels GS, 531
Michaels ML, 46, 102, 667
Michalek M, 802, 814, 835
Michalek MT, 836
Michaud C, 814, 815, 821, 838
Michaud S, 372, 381, 386, 392
Michel F, 371
Michel H, 538, 547, 554, 556,
    559
Michell RH, 428
Michelotti G, 400
Michels PAM, 350, 353
Micol V, 580
Middlebrook G, 216, 221, 225,
    226
Migliaccio G, 282
Mignotte B, 597
Miki T, 636, 678
Mikkelsen HB, 477
Miklasevics E, 522, 528, 529, 531
Miklos D, 584
Milburn MV, 612, 622
Mildvan AS, 648
Milks LC, 477, 482, 486
Milla MG, 111
Miller AD, 670, 703
Miller CA, 645, 672, 760
Miller EJ, 115, 144
Miller JD, 273, 278

Miller JH, 102, 142, 667
Miller JR, 541, 548
Miller KG, 676
Miller LP, 222
Miller RA, 254
Miller SP, 21, 54
Miller TM, 479, 485
Miller WA, 743, 744, 752, 753
Millett F, 547, 571
Milliman CL, 597
Mills JS, 676
Mills SL, 496
Millward S, 707
Milne JL, 430
Milne JLS, 430–32
Milner PF, 727
Milona N, 422
Milot E, 125, 127
Min H, 392
Min J, 581, 582
Minagawa N, 593, 595
Minami E, 522, 526, 530
Minami Y, 623
Mines GA, 540, 544
Ming ME, 679
Minkley EG Jr, 175
Minnich BN, 491
Minvielle-Sebastia L, 696, 697
Miquel J, 598
Mirabelli CK, 639, 666, 674, 675
Mirambeau G, 638, 674
Miret JJ, 111
Miriami E, 400
Mirkin SM, 659
Mirza UA, 773, 785
Misher L, 696
Miskimins R, 674
Miskimins WK, 674
Misra LM, 280, 289
Mita S, 576
Mital R, 712
Mitch WE, 840
Mitchell DL, 67, 145, 146
Mitchell EP, 452
Mitchell JLA, 746
Mitchell PJ, 714, 725, 726
Mitchison DA, 217, 221, 228,
    230, 232
Mitchison TJ, 666, 667
Mitoma JY, 297
Mitra S, 45, 137, 140
Mitsuzawa H, 249
Miura A, 669
Miura M, 151
Miura N, 55, 148
Miura S, 598
Miura Y, 258
Miura-Masuda A, 669
Miwa M, 701
Miyaji H, 840
Miyajima A, 610, 611, 620, 622
Miyajima I, 246
Miyamoto I, 55, 148

Miyazaki J, 148
Miyazaki Y, 746, 747
Mizuno S, 221
Mizushima S, 276, 278, 280,
    292–94
Mizushima T, 811
Mizutani T, 757
Mizutani Y, 538
Mizuuchi K, 636, 673
Mldozik M, 792
Mochida A, 640
Mock M, 638
Moda I, 401
MODRICH P, 101–33; 16, 44,
    102–6, 108–10, 112, 113,
    115, 120
Moerman M, 523
Moerschell RP, 810, 829
Moggs JG, 18, 37, 59, 60, 65,
    146, 147, 153
Mogi T, 589
Mohandas TK, 307, 327, 328, 331
Mohn KL, 378, 384
Mohrs K, 587
Mok M, 198, 199
Mok W, 284
Mol CD, 51, 140, 142
Molinete M, 144
Mollerup S, 643
Momand J, 72
Momburg F, 837
Monaco JJ, 803, 805, 836
Moncollin V, 19, 30–33, 37, 38,
    54, 57, 58, 73, 147, 149,
    150, 152
Mondragon A, 636, 642, 644,
    648, 652–54
Mong S, 676
Mong S-M, 676
Monnet M, 671
Monod J, 695
Monroy AF, 586
Monstein HJ, 743
Montecucco C, 570, 575
Montfort W, 444
Montreuil J, 444, 456, 458
Monzingo AF, 444
Moody PCE, 809, 813, 814
Moolenaar GF, 50, 51
Moolenaar WH, 415
Moomaw CR, 812, 818, 826,
    827, 830, 831, 833, 837
Moomaw JF, 244–49
Mooney CL, 54, 60, 148
Moorby CD, 494
Moore CW, 123
Moore DD, 595
Moore DP, 749
Moore GR, 545
Moore KJM, 170, 185, 188, 190,
    194, 195, 198, 203, 205,
    206
Moore LK, 485, 491

Moore MJ, 369, 371, 372, 374,
    375, 390
Moore ML, 307, 327, 328
Moore PA, 827
Moore R, 749
Moores SL, 244, 245, 248
Moorhead PS, 356
Moos M Jr, 61, 150
Mora C, 328
Morales F, 674
Morandi C, 378, 388, 389
Morbiducci V, 223
Moreau J, 105, 110, 123
Morel P, 668
Moreno AP, 492–94
Moretti P, 350
Morfin JP, 717
Morgan DO, 19, 20, 70
Morgan HW, 638
Morgan MM, 565, 593
Morgan WD, 185
Morgan WF, 358
Morham SG, 644, 647
Morikawa K, 636
Morimoto Y, 811
Morin GB, 339, 341, 342, 356
Morino K, 663, 665, 675
Morita EH, 148
Morita T, 45, 154
Moritz RL, 760
Moroni C, 703, 704
Morris DR, 641, 645, 657
Morris S, 218, 230
Morris SL, 219, 228
Morrison A, 412, 645
Morrison PT, 105, 112, 113, 119,
    120
Morrison SG, 662
Morrison T, 294
Morrison-Plummer J, 92
Morrissey I, 662, 663
Morse DE, 714
Morten KJ, 580
Mortensen UH, 642, 643, 655
Morton CC, 309, 316
Mosbaugh DW, 140
Mosckovitz R, 327, 328, 331
Moser CC, 549, 553, 557
Moskaitis JE, 728
Mosser SD, 244–48, 260
Mossmann H, 143
Motamedi H, 511
Mothes W, 286–88
Mott HR, 612
Mott JD, 818
Moubakak A, 571
Moudrianakis E, 786, 788
Moudrianakis EN, 785–88
Moulding TS, 221
Mount DW, 71
Mount SM, 372, 380
Moura I, 97
Moustacchi E, 27

Movva NR, 294, 623
Moyer MB, 67, 158
Moyzis RK, 356
Mroczkowska M, 142
Mu D, 31, 33, 54, 55, 57, 58, 60,
    61, 63, 65, 66, 145, 149,
    152, 153
Muckenthaler M, 704, 731
Mueckler M, 294
Mueller CGF, 776, 777
Mueller DM, 581
Mueller F, 759, 763
Mueller GC, 713
Muerhoff AS, 595
Muesch A, 297
Muhlrad D, 695–98, 702, 703,
    705, 706, 709, 710, 722,
    723, 726, 728
Muijsers AO, 576
Mukherjee SK, 640
Mulder E, 679
Mulero JJ, 583, 585
Mulhauser F, 378, 394, 395
Mulkerrin MG, 616, 617, 624,
    626, 628
Mullberg J, 621
Mullen MP, 388
Mullenders LHF, 27, 37, 54, 59,
    68, 150
Müller B, 125, 126, 177
Muller DK, 580
Müller G, 289
Muller H, 290
Muller HG, 294
Muller HJ, 338
Muller J, 525, 527, 528
Muller MT, 642–44, 647, 649,
    661
Muller S, 95, 805, 811
Muller T, 623
Muller W, 360
Muller YA, 614, 615, 627
Müller-Taubenberger A, 419, 805
Mulligan JT, 507
Mullinger AM, 664, 665, 675
Mullner EW, 709, 728, 729
Multhaup G, 808, 836, 837
Mumby SM, 257, 259
Muneses C, 713
Munn M, 57, 145, 146, 149
Munn TZ, 836
Munnich A, 586
Munroe D, 696, 706, 707, 711
Munroe SH, 392
Munster PN, 495
Muntz KH, 257, 259
Munz PL, 159
Murakami K, 817
Murakami M, 622
Murakami S, 791
Murakami Y, 720, 746, 747, 812,
    821, 832, 838
Murante RS, 143

Murdoch GH, 307, 314, 319, 320
Muriel W, 155
Murikami M, 620, 621
Murnane JP, 160, 358
Murphy C, 380
Murphy EC, 288
Murphy JB, 442
Murphy RF, 485
Murray AW, 247, 251, 252, 494,
    496, 664, 665, 675
Murray CJL, 216
Murray JM, 144, 146, 155
Murty VVVS, 327, 331
Musacchio A, 423
Müsch A, 290, 292
Musgrave DR, 638
Musha T, 245, 250
Musil LS, 487, 492, 493, 495
Musser JM, 223, 224
Mustafa S, 235
Muster-Nassal C, 107
Muto A, 760
Muto K, 822, 823
Mutzel R, 426
Myer VE, 704, 731
Myeroff L, 121
Myers LC, 246
Myers R, 356, 358
Myers RAM, 598
Mykles DL, 815, 817, 840
Mylona P, 525, 526, 528

N

Na JG, 714, 715, 718, 722, 776
Naas T, 749
Nadal M, 638, 660, 668
Nadal MS, 711, 728
Nadal-Ginard B, 378, 388, 389,
    394, 395, 696
Nadel M, 638
Nadler JV, 322
Naegeli H, 35, 54, 171, 197
Naeger LK, 714, 725
Naess V, 753
Nag B, 759
Nag DK, 122, 123
Nagabhushan TL, 612, 614
Nagai A, 148, 153
Nagai K, 377
Nagamine Y, 703
Nagashima K, 805, 837
Nagasu T, 261
Nagata A, 219
Nagata S, 617, 619, 620, 627, 631
Nagele B, 425
Nagley P, 567, 591
Nagrajan K, 217, 236
Nagy E, 703, 704
Nagy I, 810
Naider F, 241, 253, 256
Naik K, 378, 387
Nair APK, 703

Nair J, 219
Nairn AC, 492, 494, 495
Nairn RS, 145, 150
Naismith JH, 444, 450, 463
Najita L, 706
Nakabayashi T, 676
Nakabeppu Y, 45, 141
Nakafuku M, 246
Nakagawa J, 704
Nakagawa N, 621
Nakagawa T, 621, 698
Nakahara Y, 521, 522
Nakai M, 584
Nakai T, 572, 588
Nakajima N, 772
Nakamura K, 155, 157
Nakamura S, 307, 314, 317
Nakamura T, 669
Nakamura Y, 105, 112, 113
Nakane H, 161
Nakanishi H, 258
Nakanishi T, 358–60
Nakano A, 252
Nakano H, 260
Nakano M, 595
Nakari T, 568, 573
Nakase H, 576
Nakashio R, 358–60
Nakatani Y, 773–75, 784–86, 788
Nakatsu Y, 55, 57, 63, 149, 153,
    161
Nakatsuru Y, 46
Nakayama H, 192
Nakayama K, 192
Nakayama N, 185, 246
Nakayama T, 307, 314, 317
Nakazato H, 696
Naleway CA, 548
Nambi P, 676
Nambudripad R, 36
Nanbu R, 703, 704
Nance MA, 36, 54
Nandabalan K, 400
Nardo T, 35, 54
Narita T, 662–65, 675
Nash HA, 636, 649
Nash RA, 158
Nasim A, 47, 155
Nasim FH, 381
Nasmyth KA, 280, 830
Nassif N, 125, 127
Natalie D, 636
Natarajan AT, 27, 37, 54, 59, 68,
    150
Nath ST, 23
Natori S, 704
Natsuka S, 622
Naumovski L, 21, 34
Naus CC, 477
Naylor SL, 356, 358
Naysmyth K, 349
Neddermann P, 141, 142
Nedergaard M, 491

Nederlof PM, 807
Neece S, 671
Neefjes JJ, 837
Neer EJ, 36, 422
Negri C, 672, 674, 676
Negrini M, 669
Neher E, 485
Nehls P, 141
Neish AS, 819, 820, 834
Nelbock P, 827
Nellen W, 425
Nelson BD, 574, 575
Nelson EM, 636, 706
Nelson H, 307, 323, 325, 827
Nelson N, 307, 323, 325, 326, 827
Nelson WG, 674
Nemoto F, 754
Nemoto N, 46
Nerke K, 416
Nesper M, 805
Neugebauer KM, 378, 380, 382–84
Neupert B, 728, 729
Neupert W, 584, 587, 588
Nevers P, 102
Neville M, 378, 393, 394, 396, 398
New L, 105, 110, 114, 115, 123, 125, 126
Newburger PE, 88, 757
Newell PC, 412, 415, 427
Newlon CS, 347
Newman AJ, 369, 374, 376, 391, 397
Newman AP, 251, 252
Newport JW, 197, 199, 665
Newton MD, 547, 548, 550
Neyses L, 477
Neyton J, 485
Ng KK-S, 444, 453, 454, 456, 459, 460, 462
Ng L, 263
Ng S-W, 674
Ng WM, 180, 192
Nguyen DC, 108, 109, 117, 120
Nguyen HT, 696
Nguyen ST, 777
Nguyen T, 814
Nhamburo PT, 93
Niblett CL, 752
Nicchitta CV, 276, 281–83, 287, 288
Nicholls P, 568, 569, 571
Nichols AF, 61, 66, 149–51
Nichols BP, 760
Nichols LM, 754, 759
Nicholson BJ, 477, 489
Nicholson WL, 45
Nickerson JA, 402
Nickoloff JA, 123
Nicola NA, 610, 622

Nicolaides NC, 105, 108–10, 112, 113, 117, 119, 120
Nicolaou KC, 521, 522, 528
Nicolas A, 121, 125, 126, 128
Nicolas AL, 159
Niedel B, 803, 805
Nielsen FC, 728–30
Nielsen OF, 643
Nielsen PJ, 378, 384, 398
Nielson BL, 640
Nieman PE, 709
Niemczyk MP, 541
Nienhuis AW, 727
Nieuwkoop AJ, 507, 515, 518
Nigam M, 260
Nigg EA, 20
Nikaido H, 218, 231, 236
Nikaido T, 818
Niki H, 638, 657, 660
Nikolaeva OV, 752
Nikolov DB, 770, 773, 775, 777, 778, 781, 791
Nilsen TW, 376
Nilsson G, 714
Nilsson I, 281, 295
Nilsson T, 547
Nilsson-Tillgren T, 126, 128
Ning Y, 356
Ninomiya TJ, 827
Nirenberg MJ, 328
Nishida C, 66, 151, 152
Nishida E, 834
Nishigai M, 823
Nishigori C, 69
Nishikawa K, 760
Nishikawa T, 826
Nishikimi M, 591
Nishimoto T, 662, 663, 665, 675
Nishimura S, 46, 141
Nishimura Y, 638, 657, 660
Nishiyama K, 278, 292
Nishizuka Y, 494
Nitiss JL, 657, 658, 662, 670–72
Niu H, 57
Niu LM, 105, 110
Nivera NL, 643
Niwa H, 148
Noble MEM, 782
Nobrega FG, 583, 584, 589
Nobrega MP, 583, 584, 589
Nocera DG, 540, 554
Noda A, 356
Noda C, 818, 819
Noda M, 425, 488
Nodheim A, 659
Noegel AA, 416
Noel PJ, 279
Nogami M, 840
Nolan JM, 636, 679
Nonner W, 486
Nony E, 523
Noonan DJ, 595
Nordan RP, 621

Nordheim A, 661
Nordling M, 553, 554, 557
Norman C, 376
Norris JR, 546
North P, 159
North RA, 308, 313
Norton K, 46
Norwood TH, 358
Nossal NG, 193, 194, 208
Notarnicola SM, 180, 192, 194
Nöthiger R, 380
Nothwang HG, 805, 808, 812
Noumi T, 566
Nouspikel T, 59, 146
Novak DA, 307, 310–12, 320
Novick P, 244, 257
Nowell PC, 118–20
Nozaki N, 674
Nudel U, 696, 697
Nunes SL, 821
Nunes V, 327, 328
Nunn M, 217, 235
Nunnari JM, 281, 282, 584, 588
Nuss DL, 701
Nussbaum AL, 104
Nussbaum RL, 400
Nussberger S, 316, 317
Nyamuswa G, 671
Nystrom-Lahti M, 119

O

Obermaier B, 348
Obernauerova M, 640
Obeso A, 599
O'Brien DP, 614
O'Brien P, 571
O'Connell ML, 706
O'Conner PM, 38
O'Connor LB, 832
O'Connor M, 753, 758
O'Connor TR, 140
Oda H, 141
Oda T, 666, 674
O'Dea MH, 636, 645, 650
Odijk H, 29, 30, 53, 146, 147
Odom JD, 94, 95
Odom OW, 759
O'Donnell M, 52, 63, 151, 646
O'Donnell ME, 751
O'Donovan A, 18, 37, 59, 60, 65, 146, 147
Oelgeschlager T, 794
Offen D, 379, 400
Ogawa H, 160
Ogawa S, 599
Ogawa T, 160, 521, 522, 526, 530
Ogg SC, 274, 278, 282
Ogura K, 589
Ogura T, 816, 825, 840
Oguro M, 636
Oh EY, 47, 50, 175
Oh SY, 494, 496

Ohama T, 98, 757
Ohana B, 827
O'Hara BP, 444, 454
O'Hara MB, 244, 245, 248
Ohashi A, 588
Ohasi A, 572
Ohguro H, 259
Ohkura H, 663, 665, 675
Ohnishi T, 567
Ohno M, 372, 379, 399, 401
Ohori M, 119
Ohshima Y, 376
Ohsumi T, 282, 284
Ohta T, 679, 774, 775, 785, 786, 792
Ohtake Y, 707
Ohtsubo E, 752
Ohtsuka E, 46, 141
Ohya Y, 248, 249, 261
Oikawa A, 154
Oikawa T, 94
Okabe M, 260
Okada S, 674
Okaichi K, 420
Okamoto K, 412
Okamura HH, 279, 280, 290
Okker RJH, 516, 518
Okumoto DS, 24, 67
Okumura K, 840
Okuno Y, 59
Olah ME, 434
Olds PJ, 479
Oleksa L, 98
Oliff A, 260
Olink-Coux M, 839
Oliva G, 444, 454
Olivares L, 323, 324
Oliveira CC, 714, 715
Oliver D, 280, 293, 294
Oliver DB, 293, 294
Oliver J, 294
Olivera BM, 677
Oller AR, 68, 69
Olmsted MC, 173
Olovnikov AM, 339
Olsen GJ, 343
Olshevskaya E, 257
Olshevsky U, 743
Olson MOJ, 678
Oltvai ZN, 597
Omata S, 282, 284
Omer CA, 245, 247, 249, 260
Omura S, 820, 821
Ong ES, 594
Ono Y, 219, 379, 399
Onorato IM, 216, 221, 233, 234
Onrust R, 63, 646, 751
Onuchic JN, 548, 549, 554
Onura S, 821
Onyekwere O, 72
Oppliger W, 290
Orban PC, 143

Ordway DJ, 235
Orefici G, 218
Orgambide GG, 505, 511–13, 518
Orii H, 669
Orino E, 816, 825, 840
Orita S, 258
Orkin SH, 727
Orlandi R, 758
Orlinsky K, 748
Orlova EV, 759, 763
Orlow SJ, 425
Orlowski M, 802, 803, 814, 815, 821, 838
Orme IM, 235
Ornberg RL, 479
Oro AE, 263
Oroszlan S, 639, 754
Orphanides G, 636, 645, 650, 651
Orren DK, 50, 51, 70
Orr-Weaver TL, 122, 831
Orth K, 120, 836
Ortiz NV, 837
Ortiz-Navarrete V, 836
Ortlepp SA, 756
Osaka F, 809, 840
Osawa T, 591
Osborne HB, 696, 709
Osborne RS, 293
Osheim YN, 368
Osheroff N, 636, 642, 645, 646, 648–51, 666, 670, 671, 677
Oshimura M, 159
Oshino N, 570
Osslund TD, 612
Oster GF, 290
Ostermeier C, 538, 547, 559
Ostrander EO, 659
Osvath P, 540
Otori Y, 317, 318
Otrin VR, 61, 150
Ott G, 756
Ottemann KM, 278
Otter R, 641, 657
Ottersen OP, 317, 322
Otto A, 282
Otto JC, 259
Ottolenghi AC, 227
Oubridge C, 377
Oudar O, 839
Oudega B, 275
Overman LB, 173
Overton L, 246
Ovilson MT, 568, 571
Ovodov SY, 756
Owen-Hughes T, 772
Owicki JC, 479
Oxender DL, 306, 329, 332
Oyamada M, 494
Ozato K, 149
Ozer Z, 45
Ozkaynak E, 662

P

Pabich EK, 21, 54
Pablos-Mendez A, 216
Paccaly-Moulin A, 225
Pace NR, 343
Pacholczyk T, 318
Pachter JS, 695
Paddon-Row MN, 548
Padgett RA, 369, 371, 376, 391, 401
Paek I, 696
Paetkau V, 729
Page E, 477, 479, 491
Paika K, 675
Paillard S, 646
Painter A, 598
Pajic A, 584, 588
Pajor AM, 327, 328
Pal JK, 807, 839
Palacin M, 327–30
Palandjian L, 386
Palas KM, 175, 206
Palatnik CM, 696, 724
Palcic MM, 507, 510, 512, 513, 515, 522, 528
Palladino F, 349
Pallard C, 105, 110, 123
Palmer A, 802
Palmer JE, 68
Palombella VJ, 803, 819, 820, 834, 835
Palombo F, 105, 109, 111, 112, 120
Palumbo L, 223, 224
Pamnani V, 808
Pan LP, 547
Pan M, 310
Pan ZQ, 57, 63, 73, 74
Pande S, 749
Pandey NB, 695, 712
Pandit J, 612
Pandit SD, 647
Pang PP, 103, 110
Pang Q, 112, 113
Panganiban AT, 754
Panicker MM, 175
Pantoliano MW, 620
Panusz HT, 713
Panzner S, 278, 282, 289
Paonessa G, 618, 621, 622
Papa FR, 831, 833
Papadopoulos N, 105, 108–10, 112, 113, 117, 119, 120
Papadopoulou B, 585
Paparelli M, 676
Papini E, 567
Papp A, 718
Pappin DJC, 67, 144, 158
Paranjape SM, 19, 23
Pardee AB, 636, 672, 820, 839
Pardon B, 661
PARENT CA, 411–40; 416, 435

Pariat M, 832, 838
Parikh VS, 565, 593
Paris J, 696, 706, 709
Paris KJ, 245
Parish CA, 256
Parisi MA, 579
Park CH, 31, 33, 54, 55, 57, 58, 60, 63, 65, 66, 74, 145, 149, 152, 153
Park E, 21, 34
Park HW, 45, 51
Park IK, 832
Park J-G, 121
Park K, 121
Park MK, 598
Park SB, 263, 276, 583, 589
Park SJ, 444, 459
Parker BO, 103, 111
Parker KA, 372
Parker R, 729
Parker RA, 374, 390, 391, 694–98, 700, 702, 703, 705, 706, 709–11, 722, 723, 726, 729
Parkhurst K, 782
Parkhurst L, 782
Parkin N, 750
Parkin NT, 751
Parmacek MS, 330
Parsons R, 72, 73, 108, 109, 117, 119–21, 356
Parsons RE, 119, 120
Part D, 426
Partaledis JA, 593
Parvin JD, 57, 776, 782
Pas E, 679
Pascher T, 545, 553, 554, 557
Pasion SG, 640
Pasman Z, 381
Pastan I, 262
Pastor-Anglada M, 307
Pastori RL, 728
Patel DV, 260
Patel SS, 174–76, 180, 181, 184, 185, 188, 189, 192, 207, 639
Paterson H, 256, 257
Pathak RL, 245, 257
Patil CK, 159
Patil DS, 97
Pato ML, 645, 672
Patriarca EJ, 598
Patterson B, 371, 372, 388, 397, 400
Patterson C, 356
Patterson MC, 69
Patterson TE, 567–69, 571, 572
Patton JG, 377, 378, 380, 386–89, 392, 393, 401
Patton R, 260
Pattyn SR, 223
Patzelt E, 401
Paul CP, 743, 744, 752, 753
PAUL DL, 475–502; 476–78, 483, 485, 487–89, 491–94

Paul LS, 197, 199
Paul M, 477
Paul M-F, 584, 588
Paulson JC, 444, 452, 454, 465
Paulus BF, 67
Pause A, 180
Pavlovsky A, 612
Pawlowski K, 525, 526, 528, 529
Pawson T, 385
Payne A, 149
Payne MJ, 583
Pearce DA, 584
Pearl L, 51, 140
Pearlberg J, 793
Pearlstein RM, 546
Pearson A, 72, 73, 791
Pearson AJ, 223
Pearson GD, 674
Pecht I, 540, 553, 554, 557
Peck HD Jr, 97
Pederson T, 392
Peebles CL, 636, 650
Pees E, 506
Peinado MA, 118–20
Pel HJ, 577, 578, 580, 582, 586, 745
Pelegrini O, 426
Pelham HR, 743
Pelissier PP, 586
Pelletier DB, 494
Pelosi E, 761
Pelsy F, 714–16
Peltomäki P, 118–20
PELTZ SW, 693–739; 694–98, 700, 703, 709, 712–15, 718, 719, 726
Penczek P, 759, 763
Pendergrass WR, 358
Penfield KW, 541
Peng H, 638, 644, 663
Penman S, 402
Pentz ES, 831
Peppel K, 697, 703
Pepperl S, 595, 599
Per SR, 639
Peracchia C, 491
Peracchia LL, 491
Perbal B, 377, 378, 384
Pereira ME, 814
Pereira-Smith OM, 356, 358
Perez-Howard GM, 782
Perez-Sala D, 254, 261
Perkins A, 827
Perkins GA, 811
Perlman PS, 371
Permana PA, 662, 664
Perret X, 507, 525, 527
Perry RP, 712
Perucho M, 118–20
Perzova R, 490
Peskin CS, 290
Pesold-Hurt B, 808, 809, 840
Pestova TV, 389

Peters G, 749
Peters JM, 804, 807, 816, 822, 823, 825, 839
Peters NK, 519, 527, 528
Petersen LN, 70
Petersen-Bjorn S, 248
Petersheim M, 540
Peterson CA, 33, 55, 57, 58, 63, 149, 150, 153
Peterson CL, 36
Peterson GM, 119, 120
Peterson J, 102, 103
Peterson PA, 809, 836, 838, 840
Peterson SR, 159
Petes TD, 112, 114, 115, 121–23, 348, 352, 416
Petit F, 808
Petit JF, 218, 236
Petit M-A, 126
Petit PX, 597
Petrin J, 260
Petrini JHJ, 158
Petrov P, 676
Petrovics G, 507, 509, 518, 526
Petryniak B, 704
Petsko GA, 454, 455
Pette D, 574, 575
Pettersson I, 372
Petzold SJ, 671, 677
Pewitt EB, 541
Pfanner N, 587
Pfeffer SR, 258, 259
Pfeifer G, 805, 811
Pflugfelder G, 641, 661
Pflugrath JW, 452
Pfund WP, 661
Pham HD, 581
Philbrick WM, 389
Philipp W, 224
Philippe M, 696, 709
Philips MR, 254, 261, 262
Philipson L, 743
Phillipov P, 257
Phillips AM, 50, 51
Phillips BA, 696
Phillips DH, 24, 26
Phillips DR, 23
Phillips GJ, 275
Phillips JA, 623
Phillips S, 697
Phillips SL, 697
Phillips WC, 479–84, 491
Piatyszek MA, 341, 358–60
Piccoli C, 487
Pichon M, 523, 530
Pickart C, 824, 830, 832
Pickart CM, 832
Pickel VM, 328
Picking WD, 759
Pictet R, 793
Pidsley S, 671
Piechaczyk M, 695, 696, 703, 832, 838

Pieper RO, 116
Pierce RJ, 755, 757
Pieringer J, 746
Pierrat B, 695
Pierre M, 526
Piersimoni C, 223
Pietras DF, 695
Pikielny CW, 372, 374
Pil PM, 61
Pilch DR, 712
Pilkington SJ, 568
Pillinger MH, 254, 261, 262
Pinaud S, 426
Pind S, 820, 821
Pines G, 307, 315, 317, 319, 320
Piñol-Roma S, 378, 388, 389, 392, 695
Pintel DJ, 714, 725
Pinto AL, 390, 397
Pinto I, 714, 715, 718, 722, 776
Piomelli D, 491
Pious D, 574
Pitcher JA, 424, 431
Pitt GS, 419, 420, 422, 426
Pittenger C, 144
Pitts JD, 484
Pitzer F, 806, 807, 839
Piwnica-Worms H, 493, 494
Pjura P, 780
Plaitakis A, 315
Plakidou-Dymock S, 318
Planondon L, 821
Plath K, 278
Platt T, 171, 178, 208
Plazanet C, 510, 513, 518, 520
Plebanski M, 378, 380, 383, 384
Pleij CWA, 750, 751, 754
Pleij K, 343
Plevani P, 149
Ploegh H, 835
Ploegh HL, 837
Ploetz K, 94
Plug A, 105, 112, 113, 121, 123, 124
Plumpton M, 398, 399
Plunkett BS, 72, 356
Plunkett GD, 810
Pluta AF, 342
Pluznik DH, 704
Podar M, 371
Podder SK, 458
Podgorski GJ, 425, 431
Podust LM, 151
Podust VN, 66, 149, 151, 158
Poenie M, 254
Pogliano JA, 280, 293, 294
Poiret M, 526
Poirier GG, 144
Polaina J, 280
Polanyi J, 640
Polard P, 752
Politino M, 85, 89
Poljak L, 636, 666

Polyak K, 782
Pommier Y, 636, 642, 643, 645, 646, 649, 666, 669, 671, 674, 676
Pompliano DL, 247, 260
Pompon D, 125, 126, 128
Pon L, 564, 577, 578, 580, 582, 586, 587
Pon NG, 659
Pongracz K, 140
Ponte L, 105, 123, 124
Ponthus C, 505, 508, 510, 513, 523
Poole ES, 745, 753
Poole TL, 702
Poolman B, 314, 315
Poon D, 773, 775, 792
Poon RYC, 20
Popanda O, 151
Popielarz M, 377, 378, 380, 384, 400
Popoff SC, 142
Porfiri E, 251, 258
Poritz MA, 274, 275, 278
Porro EB, 378, 388
Port JD, 704
Portaels F, 222
Portemer C, 638
Porter SE, 348
Porter SJ, 759
Portillo-Gomez L, 218, 230
Posner BA, 421
Possekel S, 576, 586
Pot B, 504
Potash MJ, 714
Potashkin J, 378, 387
Poteete A, 102
Potmesil M, 636
Potter DD, 485
Pouch MN, 808
Poulet S, 227
Poulter CD, 247
Poulton J, 580
Pound A, 641, 642
Poungouras P, 727
Poupart P, 695, 708
Poupot R, 505, 508, 510, 515, 518–20, 527
Povirk LF, 669
Powelson MA, 664
Power MD, 751
Power SD, 567, 571
Powers R, 612
Powers S, 242, 243, 245–47, 249, 251
Powers T, 273
Powis SH, 836
Powls R, 540
POYTON RO, 563–607; 57, 565–69, 571–76, 585, 587, 589, 593, 596, 599
Prabhakaran K, 236
Prakash L, 16, 18, 34, 44, 47, 53,

55, 57, 59, 60, 64–66, 115, 143, 144, 146, 148, 150, 152, 153, 171, 180, 192
Prakash S, 16, 18, 34, 44, 47, 53, 55, 57, 59, 60, 64–66, 115, 144, 146, 148, 150, 152, 153, 171, 180, 192
Pramanik BC, 818, 837
Pras E, 331
Prasad J, 385
Prasad L, 444, 456, 463
Prasad R, 67, 143
Prassad PD, 330
Pratt G, 804, 813, 817–19, 822, 823, 825, 836, 838
Praznovszky T, 674
Preciado GT, 583, 589
Prehn S, 275, 276, 278, 279, 281, 286, 288, 292, 294, 831
Preisler H, 696
Prell B, 649
Prendergast GC, 245, 261
Prere MF, 752
Prescott DM, 338, 339, 341, 347, 351
Prescott J, 643
Presnell SR, 619
Preston BD, 148
Preuss U, 378, 384
Price A, 142, 143
Price CM, 347, 349, 351, 353, 355
Price DH, 22
Price NPJ, 505, 508, 509, 513, 516, 518, 520, 523, 529
Pridgen C, 696
Priebe SD, 126, 128
Priel E, 639
Priestley A, 159
Prigent C, 158
Pringle JR, 248, 249
Prioleau M-N, 780
Prives C, 38, 72
Proffitt JH, 114
Prokipcak RD, 695, 728–30
Prolla TA, 105, 112–15, 122, 123
Promé D, 505, 507–9, 511–13, 519, 520, 523, 529
PROMÉ J-C, 503–35; 505, 507, 508, 510–16, 519, 520, 522, 523, 527
Pronk SE, 445, 452
Proske RJ, 818, 825–27, 830, 831, 833
Prosperi E, 151
Protic M, 61, 149, 153
Proudfoot AEI, 612, 622
Prowse KR, 339, 341, 352, 357–60
Prufer D, 752
Pruss D, 788
Pruss GJ, 660
Prywes R, 791, 792

Pryzbyla AE, 97
Ptashne M, 387
Pueppke SG, 505, 507, 519, 520, 523
Pugh B, 773, 778, 782
Pugh F, 389
Pühler G, 805–8, 811–13
Pujol C, 510, 513, 518, 520
Pukkila PJ, 102, 103
Pulak R, 714, 718
Pulleyblank DE, 642
Pumphrey J, 809
Pupillo M, 419–21
Puppi M, 309, 310, 312, 326
Puranam RS, 307, 319, 320
Purvis IJ, 696
Putney JW Jr, 430
Puvion F, 803, 808
Puziss JW, 275
Pyke KA, 640
Pylkkänen L, 118, 119

Q

Qadota H, 249
Qian L, 714, 725
Qian Y, 746
Qian YM, 260
Quackenbush E, 307, 330
Quail JW, 444, 456, 463
Quan Y, 714, 716, 717, 722–24
Quemard A, 226–28, 230, 235
Query CC, 369, 372, 374, 375, 390, 391, 397
Qui H, 34
Quick S, 729
Quiclet-Sire B, 505, 508, 510, 512, 523, 524, 527, 530
Quinn FD, 235
Quinto C, 505, 507, 516, 519
Quiocho FA, 451, 452, 460, 461, 469

R

Rabadan-Diehl C, 485, 486, 492
Rabinovich D, 148
Rachmilewitz EA, 714, 715, 725
Radding CM, 122, 125
Radley E, 836
Radman M, 102–5, 107, 110, 113, 117, 121, 123, 125–28, 137
Raes M, 598
Ragan CI, 567
Ragnini A, 252
Rahman S, 477, 486
Rahmouni AR, 659
Rahmsdorf HJ, 703, 709
Rainey WE, 360
Raitio M, 568, 573
Raitt D, 595, 599
Rajapandi T, 293, 294

RajBhandary UL, 763
Rajewsky K, 143, 837
Raji A, 660
Ramakrishnan B, 780
Ramakrishnan T, 217, 228
Ramakrishnan V, 762, 786
Ramchatesingh J, 382, 383
Ramin VC, 728, 729
Ramirez BE, 559
Ramkumar R, 458
Rammensee HG, 837
Ramon F, 491
Ramos W, 67, 158
Ramotar D, 142
Ramsden D, 159
Randall LL, 273, 275, 276
Randerath E, 749
Randerath K, 749
Randhawa B, 236
Rando OJ, 803, 819, 820, 834, 835
Rando RR, 252–54, 256, 261
Rands E, 244, 245, 247, 248, 260, 261
Ranjeva R, 507, 515, 522, 528
Rao A, 254
Rao G, 598
Rao MR, 105, 110, 118, 119
Rao MS, 594, 595
Rao PN, 660, 667
Rao ST, 780
Rapiejko PJ, 273, 281
RAPOPORT TA, 271–303; 272, 273, 275, 276, 278, 279, 281–84, 286–90, 292, 294, 295, 297
Rappold GA, 355
Rasmussen BD, 623
Rasmussen TP, 831
Rasschaert D, 751
Rassow J, 290, 587
Rastoggi N, 218, 230, 234
Ratcliffe PJ, 599
Ratet P, 506, 507, 519
Rath VL, 278
Rathi A, 412
Rathjen PD, 751
Rathmell WK, 159
Ratnasabapathy R, 728
Ratner MA, 543
Ratrie HR III, 676
Rattray AJ, 121
Rau DC, 651
Raue HA, 698, 729
Rauen T, 316, 317
Rauscher F, 780
Ravid S, 427
Raviola E, 479, 496
Raviola G, 479
Rawlings ND, 814
Rawson TE, 260
Ray BK, 171, 180
Ray BL, 122–24, 775

Ray DS, 640
Ray P, 275
Raymond M, 262
Raymond V, 710
Rayssiguier C, 125, 128
Razin A, 114
Razin SV, 666
Read D, 102
Read GS
Read MA, 819, 820, 834
Read R, 92, 758
Reagan MS, 68, 69, 144
Realini C, 818, 819, 838
Ream W, 92, 757
Reardon JT, 19, 31, 33, 35, 45, 47, 54, 55, 57, 58, 60, 61, 64–67, 73, 145–47, 149, 150, 152, 153
Rebbert ML, 531
Rebora A, 35
Rebstein PJ, 425
Rech J, 695, 696, 703
Rechsteiner M, 802, 804, 813, 817–19, 822–25, 827, 828, 830–33, 836, 838, 839
Recillas-Targa F, 805, 812, 840
Record MT Jr, 172, 173, 200
Reddel R, 358
Reddy JK, 594, 595
Reddy MK, 640
Reddy S, 493
Redfield C, 612
Redston M, 121, 123
Reece RJ, 645
Reed MT, 217, 221
Reed R, 87, 91, 372, 377–79, 381, 382, 386, 389, 391–96, 401
Reed RR, 422
Reed SI, 21
Reeder RH, 139, 661, 770
Reenan RA, 105, 110, 114, 121–23, 125, 126, 128
Reese JC, 773, 775
Reese RJ, 636
Reese TS, 210, 479
Reeves OR, 484
Regan JD, 67
Regan JJ, 549, 552–54, 557
Reggiardo Z, 221
Reha-Krantz LJ, 175, 179
Reichelt J, 392, 393
Reid BR, 743
Reid GE, 760, 837
Rein A, 754
Reinagle P, 793
Reinauer H, 805
Reinberg D, 19, 22, 26, 30, 33, 35, 57, 58, 64, 69, 678, 770, 771, 776, 783, 791, 792
Reines D, 23, 24, 68
Reinhard P, 66, 151, 152
Reinhardt KR, 821

Reinhart GD, 283, 286, 287, 289
Reinhart U, 701
Reinheckel T, 839
Reinmann A, 570, 575, 576
Reis DJ, 599
Reiser J, 391
Reiss Y, 244–46, 248
Reitmair AH, 121, 123
Reitman ML, 456
Reizer A, 307, 309, 507, 518
Reizer J, 307, 309, 507, 518
Relic B, 504–6, 508, 513, 519, 520, 523, 525, 527, 529
Remacle J, 598
Rempel R, 349
Renard P, 598
Rendahl KG, 390
Rennie MJ, 312
Renz M, 297
Rep M, 745
Resnick MA, 126, 128
Restifo LL, 696
Restle T, 171
Reumkens J, 84, 92
Reusch P, 623
Reuter R, 380
Reveillaud I, 390
Revel JP, 477, 478, 483, 486, 487, 492
Revel V, 224
Reymond CD, 424–26
Reynolds AE, 660, 663
Reynolds P, 146
Reynolds SE, 826, 840
Rhijnsburger EH, 426
Rhode PR, 574, 575, 596, 597
Rhodes D, 349, 788
Ribas JC, 706, 707, 711
Ribi HO, 481
Ricciardi R, 791
Rice CM, 754
Rice KG, 471
Rice M, 294
Rice P, 423
Rich A, 659, 780
Richard-Foy H, 36
Richardon DC, 482
Richards E, 352
Richards JG, 326
Richards JH, 540, 549, 550, 552, 553, 557, 558
Richards SH, 115
Richards WG, 615, 622, 623
Richardson CC, 104, 178, 180, 192–94, 196, 197, 208
Richardson JP, 175, 179
Richardson JS, 482
Richardson RW, 194, 208
Richardson SMH, 660
Richter A, 661, 664, 673, 674
Richter JD, 706, 711
Richter-Ruoff B, 588, 816
Rieber P, 598

Riedl A, 694, 695
Riegler N, 228
Riemen G, 579
Rienitz A, 322
Riezman H, 821
Rigaud G, 793
Rigby WFC, 703, 704
Rigoulet M, 570
Rigual R, 599
Riley K, 412
Riley M, 567
Rilling HC, 244, 262
Rine J, 241, 242, 249, 252, 253, 255, 261, 349
Ring D, 196, 197
Ring HZ, 383
Ringe D, 454
Rini JM, 443, 444, 450, 456, 458, 460, 463
Rinkes IHMB, 311, 312
Rio DC, 368, 378, 387, 400
Riond J, 522, 528
Riordan JR, 820, 821
Ripley LS, 115
Risebrough RW, 695
Risinger JI, 108, 109, 117, 120
Risler Y, 597
Risman SS, 580
Risse B, 279, 280, 290
Ritsema T, 507, 511, 529
Rittel W, 217, 236
Ritter S, 647, 651
Rivett AJ, 802, 813, 817, 837
Rivilla R, 507, 517
Rizo J, 246
Rizzuto R, 576
Robatzek M, 105, 112, 113, 121, 123, 124
Robberson BL, 381
Robbins JH, 53, 145, 150
Robbins PW, 509, 531, 791
Robbins SM, 424
Roberge M, 349, 665, 675
Roberts E, 158
Roberts GD, 223
Roberts JD, 107–9
Roberts JM, 55
Roberts M, 639
Roberts S, 761
Roberts SGE, 791–93
Roberts TM, 496
Roberts WK, 706
Robertson AJ, 490
Robertson JD, 478
Robertus JD, 444
Robins P, 67, 144, 148, 158
Robinson D, 309–12
Robinson KM, 567, 589
Robinson RC, 612
Robinson SS, 260
Robishaw JD, 259
Robison SH, 26
Robson CN, 142

Roca J, 636, 645, 646, 649, 651, 656, 664, 667, 674
Rochaix J-D, 695
Roche P, 505–8, 510, 512–15, 518–20, 523, 526, 527, 529, 530
Roche SE, 378, 387
Rochepeau P, 504, 506
Rock KL, 802, 805, 807, 808, 814–16, 819, 820, 825, 835–37
Rodel G, 583, 589
Roder H, 545
Rodriguez-Sanchez JL, 757
Rodriquez JC, 595
Roe J-H, 172, 200
Roedel G, 585
Roeder GS, 105, 123, 124, 126, 400
ROEDER RG, 769–99; 678, 770–75, 780, 784, 791–95
Roemisch K, 296
Rogel MA, 518, 527
Rogers EJ, 759
Rogers J, 372
Rogers PJ, 575
Rogers SW, 819
Roginski RS, 485
Rogness C, 574, 592, 596
Rohde K, 297
Rohde W, 752
Rohdich F, 570, 575, 576
Röhrig H, 507, 510, 512, 515, 522, 528, 529, 531
Rohrwild M, 810, 829
Rolfe M, 637, 667, 669, 673
Rolfson DH, 674
Roll D, 676
Rolley NJ, 750, 751
Rollins RA, 584
Rom E, 746, 747
Roman LJ, 198, 199, 201
Romano LJ, 23, 197
Romanos MA, 569
Romero DP, 343
Romero MF, 316, 317
Romero-Severson J, 352
Romig H, 674
Römisch K, 273, 275, 278
Rong Z, 153
Ronne H, 122
Rook MB, 491, 492, 494
Rooney SA, 226
Roos BA, 495
Roos G, 358, 359
Ropp PA, 143
Rosa F, 531
Rosbash M, 369, 372, 374, 377, 388, 394, 696, 714, 715
Roscigno RF, 372, 381
Rose B, 494, 495
Rose D, 662
Rose IA, 455, 826, 832

Rose JK, 286, 295
Rose KM, 661
Rose MD, 279, 280, 289, 290, 292
Rosen C, 827, 828
Rosen CA, 827
Rosen N, 820
Rosenberg AH, 744
Rosenberg C, 506, 509, 511, 514, 519
Rosenberg S, 246
Rosenberg SM, 668
Rosenfeld KL, 54
Rosenfeld MG, 254, 261, 262, 717
Rosenstein BJ, 762
Rosenthal E, 834
Rosner JL, 227
Ross J, 694–98, 700, 703, 709, 712–15, 725, 728
Ross KL, 511, 514, 518
Ross R, 242
Ross WE, 636, 665, 674, 675
Rossen L, 507, 511, 516
Rossi F, 385
Rossi G, 251, 252
Rossi R, 151
Ross-Macdonald P, 105, 123, 124, 126
Rossmann MG, 228
Rossomando PC, 748
Roth M, 351, 380, 382, 383
Roth MB, 377, 378, 380, 381, 383, 384
Rothblatt J, 276, 279, 280, 289, 290
Rothblatt JA, 279, 292
Rothfield NF, 672, 676
Rothstein L, 802, 816, 819, 820, 835
Rothstein RJ, 122, 123, 125, 126, 128, 149, 160, 637, 667, 669, 673, 677
Roti R, 674
Rotig A, 586
Rotman G, 160
Rottier PJM, 286, 295
Rottman FM, 382, 383
Rotzschke O, 837
Rouault TA, 728–30
Rougé P, 444, 448, 450, 452, 456, 458, 459, 463
Rouse DA, 219, 228
Rousseau DL, 538
Roussel A, 444, 450, 459, 463
Rousset J-P, 754
Roussey G, 70
Roux J, 793
Roveri A, 84, 92, 95, 98
Rowan S, 392
Rowe AJ, 817
Rowe TC, 636, 652, 661

Rowe WP, 308
Rowell C, 260, 261
Roy AK, 696, 706, 773
Roy AL, 792, 795
Roy R, 19, 20, 30–33, 37, 38, 54, 58, 70, 73, 141, 150
Rozakis-Adock M, 619
Rozen F, 398
Rozwarski DA, 612
Rubartelli A, 297
Ruben SM, 105, 108, 112, 113, 117, 119, 120, 827
Rubin DM, 802, 822, 828, 829
Rubin EH, 678
Rubin GM, 390
Rubin JB, 488, 489
Ruby SW, 372, 388, 394, 398
Ruch BC, 619
Ruch RJ, 494, 495
Rucknagel P, 87
Rudd MD, 23
Rudert WA, 307, 315
Rudikoff S, 809
Rudner DZ, 698, 705
Rudnicki KS, 126, 128
Rudolph JA, 69
Ruegg JC, 586
Ruf B, 234
Ruigrok RWH, 452
Ruiz-Echevarria MJ, 714, 716, 717, 719, 722–24
Ruiz-Herrera J, 593
Rumessen JJ, 477
Rundell K, 703
Runge KW, 352, 353
Runswick MJ, 180
Runyon GT, 175, 176, 190, 192, 206
Rupp HL, 494, 495
Rupp WD, 47, 57, 145, 146, 149
Ruppert JM, 72, 73
Ruppert S, 773–75, 791
Ruppert T, 836–38
Rusche JR, 652
Rush MG, 261
Ruskin B, 369, 372, 374, 386, 394, 401
Russell DW, 245, 246, 670
Russell JB, 760
Russo A, 671
Russo AA, 782
Rustin P, 586
Rutenber E, 444
Ruther U, 703, 709
Ryabova LA, 756
Ryan AJ, 24, 26, 671
Rydberg B, 102, 140
Ryder U, 372, 374
Rymond BC, 369, 372, 374, 377, 388
Ryner LC, 717
Ryo H, 45, 154, 155

S

Saarma N, 753
Sabatelli P, 676
Sabatier L, 358
Sabatini DD, 284, 286
Sabbag RV, 307, 327, 328, 331
Sabesan S, 444, 463, 465, 468
Sacchettini JC, 227, 228, 230, 444, 463, 468
Sachs A, 384, 698
Sachs AB, 697, 698, 700, 705, 707, 711
Sachs CW, 262, 263
Sack JS, 452
Sadler I, 279, 280, 290
Sadlock J, 576
Sadoff BU, 669
Sadofsky M, 159
Sadowsky MJ, 507
Saez JC, 488, 492–95
Safarik R, 482
Saffitz JE, 477
Saffran WA, 148
Safir SR, 216, 232
Safran M, 93
Sage E, 27
Sage S, 176
Sagliocco FA, 727
Sahai BM, 674
Sahakian JA, 839
Saidowsky J, 804, 805
Saier MH Jr, 307, 309, 507, 518
Saijo M, 55, 57, 63, 148, 149, 153, 161, 674, 676, 679
Saiki K, 589
Sailer A, 387
Saing K, 669
Saito M, 620
Saito T, 59
Saito Y, 809, 817, 836
Saitoh N, 666, 679
Saitoh Y, 667
Sajjadi FG, 595
Sakagami Y, 242
Sakai H, 273, 274, 284, 286, 287, 293
Sakajo S, 593, 595
Sakamoto A, 73, 522
Sakamoto H, 372, 401, 717
Sakamoto M, 818
Sakker RJ, 37
Sakmann B, 485
Sakumi K, 141
Sakurai A, 242
Sakurai H, 674
Sala C, 674
Sala E, 151
Salazaar EP, 57
Salazar C, 45
Salazar EP, 31, 60, 147, 148
Salazar L, 224
Salcedo-Hernandez R, 593

Salditt-Georgieff M, 696, 697
Salfinger M, 232
Salles FJ, 706
Salmon GA, 540
Salmons S, 574, 575
Salovaara R, 119, 120
Salvat C, 832, 838
Salvati AL, 621
Sambrook J, 120
Sampson E, 668
Samson L, 45, 140
Samuel M, 641, 642
Samuels DS, 672, 676
Samuels N, 840
Sanborn BN, 575
SANCAR A, 43–81; 16, 18, 19,
    23, 25, 28–31, 33, 43–45,
    47, 49–51, 58–61, 65, 67, 71
Sancar GB, 16, 44, 45, 50, 71,
    154
Sanchez A, 161
Sanchez G, 23
Sandbaken MG, 98, 753, 757
Sander M, 644, 679
Sanders SL, 279, 280, 289, 290,
    292
Sanderson CJ, 622
Sanderson MJ, 491
Sandmeyer S, 748
Sandona D, 569
Sands AT, 161
Sanfridson A, 703
Sänger HL, 808
Sanjuan J, 505, 507, 512, 513,
    515, 516, 518–20, 523
Sanjuro JL, 216, 232
Santa-Ana AS, 242, 243, 245, 249
Santana O, 507, 516
Santi S, 676
Santiago TC, 696
Sanz P, 275, 289
Saparbaev M, 141
Sapperstein S, 255, 262
Sapperstein SK, 255
Sarafian TA, 597
Saraste M, 180, 423, 567, 571
Sarau HM, 676
Sareen M, 231
Sarfaty S, 445, 448, 452, 463,
    464, 467, 468
Sargent TD, 531
Sarkar SN, 72
Sarnecki C, 496
Sarnoff J, 763
Sarnow P, 706
Sarti P, 568, 571
Sasaki AW, 714, 725
Sasaki S, 105, 112, 113, 280, 293
Sasaki T, 151, 250
Sastry SS, 23
Satchwell SS, 780
Sato K, 662, 663
Sato M, 69, 662–65, 675

Sato N, 610, 611, 620, 622
Satoh K, 826
Satoh MS, 139, 144, 145, 158
Satokata I, 55
Saudubray JM, 586
Sauer F, 133
Sauer RT, 760, 761
Saumweber H, 378, 390, 664, 665
Saunders CA, 697
Sauter NK, 444, 449, 452, 454,
    459
Savagnac A, 505, 519, 520, 523
Savant-Bhonsale S, 704, 710, 711
Savarese TM, 432
Savino R, 618, 621, 622
Savitsky K, 160
Savitz AJ, 282, 284
Savory PJ, 802, 813
Savoure A, 526
Savva R, 51, 140
Sawa H, 372, 374
Sawada H, 822, 823, 826
Sawada MT, 822, 823
Sawadogo M, 771, 772
Sawers G, 84, 85, 92, 762
Sawicki SG, 696, 697
Sawitzke JA, 636
Sawyers G, 89
Saxe CL III, 416, 418, 419, 434
Saxena IM, 509
Sayer PJ, 675
Sayre MH, 828
Scadden D, 307, 308, 310
Scadden J, 376
Scalise G, 223
Scardaci G, 218
Scarpulla RC, 592
Schaack J, 704
Schaak J, 672
Schaal T, 377, 378
Schaap P, 419, 426, 429
Schaaper RM, 68, 69, 102, 103,
    148
Schaber MD, 244, 245, 247, 248
Schaechter F, 356, 359
Schaefer AW, 704
Schaefer TS, 754
Schaefer WB, 226
Schaeffer L, 19, 30–32, 36–38,
    54, 55, 58, 73, 150
Schafer JA, 328
Schafer WR, 241, 242, 246, 255
Schaffer LA, 620
Schaffner W, 640
Schaid D, 118, 119
Schalk-Hihi C, 615, 623
Schaller J, 477
Schaloske R, 421
Schapiro JM, 331
Schatz A, 216, 219
Schatz G, 564, 565, 572, 577,
    578, 580, 582, 584, 586–88
Schatz M, 759, 763

Schatz PJ, 272, 276, 278, 280
Schauer TM, 805
Schedl P, 672, 717
Scheer A, 262
Scheffner M, 171, 175
Scheimer C, 142
Scheja K, 576
Schekman R, 275, 276, 278–80,
    289–92
Schekman RW, 290, 292
Schell J, 507, 510, 516, 522, 529
Schellman C, 555
Schellman JA, 555
Schenk T, 704
Scheper W, 728, 729
Scherer D, 835
Scherer SS, 477, 498
Scheres B, 523
Scherly D, 37, 59, 146
Scherrer K, 707, 708, 802, 803,
    805, 807, 808, 811, 812, 839
Schiavi SC, 695, 705, 706, 709
Schiavo G, 567, 570, 575, 576
Schibler U, 368
Schiebel E, 276, 280, 292
Schiestl RH, 416
Schiller PC, 495
Schindler C, 610, 611, 630
Schindler R, 712
Schlaman HRM, 516
Schlatterer C, 419, 430
Schlenker S, 802, 803
Schlenstedt G, 279, 280, 289, 290
Schlerf A, 576
Schlessinger J, 620
Schmid HP, 802, 803, 808
Schmid M, 389
Schmid MB, 636, 639, 662, 678
Schmid SR, 171, 172, 180
Schmidt CJ, 36
Schmidt E, 419, 430
Schmidt EDL, 529
Schmidt F, 329, 330
Schmidt FH, 695
Schmidt H, 155, 385
Schmidt J, 507, 510, 512, 515,
    516, 522, 528, 529, 531
Schmidt KN, 598
Schmidt MC, 794
Schmidt PE, 525
Schmidt RA, 242
Schmidt RJ, 260
Schmidt SA, 496
Schmidt VK, 649
Schmits R, 121, 123
Schmitt B, 322, 677
Schmitt ME, 568
Schmitz J, 752
Schnabel P, 390
Schnall R, 584, 587, 588, 828
Schneider CJ, 242, 380
Schneider HC, 290
Schneider MJ, 755

Schneider RJ, 695, 710, 711, 727, 728
Schneider TD, 368
Schneiter R, 695, 704, 731
Schnipper LE, 674
Schnitzler GR, 431
Schnoes HK, 226
Schoborg RV, 714, 725
Schoenberg DR, 728
Schofield MA, 667
Scholder JC, 426
Scholz TD, 85, 89, 90
Schon EA, 576
Schonthal A, 703, 709
Schoolnik GK, 481
Schooltink H, 621, 622
Schreck R, 598
Schreiber SL, 808, 813, 821
Schripsema J, 526
Schroder KH, 230
Schroeder WE, 540
Schroer JA, 330
Schubiger K, 412, 426
Schuckelt R, 84, 92
Schulkes C, 426
Schulmeister T, 275
Schulte S, 307, 314, 317, 320
Schultes NP, 122, 123
Schultz DE, 196, 197
Schultz MC, 139, 661
Schultz PG, 225, 227, 228, 230
Schultz SC, 782
Schultz SJ, 20
Schultze M, 505–10, 512, 515, 519, 522–28, 530
Schulz GE, 180
Schulz PW, 591, 592
Schulz VP, 353
Schulze M, 583, 589
Schulze-Osthoff K, 598, 599
Schumacher TNM, 837
Schumperli D, 712
Schurmann D, 234
Schurra C, 216
Schwabe M, 621
Schwartz AL, 833
Schwartz H, 425
Schwartz JH, 803
Schwartz LM, 840
Schwartzman RA, 92, 93, 98
Schwarz E, 314
Schwarz HJ, 477
Schwarz MW, 194, 208
Schwedock JS, 507, 514
Schweizer D, 349
Schweizer E, 578
Schweizer J, 187
Schwer B, 398, 399
Schweyen RJ, 252
Schwob E, 830
Scicchitano DA, 23, 26, 27
Scidmore MA, 279, 280, 290
Scofield M, 349

Scorer CA, 569
Scott JF, 201, 205
Scott JR, 541, 558
Scott MO, 477, 498
Scott RA, 540, 565, 593
Scott RW, 820
Scovassi AI, 151
Screaton G, 378, 380, 383, 384
Scudiero DA, 116
Seabra MC, 244–46, 248, 250, 251, 254, 258
Seal BS, 309, 311
Seaver EC, 582
Sebald W, 623
Sebastian J, 154
Sebti SM, 260
Sedat J, 114, 338
Sedat JW, 667
Seddiqi I, 668
Sedgwick B, 16, 45, 137
Seeberg E, 50, 140
Seeger M, 838
Seelan RS, 576
Seelig A, 808, 809, 836, 840
Seelig HP, 297
Seemüller E, 809, 813, 814, 816
Segal S, 330, 639
Segall JE, 424, 771
Segel LA, 430
Sehgal PB, 696
Seifried SE, 174, 175, 177–79, 201, 202, 207
Seiler SR, 371, 392
Sekiguchi J, 643, 644, 652, 655
Sekiguchi M, 45, 46
Sekine Y, 752
Sela I, 695, 708
Selby CP, 23, 25, 28, 29, 48, 50, 51, 67, 68
Selby E, 105, 110
Selfridge J, 148
Selivanova G, 72
Selman MA, 540
Selva EM, 125, 126
Semino CE, 509, 531
Senger G, 836
Sen Gupta DJ, 175, 187
Senn H, 95
Sensel M, 694, 695, 728
Senshu T, 713
Sentenac A, 770, 780
Seo Y-S, 175
Seong C-M, 260
Sepp-Lorenzino L, 246, 261, 820
Séraphin B, 372, 388
Serizawa H, 19, 20, 58, 70
Seroz T, 19, 20, 31, 33, 58, 70
Serventi I, 697
Sessler A, 252
Setlow P, 45
Setlow RB, 67
Sevarino KA, 567, 571, 582, 585, 587, 588

Shaanan B, 444, 450, 463
Shadel GS, 578, 579
Shafer JA, 247
Shaffer R, 652
Shafqat S, 307, 319–22, 331
Shahinian S, 256
Shalloway D, 493
Shampay J, 341, 352
Shamu CE, 664, 665
Shan B, 575
Shanafelt AB, 622
Shaner SL, 172, 200
Shang J, 580
Shang J-G, 760
Shannon MF, 622, 623
Shao L, 180
Shao Z, 791
Shapiro RA, 696
Shapiro RI, 425
Shapiro TA, 640
Sharma A, 507, 644, 652, 653
Sharma SB, 514
Sharon N, 442–44, 450, 463, 468, 469
Sharp PA, 57, 368, 369, 371, 372, 374, 375, 378, 381, 388–90, 392, 396, 397, 401, 402, 776, 777, 782, 791
Sharpe CR, 715, 724
Shashidharan P, 315
Shatkin AJ, 701
Shatten G, 665
Shaulsky G, 419, 423, 424
Shaw AS, 286
Shaw G, 424, 695, 703, 708, 709
Shay JW, 356–58, 360
Shea JE, 379, 397
Shearman CA, 507, 511, 516
Shears SB, 428
Sheeley DM, 505, 507, 509, 511, 520, 523, 525
Sheen F, 339
Sheer D, 836
Sheetz MP, 210
Shekhtman EM, 639, 662, 663
Sheldon M, 19
Sheldrick KS, 144, 146, 155
Sheline C, 640
Shelness GS, 281, 288
Shelton ER, 651
Shelton JB, 540
Shelton JR, 540
Shen B-J, 623
Shen C, 642
Shen CKJ, 642
Shen LX, 750, 751
Shen P, 125, 126, 128
Shen Q, 88, 757
Shenk T, 672, 795
Shephard MJ, 548
Shepherd RG, 230
Sheppard JR, 493
Sheridan JD, 493

Sheridan KM, 242
Sheriff S, 466, 468
Sherman F, 123, 584, 695, 714–18, 722
Sherman M, 829, 831, 834
Shero JH, 672
Shi Q, 50
Shi Y-B, 23
Shi YQ, 254, 261
Shibata D, 118–20
Shibata M, 818
Shibuya H, 827
Shieh H-S, 612
Shiekhattar R, 19, 20, 58, 70
Shields GC, 782
Shiestl RH, 60
Shigenaga MK, 597
Shih D, 94
Shiina N, 834
Shim E-Y, 786
Shimada M, 105, 112, 113
Shimada S, 317, 318
Shimada T, 105, 110
Shimbara N, 805, 809, 828, 830, 831, 836, 840
Shimizu K, 258
Shimizu N, 672, 674, 676
Shimizu S, 488
Shimizu Y, 676, 831
Shimonishi Y, 256, 259
Shimura Y, 372, 374, 379, 382, 399, 401, 717
Shin SH, 389, 809
Shindo Y, 282, 284
Shinkai A, 280, 293
Shinnick TM, 222, 235
Shinohara A, 160
Shinshi H, 701
Shinya E, 110
Shiomi N, 59
Shiomi T, 54, 59
Shiozaki K, 663, 665, 675–77
Shipley A, 524, 529
Shippen DE, 341, 343, 353, 355
Shippen-Lentz D, 339, 341, 343, 351
Shirakawa M, 148
Shiue L, 356, 358
Shivji MKK, 33, 57, 66, 73, 146, 147, 149–52
Shobuike T, 702
Shoeb HA, 227
Shohmori T, 674
Shore DA, 338, 349, 350
Shore S, 350
Showater SD, 639
Showers M, 619
Shu L, 660
Shugar D, 701
Shum L, 121
Shuman S, 643, 644, 647–49, 652, 655, 669–71

Shure M, 642
Shuttleworth J, 20
Shyu A-B, 695, 697, 703–6, 708–10, 717, 722
Shyy S, 659
Sibghat-Ullah S, 52, 67, 116
Sibille C, 837
Sicard AM, 106, 108
Siddarth P, 549
Siddell SG, 751
Siddiqi SH, 221
Sidik K, 18, 46, 155
Sidransky D, 72
Siebel CW, 400
Siebel P, 576, 586
Sieber J, 123
Siede W, 16, 17, 27, 30, 35, 44, 46, 52, 57, 58, 67, 102, 136, 137, 144, 151, 153, 154, 159, 160
Siedlecki J, 640
Siegel NR, 826, 833
Siegel P, 360
Siegel V, 275
Sieliwanowicz B, 707
Sierakowska H, 392, 701
Sies H, 598
Sigler PB, 778, 802, 829
Sigmund CD, 714, 725
Sigrist CJA, 523
Sigrist H, 477
Sigworth FJ, 485
Sikorska M, 666
Silber R, 671
Silberman L, 674
Silbernagl S, 328
Silberstein S, 281, 282
Silcox VA, 221
Silhavy TJ, 275, 278, 293, 294
Siliciano PG, 374, 390
Silke J, 640
Sillekens PTG, 389
Silva-Pereira I, 805, 808, 812, 840
Silve G, 230
Silve S, 574, 575, 596, 597
Silver LL, 193
Silver P, 279, 280, 290
Silver PA, 279, 280, 290
Silverman N, 773, 792
Silverstein SC, 599
Silvius JR, 256
Siman R, 820
Simkovich N, 793
Simmons DL, 639
Simmons RM, 200
Simms S, 254
Simon FR, 574, 596
Simon JA, 69
Simon M, 21, 583
Simon MI, 259
Simon MN, 426
Simon SA, 491
Simon SM, 282, 283, 286, 290

Simonds WF, 254, 259
Simoni RD, 263
Simons JWIM, 27
Simons K, 257
Simonsson C, 763
Simpson L, 640
Simpson RJ, 621, 760
Sinclair P, 583, 589
Sinelnikova VV, 257
Sinensky M, 242
Singer B, 140
Singer JD, 185
Singer MS, 341, 343, 345, 347, 358
Singer RH, 711, 731
Singer SJ, 283
Singh K, 574, 576
Singh R, 368, 386–88, 392, 704
Singh SK, 829, 832
Singh WP, 85, 89
Singhal RK, 67, 143
Sinn E, 773
Sinning I, 538, 554
Siomi H, 400
Siomi MC, 400
Sioud M, 638
Sirard J-C, 638
Sirsi M, 228
Sistonen P, 118, 119
Sitaram A, 27
Sitia R, 297
Sixma TK, 445, 452, 459
Skaanild M, 126
Skaggs LA, 762
Skai H, 273, 284
Skandalis A, 119, 120
Skehel DC, 444, 452, 454
Skehel JJ, 444, 452, 465, 466
Skehel JM, 568
Skopek TR, 102
Skopp R, 349
Skorupa ES, 580
Skoultchi AI, 712
Skourtis SS, 549
Skov LK, 549, 552–54, 557
Skowyra D, 830, 834
Skuzeski JM, 754
Slade MB, 416
Slaga TJ, 487
Slaughter CA, 246, 250, 807, 812, 817, 818, 825, 827, 830, 831, 838
Slesarev AI, 637, 638, 641
Sligar SG, 541, 558
Sloat BF, 248
Slocum PC, 217, 235
Slonimski PP, 580, 583, 589, 718
Slupphaug G, 139, 140
Sly WS, 762
Smale ST, 794
Small WC, 593, 594
Smallwood-Kentro S, 674
Smeal T, 72

Smeland TE, 245, 250, 251, 254, 258
Smerdon MJ, 25, 68
Smider V, 159
Smit G, 526
Smit NPM, 318
Smith BA, 54, 355, 356
Smith CA, 18, 24, 26, 45, 46, 61, 67, 149, 155
Smith CE, 826, 833
Smith CL, 356, 645
Smith CP, 307, 314, 316
Smith CWJ, 388
Smith EO, 576
Smith FW, 45
Smith GCM, 159, 160
Smith GR, 175, 187, 188, 192, 198, 206
Smith J, 94, 149, 160, 180, 398, 788
Smith JA, 228
Smith JL, 612, 809, 813
Smith JR, 356, 358
Smith KE, 307, 323
Smith KR, 200
Smith LG, 105, 112, 113, 119
Smith ML, 38, 73, 74
Smith PP, 356, 358
Smith QR, 309, 310
Smith RA, 381, 749
Smith SJ, 491
Smith TF, 36
Smith V, 399
Smith W, 376
Smith WW, 612
Smooker PM, 583
Smrcka AV, 256
Smulson M, 144, 677
Smyrk TC, 118, 119
Snaar-Jagalska BE, 420, 421, 423, 430
Snapka RM, 662, 664
Snapper SB, 235
Sng J-H, 639
Snowden A, 18, 49
Snyder HL, 835, 837
Sobek A, 807, 816
Sobell DI, 103, 110
Soda K, 94
Soderman DD, 245
Soede RDM, 419, 426
Sofi J, 674
Sogo JM, 658
Sogo LF, 590
Solary E, 671
Soldati D, 712
Soldati S, 258, 259
Soll DR, 415
Solomon EI, 553
Solomon MJ, 21, 70, 71, 142, 663
Solski PA, 242, 255, 262
Somers TC, 260

Somers W, 612, 614, 615, 619, 627, 628
Sommer H, 523
Sommer T, 276, 278, 292
Sommers CH, 115, 144, 146
Somogyi P, 751
Son JC, 452
Sone S, 816, 825, 840
Sonenberg N, 180, 401, 706, 707, 711
Song HY, 831
Song OK, 381, 720
Song R, 820
Song SH, 538
Song X, 826
Sonnenberg MG, 235
Sonoyama T, 58, 59, 61, 150
Sontheimer EJ, 374, 376, 397, 757
Sordano C, 415, 421
Sorensen BS, 649, 676
Sorensen HB, 649
Sorenson CM, 695
Soreq H, 696, 697
Soret J, 377, 378, 384
Sorgenfrei O, 97
Soria M, 696
Soriano P, 831
Sosinsky G, 478
Sosinsky GE, 479, 481–84
Souba WW, 312
Sougakoff W, 224
Sourgen F, 671
Sout CD, 729
Southgate CD, 827
Sowadski JM, 264
Spadari S, 151
Spaink HP, 5, 505–13, 516, 518–20, 523, 525
Spaltenstein A, 821
Spangler EA, 351–53
Spanjaard RA, 748
Spann T, 665
Spatz H, 102
Spear BB, 342
Spector DL, 385
Spell RM, 662, 664, 679
Spengler SJ, 641, 657
Sperling J, 379, 400
Sperling R, 379, 400
Sperry AO, 666, 669
Spiegel AM, 254, 259
Spiegelman GB, 424, 425
Spies T, 808, 815, 836, 837
Spik G, 456
Spikes DA, 395, 396
Spira AI, 142
Spirin AS, 756
Spitzner JR, 644, 647
Spivak G, 26, 67
Sporeno E, 621
Sprague CA, 358
Sprang SR, 421, 612

Spray DC, 485, 488–95
Springate CF, 118, 119
Springer AL, 639, 678
Springer BA, 620
Sprinzl M, 756
Sproat BS, 372, 374
Spudich JA, 419, 426, 427
Spudich JL, 430
Spulino JC, 460
Squires S, 148, 671
Sramek HA, 217, 236
Srere PA, 593, 594
Srinivasan N, 444, 454
Srivastava S, 759, 763
Srivenugopal KS, 641, 657
Stacey G, 506–9, 515, 516, 518, 521–24, 526, 527, 529, 530
Stade K, 759
Stadtman ER, 597, 811, 817, 834, 839
STADTMAN TC, 83–100; 84–92, 96–98, 744, 755
Stachelin C, 507, 515, 519, 522, 525, 527, 528
Stagljar I, 282
Stahl FW, 122
Stahl G, 754
Stahl H, 171, 175, 187, 194, 208
Stahl J, 61
Stahl NA, 610, 611, 618, 620
St. Germain DL, 92, 93, 98, 755
St. John AC, 804, 834
Staknis D, 381, 386, 391–93
Stamm S, 383, 392
Stanbridge EJ, 118
Standaert RF, 808, 813, 821
Standart N, 707, 708, 711
Standera S, 831, 838
Stanewsky R, 390
Stange G, 307, 327, 328
Stansfield I, 724
Stanulis-Praeger BM, 357
Stanway CA, 674
Stanzel M, 435
Staples RR, 580, 585, 586
Stapulionis R, 756
Stark H, 759, 763
Starling JA, 352, 357
Staron K, 676
Starr DB, 782
Stasiak A, 160, 174, 175, 177, 207
Staszewski LM, 838
Staten NR, 758, 762
Staub A, 19, 31, 54, 58, 775
Stauber C, 712
Staud R, 254, 261, 262
Stauffer KA, 478, 479, 481, 485, 497
Staunton D, 612
Stayton PS, 541, 558
Stearns T, 663
Steck TR, 148, 663

Steeg CM, 123
Stefanini M, 35, 54
Steffen MA, 538
Stehle T, 444, 452, 453, 466
Steigerwalt RW, 27
Stein GS, 713
Stein JL, 713
Stein PE, 445, 468
Stein R, 802, 816, 819–21, 835
Stein RL, 821
Steinbach A, 484
Steinberg BA, 225
Steinberg TH, 490
Steiner EK, 661
Steiner R, 805
Steingrimsdotter H, 35
Steingrube VA, 236
Steitz JA, 372, 374, 376, 381, 386, 390, 397, 398, 704, 731
Steitz TA, 469, 782
Stelzer E, 257
Stelzner M, 316
Stelzner MG, 328
Stephens RM, 368
Stephenson LS, 714, 725
Stephenson RC, 254
Stepien G, 566
Stepien PP, 583
Stepinski J, 401
Sterling T, 216
Stern D, 599
Sterne R, 242, 255
Sterne RE, 262
Sternglanz R, 636, 639, 647, 658–63, 673
Sternweis PC, 257, 259
Stetter KO, 637, 638, 678
Stévenin J, 377, 378, 380, 383, 384, 400
Stevens A, 697, 698, 700–2
Stevens BR, 307, 310, 311, 320, 322
Stevens C, 703
Stevnsner T, 643, 655
Stewart AF, 661
Stewart DE, 540
Stewart LC, 590, 642
Stewart NG, 358, 359
Stewart PR, 575
Stewart SA, 31, 57
Stewart V, 755
Stewart WW, 484
Steyaert J, 678
Steyn LM, 219
Stiles CD, 695, 703
Stiles GL, 434
Stiles JI, 714, 722
Stillman B, 55, 63, 73, 143, 151, 348, 579
Stimac E, 695
Stimmel JB, 243, 253
Stingl L, 144
Stirling CJ, 276

Stitt BL, 188–90
Stivala LA, 151
Stivers JT, 648
Stock D, 805, 807, 809, 811, 813, 814, 816, 834
Stock J, 243, 253, 254
Stock JB, 254, 261, 262
Stockman L, 227
Stoeckenius W, 478, 479, 491
Stoeckle MY, 228, 728, 730
Stoecklin G, 703
Stoffel W, 307, 314, 317, 319, 320
Stohl LL, 578
Stoker M, 494
Stokkermans TJW, 519, 521, 523, 524, 527, 528, 530
Stolk JA, 377, 378, 380, 383
Stoll J, 309, 310
Stoneking M, 566
Stoner L, 489
Storck T, 307, 314, 317, 319, 320
Storm-Mathisen J, 315, 317
Storti RV, 696, 724
Stradley SJ, 245, 246
Strand M, 112, 114, 115
Strath M, 622
Strathmann M, 420
Stratton C, 223
Stratton MA, 217, 221
Straughen J, 158, 669
Strausfeld UP, 73, 664
Strauss EG, 754
Strauss EJ, 398
Strauss F, 646
Strauss JH, 754
Strauss PR, 640
Strayer JM, 664
Strecker G, 444, 456, 458, 468
Streit A, 712
Strettoi E, 496
Strickland S, 706
Strittmatter U, 623
Strnad N, 803
Strominger JL, 307, 330
Strom-Mathisen J, 322
Strong AA, 826
Strub K, 275, 278
Strubin M, 793
Struhl K, 793
Strutt HL, 671
Stryer L, 264
Strynadka NC, 460
Stuart DI, 614
Stuart RA, 587
Stubbs M, 570
Stucka R, 584, 587, 588, 828
Studier FW, 743, 744
Stukenberg PT, 751
Sturani C, 672
Sturchler-Pierrat C, 757
Stutenkemper R, 477, 485
Styles CA, 586

Su S-S, 103, 110, 112, 113
Subak-Sharpe H, 484
Subik J, 640
Suchyna T, 477, 485
Suchyna TM, 489
Suda K, 584
Suddath FL, 444, 450, 460, 463
Sudhof TC, 245, 250, 251
Suffys P, 218, 230
Sugano H, 46, 282, 284
Sugano S, 702
Sugano T, 73
Sugasawa K, 58, 59, 61, 150, 664
Sugimoto Y, 425
Sugimura T, 701
Sugino A, 143, 636, 650, 701, 702
Sullivan DM, 674, 676
Sullivan EA, 223, 224
Sullivan ML, 695
Sumida-Yasumoto C, 200
Summers DF, 696
Summers WC, 72
Sumner AT, 665
Sun D, 782
Sun H, 122
Sun J-H, 712
Sun JR, 159, 540, 545
Sun JS, 374
Sun JZ, 260
Sun L, 121
Sun MK, 599
Sun Q, 50, 51, 382, 383
Sun TJ, 416, 419
Sun Y, 282
Sundaralingam M, 780
Sundarp R, 673
Sunde RA, 86
Sundin O, 662
Sung P, 16, 18, 34, 44, 47, 53, 55, 57, 59, 60, 64–66, 146, 148, 150, 152, 153, 160, 171, 180, 192
Suo Y, 514
Suomensaari S, 66, 152
Surdej P, 694, 695
Sureau A, 377, 378, 384
Surin BP, 507
Surolia A, 458
Susi H, 555
Sussel L, 349, 350
Sussenbach JS, 728, 729
Sussman M, 415, 431
Sutherlin DP, 507, 515
Sutin N, 539, 541, 543, 544, 552
Sutton JM, 507, 517, 524
Suttorp VV, 424
Suyama K, 119
Suzuki AE, 218, 242
Suzuki CK, 584
Suzuki H, 591, 638, 662, 663
Suzuki N, 258
Suzuki T, 746

Suzuki Y, 219
Svec WA, 541
Svejstrup ABD, 643
Svejstrup JQ, 19–21, 31, 32, 37,
    58, 64, 150, 151, 153, 642,
    643, 648, 649, 670, 676
Svensson B, 589
Svoboda DL, 47, 67, 145, 146
Swaffield JC, 828
Swagemakers SMA, 27
Swaminathan S, 831
Swamy KHS, 803
Swanberg SL, 636
Swank RA, 349
Swanson M, 389
Swanson MS, 392, 698
Swanton MT, 351
Swartwout SG, 696, 703
Swat W, 837
Sweder K, 16, 30
Sweder KS, 32
Swedlow JR, 667
Sweet RM, 786
Swenson KI, 477, 478, 485, 493,
    494
Swinton D, 114
Syed R, 94
Sykes AG, 540
Symington LS, 121
Sypes MA, 795
Sytkowski AJ, 619
Syväoja J, 66, 152
Szakacs NA, 593
Szalai L, 481
Szatkowski M, 318
Szecsi J, 523
Szer W, 392
Szikora JP, 831
Szilak I, 666
Szopa J, 661
Szostak JW, 122, 123, 342, 344,
    345, 714, 722
Szumiel I, 676
Szweda LI, 811, 817, 834, 839
Szymkowski DE, 145–47

T

Taagepera S, 660, 667
Taanman JW, 567, 573, 574
Tabor S, 193, 194, 196, 197, 208
Taccioli GE, 159
Tacke E, 752
Tacke R, 378, 383, 384
Taffet S, 489
Taffet SM, 490
Tafuri SR, 673
Taga T, 620
Tager JM, 570, 576
Taggart A, 778
Tahara H, 358–60
Taillandier D, 840
Tailler D, 522, 528

Tainer JA, 51, 142
Tajima S, 278
Tajiri T, 141
Tak T, 523, 525
Takada R, 773–75, 785, 786, 792
Takagi T, 321, 805, 822
Takahashi N, 242
Takai T, 488
Takai Y, 245, 250, 254, 258, 259
Takao M, 18, 46, 61, 150, 157
Takao T, 256, 259
Takasaki Y, 151
Takashina M, 805, 809, 836, 840
Takasuga Y, 666
Takayama K, 226, 231
Takebe H, 69
Takeda T, 702
Takemori H, 45, 154, 155
Takemoto LJ, 477
Takens-Kwak BR, 491, 492
Takeuchi I, 412
Takeuchi S, 161
Takeuchi T, 226
Takiff HE, 224
Takiguchi S, 636, 678
Talerico M, 381
Talin A, 590
Talmont E, 505, 508, 513, 520,
    523, 529
Talmont F, 507, 510, 513, 514,
    518, 523, 527
Tam R, 507, 518
Tamanoi F, 245, 249, 252, 260
Tamarappoo BK, 307, 319, 320
Tambini C, 701
Tamkun JW, 36
Tamm I, 696
Tamm T, 753
Tampe R, 812
Tamura A, 325
Tamura JK, 650, 651
Tamura S, 242, 325
Tamura T, 805, 807–10, 816,
    822, 825–27, 830, 836, 840
Tan EM, 672
Tan EW, 254
Tan J, 660
Tan KB, 639
Tanabe T, 488
Tanahashi N, 811, 818, 819, 822,
    823, 826, 827, 830
Tanaka H, 94, 96
TANAKA K, 801–47; 43, 44, 55,
    148, 171, 252, 307, 314,
    317, 318, 382, 802–8, 811,
    813, 815, 816, 821–23, 825,
    826, 838, 839
Tanaka RA, 257
Tanaka T, 264
Tanaka Y, 669
Tanese N, 389, 773–75
Tang CK, 729
Tang D, 642

Tang MS, 16, 18, 43, 44, 47
Tang WJ, 423, 435
Tanguay R, 708, 711
Tanhashi N, 821
Tani T, 376
Taniguchi T, 623, 827
Tanizawa A, 642, 643
Tano K, 46, 140, 141
Tano Y, 317, 318
Taphouse C, 412
Taphouse CR, 415
Tarn W-Y, 381
Tarnai T, 481
Tarone RE, 150
Tarovaia O, 666
Tartakoff AM, 695, 704, 731
Tasaka M, 412
Tashayev VL, 832
Tassan J-P, 19, 20, 31, 33, 58, 70
Tate SS, 307, 327–29
Tate WP, 744, 745, 753
Taub R, 746
Taudou G, 674
Tauer R, 584, 587, 588, 828
Taura T, 294
Tavassoli M, 144, 146
Tavitian A, 257
Tawa NE Jr, 840
Taylor AF, 175, 187, 192, 198
Taylor DL, 485
Taylor GR, 119, 452
Taylor JH, 106, 114
Taylor JS, 18, 24, 45–47, 61, 68,
    145, 146, 149, 155
Taylor ME, 442
Taylor SS, 264
Taylor TA, 357
Tazawa S, 282, 284
Tazi J, 385, 390
Teebor G, 140
Teebor GW, 140
te Heesen S, 281, 282
Teigelkamp S, 374, 376, 397, 399
Teixeira M, 97
Teixeira S, 307, 330
Telenti A, 222
Temesgen Z, 227
Temparis S, 840
Tempst P, 378, 388, 389
Temsamani J, 390
ten Dam E, 343
ten Dam EB, 750, 751, 754
Ten Eyck LF, 264
ten Hagen-Jongman CM, 275
Teplow D, 477
te Riele H, 117, 121, 123, 127,
    137
Terleth C, 24, 26
Terranova MP, 246
Tesmer J, 159
Tewari DS, 746, 759
Tewari KK, 640
Tewey KM, 636, 639

Teyssot B, 619
Thacker J, 159
Thaler DS, 125, 668
Thalerscheer MS, 148
Thalmann E, 401
Thangada M, 26
Thanos D, 835
Thayer MM, 51, 142
Thayer R, 703
Theibert A, 423, 425, 430
Theissen H, 380
Theodor L, 714, 725
Theodorakis NG, 696, 704
Therien MJ, 540, 554
Thiagalingam S, 18, 43, 44, 47, 50
Thibodeau SN, 118, 119
Thiele D, 349
Thielmann HW, 151
Thier SO, 330
Thilly W, 107, 108
Thilly WG, 109, 116–18, 137
Thimmalapura P-R, 98
Thireos G, 510, 773
Thissen JA, 244, 245, 248, 257
Th'ng JPH, 146, 665, 675
Thoma F, 25, 68
Thomas CJ, 753
Thomas DC, 107–9, 117, 120, 141
Thomas DY, 262
Thomas EL, 330
Thomas JO, 788
Thomas JP, 230
Thomas K, 106
Thomas TE, 356, 359
Thomas W, 662, 679
Thomas-Oates JE, 505, 507, 509, 510, 512, 513, 516
Thomason LC, 668
Thommes P, 170, 172
Thompson AM, 356, 358
Thompson CB, 330, 704, 709
Thompson CM, 19
Thompson JD, 176, 400, 423
Thompson LD, 620
Thompson LH, 29, 31, 47, 54, 57, 58, 60, 147, 148, 150, 158
Thompson MW, 829, 832
Thompson S, 753, 760
Thomsen B, 643, 666
Thomson S, 837
Thorburn AM, 704, 731
Thorner J, 242, 254, 255, 262, 296, 297, 776, 777
Thornton RD, 746
Thorsness PE, 584, 588
Thresher R, 50, 160
Thrift RN, 295
Thuneberg L, 477
Thut C, 791
Thut CJ, 775, 791, 792

Tian H, 382
Tian M, 382, 383, 717
Tiao G, 840
Tibbitts TT, 481, 482, 484
Tijan R, 389
Tikhonovich I, 515, 525, 528
Tilbrook PA, 249
Tilchen EJ, 674
Tiller AA, 477, 482
Tillinghast HS, 412
Timmers HTM, 777
Tinelli S, 674
Tinoco I, 751, 763
Tinoco I Jr, 172
Tiraby J-G, 102
Tishkoff D, 701, 702
Tjian R, 19, 21, 22, 26, 33, 35, 57, 58, 64, 133, 389, 773–75, 777, 782, 791, 792, 794
Tobery T, 620
Tobey RA, 665, 675
Tobin ED, 123
Todo T, 45, 154, 155
Toh-e A, 252
Tohyama M, 317, 318
Toida I, 233
Toikkanen J, 297
Tojo T, 672
Tokino T, 72, 73, 356
Tokoro Y, 379, 401
Tokuda H, 276, 278, 280, 292–94
Tokunaga F, 805, 812, 822
Tolar LA, 339
Tolley SP, 814
Tolner B, 314, 315
Tomek W, 808
Tomizaki T, 538, 547, 559, 574
Tomizawa JI, 636
Tomkinson AE, 18, 60, 67, 141, 146, 148, 158
Tommerup H, 348
Tomonaga T, 400
Tondokoro N, 282, 284
Toney JH, 141
Tong KI, 783
Tony H-P, 623
Toogood PL, 465
Toone EJ, 470
Topal MD, 640
Topping TB, 275, 276
Tora L, 775, 791
Tormay P, 85, 90, 762
Torrey JG, 526, 529
Torri AF, 640
Toschi L, 66, 151
Tosi M, 714
Tosto LM, 159
Totsuka T, 757
Touhara K, 423, 424
Toussaint O, 598
Town GP, 586
Townsend A, 836, 837
Townsend RR, 465

Townsley K, 749
Toyn JH, 379, 397
Toyoda N, 92
Traenckner EB, 819, 820
Traicoff J, 121
Traktman P, 652
Trash C, 673
Trask DK, 642, 643, 649, 661, 672
Traub O, 477, 489, 490, 492, 494, 495
Trautmann A, 485
Travers AA, 780
Trawick JD, 565, 572, 574, 592, 593, 596
Traxler BA, 175
Treco D, 122, 123
Treiber DK, 149
Treier M, 838
Treisman R, 696, 697, 703, 709
Tremoliéres A, 511
Treutlein H, 621
Tricoli JV, 674
Trinick MJ, 529
Trivedi D, 671, 677
Trnka L, 233
Troalen F, 671
Troelstra C, 29, 36, 54, 68, 69, 136, 151, 159, 160
Troidl EM, 416
Tronchere H, 385
Tröndle N, 806
Tropschug M, 587
Trosko JE, 485, 493–95
Trotta C, 695, 714, 715, 718, 719
Trotta PP, 612
Trowsdale J, 836
Troxell M, 840
Truant R, 72, 73, 791
Trucco M, 307, 315
Truchet G, 505, 507, 508, 510, 512–14, 518, 519, 523, 526, 527, 529, 530
Trueblood CE, 246, 249, 261, 566, 568, 571, 572, 574–76, 585
Truffot-Pernot C, 216, 223, 225
Trumpower BL, 567, 568, 573
Truong O, 111
Trus BL, 811, 823
Truve E, 753
Tsai L, 85, 89
Tsai LB, 762
Tsai-Pflugfelder M, 639
Tsaneva IR, 125, 126, 174, 175, 177, 207
Tsang JSH, 349
Tsao Y-P, 659, 671
Tschochner H, 776
Tse C, 697
Tse Y-C, 647
Tse-Dinh Y-C, 641, 642, 648, 649, 652, 669, 673

Tseng L, 307–9, 310
Tsevrenis H, 727
Tsubaki M, 589
Tsubuki S, 817
Tsuchihashi Z, 751
Tsuchiya T, 171, 185, 200, 201, 203
Tsui L-C, 762
Tsukamura M, 221, 222
Tsukihara T, 538, 547, 559, 574
Tsurimoto T, 348
Tsurumi C, 808, 827, 830, 831
Tsutsui Ken, 666, 674
Tsutsui Kimiko, 666, 674
Tsuzuki T, 141
Tu C, 751
Tu G-F, 760
Tucker JD, 158
Tucker MA, 477
Tugwood JD, 595
Tuite MF, 724, 727
Tullis GE, 714, 725
Turchi JJ, 115, 143
Turin L, 485
Turk E, 321, 327, 328, 331
Turman MA, 676, 678
Turner DC, 96
Turro C, 554
Tyc K, 392, 393
Tzagoloff A, 565, 583, 584, 586, 588, 589
Tzeng TH, 751

U

Uchida I, 356, 359
Uchida T, 119
Ucla C, 59, 146
Uda Y, 282, 284
Udenfriend S, 307, 327, 328
Udvardy A, 647, 651, 674, 822, 825, 827, 828, 830, 832
Uemura T, 636, 658, 661, 663, 665, 675
Ugai S, 822, 826
Ugalde RA, 507–10
Uhl JR, 227
Ui M, 58, 59, 61, 150, 674, 676, 679
Ulery TL, 574, 579, 592
Ullrich A, 619, 620
Ullsperger CJ, 636, 658, 662
Ultsch M, 612, 615–17, 626, 627
Ultsch MH, 612, 614, 615, 619, 627, 628
Um KS, 227, 230, 235
Umar A, 108, 109, 117, 120
Umen JG, 369, 374, 388, 397, 401, 718
Umesono K, 595
Umezu K, 192
Underbrink KM. 718
Unger L, 172, 200

Unson CG, 259
Unuma M, 282, 284
Unwin N, 477, 479, 481, 482, 486, 497
Unwin PNT, 480, 481, 483
Upton T, 597
Ura T, 233
Urade R, 327
Urieli-Shoval S, 114
Urlaub G, 27, 714, 725, 726
Ursic D, 831
Ursini F, 95, 98
Usheva A, 395, 795
Ushida C, 760
Ussuf KF, 575
Ustrell V, 824, 830, 832, 836
Utans U, 379, 393, 397–99, 401
Utsumi KR, 662, 663, 665, 675
Uzawa S, 662

V

Vadgama JV, 320
Vahrson W, 659
Vaidya S, 263
Vakalopoulou E, 704
Valay J-G, 21
Valcárcel J, 368, 386–88, 391, 392
Vale RD, 210
Valencik ML, 583, 585
Valero-Guillen P, 230
Valgeirsdottir K, 339
Valkema R, 422, 423, 426
Vallano ML, 489
Vallen EA, 144
Van Beeumen JJ, 576
Van Belkum A, 343
Van Boom JH, 507, 513, 522, 780
van Brussel AAN, 505–7, 509, 511, 518, 520, 523, 525–27
Van Bun S, 576
Vancura A, 252
Vancura KL, 246
Vandekamp M, 553, 554, 557
van den Akker F, 445, 448, 452, 463, 464, 467, 468
van den Berg JDJ, 505
Vandenberg M, 318
Vandenberg RJ, 307, 316, 319, 320
Van den Bogert C, 576
Vandenbosch K, 523
Van den Brande JL, 728, 729
van den Elsen P, 102
van de Putte P, 24, 26, 27, 50, 51
van der Drift KMGM, 505, 507, 509–11, 518, 520, 527
van der Eb AJ, 30, 58, 697, 703, 709
Van der Hagen BA, 356, 358
Van der Heyden J, 622
van der Horst GTJ, 593

van de Rijn I, 509
van der Kaay J, 420, 428
Van der Kammen A, 728
van der Kemp P, 141
van der Laarse A, 492
Vanderleyden J, 506
Van der Marel GA, 507, 513, 522, 780
Van der Meer NM, 576
Van der Ploeg LTH, 350
van der Weijden CC, 275
Vandevoorde V, 598, 599
van de Wal Y, 837
Van de Wiel C, 523
van Dijken P, 428
van Dillewijn P, 507, 511
Vandonselaar M, 444, 456, 463
Van Doren K, 395, 396
Van Driel R, 425
Van Duijn B, 415
van Duin J, 748, 762
van Duin M, 53, 146
Van Dyck L, 584
Van Dyk DE, 620
Van Dyke MW, 772
van Eijk HG, 762
Vaney DI, 496
van Eyk E, 593
Vangent D, 159
Van Ginneken AC, 491, 492
van Gool AJ, 27, 29, 36, 54, 68, 69
van Haastert PJM, 412, 415, 419–23, 426–30, 434
Vanhaesebroeck B, 598
van Ham RCA, 30, 58
van Hoffen A, 37, 54, 59, 150
Van Houten B, 18, 49
Van Itallie CM, 566, 575
Van Kaer L, 805, 837
Van Kammen A, 523, 524, 526, 530
Vankelecom H, 612
van Knippenberg PH, 762
Van Kuilenburg ABP, 576
Van Leeuwen S, 507, 513, 522
van Loon APGM, 593
Van Luilenburg ABP, 576
van Melderen L, 678
Van Ments-Cohen M, 426
Van Montagu M, 505, 507–11, 513, 514, 516, 519, 522, 523, 527
Van Nocker S, 824, 830
van Noort JM, 748
Van Pel A, 831
van Rhijn P, 506
van Roermund CWT, 318
van Rooijen ML, 27
Van Schaik FMA, 728
Van Spronsen PC, 523, 525
Vanveen TAB, 495

van Venrooij WJ, 389, 672
Van Vliet DL, 540
Van Vliet TB, 529
van Vuuren AJ, 31, 36, 54, 55, 147, 150
van Vuuren H, 19, 31, 136, 151
Van Winkle LJ, 307, 309, 312, 314, 317, 329
van Zanten BAM, 445, 452, 459
van Zeeland AA, 27, 37, 54, 59, 68, 150
Varki A, 456
Varlet I, 105, 107, 110, 123
Varmus HE, 718, 744, 749–51
Varotsis C, 538
Varshavsky A, 275, 662
Varshese JN, 452
Vassalli A, 706
Vassalli J-D, 706
Vasse J, 506, 511, 514, 526, 527, 529, 530
Vassy J, 839
Vasta GR, 444, 448, 458, 468
Vaughan MH, 696
Vaughan P, 137
Vaughan R, 429
Vayssiere J-L, 597
Vaziri H, 356, 359
Vazquez E, 575
Vazquez J, 326
Vazquez M, 507, 516
Veech RL, 570
Veenstra RD, 485, 490
Velaz-Faircloth M, 307, 316, 317, 321, 322, 331
Velculescu VE, 72, 73, 356
Vellard M, 377, 378, 384
Velours GM, 586
Venable SF, 356
Venance L, 491
Vende P, 751
Vendeland SC, 92, 757
Venema J, 27, 37, 54, 59, 68, 150
Venkatesan M, 193
Venta PJ, 762
Veprek B, 85, 89, 90
Verdine GL, 246, 391, 397
Verdone L, 661
Veres Z, 85, 89, 90
Verhage R, 24, 26, 27, 59, 150
Verhee A, 622
Verma IM, 694, 695, 703
Vermeersch J, 674
Vermeesch JR, 347, 351, 353
Vermeulen W, 19, 20, 29–33, 35, 36, 54, 55, 58, 61, 68–70, 136, 144, 149–51
Vernallis AB, 612
Veron M, 426
Verrijzer CP, 774, 775, 791, 792, 794, 795
Verselis VK, 488, 489
Vessetzsky YS, 666

Vetter W, 95
Vicuna R, 200
Vie A, 695, 696, 703
Viegas-Pequignot E, 840
Viel GT, 427
Vierheilig H, 525
Vierstra RD, 824, 830
Vignais M-L, 349
Vijn I, 523, 526
Villafranca JE, 444
Villanueva GB, 642
Villemure J-F, 125, 127
Villepounteau B, 341, 343, 360
Vilpo JA, 33, 57, 147, 149, 150, 152
Vimaladithan A, 748, 749
Vimr E, 452
Vincent M, 402
Vinitsky A, 814, 815, 820, 821, 838
Vinograd J, 642
Vinuela E, 639
Violand BN, 612, 758, 762
Viprey V, 507
Virbasius CA, 592
Virbasius JV, 592
Virelizier H, 505, 507–10, 512, 523, 524, 527, 530
Virtanen A, 698
Vishvanathan N, 217, 236
Visse R, 50, 51
Visser TJ, 96
Vivas E, 636, 678
Viveros OH, 821
Voelkel K, 660, 663
Voelkel-Meiman K, 639, 658, 661
Voelkl H, 328
Voellmy R, 485
Vogel JP, 280, 289, 290, 292
Vogel U, 219
Vogelstein B, 72, 118, 119
Vogt A, 260
Vogt TF, 514
Voigt S, 294
Voith G, 416
Vojta PJ, 357
Volden G, 140
Volker C, 243, 253, 254, 261, 262
Vologodskii AV, 636, 658, 662
Von Gabain A, 714
von Heijne G, 281, 294, 295
von Hippel PH, 174, 175, 177–79, 185, 186, 188, 196, 197, 199, 201, 202, 207
von Kieckebusch-Guck A, 567, 589
von Ossterum K, 704
Vos J-M, 27
Vosberg H-P, 642, 649, 677
Vournakis JN, 510
Voytas DF, 749
Vrecken P, 698, 729

Vrensen GFJM, 480
Vrieling H, 27
Vu JH, 818, 825, 827
Vu MN, 714, 725
Vyas NK, 446, 452, 460, 461

W

Wachter E, 294
Wada C, 119
Wade M, 494
Wadhwani KC, 309, 310
Wadiche JI, 307, 314
Waechter CJ, 263
Waga S, 73, 143
Wagner BJ, 704, 814
Wagner G, 246
Wagner I, 584
Wagner K, 623
Wagner R, 102, 103, 113, 116, 661
Wagner RE, 102
Wagner TE, 627
Waguespack C, 222
Waheed A, 762
Wakasugi M, 54
Wakatsuki S, 782
Wake SA, 696, 697
Waksman SA, 216, 219
Walbot V, 706
Waldegger S, 329, 330
Walden R, 522, 528, 529, 531
Waldherr M, 252
Waldman AS, 125, 127
Waldner H, 704
Walker GC, 44, 46, 52, 57, 58, 67, 71, 72, 102, 103, 110, 112, 136, 137, 151, 153, 154, 159, 160
Walker GW, 16, 17, 27, 30, 35
Walker IG, 146
Walker JE, 180, 568
Walker JR, 748, 751
Walker LJ, 142
Walker PR, 666
Walker SS, 773, 775
Wall JS, 174, 179
Wall MA, 421
Wall R, 314, 372
Wallace B, 314, 315
Wallace BA, 482
Wallace DC, 566
Wallace E, 94
Wallace RJ, 236
Wallace SS, 16, 141
Wallach D, 831
Waller PRH, 760, 761
Wallin SA, 543
Wallis JW, 637, 667, 669, 673
Wallis MG, 580, 583
Walls L, 309, 311
Walmsley RM, 352
Walsh CT, 234

Walsh WV, 72, 356
Walter A, 485
Walter MR, 612, 614
Walter P, 272–76, 278, 279, 281, 282, 284, 286, 288, 289, 292, 295, 584, 588
Walton J, 357
Walworth N, 148
Walz G, 703
Wand AJ, 545
Wanders RJA, 570
Wandersman C, 296
Wang A-HJ, 780
Wang B-C, 785–88
Wang C, 119
Wang EH, 773, 775
Wang H, 308, 312, 313
Wang HR, 807
Wang HZ, 485, 490
Wang J, 121, 378, 381, 383, 393, 394, 396, 398, 773
WANG JC, 635–92; 223, 636–42, 644–46, 652, 653, 657, 659, 661–63, 667–69, 676
Wang K, 809, 836, 840
Wang L, 226, 260
Wang LX, 522
Wang M, 419, 429, 672, 677
Wang N, 584
Wang P, 358
Wang QW, 522
Wang S, 477, 498
Wang SH, 256
Wang SM, 223
Wang SP, 753
Wang S-S, 341, 343, 347, 593
Wang TSF, 66, 67
Wang W, 27
Wang WJ, 254, 261, 262
Wang WL, 349
Wang XW, 37, 38, 73, 727, 728
Wang YC, 158, 174, 175, 177, 179, 185, 186, 201, 202, 207, 329
Wang Y-M, 615
Wang YW, 72
Wang Z, 92, 93, 98, 382, 383
Wang ZG, 16, 19, 31, 32, 36, 37, 58, 64, 143, 151, 153
Wang Z-Q, 144
Wang ZX, 773, 775
Wang ZY, 546
Wanger A, 222
Wanker EE, 282
Ward CL, 820
Ward DC, 640
Ward LD, 618, 621, 622
Ware DE, 718, 722
Ware JA, 261
Warms JV, 826, 832
Warncke K, 549, 553, 557
Warner A, 485

Warner P, 636, 678
Warren G, 105, 112, 113, 119
Warren V, 420
Washburn BK, 176, 192
Wasielewski MR, 541
Wassarman DA, 374, 376, 398
Wasserman RA, 675
Wasserman SA, 650, 663
Wassermann K, 24, 26
Watakabe A, 382
Watanabe M, 275, 276, 674
Watanabe N, 233
Watanabe T, 760
Waterland RA, 573, 575, 576, 595
Waters MG, 275, 289
Waters R, 153
Wathelet M, 695, 708
Wathelet MG, 695, 708
Watkins JF, 59, 143, 150
Watkins ML, 328
Watowich SJ, 444, 452
Watowich SS, 619
Watson JA, 251, 252, 259
Watson JD, 339, 695
Watson JM, 507
Watson KA, 452
Watson P, 118
Watt PM, 636, 639, 653, 663, 668, 669, 676
Watt PW, 312
Watts C, 294
Watts FZ, 144, 155
Wauthier EL, 27
Wawrzynow A, 830, 834
Waxman L, 802, 804, 813, 816, 822, 838, 839
Way J, 279, 280, 290
Wayne LG, 217, 236
Weatherall DJ, 727
Weaver DT, 55, 74, 158, 662
Webb CF, 679
Webb J, 275, 278
Webb MR, 210
Weber CA, 31, 35, 57, 147
Weber ER, 583, 584, 588
Weber K, 479, 491, 743
Webster ADB, 158
Webster G, 529
Wedel A, 19
Weeda G, 21, 30, 31, 35, 54, 58, 136, 150, 151
Weeks G, 424, 425
Wefes I, 822, 828
Wei L, 356, 359
Wei N, 72, 73
Wei Y-F, 67, 105, 108, 112, 113, 117, 119, 120, 158
Weigmann C, 307, 323
Weil PA, 771, 775, 782, 792
Weiler R, 496
Wein H, 61, 149
Weinberg RA, 19, 20, 58, 70, 743

Weiner AM, 371, 372, 374, 376, 390, 394, 743
Weiner JH, 568
Weiner M, 372
Weinert TA, 356
Weinfeld M, 159
Weingart R, 495
Weinkauf S, 805, 811
Weinman J, 516, 518
Weinrich SL, 341, 343, 358, 359
Weinshank RL, 307, 323
Weinstock GM, 575
Weinstock J, 621
Weinzierl ROJ, 773–75, 791
Weis SE, 217, 235
WEIS WI, 441–73; 444, 451–54, 456–60, 462, 463, 466, 467
Weiser B, 374
Weiss H, 567, 568
Weiss IM, 703, 727, 728
Weiss J, 275
Weiss RB, 742, 745, 748, 750, 752, 753, 758–60
Weiss SR, 696
Weissbach A, 640
Weiss-Brummer B, 760
Weissenbach J, 749
Weissman JS, 802, 829
Weissmann C, 387, 391, 743
Weissmann G, 254
Weisz OA, 471
Weitman D, 817
Weitzman I, 228
Welch E, 695, 714, 718, 719
Welch J, 780
Welch MD, 775
Welch WJ, 803
Welker D, 415
Welker DL, 416, 421, 422
Weller HN, 260
Weller SK, 180
Wellinger RJ, 347, 348
Wellington CL, 695, 705, 706, 709
WELLS JA, 609–34; 610, 611, 616, 617, 619, 620, 624, 626–29, 631
Wells NJ, 676, 677
Wells RD, 659
Wells RG, 307, 327, 328, 330, 331
Wells SE, 378, 393, 394, 396, 398
Welsh K, 103, 104, 110, 112
Welsh KM, 103, 104, 110, 112, 113
Wen D, 619
Weng Y, 719
Wentz-Hunter K, 378, 387
Wenzel T, 811, 812, 815, 834
Weremowicz S, 316
Werner A, 307, 327, 328, 330
Werner D, 525
Werner M, 782

Werner R, 485, 486, 492
Werner WM, 275
Wersig C, 378, 394
Wertman KF, 591
Wessel R, 187
West MD, 341, 358, 359
West SC, 18, 37, 59, 60, 65, 121,
    125, 126, 146–48, 160, 174,
    175, 177, 207
Westergaard O, 642, 643, 645,
    648, 649, 655, 666, 670, 676
Westerhoff HV, 645
Westerveld A, 29, 30, 53, 146
Westmoreland J, 126, 128
Weston B, 623
Westphale EM, 477, 485
Wetmur JG, 173
Whanger PD, 92, 757
Whang-Pheng J, 639, 669
Wheelock MJ, 477
White CI, 122–24, 775
White CL, 460, 469
White HE, 444, 454
White JB, 477
White JH, 122, 123
White JM, 259
White MA, 123
White MF, 309, 330
White O, 280, 639, 810
White PA, 696, 697
White TW, 476–78, 485, 487–89,
    492
Whitesides GM, 449, 454
Whiteway M, 262
Whitfield KM, 290, 292
Whitfield TT, 715, 724
Whitley P, 295
Whitney KD, 321, 322, 331
Whittemore L-A, 703
Whittemore SL, 755
Wick U, 412, 419, 426
Wickens M, 384, 706
Wickner RB, 706, 707, 711, 718,
    720, 750, 753
Wickner S, 830, 834
Wickner W, 272, 276, 280, 281,
    292–94
Widen SG, 143
Widner WR, 706, 720
Wiebauer K, 102, 143
Wiedenmann B, 297
Wiedmann B, 273, 274, 284,
    286, 287, 293, 294
Wiedmann M, 273–75, 284, 286,
    287, 290, 292–94
Wiekowski M, 194, 208
Wieland C, 390
Wiemken A, 525, 527, 528
Wieneke U, 507, 510, 512, 516,
    522, 529
Wiesenberger G, 252, 581, 582,
    584
Wieslander L, 368

Wiesner RJ, 574, 575, 586
Wight DC, 627
Wight TN, 242
Wigley D, 636
Wigley DB, 650–53
Wijffelman CA, 506, 516, 518
Wijfjes AHM, 507, 516, 529
Wikstrom M, 538, 568, 570, 571,
    576
Wilcox HM, 820
Wilczynska Z, 428
Wild R, 174, 181, 184, 185, 207
Wiles M, 87, 91, 93
Wiley DC, 444, 452, 454, 465,
    466
Wilhelm H, 273, 278
Wilk S, 802, 803, 814, 815, 819,
    820
Wilkie TM, 420
Wilkinson A, 647
Wilkinson G, 89
Wilkinson KD, 455, 832
Wilkinson MF, 307, 309, 311,
    714, 725
Wilkinson RG, 230
Will CL, 381, 383, 385, 390, 399
Willardgallo K, 837
Willecke K, 477, 485, 492
Willems A, 504
Willett WC, 75
Williams AS, 712
Williams DB, 820, 821
Williams DL, 222, 236, 351, 694,
    695, 728, 755, 757
Williams J, 412, 426
Williams JG, 412, 424, 426
Williams JR, 669
Williams KL, 416
Williams LD, 148
Williams RJP, 545
Williams RS, 574, 575, 586
Williams SA, 554
Williamson MP, 389
Williamson MS, 105, 106, 112,
    114, 122, 123, 125, 127
Willie A, 541, 558
Willis A, 837
Willis MC, 349
Willis NM, 753
Willmore E, 675
Wills NM, 747, 748, 751, 752,
    754
Willson JKV, 120
Willy PJ, 818
Wilson AL, 251
Wilson C, 294
Wilson DB, 760
Wilson DF, 570
Wilson GL, 26
Wilson IA, 466
Wilson KE, 505, 507, 515, 518
Wilson KL, 246
Wilson KS, 444, 456, 463

Wilson MC, 696, 697
Wilson PG, 753
Wilson SH, 67, 143
Wilson SJ, 70
Wilson T, 696, 697, 703, 709
Wilson W, 743, 749, 751
Wilting R, 762
Wiltshire S, 597
Wimmer MJ, 455
Winchester E, 72, 73
Winder FG, 226
Windsor WT, 614
Wing SS, 840
Wingender R, 516, 529
Winkelmann G, 396, 397
WINKLER JR, 537–61; 538,
    540, 541, 543, 545–50, 552–
    54, 557–59
Winkler MM, 706
Winsor B, 696, 697
Winstall E, 710
Winston F, 36
Winter M, 576
Wisdom R, 694, 695, 730
Wise JA, 371
Wishart JF, 540, 545
Witherell GW, 389
Witkin EM, 28, 68
Witte LD, 242
Wittenberg I, 315
Wittig B, 659
Wittwer CT, 674
Wlodawer A, 612, 615, 623, 645
Wodnar-Filipowicz A, 703
Woessner RD, 666, 674, 675
Wojtkowiak D, 830, 834
Wold MS, 55, 57, 63, 65, 67, 74,
    149, 153, 192, 658, 664
Wolda SL, 242, 243
Woldringh CL, 679
Wolf AJ, 347–50
Wolf DH, 588, 802–6, 808, 814,
    816, 840
Wolfe PB, 281, 294
Wolff S, 659
Wolff T, 374
Wolffe AP, 19, 673, 788
Wolfram S, 478, 487, 489
Wolin SL, 284, 286, 372
Wolucka BA, 224, 230, 231
Wong A, 228
Wong CH, 510
Wong I, 171, 174–76, 181, 186,
    190, 195, 196, 201, 203, 206
Wong JM, 782
Wong K, 619
Wong SC, 611, 620
Wong SYC, 465
Wong TC, 465
Wonigeit K, 837
Woo KM, 829
Woo SK, 832
Wood EA, 125

Wood ER, 665, 666, 679
Wood JM, 714, 718, 719, 721, 722, 753
Wood KW, 496
WOOD RD, 135–67; 16, 18, 37, 43, 44, 55, 57, 59, 60, 63, 65, 66, 73, 144–51, 153
Wood WI, 619
Woodard MH, 310, 311
Woodbury CP, 173
Woodcock J, 622
Woodley CL, 221
Woodruff TK, 698
Woods RJ, 466
Wooton JC, 61
Wootton JC, 150, 774, 775, 784
Woppmann A, 385
Workman JL, 772, 780, 793
Worland PJ, 84, 86
Worland S, 647
Wormington M, 706
Worosila G, 540
Worrell VE, 283, 286, 287
Worth L Jr, 126, 128
Wrede A, 219
Wreden C, 706
Wright A, 660, 663
Wright CS, 444, 450, 451, 458, 459, 469
Wright EM, 321, 322, 327, 328, 331
Wright JA, 703
Wright JH, 349, 350
Wright RM, 566, 568, 572, 574–76, 585, 591, 596
Wright SJ, 665
Wright TR, 831
Wright WE, 356–58, 360
Wu H-Y, 659–61, 674
Wu JA, 374, 394
Wu JY, 309, 374, 381, 383, 387, 390
Wu L, 420–23, 425, 426, 430, 431
Wu M, 583, 589
Wu RS, 713, 714, 722
Wu W-H, 776
Wu XH, 19, 31, 32, 37, 58, 64, 143, 151, 153
Wu Y-N, 307, 312, 319, 320
Wu Z-P, 94, 95
Wuarin J, 368
Wurster B, 412, 419, 421, 426
Wurzburg BA, 449, 454
Wuttke DS, 540, 558
Wyatt JR, 397
Wyckoff E, 636
Wydro RM, 696
Wylie CC, 715, 724
Wynford-Thomas D, 73
Wyns L, 678

X

Xanthoudakis S, 142
Xiao H, 72, 73, 791, 793
Xiao W, 140
Xiao Y, 158
Xie X, 773, 775, 785, 786, 788
Xie ZP, 525
Xu B, 579
Xu LX, 485, 489
Xu N, 705, 710
Xue CB, 256
Xue F, 753
Xuong NG, 264

Y

Yadeau JT, 281
Yaffe MP, 283, 590
Yager TD, 178, 179, 185
Yagi A, 811
Yagi T, 69
Yagura T, 666
Yajima H, 18, 46, 154, 157
Yakovleva T, 72
Yale HL, 225
Yamada F, 713, 808
Yamada K, 623
Yamada T, 219
Yamaguchi H, 538, 547, 559, 574
Yamaizumi M, 54, 73
Yamamoto K, 141
Yamamoto M, 495
Yamamoto S, 233
Yamamoto T, 773
Yamano S, 93
Yamanouchi T, 219
Yamasaki H, 494
Yamasaki K, 622
Yamashita E, 538, 547, 559, 574
Yamashita J, 666
Yamashita K, 20
Yamashita S, 773–75
Yamashita T, 669
Yamazaki S, 97
Yan D, 393, 394, 398
Yan N, 307, 327, 331
Yan Y, 444, 452, 453, 466
Yanagida M, 636, 658, 661–63, 665, 673, 675–77
Yanagisawa J, 150
Yanagisawa K, 669
Yancey JE, 200
Yancopoulos GD, 610, 611, 618, 620
Yang C-C, 246
Yang G, 673
Yang J-G, 92, 758
Yang L, 636, 658, 660, 664
Yang WC, 523, 526
Yang XM, 392
Yang Y, 809, 836, 838, 840
Yang-Feng TL, 307, 321–23, 331

Yangisawa J, 58, 59, 61
Yanofsky C, 714
Yapo A, 218, 236
Yarema KJ, 146
Yariv J, 444, 463
Yarmush ML, 311, 312
Yarranton GT, 175, 182, 185, 186, 193, 200, 201, 203
Yarus M, 745
Yasuhara T, 588
Yasuhira S, 18, 46, 154, 157
Yasui A, 53, 146, 147, 154
Yasui M, 226
Yasukawa K, 618, 621, 622
Yawata H, 622
Ycas M, 595
Ye T-Z, 158, 669
Yeager M, 477, 479, 481
Yeakley JM, 717
Yean SL, 374, 376
Yeh H, 37, 38, 72, 73
Yeh J-Y, 92, 757
Yeh RP, 426
Yeh Y-C, 142, 649, 670
Yen J-YJ, 105, 110
Yen TJ, 666, 695, 709
Yenofsky R, 696
Yeung AT, 47
Yewdell J, 837
Yewdell JW, 835, 837
Yi MK, 791
Yin S, 24, 68, 763
Yin XM, 597
Ying W, 246
Yingfang Z, 669
Yocom KM, 540
Yoder BL, 67, 171
Yokogawa T, 760
Yokomori K, 774, 775, 791, 792, 794, 795
Yokosawa H, 807, 826, 839, 840
Yokota A, 512
Yokota K, 809, 828, 831, 836
Yokoyama K, 244, 245, 248, 249, 252, 253
Yokoyama S, 756
Yoo SJ, 810, 829
Yoo Y, 839
Yoo YJ, 832
Yoon C, 780
Yoon H, 21, 54
Yorgey P, 636, 678
York WS, 505, 507, 509, 511, 520, 523, 525
Yoshida MC, 55, 141
Yoshida Y, 245, 250, 254
Yoshikura H, 754
Yoshimatsu K, 261
Yoshimoto A, 593, 595
Yoshimoto E, 309–12
Yoshimoto T, 309–12
Yoshimura A, 619
Yoshimura K, 279

Yoshimura T, 805, 807, 816, 822
Yoshimura Y, 619
Yoshinaka Y, 754
Yoshizawa T, 259
Yosuhara T, 572
You Y, 704, 706
Youatt J, 225, 228
Youdale T, 666
Young AT, 714, 722
Young CS, 159
Young D, 219, 227, 228
Young JD, 485
Young JPW, 504, 510
Young JW, 359
Young MC, 196, 197
Young MG, 260
Young PA, 86
Young RA, 19, 20, 58, 70, 153,
    776, 792, 822, 828
Younglai EV, 356
Yu BP, 597, 814
Yu CA, 670
Yu G-L, 339, 343, 345–47
Yu K, 478, 487
Yu L, 652
Yu WH, 569
Yu X, 160, 174, 175, 177, 181,
    184, 185, 207
Yu YT, 376, 418
Yuan H, 754
Yuan JG, 275, 280
Yuan R, 660
Yuba S, 161
Yue J, 584, 588
Yuen IS, 412, 415
Yun DF, 716, 717

Z

Zaat SAJ, 506
Zabel DJ, 660
Zaccai G, 645
Zachar Z, 378, 380, 395, 396
Zafra F, 318, 323, 324
Zagnitko OP, 827, 830, 831, 833
Zahler AM, 339, 341, 347, 377,
    378, 380–84
Zaidi SHE, 703, 704
Zaitzev DA, 638
Zak R, 574, 575, 586
Zakian VA, 339, 342, 344, 345,
    347–50, 352, 353

Zakin MM, 389
Zalensky A, 523
Zamble DB, 47, 61, 65, 145
Zambronicz EB, 569
Zamore PD, 378, 380, 386, 387,
    394
Zampighi G, 480, 481, 483
Zanker B, 703
Zaret KS, 786, 791
Zassenhaus HP, 581, 582
Zaug AJ, 343
Zawel L, 19, 22, 26, 31, 33, 35,
    54, 57, 58, 64, 69, 150, 770–
    72, 776
Zdanov A, 615, 623
Zdzienicka MZ, 27
Zechiedrich EL, 636, 639, 642,
    645, 648, 649, 658, 662
Zeeman AM, 59, 150
Zehelein E, 85, 86, 89, 90, 762
Zehelin E, 85
Zeidler MP, 792
Zeikus R, 758
Zeng C, 402
Zeng XR, 66, 152
Zenke G, 623
Zensen M, 322
Zerangue N, 318
Zerfass K, 754
Zerial M, 257
Zerivitz K, 381
Zernik-Kobak M, 57, 74
Zetter BR, 279
Zgaga Z, 126
Zghal M, 69
Zhan Q, 38, 72–74
Zhang BW, 595
ZHANG FL, 241–69; 248, 249
Zhang H, 159, 160, 540, 660, 661
Zhang HL, 641, 642
Zhang JG, 621
Zhang J-T, 477
Zhang L, 105, 112, 113, 121,
    123, 124, 262, 263
Zhang M, 378, 380, 387
Zhang SJ, 66, 72, 73, 152, 714–
    18, 722, 724, 791
Zhang WB, 312, 704
Zhang Y, 160, 219, 227, 228,
    236, 307, 315, 317, 319,
    320, 538
Zhang YX, 831

Zhang Y-Z, 576
Zhang Z, 312
Zhao BG, 612
Zhao H, 748, 749
Zhao JH, 18, 46, 157
Zhao QH, 714, 725
Zhao X, 61, 571, 573, 595, 599
Zheng J, 264
Zhou KM, 420
Zhou Q, 773, 775, 794
Zhou S, 774, 775, 791, 792
Zhou W, 23
Zhou X, 714, 725, 726
Zhou Y, 477, 482
Zhu C-X, 641, 642, 648, 652, 669
Zhu DL, 727
Zhu J, 759, 763
Zhu L, 180
Zhu XM, 356, 359
Zhu YJ, 594, 595
Zhuang Y, 372, 374, 390
Ziaja K, 583
Zielenski J, 762
Zimmer DB, 477
Zimmerberg J, 485
Zimmerman DL, 282
Zimmerman JP, 639
Zimmermann R, 289, 294
Zimmermann W, 640
Zini N, 676
Zinoni F, 85, 88, 89, 97, 744,
    755
Ziolkowski CHJ, 116
Zipkin ID, 728–30
Zitomer RS, 566, 574, 585
Ziv Y, 160
Zlotnick A, 287
Zlotnik A, 87, 91, 93
Zomerdijk J, 777
Zopf D, 273
Zubay G, 744
Zubiaga AM, 703
Zufferey R, 282
Zühl F, 813
Zumstein L, 642, 660
Zunino F, 636, 674
Zuo P, 378, 380, 381, 383–85,
    390, 399
Zurawski G, 623
Zurawski SM, 623
Zwelling LA, 669, 674
Zwickl P, 805–8, 810–13, 834

SUBJECT INDEX

A

ABC-transporters, 296
Acetylaminofluorene
  adducts
    nucleotide excision repair
    and, 145
Acetyl-leu-leu-norleucinal
  20S proteasome and, 819
Acyclovir
  viral mutants resistant, 761
Adenylyl cyclase
  activation during *Dictyostelium* development,
    422-25
Adherens junctions
  blocking assembly with antisera, 487
a-factor
  CaaX motif, 242-44
African swine fever virus
  DNA topoisomerase, 639
Aging and
  telomeres and, 355-57
Alkylating agents
  lethal and mutagenic effects
    O^6-methylguanine-DNA
      methyltransferase and,
      137
All-rich elements (AREs)
  eukaryotic gene expression
    and, 703-5
American Association for the Advancement of Science
  (AAAS), 10-11
Amikacin
  structure, 220
Amino acid transporters, 305-32
  brain-specific, 321-22
  CAT family, 308-14
    cloning, 308-10
    regulation of expression,
      310-13
    structure-function analysis,
      313-14
  glutamate family, 314-20
    cloning, 314-16
    distribution of expression,
      316-18
    regulation, 318
    structure-function analysis,
      319
  glycine-specific, 322-27
  GLYT family
    cloning, 323
    GLYT1 isoforms, 323-25

GLYT2, 325-27
rBAT/4F2hc family, 327-31
  cloning, 327-28
  cystinuria and, 330-31
  structure-function analysis,
    328-30
Amphibians
  photoreactivating activity,
    154
Anandamide
  gap-junctional intercellular
    communication and, 491
Animal lectins
  C-type, 462-63
Antibiotic resistance
  tuberculosis treatment and,
    233-36
Antigen presentation
  proteasomes and, 835-38
Antizyme
  recoding, 746-48
AP endonucleases
  base excision repair and, 142
Aphidicolin
  DNA mismatch repair and,
    108
  nucleotide excision repair and,
    151
Apo-TBP, 777-78
Arabidopsis
  telomere length
    strain-specific differences,
      352
Arabidopsi thaliana
  TBP isoform 2, 777-78
AREs
  See All-rich elements
ASCT1 glutamate transporter,
  319-20
Astrocytes
  gap-junctional intercellular
    communication
    inhibition, 491
Atkins, J. F., 741-63
ATP
  binding to type II DNA topoisomerases, 650-51
  protein breakdown in cells
    and, 804
ATPase
  26S proteasome and, 826-30
ATP synthase
  assembly, 589
Azorhizobium, 504
Azurin
  electron-transfer reactions, 551-54

B

Bacillus subtilis
  DNA photolyase and, 45
  nucleotide excision repair, 47
  proteasomes, 810
Bacteria
  cytoplasmic membrane
    protein translocation and,
      271-98
  homeologous recombination,
    125-26
  insertion sequence elements,
    752
  Sec61/SecYEGp complex, 276-78
  selenocysteine insertion, 88
Bacteriophages
  DNA helicases, 170
Bacteriophage T4
  gene 60
    translational bypass, 758-59
Bacteroides ruminicola
  ribosomal bypassing and, 760
Barley yellow dwarf virus
  frameshifting and, 762
Base excision repair (BER), 17,
  46, 137-44
  completion
    pathways, 142-43
  poly(ADP-ribose) polymerase
    and, 144
BER
  See Base excision repair
BIDS
  nucleotide excision repair
    genes and, 35
Bjornson, K. P., 169-210
Bladder cancer
  microsatellite instability and,
    119
Blanchard, J. S., 215-36
Bradyrhizobium, 504
Breast cancer
  microsatellite instability and,
    119
Brookhaven National Laboratory,
  7-9
Burley, S. K., 769-96

C

CaaX motif, 242-44
CaaX proteins
  C-terminal processing, 253-55
  prenylation, 245-50
Caenorhabditis elegans

901

G protein β-subunit, 421
mRNA decay pathways
    nonsense-mediated, 717-18
    recessive heavy chain mutants,
        714-15
    translocating-chain association
        membrane, 279
cAMP
    intracellular
        gene expression in Dic-
            tyostelium and, 426
Camptothecin
    genome stability and, 670
Cancer
    DNA mismatch repair defi-
        ciency and, 118-21
    p53 gene and, 72-73
    See also specific type
Cancer cells
    telomerase activation and, 357-
        61
Cap-binding proteins
    pre-mRNA splicing and, 401
Capreomycin
    structure, 220
Carassius auratusa
    DNA photolyase gene, 154
Carbohydrate-binding proteins
    lectins compared, 460-61
Carbohydrate-lectin complexes
    three-dimensional structure,
        444-45
Carbohydrate-lectin hydrogen
    bonds, 447
    water molecules mediating,
        451-52
Carbohydrate-lectin recognition,
    441-71
    divalent cations and, 453-54
Carbonic anhydrase II
    osteopetrosis and, 762
Carcinogens
    adducts
        nucleotide excision repair
            and, 145
Casey, P. J., 241-65
CAT transporter family, 308-14
    cloning, 308-10
    regulation of expression, 310-
        13
    structure-function analysis,
        313-14
CBZ-leu-leu-leucinal
    20S proteasome and, 819-20
Cell cycle
    DNA excision repair and, 69-
        71
    DNA topoisomerases and, 675
Cellular energy production
    regulation by cytochrome c oxi-
        dase, 569-71
Cellular senescence
    telomeres and, 355-57

Cervical cancer
    microsatellite instability and,
        118-19
cGMP
    chemotaxis in Dictyostelium
        and, 426-28
Chaetomellic acid
    protein prenylation inhibition
        and, 260
Chaperones
    26S proteasome and, 834
Chaperonins
    mitochondrial assembly and,
        587
Chironomus tentans
    RNA synthesis
        DNA topoisomerases and,
            661
Choroideremia
    defective Rep1 function and,
        251
Chromatin
    eukaryotic DNA topoisom-
        erase type II and, 666-
        67
Chromosome condensation
    eukaryotic DNA topoisom-
        erase type II and, 665-67
Chromosome segregation
    DNA topoisomerases and, 663-
        64
Chymostatin
    20S proteasome and, 819
Ciliates
    telomeric DNA, 338-39
Ciprofloxacin
    structure, 223
Cisplatin-purine adducts
    nucleotide excision repair and,
        145
Closteroviruses
    frameshifting and, 752
Clostridial glycine reductase se-
    lenoprotein A
    catalytic activity, 98
Clostridium sticklandii
    selenocysteine insertion, 88
CMF
    See Conditioned medium factor
Cockayne syndrome (CS)
    defective nucleotide excision
        repair and, 54
    genetic complementation
        groups, 29-30
    nucleotide excision repair
        genes and, 36-38
    strand-specific repair of tran-
        scriptionally active genes
        and, 27
    transcription-repair coupling
        and, 68-69
Cocksfoot mottle sobemovirus
    frameshifting and, 753

Codons
    nonsense
        rapid mRNA decay and, 716-
            17
    redefinition, 753-60
Coenzyme Q cytochrome c oxi-
    doreductase
    assembly, 589
Colon cancer
    Ras proteins and, 260
Colorectal cancer
    hereditary nonpolyposis
        DNA mismatch repair defi-
            ciency and, 118-20
Conditioned medium factor
    (CMF)
    Dictyostelium cells and, 413-15
Connexins, 475-98
    function
        phosphorylation and, 492-97
    functional anatomy, 485-90
    hydrophobic domains, 477
    oligomerization, 478
    selective permeabilities, 490
    structure in intercellular chan-
        nels, 476-84
Connexons, 475-98
    formation, 478
Copper proteins
    electron-transfer reactions, 551-
        54
Coux, O., 801-41
Crithidia fasciculata
    DNA topoisomerase, 640
CS
    See Cockayne syndrome
C-terminal methyltransferase
    prenyl protein processing and,
        261-62
Cyclin-dependent kinase-activat-
    ing enzyme
    DNA excision repair and, 70-
        71
Cyclophilin
    mitochondrial assembly and,
        587
Cysteine-containing enzymes
    catalytic activities, 97-98
Cystic fibrosis transmembrane
    conductance regulator
    UGA mutants, 761-62
Cystinuria
    rBAT/4F2hc transporter family
        and, 330-31
Cytochrome c
    electron-transfer reactions, 543-
        45
    structure, 557
Cytochrome c oxidase
    assembly, 589
    cellular energy production and,
        569-71
    electron-transfer reactions, 538

levels
 environmental control, 574-75
 subunits
  function, 571-74
  nuclear-coded, 575-76
Cytokines
 receptor subunits, 620-21
Cytoplasm
 nonsense-mediated mRNA decay and, 715-16
Cytoplasmic membrane
 bacterial
  protein translocation and, 271-98

D

Debelle, F., 503-31
Deinococcus radiodurans
 nucleotide excision repair, 47
Denarie, J., 503-31
Desulfurococcus amylolyticus
 DNA topoisomerase, 638
de Vos, A. M., 609-31
Devreotes, P. N., 411-37
Diabetes mellitus
 insulin-dependent
  autoantibodies against DNA topoisomerase II and, 672
Dictyostelium
 adenylyl cyclase activation, 422-24
  regulation, 424-25
 ambient levels of cAMP
  regulation, 425-26
 cell culture and mutant handling, 415
 chemotaxis, 414
  cGMP and, 426-28
 desensitization of receptor-mediated responses, 428-30
 gene expression
  intracellular cAMP and, 426
 genome, 415
 G protein α-subunits, 420-21
 G protein β-subunit, 421-22
 G protein-independent processes
  receptor-mediated, 430-31
 growth and development
  intercellular signals and, 412-15
 high-efficiency transformation, 416
 homologous recombination, 415
 morphological development, 414
 random mutagenesis
  rapid phenotypic screens, 431-36

receptor subtype switching programs development, 416-19
responses mediated by chemoattractant stimulation, 419-20
restriction-mediated integration, 416
signal transduction, 411-37
Dictyostelium discoideum
 cytochrome c oxidase
  nuclear-coded subunits, 575-76
DNA
 basal transcription
  initiation, 19-21
  properties, 172-74
 repair mechanisms
  nucleotide excision, 46-47
  repair/transcription relationships, 24-32
 single-stranded
  DNA helicase translocation and, 196-98
  duplex DNA unwinding and, 193-96
  polarity of binding, 180
  proteins binding, 148-50
  single-stranded vs duplex, 172-73
 transcriptionally silent
  nucleotide excision repair and, 32-33
 UV light-induced damage
  pathways for repair, 154-57
DNA binding, 180-88
 single-stranded/double-stranded
  stoichiometries and energetics, 180-85
DNA duplexes
 intertwined
  unlinking, 662-64
DNA-enzyme complexes
 covalent
  active-site tyrosines and, 647-49
DNA excision repair, 43-76
 cell cycle and, 69-71
 in humans, 52-67
  polypeptides required, 56
 in prokaryotes, 47-52
 SOS response and, 71-74
 transcription-repair coupling, 67-69
DNA glycosylases
 base excision repair and, 17, 138-42
DNA helicases, 170-72
 hexameric, 207-8
 structural features, 174-80
DNA lesions
 proteins binding, 148-50

DNA ligases
 DNA repair and, 157-58
DNA mismatch repair, 101-29
 deficient
  cancer predisposition and, 118-21
 in Escherichia coli
  methyl-directed, 102-5
 in eukaryotic cells
  long-patch, 105-14
 genetic recombination and, 121-29
DNA mismatch repair mutants
 mitotic phenotype, 114-21
DNA photolyase
 direct DNA repair and, 44-45
 UV light-induced DNA damage and, 154-57
DNA polymerase
 nucleotide excision repair and, 151-52
DNA polymerase β
 base excision repair and, 143
DNA polymerase III holoenzyme
 repair DNA synthesis and, 109
DNA-protein interactions
 energetics and kinetics
  solution conditions and, 173-74
DNA repair, 44-47. See also Nucleotide excision repair
 direct, 44-46
 DNA ligases and, 157-58
 double-strand break, 158-60
 in eukaryotes, 135-61
 See also Base excision repair; Nucleotide excision repair
DNAse IV
 homologous activities in yeast and, 143-44
DNA supercoiling
 consequences, 660-62
 replication and, 656-58
 transcription and, 658-59
DNA topoisomerases, 635-81
 active-site tyrosines, 647-50
 biological functions, 656-79
 cellular levels
  regulation, 673-77
 DNA binding, 641-47
 genome stability and, 667-72
 mechanistic studies, 640-56
 structures and classification, 637-40
 three-dimensional structures, 652-54
 type I
  association with other proteins, 677-78
 type IA, 641-42
 type IB, 642-44
 type II, 644-47

association with other proteins, 678-79
eukaryotic, 665-67
nucleotides binding, 650-52
DNA unwinding
duplex
flanking single-stranded
DNA and, 193-96
helicase-catalyzed
mechanisms, 169-210
initiation in vitro, 192-93
mechanisms, 201-9
active vs passive, 202-3
rates and processivities, 198-200
thermodynamic efficiency, 200-1
Drickamer, K., 441-71
Drosophila
G protein β-subunit, 421
homeologous recombination, 127
Drosophila melanogaster
DNA mismatch repair, 107
DNA photolyase, 154
MutS/MutL homologs, 105

E

EAAC glutamate transporter, 315-16
distribution of expression, 317-18
regulation, 317-18
Electronic coupling
electron-transfer reactions and, 547-59
Electron transfer
in proteins, 537-59
Electron-transfer reactions
azurin and, 551-54
conformational changes and, 543
coupling zones, 556-59
cytochrome c and, 543-45
electronic coupling and, 547-59
α-helical proteins and, 554-56
homogeneous-barrier models, 547
myoglobin and, 545-47
nuclear factor and, 541-47
reorganization energy and, 541-42
self-exchange, 542
superexchange-coupling models, 547-49
tunneling-pathway model, 549-51
Electron-transfer theory, 538-40
Electron tunneling, 539
Endometrial cancer
microsatellite instability and, 118-19

Endoplasmic reticulum (ER)
protein translocation and, 271-98
Energy production
cellular
regulation by cytochrome c oxidase, 569-71
Enoyl-CoA hydratase
ARE-binding activity, 704
Enzyme-DNA complexes
covalent
active-site tyrosines and, 647-49
Enzymes
biochemical roles of selenocysteine residues, 95-97
protein translocation site and, 281-82
selenocysteine-/cysteine-containing
catalytic activities, 97-98
ER
See Endoplasmic reticulum
Escherichia coli
codon redefinition and, 753
DnaB protein, 179, 184-85
DNA binding and, 189
DNA excision repair
SOS regulation, 71
DNA helicases, 170-71
DNA mismatch repair
methyl-directed, 102-5
strand signals and, 113-14
DNA photolyase, 45
DNA topoisomerases, 637-38
regulation, 673
dnaX gene, 751-52
membrane protein insertion
post-translational, 294-95
nucleotide excision repair, 17-18, 47
coupling of transcription and, 28-29
proteins required, 48
protein translocation
post-translational, 292-94
RecBCD helicase, 187-88
release factor 2 gene, 745-46
Rep helicase
DNA binding and, 185-89
DNA unwinding and, 203-6
Rep monomer
ATP binding and hydrolysis and, 191-92
Rep protein, 182-84
Rho protein, 178-79, 185
DNA binding and, 189-90
DNA unwinding and, 208-9
ribosomal bypassing and, 760
RuvB protein, 177-78
Sec61/SecYEGp complex, 278
SecA protein, 280
SecD/SecFp complex, 280

selenocysteine synthesis, 86
selenophosphate synthetase, 91
transcription-repair coupling, 68
UvrD protein
DNA unwinding and, 206-7
Esophageal adenoma
microsatellite instability and, 118-19
Ethambutol
drug-resistant tuberculosis and, 230-32
structure, 230
Ethionamide
drug-resistant tuberculosis and, 225-30
structure, 225
Eubacteria
proteasomes, 810-11
Eukaryotes
DNA helicases, 170
DNA repair, 135-61
DNA topoisomerase type II
chromosome condensation and, 665-67
gene expression
all-rich elements and, 703-5
homeologous recombination, 127
membrane protein insertion
post-translational, 294
nucleotide excision repair
coupling of transcription and, 29-30
nucleotide excision repair
genes and proteins, 156
photoreactivating activity, 154
protein translocation pathways, 272-76
RNA polymerases, 770-71
selenocysteine insertion, 87-88, 757-58
telomeric DNA, 338
Eukaryotic cells
DNA mismatch repair
long-patch, 105-14
mRNA decay, 693-731
endonucleolytic cleavage initiating, 728-30
impaired translation initiating, 28
nonsense-mediated, 714-26
poly(A) shortening and, 696-713
proteasomes, 802
Euplotes
telomere processing, 351-52
telomeric DNA, 339
Excision nuclease
human
structure and function, 55-61
nucleotide excision repair and, 46-47

Exons
  SR proteins and, 381-83

F

Farnesyltransferase (FTase)
  isoprenoid addition to proteins
    and, 244-45
  prenylation, 245-50
FFP analog
  protein prenylation inhibition
    and, 260
Fish
  photoreactivating activity, 154
Flagellates
  telomeric DNA, 338
Fluoroquinolones
  drug-resistant tuberculosis and,
    223-25
Formamidopyrimidine-DNA glycosylase
  base excision repair and, 141
Formyluracil
  elimination from DNA
    DNA glycosylase and, 140
Frameshifting
  ribosomal, 745-53
Franck-Condon principle
  electron-transfer reactions and,
    539
Friedberg, E. C., 15-39
Fungi
  photoreactivating activity, 154

G

Galactose
  packing interactions unique,
    462
Gap-junctional intercellular communication (GJIC), 476
  functional assays, 484-85
  inhibition
    anandamide and, 491
Gap junctions
  blocking assembly with antis-
    era, 487
  isolation from liver and heart,
    479
  lipid composition, 491-92
  structural model, 480
Gene conversion
  DNA mismatch repair and,
    121-24
Gene expression
  eukaryotic
    all-rich elements and, 703-5
  mitochondrial, 577-87
    coordination, 591-92
    differential regulation, 586-
      87
    global regulation, 585-86
    RNA processing, 580-82

transcription, 578-80
  translation, 582-85
  nuclear
    coordination, 591-92
Genes
  transcriptionally active
    nucleotide excision repair,
      33-34
    preferential excision repair,
      24-26
    strand-specific repair, 26-28
Genetic recombination
  DNA mismatch repair and,
    121-29
Geranylgeranyltransferase type I
  (GGTase-I)
  isoprenoid addition to proteins
    and, 244-45
  prenylation, 245-50
Geranylgeranyltransferase type II
  (GGTase-II)
  isoprenoid addition to proteins
    and, 244-45
  prenylation, 250-52
Gesteland, R. F., 741-63
GJIC
  See Gap-junctional intercellu-
    lar communication
GLAST glutamate transporter,
  314-16
  distribution of expression, 316-
    18
  regulation, 317-18
Gliotoxon
  protein prenylation inhibition
    and, 260
Glutamate transporter family
  cloning, 314-16
  distribution of expression, 316-
    18
  regulation, 318
  structure-function analysis, 319
Glutathione peroxidases
  sulfur analogs, 98
Glyceraldehyde-3-phosphate de-
  hydrogenase
  ARE-binding activity, 704
Glycine-specific transporters,
  322-27
Glycolipid-binding toxins, 463-
  64
GLYT1 isoforms, 323-25
GLYT2 transporter, 325-27
GLYT transporter family
  cloning, 323
GM-CSF
  See Granulocyte-monocyte col-
    ony-stimulating factor
Goldberg, A. L., 801-41
Goliger, J. A., 475-98
Goodenough, D. A., 475-98
G protein α-subunits, 420-21
G protein β-subunit, 421-22

Granulocyte-monocyte colony-
  stimulating factor (GM-CSF)
  structure, 612-14
Gray, H. B., 537-59
Greider, C. W., 337-62

H

Haemophilus influenzae
  DNA topoisomerase, 639
  proteasomes, 810
Halobacterium salinarium
  G protein-independent proc-
    esses, 430
Heart
  isolation of gap junctions,
    479
α-Helical proteins
  electron-transfer reactions, 554-
    56
Hematopoietic ligand-receptor superfamily
  class I, 611
Hematopoietic ligands
  structure, 612-14
Hematopoietic receptor
  ligand binding domain
    structure, 614-16
Hematopoietic receptor complexes, 609-31
  hetero-oligomeric group, 620-
    23
  homodimeric group, 616-20
Hereditary disease
  transcription hypothesis, 35-38
Hereditary nonpolyposis colorectal cancer (HNPCC)
  DNA mismatch repair defi-
    ciency and, 118-20
Herpes virus thymidine kinase
  gene
  framing errors and, 761
hGH
  See Human growth hormone
Histone mRNA decay, 712-13
HIV infection
  tuberculosis and, 216
HNPCC
  See Hereditary nonpolyposis
    colorectal cancer
HNRNP proteins
  pre-mRNA splicing and,
    392
Homeologous recombinants
  structure, 128
Homeologous recombination,
  125-29
Homologous recombination
  Dictyostelium and, 415
Human excision nuclease
  structure and function, 55-61
Human growth hormone (hGH)
  structure, 612-14

Human growth hormone (hGH)-
    prolactin receptor complex,
    627-29
Human growth hormone (hGH)
    receptor
    oligomerization mechanism,
    618
Human growth hormone (hGH)
    receptor complexes
    Site 1
        structure and function, 624-
        26
    Site 2
        structure and function, 626-
        27
Hydrogen bonding
    saccharide-lectin interactions
        and, 443-52
Hydroxymethyluracil
    elimination from DNA
        DNA glycosylase and, 140
Hypoxanthine
    elimination from DNA
        3-methyladenine-DNA gly-
        cosylase and, 140-41

I

IBIDS
    nucleotide excision repair
        genes and, 35
Ichthyosis
    nucleotide excision repair
        genes and, 35
Induced-fit theory, 8
Intercellular channels
    structure of connexins, 476-84
Intercellular communication
    gap-junctional, 476
        functional assays, 484-85
        inhibition, 491
Interferon-γ
    proteasome subunit composi-
        tion and, 836
Interleukin-2 (IL-2)
    receptor subunits, 623
Interleukin-4 (IL-4)
    oligomerization mechanism,
        623
Interleukin-5 (IL-5)
    receptor subunits, 622-23
Interleukin-6 (IL-6)
    receptor subunits, 620-22
Isoniazid
    drug-resistant tuberculosis and,
        225-30
    mechanism of action, 229
    structure, 225
Isoprenoids
    polypeptide modification and,
        242

J

Jacobson, A., 693-731
Jungnickel, B., 271-98

K

Kanamycin
    structure, 220
Kar2p
    protein translocation and, 280,
        290-91
Ketomycolates
    structure, 226
Kilberg, M. S., 305-32
Kluveromyces lactis
    telomerase RNA component,
        346
Koshland, E. E., Jr., 1-13
Krämer, A., 367-403
Kutay, U., 271-98

L

Lactacystin
    20S proteasome and, 821
Lahue, R., 101-29
Lamin B
    CaaX motif, 242-44
Lectin-carbohydrate complexes
    three-dimensional structure,
        444-45
Lectin-carbohydrate hydrogen
        bonds, 447
    water molecules mediating,
        451-52
Lectin-carbohydrate recognition,
        441-71
    divalent cations and, 453-54
Lectin-oligosaccharide recogni-
        tion
    multivalency and importance
        of geometry, 464-69
Lectins
    bridging
        lattice formation and, 468-69
    carbohydrate-binding proteins
        compared, 460-61
    conformational changes upon
        sugar binding, 459-60
    C-type, 462-63
    extended binding sites, 456-58
    legume, 463
    nonpolar interactions, 454-56
    oligomeric
        arrangement of binding
            sites, 466-67
    recognition of charged sugars
        and, 452-53
    secondary binding sites, 458-59
Lectin-saccharide recognition
    structural basis, 443-61
Legume lectins, 463

Legumes
    rhizobial infection and nodula-
        tion, 504-6
Li-Fraumeni syndrome
    nucleotide excision repair and,
        29-30, 38
Lipids
    gap junctions and, 491-92
Lipo-chitooligosaccharide nodula-
        tion factors, 503-31
Liver
    isolation of gap junctions, 479
Liver cancer
    autoantibodies against DNA
        topoisomerase II and, 672
Lohman, T. M., 169-210
LPA
    See Lysophosphatidic acid
Lung cancer
    microsatellite instability and,
        118-19
Lynch syndrome
    DNA mismatch repair defi-
        ciency and, 118-20
Lysophosphatidic acid (LPA)
    Dictyostelium cells and, 415

M

Maize
    telomere length
        strain-specific differences,
            352
Malandro, M. S., 305-32
Mammals
    amino acid transporters, 305-32
    G protein β-subunit, 421
    MutS/MutL homologs, 105
    photoreactivating activity, 154
    poly(A) shortening, decapping,
        and exonucleolytic decay
        enzymology and genetics,
            699-701
    Sec61/SecYEGp complex, 276-
        78
    Sec62/Sec63p complex, 279-80
Manhattan Project, 3-5
Manumycin
    protein prenylation inhibition
        and, 260
MAP kinase
    adenylyl cyclase activation
        and, 424-25
McEwen, J. A., 563-601
Membrane proteins
    insertion, 294-95
    protein translocation and,
        282
Methanococcus vannielii
    selenophosphate synthetase, 91
Methanopyrus kandleri
    DNA topoisomerase, 638
    subunits, 678

Methoxymycolates
    structure, 226
5-Methyladenine-DNA glycosy-
    lase
    base excision repair and, 140-
        41
Methylating agents
    mammalian cell lines without
        MGMT activity and, 137
Methylation tolerance
    DNA mismatch mutations and,
        117-18
O^6-Methylguanine DNA methyl-
    transferase (MGMT)
    direct DNA repair and, 45-46,
        137
Methylnitrosourea
    mammalian cell lines without
        MGMT activity and, 137
Mice
    telomerase regulation, 360-61
    telomere length
        strain-specific differences,
            352
Micrococcus luteus
    nucleotide excision repair, 18,
        47
Mitochondria
    signaling to nucleus, 592-99
Mitochondrial assembly
    chaperonins and, 587
    nuclear control, 587-91
    pathways, 587
    proteases and, 588
Mitochondrial enzymes
    functioning
        electron-transfer reactions
            and, 538
Mitochondrial gene expression,
    577-87
    coordination, 591-92
    differential regulation, 586-87
    global regulation, 585-86
    RNA processing, 580-82
    transcription, 578-80
    translation, 582-85
Mitochondrial genes
    synthesis/assembly
        nuclear PET genes and, 583-
            84
Mitochondrial-nuclear interac-
    tions, 563-601
Mitochondrial transcription
    initiation
        nuclear-coded protein fac-
            tors and, 578-80
    RNA processing and, 580-82
Mitochondrial translation, 582-85
Mitochondrial transmission
    nuclear genes required, 590-91
Mitotic recombination
    DNA topoisomerases and, 667-
        69

MMTV
    See Mouse mammary tumor vi-
        rus
Modrich, P., 101-29
Molecular matchmaker
    DNA damage recognition and,
        49-51
Moloney murine leukemia virus
    (MuLV)
    codon redefinition and, 754
Monodelphus domesticus
    DNA photolyase, 45, 154
Monosaccharides
    lectins binding, 456-58
Mouse mammary tumor virus
    (MMTV)
    frameshifting and, 749-51
mRNA
    recoding, 741-63
        classes, 744-45
        history, 743-44
mRNA decay, 693-731
    endonucleolytic cleavage initi-
        ating, 728-30
    histone, 712-13
    impaired translation initiating,
        713-28
    nonsense-mediated, 714-26
    pathway, 697-99
    poly(A) shortening and, 696-
        713
mRNA translation
    poly(A)-binding proteins and,
        707-8
    poly(A) shortening and, 711-12
    poly(A) tails and, 706-7
    3′-UTR instability elements
        and, 708
MT1 cell line
    methylation-tolerant, 117
MuLV
    See Moloney murine leukemia
        virus
Murine leukemia virus
    codon redefinition and, 754
Mus spretus
    telomerase regulation, 360-61
Mut homologs
    DNA mismatch repair and,
        106-10
MutL homologs
    molecular nature, 112-13
MutS homologs
    molecular nature, 110-12
Mycobacterium avium-intracellu-
    lare complex (MAC)
    susceptibility to isoniazid,
        218
Mycobacterium bovis BCG
    susceptibility to isoniazid,
        218
Mycobacterium leprae
    proteasomes, 810

Mycobacterium smegmatis
    doubling time, 217
    susceptibility to isoniazid, 218
Mycobacterium tuberculosis
    doubling time, 217
    drug resistance, 215-36
    susceptibility to isoniazid, 218
α-Mycolates
    structure, 226
Mycoplasma genitalium
    nucleotide excision repair, 47
    DNA topoisomerase, 639
Myoglobin
    electron-transfer reactions, 545-
        47

N

Nalidixic acid
    structure, 223
NER
    See Nucleotide excision repair
Neurospora crassa
    DNA photolyase, 154
    nuclear-coded mitochondrial
        proteins, 592-93
    nucleotide excision repair, 18,
        46, 155-57
Neurotransmission
    glycine and, 322-23
Neurotransmitter transporter fam-
    ily, 321
Nicotinamide
    structure, 233
Nod factors, 503-31
    chemical and enzymatic syn-
        thesis, 521-22
    chitin oligomer backbone, 508-
        10
    diversity within strains, 520
    fatty acid substituents, 511-12
    identification and charac-
        terization, 506-8
    N-acyl transfer, 512-13
    N-deacetylation at nonreduc-
        ing end, 510
    N-methylation, 510-11
    O-acetylation, 513
    6-O-acetylation, 515
    O-carbamoylation, 513-14
    1-O-glycerol substitution, 516
    6-O-glycosylation, 515-16
    6-O-sulfation, 514-15
    perception and transduction,
        527-29
    plant growth regulation and,
        522-31
    structural variation at both
        ends, 520-21
    structural variation between
        species, 519-20
    structure, 505

structure and host specificity, 518-19
transport, 516-18
Nod genes, 506
Nodule primordia
  nod factors eliciting, 527
Nonsense codons
  rapid mRNA decay and, 716-17
Nuclear factor, 541-47
Nuclear gene expression
  coordination, 591-92
Nuclear-matrix associated proteins
  pre-mRNA splicing and, 402
Nuclear-mitochondrial interactions, 563-601
Nucleotide excision repair (NER), 46-47, 144-53
  DNA polymerases and, 151-52
  in Escherichia coli
    coupling of transcription and, 28-29
  in eukaryotes
    coupling of transcription and, 29-30
    RNAP II basal transcription proteins and, 30-32
  incision during, 146-48
  pathways, 17-18
  reconstitution with purified components, 152-53
  regulation, 153
  transcription factor IIH and, 19-21, 32-35, 57-58, 150-51
Nucleotide excision repair (NER) genes
  human hereditary diseases and, 35-38
Nucleotide excision repair (NER) proteins
  eukaryotic, 156
  interactions, 153
Nucleotides
  binding to type II topoisomerases, 650-52
  DNA binding to dimeric Escherichia coli Rep helicase and, 185-86

O

Ofloxacin
  structure, 223
Oligomeric lectins
  arrangement of binding sites, 466-67
Oligosaccharide-lectin recognition
  multivalency and importance of geometry, 464-69
Oligosaccharides
  lectins binding, 456-58
Oligosaccharyltransferase

protein translocation site and, 281-82
Open reading frame (ORF)
  recoding and, 743
ORF
  See Open reading frame
Ornithine decarboxylase antizyme gene
  recoding, 746-48
Oryzias latipes
  DNA photolyase, 154
Osteopetrosis
  carbonic anhydrase II and, 762
Ovarian cancer
  microsatellite instability and, 118-19
8-Oxoguanine endonuclease
  nucleotide excision repair and, 46
Oxygen
  reactive species
    as second messengers in retrograde regulation, 597-99
Oxytricha
  telomere protein, 348-49
  telomeric DNA, 339

P

p53 protein
  cell cycle regulation and, 72-73
PA28 activator
  20S proteasome and, 817-19
PABPs
  See Poly(A)-binding proteins
PAF
  See Platelet activating factor
Pancreatic cancer
  microsatellite instability and, 118-19
  Ras proteins and, 260
Paracoccus denitrificans
  cytochrome c oxidase subunit III, 573-74
Paramecium
  telomeric DNA, 338
Parent, C. A., 411-37
Pasteurella haemolytica
  proteasomes, 810
Paul, D. L., 475-98
PCNA
  nucleotide excision repair and, 151-52
Peltz, S. W., 693-731
Pepticiinnamins
  protein prenylation inhibition and, 260
Phage T4 gene 41 protein, 178
Phage T7 gene 10
  frameshifting and, 762
Phage T7 gene 4 helicase, 184
Phage T7 gene 4 proteins, 178

DNA binding and, 189
Phosphodiesterase
  ambient levels of cAMP in Dictyostelium and, 425-26
α-hydroxyfarnesyl)Phosphonic acid
  protein prenylation inhibition and, 260
Phosphorylation
  connexin function and, 492-97
6-4 Photoproduct lyase
  direct DNA repair and, 45, 155
Photosensitivity
  nucleotide excision repair genes and, 35
PIBIDS
  nucleotide excision repair genes and, 35
Plant RNA viruses
  codon redefinition and, 754
  frameshifting and, 752-53
Plants
  telomeric DNA, 338
Platelet activating factor (PAF)
  Dictyostelium cells and, 415
PMS
  See Postmeiotic segregation
Poly(A)-binding proteins (PABPs), 699-700
  mRNA translation and, 707-8
Poly(ADP-ribose) polymerase
  base excision repair and, 144
Polyamine biosynthesis
  ornithine decarboxylase antizyme and, 746
Polypyrimidine tract-binding proteins
  pre-mRNA splicing and, 386-90
Postmeiotic segregation (PMS), 122-24
Potato virus M
  frameshifting and, 752-53
Potorous tridactylis
  DNA photolyase, 154
Poyton, R. O., 563-601
Pre-mRNA splicing, 367-403
  ATP-dependent RNA helicases and, 398-99
  branch site-binding proteins and, 390-91
  Cap-binding proteins and, 401
  catalysis, 369-71
    protein factors required, 400-1
  HNRNP proteins and, 392
  nuclear-matrix associated proteins and, 402
  polypyrimidine tract-binding proteins and, 386-90
  spliceosome assembly, 371-77
    protein factors required, 399-400

SR proteins and, 377-86
Prenylated C-termini
  proteolytic maturation, 252-53
Prenylated proteins
  cellular localization, 255-58
  membrane targeting, 255-58
  metabolism, 262-63
  protein-protein interactions,
    258-59
Prenylation
  protein, 241-65
    enzymology, 244-55
    functional consequences,
      255-63
    inhibitor studies, 259-62
Prenylcysteines
  C-terminal
    methylation, 253-55
Prestarvation factor (PSF)
  Dictyostelium cells and, 413
Prevotella loescheii
  ribosomal bypassing and, 760
Prokaryotes
  DNA excision repair, 47-52
  DNA helicases, 170
  selenocysteine insertion, 755-
    57
Prolactin-human growth hormone
  receptor complex, 627-29
Proline transporter
  brain-specific, 321-22
Prome, J.-C., 503-31
Prostatic cancer
  microsatellite instability and,
    119
Proteases
  mitochondrial, 588
Proteasomes, 801-41
  antigen presentation and, 835-
    38
  content and composition
    regulation, 839-40
  eubacterial, 810-11
  proteolytic processing of pre-
    cursor proteins and, 834-
    35
  20S, 805-21
    assembly, 808-9
    catalytic mechanism, 812-14
    components, 806-8
    peptidase activities, 814-16
    regulators of activity, 816-19
    selective inhibitors, 819-21
    X-ray analysis, 811-12
  26S, 821-34
    functions and cofactors, 831-
      34
    regulatory complex, 824-26
    subunits, 826-31
Protein-DNA interactions
  energetics and kinetics
    solution conditions and, 173-
      74

Protein kinases
  cyclin-dependent
    cell division cycle and, 70-
      71
Protein prenylation, 241-65
  enzymology, 244-55
  functional consequences, 255-
    63
  inhibitor studies, 259-62
Protein processing
  proteasomes and, 834-35
Protein-protein interactions
  protein prenylation and, 258-
    59
Proteins
  electron transfer, 537-59
Protein translocation, 271-98
  cotranslational, 283-88
  enzymes at site, 281-82
  Kar2p and, 280
  post-translational in *Es-
    cherichia coli*, 292-94
  post-translational in yeast, 289-
    92
  protein-conducting channel
    and, 282-83
  Sec61/SecYEGp complex and,
    276-78
  Sec62/Sec63p complex and,
    279-80
  SecA protein and, 280
  SecD/SecFp complex and,
    280
Protein unfolding
  26S proteasome and, 834
Proteoliposomes
  protein translocation and, 283-
    84
Proteolysis
  ubiquitin-independent
    26S proteasome and, 838-
      39
Protozoa
  telomeric DNA, 338
PSF
  See Prestarvation factor
Psoralen derivatives
  adducts
    nucleotide excision repair
      and, 144-45
PTB-associated splicing factor
  pre-mRNA splicing and, 389-
    90
Puromycin
  protein-conducting channel
    and, 282
Pyrazinamide
  drug-resistant tuberculosis and,
    232-33
  structure, 233
Pyrimidine adducts
  nucleotide excision repair and,
    144-45

Pyrimidine hydrate-DNA glyco-
    sylase
  base excision repair and, 141
Pyrococcus furiosus
  DNA topoisomerase, 638

R

Rab proteins
  membrane association, 258
  prenylation, 250-52
RAD4
  DNA excision repair and, 150
RAD14
  DNA excision repair and, 148-
    49
RAD23
  DNA excision repair and, 150
Rap1 protein
  telomeric DNA and, 349-50
Rapoport, T. A., 271-98
Ras proteins
  adenylyl cyclase activation
    and, 424-25
  CaaX motif, 242-44
rBAT/4F2hc transporter family,
    327-31
  cloning, 327-28
  cystinuria and, 330-31
  structure-function analysis,
    328-30
Reactive oxygen species
  as second messengers in retro-
    grade regulation, 597-99
Recoding, 741-63
  classes, 744-45
  history, 743-44
REMI
  See Restriction enzyme-medi-
    ated integration
Reptiles
  photoreactivating activity,
    154
Respiratory proteins
  function
    nuclear-mitochondrial contri-
      butions, 566-69
Restriction enzyme-mediated inte-
    gration (REMI)
  Dictyostelium gene products
    and, 416
Retroviruses
  codon redefinition and, 754
  frameshifting and, 749-51
Rhizobium
  infection and nodulation of leg-
    umes, 504-6
  lipo-chitooligosaccharide nodu-
    lation factors, 503-31
  nodulation genes, 506
Rhodobacter sphaeroides
  photosynthetic reaction center
    structure, 556

Rhodococcus
  proteasomes, 810-11
Rhodopsin kinase
  farnesylation, 257-58
Ribosomal frameshifting, 745-53
Rifampicin
  drug-resistant tuberculosis and,
    221-23
RNA helicases, 172
  ATP-dependent
    pre-mRNA splicing and,
      398-99
RNA polymerase
  eukaryotic, 770-71
  telomerase and, 342
RNA polymerase II
  basal transcription proteins
    nucleotide excision repair in
      eukaryotes and, 30-32
  DNA transcription and, 19-23
  transcriptional arrest on dam-
    aged templates, 23-24
  transcription factor IIH and,
    150-51
  transcription initiation
    regulated, 789-94
RNA processing
  mitochondrial transcription
    and, 580-82
RNA synthesis
  DNA topoisomerases and, 661
RNA viruses
  codon redefinition and, 754
  frameshifting and, 752-53
Roeder, R. G., 769-96
Root cortex
  cell cycle activation
    nod factors and, 525-26
Root epidermis
  responses to nod factors, 522-
    25
Rous sarcoma virus (RSV)
  frameshifting and, 750
RPA (HSSB)
  DNA excision repair and, 55-
    57, 74, 149
RSV
  See Rous sarcoma virus
Ruthenium complexes
  electron-transfer reactions and,
    540-41

S

Saccharide-lectin recognition
  structural basis, 443-61
Saccharomyces cerevisiae
  CaaX proteins, 254-55
  cytochrome *c* oxidase
    environmental regulation,
      574-75
    nuclear-coded subunits, 575-
      76

DNA helicases, 170-71
DNA ligase, 157
DNA mismatch repair, 106-7
  gene conversion and, 122
DNA photolyase, 154
DNA topoisomerases
  active-site tyrosines, 647
  regulation, 673
double-strand break repair,
  159-60
intergenomic signaling, 596-
  97
mitochondrial transcription
  units, 578-80
mitochondrial transmission
  nuclear genes required, 590-
    91
mitotic mutability, 114-16
mitotic recombination
  homology and, 125
mRNA decay pathways
  nonsense-mediated, 717-18
nucleotide excision repair, 18,
  47
  incision during, 146-48
postmeiotic segregation, 122
prenylpeptide endoprotease activity, 253
protein translocation
  post-translational, 289-92
respiratory capacity, 585-86
respiratory chain
  composition, 567-69
retrograde regulation, 594-96
ribonuclease, 701-2
Sec61/SecYEGp complex, 276-
  78
Sec62/Sec63p complex, 279-80
transcription factor IIH, 20
transposable elements, 748-49
UPF1 gene, 718-20
UPF2/NMD2 gene, 720-22
UPF3 gene, 722
Salmonella typhimurium
  DNA topoisomerase, 639
  selenocysteine synthesis, 86
Sancar, A., 43-76
Schizosaccharomyces pombe
  DNA ligase, 157
  DNA topoisomerases
    active-site tyrosines, 647
    regulation, 673
  nucleotide excision repair, 18,
    46-47, 155
  rad2 mutants
    spontaneous chromosome
      loss and, 144
Scleroderma
  autoantibodies against DNA
    topoisomerase I and,
      672
Sec61p complex
  protein targeting, 284-88

Sec62/Sec63p complex
  protein translocation and, 279-
    80
SecA protein
  protein translocation and, 280
SecD/SecFp complex
  protein translocation and,
    280
Selenocysteine, 83-98, 755-58
  biosynthesis, 85-87
  enzymes and proteins contain-
    ing residues, 91-94
  eukaryotic insertion, 757-58
  prokaryotic insertion, 755-57
Selenocysteine-containing enzymes
  catalytic activities, 97-98
Selenocysteine residues
  biochemical roles in enzymes,
    95-97
Selenocysteyl-tRNA^{Sec}
  UGA decoding and, 87
Selenophosphate synthetase, 89-
  90
  distribution, 91
  properties, 90
Selenoprotein P, 92
  UGA codons, 757-58
Selenoprotein W, 92
Selenosubtilisin
  reacting with hydroperoxides,
    94-95
SGS1 protein
  chromosome segregation, 668-
    69
Signal peptidase
  protein translocation site and,
    281-82
Signal recognition particle (SRP),
  272
  protein targeting, 284-86
Signal recognition particle (SRP)
  receptor, 278
Signal transduction
  in *Dictyostelium*, 411-37
  prenylated proteins and, 258
Sinorhizobium, 504
Skin cancer
  microsatellite instability and,
    118-19
Small-cell lung cancer
  microsatellite instability and,
    118-19
Small nuclear ribonucleoprotein
  particles (snRNPs)
  17S U2, 392-96
  20S U5, 396-97
  25S U4/U6.U5, 397-98
  pre-mRNA splicing and, 371-
    77
snRNPs
  See Small nuclear ribonucleo-
    protein particles

Sololobus shibatae
  DNA topoisomerase, 639
SOS response
  DNA excision repair and, 71-74
Spiroplasma
  DNA topoisomerase, 640
Spliceosome assembly
  pre-mRNA splicing and, 371-77
  protein factors required, 399-400
Spore photoproduct lyase
  direct DNA repair and, 45
Squamous cell skin cancer
  microsatellite instability and, 118-19
SRP
  See Signal recognition particle
SR proteins, 377-86
  functions, 380-84
  regulation
    reversible phosphorylation and, 385-86
    transcriptional and post-transcriptional, 384
  structural organization, 377-80
Stadtman, T. C., 83-98
Staphylococcus aureus
  DNA topoisomerase, 639
  drug-resistant strains
    confined populations and, 233
Stele
  responses to nod factors, 526-27
Steric exclusion
  C-type lectin selectivity and, 463
Stomach cancer
  microsatellite instability and, 118-19
Streptococcus pneumoniae
  DNA mismatch repair
    genetic recombination and, 122
  strand signals and, 113
  nucleotide excision repair, 47
Streptolysin O
  protein-conducting channel and, 283
Streptomycin
  drug-resistant tuberculosis and, 219-21
  structure, 220
Subtilisin
  selenium analog, 94
Sugars
  charged
    recognition by lectins, 452-53
Sulfolobus acidocaldarius

DNA topoisomerase, 638
reverse gyrase
  DNA binding, 641
SV40 large T antigen
  binding with synthetic DNA fork, 187
  hexameric helicases and, 177
Systemic lupus erythematosus (SLE)
  autoantibodies against DNA topoisomerase II and, 672

T

Tanaka, K., 801-41
TATA box binding protein
  transcription factor IIB and, 776-84
TBP-DNA complex, 778-84
Telomerase, 339-46
  activation in cancer cells, 357-61
  alteration
    in vivo effects, 345-46
  biochemistry, 341-42
  elongation model, 340
  protein components, 344-45
  regulation
    mouse models, 360-61
  RNA components, 341, 343-44
Telomeres, 337-62
  aging and, 355-57
  binding proteins, 348-50
  cellular senescence and, 355-57
  length equilibrium
    regulation, 352-55
  length regulation
    models, 353-55
  processing, 351-52
  replication, 346-48
  structure and function, 338-39
Teniposide
  genome stability and, 670
Testicular cancer
  microsatellite instability and, 119
Tetrahymena
  nuclear-coded mitochondrial proteins, 592
  telomerase, 339-45
  telomere processing, 351-52
  telomeric DNA, 338
Thermoplasma acidophilum
  DNA topoisomerase, 638
  20S proteasome, 805
    assembly, 808-9
    subunits, 808
Thiolase
  ARE-binding activity, 704
Thyroid hormone deiodinases, 92-93

TMV
  See Tobacco mosaic virus
Tobacco mosaic virus (TMV)
  codon redefinition and, 754
Toxins
  glycolipid-binding, 463-64
Transcription
  DNA supercoiling and, 658-59
  nucleotide excision repair in Escherichia coli and, 28-29
  nucleotide excision repair in eukaryotes and, 29-30
Transcription factor IIB
  TATA box binding protein and, 776-84
Transcription factor IID, 769-96
  molecular architecture, 771-76
  RNA polymerase II transcription and, 791
  TATA-less promoters and, 794-95
Transcription factor IIH
  DNA excision repair and, 70-71
  nucleotide excision repair and, 19-21, 32-35, 57-58, 150-51
Transcription-repair coupling, 67-69
  in Escherichia coli, 68
  in humans, 68-69
Translational bypassing, 758-60
Translocating-chain associating membrane (TRAM) protein, 278-79
Trichothiodystrophy (TTD)
  defective nucleotide excision repair and, 54-55
  nucleotide excision repair genes and, 35-36
Trypanosoma equiperdum
  DNA-DNA topoisomerase complexes, 640
TTD
  See Trichothiodystrophy
Tuberculosis
  drug-resistant, 215-36
    ethambutol and, 230-32
    ethionamide and, 225-30
    fluoroquinolones and, 223-25
    isoniazid and, 225-30
    pyrazinamide and, 232-33
    rifampicin and, 221-23
    streptomycin and, 219-21
Tumor necrosis factor-α
  proteasome subunit composition and, 836
Tyrosines
  active-site
    covalent enzyme-DNA complexes and, 647-49

U

U2 auxiliary factor
  pre-mRNA splicing and, 386-
    88
Ubiquitin conjugation
  protein degradation in cells
    and, 804
Ubiquitin recycling
  26S proteasome and, 832-34
uORFs
  See Upstream open reading
    frames
Upstream open reading frames
  (uORFs)
  nonsense-mediated mRNA
    pathway and, 715
Uracil-DNA glycosylase
  base excision repair and, 138-
    40

V

Vaccinia virus
  DNA topoisomerase, 647-48
Vertebrates
  photoreactivating activity, 154
Viruses
  DNA helicases, 170

W

Wang, J. C., 635-81
Water molecules
  lectin-carbohydrate hydrogen
    bonds and, 451-52
Weis, W. I., 441-71

Wells, J. A., 609-31
Winkler, J. R., 537-59
Wood, R. D., 135-61

X

Xenopus laevis
  DNA mismatch repair, 107
  MutS/MutL homologs, 105
  nucleotide excision repair, 47
Xeroderma pigmentosum (XP)
  defective nucleotide excision
    repair and, 30-31, 35-37,
    53-54, 145
  preferential repair of transcrip-
    tionally active genes and,
    27
XP
  See Xeroderma pigmentosum
XPA
  DNA excision repair and, 55,
    148-49
XPC
  DNA excision repair and, 150
XPC-HHR23B
  DNA excision repair and, 58-
    59
XPE
  DNA excision repair and, 60-
    61, 149-50
XPF-ERCC1
  DNA excision repair and, 60
XPG
  DNA excision repair and, 59-
    60
XPG protein

nucleotide excision repair and,
  146
X-ray diffraction
  20S proteasome and, 811-12

Y

Yeast
  DNA mismatch repair
    strand-specific, 114-16
  DNA topoisomerases, 637-38
  homeologous recombination,
    126-27
  homologous activities
    DNAse IV and, 143-44
  MutS/MutL homologs, 105
  poly(A) shortening, decapping,
    and exonucleolytic decay
    enzymology and genetics,
    699-701
  protein translocation
    post-translational, 289-92
  Ste6 transporter, 262
Yeast extract
  Dictyostelium cells and, 415
Yeast LA-virus
  frameshifting and, 753
Yeast Ras2
  farnesylation, 258

Z

Zaragozic acid
  protein prenylation inhibition
    and, 260
Zhang, F. L., 241-65

CUMULATIVE INDEXES

CONTRIBUTING AUTHORS, VOLUMES 61–65

A

Adams, M. E., 63:823–67
Adhya, S., 62:749–95
Affolter, M., 63:487–526
Alberts, B. M., 63:639–74
Anderson, W. F., 62:191–217
Arnheim, N., 61:131–56
Atkins, J. F., 65:741–68

B

Balch, W. E., 63:949–90
Barnes, D. E., 61:251–81
Belfort, M., 62:587–622
Benkovic, S. J., 61:29–54
Bennett, M. K., 63:63–100
Berg, O. G., 64:652–87
Bhatnagar, R. S., 63:869–914
Biemann, K., 61:977–1010
Bjornson, K. P., 65:169–214
Blackburn, E. H., 61:113–29
Blanchard, J. S., 65:215–39
Blau, H., 61:1213–30
Bossone, S. A., 61:809–60
Bradshaw, R. A., 62:823–50
Bredt, D. S., 63:175–95
Brennan, P. J., 64:29–63
Brun, Y. V., 63:419–50
Brunner, J., 62:483–514
Buc, H., 62:749–95
Burd, C. G., 62:289–321
Bürglin, T., 63:487–526
Burley, S., 65:769–99
Busby, S., 62:749–95
Buxser, S. E., 62:823–50

C

Carreras, C. W., 64:721–62
Carter, C. W. Jr., 62:715–48
Casey, P. J., 65:241–69
Catterall, W. A., 64:493–531
Chaudhary, V., 61:331–54
Ciechanover, A., 61:761–807
Clarke, S. G., 61:355–86
Claus, T. H., 64:799–835
Clipstone, N. A., 63:1045–83
Coleman, J. E., 61:897–946
Conaway, J. W., 62:161–90
Conaway, R. C., 62:161–90
Coux, O., 65:801–47
Coverley, D., 63:745–76
Crabtree, G. R., 63:1045–83

D

Darnell, J. E., 64:621–51
Das, A., 62:893–930
Davis, L. I., 64:865–96
Debelle, F., 65:503–35
Demple, B., 63:915–48
Denarie, J., 65:503–35
DePamphilis, M. L., 62:29–63
de Vos, A. M., 65:609–34
Devreotes, P. N., 65:411–40
Dixon, J. E., 62:101–20
Dixon, R. A. F., 63:101–32
Donkersloot, J. A., 61:517–57
Doolittle, R. F., 64:287–314
Draper, D. E., 64:593–620
Dreyfuss, G., 62:289–321
Drickamer, K., 65:441–73
Dwek, R. A., 62:65–100

E

Eaton, B. E., 64:837–63
Edge, C. J., 62:65–100
Eggleston, A. K., 63:991–1043
Englund, P. T., 62:121–38
Erickson, J. W., 62:543–85
Erlich, H., 61:131–56

F

Fanning, E., 61:55–85
Fantl, W. J., 62:453–81
Feher, G., 61:861–96
Ferat, J., 64:435–61
Ferrari, M. E., 63:527–70
FitzGerald, D. J., 61:331–54
Fong, T. M., 63:101–32
Ford-Hutchinson, A. W., 63:383–417
Fournier, M. J., 64:897–933
Frank-Kamenetskii, M. D., 64:65–95
Fridovich, I., 64:97–112
Friedberg, E. C., 65:15–42
Fuchs, E., 63:345–82

G

Garges, S., 62:749–95
Gehring, W. J., 63:487–526
Gelb, M. H., 64:652–87
Gennis, R. B., 63:675–716
Gesteland, R. F., 65:741–68

Gierasch, L., 61:387–418
Gold, L., 64:763–97
Goldberg, A. L., 65:801–47
Goliger, J. A., 65:475–502
Goodenough, D. A., 65:475–502
Gordon, J. I., 63:869–914
Gottesman, M. M., 62:385–427
Granner, D. K., 61:1131–73
Gray, H. B., 65:537–61
Greider, C. W., 65:337–65
Gresser, M., 63:383–417
Gudas, L. J., 64:201–33

H

Hanel, A. M., 64:652–87
Hanson, P. I., 61:559–601
Harpel, M. R., 63:197–234
Harrison, L., 63:915–48
Hartl, F.-U., 62:349–84
Hartman, F. C., 63:197–234
Harvey, D. J., 62:65–100
Hayaishi, O., 63:1–24
Heemels, M., 64:463–91
Hendrick, J. P., 62:349–84
Hershko, A., 61:761–807
Herz, J., 63:601–37
Hiraga, S., 61:283–306
Howard, J. B., 63:235–64
Hu, P. Y., 64:375–401

J

Jacobson, A., 65:693–739
Jaeger, J. A., 62:255–87
Jain, M. K., 64:652–87
Johnson, D. E., 62:453–81
Johnson, D. R., 63:869–914
Johnson, K. A., 62:685–713
Jones, K. A., 63:717–43
Joyce, C. M., 63:777–822
Jungnickel, B., 65:271–303

K

Kadonaga, J. T., 63:265–97
Kamakaka, R. T., 63:265–97
Katz, R. A., 63:133–73
Keller, W., 61:419–40
Kellogg, D. R., 63:639–74
Kelman, Z., 64:171–200
Kennedy, E. P., 61:1–28
Kennedy, M. B., 63:571–600

Kent, C., 64:315–43
Kilberg, M. S., 65:305–36
Kivirikko, K. I., 64:403–34
Klinman, J. P., 63:299–344
Knippers, R., 61:55–85
Knoll, L. J., 63:869–914
Kodukula, K., 64:563–91
Kolb, A., 62:749–95
Kornfeld, S., 61:307–30
Koshland, D. E., 65:1–13
Kowalczykowski, S. C., 63:991–1043
Kramer, A., 65:367–409
Krieger, M., 63:601–37
Kurjan, J., 61:1097–129
Kurland, I. J., 64:799–835
Kutay, U., 65:271–303

L

Lahue, R., 65:101–33
Lai, M., 64:259–86
Lambowitz, A. M., 62:587–622
Lane, M. D., 64:345–73
Lange, A. J., 64:799–835
Lanka, E., 64:141–69
Laskey, R. A., 63:745–76
Lasky, L. A., 64:113–39
Law, J. H., 61:87–111
Levine, A. J., 62:623–51
Lindahl, T., 61:251–81
Lohman, T. M., 63:527–70; 65:169–214
Lowy, D. R., 62:851–91
Lucas, P. C., 61:1131–73

M

MacDougald, O. A., 64:345–73
Majerus, P. W., 61:225–50
Malandro, M. S., 65:305–36
Marcu, K. B., 61:809–60
Marczynski, G., 63:419–50
Marians, K. J., 61:673–719
Massagué, J., 62:515–41
Matthews, B. W., 62:139–60
Matthews, C. R., 62:653–83
Matunis, M. J., 62:289–321
Maxwell, E. S., 64:897–933
McEwen, J. E., 65:563–607
McGarry, J. D., 64:689–719
Means, A. L., 64:201–33
Michel, F., 64:435–61
Miljanich, G. P., 63:823–67
Mirkin, S. M., 64:65–95
Mizuuchi, K., 61:1011–51
Modrich, P., 65:101–33
Morgan, R. A., 62:191–217
Moritz, M., 63:639–74
Mu, D., 63:299–344

N

Newgard, C. B., 64:689–719
Nikaido, H., 64:29–63

Norbury, C., 61:441–70
Nuoffer, C., 63:949–90
Nurse, P., 61:441–70

O

O'Donnell, M., 64:171–200
Okamura, M. Y., 61:861–96
Olivera, B. M., 63:823–67
O'Malley, B. W., 63:451–86
Osborne, M. A., 62:219–54

P

Pabo, C. O., 61:1053–95
Pandiella, A., 62:515–41
Paranjape, S. M., 63:265–97
Parekh, R. B., 62:65–100
Parent, C. A., 65:411–40
Pastan, I. H., 61:331–54; 62:385–427
Patel, A. J., 61:809–60
Paul, D. L., 65:475–502
Peltz, S. W., 65:693–739
Peterlin, B. M., 63:717–43
Pieken, W. A., 64:837–63
Pilkis, S. J., 64:799–835
Piñol-Roma, S., 62:289–321
Ploegh, H., 64:463–91
Polisky, B., 64:763–97
Poyton, R. O., 65:563–607
Prockop, D. J., 64:403–34
Promé, J.-C., 65:503–35
Pryer, N. K., 61:471–516

R

Raffioni, S., 62:823–50
Ramachandran, J., 63:823–67
Rapoport, T. A., 65:271–303
Ravid, S., 61:721–59
Rees, D. C., 63:235–64
Reichard, P., 64:1–28
Reinberg, D., 64:533–61
Ribeiro, J. M. C., 61:87–111
Rizo, J., 61:387–418
Roeder, R., 65:769–99
Russell, D. W., 63:25–61

S

Sancar, A., 65:43–81
SantaLucia, J., 62:255–87
Santi, D., 64:721–62
Sauer, R. T., 61:1053–95
Schekman, R., 61:471–516
Scheller, R. H., 63:63–100
Schindler, C., 64:621–51
Schleif, R., 61:199–223
Schulman, H., 61:559–601
Schutgens, R. B. H., 61:157–97
Shapiro, L., 63:419–50
Sheetz, M. P., 62:429–51
Silver, P. A., 62:219–54
Sipe, J. D., 61:947–75
Skalka, A. M., 63:133–73

Sly, W., 64:375–401
Snell, E. E., 62:1–27
Snyder, S. H., 63:175–95
Spudich, J. A., 61:721–59
Stadtman, E. R., 62:797–821
Stadtman, T., 65:83–100
Steitz, T. A., 63:777–822
Strader, C. D., 63:101–32
Symons, R. H., 61:641–71

T

Tager, J. M., 61:157–97
Tall, A., 64:235–57
Tan, J. L., 61:721–59
Tanaka, K., 65:801–47
Thompson, J., 61:517–57
Tinoco, I. Jr., 62:255–87
Tota, M. R., 63:101–32
Trumpower, B. L., 63:675–716
Tsai, M.-J., 63:451–86

U

Udenfriend, S., 64:563–91
Uhlenbeck, O., 64:763–97
Underwood, D., 63:101–32

V

van den Bosch, H., 61:157–97

W

Wahle, E., 61:419–40
Walker, R. A., 62:429–51
Wallace, D., 61:1175–212
Walton, K. M., 62:101–20
Wanders, R. J. A., 61:157–97
Wang, J. C., 65:635–92
Warren, G., 62:323–48
Weber, K., 63:345–82
Weis, W. I., 65:441–73
Wells, J. A., 65:609–34
Wells, M., 61:87–111
West, S. C., 61:603–40
Wilkins, B. M., 64:141–69
Williams, L. T., 62:453–81
Willumsen, B. M., 62:851–91
Wilson, J. D., 63:25–61
Winkler, J. R., 65:537–61
Wlodawer, A., 62:543–85
Wood, R. D., 65:135–67
Wormald, M. R., 62:65–100
Wuestehube, L., 61:471–516

Y

Yarus, M., 64:763–97
Young, R. N., 63:383–417

Z

Zawel, L., 64:533–61
Zhang, F. L., 65:241–69

CHAPTER TITLES, VOLUMES 61–65

PREFATORY
    Sailing to Byzantium   E. P. Kennedy   61:1–28
    From Bacterial Nutrition to Enzyme Structure:
        A Personal Odyssey   E. E. Snell   62:1–27
    Tryptophan, Oxygen, and Sleep   O. Hayaishi   63:1–24
    To Be There When the Picture Is Painted   P. Reichard   64:1–28
    How to Get Paid for Having Fun   D. E. Koshland, Jr.   65:1–13

AMINO ACIDS
    N-(Carboxyalkyl) Amino Acids: Occurrence,
        Synthesis, and Functions   J. Thompson, J. A. Donkersloot   61:517–57
    From Bacterial Nutrition to Enzyme Structure:
        A Personal Odyssey   E. E. Snell   62:1–27
    Oxidation of Free Amino Acids and Amino
        Acid Residues in Proteins by Radiolysis and
        by Metal-Catalyzed Reactions   E. R. Stadtman   62:797–821
    Nitric Oxide: A Physiologic Messenger
        Molecule   D. S. Bredt, S. H. Snyder   63:175–95
    Selenocysteine   T. C. Stadtman   65:83–100
    Molecular Biology of Mammalian Amino Acid
        Transporters   M. S. Malandro, M. S. Kilberg   65:305–36

BIOENERGETICS (See also Contractile Proteins, Membranes, and Transport)
    Control of Nonmuscle Myosins by
        Phosphorylation   J. L. Tan, S. Ravid, J. A. Spudich   61:721–59
    Proton Transfer in Reaction Centers from
        Photosynthetic Bacteria   M. Y. Okamura, G. Feher   61:861–96
    Diseases of the Mitochondrial DNA   D. C. Wallace   61:1175–212
    Energy Transduction by Cytochrome
        Complexes in Mitochondrial and Bacterial
        Respiration: The Enzymology of Coupling
        Electron Transfer Reactions to
        Transmembrane Proton Translocation   B. L. Trumpower, R. B. Gennis   63:675–716
    Electron Transfer in Proteins   H. B. Gray, J. R. Winkler   65:537–61
    Crosstalk Between Nuclear and Mitochondrial
        Genomes   R. O. Poyton, J. E. McEwen   65:563–607

CANCER (See Disease, Biochemistry of)

CARBOHYDRATES
    Analysis of Glycoprotein-Associated
        Oligosaccharides   R. A. Dwek, C. J. Edge, D. J. Harvey, M. R. Wormald, R. B. Parekh   62:65–100
    Selectin-Carbohydrate Interactions and the
        Initiation of the Inflammatory Response   L. A. Lasky   64:113–39
    Structural Basis of Lectin-Carbohydrate
        Recognition   W. I. Weis, K. Drickamer   65:441–73

CELL ORGANELLES
    Biochemistry of Peroxisomes   H. van den Bosch, R. B. H. Schutgens, R. J. A. Wanders, J. M. Tager   61:157–97

Structure and Function of the Mannose
  6-Phosphate/Insulinlike Growth Factor II
  Receptors   S. Kornfeld   61:307–30
Vesicle-Mediated Protein Sorting   N. K. Pryer, L. J. Wuestehube, R.   61:471–516
     Schekman
Diseases of the Mitochondrial DNA   D. C. Wallace   61:1175–
     212

Nucleocytoplasmic Transport in the Yeast
  Saccharomyces cerevisiae   M. A. Osborne, P. A. Silver   62:219–54
Membrane Partitioning During Cell Division   G. Warren   62:323–48
Cytoplasmic Microtubule-Associated Motors   R. A. Walker, M. P. Sheetz   62:429–51
The Centrosome and Cellular Organization   D. R. Kellogg, M. Moritz, B. M.   63:639–74
     Alberts
GTPases: Multifunctional Molecular Switches
  Regulating Vesicular Traffic   C. Nuoffer, W. E. Balch   63:949–90
The Nuclear Pore Complex   L. I. Davis   64:865–96
Protein Transport Across the Eukaryotic
  Endoplasmic Reticulum and Bacterial Inner
  Membranes   T. A. Rapoport, B. Jungnickel, U.   65:271–303
     Kutay
Connexins, Connexons, and Intercellular
  Communication   D. A. Goodenough, J. A. Goliger, D.   65:475–502
     L. Paul

CELL WALLS
Chromosome and Plasmid Partition in
  Escherichia coli   S. Hiraga   61:283–306
The Envelope of Mycobacteria   P. J. Brennan, H. Nikaido   64:29–63

DEVELOPMENT AND DIFFERENTIATION
Structure and Function of the Mannose
  6-Phosphate/Insulinlike Growth Factor II
  Receptors   S. Kornfeld   61:307–30
Animal Cell Cycles and Their Control   C. Norbury, P. Nurse   61:441–70
Pheromone Response in Yeast   J. Kurjan   61:1097–
     129
Diseases of the Mitochondrial DNA   D. C. Wallace   61:1175–
     212
Differentiation Requires Continuous Active
  Control   H. M. Blau   61:1213–30
Human Gene Therapy   R. A. Morgan, W. F. Anderson   62:191–217
Membrane-Anchored Growth Factors   J. Massagué, A. Pandiella   62:515–41
Intermediate Filaments: Structure, Dynamics,
  Function, and Disease   E. Fuchs, K. Weber   63:345–82
The Expression of Asymmetry During
  Caulobacter Cell Differentiation   Y. V. Brun, G. Marczynski, L.   63:419–50
     Shapiro
Homeodomain Proteins   W. J. Gehring, M. Affolter, T. Bürglin   63:487–526
Signal Transmission between the Plasma
  Membrane and Nucleus of T Lymphocytes   G. R. Crabtree, N. A. Clipstone   63:1045–83
The Roles of Retinoids in Vertebrate
  Development   A. L. Means, L. J. Gudas   64:201–33
Rhizobium Lipo-Chitooligosaccharide
  Nodulation Factors: Signaling Molecules
  Mediating Recognition and Morphogenesis   J. Dénarié, F. Debellé, J.-C. Promé   65:503–35

DISEASE, BIOCHEMISTRY OF
Mammalian DNA Ligases   T. Lindahl, D. E. Barnes   61:251–81
Amyloidosis   J. D. Sipe   61:947–75
Diseases of the Mitochondrial DNA   D. C. Wallace   61:1175–
     212
Human Gene Therapy   R. A. Morgan, W. F. Anderson   62:191–217

The Tumor Suppressor Genes | A. J. Levine | 62:623–51
Steroid 5α-Reductase: Two Genes/Two Enzymes | D. W. Russell, J. D. Wilson | 63:25–61
The Retroviral Enzymes | R. A. Katz, A. M. Skalka | 63:133–73
Control of RNA Initiation and Elongation at the HIV-1 Promoter | K. A. Jones, B. M. Peterlin | 63:717–43
Plasma Lipid Transfer Proteins | A. Tall | 64:235–57
Human Carbonic Anhydrases and Carbonic Anhydrase Deficiencies | W. Sly, P. Y. Hu | 64:375–401
Mismatch Repair in Replication Fidelity, Genetic Recombination, and Cancer Biology | P. Modrich, R. Lahue | 65:101–33

DNA

General
Polymerase Chain Reaction Strategy | N. Arnheim, H. Erlich | 61:131–56
DNA Looping | R. Schleif | 61:199–223
Chromosome and Plasmid Partition in *Escherichia coli* | S. Hiraga | 61:283–306
Transcription Factors: Structural Families and Principles of DNA Recognition | C. O. Pabo, R. T. Sauer | 61:1053–95
Function and Regulation of Ras | D. R. Lowy, B. M. Willumsen | 62:851–91
Role of Chromatin Structure in the Regulation of Transcription by RNA Polymerase II | S. M. Paranjape, R. T. Kamakaka, J. T. Kadonaga | 63:265–97

Recombination
Mammalian DNA Ligases | T. Lindahl, D. E. Barnes | 61:251–81
Enzymes and Molecular Mechanisms of Genetic Recombination | S. C. West | 61:603–40
Transpositional Recombination: Mechanistic Insights from Studies of Mu and Other Elements | K. Mizuuchi | 61:1011–51
Diseases of the Mitochondrial DNA | D. C. Wallace | 61:1175–212
Human Gene Therapy | R. A. Morgan, W. F. Anderson | 62:191–217
Introns as Mobile Genetic Elements | A. M. Lambowitz, M. Belfort | 62:587–622
Escherichia coli Single-Stranded DNA-Binding Protein: Multiple DNA-Binding Modes and Cooperativities | T. M. Lohman, M. E. Ferrari | 63:527–70
Repair of Oxidative Damage to DNA: Enzymology and Biology | B. Demple, L. Harrison | 63:915–48
Homologous Pairing and DNA Strand-Exchange Proteins | S. C. Kowalczykowski, A. K. Eggleston | 63:991–1043
DNA Processing Reactions in Bacterial Conjugation | E. Lanka, B. M. Wilkins | 64:141–69

Repair
Mammalian DNA Ligases | T. Lindahl, D. E. Barnes | 61:251–81
Escherichia coli Single-Stranded DNA-Binding Protein: Multiple DNA-Binding Modes and Cooperativities | T. M. Lohman, M. E. Ferrari | 63:527–70
Function and Structure Relationships in DNA Polymerases | C. M. Joyce, T. A. Steitz | 63:777–822
Repair of Oxidative Damage to DNA: Enzymology and Biology | B. Demple, L. Harrison | 63:915–48
Relationships Between DNA Repair and Transcription | E. C. Friedberg | 65:15–42
DNA Excision Repair | A. Sancar | 65:43–81
DNA Repair in Eukaryotes | R. D. Wood | 65:135–67

Replication

Structure and Function of Simian Virus 40 Large Tumor Antigen	E. Fanning, R. Knippers	61:55–85
Telomerases	E. H. Blackburn	61:113–29
Mammalian DNA Ligases	T. Lindahl, D. E. Barnes	61:251–81
Chromosome and Plasmid Partition in *Escherichia coli*	S. Hiraga	61:283–306
Prokaryotic DNA Replication	K. J. Marians	61:673–719
Zinc Proteins: Enzymes, Storage Proteins, Transcription Factors, and Replication Proteins	J. E. Coleman	61:897–946
Eukaryotic DNA Replication: Anatomy of an Origin	M. L. DePamphilis	62:29–63
Conformational Coupling in DNA Polymerase Fidelity	K. A. Johnson	62:685–713
Escherichia coli Single-Stranded DNA-Binding Protein: Multiple DNA-Binding Modes and Cooperativities	T. M. Lohman, M. E. Ferrari	63:527–70
The Centrosome and Cellular Organization	D. R. Kellogg, M. Moritz, B. M. Alberts	63:639–74
Regulation of Eukaryotic DNA Replication	D. Coverley, R. A. Laskey	63:745–76
Function and Structure Relationships in DNA Polymerases	C. M. Joyce, T. A. Steitz	63:777–822
DNA Polymerase III Holoenzyme: Structure and Function of a Chromosomal Replicating Machine	Z. Kelman, M. O'Donnell	64:171–200
Mechanisms of Helicase-Catalyzed DNA Unwinding	T. M. Lohman, K. P. Bjornson	65:169–214
DNA Topoisomerases	J. C. Wang	65:635–92

Structure

Telomerases	E. H. Blackburn	61:113–29
Transcriptional Regulation by cAMP and its Receptor Protein	A. Kolb, S. Busby, H. Buc, S. Garges, S. Adhya	62:749–95
Homologous Pairing and DNA Strand-Exchange Proteins	S. C. Kowalczykowski, A. K. Eggleston	63:991–1043
Triplex DNA Structures	M. D. Frank-Kamenetskii, S. M. Mirkin	64:65–95
Telomere Length Regulation	C. W. Greider	65:337–65

DRUGS, ANTIBIOTICS, AND ANTIMETABOLITES

Biochemistry of Multidrug Resistance Mediated by the Multidrug Transporter	M. M. Gottesman, I. Pastan	62:385–427
Molecular Mechanisms of Drug Resistance in *Mycobacterium Tuberculosis*	J. S. Blanchard	65:215–39

ENZYMES

Mechanisms and Kinetics

Catalytic Antibodies	S. J. Benkovic	61:29–54
Protein Tyrosine Phosphatases	K. M. Walton, J. E. Dixon	62:101–20
Structure-Based Inhibitors of HIV-1 Protease	A. Wlodawer, J. W. Erickson	62:543–85
Conformational Coupling in DNA Polymerase Fidelity	K. A. Johnson	62:685–713
Structure, Function, Regulation, and Assembly of D-Ribulose-1,5-Bisphosphate Carboxylase/Oxygenase	F. C. Hartman, M. R. Harpel	63:197–234
5-Lipoxygenase	A. W. Ford-Hutchinson, M. Gresser, R. N. Young	63:383–417

Function and Structure Relationships in DNA
  Polymerases   C. M. Joyce, T. A. Steitz   63:777–822
The Catalytic Mechanism and Structure of
  Thymidylate Synthase   C. W. Carreras, D. Santi   64:721–62
Protein Prenylation: Molecular Mechanisms
  and Functional Consequences   F. L. Zhang, P. J. Casey   65:241–69

Regulation
Structure, Function, Regulation, and Assembly
  of D-Ribulose-1,5-Bisphosphate
  Carboxylase/Oxygenase   F. C. Hartman, M. R. Harpel   63:197–234

Specific Enzymes and Classes
Telomerases   E. H. Blackburn   61:113–29
Inositol Phosphate Biochemistry   P. W. Majerus   61:225–50
Mammalian DNA Ligases   T. Lindahl, D. E. Barnes   61:251–81
Protein Isoprenylation and Methylation at
  Carboxy-Terminal Cysteine Residues   S. Clarke   61:355–86
Enzymes and Molecular Mechanisms of
  Genetic Recombination   S. C. West   61:603–40
Zinc Proteins: Enzymes, Storage Proteins,
  Transcription Factors, and Replication
  Proteins   J. E. Coleman   61:897–946
Analysis of Glycoprotein-Associated
  Oligosaccharides   R. A. Dwek, C. J. Edge, D. J. Harvey, 62:65–100
                                                  M. R. Wormald, R. B. Parekh
Conformational Coupling in DNA Polymerase
  Fidelity   K. A. Johnson   62:685–713
Cognition, Mechanism, and Evolutionary
  Relationships in Aminoacyl-tRNA
  Synthetases   C. W. Carter Jr.   62:715–48
Steroid 5α-Reductase: Two Genes/Two
  Enzymes   D. W. Russell, J. D. Wilson   63:25–61
The Retroviral Enzymes   R. A. Katz, A. M. Skalka   63:133–73
Nitric Oxide: A Physiologic Messenger
  Molecule   D. S. Bredt, S. H. Snyder   63:175–95
Quinoenzymes in Biology   J. P. Klinman, D. Mu   63:299–344
GTPases: Multifunctional Molecular Switches
  Regulating Vesicular Traffic   C. Nuoffer, W. E. Balch   63:949–90
Superoxide Radical and Superoxide
  Dismutases   I. Fridovich   64:97–112
Interfacial Enzymology of Glycerolipid
  Hydrolases: Lessons from Secreted
  Phospholipases A2   M. M. H. Gelb, M. K. Jain, A. M.   64:652–87
                                                  Hanel, O. G. Berg
6-Phosphofructo-2-Kinase/Fructose-2,6-Bisphosp
  hatase: A Metabolic Signaling Enzyme   S. J. Pilkis, T. H. Claus, I. J. Kurland, 64:799–835
                                                  A. J. Lange

Structure (Protein)
From Bacterial Nutrition to Enzyme Structure:
  A Personal Odyssey   E. E. Snell   62:1–27
Signaling by Receptor Tyrosine Kinases   W. J. Fantl, D. E. Johnson, L. T.   62:453–81
                                                  Williams
Structure-Based Inhibitors of HIV-1 Protease   A. Wlodawer, J. W. Erickson   62:543–85
Nitrogenase: A Nucleotide-Dependent
  Molecular Switch   J. B. Howard, D. C. Rees   63:235–64
Function and Structure Relationships in DNA
  Polymerases   C. M. Joyce, T. A. Steitz   63:777–822
How Glycosylphosphatidylinositol-Anchored
  Membrane Proteins Are Made   S. Udenfriend, K. Kodukula   64:563–91
Biochemistry and Structural Biology of
  Transcription Factor IID (TFIID)   S. K. Burley, R. G. Roeder   65:769–99
Structure and Functions of the 20S and 26S
  Proteasomes   O. Coux, K. Tanaka, A. L. Goldberg   65:801–47

GENES AND BIOCHEMICAL GENETICS (See also DNA and RNA)

Polymerase Chain Reaction Strategy | N. Arnheim, H. Erlich | 61:131–56
Animal Cell Cycles and Their Control | C. Norbury, P. Nurse | 61:441–70
Enzymes and Molecular Mechanisms of
 Genetic Recombination | S. C. West | 61:603–40
myc Function and Regulation | K. B. Marcu, S. A. Bossone, A. J. Patel | 61:809–60

Transcription Factors: Structural Families and
 Principles of DNA Recognition | C. O. Pabo, R. T. Sauer | 61:1053–95
Hormone Response Domains in Gene
 Transcription | P. C. Lucas, D. K. Granner | 61:1131–73
Diseases of the Mitochondrial DNA | D. C. Wallace | 61:1175–212

Human Gene Therapy | R. A. Morgan, W. F. Anderson | 62:191–217
The Tumor Suppressor Genes | A. J. Levine | 62:623–51
Transcriptional Regulation by cAMP and its
 Receptor Protein | A. Kolb, S. Busby, H. Buc, S. Garges, S. Adhya | 62:749–95

Control of Transcription Termination by
 RNA-Binding Proteins | A. Das | 62:893–930
Steroid 5α-Reductase: Two Genes/Two
 Enzymes | D. W. Russell, J. D. Wilson | 63:25–61
The Expression of Asymmetry During
 Caulobacter Cell Differentiation | Y. V. Brun, G. Marczynski, L. Shapiro | 63:419–50

Control of RNA Initiation and Elongation at the
 HIV-1 Promoter | K. A. Jones, B. M. Peterlin | 63:717–43
Regulation of Eukaryotic DNA Replication | D. Coverley, R. A. Laskey | 63:745–76

HORMONES

Biochemical Insights Derived from Insect
 Diversity | J. H. Law, J. M. C. Ribeiro, M. A. Wells | 61:87–111

Pheromone Response in Yeast | J. Kurjan | 61:1097–129

Hormone Response Domains in Gene
 Transcription | P. C. Lucas, D. K. Granner | 61:1131–73
Steroid 5α-Reductase: Two Genes/Two
 Enzymes | D. W. Russell, J. D. Wilson | 63:25–61
Molecular Mechanisms of Action of
 Steroid/Thyroid Receptor Superfamily
 Members | M.-J. Tsai, B. W. O'Malley | 63:451–86
Generation, Translocation, and Presentation of
 MHC Class I-Restricted Peptides | M. Heemels, H. Ploegh | 64:463–91
Transcriptional Responses to Polypeptide
 Ligands: The JAK-STAT Pathway | C. Schindler, J. E. Darnell | 64:621–51
Hematopoietic Receptor Complexes | J. A. Wells, A. M. de Vos | 65:609–34

IMMUNOBIOCHEMISTRY

Signal Transmission between the Plasma
 Membrane and Nucleus of T Lymphocytes | G. R. Crabtree, N. A. Clipstone | 63:1045–83

LIPIDS

Inositol Phosphate Biochemistry | P. W. Majerus | 61:225–50
Structures and Functions of Multiligand
 Lipoprotein Receptors: Macrophage
 Scavenger Receptors and LDL
 Receptor-Related Proteins (LRP) | M. Krieger, J. Herz | 63:601–37
Genetic and Biochemical Studies of Protein
 N-Myristoylation | D. R. Johnson, R. S. Bhatnagar, L. J. Knoll, J. I. Gordon | 63:869–914

Eukaryotic Phospholipid Biosynthesis | C. Kent | 64:315–43

MEMBRANES

Inositol Phosphate Biochemistry	P. W. Majerus	61:225–50
Chromosome and Plasmid Partition in *Escherichia coli*	S. Hiraga	61:283–306
Structure and Function of the Mannose 6-Phosphate/Insulinlike Growth Factor II Receptors	S. Kornfeld	61:307–30
Protein Isoprenylation and Methylation at Carboxy-Terminal Cysteine Residues	S. Clarke	61:355–86
Vesicle-Mediated Protein Sorting	N. K. Pryer, L. J. Wuestehube, R. Schekman	61:471–516
Neuronal Ca^{2+}/Calmodulin-Dependent Protein Kinases	P. I. Hanson, H. Schulman	61:559–601
The Structure and Biosynthesis of Glycosyl Phosphatidylinositol Protein Anchors	P. T. Englund	62:121–38
Membrane Partitioning During Cell Division	G. Warren	62:323–48
Signaling by Receptor Tyrosine Kinases	W. J. Fantl, D. E. Johnson, L. T. Williams	62:453–81
Membrane-Anchored Growth Factors	J. Massagué, A. Pandiella	62:515–41
The Receptors for Nerve Growth Factor and Other Neurotrophins	S. Raffioni, R. A. Bradshaw, S. E. Buxser	62:823–50
Genetic and Biochemical Studies of Protein N-Myristoylation	D. R. Johnson, R. S. Bhatnagar, L. J. Knoll, J. I. Gordon	63:869–914
GTPases: Multifunctional Molecular Switches Regulating Vesicular Traffic	C. Nuoffer, W. E. Balch	63:949–90

METABOLISM

Diseases of the Mitochondrial DNA	D. C. Wallace	61:1175–212
From Bacterial Nutrition to Enzyme Structure: A Personal Odyssey	E. E. Snell	62:1–27
Transcriptional Regulation of Gene Expression During Adipocyte Differentiation	O. A. MacDougald, D. Lane	64:345–73
Metabolic Coupling Factors in Pancreatic Beta-Cell Signal Transduction	C. B. Newgard, J. D. McGarry	64:689–719

METHODOLOGY

Polymerase Chain Reaction Strategy	N. Arnheim, H. Erlich	61:131–56
Mass Spectrometry of Peptides and Proteins	K. Biemann	61:977–1010
New Photolabeling and Crosslinking Methods	J. Brunner	62:483–514

MUSCLE AND CONTRACTILE PROTEINS

Neuronal Ca^{2+}/Calmodulin-Dependent Protein Kinases	P. I. Hanson, H. Schulman	61:559–601
Control of Nonmuscle Myosins by Phosphorylation	J. L. Tan, S. Ravid, J. A. Spudich	61:721–59

NUCLEOTIDES, NUCLEOSIDES, PURINES, AND PYRIMIDINES

Repair of Oxidative Damage to DNA: Enzymology and Biology	B. Demple, L. Harrison	63:915–48

NEUROBIOLOGY AND NEUROCHEMISTRY

Neuronal Ca^{2+}/Calmodulin-Dependent Protein Kinases	P. I. Hanson, H. Schulman	61:559–601
Diseases of the Mitochondrial DNA	D. C. Wallace	61:1175–212
The Receptors for Nerve Growth Factor and Other Neurotrophins	S. Raffioni, R. A. Bradshaw, S. E. Buxser	62:823–50

A Molecular Description of Synaptic Vesicle
  Membrane Trafficking — M. K. Bennett, R. H. Scheller — 63:63–100
Nitric Oxide: A Physiologic Messenger
  Molecule — D. S. Bredt, S. H. Snyder — 63:175–95
The Biochemistry of Synaptic Regulation in the
  Central Nervous System — M. B. Kennedy — 63:571–600
Calcium Channel Diversity and
  Neurotransmitter Release: The ω-Conotoxins
  and ω-Agatoxins — B. M. Olivera, G. P. Miljanich, J. Ramachandran, M. E. Adams — 63:823–67

NITROGEN FIXATION
Nitrogenase: A Nucleotide-Dependent
  Molecular Switch — J. B. Howard, D. C. Rees — 63:235–64

NUTRITION (See Vitamins, Growth Factors, and Essential Metabolites)

PEPTIDES
Constrained Peptides: Models of Bioactive
  Peptides and Protein Substructures — J. Rizo, L. M. Gierasch — 61:387–418
Mass Spectrometry of Peptides and Proteins — K. Biemann — 61:977–1010

PHOTOBIOLOGY AND PHOTOSYNTHESIS (See also Bioenergetics)
Proton Transfer in Reaction Centers from
  Photosynthetic Bacteria — M. Y. Okamura, G. Feher — 61:861–96
Repair of Oxidative Damage to DNA:
  Enzymology and Biology — B. Demple, L. Harrison — 63:915–48

PROTEINS

Binding and Transport Proteins
Molecular Chaperone Functions of Heat-Shock
  Proteins — J. P. Hendrick, F.-U. Hartl — 62:349–84
Biochemistry of Multidrug Resistance Mediated
  by the Multidrug Transporter — M. M. Gottesman, I. Pastan — 62:385–427
The Receptors for Nerve Growth Factor and
  Other Neurotrophins — S. Raffioni, R. A. Bradshaw, S. E. Buxser — 62:823–50
Function and Regulation of Ras — D. R. Lowy, B. M. Willumsen — 62:851–91
Structure and Function of Voltage-Gated Ion
  Channels — W. A. Catterall — 64:493–531

Biosynthesis
The Biochemistry of 3'-End Cleavage and
  Polyadenylation of Messenger RNA
  Precursors — E. Wahle, W. Keller — 61:419–40
Cognition, Mechanism, and Evolutionary
  Relationships in Aminoacyl-tRNA
  Synthetases — C. W. Carter Jr. — 62:715–48

Contractile Proteins
Cytoplasmic Microtubule-Associated
  Motors — R. A. Walker, M. P. Sheetz — 62:429–51

Post-Translational Modification
Protein Isoprenylation and Methylation at
  Carboxy-Terminal Cysteine Residues — S. Clarke — 61:355–86
Vesicle-Mediated Protein Sorting — N. K. Pryer, L. J. Wuestehube, R. Schekman — 61:471–516
The Ubiquitin System for Protein Degradation — A. Hershko, A. Ciechanover — 61:761–807
Protein Tyrosine Phosphatases — K. M. Walton, J. E. Dixon — 62:101–20
Oxidation of Free Amino Acids and Amino
  Acid Residues in Proteins by Radiolysis and
  by Metal-Catalyzed Reactions — E. R. Stadtman — 62:797–821

Genetic and Biochemical Studies of Protein
  N-Myristoylation   D. R. Johnson, R. S. Bhatnagar, L. J.   63:869–914
     Knoll, J. I. Gordon

Special Classes
  Catalytic Antibodies   S. J. Benkovic   61:29–54
  Inositol Phosphate Biochemistry   P. W. Majerus   61:225–50
  Structure and Function of the Mannose
    6-Phosphate/Insulinlike Growth Factor II
    Receptors   S. Kornfeld   61:307–30
  Neuronal Ca^{2+}/Calmodulin-Dependent Protein
    Kinases   P. I. Hanson, H. Schulman   61:559–601
  Control of Nonmuscle Myosins by
    Phosphorylation   J. L. Tan, S. Ravid, J. A. Spudich   61:721–59
  Zinc Proteins: Enzymes, Storage Proteins,
    Transcription Factors, and Replication
    Proteins   J. E. Coleman   61:897–946
  Amyloidosis   J. D. Sipe   61:947–75
  Transcription Factors: Structural Families and
    Principles of DNA Recognition   C. O. Pabo, R. T. Sauer   61:1053–95
  Analysis of Glycoprotein-Associated
    Oligosaccharides   R. A. Dwek, C. J. Edge, D. J. Harvey,   62:65–100
     M. R. Wormald, R. B. Parekh
  hnRNP Proteins and the Biogenesis of mRNA   G. Dreyfuss, M. J. Matunis, S.   62:289–321
     Piñol-Roma, C. G. Burd
  Signalling by Receptor Tyrosine Kinases   W. J. Fantl, D. E. Johnson, L. T.   62:453–81
     Williams
  Control of Transcription Termination by
    RNA-Binding Proteins   A. Das   62:893–930
  The Expression of Asymmetry During
    Caulobacter Cell Differentiation   Y. V. Brun, G. Marczynski, L.   63:419–50
     Shapiro
  Homeodomain Proteins   W. J. Gehring, M. Affolter, T. Bürglin   63:487–526
  Escherichia coli Single-Stranded DNA-Binding
    Protein: Multiple DNA-Binding Modes and
    Cooperativities   T. M. Lohman, M. E. Ferrari   63:527–70
  Structures and Functions of Multiligand
    Lipoprotein Receptors: Macrophage
    Scavenger Receptors and LDL
    Receptor-Related Proteins (LRP)   M. Krieger, J. Herz   63:601–37
  Genetic and Biochemical Studies of Protein
    N-Myristoylation   D. R. Johnson, R. S. Bhatnagar, L. J.   63:869–914
     Knoll, J. I. Gordon
  Homologous Pairing and DNA
    Strand-Exchange Proteins   S. C. Kowalczykowski, A. K.   63:991–
     Eggleston   1043
  Collagens: Molecular Biology, Diseases, and
    Potentials for Therapy   D. J. Prockop, K. I. Kivirikko   64:403–34

Structure
  The Structure and Biosynthesis of Glycosyl
    Phosphatidylinositol Protein Anchors   P. T. Englund   62:121–38
  Structural and Genetic Analysis of Protein
    Stability   B. W. Matthews   62:139–60
  Determination of RNA Structure and
    Thermodynamics   J. A. Jaeger, J. SantaLucia Jr., I.   62:255–87
     Tinoco Jr.
  hnRNP Proteins and the Biogenesis of mRNA   G. Dreyfuss, M. J. Matunis, S.   62:289–321
     Piñol-Roma, C. G. Burd
  Pathways of Protein Folding   C. R. Matthews   62:653–83
  Cognition, Mechanism, and Evolutionary
    Relationships in Aminoacyl-tRNA
    Synthetases   C. W. Carter Jr.   62:715–48

Function and Structure Relationships in DNA
    Polymerases   C. M. Joyce, T. A. Steitz   63:777–822
The Multiplicity of Domains in Proteins   R. F. Doolittle   64:287–314

RECEPTORS
    Protein Isoprenylation and Methylation at
        Carboxy-Terminal Cysteine Residues   S. Clarke   61:355–86
    Pheromone Response in Yeast   J. Kurjan   61:1097–
                                                                                                    129
    Signaling by Receptor Tyrosine Kinases   W. J. Fantl, D. E. Johnson, L. T.   62:453–81
                                                            Williams
    Membrane-Anchored Growth Factors   J. Massagué, A. Pandiella   62:515–41
    Transcriptional Regulation by cAMP and its
        Receptor Protein   A. Kolb, S. Busby, H. Buc, S.   62:749–95
                                                            Garges, S. Adhya
    The Receptors for Nerve Growth Factor and
        Other Neurotrophins   S. Raffioni, R. A. Bradshaw, S. E.   62:823–50
                                                            Buxser
    Structure and Function of G Protein-Coupled
        Receptors   C. D. Strader, T. M. Fong, M. R.   63:101–32
                                                            Tota, D. Underwood, R. A. F.
                                                            Dixon
    Nitric Oxide: A Physiologic Messenger
        Molecule   D. S. Bredt, S. H. Snyder   63:175–95
    Molecular Mechanisms of Action of
        Steroid/Thyroid Receptor Superfamily
        Members   M.-J. Tsai, B. W. O'Malley   63:451–86
    Structures and Functions of Multiligand
        Lipoprotein Receptors: Macrophage
        Scavenger Receptors and LDL
        Receptor-Related Proteins (LRP)   M. Krieger, J. Herz   63:601–37
    Genetic and Biochemical Studies of Protein
        N-Myristoylation   D. R. Johnson, R. S. Bhatnagar, L. J.   63:869–914
                                                            Knoll, J. I. Gordon
    Molecular Genetics of Signal Transduction in
        Dictyostelium   C. A. Parent, P. N. Devreotes   65:411–40

RNA
    Small Catalytic RNAs   R. H. Symons   61:641–71
    Hormone Response Domains in Gene
        Transcription   P. C. Lucas, D. K. Granner   61:1131–73
    General Initiation Factors for RNA Polymerase
        II   R. C. Conaway, J. W. Conaway   62:161–90
    Introns as Mobile Genetic Elements   A. M. Lambowitz, M. Belfort   62:587–622
    Transcriptional Regulation by cAMP and its
        Receptor Protein   A. Kolb, S. Busby, H. Buc, S.   62:749–95
                                                            Garges, S. Adhya
    Control of Transcription Termination by
        RNA-Binding Proteins   A. Das   62:893–930
    The Retroviral Enzymes   R. A. Katz, A. M. Skalka   63:133–73
    Role of Chromatin Structure in the Regulation
        of Transcription by RNA Polymerase II   S. M. Paranjape, R. T. Kamakaka, J.   63:265–97
                                                            T. Kadonaga
    Molecular Mechanisms of Action of
        Steroid/Thyroid Receptor Superfamily
        Members   M.-J. Tsai, B. W. O'Malley   63:451–86
    Control of RNA Initiation and Elongation at the
        HIV-1 Promoter   K. A. Jones, B. M. Peterlin   63:717–43
    Common Themes in Assembly and Function of
        Eukaryotic Transcription Complexes   D. Reinberg, L. Zawel   64:533–61
    Diversity of Oligonucleotide Functions   L. Gold, B. Polisky, O. Uhlenbeck,   64:763–97
                                                            M. Yarus
    Structure and Activities of Group II Introns   F. Michel, J. Ferat   64:435–61

Protein-RNA Recognition | D. E. Draper | 64:593–620
Ribonucleosides and RNA | B. E. Eaton, W. A. Pieken | 64:837–63
The Small Nucleolar RNAs | E. S. Maxwell, M.J. Fournier | 64:897–933
The Structure and Function of Proteins
    Involved in Mammalian Pre-mRNA Splicing | A. Krämer | 65:367–409
Interrelationships of the Pathways of mRNA
    Decay and Translation | A. Jacobson, S. W. Peltz | 65:693–739
Recoding: Dynamic Reprogramming of
    Translation | R. F. Gesteland, J. F. Atkins | 65:741–68

TOXINS AND TOXIC AGENTS
Recombinant Toxins as Novel Therapeutic
    Agents | I. Pastan, V. Chaudhary, D. J. FitzGerald | 61:331–54

Calcium Channel Diversity and
    Neurotransmitter Release: The ω-Conotoxins
    and ω-Agatoxins | B. M. Olivera, G. P. Miljanich, J. Ramachandran, M. E. Adams | 63:823–67

Repair of Oxidative Damage to DNA:
    Enzymology and Biology | B. Demple, L. Harrison | 63:915–48

TRANSPORT
Biochemistry of Multidrug Resistance Mediated
    by the Multidrug Transporter | M. M. Gottesman, I. Pastan | 62:385–427

VIRUSES AND BACTERIOPHAGES
Structure and Function of Simian Virus 40
    Large Tumor Antigen | E. Fanning, R. Knippers | 61:55–85
Human Gene Therapy | R. A. Morgan, W. F. Anderson | 62:191–217
Function and Regulation of RAS | D. R. Lowy, B. M. Willumsen | 62:851–91
The Retroviral Enzymes | R. A. Katz, A. M. Skalka | 63:133–73
Control of RNA Initiation and Elongation at the
    HIV-1 Promoter | K. A. Jones, B. M. Peterlin | 63:717–43
The Molecular Biology of Hepatitis Delta Virus | M. Lai | 64:259–86

VITAMINS, GROWTH FACTORS, AND ESSENTIAL METABOLITES
Zinc Proteins: Enzymes, Storage Proteins,
    Transcription Factors, and Replication
    Proteins | J. E. Coleman | 61:897–946
From Bacterial Nutrition to Enzyme Structure:
    A Personal Odyssey | E. E. Snell | 62:1–27
Membrane-Anchored Growth Factors | J. Massagué, A. Pandiella | 62:515–41
Function and Regulation of RAS | D. R. Lowy, B. M. Willumsen | 62:851–91

ANNUAL REVIEWS INC.
4139 El Camino Way • P.O. Box 10139
Palo Alto, CA 94303-0139 • USA

BB96

Step 1 *Ordered by:*

Name _____

Address _____

_____ Zip Code _____

Please Mention
Priority Code
BB96
when placing
orders by phone.

Call from USA or Canada
1.800.523.8635
FAX orders 24 hours a day
1.415.424.0910

Today's Date _____ Day Phone: (____) ____

Fax (____) ____ _____ e-mail _____

Step 4 *Payment Method*

☐ Check or money order enclosed. Make checks payable
to "Annual Reviews Inc."
or charge

☐ VISA ☐ M/C ☐ AMEX

Account Number _____
Mo __/__ Yr __/__
Expiration Date _____ Print name exactly as it appears on credit card.
Signature _____

Qty	Annual Review of	Vol.	Place on Standing Order? Save 10% now with payment	Price	Total
			☐ Yes, save 10% ☐ No	$	
			☐ Yes, save 10% ☐ No	$	
			☐ Yes, save 10% ☐ No	$	
			☐ Yes, save 10% ☐ No	$	
			☐ Yes, save 10% ☐ No	$	

Step 2
Enter Order

☐ Student / Recent Graduate (past three years) discount 30% off. Not applicable to standing orders. Proof of status enclosed.

☐ California customers. Add applicable California sales tax for your location.

☐ Canadian customers. Add 7% Canadian GST. **(Reg. # 121449029 RT)**

Step 3
Shipping and Handling

✓ Handling Charges. Add $3 per volume. Applies to all orders.

☐ Standard shipping, US Mail 4th class bookrate (surface). No extra charge. | N/C

☐ Optional UPS Ground service, $3 extra per volume in 48 contiguous states only. UPS not available to PO boxes.

UPS Next Day Air ☐ UPS Second Day Air ☐ US Airmail ☐   Note option at left. We will calculate amount and add to your total.
Optional shipping to anywhere. Charged at actual cost and added to total. Prices vary by weight of volumes. **Total**

 Call Toll Free 1.800.523.8635 from USA or Canada 8am-4pm, M-F, Pacific Time. From elsewhere call 1.415.493.4400 ext. 1

 Mail Orders, fill in form, send in attached envelope. **e** e-mail **service@annurev.org**

Orders may also be placed through booksellers or subscription agents or through our Authorized Stockists
From Europe, the UK, the Middle East, and Africa contact: Gazelle Book Service Ltd., Fax 44 (0) 1524-63232
From India, Pakistan, Bangladesh or Sri Lanka contact: SARAS Books, Fax 91-11-941111.

ANNUAL REVIEWS INC. on the Web **http://www.annurev.org**